AF788613

HANDBOOK OF METAL LIGAND HEATS

HANDBOOK OF METAL LIGAND HEATS

and Related Thermodynamic Quantities

Third Edition, Revised and Expanded

JAMES J. CHRISTENSEN
DEPARTMENT OF CHEMICAL ENGINEERING SCIENCE
BRIGHAM YOUNG UNIVERSITY
PROVO, UTAH

REED M. IZATT
DEPARTMENT OF CHEMISTRY
BRIGHAM YOUNG UNIVERSITY
PROVO, UTAH

MARCEL DEKKER, INC. New York

Library of Congress Cataloging in Publication Data

Christensen, James J.
Handbook of metal ligand heats and related thermodynamic quantities.

Includes index.
1. Thermodynamics--Tables. 2. Complex compounds--Tables. I. Izatt, Reed McNeil . II. Title.
QD504.C528 1983 541.3'69 83-1908
ISBN 0-8247-7030-7

COPYRIGHT © 1983 by MARCEL DEKKER, INC. ALL RIGHTS RESERVED

Neither this book nor any part may be reproduced or transmitted in any form or by any means, electronic or mechanical, including photocopying, microfilming, and recording, or by any information storage and retrieval system, without permission in writing from the publisher.

MARCEL DEKKER, INC.
270 Madison Avenue, New York, New York 10016

Current printing (last digit):
10 9 8 7 6 5 4 3 2 1

PRINTED IN THE UNITED STATES OF AMERICA

To

V. B. C. and H. F. I.

PREFACE

We have undertaken the compilation of heats of metal*-ligand interactions in solution to aid those working in the fields of chemistry, physics, biology, medicine, engineering, etc., where these data are necessary and useful. We have found that an enormous amount of such data has been collected and in some cases partly tabulated, but that up to now there has been no one reasonably complete source of this information.

This book is a compilation of heats of interaction, ΔH, together with the related thermodynamic quantities, log K, ΔS, and ΔCp, where available. The book consists of a Table in which are summarized the published literature values through 1980 under the headings of the various ligands and four indexes. The first edition published in 1970 covered the literature up to mid-1969 and the second edition published in 1975 covered up through 1973. This edition contains 1547 separate ligand entries, more than triple the 514 entries in the first edition and 464 more than the second edition. The criterion for including a metal-ligand pair in the Table was that ΔH values must have been reported in the published literature. Values appearing in theses, technical reports, books, or other *nonrefereed* sources were not included. In addition to the ΔH, log K, ΔS, and ΔCp values, the following information is included in the Table: the appropriate reaction, temperature, method, and conditions of measurement of ΔH; original and additional literature references; and pertinent supplemental information.

Four indexes -- Empirical Formula, Element, Synonym, and Reference -- are included in the book. The Empirical Formula Index contains the empirical formula of each ligand together with the location of the ligand in the Table. The Element Index contains an alphabetical listing of all elements appearing in the Metal Ion column of the Table together with their valence states and locations in the Table. The Synonym Index is an alphabetical listing of the common synonyms of the ligands appearing in the Table together with their locations in the Table. The Reference Index includes a year by year alphabetical listing of all references cited in the Table.

The compilers have read the original of each paper, if possible, and data from other sources are reported only if the original paper could not be obtained through any of the means available including interlibrary loans. All values are those reported by the original investigators, except when only log K or ΔS was reported in addition to ΔH and in such cases the third quantity, either log K or

---

*In the compilation the *word* metal includes many elements from the right side of the periodic table, i.e., I, Br, Se, etc.

ΔS, was calculated from the other two. Values calculated in this manner are indicated by being underlined. Where sufficient data for a given reaction were available a critical evaluation was made by the compilers and the value or values reported are those judged to be most reliable. The authors do not claim to be infallible with respect to the selection of the most correct value in these cases and are solely responsible for any errors which may have been made in either compiling or critically evaluating the data. Comments concerning errors and omitted data would be most appreciated.

This book is intended as a companion to the following published Tables.

Handbook of Proton Ionization Heats and Related Thermodynamic Quantities, John Wiley and Sons, New York, N.Y. 1976, by J. J. Christensen, L. D. Hansen, and R. M. Izatt.

Critical Stability Constants, Volumes 1-5, Plenum Press, New York, N.Y. 1974-1982, by A. E. Martell and R. M. Smith.

Stability Constants of Metal-Ion Complexes, Special Publication No. 17, The Chemical Society, Burlington House, London, 1964, by L. G. Sillen and A. E. Martell.

Stability Constants of Metal-Ion Complexes, Supplement No. 1, Special Publication No. 25, The Chemical Society, Burlington House, London, 1971, by L. G. Sillen and A. E. Martell.

Instability Constants of Complex Compounds, Translated from Russian, Consultants Bureau, New York, 1960, by K. B. Yatsimirskii and V. P. Vasil'ev.

The authors express their gratitude to Ted Izatt for editorial assistance and to Kate Toepke for technical typing of the manuscript.

James J. Christensen
Reed M. Izatt

CONTENTS

HANDBOOK OF METAL LIGAND HEATS

USE OF TABLE AND INDEXES

A. Ligands

1. The ligands are ordered alphabetically and each is assigned a letter-number combination, i.e., A1, A2, A3, etc. In general, the nomenclature follows that in the 53rd Edition of the Handbook of Chemistry and Physics, The Chemical Rubber Co., 1972-1973. One exception is the naming of the synthetic macrocyclic compounds containing O and S atoms for which the term "crown" is used to represent the central ring following the nomenclature devised by C. J. Pedersen, J. Am. Chem. Soc., 89, 7017 (1967).

2. The charge on the ligand is indicated by a superscript on the symbol which is located directly after the formula of the compound/ion. In those cases where the coordinating form of the ligand is variable, the form is noted in the Conditions column if it has been indicated by the author.

B. Formula, Synonyms

1. Empirical and structural formulas of each ligand are given. Empirical formulas are also given in the Empirical Formula Index where they are arranged with the elements in the order C, H, O, N, Cl, Br, I, F, S, P followed by other elements in the alphabetical order of their symbols according to Beilstein's system.

2. The neutral form of each ligand is given except in a few cases, i.e., Cl^-, Br^-, I^-, CrO_4^{2-}, etc.

3. Synonyms for each ligand are given alphabetically in the Synonym Index.

C. Metals

1. The metals under each ligand follow the order given in L. G. Sillen and A. E. Martell, Stability Constants, Chem. Soc. Special Publication No. 17, 1964, which is given below. Where an element has several valencies, a lower oxidation state precedes a higher one.

group IA:	H, Li, Na, K, Rb, Cs, Fr
group IIA:	Be, Mg, Ca, Sr, Ba, Ra
group IIIB + 4f:	Sc, Y, La, Ce, Pr, Nd, Pm, Sm, Eu, Gd, Tb, Dy, Ho, Er, Tm, Yb, Lu, Ac

group IVB:	Ti, Zr, Hf, Th
group VB:	V, Nb, Ta, Pa
group VIB + 5f:	Cr, Mo, W, U, Np, Pu, Am, Cm, Bk, Cf, Es, Fm, Md, No, Lw
group VIIB:	Mn, Tc, Re
group VIII:	Fe, Co, Ni, Ru, Rh, Pd, Os, Ir, Pt
group IB:	Cu, Ag, Au
group IIB:	Zn, Cd, Hg
group IIIA:	B, Al, Ga, In, Tl
groups IVA:	C, Si, Ge, Sn, Pb
group VA:	N, P, As, Sb, Bi
group VIA:	O, S, Se, Te, Po
group VIIA:	F, Cl, Br, I, At

2. When the metal is part of a complex molecule or ion consisting of the metal and one or more ligands, the formula of the complex is given in the Metal Ion column. Ligands attached to the metal, except oxygen, are indicated by letters of the alphabet, e.g., A, B, etc., and each ligand is identified in the Conditions column, i.e., A = Cl, B = C_2H_4, etc.

3. Enclosure of the metal or metal ion in brackets, [ ], indicates that the metal is part of a complex molecule or ion which is named in the Conditions column.

4. All elements in the Metal Ion column are listed alphabetically in the Metal Index together with their oxidation states and locations in the Table.

D. ΔH

1. The first criterion for construction of the Table was to include all reliable ΔH data. When only one ΔH value was reported for a given reaction, this value is given in the Table irrespective of its apparent validity and can be identified by the single reference. Where sufficient data for a given reaction were available to make a critical evaluation the value or values judged by the authors to be most reliable are the ones reported. In those cases where values reported by different workers disagree widely with no apparent reason all values are given.

2. The order of the thermodynamic quantities is as follows. All values for the first consecutive reaction are presented before the second, etc. Values valid for aqueous solutions are presented first with those for other solvents following, the solvents being arranged alphabetically. For a given solvent, values are arranged in order of increasing ionic strength. For a given ionic strength, values are arranged in order of increasing temperature. This order is given below in outline form.

   a. Reaction

      (1) Consecutive reactions beginning with first
      (2) Overall reactions

   b. Solvent

      (1) Water
      (2) Other solvents arranged alphabetically

c. Ionic Strength, μ

(1) $\mu = 0$
(2) Increasing μ values

d. Temperature - lowest temperature first

3. The ΔH values are valid either at the temperature or over the temperature range listed in the T, °C column.

4. Uncertainties of the ΔH values are not reported when less that 10%; however, the number of significant figures given is that reported in the original article.

E. Log K

1. Each log K value is valid for the reaction designated by the italicized number(s) which is in parentheses immediately following the log K value. These numbers have the following meanings.

Consecutive reactions -- (1) (M + L = ML): (2) (ML + L = ML_2): etc.

Overall reactions -- (1-2) (M + 2L = ML_2): (1-3) (M + 3L = ML_3): (2-3) (2M + 3L = M_2L_3): etc.

If no number appears after the log K value see Conditions or Remarks column for the reaction.

2. Each log K value is valid at the temperature listed in the T, °C column, unless otherwise specified in parentheses immediately following the log K value.

3. Uncertainties of the log K values are not reported. However, the number of significant figures given is that reported in the original article if log K is reported directly. If ΔG or K is reported in the original article then log K is usually reported with one additional significant figure.

F. ΔS

1. The ΔS values are valid at the temperatures listed in the T, °C column if the ΔH and log K values are valid at the same temperature. If the ΔH and log K values are not valid at the same temperature, no attempt has been made to designate the temperature at which ΔS is valid.

2. Uncertainties of the ΔS values are not reported. However, the number of significant figures given is that reported in the original article.

G. Values Calculated by Compilers

1. In those cases where ΔH and only one of the other two thermodynamic quantities, log K or ΔS, are given, the compilers have taken the liberty to calculate the third quantity. This calculated value is underlined in each case.

H. T,°C

1. Where ΔH values were determined calorimetrically, they are valid at the stated temperatures.

2. Where ΔH values were determined by the temperature variation method, the temperature or temperature range is that given by the author(s).

I. Method

1. The method of ΔH determination is designated by C (calorimetric) or T (temperature variation).

J. Conditions

1. The experimental conditions under which ΔH was determined are given. Supporting electrolyte, when present, is identified by its composition and molarity (M). Composition of non-aqueous solvents, when present, is indicated by the symbol, S: followed by the composition of the solvent. In a solvent mixture if the second component is not stipulated it is understood to be water. The letters A, B, etc., indicate ligands attached to the metal as described under heading C.2.

2. If the reaction is not stipulated in parentheses immediately following the log K values (see heading E.1.), it is indicated either in the Conditions column by R: followed by the chemical reaction or in the Remarks column by an alphabetic letter.

3. Abbreviations used are defined under heading N.

4. The units of ΔCp are joule/°K-gram-mole.

K. References

1. The references given are those from which the data were taken.

2. All references in the Table are listed in the Reference Index according to the year of publication. Under each year the references are in alphabetical order according to the authors. In the Reference Index all journal titles are abbreviated in accordance with Chemical Abstracts usage as given in ACCESS, 1969 edition.

L. Additional References

1. These references also contain ΔH values, but the values were not included in the Table since in the opinion of the compilers more reliable, or more extensive data were available for a given set of conditions. However, the compilers do not consider their opinions to be infallible in every case. These references are underlined and appear with the first data listings for each metal. No attempt was made to indicate the reactions for which data are valid in these cases.

M. Remarks

1. When the same comment was applicable to several data entries that comment was given an alphabetic letter. These letters have the following meanings.

   a. ΔH values are calculated from log K values determined at only two temperatures.

   b. ΔH values are average values over the indicated temperature range, resulting in a combination of thermodynamic data valid at different temperatures.

   c. No correction has been made for other species prior to calculation of ΔH. Species, if known, are given in parentheses.

   d. Numbers in the Conditions column refer to the pH of the solution at the ionic strength indicated by the numbers following the letter, e.g., if all the pH measurements were made at $\mu = 0.05$ then d:0.05 appears in the Remarks column.

   e. Questionable whether ΔH value refers to overall or stepwise reaction.

   f. The ΔH value for the stepwise reaction was obtained by dividing the ΔH value for the overall reaction by the number of consecutive reactions, e.g., f:6 where 6 is the number of consecutive reactions.

   g. The salt of the ligand was used. This is indicated by placing the cation formula after the letter, e.g., if the sodium salt of the ligand was used then g:Na appears in the Remarks column.

   h. Indicates temperature range (°C) over which ΔH was assumed constant, e.g., if temperature range was 2-45°C then h:2-45° appears in the Remarks column.

   i. ΔH value was computed from earlier work, e.g., if earlier work has a reference number of 50Ha then i:50Ha appears in the Remarks column.

   j. Reaction not specified or some question as to what reaction applies to ΔH.

   k. ΔH value was calculated from log K values where K was in mole fraction units.

   l. K is in mole fraction units.

   m. $K = [I_2L]/[I_2]L$ where $[I_2L]$ and $[I_2]$ are in concentration units and L is in mole fraction units.

   n. ΔH value appears questionable.

   o. The reaction took place in a two phase system, CCl_4 and H_2O. The reaction was $Ag^+(H_2O) + L(CCl_4) = complex^+(H_2O)$ where the phase is indicated in parentheses after each reactant and product.

p. Data are not internally consistent.

q. The author has placed a large uncertainty of ΔH, e.g., if the uncertainty was reported as ± 3 kjoules/mole then q:±3 appears in the Remarks column.

r. Temperature at which log K is valid is not given or is uncertain.

s. Reaction consists of the complex changing from a tetrahedral to an octahedral configuration.

t. Reaction consists of the complex changing from a planar to a tetrahedral configuration.

u. Temperature at which ΔH is valid is not given. Most probable temperature as estimated by the compilers is given, e.g., if the most probable temperature is 25°C then u:25° appears in the Remarks column.

v. The author has placed a large uncertainty on log K, e.g., if the uncertainty was reported as ± 0.1 log K unit then v:0.1 appears in the Conditions column.

w. Indicates counter ion accompanying positvely charged ligand, e.g., if the counter ion was I^- then w:I appears in the Remarks column.

x. The author has calculated the value of ΔH by combining his data with data in the literature. If other work has a reference number of 50Ha then x:50Ha appears in the Remarks column.

y. The sign on ΔH was changed to make value conform to other reported values.

z. Product of reaction is a solvent separated complex.

aa. Reaction consists of the complex changing from an octahedral to a tetrahedral configuration.

bb. Reaction consists of the complex changing from a <u>cis</u> to a <u>trans</u> configuration.

cc. ΔH value was calculated from K values where K was in pressure units (mm^{-1}).

dd. Data were not internally consistent. Value of log K or ΔS was recalculated from reported value of ΔH and either log K or ΔS. Calculated value is underlined.

ee. The sign on ΔS was changed to make value conform to other reported values.

ff. The sign on ΔG was changed to make value conform to other reported values.

gg. K has the units kg/mole (molality concentration scale).

N. Abbreviations

1. The following abbreviations were used in the Table primarily in the Conditions column.

L = ligand	$\bar{n}$ = average number of ligands bound per metal atom
S = solvent	CM = concentration of metal in moles/liter
R = reaction	CL = concentration of ligand in moles/liter
v = volume	μ = ionic strength
w = weight	D = dielectric constant
M = moles/liter	P = product of reaction
P = pressure	Cp = heat capacity in joules/K mole
	ΔV = volume change in cc/mole

O. Heats of Proton Ionization and Related Thermodynamic Quantities

1. In many cases it is desirable to know related thermodynamic values for proton ionization. A book containing heat of proton ionization data and related thermodynamic quantities for many of the ligands contained in the present Table has been published as the Handbook of Proton Ionization Heats and Related Thermodynamic Quantities, by J. J. Christensen, L. D. Hansen, and R. M. Izatt, John Wiley and Sons, New York, NY, 1976.

A

A1 ACETIC ACID $C_2H_4O_2$ CH_3CO_2H L^-

Metal Ion	ΔH, kJ /mol	Log K	ΔS, J /mol K	T, °C	M	Conditions	Ref.	Remarks
Mg^{2+}	-6.36	1.24(25°,1)	2.5	25-35	T	μ=0	56Na	b
Ca^{2+}	3.81	1.24(25°,1)	36.4	25-35	T	μ=0	56Na	b
Sr^{2+}	3.10	1.16(25°,1)	32.6	25-35	T	μ=0	56Na	b
Ba^{2+}	-9.71	1.15(25°,1)	-10.5	25-35	T	μ=0	56Na	b
Y^{3+}	13.648	1.57(1)	75.7	25	C	μ=2.0($NaClO_4$)	64Gb	
	22.552	2.73(1-2)	128.0	25	C	μ=2.0($NaClO_4$)	64Gb	
	21.937	3.44(1-3)	139.3	25	C	μ=2.0($NaClO_4$)	64Gb	
La^{3+}	9.125	1.58(1)	61.1	25	C	μ=2.0($NaClO_4$)	64Gb	
	13.72	4.733(1)	136.8	25	C	0.5M($NaClO_4$) S:95v%CH_3OH	67Gd	
	10.88	3.571(2)	104.6	25	C	0.5M($NaClO_4$) S:95v%CH_3OH	67Gd	
	4.52	2.589(3)	64.8	25	C	0.5M($NaClO_4$) S:95v%CH_3OH	67Gd	
	1.72	0.85(4)	21.8	25	C	0.5M($NaClO_4$) S:95v%CH_3OH	67Gd	
	15.841	2.52(1-2)	101.2	25	C	μ=2.0($NaClO_4$)	64Gb	
	19.33	3.03(1-3)	123.0	25	C	μ=2.0($NaClO_4$)	64Gb	
Ce^{3+}	9.2	1.708(25°,1)	63.6	2.1-46.6	T	μ=2.0M($NaClO_4$)	70Cg	b,q: ±0.8
	8.753	1.71(1)	61.9	25	C	μ=3.0($NaClO_4$)	64Gb	
	15.330	2.73(1-2)	103.8	25	C	μ=3.0($NaClO_4$)	64Gb	
	21.38	3.20(1-3)	133.1	25	C	μ=3.0($NaClO_4$)	64Gb	
Pr^{3+}	7.192	1.83(1)	59.4	25	C	μ=2.0($NaClO_4$)	64Gb	
	12.09	5.114(1)	138.5	25	C	0.5M($NaClO_4$) S:95v%CH_3OH	67Gd	
	11.04	3.821(2)	110.0	25	C	0.5M($NaClO_4$) S:95v%CH_3OH	67Gd	
	6.53	2.609(3)	72.0	25	C	0.5M($NaClO_4$) S:95v%CH_3OH	67Gd	
	2.76	0.92(4)	27.2	25	C	0.5M($NaClO_4$) S:95v%CH_3OH	67Gd	
	17.380	2.86(1-2)	113.0	25	C	μ=2.0($NaClO_4$)	64Gb	
	15.10	3.33(1-3)	114.2	25	C	μ=2.0($NaClO_4$)	64Gb	
Nd^{3+}	7.146	1.93(1)	61.1	25	C	μ=2.0($NaClO_4$)	64Gb	
	9.20	5.230(1)	131.0	25	C	0.5M($NaClO_4$) S:95v%CH_3OH	67Gd	
	9.25	3.963(2)	107.1	25	C	0.5M($NaClO_4$) S:95v%CH_3OH	67Gd	
	8.12	2.601(3)	77.0	25	C	0.5M($NaClO_4$) S:95v%CH_3OH	67Gd	
	7.15	1.30(4)	1.2	25	C	0.5M($NaClO_4$) S:95v%CH_3OH	67Gd	
	14.585	3.06(1-2)	107.5	25	C	μ=2.0($NaClO_4$)	64Gb	
	18.24	3.52(1-3)	128.4	25	C	μ=2.0($NaClO_4$)	64Gb	
Sm^{3+}	6.079	2.03(1)	59.4	25	C	μ=2.0($NaClO_4$)	64Gb	
	11.59	5.137(1)	137.2	25	C	0.5M($NaClO_4$)	67Gd	

Metal Ion	ΔH, kJ /mol	Log K	ΔS, J /mol K	T,°C	M	Conditions	Ref.	Remarks
Al, continued								
Sm^{3+}, cont.						S:95v%CH_3OH		
	6.44	4.059(2)	99.2	25	C	0.5M($NaClO_4$) S:95v%CH_3OH	67Gd	
	8.08	2.694(3)	78.6	25	C	0.5M($NaClO_4$) S:95v%CH_3OH	67Gd	
	-1.84	1.11(4)	15.1	25	C	0.5M($NaClO_4$) S:95v%CH_3OH	67Gd	
	12.033	3.29(1-2)	103.3	25	C	μ=2.0($NaClO_4$)	64Gb	
	15.786	3.90(1-3)	127.6	25	C	μ=2.0($NaClO_4$)	64Gb	
Eu^{3+}	5.8	1.903(25°,1)	56.1	0-55	T	μ=2.0M($NaClO_4$)	70Cg	b,q:±0.4
Gd^{3+}	78.157	1.86(1)	61.9	25	C	μ=2.0($NaClO_4$)	64Gb	
	14.14	4.977(1)	142.7	25	C	0.5M($NaClO_4$) S:95v%CH_3OH	67Gd	
	7.15	4.000(2)	100.4	25	C	0.5M($NaClO_4$) S:95v%CH_3OH	67Gd	
	6.65	2.687(3)	74.0	25	C	0.5M($NaClO_4$) S:95v%CH_3OH	67Gd	
	-1.42	1.16(4)	17.2	25	C	0.5M($NaClO_4$) S:95v%CH_3OH	67Gd	
	13.577	3.16(1-2)	105.8	25	C	μ=2.0($NaClO_4$)	64Gb	
	14.824	3.76(1-3)	121.8	25	C	μ=2.0($NaClO_4$)	64Gb	
Dy^{3+}	12.238	1.71(1)	73.6	25	C	μ=2.0($NaClO_4$)	64Gb	
	17.11	5.026(1)	153.6	25	C	0.5M($NaClO_4$) S:95v%CH_3OH	67Gd	
	11.38	3.987(2)	114.6	25	C	0.5M($NaClO_4$) S:95v%CH_3OH	67Gd	
	4.18	2.777(3)	66.9	25	C	0.5M($NaClO_4$) S:95v%CH_3OH	67Gd	
	1.09	1.39(4)	30.5	25	C	0.5M($NaClO_4$) S:95v%CH_3OH	67Gd	
	18.573	3.03(1-2)	120.1	25	C	μ=2.0($NaClO_4$)	64Gb	
	18.028	3.84(1-3)	133.9	25	C	μ=2.0($NaClO_4$)	64Gb	
Ho^{3+}	13.251	1.67(1)	76.6	25	C	μ=3.0($NaClO_4$)	64Gb	
	20.958	2.92(1-2)	126.36	25	C	μ=3.0($NaClO_4$)	64Gb	
	18.966	3.80(1-3)	136.4	25	C	μ=3.0($NaClO_4$)	64Gb	
Er^{3+}	13.707	1.64(1)	77.4	25	C	μ=2.0($NaClO_4$)	64Gb	
	19.16	3.150(1)	162.8	25	C	0.5M($NaClO_4$) S:95v%CH_3OH	67Gd	
	12.21	3.965(2)	116.7	25	C	0.5M($NaClO_4$) S:95v%CH_3OH	67Gd	
	7.86	2.760(3)	79.5	25	C	0.5M($NaClO_4$) S:95v%CH_3OH	67Gd	
	-14.6	1.19(4)	-25.9	25	C	0.5M($NaClO_4$) S:95v%CH_3OH	67Gd	
	23.050	2.90(1-2)	133.0	25	C	μ=2.0($NaClO_4$)	64Gb	
	21.614	3.72(1-3)	143.5	25	C	μ=2.0($NaClO_4$)	64Gb	
Yb^{3+}	14.673	1.68(1)	81.6	25	C	μ=2.0($NaClO_4$)	64Gb	
	20.25	5.530(1)	173.6	25	C	0.5M($NaClO_4$) S:95v%CH_3OH	67Gd	
	14.64	4.472(2)	134.7	25	C	0.5M($NaClO_4$) S:95v%CH_3OH	67Gd	

Metal Ion	ΔH, kJ /mol	Log K	ΔS, J /mol K	T,°C	M	Conditions	Ref.	Re-marks
A1, continued								
Yb^{3+}, cont.	17.45	3.368(3)	123.0	25	C	0.5M($NaClO_4$) S:95v%CH_3OH	67Gd	
	-0.46	2.15(4)	39.3	25	C	0.5M($NaClO_4$) S:95v%CH_3OH	67Gd	
	25.376	2.91(1-2)	141.0	25	C	μ=2.0($NaClO_4$)	64Gb	
	27.53	3.63(1-3)	161.9	25	C	μ=2.0($NaClO_4$)	64Gb	
Th^{4+}	11.3	3.85(1)	112	25	C	μ=1.0M($NaClO_4$)	75Pe	
	4.48	3.12(2)	74.5	25	C	μ=1.0M($NaClO_4$)	75Pe	
	13.7	1.97(3)	83.7	25	C	μ=1.0M($NaClO_4$)	75Pe	
	5.19	1.36(4)	44.6	25	C	μ=1.0M($NaClO_4$)	75Pe	
	3.72	0.71(5)	25.9	25	C	μ=1.0M($NaClO_4$)	75Pe	
$UO_2{}^{2+}$	10.539	2.420(1)	81.6	25	C	μ=1.0($NaClO_4$)	71Ac	
	11.8	2.46(1)	86.6	25	C	μ=1.0M($NaClO_4$)	74Ph	
	9.66	1.995(2)	70.7	25	C	μ=1.0($NaClO_4$)	71Ac	
	6.07	1.92(2)	57.3	25	C	μ=1.0M($NaClO_4$)	74Ph	
	-4.02	1.983(3)	24.52	25	C	μ=1.0($NaClO_4$)	71Ac	
	-1.21	2.14(3)	36.8	25	C	μ=1.0M($NaClO_4$)	74Ph	
Am^{3+}	18.0	1.954(25°,1)	97.9	0-55	T	μ=2.0M($NaClO_4$)	70Cg	b,q:±1.2
Cm^{3+}	18.0	2.033(25°,1)	99.6	0-55	T	μ=2.0M($NaClO_4$)	70Cg	b
Bk^{3+}	18.4	2.029(25°,1)	100.8	0.7-39.8	T	μ=2.0M($NaClO_4$)	70Cg	b
Cf^{3+}	15.9	2.124(25°,1)	93.7	0-55.2	T	μ=2.0M($NaClO_4$)	70Cg	b,q:±1.2
CoA_5B^{2+}	15.9	-0.49(1)	44	25	C	μ=3.0M($NaClO_4$) A=NH_3 B=Br	74Mb	
	20.9	-0.65(1)	63	25	C	μ=3.0M($NaClO_4$) A=NH_3 B=Cl	74Mb	
	-15.9	-0.04(2)	-54	25	C	μ=3.0M($NaClO_4$) A=NH_3 B=Br	74Mb	
	-20.9	-0.04(2)	-75	25	C	μ=3.0M($NaClO_4$) A=NH_3 B=Cl	74Mb	
Cu^{2+}	5.19	1.33(1)	42.7	25	C	μ=1M($NaNO_3$)	74Ad	
	4.35	1.88(1)	50.6	25	C	3M($NaClO_4$)	67Gc	e
	13.6	2.82(30°,1)	97.1	20&30	T	μ=0 S:50v%Ethanol	76Ca	a,b
	3.39	0.98(2)	30.1	25	C	μ=1M($NaNO_3$)	74Ad	
	1.72	1.24(2)	29.7	25	C	3M($NaClO_4$)	67Gc	e
	47.6	2.36(30°,2)	205.6	20&30	T	μ=0 S:50v%Ethanol	76Ca	a,b
	0.29	0.47(3)	10.0	25	C	3M($NaClO_4$)	67Gc	e
CuL_2	-21	5.00(25°)	25	25&70	T	S:HL with 0.2M(H_2O) R:2CuL_2=Cu_2L_4	71Gj	a,b,q: ±4
	-29	3.00(25°)	-46	25&70	T	S:Ethanol with 0.1M(H_2O) R:2CuL_2=Cu_2L_4	71Gj	a,b,q: ±8
Ag^{+}	5.0	0.683(1)	30	25	C	μ=0.5M($NaClO_4$)	75Mb	q:±2.7
Zn^{2+}	8.54	0.946(1)	46.9	25	C	3M($NaClO_4$)	67Gc	e
	13.47	0.396(2)	52.3	25	C	3M($NaClO_4$)	67Gc	e
	4.60	0.506(3)	25.5	25	C	3M($NaClO_4$)	67Gc	e
Cd^{2+}	7.6	1.772(25°,1)	59	5-35	T	μ=0	75Cc	b

Metal Ion	ΔH, kJ /mol	Log K	ΔS, J /mol K	T, °C	M	Conditions	Ref.	Re- marks
A1, continued								
Cd^{2+}, cont.	7.07	1.27(1)	48.1	25	C	μ=0.25($NaClO_4$)	68Gd 67Gc	
	7.36	1.19(1)	47.3	25	C	μ=0.5($NaClO_4$)	68Gd	
	6.78	1.17(1)	45.2	25	C	μ=1($NaClO_4$)	68Gd	
	6.28	1.24(1)	44.8	25	C	μ=2($NaClO_4$)	68Gd	
	6.12	1.30(1)	45.6	25	C	μ=3($NaClO_4$)	68Gd	
	11.7	0.592(25°,2)	50.6	5-35	T	μ=0	75Cc	b
	6.15	0.71(2)	34.3	25	C	μ=0.5($NaClO_4$)	68Gd	
	5.15	0.64(2)	29.7	25	C	μ=1($NaClO_4$)	68Gd	
	4.52	0.74(2)	29.3	25	C	μ=2($NaClO_4$)	68Gd	
	3.51	0.97(2)	30.5	25	C	μ=3($NaClO_4$)	68Gd	
	-2.51	0.22(3)	-4.2	25	C	μ=1($NaClO_4$)	68Gd	
	-0.29	0.15(3)	2.1	25	C	μ=2($NaClO_4$)	68Gd	
	1.72	0.14(3)	8.4	25	C	μ=3($NaClO_4$)	68Gd	
CdA	-75	3.63(25°)	-184	25-45	T	L=0.02M S:Ethanol A=Phaeophytin R:CdA+2HL= CdL_2+H_2A	71Bh	b
Hg^{2+}	7.6	5.52(1)	131	25	C	μ=0.5M($NaClO_4$)	75Mb	
	7.5	2.54(2)	74	25	C	μ=0.5M($NaClO_4$)	75Mb	
Pb^{2+}	-1.7	1.83(25°,1)	29.41	15-35	T	μ=0.75	67Hi	b
	-3.3	2.037(25°,1)	27.78	15-35	T	μ=1.0	67Hi	b
	-4.6	2.238(25°,1)	27.40	15-35	T	μ=2.0	67Hi	b
	-0.25	2.33(1)	43.9	25	C	3M($NaClO_4$)	67Gc	e
	-4.2	2.740(25°,2)	38.41	15-35	T	μ=0.75	67Hi	b
	-3.3	2.914(25°,2)	44.4	15-35	T	μ=1.0	67Hi	b
	-2.9	3.299(25°,2)	53.1	15-35	T	μ=2.0	67Hi	b
	-0.38	1.27(2)	23.0	25	C	3M($NaClO_4$)	67Gc	e
	-5.8	3.243(25°,3)	42.2	15-35	T	μ=0.75	67Hi	b
	-8.4	3.581(25°,3)	40.38	15-35	T	μ=1.0	67Hi	b
	-8.8	3.956(25°,3)	46.4	15-35	T	μ=2.0	67Hi	b
	-4.02	-0.015(3)	-13.8	25	C	3M($NaClO_4$)	67Gc	e
A2	ACETIC ACID, amide		C_2H_5ON			CH_3CONH_2		L
I_2	-19.29	-0.215(22°,1)	-69.4	0&22	T	CM=0.35 S:CH_2Cl_2	61Ta	a,b
A3	ACETIC ACID, amide, N-acetyl-N-phenyl-		$C_{10}H_{12}O_2N$			$(CH_3CO)_2NC_6H_5$		L^-
Be^{2+}	-7.1	8.69(25°,1)	146	10-40	T	S:50% Dioxane	69Ha	b
	0	7.23(25°,2)	155	10-40	T	S:50% Dioxane	69Ha	b
A4	ACETIC ACID, amide, N-benzoyl-N-(m-nitrophenyl)-		$C_{15}H_{11}O_4N_2$			$C_6H_5CON(C_6H_5NO_2)COCH_3$		HL
Fe^{3+}	-20.79	6.97(25°)	63.68	15-35	T	S:60w%Ethanol μ=0.1 pH=1.8-4.2	72Ta	b,j

Metal Ion	ΔH, kJ /mol	Log K	ΔS, J /mol K	T,°C	M	Conditions	Ref.	Re-marks
A4, continued								
Fe^{3+}, cont.						R:Fe^{3+}+2HL= FeL_2^{+}+$2H^{+}$		
A5 ACETIC ACID, amide, N,N-diethyl $C_6H_{13}ON$						$CH_3CON(C_2H_5)_2$		L
I_2	-22.6	1.27(25°,1)	-51.5	5-25	T	S:Heptane	71Td	b
A6 ACETIC ACID, amide, N,N-dimethyl C_4H_9ON						$CH_3CON(CH_3)_2$		L
$Rh_2A_4B_2$	-22.2	1.73	-41	24	C	S:Benzene $Rh_2A_4B_2$+2L= $2RhA_2BL$ A=CO B=Cl	73Pd	
CuA_2	-33.5	3.217(1)	-53.6	25	C	S:CCl_4 A=Hexafluoro-acetylacetone	70Pa	
	-33.5	3.23(1)	--	--	C	S:CCl_4 A=Hexafluoro-acetylacetone	76Mc	μ:25°
	-27.6	(1)	--	25	C	S:O-Dichloro-benzene A=Hexafluoro-acetylacetone	70Pa	
	-24.7	(1)	--	25	C	S:Methylene chloride A=Hexafluoro-acetylacetone	74Dc	
	-21.8	0.964(2)	-54.4	25	C	S:CCl_4 A=Hexafluoro-acetylacetone	70Pa	
CuA_2B	-8.8	--	--	25	C	S:CCl_4 A=Hexafluoro-acetylacetone B=Ethyl acetate R:CuA_2B+L=CuA_2L+B	72Nd	
	-7.9	--	--	25	C	S:o-$Cl_2C_6H_4$ A=Hexafluoro-acetylacetone B=Ethyl acetate R:CuA_2B+L=CuA_2L+B	72Nd	
ZnA_2	-39.3	2.33(1)	-87.2	25	C	S:Benzene A=N[Si$(CH_3)_3$]$_2$	75Fc	
CdA_2	-33.9	2.08(1)	-73.9	25	C	S:Benzene A=N[Si$(CH_3)_3$]$_2$	75Fc	
SnA_3B	-33.0	0.59(1)	-100	26	C	S:CCl_4 A=CH_3 B=Cl	66Be	
	-32.2	0.54(1)	-96	27	C	S:CCl_4 A=CH_3 B=Cl	65Bb	

Metal Ion	ΔH, kJ /mol	Log K	ΔS, J /mol K	T,°C	M	Conditions	Ref.	Remarks
A6, continued								
SnA_3B, cont.	-39.3	0.193(1)	-126	26	C	S:N,N-Dimethyl-acetamide A=CH_3 B=Cl	66Bf	
SbA_5	-116.32	(1)	--	25	C	CM<0.01 S:Ethylene chloride A=Cl	64Ob	
Br_2	-6.7	0.255(25°,1)	-17	5-40	T	S:CCl_4	62Db	b
I_2	-17.32	0.710(38°,1)	42	25.2-39.5	T	S:Benzene	66Dc	b
	-16.7	0.839(25°,1)	39.3	29.4-46.7	T	S:CCl_4	60Sf 61Db	b
	-22.2	1.18(25°,1)	-51.9	5-25	T	S:Heptane	71Td	b
IA	-39.7	3.042(25°,1)	-71	28-46	T	CM=10^{-2}M CL=5x10^{-2}M S:CCl_4 A=Cl	62Db	b
A7 ACETIC ACID, amide, N-phenyl-N-benzoyl			$C_{15}H_{13}O_2N$			$C_6H_5CON(C_6H_5)COCH_3$		L
Fe^{3+}	-10.38	8.67(25°,1-3)	131.13	20-35	T	S:60% Ethanol pH=3.0	71Tb	b
A8 ACETIC ACID, amide, thiono			C_2H_5NS			CH_3CSNH_2		L
I_2	-34.3	4.398(25°,1)	-39.7	2.5 &17	T	CM=1x10^{-4}M CL=10^{-4}-3x10^{-4}M S:CH_2Cl_2	62La	a,b
A9 ACETIC ACID, amide, thiono N,N-dimethyl			C_4H_9NS			$CH_3CSN(CH_3)_2$		L
SnA_3B	-24.7	-0.322(1)	-88	26	C	S:CCl_4 A=CH_3 B=Cl	66Be	
I_2	-39.7	3.076(25°,1)	-74.5	25-40	T	CM=10^{-3}M CL=10^{-3}M S:CCl_4	64Na	b
	-38.67	3.18(25°,1)	-69.4	--	T	S:CCl_4	70Gj	b,u:25°
A10 ACETIC ACID, amide, thiono, N-methyl-			C_3H_7NS			$CH_3C(S)NHCH_3$		
I_2	-33.5	2.732(25°,1)	-65.3	0-45	T	S:CH_2Cl_2	66Bc	b,q:±4
A11 ACETIC ACID, bromide			C_2H_3OBr			CH_3COBr		L
I_2	-3.8	-0.89(25°,1)	-29	5-25	T	S:Heptane	71Td	b

Metal Ion	ΔH, kJ /mol	Log K	ΔS, J /mol K	T,°C	M	Conditions	Ref.	Re-marks
A12	ACETIC ACID, tert-butyl ester		$C_6H_{12}O_2$			$CH_3CO_2C(CH_3)_3$		L
SbA_5	-150.2	(1)	--	25	C	Excess $SbCl_5$ CL low, variable, ~0.02M S:1,2-Dichloroethane A=Cl	640a	
A13	ACETIC ACID, chloride		C_2H_3OCl			CH_3COCl		L
I_2	-4.6	-0.92(25°,1)	-32.6	5-25	T	S:Heptane	71Td	b
A14	ACETIC ACID, ethenyl ester		$C_4H_6O_2$			$CH_3CO_2CH:CH_2$		
NiAB	-57.3	2.49(27°)	-143.5	18-42	T	S:Tetrahydrofuran A=Dipyridyl B=Tetrahydrofuran R:NiAB+L=NiAL+B	71Ya	b
A15	ACETIC ACID, ethyl ester		$C_4H_8O_2$			$CH_3CO_2CH_2CH_3$		L
$[NiA_2]_2$	-17.4	1.35(100°,1)	-21	75-125	T	S:Squalane A=3-Trifluoro-acetylcamphorate	74Sj	b,gg
$Rh_2A_4B_2$	-1.2	--	--	24	C	S:Benzene R:$Rh_2A_4B_2$+2L= $2RhA_2BL$ A=CO B=Cl	74Pd	
CuA_2	-24.7	2.290(1)	-38.9	25	C	S:CCl_4 A=Hexafluoro-acetylacetone	70Pa	
	-24.7	2.28(1)	--	--	C	S:CCl_4 A=Hexafluoro-acetylacetonato	76Mc	u:25°
	-19.7	1.545(1)	-36.4	25	C	S:o-Dichloro-benzene A=Hexafluoro-acetylacetone	72Nd	q:±1.3
	-27.6	(1)	--	--	C	S:o-$Cl_2C_6H_4$ A=Hexafluoro-acetylacetonato	76Mc	u:25°
SbA_5	-64.8	(1)	--	25	C	S:CCl_4 A=Cl	72Lb	
	-66.5	(1)	--	25	C	S:CCl_4 A=Cl	730a	
	-69.4	(1)	--	25	C	S:1,2-Dichloro-ethane A=Cl	72Lb	
	-71.46	(1)	--	25	C	S:1,2-Dichloro-ethane A=Cl	630a	
I_2	-8.58	(1)	--	--	C	S:Ethyl acetate	50Ha 09Ha	
	-12.6	-0.03(25°,1)	-43.9	5-25	T	S:Heptane	71Td	b
IA	-25.5	0.769(1)	-70.62	26	C	CM=0.03M	68Bb	

Metal Ion	ΔH, kJ /mol	Log K	ΔS, J /mol K	T,°C	M	Conditions	Ref.	Remarks
A15, continued								
IA, cont.						CL=0.1-0.9M S:CCl_4 A=Cl		
A16 ACETIC ACID, isopropenyl ester $C_5H_8O_2$						$CH_3CO_2C(CH_3):CH_2$		L
NiAB	-36.4	1.07(27°)	-100.8	17-42	T	S:Tetrahydrofuran A=Dipyridyl B=Tetrahydrofuran R:NiAB+L=NiAL+B	71Ya	b
A17 ACETIC ACID, methyl ester $C_3H_6O_2$						$CH_3CO_2CH_3$		L
$[NiA_2]_2$	-19.2	1.22(100°,1)	-28	75-125	T	S:Squalane A=3-Trifluoro-acetylcamphorate	74Sj	b,gg
SbA_5	-68.53	(1)	--	25	C	S:1,2-Dichloro-ethane A=Cl	63Oa	
I_2	-10.5	-0.284(25°,1)	-40.71	25-45	T	CM=1.9x10^{-3}M CL=0.5M S:CCl_4	64Ma	b
	-13.0	-0.12(25°,1)	-46	5-25	T	S:Heptane	71Td	b
	-10.0	(1)	--	--	C	S:Methyl acetate	50Ha	
A18 ACETIC ACID, nitrile C_2H_3N						CH_3CN		L
CoA_2L_2	-61.1	-1.506	--	-38-29	T	S:Acetonitrile CM=0.2-0.4M A=Cl R:CoL_2A_2+2L=CoL_4A_2	72Bf	b,l
	-76.1	-2.821	--	-34-49	T	S:Acetonitrile CM=0.2-0.4M A=Br R:CoL_2A_2+2L=CoL_4A_2	72Bf	b,l
	-74.9	-2.966	--	-33-50	T	S:Acetonitrile CM=0.2-0.4M A=I R:CoL_2A_2+2L=CoL_4A_2	72Bf	b,l
	61.1	-0.420	--	-38-29	T	S:Acetonitrile CM=0.2-0.4M A=Cl R:2CoL_2A_2+2L=CoL_6^{2+}+CoA_4^{2-}	72Bf	b,l
	46.9	-0.468	--	-34-49	T	S:Acetonitrile CM=0.2-0.4M A=Br R:2CoL_2A_2+2L=CoL_6^{2+}+CoA_4^{2-}	72Bf	b,l
	32.01	-0.409	--	-33-50	T	S:Acetonitrile CM=0.2-0.4M A=I R:2CoL_2A_2+2L=CoL_6^{2+}+CoA_4^{2-}	72Bf	b
NiA	-29.01	0.947(25°,1-2)	-79.1	5-110	T	S:Acetonitrile	79Vb	

Metal Ion	ΔH, kJ /mol	Log K	ΔS, J /mol K	T,°C	M	Conditions	Ref.	Re-marks
A18, continued								
NiA, cont.						A=1,4,8,11-Tetraazacyclo-tetradecane		
NiA_2B	-18.0	-1.097(22°,1)	-84	-50-50	T	S:Acetonitrile (L) $A=C_2H_5$ B=2,2'-Bipyridine	76Yc	b
IrA_2BC_2	-70.7	1.74(1)	-209	25	C	$S:CH_2Cl_2$ A=Cl $B=C(O)CF_3$ $C=P(C_6H_5)_3$	77La	
BA_3	-88	(1)	--	25	C	S:Acetonitrile A=I	73Bl	q:±10
SnA_4	-25.1	2.20(1)	-42.2	25	C	S:Benzene A=Cl	70Gh	
	-25.5	(1)	--	--	C	S:Benzene CM=0.05-0.08M A=Cl	68Ge	
$SnAB_3$	-20.1	-0.260(1)	-72.0	26	C	$S:CCl_4$ A=Cl $B=CH_3$	66Be	
SbA_5	-61.1	(1)	--	25	C	$S:C_2H_4Cl_2$ A=Cl	72Lb	
IA	-20.5	0.924(25°,1)	-51.5	0-40	T	$S:CCl_4$ A=Cl	63Pc	b
	-17.2	0.170(25°,1)	-53.6	0-40	T	$S:CCl_4$ A=Br	63Pc	b
I_2	-7.9	-0.38(25°,1)	-34.7	0-40	T	$S:CCl_4$	63Pc	b
	-12.22	-0.336(25°)	-47.28	--	T	$S:CCl_4$ CL=0.2-1.0M $CM=2x10^{-3}M$	72Ma	b,j
A19 ACETIC ACID, 2-propyl ester		$C_5H_{10}O_2$				$CH_3CO_2CH_2CH_2CH_3$		L
SbA_5	-73.34	(1)	--	25	C	S:1,2-Dichloro-ethane A=Cl	63Oa	
A20 ACETIC ACID, 4-tolyl ester		$C_9H_{10}O_2$				$(CH_3)C_6H_4(CO_2CH_3)$		L
TiA_4	-36.0	2.602(25°,1)	-71	25&60	T	μ=0 A=Cl	66Gc	a,b,q ±7.5
A21 ACETIC ACID, n-butyl-(hydroxy)-(thio)-		$C_6H_{12}O_3S$				$(C_4H_9)(OH)C(SH)CO_2H$		L^-
Tl^+	-45.2	1.26(30°,1)	-121	30&40	T	μ=0.1M S:50v%Ethanol	76Sm	a,b
	6.90	0.36(30°,2)	29.2	30&40	T	μ=0.1M S:50v%Ethanol	76Sm	a,b
	22.9	0.79(30°,3)	88.4	30&40	T	μ=0.1M S:50v%Ethanol	76Sm	a,b
Pb^{2+}	2.59	1.94(27°,1)	66.9	27&37	T	$μ=0.1M(NaClO_4)$ S:50v%Dimethyl formamide	76Sn	a,b,n
	-21.4	2.62(27°,2)	-65.6	27&37	T	$μ=0.1M(NaClO_4)$	76Sn	a,b,n

Metal Ion	ΔH, kJ /mol	Log K	ΔS, J /mol K	T, °C	M	Conditions	Ref.	Remarks
A21, continued								
Pb^{2+}, cont.						S:50v%Dimethyl formamide		
	12.3	2.80(27°,3)	80.7	27&37	T	μ=0.1M($NaClO_4$) S:50v%Dimethyl formamide	76Sn	a,b,n
A22 ACETIC ACID, chloro			$C_2H_3O_2Cl$			$ClCH_2CO_2H$		L^-
UO_2^{2+}	8.08	1.44(1)	54.4	25	C	μ=1.0M($NaClO_4$)	74Ph	
	7.99	0.80(2)	42.2	25	C	μ=1.0M($NaClO_4$)	74Ph	
	8.28	0.33(3)	34.3	25	C	μ=1.0M($NaClO_4$)	74Ph	
Th^{4+}	12.22	2.749(1)	93.7	25	C	μ=1.0M($NaClO_4$)	78Dc	
	13.01	1.884(2)	79.5	25	C	μ=1.0M($NaClO_4$)	78Dc	
	10.71	1.151(3)	58.2	25	C	μ=1.0M($NaClO_4$)	78Dc	
	7.99	0.748(4)	41.0	25	C	μ=1.0M($NaClO_4$)	78Dc	
A23 ACETIC ACID, chloro-, amide, N,N-dimethyl-			C_4H_8ONCl			$ClCH_2CON(CH_3)_2$		L
I_2	-13.8	-0.114(25°,1)	-43.9	25-43	T	CM=1.7x10^{-3}M CL=0.1-0.2M S:CCl_4	62Dc	b
A24 ACETIC ACID, chloro-, ethyl ester			$C_4H_7O_2Cl$			$ClCH_2CO_2CH_2CH_3$		L
SbA_5	-53.6	(1)	--	25	C	S:CCl_4 A=Cl	73Oa	
	-53.56	2.18(1)	-137.6	25	C	S:Ethylene chloride A=Cl	67Od	
A25 ACETIC ACID, chloro-, nitrile			C_2H_2NCl			$ClCH_2CN$		L
SbA_5	-40.2	(1)	--	25	C	S:CCl_4 A=Cl	72Lb	
IA	-22.2	0.382(25°,1)	-67	0-40	T	S:CCl_4 A=Cl	63Pc	b
	-13.0	-0.260(25°,1)	-47.7	0-40	T	S:CCl_4 A=Br	63Pc	b
I_2	-6.3	-0.72(25°,1)	-35.1	0-40	T	S:CCl_4 A=Br	63Pc	b
A26 ACETIC ACID, dichloro- ethyl ester			$C_4H_6O_2Cl_2$			$Cl_2CHCO_2CH_2CH_3$		L
SbA_5	-39.1	0.30(1)	-125.5	25	C	S:Ethylene chloride A=Cl	67Od	
A27 ACETIC ACID, dichloro-, nitrile			C_2HNCl_2			Cl_2CHCN		L
IA	-15.5	-0.161(25°,1)	-55.6	0-40	T	S:CCl_4 A=Cl	63Pc	b
	-8.4	-0.638(25°,1)	-40.6	0-40	T	S:CCl_4 A=Br	63Pc	b

Metal Ion	ΔH, kJ /mol	Log K	ΔS, J /mol K	T,°C	M	Conditions	Ref.	Remarks
A28	ACETIC ACID, 2,2-dimercapto-2-glycolyl		$C_4H_6O_4S_2$			$HOCH_2COC(SH)_2COOH$		
Cu^{2+}	-24.8	7.56(30°,1)	60.5	20-40	T	μ=0.1M($NaClO_4$) S:25v%Ethanol	76Sk	b
Ag^{+}	-124.7	8.33(30°,1-2)	-254	20-40	T	μ=0.1M($NaClO_4$) S:25v%Ethanol	76Sk	b
A29	ACETIC ACID, ethoxy		$C_4H_8O_3$			$C_2H_5OCH_2CO_2H$		L^-
Ni^{2+}	4.06	1.02(1)	33.0	25	C	μ=1.0($NaNO_3$)	77Ao	
	4.98	0.48(2)	25.9	25	C	μ=1.0($NaNO_3$)	77Ao	q:±0.5
Cu^{2+}	2.09	1.80(1)	41.4	25	C	μ=1.0($NaNO_3$)	77Ao	
	4.98	1.09(2)	37.6	25	C	μ=1.0($NaNO_3$)	77Ao	q:±0.5
	3.72	0.34(3)	18.8	25	C	μ=1.0($NaNO_3$)	77Ao	q:±2.7
Zn^{2+}	4.56	1.13(1)	36.8	25	C	μ=1.0($NaNO_3$)	77Ao	
	4.94	0.74(2)	30.5	25	C	μ=1.0($NaNO_3$)	77Ao	q:±2.3
	15.98	-0.09(3)	51.9	25	C	μ=1.0($NaNO_3$)	77Ao	q:±29.5
Cd^{2+}	4.73	1.07(1)	36.4	25	C	μ=1.0($NaNO_3$)	77Ao	q:±0.54
	0.75	0.62(2)	14.2	25	C	μ=1.0($NaNO_3$)	77Ao	q:±4.8
	43.97	-0.15(3)	144.8	25	C	μ=1.0($NaNO_3$)	77Ao	q:±58
A30	ACETIC ACID, (ethylthio)		$C_4H_8O_2S$			$C_2H_5SCH_2CO_2H$		L^-
Ni^{2+}	1.38	1.04(1)	24.7	25	C	μ=1.0($NaNO_3$)	77Ao	
	1.38	1.04(1)	24.5	25	C	μ=1.0($NaNO_3$)	79Ac	
	1.30	0.77(2)	19.2	25	C	μ=1.0($NaNO_3$)	77Ao	
	-1.55	0.46(3)	3.8	25	C	μ=1.0($NaNO_3$)	77Ao	q:±2.3
Cu^{2+}	2.30	2.56(1)	56.9	25	C	μ=1.0($NaNO_3$)	77Ao	
	2.30	2.57(1)	56.9	25	C	μ=1.0($NaNO_3$)	79Ac	
	0.17	2.20(2)	42.7	25	C	μ=1.0($NaNO_3$)	77Ao	
	-8.16	0.06(3)	-26.4	25	C	μ=1.0($NaNO_3$)	77Ao	
Zn^{2+}	5.10	0.74(1)	31.4	25	C	μ=1.0($NaNO_3$)	77Ao	
	5.10	0.74(1)	31.3	25	C	μ=1.0($NaNO_3$)	79Ac	
	-0.25	0.46(2)	7.9	25	C	μ=1.0($NaNO_3$)	77Ao	q:±1.2
	40.88	-0.06(3)	136.0	25	C	μ=1.0($NaNO_3$)	77Ao	q:±11.7
Cd^{2+}	3.05	1.27(1)	34.7	25	C	μ=1.0($NaNO_3$)	77Ao	
	3.05	1.27(1)	34.5	25	C	μ=1.0($NaNO_3$)	79Ac	
	0.92	0.85(2)	19.2	25	C	μ=1.0($NaNO_3$)	77Ao	q:±0.8
	-0.08	0.38(3)	7.1	25	C	μ=1.0($NaNO_3$)	77Ao	q:±3.2
A31	ACETIC ACID, (ethylthio) hydroxide-		$C_4H_9O_3S$			$CH_3CH_2SCH_2(OH)CO_2H$		L^-
Tl^{+}	-6.7	1.708(30°,1)	10.0	20&30	T	μ=0.5M($NaClO_4$)	72Sd	a,b,j
	-3.97	1.67(30°,1)	18.91	20&30	T	μ=2M($NaClO_4$)	74Ma	a,b
	-0.8	0.246(30°,2)	-0.4	20&30	T	μ=0.5M($NaClO_4$)	72Sd	a,b,j
	-1.46	0.72(30°,2)	9.08	20&30	T	μ=2M($NaClO_4$)	74Ma	a,b,i: 72Sd

Metal Ion	ΔH, kJ /mol	Log K	ΔS, J /mol K	T,°C	M	Conditions	Ref.	Re-marks
A31, continued								
Tl^{+}, cont.	-0.8	1.350(30°,3)	23.8	20&30	T	μ=0.5M($NaClO_4$)	74Ma	a,b,i: 72Sd
	-1.30	0.55(30°,3)	5.86	20&30	T	μ=2M($NaClO_4$)	74Ma	a,b,i: 72Sd
Pb^{2+}	22.2	2.00(30°,1)	113	20&30	T	μ=0.5M($NaClO_4$)	71Sa 72Sc	a,b,j
	13.4	4.477(30°,1-2)	130	20&30	T	μ=0.5M($NaClO_4$)	71Sa 72Sc	a,b,j
	11.7	6.832(30°,1-3)	172	20&30	T	μ=0.5M($NaClO_4$)	71Sa 72Sc	a,b,j
A32 ACETIC ACID, hydroxy-			$C_2H_4O_3$			$HOCH_2CO_2H$		L^-
Ca^{2+}	35.52	1.65(25°,1)	150.8	25-45	T	μ=0 pH=3.6-3.8	75Da	b
Y^{3+}	-0.63	2.36(1)	43.1	25	C	μ=2.0($NaClO_4$)	66Cc	
	-0.305	2.47(1)	46.4	25	C	μ=2.0($NaClO_4$)	64Gb	
	-1.05	1.99(2)	34.7	25	C	μ=2.0($NaClO_4$)	66Cc	
	-0.418	1.93(2)	35.6	25	C	μ=2.0($NaClO_4$)	64Gb	
	-5.10	1.23(3)	6.3	25	C	μ=2.0($NaClO_4$)	66Cc	
	-3.021	1.29(3)	14.2	25	C	μ=2.0($NaClO_4$)	64Gb	
	-0.105	0.55(4)	10.5	25	C	μ=2.0($NaClO_4$)	64Gb	
La^{3+}	-3.56	2.14(1)	29	25	C	μ=2.0($NaClO_4$)	66Cc	
	-2.648	2.18(1)	33.0	25	C	μ=2.0($NaClO_4$)	64Gb	
	-0.96	1.80(2)	38	25	C	μ=2.0($NaClO_4$)	66Cc	
	-1.766	1.56(2)	23.8	25	C	μ=2.0($NaClO_4$)	64Gb	
	-2.615	1.05(3)	11.3	25	C	μ=2.0($NaClO_4$)	64Gb	
	-2.13	0.27(4)	-2.1	25	C	μ=2.0($NaClO_4$)	64Gb	
Ce^{3+}	-3.389	2.34(1)	33.5	25	C	μ=2.0($NaClO_4$)	64Gb 61Ca	
	-3.301	1.66(2)	20.5	25	C	μ=2.0($NaClO_4$)	64Gb	
	-2.799	1.12(3)	12.1	25	C	μ=2.0($NaClO_4$)	64Gb	
	-3.01	0.38(4)	-2.9	25	C	μ=2.0($NaClO_4$)	64Gb	
Nd^{3+}	-4.02	2.40(1)	32.6	25	C	μ=2.0($NaClO_4$)	66Cc	
	-4.992	2.49(1)	31.0	25	C	μ=2.0($NaClO_4$)	64Gb	
	-2.01	1.96(2)	31.0	25	C	μ=2.0($NaClO_4$)	66Cc	
	-4.163	1.83(2)	20.9	25	C	μ=2.0($NaClO_4$)	64Gb	
	-5.402	1.21(3)	5.0	25	C	μ=2.0($NaClO_4$)	64Gb	
	-2.18	0.47(4)	1.7	25	C	μ=2.0($NaClO_4$)	64Gb	
Sm^{3+}	-3.89	2.40(1)	33.0	25	C	μ=2.0($NaClO_4$)	66Cc	
	-4.347	2.54(1)	33.9	25	C	μ=2.0($NaClO_4$)	64Gb	
	-3.35	2.07(2)	28.4	25	C	μ=2.0($NaClO_4$)	66Cc	
	-1.527	1.96(2)	18.4	25	C	μ=2.0($NaClO_4$)	64Gb	
	-5.364	1.31(3)	7.1	25	C	μ=2.0($NaClO_4$)	64Gb	
	-5.23	0.57(4)	-6.7	25	C	μ=2.0($NaClO_4$)	64Gb	
Eu^{3+}	-3.51	2.38(1)	33.9	25	C	μ=2.0($NaClO_4$)	66Cc	
	-2.22	2.52(25°,1)	40.6	0.5-52	T	μ=2.0($NaClO_4$)	72Cd	b
	-3.76	2.08(2)	27.2	25	C	μ=2.0($NaClO_4$)	66Cc	
Gd^{3+}	-2.59	2.38(1)	36.8	25	C	μ=2.0($NaClO_4$)	66Cc	
	-2.569	2.47(1)	38.9	25	C	μ=2.0($NaClO_4$)	64Gb	

Metal Ion	ΔH, kJ /mol	Log K	ΔS, J /mol K	T,°C	M	Conditions	Ref.	Re- marks
A32, continued								
Gd^{3+},	-4.98	2.01(2)	21.8	25	C	μ=2.0($NaClO_4$)	66Cc	
cont.	-4.686	1.94(2)	21.3	25	C	μ=2.0($NaClO_4$)	64Gb	
	-7.452	1.34(3)	0.4	25	C	μ=2.0($NaClO_4$)	64Gb	
	-3.14	0.61(4)	1.2	25	C	μ=2.0($NaClO_4$)	64Gb	
Tb^{3+}	-0.38	2.40(1)	44.8	25	C	μ=2.0($NaClO_4$)	66Cc	
	-3.47	2.05(2)	27.6	25	C	μ=2.0($NaClO_4$)	66Cc	
Dy^{3+}	-0.33	2.48(1)	46.4	25	C	μ=2.0($NaClO_4$)	66Cc	
	-0.686	2.53(1)	46.0	25	C	μ=2.0($NaClO_4$)	64Gb	
	-4.06	1.96(2)	24.3	25	C	μ=2.0($NaClO_4$)	66Cc	
	-1.858	1.94(2)	31.0	25	C	μ=2.0($NaClO_4$)	64Gb	
	-4.724	1.41(3)	11.3	25	C	μ=2.0($NaClO_4$)	64Gb	
	-3.51	0.57(4)	0.8	25	C	μ=2.0($NaClO_4$)	64Gb	
Ho^{3+}	-0.33	2.42(1)	45.2	25	C	μ=2.0($NaClO_4$)	66Cc	
	-1.67	2.12(2)	35.1	25	C	μ=2.0($NaClO_4$)	66Cc	
	-10.04	1.22(3)	-10.5	25	C	μ=2.0($NaClO_4$)	66Cc	
Er^{3+}	-0.38	2.64(1)	49.4	25	C	μ=2.0($NaClO_4$)	66Cc	
	-0.79	2.60(1)	47.3	25	C	μ=2.0($NaClO_4$)	64Gb	
	-1.63	1.88(2)	30.5	25	C	μ=2.0($NaClO_4$)	66Cc	
	-1.72	1.98(2)	31.8	25	C	μ=2.0($NaClO_4$)	64Gb	
	-2.84	1.42(3)	10.0	25	C	μ=2.0($NaClO_4$)	64Gb	
	-0.63	0.48(4)	16.3	25	C	μ=2.0($NaClO_4$)	64Gb	
Tm^{3+}	-1.21	2.62(1)	46.4	25	C	μ=2.0($NaClO_4$)	66Cc	
	-2.38	2.03(2)	31.0	25	C	μ=2.0($NaClO_4$)	66Cc	
	-3.51	1.46(3)	16.3	25	C	μ=2.0($NaClO_4$)	66Cc	
Yb^{3+}	-1.51	2.59(1)	44.4	25	C	μ=2.0($NaClO_4$)	66Cc	
	-1.205	2.71(1)	47.7	25	C	μ=2.0($NaClO_4$)	64Gb	
	-3.14	2.18(2)	31.0	25	C	μ=2.0($NaClO_4$)	66Cc	
	-2.008	2.10(2)	33.5	25	C	μ=2.0($NaClO_4$)	64Gb	
	-4.81	1.43(3)	11.3	25	C	μ=2.0($NaClO_4$)	66Cc	
	-3.72	1.49(3)	16.3	25	C	μ=2.0($NaClO_4$)	64Gb	
	4.48	0.47(4)	23.8	25	C	μ=2.0($NaClO_4$)	64Gb	
Lu^{3+}	-2.72	2.67(1)	41.8	25	C	μ=2.0($NaClO_4$)	66Cc	
	-2.47	2.09(2)	31.8	25	C	μ=2.0($NaClO_4$)	66Cc	
Ti^{3+}	-15.4	3.85(30°,1)	21.5	30-50	T	μ=0.1M($NaClO_4$)	75Cb	
	-39.6	11.22(30°,1-3)	83.5	30-50	T	μ=0.1M($NaClO_4$)	75Cb	
$UO_2{}^{2+}$	5.40	2.35(1)	63.2	25	C	μ=1M($NaClO_4$)	76De	
	7.53	1.62(2)	56.1	25	C	μ=1M($NaClO_4$)	76De	
	-0.79	1.19(3)	20.1	25	C	μ=1M($NaClO_4$)	76De	
Am^{3+}	-5.56	2.57(25°,1)	30.5	0.5-52.6	T	μ=2.0($NaClO_4$)	72Cd	b
Cm^{3+}	-3.89	2.57(25°,1)	36.0	0.5-52	T	μ=2.0($NaClO_4$)	72Cd	b
Bk^{3+}	-5.06	2.65(25°,1)	33.9	0.5-52.8	T	μ=2.0($NaClO_4$)	72Cd	b
Cu^{2+}	24.52	3.10(25°,1)	141.6	25-45	T	μ=0 pH=3.6-3.8	75Da	b
	-1.46	2.17(1)	36	25	C	μ=1M($NaNO_3$)	74Ad	
	-1.55	1.13(2)	16	25	C	μ=1M($NaNO_3$)	74Ad	

Metal Ion	ΔH, kJ /mol	Log K	ΔS, J /mol K	T,°C	M	Conditions	Ref.	Remarks
A32, continued								
In^{3+}	13.0	3.00(35°,1)	100	25-45	T	μ=2.0M($NaClO_4$)	73Sc	b,q:±0.5
	9.2	2.58(35°,2)	79	25-45	T	μ=0.2M($NaClO_4$)	73Sc	b,q;±0.5
Tl^{3+}	-7.7	4.04(30°,1)	50.2	30-50	T	μ=0.1M($NaClO_4$)	75Ca	b
	-19.2	11.69(30°,1-3)	161	30-50	T	μ=0.1M($NaClO_4$)	75Ca	b
A33	ACETIC ACID, hydroxy- (n-butylthio)		$C_6H_{12}O_3S$			$C_4H_9SCH(OH)CO_2H$		
La^{3+}	-13.35	7.92(30°,1-3)	107.4	20-40	T	μ=0.1M($NaClO_4$) S:40v%Ethanol	79Sj	
Ce^{3+}	-16.61	8.57(30°,1-3)	109.3	20-40	T	μ=0.1M($NaClO_4$) S:40v%Ethanol	79Sj	
Pr^{3+}	-14.06	8.82(30°,1-3)	122.4	20-40	T	μ=0.1M($NaClO_4$) S:40v%Ethanol	79Sj	
Sm^{3+}	-13.39	9.40(30°,1-3)	108.4	20-40	T	μ=0.1M($NaClO_4$) S:40v%Ethanol	79Sj	
A34	ACETIC ACID, hydroxy (phenyl)-		$C_8H_8O_3$			$C_6H_5CH(OH)CO_2H$		L^-
Ca^{2+}	32.05	1.46(25°,1)	135.6	25-45	T	μ=0 pH=3.6-3.8	75Da	b
La^{3+}	-5.86	1.93(1)	17.2	25	C	μ=2.0M($NaClO_4$)	72Da	
Ce^{3+}	-5.27	2.17(1)	23.8	25	C	μ=2.0M($NaClO_4$)	72Da	
Pr^{3+}	-5.70	2.30(1)	23.0	25	C	μ=2.0M($NaClO_4$)	72Da	
Nd^{3+}	-5.98	2.43(1)	26.4	25	C	μ=2.0M($NaClO_4$)	72Da	
Sm^{3+}	-6.11	2.47(1)	26.8	25	C	μ=2.0M($NaClO_4$)	72Da	
Eu^{3+}	-5.82	2.25(1)	23.4	25	C	μ=2.0M($NaClO_4$)	72Da	
Gd^{3+}	-4.35	2.42(1)	31.8	25	C	μ=2.0M($NaClO_4$)	72Da	
Tb^{3+}	-3.89	2.52(1)	35.1	25	C	μ=2.0M($NaClO_4$)	72Da	
Dy^{3+}	-4.39	2.57(1)	34.7	25	C	μ=2.0M($NaClO_4$)	72Da	
Ho^{3+}	-4.27	2.54(1)	34.3	25	C	μ=2.0M($NaClO_4$)	72Da	
Er^{3+}	-5.23	2.68(1)	33.9	25	C	μ=2.0M($NaClO_4$)	72Da	
Tm^{3+}	-4.81	2.71(1)	36.0	25	C	μ=2.0M($NaClO_4$)	72Da	
Yb^{3+}	-4.31	2.72(1)	37.6	25	C	μ=2.0M($NaClO_4$)	72Da	
Lu^{3+}	-5.15	2.77(1)	36.0	25	C	μ=2.0M($NaClO_4$)	72Da	
Ti^{3+}	-15.4	3.25(30°,1)	8.6	30-50	T	μ=0.1M($NaClO_4$)	75Cb	
	-39.6	9.22(30°,1-3)	58.4	30-50	T	μ=0.1M($NaClO_4$)	75Cb	
A35	ACETIC ACID, iminodi-		$C_4H_7O_4N$			$HN(CH_2CO_2H)_2$		L^{2-}
Mg^{2+}	12.30	2.94(1)	98.3	20	C	0.1M(KNO_3)	64Ab	
Ca^{2+}	1.26	2.59(1)	53.1	20	C	0.1M(KNO_3)	64Ab	
Sr^{2+}	0.4	2.23(1)	43.9	20	C	0.1M(KNO_3)	64Ab	

Metal Ion	ΔH, kJ /mol	Log K	ΔS, J /mol K	T, °C	M	Conditions	Ref.	Re-marks
A35, continued								
Ba^{2+}	0.4	1.67(1)	33.5	20	C	0.1M(KNO_3)	64Ab	
Y^{3+}	-27.32	21.32	306.7	20	C	μ=0.1M(KNO_3) R:Y^{3+}+$EDTA^{4-}$+ L^{2-}=Y(EDTA)L^{3-}	71Ga	x:62Ma
YA^-	-27.32	3.24(1)	-31.0	20	C	μ=0.1M(KNO_3) A=EDTA	71Ga	
La^{3+}	0.71	5.70(1)	107.1	20	C	0.1M(KNO_3)	64Ab	
	-1.38	3.97(2)	75.3	20	C	0.1M(KNO_3)	64Ab	
	-18.79	18.31	286.6	20	C	μ=0.1M(KNO_3) R:La^{3+}+$EDTA^{4-}$+ L^{2-}=La(EDTA)L^{3-}	71Ga	x:62Ma
LaA^-	-6.53	2.80(1)	31.8	20	C	μ=0.1M(KNO_3) A=EDTA	71Ga	
Ce^{3+}	-19.00	19.08	300.4	20	C	μ=0.1M(KNO_3) R:Ce^{3+}+$EDTA^{4-}$+ L^{2-}=Ce(EDTA)L^{3-}	71Ga	x:62Ma
CeA^-	-6.69	3.11(1)	36.8	20	C	μ=0.1M(KNO_3) A=EDTA	71Ga	
Pr^{3+}	-3.628	6.074(1)	100.8	25	C	μ=1.0M($NaClO_4$)	72Ge	q:±0.27
	-7.40	4.599(2)	62.8	25	C	μ=1.0M($NaClO_4$)	72Ge	
	-9.92	3.099(3)	26.8	25	C	μ=1.0M($NaClO_4$)	72Ge	
	-31.46	10.741	100.0	25	C	μ=1.0M($NaClO_4$) R:M^{3+}+H^++L^{2-}= MHL^{2+}	72Ge	
	-46.02	12.873	92.0	25	C	μ=1.0M($NaClO_4$) R:M^{3+}+2H^++L^{2-}= MH_2L^{3+}	72Ge	
	-26.15	19.88	291.2	20	C	μ=0.1M(KNO_3) R:Pr^{3+}+$EDTA^{4-}$+ L^{2-}=Pr(EDTA)L^{3-}	71Ga	x:62Ma
PrA^-	-12.76	3.48(1)	23.0	20	C	μ=0.1M(KNO_3) A=EDTA	71Ga	
Nd^{3+}	-26.28	20.30	298.7	20	C	μ=0.1M(KNO_3) R:Nd^{3+}+$EDTA^{4-}$+ L^{2-}=Nd(EDTA)L^{3-}	71Ga	x:62Ma
NdA^-	-11.13	3.68(1)	32.6	20	C	μ=0.1M(KNO_3) A=EDTA	71Ga	
Sm^{3+}	-17.82	4.11(1)	18.0	20	C	μ=0.1M(KNO_3) A=EDTA	71Ga	
SmA^-	-31.84	21.26	298.3	20	C	μ=0.1M(KNO_3) R:Sm^{3+}+$EDTA^{4-}$+ L^{2-}=Sm(EDTA)L^{3-}	71Ga	x:62Ma
Eu^{3+}	-3.720	6.458(1)	110.9	25	C	μ=1.0M($NaClO_4$)	62Ge	q:±0.39

Metal Ion	ΔH, kJ /mol	Log K	ΔS, J /mol K	T,°C	M	Conditions	Ref.	Re-marks
A35, continued								
Eu^{3+}, cont.	-12.790	5.200(2)	56.9	25	C	μ=1.0M($NaClO_4$)	72Ge	
	-16.640	4.129(3)	23.0	25	C	μ=1.0M($NaClO_4$)	72Ge	
	-30.819	10.814	103.8	25	C	μ=1.0M($NaClO_4$) R:M^{2+}+H^++L^{2-}= MHL^{2+}	72Ge	
	-44.06	12.908	99.2	25	C	μ=1.0M($NaClO_4$) R:M^{2+}+2H^++L^{2-}= MH_2L^{3+}	72Ge	
	-34.02	21.60	297.1	20	C	μ=0.1M(KNO_3) R:Eu^{3+}+$EDTA^{4-}$+ L^{2-}=Eu(EDTA)L^{3-}	71Ga	x:62Ma
EuA^-	-23.30	4.23(1)	1.2	20	C	μ=0.1M(KNO_3) A=EDTA	71Ga	
Gd^{3+}	-34.89	21.69	295.8	20	C	μ=0.1M(KNO_3) R:Gd^{3+}+$EDTA^{4-}$+ L^{2-}=Gd(EDTA)L^{3-}	71Ga	x:62Ma
GdA^-	-27.66	4.30(1)	-12.1	20	C	μ=0.1M(KNO_3) A=EDTA	71Ga	
Tb^{3+}	-38.53	21.92	287.8	20	C	μ=0.1M(KNO_3) R:Tb^{3+}+$EDTA^{4-}$+ L^{2-}=Tb(EDTA)L^{3-}	71Ga	
TbA^-	-33.89	3.98(1)	-39.3	20	C	μ=0.1M(KNO_3) A=EDTA	71Ga	
Dy^{3+}	-36.28	22.14	300.0	20	C	μ=0.1M(KNO_3) R:Dy^{3+}+$EDTA^{4-}$+ L^{2-}=Dy(EDTA)L^{3-}	71Ga	x:62Ma
DyA^-	-31.21	3.84(1)	33.0	20	C	μ=0.1M(KNO_3) A=EDTA	71Ga	
Ho^{3+}	1.46	6.642(1)	131.8	25	C	μ=1.0M($NaClO_4$)	72Ge	q:±0.4
	-7.61	5.325(2)	76.1	25	C	μ=1.0M($NaClO_4$)	72Ge	
	-28.12	4.316(3)	-10.9	25	C	μ=1.0M($NaClO_4$)	72Ge	
	-28.58	10.677	107.9	25	C	μ=1.0M($NaClO_4$) R:M^{3+}+H^++L^{2-}= MHL^{2+}	72Ge	
	-46.36	12.829	90.0	25	C	μ=1.0M($NaClO_4$) R:M^{3+}+2H^++L^{2-}= MH_2L^{3+}	72Ge	q: ±3.9
	-35.56	22.28	305.4	20	C	μ=0.1M(KNO_3) R:Ho^{3+}+$EDTA^{4-}$+ L^{2-}=Ho(EDTA)L^{3-}	71Ga	x:62Ma
HoA^-	-29.87	3.54(1)	-34.3	20	C	μ=0.1M(KNO_3) A=EDTA	71Ga	
Er^{3+}	-36.40	22.11	298.7	20	C	μ=0.1M(KNO_3)	71Ga	x:62Ma

Metal Ion	ΔH, kJ /mol	Log K	ΔS, J /mol K	T,°C	M	Conditions	Ref.	Re-marks
A35, continued								
Er^{3+}, cont.						R:Er^{3+}+$EDTA^{4-}$+ L^{2-}=Er(EDTA)L^{3-}		
ErA^{-}	-29.25	3.25(1)	-37.6	20	C	μ=0.1M(KNO_3) A=EDTA	71Ga	
Tm^{3+}	-35.02	22.17	305.0	20	C	μ=0.1M(KNO_3) R:Tm^{3+}+$EDTA^{4-}$+ L^{2-}=Tm(EDTA)L^{3-}	71Ga	x:62Ma
TmA^{-}	-25.10	2.86(1)	-38.1	20	C	μ=0.1M(KNO_3) A=EDTA	71Ga	
Yb^{3+}	-32.84	22.08	310.4	20	C	μ=0.1M(KNO_3) R:Yb^{3+}+$EDTA^{4-}$+ L^{2-}=Yb(EDTA)L^{3-}	71Ga	x:62Ma
YbA^{-}	-22.59	2.55(1)	-30.1	20	C	μ=0.1M(KNO_3) A=EDTA	71Ga	
Lu^{3+}	-0.54	7.064(1)	133.0	25	C	μ=1.0M($NaClO_4$)	72Ge	q:±0.38
	-8.37	5.810(2)	84.1	25	C	μ=1.0M($NaClO_4$)	72Ge	
	-23.43	3.275(3)	-17.2	25	C	μ=1.0M($NaClO_4$)	72Ge	
	-27.99	10.601	108.8	25	C	μ=1.0M($NaClO_4$) R:M^{3+}+H^{+}+L^{2-}= MHL^{2+}	72Ge	
	-49.45	12.707	77.0	25	C	μ=1.0M($NaClO_4$) R:M^{3+}+2H^{+}+L^{2-}= MH_2L^{3+}	72Ge	q:±4.6
	-27.24	22.34	334.7	20	C	μ=0.1M(KNO_3) R:Lu^{3+}+$EDTA^{4-}$+ L^{2-}=Lu(EDTA)L^{3-}	71Ga	x:62Ma
LuA^{-}	-16.74	2.51(1)	-9.2	20	C	μ=0.1M(KNO_3) A=EDTA	71Ga	
Th^{4+}	-28.6	3.81(30°)	-21.4	30&35	T	μ=0.1M(KNO_3) R:Th+A+L=ThAL A=3-Aza-1,5-diamino-N,N,N',-N",N"-pentaacetic acid	75Pa	a,b
ThA	3.51	6.11(1)	128.9	25	C	μ=0.10(KNO_3) A=1,2-Diamino-cyclohexane-N,N,N',N'-tetraacetic acid	70Kb	
	10.00	6.71(1)	161.9	25	C	μ=0.10(KNO_3) A=EDTA	70Kb	
	36.61	--	--	25	C	μ=0.10(KNO_3) A=1,2-Diamino-cyclohexane-N,N,N',N'-tetraacetic acid R:ThA+H_2A=	70Kb	

Metal Ion	ΔH, kJ /mol	Log K	ΔS, J /mol K	T,°C	M	Conditions	Ref.	Re-marks
A35, continued								
ThA, cont.	43.1	--	--	25	C	$ThAL^{2-}+2H^+$ μ=0.10(KNO_3) A=EDTA R:ThA+H_2A =$ThAL^{2-}+2H^+$	70Kb	
Am^{3+}	25.1	(1)	54.8	15-90	T	CM=8.4x10^{-4}M CL=5x10^{-3}M	71Bg	b
Co^{2+}	-8.95	6.97(1)	102.9	20	C	0.1M(KNO_3)	64Ab	
	-16.15	5.34(2)	46.9	20	C	0.1M(KNO_3)	64Ab	
Ni^{2+}	-21.13	8.19(1)	83.7	20	C	0.1M(KNO_3)	64Ab	
	-22.6	4.11(1)	2.9	25	C	μ=0.1M(KNO_3)	71Ld	
	-18.62	6.11(2)	54.8	20	C	0.1M(KNO_3)	64Ab	
	-17.61	3.21(2)	2.5	25	C	μ=0.1M(KNO_3)	71Ld	
Cu^{2+}	-18.8	10.63(1)	139.3	20	C	0.1M(KNO_3)	64Ab	
	-16.6	10.42(1)	144	25	C	μ=0.1M($NaClO_4$)	79Be	
	-26.8	6.05(2)	24.7	20	C	0.1M(KNO_3)	64Ab	
	-26.5	5.60(2)	18	25	C	μ=0.1M($NaClO_4$)	79Be	
	-19	12.35	172	25	C	μ=0.1M($NaClO_4$) R:$Cu^{2+}+L^{2-}+H^+$= $CuHL^+$	79Be	
	-61.5	16.63	112	25	C	μ=0.1M($NaClO_4$) R:$Cu^{2+}+L^{2-}$+A=CuAL A=2,2'-Bipyridyl	79Be	
Ag^+	-21.5	2.69(1)	-20.8	25	C	μ=0.5M(ClO_4^- or NO_3^-)	78Mc	q:±1.8
	-27.6	2.79(2)	-38.9	25	C	μ=0.5M(ClO_4^- or NO_3^-)	78Mc	
Zn^{2+}	-9.2	7.27(1)	107.5	20	C	0.1M(KNO_3)	64Ab	
	-15.5	5.33(2)	49.4	20	C	0.1M(KNO_3)	64Ab	
	-17.6	6.74(25°)	70.0	25&35	T	R:Zn+L+A=ZnLA A=Oxalic acid	78Ka	a,b
	-29.9	11.69(25°)	123.4	25&35	T	R:Zn+L+A=ZnLA A=Malonic acid	78Ka	a,b
	-29.9	11.08(25°)	111.8	25&35	T	R:Zn+L+A=ZnLA A=Phthalic acid	78Ka	a,b
Cd^{2+}	-6.11	5.73(1)	89.1	20	C	0.1M(KNO_3)	64Ab	
	-9.3	2.40(1)	14.8	25	C	μ=0.5M(ClO_4^- or NO_3^-)	78Mc	q:±1.2
	-16.82	4.46(2)	27.6	20	C	0.1M(KNO_3)	64Ab	
	-8.7	2.12(2)	11.4	25	C	μ=0.5M(ClO_4^- or NO_3^-)	78Mc	
	-19.3	6.04(25°)	50.7	25&35	T	R:Zn+L+A=ZnLA A=Oxalic acid	78Ka	a,b
	-35.1	10.92(25°)	91.0	25&35	T	R:Zn+L+A=ZnLA A=Malonic acid	78Ka	a,b
	-29.9	10.76(25°)	105.6	25&35	T	R:Zn+L+A=ZnLA A=Phthalic acid	78Ka	a,b

Metal Ion	ΔH, kJ /mol	Log K	ΔS, J /mol K	T,°C	M	Conditions	Ref.	Remarks
A35, continued								
Hg^{2+}	-80	7.84(1)	-117	25	C	μ=0.5M(ClO_4^- or NO_3^-)	78Mc	q:±24
	-35	7.82(2)	30	25	C	μ=0.5M(ClO_4^- or NO_3^-)	78Mc	q:±15
Pb^{2+}	-13.97	7.45(1)	95.0	20	C	0.1M(KNO_3)	64Ab	
A36	ACETIC ACID, iminodi, N-(2-carboxyethyl)-		$C_7H_{11}O_6N$			$CO_2HCH_2CH_2N(CH_2CO_2H)_2$		L^{3-}
Mg^{2+}	14.02	5.32(1)	149.0	25	C	μ=0.10(KNO_3)	69Ja	
Ca^{2+}	-1.05	5.04(1)	92.9	25	C	μ=0.10(KNO_3)	69Ja	
Sr^{2+}	-1.92	3.87(1)	67.8	25	C	μ=0.10(KNO_3)	69Ja	
Ba^{2+}	-0.46	3.40(1)	63.6	25	C	μ=0.10(KNO_3)	69Ja	
Mn^{2+}	4.6	7.33(1)	155.6	25	C	μ=0.1M(KNO_3)	70Mb	
Co^{2+}	-7.1	10.00(1)	167.8	25	C	μ=0.1M(KNO_3)	70Mb	
Ni^{2+}	-15.1	11.37(1)	166.9	25	C	μ=0.1M(KNO_3)	70Mb	
	-15.1	11.37(1)	166.9	25	C	μ=0.1M(KNO_3)	69Ja	
Cu^{2+}	-17.2	13.22(1)	195.4	25	C	μ=0.1M(KNO_3)	70Mb	
Zn^{2+}	-4.6	9.98(1)	176.1	25	C	μ=0.1M(KNO_3)	70Mb	
A37	ACETIC ACID, iminodi, N-(o-carboxyphenyl)		$C_{11}H_{11}O_6N$			$(CO_2HC_6H_4)N(CH_2CO_2H)_2$		L^{3-}
Mg^{2+}	31.96	4.01(1)	184.1	25	C	μ=0.1(KNO_3)	69Ua	
Ca^{2+}	5.61	5.08(1)	115.9	25	C	μ=0.1(KNO_3)	69Ua	
Sr^{2+}	5.69	3.93(1)	94.1	25	C	μ=0.1(KNO_3)	69Ua	
Ba^{2+}	5.36	3.59(1)	86.6	25	C	μ=0.1(KNO_3)	69Ua	
Mn^{2+}	19.7	3.85(1)	177.4	25	C	μ=0.1M(KNO_3)	70Mb	
Co^{2+}	13.8	8.42(1)	207.5	25	C	μ=0.1M(KNO_3)	70Mb	
Ni^{2+}	6.3	9.48(1)	202.1	25	C	μ=0.1M(KNO_3)	70Mb	
	6.11	9.48(1)	202.1	25	C	μ=0.1(KNO_3)	69Ua	
Cu^{2+}	7.5	10.93(1)	234.7	25	C	μ=0.1M(KNO_3)	70Mb	
Zn^{2+}	13.0	8.42(1)	204.6	25	C	μ=0.1M(KNO_3)	70Mb	
A38	ACETIC ACID, iminodi, N-(2,5-dicarboxyphenyl)-		$C_{12}H_{11}O_8N$			$(CO_2HC_6H_3CO_2H)N(CH_2CO_2H)_2$		
Mg^{2+}	24.7	4.14(1)	195.8	25	C	μ=0.1M(KNO_3)	70Mc	
Ca^{2+}	6.7	5.05(1)	119.7	25	C	μ=0.1M(KNO_3)	70Mc	
Sr^{2+}	5.8	4.12(1)	99.2	25	C	μ=0.1M(KNO_3)	70Mc	
Co^{2+}	15.9	8.83(1)	223.0	25	C	μ=0.1M(KNO_3)	70Mc	
Ni^{2+}	7.5	9.95(1)	210.4	25	C	μ=0.1M(KNO_3)	70Mc	

Metal Ion	ΔH, kJ /mol	Log K	ΔS, J /mol K	T,°C	M	Conditions	Ref.	Re-marks
A38, continued								
Cu^{2+}	9.2	10.85(1)	238.1	25	C	μ=0.1M(KNO_3)	70Mc	
Zn^{2+}	14.2	8.86(1)	218.0	25	C	μ=0.1M(KNO_3)	70Mc	
A39 ACETIC ACID, iminodi, N-hydroxethyl-			$C_6H_{11}O_5N$			$HOCH_2CH_2N(CH_2COOH)_2$		L^{2-}
UO_2^{2+}	9.2	8.34(25°,1)	192	20-40	T	μ=0.1M(KNO_3)	69Sg	b
A40 ACETIC ACID, iminodi, N-methyl			$C_5H_9O_4N$			$CH_3N(CH_2CO_2H)_2$		L^{2-}
VO_2^{+}	-25	10.16(25°,1)	117	15-35	T	μ=1.0M($NaClO_4$ & $HClO_4$)	76Yb	q:±17
A41 ACETIC ACID, iminotri-			$C_6H_9O_6N$			$N(CH_2CO_2H)_3$		L^{3-}
Be^{2+}	25.9	7.86(25°,1)	238	20-40	T	μ=0.1M(KNO_3)	77Fb	q:±2.5
Mg^{2+}	10.5	6.310(1)	159.0	0	T	μ=0	56Hb, 56Ma	
	14.2	6.396(1)	172.4	10	T	μ=0	56Hb	
	18.0	6.500(1)	185.4	20	T	μ=0	56Hb	
	20.1	6.612(1)	192.5	30	T	μ=0	56Hb	
	18.58	5.41(1)	166.5	20	C	μ=0.1(KNO_3)	64Aa	
	18.6	5.41(1)	166	20	C	μ=0.1M(KNO_3)	76Ah	
	17.029	(1)	--	20	C	μ=0.1	64Db	
	8.24	5.58(1)	134	25	T	μ=0.10(KNO_3)	65Ma	
	12.6	(1)	142	20-40	T	μ=0.1M(KNO_3)	77Fb	
Ca^{2+}	-8.8	7.704(1)	115.1	0	T	μ=0	56Hb, 56Ma	
	-7.1	7.652(1)	121.3	10	T	μ=0	56Hb	
	-4.2	7.608(1)	131.4	20	T	μ=0	56Hb	
	0	7.595(1)	144.8	30	T	μ=0	56Hb	
	-5.69	6.30(1)	103.3	20	C	μ=0.1	64Hc	
	-5.69	6.41(1)	101.2	20	C	μ=0.1(KNO_3)	64Aa	
	-5.69	6.41(1)	101	20	C	μ=0.1M(KNO_3)	76Ah	
	-3.39	6.57(1)	114.6	25	T	μ=0.1(KNO_3)	62Mf	
	-8.4	(1)	92	20-40	T	μ=0.1M(KNO_3)	77Fb	
	-26.94	2.49(2)	-42.7	20	C	μ=0.1(KNO_3)	64Aa	
Sr^{2+}	-2.22	4.98(1)	73.2	20	C	μ=0.1(KNO_3)	64Aa	
	-2.26	6.68(1)	87.4	20	C	μ=0.1	64Hc	
	-2.2	4.98(1)	73.2	20	C	μ=0.1M(KNO_3)	76Ah	
	-11.13	5.16(1)	63	25	T	μ=0.10(KNO_3)	65Ma	
	0	(1)	100	20-40	T	μ=0.1M(KNO_3)	77Fb	
Ba^{2+}	-9.41	5.968(1)	79.9	0	T	μ=0	56Hb	
	-7.11	5.914(1)	87.9	10	T	μ=0	56Hb	
	-5.02	5.875(1)	95.4	20	T	μ=0	56Hb	
	-4	5.86(1)	100	25	T	μ=0	56Ma	
	-2.1	5.587(1)	105.0	30	T	μ=0	56Hb	
	-6.02	4.82(1)	71.5	20	C	μ=0.1(KNO_3)	64Aa	
	-6.02	4.82(1)	71.5	20	C	μ=0.1M(KNO_3)	76Ah	

Metal Ion	ΔH, kJ /mol	Log K	ΔS, J /mol K	T,°C	M	Conditions	Ref.	Remarks
A41, continued								
Ba^{2+}, cont.	-6.07	4.94(1)	75	25	T	μ=0.10(KNO_3)	65Ma	
	-8.4	(1)	63	20-40	T	μ=0.1M(KNO_3)	77Fb	
Y^{3+}	4.297	11.458(1)	21	20	C	μ=0.1	64Db	
	11.25	11.48(1)	257.7	25	T	μ=0.1(KNO_3)	62Mf	
	-17.53	9.02(2)	112.5	25	T	μ=0.1(KNO_3)	62Mf	
	-31.76	21.82	309.6	20	C	μ=0.1M(KNO_3) R: Y^{3+}+$EDTA^{4-}$+L^{3-}= $Y(EDTA)L^{4-}$	71Ga	x:62Ma
YA^-	-29.29	3.74(1)	-28.4	20	C	μ=0.1M(KNO_3) A=EDTA	71Ga	
La^{3+}	1.339	10.370(1)	204.2	20	C	μ=0.1	64Db 61La	
	8.6	10.36(1)	227.2	25	T	μ=0.1(KNO_3)	62Mf	
	-10.17	7.24(2)	104.6	25	T	μ=0.1(KNO_3)	62Mf	
	-37.70	20.30	259.8	20	C	μ=0.1M(KNO_3) R: La^{3+}+$EDTA^{4-}$+ L^{3-}=$La(EDTA)L^{4-}$	71Ga	x:62Ma
LaA^-	-25.44	4.79(1)	5.0	20	C	μ=0.1M(KNO_3) A=EDTA	71Ga	
Ce^{3+}	-0.900	10.824(1)	204.2	20	C	μ=0.1	64Db 61La	
	5.23	10.83(1)	225.1	25	T	μ=0.1(KNO_3)	62Mf	
	1.67	10.26(1)	200.8	25	C	μ=0.2	69Mc	g:Na, q:±0.4
	-12.80	7.84(2)	107.1	25	T	μ=0.1(KNO_3)	62Mf	
	-12.55	7.99(2)	110.0	25	C	μ=0.2	69Mc	g:Na
	-36.23	20.70	272.4	20	C	μ=0.1M(KNO_3) R: Ce^{3+}+$EDTA^{4-}$+ L^{3-}=$Ce(EDTA)L^{4-}$	71Ga	x:62Ma
CeA^-	-23.93	4.73(1)	8.8	20	C	μ=0.1M(KNO_3) A=EDTA	71Ga	
Pr^{3+}	-2.100	11.063(1)	204.6	20	C	μ=0.1	64Db 61La	
	1.88	11.07(1)	218.4	25	T	μ=0.1(KNO_3)	62Mf	
	1.67	10.47(1)	205.0	25	C	μ=0.2	69Mc	g:Na, q:±0.4
	-15.56	8.18(2)	104.2	25	T	μ=0.1(KNO_3)	62Mf	
	-14.22	8.15(2)	113.4	25	C	μ=0.2	69Mc	g:Na
	-32.84	21.08	291.6	20	C	μ=0.1M(KNO_3) R: Pr^{3+}+$EDTA^{4-}$+ L^{3-}=$Pr(EDTA)L^{4-}$	71Ga	x:62Ma
PrA^-	-19.46	4.67(1)	23.0	20	C	μ=0.1M(KNO_3) A=EDTA	71Ga	
Nd^{3+}	-3.36	11.249(1)	204.2	20	C	μ=0.1	64Db 61La	
	2.84	11.26(1)	225.1	25	T	μ=0.1(KNO_3)	62Mf	

Metal Ion	ΔH, kJ /mol	Log K	ΔS, J /mol K	T,°C	M	Conditions	Ref.	Remarks
A41, continued								
Nd^{3+}, cont.	0.84	10.67(1)	205.9	25	C	μ=0.2	69Mc	g:Na, q:±0.21
	-15.82	8.47(2)	109.2	25	T	μ=0.1(KNO_3)	62Mf	
	-11.72	8.47(2)	122.2	25	C	μ=0.2	69Mc	g:Na
	-32.51	21.38	298.7	20	C	μ=0.1M(KNO_3) R:Nd^{3+}+$EDTA^{4-}$+ L^{3-}=Nd(EDTA)L^{4-}	71Ga	x:62Ma
NdA^-	-17.36	4.77(1)	32.2	20	C	μ=0.1M(KNO_3) A=EDTA	71Ga	
Sm^{3+}	-4.381	11.50(1)	205.4	20	C	μ=0.1	64Db	
							61La	
	-1.67	11.53(1)	226.4	25	T	μ=0.1(KNO_3)	62Mf	
	-1.67	10.91(1)	200.8	25	C	μ=0.2	69Mc	g:Na, q:±0.42
	-18.28	9.00(2)	110.9	25	T	μ=0.1(KNO_3)	62Mf	
	-16.32	9.18(2)	119.7	25	C	μ=0.2	69Mc	g:Na
	-40.08	22.15	287.0	20	C	μ=0.1M(KNO_3) R:Sm^{3+}+$EDTA^{4-}$+ L^{3-}=Sm(EDTA)L^{4-}	71Ga	x:62Ma
SmA^-	-26.07	5.00(1)	6.7	20	C	μ=0.1M(KNO_3) A=EDTA	71Ga	
Eu^{3+}	-4.305	11.49(1)	205.4	20	C	μ=0.1	64Db	
							61La	
	3.89	11.52(1)	233.5	25	T	μ=0.1(KNO_3)	62Mf	
	-21.25	9.18(2)	104.6	25	T	μ=0.1(KNO_3)	62Mf	
	-40.75	22.40	289.5	20	C	μ=0.1M(KNO_3) R:Eu^{3+}+$EDTA^{4-}$+ L^{3-}=Eu(EDTA)L^{4-}	71Ga	x:62Ma
EuA^-	-30.04	5.03(1)	-6.3	20	C	μ=0.1M(KNO_3) A=EDTA	71Ga	
Gd^{3+}	-2.619	11.54(1)	212.13	20	C	μ=0.1	64Db	
							61La	
	4.27	11.54(1)	235.1	25	T	μ=0.1(KNO_3)	62Mf	
	-1.67	11.00(1)	203.8	25	C	μ=0.2	69Mc	g:Na, q:±0.42
	-24.27	9.27(2)	95.8	25	T	μ=0.1(KNO_3)	62Mf	
	-20.08	9.31(2)	110.4	25	C	μ=0.2	69Mc	g:Na
	-40.38	22.25	287.8	20	C	μ=0.1M(KNO_3) R:Gd^{3+}+$EDTA^{4-}$+ L^{3-}=Gd(EDTA)L^{4-}	71Ga	x:62Ma
GdA^-	-33.14	4.86(1)	-19.7	20	C	μ=0.1M(KNO_3) A=EDTA	71Ga	
Tb^{3+}	-0.255	11.58(1)	220.9	20	C	μ=0.1	64Db	
							61La	
	7.15	11.60(1)	246.0	25	T	μ=0.1(KNO_3)	62Mf	
	-21.80	9.38(2)	106.3	25	T	μ=0.1(KNO_3)	62Mf	
	-40.17	22.58	295.4	20	C	μ=0.1M(KNO_3)	71Ga	x:62Ma

Metal Ion	ΔH, kJ /mol	Log K	ΔS, J /mol K	T,°C	M	Conditions	Ref.	Re-marks
A41, continued								
Tb^{3+}, cont.						R: $Tb^{3+}+EDTA^{4-}+L^{3-}=Tb(EDTA)L^{4-}$		
TbA^{-}	-35.52	4.65(1)	-32.2	20	C	μ=0.1M(KNO_3) A=EDTA	71Ga	
Dy^{3+}	1.464	11.70(1)	229.3	20	C	μ=0.1	64Db	
	9.41	11.74(1)	256.5	25	T	μ=0.1(KNO_3)	61La/62Mf	
	-23.47	9.41(2)	101.2	25	T	μ=0.1(KNO_3)	62Mf	
	-40.42	22.59	294.6	20	C	μ=0.1M(KNO_3)	71Ga	x:62Ma
						R: $Dy^{3+}+EDTA^{4-}+L^{3-}=Dy(EDTA)L^{4-}$		
DyA^{-}	-35.35	4.29(1)	-38.5	20	C	μ=0.1M(KNO_3) A=EDTA	71Ga	
Ho^{3+}	2.272	11.846(1)	234.7	20	C	μ=0.1	64Db	
	11.00	11.90(1)	264.8	25	T	μ=0.1(KNO_3)	61La/62Mf	
	-21.55	9.35(2)	106.7	25	T	μ=0.1(KNO_3)	62Mf	
	-39.41	22.70	300.0	20	C	μ=0.1M(KNO_3)	71Ga	x:62Ma
						R: $Ho^{3+}+EDTA^{4-}+L^{3-}=Ho(EDTA)L^{4-}$		
HoA^{-}	-33.72	3.95(1)	-39.3	20	C	μ=0.1M(KNO_3) A=EDTA	71Ga	
Er^{3+}	2.481	11.995(1)	238.1	20	C	μ=0.1	64Db	
	10.50	12.04(1)	265.7	25	T	μ=0.1(KNO_3)	61La/62Mf	
	-16.19	9.26(2)	123.0	25	T	μ=0.1(KNO_3)	62Mf	
	-37.40	22.38	300.8	20	C	μ=0.1M(KNO_3)	71Ga	x:62Ma
						R: $Er^{3+}+EDTA^{4-}+L^{3-}=Er(EDTA)L^{4-}$		
ErA^{-}	-30.63	3.52(1)	-35.6	20	C	μ=0.1M(KNO_3) A=EDTA	71Ga	
Tm^{3+}	2.448	12.196(1)	241.8	20	C	μ=0.1	64Db	
	7.95	12.23(1)	260.7	25	T	μ=0.1(KNO_3)	61La/62Mf	
	-9.87	9.24(2)	143.5	25	T	μ=0.1(KNO_3)	62Mf	
	-36.07	22.44	306.7	20	C	μ=0.1M(KNO_3)	71Ga	x:62Ma
						R: $Tm^{3+}+EDTA^{4-}+L^{3-}=Tm(EDTA)L^{4-}$		
TmA^{-}	-28.24	3.13(1)	-36.40	20	C	μ=0.1M(KNO_3) A=EDTA	71Ga	
Yb^{3+}	1.674	12.368(1)	242.7	20	C	μ=0.1	64Db	
	8.74	12.40(1)	266.5	25	T	μ=0.1(KNO_3)	61La/62Mf	
	-7.78	9.29(2)	151.9	25	T	μ=0.1(KNO_3)	62Mf	
	-31.63	22.38	320.1	20	C	μ=0.1M(KNO_3)	71Ga	x:62Ma

Metal Ion	ΔH, kJ /mol	Log K	ΔS, J /mol K	T,°C	M	Conditions	Ref.	Remarks
A41, continued								
Yb^{3+}, cont.						R:Yb^{3+}+$EDTA^{4-}$+ L^{3-}=Yb(EDTA)L^{4-}		
YbA^-	-21.97	2.85(1)	-20.5	20	C	μ=0.1M(KNO_3) A=EDTA	71Ga	
Lu^{3+}	0.753	12.465(1)	241.4	20	C	μ=0.1	64Db 61La	
	8.28	12.49(1)	266.9	25	T	μ=0.1(KNO_3)	62Mf	
	-4.64	9.42(2)	164.8	25	T	μ=0.1(KNO_3)	62Mf	
	-25.56	22.67	346.4	20	C	μ=0.1M(KNO_3)	71Ga	x:62Ma
						R:Lu^{3+}+$EDTA^{4-}$+ L^{3-}=Lu(EDTA)L^{4-}		
LuA^-	-15.06	2.81(1)	2.5	20	C	μ=0.1M(KNO_3) A=EDTA	71Ga	
Th^{4+}	-17.9	3.94(30°)	16.4	30&35	T	μ=0.1M(KNO_3) R:Th+A+L=ThAL A=3-Aza-1,5-diamino-N,N',-N",N"-pentaacetic acid	75Pa	a,b
VO_2^+	~0	13.8(25°,1)	263	15-35	T	μ=1.0M($NaClO_4$ & $HClO_4$)	76Yb	
Am^{3+}	2.9	11.95(25°,1)	234	15-50	T	0.10M($HClO_4$)	72Eb	b,q:±2
	-23.8	9.03(25°,2)	93	15-50	T	0.10M($HClO_4$)	72Eb	b,q:±3
Mn^{2+}	~0	8.527(1)	163.6	0	T	μ=0	56Hb	a
	3.3	8.534(1)	175.3	10	T	μ=0	56Hb	a
	8.4	8.573(1)	192.5	20	T	μ=0	56Hb	a
	14.6	8.644(1)	213.8	30	T	μ=0	56Hb	a
	4.77	7.44(1)	158.6	20	C	μ=0.1	64Hc	
	6.02	7.44(1)	162.8	20	C	μ=0.1(KNO_3)	64Aa	
	-23.35	3.55(2)	-7.1	20	C	μ=0.1(KNO_3)	64Aa	
Fe^{2+}	6.3	6.06	-96	25	T	μ=1.0($NaClO_4$)	74Ca	
						A=NH_3 R:Fe^{2+}+ CoA_5L=CoA_5LFe^{2+}		
Fe^{3+}	13.4	15.87(1)	349.8	20	C	μ=0.1(KNO_3)	64Aa	
Co^{2+}	-0.63	10.40(1)	197.1	20	C	μ=0.1	64Hc	
	-0.29	10.38(1)	197.5	20	C	μ=0.1(KNO_3)	64Aa	
	-19.92	4.01(2)	8.8	20	C	μ=0.1(KNO_3)	64Aa	
Ni^{2+}	-10.58	11.53(1)	184.5	20	C	μ=0.1	64Hc	
	-10.71	11.53(1)	183.7	20	C	μ=0.1(KNO_3)	64Aa	
	-23.30	4.88(2)	29.3	20	C	μ=0.1(KNO_3)	64Aa	
Cu^{2+}	-7.70	12.96(1)	221.8	20	C	μ=0.1	64Hc	
	-7.82	12.96(1)	220.9	20	C	μ=0.1(KNO_3)	64Aa	
	-4.6	13.10(1)	235.6	25	T	μ=0.1(KNO_3)	62Mf	
	-25.23	4.47(2)	-14.6	20	C	μ=0.1(KNO_3)	64Aa	
Zn^{2+}	-3.51	10.67(1)	192.5	20	C	μ=0.1	64Hc	

Metal Ion	ΔH, kJ /mol	Log K	ΔS, J /mol K	T,°C	M	Conditions	Ref.	Re-marks
A41, continued								
Zn^{2+},	-3.64	10.67(1)	190.4	20	C	μ=0.1(KNO_3)	64Aa	
cont.	-11.76	3.62(2)	31.0	20	C	μ=0.1(KNO_3)	64Aa	
Cd^{2+}	16.61	9.82(1)	131.0	20	C	μ=0.1(KNO_3)	64Aa	
	-54.48	4.79(2)	19.7	20	C	μ=0.1(KNO_3)	64Aa	
Pb^{2+}	-15.94	11.39(1)	163.6	20	C	μ=0.1(KNO_3)	64Aa	
A42 ACETIC ACID, mercapto-			$C_2H_4O_2S$			$HSCH_2CO_2H$		L^-
La^{3+}	6.217	1.41(1)	48.1	25	C	μ=2.0($NaClO_4$)	64Gb	
	10.0	0.70(2)	46.9	25	C	μ=2.0($NaClO_4$)	64Gb	
Pr^{3+}	12.18	2.742(1)	93.3	25	C	μ=1.00M($NaClO_4$)	73Dd	
	7.53	1.643(2)	56.9	25	C	μ=1.00M($NaClO_4$)	73Dd	q:±1.2
	6.40	5.447	125.9	25	C	μ=1.00M($NaClO_4$) R:$M^{3+}+L^{2-}+H^+=MHL^{2+}$	73Dd	q:±2.5
	9.58	2.190	74.0	25	C	μ=1.00M($NaClO_4$) R:$MHL^{2+}+L^{2-}=MHL_2$	73Dd	q:±5.0
Sm^{3+}	11.38	2.903(1)	93.7	25	C	μ=1.00M($NaClO_4$)	73Dd	
	5.586	1.80(1)	53.1	25	C	μ=2.0($NaClO_4$)	64Gb	
	9.25	1.773(2)	65.3	25	C	μ=1.00M($NaClO_4$)	73Dd	
	8.24	0.91(2)	45.2	25	C	μ=2.0($NaClO_4$)	64Gb	
	5.40	5.592	125.1	25	C	μ=1.00M($NaClO_4$) R:$M^{3+}+L^{2-}+H^+=MHL^+$	73Dd	q:±0.58
	9.08	2.225	72.8	25	C	μ=1.00M($NaClO_4$) R:$MHL^{2+}+L^{2-}=MHL_2$	73Dd	q:±1.46
Tb^{3+}	17.20	2.521(1)	105.8	25	C	μ=1.00M($NaClO_4$)	73Dd	
	9.08	1.606(2)	61.1	25	C	μ=1.00M($NaClO_4$)	73Dd	
	12.89	5.205	143.1	25	C	μ=1.00M($NaClO_4$) R:$M^{3+}+L^{2-}+H^+=MHL^{2+}$	73Dd	q:±2.5
	7.11	2.085	64.02	25	C	μ=1.00M($NaClO_4$) R:$MHL^{2+}+L^{2-}=MHL_2$	73Dd	q:±5.8
Er^{3+}	20.96	2.360(1)	115.5	25	C	μ=1.00M($NaClO_4$)	73Dd	
	9.958	1.28(1)	57.7	25	C	μ=2.0($NaClO_4$)	64Gb	
	12.13	1.489(2)	69.4	25	C	μ=1.00M($NaClO_4$)	73Dd	
	12.80	0.90(2)	60.2	25	C	μ=2.0($NaClO_4$)	64Gb	
	15.31	5.139	149.8	25	C	μ=1.00M($NaClO_4$) R:$M^{3+}+L^{2-}+H^+=MHL^{2+}$	73Dd	q:2.3
	9.92	2.062	72.0	25	C	μ=1.00M($NaClO_4$) R:$MHL^{2+}+L^{2-}=MHL_2$	73Dd	q:±3.8
Yb^{3+}	22.38	2.360(1)	120.0	25	C	μ=1.00M($NaClO_4$)	73Dd	
	13.72	1.400(2)	72.8	25	C	μ=1.00M($NaClO_4$)	73Dd	
	14.68	5.095	146.8	25	C	μ=1.00M($NaClO_4$) R:$M^{3+}+L^{2-}+H^+=MHL^{2+}$	73Dd	q:±1.1

Metal Ion	ΔH, kJ /mol	Log K	ΔS, J /mol K	T,°C	M	Conditions	Ref.	Re-marks
A41, continued								
Yb^{3+}, cont.	11.88	2.176	81.2	25	C	μ=1.00M($NaClO_4$) R:MHL^{2+}+L^{2-}=MHL_2	73Dd	q:±1.6
UO_2^{2+}	-74.9	16.15(30°,1-2)	62	20-40	T	μ=0.1M($NaClO_4$)	74Si	b
Ni^{2+}	-14.6	13.041(1-2)	201	25	T	μ=0.11(KCl) L=L^{2-}	60Lb	
	-12.6	12.759(1-2)	201	25	C	μ=0.5(KNO_3)	74Da	
	-87.9	32.219(3-4)	314	25	C	μ=0.5(KNO_3)	74Da	
	-130	49.845(4-6)	519	25	T	μ=0.11(KCl) L=L^{2-}	60Lb	
	-129.7	49.157(4-6)	494	25	C	μ=0.5(KNO_3)	74Da	
Zn^{2+}	-3.4	15.21(30°,1-2)	280	20-40	T	μ=0.1M($NaClO_4$)	74Si	b
In^{3+}	-25.5	11.98(35°,1)	146	25-45	T	μ=0.2M($NaClO_4$)	73Sc	b,q:±2.1
	-26.4	10.19(35°,2)	109	25-45	T	μ=0.2M($NaClO_4$)	73Sc	b,q:±2.1
	-29.3	6.16(35°,3)	21	25-45	T	μ=0.2M($NaClO_4$)	73Sc	b,q:±2.1
A43 ACETIC ACID, (2-mercaptophenyl)-			$C_8H_8O_2S$			$C_6H_5SCH_2CO_2H$		
Ni^{2+}	2.9	0.5(1)	19.7	25	C	μ=1($NaNO_3$)	78Ag	q:±0.8
Cu^{2+}	6.7	1.3(1)	46.4	25	C	μ=1($NaNO_3$)	78Ag	q:±1.7
Cd^{2+}	4.2	1.0(1)	33.9	25	C	μ=1($NaNO_3$)	78Ag	q:±1.3
A44 ACETIC ACID, methoxy			$C_3H_6O_3$			$CH_3OCH_2CO_2H$		
Cu^{2+}	3.10	1.83(1)	45.2	25	C	μ=1($NaNO_3$)	74Ad	
	5.82	1.01(2)	38.9	25	C	μ=1($NaNO_3$)	74Ad	
A45 ACETIC ACID, methyliminodi-			$C_5H_9O_4N$			$CH_3N(CH_2CO_2H)_2$		L^{2-}
Mg^{2+}	11.92	3.48(1)	107.1	25	C	μ=0.1	68Na	
	-2.80	2.32(2)	35.1	25	C	μ=0.1	68Na	
Ca^{2+}	-4.98	3.85(1)	56.9	25	C	μ=0.1	68Na	
	-3.10	2.81(2)	41.8	25	C	μ=0.1	68Na	
Sr^{2+}	-3.47	2.95(1)	44.8	25	C	μ=0.1	68Na	
Ba^{2+}	-4.44	2.61(1)	35.1	25	C	μ=0.1	68Na	
UO_2^{2+}	4.2	9.71(25°,1)	201	20-40	T	μ=0.1M(KNO_3)	69Sg	b
A46 ACETIC ACID, oxydi-			$C_4H_6O_5$			$O(CH_2CO_2H)_2$	L^{2-}	
Sc^{3+}	-1.97	8.28(1)	151.9	25	C	μ=1.0($NaClO_4$)	69Gf	
	-10.67	4.48(2)	50.2	25	C	μ=1.0($NaClO_4$)	69Gf	
Y^{3+}	7.247	5.26(1)	125.5	25	C	μ=1.0	63Gd	
	-5.201	4.51(2)	68.6	25	C	μ=1.0	63Gd	
	-17.564	3.23(3)	3.3	25	C	μ=1.0	63Gd	
La^{3+}	-0.293	4.93(1)	93.3	25	C	μ=1.0	63Gd	
	-3.171	3.47(2)	55.6	25	C	μ=1.0	63Gd	
	1.481	1.84(3)	40.2	25	C	μ=1.0	63Gd	

Metal Ion	ΔH, kJ /mol	Log K	ΔS, J /mol K	T, °C	M	Conditions	Ref.	Remarks
A46, continued								
Ce^{3+}	-1.678	5.15(1)	92.9	25	C	μ=1.0	63Gd	
	-3.636	3.76(2)	59.8	25	C	μ=1.0	63Gd	
	-2.067	2.30(3)	37.2	25	C	μ=1.0	63Gd	
Pr^{3+}	-5.356	(1)	--	5	C	μ=1.0M($NaClO_4$)	72Gf	
	-3.167	(1)	--	20	C	μ=1.0M($NaClO_4$)	72Gf	
	-2.845	5.33(1)	92.5	25	C	μ=1.0	63Gd	
	-1.188	(1)	--	35	C	μ=1.0M($NaClO_4$)	72Gf	
	0.979	(1)	--	50	C	μ=1.0M($NaClO_4$)	72Gf	
		(1)				$\Delta H=A+BT+CT^2+DT^3$		
						A=-137.928	72Gf	
						B=1.306856		
						C=-4.237067x10^{-3}		
						D=4.691358x10^{-6}		
						ΔCp(25°,1)=131.4		
	-5.699	(2)	--	5	C	μ=1.0M($NaClO_4$)	72Gf	
	-4.347	(2)	--	20	C	μ=1.0M($NaClO_4$)	72Gf	
	-4.318	3.88(2)	59.8	25	C	μ=1.0	63Gd	
	-3.310	(2)	--	35	C	μ=1.0M($NaClO_4$)	72Gf	
	-2.067	(2)	--	50	C	μ=1.0M($NaClO_4$)	72Gf	
		(2)				$\Delta H=A+BT+CT^2+DT^3$		
						A=-174.819	72Gf	
						B=1.694176		
						C=-5.5552124x10^{-3}		
						D=6.123457x10^{-6}		
						ΔCp(25°,2)=68.6		
	-1.402	(3)	--	5	C	μ=1.0M($NaClO_4$)	72Gf	
	-1.385	(3)	--	20	C	μ=1.0M($NaClO_4$)	72Gf	
	-3.297	2.39(3)	34.7	25	C	μ=1.0	63Gd	
	-1.891	(3)	--	35	C	μ=1.0M($NaClO_4$)	72Gf	
	-2.280	(3)	--	50	C	μ=1.0M($NaClO_4$)	72Gf	
		(3)				$\Delta H=A+BT+CT^2+DT^3$		
						A=-212.924	72Gf	
						B=2.105303		
						C=-6.922738x10^{-3}		
						D=7.555556x10^{-6}		
						ΔCp(25°,3)=-32.6		
Nd^{3+}	-3.548	5.44(1)	92.5	25	C	μ=1.0	63Gd	
	-5.251	4.04(2)	59.8	25	C	μ=1.0	63Gd	
	-3.753	2.65(3)	37.6	25	C	μ=1.0	63Gd	
Sm^{3+}	-7.000	(1)	--	5	C	μ=1.0M($NaClO_4$)	72Gf	
	-4.853	(1)	--	20	C	μ=1.0M($NaClO_4$)	72Gf	
	-4.385	5.53(1)	91.2	25	C	μ=1.0	63Gd	
	-2.753	(1)	--	35	C	μ=1.0M($NaClO_4$)	72Gf	
	-0.339	(1)	--	50	C	μ=1.0M($NaClO_4$)	72Gf	
		(1)				$\Delta H=A+BT+CT^2+DT^3$		
						A=-119.900	72Gf	q:
						B=1.142185		±0.067
						C=-3.759520x10^{-3}		
						D=4.246914x10^{-6}		
						ΔCp(25°,1)=138.1		
	-9.050	(2)	--	5	C	μ=1.0M($NaClO_4$)	72Gf	
	-7.678	(2)	--	20	C	μ=1.0M($NaClO_4$)	72Gf	

Metal Ion	ΔH, kJ /mol	Log K	ΔS, J /mol K	T,°C	M	Conditions	Ref.	Re-marks
A46, continued								
Sm^{3+}, cont.	-7.657	4.33(2)	57.3	25	C	μ=1.0	63Gd	
	-6.665	(2)	--	35	C	μ=1.0M($NaClO_4$)	72Gf	
	-5.694	(2)	--	50	C	μ=1.0M($NaClO_4$)	72Gf	
		(2)				$\Delta H=A+BT+CT^2+DT^3$		
						A=-159.179	72Gf	
						B=1.505553		
						$C=-4.839253x10^{-3}$		
						$D=5.234568x10^{-6}$		
						ΔCp(25°,2)=66.5		
	-4.322	(3)	--	5	C	μ=1.0M($NaClO_4$)	72Gf	
	-4.477	(3)	--	20	C	μ=1.0M($NaClO_4$)	72Gf	
	-5.912	2.89(3)	35.1	25	C	μ=1.0	63Gd	
	-5.125	(3)	--	35	C	μ=1.0M($NaClO_4$)	72Gf	
	-5.732	(3)	--	50	C	μ=1.0M($NaClO_4$)	72Gf	
		(3)				$\Delta H=A+BT+CT^2+DT^3$		
						A=-180.570	72Gf	
						B=1.775654		
						$C=-5.821404x10^{-3}$		
						$D=6.320988x10^{-6}$		
						ΔCp(25°,3)=-41.8		
Eu^{3+}	-3.268	5.52(1)	94.6	25	C	μ=1.0	63Gd	
	9.046	4.48(2)	55.6	25	C	μ=1.0	63Gd	
	6.544	3.15(3)	38.1	25	C	μ=1.0	63Gd	
Gd^{3+}	-4.084	(1)	--	5	C	μ=1.0M($NaClO_4$)	72Gf	
	-1.992	5.490(1)	97.9	20	C	μ=1.0M($NaClO_4$)	72Gf	
	-1.506	5.40(1)	98.3	25	C	μ=1.0	63Gd	
	0.142	(1)	--	35	C	μ=1.0M($NaClO_4$)	72Gf	
	2.519	(1)	--	50	C	μ=1.0M($NaClO_4$)	72Gf	
		(1)				$\Delta H=A+BT+CT^2+DT^3$		
						A=-68.001	72Gf	
						B=0.631253		
						$C=-2.062471x10^{-3}$		q:
						$D=2.370370x10^{-6}$		±0.067
						ΔCp(25°,1)=140.2		
	-11.322	(2)	--	5	C	μ=1.0M($NaClO_4$)	72Gf	
	-9.878	4.543(2)	53.1	20	C	μ=1.0M($NaClO_4$)	72Gf	
	-9.673	4.50(2)	53.5	25	C	μ=1.0	63Gd	
	-8.900	(2)	--	35	C	μ=1.0M($NaClO_4$)	72Gf	
	-7.360	(2)	--	50	C	μ=1.0M($NaClO_4$)	72Gf	
		(2)				$\Delta H=A+BT+CT^2+DT^3$		
						A=-333.248	72Gf	
						B=3.280608		
						$C=-10.887279x10^{-3}$		
						$D=12.098765x10^{-6}$		
						ΔCp(25°,2)=62.8		
	-5.602	(3)	--	5	C	μ=1.0M($NaClO_4$)	72Gf	
	-6.201	3.187(3)	15.5	20	C	μ=1.0M($NaClO_4$)	72Gf	
	-8.025	3.08(3)	32.2	25	C	μ=1.0	63Gd	
	-7.171	(3)	--	35	C	μ=1.0M($NaClO_4$)	72Gf	
	-8.180	(3)	--	50	C	μ=1.0M($NaClO_4$)	72Gf	
		(3)				$\Delta H=A+BT+CT^2+DT^3$		
						A=-114.090	72Gf	
						B=1.121153		

Metal Ion	ΔH, kJ /mol	Log K	ΔS, J /mol K	T,°C	M	Conditions	Ref.	Re-marks
A46, continued								
Gd^{3+}, cont.						$C=-3.672267x10^{-3}$		
						$D=3.950617x10^{-6}$		
						$\Delta Cp(25°,3)=-63.2$		
Tb^{3+}	0.393	(1)	--	5	C	$\mu=1.0M(NaClO_4)$	72Gf	
	2.276	5.422(1)	96.2	20	C	$\mu=1.0M(NaClO_4)$	72Gf	
	3.201	5.34(1)	113.0	25	C	$\mu=1.0$	63Gd	
	4.117	(1)	--	35	C	$\mu=1.0M(NaClO_4)$	72Gf	
	6.188	(1)	--	50	C	$\mu=1.0M(NaClO_4)$	72Gf	
		(1)				$\Delta H=A+BT+CT^2+DT^3$		
						$A=-90.724$	72Gf	q:
						$B=0.869571$		±0.038
						$C=-2.845244x10^{-3}$		
						$D=3.209877x10^{-6}$		
						$\Delta Cp(25°,1)=121.3$		
	-12.757	(2)	--	5	C	$\mu=1.0M(NaClO_4)$	72Gf	
	-10.983	4.657(2)	51.9	20	C	$\mu=1.0M(NaClO_4)$	72Gf	
	-11.096	4.61(2)	51.0	25	C	$\mu=1.0$	63Gd	
	-9.305	(2)	--	35	C	$\mu=1.0M(NaClO_4)$	72Gf	
	-7.159	(2)	--	50	C	$\mu=1.0M(NaClO_4)$	72Gf	
		(2)				$\Delta H=A+BT+CT^2+DT^3$		
						$A=-182.606$	72Gf	
						$B=1.774823$		
						$C=-5.914311x10^{-3}$		
						$D=6.666667x10^{-6}$		
						$\Delta Cp(25°,2)=108.8$		
	-8.293	(3)	--	5	C	$\mu=1.0M(NaClO_4)$	72Gf	
	-9.418	3.329(3)	31.8	20	C	$\mu=1.0M(NaClO_4)$	72Gf	
	-10.782	3.25(3)	25.9	25	C	$\mu=1.0$	63Gd	
	-11.092	(3)	--	35	C	$\mu=1.0M(NaClO_4)$	72Gf	
	-12.790	(3)	--	50	C	$\mu=1.0M(NaClO_4)$	72Gf	
		(3)				$\Delta H=A+BT+CT^2+DT^3$		
						$A=-175.850$	72Gf	
						B-1.738528		
						$C=-5.720000x10^{-3}$		
						$D=6.172840x10^{-6}$		
						$\Delta Cp(25°,3)=-109.2$		
Dy^{3+}	2.402	(1)	--	5	C	$\mu=1.0M(NaClO_4)$	72Gf	
	4.535	(1)	--	20	C	$\mu=1.0M(NaClO_4)$	72Gf	
	5.523	5.33(1)	120.5	25	C	$\mu=1.0$	63Gd	
	6.142	(1)	--	35	C	$\mu=1.0M(NaClO_4)$	72Gf	
	8.535	(1)	--	50	C	$\mu=1.0M(NaClO_4)$	72Gf	
		(1)				$\Delta H=A+BT+CT^2+DT^3$		
						$A=-63.978$	72Gf	
						$B=0.604455$		
						$C=-1.956622x10^{-3}$		
						$D=2.222222x10^{-6}$		
						$\Delta Cp(25°,1)=126.8$		
	-11.853	(2)	--	5	C	$\mu=1.0M(NaClO_4)$	72Gf	
	-10.029	(2)	--	20	C	$\mu=1.0M(NaClO_4)$	72Gf	
	-10.058	4.64(2)	55.2	25	C	$\mu=1.0$	63Gd	
	-7.657	(2)	--	35	C	$\mu=1.0M(NaClO_4)$	72Gf	
	-5.958	(2)	--	50	C	$\mu=1.0M(NaClO_4)$	72Gf	
		(2)				$\Delta H=A+BT+CT^2+DT^3$		

Metal Ion	ΔH, kJ /mol	Log K	ΔS, J /mol K	T,°C	M	Conditions	Ref.	Remarks
A46, continued								
Dy^{3+}, cont.						A=-14.249	72Gf	
						B=0.052939		
						C=-0.056764x10^{-3}		
						D=0.049383x10^{-6}		
						ΔCp(25°,2)=135.1		
	-10.870	(3)	--	5	C	μ=1.0M($NaClO_4$)	72Gf	
	-12.452	(3)	--	20	C	μ=1.0M($NaClO_4$)	72Gf	
	-13.941	3.34(3)	17.2	25	C	μ=1.0	63Gd	
	-14.435	(3)	--	35	C	μ=1.0M($NaClO_4$)	72Gf	
	-16.288	(3)	--	50	C	μ=1.0M($NaClO_4$)	72Gf	
		(3)				ΔH=A+BT+CT^2+DT^3		
						A=-170.854	72Gf	
						B=1.712268		
						C=-5.729084x10^{-3}		
						D=6.271605x10^{-6}		
						ΔCp(25°,3)=-131.8		
Ho^{3+}	3.987	(1)	--	5	C	μ=1.0M($NaClO_4$)	72Gf	
	5.782	5.396(1)	123.0	20	C	μ=1.0M($NaClO_4$)	72Gf	
	6.657	5.30(1)	123.8	25	C	μ=1.0	63Gd	
	7.481	(1)	--	35	C	μ=1.0M($NaClO_4$)	72Gf	
	9.297	(1)	--	50	C	μ=1.0M($NaClO_4$)	72Gf	
		(1)				ΔH=A+BT+CT^2+DT^3		
						A=-74.458	72Gf	
						B=0.706580		
						C=-2.266098x10^{-3}		
						D=2.518519x10^{-6}		
						ΔCp(25°,1)=112.5		
	-10.326	(2)	--	5	C	μ=1.0M($NaClO_4$)	72Gf	
	-7.807	4.676(2)	62.8	20	C	μ=1.0M($NaClO_4$)	72Gf	
	-7.749	4.64(2)	62.8	25	C	μ=1.0	63Gd	
	-5.606	(2)	--	35	C	μ=1.0M($NaClO_4$)	72Gf	
	-3.615	(2)	--	50	C	μ=1.0M($NaClO_4$)	72Gf	
		(2)				ΔH=A+BT+CT^2+DT^3		
						A=-59.668	72Gf	
						B=0.467373		
						C=-1.298098x10^{-3}		
						D=1.283951x10^{-6}		
						ΔCp(25°,2)=149.4		
	-14.016	(3)	--	5	C	μ=1.0M($NaClO_4$)	72Gf	
	-15.870	3.392(3)	10.9	20	C	μ=1.0M($NaClO_4$)	72Gf	
	-17.251	3.32(3)	5.8	25	C	μ=1.0	63Gd	
	-17.824	(3)	--	35	C	μ=1.0M($NaClO_4$)	72Gf	
	-19.539	(3)	--	50	C	μ=1.0M($NaClO_4$)	72Gf	
		(3)				ΔH=A+BT+CT^2+DT^3		
						A=-100.000	72Gf	
						B=1.031351		
						C=-3.571253x10^{-3}		
						D=4.000000x10^{-6}		
						ΔCp(25°,3)=131.8		
Er^{3+}	4.297	(1)	--	5	C	μ=1.0M($NaClO_4$)	72Gf	
	6.134	5.470(1)	125.5	20	C	μ=1.0M($NaClO_4$)	72Gf	
	6.945	5.36(1)	125.9	25	C	μ=1.0	63Gd	
	7.816	(1)	--	35	C	μ=1.0M($NaClO_4$)	72Gf	

Metal Ion	ΔH, kJ /mol	Log K	ΔS, J /mol K	T,°C	M	Conditions	Ref.	Remarks
A46, continued								
Er^{3+}, cont.	9.531	(1)	--	50	C	μ=1.0M($NaClO_4$)	72Gf	
		(1)				ΔH=A+BT+CT^2+DT^3		
						A=-69.661	72Gf	
						B=0.648694		
						C=-2.036622x10^{-3}		
						D=2.222222x10^{-6}		
						ΔCp(25°,1)=112.5		
	-6.937	(2)	--	5	C	μ=1.0M($NaClO_4$)	72Gf	
	-4.686	4.688(2)	73.6	20	C	μ=1.0M($NaClO_4$)	72Gf	
	-4.021	4.67(2)	75.7	25	C	μ=1.0	63Gd	
	-3.054	(2)	--	35	C	μ=1.0M($NaClO_4$)	72Gf	
	-1.565	(2)	--	50	C	μ=1.0M($NaClO_4$)	72Gf	
		(2)				ΔH=A+BT+CT^2+DT^3		
						A=-179.921	72Gf	
						B=1.673979		
						C=-5.280036x10^{-3}		
						D=5.629630x10^{-6}		
						ΔCp(25°,2)=112.1		
	-17.322	(3)	--	5	C	μ=1.0M($NaClO_4$)	72Gf	
	-18.615	3.248(3)	-1.30	20	C	μ=1.0M($NaClO_4$)	72Gf	
	-20.506	3.15(3)	-8.4	25	C	μ=1.0	63Gd	
	-20.146	(3)	--	35	C	μ=1.0M($NaClO_4$)	72Gf	
	-21.384	(3)	--	50	C	μ=1.0M($NaClO_4$)	72Gf	
		(3)				ΔH=A+BT+CT^2+DT^3		
						A=-166.338	72Gf	
						B=1.667354		
						C=-5.642418x10^{-3}		
						D=6.271605x10^{-6}		
						ΔCp(25°,3)=-103.3		
Tm^{3+}	6.586	5.51(1)	127.6	25	C	μ=1.0	63Gd	
	-1.711	4.73(2)	84.9	25	C	μ=1.0	63Gd	
	-20.853	3.00(3)	-12.6	25	C	μ=1.0	63Gd	
Yb^{3+}	3.414	(1)	--	5	C	μ=1.0M($NaClO_4$)	72Gf	
	5.485	(1)	--	20	C	μ=1.0M($NaClO_4$)	72Gf	
	5.954	5.57(1)	126.4	25	C	μ=1.0	63Gd	
	7.042	(1)	--	35	C	μ=1.0M($NaClO_4$)	72Gf	
	9.029	(1)	--	50	C	μ=1.0M($NaClO_4$)	72Gf	
		(1)				ΔH=A+BT+CT^2+DT^3		
						A=-101.812	72Gf	
						B=0.984906		
						C=-3.217138x10^{-3}		
						D=3.604938x10^{-6}		
						ΔCp(25°,1)=116.7		
	-2.251	(2)	--	5	C	μ=1.0M($NaClO_4$)	72Gf	
	-1.506	(2)	--	20	C	μ=1.0M($NaClO_4$)	72Gf	
	-1.577	4.81(2)	87.0	25	C	μ=1.0	63Gd	
	-0.569	(2)	--	35	C	μ=1.0M($NaClO_4$)	72Gf	
	0.167	(2)	--	50	C	μ=1.0M($NaClO_4$)	72Gf	
		(2)				ΔH=A+BT+CT^2+DT^3		
						A=-88.148	72Gf	
						B=0.836912		
						C=-2.686880x10^{-3}		q:
						D=2.913580x10^{-6}		±0.063

Metal Ion	ΔH, kJ /mol	Log K	ΔS, J /mol K	T,°C	M	Conditions	Ref.	Remarks
A46, continued								
Yb^{3+}, cont.						$\Delta Cp(25°,2)=50.0$		q: ±0.079
	-18.857	(3)	--	5	C	$\mu=1.0M(NaClO_4)$	72Gf	
	-18.782	(3)	--	20	C	$\mu=1.0M(NaClO_4)$	72Gf	
	-20.514	2.75(3)	-16.3	25	C	$\mu=1.0$	63Gd	
	-20.050	(3)	--	35	C	$\mu=1.0M(NaClO_4)$	72Gf	
	-21.121	(3)	--	50	C	$\mu=1.0M(NaClO_4)$	72Gf	
		(3)				$\Delta H=A+BT+CT^2+DT^3$		
						$A=-519.676$	72Gf	
						$B=5.090126$		
						$C=-16.695982\times10^{-3}$		
						$D=18.172840\times10^{-6}$		
						$\Delta Cp(25°,3)=-81.2$		
Lu^{3+}	3.151	(1)	--	5	C	$\mu=1.0M(NaClO_4)$	72Gf	
	4.920	5.793(1)	127.6	20	C	$\mu=1.0M(NaClO_4)$	72Gf	
	5.146	5.65(1)	125.5	25	C	$\mu=1.0$	63Gd	
	6.690	(1)	--	35	C	$\mu=1.0M(NaClO_4)$	72Gf	
	8.577	(1)	--	50	C	$\mu=1.0M(NaClO_4)$	72Gf	
		(1)				$\Delta H=A+BT+CT^2+DT^3$		
						$A=-41.837$	72Gf	
						$B=0.384392$		
						$C=-1.216071\times10^{-3}$		
						$D=1.382716\times10^{-6}$		
						$\Delta Cp(25°,1)=117.2$		
	-3.071	(2)	--	5	C	$\mu=1.0M(NaClO_4)$	72Gf	
	-1.799	4.960(2)	--	20	C	$\mu=1.0M(NaClO_4)$	72Gf	
	-1.879	4.91(2)	87.4	25	C	$\mu=1.0$	63Gd	
	-0.962	(2)	--	35	C	$\mu=1.0M(NaClO_4)$	72Gf	
	-0.423	(2)	--	50	C	$\mu=1.0M(NaClO_4)$	72Gf	
		(2)				$\Delta H=A+BT+CT^2+DT^3$		
						$A=-66.168$	72Gf	
						$B=0.572103$		
						$C=-1.664338\times10^{-3}$		
						$D=1.629630\times10^{-6}$		
						$\Delta Cp(25°,2)=59.4$		q: ±0.067
	-17.117	(3)	--	5	C	$\mu=1.0M(NaClO_4)$	72Gf	
	-17.163	2.669(3)	--	20	C	$\mu=1.0M(NaClO_4)$	72Gf	
	-19.246	2.55(3)	-11.3	25	C	$\mu=1.0$	63Gd	
	-17.615	(3)	--	35	C	$\mu=1.0M(NaClO_4)$	72Gf	
	-17.284	(3)	--	50	C	$\mu=1.0M(NaClO_4)$	72Gf	
		(3)				$\Delta H=A+BT+CT^2+DT^3$		
						$A=-373.891$	72Gf	
						$B=3.735225$		
						$C=-12.549991\times10^{-3}$		
						$D=14.024691\times10^{-6}$		
						$\Delta Cp(25°,3)=-34.3$		
Ni^{2+}	11.92	2.3(1)	84.1	25	C	$\mu=1(NaNO_3)$	78Ah	
Cu^{2+}	7.95	3.6(1)	95.4	25	C	$\mu=1(NaNO_3)$	78Ah	
Zn^{2+}	7.95	3.3(1)	90.0	25	C	$\mu=1(NaNO_3)$	78Ah	
Cd^{2+}	-0.25	3.0(1)	56.5	25	C	$\mu=1(NaNO_3)$	78Ah	

Metal Ion	ΔH, kJ /mol	Log K	ΔS, J /mol K	T,°C	M	Conditions	Ref.	Re-marks
A47	ACETIC ACID (phenylthio)		$C_8H_8O_2S$			$C_6H_5SCH_2CO_2H$		
Ni^{2+}	2.9	0.5(1)	19.5	25	C	μ=1.0($NaNO_3$)	79Ac	
Cu^{2+}	6.7	1.3(1)	47.3	25	C	μ=1.0($NaNO_3$)	79Ac	
Cd^{2+}	4.2	1.0(1)	33.2	25	C	μ=1.0($NaNO_3$)	79Ac	
A48	ACETIC ACID, thio-, S-(3-butenyl)-		$C_6H_{10}O_2S$			$CH_2=CH(CH_2)_2SCH_2CO_2H$		L^-
Ag^+	-46.11	4.77(25°,1)	-63.39	0-40	T	μ=0.20M pH=5.8	71Bd	b,q:±5.8
	-79.9	2.26(25°,2)	-224.7	0-40	T	μ=0.20M pH=5.8	71Bc	b,q:±10
	7.9	2.07(25°)	66.9	0-40	T	μ=0.20M R:$AgL^+ + Ag^+ = Ag_2L^+$ pH=5.8	71Bd	b,q: ±5.8
A49	ACETIC ACID, thio, S-(n-butyl)		$C_6H_{12}O_2S$			$CH_3(CH_2)_3SCH_2CO_2H$		L^-
Ag^+	-28.49	3.92(25°,1)	-20.46	0-40	T	μ=0.20M pH=5.8	71Bd 71Bc	b,q: ±5.0
	-30.00	3.56(25°,2)	-33.56	0-40	T	μ=0.20M pH=5.8	71Bd 71Bc	b,q: ±2.9
A50	ACETIC ACID, thio, S-(4-bentyl)-		$C_7H_{12}O_2S$			$CH_2:CH(CH_2)_3SCH_2CO_2H$		
Ag^+	-35.10	4.19(25°,1)	-37.57	0-40	T	μ=0.20M pH=5.8	71Bd 71Bc	b,q: ±2.9
	-20.00	2.82(25°,2)	-13.43	0-40	T	μ=0.20M pH=5.8	71Bd 71Bc	b,q: ±10.0
	-31.00	1.91(25°)	-67.07	0-40	T	μ=0.20M R:$AgL+Ag^+=Ag_2L^+$ pH=5.8	71Bd	b,q: ±2.9
A51	ACETIC ACID, thio-S-(allyl)-		$C_5H_8O_2S$			$CH_2:CHCH_2SCH_2CO_2H$		L^-
Ag^+	-28.91	3.75(25°,1)	-25.14	0-40	T	μ=0.20M pH=5.8	71Ed	b,q:±4
	-20.00	2.96(25°,2)	-10.04	0-40	T	μ=0.20M pH=5.8	71Ed	b,q: ±2.9
A52	ACETIC ACID, thiolo- S-methyl ester		C_3H_6OS			CH_3COSCH_3		L
I_2	-13.18	-0.143(25°,1)	-46.9	24-41	T	CM=$3.5x10^{-3}$M CL=0.33M S:CCl_4	64Ma	b
A53	ACETIC ACID, 2,2,2-trichloro-, amide, N,N-dimethyl-		$C_4H_6ONCl_3$			$CCl_3CON(CH_3)_2$		L
I_2	-10.5	-0.523(25°,1)	-44.8	14-43	T	CM=$3x10^{-3}$-$7x10^{-3}$	62Dc	b

Metal Ion	ΔH, kJ /mol	Log K	ΔS, J /mol K	T,°C	M	Conditions	Ref.	Remarks
A53, continued								
I_2, cont.						CL=4x10^{-2}-0.18 S:CCl_4		
A54	ACETIC ACID, 2,2,2-trichloro-, ethyl ester	$C_4H_5O_2Cl_3$				$CCl_3CO_2CH_2CH_3$		L
SbA_5	-12.76	-1.17(1)	-65.3	25	C	S:Ethylene chloride A=Cl	670d	
A55	ACETIC ACID, trichloro-, nitrile	C_2NCl_3				Cl_3CCN		L
IA	-22.6	-0.70(25°,1)	-89.1	0-40	T	S:CCl_4 A=Cl	63Pc	b
A56	ACETIC ACID, 2,2,2-trifluoro-, amide, N,N-dimethyl-	$C_4H_6ONF_3$				$CF_3CON(CH_3)_2$		L
SbA_5	-69.37	(1)	--	25	C	S:Ethylene chloride A=Cl	670a	
A57	ACETIC ACID, 2,2,2-trifluoro-, methyl ester	$C_3H_3O_2F_3$				$CF_3CO_2CH_3$		L
SbA_5	-11.3	-0.222(1)	-68.6	25	C	S:Ethylene chloride A=Cl	670e	
A58	ACETIC ACID, vinyl ester	$C_4H_6O_2$				$CH_3CO_2CH:CH_2$		
$[NiA_2]_2$	-15.3	0.67(100°,1)	-28	75-125	T	S:Squalane A=3-Trifluoro-acetylcamphorate	74Sj	b,gg
	-12.6	1.28(100°,1)	-9.2	75-125	T	S:Squalane A=3-Trifluoro-acetylcamphorate	74Sj	b,gg
A59	ACETALDEHYDE	C_2H_4O				CH_3CHO		
$[NiA_2]_2$	-12.6	-1.28(100°,1)	-9.2	75-125	T	S:Squalane A=3-Trifluoro-acetylcamphorate	74Sj	b,gg
A60	ACETOPHENONE	C_8H_8O				$CH_3COC_6H_5$		L
I_2	-14.2	0.060(25°,1)	-46	5-25	T	S:Heptane	71Td	b
A61	ACETOPHENONE, 2-hydroxy	$C_8H_8O_2$				$CH_3COC_6H_4OH$		
Cu^{2+}	-56.11	22.14(25°,1-2)	209.2	15&40	T	S:75v%Dioxane	79Sb 78Sg	a,b,p,y

Metal Ion	ΔH, kJ /mol	Log K	ΔS, J /mol K	T,°C	M	Conditions	Ref.	Re-marks
A62	ACRIDINE		$C_{13}H_9N$			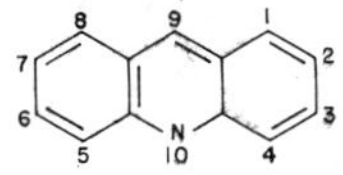		L
CoA_2	-4.2	2.27(25°,1)	29.3	25	C	S:Acetone A=Cl	73Lb	q:±0.4
A63	ACROLEIN							
$[NiA_2]_2$	-11.1	1.20(100°,1)	-6.9	75-125	T	S:Squalane A=3-Trifluoro-acetylcamphorate	74Sj	b,gg
A64	ADENOSINE-5'-diphosphoric acid		$C_{10}H_{15}O_{10}N_5P_2$			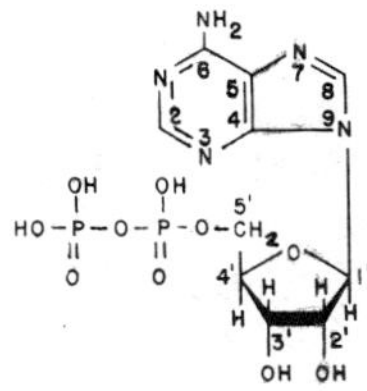		L^{3-}
Mg^{2+}	18.0	4.27(1)	141.8	25	T	μ=0	66Pe 67Ka	
	24.27	4.10(1)	159	25	T	μ=0	63Ga	i:K. Burton Biochem. J.,71, 388 (1959)
	3.8	2.45(1)	59.8	25	T	μ=0 $L=HL^{2-}$	66Pe 63Ga	
	8.4	2.00(1)	66.5	25	T	μ=0.1 $L=HL^{2-}$	66Pe	
	15.1	3.17(25°,1)	111.3	5-35	T	μ=0.1M(KNO_3)	73Bb	b,g:Na
	16.3	1.64(25°,1)	86.2	5-35	T	μ=0.1M(KNO_3) $L=HL^{3-}$	73Bb	b,g:Na
	13.18	3.695(1)	114.2	30	C	μ=0.2 (tetra-methylammonium chloride) pH=8.5 Conc. triethanol-amine buffer =0.15M	69Bf	g:Na
	13.30	3.66(1)	114.2	30	C	μ=0.2 (tetra-methylammonium chloride) pH=8.50	73Sa	
	14.06	3.48(1)	113	30	C	μ=0.2 (Tetra-methylammonium chloride) pH=8.5	74Sb	
Ca^{2+}	-5.0	2.86(25°,1)	38.1	0-40	T	μ=0.1(KNO_3)	67Ka	b,c,g:Na
	-2.5	1.58(25°,1)	21.8	0-40	T	μ=0.1(KNO_3) $L=HL^{2-}$	67Ka	b,c,g: Na

Metal Ion	ΔH, kJ /mol	Log K	ΔS, J /mol K	T,°C	M	Conditions	Ref.	Re-marks
A64, continued								
Sr^{2+}	-11.21	2.54(25°,1)	23.0	0-40	T	μ=0.1(KNO_3)	67Ka	b,c,g:Na
	-5.0	1.52(25°,1)	12.5	0-40	T	μ=0.1(KNO_3) L=HL^{2-}	67Ka	b,c,g: Na
Ba^{2+}	-12.1	2.36(25°,1)	4.6	0-40	T	μ=0.1(KNO_3)	67Ka	b,c,g: Na
	-7.5	1.44(25°,1)	2.5	0-40	T	μ=0.1(KNO_3) L=HL^{2-}	67Ka	b,c,g: Na
Mn^{2+}	10.9	4.982(24.8°,1)	130	1.5-47.2	T	CL=$2x10^{-5}$M - $1.0x10^{-4}$M CM≅$1x10^{-4}$M 50mM n-Ethyl morpholine buffer	70Ja	b
	-10.0	4.16(25°,1)	46.0	0-40	T	μ=0.1(KNO_3)	67Ka	b,c,g:Na
	-7.9	1.88(25°,1)	9.2	0-40	T	μ=0.1(KNO_3) L=HL^{2-}	67Ka	b,c,g: Na
Co^{2+}	-8.4	4.19(25°,1)	52.3	0-40	T	μ=0.1(KNO_3)	67Ka	b,c,g:Na
	-7.9	2.01(25°,1)	11.7	0-40	T	μ=0.1(KNO_3) L=HL^{2-}	67Ka	b,c,g: Na
Ni^{2+}	-7.9	4.49(25°,1)	59.0	0-40	T	μ=0.1(KNO_3)	67Ka	b,c,g:Na
	-8.8	2.29(25°,1)	13.8	0-40	T	μ=0.1(KNO_3) L=HL^{2-}	67Ka	b,c,g: Na
	-6.3	3.99(1)	80.4	5	T	0.2M(Tetramethyl-ammonium bromide) ΔCp=845	79Me	
	0.8	3.98(1)	77.0	15	T	0.2M(Tetramethyl-ammonium bromide) ΔCp=582	79Me	
	5.4	4.00(1)	94.6	25	T	0.2M(Tetramethyl-ammonium bromide) ΔCp=343	79Me	
	6.3	4.2(1)	93.8	25	C	0.2M(Tetramethyl-ammonium bromide)	79Me	
	7.9	4.04(1)	103.0	37	T	0.2M(Tetramethyl-ammonium bromide) ΔCp=84	79Me	
Cu^{2+}	-17.2	5.89(25°,1)	54.4	0-40	T	μ=0.1(KNO_3)	67Ka	b,c,g:Na
	-11.3	2.62(25°,1)	12.6	0-40	T	μ=0.1(KNO_3) L=HL^{2-}	67Ka	b,c,g: Na
Ag^{+}	-22.97	2.02(25°,1)	-38.1	10-40	T	μ=0	68Pb	b,q:±0.5
	-15.31	1.84(25°,2)	-15.9	10-40	T	μ=0	68Pb	b,q:±1.8
Zn^{2+}	-8.4	4.27(25°,1)	52.3	0-40	T	μ=0.1(KNO_3)	67Ka	b,c,g:Na
	-7.9	2.04(25°,1)	12.6	0-40	T	μ=0.1(KNO_3) L=HL^{2-}	67Ka	b,c,g: Na

Metal Ion	ΔH, kJ /mol	Log K	ΔS, J /mol K	T,°C	M	Conditions	Ref.	Re-marks
A65	ADENOSINE-2'-monophosphoric acid		$C_{10}H_{14}O_7N_5P$			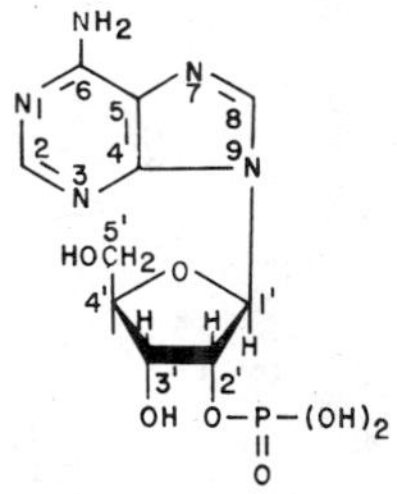		L^{2-}
Mg^{2+}	14.6	1.93(25°,1)	85.8	0-40	T	μ=0.1(KNO_3)	67Ka	b
Ca^{2+}	-2.5	1.82(25°,1)	27.2	0-40	T	μ=0.1(KNO_3)	67Ka	b
Sr^{2+}	-4.2	1.74(25°,1)	18.8	0-40	T	μ=0.1(KNO_3)	67Ka	b
Ba^{2+}	-8.4	1.71(25°,1)	5.0	0-40	T	μ=0.1(KNO_3)	67Ka	b
Mn^{2+}	-4.2	2.37(25°,1)	31.4	0-40	T	μ=0.1(KNO_3)	67Ka	b
Co^{2+}	-2.9	2.34(25°,1)	35.6	0-40	T	μ=0.1(KNO_3)	67Ka	b
Ni^{2+}	-4.2	2.81(25°,1)	39.7	0-40	T	μ=0.1(KNO_3)	67Ka	b
Cu^{2+}	-7.9	3.16(25°,1)	33.5	0-40	T	μ=0.1(KNO_3)	67Ka	b
Zn^{2+}	-5.0	2.64(25°,1)	33.5	0-40	T	μ=0.1(KNO_3)	67Ka	b
A66	ADENOSINE-3'-monophosphoric acid		$C_{10}H_{14}O_7N_5P$					L^{2-}
Mg^{2+}	14.6	1.89(25°,1)	85.8	0-40	T	μ=0.1(KNO_3)	67Ka	b
Ca^{2+}	-2.5	1.80(25°,1)	27.2	0-40	T	μ=0.1(KNO_3)	67Ka	b
Sr^{2+}	-3.8	1.71(25°,1)	18.8	0-40	T	μ=0.1(KNO_3)	67Ka	b
Ba^{2+}	-7.9	1.68(25°,1)	5.0	0-40	T	μ=0.1(KNO_3)	67Ka	b
Mn^{2+}	-3.8	2.28(25°,1)	31.8	0-40	T	μ=0.1(KNO_3)	67Ka	b
Co^{2+}	-2.5	2.24(25°,1)	35.6	0-40	T	μ=0.1(KNO_3)	67Ka	b
Ni^{2+}	-4.2	2.78(25°,1)	40.2	0-40	T	μ=0.1(KNO_3)	67Ka	b
Cu^{2+}	-7.1	2.95(25°,1)	33.5	0-40	T	μ=0.1(KNO_3)	67Ka	b
Zn^{2+}	-4.6	2.59(25°,1)	34.3	0-40	T	μ=0.1(KNO_3)	67Ka	b
A67	ADENOSINE-5'-monophosphoric acid		$C_{10}H_{14}O_7N_5P$					L^{2-}
Mg^{2+}	14.2	1.96(25°,1)	85.4	0-40	T	μ=0.1(KNO_3)	67Ka	b
	7.40	1.79(1)	58.6	30	C	μ=0.2 $(CH_3)_4NCl$ pH=8.50	73Sa	
	7.45	1.806(1)	59.0	30	C	μ=0.20 $(CH_3)_4NCl$ pH=8.5 Conc. triethanolamine buffer=0.15M	69Bf	
Ca^{2+}	-2.5	1.85(25°,1)	26.8	0-40	T	μ=0.1(KNO_3)	67Ka	b
Sr^{2+}	-5.8	1.79(25°,1)	18.4	0-40	T	μ=0.1(KNO_3)	67Ka	b

Metal Ion	ΔH, kJ /mol	Log K	ΔS, J /mol K	T,°C	M	Conditions	Ref.	Remarks
A67, continued								
Ba^{2+}	-8.4	1.73(25°,1)	5.0	0-40	T	μ=0.1(KNO_3)	67Ka	b
Mn^{2+}	-4.2	2.40(25°,1)	31.8	0-40	T	μ=0.1(KNO_3)	67Ka	b
Co^{2+}	-4.6	2.40(25°,1)	35.1	0-40	T	μ=0.1(KNO_3)	67Ka	b
Ni^{2+}	-4.2	2.84(25°,1)	40.2	0-40	T	μ=0.1(KNO_3)	67Ka	b
	-23.4	2.67(1)	-33.1	5	T	0.2M $(CH_3)_4NBr$ ΔCp=933	79Me	
	-15.1	2.54(1)	-3.4	15	T	0.2M $(CH_3)_4NBr$ ΔCp=732	79Me	
	-8.8	2.46(1)	14.6	25	T	0.2M $(CH_3)_4NBr$ ΔCp=515	79Me	
	-10.4	2.48(1)	12.7	25	C	0.2M $(CH_3)_4NBr$	79Me	
	-4.2	2.42(1)	32.6	37	T	0.2M $(CH_3)_4NBr$ ΔCp=243	79Me	
Rh_2A_4	-22.7	3.45(1)	-10.0	25	C	A=MeO_2Ac	77Da	
	-21.8	3.66(1)	-5.8	25	C	A=OAc	77Da	
	-22.8	3.78(1)	-4.2	25	C	A=Propionate	77Da	
	-19.2	2.70(2)	-13.0	25	C	A=MeO_2Ac	77Da	q:±1.7
	-11.7	2.66(2)	11.3	25	C	A=OAc	77Da	q:±0.8
	-8.4	2.57(2)	20.9	25	C	A=Propionate	77Da	q:±0.8
Cu^{2+}	-8.4	3.17(25°,1)	33.5	0-40	T	μ=0.1(KNO_3)	67Ka	b
Zn^{2+}	-5.0	2.71(25°,1)	34.3	0-40	T	μ=0.1(KNO_3)	67Ka	b
A68	ADENOSINE-5'-triphosphoric acid		$C_{10}H_{16}O_{13}N_5P_3$					L^{4-}
Mg^{2+}	21.3	5.83(1)	182.8	25	T	μ=0	66Pe 66Ka	
	20.1	5.72(1)	176	25	T	μ=0	63Ga	i: K. Burton Biochem. J. 71, 388 (1959)
	9.2	3.59(1)	99.6	25	T	μ=0 L=HL^{3-}	66Pe 63Ga 57Nc	
	7.9	2.86(1)	81.2	25	T	μ=0.1 L=HL^{3-}	66Pe	
	10.9	4.22(25°,1)	117.2	5-35	T	μ=0.1M(KNO_3)	73Bb	b,g:Na
	14.2	2.24(25°,1)	90.4	5-35	T	μ=0.1M(KNO_3)	73Bb	b,g:Na
	17.238	3.343(25°,1)	121.8	1-43	T	μ=0.15(NaCl) pH=8.8	57Nc	b
	18.70	4.693(1)	151.5	30	C	μ=0.2 $(CH_3)_4NCl$	69Bf	g:Na

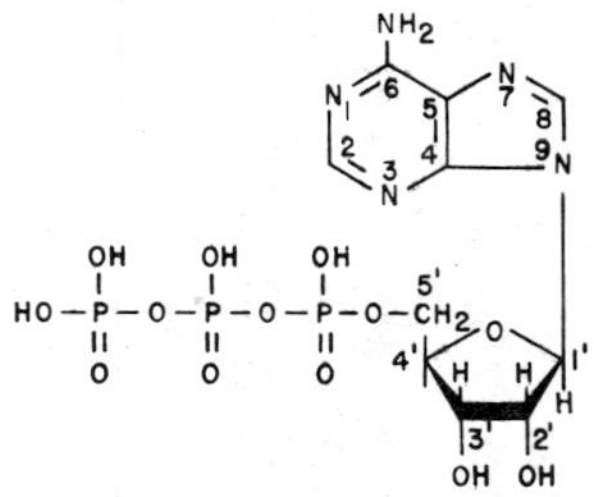

Metal Ion	ΔH, kJ /mol	Log K	ΔS, J /mol K	T,°C	M	Conditions	Ref.	Remarks
A68, continued								
Mg^{2+}, cont.						pH=8.5 Conc. triethanolamine buffer=0.15M		
	18.66	4.67(1)	151.0	30	C	μ=0.2 $(CH_3)_4NCl$ pH=8.50	73Sa	
Ca^{2+}	-3.8	3.97(25°,1)	50	0-40	T	0.1M(KNO_3)	66Ka 57Nc	b,c
	-1.3	2.17(25°,1)	38	0-40	T	0.1M(KNO_3) L=HL^{3-}	66Ka	b,c
	19.079	2.932(25°,1)	120.7	1-43	T	μ=0.15(NaCl) pH=8.8	57Nc	b
Sr^{2+}	-12.6	3.54(25°,1)	25	0-40	T	0.1M(KNO_3)	66Ka	b,c
	-6.7	2.05(25°,1)	17	0-40	T	0.1M(KNO_3) L=HL^{3-}	66Ka	b,c
Ba^{2+}	-16.3	3.29(25°,1)	8	0-40	T	0.1M(KNO_3)	66Ka	b,c
	-8.8	1.85(25°,1)	8	0-40	T	0.1M(KNO_3) L=HL^{3-}	66Ka	b,c
YA^-	-18.8	5.3(25°,1)	38	2-45	T	μ=0.10M(KNO_3) A=$EDTA^{2-}$	73Kg	b,g:Na q:±8.4
LaA^-	-14.2	4.5(25°,1)	38	2-45	T	μ=0.10M(KNO_3) A=$EDTA^{2-}$	73Kg	b,g:Na q:±5.8
CeA^-	-10.0	4.3(25°,1)	46	2-45	T	μ=0.10M(KNO_3) A=$EDTA^{2-}$	73Kg	b,g:Na q:±2.1
PrA^-	-10.0	4.4(25°,1)	50	2-45	T	μ=0.10M(KNO_3) A=$EDTA^{2-}$	73Kg	b,g:Na q:±1.7
NdA^-	-10.0	4.4(25°,1)	54	2-45	T	μ=0.10M(KNO_3) A=$EDTA^{2-}$	73Kg	b,g:Na q:±1.7
SmA^-	-14.6	4.9(25°,1)	46	2-45	T	μ=0.10M(KNO_3) A=$EDTA^{2-}$	73Kg	b,g:Na q:±4.6
EuA^-	-10.0	4.7(25°,1)	54	2-45	T	μ=0.10M(KNO_3) A=$EDTA^{2-}$	73Kg	b,g:Na q:±2.1
GdA^-	-11.3	4.3(25°,1)	46	2-45	T	μ=0.10M(KNO_3) A=$EDTA^{2-}$	73Kg	b,g:Na q:±2.5
TbA^-	-13.8	5.4(25°,1)	58	2-45	T	μ=0.10M(KNO_3) A=$EDTA^{2-}$	73Kg	b,g:Na q:±4.2
DyA^-	-19.7	5.7(25°,1)	42	2-45	T	μ=0.10M(KNO_3) A=$EDTA^{2-}$	73Kg	b,g:Na q:±7.5
HoA^-	-19.2	5.6(25°,1)	42	2-45	T	μ=0.10M(KNO_3) A=$EDTA^{2-}$	73Kg	b,g:Na q:±3.3
ErA^-	-15.0	6.1(25°,1)	67	2-45	T	μ=0.10M(KNO_3) A=$EDTA^{2-}$	73Kg	b,g:Na q:±2.1
TmA^-	-18.8	6.5(25°,1)	58	2-45	T	μ=0.10M(KNO_3) A=$EDTA^{2-}$	73Kg	b,g:Na q:±8.4
Mn^{2+}	-20.5	5.518(24.8°,1)	38	1.5-47.2	T	CL=$2x10^{-5}$-$1.0x10^{-4}$M	70Ja	b

Metal Ion	ΔH, kJ /mol	Log K	ΔS, J /mol K	T,°C	M	Conditions	Ref.	Remarks
A68, continued								
Mn^{2+}, cont.						CM≅1x10^{-4}M 50mM N-Ethyl morpholine buffer		
	-12.6	4.78(25°,1)	50	0-40	T	0.1M(KNO_3)	66Ka	b,c
	-9.6	2.39(25°,1)	13	0-40	T	0.1M(KNO_3) L=HL^{3-}	66Ka	b,c
[Fe]	-43.5	3.778(1)	-84	5	C	CCl^-=0 pH(initial)=7.32 [Fe]=Horse oxyhemoglobin	72He	q:±5.8
	-21.8	4.799(1)	18.8	25	C	CCl^-=0 pH(initial)=6.93 [Fe]=Horse oxyhemoglobin	72He	
	-43.1	5.114(1)	-58	5	C	CCl^-=0.005 pH(initial)=6.50 [Fe]=Horse oxyhemoglobin	72He	
	-18.8	2.903(1)	-6.3	25	C	CCl^-=0.10 pH(initial)=7.05 [Fe]=Horse oxyhemoglobin	72He	q:±2.1
	-28.2	2.167(14°)	-56.9	5-21	T	pH=6.2 [Fe]=Human methaemoglobin R:[Fe]+10L=[Fe]L_{10}	70Jb	b,j: Probably per mole L bound
	-28.66	2.201(17°)	-60.2	5&17	T	pH=7.2 [Fe]=Human methaemoglobin R:[Fe]+10L=[Fe]L_{10}	70Jb	b,j: Probably per mole L bound
Co^{2+}	-9.2	4.66(25°,1)	58	0-40	T	0.1M(KNO_3)	66Ka	b,c
	-8.8	2.32(25°,1)	17	0-40	T	0.1M(KNO_3) L=HL^{3-}	66Ka	b,c
Ni^{2+}	-10.5	5.02(25°,1)	63	0-40	T	0.1M(KNO_3)	66Ka	b,c
	-10.0	2.72(25°,1)	17	0-40	T	0.1M(KNO_3) L=HL^{3-}	66Ka	b,c
Cu^{2+}	-18.0	6.13(25°,1)	58	0-40	T	0.1M(KNO_3)	66Ka	b,c
	-12.6	3.12(25°,1)	17	0-40	T	0.1M(KNO_3) L=HL^{3-}	66Ka	b,c
Zn^{2+}	-11.3	4.85(25°,1)	54	0-40	T	0.1M(KNO_3)	66Ka	b,c
	-10.0	2.67(25°,1)	17	0-40	T	0.1M(KNO_3) L=HL^{3-}	66Ka	b,c
A69 ADRENALINE			$C_9H_{13}O_3N$					L
Be^{2+}	-104.6	9.65(25°,1)	-167	0-45	T	μ=0.12M(KCl) CM≅2x10^{-3}M CL≅8x10^{-3}M	69Cb	b

Metal Ion	ΔH, kJ /mol	Log K	ΔS, J /mol K	T,°C	M	Conditions	Ref.	Re-marks
A69, continued								
Be^{2+}, cont.	-99.6	6.31(25°,2)	-213	0-45	T	μ=0.12M(KCl) CM≅2x10^{-3}M CL≅8x10^{-3}M	69Cb	b
A70 α-ALANINE			$C_3H_7O_2N$			$CH_3CH(NH_2)CO_2H$		L^-
Eu^{3+}	5.02	0.743	31.0	25	C	μ=2.0($NaClO_4$) R:HL^++Eu^{3+}=$EuHL^{3+}$	71Aj	
U^{4+}	-192.5	10.3(22°,1)	-456	0-40	T	CM=1-8x10^{-3}M pH=2.3	76Nb	
	77.4	3.1(22°,2)	323	0-40	T	CM=1-8x10^{-3}M pH=2.3	76Nb	
	-18.8	18.8(22°,1-4)	295	0-40	T	CM=1-8x10^{-3}M pH=2.3	76Nb	
Mn^{2+}	-10.5	3.08(20°,1)	21	20-60	T	μ=0.1(KNO_3)	73Bh	b
	-5.0	3.00(20°,2)	37	20-60	T	μ=0.1(KNO_3)		b
Co^{2+}	-5.4	4.354(25°,1)	63	27	C	μ=0.05M(KCl)	71Gb	
	-8.4	4.354(25°,1)	54	20-35	T	μ=0.05(KCl)	71Gb 70Ga 70Gb	b,q: ±2.1
	-10.9	4.41(15°,1)	46.0	15-40	T	μ=0.2	65Sb 64Sd	b
	-9.6	3.505(25°,2)	38	27	C	μ=0.05M(KCl)	71Gb	
	-9.6	3.505(25°,2)	33	20-35	T	μ=0.05(KCl)	71Gb 70Ga 70Gb	b,q: ±3.3
	-13.0	3.27(15°,2)	16.7	15-40	T	μ=0.2	65Sb	b
	-24.7	7.7(1-2)	83.7	21.8	C	0.1M(KNO_3)	67Sc	
	-23.8	7.68(15°,1-2)	62.8	15-40	T	μ=0.2	65Sb	b
Ni^{2+}	-14.2	5.93(1)	63.2	10	C	μ=0	67Aa	
	-13.8	5.81(1)	64.8	25	C	μ=0	67Aa	
	-15.1	5.69(1)	64.4	40	C	μ=0	67Aa	
	-15.1	5.463(25°,1)	54	27	C	μ=0.05M(KCl)	71Gb	
	-14.2	5.463(25°,1)	58	20-35	T	μ=0.05(KCl)	70Ga 70Gb	b,q: ±0.84
	-14.02	3.09(1)	11.92	25	C	0.1M(KNO_3)	62Sc 67Sc	
	-16.9	5.35(1)	46	25	C	μ=0.2M(KCl)	77Gc	
	-16.9	5.32(1)	45.2	25	C	μ=0.2M(KCl)	74Sm	
	-18.4	5.65(15°,1)	43.5	15-40	T	μ=0.2	65Sb 64Sd	b
	-16.82	22.566(1)	46.82	25	C	μ=1M($NaClO_4$)	79Ea	
	-17.8	6.01(1)	56	25	C	μ=0.2M(KCl) S:35w%Dioxane	77Gc	
	-20.5	6.48(1)	56	25	C	μ=0.2M(KCl) S:50w%Dioxane	77Gc	
	-21.3	6.96(1)	62	25	C	μ=0.1M(KNO_3) S:65w%Dioxane	77Gc	
	-21.9	6.85(1)	58	25	C	μ=0.2M(KCl) S:65w%Dioxane	77Gc	
	-17.0	5.62(1)	51	25	C	μ=0.2M(KCl)	77Gc	

Metal Ion	ΔH, kJ /mol	Log K	ΔS, J /mol K	T,°C	M	Conditions	Ref.	Remarks
A70, continued								
Ni^{2+},						S:20w%Methanol	76Gc	
cont.	-17.0	5.92(1)	56	25	C	μ=0.2M(KCl) S:40w%Methanol	77Gc 76Gc	
	-18.3	6.27(1)	59	25	C	μ=0.2M(KCl) S:60w%Methanol	77Gc 76Gc	
	-17.6	6.57(1)	67	25	C	μ=0.1M(KNO_3) S:75w%Methanol	77Gc 76Gc	
	-17.8	6.54(1)	65	25	C	μ=0.2M(KCl) S:75w%Methanol	77Gc 76Gc	
	-19.2	4.87(2)	24.7	10	C	μ=0	67Aa	
	-16.3	4.73(2)	36.4	25	C	μ=0	67Aa	
	-10.9	4.50(2)	47.7	40	C	μ=0	67Aa	
	-15.9	4.467(25°,2)	33	27	C	μ=0.05M(KCl)	71Gb	
	-16.3	4.467(25°,2)	29	20-35	T	μ=0.05(KCl)	70Ga 70Gb	b,q: ±1.2
	-16.57	2.47(2)	-8.28	25	C	0.1M(KNO_3)	62Sc	
	-19.4	4.42(2)	17.6	25	C	μ=0.2M(KCl)	75Sn	
	-20.1	4.57(15°,2)	18.0	15-40	T	μ=0.2	65Sb	b
	-20.50	18.924(2)	17.82	25	C	μ=1M($NaClO_4$)	79Ea	
	-19.4	4.44(2)	20	25	C	μ=0.2M(KCl)	77Gc	
	-20.8	4.80(2)	22	25	C	μ=0.2M(KCl) S:20w%Dioxane	77Gc	
	-21.1	5.05(2)	25	25	C	μ=0.2M(KCl) S:35w%Dioxane	77Gc	
	-23.6	5.50(2)	26	25	C	μ=0.2M(KCl) S:50w%Dioxane	77Gc	
	-25.1	5.96(2)	30	25	C	μ=0.1M(KNO_3) S:65w%Dioxane	77Gc	
	-25.0	5.86(2)	28	25	C	μ=0.2M(KCl) S:65w%Dioxane	77Gc	
	-19.4	4.44(2)	20	25	C	μ=0.2M(KCl)	77Gc 76Gc	
	-19.2	4.68(2)	24	25	C	μ=0.2M(KCl) S:20w%Methanol	77Gc 76Gc	
	-20.1	4.95(2)	28	25	C	μ=0.2M(KCl) S:40w%Methanol	77Gc 76Gc	
	-20.3	5.26(2)	32	25	C	μ=0.2M(KCl) S:60w%Methanol	77Gc 76Gc	
	-21.4	5.54(2)	37	25	C	μ=0.1M(KNO_3) S:75w%Methanol	77Gc 76Gc	
	-20.6	5.52(2)	37	25	C	μ=0.2M(KCl) S:75w%Methanol	77Gc 76Gc	
	-10.9	1.44(3)	-9.00	25	C	0.1M(KNO_3)	62Sc	
	-19.29	13.146(3)	-4.54	25	C	μ=1M($NaClO_4$)	79Ea	
	-38.4	10.22(15°,1-2)	61.5	15-40	T	μ=0.2	65Sb	b
	-37.64	44.671	78.14	25	C	μ=1M($NaClO_4$) R:Ni^{2+}+L+A=$NiLA^{2+}$ A=Glycine	79Ea	
	-57.21	59.031	78.20	25	C	μ=1M($NaClO_4$) R:Ni^{2+}+2L+A= $NiAL_2{}^{2+}$ A=Glycine	79Ea	

Metal Ion	ΔH, kJ /mol	Log *K*	ΔS, J /mol K	T,°C	M	Conditions	Ref.	Re-marks
A70, continued								
Ni^{2+}, cont.	-59.94	59.659	71.92	25	C	μ=1M($NaClO_4$) R:Ni^{2+}+2A+L= NiA_2L^{2+} A=Glycine	79Ea	
Cu^{2+}	-22.51	8.70(1)	87.0	10	C	μ=0	66Ab 65Ib	
	-18.87	8.54(1)	100.0	25	C	μ=0	66Ab 64Sd	
	-16.69	8.32(1)	105.8	40	C	μ=0	66Ab	
	-23.4	8.51(25°,1)	85.0	0-40	T	μ=0	61Ib	b
	-20.5	8.174(25°,1)	88	27	C	μ=0.05M(KCl)	71Gb	
	-19.7	8.174(25°,1)	92	20-35	T	μ=0.05(KCl)	70Ga 70Gb	b,q: ±1.7
	-23.4	8.15(1)	77.4	25	C	μ=0.1M(KNO_3)	72Ia	
	-26.4	8.19(1)	68.2	25	C	μ=0.1M($NaClO_4$)	72Ia	
	-20.67	8.07(1)	84	25	C	μ=0.2M(KCl)	73Ga	
	-20.7	8.07(1)	85	25	C	μ=0.2M(KCl)	77Gc 75Gd	
	-20.5	8.40(15°,1)	88.7	15-40	T	μ=0.2	65Sb	b
	-23.0	8.45(1)	84	25	C	μ=0.2M(KCl) S:20w%Dioxane	77Gc 75Gd	
	-27.6	8.74(1)	75	25	C	μ=0.2M(KCl) S:w35%Dioxane	77Gc 75Gd	
	-31.3	9.27(1)	71	25	C	μ=0.2M(KCl) S:50w%Dioxane	77Gc 75Gd	
	-33.5	10.06(1)	81	25	C	μ=0.1M(KNO_3) S:65w%Dioxane	77Gc 75Gd	
	-35.1	9.64(1)	67	25	C	μ=0.2M(KNO_3) S:65w%Dioxane	77Gc 75Gd	
	-25.2	8.38(1)	76	25	C	μ=0.2M(KCl) S:20w%Methanol	77Gc 76Gc	
	-30.5	8.76(1)	65	25	C	μ=0.2M(KCl) S:40w%Methanol	77Gc 76Gc	
	-34.7	9.17(1)	59	25	C	μ=0.2M(KCl) S:60w%Methanol	77Gc 76Gc	
	-34.9	9.74(1)	70	25	C	μ=0.1M(KNO_3) S:75w%Methanol	77Gc 76Gc	
	-37.4	9.47(1)	56	25	C	μ=0.2M(KCl) S:75w%Methanol	77Gc 76Gc	
	-22.89	7.26(2)	54.0	10	C	μ=0	66Ab	
	-21.92	6.98(2)	37.6	25	C	μ=0	66Ab	
	-23.64	6.76(2)	44.8	40	C	μ=0	66Ab	
	-25.9	6.86(25°,2)	45.6	0-40	T	μ=0	61Ib	b
	-20.9	6.779(25°,2)	54	27	C	μ=0.05M(KCl)	71Gb	
	-25.1	6.779(25°,2)	42	20-35	T	μ=0.05(KCl)	70Ga 70Gb	b,q: ±2.9
	-25.73	6.67(2)	41.4	25	C	μ=0.1M(KNO_3)	72Ia	
	-22.6	6.75(2)	53.6	25	C	μ=0.1M($NaClO_4$)	72Ia	
	-29.54	6.72(2)	29	25	C	μ=0.2M(KCl)	73Ga	
	-29.5	6.72(2)	29	25	C	μ=0.2M(KCl)	77Gc 75Gd	
	-24.3	6.86(15°,2)	47.3	15-40	T	μ=0.2	65Sb	b
	-31.4	6.88(2)	26	25	C	μ=0.2M(KCl) S:20w%Dioxane	77Gc 75Gd	

Metal Ion	ΔH, kJ /mol	Log K	ΔS, J /mol K	T, °C	M	Conditions	Ref.	Remarks
A70, continued								
Cu^{2+}, cont.	-33.0	7.13(2)	25	25	C	μ=0.2M(KCl) S:35w%Dioxane	77Gc 76Gf	
	-36.1	7.46(2)	25	25	C	μ=0.2M(KCl) S:50w%Dioxane	77Gc 76Gf	
	-35.3	8.12(2)	35	25	C	μ=0.1M(KNO_3) S:65w%Dioxane	77Gc	
	-37.7	7.69(2)	21	25	C	μ=0.2M(KCl) S:65w%Dioxane	77Gc 76Gf	
	-31.0	6.94(2)	29	25	C	μ=0.2M(KCl) S:20w%Methanol	77Gc 76Gc	
	-32.5	7.19(2)	29	25	C	μ=0.2M(KCl) S:40w%Methanol	77Gc 76Gc	
	-35.3	7.46(2)	25	25	C	μ=0.2M(KCl) S:60w%Methanol	77Gc 76Gc	
	-35.6	7.96(2)	37	25	C	μ=0.1M(KNO_3) S:75w%Methanol	77Gc 76Gc	
	-37.7	7.66(2)	20	25	C	μ=0.2M(KCl) S:75w%Methanol	77Gc 76Gc	
	-49.8	15.0(1-2)	118.8	22.3	C	0.1M(KNO_3)	67Sc	
	-51.9	14.9(1-2)	112.1	25	C	μ=0.1M	71Be	
	-51.5	14.9(1-2)	113.0	25	C	μ=0.1M	71Be	
	-44.8	15.26(15°,1-2)	136.0	15-40	T	μ=0.2	65Sb	
	-50.29	15.77	133.0	25	C	μ=0 A=α-Amino-isobutric acid R:Cu^{2+}+L^-+A^-=CuLA	72Yb	
	-29.3	16.41	216	25	C	μ=0.50($NaClO_4$) R:Cu^{2+}+L^-+A^{3-}=$CuLA^{2-}$ A=2-Hydroxy-5-sulfobenzoic acid	75Mc	
	-49.0	14.91	121.3	25	C	μ=0.1M(KNO_3) R:Cu+L+Serine=[CuL(Serine)]	72Ia	
	-49.29	15.20	125.5	25	C	μ=0.1M(KNO_3) R:Cu+L+Valine=[CuL(Valine)]	72Ia	
	3.85	1.77(25°)	46.9	-75&25	T	S:50v%CH_3OH R:Cu+CuL_2=2CuL	73Ya	a,b,q: ±5.3
Zn^{2+}	-6.3	4.604(25°,1)	71	27	C	μ=0.05M(KCl)	71Gb	
	-9.6	4.604(25°,1)	54	20-35	T	μ=0.05(KCl)	70Ga 70Gb	b,q: ±1.2
	-12.6	4.98(15°,1)	51.9	15-40	T	μ=0.2	65Sb 64Sd	b
	-11.7	4.069(25°,2)	42	27	C	μ=0.05M(KCl)	71Gb	
	-7.1	4.069(25°,2)	54	20-35	T	μ=0.05(KCl)	70Ga 70Gb	b,q: ±0.8
	-18.0	4.12(15°,2)	16.7	15-40	T	μ=0.2	65Sb	b
	-30.5	9.10(15°,1-2)	68.6	15-40	T	μ=0.2	65Sb	b
In^{3+}	14.6	2.57(25°,1)	96	25-45	T	μ=0.2M($NaClO_4$)	73Sc	b,q: ±2.1

Metal Ion	ΔH, kJ /mol	Log K	ΔS, J /mol K	T,°C	M	Conditions	Ref.	Remarks
A71	β-ALANINE		$C_3H_7O_2N$			$H_2NCH_2CH_2CO_2H$		L^-
Mn^{2+}	-8.8	2.13(20°,1)	13	20-60	T	μ=0.1(KNO_3)	73Bh	b
Co^{2+}	-13.89	4.21(1)	33.9	25	C	μ=0	67Bd 64Sd	
	-10.9	3.69(15°,1)	32.6	15-40	T	μ=0.2	65Sb	b
	-9.6	2.59(15°,2)	15.9	15-40	T	μ=0.2	65Sb	b
	-20.5	6.28(15°,1-2)	48.5	15-40	T	μ=0.2	65Sb	b
Ni^{2+}	-15.94	5.01(1)	42.7	25	C	μ=0	67Bd 64Sd	
	-16.7	4.80(15°,1)	34.3	15-40	T	μ=0.2	65Sb	b
	-20.9	3.54(15°,2)	-4.2	15-40	T	μ=0.2	65Sb	b
	-25.5	8.0(1-2)	66.5	21.7	C	0.1M(KNO_3)	67Sc	
	-37.6	8.34(15°,1-2)	30.1	15-40	T	μ=0.2	65Sb	b
Cu^{2+}	-16.7	7.16(15°,1)	82.0	15-40	T	μ=0.2	67Bd 64Sd	b
	-25.1	5.59(15°,2)	18.4	15-40	T	μ=0.2	65Sb	b
	-45.6	12.6(1-2)	84.9	22.0	C	0.1M(KNO_3)	67Sc	
	-41.8	12.75(15°,1-2)	100.4	15-40	T	μ=0.2	65Sb	b
	-0.46	1.36(25°)	24.7	-75&25	T	S:50v%CH_3OH R:Cu+CuL_2=2CuL	73Ya	a,b,q: ±1.9
Ag^+	-22.6	3.35(1)	-11.7	25	C	0.5M(KNO_3)	70Va	
	-21.9	3.35(25°,1)	-8.8	15-40	T	0.5M(KNO_3)	70Va	b
	-29.4	5.14(2)	-14.6	25	C	0.5M(KNO_3)	70Va	
	-28.2	5.14(25°,2)	-23.0	15-40	T	0.5M(KNO_3)	70Va	b
Zn^{2+}	-4.90	4.274(1)	64.4	5	T	μ=0.1(KCl)	73Rb	q:±3.0
	-6.15	4.205(1)	59.8	25	T	μ=0.1(KCl)	73Rb	q:±3.0
	-7.61	4.130(1)	55.2	45	T	μ=0.1(KCl)	73Rb	q:±3.0
	-17.2	(1-2)	--	21.4	C	0.1M(KNO_3)	67Sc	
In^{3+}	33.5	5.51(35°,1-2)	213	25-45	T	μ=0	72Sa	b,q:±6.3
	30.5	5.42(35°,1-2)	201	25-45	T	μ=0.01M($NaClO_4$)	72Sa	b,q:±6.3
	12.6	2.77(35°,1)	92	25-45	T	μ=0.20M($NaClO_4$)	73Sc	b
A72	DL-α-ALANINE, N-DL-α-alanyl		$C_6H_{12}O_3N_2$			$CH_3CH(NH_2)CONHCH(CH_3)CO_2H$		
Cu^{2+}	-28	5.33(1)	7	25	C	μ=0.2M(KCl)	77Gd 76Gd	q:±2
	5	1.43	44	25	C	μ=0.2M(KCl) ΔH for formation of[CuLH-1]	77Gd 76Gd	q:±1
	49	-8.01	11	25	C	μ=0.2M(KCl) ΔH for formation of$[CuLH-2]^-$	77Gd 76Gd	
	-22	4.13	6	25	C	μ=0.2M(KCl) ΔH for formation of$[CuL_2H-1]^-$	77Gd 76Gd	
	41	-4.39	53	25	C	μ=0.2M(KCl) ΔH for formation of$[Cu_2L_2H-3]^-$	77Gd 76Gd	
	-27	2.70	-38	25	C	μ=0.2M(KCl) R:[CuLH-1]+L^-= $[CuL_2H-1]^-$	77Gd	

Metal Ion	ΔH, kJ /mol	Log K	ΔS, J /mol K	T, °C	M	Conditions	Ref.	Remarks
A72, continued								
Cu^{2+}, cont.	-13	4.38	-41	25	C	μ=0.2M(KCl) R:$(OH)^-$+CuLH-1= $[CuLH\text{-}1(OH)]^-$	77Gd	q:±2
	-13	2.18	-2	25	C	μ=0.2M(KCl) R:[CuLH-1(OH)]+ [CuLH-1]= $[(CuLH\text{-}1)_2(OH)]^-$	77Gd	q:±2
A73	ALANINE, dehydro, N-acetyl- (polymer)		$(C_5H_7O_3N)_n$			$[H_2C:C(CO_2H)NHCOCH_3]_n$		
Co^{2+}	-5.573	--	--	25	C	μ<0.1M R:$(C_5H_7O_3N)_n$+ $0.5nCoCl_2$= $[C_5H_6O_3N(0.5Co)]_n$ +nHCl	67Ia	
A74	ALANINE, 3-(p-fluorophenyl)		$C_9H_{10}O_2NF$			$FC_6H_4CH_2CH(NH_2)CO_2H$		L^-
Cu^{2+}	-28.4	9.41(1)	79.5	25	C	μ=0.2M(KCl) S:50w%Dioxane	75Gd	
	-34.7	6.97(2)	20.9	25	C	μ=0.2M(KCl) S:50w%Dioxane	75Gd	
A75	ALANINE, N-glycyl-		$C_5H_{10}O_3N_2$			$H_2NCH_2CONHCH(CH_3)CO_2H$		L^-
Cu^{2+}	-27.6	5.80(1)	18.4	25	C	μ=0.1M(KNO_3)	72Bl	
	-27.0	5.76(1)	20	25	C	μ=0.2(KCl)	77Gd 76Gd	q:±2
	-3.2	1.55	40	25	C	μ=0.2(KCl) R:$Cu^{2+}+L^-=CuL+H^+$	77Gd 76Gd	j,q:±1
	47.6	7.94	8	25	C	μ=0.2(KCl) R:$Cu^{2+}+L^-=CuL^-$ $+2H^+$	77Gd 76Gd	j
	-24.1	4.63	8	25	C	μ=0.2(KCl) R:$Cu^{2+}+2L^-=$ $CuL_2^-+H^+$	77Gd 76Gd	j
	36.8	-4.18	43	25	C	μ=0.2(KCl) R:$2Cu^{2+}+2L^-=$ $Cu_2L_2^-+3H^+$	77Gd 76Gd	j
	-27	3.08	-32	25	C	μ=0.2M(KCl) R:[CuLH-1]+L^-= $[CuL_2H\text{-}1]^-$	77Gd	
	-12	4.33	-42	25	C	μ=0.2M(KCl) R:$(OH)^-$+CuLH-1= $[CuLH\text{-}1(OH)]^-$	77Gd	q:±2
	-14	2.21	-5	25	C	μ=0.2M(KCl) R:$[CuLH\text{-}1(OH)]^-$+	77Gd	q:±2

Metal Ion	ΔH, kJ /mol	Log K	ΔS, J /mol K	T,°C	M	Conditions	Ref.	Re-marks
A75, continued								
Cu^{2+}, cont.						[CuLH-1]= [(CuLH-1)$_2$(OH)]$^-$		
A76 α-ALANINE, 3-phenyl-			$C_9H_{11}O_2N$			$C_6H_5CH_2CH(NH_2)CO_2H$		L^-
Mn^{2+}	-4.6	2.39(20°,1)	29	20-60	T	μ=0.1(KNO_3)	73Bh	b
Co^{2+}	-6.3	4.02(25°,1)	58.6	27	C	μ=0.05M(KCl)	71Gb	
	-4.6	4.02(25°,1)	63	20-35	T	μ=0.05M(KCl)	71Gb	b
	-5.31	4.450(1)	67.4	25	C	μ=3.0($NaClO_4$)	72Wd	
	-1.3	3.43(25°,2)	63	27	C	μ=0.05M(KCl)	71Gb	
	-9.2	3.43(25°,2)	33	20-35	T	μ=0.05M(KCl)	71Gb	b
	-8.08	3.991(2)	49.4	25	C	μ=3.0($NaClO_4$)	72Wd	
Ni^{2+}	-14.2	5.61(1)	57.7	10	C	μ=0	67Aa	
	-13.4	5.56(1)	61.9	25	C	μ=0	67Aa	
	-10.9	5.52(1)	71.1	40	C	μ=0	67Aa	
	-11.3	5.13(25°,1)	63	27	C	μ=0.05M(KCl)	71Gb	b
	-9.2	5.13(25°,1)	67	20-35	T	μ=0.05M(KCl)	71Gb	
	-9.83	5.356(1)	69.4	25	C	μ=3.0($NaClO_4$)	72Wd	
	-15.1	4.95(2)	41.8	10	C	μ=0	67Aa	
	-13.8	4.66(2)	43.1	25	C	μ=0	67Aa	
	-12.1	4.39(2)	44.8	40	C	μ=0	67Aa	
	-7.9	4.36(25°,2)	59	27	C	μ=0.05M(KCl)	71Gb	
	-9.6	4.36(25°,2)	50	20-35	T	μ=0.05M(KCl)	71Gb	b
	-14.73	5.126(2)	49.0	25	C	μ=3.0($NaClO_4$)	72Wd	
Cu^{2+}	-25.1	8.48(1)	74.0	10	C	μ=0	66Aa	
	-22.2	8.25(1)	82.8	25	C	μ=0	66Aa	
	-21.3	8.30(25°,1)	86.6	0-40	T	μ=0	65Ib	b
	-21.3	7.70(30°,1)	86.6	0-40	T	μ=0	61Ib	b
	-20.29	8.13(1)	91.2	40	C	μ=0	66Aa	
	-19.7	7.90(25°,1)	84	27	C	μ=0.05M(KCl)	71Gb	
	-19.2	7.90(25°,1)	88	20-35	T	μ=0.05M(KCl)	71Gb	b
	-21.9	7.51(1)	70.3	25	C	μ=0.16(KNO_3)	70Md	
	-23.0	7.70(1)	71.1	25	C	μ=0.2M(KCl)	75Gd	
	-19.16	8.250(1)	93.7	25	C	μ=3.0($NaClO_4$)	72Wd	
	-29.3	9.07(1)	75.3	25	C	μ=0.2M(KCl) S:50w%Dioxane	75Gd	
	-26.8	7.43(2)	43.1	10	C	μ=0	66Aa	
	-26.8	7.13(2)	57.3	25	C	μ=0	66Aa	
	-26.8	7.01(25°,2)	44.8	0-40	T	μ=0	65Ib	b
	-26.8	6.94(30°,2)	44.8	0-40	T	μ=0	61Ib	b
	-23.0	6.94(3)	59.4	40	C	μ=0	66Aa	
	-28.9	6.91(25°,2)	38	27	C	μ=0.05M(KCl)	71Gb	
	-26.8	6.91(25°,2)	42	20-35	T	μ=0.05M(KCl)	71Gb	b
	-24.1	6.74(2)	48.1	25	C	μ=0.16(KNO_3)	70Md	
	-29.3	6.78(2)	29.3	25	C	μ=0.2M(KCl)	75Gd	
	-39.16	7.304(2)	8.4	25	C	μ=3.0($NaClO_4$)	72Wd	
	-36.4	7.44(2)	20.9	25	C	μ=0.2M(KCl) S:50w%Dioxane	75Gd	
Zn^{2+}	-3.8	4.31(25°,1)	71	27	C	μ=0.05M(KCl)	71Gb	
	-7.5	4.31(25°,1)	54	20-35	T	μ=0.05M(KCl)	71Gb	b
	-1.00	4.477(1)	82.0	5	T	μ=0.1(KCl)	73Rb	q:±3.0

Metal Ion	ΔH, kJ /mol	Log K	ΔS, J /mol K	T,°C	M	Conditions	Ref.	Remarks
A76, continued								
Zn^{2+}, cont.	-2.9	4.453(1)	75.3	25	T	μ=0.1(KCl)	73Rb	q:±3.0
	-5.4	4.408(1)	67.4	45	T	μ=0.1(KCl)	73Rb	q:±3.0
	-7.1	4.09(25°,2)	54	27	C	μ=0.05M(KCl)	71Gb	
	-11.7	4.09(25°,2)	38	20-35	T	μ=0.05M(KCl)	71Gb	b
A77 β-ALANINE, 3-phenyl-			$C_9H_{11}O_2N$			$C_6H_5CH(NH_2)CH_2CO_2H$		L^-
Mn^{2+}	-7.1	2.06(20°,1)	17	20-60	T	μ=0.1(KNO_3)	73Sx	b
A78 AMINE, allyl-			C_3H_7N			$CH_2:CHCH_2NH_2$		L^+
PtA_4^{2-}	-29.7	3.540(25°)	-31.8	25-45	T	1.90M(NaCl) 0.1M(HCl) A^-=Cl R:PtA_4^{2-}+L^+= PtA_3L+A^-	68Me	
	-29.7	3.240(44°)	-31.8	30-60	T	μ=2M(NaCl+HCl) A^-=Cl R:PtA_4^{2-}+ L^+=PtA_3L+A^-	67Db	b
A79 AMINE, allyl diethyl-			$C_7H_{15}N$			$CH_2:CHCH_2NH(C_2H_5)_2$		L^+
PtA_4^{2-}	-23.4	2.740(45.3°)	-21	30-60	T	μ=2M(NaCl+HCl) A^-=Cl R:PtA_4^{2-}+ L^+=PtA_3L+A^-	67Db	b
A80 AMINE, allyl ethyl-			$C_5H_{12}N$			$CH_2:CHCH_2NHC_2H_5$		
PtA_4^{2-}	-20.1	2.382(25°)	-22.2	0-35	T	μ=2M(KBr+HBr) A^-=Br R:PtA_4^{2-}+ L^+=PtA_3L+A^-	67Dd	b
	-24.7	3.091(44°)	-18.4	30-60	T	μ=2M(NaCl+HCl) A^-=Cl R:PtA_4^{2-}+ L^+=PtA_3L+A^-	67Db	b
A81 AMINE, allyl methyl-			C_4H_9N			$CH_2:CHCH(CH_3)NH_2$		L^+
PtA_4^{2-}	-28.0	3.108(45.3°)	-28.9	30-60	T	μ=2M(NaCl+HCl) A^-=Cl R:PtA_4^{2-}+ L^+=PtA_3L+A^-	67Dc	b
A82 AMINE, bis(2-pyridylmethyl)			$C_{12}H_{13}N_3$			$NH[CH_2(C_5H_4N)]_2$		
Mn^{2+}	-10.5	3.52(1)	32.2	25	C	μ=0.1(KNO_3)	77Ag	
	-10.5	2.53(2)	5.4	25	C	μ=0.1(KNO_3)	77Ag	q:±4.2
Fe^{2+}	-25.5	6.15(1)	30	25	C	μ=0.1(KNO_3)	77Ag	
	-46.0	6.07(2)	-36.4	25	C	μ=0.1(KNO_3)	77Ag	
Co^{2+}	-36.2	8.05(1)	32.6	25	C	μ=0.1(KNO_3)	77Ag	
	-7.65	5.80(2)	2.9	25	C	μ=0.1(KNO_3)	77Ag	
Ni^{2+}	-46.44	9.30(1)	22.6	25	C	μ=0.1(KNO_3)	77Ag	
	-42.00	8.15(2)	15.1	25	C	μ=0.1(KNO_3)	77Ag	

Metal Ion	ΔH, kJ /mol	Log K	ΔS, J /mol K	T, °C	M	Conditions	Ref.	Remarks
A82, continued								
Cu^{2+}	-66.1	13.85(1)	43.9	25	C	μ=0.1(KNO_3)	77Ag	
	-25.9	4.65(2)	0.8	25	C	μ=0.1(KNO_3)	77Ag	
Zn^{2+}	-32.0	7.63(1)	37.6	25	C	μ=0.1(KNO_3)	77Ag	
	-22.8	4.52(2)	10.9	25	C	μ=0.1(KNO_3)	77Ag	
Cd^{2+}	-24.7	6.40(1)	39.7	25	C	μ=0.1(KNO_3)	77Ag	
	-33.1	5.36(2)	-6.7	25	C	μ=0.1(KNO_3)	77Ag	
Pb^{2+}	-31.0	6.00(1)	11.3	25	C	μ=0.1(KNO_3)	77Ag	
	-23.4	2.55(2)	-48.9	25	C	μ=0.1(KNO_3)	77Ag	q:±8.4
Hg^{2+}	-111.3	22.25(1-2)	52.5	25	C	μ=0.1(KNO_3)	77Ag	
A83 AMINE, 2-butenyl-(trans)			C_4H_9N			$CH_3CH:CHCH_2NH_2$		L^+
PtA_4^{2-}	-21.3	2.483(44.5°)	-19.2	30-60	T	μ=2M(NaCl+HCl) $A=Cl^-$ R:PtA_4^{2-}+ L^+=PtA_3L+A^-	67Dc	b
A84 AMINE, 3-butenyl-			C_4H_9N			$CH_2:CHCH_2CH_2NH_2$		L^+
PtA_4^{2-}	-21.3	3.480(44.5°)	-0.8	30-60	T	μ=2M(NaCl+HCl) $A=Cl^-$ R:PtA_4^{2-}+ L^+=PtA_3L+A^-	67Db	b
A85 AMINE, cyclohexyl dimethyl			$C_8H_{17}N$			$C_6H_{13}N(CH_3)_2$		L
BA_3C	-71.1	--	--	25	C	S:Benzene A=F C=$(CH_3CH_2CH_2CH_2)_2O$ R:Ba_3C+L=BA_3L+C	68Rd	
	-84.5	--	--	25	C	S:Benzene CM=0.03-0.04M A=F C=Tributyl- phosphine R:BA_3C+L=BA_3L+C	73Rd	
	-81.6	--	--	25	C	S:Benzene CM=0.03-0.04M A=F C=Trihexyl- phosphine R:BA_3C+L=BA_3L+C	73Rd	
	-32.2	--	--	25	C	S:Benzene CM=0.03-0.04M A=F C=Triphenyl- phosphine R:BA_3C+L=BA_3L+C	73Rd	
Al_2A_6	-136.8	--	--	25	C	S:Benzene CM=0.07M A=Br R:0.5(Al_2A_6)+L= AlA_3L	68Rc	

Metal Ion	ΔH, kJ /mol	Log K	ΔS, J /mol K	T,°C	M	Conditions	Ref.	Remarks
A86	AMINE, 2,2'-diacetic acid methyl-		C_5H_9N			$CH_3N(CH_2CO_2H)_2$		L^{2-}
Mg^{2+}	-13.05	3.44(1)	110.4	20	C	0.1M(KNO_3)	65Ab 56Ma	
Ca^{2+}	-6.86	3.75(1)	48.45	20	C	0.1M(KNO_3)	65Ab	
Sr^{2+}	-5.15	2.85(1)	37.03	20	C	0.1M(KNO_3)	65Ab 56Ma	
Ba^{2+}	-3.30	2.59(1)	38.37	20	C	0.1M(KNO_3)	65Ab	
Mn^{2+}	2.34	5.4(1)	111.3	20	C	0.1M(KNO_3)	65Ab 56Ma	
	0.96	9.56(1-2)	186.4	20	C	0.1M(KNO_3)	65Ab	
Co^{2+}	-7.74	7.62(1)	119.58	20	C	0.1M(KNO_3)	65Ab	
	-22.93	13.91(1-2)	188.3	20	C	0.1M(KNO_3)	65Ab	
Ni^{2+}	-19.7	8.73(1)	100.0	20	C	0.1M(KNO_3)	65Ab	
	-32.0	15.95(1-2)	196.2	20	C	0.1M(KNO_3)	65Ab	
Cu^{2+}	-16.07	11.09(1)	157.7	20	C	0.1M(KNO_3)	65Ab	
	-50.67	17.92(1-2)	170.50	20	C	0.1M(KNO_3)	65Ab	
Zn^{2+}	-9.08	7.66(1)	115.9	20	C	0.1M(KNO_3)	65Ab	
	-24.39	14.09(1-2)	185.77	20	C	0.1M(KNO_3)	65Ab	
Cd^{2+}	-7.91	6.77(1)	102.59	20	C	0.1M(KNO_3)	65Ab	
	-30.42	12.52(1-2)	136.0	20	C	0.1M(KNO_3)	65Ab	
Pb^{2+}	-14.90	8.02(1)	102.9	20	C	0.1M(KNO_3)	65Ab	
A87	AMINE, di(2-aminoethyl)-		$C_4H_{13}N$			$HN(CH_2CH_2NH_2)_2$		L
La^{3+}	-99.6	(1)	--	23	C	μ=0.024MLa(ClO_4)$_3$	71Fd	
	-82.8	(2)	--	23	C	μ=0.024MLa(ClO_4)$_3$	71Fd	
	-54.8	(3)	--	23	C	μ=0.024MLa(ClO_4)$_3$	71Fd	
Pr^{3+}	-102.9	(1)	--	23	C	μ=0.024MLa(ClO_4)$_3$	71Fd	
	-87.0	(2)	--	23	C	μ=0.024MLa(ClO_4)$_3$	71Fd	
	-52.7	(3)	--	23	C	μ=0.024MLa(ClO_4)$_3$	71Fd	
Nd^{3+}	-107.1	(1)	--	23	C	μ=0.024MLa(ClO_4)$_3$	71Fd	
	-87.4	(2)	--	23	C	μ=0.024MLa(ClO_4)$_3$	71Fd	
	-48.1	(3)	--	23	C	μ=0.024MLa(ClO_4)$_3$	71Fd	
Sm^{3+}	-107.5	(1)	--	23	C	μ=0.024MLa(ClO_4)$_3$	71Fd	
	-88.7	(2)	--	23	C	μ=0.024MLa(ClO_4)$_3$	71Fd	
	-44.8	(3)	--	23	C	μ=0.024MLa(ClO_4)$_3$	71Fd	
Gd^{3+}	-109.2	(1)	--	23	C	μ=0.024MLa(ClO_4)$_3$	71Fd	
	-90.0	(2)	--	23	C	μ=0.024MLa(ClO_4)$_3$	71Fd	
	-41.4	(3)	--	23	C	μ=0.024MLa(ClO_4)$_3$	71Fd	
Dy^{3+}	-113.8	(1)	--	23	C	μ=0.024MLa(ClO_4)$_3$	71Fd	
	-92.9	(2)	--	23	C	μ=0.024MLa(ClO_4)$_3$	71Fd	
	-28.9	(3)	--	23	C	μ=0.024MLa(ClO_4)$_3$	71Fd	
Er^{3+}	-117.6	(1)	--	23	C	μ=0.024MLa(ClO_4)$_3$	71Fd	
	-98.3	(2)	--	23	C	μ=0.024MLa(ClO_4)$_3$	71Fd	
	-25.9	(3)	--	23	C	μ=0.024MLa(ClO_4)$_3$	71Fd	

Metal Ion	ΔH, kJ /mol	Log K	ΔS, J /mol K	T,°C	M	Conditions	Ref.	Remarks
A87, continued								
Yb^{3+}	-114.6	(1)	--	23	C	μ=0.024MLa$(ClO_4)_3$	71Fd	
	-95.0	(2)	--	23	C	μ=0.024MLa$(ClO_4)_3$	71Fd	
	-25.5	(3)	--	23	C	μ=0.024MLa$(ClO_4)_3$	71Fd	
Mn^{2+}	-17	3.99(30°,1)	21.3	30-40	T	$\mu\cong2$(KCl, KNO_3)	52Ja	a,b
	-21	2.83(30°,2)	-14.6	30-40	T	$\mu\cong2$(KCl, KNO_3)	52Ja	a,b
	-29.08	(1)	--	25	C	0.1(KCl)	61Cb	
Fe^{2+}	-38	6.23(30°,1)	-5.0	30-40	T	$\mu\cong2$(KCl, KNO_3)	52Ja	a,b
	-33	4.13(30°,2)	-31.4	30-40	T	$\mu\cong2$(KCl, KNO_3)	52Ja	a,b
	-54.18	(1-2)	--	25	C	0.1(KCl)	61Cb	
Co^{2+}	-34.10	7.99(1)	37.6	25	C	μ=0.1(KCl)	61Cb	
	-38	8.47(30°,1)	38.1	30-40	T	$\mu\cong2$(KCl, KNO_3)	52Ja	a,b
	-42.89	5.86(2)	-31.4	25	C	μ=0.1(KCl)	61Cb	
	-43	6.07(30°,2)	-21.8	30-40	T	$\mu\cong2$(KCl, KNO_3)	52Ja	a,b
	-76.98	13.85(1-2)	6.3	25	C	μ=0.1(KCl)	61Cb	
Ni^{2+}	-49.58	10.59(1)	35.6	25	C	μ=0.1(KCl)	61Cb	
	-50	10.81(30°,1)	41.4	30-40	T	$\mu\cong2$(KCl, KNO_3)	50Jb 52Ja	a,b
	-56.27	7.99(2)	-35.6	25	C	μ=0.1(KCl)	61Cb	
	-54	8.14(30°,2)	-23.4	30-40	T	$\mu\cong2$(KCl, KNO_3)	52Ja	a,b
	-105.8	18.58(1-2)	0.0	25	C	μ=0.1(KCl)	61Cb	
Cu^{2+}	-79.1	15.43(30°,1)	33	10-40	T	μ=0	59Ma	b
	-75.31	15.80(1)	50.2	25	C	μ=0.1(KCl)	50Jb 61Cb	
	-84	16.11(30°,1)	32.6	30-40	T	$\mu\cong2$(KCl, KNO_3)	52Ja	a,b
	-26.4	4.68(30°,2)	4	10-40	T	μ=0	59Ma	b
	-34.10	5.20(2)	-14.6	25	C	μ=0.1(KCl)	61Cb	
	-109.41	21.00(1-2)	35.6	25	C	μ=0.1(KCl)	61Cb	
Zn^{2+}	-26.8	8.57(30°,1)	75	10-40	T	μ=0	59Ma	b
	-26.99	8.80(1)	77.4	25	C	μ=0.10(KCl)	61Cb	
	-33	9.14(30°,1)	64.4	30-40	T	$\mu\cong2$(KCl, KNO_3)	52Ja	a,b
	-42.47	5.50(2)	-37.6	25	C	μ=0.10(KCl)	61Cb	
	-69.45	14.29(1-2)	39.7	25	C	μ=0.10(KCl)	61Cb	
Hg^{2+}	-151	25.03(1-2)	-29	25	T	μ=0.1(KNO_3)	61Rc	
A88	AMINE, di(3-aminopropyl)-		$C_6H_{17}N$			$HN(CH_2CH_2CH_2NH_2)_2$		L
Co^{2+}	-32.51	6.92(1)	23.4	25	C	0.1M(KCl)	66Pa	
	-37.6	6.63(1)	4	30	T	1M(KNO_3)	56Ha	
	-27.45	6.73(25°,1)	36.8	25-55	T	μ=1.0M(KNO_3)	76Dc	
Ni^{2+}	-44.18	9.18(1)	27.6	25	C	0.1M(KCl)	66Pa	
	-42	9.09(1)	38	30	T	1M(KNO_3)	56Ha	
	-45.65	9.53(25°,1)	29.3	25-55	T	μ=1.0M(KNO_3)	76Dc	
	-29.62	3.55(2)	-31.4	25	C	0.1M(KCl)	66Pa	
	-28.07	4.28(25°,2)	-12.1	25-55	T	μ=1.0M(KNO_3)	76Dc	
Cu^{2+}	-67.32	14.20(1)	46.0	25	C	0.1M(KCl)	66Pa	
	-67	14.25(1)	50	30	T	1M(KNO_3)	56Ha	
Zn^{2+}	-22.76	7.92(1)	75	25	C	0.1M(KCl)	66Pa	

Metal Ion	ΔH, kJ /mol	Log K	ΔS, J /mol K	T, °C	M	Conditions	Ref.	Re-marks
A89	AMINE, dibutyl-		$C_8H_{19}N$			$HN(CH_2CH_2CH_2CH_3)_2$		L
VOA_2	-40.79	-1.47(1)	-116.3	25	C	μ≅0 S:Nitrobenzene A=Acetyl-acetonate ion	65Ca	
NiA	-54.8	3.03(1)	-125.5	25	T	$CM=10^{-4}M$ S:Benzene A=Di-acetylbisbenzoyl hydrazine	60Sa	
CuA_2	-37	3.35(1)	-61	20	T	S:Benzene A=Trifluoro-acetylacetone	79Si	b,q:±3
ZnA_2	-37.2	2.463(25°,1)	-79	18.5-65	T	S:Toluene $A=S_2CH(CH_3)_2$	65Ch	b
A90	AMINE, di(2-butyl)-		$C_8H_{19}N$			$HN[CH(CH_3)CH_2CH_3]_2$		
NiA	-52.7	0.56(1-2)	-166.1	25	T	$CM=10^{-4}M$ S:Benzene A=Di-acetylbisbenzoyl hydrazine	60Sa	
A91	AMINE, diethyl		$C_4H_{11}N$			$HN(CH_2CH_3)_2$		L
VOA_2	-39.75	-1.36(1)	-114.2	25	C	μ≅0 S:Nitrobenzene A=Acetyl-acetonate ion	65Ca	
NiA	-56.5	3.72(1-2)	-118.8	25	T	$CM=10^{-4}M$ S:Benzene A=Di-acetylbisbenzoyl hydrazine	60Sa	
NiA_2	-31.51	(1)	--	25	C	S:Benzene A=2,2,6,6-Tetramethyl-3,5-heptanedione	73Ce	
	-48.78	1.569(2)	-134.7	25	C	S:Benzene A=2,2,6,6-Tetramethyl-3,5-heptanedione	73Ce	
[Ni]	-22.2	1.176(20°,1)	-53	-12-32	T	S:Toluene [Ni]=Bis(0,0'-diethyldithiophosphato) nickel(II)	71Cc	b,q: ±2.1
	-20.1	-0.377(20°,2)	-75	-12-32	T	S:Toluene [Ni]=Bis(0,0'-diethyldithiophosphato) nickel(II)	71Cc	b,q: ±1.2
IrA_2BC_2	-15.5	4.46(1)	33	25	C	$S:CH_2Cl_2$ A=Cl	77La	

Metal Ion	ΔH, kJ /mol	Log K	ΔS, J /mol K	T,°C	M	Conditions	Ref.	Re-marks
A91, continued								
IrA_2BC_2, cont.						$B=C(O)CF_3$ $C=P(C_6H_5)_3$		
Ag^+	-44.56	-8.70(1-2)	-27.2	25	C	--	55Fa	
AlA_3	-114.35	(1)	--	25-28	C	S:n-Hexane $A=CH_3$	68Hc	
I_2	-36.8	3.853(20°,1)	-5.0	22	C	low μ S:n-Heptane	64Sc 60Ya	
A92	AMINE, diethyl, 2,2'-dihydroxy	$C_4H_{11}O_2N$				$NH(CH_2CH_2OH)_2$		
Ag^+	-21.6	2.68(1)	-21	25	C	μ=0.5M(Di-ethanolammonium perchlorate	75Mb	
	-21.5	2.69(1)	-21	25	C	μ=0.5M($NaClO_4$)	77Mc	q:±1.8
	-27.6	2.80(2)	-39	25	C	μ=0.5M(Di-ethanolammonium perchlorate	75Mb	
	-27.6	2.79(2)	-39	25	C	μ=0.5M($NaClO_4$)	77Mc	
Cd^{2+}	-6.4	2.40(1)	24	25	C	μ=0.5M($NaClO_4$)	75Mb	q:±1.3
	-9.3	2.40(1)	15	25	C	μ=0.5M($NaClO_4$)	77Mc	
	-11.1	2.12(2)	3	25	C	μ=0.5M($NaClO_4$)	75Mb	
	-8.7	2.12(2)	11	25	C	μ=0.5M($NaClO_4$)	77Mc	
Hg^{2+}	-48	7.85(1)	-10	25	C	μ=0.5M($NaClO_4$)	75Mb	
	-80	7.84(1)	-117	25	C	μ=0.5M($NaClO_4$)	77Mc	q:±24
	-41	7.81(2)	11	25	C	μ=0.5M($NaClO_4$)	75Mb	
	-35	7.82(2)	30	25	C	μ=0.5M($NaClO_4$)	77Mc	q:±15
A93	AMINE, diisobutyl-		$C_8H_{19}N$			$HN[CH_2CH(CH_3)_2]_2$		L
NiA	-55.2	1.69(1-2)	-150.6	25	T	CM=10^{-4}M S:Benzene A=Di-acetylbisbenzoyl hydrazine	60Sa	
A94	AMINE, dimethyl-		C_2H_7N			$HN(CH_3)_2$		L
NiA	-60.2	6.46(1-2)	-79.9	25	T	CM=10^{-4}M S:Benzene A=Di-acetylbisbenzoyl hydrazine	60Sa	
Ag^+	-47.07	5.77(1-2)	-60.7	2	T	μ=0	65Le	h:2-45°
	-47.07	5.22(1-2)	-58	25	T	μ=0	65Le	h:2-45°
	-47.07	4.94(1-2)	-58.6	35	T	μ=0	65Le	h:2-45°
	-47.07	4.68(1-2)	-58.6	45	T	μ=0	65Le	h:2-45°
	-40.6	5.36(1-2)	-33.5	25	C	--	55Fa	
	-47.07	6.28(1-2)	-51.0	2	T	μ=0 S:50v%CH_3OH	65Le	h:2-45°
	-47.07	5.54(1-2)	-51.9	25	T	μ=0 S:50v%CH_3OH	65Le	h:2-45°
	-47.07	5.30(1-2)	-51.0	35	T	μ=0 S:50v%CH_3OH	65Le	h:2-45°
	-47.07	4.97(1-2)	-53.1	45	T	μ=0 S:50v%CH_3OH	65Le	h:2-45°
AlA_3	-129.03	(1)	--	25-28	C	S:n-Hexane $A=CH_3$	68Hc	

Metal Ion	ΔH, kJ /mol	Log K	ΔS, J /mol K	T,°C	M	Conditions	Ref.	Remarks
A95	AMINE, dioctyl		$C_{16}H_{35}N$			$NH(C_8H_{17})_2$		
CuA_2	-36	3.50(1)	-57	20	T	S:Benzene A=Trifluoro-acetylacetone	79Si	b,q:±3
A96	AMINE, diphenyl methyl-		$C_{13}H_{13}N$			$(C_6H_5)_2NCH_3$		
Al_2A_6	-72.8	--	--	25	C	S:Benzene CM=0.07M A=Br R:$0.5(Al_2A_6)+L=AlA_3L$	68Rc	
A97	AMINE, dipropyl-		$C_6H_{15}N$			$HN(CH_2CH_2CH_3)_2$		L
NiA	-55.6	3.36(1-2)	-122.6	25	T	$CM=10^{-4}M$ S:Benzene A=Di-acetylbisbenzoyl hydrazine	60Sa	
A98	AMINE, di(2-propyl)-		$C_6H_{15}N$			$HN[CH(CH_3)_2]_2$		L
NiA	-52.7	1.80(1-2)	-144.3	25	T	$CM=10^{-4}M$ S:Benzene A=Di-acetylbisbenzoyl hydrazine	60Sa	
A99	AMINE, di(2-pyridyl)		$C_{10}H_9N_3$			$C_5H_4NNHNC_5H_4$		
Mn^{2+}	-45.6	1.78(25°,1)	-118.4	25-40	T	$\mu=0.1M(KNO_3+HNO_3)$	76Bb	
	31.8	4.16(25°,2)	186.2	25-40	T	$\mu=0.1M(KNO_3+HNO_3)$	76Bb	
Co^{2+}	-11.3	4.98(25°,1)	58.2	25-40	T	$\mu=0.1M(KNO_3+HNO_3)$	76Bb	
	-24.3	3.93(25°,2)	-6.3	25-40	T	$\mu=0.1M(KNO_3+HNO_3)$	76Bb	
Ni^{2+}	-18.4	6.76(25°,1)	68.2	25-40	T	$\mu=0.1M(KNO_3+HNO_3)$	76Bb	
	-19.2	6.29(25°,2)	52.7	25-40	T	$\mu=0.1M(KNO_3+HNO_3)$	76Bb	
Cu^{2+}	-10.5	7.60(25°,1)	111.3	25-40	T	$\mu=0.1M(KNO_3+HNO_3)$	76Bb	
	-9.6	6.78(25°,2)	98.3	25-40	T	$\mu=0.1M(KNO_3+HNO_3)$	76Bb	
Zn^{2+}	-23.4	3.70(25°,1)	-7.5	25-40	T	$\mu=0.1M(KNO_3+HNO_3)$	76Bb	
	-7.1	3.57(25°,2)	42.7	25-40	T	$\mu=0.1M(KNO_3+HNO_3)$	76Bb	
Cd^{2+}	-8.8	2.89(25°,1)	25.9	25-40	T	$\mu=0.1M(KNO_3+HNO_3)$	76Bb	
	-33.0	2.09(25°,2)	-70.7	25-40	T	$\mu=0.1M(KNO_3+HNO_3)$	76Bb	
A100	AMINE, methyl bis-(3-aminopropyl)-		$C_7H_{19}N_3$			$CH_3N(CH_2CH_2CH_2NH_2)_2$		L
Ni^{2+}	-32.8	7.05(30°,1)	25	10-40	T	$\mu=0$	59Gc	b
Cu^{2+}	-59.0	12.28(30°,1)	42	10-40	T	$\mu=0$	59Gc	b
Cd^{2+}	-19.4	5.87(30°,1)	50	10-40	T	$\mu=0$	59Gc	b

Metal Ion	ΔH, kJ /mol	Log K	ΔS, J /mol K	T,°C	M	Conditions	Ref.	Remarks
A101	AMINE, N,N,N',N'-tetrakis-(2-pyridylmethyl)		$C_{26}H_{28}N_6$			$[(C_5H_4N)CH_2]_2NCH_2CH_2N[CH_2(C_5H_4N)]_2$		
Mn^{2+}	-47.82	10.27(1)	32.6	20	C	μ=0.1(KNO_3)	77Ag	
Fe^{2+}	-75.52	14.61(1)	22.2	20	C	μ=0.1(KNO_3)	77Ag	
Co^{2+}	-72.09	16.59(1)	69.4	20	C	μ=0.1(KNO_3)	77Ag	
Ni^{2+}	-98.74	21.55(1)	75.7	20	C	μ=0.1(KNO_3)	77Ag	
Cu^{2+}	-95.52	20.54(1)	67.8	20	C	μ=0.1(KNO_3)	77Ag	
Zn^{2+}	-61.84	15.58(1)	87.0	20	C	μ=0.1(KNO_3)	77Ag	
Cd^{2+}	-78.7	16.33(1)	44.3	20	C	μ=0.1(KNO_3)	77Ag	
Pb^{2+}	-80.3	13.98(1)	7.9	20	C	μ=0.1(KNO_3)	77Ag	
Ag^{+}	-77.0	11.29(1)	-46.4	20	C	μ=0.1(KNO_3)	77Ag	
	-126.1	17.40(2-1)	-97.1	20	C	μ=0.1(KNO_3)	77Ag	
	-94.1	15.45	-25.5	20	C	μ=0.1(KNO_3) R:$Ag^{+}+H^{+}+L=AgHL^{2+}$	77Ag	
Hg^{2+}	-124.3	25.05(1)	55.6	20	C	μ=0.1(KNO_3)	77Ag	
A102	AMINE, tri(2-aminoethyl)-		$C_6H_{18}N_4$			$N(CH_2CH_2NH_2)_3$		L
Ca^{2+}	-23.18	8.43(25°,1)	83.3	15-40	T	μ=0.11(KNO_3)	61Mf	b
Pr^{3+}	-261.5	(1-2)	--	23	C	S:Acetonitrile	71Fc	
Gd^{3+}	-252.3	(1-2)	--	23	C	S:Acetonitrile	71Fc	
Mn^{2+}	-12.55	5.80(1)	69.0	25	C	μ=0.1(KCl)	63Pb	
Fe^{2+}	-26.36	8.6(1)	77.4	25	C	μ=0.1(KCl)	63Pb 60Pa	
Co^{2+}	-44.56	12.61(1)	92.0	25	C	μ=0.1(KCl)	63Pb 60Pa	
Ni^{2+}	-63.39	14.59(1)	66.9	25	C	μ=0.1(KCl)	63Pb 60Pa	
Cu^{2+}	-65.7	18.40(30°,1)	134	10-40	T	μ=0	58Ba 60Pa	b
	-85.35	18.84(1)	75.3	25	C	μ=0.1(KCl)	63Pb 61Mf	
Zn^{2+}	-57.95	14.48(1)	81.6	25	C	μ=0.1(KCl)	63Pb 60Pa	
A103	AMINE, tri(3-aminopropyl)-		$C_9H_{24}N_4$			$N(CH_2CH_2CH_2NH_2)_3$		L
Ni^{2+}	-38.45	8.702(1)	37.6	25	C	0.1M(KCl)	68Va	
Cu^{2+}	-61.38	13.113(1)	45.2	25	C	0.1M(KCl)	68Va	
	-53.68	10.752(1)	25.5	25	C	0.1M(KCl) L=HL^{+}	68Va	
Zn^{2+}	-37.20	10.696(1)	79.9	25	C	0.1M(KCl)	68Va	

Metal Ion	ΔH, kJ /mol	Log K	ΔS, J /mol K	T,°C	M	Conditions	Ref.	Re- marks
A104	AMINE, tribenzyl-		$C_{21}H_{21}N$			$(C_6H_5CH_2)_3N$		L
I_2	-9.6	(1)	--	25	C	S:Benzene	66Ad	
A105	AMINE, tribenzyl-N-oxide		$C_{21}H_{21}NO$			$(C_6H_5CH_2)_3NO$		L
I_2	-43.9	3.540(20°,1)	-82.0	4-20	T	CM=4.6x10^{-4}M CL=4x10^{-4}- 1.4x10^{-3}M S:CH_2Cl_2	65Kc	b
A106	AMINE, tributyl-		$C_{12}H_{27}N$			$(CH_3CH_2CH_2CH_2)_3N$		L
NiA	-54.8	0.51(1-2)	-174.9	25	T	CM=10^{-4}M S:Benzene A=Di-acetylbisbenzoyl hydrazine	60Sd	
HgA_2	-43.30	2.322(1)	-99.2	30	C	S:Benzene A=Cl CM=1-5x10^{-3}M	73Fa	
	-43.81	2.539(1)	-95.8	30	C	S:Benzene A=Br CM=1-5x10^{-3}M	73Fa	
	-44.69	2.348(1)	-102.9	30	C	S:Benzene A=I CM=1-5x10^{-3}M	73Fa	
I_2	-48.5	3.20(20°,1)	-105	22	C	low μ S:n-Heptane	64Sc 60Ya	b
A107	AMINE, triethyl-		$C_6H_{15}N$			$(CH_3CH_2)_3N$		L
NiA	-54.8	1.36(1-2)	-159.8	25	T	CM=10^{-4}M S:Benzene A=Di-acetylbisbenzoyl hydrazine	60Sd	
NiA_2	-31.34	1.918(1)	-68.2	25	C	S:Benzene A=2,2,6,6-Tetramethyl-3,5-heptanedione	73Ce	
	-10.33	(2)	--	25	C	S:Benzene A=2,2,6,6-Tetramethyl-3,5-heptanedione	73Ce	
AlA_3	-110.75	(1)	--	25-28	C	S:n-Hexane A=CH_3	68Hc	
Al_2A_6	-131.8	--	--	25	C	S:Benzene CM=0.07M A=Br R:0.5(Al_2A_6)+L= AlA_3L	68Rc	
CuA_2	-20	0.591(1)	-55	20	T	S:Benzene A=Trifluoro-acetylacetone	79Si	b,q:±2
	-70.7	3.699(1)	-165.7	25	C	S:CCl_4 A=Hexafluoro-acetylacetone	70Pa	

Metal Ion	ΔH, kJ /mol	Log K	ΔS, J /mol K	T,°C	M	Conditions	Fef.	Re-marks
A107, continued								
CuA_2, cont.	-63.2	4.400(1)	-127.6	25	C	S:Cyclohexane A=Hexafluoro-acetylacetone	72Nd	
	-56.1	3.201(1)	-126.4	25	C	S:o-Dichloro-benzene A=Hexafluoro-acetylacetone	72Nd	
CuA_2B	-28.4	--	--	25	C	S:o-$Cl_2C_6H_4$ A=Hexafluoro-acetylacetone B=Dimethyl-acetate R:CuA_2B+L=CuA_2L+B	72Nd	
	-25.9	--	--	25	C	S:o-$Cl_2C_6H_4$ A=Hexafluoro-acetylacetone B=Dimethyl sulfoxide R:CuA_2B+L=CuA_2L+B	72Nd	
	-36.4	--	--	25	C	S:o-$Cl_2C_6H_4$ A=Hexafluoro-acetylacetone B=Ethyl acetate R:CuA_2B+L=CuA_2L+B	72Nd	
	-29.7	--	--	25	C	S:Cyclohexane A=Hexafluoro-acetylacetone B=Dimethyl-acetate R:CuA_2B+L=CuA_2L+B	72Nd	
	-27.6	--	--	25	C	S:Cyclohexane A=Hexafluoro-acetylacetone B=Dimethyl sulfoxide R:CuA_2B+L=CuA_2L+B	72Nd	
	-38.5	--	--	25	C	S:Cyclohexane A=Hexafluoro-acetylacetone B=Ethyl acetate R:CuA_2B+L=CuA_2L+B	72Nd	
BA_3C	-61.1	(1)	--	25	C	S:Benzene A=F C=$(CH_3CH_2)_2O$	68Rd	
	-72.0	(1)	--	25	C	S:Benzene A=F C=$(CH_3CH_2CH_2CH_2)_2O$	68Rd	
GaA_3	-126.4	4.91(1)	-329.7	25	C	S:Benzene A=Cl	70Ga	

Metal Ion	ΔH, kJ /mol	Log K	ΔS, J /mol K	T,°C	M	Conditions	Ref.	Re-marks
A107, continued								
SnA_2B_2	-14.8	1.199(1)	-26	25	C	S:Benzene A=Bu B=(OCOMe)	77Ge	
	-15.8	1.013(1)	-33	25	C	S:Benzene A=Bu B=(OCOEt)	77Ge	
	-14.2	1.905(1)	-11	25	C	S:Benzene A=Ph B=(OCOMe)	77Ge	
	-133	(2-1)	--	30	C	S:Benzene A=Phenyl B=Cl	75Fa	
	-115	(2-1)	--	30	C	S:Benzene A=Methyl B=Cl	75Fa	
	-116	(2-1)	--	30	C	S:Benzene A=Butyl B=Cl	75Fa	
	-142	(2-1)	--	30	C	S:Benzene A=Butyl B=NCS	75Fa	
SnA_4	-110.9	7.7(1)	-307.5	25	C	S:Benzene A=Cl	79Kc	
	-64	2.35(2)	-167	30	C	S:Benzene A=Cl	75Fa	
	-147	3.54(2)	-418	30	C	S:Benzene A=NCS	75Fa	
	-62	2.10(2)	-165	30	C	S:Benzene A=Cl	75Fa	
	-85	2.54(2)	-232	30	C	S:Benzene A=Cl	75Fa	
I_2	-49.4	3.671(25°,1)	-95.4	22	C	low μ S:n-Heptane	64Sc 58Na	b
A108	AMINE, tri(2-hydroxyethyl)-	$C_6H_{15}O_3N$				$(CH_2CH_2OH)_3N$		L
Ni^{2+}	-16.15	2.27(1)	-10.5	25	C	μ=0.5	63Sa	
	-2.13	0.82(2)	8.8	25	C	μ=0.5	63Sa	
	-18.28	3.09(1-2)	-1.7	25	C	μ=0.5	63Sa	
A109	AMINE, trimethyl-	C_3H_9N				$(CH_3)_3N$		L
NiA	-58.2	3.39(1-2)	-129.7	25	T	CM=10^{-4}M S:Benzene A=Diacetylbisbenzoyl hydrazine	60Sd	
AlA_3	-125.35	(1)	--	25-28	C	S:n-Hexane A=CH_3	68Hc	
I_2	-50.6	4.083(20°,1)	-94.6	20-40	T	CM=10^{-5}M CL=10^{-4}-10^{-3}M S:n-Heptane	60Ya	b
A110	AMINE, trimethyl, oxide	C_3H_9ON				$(CH_3)_3NO$		L
VAB_2	-157.3	(1)	--	25	C	S:CH_2Cl_2 A=O B=2,4-Pentanedione	69Pd	
CuA_2	-47.7	(1)	--	25	C	S:o-Dichlorobenzene A=Hexafluoroacetylacetone	70Pa	
	-47.7	(1)	--	25	C	S:o-Dichlorobenzene	72Nd 72Nd	

Metal Ion	ΔH, kJ /mol	Log K	ΔS, J /mol K	T,°C	M	Conditions	Ref.	Remarks
A110, continued								
CuA_2, cont.						A=Hexafluoro-acetylacetone		
I_2	-41.8	3.753(22°,1)	-70.7	6.5-22	T	CM=$5.2x10^{-4}$M CL=$1.8x10^{-4}$-$6.5x10^{-4}$M S:CH_2Cl_2	65Kc	b
A111	AMINE, trioctyl-		$C_{24}H_{51}N$			$[CH_2(CH_2)_6CH_3]_3N$		
CuA_2	-14	0.544(1)	-39	20	T	S:Benzene A=Trifluoro-acetylacetone	79Si	b,q:±2
I_2	-51.0	(1)	--	25	C	S:Benzene	66Ad	
A112	AMINE, tripentyl		$C_{15}H_{33}N$			$[CH_3(CH_2)_4]_3N$		L
BA_3C	-50.6	(1)	--	25	C	S:Benzene A=F C=$(CH_3CH_2)_2O$	68Rd	
Al_2A_6	-131.8	--	--	25	C	S:Benzene CM=0.07M A=Br R:0.5(Al_2A_6)+L=AlA_3L	68Rc	
I_2	-50.6	(1)	--	22	C	low μ S:n-Heptane	64Sc	
A113	AMINE, triphenyl-		$C_{18}H_{15}N$			$(C_6H_5)_3N$		L
I_2	-4.6	(1)	--	25	C	S:Benzene	66Ad	
A114	AMINE,tripropyl		$C_9H_{21}N$			$(CH_2CH_2CH_3)_3N$		L
NiA	-55.2	0.88(1-2)	-168.6	25	T	CM=10^{-4}M S:Benzene A=Di-acetylbisbenzoyl hydrazine	60Sd	
BA_3C	-60.2	(1)	--	25	C	S:Benzene A=F C=$(CH_3CH_2)_2O$	68Rd	
	-70.3	(1)	--	25	C	S:Benzene A=F C=$(CH_3CH_2CH_2CH_2)_2O$	68Rd	
Al_2A_6	-139.7	--	--	25	C	S:Benzene CM=0.07M A=Br R:0.5(Al_2A_6)+L=AlA_3L	68Rc	
I_2	-50.2	3.143(25°,1)	-107.9	22	C	low μ S:n-Heptane	64Sc 60Ya	

Metal Ion	ΔH, kJ /mol	Log K	ΔS, J /mol K	T,°C	M	Conditions	Ref.	Re-marks
A115	AMINE, tri(8-quinolyloxy-ethane)		$C_{33}H_{30}O_3N_4$			$[(C_9H_6N)OCH_2CH_2]_3N$		
Li^+	-19	2.18(25°,1)	-20	25	T	μ=0.08M$[N(CH_3)_4Br]$ S:Methanol	79Tb	b,q:±3
Na^+	-35	3.69(25°,1)	-46	25	T	μ=0.08M$[N(CH_3)_4Br]$ S:Methanol	79Tb	b
K^+	-50	2.58(25°,1)	-119	25	T	μ=0.08M$[N(CH_3)_4Br]$ S:Methanol	79Tb	b
Rb^+	-66	2.05(25°,1)	-184	25	T	μ=0.08M$[N(CH_3)_4Br]$ S:Methanol ΔCp=600	79Tb	b
Cs^+	-50	1.57(25°,1)	-140	25	T	μ=0.08M$[N(CH_3)_4Br]$ S:Methanol ΔCp=800	79Tb	b,q:±8
A116	AMINE, tris(6-methyl-2-pyridylmethyl)		$C_{21}H_{24}N_4$			$N[CH_2(C_5H_3N)CH_3]_3$		
Co^{2+}	-10.96	5.55(1)	68.6	20	C	μ=0.1(KNO_3)	77Ag	
Cu^{2+}	-33.22	10.52(1)	87.9	20	C	μ=0.1(KNO_3)	77Ag	
Zn^{2+}	-4.81	5.45(1)	87.9	20	C	μ=0.1(KNO_3)	77Ag	
Cd^{2+}	-16.15	6.77(1)	74.5	20	C	μ=0.1(KNO_3)	77Ag	
Hg^{2+}	-65.7	15.55(1)	72.8	20	C	μ=0.1(KNO_3)	77Ag	
Pb^{2+}	-21.76	6.80(1)	55.6	20	C	μ=0.1(KNO_3)	77Ag	
A117	AMINE, tris (2-pyridyl-methyl)		$C_{18}H_{18}N_4$			$N[CH_2(C_5H_4N)]_3$		
Mn^{2+}	-26.0	5.62(1)	18.8	20	C	μ=0.1(KNO_3)	77Ag 70Wb	
Fe^{2+}	-32.7	8.65(1)	54.0	20	C	μ=0.1(KNO_3)	77Ag 70Wb	
Co^{2+}	-46.82	11.38(1)	58.2	20	C	μ=0.1(KNO_3)	77Ag 70Wb	
Ni^{2+}	-67.95	14.60(1)	47.3	20	C	μ=0.1(KNO_3)	77Ag 70Wb	
Cu^{2+}	-69.87	16.15(1)	69.4	20	C	μ=0.1(KNO_3)	77Ag	
	-73.2	16.2(1)	59.8	20	C	μ=0.1	70Wb	
Zn^{2+}	-40.88	11.00(1)	70.3	20	C	μ=0.1(KNO_3)	77Ag 70Wb	
Cd^{2+}	-58.6	9.9(1)	-10.0	20	C	μ=0.1(KNO_3)	77Ag	
Pb^{2+}	-43.81	8.58(1)	14.2	20	C	μ=0.1(KNO_3)	77Ag	
Ag^+	-68.62	7.92(1)	-82.8	20	C	μ=0.1(KNO_3)	77Ag	
Hg^{2+}	-89.87	17.15(1)	16.7	20	C	μ=0.1(KNO_3)	77Ag	

Metal Ion	ΔH, kJ /mol	Log K	ΔS, J /mol K	T,°C	M	Conditions	Ref.	Re-marks
A118	AMMONIA		NH_3			NH_3		L
Li^+	-2.1	2.47(15°)	-52.7	14-17	C	Very qualitative calorimetry. Not known how many NH_3 coordinate-thought to be 3	52Fa	
Mg^{2+}	-0.4	3.40(15°)	-79.5	14-17	C	Very qualitative calorimetry. Not known how many NH_3 coordinate-thought to be 6	52Fa	
Co^{2+}	-22.01	(1-3)	--	--	C	--	36Ca	
Co^{3+}	-237.6	29.68(1-6)	-225.5	25	C	--	61Kb	
	-204.2	32.80	-57.3	25	C	R:$Co^{3+}+5NH_3+H_2O$ $=[Co(NH_3)_5H_2O]^{3+}$		
Ni^{2+}	-14.6	2.77(1)	3.8	25	C	1M(KNO_3)	55Pb 57Pa 46Ca 52Fa	f:6
	-16.7	2.80(1)	-2.1	26.8	C	--	57Yc	
	-14.6	2.22(2)	-6.7	25	C	1M(KNO_3)	55Pb	f:6
	-16.7	2.24(2)	-13.0	26.8	C	--	57Yc	
	-14.6	1.71(3)	-16.3	25	C	1M(KNO_3)	55Pb	f:6
	-16.7	1.73(3)	-22.6	26.8	C	--	57Yc	
	-14.6	1.17(4)	-26.8	25	C	1M(KNO_3)	55Pb	f:6
	-16.7	1.19(4)	-33.0	26.8	C	--	57Yc	
	-14.6	0.73(5)	-35.1	25	C	1M(KNO_3)	55Pb	f:6
	-18.0	0.75(5)	-45.6	26.8	C	--	57Yc	
	-14.6	0.01(6)	-49.0	25	C	1M(KNO_3)	55Pb	f:6
	-18.0	0.03(6)	-59.4	26.8	C	--	57Yc	
	-87.9	8.61(1-6)	-130.1	25	C	1M(KNO_3)	55Pb	f:6
NiA	-61.1	2.10(25°,1-2)	-164.0	10-45	T	CM=10^{-4}M S:Benzene A=Di-acetylbisbenzoyl hydrazine	60Sb	b
PtL_2A_2	-12.6	--	--	25	C	9.4%NH_3 solution A=I R:PtL_2A_2(cis) =PtL_2A_2(trans)	73Pa	
	-15.9	--	--	25	C	9.4%NH_3 solution A=I R:PtL_2A_2(cis) =PtL_2A_2(trans)	73Pa	
	-30.1	--	--	25	C	9.4%NH_3 solution A=I R:PtL_2A_2(cis) =PtL_2A_2(trans)	73Pa	
Cu^{2+}	-20.9	4.14(1)	8.8	25	C	1M(KNO_3) (Bouzat Ann.Chim.Phys., 29,312(1903))	55Pb 52Fa	i:see cond. col.
	-23.4	4.15(1)	1.25	26.8	C	--	57Yc	
	-20.9	3.49(2)	-3.3	25	C	1M(KNO_3) (Bouzat Ann.Chim.Phys., 29,312(1903))	55Pb	i:see cond. col.

A118, continued

Metal Ion	ΔH, kJ /mol	Log K	ΔS, J /mol K	T,°C	M	Conditions	Ref.	Remarks
Cu^{2+}, cont.	-23.0	3.50(2)	-9.6	26.8	C	--	57Yc	
	-20.9	2.88(3)	-15.1	25	C	1M(KNO_3) (Bouzat Ann.Chim.Phys., 29,312(1903))	55Pb	i:see cond. col.
	-23.4	2.89(3)	-22.6	26.8	C	--	57Yc	
	-20.9	2.12(4)	-29.7	25	C	1M(KNO_3) (Bouzat Ann.Chim.Phys., 29,312(1903))	55Pb	i:see cond. col.
	-22.2	2.13(4)	-33.0	26.8	C	--	57Yc	
	-13.4	-0.55(5)	-55.6	25	C	1M(KNO_3) (Bouzat Ann.Chim.Phys., 29,312(1903))	55Pb	i:see cond. col.
	-21.3	-0.52(5)	-81.2	26.8	C	--	57Yc	
	-50.2	7.865(1-2)	-18.24	25	T	μ=2	53Sb	
	-87.9	12.09(1-4)	-64	20	T	μ=1	52Sa	
	-83.7	12.63(1-4)	-39.3	25	C	1M(KNO_3) (Bouzat Ann.Chim.Phys., 29,312(1903))	55Pb	i:see cond. col.
	-98.7	13.05(1-4)	-82.8	25	T	μ=2	53Sb	
Ag^{+}	-56.1	7.225(1-2)	49.8	25	C	μ=0	37Sa 52Fa	
	-54	7.204(1-2)	-44.8	0-45	T	μ≅0.15	49Ja 55Fa 66Ma 57Sb	b
Zn^{2+}	-10.9	2.35(1)	8.8	26.8	C	CL=0.1-10M	57Yb 61Me 52Fa	
	-13.0	2.45(2)	3.8	26.8	C	CL=0.1-10M	57Yb	
	-16.3	2.51(3)	-6.3	26.8	C	CL=0.1-10M	57Yb	
	-21.8	2.15(4)	-31.4	26.8	C	CL=0.1-10M	57Yb	
ZnA_4^{2-}	-87.9	(1-4)	--	--	C	S:25w%NH_4Cl A=Cl	57Ta	
Cd^{2+}	-14.6	2.66(1)	2.1	26.8	C	CL=0.1-10M	57Yb 52Fa	f:6
	-14.6	2.09(2)	-8.8	26.8	C	CL=0.1-10M	57Yb	f:6
	-14.6	1.43(3)	-21.3	26.8	C	CL=0.1-10M	57Yb	f:6
	-14.6	0.93(4)	-31.0	26.8	C	CL=0.1-10M	57Yb	f:6
	-14.6	-0.29(5)	-54.5	26.8	C	CL=0.1-10M	57Yb	f:6
	-14.6	-2.42(6)	-81.2	26.8	C	CL=0.1-10M	57Yb	f:6
	-29.79	4.95(1-2)	-5.19	25	T	μ=2	53Sa	
	-53.1	7.44(1-4)	-142.6	25	T	μ=2	53Sa	
Hg^{2+}	-13.8	1.00(3)	-26.8	26.8	C	CL=0.1-10M	57Yb 52Fa	
	-15.1	0.80(4)	-34.7	26.8	C	CL=0.1-10M	57Yb	
	-103.3	17.5(1-2)	-9.6	26.8	C	CL=0.1-10M	57Yb	
AlA_3	-115.27	(1)	--	25-28	C	S:n-Hexane A=CH_3	68Hc	
I_2	-20.1	1.83(20°,1)	-33.5	10&20	T	CM=10^{-5}M CL=10^{-4}M-10^{-3}M S:n-Heptane	60Ya	a,b

Metal Ion	ΔH, kJ /mol	Log K	ΔS, J /mol K	T,°C	M	Conditions	Ref.	Remarks
A119	AMMONIUM ION, allyl triethyl- $C_9H_{20}N$					$CH_2:CHCH_2N(Et)_3$		L^+
PtA_4^{2-}	-20.5	2.185(45°)	-19.2	30-60	T	μ=2M(NaCl+HCl) A=Cl R:PtA_4^{2-}+ L^+=PtA_3L+A^-	67Db	b
A120	AMMONIUM ION, allyl trimethyl- $C_6H_{14}N$					$CH_2:CHCH_2N(Me)_3$		L^+
PtA_4^2	-22.2	2.238(44.5°)	-27.6	30-60	T	μ=2M(NaCl+HCl) A=Cl R:PtA_4^{2-}+ L^+=PtA_3L+A^-	67Db	b
A121	AMMONIUM ION, 3-butenyl triethyl-		$C_{10}H_{22}N$			$CH_2:CHCH_2CH_2N(Et)_3$		L^+
PtA_4^{2-}	-15.9	3.772(44.8°)	22.6	30-60	T	μ=2M(NaCl+HCl) A=Cl R:PtA_4^{2-}+ L^+=PtA_3L+A^-	67Db	b,q: ±1.7
A122	AMYLOSE		$(C_6H_{10}O_5)_n$			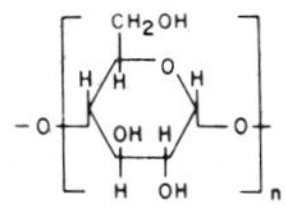		L
I_2	-64.9	9.34	-46.0	16	T	0.05M(KI)	60Ka	j
	-87.0	--	--	50	T	0.05M(KI)	60Ka	j
A123	ANILINE		C_6H_7N			$C_6H_5NH_2$		
CoA_2	-63.6	1.83(25°,1-2)	-178	--	C	CM=1x10^{-3}M CL=1.5x10^{-1}M A=Cl	72Zb	u:25°
	-31.30	4.897(1-2)	-25.27	25	C	S:Acetone CM=1x10^{-3}M CL=0.2M A=Cl	71Zc	q:±2.5
	-31.30	4.163(1-2)	-25.27	25	C	S:Acetone A=Cl	71Ze	q:±2.5
	-57.32	3.577(1-2)	-123.85	25	C	S:Acetonitrile A=Cl	71Ze	q:±6.3
	-41.55	3.188(1-2)	-78.24	25	C	S:n-Butanol A=Cl	71Ze	
	-63.6	1.832(1-2)	-178.2	25	C	S:t-Butyl alcohol CM=1x10^{-3}M CL=0.15M A=Cl	71Zd	
	-62.01	4.354(1-2)	-124.68	25	C	S:Cyclohexanone A=Cl	71Ze	
	-17.8	4.178(1-2)	19.92	25	C	S:Dimethylform-amide A=Cl	71Ze	
	-54.31	2.030(1-2)	-143.39	25	C	S:CH_3OH A=Cl	71Ze	
CuA_2	-20.1	1.18(10°,1)	-49.4	-60-60	T	S:Toluene A=Thenoyltri-fluoroacetonato	78Mb	
Ag^+	-16.5	1.82(1)	-20	25	C	μ=0.5M($NaClO_4$)	75Mb	
Hg^{2+}	-20	4.61(1)	21	25	C	μ=0.15M($NaClO_4$)	75Mb	q:±6
	-37	4.61(2)	-34	25	C	μ=0.15M($NaClO_4$)	75Mb	q:±6

Metal Ion	ΔH, kJ /mol	Log K	ΔS, J /mol K	T, °C	M	Conditions	Ref.	Re-marks
A124	ANILINE, 4-bromo-N,N-dimethyl-N-oxide-		$C_8H_{10}ONBr$			$C_6H_4N(Br)(CH_3)_2(\rightarrow O)$		
VAB_2	-141.8	(1)	--	25	C	S:CH_2Cl_2 A=O B=2,4-Pentane-dione	69Pd	
A125	ANILINE, 2-bromo-N-methylazoxy		$C_7H_7ON_3Br$			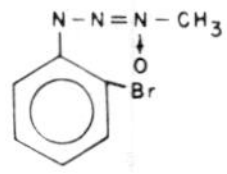		
NiL_2	4	--	1.7	21-50	T	S:Benzene R:NiL_2(planar, S=0)=NiL_2(pseudo-octahedral,S=1)	71Za	b,q: ±1.2
A126	ANILINE, 4-chloro-		C_6H_6NCl			C_6H_5NHCl		
CoA_2	-40.8	3.848(1-2)	-63.60	25	C	S:Acetone CM=$1x10^{-3}$M CL=0.2M A=Cl	71Zc	
A127	ANILINE, 2-chloro-N-methylazo, 3'-N-oxide		$C_7H_7ON_3Cl$			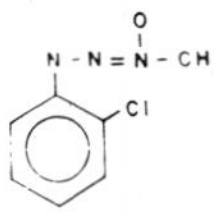		
NiL_2	8.4	--	13	20-60	T	S:Benzene R:NiL_2(planar, S=0)=NiL_2(pseudo-octahedral,S=1)	71Za	b,q: ±0.8
A128	ANILINE, N,N-diethyl-		$C_{10}H_{15}N$			$C_6H_5N(C_2H_5)_2$		L
CoA_2	-15.98	4.545(1-2)	33.26	25	C	S:Acetone CM=$1x10^{-3}$M CL=0.2M A=Cl	71Zc	q:±3.8
A129	ANILINE, N,N-dimethyl-		$C_8H_{11}N$			$C_6H_5N(CH_3)_2$		L
CoA_2	-26.86	4.347(1-2)	-6.86	25	C	S:Acetone CM=$1x10^{-3}$M CL=0.02M A=Cl	71Zc	q:±3.8
BA_3C	-30.5	(1)	--	25	C	S:Benzene A=F C=$(CH_3CH_2CH_2CH_3)_2O$	68Rd	
AlA_6	-101.3	--	--	25	C	S:Benzene CM=0.07M A=Br R:0.5(Al_2A_6)+L= AlA_3L	68Rc	
I_2	-34.3	1.24(27°,1)	-90.4	11-40	T	CM=$2.9x10^{-5}$M	60Tc	b

Metal Ion	ΔH, kJ /mol	Log K	ΔS, J /mol K	T,°C	M	Conditions	Ref.	Remarks
A129, continued								
I_2, cont.						CL=0.01-0.04M S:n-Heptane		
A130	ANILINE, 2,3-dimethyl-N-(2'-hydroxybenzylidine)-		$C_{15}H_{15}ON$					
Co^{2+}	-48.45	11.77(25°,<u>1-2</u>)	61.17	--	T	S:80v%CH_3OH	71De	b,j,u:25
Cu^{2+}	-54.4	15.91(25°,<u>1-2</u>)	121	15-45	T	S:80v%CH_3OH μ=0.1 CM=4x10^{-4}M	71Kb	b
	-56.78	15.86(25°,<u>1-2</u>)	109.96	--	T	S:80v%CH_3OH	71De	b,j,u:25
A131	ANILINE, 1,2-dimethyl-N-(2'-hydroxy-3'-methoxy-benzylidene)		$C_{16}H_{17}O_2N$			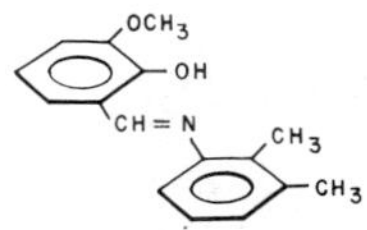		
Co^{2+}	-41.7	11.16(25°,<u>1-2</u>)	88.91	--	T	S:80v%CH_3OH	71De	b,j,u:25
Cu^{2+}	-46.9	16.86(25°,<u>1-2</u>)	159	15-45	T	S:80v%CH_3OH μ=0.1 CM=4x10^{-4}M	71Kb	b
	-43.89	16.52(25°,<u>1-2</u>)	161.46	--	T	S:80v%CH_3OH	71De	b,j,u:25
A132	ANILINE, 2-fluoro-N-methylazo, 3'-N-oxide		$C_7H_7ON_3F$			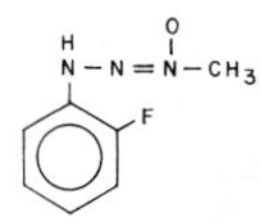		
NiL_2	10.0	--	6.3	30-70	T	S:Benzene R:NiL_2(planar, S=0)=NiL_2(pseudo-octahedral,S=1)	71Za	b,q: ±0.8
A133	ANILINE, N-(2'-hydroxy-benzylidene)-		$C_{13}H_{11}ON$			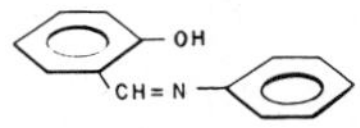		L
Co^{2+}	-51.63	10.75(25°,<u>1-2</u>)	31.04	--	T	S:80v%CH_3OH	71De	b,j,u:25
Cu^{2+}	-66.1	14.07(25°,<u>1-2</u>)	46	15-45	T	S:80v%CH_3OH μ=0.1 CM=4x10^{-4}M	71Kb	b
	-66.23	14.10(25°,<u>1-2</u>)	47.95	--	T	S:80v%CH_3OH	71De	b,j, u:25
A134	ANILINE, N-(2-hydroxy-5-chlorodiphenylmethane)		$C_{19}H_{16}ONCl$			$(ClC_6H_3OH)(C_6H_5)CHNC_6H_5$		HL
Zn^{2+}	-84.14	13.36(25°,<u>1-2</u>)	-26.53	25-35	T	μ=0.1M S:60v%Dioxane	79Pa	b

Metal Ion	ΔH, kJ /mol	Log K	ΔS, J /mol K	T, °C	M	Conditions	Ref.	Remarks
A134, continued								
Cd^{2+}	-89.29	13.12(25°,1-2)	-48.28	25-35	T	μ=0.1M S:60v%Dioxane	79Pa	b
A135	ANILINE, N-(2'-hydroxy-3'methoxybenzylidene)		$C_{14}H_{13}O_2N$			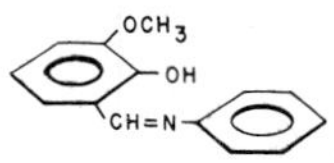		
Co^{2+}	-46.69	10.94(25°,1-2)	51.80	--	T	S:80v%CH_3OH	71De	b,j,u:25
Cu^{2+}	-59.8	15.32(25°,1-2)	88	15-45	T	S:80v%CH_3OH μ=0.1 CM=4x10^{-4}M	71Kb	b
	-59.96	15.36(25°,1-2)	89.45	--	T	S:80v%CH_3OH	71De	b,j,u:25
A136	ANILINE, 2-methoxy-		C_7H_9ON			$C_6H_4NH_2(OCH_3)$		
CoA_2	-15.36	3.878(1-2)	23.01	25	C	S:Acetone CM=1x10^{-3}M CL=0.2M A=Cl	71Zc	q:±2.0
A137	ANILINE, 4-methoxy-		C_7H_9ON			$C_6H_4NH_2(OCH_3)$		
CoA_2	-65.06	4.083(1-2)	-142.92	25	C	S:Acetone CM=1x10^{-3}M CL=0.2M A=Cl	71Zc	q:±6.3
A138	ANILINE, 4-methyl-N-(2'-hydroxybenzylidene)-		$C_{14}H_{13}ON$			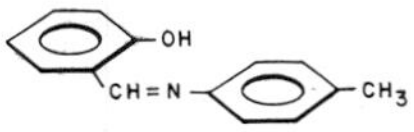		
Co^{2+}	-49.92	11.14(25°,1-2)	62.38	--	T	S:80v%CH_3OH	71De	b,j,u:25
Cu^{2+}	-64.0	15.98(25°,1-2)	84	15-45	T	S:80v%CH_3OH μ=0.1 CM=4x10^{-4}M	71Kb	b
	-64.10	16.00(25°,1-2)	86.78	--	T	S:80v%CH_3OH	71De	b,j,u:25
A139	ANILINE, 4-methyl-N-(2'-hydroxy-3'-methoxy-benzylidene)-		$C_{15}H_{15}O_2N$			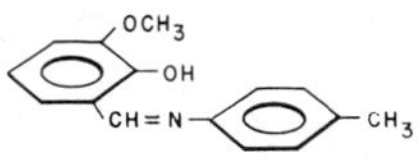		
Co^{2+}	-52.22	11.96(25°,1-2)	90.42	--	T	S:80v%CH_3OH	71De	b,j,u:25
Cu^{2+}	-51.9	16.20(25°,1-2)	138	15-45	T	S:80v%CH_3OH μ=0.1 CM=4x10^{-4}M	71Kb	b
	-52.09	16.19(25°,1-2)	140.20	--	T	S:80v%CH_3OH	71De	b,j,u:25
A140	ANILINE, 3-nitro-		$C_6H_6O_2N_2$			$C_6H_4NH_2(NO_2)$		
CoA_2	-58.03	2.566(1-2)	-145.60	25	C	S:Acetone CM=5x10^{-4}M CL=7.5x10^{-4}M A=Cl	71Zc	q:±4

Metal Ion	ΔH, kJ /mol	Log K	ΔS, J /mol K	T,°C	M	Conditions	Ref.	Re- marks
A141	ANILINE, 4-nitro-		$C_6H_6O_2N_2$			$C_6H_4NH_2(NO_2)$		
CoA_2	-61.59	2.155(1-2)	-165.39	25	C	S:Acetone CM=5x10^{-4}M CL=7.5x10^{-4}M A=Cl	71Zc	q:±4.6
A142	ANILINE, 2,4,6-trimethyl- N-(2'-hydroxybenzylidene)-		$C_{16}H_{17}ON$			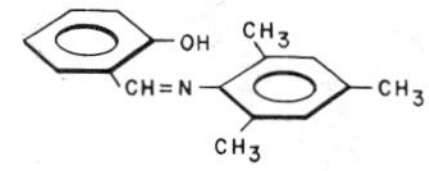		
Cu^{2+}	-77.4	15.83(25°,1-2)	109	15-45	T	S:80v%CH_3OH μ=0.1 CM=4x10^{-4}M	71Kb	b
A143	ANILINE, 2,4,6-trimethyl- N-(2'hydroxy-3'-methoxy- benzilidene)-		$C_{17}H_{19}O_2N$			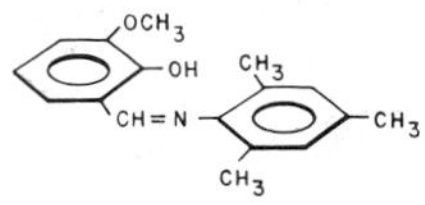		
Cu^{2+}	-43.9	16.49(25°,1-2)	159	15-45	T	S:80v%CH_3OH μ=0.1 CM=4x10^{-4}M	71Kb	b
A144	ANTAZOLINE		$C_{17}H_{19}N_3$			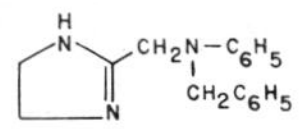		L
Be^{2+}	-23.8	7.44(25°,1)	63	0-45	T	μ=0.12M(KCl) CM≅2x10^{-3}M CL≅8x10^{-3}M	69Cb	b
	-13.8	6.20(25°,2)	71	0-45	T	μ=0.12M(KCl) CM≅2x10^{-3}M CL≅8x10^{-3}M	69Cb	b
Co^{2+}	46.0	7.54(25°,1-2)	301	0&25	T	0.06M(KCl)	62Aa	a,b
Ni^{2+}	38.1	7.70(25°,1-2)	276	0&25	T	0.06M(KCl)	62Aa	a,b
Cu^{2+}	-52.3	12.30(25°,1-2)	60.2	0&25	T	0.06M(KCl)	62Aa	a,b
Cd^{2+}	-74.0	8.73(25°,1-2)	-79	0-45	T	μ=0.1(KNO_3)	64Ac	b
A145	ANTHRACENE		$C_{14}H_{10}$			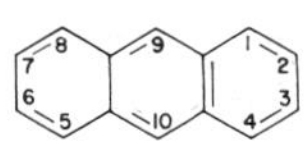		L^-
Li^+	0.0	5.149(1)	100	-65	T	S:Tetrahydrofuran	68Nb	q:±1.7
	7.9	5.305(1)	129.7	15	T	S:Tetrahydrofuran	68Nb	q:±1.7
Na^+	-19.2	--	-96	-92-12	T	S:2-Methyltetra- hydrofuran CL=10^{-5}-10^{-4}M R:Tight ion pair =Loose ion pair	67He	b,q: ±1.2
	-25.9	--	-105	-40-30	T	S:50%Tetrahydro- furan50%2-Methyl- tetrahydrofuran CL=10^{-5}-10^{-4}M	67He	b,q: ±2.1

Metal Ion	ΔH, kJ /mol	Log K	ΔS, J /mol K	T,°C	M	Conditions	Ref.	Re-marks
A145, continued								
Na^{+}, cont.						R:Tight ion pair =Loose ion pair		
	0	4.54(1)	88	-65	T	CM=CL=10^{-6}-10^{-3}M S:Tetrahydrofuran	66Cb	q:±1.7
	25.5	5.19(1)	188	15	T	CM=CL=10^{-6}-10^{-3}M S:Tetrahydrofuran	66Cb	q:±1.7
A146	2-ANTHRACENESULFONIC ACID, 9,10-dihydro-3,4-dihydroxy-9,10-dioxo-, sodium salt	$C_{14}H_7O_7SNa$				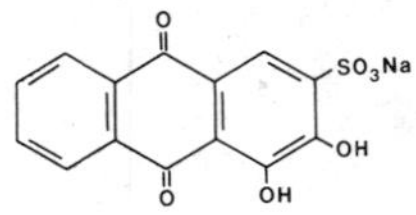		
Ni^{2+}	-33.8	9.30(30°,1)	65.3	10-50	T	μ>0.025($NaClO_4$)	73Gc	b
	15.2	6.10(30°,2)	167	10-50	T	μ>0.025($NaClO_4$)	73Gc	b
Co^{2+}	-29.6	8.90(30°,1)	71.8	10-50	T	μ>0.025($NaClO_4$)	73Gc	b
	15.5	6.70(30°,2)	179	10-50	T	μ>0.025($NaClO_4$)	73Gc	b
Be^{2+}	22.1	9.90(30°,1)	262	10-50	T	μ>0.025($NaClO_4$)	73Gc	b
	-14.8	6.82(30°,2)	82.1	10-50	T	μ>0.025($NaClO_4$)	73Gc	b
UO_2^{2+}	-31.4	11.80(30°,1)	121	10-50	T	μ>0.025($NaClO_4$)	73Gc	b
	-23.0	5.60(30°,2)	30.5	10-50	T	μ>0.025($NaClO_4$)	73Gc	b
A147	ANTHRANIL, N,N-diacetic acid	$C_{11}H_{13}O_5N$				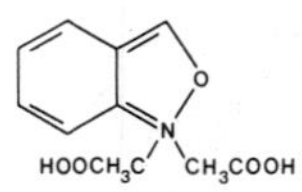		
Ga^{3+}	35.7	6.146(1)	237.8	25	C	--	77Dc	
In^{3+}	28.6	6.255(1)	215.3	25	C	--	77Dc	
Sc^{3+}	32.2	5.820(1)	222.7	25	C	--	77Dc	
Y^{3+}	28.0	5.477(1)	198.7	25	C	--	77Dc	
La^{3+}	20.6	5.531(1)	174.7	25	C	--	77Dc	
A148	9,10-ANTHRAQUINONE, 1,2,5,8-tetrahydroxy	$C_{14}H_8O_6$				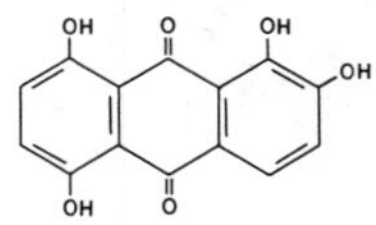		
CrO_4^{2-}	-123.6	8.59(25°,1-2)	-249	--	T	μ=0.08(NaCl) S:50v%Ethanol pH=6.0	71Ra	b
MoO_4^{2-}	-42.3	4.67(25°,1)	-28.8	--	T	μ=0.08(NaCl) S:50v%Ethanol pH=5.0	71Ra	b,p
WO_4^{2-}	-44.2	8.92(25°,1-2)	-58.7	--	T	μ=0.08(NaCl) S:50v%Ethanol pH=5.5	71Ra	b,p
UO_4^{2-}	12.1	5.26(25°,1)	142	--	T	μ=0.08(NaCl) S:50v%Ethanol pH=5.5	71Ra	b,p

Metal Ion	ΔH, kJ /mol	Log K	ΔS, J /mol K	T,°C	M	Conditions	Ref.	Remarks
A149	APOCARBONIC ANHYDRASE B, bovine							
Zn^{2+}	16.15	10.48	254.8	25	C	pH=5.5	69Hc	j
	41.17	12.02	368.2	25	C	pH=7.0	69Hc	j
A150	L-(+)-ARABINOSE		$C_5H_{10}O_5$					L
$H_5TeO_6^-$	-5.4	1.67	13	25	T	μ=0.1 R:$H_5TeO_6^-$ +L=$HTeO_4L^-$+$2H_2O$	62Ea	
A151	ARGININE		$C_6H_{14}O_2N_4$			$H_2NC(:NH)NH(CH_2)_3CH(NH_2)CO_2H$		L^-
Ni^{2+}	-15.69	4.92(25°,1)	41.4	17-40	T	--	60Pe	b
	-19.87	4.20(25°,2)	13.8	17-40	T	--	60Pe	b
	-27.20	3.08(25°,3)	-32.2	17-40	T	--	60Pe	b
Cu^{2+}	-38.70	7.34(25°,1)	10.5	17-40	T	--	60Pe	b
	-32.84	6.42(25°,2)	12.6	17-40	T	--	60Pe	b
Ag^+	-24.10	3.18(1)	-20.00	25	T	μ=0.024	59Da	
	-30.29	3.65(2)	-31.97	25	T	μ=0.024	59Da	
Zn^{2+}	-5.86	4.19(25°,1)	60.7	17-40	T	--	60Pe	b
	-14.02	3.93(25°,2)	28.0	17-40	T	--	60Pe	b
Cd^{2+}	-8.79	3.27(25°,1)	33.0	17-40	T	--	60Pe	b
	-26.57	3.18(25°,2)	-28.0	17-40	T	--	60Pe	b
A152	ARSINE, dimethyl phenyl		$C_8H_{11}As$			$(CH_3)_2As(C_6H_5)$		
HgA_2	-75.1	>10^5(1)	--	30	C	S:Benzene CM=5-10x10^{-4}M A=Cl	76Gb	
	-66.7	>10^5(1)	--	30	C	S:Benzene CM=7-14x10^{-4}M A=Br	76Gb	
	-55.0	3.607(1)	-113	30	C	S:Benzene CM=1-1.4x10^{-4}M A=I	76Gb	
A153	ARSINE, diphenyl methyl		$C_{13}H_{13}As$			$(C_6H_5)_2AsCH_3$		
HgA_2	-72.4	3.473(1)	-173	30	C	S:Benzene CM=4-6x10^{-4}M A=Cl	76Gb	
	-77.3	3.045(1)	-198	30	C	S:Benzene CM=4-6x10^{-4}M A=Br	76Gb	
	-71.7	2.049(1)	-198	30	C	S:Benzene CM=6-12x10^{-4}M A=I	76Gb	
A154	ARSINE, trimethyl		C_3H_9As			$As(CH_3)_3$		
$PtAB_2^+$	-64.4	(1)	--	25	C	S:THF A=CH_3 B=$[P(CH_3)_2(C_6H_5)]$	75Ma	

Metal Ion	ΔH, kJ /mol	Log K	ΔS, J /mol K	T,°C	M	Conditions	Ref.	Remarks
A154, continued								
HgA_2	-74.8	(1)	--	30	C	S:Benzene $CM=1\text{-}3x10^{-4}$ A=Cl	76Gb	
	-71.9	(1)	--	30	C	S:Benzene $CM=2\text{-}4x10^{-4}$ A=Br	76Gb	
	-70.0	5.778(1)	-120	30	C	S:Benzene $CM=3\text{-}5x10^{-4}$ A=I, $K=[(HgI_2AsMe_3)_2]^{1/2}/[HgI_2][AsMe_3]$	76Gb	
A155	ARSINE, triphenyl-		$C_{18}H_{15}As$			$(C_6H_5)_3As$		L
$PtAB_2^+$	-53.6	(1)	--	25	C	S:THF $A=CH_3$ $B=[P(CH_3)_2(C_6H_5)]$	75Ma	
Cu^+	-23	2.65(1)	-27	25	C	$\mu=0.1M(NH_4ClO_4)$ S:Dimethyl sulfoxide	78Ab	
	-32	1.40(2)	-80	25	C	$\mu=0.1M(NH_4ClO_4)$ S:Dimethyl sulfoxide	78Ab	
Hg^{2+}	-34	6.77(1)	15	25	C	$\mu=0.1M(NH_4ClO_4)$ S:Dimethyl sulfoxide	78Ab	
	-27	2.20(2)	-49	25	C	$\mu=0.1M(NH_4ClO_4)$ S:Dimethyl sulfoxide	78Ab	
HgA_2	-54.0	2.583(1)	-128.9	30	C	S:Benzene A=Cl $CM=1\text{-}5x10^{-3}M$	73Fa	
	-54.0	2.583(1)	-129	30	C	S:Benzene A=Cl	76Gb	
	-66.9	2.120(1)	-180.7	30	C	S:Benzene A=Br $CM=1\text{-}5x10^{-3}M$	73Fa	
	-67.0	2.120(1)	-181	30	C	S:Benzene A=Br	76Gb	
A156	ARSINE, triphenyl, sulfide		$C_{18}H_{15}SAs$			$(C_6H_5)_3AsS$		
HgA_2	-35.3	2.942(1)	-61	30	C	S:Benzene $CM=3\text{-}5x10^{-4}M$ A=Cl	76Gb	
	-25.0	3.699(1)	-11	30	C	S:Benzene $CM=4\text{-}6x10^{-4}M$ A=Br	76Gb	
	-29.5	2.989(1)	-40	30	C	S:Benzene $CM=3\text{-}5x10^{-4}M$ A=I	76Gb	q:±4.4
A157	ARSINE, tri(3-sulphophenyl)		$C_{18}H_{15}O_9S_3As$			$(HO_3SC_6H_4)_3As$		L^{-3}
Ag^+	-31.2	4.92(1)	-10	25	C	$\mu=0.5M(NaClO_4)$	75Mb	
	-31	4.96(1)	-3	25	C	$\mu=0.5M(NaClO_4)$	76Ha	
	-10.8	2.03(2)	-6.7	25	C	$\mu=0.5M(NaClO_4)$	75Mb	q:±2
	-17	1.74(2)	-7	25	C	$\mu=0.5M(NaClO_4)$	76Ha	
Hg^{2+}	-43	9.28(1)	33.5	25	C	$\mu=0.5M(NaClO_4)$	75Mb	q:±12
	-43	9.3(1)	10	25	C	$\mu=0.5M(NaClO_4)$	76Ha	
	-30	5.04(2)	-7	25	C	$\mu=0.5M(NaClO_4)$	75Mb	q:±8

Metal Ion	ΔH, kJ /mol	Log K	ΔS, J /mol K	T,°C	M	Conditions	Ref.	Re-marks
A157, continued								
Hg^{2+},	-30	5.06(2)	-1	25	C	μ=0.5M($NaClO_4$)	76Ha	
cont.	-12	2.89(3)	17	25	C	μ=0.5M($NaClO_4$)	75Mb	q:±6
A158	ARSONIUM ION, allyl triethyl- $C_9H_{20}As$					$CH_2:CHCH_2As(C_2H_5)_3$		L^+
PtA_4^{2-}	-24.7	3.121(45°)	-18.0	30-60	T	μ=2M(NaCl+HCl) A^-=Cl R:PtA_4^{2-}+L^+=PtA_3L+A^-	67Db	b,q: ±2.5
A159	ARSONIUM ION, 3-butenyl triethyl $C_{10}H_{22}As$					$CH_2:CHCH_2CH_2As(C_2H_5)_3$		L^+
PtA_4^{2-}	-13.8	3.846(44.8°)	30.1	30-60	T	μ=2M(NaCl+HCl) A^-=Cl R:PtA_4^{2-}+L^+=PtA_3L+A^-	67Db	b,q: ±1.5
A160	ASPARAGINE $C_4H_8O_3N_2$					$H_2NCOCH_2CH(NH_2)CO_2H$		L^-
Ni^+	-16.3	5.59(1)	42	25	C	μ=0.2(KCl)	75Gc	
	-20.5	4.54(2)	17	25	C	μ=0.2(KCl)	75Gc	
Cu^{2+}	-26.3	7.79(1)	60.6	25	C	μ=0.2(KCl)	75Ce	
	-47.49	14.4(2)	117.2	25	C	μ=0.1M	70Bd 71Be	y:70Bd
	-47.15	14.4(2)	118.0	25	C	μ=0.1M	70Bd 71Be	y:70Bd
	-27.6	6.50(2)	31.8	25	C	μ=0.2(KCl)	75Ce	
	-67.5	18.60	129.7	25	C	μ=3.0M($NaClO_4$) R:Cu^{2+}+L^-+A^-=CuLA A=Histidine	75Bd	
	-88.2	23.32	150.7	25	C	μ=3.0M($NaClO_4$) R:Cu^{2+}+L^-+HA=$CuLHA^+$ HA=Histidine	75Bd	
	-50.0	16.48	147.7	25	C	μ=3.0M($NaClO_4$) R:Cu^{2+}+L^-+A^-=CuLA A=Threonine	75Bd	
Zn^{2+}	-2.22	6.000(1)	106.7	5	T	μ=0.1(KCl)	73Rb	q:±3.0
	-7.65	5.940(1)	87.9	25	T	μ=0.1(KCl)	73Rb	q:±3.0
	-14.48	5.820(1)	66.1	45	T	μ=0.1(KCl)	73Rb	q:±3.0
	21.3	3.67(2)	141	25	C	μ=0.2(KCl)	75Gc	
A161	ASPARTIC ACID $C_4H_7O_4N$					$HO_2CCH_2CH(NH_2)CO_2H$		L^{2-}
Y^{3+}	8.8	8.07(25°,1-2)	184	25-45	T	μ=0.2M($NaClO_4$)	75Pb	b
Pr^{3+}	-196	5.62(25°,1-2)	-529	35-55	T	μ=0.1M($NaClO_4$)	74Ab	b,j
Nd^{3+}	-382.4	5.66(30°)	-1,153.9	30-50	T	μ=0.1M($NaClO_4$) R:Nd^{3+}+H_2L=$NdHL^{2+}$+H^+	71Td	b

Metal Ion	ΔH, kJ /mol	Log K	ΔS, J /mol K	T,°C	M	Conditions	Ref.	Re-marks
A161, continued								
Sm^{3+}	-238.99	7.83(30°)	-639.44	30-50	T	μ=0.1M($NaClO_4$) R:Sm^{3+}+H_2L= $SmHL^{2+}$+H^+	71Td	b
Gd^{3+}	-218	5.62(25°,1-2)	-600	35-55	T	μ=0.1M($NaClO_4$)	74Ab	b,j
$UO_2{}^{2+}$	-108.642	8.34(30°)	-199.45	30-50	T	μ=0.1M($NaClO_4$) R:$UO_2{}^{2+}$+HL^-= UO_2L+H^+	71Tc	b
Ni^{2+}	-17.2	7.14(1)	79	25	C	μ=0.2(KCl)	75Gc	
	-21.3	5.29(2)	29	25	C	μ=0.2(KCl)	75Gc	
Cu^{2+}	-25.5	8.80(1)	84	25	C	μ=0.2(KCl)	74Na	
	-25.5	8.70(1)	84	25	C	μ=0.2(KCl)	73Na	
	-25.5	8.80(25°,1)	84	20-35	T	μ=0.2(KCl)	74Na	b
	-26.4	6.96(2)	46	25	C	μ=0.2(KCl)	74Na	
	-26.4	6.96(2)	46	25	C	μ=0.2(KCl)	73Na	
	-25.1	6.96(25°,2)	50	20-35	T	μ=0.2(KCl)	74Na	b
Zn^{2+}	19.7	4.47(2)	130	25	C	μ=0.2(KCl)	75Gc	
In^{3+}	20.5	3.28(35°,1)	130	25-45	T	μ=0.20M($NaClO_4$)	73Sc	b
	41.80	6.40(35°,1-2)	258.2	25-45	T	μ=0	72Sb	b
A162	ASPARTIC ACID, N(2-hydroxybenzyl)		$C_{11}H_{13}O_5N$			$C_6H_4(OH)[CH_2NHCH(CO_2H)CH_2CO_2H]$		
$CoA_4{}^{3+}$	43.93	3.50(25°,1)	214.64	15-30	T	0.1M(Sodium acetate) A=NH_3	79Mb	b
$CoA_5{}^{3+}$	15.40	2.62(25°,1)	102.42	15-30	T	0.1M(Sodium acetate) A=NH_3	79Mb	b
A163	ASPARTIC ACID, N(3-methoxy-4-hydroxybenzyl)		$C_{12}H_{15}O_6N$			$C_6H_3(OCH_3)(OH)[CH_2NHCH(CO_2H)$-$CH_2CO_2H]$		
$CoA_4{}^{3+}$	15.77	3.62(25°,1)	122.76	15-30	T	0.1M(Sodium acetate) A=NH_3	79Mb	b
$CoA_5{}^{3+}$	16.07	2.60(25°,1)	104.10	15-30	T	0.1M(Sodium acetate) A=NH_3	79Mb	b
A164	ATPase							L
Mg^{2+}	-177.8	-3.10(1)	-536	24.5	C	0.04M Imidazole hydrochloride pH=6.2 L=ATPase (Enzyme purified from electric eel)	76Kn	

Metal Ion	ΔH, kJ /mol	Log K	ΔS, J /mol K	T,°C	M	Conditions	Ref.	Remarks
A165	1,2-AZABOROLIDINE, 2-butyl-		$C_7H_{16}NB$			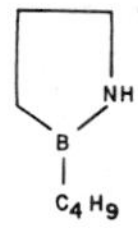		
[B]	63.6	0.535(100°)	180.7	30-110	T	[B]=L R:2L=L_2	71Td	b
A166	4-AZAOCTANE, 1,8-diamine		$C_7H_{19}N_3$			$NH_2(CH_2)_3NH(CH_2)_4NH_2$		L
Cu^{2+}	-59.12	11.60(1)	23.8	25	C	0.1M(NaCl) pH:5-9	77Al	
	-46.48	6.83	-25.1	25	C	0.1M(NaCl) pH:5-9 R:Cu^{2+}+HL^+= $[Cu(HL)]^{3+}$	77Al	
A167	3-AZAPENTANE, 1,5-diamine		$C_4H_{12}N_3$			$H_2NCH_2CH_2NCH_2CH_2NH_2$		L^3
Cu^{2+}	-75.3	15.8(25°,1)	50	15-35	T	μ=0.2M($NaClO_4$)	78Kd	b
Zn^{2+}	-27.2	8.8(25°,1)	79	15-35	T	μ=0.2M($NaClO_4$)	78Kd	b
Cd^{2+}	-42.2	8.2(25°,1)	17	15-35	T	μ=0.2M($NaClO_4$)	78Kd	b
Pb^{2+}	-35.6	7.4(25°,1)	21	15-35	T	μ=0.2M($NaClO_4$)	78Kd	b
A168	AZIDE ION		N_3			N_3^-		L^-
MnA	5.36	1.473(1)	46.9	25	T	μ=0.25 A=EDTA	67Sd	
	2.9	1.466(1)	38.5	25	T	μ=0.25 ΔH calculated from rate data A=EDTA	67Sd	
Fe^{3+}	-4.6	4.84(25°,1)	76.6	15-35	T	μ=0.25	61Ba 52Ua	b
$[Fe^{3+}]$	-63.6	4.398	-130	25	C	Tris-cacodylate buffer(≅0.1M Tris)[Fe]=G gouldii Hemerythrin=$HrFe_2$ R:$HrFe_2(H_2O)_m$+L =$HrFe_2L$+mH_2O	71Lb	g:Na q:±6.3
$[Fe^{3+}]$	-54.0	5.5(1)	-77.0	25	-	Hemoglobin	60Gb	
$[Fe^{3+}]$	-42.7	4.8(1)	-51.0	25	-	Chironomus hemoglobin	60Gb	
						$[Fe^{3+}]$=Ferri-hemoglobin μ=0.05(NaCl)		d:0.05
	-59.0	(1)	--	25.9	C	5.65	73Ah	
	-59.0	(1)	--	25.9	C	5.99	73Ah	
	-59.0	5.290(1)	-97.03	25.9	C	6.00	73Ah	
	-61.21	5.209(1)	-105.44	25.9	C	6.40	73Ah	
	-61.9	(1)	--	25.9	C	6.41	73Ah	
	-71.1	(1)	--	25.9	C	6.82	73Ah	

Metal Ion	ΔH, kJ /mol	Log K	ΔS, J /mol K	T,°C	M	Conditions	Ref.	Remarks
A168, continued								
$[Fe^{3+}]$, cont.	-74.18	5.055(1)	-151.75	25.9	C	7.00	73Ah	
	-70.3	(1)	--	25.9	C	7.24	73Ah	
	-70.3	(1)	--	25.9	C	7.60	73Ah	
	-68.2	(1)	--	25.9	C	7.82	73Ah	
	-66.9	(1)	--	25.9	C	8.01	73Ah	
	-64.5	5.231(1)	-115.94	25.9	C	8.47	73Ah	
	-60.7	(1)	--	25.9	C	8.83	73Ah	
	-52.7	(1)	--	25.9	C	9.06	73Ah	
	-48.1	(1)	--	25.9	C	9.21	73Ah	
						μ=0.05(NaCl) S:5%Tert-butyl alcohol		d:0.05
	-57.7	(1)	--	25.9	C	5.8	73Ah	
	-61.9	(1)	--	25.9	C	6.0	73Ah	
	-61.5	(1)	--	25.9	C	6.41	73Ah	
	-63.05	4.791(1)	-119.58	25.9	C	6.50	73Ah	
	-63.6	(1)	--	25.9	C	6.64	73Ah	
	-63.2	(1)	--	25.9	C	6.80	73Ah	
	-65.02	4.696(1)	-127.95	25.9	C	7.0	73Ah	
	-69.4	(1)	--	25.9	C	7.25	73Ah	
	-69.45	4.549(1)	-145.60	25.9	C	7.52	73Ah	
	-69.4	(1)	--	25.9	C	7.55	73Ah	
	-61.50	4.630(1)	-117.44	25.9	C	7.94	73Ah	
	-63.6	(1)	--	25.9	C	8.02	73Ah	
	-59.4	(1)	--	25.9	C	8.40	73Ah	
	-59.83	4.813(1)	-108.36	25.9	C	8.52	73Ah	
	-59.8	(1)	--	25.9	C	8.56	73Ah	
	-57.3	(1)	--	25.9	C	8.80	73Ah	
	-51.92	4.769(1)	-82.76	25.9	C	9.00	73Ah	
	-51.9	(1)	--	25.9	C	9.01	73Ah	
$[Fe^{3+}]$						$[Fe^{3+}]$=Sperm whale melmyoglobin μ=0.05		b,d: 0.05
	-20.0	4.79(20°,1)	-83.7	8.7-27.2	T	6.0	69Ba	
	-50.6	4.73(20°,1)	-83.7	8.7-27.2	T	6.4	69Ba	b
	-46.9	4.66(20°,1)	-70.3	8.7-27.2	T	6.8	69Ba	b
	-46.9	4.60(20°,1)	-72.0	8.7-27.2	T	7.2	69Ba	b
	-49.8	4.56(20°,1)	-82.4	8.7-27.2	T	7.6	69Ba	b
	-53.6	4.52(20°,1)	-96.2	8.7-27.2	T	8.0	69Ba	b
	-54.8	4.50(20°,1)	-100.8	8.7-27.2	T	8.4	69Ba	b
	-54.8	4.50(20°,1)	-100.8	8.7-27.2	T	8.8	69Ba	b
	-53.6	4.51(20°,1)	-96.2	8.7-27.2	T	9.2	69Ba	b
	-51.9	4.52(20°,1)	-90.4	8.7-27.2	T	9.6	69Ba	b

Metal Ion	ΔH, kJ /mol	Log K	ΔS, J /mol K	T,°C	M	Conditions	Ref.	Remarks
A168, continued								
[Fe^{3+}], cont.	-45.6	4.57(20°,1)	-68.2	8.7-27.2	T	10.0	69Ba	b
	-9.6	1.03(20°,1)	-6.7	14.4-30.9	T	0.1M(K-phos) pH=7.0 [Fe^{3+}]= Horse heart ferric cytochrome c	75Ia	
[Fe^{3+}]	Reaction: Pigeon methaemoglobin + L. ΔH values reported as a function of pH in graphical form.						68Ac	
[Fe^{3+}]	Reaction: Dog methaemoglobin + L. ΔH values reported as a function of pH in graphical form.						68Ac	
[Fe^{3+}]	Reaction: Guinea pig methaemoglobin + L. ΔH values reported as a function of pH in graphical form.						68Ac	
[Fe^{3+}]	Reaction: Human methaemoglobin + L. ΔH values reported as a function of pH in graphical form.						66Ac	
Co^{2+}	5.73	1.279(25°,1)	76.06	20.5-36.5	T	CL=CM=1.8x10^{-3}M	73Aa	b
	17.6	3.743(27°,1-4)	131.8	23-33	T	S:CH_3OH	73Aa	b
$CoA_3{}^{3+}$	-21.8	1.06(1)	-54	25	T	μ=0.054 A=Ethylenediamine	53Ea	
$CoA_6{}^{3+}$	-12.59	2.01(1)	-4.2	25	T	μ=0 A=NH_3	55Na 53Ea	
CoA_2BC	-13.5	3.50(25°,1)	23	32.7-48	T	S:DMSO R:trans-CoA_2BC+L^-=trans-CoA_2BL^-+C A=Methylglyoximato B=CH_3 C=DMSO	74Lb	b
Ni^{2+}	-0.96	0.87(1)	13.4	25	C	μ=1.0M($NaClO_4$)	75Aa	q:±0.14
	-3.3	0.38(2)	-3.7	25	C	μ=1.0M($NaClO_4$)	75Aa	q:±0.8
	-10.3	0.052(3)	-34	25	C	μ=1.0M($NaClO_4$)	75Aa	q:±1.7
$PtA_2{}^{2+}$	≅0	1.59(23°,1)	30.5	23-48	T	μ=0 S:CH_3OH A=o-Phenylenebisdimethyl arsine	66Db	b
CuA^{2+}	13.89	2.48(1)	93.7	25	C	S:CH_3OH A=N,N'-bis(2-Aminoethyl) propane-1,3-diamine	72Bc	
	6.65	3.07(1)	83.3	25	C	S:CH_3OH A=N,N'-bis(3-Aminopropyl) propane-1,3-diamine	72Bc	
	12.93	2.79(1)	96.6	25	C	S:CH_3OH A=Triethylenetetramine	72Bc	
$CuA_2{}^{2+}$	14.56	2.48(1)	96.2	25	C	S:CH_3OH A=Ethylenediamine	72Bc	
	10.08	2.57(1)	82.8	25	C	S:CH_3OH A=Trimethylenediamine	72Bc	

Metal Ion	ΔH, kJ /mol	Log K	ΔS, J /mol K	T,$^{\circ}$C	M	Conditions	Ref.	Re-marks
A168, continued								
Zn^{2+}	2.57	0.759(1)	23.1	25	C	μ=1.0M($NaClO_4$)	75Ab	q:±0.3
	2.3	0.561(2)	19	25	C	μ=1.0M($NaClO_4$)	75Ab	q:±1.7
	7.6	0.841(3)	42	25	C	μ=1.0M($NaClO_4$)	75Ab	q:±2.3
	-6.5	0.280(4)	-16	25	C	μ=1.0M($NaClO_4$)	75Ab	q:±3.8
Cd^{2+}	-4.81	1.62(1)	14.6	25	C	3.0M($NaClO_4$)	66Gb	
	-5.90	1.13(2)	0.8	25	C	3.0M($NaClO_4$)	66Gb	
	-6.94	0.53(3)	-13.0	25	C	3.0M($NaClO_4$)	66Gb	
	-5.31	0.56(4)	-7.1	25	C	3.0M($NaClO_4$)	66Gb	
Hg^{2+}	-30.4	6.978(1)	32	25	C	μ=1.0($NaClO_4$)	76Ad	
	-37.1	7.410(2)	17	25	C	μ=1.0($NaClO_4$)	76Ad	
Tl^{+}	-5.56	0.39(1)	-11.3	25	T	μ=0	57Na	
A169	4-AZOBENZENESULFONIC ACID, 2-carboxy-3',5'-dimethyl-2'-hydroxy		$C_{15}H_{13}O_6N_2S$			$(CH_3)_2(OH)C_6H_2N_2C_6H_3(COOH)(SO_3)$		
Fe^{3+}	-50.4	24.80(25°,1-2)	305	25-45	T	μ=0.2M(KNO_3) S:50v%Ethanol	76Bf	
A170	4-AZOBENZENESULFONIC ACID, 2-carboxy-3',5'-dichloro-2'-hydroxy		$C_{13}H_7O_6N_2Cl_2S$			$(Cl_2)(OH)C_6H_2N_2C_6H_3(COOH)(SO_3)$		
Fe^{3+}	-104.8	18.75(25°,1-2)	6.3	25-45	T	μ=0.2M(KNO_3) S:50v%Ethanol	76Bf	

B

Metal Ion	ΔH, kJ /mol	Log K	ΔS, J /mol K	T,$^{\circ}$C	M	Conditions	Ref.	Re-marks
B1	BARBITURIC ACID, 5-amino-N,N-diacetic acid		$C_8H_9O_7N_3$					L^{3-}
Li^{+}	-7.61	4.9(1)	67.8	20	C	μ=0.1M[$(CH_3)_4NCl$]	76Ah	
	-29.3	4.70(27°,1)	-4	20-39	T	μ=0.1M[$(CH_3)_4$-NNO_3]	63Ia	b
	-29.3	(1)	-4	20-40	T	μ=0.1M(KNO_3)	77Fb	
Na^{+}	-4.39	2.72(1)	37.2	20	C	μ=0.1M[$(CH_3)_4NCl$]	76Ah	
	-36.4	2.54(27°,1)	-75	20-34	T	μ=0.1M[$(CH_3)_4$-NNO_3]	63Ia	b
	-36.4	(1)	-75	20-40	T	μ=0.1M(KNO_3)	77Fb	
K^{+}	-1.7	1.23(1)	18.0	20	C	μ=0.1M[$(CH_3)_4NCl$]	76Ah	
	-49.4	1.00(27°,1)	-146	20-39	T	μ=0.1M[$(CH_3)_4$-NNO_3]	63Ia	b
	-49.4	(1)	-146	20-40	T	μ=0.1M(KNO_3)	77Fb	
Be^{2+}	-28.9	10.13(25°,1)	96	20-40	T	μ=0.1M(KNO_3)	77Fb	
Mg^{2+}	2.3	8.19(1)	164.8	20	C	μ=0.1M[$(CH_3)_4NCl$]	76Ah	
	-13.4	8.09(25°,1)	109	20-40	T	μ=0.1M(KNO_3)	77Fb	q:±0.8

Metal Ion	ΔH, kJ /mol	Log K	ΔS, J /mol K	T, °C	M	Conditions	Ref.	Remarks
B1, continued								
Ca^{2+}	-13.6	8.31(1)	117.2	20	C	μ=0.1M[$(CH_3)_4NCl$]	76Ah	
	-28.0	8.15(25°,1)	63	20-40	T	μ=0.1M(KNO_3)	77Fb	
Sr^{2+}	-12.1	6.93(1)	90.4	20	C	μ=0.1M[$(CH_3)_4NCl$]	76Ah	
	-25.1	6.82(25°,1)	46	20-40	T	μ=0.1M(KNO_3)	77Fb	
Ba^{2+}	-11.3	6.13(1)	79.1	20	C	μ=0.1M[$(CH_3)_4NCl$]	76Ah	
	-29.3	6.02(25°,1)	17	20-40	T	μ=0.1M(KNO_3)	77Fb	
Tl^{+}	-24.3	5.99(1)	31.8	20	C	μ=0.1M[$(CH_3)_4NCl$]	76Ah	
Th^{+}	-64.4	5.76(27°,1)	-105	20-39	T	μ=0.1M[$(CH_3)_4$-NNO_3]	63Ia	b
B2 BENZALDEHYDE			C_7H_6O			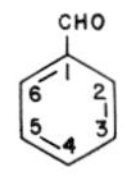		L
FeA_3	-53.6	1.48(1)	-151	25	C	S:Chlorobenzene A=Cl	49Db	
AlA_3	-103.3	1.78(1)	-314	25	C	S:Chlorobenzene A=Cl	49Db	
SnA_4	-117.2	5.56(1-2)	-285	25	C	S:Chlorobenzene A=Cl	49Db	
SbA_3	-16.3	0.95(1)	-38	25	C	S:Chlorobenzene A=Cl	49Db	
I_2	-12.6	-0.28(25°,1)	-47.3	5-25	T	S:Heptane	71Td	
B3 BENZALDEHYDE, oxime, N-methyl-			C_8H_9ON			$C_6H_5CH:N(CH_3)OH$		L
I_2	-21.97	1.554(15°,1)	-46.9	5-28	T	CM=$7.5x10^{-5}$M CL=$3.8x10^{-3}$-$1.5x10^{-2}$M S:CCl_4	65Kc	b
B4 BENZALDEHYDE, 5-bromo-2-hydroxy-			$C_7H_5O_2Br$			$C_6H_4O(Br)(OH)$		
Mg^{2+}	-11.7	3.18(25°,1)	23.4	15-50	T	S:50% Dioxane μ=0.3M($NaClO_4$)	73Ca	b
	-13.8	2.68(25°,2)	5.4	15-50	T	S:50% Dioxane μ=0.3M($NaClO_4$)	73Ca	b
Sr^{2+}	-5.4	3.09(25°,1)	41.4	15-50	T	S:50% Dioxane μ=0.3M($NaClO_4$)	73Ca	b
	-9.6	2.65(25°,2)	17.2	15-50	T	S:50% Dioxane μ=0.3M($NaClO_4$)	73Ca	b
Ni^{2+}	-20.5	4.28(25°,1)	13.8	15-50	T	S:50% Dioxane μ=0.3M($NaClO_4$)	73Ca	b
	-14.6	3.36(25°,2)	14.6	15-50	T	S:50% Dioxane μ=0.3M($NaClO_4$)	73Ca	b

Metal Ion	ΔH, kJ /mol	Log *K*	ΔS, J /mol K	T,°C	M	Conditions	Ref.	Re-marks
B4, continued								
Cu^{2+}	-26.4	6.37(25°,1)	32.6	15-50	T	S:50% Dioxane μ=0.3M($NaClO_4$)	73Ca	b
	-27.2	4.75(25°,2)	0.0	15-50	T	S:50% Dioxane μ=0.3M($NaClO_4$)	73Ca	b
Zn^{2+}	-10.5	3.81(25°,1)	38.1	15-50	T	S:50% Dioxane μ=0.3M($NaClO_4$)	73Ca	b
	-10.5	3.13(25°,2)	23.8	15-50	T	S:50% Dioxane μ=0.3M($NaClO_4$)	73Ca	b
B5 BENZALDEHYDE, 5-chloro-2-hydroxy-			$C_7H_5O_2Cl$			$C_6H_4O(Cl)(OH)$		
Mg^{2+}	-16.7	3.30(25°,1)	6.7	15-50	T	S:50% Dioxane μ=0.3M($NaClO_4$)	73Ca	b
	-17.6	2.82(25°,2)	-4.6	15-50	T	S:50% Dioxane μ=0.3M($NaClO_4$)	73Ca	b
Sr^{2+}	-5.4	3.06(25°,1)	40.6	15-50	T	S:50% Dioxane μ=0.3M($NaClO_4$)	73Ca	b
	-2.9	2.59(25°,2)	-39.7	15-50	T	S:50% Dioxane μ=0.3M($NaClO_4$)	73Ca	b
Ni^{2+}	-24.3	4.56(25°,1)	5.8	15-50	T	S:50% Dioxane μ=0.3M($NaClO_4$)	73Ca	b
	-22.6	3.52(25°,2)	-8.4	15-50	T	S:50% Dioxane μ=0.3M($NaClO_4$)	73Ca	b
Cu^{2+}	-23.8	6.21(25°,1)	39.7	15-50	T	S:50% Dioxane μ=0.3M($NaClO_4$)	73Ca	b
	-20.5	4.66(25°,2)	20.1	15-50	T	S:50% Dioxane μ=0.3M($NaClO_4$)	73Ca	b
Zn^{2+}	-13.8	3.65(25°,1)	24.3	15-50	T	S:50% Dioxane μ=0.3M($NaClO_4$)	73Ca	b
	-2.5	3.03(25°,2)	49.8	15-50	T	S:50% Dioxane μ=0.3M($NaClO_4$)	73Ca	b
B6 BENZALDEHYDE, 3,4-dihydroxy			$C_7H_6O_3$			$OHCC_6H_3(OH)_2$		
Pr^{3+}	-13.5	9.39(25°,1)	148.5	25-45	T	μ=0.2M($NaClO_4$) S:50v%Dioxane	76Md	
	-18.3	5.18(25°)	37.2	25-45	T	μ=0.2M($NaClO_4$) S:50v%Dioxane R:Pr+L+A=PrLA A=1,2-Diamino-N,N,N',N'-tetra-acetic acid	76Md	
B7 BENZALDEHYDE, 2-hydroxy-			$C_7H_6O_2$			$C_6H_5O(OH)$		L
Fe^{3+}	8.4	8.8(25°,1)	197	15-35	T	μ=3.0	56Aa	b
Cu^{2+}	-34.64	17.89(25°,1-2)	200.8	15&40	T	S:75v%Dioxane	78Sh 78Sg	a,b,p,y

Metal Ion	ΔH, kJ /mol	Log K	ΔS, J /mol K	T,°C	M	Conditions	Ref.	Re- marks
B8	BENZALDEHYDE, 4-hydroxy		$C_7H_6O_2$			$C_6H_5O(OH)$		
Fe^{3+}	2.1	5.35(1)	109	25	T	μ=0.1	69Da	
B9	BENZALDEHYDE, 2-hydroxy-5-nitro-		$C_7H_5O_4N$			$C_6H_3O(OH)(NO_2)$		
Mg^{2+}	-30.1	3.18(25°,1)	-49.8	15-50	T	S:50% Dioxane μ=0.3M($NaClO_4$)	73Ca	b
	-18.0	2.63(25°,2)	-12.6	15-50	T	S:50% Dioxane μ=0.3M($NaClO_4$)	73Ca	b
Sr^{2+}	-43.1	3.16(25°,1)	-87.9	15-50	T	S:50% Dioxane μ=0.3M($NaClO_4$)	73Ca	b
	-25.9	2.65(25°,2)	-38.9	15-50	T	S:50% Dioxane μ=0.3M($NaClO_4$)	73Ca	b
Ni^{2+}	-42.7	3.80(25°,1)	-71.5	15-50	T	S:50% Dioxane μ=0.3M($NaClO_4$)	73Ca	b
	-23.0	3.13(25°,2)	-18.0	15-50	T	S:50% Dioxane μ=0.3M($NaClO_4$)	73Ca	b
Cu^{2+}	-33.0	4.39(25°,1)	-27.2	15-50	T	S:50% Dioxane μ=0.3M($NaClO_4$)	73Ca	b
	-16.7	3.34(25°,2)	9.6	15-50	T	S:50% Dioxane μ=0.3M($NaClO_4$)	73Ca	b
Zn^{2+}	-15.1	3.27(25°,1)	11.7	15-50	T	S:50% Dioxane μ=0.3M($NaClO_4$)	73Ca	b
	-12.1	2.69(25°,2)	10.9	15-50	T	S:50% Dioxane μ=0.3M($NaClO_4$)	73Ca	b
B10	BENZALDEHYDE, 4-nitro		$C_7H_5O_3N$			$(O_2N)C_6H_4CHO$		
K^+	0	1.22(25°)	23	?	T	S:Hexamethyl-phosphoramide R:$K^+ + A^- = KA$ A=Free anion radical of L	75Ac	
B11	BENZENE		C_6H_6					
VAB_3	-0.586	(1)	--	5-50	T	S:CCl_4 A=O B=Cl	71Ha	b
	-1.255	(1)	--	5-50	T	S:Cyclohexane A=O B=Cl	71Ha	b
NbA_5	-8.786	(1)	--	-30-0	T	S:CCl_4 A=Cl	71Ha	b,q: ±1.7
	-10.042	(1)	--	-30-0	T	S:Cyclohexane A=Cl	71Ha	b,q: ±1.7
TaA_5	-4.602	(1)	--	-30-0	T	S:CCl_4 A=Cl	71Ha	b
	-10.460	(1)	--	-30-0	T	S:Cyclohexane A=Cl	71Ha	b,q: ±1.7
WA_6	-0.628	(1)	--	-24-26	T	S:CCl_4 A=Cl	71Hd	b

Metal Ion	ΔH, kJ /mol	Log K	ΔS, J /mol K	T,°C	M	Conditions	Ref.	Re-marks
B11, continued								
WA_6, cont.	-0.209	(1)	--	-24-26	T	S:Perfluoro-methylcyclo-hexane A=Cl	71Hd	b
WAB_4	0.544	(1)	--	-14-26	T	S:CCl_4 A=O B=Cl	71Hd	b
	0.251	(1)	--	-14-26	T	S:Cyclohexane A=O B=Cl	71Hd	b
OsA_4	-2.197	(1)	--	0-31.5	T	S:Cyclohexane A=O	71Hc	b,q: ±0.73
Ag^+	-11.97	0.041(1)	-39.33	25	T	μ=0.5 S:Equi-molar H_2O-CH_3OH solution	56Oa	
AgA	16.7	0.220(20°,1)	61.1	1.4&20	T	S:CH_3OH A=ClO_4^-	64Tb	a,b
GaA_3	-47.3	(1)	--	25	C	S:Benzene A=Cl	72Rd	
SO_2	-32.6	-0.252(20°,1)	-115.81	0-30	T	0.4M(NaOH) 0.2M(HCl) S:CCl_4	57Da	b
	-32.6	-0.252(20°,1)	-115.81	11-25	T	low μ S:Ethanol	57Db	b
I_2	-5.86	(1)	--	--	C	S:Benzene	61Aa 55Kb 61Ja	i: 50Ha
	-3.56	(1)	--	--	C	S:Benzene	50Ha	
	-6.075	0.336(25°,1)	-13.8	0-44.5	T	Mole fraction of Benzene in CCl_4:1.00	50Ca	b
	-5.924	0.282(25°,1)	-14.6	0-44.5	T	Mole fraction of Benzene in CCl_4:0.620	50Ca	b
	-5.644	0.265(25°,1)	-13.8	0-44.5	T	Mole fraction of Benzene in CCl_4:0.0434	50Ca	b
	-5.510	0.281(25°,1)	-13.0	0-44.5	T	Mole fraction of Benzene in CCl_4:0.0217	50Ca	b
	-6.28	0.196(25°,1)	-17.3	25-60	T	S:CCl_4	76Ko	b
	-5.4	(1)	--	--	-	S:Cyclohexane	60Bb	
	-5.439	0.083(1)	-16.7	25	T	S:n-Hexane	54Ka	1
	-4.77	0.649(25°)	-3.56	-20-40	T	S:Hexane A=Hexane R:$I_2A+L=I_2L+A$	73Sn	b,q: ±0.63
IA	-10.63	0.74(25°,1)	-21.50	1.6&25	T	CL=0.1-1M C=10^{-4}-10^{-3}M S:CCl_4 A=Cl	55Oa	a,b
	-13.14	0.74(25°,1)	-29.87	25& 45.8	T	CL=0.1-1M C=10^{-4}-10^{-3}M S:CCl_4 A=Cl	55Oa	a,b

B12 BENZENE, 1,2-bis(carboxy-methoxy) $C_{10}H_{10}O_6$ $C_6H_4(OCH_2CO_2H)_2$

Y^{3+}	24.3	3.78(1)	154	25	C	μ=0.1M($NaClO_4$)	77Hb	

Metal Ion	ΔH, kJ /mol	Log K	ΔS, J /mol K	T,°C	M	Conditions	Ref.	Re-marks
B12, continued								
La^{3+}	16.3	3.81(1)	127	25	C	μ=0.1M($NaClO_4$)	77Hb	
	10.0	3.49(2)	100	25	C	μ=0.1M($NaClO_4$)	77Hb	
Pr^{3+}	14.4	4.36(1)	132	25	C	μ=0.1M($NaClO_4$)	77Hb	
	6.32	3.33(2)	85	25	C	μ=0.1M($NaClO_4$)	77Hb	
Nd^{3+}	17.2	4.45(1)	141	25	C	μ=0.1M($NaClO_4$)	77Hb	
Sm^{2+}	11.3	4.57(1)	126	25	C	μ=0.1M($NaClO_4$)	77Hb	
	3.9	2.91(2)	69	25	C	μ=0.1M($NaClO_4$)	77Hb	
Eu^{3+}	14.7	4.60(1)	138	25	C	μ=0.1M($NaClO_4$)	77Hb	
	1.8	2.85(2)	61	25	C	μ=0.1M($NaClO_4$)	77Hb	
Gd^{3+}	13.6	4.48(1)	131	25	C	μ=0.1M($NaClO_4$)	77Hb	
Tb^{3+}	19.7	4.30(1)	149	25	C	μ=0.1M($NaClO_4$)	77Hb	
Dy^{3+}	20.2	4.26(1)	149	25	C	μ=0.1M($NaClO_4$)	77Hb	
Ho^{3+}	23.9	4.11(1)	159	25	C	μ=0.1M($NaClO_4$)	77Hb	
Er^{3+}	26.9	3.92(1)	165	25	C	μ=0.1M($NaClO_4$)	77Hb	
Tm^{3+}	35	3.94(1)	179	25	C	μ=0.1M($NaClO_4$)	77Hb	
Lu^{3+}	32.4	4.04(1)	186	25	C	μ=0.1M($NaClO_4$)	77Hb	
B13	BENZENE, n-butyl		$C_{10}H_{14}$			$C_6H_5(C_4H_9)$		
I_2	-8.87	0.413(25°,1)	-21.8	25-60	T	S:CCl_4	76Ko	b
B14	BENZENE, tert-butyl		$C_{10}H_{14}$			$C_6H_5(C_4H_9)$		
I_2	-8.45	0.364(25°,1)	-21.4	25-60	T	S:CCl_4	76Ko	b
B15	BENZENE, 1-tert-butyl-4-nitro-		$C_{10}H_{13}O_2N$			$C_6H_4(NO_2)[C(CH_3)_3]$		L^-
Li^+	55.6	--	--	25	T	S:Hexamethyl-phosphoramide R:$Li^+ + L^- = LiL$	73Sw	q:3.3
B16	BENZENE, 1-carbamyl-4-nitro-		$C_7H_6O_3N_2$			$C_6H_4(NO_3)(CONH_2)$		
Li^+	29.3	--	--	25	T	S:Hexamethyl-phosphoramide R:$Li^+ + L^- = LiL$	73Sw	q:±2.5
B17	BENZENE, chloro-		C_6H_5Cl			$C_6H_5(Cl)$		
I_2	-4.60	-0.174(1)	-18.4	25	T	S:n-Hexane	54Ka	1

Metal Ion	ΔH, kJ /mol	Log K	ΔS, J /mol K	T,°C	M	Conditions	Ref.	Remarks
B18	BENZENE, 1-chloro-2-cyclopropyl-		C_9H_9Cl			$C_6H_4(Cl)(CHCH_2CH_2)$		L
HgA_2	40.88	-0.777(27°)	121.3	35-20	T	Large excess of L S:CH_3OH R:HgA_2+L=HgAl+A A=$OCOCH_3$	64Pb	b
B19	BENZENE, 4-chloro-1,2-dihydroxy-		$C_6H_5O_2Cl$			$C_6H_3(Cl)(OH)_2$		L^{2-}
Cu^{2+}	-16.7	0.92	-38.1	25	C	μ=0.100M(KNO_3) R:$Cu^{2+}+HL^-=CuL+H^+$	72Jb	
	11.9	-7.602	-105.0	25	C	μ=0.100M(KNO_3) R:$Cu^{2+}+H_2L=CuL+2H^+$	72Jb	
B20	BENZENE, 1-chloro-4-nitro-		$C_6H_4O_2NCl$			$C_6H_4(NO_2)(Cl)$		L^-
Li^+	45.2	(1)	--	25	T	S:Hexamethyl-phosphoramide	73Sw	
B21	BENZENE, cyclopropyl		C_9H_{10}			$C_6H_5(CHCH_2CH_2)$		L
HgA_2	23.8	-0.356(11°)	77.0	-14-20	T	Large excess of L S:CH_3OH R:HgA_2+L=HgAL+A A=$OCOCH_3$	64Pb	b
B22	BENZENE, 1-cyclopropyl-2-nitro-		$C_9H_9O_2N$			$C_6H_4(CHCH_2CH_2)(NO_2)$		L
HgA_2	44.39	-0.914(27°)	130.5	15-35	T	Large excess of L S:CH_3OH R:HgA_2+L=HgAl+A A=$OCOCH_3$	64Pb	b
B23	BENZENE, 1,2-diamino		$C_6H_8N_2$			$C_6H_4(NH_2)_2$		L
Co^{2+}	-20.48	3.03(26°,1)	-10.82	20-32	T	μ=0.1M(KNO_3)	80Aa	b
Ni^{2+}	-89.00	2.88(26°,1)	-250.71	20-32	T	μ=0.1M(KNO_3)	80Aa	b
Cu^{2+}	-36.24	4.70(26°,1)	-30.19	20-32	T	μ=0.1M(KNO_3)	80Aa	b
	-64.04	5.05(26°,1)	-117.78	20-32	T	μ=0.1M(KNO_3)	80Aa	b
	-38.96	3.98(26°,1)	-53.95	20-32	T	μ=0.1M(KNO_3)	80Aa	b
B24	BENZENE, 1,3-diamino		$C_6H_8N_2$			$C_6H_4(NH_2)_2$		L
Co^{2+}	-19.60	3.02(26°,1)	-7.36	20-32	T	μ=0.1M(KNO_3)	80Aa	b
Ni^{2+}	-83.14	3.09(26°,1)	-220.60	20-32	T	μ=0.1M(KNO_3)	80Aa	b
Cu^{2+}	-57.85	3.77(26°,1)	-140.92	20-32	T	μ=0.1M(KNO_3)	80Aa	b

Metal Ion	ΔH, kJ /mol	Log K	ΔS, J /mol K	T,°C	M	Conditions	Ref.	Re-marks
B24, continued								
Cu^{2+},	-29.75	3.06(26°,1)	-56.76	20-32	T	μ=0.1M(KNO_3)	80Aa	b
cont.	-32.10	2.85(26°,1)	-71.18	20-32	T	μ=0.1M(KNO_3)	80Aa	b
B25	BENZENE, 1,4-diamino		$C_6H_8N_2$			$C_6H_4(NH_2)_2$		
Co^{2+}	-38.54	3.04(26°,1)	-70.84	20-32	T	μ=0.1M(KNO_3)	80Aa	
Ni^{2+}	-38.20	3.25(26°,1)	-66.11	20-32	T	μ=0.1M(KNO_3)	80Aa	
Cu^{2+}	-38.92	3.84(26°,1)	-56.99	20-32	T	μ=0.1M(KNO_3)	80Aa	
	-36.87	3.36(26°,1)	-59.36	20-32	T	μ=0.1M(KNO_3)	80Aa	
	-40.63	2.88(26°,1)	-80.91	20-32	T	μ=0.1M(KNO_3)	80Aa	
B26	BENZENE, 1,2-diamino-4-methyl		$C_7H_{10}N_2$			$C_6H_3(NH_2)_2(CH_3)$		
Co^{2+}	-27.42	3.16(26°,1)	-30.43	20-32	T	μ=0.1M(KNO_3)	80Aa	
Ni^{2+}	-59.10	3.53(26°,1)	-129.88	20-32	T	μ=0.1M(KNO_3)	80Aa	
Cu^{2+}	-34.28	5.41(26°,1)	-38.99	20-32	T	μ=0.1M(KNO_3)	80Aa	
	-34.15	5.36(26°,1)	-38.61	20-32	T	μ=0.1M(KNO_3)	80Aa	
	-32.56	4.36(26°,1)	-47.64	20-32	T	μ=0.1M(KNO_3)	80Aa	
B27	BENZENE, 1,3-diamino-4-methyl		$C_7H_{10}N_2$			$C_6H_3(NH_2)_2(CH_3)$		
Co^{2+}	-38.25	3.32(26°,1)	-63.92	20-32	T	μ=0.1M(KNO_3)	80Aa	
Ni^{2+}	-46.19	2.87(26°,1)	-98.73	20-32	T	μ=0.1M(KNO_3)	80Aa	
Cu^{2+}	-58.10	5.02(26°,1)	-125.17	20-32	T	μ=0.1M(KNO_3)	80Aa	
	-29.09	3.58(26°,1)	-47.64	20-32	T	μ=0.1M(KNO_3)	80Aa	
	-29.01	3.14(26°,1)	-53.37	20-32	T	μ=0.1M(KNO_3)	80Aa	
B28	BENZENE, 1,2-dichloro-		$C_6H_4Cl_2$			$C_6H_4(Cl)_2$		L
I_2	-4.18	-0.180(1)	-18.0	25	T	S:n=Hexane	54Ka	1
B29	BENZENE, 1,2-dihydroxy-		$C_6H_6O_2$			$C_6H_4(OH)_2$		L^{2-}
Be^{2+}	-158.1	14.21(30°,1)	-249.8	30&40	T	μ=0	79Bb	a,b
	-216.3	10.43(30°,2)	-512.8	30&40	T	μ=0	79Bb	a,b
Ce^{3+}	-152.8	18.26(25°,1-2)	-117	25&45	T	--	79Na	a,b,j,n
Pr^{3+}	-42.8	8.72(25°,1)	23.4	25-45	T	μ=0.2M($NaClO_4$)	76Md	
	-155.3	18.80(25°,1-2)	-159	25&45	T	--	79Na	a,b,j,n
	-37.6	5.26(25°)	-26.4	25-45	T	μ=0.2M($NaClO_4$) R:Pr+L+A=PrLA A=1,2-Diamino-N,N,N',N'-tetra-acetic acid	76Md	
Nd^{3+}	-111.1	19.25(25°,1-2)	-161	25&45	T	--	79Na	a,b,j,n
Sm^{3+}	-113.4	19.66(25°,1-2)	-4	25&45	T	--	79Na	a,b,j,n
Gd^{3+}	-101.0	20.08(25°,1-2)	45	25&45	T	--	79Na	a,b,j,n

Metal Ion	ΔH, kJ /mol	Log K	ΔS, J /mol K	T,°C	M	Conditions	Ref.	Re-marks
B29, continued								
Tb^{3+}	-93.5	20.43(25°,1-2)	77	25&45	T	--	79Na	a,b,j,n
Dy^{3+}	-91.4	20.83(25°,1-2)	92	25&45	T	--	79Na	a,b,j,n
Ho^{3+}	-121.9	21.22(25°,1-2)	2.5	25&45	T	--	79Na	a,b,j,n
ThA	53.6	--	--	25	C	μ=0.10(KNO_3) A=1,2-Diamino-cyclohexane-N,N',N'-tetra-acetic acid R:ThA+H_2L= $ThAL^{2-}$+$2H^+$	70Kb	
	51.0	--	--	25	C	μ=0.10(KNO_3) A=EDTA R:ThA+ H_2L=$ThAL^{2-}$+$2H^+$	70Kb	
Co^{2+}	-177.6	9.67(30°,1-2)	-398.9	30-50	T	μ=0.2M($NaClO_4$)	76Sg	
Ni^+	82.0	8.56(30°,1-2)	439.1	30-50	T	μ=0.2M($NaClO_4$)	76Sg	
Cu^{2+}	-32.72	14.11(30°,1)	155.2	30&40	T	μ=0	79Bb	a,b
	-41.8	13.827(1)	-129.7	25	C	μ=0.100M(KNO_3)	72Ja	
	-21.8	13.81(25°,1)	191.6	15-35	T	μ=0.2M(KNO_3) CM=2.0x10^{-3}	79Mb	b
	-154.5	13.36(30°,2)	-253.8	30&40	T	μ=0	79Bb	a,b
	-23.0	11.09(25°,2)	135.1	15-35	T	μ=0.2M(KNO_3) CM=2.0x10^{-3}	79Mb	b
	-218.6	19.05(30°,1-2)	-364.2	30-50	T	μ=0.2M($NaClO_4$)	76Sg	
	-20.9	0.85	-54.0	25	C	μ=0.100M(KNO_3) R:Cu^{2+}+HL^-= CuL+H^+	72Ja	
	-13.4	-8.345	-115.1	25	C	μ=0.100M(KNO_3) R:Cu^{2+}+H_2L= CuL+$2H^+$	72Ja	
CuA^{2+}	-29.7	14.12(25°,1)	171.1	15-35	T	0.2M(KNO_3) CM=2.0x10^{-3} A=2,2'-Bipyridyl	79Mg	b
Al^{3+}	-69.1	15.11(30°,1)	69.1	30&40	T	μ=0	79Bb	a,b
	-41.8	13.86(30°,2)	121.1	30&40	T	μ=0	79Bb	a,b
B30	BENZENE, 1,4-dihydroxy		$C_6H_6O_2$			$C_6H_4(OH)_2$		L
SO_2	-58.6	0.121(14.5°,1)	-177.69	4.5-20	T	Low μ S:C_2H_5OH	57Db	b
B31	BENZENE, (1,2-dihydroxyethyl)		$C_8H_{10}O_2$			$C_6H_5[CH(OH)(CH_2OH)]$		L
$H_5TeO_6^-$	-15.1	1.63	-21	25	T	μ=0.1 R:$H_5TeO_6^-$ +L=$HTeO_4L^-$+$2H_2O$	62Ea	

Metal Ion	ΔH, kJ /mol	Log K	ΔS, J /mol K	T,°C	M	Conditions	Ref.	Re-marks
B32	BENZENE, 1,3-dihydroxy-4-(2-pyridylazo)		$C_{11}H_9O_2N_3$			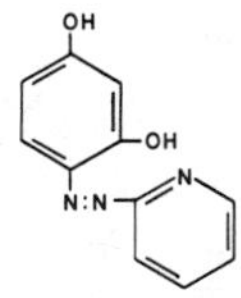		
Y^{3+}	-24.7	24.03(25°,1-2)	377	25-45	T	μ=0.1M($NaClO_4$) S:50v%Dioxane	74Pb	b
B33	BENZENE, 1,3-dimethyl-		C_8H_{10}			$C_6H_4(CH_3)_2$		L
AgA	18.8	0.783(20°,1)	79.1	1.4&20	T	S:CH_3OH A=ClO_4^-	64Tb	a,b
B34	BENZENE, 1,4-dimethyl-		C_8H_{10}			$C_6H_4(CH_3)_2$		L
I_2	-8.37	0.471(1)	-20.5	25	T	S:n-Hexane Orientation (o,m,or p) not specified	54Ka	1
IA	-15.06	1.29(25°,1)	-25.86	1.6&25	T	CM=10^{-4}-10^{-3}M A=Cl CL=0.1-1M S:CCl_4	55Oa 60Bb 55Kb	a,b,k
	-15.98	1.29(25°,1)	-28.91	25& 45.8	T	CM=10^{-4}M-10^{-3}M A=Cl CL=0.1-1M S:CCl_4	55Oa	a,b,k
B35	BENZENE, 1,3-dimethyl-2-(N,N'dimethylamino)-		$C_{10}H_{15}N$			$C_6H_3(CH_3)_2[N(CH_3)_2]$		L
I_2	-7.1	0.42(33°,1)	-15.1	16&33	T	CM≅10^{-5}M CL≅0.05-0.13M S:n-Heptane	60Tc	a,b
B36	BENZENE, 1,4-dinitro		$C_6H_4O_4N$			$C_6H_4(NO_2)_2$		
K^+	7.9	0.15(25°)	39	?	T	S:Hexamethyl-phosphoramide R:K^++A^-=KA A=Free anion radical of L	75Ac	
B37	BENZENE, ethoxy		$C_8H_{10}O$			$C_6H_5OCH_2CH_3$		L
I_2	-2.1	(1)	--	25	C	S:Octane	66Ad	
B38	BENZENE, ethyl		C_8H_{10}			$C_6H_5(C_2H_5)$		
I_2	-7.82	0.308(25°,1)	-20.4	25-60	T	S:CCl_4	76Ko	b

Metal Ion	ΔH, kJ /mol	Log K	ΔS, J /mol K	T,°C	M	Conditions	Ref.	Remarks
B39	BENZENE, ethynyl-		C_8H_6			$C_6H_5C{:}CH$		L
AlA_3	3.10	-1.19(27°,1)	--	-5-27	T	S:Cyclohexane R:½Al_2A_6+L= AlA_3L A=Et	73Ae	b,q: ±0.84
B40	BENZENE, hexaethyl-		$C_{18}H_{30}$			$C_6(CH_2CH_3)_6$		L
I_2	-7.49	0.58(25°,1)	-14.2	1.6-45.8	T	CL=0.01-0.15mole fraction S:CCl_4	55Kb	b,m
IA	-15.02	1.25(25°,1)	-26.36	1.6&25	T	CM=10^{-4}-10^{-3}M CL=0.1-1M S:CCl_4 A=Cl	55Oa	a,b,k
	-14.02	1.25(25°,1)	-23.0	25&45.8	T	CM=10^{-4}-10^{-3}M CL=0.1-1M S:CCl_4 A=Cl	55Oa	a,b,k
B41	BENZENE, hexamethyl-		$C_{12}H_{18}$			$C_6(CH_3)_6$		L
Ag^+	-14.43	-0.201(1)	-52.3	25	T	μ=0.5 S:Equimolar H_2O-CH_3OH soln.	56Oa	
PA_5	2.468	(1)	--	-30-31	T	S:Cyclohexane A=Cl	71Hb	b
I_2	-15.61	1.20(25°,1)	-29.7	1.6-45.8	T	CL=0.01-0.15M mole fraction S:CCl_4	55Kb	b,m
IA	-21.97	2.13(25°,1)	-32.89	1.6&25	T	CM=10^{-4}-10^{-3}M CL=0.1-1M S:CCl_4 A=Cl	55Oa	a,b,k
	-22.55	2.13(25°,1)	-34.81	25&45.8	T	CM=10^{-4}-10^{-3}M CL=0.1-1M S:CCl_4 A=Cl	55Oa	a,b,k
B42	BENZENE, isobutyl		$C_{10}H_{14}$			$C_6H_5(C_4H_9)$		
I_2	-8.37	0.364(25°,1)	-21.1	25-60	T	S:CCl_4	76Ko	b
B43	BENZENE, isopropyl		C_9H_{12}			$C_6H_5(C_3H_7)$		
I_2	-8.33	0.344(25°,1)	-21.3	25-60	T	S:CCl_4	76Ko	b
B44	BENZENE, 1-isopropyl-4-nitro-		$C_9H_{11}O_2N$			$(CH_3)_2CHC_6H_5NO_2$		L^-
Li^+	52.3	(1)	--	25	T	S:Hexamethylphosphoramide	73Sw	q:±5.0

Metal Ion	ΔH, kJ /mol	Log K	ΔS, J /mol K	T,°C	M	Conditions	Ref.	Remarks
B45	BENZENE, α-ketoethyl-2-hydroxy-		$C_8H_8O_2$			$C_6H_4(COCH_3)(OH)$		L
Fe^{3+}	3.3	10.6(25°,1)	213	15-35	T	μ=3.0	56Aa	b
B46	BENZENE, methoxy-		C_7H_8O			$C_6H_5OCH_3$		L
Al_2A_6	-64.8	--	--	25	C	S:Benzene CM=0.07M A=Br R:0.5(Al_2A_6)+L= AlA_3L	68Rc	
	-61.1	--	--	25	C	S:Cyclohexane CM=0.07M A=Br R:0.5(Al_2A_6)+L= AlA_3L	68Rc	
GaA_3	-88.3	(1)	--	25	C	S:Benzene A=Cl	72Rd	
	-81.6	(1)	--	25	C	S:Cyclohexane A=Cl	72Rd	
	-91.6	(1)	--	25	C	S:Mesitylene A=Cl	72Rd	
B47	BENZENE, 1-methoxy-4-(S-methylmercapto)		$C_8H_{10}OS$			$C_6H_4(OCH_3)(SCH_3)$		L
I_2	-25.9	2.339(25°,1)	-32.6	25-55	T	CL<0.02 mole fraction S:CCl_4	65Lb	b,l
B48	BENZENE, methylthio-		C_7H_8S			$C_6H_5(SCH_3)$		L
PtA_2L_2	15.1	0.398(28°)	58	30-60	T	S:Chloroform A=Cl R:PtL_2A_2(cis)= PtL_2A_2(trans)	73Re	b
Al_2A_6	-54.8	--	--	25	C	S:Benzene CM=0.07M A=Br R:0.5(Al_2A_6)+L= AlA_3L	68Rc	
I_2	-25.5	0.964(25°,1)	-67.4	25-40	T	CM=10^{-3}M CL=0.16M S:CCl_4	64Na	b
	-23.0	1.083(25°,1)	-54	15-35	T	S:Cyclohexane	63Va	b
B49	BENZENE, methylthio-4-bromo-		C_7H_7BrS			$C_6H_4(SCH_3)(Br)$		L
I_2	-24.3	0.771(25°,1)	-67	15-35	T	S:Cyclohexane	63Va	b
B50	BENZENE, methylthio-3-chloro-		C_7H_7ClS			$C_6H_5(SCH_3)(Cl)$		L
I_2	-24.3	0.663(25°,1)	-67	15-35	T	S:Cyclohexane	63Va	b

Metal Ion	ΔH, kJ /mol	Log K	ΔS, J /mol K	T,°C	M	Conditions	Ref.	Re-marks
B51	BENZENE, methylthio-4-chloro-		C_7H_7ClS			$C_6H_4(SCH_3)(Cl)$		L
I_2	-24.2	0.748(25°,1)	-67	15-35	T	S:Cyclohexane	63Va	b
B52	BENZENE, methylthio-3-fluoro-		C_7H_7FS			$C_6H_4(SCH_3)(F)$		L
I_2	-22.2	0.633(25°,1)	-63	15-35	T	S:Cyclohexane	63Va	b
B53	BENZENE, methylthio-4-fluoro-		C_7H_7FS			$C_6H(SCH_3)(F)$		L
I_2	-27.2	0.949(25°,1)	-71	15-35	T	S:Cyclohexane	73Va	b
B54	BENZENE, methylthio-4-methyl-		$C_8H_{10}S$			$C_6H_4(SCH_3)(CH_3)$		L
I_2	-26.4	1.292(25°,1)	-63	15-35	T	S:Cyclohexane	63Va	b
B55	BENZENE, nitro-		$C_6H_5O_2N$			$C_6H_5(NO_2)$		L
Li^+	37.6	(1)	--	25	T	S:Hexamethyl-phosphoramide	73Sw	q:±3.3
SbA_5	-36.0	(1)	--	25	C	S:CCl_4 A=Cl	73Oa	q:±8.4
I_2	-11.51	(1)	--	--	C	S:Nitrobenzene	50Ha	
B56	BENZENE, 4-nitro-1,2-dihydroxy-		$C_6H_5O_5N$			$C_6H_3(NO_3)(OH)_2$		L^{2-}
Cu^{2+}	-16.7	0.81	-41.8	25	C	μ=0.100M(KNO_3) R:$Cu^{2+}+HL^-=CuL+H^+$	72Jb	
	7.15	-5.888	-88.7	25	C	μ=0.100M(KNO_3) R:$Cu^{2+}+H_2L=CuL+2H^+$	72Jb	
B57	BENZENE, pentamethyl-		$C_{11}H_{16}$			$C_6H(CH_3)_5$		
IA	-21.76	1.82(25°,1)	-38.07	1.6&25	T	CM=10^{-4}-10^{-3}M CL=0.1-1.0M S:CCl_4 A=Cl	55Oa	a,b,k
	-20.79	1.82(25°,1)	-34.81	25& 45.8	T	CM=10^{-4}-10^{-3}M CL=0.1-1M S:CCl_4 A=Cl	55Oa	a,b,k
B58	BENZENE, n-propyl		C_9H_{12}			$C_6H_5(C_3H_7)$		
I_2	-8.62	0.369(25°,1)	-21.8	25-60	T	S:CCl_4	76Ko	b
B59	BENZENE, 1,2,4,5-tetramethyl-		$C_{10}H_{14}$			$C_6H_2(CH_3)_4$		L
I_2	-11.63	0.814(25°,1)	-23.4	1.6-	T	CL=0.01-0.15	55Kb	b,m

Metal Ion	ΔH, kJ /mol	Log K	ΔS, J /mol K	T,°C	M	Conditions	Ref.	Remarks
B59, continued								
I_2, cont.				45.8		mole fraction S:CCl_4		
B60	BENZENE, 1,2,4-triethyl-		$C_{12}H_{18}$			$C_6H_3(CH_2CH_3)_4$		L
IA	-17.66	1.40(25°,1)	-32.47	1.6&25	T	$CM=10^{-4}-10^{-3}M$ CL=0.1-1M S:CCl_4 A=Cl	55Oa	a,b,k
	-18.67	1.40(25°,1)	-35.77	25& 45.8	T	$CM=10^{-4}-10^{-3}M$ CL=0.1-1M S:CCl_4 A=Cl	55Oa	a,b,k
B61	BENZENE, 1,3,5-triethyl		$C_{12}H_{18}$			$C_6H_2(CH_2CH_3)_3$		L
I_2	-11.04	0.72(25°,1)	-23.4	1.6- 45.8	T	CL=0.01-0.15 mole fraction S:CCl_4	55Kb	b,m
B62	BENZENE, 1,2,3-trihydroxy		$C_6H_6O_3$			$C_6H_3(OH)_3$		
Pr^{3+}	-24.1	9.78(25°,1)	114.6	25-45	T	μ=0.2M($NaClO_4$)	76Md	
	31.6	4.27(25°)	188.3	25-45	T	μ=0.2M($NaClO_4$) R:Pr+L+A=PrLA A=1,2-Diamino-N,N,N',N'-tetra-acetic acid	76Md	
In^{3+}	-40.6	16.68(25°,1)	184	25-45	T	μ=0	78Sc	
	-38.9	14.94(25°,2)	155	25-45	T	μ=0	78Sc	
	-30.1	6.38(25°,3)	21	25-45	T	μ=0	78Sc	
B63	BENZENE, 1,3,5-trihydroxy		$C_6H_6O_3$			$C_6H_3(OH)_3$		
Sc^{3+}	21.03	15.27(30°,1)	363.7	20-40	T	μ=0.02M($NaClO_4$)	77Ma	b
Y^{3+}	21.39	8.10(30°,1)	226.7	20-40	T	μ=0.02M($NaClO_4$)	77Ma	b
La^{3+}	30.63	6.25(30°,1)	214.5	20-40	T	μ=0.02M($NaClO_4$)	77Ma	b
Ce^{3+}	30.84	6.37(30°,1)	224.5	20-40	T	μ=0.02M($NaClO_4$)	77Ma	b
Pr^{3+}	24.99	6.85(30°,1)	214.3	20-40	T	μ=0.02M($NaClO_4$)	77Ma	b
Nd^{3+}	33.16	7.12(30°,1)	246.6	20-40	T	μ=0.02M($NaClO_4$)	77Ma	b
B64	BENZENE, 1,3,5-trimethyl-		C_9H_{12}			$C_6H_3(CH_3)_3$		L
NbA_5	-5.021	(1)	--	-30-0	T	S:CCl_4 A=Cl	71Ha	b
OsA_4	-3.222	(1)	--	0-31.5	T	S:CCl_4 A=O	71Hc	b,q: ±0.46
Ag^+	-12.26	-0.097(1)	-43.1	25	T	μ=0.5 S:Equimolar	56Oa	

Metal Ion	ΔH, kJ /mol	Log K	ΔS, J /mol K	T,°C	M	Conditions	Ref.	Re-marks
B64, continued								
Ag^+, cont.	2.93	-0.292(20°,$\underline{1}$)	--	1.4&20	T	H_2O-CH_3OH soln. S:CH_3OH A=ClO_4	64Tb	a,b
AlA_3	-64.4	($\underline{1}$)	--	25	C	S:Mesitylene A=Cl	72Rd	
GaA_3	-63.6	($\underline{1}$)	--	25	C	S:Mesitylene A=Cl	72Rd	
I_2	-11.97	0.780(25°,$\underline{1}$)	-25.1	1.6-45.8	T	CL=0.01-0.15M mole fraction S:CCl_4	55Kb	b,m
IA	-18.62	0.936($\underline{1}$)	-44.4	25	C	S:CCl_4 A=Cl	73Sk	
	-19.87	1.58(25°,$\underline{1}$)	$\underline{-36.36}$	1.6&45.8	T	CM=10^{-4}-10^{-3}M CL=0.1-1M S:CCl_4 A=Cl	55Oa	a,b,k
	-19.58	1.58(25°,$\underline{1}$)	$\underline{-35.35}$	25&45.8	T	CM=10^{-4}-10^{-3}M CL=0.1-1M S:CCl_4 A=Cl	55Oa	a,b,k
B65	BENZENE, 1,3,5-tris-(1,1-dimethylethyl)-		$C_{18}H_{30}$			$C_6H_3[C(CH_3)_3]_3$		L
Ag^+	-13.76	-0.495($\underline{1}$)	-55.6	25	T	μ=0.5 S:Equimolar H_2O-CH_3OH soln.	56Oa	a
I_2	-9.12	0.46(25°,$\underline{1}$)	-21.8	1.6-45.8	T	CL=0.01-0.15M mole fraction	55Kb	b,m
B66	BENZENEARSONIC ACID, o'(1,8-dihydroxy-3,6-disulpho-2-napthylazo)		$C_{16}H_{13}O_{11}N_2S_2As$			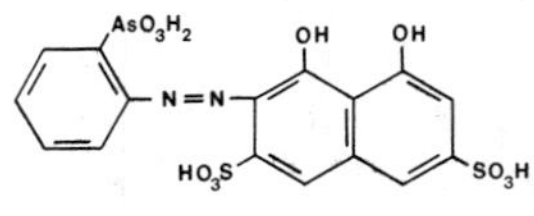		
Y^{3+}	-24.5	14.2(30°,$\underline{1}$)	$\underline{187}$	30&45	T	0.5M(KCl)	77Mb	a,b,n
	85.8	9.8(30°,$\underline{2}$)	$\underline{467}$	30&45	T	0.5M(KCl)	77Mb	a,b,n
Ce^{4+}	-36.7	19.8(30°,$\underline{1}$)	$\underline{232}$	30&45	T	0.5M(KCl)	77Mb	a,b,n
	-85.8	9.2(30°,$\underline{2}$)	$\underline{-110}$	30&45	T	0.5M(KCl)	77Mb	a,b,n
Zr^{4+}	-12.2	17.9(30°,$\underline{1}$)	$\underline{297}$	30&45	T	0.5M(KCl)	77Mb	a,b,n
	-36.7	13.3(30°,$\underline{2}$)	$\underline{129}$	30&45	T	0.5M(KCl)	77Mb	a,b,n
B67	1,3-BENZENEDISULFONIC ACID, 4,5-dihydroxy-,		$C_6H_4O_8S_2Na_2$			$C_6H_2(SO_3Na)_2(OH)_2$		L^{2-}
La^{3+}	-67.49	12.76(30°,$\underline{1}$)	22.9	20-40	T	μ=0.1M($NaClO_4$)	76Cb	
Pr^{3+}	-69.08	13.03(30°,$\underline{1}$)	22.5	20-40	T	μ=0.1M($NaClO_4$)	76Cb	
Nd^{3+}	-70.25	13.26(30°,$\underline{1}$)	24.0	20-40	T	μ=0.1M($NaClO_4$)	76Cb	
Sm^{3+}	-73.81	13.59(30°,$\underline{1}$)	20.4	20-40	T	μ=0.1M($NaClO_4$)	76Cb	
Y^{3+}	-117.65	14.10(30°,$\underline{1}$)	114.6	20-40	T	μ=0.1M($NaClO_4$)	76Cb	

Metal Ion	ΔH, kJ /mol	Log K	ΔS, J /mol K	T,°C	M	Conditions	Ref.	Remarks
B67, continued								
ThA	15.81	--	--	25	C	μ=0.10(KNO_3) A=EDTA R:ThA+ H_2L=$ThAL^{2-}$+$2H^+$	70Kb	g:Na
	28.41	--	--	25	C	μ=0.10(KNO_3) A=1,2=Diamino-cyclohexane-N,N-N',N'-tetra-acetic acid R:ThA+H_2L= $ThAL^{2-}$+$2H^+$	70Kb	g:Na
VO^{2+}	-10.5	16.74(25°,1)	280	25-35	T	μ=0.10(KNO_3)	66Mg	b,q:±2.1
Ga^{3+}	1.640	4.62(1)	89.1	25	T	μ=0.1	65Da	g:Na
In^{3+}	5.98	3.77(29°,1)	92.9	20-45	T	μ=0.1	65Nb	b,g:Na
B68 BENZENESULFONIC ACID, 3-(diphenylphosphino)		$C_{18}H_{15}O_3SP$				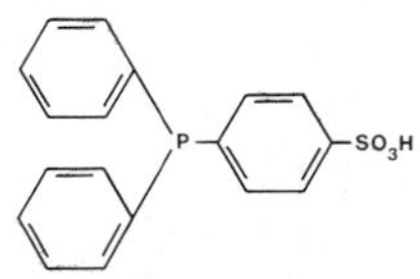		
Ag^+	-71	8.15(1)	-80	25	C	μ=0.1M($NaClO_4$)	75Mb	
	-60	5.95(2)	-91	25	C	μ=0.1M($NaClO_4$)	75Mb	
	-64	5.40(3)	-111	25	C	μ=0.1M($NaClO_4$)	75Mb	
	-30	3.77(4)	-27	25	C	μ=0.1M($NaClO_4$)	75Mb	
Hg^{2+}	-86	14.51(1)	-10	25	C	μ=0.1M($NaClO_4$)	75Mb	q:±10
	-88	10.37(2)	-97	25	C	μ=0.1M($NaClO_4$)	75Mb	
	-43	5.08(3)	-47	25	C	μ=0.1M($NaClO_4$)	75Mb	
	-19	3.33(4)	0	25	C	μ=0.1M($NaClO_4$)	75Mb	q:±35
B69 BENZENESULFONIC ACID, 4-(diphenylphosphino)		$C_{18}H_{15}O_3SP$				$(C_6H_5)_2PC_6H_4SO_3H$		L^-
Ag^+	-71	8.15(1)	-25	25	C	μ=0.1M($NaClO_4$)	76Ha	
	-60.6	5.95(2)	-25	25	C	μ=0.1M($NaClO_4$)	76Ha	
	-63.4	5.40(3)	-34	25	C	μ=0.1M($NaClO_4$)	76Ha	
	-30.0	3.80(4)	-8	25	C	μ=0.1M($NaClO_4$)	76Ha	
Hg^{2+}	-86	14.5(1)	-3	25	C	μ=0.1M($NaClO_4$)	76Ha	
	-88	10.3(2)	-29	25	C	μ=0.1M($NaClO_4$)	76Ha	
	-43	5.1(3)	-14	25	C	μ=0.1M($NaClO_4$)	76Ha	
	-19	3.3(4)	-1	25	C	μ=0.1M($NaClO_4$)	76Ha	
B70 BENZENESULFONIC ACID, o-(N-α-methyl-2-hydroxybenzylideneimino)		$C_{14}H_{15}O_4NS$				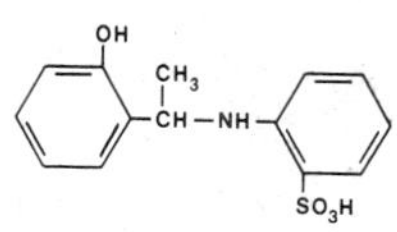		
VO^{2+}	-18.10	14.935(25°,1-2)	225.1	25&35	T	μ=1.0($NaClO_4$)	77Sf	a,b,j
Cu^{2+}	-79.08	10.739(25°,1-2)	-59.8	25&35	T	μ=1.0($NaClO_4$)	77Sf	a,b,j

Metal Ion	ΔH, kJ /mol	Log K	ΔS, J /mol K	T,°C	M	Conditions	Ref.	Remarks
B70, continued								
UO_2^{2+}	40.24	14.068(25°,1-2)	404.7	25&35	T	μ=1.0($NaClO_4$)	77Sf	a,b,j
B71 BENZENESULFONIC ACID, o-(α-2-oxopropylbenzylidene-imino)		$C_{16}H_{19}O_4NS$				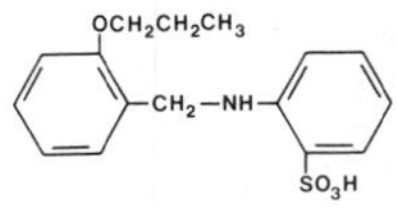		
La^{3+}	196.8	10.53(25°,1-2)	855.6	25&35	T	--	76So	a,b
Ce^{3+}	189.8	12.43(25°,1-2)	867.8	25&35	T	--	76So	a,b
Pr^{3+}	224.9	13.33(25°,1-2)	1002	25&35	T	--	76So	a,b
Nd^{3+}	232.0	14.41(25°,1-2)	858.1	25&35	T	--	76So	a,b
Sm^{3+}	251.3	15.42(25°,1-2)	1130.1	25&35	T	--	76So	a,b
Gd^{3+}	223.2	16.18(25°,1-2)	1051.2	25&35	T	--	76So	a,b
B72 BENZIMIDAZOLE, 2-acetic acid		$C_9H_8O_2N_2$				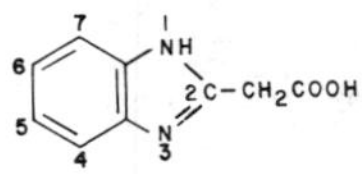		L^-
Mn^{2+}	-18.8	5.40(25°,1)	40.2	3-40	T	S:50v% Dioxane	59La	b
	-35.2	3.95(25°,2)	42.3	3-40	T	S:50v% Dioxane	59La	b
Co^{2+}	-30.5	5.09(25°,1)	-5.0	3-40	T	S:50v% Dioxane	59La	b
	-26.8	4.37(25°,1)	-6.3	3-40	T	S:50v% Dioxane	59La	b
Ni^{2+}	-40.6	6.98(25°,1)	-2.51	25&40	T	S:50v% Dioxane	59La	b
	-34.7	6.50(25°,2)	7.99	3-40	T	S:50v% Dioxane	59La	b
Cu^{2+}	-49.0	7.85(25°,1)	-14.2	25&40	T	S:50v% Dioxane	59La	b
	-37.2	6.72(25°,2)	3.81	3-40	T	S:50v% Dioxane	59La	b
Zn^{2+}	-41.0	7.59(25°,1)	7.87	3-40	T	S:50v% Dioxane	59La	b
	-38.9	6.95(25°,2)	2.51	3-40	T	S:50v% Dioxane	59La	b
Pb^{2+}	-12.6	2.76(25°,1)	10.9	3&25	T	S:50v% Dioxane	59La	b
	-13.4	2.45(25°,2)	1.97	3&25	T	S:50v% Dioxane	59La	b
B73 BENZIMIDAZOLE, 2-aminomethyl-		$C_8H_9N_3$				$C_7H_5N_2(CH_2NH_2)$		L
Cu^{2+}	-80.3	14.23(25°,1-2)	2.80	15-45	T	S:50v% Dioxane	63La	b
B74 BENZIMIDAZOLE, 2-(butylaminomethyl)-		$C_{12}H_{17}N_3$				$C_7H_5N_2[CH_2NH(CH_2)_3CH_3]$		L
Cu^{2+}	-48.5	11.45(25°,1-2)	56.5	15-45	T	S:50v% Dioxane	63La	b
B75 BENZIMIDAZOLE, 2-(ethylaminomethyl)-		$C_{10}H_{13}N_3$				$C_7H_5N_2(CH_2NHCH_2CH_2)_3$		L
Cu^{2+}	-53.6	11.56(25°,1-2)	41.8	15-45	T	S:50v% Dioxane	63La	b

Metal Ion	ΔH, kJ /mol	Log K	ΔS, J /mol K	T,°C	M	Conditions	Ref.	Re-marks
B76	BENZIMIDAZOLE, 2-(1-hydroxy-butyl)-		$C_{11}H_{14}ON_2$			$C_7H_5N_2[CHOH(CH_2)_2CH_3]$		L
Cu^{2+}	-46.4	17.54(25°,1-2)	179.9	15-45	T	S:50v% Dioxane	63La	b
B77	BENZIMIDAZOLE, 2-(1-hydroxy-ethyl)-		$C_9H_{10}ON_2$			$C_7H_5N_2(CHOHCH_3)$		L
Cu^{2+}	-56.5	17.34(25°,1-2)	142.7	15-45	T	S:50v% Dioxane	63La	b
B78	BENZIMIDAZOLE, 2-hydroxy-methyl-		$C_8H_8ON_2$			$C_7H_5N_2(CH_2OH)$		L
Cu^{2+}	-69.9	17.68(25°,1-2)	104.2	15-45	T	S:50v% Dioxane	63La	b
B79	BENZIMIDAZOLE,2-(1-hydroxy-pentyl)-		$C_{12}H_{16}ON_2$			$C_7H_5N_2[CHOH(CH_2)_3CH_3]$		L
Cu^{2+}	-42.3	17.86(1-2)	200.0	15-45	T	S:50v% Dioxane	63La	b
B80	BENZIMIDAZOLE,2-(1-hydroxy-propyl)-		$C_{10}H_{12}ON_2$			$C_7H_5N_2(CHOHCH_2CH_3)$		L
Cu^{2+}	-47.3	17.60(25°,1-2)	178.2	15-45	T	S:50v% Dioxane	63La	b
B81	BENZIMIDAZOLE, 2-(N-methyl-aminoethyl)		$C_{10}H_{13}N_3$			$C_7H_5N_2(CH_2CH_2NHCH_3)$		L
Cu^{2+}	-57.3	11.68(25°,1-2)	31.4	15-45	T	S:50v% Dioxane	63La	b
B82	BENZIMIDAZOLE, 2-propanoic acid-		$C_{10}H_{10}O_2N_2$			$C_7H_5N_2(CH_2CH_2CO_2H)$		L^-
Mn^{2+}	-18.8	3.61(25°,1)	5.9	3-40	T	S:50v% Dioxane	59La	b
	-16.7	3.21(25°,2)	5.4	3-40	T	S:50v% Dioxane	59La	b
Co^{2+}	-30.5	3.79(25°,1)	-29.7	3-40	T	S:50v% Dioxane	59La	b
	-17.6	3.37(25°,2)	5.4	3-40	T	S:50v% Dioxane	59La	b
Ni^{2+}	-27.2	4.11(25°,1)	-12.6	3-40	T	S:50v% Dioxane	59La	b
	-17.6	3.27(25°,2)	3.64	3-40	T	S:50v% Dioxane	59La	b
Cu^{2+}	-33.9	5.25(25°,1)	-13.4	3-40	T	S:50v% Dioxane	59La	b
	-8.8	4.15(25°,2)	49.8	3-40	T	S:50v% Dioxane	59La	b
Zn^{2+}	-75.3	4.65(25°,1)	-163.6	3-40	T	S:50v% Dioxane	59La	b
	-10.5	3.95(25°,2)	40.6	3-40	T	S:50v% Dioxane	59La	b
B83	BENZIMIDAZOLE, 2-(N-propyl-aminomethyl)-		$C_{11}H_{15}N_3$			$C_7H_5N_2(CH_2NHCH_2CH_2CH_3)$		L
Cu^{2+}	-47.7	12.15(25°,1-2)	72.8	15-45	T	S:50v% Dioxane	63La	b

Metal Ion	ΔH, kJ /mol	Log K	ΔS, J /mol K	T,°C	M	Conditions	Ref.	Re-marks
B84	1,8-BENZODIAZINE		$C_8H_6N_2$			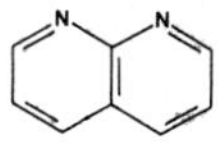		
YA_3	-39.8	(1)	--	25	C	S:Benzene A=Dipi-valoylmethanato	76Da	
SmA_3	-45.0	(1)	--	25	C	S:Benzene A=Dipi-valoylmethanato	76Da	
GdA_3	-51.0	(1)	--	25	C	S:Benzene A=Dipi-valoylmethanato	76Da	
TbA_3	-52.0	(1)	--	25	C	S:Benzene A=Dipi-valoylmethanato	76Da	
HoA_3	-39.8	(1)	--	25	C	S:Benzene A=Dipi-valoylmethanato	76Da	q:±3.5
ErA_3	-38.1	(1)	--	25	C	S:Benzene A=Dipi-valoylmethanato	76Da	q:±3.5
TmA_3	-37.3	(1)	--	25	C	S:Benzene A=Dipi-valoylmethanato	76Da	
YbA_3	-39.6	(1)	--	25	C	S:Benzene A=Dipi-valoylmethanato	76Da	
B85	5,6-BENZO-4,7,13,16,21,24-hexaoxa-1,10-diazabicyclo[8.8.8]-hexacosane($2_B2,2$)							L
Na^+	-39.7	7.50(1)	10	25	C	S:Methanol	80Cb	
K^+	-65.3	9.21(1)	-43	25	C	S:Methanol	80Cb	
Rb^+	-57.7	7.19(1)	-56	25	C	S:Methanol	80Cb	
Cs^+	-31.8	2.99(1)	-50	25	C	S:Methanol	80Cb	
B86	BENZOHYDROXAMIC ACID, 2-bromo-N-phenyl		$C_{13}H_{10}O_2NBr$			$(BrC_6H_4)C(=O)N(OH)C_6H_5$		
Mn^{2+}	-31.7	6.25(25°,1)	13.5	25&35	T	S:50v%Dioxane CL/CM=10	77Ac	a,b
	-31.7	4.94(25°,2)	-8.5	25&35	T	S:50v%Dioxane CL/CM=10	77Ac	a,b
Ni^{2+}	-24.6	7.35(25°,1)	58.1	25&35	T	S:50v%Dioxane CL/CM=10	77Ac	a,b
	-21.1	5.45(25°,2)	33.6	25&35	T	S:50v%Dioxane CL/CM=10	77Ac	a,b
Cu^{2+}	-36.9	10.11(25°,1)	69.7	25&35	T	S:50v%Dioxane CL/CM=10	77Ac	a,b
	-26.4	8.12(25°,2)	66.9	25&35	T	S:50v%Dioxane CL/CM=10	77Ac	a,b
Zn^{2+}	-29.9	7.65(25°,1)	32.1	25&35	T	S:50v%Dioxane CL/CM=10	77Ac	a,b

Metal Ion	ΔH, kJ /mol	Log K	ΔS, J /mol K	T,°C	M	Conditions	Ref.	Re-marks
B86, continued								
Zn^{2+}, cont.	-28.2	6.03(25°,2)	35.1	25&35	T	S:50v%Dioxane CL/CM=10	77Ac	a,b
B87	BENZOHYDROXAMIC ACID, 2-chloro-N-phenyl		$C_{13}H_{10}O_2NCl$			$(ClC_6H_4)C(=O)N(OH)C_6H_5$		
Mn^{2+}	-31.7	6.22(25°,1)	12.9	25&35	T	S:50v%Dioxane CL/CM=10	77Ac	a,b
	-31.7	5.11(25°,2)	-5.1	25&35	T	S:50v%Dioxane CL/CM=10	77Ac	a,b
Ni^{2+}	-12.3	7.30(25°,1)	98.5	25&35	T	S:50v%Dioxane CL/CM=10	77Ac	a,b
	-19.4	5.58(25°,2)	42.0	25&35	T	S:50v%Dioxane CL/CM=10	77Ac	a,b
Cu^{2+}	-31.7	10.14(25°,1)	88.0	25&35	T	S:50v%Dioxane CL/CM=10	77Ac	a,b
	-19.4	8.15(25°,2)	91.1	25&35	T	S:50v%Dioxane CL/CM=10	77Ac	a,b
Zn^{2+}	-26.4	7.67(25°,1))	58.4	25&35	T	S:50v%Dioxane CL/CM=10	77Ac	a,b
	-26.4	6.07(25°,2)	27.8	25&35	T	S:50v%Dioxane CL/CM=10	77Ac	a,b
B88	BENZOHYDROXAMIC ACID, 2-fluoro-N-phenyl		$C_{13}H_{10}O_2NF$			$FC_6H_4C(=O)N(OH)C_6H_5$		
Mn^{2+}	-35.2	6.30(25°,1)	2.7	25&35	T	S:50v%Dioxane CL/CM=10	77Ac	a,b
	-35.2	5.08(25°,2)	-17.4	25&35	T	S:50v%Dioxane CL/CM=10	77Ac	a,b
Ni^{2+}	-33.4	7.55(25°,1)	32.6	25&35	T	S:50v%Dioxane CL/CM=10	77Ac	a,b
	-10.5	5.60(25°,2)	71.8	25&35	T	S:50v%Dioxane CL/CM=10	77Ac	a,b
Cu^{2+}	-29.9	10.20(25°,1)	95.0	25&35	T	S:50v%Dioxane CL/CM=10	77Ac	a,b
	-21.1	8.20(25°,2)	86.1	25&35	T	S:50v%Dioxane CL/CM=10	77Ac	a,b
Zn^{2+}	-36.9	7.61(25°,1)	21.9	25&35	T	S:50v%Dioxane CL/CM=10	77Ac	a,b
	-19.4	5.80(25°,2)	46.1	25&35	T	S:50v%Dioxane CL/CM=10	77Ac	a,b
B89	BENZOHYDROXAMIC ACID, 2-iodo-N-phenyl		$C_{13}H_{10}O_2NI$			$(IC_6H_4)C(=O)N(OH)C_6H_5$		
Mn^{2+}	-29.9	6.19(25°,1)	18.2	25&35	T	S:50v%Dioxane CL/CM=10	77Ac	a,b

Metal Ion	ΔH, kJ /mol	Log K	ΔS, J /mol K	T,°C	M	Conditions	Ref.	Remarks
B89, continued								
Mn^{2+}, cont.	-28.2	4.84(25°,2)	1.42	25&35	T	S:50v%Dioxane CL/CM=10	77Ac	a,b
Ni^{2+}	-26.4	7.38(25°,1)	52.8	25&35	T	S:50v%Dioxane CL/CM=10	77Ac	a,b
	-18.7	5.43(25°,2)	27.2	25&35	T	S:50v%Dioxane CL/CM=10	77Ac	a,b
Cu^{2+}	-33.4	10.15(25°,1)	82.2	25&35	T	S:50v%Dioxane CL/CM=10	77Ac	a,b
	-7.03	8.16(25°,2)	132.3	25&35	T	S:50v%Dioxane CL/CM=10	77Ac	a,b
Zn^{2+}	-31.7	7.58(25°,1)	39.0	25&35	T	S:50v%Dioxane CL/CM=10	77Ac	a,b
	-56.3	5.96(25°,2)	-74.8	25&35	T	S:50v%Dioxane CL/CM=10	77Ac	a,b
B90 BENZOHYDROXAMIC ACID, 2-methoxy-N-phenyl			$C_{14}H_{13}O_3N$			$(CH_3OC_6H_4)C(=O)N(OH)C_6H_5$		
Mn^{2+}	-31.7	6.98(25°,1)	27.5	25&35	T	S:50v%Dioxane CL/CM=10	77Ac	a,b
	-31.7	5.85(25°,2)	23.4	25&35	T	S:50v%Dioxane CL/CM=10	77Ac	a,b
Ni^{2+}	-42.3	8.01(25°,1)	11.6	25&35	T	S:50v%Dioxane CL/CM=10	77Ac	a,b
	-7.03	5.72(25°,2)	86.0	25&35	T	S:50v%Dioxane CL/CM=10	77Ac	a,b
Cu^{2+}	-35.2	10.55(25°,1)	84.0	25&35	T	S:50v%Dioxane CL/CM=10	77Ac	a,b
	-38.7	8.90(25°,2)	40.5	25&35	T	S:50v%Dioxane CL/CM=10	77Ac	a,b
Zn^{2+}	-36.9	8.84(25°,1)	45.5	25&35	T	S:50v%Dioxane CL/CM=10	77Ac	a,b
	-29.9	6.82(25°,2)	30.3	25&35	T	S:50v%Dioxane CL/CM=10	77Ac	a,b
B91 BENZOHYDROXAMIC ACID, 2-methyl-N-phenyl			$C_{14}H_{13}O_2N$			$(CH_3C_6H_4)C(=O)N(OH)C_6H_5$		
Mn^{2+}	-26.4	6.61(25°,1)	24.0	25&35	T	S:50v%Dioxane CL/CM=10	77Ac	a,b
	-26.4	4.92(25°,2)	9.0	25&35	T	S:50v%Dioxane CL/CM=10	77Ac	a,b
Ni^{2+}	-29.9	7.99(25°,1)	52.8	25&35	T	S:50v%Dioxane CL/CM=10	77Ac	a,b
	-24.6	6.00(25°,2)	32.2	25&35	T	S:50v%Dioxane CL/CM=10	77Ac	a,b
Cu^{2+}	-35.2	10.45(25°,1)	82.1	25&35	T	S:50v%Dioxane CL/CM=10	77Ac	a,b

Metal Ion	ΔH, kJ /mol	Log K	ΔS, J /mol K	T,°C	M	Conditions	Ref.	Re- marks
B91, continued								
Cu^{2+}, cont.	-33.4	8.75(25°,2)	55.4	25&35	T	S:50v%Dioxane CL/CM=10	77Ac	a,b
Zn^{2+}	-31.7	8.04(25°,1)	47.7	25&35	T	S:50v%Dioxane CL/CM=10	77Ac	a,b
	-33.4	6.40(25°,2)	10.5	25&35	T	S:50v%Dioxane CL/CM=10	77Ac	a,b
B92	BENZOHYDROXAMIC ACID, 4-methyl-N-(m-tolyl)		$C_{15}H_{15}O_2N$			$CH_3C_6H_4C(=O)N(OH)C_6H_4CH_3$		
Mn^{2+}	-17.57	6.50(25°,1)	65.52	25&35	T	S:50v%Dioxane	77Ab	a,b,n
	-17.57	7.45(25°,1)	83.76	25&35	T	S:60v%Dioxane	77Ab	a,b,n
	-17.57	8.65(25°,1)	106.8	25&35	T	S:70v%Dioxane	77Ab	a,b,n
	-17.57	5.31(25°,2)	42.5	25&35	T	S:50v%Dioxane	77Ab	a,b,n
	-10.54	6.68(25°,2)	92.6	25&35	T	S:60v%Dioxane	77Ab	a,b,n
	-10.54	6.88(25°,2)	96.4	25&35	T	S:70v%Dioxane	77Ab	a,b,n
Ni^{2+}	-17.57	7.71(25°,1)	88.7	25&35	T	S:50v%Dioxane	77Ab	a,b,n
	-17.57	8.60(25°,1)	105.8	25&35	T	S:60v%Dioxane	77Ab	a,b,n
	-17.57	9.80(25°,1)	128.8	25&35	T	S:70v%Dioxane	77Ab	a,b,n
	-10.54	6.64(25°,2)	106.1	25&35	T	S:50v%Dioxane	77Ab	a,b,n
	-10.54	6.70(25°,2)	95.9	25&35	T	S:60v%Dioxane	77Ab	a,b,n
	-10.54	7.90(25°,2)	115.9	25&35	T	S:70v%Dioxane	77Ab	a,b,n
Cu^{2+}	-15.86	10.67(25°,1)	151.1	25&35	T	S:50v%Dioxane	77Ab	a,b,n
	-15.86	11.73(25°,1)	171.5	25&35	T	S:60v%Dioxane	77Ab	a,b,n
	-15.86	12.85(25°,1)	193.2	25&35	T	S:70v%Dioxane	77Ab	a,b,n
	-28.11	8.85(25°,2)	75.1	25&35	T	S:50v%Dioxane	77Ab	a,b,n
	-28.11	9.85(25°,2)	94.4	25&35	T	S:60v%Dioxane	77Ab	a,b,n
	-28.11	11.05(25°,2)	117.3	25&35	T	S:70v%Dioxane	77Ab	a,b,n
Zn^{2+}	-7.03	8.20(25°,1)	133.3	25&35	T	S:50v%Dioxane	77Ab	a,b,n
	-7.03	9.31(25°,1)	154.8	25&35	T	S:60v%Dioxane	77Ab	a,b,n
	-7.03	10.51(25°,1)	177.6	25&35	T	S:70v%Dioxane	77Ab	a,b,n
	-17.57	6.74(25°,2)	83.8	25&35	T	S:50v%Dioxane	77Ab	a,b,n
	-17.57	7.83(25°,2)	91.1	25&35	T	S:60v%Dioxane	77Ab	a,b,n
	-17.57	9.03(25°,2)	114.1	25&35	T	S:70v%Dioxane	77Ab	a,b,n
B93	BENZOHYDROXAMIC ACID, 4-methyl-N-(p-tolyl)		$C_{15}H_{15}O_2N$			$CH_3C_6H_4C(=O)N(OH)C_6H_4CH_3$		
Mn^{2+}	-21.3	6.16(25°,1)	46.6	25&35	T	S:50v%Dioxane	76Ac	a,b,p,y
	-16.9	7.43(25°,1)	85.6	25&35	T	S:60v%Dioxane	76Ac	a,b,p,y
	-16.9	8.93(25°,1)	114.4	25&35	T	S:70v%Dioxane	76Ac	a,b,p,y
	-8.8	5.21(25°,2)	70.3	25&35	T	S:50v%Dioxane	76Ac	a,b,p,y
	-16.9	6.66(25°,2)	70.9	25&35	T	S:60v%Dioxane	76Ac	a,b,p,y
	-16.9	8.15(25°,2)	99.5	25&35	T	S:70v%Dioxane	76Ac	a,b,p,y
Ni^{2+}	-33.4	7.40(25°,1)	29.6	25&35	T	S:50v%Dioxane	76Ac	a,b,p,y
	-17.6	8.58(25°,1)	105.1	25&35	T	S:60v%Dioxane	76Ac	a,b,p,y
	-17.6	10.08(25°,1)	133.9	25&35	T	S:70v%Dioxane	76Ac	a,b,p,y
	-36.9	6.33(25°,2)	-2.59	25&35	T	S:50v%Dioxane	76Ac	a,b,p,y
	-9.4	6.88(25°,2)	96.3	25&35	T	S:60v%Dioxane	76Ac	a,b,p,y
	-10.5	8.38(25°,2)	125.0	25&35	T	S:70v%Dioxane	76Ac	a,b,p,y

Metal Ion	ΔH, kJ /mol	Log K	ΔS, J /mol K	T,°C	M	Conditions	Ref.	Remarks
B93, continued								
Cu^{2+}	-15.8	10.66(25°,1)	150.9	25&35	T	S:50v%Dioxane	76Ac	a,b,p,y
	-19.4	11.71(25°,1)	159.2	25&35	T	S:60v%Dioxane	76Ac	a,b,p,y
	-19.4	13.22(25°,1)	188.0	25&35	T	S:70v%Dioxane	76Ac	a,b,p,y
	-15.8	8.83(25°,2)	115.9	25&35	T	S:50v%Dioxane	76Ac	a,b,p,y
	-19.4	9.84(25°,2)	155.3	25&35	T	S:60v%Dioxane	76Ac	a,b,p,y
	-19.4	11.34(25°,2)	152.1	25&35	T	S:70v%Dioxane	76Ac	a,b,p,y
Zn^{2+}	-8.8	8.25(25°,1)	128.5	25&35	T	S:50v%Dioxane	76Ac	a,b,p,y
	-7.0	9.29(25°,1)	154.2	25&35	T	S:60v%Dioxane	76Ac	a,b,p,y
	-7.0	10.78(25°,1)	182.8	25&35	T	S:70v%Dioxane	76Ac	a,b,p,y
	-10.5	6.71(25°,2)	93.0	25&35	T	S:50v%Dioxane	76Ac	a,b,p,y
	-15.8	7.81(25°,2)	96.4	25&35	T	S:60v%Dioxane	76Ac	a,b,p,y
	-15.5	9.32(25°,2)	125.1	25&35	T	S:70v%Dioxane	76Ac	a,b,p,y
B94	BENZOHYDROXAMIC ACID, 2-nitro-N-phenyl		$C_{13}H_{10}O_4N_2$			$(NO_2C_6H_4)C(=O)N(OH)C_6H_5$		
Be^{2+}	-21.1	7.78(25°,1)	78.2	25&35	T	S:50v%Dioxane	77Mf	a,b
	-24.0	5.87(25°,2)	31.7	25&35	T	S:50v%Dioxane	77Mf	a,b
UO_2^{2+}	-35.1	9.80(25°,1)	69.9	25&35	T	S:50v%Dioxane	77Mf	a,b
	-33.4	7.69(25°,2)	49.2	25&35	T	S:50v%Dioxane	77Mf	a,b
Mn^{2+}	-28.2	5.90(25°,1)	18.5	25&35	T	S:50v%Dioxane CL/CM=10	77Ac	a,b
	-28.2	4.54(25°,2)	-4.48	25&35	T	S:50v%Dioxane CL/CM=10	77Ac	a,b
Ni^{2+}	-14.1	7.15(25°,1)	90.0	25&35	T	S:50v%Dioxane CL/CM=10	77Ac	a,b
	-3.51	5.06(25°,2)	85.2	25&35	T	S:50v%Dioxane CL/CM=10	77Ac	a,b
Cu^{2+}	-44.0	9.70(25°,1)	38.2	25&35	T	S:50v%Dioxane CL/CM=10	77Ac	a,b
	-36.9	7.70(25°,2)	23.6	25&35	T	S:50v%Dioxane CL/CM=10	77Ac	a,b
Zn^{2+}	-17.6	7.22(25°,1)	79.2	25&35	T	S:50v%Dioxane CL/CM=10	77Ac	a,b
	-35.2	5.64(25°,2)	-10.0	25&35	T	S:50v%Dioxane CL/CM=10	77Ac	a,b
B95	BENZOHYDROXAMIC ACID, N-o-tolyl		$C_{13}H_{11}O_2N$			$OHC_6H_4(NOH)COC_6H_5$		L^-
Be^{2+}	-24.64	11.26(25°,1)	75.35	25-35	T	S:50v%Dioxane	80Sa	a,b
	-28.16	9.27(25°,2)	35.65	25-35	T	S:50v%Dioxane	80Sa	a,b
UO_2^{2+}	-31.67	14.98(25°,1)	103.97	25-35	T	S:50v%Dioxane	80Sa	a,b
	-15.86	12.28(25°,2)	119.11	25-35	T	S:50v%Dioxane	80Sa	a,b
Cd^{2+}	-26.40	10.07(25°,1)	52.76	25-35	T	S:50v%Dioxane	80Sa	a,b
	-14.10	8.34(25°,2)	69.75	25-35	T	S:50v%Dioxane	80Sa	a,b
Hg^{2+}	-45.77	8.38(25°,1)	-35.94	25-35	T	S:50v%Dioxane	80Sa	a,b
	-40.46	6.81(25°,2)	-40.12	25-35	T	S:50v%Dioxane	80Sa	a,b

Metal Ion	ΔH, kJ /mol	Log K	ΔS, J /mol K	T,°C	M	Conditions	Ref.	Re-marks
B96	BENZOIC ACID		$C_7H_6O_2$			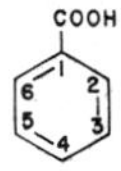		L^-
$[Fe^{3+}]$	-8.8	2.78(1)	23.8	25	T	Myoglobin	60Gb	
Cu^{2+}	0.0	4.04(30°,1)	76.2	25-35	T	≅0.03 S:50v% Dioxane	62Mb	b,j
Pb^{2+}	29.79	3.299(30°,1-2)	161.5	30&40	T	μ=1.0($NaClO_4$)		a,b
B97	BENZOIC ACID, amide, N,N-dimethyl-		C_9H_4ON			$C_6H_5[CON(CH_3)_2]$		L
I_2	-16.7	0.58(25°,1)	-44.8	25-45	T	CM=$1.6x10^{-3}$M CL=0.038-0.18M S:CCl_4	62Ce	b
	-20.9	0.91(25°,1)	-49.8	5-25	T	S:Heptane	71Td	b
B98	BENZOIC ACID, bromide		C_7H_5OBr			C_6H_5COBr		L
I_2	-3.8	-1.0(25°,1)	-30.5	5-25	T	S:Heptane	71Td	b
B99	BENZOIC ACID, chloride		C_7H_5OCl			C_6H_5COCl		L
I_2	-5.4	-1.0(25°,1)	-38.1	5-25	T	S:Heptane	71Td	b
B100	BENZOIC ACID, ethyl ester		$C_9H_{10}O_2$			$C_6H_5COOC_2H_5$		L
I_2	-10.9	-0.15(25°,1)	-38.5	5-25	T	S:Heptane	71Td	b
B101	BENZOIC ACID, methyl ester		$C_8H_8O_2$			$C_6H_5COOCH_3$		L
I_2	-10.5	-0.21(25°,1)	-40.2	5-25	T	S:Heptane	71Td	b
B102	BENZOIC ACID, nitrile		C_7H_5N			C_6H_5CN		L
TiA_4	-47.3	(1)	--	--	C	S:Benzene CM=0.05-0.08m A=Cl	68Ge	μ:25
SnA_4	-21.3	(1)	--	--	C	S:Benzene CM=0.05-0.08M A=Cl	68Ge	μ:25
SbA_5	-53.1 -54.4	(1) (1)	-- --	25 25	C C	S:CCl_4 A=Cl S:$C_2H_4Cl_2$ A=Cl	72Lb 72Lb	
B103	BENZOIC ACID, 2-amino-		$C_7H_7O_2N$			$C_6H_4(NH_2)(CO_2H)$		L^-
Pr^{3+}	-9.20	2.86(25°,1)	23.8	25-51	T	μ=1M($LiClO_4$) S:50v%Propanol	76Kk	j,q: ±1.7

Metal Ion	ΔH, kJ /mol	Log K	ΔS, J /mol K	T,°C	M	Conditions	Ref.	Re-marks
B103, continued								
In^{3+}	-8.0	5.24(30°,1)	71	20-40	T	S:75v% Ethanol μ=0.2($NaClO_4$)	73Sb	b,j,q: ±2.1
	-13.4	4.27(30°,2)	37.7	20-40	T	S:75v% Ethanol μ=0.2($NaClO_4$)	73Sb	b,j,q: ±4.2
B104	BENZOIC ACID, 4-amino-2-hydroxy-, amide		$C_7H_8O_2N_2$			$C_6H_3(CONH_2)(OH)(NH_2)$		L
Fe^{3+}	16.7	10.8(25°,1)	213	15-35	T	μ=3.0	56Aa	b
B105	BENZOIC ACID, 2-amino-5-sulpho-		$C_7H_7O_5NS$			$C_6H_3(CO_2H)(NH_2)(HSO_3)$		
Fe^{3+}	9.58	3.647(22°,1)	102.1	22-41	T	μ=0.1M(KCl) pH<2	69Bi	b
B106	BENZOIC ACID, 3-bromo-		$C_7H_5O_2Br$			$C_6H_4(CO_2H)(Br)$		L^-
Cu^{2+}	89.5	3.72(30°,1)	365.7	25-35	T	μ≅0.03 S:50v% Dioxane	62Mb	b,j
B107	BENZOIC ACID, 2-chloro-, amide, N,N-dimethyl-		$C_9H_{10}ONCl$			$C_6H_4[CON(CH_3)_2](Cl)$		L
I_2	-17.2	0.481(25°,1)	-46.0	25-45	T	CM=$1.5x10^{-3}$M CL=0.08-0.13M S:CCl_4	63Ca	b
B108	BENZOIC ACID, 3-chloro-		$C_7H_5O_2Cl$			$C_6H_4(CO_2H)(Cl)$		L^-
Cu^{2+}	45.6	3.88(30°,1)	223.4	25-35	T	μ≅0.03 S:50v% Dioxane	62Mb	b,j
B109	BENZOIC ACID, 4,-chloro-, amide, N,N-dimethyl-		$C_9H_{10}ONCl$			$C_6H_4[CON(CH_3)_2](Cl)$		L
I_2	-15.9	0.447(25°,1)	-44.8	25-45	T	CM=$1.4x10^{-3}$M CL=0.08-0.15M S:CCl_4	63Ca	b
B110	BENZOIC ACID, 4-chloro-nitrile		C_7H_4NCl			$C_6H_4(CN)(Cl)$		L
SnA_4	-29.7	0.556(25°,1)	-89.1	20-35	T	S:Benzene A=Cl	64La	b
B111	BENZOIC ACID, 3,4-dichloro-, amide, N,N-dimethyl-		$C_9H_9ONCl_2$			$C_6H_3[CON(CH_3)_2](Cl)_2$		L
I_2	-14.6	0.230(25°,1)	-45.6	25-45	T	CM=$1.2x10^{-3}$M CL=0.09-0.43M S:CCl_4	63Ca	b

Metal Ion	ΔH, kJ /mol	Log K	ΔS, J /mol K	T,°C	M	Conditions	Ref.	Remarks
B112	BENZOIC ACID, 2,3-dihydroxy-	$C_7H_6O_4$				$C_6H_3(CO_2H)(OH)_2$		L^-
UO_2^{2+}	-45.35	5.76(30°,1)	-39.37	20-50	T	μ=0.1 pH=5.5	73Pc	b
B113	BENZOIC ACID, 2,4-dihydroxy-	$C_7H_6O_4$				$C_6H_3(CO_2H)(OH)_2$		L^-
Be^{2+}	6.23	4.330(30°,1)	103.3	10-60	T	μ=0.01 pH=5.0	69Gg	b
Cu^{2+}	6.3	2.770(30°,1)	74.1	10-60	T	μ=0.01 pH=5.5	69Gi	b
B114	BENZOIC ACID, 2,5-dihydroxy-	$C_7H_6O_4$				$C_6H_3(CO_2H)(OH)_2$		L^-
UO_2^{2+}	-9.12	3.20(30°,1)	20.25	20-50	T	μ=0.1 pH=4.7	73Pc	b
B115	BENZOIC ACID, 2,6-dihydroxy-	$C_7H_6O_4$				$C_6H_3(CO_2H)(OH)_2$		L^-
UO_2^{2+}	-10.42	3.02(30°,1)	23.43	20-50	T	μ=0.1 pH=4.9	73Pc	b
B116	BENZOIC ACID, 3,4-dihydroxy	$C_7H_6O_4$				$HO_2CC_6H_3(OH)_2$		
Pr^{3+}	-38.3	5.74(25°,1)	-18.0	25-45	T	μ=0.2M($NaClO_4$)	76Md	
	-30.6	3.40(25°)	-37.6	25-45	T	μ=0.2M($NaClO_4$) R:Pr+L+A=PrLA A=1,2-Diamino-N,N,N',N',-tetra-acetic acid	76Mc	
B117	BENZOIC ACID, 3,5-dinitro-2-hydroxy	$C_7H_4O_7N_2$				$C_6H_2(CO_2H)(OH)(NO_2)_2$		
YA	-35.1	4.73(30°,1)	-25.5	30-50	T	μ=0.2M(KNO_3) A=Nitrilotriacetic acid	75Pg	b
Mn^{2+}	14.05	4.36(25°,1)	130.6	25-45	T	μ=0	75Dd	b
Co^{2+}	15.85	4.89(25°,1)	146.8	25-45	T	μ=0	75Dd	b
Ni^{2+}	17.23	5.29(25°,1)	159.0	25-45	T	μ=0	75Dd	b
Cu^{2+}	21.75	7.69(25°,1)	220.2	25-45	T	μ=0	75Dd	b
In^{3+}	-18.8	6.77(25°,1)	67	25-45	T	μ=0	78Sc	
	-16.7	5.86(25°,2)	54	25-45	T	μ=0	78Sc	
B118	BENZOIC ACID, 2-hydroxy-	$C_7H_6O_3$				$C_6H_4(CO_2H)(OH)$		L^-
Be^{2+}	-6.99	3.47(25)	90.0	20-45	T	μ=0.2 pH=4.0 R:Be+HL=BeL+H	64Da	b,q: ±0.63
ThA	-19.41	10.7(1)	140.2	25	C	μ=0.10(KNO_3) A=EDTA	70Kb	
	35.98	--	--	25	C	μ=0.10(KNO_3) A=1,2-Diamino-cyclohexane-	70Kb	

Metal Ion	ΔH, kJ /mol	Log K	ΔS, J /mol K	T,°C	M	Conditions	Ref.	Re-marks
B118, continued								
ThA, cont.						N,N,-N',N'-tetra-acetic acid R:ThA+H_2L=		
	35.82	--	--	25	C	$ThAL^{2-}+2H^+$ μ=0.10(KNO_3) A=EDTA R:ThA+ $H_2L=ThAL^{2-}+2H^+$	70Kb	
VO^{2+}	-8.8	13.38(25°,1)	226	25-35	T	μ=0.10(KNO_3)	66Mg	b,q:±2.1
Fe^{3+}	-33.263	17.80(1)	223.4	10	T	μ=0	73Vb	
	-28.284	17.57(1)	241.4	25	T	μ=0	73Vb	Cp=85
	-24.476	17.46(1)	254.8	35	T	μ=0	73Vb	
	-19.121	17.18(1)	269.9	50	T	μ=0	73Vb	
	-37.865	14.44(1)	142.7	10	T	μ=1.0	73Vb	
	-33.681	14.13(1)	157.3	25	T	μ=1.0	73Vb	Cp=69
	-30.794	13.97(1)	167.4	35	T	μ=1.0	73Vb	
	-26.359	13.70(1)	180.7	50	T	μ=1.0	73Vb	
	-39.288	14.32(1)	135.1	10	T	μ=1.5	73Vb	
	-35.229	13.98(1)	149.4	25	T	μ=1.5	73Vb	Cp=66
	-32.468	13.81(1)	159.0	35	T	μ=1.5	73Vb	
	-28.242	13.55(1)	172.0	50	T	μ=1.5	73Vb	
	-40.501	14.17(1)	128.0	10	T	μ=2.0	73Vb	
	-36.568	13.83(1)	141.8	25	T	μ=2.0	73Vb	Cp=64
	-33.890	13.64(1)	151.0	35	T	μ=2.0	73Vb	
	-29.832	13.37(1)	163.6	50	T	μ=2.0	73Vb	
	-22.2	15.9(25°,1)	230	15-35	T	μ=3.0	56Aa	b
	-46.74	5.73(25°)	-46.9	15-35	T	S:70w% Dioxane R:$Fe^{3+}+HL^-$= FeL^++H^+	69Bj	b
	-58.24	7.29(25°)	-55.6	15-35	T	S:82w% Dioxane R:$Fe^{3+}+HL^-$= FeL^++H^+	69Bj	b
Al^{3+}	32.6	0.183(25°)	113	23-45	T	μ≃0 R:$Al^{3+}+HL^-$= AlA^++H^+	75Sh	q:±2.1
In^{3+}	8.8	2.59(30°,1)	79	20-40	T	S:75v% Ethanol μ=0.2($NaClO_4$)	73Sb	b,j,q: ±2.1
B119	BENZOIC ACID, 2-hydroxy-, amide		$C_7H_7O_2H$			$C_6H_4(CONH_2)(OH)$		L
Fe^{3+}	0.8	10.1(25°,1)	197	15-35	T	μ=3.0	56Aa	b
B120	BENZOIC ACID, 2-hydroxy-, ethyl ester		$C_9H_{10}O_3$			$C_6H_4(CO_2CH_2CH_3)(OH)$		L
Mn^{2+}	-248.86	6.15(25°,1)	-717.56	25-45	T	S:50v%Ethanol μ=0.05($NaClO_4$)	79Gd	b
	-134.01	4.85(25°,2)	-356.85	25-45	T	S:50v%Ethanol μ=0.05($NaClO_4$)	79Gd	b

Metal Ion	ΔH, kJ /mol	Log K	ΔS, J /mol K	T,°C	M	Conditions	Ref.	Remarks
B121	BENZOIC ACID, 2-hydroxy-, methyl ester		$C_8H_8O_3$			$C_6H_4(CO_2CH_3)(OH)$		L
Mn^{2+}	-269.78	6.75(25°,1)	-773.62	25-45	T	S:50v%Ethanol μ=0.05($NaClO_4$)	79Gd	b
	-135.73	4.98(25°,2)	-372.92	25-45	T	S:50v%Ethanol μ=0.05($NaClO_4$)	79Gd	b
Fe^{3+}	6.7	9.82(25°,1)	213	15-35	T	μ=3.0	56Aa	b
B122	BENZOIC ACID, 2-hydroxy-5-sulfo-		$C_7H_6O_6S$			$C_6H_3(CO_2H)(OH)(SO_3H)$		L^{3-}
Be^{2+}	5.86	3.56(25)	88	20-45	T	μ=0.2 pH=4.0 R:Be+HL=BeL+H	64Da	b,q: ±0.63
YA	16.3	6.61(25°,1)	180	25-45	T	μ=0.2M(KNO_3) A=Nitrilotriacetic acid	75Pg	b
ThA	31.67	--	--	25	C	μ=0.10(KNO_3) A=EDTA R:ThA+ $H_3L=ThAL^{3-}+3H^+$	70Kb	
	34.98	--	--	25	C	μ=0.10(KNO_3) A=1,2-Diamino-cyclohexane-N,N-N',N'-tetra-acetic acid R:ThA+H_3L= $ThAL^{3-}+3H^+$	70Kb	
	-7.1	11.71(25°,1)	188	25-35	T	μ=0.10(KNO_3)	66Mg	b,q:±2.1
Fe^{3+}	-11.3	14.5(25°,1)	243	15-35	T	μ=3.0	56Aa	b
	-30.00	5.85(25°)	11.3	15-35	T	S:70w% Dioxane R:$Fe^{3+}+HL^-$= FeL^++H^+	69Bj	b
	-51.46	6.55(25°)	-47.28	15-35	T	S:82w% Dioxane R:$Fe^{3+}+HL^-$= FeL^++H^+	69Bj	b
Cu^{2+}	-16.7	9.02(1)	117	25	C	μ=0.50($NaClO_4$)	75Mc	
	-33.5	16.05	195	25	C	μ=0.50($NaClO_4$) R:$Cu^{2+}+A^-+L^{3-}$= $CuAL^{2-}$ A=Serine	75Mc	
	-30.1	16.13	208	25	C	μ=0.50($NaClO_4$) R:$Cu^{2+}+A^-+L^{3-}$= $CuAL^{2-}$ A=2-Amino-butanoic acid	75Mc	
	-29.3	16.41	216	25	C	μ=0.50($NaClO_4$) R:$Cu^{2+}+A^-+L^{3-}$= $CuAL^{2-}$ A=α-Alanine	75Mc	

Metal Ion	ΔH, kJ /mol	Log K	ΔS, J /mol K	T,°C	M	Conditions	Ref.	Remarks
B122, continued								
Cu^{2+}, cont.	-27.2	17.08	236	25	C	μ=0.50($NaClO_4$) R:Cu^{2+}+A^-+L^{3-}= $CuAL^{2-}$ A=Valine	75Mc	
B123	BENZOIC ACID, 2-hydroxy-3-nitro		$C_7H_5O_5N$			$C_6H_3(CO_2H)(OH)(NO_2)$		
Tl^+	-120.8	7.87(30°,1)	-246	30-40	T	μ=0.1M($HClO_4$) S:50v% Ethanol	76Kc	b,dd
	-109.7	5.05(30°,2)	-62.7	30-40	T	μ=0.1M($HClO_4$) S:50v% Ethanol	76Kc	b,dd
B124	BENZOIC ACID, 2-hydroxy-5-nitro-		$C_7H_5O_5N$			$C_6H_3(CO_2H)(OH)(NO_2)$		
Mn^{2+}	18.15	5.83(25°,1)	172.5	25-45	T	μ=0	75Db	b
Co^{2+}	19.95	6.41(25°,1)	189.6	25-45	T	μ=0	75Db	b
Ni^{2+}	20.87	6.71(25°,1)	198.4	25-45	T	μ=0	75Db	b
Cu^{2+}	22.67	9.24(25°,1)	252.9	25-45	T	μ=0	75Db	b
	18.07	6.19(25°,2)	179.2	25-45	T	μ=0	75Db	b
B125	BENZOIC ACID, 2-hydroxy-, phenyl ester		$C_{13}H_{10}O_3$			$C_6H_4(OH)[C(=O)OC_6H_5]$		L
Mn^{2+}	-268.03	6.04(25°,1)	-784.50	25-45	T	S:50v%Ethanol μ=0.05($NaClO_4$)	79Gd	b
	-179.83	4.61(25°,2)	-515.05	25-45	T	S:50v%Ethanol μ=0.05($NaClO_4$)	79Gd	b
B126	BENZOIC ACID, 3-iodo-		$C_7H_5O_2I$			$C_6H_4(CO_2H)(I)$		L^-
Cu^{2+}	102.1	3.88(30°,1)	410.0	25-35	T	μ=0.03 S:50v% Dioxane	62Mb	b,j
B127	BENZOIC ACID, 2-mercapto-		$C_7H_6O_2S$			$C_6H_4(CO_2H)(SH)$		L^{2-}
In^{3+}	-9.6	11.10(30°,1)	180	20-40	T	S:75v% Ethanol μ=0.2($NaClO_4$)	73Sb	b,j,q: ±2.1
	-19.2	8.90(30°,2)	105	20-40	T	S:75v% Ethanol μ=0.2($NaClO_4$)	73Sb	b,j,q: ±4.2
	-28.4	5.96(30°,3)	21	20-40	T	S:75v% Ethanol μ=0.2($NaClO_4$)	73Sb	b,j,q: ±4.2
B128	BENZOIC ACID, 4-methoxy-, amide, N,N-dimethyl-		$C_{10}H_{13}O_2N$			$C_6H_4[CON(CH_3)_2](OCH_3)$		L
I_2	-19.2	0.724(25°,1)	-50.2	25-45	T	CM=1.3x10^{-3}M CL=0.05-0.25M S:CCl_4	63Ca	b

Metal Ion	ΔH, kJ /mol	Log K	ΔS, J /mol K	T,°C	M	Conditions	Ref.	Remarks
B129	BENZOIC ACID, 4-methoxy-, nitrile		C_8H_7ON			$C_6H_4(CN)(OCH_3)$		L
SnA_4	-52.3	1.19(30°,1)	-149.8	30-40	T	S:Benzene A=Cl	64La	b
B130	BENZOIC ACID, 4-methyl-, amide, N,N-dimethyl-		$C_{10}H_{13}ON$			$C_6H_4[CON(CH_3)_2](CH_3)$		
I_2	-18.8	0.681(25°,1)	-50.2	25-45	T	CM=1.5×10^{-3}M CL=0.069M S:CCl_4	63Ca	b
B131	BENZOIC ACID, 4-nitro, methyl ester		$C_8H_7O_4N$			$O_2NC_6H_4CO_2CH_3$		L^-
Li^+	4.6	(1)	--	25	T	S:Hexamethyl-phosphoramide	73Sw	q:±0.84
B132	BENZOIC ACID, 4-nitro-, nitrile		$C_7H_4O_2N_2$			$C_6H_4(NO_2)(CN)$		L^-
Li^+	0	(1)	--	25	T	S:Hexamethyl-phosphoramide	73Sw	
B133	BENZOIC ACID, 2-phosphonooxy-		$C_7H_7O_6P$			$C_6H_4(PO_4H_2)(CO_2H)$		L^{3-}
VO^{2+}	-2.1	5.81(25°,1)	105	25-35	T	μ=0.10(KNO_3)	66Mg	b,q:±2.1
B134	9,10-BENZOPHENANTHRENE		$C_{18}H_{12}$					L
Na^+	0	4.830(1)	92	-55	T	CM=CL=10^{-6}-10^{-3}M S:Dimethoxyethane	66Cb	
	10.0	5.201(15°,1)	136.0	20	T	CM=CL=10^{-6}-10^{-3}M S:Dimethoxyethane	66Cb	
	1.7	4.534(1)	96	-65	T	CM=CL=10^{-6}-10^{-3}M S:Tetrahydrofuran	66Cb	
	21.8	5.130(1)	176	15	T	CM=CL=10^{-6}-10^{-3}M S:Tetrahydrofuran	66Cb	
B135	BENZOPHENONE		$C_{13}H_{10}O$			$C_6H_5COC_6H_5$		L
I_2	-11.7	-0.06(25°,1)	-42	5-25	T	S:Heptane	71Td	b
B136	BENZOPHENONE, 2,6-di-tert-butyl		$C_{21}H_{26}O$			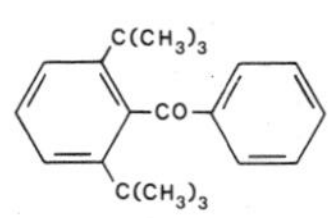		
K^+	6.4	0.98(20°)	40	-5-20	T	S:Hexamethyl-	75Ac	

Metal Ion	ΔH, kJ /mol	Log K	ΔS, J /mol K	T,°C	M	Conditions	Ref.	Re-marks
B136, continued								
K^+, cont.						phosphoramide R:$K^+ + A^- = KA$ A=Free anion radical of L A= 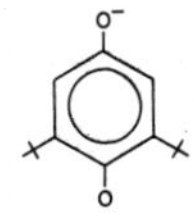		
B137	BENZOPHENONE, 4-nitro		$C_{13}H_9O_3N$			$(NO_2)C_6H_5COC_6H_4$		
K^+	3.7	1.23(25°)	36	?	T	S:Hexamethyl-phosphoramide R:$K^+ + A^- = KA$ A=Free anion radical of L	75Ac	
B138	1,4-BENZOQUINONE		$C_6H_4O_2$					L^-
Na^+	4.39	0.824	30.5	0-35	T	S:Hexamethyl-phosphoramide	73Sv	b,q: ±0.33
K^+	15.40	1.444	79	-10-30	T	S:Hexamethyl-phosphoramide	73Sv	b,z
B139	1,4-BENZOQUINONE, 2,5-dichloro-3,6-dihydroxy-		$C_6H_2O_4Cl_2$			$C_6H_4O_2(OH)_2(Cl)_2$		L^{2-}
Fe^{3+}	-14.06	5.813(25°,<u>1</u>)	-64.4	15-35	T	μ=0.15 pH=4.77 Complex in configuration 1	67Ca	b
	-61.5	5.513(25°,<u>1</u>)	102.1	15-35	T	μ=0.15 pH=4.77 Complex in configuration 2	67Ca	b,p
	2.9	4.030(25°,<u>2</u>)	-87.0	15-35	T	μ=0.15 pH=4.77 Complex in configuration 1	67Ca	b,p
	18.20	3.954(25°,<u>2</u>)	-133.0	15-35	T	μ=0.15 pH=4.77 Complex in configuration 2	67Ca	b,p
Ni^{2+}	-6.94	4.041(25°,<u>1</u>)	-54.0	15-35	T	μ=0.15 pH=7.56	67Ca	b
B140	1,2-BENZOQUINONE, tetrachloro		$C_6O_2Cl_4$			$C_6(=O)_2(Cl)_4$		
$RhABC_2$ (trans)	-102.5	(<u>1</u>)	--	25	C	S:Benzene A=Cl B=CO C=PPh_3	79Bf	
	-264.0	(<u>1</u>)	--	25	C	S:Benzene A=Cl B=CO C=$P(OMe)_3$	79Bf	
	-256.0	(<u>1</u>)	--	25	C	S:Benzene A=Cl B=CO C=PMe_3	79Bf	
	-123.4	(<u>1</u>)	--	25	C	S:Benzene A=Cl B=CO C=PMe_2Ph	79Bf	

Metal Ion	ΔH, kJ /mol	Log K	ΔS, J /mol K	T,°C	M	Conditions	Ref.	Remarks
B140, continued								
$RhABC_2$ (trans), cont.	-127.6	(1)	--	25	C	S:Benzene A=Cl B=CO $C=P(OPh)_2$	79Bf	
	-97.9	(1)	--	25	C	S:Benzene A=Cl B=CO $C=PMePh_2$	79Bf	
	-108.4	(1)	--	25	C	S:Benzene A=Cl B=CO $C=PEt_2Ph$	79Bf	
	-102.5	(1)	--	25	C	S:Benzene A=Cl B=CO $C=PPh_3$	79Bf	
	-87.9	(1)	--	25	C	S:Benzene A=Cl B=CO $C=P(t\text{-}Bu)Ph_2$	79Bf	
	-97.5	(1)	--	25	C	S:Benzene A=Cl B=CO $C=AsPh_3$	79Bf	
	-102.1	(1)	--	25	C	S:Benzene A=Cl B=CS $C=PPh_3$	79Bf	
	-105.0	(1)	--	25	C	S:Benzene A=Br B=CO $C=PPh_3$	79Bf	
	-105.0	(1)	--	25	C	S:Benzene A=I B=CO $C=PPh_3$	79Bf	
$IrABC_2$ (trans)	-139.3	(1)	--	25	C	S:Benzene A=F B=CO $C=PPh_3$	79Bf	
	-172.4	(1)	--	25	C	S:Benzene A=Cl B=CO $C=PMe_2Ph$	79Bf	
	-127.6	(1)	--	25	C	S:Benzene A=Cl B=CO $C=P(OPh)_3$	79Bf	
	-133.5	(1)	--	25	C	S:Benzene A=Cl B=CO $C=PMePh_2$	79Bf	
	-165.3	(1)	--	25	C	S:Benzene A=Cl B=CO $C=PEt_2Ph$	79Bf	
	-120.9	(1)	--	25	C	S:Benzene A=Cl B=CO $C=PPh_3$	79Bf	
	-123.0	(1)	--	25	C	S:Benzene A=Cl B=CO $C=P(t\text{-}Bu)Ph_2$	79Bf	
	-134.3	(1)	--	25	C	S:Benzene A=Cl B=CO $C=AsPh_3$	79Bf	
	-143.1	(1)	--	25	C	S:Benzene A=Cl B=CS $C=PPh_3$	79Bf	
	-120.9	(1)	--	25	C	S:Benzene A=Cl B=CO $C=PPh_3$	79Bf	
	-138.5	(1)	--	25	C	S:Benzene A=Br B=CO $C=PPh_3$	79Bf	
	-123.0	(1)	--	25	C	S:Benzene A=I B=CO $C=PPh_3$	79Bf	
	-115.9	(1)	--	25	C	S:Benzene A=NCS B=CO $C=PPh_3$	79Bf	
	-135.1	(1)	--	25	C	S:Benzene $A=N_3$ B=CO $C=PPh_3$	79Bf	
B141	1,4-BENZOQUINONE, tetrachloro		$C_6O_2Cl_4$			$C_6(=O)_2(Cl)_4$		
[Te]	-5.0	0.04(20°,1)	-17	10-30	T	S:Dichloromethane	74Hc	b,q:±4

Metal Ion	ΔH, kJ /mol	Log K	ΔS, J /mol K	T,°C	M	Conditions	Ref.	Re-marks
B141, continued								
[Te], cont.						[Te]=Phenoxatel-lurine 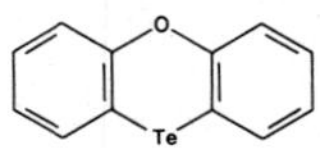		
B142	BICYCLO[2,2,1] HEPTANE, 7-oxy $C_6H_{10}O$							
ZnA	-29.7	2.677(25°,1)	-49.0	15-35	T	S:Cyclohexane A=α,β,γ,δ-Tetra-phenylporphine	77Vb	
B143	BICYCLO [2,2,1]-hepta-2,5-diene C_7H_8					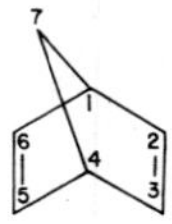		L
PdA_2B_2	-55.6	--	--	25	C	S:CH_2Cl A=Cl B=C_6H_5CN R:PdA_2B_2+L= PdA_2L+2B CL=$3.5x10^{-3}$M CM=$1x10^{-4}$M	72Pd	
Cu^+	-63	4.11(1)	-128.4	30	T	S:1M($LiClO_4$) in 2-propanol	69Hb	
B144	BICYCLO[2,2,1]-heptadiene C_7H_8							L
Ag^+	-46.78	-0.243	-161.9	25	T	1M(KNO_3) Location of double bonds not specified Two phase system: (CCl_4&H_2O) R:Ag^+(H_2O)+ L(CCl_4)⇌ $Complex^+$(H_2O)	59Ta	
B145	BICYCLO [2,2,1]-2-heptene C_7H_{10}					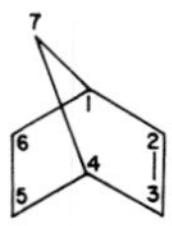		L
Cu^+	-45.2	4.26(1)	-67.4	30	T	S:1M($LiClO_4$) in 2-propanol	69Hb	
Ag^+	-27.95	-0.572	-104.6	25	T	1M(KNO_3) Two phase system: (CCl_4&H_2O) R:Ag^+(H_2O)+ L(CCl_4)⇌ $Complex^+$(H_2O)	59Ta	

Metal Ion	ΔH, kJ /mol	Log K	ΔS, J /mol K	T,°C	M	Conditions	Ref.	Remarks
B146	BICYCLO [2,2,2] OCTANE, 1-aza $C_7H_{13}N$							
CuA_2	-82.4	3.61(1)	--	--	C	S:C_6H_{12} A=Hexafluoro-acetylacetonato	76Mc	u:25°
B147	BICYCLO [2,2,2]-2-octene C_8H_{12}					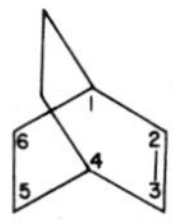		L
Ag^+	-20.67	-1.02(1)	-88.7	25	T	1M(KNO_3) Two phase system: (CCl_4&H_2O) R:$Ag^+(H_2O)$+ L(CCl_4)$\rightleftharpoons$ Complex$^+$(H_2O)	59Ta	
B148	BICYCLO [2,2,2] OCTANE, 1-phospho-, 4-ethyl-2,6,7-trioxo $C_{10}H_{11}O_5P$							
CuA_2	-27.2	1.72(1)	--	--	C	S:CCl_4 A=Hexafluoro-acetylacetonato	76Mc	u:25°
B149	BICYCLO [2,2,2] OCTANE, 1-phospho-2,6,7-trioxa-4-ethyl							
PdAB	-64.8	--	--	24	C	S:Benzene R:$[PdAB]_2$+2L= 2[PdABL] A=(π-MeAllyl) B=Cl	77Lb	
B150	BIPHENYL $C_{12}H_{10}$					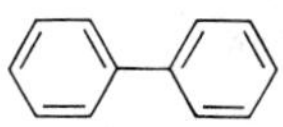		L
Li^+	1.7	5.119(1)	109	-65	T	S:Tetrahydrofuran	68Nb	q:±1.7
	10.5	5.328(1)	138	15	T	S:Tetrahydrofuran	68Nb	q:±1.7
Na^+	0	4.769(1)	92.0	-55	T	CM=CL=10^{-6}-10^{-3}M S:Dimethoxyethane	66Cb	
	8.8	5.207(15°,1)	131.8	20	T	CM=CL=10^{-6}-10^{-3}M S:Dimethoxyethane	66Cb	
	14.2	9.75(25°,1)	234	-70-25	T	low μ ~10^{-4} S:2-Methyltetra-hydrofuran	67Sb	b
	6.7	4.530(1)	117	-65	T	CM=CL=10^{-6}-10^{-3}M S:Tetrahydrofuran	66Cb 68Nb	
	30.5	5.824(1)	218	15	T	CM=CL=10^{-6}-10^{-3}M S:Tetrahydrofuran	66Cb 68Nb	

Metal Ion	ΔH, kJ /mol	Log K	ΔS, J /mol K	T,°C	M	Conditions	Ref.	Re-marks
B151	2,2'-BIPHENYLDICARBOXYLIC ACID		$C_{14}H_{10}O_4$			$C_{12}H_8(CO_2H)_2$		
Mn^{2+}	-34.8	5.74(30°)	-5.0	30&40	T	0.1M($NaClO_4$) S:50v%Dioxane R:Mn^{2+}+L=MnL^++H^+	78Sf	a,b
	-31.3	2.18(30°)	-74.0	30&40	T	0.1M($NaClO_4$) S:50v%Dioxane R:MnL^++L=MnL_2+H^+	78Sf	a,b,j
Co^{2+}	-40.0	5.88(30°)	-19.5	30&40	T	0.1M($NaClO_4$) S:50v%Dioxane R:Mn^{2+}+L=MnL^++H^+	78Sf	a,b
	-38.3	2.39(30°)	-79.1	30&40	T	0.1M($NaClO_4$) S:50v%Dioxane R:MnL^++L=MnL_2+H^+	78Sf	a,b,j
Ni^{2+}	-43.5	7.25(30°)	-6.2	30&40	T	0.1M($NaClO_4$) S:50v%Dioxane R:Mn^{2+}+L=MnL^++H^+	78Sf	a,b
	-45.2	3.85(30°)	-73.2	30&40	T	0.1M($NaClO_4$) S:50v%Dioxane R:MnL^++L=MnL_2+H^+	78Sf	a,b,j
Pd^{2+}	-83.6	7.00(30°)	-141.7	30&40	T	0.1M($NaClO_4$) S:50v%Dioxane R:Mn^{2+}+L=MnL^++H^+	78Sf	a,b
	-125.3	6.24(30°)	-290.2	30&40	T	0.1M($NaClO_4$) S:50v%Dioxane R:MnL^++L=MnL_2+H^+	78Sf	a,b,j
Cu^{2+}	-47.0	7.51(30°)	-11.3	30&40	T	0.1M($NaClO_4$) S:50v%Dioxane R:Mn^{2+}+L=MnL^++H^+	78Sf	a,b
	-26.1	3.76(30°)	-13.8	30&40	T	0.1M($NaClO_4$) S:50v%Dioxane R:MnL^++L=MnL_2+H^+	78Sf	a,b
B152	2,2'-BIPYRIDYL		$C_{10}H_8N_2$			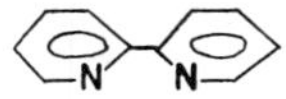		L
LaA_3	-29.8	(1)	--	25	C	S:Benzene A=Dipivaloylmethanato	76Da	q:±3
	-27.70	3.25(1)	-31.21	25	C	S:Chloroform A=Thenoyltrifluoroacetone	72Kb	
	-24.48	2.25(2)	-38.45	25	C	S:Chloroform A=Thenoyltrifluoroacetone	72Kb	
PrA_3	-42.4	(1)	--	25	C	S:Benzene A=Dipivaloylmethanato	76Da	

Metal Ion	ΔH, kJ /mol	Log K	ΔS, J /mol K	T,°C	M	Conditions	Ref.	Re-marks
B152, continued								
NdA_3	-49.2	(1)	--	25	C	S:Benzene A=Dipi-valoylmethanato	76Da	
	-33.05	3.61(1)	-41.71	25	C	S:Chloroform A=Thenoyltri-fluoroacetone	72Kb	
	-44.10	3.26(2)	-85.56	25	C	S:Chloroform A=Thenoyltri-fluoroacetone	72Kb	
YA_3	-67.3	(1)	--	25	C	S:Benzene A=Dipi-valoylmethanato	76Da	
SmA_3	-63.4	(1)	--	25	C	S:Benzene A=Dipi-valoylmethanato	76Da	
EuA_3	-70.5	(1)	--	25	C	S:Benzene A=Dipi-valoylmethanato	76Da	
GdA_3	-64.9	(1)	--	25	C	S:Benzene A=Dipi-valoylmethanato	76Da	q:±5
	-40.92	3.71(1)	-65.40	25	C	S:Chloroform A=Thenoyltri-fluoroacetone	72Kb	
	-56.06	3.79(2)	-116.27	25	C	S:Chloroform A=Thenoyltri-fluoroacetone	72Kb	
TbA_3	-66.6	(1)	--	25	C	S:Benzene A=Dipi-valoylmethanato	76Da	
DyA_3	-63.0	(1)	--	25	C	S:Benzene A=Dipi-valoylmethanato	76Da	
HoA_3	-65.5	(1)	--	25	C	S:Benzene A=Dipi-valoylmethanato	76Da	
ErA_3	-63.0	(1)	--	25	C	S:Benzene A=Dipi-valoylmethanato	76Da	
TmA_3	-65.5	(1)	--	25	C	S:Benzene A=Dipi-valoylmethanato	76Da	
Yb	-52.9	(1)	--	25	C	S:Benzene A=Dipi-valoylmethanato	76Da	
LuA_3	-59.91	4.75(1)	-110.46	25	C	S:Chloroform A=Thenoyltri-fluoroacetone	72Kb	
	-62.01	3.00(2)	-150.12	25	C	S:Chloroform A=Thenoyltri-fluoroacetone	72Kb	
VO^{2+}	-50.58	23.26(30°)	280.79	35&45	T	μ=0.1(KNO_3) CM=0.025M CL=0.025M R:$VO^{2+}+L^{-}+A^{-}=VOLA$ A=Glycollic acid	78Ja	
	-82.47	11.44(30°)	-48.87	35&45	T	μ=0.1(KNO_3) CM=0.025M	78Ja	a&b

Metal Ion	ΔH, kJ /mol	Log K	ΔS, J /mol K	T,°C	M	Conditions	Ref.	Remarks
B152, continued								
VO^{2+}, cont.						CL=0.025M		
	-59.54	11.34(35°)	23.64	35&45	T	R:$VO^{2+}+L^-+A^-$=VOLA A=Maleic acid μ=0.1(KNO_3) CM=0.025M CL=0.025M	78Ja	a&b
	-74.98	8.88(35°)	-73.47	35&45	T	R:$VO^{2+}+L^-+A^-$=VOLA A=Malonic acid μ=0.1(KNO_3) CM=0.025M CL=0.025M	78Ja	a&b
	-50.58	10.97(35°)	45.60	35&45	T	R:$VO^{2+}+L^-+A^-$=VOLA A=Oxalic acid μ=0.1(KNO_3) CM=0.025M CL=0.025M	78Ja	a&b
	-67.49	27.78(30°)	312.54	35&45	T	R:$VO^{2+}+L^-+A^-$=VOLA A=Phthalic acid μ=0.1(KNO_3) CM=0.025M CL=0.025M	78Ja	a&b
	-58.07	22.93(30°)	250.20	35&45	T	R:$VO^{2+}+L^-+A^-$=VOLA A=Salicylic acid μ=0.1(KNO_3) C_M=0.025M C_L=0.025M	78Ja	a&b
						R:$VO^++L^-+A^-$=VoLA A=5-Sulfosalicylic acid		
Mn^{2+}	-14.6	2.6(1)	0	20	C	μ=0.1($NaNO_3$)	63Ac 62Ab	
	-8	2.7(25°,1)	25	-46-0	T	μ=0.2($NaClO_4$) S:CH_3OH	73Bg	b
	-23.97	2.54(1)	-30.5	30.3	C	1M(KNO_3)	65Db	
	-25.56	4.40(1-2)	0	30.3	C	1M(KNO_3)	65Db	
	-26.07	5.90(1-3)	27.2	30.3	C	1M(KNO_3)	65Db	
Fe^{2+}	-131.17	17.45(1-3)	-113.0	20	C	μ=0.1($NaNO_3$)	63Ac 50Bb 50Ba 64Lc 49Kc	
	-117.2	17.16(1-3)	-57.7	30.3	C	μ=1.0(KNO_3)	65Db	
	-104.6	17.777(1-3)	-10.5	25	C	μ=∼10^{-3}M	77Bb	
	-121.9	17.850(1-3)	-65.7	25	C	μ∼10^{-3}M S:8w%Methanol	77Bb	
	-147.2	17.879(1-3)	-151.6	25	C	μ∼10^{-3}M S:16.4w%Methanol	77Bb	
	-167.6	17.769(1-3)	-222.0	25	C	μ∼10^{-3}M S:25.2w%Methanol	77Bb	

Metal Ion	ΔH, kJ /mol	Log K	ΔS, J /mol K	T,°C	M	Conditions	Ref.	Re-marks
B152, continued								
Fe^{2+}, cont.	-152.0	17.205(1-3)	-180.4	25	C	$\mu \sim 10^{-3}$M S:34.4w%Methanol	77Bb	
	-137.4	16.838(1-3)	-138.6	25	C	$\mu \sim 10^{-3}$M S:44.1w%Methanol	77Bb	
	-132.4	16.157(1-3)	-134.9	25	C	$\mu \sim 10^{-3}$M S:54.2w%Methanol	77Bb	
	-129.1	15.490(1-3)	-136.5	25	C	$\mu \sim 10^{-3}$M S:64.7w%Methanol	77Bb	
	-136.40	17.18(1-3)	-128.6	25	C	$\mu = 10^{-3}$M S:8w%Ethanol	78Bc	
	-159.95	16.92(1-3)	-212.7	25	C	$\mu = 10^{-3}$M S:16.4w%Ethanol	78Bc	
	-193.30	16.88(1-3)	-323.2	25	C	$\mu = 10^{-3}$M S:25.2w%Ethanol	78Bc	
	-182.17	16.24(1-3)	-300.3	25	C	$\mu = 10^{-3}$M S:34.4w%Ethanol	78Bc	
	-174.22	15.64(1-3)	-285.1	25	C	$\mu = 10^{-3}$M S:44.1w%Ethanol	78Bc	
	-153.6	15.14(1-3)	-225.4	25	C	$\mu = 10^{-3}$M S:54.2w%Ethanol	78Bc	
	-144.47	14.69(1-3)	-203.5	25	C	$\mu = 10^{-3}$M S:64.7w%Ethanol	78Bc	
	-74.48	4.354	-166.5	25	C	$\mu \sim 10^{-3}$M R:$Fe^{2+}+3HL^{+}=FeL_3^{2+}+3H^{+}$	77Bb	
	-75.56	4.670	-163.2	25	C	$\mu \sim 10^{-3}$M S:8.0w%Methanol R:$Fe^{2+}+3HL^{+}=FeL_3^{2+}+3H^{+}$	77Bb	
	-92.51	5.139	-212.1	25	C	$\mu \sim 10^{-3}$M S:16.4w%Methanol R:$Fe^{2+}+3HL^{+}=FeL_3^{2+}+3H^{+}$	77Bb	
	-88.12	5.322	-197.9	25	C	$\mu \sim 10^{-3}$M S:25.2w%Methanol R:$Fe^{2+}+3HL^{+}=FeL_3^{2+}+3H^{+}$	77Bb	
	-82.17	5.249	-175.9	25	C	$\mu \sim 10^{-3}$M S:34.4w%Methanol R:$Fe^{2+}+3HL^{+}=FeL_3^{2+}+3H^{+}$	77Bb	
	-76.40	5.359	-154.0	25	C	$\mu \sim 10^{-3}$M S:44.1w%Methanol R:$Fe^{2+}+3HL^{+}=FeL_3^{2+}+3H^{+}$	77Bb	
	-82.84	5.366	-179.9	25	C	$\mu \sim 10^{-3}$M S:54.2w%Methanol	77Bb	

Metal Ion	ΔH, kJ /mol	Log K	ΔS, J /mol K	T,°C	M	Conditions	Ref.	Re-marks
B152, continued								
Fe^{2+}, cont.						R:Fe^{2+}+3HL^{+}= FeL_3^{2+}+3H^{+}		
	-86.65	5.556	-184.9	25	C	μ~10^{-3}M S:64.7w%Methanol R:Fe^{2+}+3HL^{+}= FeL_3^{2+}+3H^{+}	77Bb	
	-88.07	4.596	-207.2	25	C	S:8w%Ethanol R:Fe^{2+}+3HL^{+}= FeL_3^{2+}+3H^{+}	78Bc	
	-27.45	4.809	-209.3	25	C	S:16.4w%Ethanol R:Fe^{2+}+3HL^{+}= FeL_3^{2+}+3H^{+}	78Bc	
	-29.37	5.146	-266.0	25	C	S:25.2w%Ethanol R:Fe^{2+}+3HL^{+}= FeL_3^{2+}+3H^{+}	78Bc	
	-29.08	5.095	-266.2	25	C	S:34.4w%Ethanol R:Fe^{2+}+3HL^{+}= FeL_3^{2+}+3H^{+}	78Bc	
	-29.79	5.219	-320.5	25	C	S:44.1w%Ethanol R:Fe^{2+}+3HL^{+}= FeL_3^{2+}+3H^{+}	78Bc	
	-31.67	5.549	-169.8	25	C	S:54.2w%Ethanol R:Fe^{2+}+3HL^{+}= FeL_3^{2+}+3H^{+}	78Bc	
	-30.79	5.394	-160.6	25	C	S:64.7w%Ethanol R:Fe^{2+}+3HL^{+}= FeL_3^{2+}+3H^{+}	78Bc	
Fe^{3+}	-5.0	11.3(1-3)	198.7	25	T	μ=0	59Ga	
Co^{2+}	-34.3	6.06(1)	-1.46	20	C	μ=0.1($NaNO_3$)	63Ac	
	-30.12	-7.95(1)	10.5	30.3	C	1M(KNO_3)	65Db	
	-63.6	11.42(1-2)	1.46	20	C	μ=0.1($NaNO_3$)	63Ac	
	-60.25	-15.46(1-2)	14.6	30.3	C	1M(KNO_3)	65Db	
	-89.1	16.02(1-3)	5.86	20	C	μ=0.1($NaNO_3$)	63Ac	
	-82.26	-22.13(1-3)	33.9	30.3	C	1M(KNO_3)	65Db	
CoA_2	-36.8	2.188(1)	-80	30	C	S:Benzene CM=1.3-4.0x10^{-3}M A= 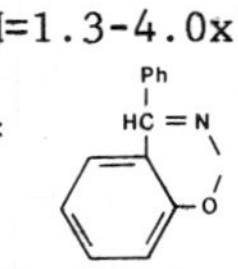	76Ai	
	-39.3	2.734(1)	-78	30	C	S:Benzene CM=1.5-5x10^{-3}M A= 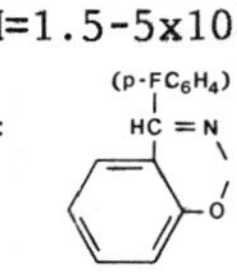	76Ai	

Metal Ion	ΔH, kJ /mol	Log K	ΔS, J /mol K	T,°C	M	Conditions	Ref.	Remarks
B152, continued								
CoA_2, cont.	-43.4	1.568(1)	-113	30	C	S:Benzene $CM=4-6x10^{-3}M$ A= 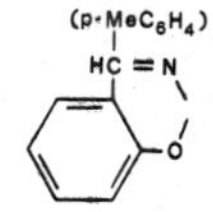	76Ai	q:±3.0
	-51.9	1.556(1)	-143	30	C	S:Benzene $CM=1.3-4.0x10^{-3}M$ A= 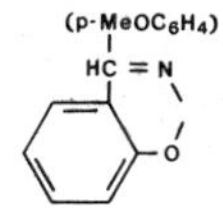	76Ai	q:±3.4
Ni^{2+}	-40.2	7.13(1)	0	20	C	μ=0.1($NaNO_3$)	63Ac	
	-37.24	9.65(1)	10.5	30.3	C	1M(KNO_3)	62Ab 65Db	
	-79.5	14.01(1-2)	-2.9	20	C	μ=0.1($NaNO_3$)	63Ac	
	-74.48	19.14(1-2)	18.4	30.3	C	1M(KNO_3)	65Db	
	-118.0	20.54(1-3)	-8.8	20	C	μ=0.1($NaNO_3$)	63Ac	
	-111.71	27.95(1-3)	17.2	30.3	C	1M(KNO_3)	65Db	
NiA_2	-76.1	(1)	--	30	C	S:Benzene A=Diethyldithio-phosphato	71Dc	q:±8.4
	-73.7	(1)	--	30	C	$CM=1-4x10^{-3}M$ S:Benzene A=CH_3OCS_2	74Gd	
	-75.5	(1)	--	30	C	$CM=1-4x10^{-3}M$ S:Benzene A=$CH_3(CH_2)_5OCS_2$	74Gd	
	-70.2	(1)	--	30	C	$CM=1-6x10^{-3}M$ S:Benzene A=$(CH_3)_2CHOCS_2$	74Gd	
	-84.1	~5(1-2)	~186	25	C	S:Nitrobenzene K determined in benzene A=Ethylxanthato	70Cc	
Cu^{2+}	-44.56	8.11(1)	5.4	25	C	μ=0.1M($NaNO_3$-$NaClO_4$)	76Al	
	-44.6	8.11(1)	5.4	25	C	μ=0.1M($NaClO_4$)	79Be	
	-49.8	8.0(1)	-17.2	20	C	μ=0.1($NaNO_3$)	63Ac	
	-42.51	8.39(1)	20.5	30.3	C	1M(KNO_3)	62Ab 65Db	
	-25.48	5.55(2)	20.9	25	C	μ=0.1M($NaNO_3$-$NaClO_4$)	62Cb 76Al	
	-25.4	5.55(2)	21	25	C	μ=0.1M($NaClO_4$)	79Be	
	-27.2	5.30(25°,2)	10.0	15-35	T	μ=0.2M(KNO_3) $CM=2.0x10^{-3}$	79Mg	b
	-33.0	3.4(3)	-46.0	25	C	μ=0.1M($NaNO_3$-$NaClO_4$)	76Al	
	-72.4	13.6(1-2)	13.0	20	C	μ=0.1($NaNO_3$)	63Ac	
	-79.58	14.03(1-2)	6.3	30.3	C	1M(KNO_3)	65Db	
	-99.6	17.08(1-3)	-13.0	20	C	μ=0.1($NaNO_3$)	63Ac	

Metal Ion	ΔH, kJ /mol	Log K	ΔS, J /mol K	T,°C	M	Conditions	Ref.	Remarks
B152, continued								
Cu^{2+},	-90.46	17.66(1-3)	39.7	30.3	C	1M(KNO_3)	65Db	
cont.	-36.4	14.209	150.2	25	C	μ=0.1M($NaClO_4$) R:$Cu^{2+}+L+A^{2-}=CuLA$ A=1,1-Cyclobutane-dicarboxylic acid	77Am	
	-35.6	13.018	129.7	25	C	μ=0.1M($NaClO_4$) R:$Cu^{2+}+L+A^{2-}=CuLA$ A=1,1-Cyclohexane-dicarboxylic acid	77Am	
	-36.0	13.263	130.1	25	C	μ=0.1M($NaClO_4$) R:$Cu^{2+}+L+A^{2-}=CuLA$ A=1,1-Cyclopentane-dicarboxylic acid	77Am	
	-46.0	15.471	141.8	25	C	μ=0.1M($NaClO_4$) R:$Cu^{2+}+L+A^{2-}=CuLA$ A=1,1-Cyclopropane-dicarboxylic acid	77Am	
	-61.5	16.63	112	25	C	μ=0.1M($NaClO_4$) R:$Cu^{2+}+L+A^{2-}=CuLA$ A=Iminodiacetic acid	79Be	
	-34.7	11.879	112.1	25	C	μ=0.1M($NaClO_4$) R:$Cu^{2+}+L+A^{2-}=CuLA$ A=Maleate	78Ae	
	-42.7	13.403	113.8	25	C	μ=0.1M($NaClO_4$) R:$Cu^{2+}+L+A^{2-}=CuLA$ A=Malonate	78Ae	
	-33.8	11.861	113.8	25	C	μ=0.1M($NaClO_4$) R:$Cu^{2+}+L+A^{2-}=CuLA$ A=Phthalate	78Ae	
	-43.9	15.313	146.0	25	C	μ=0.1M($NaClO_4$) R:$Cu^{2+}+L+A^{2-}+H^+=[CuLAH]^+$ A=Phthalate	78Ae	
	-74.1	16.11	60	25	C	μ=0.1M($NaClO_4$) R:$Cu^{2+}+L+A^{2-}=CuLA$ A=2,6-Pyridine-dicarboxylic acid	79Be	
	-42.7	10.915	65.7	25	C	μ=0.1M($NaClO_4$) R:$Cu^{2+}+L+A^{2-}=CuLA$ A=Succinate	78Ae	
	-33.0	15.979	195.8	25	C	μ=0.1M($NaClO_4$) R:$Cu^{2+}+L+A^{2-}+H^+=[CuLAH]^+$ A=Succinate	78Ae	
	-84	14.1	-12	25	C	μ=0.1M($NaNO_3$-	76Al	

Metal Ion	ΔH, kJ /mol	Log K	ΔS, J /mol K	T,°C	M	Conditions	Ref.	Re-marks
B152, continued								
Cu^{2+}, cont.						NaClO$_4$) R:Cu^{2+}+L+ OH^-=$CuLOH^+$		
	-133.0	33.1	186.6	25	C	μ=0.1M(NaNO$_3$-NaClO$_4$) R:2Cu^{2+}+2L+2OH^-= $Cu_2L_2(OH)_2{}^{2+}$	76Al	
	-27.2	0.4	85	25	C	μ=0.1M(NaNO$_3$-NaClO$_4$) R:Cu^{2+}+L+ H_2O=$CuL(OH)^+$+H^+	76Al	
	-20.5	5.65	39.3	25	C	μ=0.1M(NaNO$_3$-NaClO$_4$) R:2Cu^{2+}+2L+2H_2O= $Cu_2L_2(OH)_2{}^{2+}$+2H^+	76Al	
	33	4.9	205	25	C	μ=0.1M(NaNO$_3$-NaClO$_4$ R:2$CuLOH^+$= $Cu_2L_2(OH)_2{}^{2+}$	76Al	
CuA_2	-79.9	(1)	--	30	C	S:Benzene A=1,1,1,5,5,5-Hexafluoropentane-2,4-dione	74Gc	
	-50.00	3.004(1)	-107.9	30	C	S:Benzene A=1,1,1-Trifluoropentane-2,4-dione	74Gc	
	-56.9	3.164(1)	--	30	C	S:Benzene A=1,1,1-Trifluoro-4-phenylbutane-2,4-dione	74Gc	
	-54.4	(1)	--	30	C	S:Benzene A=1,1,1-Trifluoro-4-(2-thienyl) butane-2,4-dione	74Gc	
	-54.77	6.07(1)	-67.53	25	C	S:Chloroform A=1,1,1,5,5,5-Hexafluoro-2,4-pentanedione	72Ka	
	-1.46	1.02(1)	14.73	25	C	S:Chloroform A=2,4-Pentanedione	72Ka	
	-40.38	2.65(1)	-85.48	25	C	S:Chloroform A=1,1,1-Trifluoro-2,4-pentanedione	72Ka	
	-43.18	3.67(1)	-74.39	25	C	S:Chloroform A=4,4,4-Trifluoro-1-(2-thienyl)-1,3-butanedione	72Ka	

Metal Ion	ΔH, kJ /mol	Log K	ΔS, J /mol K	T, °C	M	Conditions	Ref.	Re-marks
B152, continued								
Ag^{+}	-32.024	3.70(25°,1)	-36.568	20-40	T	0.1M(KNO_3) S:50 v%(95%C_2H_5OH)	58Ca 58Cc	b
	-17.096	3.52(25°,2)	10.04	20-40	T	0.1M(KNO_3) S:50 v%(95%C_2H_5OH)	58Ca	b
	-49.120	7.22(25°,1-2)	-26.53	20-40	T	0.1M(KNO_3) S:50 v%(95%C_2H_5OH)	58Ca	b
Zn^{2+}	-29.3	5.3(1)	0	20	C	μ=0.1($NaNO_3$)	63Ac 62Ab	
	-29.3	5.16(25°,1)	0.4	10-40	T	μ=0.1M(KNO_3)	62Ca	b
	-26.15	5.26(1)	14.6	30.3	C	1M(KNO_3)	65Db	
	-31.8	4.88(25°,1)	-13.0	10-40	T	μ=0.1M(KNO_3) S:41.5w%EtOH	62Ca	b
	-28.0	4.46(25°,2)	-8.8	10-40	T	μ=0.1M(KNO_3)	62Ca	b
	-31.0	4.08(25°,2)	-25.5	10-40	T	μ=0.1M(KNO_3) S:41.5w%EtOH	62Ca	b
	-24.3	3.74(25°,3)	-9.6	10-40	T	μ=0.1M(KNO_3)	62Ca	b,q:±2.5
	-21.8	3.05(25°,3)	-14.6	10-40	T	μ=0.1M(KNO_3) S:41.5w%EtOH	62Ca	b
	-52.3	9.83(1-2)	10.0	20	C	μ=0.1M($NaNO_3$)	63Ac	
	-49.16	9.83(1-2)	25.9	30.3	C	1M(KNO_3)	65Db	
	-73.2	13.63(1-3)	11.3	20	C	μ=0.1($NaNO_3$)	63Ac	
	-66.61	13.78(1-3)	44.4	30.3	C	1M(KNO_3)	65Db	
ZnA_2	-45	(1)	--	30	C	S:Acetonitrile A=I	76Ak	
	-45	(2)	--	30	C	S:Acetonitrile A=I	76Ak	
	-45	(3)	--	30	C	S:Acetonitrile A=I	76Ak	q:±4
	-40.0	0.785(1)	-117	30	C	S:Benzene CM=0.5-5x10^{-3}M A=[$(Bu)_2NCS_2$]	76Aj	
	-53.3	1.602(1)	-145	30	C	S:Benzene CM=2.4-5x10^{-3}M A=[$(PhCH_2)_2NCS_2$]	76Aj	
	-62.3	-2.572(1)	-155.6	30	C	S:Benzene A=Diisobutyl-dithiophosphate	71Db	q:±5.8
	-24.68	1.95(1)	-42.55	25	C	S:Chloroform A=4,4,4-Tri-fluoro-1-(2-thienyl)-1,3-butanedione	72Ka	
	-73.6	2.866(1)	-192	25	T	S:Toluene A=CH_3	67Ra	
	-146.8	3.651(1)	-431	25	T	S:Toluene A=C_2H_5	67Ra	
	-65.7	3.269(1)	-247	25	T	S:Toluene A=C_3H_7	67Ra	
	-45.2	2.925(1)	-96	25	T	S:Toluene A=C_4H_9	67Ra	
	-54.4	3.028(1)	-126	25	T	S:Toluene A=C_5H_{11}	67Ra	
	-52.7	3.079(1)	-117	25	T	S:Toluene A=C_6H_{13}	67Ra	
	-65.3	3.137(1)	-159	25	T	S:Toluene A=C_7H_{15}	67Ra	

Metal Ion	ΔH, kJ /mol	Log K	ΔS, J /mol K	T,°C	M	Conditions	Ref.	Remarks
B152, continued								
ZnA_2, cont.	-62.8	3.064(1)	-151	25	T	S:Toluene $A=C_8H_{17}$	67Ra	
	-66.9	3.079(1)	-163	25	T	S:Toluene $A=C_9H_{19}$	67Ra	
Cd^{2+}	-21.25	4.28(25°,1)	10.63	11-40	T	$0.04M(KNO_3)$	58Cb 58Cc	b
	-21.3	4.25(1)	8.8	20	C	$\mu=0.1(NaNO_3)$	63Ac	
	24.52	3.99(25°,1)	-5.8	20-40	T	$0.05M(KNO_3)$ S:50v%(95%C_2H_5OH)	58Cb	b
	-15.86	3.51(25°,2)	13.97	11-40	T	$0.04M(KNO_3)$	58Cb	b
	-17.57	3.05(25°,2)	-0.4	20-40	T	$0.05M(KNO_3)$ S:50v%(95%C_2H_5OH)	58Cb	b
	-12.84	2.69(25°,3)	8.79	11-40	T	$0.04M(KNO_3)$	58Cb	b
	-11.38	2.10(25°,3)	2.1	20-40	T	$0.05M(KNO_3)$ S:50v%(95%C_2H_5OH)	58Cb	b
	-39.3	7.85(1-2)	15.5	20	C	$\mu=0.1(NaNO_3)$	63Ac	
	-49.96	10.48(25°,1-3)	33.01	11-40	T	$0.04M(KNO_3)$	58Cb	b
	-58.6	10.55(1-3)	1.2	20	C	$\mu=0.1(NaNO_3)$	63Ac	
	-53.47	9.14(25°,1-3)	-5.8	20-40	T	$0.05M(KNO_3)$ S:50v%(95%C_2H_5OH)	58Cb	b
CdA_2	-72	3.00(1)	-184	30	C	S:Acetonitrile A=I	76Ak	
	-43	2.380(1)	-97	30	C	S:Benzene $CM=5.8\text{-}32x10^{-3}M$ $A=[(Bu)_2NCS_2]$	76Aj	
	-61.1	(1)	--	30	C	S:Benzene A=Diisobutyldi-thiophosphate	71Db	q:±5.8
Cd_2A_4	-45.2	-2.308(1)	-105.0	30	C	S:Benzene A=Diisobutyldi-thiophosphato R:$\frac{1}{2}Cd_2A_4+L=CdA_2L$	71Db	q:±4
HgA_2	-43	~4(1)	~-76	30	C	S:Acetonitrile A=I	76Ak	
	-17.2	1.15(1)	-36	30	C	$CM=1\text{-}20x10^{-3}M$ S:Benzene $A=C_6F_5$	74Fc	
	-32.8	(1)	--	30	C	$CM=0.63\text{-}1.5x10^{-3}M$ S:Benzene A=Cl	74Fb	
	-40.1	(1)	--	30	C	$CM=0.97\text{-}1.86x10^{-3}$ S:Benzene A=Br	74Fb	
	-48.8	3.63(1)	-91	30	C	$CM=0.7\text{-}2.1x10^{-3}M$ S:Benzene A=I	74Fb	
	-17.53	1.283(1)	-34.3	25	C	$S:C_6H_6$ $A=C_6H_5$	73Pm	
	-30.46	1.744(1)	-64.4	25	C	$S:CCl_4$ $A=C_6F_5$	73Pm	
Tl^{3+}	-42.2	9.38(1)	-38	25	C	$\mu=1.0$	62Kc	
	-84.1	16.1(1-2)	-25	25	C	$\mu=1.0$	62Kc	
	-25.5	4.69	-4	25	C	$\mu=1.0$ R:$Tl^{3+}+HL^+=TlL_3^{+}+2H^+$	62Kc	
	-50.6	6.74	42	25	C	$\mu=1.0$ R:$Tl^{3+}+2HL^+=TlL_2^{3+}+2H^+$	62Kc	

Metal Ion	ΔH, kJ /mol	Log K	ΔS, J /mol K	T,°C	M	Conditions	Ref.	Re-marks
B152, continued								
SnA_4	-25.86	>6.96(1)	>47.3	25	C	μ≅0 S:Acetonitrile A=Cl	66Mb	
$SnAB_3$	-85.48	>6.96(1)	>-153.1	25	C	μ≅0 S:Acetonitrile A=n-C_4H_9 B=Cl	66Mb	
	-112.3	(1)	--	25	C	μ≅0 S:Acetonitrile A=n-C_4H_9 B=Cl $SnAB_3$ was initially dissolved in CCl_4	66Mb	
SnA_2B_2	-49.70	3.30(1)	-103.7	25	C	μ≅0 S:Acetonitrile A=CH_3 B=Cl	66Mb	
	-72.17	(1)	--	25	C	μ≅0 S:Acetonitrile A=CH_3 B=Cl SnA_2B_2 was initially dissolved in CCl_4	66Mb	
	-52.01	3.47(1)	-107.9	25	C	μ≅0 S:Acetonitrile A=C_2H_5 B=Cl	66Mb	
	-71.17	(1)	--	25	C	μ≅0 S:Acetonitrile A=C_2H_5 B=Cl SnA_2B_2 was initially dissolved in CCl_4	66Mb	
	-54.89	3.07(1)	-125.5	25	C	μ≅0 S:Acetonitrile A=n-C_3H_7 B=Cl	66Mb	
	-54.56	3.03(1)	-125.1	25	C	μ≅0 S:Acetonitrile A=n-C_4H_9 B=Cl	66Mb	
	-66.02	(1)	--	25	C	μ≅0 S:Acetonitrile A=n-C_4H_9 B=Cl SnA_2B_2 was initially dissolved in CCl_4	66Mb	
	-81.8	3.81(1)	-198	30	C	S:Benzene A=Phenyl B=Cl	75Fa	
	-64.8	2.09(1)	-174	30	C	S:Benzene A=Methyl B=Cl	75Fa	
	-62.1	1.60(1)	-174	30	C	S:Benzene A=Butyl B=Cl	75Fa	
	-70.7	>4(1)	<-158	30	C	S:Benzene A=Butyl B=NCS	75Fa	
	-104	5.0(1)	<-250	30	C	S:Benzene A=Butyl B=Cl	75Fa	
	-130	>5(1)	<-340	30	C	S:Benzene A=Phenyl B=Cl	75Fa	
	-70.3	3.96(35°,1)	-159.8	30-40	T	S:Benzene A=C_4H_9 B=SCN	71Wb	b
	-53.1	1.32(35°,1)	-145.6	30-40	T	S:Benzene A=C_8H_{17} B=Cl	71Wb	b

Metal Ion	ΔH, kJ /mol	Log K	ΔS, J /mol K	T,°C	M	Conditions	Ref.	Re-marks
B152, continued								
SnA_2B_2, cont.	-72.0	4.01(35°,1)	-160.7	30-40	T	S:Benzene $A=C_8H_{17}$ B=SCN	71Wb	b
	-52.7	1.24(35°,1)	-146.8	30-40	T	S:Toluene $A=C_8H_{17}$ B=Cl	71Wb	b
B153 4,4'-BIPYRIDYL			$C_{10}H_8N_2$			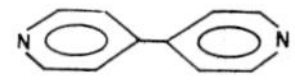		L
CrA_3^{2+}	50	-2.149(25°)	138	13.5-31.5	T	μ=0.1(NaCl) A=α,α'-Bipyridyl R:$CrA_3^{2+}+HL^+=CrA_2HL^{3+}+A$	71Ua	b
CuA_2	-28.07	2.51(1)	-46.15	25	C	S:Chloroform A=4,4,4-Trifluoro-1-(2-thienyl)-1,3-butanedione	72Ka	
ZnA_2	-20.46	1.71(1)	-35.86	25	C	S:Chloroform A=4,4,4-Trifluoro-1-(2-thienyl)-1,3-butanedione	72Ka	
B154 BISMUTHATE(III) ION, hexachloro			Cl_6Bi			$BiCl_6^{3-}$	L^{3-}	
Na^+	≅-3.8	2.64(25°)	-20.9	25-65	T	3M(Li^+,H^+, Na^+, Cl^-) R:xNa+L=Na_xL	63Mf	b
K^+	-5.0	0.19(25°)	-13.4	25-65	T	3M(Li^+,H^+,K^+, Cl^-) R:xK+L=K_xL	63Mb	b
Rb^+	-5.4	0.30(25°)	-12.6	25-65	T	3M(Li^+,H^+,Rb^+, Cl^-) R:xRb+L=Rb_xL	63Mf	b
Cs^+	-5.8	0.40(25°)	-12.1	25-65	T	3M(Li^+,H^+,Cs^+, Cl^-) R:xCs+L=Cs_xL	63Mf	b
NH_4^+	22.2	0.26(25°,1)	74	15-65	T	μ=1	63Md	b
B155 BORIC ACID, TRIS (pyrazoly)			$Fe[HB(pz)_3]_2$					L^-
FeL_2	16.1	-0.328(25°)	47.7	--	T	S:Methanoldi-chloromethane R:Transition from high to low spin isomer. ^1A ^5A	73Be	u:25
B156 BOVINE SERUM ALBUMIN								
Cu^{2+}	16.7	3.739(1)	130	20	C	μ=0.1(NaAc)	73Rc	

Metal Ion	ΔH, kJ /mol	Log K	ΔS, J /mol K	T,°C	M	Conditions	Ref.	Re-marks
B156, continued								
Cu^{2+}, cont.						pH=4.80 Note: For the reactions with Cu^{2+} one or more hydrogen ions may have been released during binding of the Cu^{2+}		
	11.632	4.331(25°,1)	122.1	0&25	T	Buffer:0.036M-(HAc) 0.064M(NaAc) CL=2% protein	48Ka	a,b
	15.5	3.313	117	20	C	μ=0.1M(NaAc) pH=4.80	73Rc	
						R:CuL+Cu=Cu_2L		
	11.882	3.958(25°)	115.5	0&25	T	Buffer:0.036M-(HAc) 0.064M(NaAc) CL=2% protein	48Ka	a,b
						R:CuL+Cu=Cu_2L		
	14.2	2.961	105	20	C	μ=0.1M(NaAc) pH=4.80	73Rc	
						R:Cu_2L+Cu=Cu_3L		
	12.134	3.708(25°)	111.7	0&25	T	Buffer:0.036M-(HAc) 0.064M(NaAc) CL=2% protein	48Ka	a,b
						R:Cu_2L+Cu=Cu_3L		
	13.4	2.940	100	20	C	μ=0.1M(NaAc) pH=4.80	73Rc	
						R:Cu_3L+Cu=Cu_4L		
	12.385	3.507(25°)	108.8	0&25	T	Buffer:0.036M-(HAc) 0.064M(NaAc) CL=2% protein	48Ka	a,b
						R:Cu_3L+Cu=Cu_4L		
	12.6	2.616	92	20	C	μ=0.1M(NaAc) pH=4.80	73Rc	
						R:Cu_4L+Cu=Cu_5L		
	12.636	3.331(25°)	106.3	0&25	T	Buffer:0.036M-(HAc) 0.064M(NaAc) CL=2% protein	48Ka	a,b
						R:Cu_4L+Cu=Cu_5L		
	11.7	2.535	88	20	C	μ=0.1M(NaAc) pH=4.80	73Rc	
						R:Cu_5L+Cu=Cu_6L		
	12.887	3.170(25°)	103.8	0&25	T	Buffer:0.036M-(HAc) 0.065M(NaAc) CL=2% protein	48Ka	a,b
						R:Cu_5L+Cu=Cu_6L		
	10.9	2.417	84	20	C	μ=0.1M(NaAc) pH=4.80	73Rc	
						R:Cu_6L+Cu=Cu_7L		
	13.138	3.018(25°)	101.7	0&25	T	Buffer:0.036M-(HAc) 0.064M(NaAc) CL=2% protein	48Ka	a,b
						R:Cu_6L+Cu=Cu_7L		
	10.0	2.267	79	20	C	μ=0.1M(NaAc) pH=4.80	73Rc	

Metal Ion	ΔH, kJ /mol	Log K	ΔS, J /mol K	T,°C	M	Conditions	Ref.	Re-marks
B156, continued								
Cu^{2+}, cont.	13.389	2.870(25°)	99.6	0&25	T	R:$Cu_7L+Cu=Cu_8L$ Buffer:0.036M-(HAc) 0.064M(NaAc) CL=2% protein	48Ka	a,b
	9.6	2.144	75	20	C	R:$Cu_7L+Cu=Cu_8L$ μ=0.1M(NaAc) pH=4.80	73Rc	
	13.640	2.723(25°)	97.9	0&25	T	R:$Cu_8L+Cu=Cu_9L$ Buffer:0.036M-(HAc) 0.064M(NaAc) CL=2% protein	48Ka	a,b
	9.2	2.037	71	20	C	R:$Cu_8L+Cu=Cu_9L$ μ=0.1M(NaAc) pH=4.80	73Rc	
	13.891	2.575(25°)	95.8	0&25	T	R:$Cu_9L+Cu=Cu_{10}L$ Buffer:0.036M-(HAc) 0.064M(NaAc) CL=2% protein	48Ka	a,b
	14.142	2.423(25°)	93.7	0&25	T	R:$Cu_9L+Cu=Cu_{10}L$ Buffer:0.036M-(HAc) 0.064M(NaAc) CL=2% protein	48Ka	a,b
	14.393	2.262(25°)	91.2	0&25	T	R:$Cu_{10}L+Cu=Cu_{11}L$ Buffer:0.036M-(HAc) 0.064M(NaAc) CL=2% protein	48Ka	a,b
	14.6	2.086(25°)	88.7	0&25	T	R:$Cu_{11}L+Cu=Cu_{12}L$ Buffer:0.036M-(HAc) 0.064M(NaAc) CL=2% protein	48Ka	a,b
	14.895	1.885(25°)	85.8	0&25	T	R:$Cu_{12}L+Cu=Cu_{13}L$ Buffer:0.036M-(HAc) 0.064M(NaAc) CL=2% protein	48Ka	a,b
	15.146	1.635(25°)	82.0	0&25	T	R:$Cu_{13}L+Cu=Cu_{14}L$ Buffer:0.036M-(HAc) 0.064M(NaAc) CL=2% protein	48Ka	a,b
	15.397	1.262(25°)	75.3	0&25	T	R:$Cu_{14}L+Cu=Cu_{15}L$ Buffer:0.036M-(HAc) 0.064M(NaAc) CL=2% protein R:$Cu_{15}L+Cu=Cu_{16}L$	48Ka	a,b
B157	BROMATE ION		O_3Br			BrO_3^-		L^-
K^+	-39.915	1.24(<u>1</u>)	-110.0	25	T	μ=0	34La	
B158	BROMIDE ION		Br			Br^-		L^-
Li^+	19.66	14.69(<u>1</u>)	376	25	T	CM≅10^{-4}M S:12.65w%Butanol, 84.35w%Hexane	61Rd	

Metal Ion	ΔH, kJ /mol	Log K	ΔS, J /mol K	T,°C	M	Conditions	Ref.	Remarks
B158, continued								
Li^+, cont.	64.02	11.65(1)	439	25	T	D=2.16 CM≅10^{-4}M S:25w%Butanol, 75w%Hexane	61Rd	
	38.49	13.44(1)	385	25	T	D=2.85 CM≅10^{-4}M S:80w%Butanol, 20w%Hexane D=2.51	61Rd	
Na^+	0.184	-3.101(1)	-59	25	T	μ=0	36La 34La	
K^+	-20.543	-1.133(1)	-90.8	25	T	μ=0	34La	
	-31.0	-0.824	-121	25&35	T	μ=4M(Li,K,ClO_4, Br) R:K^++ CdL_4^{2-}=$K^+CdL_4^{2-}$	64Mc	a,b
Rb^+	-29.3	-0.658	-113	25&35	T	μ=4M(Li,Rb,ClO_4, Br) R:Rb^++ CdL_4^{2-}=$Rb^+CdL_4^{2-}$	64Mc	a,b
Cs^+	-25.9	-0.456	-96	25&35	T	μ=4M(Li,Rb,ClO_4, Br) R:Cs^++ CdL_4^{2-}=$Cs^+CdL_4^{2-}$	64Mc	a,b
Sc^{3+}	-2.1	1.17(1)	17	25	T	μ=0.5($NaClO_4$)	62Pa	
	-2.5	0.5(2)	0	25	T	μ=0.5($NaClO_4$)	62Pa	
Y^{3+}	-3.8	0.45(1)	-4	25	T	μ=0.5($NaClO_4$)	62Pa	
Sm^{3+}	-9.6	0.20(25°,1)	-28.4	0-50	T	μ=1M(ClO_4)	75Mf	
Eu^{3+}	-12.1	0.21(25°,1)	-36.8	0-50	T	μ=1M(ClO_4)	75Mf	
Gd^{3+}	-13.0	0.23(25°,1)	-38.9	0-50	T	μ=1M(ClO_4)	75Mf	
Tb^{3+}	-11.7	0.23(25°,1)	-34.7	0-50	T	μ=1M(ClO_4)	75Mf	
Dy^{3+}	-9.2	0.16(25°,1)	-27.6	0-50	T	μ=1M(ClO_4)	75Mf	
Cr^{3+}	37.36	(1)	--	25	C	0.5M($HClO_4$)	76Dd	
	23.3	-2.648(25°,1)	20.5	0-45	T	μ=2.00($NaClO_4$)	60Ea	b
	31.4	--	--	45-75	T	μ=0.50-2.0M (Hydrobromic acid solution) A=H_2O R:CrA_6^{3+}+ L^-=CrA_5L^{2+}+A	71Zf	b,q: ±2.1
	27.2	--	--	45-75	T	μ=4.0M (Hydro- bromic acid solution) A=H_2O R:CrA_6^{3+}+L^-= CrA_5L^{2+}+A	71Zf	b,q: ±2.1
	24.7	--	--	45-75	T	μ=8.0M (Hydro- bromic acid solution) A=H_2O	71Zf	b,q: ±2.1

Metal Ion	ΔH, kJ /mol	Log K	ΔS, J /mol K	T,°C	M	Conditions	Ref.	Re-marks
B158, continued								
Cr^{3+}, cont.						R: $CrA_6^{3+}+L^-=$ $CrA_5L^{2+}+A$		
	12.6	--	--	45-75	T	μ=4.0M (Hydrobromic acid solution) A=H_2O	71Zf	b,q: ±2.1
						R: $CrA_5L^{2+}+L=$ $CrA_4L_2^{+}+A$		
	11.3	--	--	45-75	T	μ=8.0M (Hydrobromic acid solution) A=H_2O	71Zf	b,q: ±2.1
						R: $CrA_5L^{2+}+L=$ $CrA_4L_2^{+}1+A$		
$HCrO_4^-$	-9.41	1.139(26°)	-9.6	26&38	T	μ=2.0M($NaClO_4$) R: $HCrO_4^-+2H^++Br^-=HCrO_3Br+H_2O$	71Yc	a,b
CrA_3^{3+}	17.97	1.54(35°,3)	87.90	25-45	T	μ=0 R: $CrA_3^{3+}+L^-=$ CrA_3L^{2+} A=2-Guanidinobenzimidazole C_6H_4-NH-C(=N)-NH-C(=NH)-NH_2	79Mf	b
Fe^{3+}	25.5	0.60(1)	96	25	T	μ=0	42Ra 52Ua	
	3.3	1.15(1)	33.9	25	T	μ=0	77Nb	q:±2
	10.5	1.27(1)	56.9	50	T	μ=0	77Nb	q:±2
	15.9	1.41(1)	73.6	70	T	μ=0	77Nb	q:±2
	22.6	1.53(1)	91.6	90	T	μ=0	77Nb	
	26.4	1.62(1)	101.7	100	T	μ=0	77Nb	
	36.8	1.93(1)	129.7	125	T	μ=0	77Nb	
	3.3	0.74(1)	25.1	25	T	μ=1M HBr	77Nb	q:±2
	6.7	0.81(1)	36.0	50	T	μ=1M HBr	77Nb	q:±2
	10.0	0.90(1)	46.4	70	T	μ=1M HBr	77Nb	q:±2
	14.2	0.98(1)	57.7	90	T	μ=1M HBr	77Nb	q:±2
	16.3	1.05(1)	64.0	100	T	μ=1M HBr	77Nb	q:±2
	23.0	1.22(1)	80.8	125	T	μ=1M HBr	77Nb	q:±2
Co^{2+}	0.58	-0.114(1)	-0.29	25	T	μ=2.0	66Kb	
	0.63	-0.108(1)	0.0	40	C	μ=2.0	66Kb	
	9.2	-0.72(1)	14.6	25	C	μ=3.0N($LiClO_4$) R: $Co(H_2O)_6^{2+}+L^-=$ $Co(H_2O)_6L^+$	74Bf	
	46.94	2.39(20°,1)	60.08	20-40	T	CM=$5x10^{-4}$-$5x10^{-3}$M CL=$1x10^{-3}$-$3.5x10^{-1}$M S:C_2H_5OH	61Sd	b
	48.95	1.49(20°,2)	57.61	20-40	T	CM=$5x10^{-4}$-$5x10^{-3}$M CL=$1x10^{-3}$-$3.5x10^{-1}$M	61Sd	b

Metal Ion	ΔH, kJ /mol	Log K	ΔS, J /mol K	T,°C	M	Conditions	Ref.	Remarks
B158, continued								
Co^{2+}, cont.	61.00	1.04(20°,3)	66.19	20-40	T	S:C_2H_5OH CM=$5x10^{-4}$-$5x10^{-3}$M CL=$1x10^{-3}$-$3.5x10^{-1}$M	61Sd	b
	31.96	0.865(20°,4)	40.46	20-40	T	S:C_2H_5OH CM=$5x10^{-4}$-$5x10^{-3}$M CL=$1x10^{-3}$-$3.5x10^{-1}$M	61Sd	b
	57.66	-0.34(20°,5)	55.77	20-40	T	S:C_2H_5OH CM=$5x10^{-4}$-$5x10^{-3}$M CL=$1x10^{-3}$-$3.5x10^{-1}$M S:C_2H_5OH	61Sd	b
$CoA_6{}^{3+}$	11.84	2.38(1)	84	25	T	μ=0 A=NH_3	55Na 53Ea	a
CoA_2	2.5	(1)	--	--	C	A=1,2-Bis(Diphenylphosphino) Ethane	72Mi	q:±0.8 u:25
$CoA_3{}^{3+}$	8.20	1.33(1)	54	25	T	μ=0.054 A=Ethylenediamine	53Ea	a
cis-$CoABC_2$	-5.31	2.839(1)	36.8	30	T	μ=0 A=Br B=Cl C=Ethylenediamine	66La	
	30.1	3.964(1)	175.7	30	T	μ=0 A=Dimethylformamide B=Cl C=Ethylenediamine	66La	
Ni^{2+}	0.33	-0.119(1)	-1.130	25	C	μ=2.0	66Kb	
	11.7	-0.81(1)	23	25	C	μ=3.0N($LiClO_4$) R:$Ni(H_2O)_6{}^{2+}+L^-$= $Ni(H_2O)_6L^+$	74Bf	
	0.29	-0.114(2)	-1.21	40	C	μ=2.0	66Kb	
NiA_2	0.0	(1)	--	--	C	A=1,2-Bis(Diphenylphosphino) Ethane	72Mi	q:±0.8 u:25
	2.669	0.155(25°)	11.92	10-59	T	S:Chloroform A=[(p-C_6H_5)(p-C_6H_5)MeP] R:NiA_2L_2(planar)= NiA_2L_2(tetrahedral)	70Pc	b
	4.812	-0.119(25°)	13.849	4-60	T	S:Chloroform A= A=[(p-C_6H_5)(p-ClC_6H_4)MeP] R:NiA_2L_2(planar)= NiA_2L_2(tetrahedral)	70Pc	b
	2.079	0.310(25°)	12.928	9-59	T	S:Chloroform A=[(p-C_6H_5)(p-MeC_6H_4)MeP] R:NiA_2L_2(planar)= NiA_2L_2(tetrahedral)	70Pc	b
	1.602	0.375(25°)	12.55	13-70	T	S:Chloroform A=[(p-C_6H_5)	70Pc	b

Metal Ion	ΔH, kJ /mol	Log K	ΔS, J /mol K	T,°C	M	Conditions	Ref.	Remarks
B158, continued								
NiA_2, cont.						$(p\text{-}MeOC_6H_4)MeP]$ R:NiA_2L_2(planar)= NiA_2L_2(tetrahedral)		
	0.569	0.675(25°)	14.85	-5-59	T	S:Chloroform A=[$(p\text{-}C_6H_5)$ $(p\text{-}Me_2NC_6H_4)MeP]$ R:NiA_2L_2(planar)= NiA_2L_2(tetrahedral)	70Pc	b
	7.782	-0.548(25°)	15.61	-8-56	T	S:Chloroform A=[$(p\text{-}ClC_6H_4)$ $(p\text{-}ClC_6H_4)MeP]$ R:NiA_2L_2(planar)= NiA_2L_2(tetrahedral)	70Pc	b
	2.058	0.336(25°)	13.35	10-59	T	S:Chloroform A=[$(p\text{-}MeC_6H_4)$ $(p\text{-}MeC_6H_4)MeP]$ R:NiA_2L_2(planar)= NiA_2L_2(tetrahedral)	70Pc	b
	0.0264	0.516(25°)	9.96	-8-59	T	S:Chloroform A=[$(P\text{-}MeOC_6H_4)$ $(p\text{-}MeOC_6H_4)MeP]$ R:NiA_2L_2(planar)= NiA_2L_2(tetrahedral)	70Pc	b
$NiAL_2$	14.22	0.00(27°)	48.1	17-57	T	S:Nitrobenzene A:$(C_6H_5)_2P(CH_2)_3$ $P(C_6H_5)_2$ R:$NiAL_2$(planar)= $NiAL_2$(tetrahedral)	66Va	b
NiA_4L_2	50.75	--	--	25-69	T	μ=4.30M(LiBr) S:CH_3OH A=CH_3OH R:$NiL_2A_4+L^-$= NiL_3A^-+3A (octa-hedral→tetrahedral)	67Sa	b
	71.59	--	--	25-69	T	μ=2.3M(LiBr) S:C_2H_5OH A=C_2H_5OH R:$NiL_2A_4+L^-$= NiL_3A^-+3A (octa-hedral→tetrahedral)	67Sa	b
	71.25	--	--	25-69	T	μ=1.152M(LiBr) S:n-C_4H_9OH A=n-C_4H_9OH R:NiL_2A_4=NiL_2A_2 +2A (octahedral →tetrahedral)	67Sa	b
	61.9	--	--	54-111	T	μ=1.152M(LiBr) S:n-C_4H_9OH A=n-C_4H_9OH R:$NiL_2A_4+L^-$= NiL_3A^-+3A (octa-hedral→tetrahedral)	67Sa	b

Metal Ion	ΔH, kJ /mol	Log K	ΔS, J /mol K	T,°C	M	Conditions	Ref.	Re-marks
B158, continued								
RhA_2BC	-13.4	3.222	21	50	T	μ=0.2M A=Ethyl-enediamine B=Cl $C=H_2O$ R:trans-$RhA_2BC^{2+}+L^-$=trans-RhA_2BL^++C	67Bc	q:±1.7
	-12.1	3.000	21	50	T	μ=0.2M A=Ethyl-enediamine B=Br $C=H_2O$ R:trans-$RhA_2BC^{2+}+L^-$=trans-RhA_2BL^++C	67Bc	q:±1.7
	-18.8	3.222	8	50	T	μ=0.2M A=Ethyl-enediamine B=I $C=H_2O$ R:trans-$RhA_2BC^{2+}+L^-$=trans-RhA_2BL^++C	67Bc	
RhA_5B	2.9	--	--	--	-	$A=NH_3$ $B=H_2O$ R:$RhA_5B^{3+}+L^-$=$RhA_5L^{2+}+B$	67Pa	q:±1.7 u:25
Pd^{2+}	-21.34	5.17(1)	27.2	25	C	μ=1M($HClO_4$)	72Re	
	-18.0	2.3(25°,4)	-14.6	10-45	T	μ=1.0	66Sb	b
	-54.85	(1-4)	--	25	C	μ=0.10	67Ic	
PdA_2	-9.2	2.283(1)	13.8	25	T	μ=0 A=o-Phenyl-lenebisdimethyl-arsine Thermo-dynamic values are average values as reported in Ref.67Ed	67Ec 67Ed	
	-10.5	2.057(35°,1)	5.4	25-55	T	μ=0.01($NaNO_3$) A=o-Phenylene-bisdimethylarsine	67Ed	b
	-9.2	1.832(35°,1)	5.4	25-55	T	μ=0.06($NaNO_3$) A=o-Phenylene-bisdimethylarsine	67Ed	b
$PdAB^+$	-5.0	0.471(27°)	-7.5	20-63	T	μ=0.5 A=Diethyl-enetriamine B=Cl R:$PdAB^++L^-=PdAL^++B^-$	67Hd	b,q: ±0.4
	2.5	0.114(27°)	12.6	20-63	T	μ=0.5 A=1,1,7,7-Tetraethyldiethyl-enetriamine B=Cl R:$PdAB^++L^-=PdAL^++B^-$	67Hd	b,q: ±0.4
$IrABC_2$	-246.8	(1)	--	25	C	S:Benzene A=Cl B=CO $C=PPh_3$	79Bf	
PtA_2^{2+}	-13.0	1.732(1)	-10.9	25	T	μ=0 A=o-Phenyl-enebisdimethyl-arsine	67Ec 67Ed	
	-4.27	3.81(23°,1)	58	23-56	T	μ=0 S:CH_3OH A=o-Phenylene-	66Dc	b

Metal Ion	ΔH, kJ /mol	Log K	ΔS, J /mol K	T,°C	M	Conditions	Ref.	Remarks
B158, continued								
PtA_2^{2+}, cont.	-14.2	1.152(49°,1)	-22.2	25-55	T	bisdimethylarsine μ=0.06($NaNO_3$)	67Ed	b
	-2.18	0.663(1)	5.0	25	C	A=o-Phenylene-bisdimethylarsine μ=1M(NaF) CM=0.040-0.050M	73Ma	q: ±0.21
PtA_3^{4+}	5.4	1.18(1)	40.6	25	T	A=Ethylenediamine μ→0	63Gb	
	12.6	-0.95(1)	25	25	C	A=Ethylenediamine 3.0M($NaClO_4$+NaCl)	72Mf	q: ±0.08
PtA_4^{2+}	-3.51	0.505(1)	-12.1	25	C	A=Ethylenediamine μ=1M(NaF) CM=0.040-0.050M	73Ma	q: ±0.63
PtA_6^{2-}	17	1.90(1-6)	92	25	T	A=NH_3 μ=0.5 A=Cl	60Pf	
Cu^{2+}	3.76	-0.071(1)	11.51	25	C	μ=2.0	66Kb 60Ld	
	4.18	-0.036(1)	12.68	40	C	μ=2.0	66Kb	
	12.6	-0.51(1)	31	25	C	μ=3.0N($LiClO_4$) R:$Cu(H_2O)_6^{2+}+L^-=Cu(H_2O)_6L^+$	74B[illegible]	
	57.7	3.20(1)	251	20	C	S:Methanol	76Ra	q:±5
	44.8	2.41(2)	192	20	C	S:Methanol	76Ra	q:±3.3
	41.8	1.00(3)	167	20	C	S:Methanol	76Ra	q:±3.3
CuA^{2+}	-8.37	1.033(25°,1)	-8.8	25-45	T	μ=0.4M(KNO_3) A=Macrocyclic ligand, rac-5,7,7,12,14,14-hexamethyl-1,4,-8,11-tetraaza-cyclotetradecane (tet b), (blue form)	76Cc	
CuA_2^{2+}	3.47	3.04(1)	69.9	25	C	S:CH_3OH A=N,N-Dimethylethylene-diamine	72Be	
	5.77	3.13(1)	79.1	25	C	S:CH_3OH A=N,N'-Dimethylethylene-diamine	72Be	
	11.97	1.91(1)	76.6	25	C	S:CH_3OH A=Ethyl-enediamine	72Be	
	6.74	2.68(1)	73.6	25	C	S:CH_3OH A=N-Methylethylene-diamine	72Be	
AuA_4^-	-46.0	--	--	25	C	μ=2.4M R:$AuA_4^-+4L^-=4A^-+AuL_4^-$	69Pb	

Metal Ion	ΔH, kJ /mol	Log K	ΔS, J /mol K	T,°C	M	Conditions	Ref.	Remarks
B158, continued								
Ag^{+}	-24.85	1.97(376°,1)	-0.58	376& 415	T	μ=0 S:$NaNO_3$-KNO_3 Eutectic mixture	61Dc 60Ja 50Fa	a,b
	-58.2	1.52(376°,2)	-60.2	376& 415	T	μ=0 S:$NaNO_3$-KNO_3 Eutectic mixture	61Dc 38Oa	a,b
	-10.5	0.41(25°,3)	-27	25-70	T	μ=4.6M($NaNO_3$,KBr)	75Pf	b
	-12.6	0.64(25°,4)	-30	25-70	T	μ=4.6M($NaNO_3$,KBr)	75Pf	b
	-25	0.123(4)	-82	25	C	μ=5M($NaClO_4$)	77Ad	q:±6
	-43.9	7.47(25°,1-2)	-4	25-70	T	μ=4.6M($NaNO_3$,KBr)	75Pf	b
	-54.6	9.076(1-3)	-9	25	C	μ=5M($NaClO_4$)	77Ad	q:±5.2
	-52.7	9.021(20°)	-7.1	20-70	T	R:AgBr+2Br^-= $AgBr_3^{2-}$	54Pc	b
	-85.4	9.365(20°)	-111.7	20-70	T	R:AgBr+4Br^-= $AgBr_5^{4-}$	54Pc	b
Zn^{2+}	1.46	-0.57(1)	-5.8	25	C	μ=3M($NaClO_4$)	69Gc	q:±0.12
	21.3	(1)	36.4	20-95	T	CM+CL=4M	70Ma	b
	0	0.601(1)	-11.51	25	C	μ=4.5	56Sa	
	27.8	0.841(1)	110	25	C	μ=1M(NH_4ClO_4) S:DMSO	76Af	
	42	-0.83(2)	126	25	C	μ=3M($NaClO_4$)	69Gc	q:±8
	-13.0	(2)	15.9	20-95	T	CM+CL=4M	70Ma	b
	9.1	2.891(2)	85	25	C	μ=1M(NH_4ClO_4) S:DMSO	76Af	
	-8	-0.51(3)	-17	25	C	μ=3M($NaClO_4$)	69Gb	q:±8
	-3.3	(3)	-14.6	20-95	T	CM+CL=4M	70Ma	b
	-4.2	1.349(3)	12	25	C	μ=1M(NH_4ClO_4) S:DMSO	76Af	
	-1.3	(4)	-10.5	20-95	T	CM+CL=4M	70Ma	b
	0	1.026(1-2)	-19.66	25	C	μ=4.5	56Sa	
	110.457	1.680(1-3)	338.1	25	C	μ=4.5	56Sa	n
	-8.368	1.320(1-4)	-53.1	25	C	μ=4.5	56Sa	
Cd^{2+}	-1.339	(1)	--	25	T	μ=0	59Aa	a
	-4.2	2.08(1)	26	25	C	μ=0	74Fd	q:±0.4
	-3.347	2.054(25°,1)	28.0	5-35	T	μ=0	73Sl	b,q:±1.3
	-0.8	2.20(25°,1)	36	0-50	T	μ=0	74Ra	b
	-3.3	1.51(1)	18	25	C	μ=0.5($LiClO_4$)	74Fd	q:±0.4
	-3.8	1.57(1)	17	25	C	μ=1.0($LiClO_4$)	74Fd	q:±0.4
	-4.6	1.56(25°,1)	14	0-50	T	μ=1M($NaClO_4$)	74Ra	b
	-3.8	1.68(1)	20	25	C	μ=2.0($LiClO_4$)	74Fd	q:±0.4
	-6.3	1.64(25°,1)	12	0-50	T	μ=2M($NaClO_4$)	74Ra	b
	-4.2	1.82(1)	21	25	C	μ=3.0($LiClO_4$)	74Fd	q:±0.4
	-4.10	1.76(1)	19.7	25	C	μ=3.0($NaClO_4$)	66Ga	
	-7.1	1.77(25°,1)	11	0-50	T	μ=3M($NaClO_4$)	74Ra	b
	-5.8	1.95(1)	18	25	C	μ=4.0($LiClO_4$)	74Fd	
	11.3	(1)	15.5	20-95	T	CM+CL=4M	70Ma	b
	9.6	1.68(1)	64.4	25	T	μ=4.5	57Sc	
	-3.9	2.919(1)	43	25	C	μ=1M(NH_4ClO_4) S:DMSO	76Ae	
	-16.7	2.75(40°,1)	-0.8	40-70	T	S:Melt of NH_4NO_3	72Nb	b
	0.8	0.87(2)	19	25	C	μ=0	74Fd	q:±0.8
	92.05	-0.110(25°,2)	305	5-35	T	μ=0	73Sq	b,q:±8

Metal Ion	ΔH, kJ /mol	Log K	ΔS, J /mol K	T,°C	M	Conditions	Ref.	Re-marks
B158, continued								
Cd^{2+}, cont.	6.3	0.78(25°,2)	44	0-50	T	μ=0	74Ra	b,n
	0	0.75(2)	14	25	C	μ=0.5($LiClO_4$)	74Fd	q:±0.4
	-2.1	0.69(2)	6.3	25	C	μ=1.0($LiClO_4$)	74Fd	q:±0.4
	6.7	0.56(25°,2)	32	0-50	T	μ=1M($NaClO_4$)	74Ra	b,n
	-7.1	0.43(2)	-15	25	C	μ=2.0($LiClO_4$)	74Fd	q:±0.8
	4.2	0.63(25°,2)	26	0-50	T	μ=2M($NaClO_4$)	74Ra	b,n
	-6.3	0.75(2)	-4	25	C	μ=3.0($LiClO_4$)	74Fd	q:±0.8
	-2.38	0.58(2)	3.3	25	C	μ=3.0($NaClO_4$)	66Ga	
	1.2	0.79(25°,2)	26	0-50	T	μ=3M($NaClO_4$)	74Ra	b,n
	-4.6	1.05(2)	4.6	25	C	μ=4.0($LiClO_4$)	74Fd	q:±1.3
	-0.4	(2)	25.1	20-95	T	CM+CL=4M	70Ma	b
	-18.0	0.733(2)	-46.4	25	T	μ=4.5	57Sc	
	17	1.913(2)	94	25	C	μ=1M(NH_4ClO_4)	76Ae	q:±2
	3.3	0.05(3)	12	25	C	μ=0	74Fd	q:±0.8
	-19.2	0.26(25°,3)	-67	0-50	T	μ=0	74Ra	b,n
	3.3	-0.15(3)	8.8	25	C	μ=0.5($LiClO_4$)	74Fd	q:±0.4
	5.8	0.67(3)	31	25	C	μ=1.0($LiClO_4$)	74Fd	q:±0.4
	-14.2	0.37(25°,3)	-38	0-50	T	μ=1M($NaClO_4$)	74Ra	b,n
	10.5	0.89(3)	50	25	C	μ=2.0($LiClO_4$)	74Fd	q:±0.4
	-1.7	0.40(25°,3)	-8	0-50	T	μ=2M($NaClO_4$)	74Ra	b,n
	7.1	0.95(3)	38	25	C	μ=3.0($LiClO_4$)	74Fd	q:±0.4
	7.20	0.98(3)	42.7	25	C	μ=3.0($NaClO_4$)	66Ga	
	5.8	0.52(25°,3)	17	0-50	T	μ=3M($NaClO_4$)	74Ra	b,n
	4.6	1.30(3)	40	25	C	μ=4.0($LiClO_4$)	74Fd	q:±0.4
	15.9	(3)	35.1	20-95	T	CM+CL=4M	70Ma	b
	8.4	0.806(3)	43.5	25	T	μ=4.5	57Sc	
	2	2.747(3)	59	25	C	μ=1M(NH_4ClO_4) S:DMSO	76Ae	q:±3
	-4.6	-0.35(4)	-22	25	C	μ=0	74Fd	q:±1.6
	28	-0.28(25°,4)	96	0-50	T	μ=0	74Ra	b,n
	19	0.26(25°,4)	75	0-50	T	μ=1M($NaClO_4$)	74Ra	b,n
	-1.7	0.57(4)	8.4	25	C	μ=2.0($LiClO_4$)	74Fd	q:±0.4
	7	0.64(25°,4)	50	0-50	T	μ=2M($NaClO_4$)	74Ra	b,n
	-1.7	0.66(4)	8.8	25	C	μ=3.0($LiClO_4$)	74Fd	q:±0.4
	1.26	0.35(4)	11.3	25	C	μ=3.0($NaClO_4$)	66Ga	
	-1.2	0.79(25°,4)	12	0-50	T	μ=3M($NaClO_4$)	74Ra	b,n
	-3.3	0.88(4)	28	25	C	μ=4.0($LiClO_4$)	74Fd	q:±0.4
	-1.7	(4)	2.9	20-95	T	CM+CL=4M	70Ma	b
	-13.0	1.675(4)	-12	25	C	μ=1M(NH_4ClO_4) S:DMSO	76Ae	
	10.9	0.513(4)	46.4	25	T	μ=4.5	57Sc	
Hg^{2+}	-46.4	9.48(1)	16.7	8	C	μ=0	64Cb 58Sd	
	-42.7	9.01(1)	29.7	25	C	μ=0	64Cb 61Ma 63Bb	
	-41.8	8.75(1)	33.0	40	C	μ=0	64Cb	
	-42.2	(1)	--	25	C	0.5M($NaClO_4$)	65Ae	
	-40.04	9.40(1)	45.6	25	C	3M($NaClO_4$)	65Ae	
	-48.1	8.78(2)	-2.5	8	C	μ=0	64Cb	
	-46.0	8.28(2)	4.6	25	C	μ=0	64Cb	
	-47.7	7.77(2)	-3.8	40	C	μ=0	64Cb	
	-44.8	(2)	--	25	C	0.5M($NaClO_4$)	65Ae	

Metal Ion	ΔH, kJ /mol	Log K	ΔS, J /mol K	T, °C	M	Conditions	Ref.	Remarks
B158, continued								
Hg^{2+}, cont.	-40.21	8.58(2)	29.7	25	C	$3M(NaClO_4)$	65Ae	
	-8.70	2.26(3)	14.2	25	C	Dilute solution	63Ba	
	-12.6	(3)	--	25	C	$0.5M(NaClO_4)$	65Ae	
	-26.9	5.6(3)	17	25	C	$\mu=1M(NaClO_4)$ S:Dimethyl sulfoxide	76Am	
	-10.79	2.76(3)	16.7	25	C	$3M(NaClO_4)$	65Ae	
	-14.48	1.38(4)	-22.2	25	C	Dilute solution	63Ba	
	-17.2	(4)	--	25	C	$0.5M(NaClO_4)$	65Ae	
	-22.7	2.65(4)	-25	25	C	$\mu=1M(NaClO_4)$ S:Dimethyl sulfoxide	76Am	
	-18.58	1.49(4)	-33.9	25	C	$3M(NaClO_4)$	65Ae	
	-94.56	18.26(1-2)	14.2	8	C	$\mu=0$	64Cb	
	-88.83	17.29(1-2)	34.3	25	C	$\mu=0$	64Cb	
	-89.54	16.52(1-2)	29.3	40	C	$\mu=0$	64Cb	
	-87.0	(1-2)	--	25	C	$0.5M(NaClO_4)$	65Ae	
	-80.25	(1-2)	--	25	C	$3M(NaClO_4)$	65Ae	
	-116.7	(1-4)	--	25	C	$0.5M(NaClO_4)$	65Ae	
	-115.9	20.95(1-4)	12.6	25	C	$\mu=0.5$	60Ga	
	-108.8	(1-4)	--	25	C	$3M(NaClO_4)$	65Ae	
	-23.18	3.64	-7.9	25	C	Dilute solution R:$HgBr_2+2Br^-=HgBr_4^{2-}$	63Ba	
	268.2	1.16(42.1°)	874	42& 55.6	T	S:Benzene R:$2(HgL_2)=(HgL_2)_2$	74Ea	a,b
	31.4	2.58(55.6°)	146	55&65	T	S:Mesitylene R:$2(HgL_2)=(HgL_2)_2$	74Ea	a,b
	-79.0	2.27(42.1°)	-209	42& 55.6	T	S:Toluene R:$2(HgL_2)=(HgL_2)_2$	74Ea	a,b
	-110.0	2.95(55.6°)	-280	55&65	T	S:p-Xylene R:$2(HgL_2)=(HgL_2)_2$	74Ea	a,b
Hg(II)	-2.1	0.60(25°)	4.6	5-60	T	$\mu\cong10^{-3}$ R:$\frac{1}{2}HgBr_2+\frac{1}{2}HgCl_2=HgBrCl$	64Ea	b,r
	-1.7	0.55(25°)	5.0	5-60	T	$\mu\cong10^{-3}$ R:$\frac{1}{2}HgBr_2+\frac{1}{2}HgI_2=HgBrI$	64Ea	b,r
HgA^+	-41.4	6.635(1)	-15.1	20	C	$\mu=0.1(KNO_3)$ $A=CH_3$	65Sa	
HgA^{2-}	-42.7	5.20(1)	-42	25	C	$\mu=2M(NaClO_4)$ $A=EDTA^{4-}$	75Bc	
$HgAB^{3-}$	23.4	-4.03	0	25	C	$\mu=2M(NaClO_4)$ R:$HgAB^{3-}+L^-=HgLB^{3-}+A^-$ $A=I^-$ $B=EDTA^{4-}$	75Bc	
Ga^{3+}	54.4	-2.01(1-4)	115.9	80	T	$\mu=6.0$	62Nb	
	47.7	-3.00(1-4)	77.0	80	T	$\mu=7.5$	62Nb	
	43.5	-3.08(1-4)	64.0	80	T	$\mu=9.0$	62Nb	
	40.2	-3.11(1-4)	52.7	80	T	$\mu=9.9$	62Nb	
	38.9	-3.10(1-4)	51.0	80	T	$\mu=10.4$	62Nb	
	36.8	-3.36(1-4)	40.6	80	T	$\mu=11.0$	62Nb	
	38.4	-3.40(1-4)	43.9	80	T	$\mu=11.5$	62Nb	

Metal Ion	ΔH, kJ /mol	Log K	ΔS, J /mol K	T,°C	M	Conditions	Ref.	Re- marks
B158, continued								
Ga^{3+}, cont.	33.0	-3.30(1-4)	29.7	80	T	μ=13.6	62Nb	
	35.1	-3.51(1-4)	32.2	80	T	μ=14.9	62Nb	
In^{3+}	1.97	1.98(1)	44.4	25	C	μ=2.00M($NaClO_4$)	69Rd	
	5.65	0.60(2)	30.5	25	C	μ=2.00M($NaClO_4$)	69Rd	
Tl^{+}	-10.25	0.88(1)	-17.6	25	T	μ=0	57Na	
	-14.6	1.10(25°,1)	-29	10-60	T	μ=0	74Fe	q:±1.2
	-13.0	0.73(25°,1)	-29	10-60	T	μ=0.5($LiClO_4$)	74Fe	q:±1.2
	-15.9	0.66(25°,1)	-42	10-60	T	μ=1.0($LiClO_4$)	74Fe	q:±1.2
	-18.4	0.66(25°,1)	-50	10-60	T	μ=2.0($LiClO_4$)	74Fe	q:±2
	-21.8	0.59(25°,1)	-64	10-60	T	μ=3.0($LiClO_4$)	74Fe	q:±2
	-23.0	0.59(25°,1)	-67	10-60	T	μ=4.0($LiClO_4$)	74Fe	q:±3.3
	-10.5	0.59(25°,2)	-25	10-60	T	μ=0	74Fe	q:±1.2
	-5.8	0.29(25°,2)	-12	10-60	T	μ=0.5($LiClO_4$)	74Fe	q:±0.4
	-12.6	0.37(25°,2)	-33	10-60	T	μ=1.0($LiClO_4$)	74Fe	q:±2
	-0.84	0.15(25°,2)	0	10-60	T	μ=2.0($LiClO_4$)	74Fe	q:±0.4
	-5.0	0.37(25°,2)	-8.4	10-60	T	μ=3.0($LiClO_4$)	74Fe	q:±0.8
	0.0	0.29(25°,2)	4	10-60	T	μ=4.0($LiClO_4$)	74Fe	q:±3.3
Tl^{3+}	-37.49	9.50(1)	56.1	25	C	3M($HClO_4$)+ 1M($NaClO_4$)	64Lf	
	-25.48	7.37(2)	55.6	25	C	3M($HClO_4$)+ 1M($NaClO_4$)	64Lf	
	-19.16	5.44(3)	40.6	25	C	3M($HClO_4$)+ 1M($NaClO_4$)	64Lf	
	-8.95	4.12(4)	49.0	25	C	3M($HClO_4$)+ 1M($NaClO_4$)	64Lf	
Sn^{2+}	5.774	0.729(1)	33.5	25	T	3M($NaClO_4$)	52Va	
	11.422	1.140(1-2)	60.2	25	T	3M($NaClO_4$)	52Va	
	9.895	1.338(1-3)	59.0	25	T	3M($NaClO_4$)	52Va	
Pb^{2+}	1.255	(1)	--	25	C	μ=0	59Aa 55Na	
	0.201	(1)	--	25	C	μ=0.15	59Aa	
	-5.65	1.342(1)	5.4	5	T	μ=3M($LiClO_4$)	68Fb	
	-3.68	1.310(1)	12.6	15	T	μ=3M($LiClO_4$)	68Fb	
	-2.18	1.292(1)	17.2	25	T	μ=3M($LiClO_4$)	68Fb	
	-0.84	1.286(1)	21.8	35	T	μ=3M($LiClO_4$)	68Fb	q:±0.21
	0.33	1.283(1)	25.5	45	T	μ=3M($LiClO_4$)	68Fa	q:±0.21
	2.26	1.290(1)	31.8	55	T	μ=3M($LiClO_4$)	68Fb	q:±0.42
	3.14	1.305(1)	34.3	65	T	μ=3M($LiClO_4$)	68Fb	q:±0.84
	-2.18	1.283(25°,1)	17	18-65	T	μ=3M($LiClO_4$)	68Fa	b
	3.347	(1)	--	25	C	μ=0 S:20.86 mole % CH_3OH	59Aa	
	2.669	(1)	--	25	C	μ=0.015 S:20.86 mole % CH_3OH	59Aa	
	6.067	(1)	--	25	C	μ=0 S:40.16 mole % CH_3OH	59Aa	
	4.929	(1)	--	25	C	μ=0.015 S:40.16 mole % CH_3OH	59Aa	
	8.661	(1)	--	25	C	μ=0 S:61.01 mole % CH_3OH	59Aa	
	7.439	(1)	--	25	C	μ=0.009 S:61.01 mole % CH_3OH	59Aa	

Metal Ion	ΔH, kJ /mol	Log K	ΔS, J /mol K	T,°C	M	Conditions	Ref.	Re-marks
B158, continued								
Pb^{2+}, cont.	-10.9	2.09(40°,1)	5.0	40-70	T	S:Melt of NH_4NO_3	72Nb	b
	-6.78	0.984(2)	-5.4	5	T	μ=3M($LiClO_4$)	68Fb	
	-5.82	0.940(2)	-2.5	15	T	μ=3M($LiClO_4$)	68Fb	
	-3.60	0.918(2)	5.8	25	T	μ=3M($LiClO_4$)	68Fb	q:±0.50
	-1.13	0.896(2)	13.8	35	T	μ=3M($LiClO_4$)	68Fb	q:±0.63
	2.89	0.905(2)	25.5	45	T	μ=3M($LiClO_4$)	68Fb	q:±0.84
	5.94	0.925(2)	36.0	55	T	μ=3M($LiClO_4$)	68Fb	q:±0.50
	10.63	0.962(2)	49.4	65	T	μ=3M($LiClO_4$)	68Fb	q:±1.7
	-3.60	0.931(25°,2)	6.3	18-65	T	μ=3M($LiClO_4$)	68Fa	b
	-3.47	0.593(3)	-1.2	5	T	μ=3M($LiClO_4$)	68Fb	q:±2.1
	-1.67	0.582(3)	5.4	15	T	μ=3M($LiClO_4$)	68Fb	q:±1.7
	-0.38	0.568(3)	9.6	25	T	μ=3M($LiClO_4$)	68Fb	q:±0.71
	0.84	0.581(3)	14.6	35	T	μ=3M($LiClO_4$)	68Fb	q:±0.84
	1.80	0.583(3)	17.6	45	T	μ=3M($LiClO_4$)	68Fb	q:±0.54
	2.68	0.598(3)	19.2	55	T	μ=3M($LiClO_4$)	68Fb	q:±0.75
	4.10	0.614(3)	25.1	65	T	μ=3M($LiClO_4$)	68Fb	q:±1.7
	-1.51	0.579(25°,3)	10.5	18-65	T	μ=3M($LiClO_4$)	68Fa	b
	-9.2	0.271(4)	-28.0	5	T	μ=3M($LiClO_4$)	68Fb	q:±2.5
	-8.91	0.209(4)	-28.0	15	T	μ=3M($LiClO_4$)	68Fb	q:±1.7
	-9.33	0.151(4)	-28.4	25	T	μ=3M($LiClO_4$)	68Fb	q:±0.96
	-11.00	0.088(4)	-33.5	35	T	μ=3M($LiClO_4$)	68Fb	q:±1.2
	-13.8	0.007(4)	-43.5	45	T	μ=3M($LiClO_4$)	68Fb	q:±0.84
	-15.9	-0.050(4)	-49.4	55	T	μ=3M($LiClO_4$)	68Fb	q:±1.7
	-21.21	-0.149(4)	-67	65	T	μ=3M($LiClO_4$)	68Fb	
	-8.20	0.132(25°,4)	-29	18-65	T	μ=3M($LiClO_4$)	68Fa	b
Bi^{3+}	13.807	3.28(1)	109.2	25	C	μ=0	67Va	
	-2.330	2.64(1)	41.8	10	C	μ=4.0($HClO_4$) ΔCp=602.1	67Va	
	-1.485	2.64(1)	46.0	18	C	μ=4.0($HClO_4$) ΔCp=593.3	67Va	
	-0.079	2.64(1)	49.4	25	C	μ=4.0($HClO_4$) ΔCp=584.5	67Va	
	0.406	2.64(1)	51.9	30	C	μ=4.0($HClO_4$) ΔCp=579.5	67Va	
	1.046	2.64(1)	54.0	35	C	μ=4.0($HClO_4$) ΔCp=574.0	67Va	
	3.142	2.64(1)	60.2	50	C	μ=4.0($HClO_4$) ΔCp=556.9 ΔCp=219.4-0.267T	67Va	
	-6.360	2.80(1)	45.2(1)	25	C	μ=5.0($HClO_4$)	67Va	
TeA_4	-166.9	--	--	25	C	S:Benzene A=Cl R:TeA_4+AlL_3= $TeA_4 \cdot AlL_3$	73Gb	
	-110.0	--	--	25	C	S:Benzene A=Cl R:TeA_4+GaL_3= $TeA_4 \cdot GaL_3$	73Gb	
Br_2	~0	1.22(1)	~0	25	T	μ=0.5 63Db reports ΔH=12.8 as the value given in 58Sb	58Sc	
	-15.69	1.373(25°,1)	26.4	18-39	T	--	60Db	b
	-14.2	2.00(2°,1)	13.8	2&25	T	μ=0.1 S:75%	60Db	a,b,n

Metal Ion	ΔH, kJ /mol	Log K	ΔS, J /mol K	T,°C	M	Conditions	Ref.	Remarks
B158, continued								
Br^2, cont.	-24.2	2.31(18°,1)	39.3	-15-25	T	Acetic acid S:CH_3OH	63Cb	b
B159 BROMINE			Br_2			Br_2		L
KA	-29.748	1.435(1)	64.4	50.00	T	μ=0.2(Lithium ethane sulfate) S:Glacial acetic acid A=Br	42Na	a
PbA_4	-362.12	--	--	--	C	S:$CHCl_3$ A=Phenyl R:PbA_4+$2L_2$= PbL_2A_2+2LA	70Bg	u
PbA_3L	-170.96	--	--	--	C	S:$CHCl_3$ A=Phenyl R:PbA_3L+L_2= PbA_2L_2+LA	70Bg	u
SeA	-54.8	3.04(1)	-126	25	T	Dilute solution CM→0M S:CCl_4 A= 4-chlorodiphenyl	60Ma 49Ma	n
	-40.6	3.44(1)	-71	25	T	Dilute solution CM→0M S:CCl_4 A=Diphenyl	60Ma	n
	-47.3	3.65(1)	-88	25	T	Dilute solution CM→0M S:CCl_4 A= 3-Methoxydiphenyl	60Ma	n
	-60.2	4.21(1)	-121	25	T	Dilute solution CM→0M S:CCl_4 A= 4-Methoxydiphenyl	60Ma	n
SeA_2	-32.6	2.68(1)	-58	25	T	S:CCl_4 A=p-Chlorophenyl	45Ma	
	-40.6	2.55(1)	-75	25	T	S:CCl_4 A=p-Bromophenyl	45Ma	
	-35.1	4.44(1)	-33	25	T	S:CCl_4 A=o-Methylphenyl	45Ma	
B160 1,3-BUTADIENE			C_4H_6			CH_2=CHCH=CH_2		L
Ag^+	13.4	0.653(25°,1)	-32.2	0-40	T	S:Ethylene glycol	65Ci	b
B161 1,3-BUTADIENE, 2,3-dimethyl-			C_6H_{10}			CH_2:CC(CH_3)C(CH_3):CH_2		
NiA_2	-71.5	4.041(25°,1)	-159	--	T	S:Hydrocarbons CCl_4 CL=CM= $2x10^{-4}$M A=cis (1,2-perfluoro-methylethene-1,2-dithiol)	71Bb	u:25
B162 1,3-BUTADIENE, 2-methyl			C_5H_8			CH_2:CHC(CH_3):CH_2		
Cu^+	-36.0	0.544(30°,1)	-108	18-63	T	S:Hexamethyl-	75Pd	

Metal Ion	ΔH, kJ /mol	Log K	ΔS, J /mol K	T,°C	M	Conditions	Ref.	Remarks
B162, continued								
Cu^{+}, cont.						phosphoramide Copper is bonded only to the vinyl group in L		
	-38.1	0.633(30°)	-114	18-63	T	S:Hexamethyl-phosphoramide R:Cu+L=[CuL+Cu_2L] Copper in CuL is bonded only to vinyl group in L	75Pd	
	-23.0	-0.398(30°)	-83	18-63	T	S:Hexamethyl-phosphoramide R:Cu+L=[CuL+Cu_2L] Copper in CuL is bonded only to isopropenyl group in L	75Pd	q:±3.3
B163 BUTANAL			C_4H_8O			$CH_3CH_2CH_2CHO$		
$[NiA_2]_2$	-17.4	1.20(100°,1)	-24	75-125	T	S:Squalane A=3-Trifluoro acetylcamphorate	74Sj	b,gg
B164 BUTANE, 1-amino-			$C_4H_{11}N$			$CH_3(CH_2)_3NH_2$		L
VOA_2	-43.60	(1)	--	25	C	μ→0 S:Nitro-benzene A=Acetylacetone	65Ca	
NiA	-70.7	6.01(1-2)	-121.8	25	T	CM=10^{-4}M S:Ben-zene A=Diacetyl-bisbenzoyl hydrazine	60Sb	
NiA_2	-25	2.700(35°,1)	-29	25-45	T	S:Toluene A=0,0' -diethyldithio-phosphate	71Cc	b,q:±4
	-21	2.725(35°,2)	-17	25-45	T	S:Toluene A=0,0' -diethyldithio-phosphate	71Cc	b,q:±4
CuA_2	-54	4.64(1)	-96	20	T	S:Benzene A=Trifluoro-acetylacetone	79Si	b,q:±6
Ag^{+}	-17	3.43(1)	14.2	25	C	μ=0	71He	q:±8
	-37.6	4.05(2)	-41.8	25	C	μ=0	71He	q:±8
	-52.59	7.48(1-2)	-33.5	25	C	μ=0	71He	
ZnA_2	-31.4	3.159(25°,1)	-46	18.5-65	T	S:Toluene A=$[S_2CN(CH_3)_2]$	65Ch	b
[B]	-49.8	-0.628(100°,1)	-149.0	40-120	T	[B]=2-Hexyl-7-methyl-1,2-oxa-	72Vc	b

Metal Ion	ΔH, kJ /mol	Log K	ΔS, J /mol K	T,°C	M	Conditions	Ref.	Re-marks
B164, continued								
[B], cont.						borepane 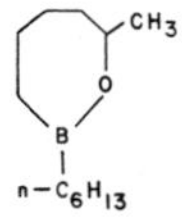		
I_2	-6.19	1.173(25°,1)	1.67	15-40	T	μ<0.02M	60Da	b
	-5.73	1.021(1)	0	20	T	--	64Eb	
	-35.1	3.090(20°,1)	-61.9	20-40	T	CM=10^{-5}M CL=10^{-4}-10^{-3}M S:n-Heptane	60Ya	b
B165	BUTANE, 2-amino-		$C_4H_{11}N$			$CH_3CH_2CH(NH_2)CH_3$		L
NiA	-69.0	4.08(1-2)	-152.7	25	T	CM=10^{-4}M S:Benzene A=Diacetyl-bisbenzoyl hydrazine	60Sb	
B166	BUTANE, 1-amino-3-methyl-		$C_5H_{13}N$			$(CH_3)_2CHCH_2CH_2NH_2$		L
NiA	-72.0	5.96(1-2)	-126.8	25	T	CM=10^{-4}M S:Benzene A=Diacetyl-bisbenzoyl hydrazine	60Sb	
B167	BUTANE, 1-amino-3-thia-		C_3H_9NS			$CH_3SCH_2CH_2NH_2$		L
Ni^{2+}	-23.30	3.299(1)	-15.1	25	C	0.5M(KNO_3)	77Hd	
	-21	3.23(30°,1)	-12	0-50	T	μ=1.0(KClorKNO$_3$)	54Gb	b
	-25.31	2.805(2)	-31.0	25	C	0.5M(KNO_3)	77Hd	
	-25	2.79(30°,2)	-33	0-50	T	μ=1.0(KClorKNO$_3$)	54Gb	b
	-25.52	1.63(3)	54.4	25	C	0.5M(KNO_3)	77Hd	
	-17	1.73(30°,3)	-29	0-50	T	μ=1.0(KClorKNO$_3$)	54Gb	b
	-63	7.75(30°,1-3)	-71	0-50	T	μ=1.0(KClorKNO$_3$)	54Gb	b
Cu^{2+}	-30.1	5.41(30°,1)	4	10-40	T	μ=0	59Ma	b
	-33.26	5.572(1)	-5.0	25	C	0.5M(KNO_3)	77Hd	
	-38.07	5.082(2)	-30.5	25	C	0.5M(KNO_3)	77Hd	
B168	BUTANE, 1,2-diamino-		$C_4H_{12}N_2$			$NH_2CH_2CH(NH_2)CH_2CH_3$		L
Cu^{2+}	-49.20	10.047(1)	27.2	25	C	μ=0	67Pc	
	-50.38	9.069(2)	4.6	25	C	μ=0	67Pc	
	-99.6	19.12(1-2)	31.8	25	C	μ=0	67Pc	
B169	BUTANE, 1,4-diamino		$C_4H_{12}N_2$			$NH_2CH_2CH_2CH_2CH_2NH_2$		
CuA_2	-91	(1)	--	30	C	S:Benzene A=1,1,1,5,5,5-	75Gg 76Gh	

Metal Ion	ΔH, kJ /mol	Log K	ΔS, J /mol K	T,°C	M	Conditions	Ref.	Remarks
B169, continued								
CuA_2, cont.						Hexafluoropentane-2,4-dionato		
	-39	(2)	--	30	C	S:Benzene A=1,1,1,5,5,5-Hexafluoropentane-2,4-dionato	75Gg 76Gh	
B170	(±)-BUTANE, 2,3-diamino		$C_4H_{12}N_2$			$CH_3CH(NH_2)CH(NH_2)CH_3$		L
Ni^{2+}	-32.5	7.639(1)	37.2	25	C	μ=0.1M(KNO_3)	77Pb	
	-29.5	6.360(2)	22.8	25	C	μ=0.1M(KNO_3)	77Pb	
	-19.7	3.960(3)	9.73	25	C	μ=0.1M(KNO_3)	77Pb	
Cu^{2+}	-52.7	11.266(1)	38.9	25	C	μ=0.1M(KNO_3)	77Pb	
	-47.3	9.601(2)	25.2	25	C	μ=0.1M(KNO_3)	77Pb	
Zn^{2+}	-18.0	5.939(1)	53.3	25	C	μ=0.1M(KNO_3)	77Pb	
	-11.9	5.326(2)	62.0	25	C	μ=0.1M(KNO_3)	77Pb	
B171	meso-BUTANE, 2,3-diamino		$C_4H_{12}N_2$			$CH_3CH(NH_2)CH(NH_2)CH_3$		L
Ni^{2+}	-28.2	6.728(1)	34.2	25	C	μ=0.1M(KNO_3)	77Pb	
	-23.7	5.572(2)	27.2	25	C	μ=0.1M(KNO_3)	77Pb	
	-11.7	2.558(3)	9.7	25	C	μ=0.1M(KNO_3)	77Pb	
Cu^{2+}	-46.9	10.530(1)	44.3	25	C	μ=0.1M(KNO_3)	77Pb	
	-49.9	9.251(2)	9.7	25	C	μ=0.1M(KNO_3)	77Pb	
Zn^{2+}	-13.9	5.449(1)	57.7	25	C	μ=0.1M(KNO_3)	77Pb	
	-16.4	5.326(2)	47.0	25	C	μ=0.1M(KNO_3)	77Pb	
B172	BUTANE, 2,3-diamino-2,3-dimethyl-		$C_6H_{16}N_2$			$CH_3C(CH_3)(NH_2)C(CH_3)(NH_2)CH_3$		L
Ni^{2+}	-59.0	14.7(1-2)	84	25	C	μ=0.5(KNO_3)	54Ba	
Cu^{2+}	-102.1	21.9(1-2)	79	25	C	μ=0.5(KNO_3)	54Ba	
Zn^{2+}	6.063	-1.083(20°,1-2)	0	0-40	T	μ=0	68Pc	b
Cd^{2+}	5.288	-1.039(20°,1-2)	1.7	0-40	T	μ=0	68Pc	b
B173	BUTANE, 1,3-diamino-N,N'-(2-pyridylmethyl) (bis)		$C_{16}H_{24}N_4$			$(C_5H_5N)CH_2NHCH_2CH_2CH_2CH_2NHCH_2(NC_5H_5)$		
Mn^{2+}	-0.84	2.64(1)	47.7	25	C	μ=0.1(KNO_3)	75Ae	
Co^{2+}	-35.1	7.99(1)	35.1	25	C	μ=0.1(KNO_3)	75Ae	
Ni^{2+}	-51.9	11.14(1)	39.3	25	C	μ=0.1(KNO_3)	75Ae	
Cu^{2+}	-70.3	15.54(1)	61.9	25	C	μ=0.1(KNO_3)	75Ae	
Ag^{+}	-48.1	5.79(1)	-50.6	25	C	μ=0.1(KNO_3)	75Ae	
Zn^{2+}	-24.7	7.70(1)	64.4	25	C	μ=0.1(KNO_3)	75Ae	
Cd^{2+}	-28.0	7.04(1)	40.6	25	C	μ=0.1(KNO_3)	75Ae	

Metal Ion	ΔH, kJ /mol	Log K	ΔS, J /mol K	T,°C	M	Conditions	Ref.	Re- marks
B173, continued								
Hg^{2+}	-87.0	17.52(1)	43.1	25	C	μ=0.1(KNO_3)	75Ae	
B174	BUTANE, 1,4-diamino-N,N,N',N'-tetramethyl		$C_8H_{20}N_2$			$(CH_3)_2NCH_2CH_2CH_2CH_2N(CH_3)_2$		
CuA_2	-49	(1)	--	30	C	S:Benzene A=1,1,1,5,5,5-Hexafluoropentane-2,4-dionato	75Gh 75Gg	
HgA_2	-85.3	5.30(1)	-179	30	C	CM=(0.33-0.98) $x10^{-3}$ S:Benzene A=Cl	74Fb	
	-74.3	5.30(1)	-143	30	C	CM=(0.5-1.7)$x10^{-3}$ S:Benzene A=Br	74Fb	
	-78.7	4.30(1)	-177	30	C	CM=(0.5-1.4)$x10^{-3}$ S:Benzene A=I	74Fb	
	-28.8	1.12(1)	-74	30	C	CM=(1-20)$x10^{-3}$ S:Benzene A=C_6F_5	74Fc	
B175	BUTANE, 1,4-diamino-N,N,-N',N'-tetraacetic acid		$C_{12}H_{20}O_8N_2$			$(CH_2CO_2H)_2N(CH_2)_4N(CH_2CO_2H)_2$		L^{4-}
Mg^{2+}	35.6	6.22(1)	225.9	20	C	μ=0.1(KNO_3)	64Ab	
Ca^{2+}	-3.8	5.66(1)	124.3	20	C	μ=0.1(KNO_3)	64Ab	
La^{3+}	7.86	9.13(1)	201.7	20	C	μ=0.1(KNO_3)	64Ab	
Mn^{2+}	14.27	9.53(1)	231.0	20	C	μ=0.1(KNO_3)	64Ab	
Co^{2+}	-6.7	15.66(1)	277.4	20	C	μ=0.1(KNO_3)	64Ab	
Ni^{2+}	-29.08	17.36(1)	232.2	20	C	μ=0.1(KNO_3)	64Ab	
Cu^{2+}	-27.28	17.33(1)	238.5	20	C	μ=0.1(KNO_3)	64Ab	
Zn^{2+}	-14.56	15.01(1)	237.6	20	C	μ=0.1(KNO_3)	64Ab	
Cd^{2+}	-12.05	12.02(1)	189.1	20	C	μ=0.1(KNO_3)	64Ab	
Hg^{2+}	-79.9	20.99(1)	128.9	20	C	μ=0.1(KNO_3)	64Ab	
Pb^{2+}	-20.29	10.53(1)	132.2	20	C	μ=0.1(KNO_3)	64Ab	
B176	BUTANE, 1,2-diethoxy-		$C_8H_{18}O_2$			$CH_3CH_2CH(OC_2H_5)CH_2OC_2H_5$		L
SnA_4	-51.5	2.71(1)	-120.9	25	C	S:Benzene A=Cl	70Gh 68Ge	
B177	BUTANE, 2,3-dihydroxy-(ℓ)		$C_4H_{10}O_2$			$CH_3CH(OH)CH(OH)CH_3$		L
BA_4^-	-18.0	1.568(25°,1)	-30.1	0-35	T	CL=0-0.9M A=OH^-	67Cb	b,q:±1.7
	-22.93	2.215(25°,1-2)	-34.3	0-35	T	CL=0-0.9M A=OH^-	67Cb	b,q:±2.5

Metal Ion	ΔH, kJ /mol	Log K	ΔS, J /mol K	T,°C	M	Conditions	Ref.	Re-marks
B178	BUTANE, 2,3-dihydroxy-,(meso) $C_4H_{10}O_2$					$CH_3CH(OH)CH(OH)CH_3$		L
BA_4^-	-14.35	0.431(25°,1)	-39.7	0-35	T	CL=0-0.9M A=OH^-	67Cb	b,q:±1.7
	-30.08	0.663(25°,1-2)	-88	0-35	T	CL=0-0.9M A=OH^-	67Cb	b,q:±2.9
H_5TeO_6	-13.4	1.22	-21	25	T	μ=0.1 R:$H_5TeO_6^-$ +L=$HTeO_4L$+$2H_2O$	62Ea	
B179	BUTANE, 2,3-dihydroxy-2,3-dimethyl $C_6H_{14}O_2$					$CH_3C(OH)(CH_3)C(OH)(CH_3)CH_3$		L
BA_4^-	-29.66	1.041(25°,1)	-79	0-35	T	CL=0-0.9M A=OH^-	67Cb	b,q:±2.9
	-37.66	2.334(25°,1-2)	-84	0-35	T	CL=0-0.9M A=OH^-	67Cb	b,q:±3.8
H_5TeO_6	-12.1	0.59	-29	25	T	μ=0.1 R:$H_5TeO_6^-$+ L=$HTeO_4L^-$+$2H_2O$	62Ea	
B180	1,4-BUTANEDIOL, diethyl ether $C_8H_{18}O_2$					$C_2H_5O(CH_2)_4OC_2H_5$		
SnA_4	-51.5	(1)	--	--	C	S:Benzene CM= 0.05-0.08M A=Cl	68Ge	u
B181	2,3-BUTANEDIOL, 2-methyl- $C_5H_{12}O_2$					$CH_3CH(OH)C(OH)(CH_3)_2$		L
BA_4^-	-20.50	1.255(25°,1)	-46	0-35	T	CL=0-0.9M A=OH^-	67Cb	b,q:±2.1
	-33.47	2.322(25°,1-2)	-67	0-35	T	CL=0-0.9M A=OH^-	67Cb	b,q:±3.3
B182	1,3-BUTANEDIONE, 1-phenyl- $C_{10}H_{10}O_2$					$C_6H_5C(O)CH_2C(O)CH_3$		
Co^{3+}	-0.71	-0.421(56°)	-10.5	56-104	T	S:Chlorobenzene R:CoL_3(trans)= CoL_3(cis)	70Gg	b,q: ±1.4
Ni^{2+}	-10.9	9.13(25°,1)	138	15-40	T	S:75v% Dioxane μ=0	71Rc	b,q: ±2.1
	-10.5	7.76(25°,2)	113	15-40	T	S:75v% Dioxane μ=0	71Rc	b,q: ±2.1
NiL_2	225.02	6.0	854	30	C	S:Benzene CM=0.006-0.062M R:$3NiL_2$=Ni_3L_6	73Ag	q
Cu^{2+}	-28.4	12.42(25°,1)	142	15-40	T	S:75v% Dioxane μ=0	71Rc	b
	-27.6	10.81(25°,2)	113	15-40	T	S:75v% Dioxane μ=0	71Rc	b
B183	1-BUTANESULPHONIC ACID, 4-amino- $C_4H_{11}O_3NS$					$NH_2(CH_2)_4SO_3H$		L^-
Ag^+	-23.05	3.379(1)	-12.6	25	C	μ=0.5M(KNO_3)	72Va	
	-29.96	3.702(2)	-29.7	25	C	μ=0.5M(KNO_3)	72Va	

Metal Ion	ΔH, kJ /mol	Log K	ΔS, J /mol K	T, °C	M	Conditions	Ref.	Re-marks
B184	2-BUTANETHIOL		$C_4H_{10}S$			$CH_3CH_2CHSHCH_3$		
Cu^{2+}	-57.36	7.565(25°,1-2)	-50.96	10-30	T	0.2M($NaClO_4$) S:25v%Ethanol	78Sd	
Ag^{+}	-28.20	7.045(25°,1)	38.70	10-30	T	0.2M($NaClO_4$) S:25v%Ethanol	78Sd	
Hg^{2+}	-56.11	7.842(25°,1-2)	-41.25	10-30	T	0.2M($NaClO_4$) S:25v%Ethanol	78Sd	
B185	BUTANOIC ACID, amide, 3-oxo- N-(o-chlorophenyl)-		$C_{10}H_{10}O_2NCl$			$C_6H_4(Cl)[NHC(O)CH_2C(O(CH_3]$		L
Fe^{3+}	-11.97	5.91(25°)	73.09	15-35	T	S:50w% Ethanol μ=0.1, pH=1.5-3.8 R:Fe^{3+}+2HL= FeL_2^{+}+$2H^{+}$	72Ta	b,j
B186	BUTANOIC ACID, amide, 3-oxo- 4-(2,4-dimethylphenyl)-		$C_{12}H_{15}O_2N$			$C_6H_3(CH_3)_2[NHC(O)CH_2C(O)CH_3]$		L
Fe^{3+}	-16.11	5.86(25°)	58.12	15-35	T	S:50w% Ethanol μ=0.1 pH=1.8-3.8 R:Fe^{3+}+2HL= FeL_2^{+}+$2H^{+}$	72 Ta	b,j
B187	BUTANOIC ACID, amide, 3-oxo- N-(o-methoxyphenyl)-		$C_{11}H_{13}O_3N$			$C_6H_4(OCH_3)[NHC(O)CH_2C(O)CH_3]$		L
Fe^{3+}	-18.53	5.81(25°)	48.99	15-35	T	S:50w% Ethanol μ=0.1 pH=1.5-3.8 R:Fe^{3+}+2HL= FeL_2^{+}+$2H^{+}$	72Ta	b,j
B188	BUTANOIC ACID, amide, 3-oxo- N-(o-tolyl)-		$C_{11}H_{13}O_3N$			$C_6H_4(CH_3)[NHC(O)CH_2C(O)CH_3]$		L
Fe^{3+}	-6.86	2.31(25°)	21.21	15-35	T	S:50w% Ethanol μ=0.1 pH=1.5-3.8 R:Fe^{3+}+HL= FeL^{2+}+H^{+}	72Ta	b,j
B189	BUTANOIC ACID, ethyl ester		$C_6H_{12}O_2$			$CH_3CH_2CH_2CO_2CH_2CH_3$		L
$[NiA_2]_2$	-18.0	1.15(100°,1)	-26	75-125	T	S:Squalane A=3-Trifluoro-acetylcamphorate	74Sj	b,gg
SbA_5	-70.12	(1)	--	25	C	S:1,2-Dichloro-ethane A=Cl	63Oa	

Metal Ion	ΔH, kJ /mol	Log K	ΔS, J /mol K	T,°C	M	Conditions	Ref.	Remarks
B190	BUTANOIC ACID, methyl ester		$C_5H_{10}O_2$			$CH_3CH_2CH_2CO_2CH_3$		
$[NiA_2]_2$	-17.4	1.01(100°,1)	-27	75-125	T	S:Squalane A=3-Trifluoro-acetylcamphorate	74Sj	b,gg
B191	BUTANOIC ACID, nitrile		C_4H_7N			$CH_3CH_2CH_2CN$		L
I_2	-7.20	-0.284(25°,1)	-29.7	--	T	S:Carbon disulfide	72Ma	b,j, u:25
	-11.84	-0.294(25°,1)	-45.60	--	T	S:CCl_4 CL=0.2-1.0M CM=$2x10^{-3}$M Data analyzed by method 1.	72Ma	j,u:25
	-12.43	-0.294(25°,1)	-47.32	--	T	S:CCl_4 CL=0.2-1.0M CM=$2x10^{-3}$M Data analyzed by method 2.	72Ma	j,u:25
	-12.09	-0.032(25°,1)	-41.0	--	T	S:Cyclohexane	72Ma	b,j,u:25
	-13.39	0.005(25°,1)	-44.77	--	T	S:n-Heptane CL=0.2-1.0M CM=$2x1^{-3}$M Data analyzed by method 1.	72Ma	j,u:25
	-16.36	0.005(25°,1)	-54.68	--	T	S:n-Heptane CL=0.2-1.0M CM=$2x10^{-3}$M Data analyzed by method 2.	72Ma	j,u:25
	-17.74	-0.032(25°,1)	-60.2	--	T	S:n-Hexane	72Ma	b,j,u:25
	-15.44	-0.387(25°,1)	-59.0	--	T	S:Tetrachloro-ethylene	72Ma	b,j, u:25
B192	BUTANOIC ACID, 2-amino		$C_4H_9O_2N$			$CH_3CH_2CHNH_2CO_2H$		L^-
Be^{2+}	-32.2	6.80(25°,1)	21	20-40	T	μ=0.1M($NaClO_4$)	77Sg	b
	-15.5	5.90(25°,2)	63	20-40	T	μ=0.1M($NaClO_4$)	77Sg	b
Y^{3+}	25.9	5.04(25°,1)	184	20-35	T	μ=0.1M(KNO_3)	78Sj	b,q:±2.1
La^{3+}	24.7	4.27(25°,1)	163	20-35	T	μ=0.1M(KNO_3)	78Sj	b,q:±2.5
Ce^{3+}	17.2	4.67(25°,1)	146	20-35	T	μ=0.1M(KNO_3)	78Sj	b,q:±2.1
Pr^{3+}	22.6	4.86(25°,1)	167	20-35	T	μ=0.1M(KNO_3)	78Sj	b,q:±1.7
Nd^{3+}	27.6	5.01(25°,1)	188	20-35	T	μ=0.1M(KNO_3)	78Sj	b,q:±2.1
Sm^{3+}	18.0	5.26(25°,1)	159	20-35	T	μ=0.1M(KNO_3)	78Sj	b,q:±2.1
Co^{2+}	-5.0	4.21(1)	63	25	C	μ=0.05M(KCl)	72Gb 72Ga	
	-10.5	4.31(15°,1)	46.4	15&40	T	μ=0.2	65Sb 64Sd	a,b
	-9.2	3.50(2)	29	25	C	μ=0.05M(KCl)	72Gb 72Ga	
	-12.6	3.19(15°,2)	18.0	15&40	T	μ=0.2	65Sb	a,b
	-23.0	7.50(15°,1-2)	64.4	15&40	T	μ=0.2	65Sb	a,b

Metal Ion	ΔH, kJ /mol	Log K	ΔS, J /mol K	T,°C	M	Conditions	Ref.	Remarks
B192, continued								
Ni^{2+}	-11.3	5.35(1)	63	25	C	μ=0.05M(KCl)	72Gb 72Ga	
	-17.2	5.22(1)	42.2	25	C	μ=0.2M(KCl)	74Sn	
	-20.1	5.46(25°,1)	37.2	15-40	T	μ=0.2	65Sb 64Sd	b
	-11.7	4.41(2)	46	25	C	μ=0.05M(KCl)	72Gb 72Ga	
	-16.7	4.31(2)	26.4	25	C	μ=0.2M(KCl)	74Sn	
	-20.1	4.36(25°,2)	16.3	15-40	T	μ=0.2	65Sb	b
	-40.2	9.82(25°,1-2)	53.1	15-40	T	μ=0.2	65Sb	b
Cu^{2+}	-18.4	8.01(1)	92	25	C	μ=0.05M(KCl)	72Gb 72Ga	
	-22.59	8.02(1)	79	25	C	μ=0.2M(KCl)	73Ga	
	-22.6	8.21(25°,1)	80.3	15-40	T	μ=0.2	65Sb 64Sd	b
	-22.6	6.75(1)	54	25	C	μ=0.05M(KCl)	72Gb 72Ga	
	-27.6	6.70(2)	38	25	C	μ=0.2M(KCl)	73Ga	
	-22.2	6.72(25°,2)	54.4	15-40	T	μ=0.2	65Sb	b
	-44.8	14.93(25°,1-2)	134.3	15-40	T	μ=0.2	65Sb	b
	-30.1	16.13	208	25	C	μ=0.50($NaClO_4$) R: $Cu^{2+}+L^{-}+A^{3-}=$ $CuLA^{2-}$ A=2-Hydroxy-5-sulfobenzoic acid	75Mc	
Ag^{+}	-10.0	3.30(25°,1)	29	20-40	T	μ=0.1M(KNO_3)	77Sh	b
	-12.1	4.24(25°,2)	42	20-40	T	μ=0.1M(KNO_3)	77Sh	b
Zn^{2+}	-8.8(1)	4.50	58	25	C	μ=0.05M(KCl)	72Gb 72Ga	
	-13.0	4.78(25°,1)	47.7	15-40	T	μ=0.2	65Sb 65Se	b
	-6.7(2)	4.15	54	25	C	μ=0.05M(KCl)	72Gb 72Ga	
	-13.8	3.90(25°,2)	28.0	15-40	T	μ=0.2	65Sb	b
	-26.8	8.68(25°,1-2)	75.3	15-40	T	μ=0.2	65Sb	b
Cd^{2+}	-13.0	3.46(25°,1)	21	20-40	T	μ=0.1M($NaClO_4$)	77Sg	b
	-14.2	3.40(25°,2)	21	20-40	T	μ=0.1M($NaClO_4$)	77Sg	b
Hg^{2+}	-18.8	6.55(25°,1)	63	20-40	T	μ=0.1M($NaClO_4$)	77Sg	b
	-11.7	5.47(25°,2)	67	20-40	T	μ=0.1M($NaClO_4$)	77Sg	b
In^{3+}	-27.2	16.24(25°,1-2)	222	20-40	T	μ=0.1M($NaClO_4$)	77Sh	b
B193	BUTANOIC ACID, 3-amino-		$C_4H_9O_2N$			$CH_3CH(NH_2)CH_2CO_2H$		L^-
Co^{2+}	-7.1	3.52(25°,1)	43.9	15-40	T	μ=0.2	65Sb 64Sd	b
	-13.8	2.36(25°,1)	-0.8	15-40	T	μ=0.2	65Sb	b
	-20.9	5.88(25°,1-2)	42.7	15-40	T	μ=0.2	65Sb	b
Ni^{2+}	-11.30	4.56(1)	49.4	25	C	μ=0.16	70Lb	

Metal Ion	ΔH, kJ /mol	Log K	ΔS, J /mol K	T,°C	M	Conditions	Ref.	Remarks
B193, continued								
Ni^{2+}, cont.	-15.9	4.60(25°,1)	35.1	15-40	T	μ=0.2	65Sb	b
							64Sd	
	-12.09	3.30(2)	22.6	25	C	μ=0.16	70Lb	
	-20.1	3.32(25°,2)	4.2	15-40	T	μ=0.2	65Sb	b
	-36.0	7.92(25°,1-2)	30.5	15-40	T	μ=0.2	65Sb	b
Cu^{2+}	-20.67	7.12(1)	66.9	25	C	μ=0.16	70Lb	
	-21.9	7.18(25°,1)	67.8	15-40	T	μ=0.2	65Sb	b
							64Sd	
	-23.35	4.73(2)	31.4	25	C	μ=0.16	70Lb	
	-19.2	5.59(25°,1)	41.0	15-40	T	μ=0.2	65Sb	b
	-40.2	12.77(25°,1-2)	108.4	15-40	T	μ=0.2	65Sb	b
B194	BUTANOIC ACID, 4-amino		$C_4H_9O_2N$			$H_2NCH_2CH_2CH_2CO_2H$		
Ag^{+}	-22.8	3.42(1)	-10.5	25	C	0.5M(KNO_3)	70Va	
	-22.8	3.42(25°,1)	-10.5	15-40	T	0.5M(KNO_3)	70Va	b
	-29.8	5.17(2)	-18.0	25	C	0.5M(KNO_3)	70Va	
	-26.8	5.17(25°,2)	-18.8	15-40	T	0.5M(KNO_3)	70Va	b
B195	BUTANOIC ACID, 4-amino-N-glycyl-		$C_6H_{12}O_3N_2$			$NH_2CH_2C(O)NHCH_2CH_2CH_2CO_2H$		
Ni^{2+}	-19.2	4.11(1)	14.2	25	C	μ=0.1M(KNO_3)	71Ld	
	-17.6	3.38(2)	7.1	25	C	μ=0.1M(KNO_3)	71Ld	
Cu^{2+}	-25.5	5.66(1)	22.6	25	C	μ=0.1M(KNO_3)	72Bl	
B196	BUTANOIC ACID, 2-amino-4-hydroxy-		$C_4H_9O_3N$			$HOCH_2CH_2CH(NH_2)CO_2H$		L^-
Ni^{2+}	-16.40	5.51(1)	50.6	25	C	μ=0.16	70Lb	
	-18.74	4.60(2)	25.1	25	C	μ=0.16	70Lb	
	-22.97	3.26(3)	-14.6	25	C	μ=0.16	70Lb	
Cu^{2+}	-22.80	8.00(1)	76.6	25	C	μ=0.16	70Lb	
	-24.85	6.69(2)	44.8	25	C	μ=0.16	70Lb	
B197	BUTANOIC ACID, 4-amino-3-hydroxy-		$C_4H_9O_3N$			$NH_2CH_2CH(OH)CH_2CO_2H$		L^-
Ni^{2+}	-10.9	3.99(1)	39.3	25	C	μ=0.16	70Lb	
	-11.34	3.18(2)	23.0	25	C	μ=0.16	70Lb	
Cu^{2+}	-20.38	6.48(1)	55.6	25	C	μ=0.16	70Lb	
	-23.22	6.06(2)	38.1	25	C	μ=0.16	70Lb	
B198	BUTANOIC ACID, 2-(dihydroxyamino)-		$C_4H_9O_4N$			$CH_3CH_2CH[N(OH)_2]CO_2H$		L^-
Ce^{3+}	-85.8	12.26(30°,1-3)	-42	30-50	T	μ=0.1M($NaClO_4$)	73Kb	b
Ce^{4+}	-102.9	14.35(30°,1-3)	-63	30-50	T	μ=0.1M($NaClO_4$)	73Kb	b

Metal Ion	ΔH, kJ /mol	Log K	ΔS, J /mol K	T,°C	M	Conditions	Ref.	Remarks
B199	BUTANOIC ACID, 3-oxo, amide, N-(4-bromophenyl)			$C_{10}H_{10}O_2NBr$		$CH_3CO_2CH_2CONHC_6H_4Br$		
SiA_4	-48.356	(1-3)	--	25	C	S:1,2-Dichloro-ethane A=Cl	79Ka	
SbA_5	-51.5	(1)	--	25	C	S:1,2-Dichloro-ethane A=Cl	76Kb	
B200	BUTANOIC ACID, 3-oxo, amide, N-(2-chlorophenyl)			$C_{10}H_{10}O_2NCl$		$CH_3COCH_2CONHC_6H_4Cl$		L
SiA_4	-45.978	(1-3)	--	25	C	S:1,2-Dichloro-ethane A=Cl	79Ka	
SbA_5	-33.6	(1)	--	25	C	S:1,2-Dichloro-ethane A=Cl	76Kb	
B201	BUTANOIC ACID, 3-oxo, amide, N-(3-chlorophenyl)			$C_{10}H_{10}O_2NCl$		$CH_3COCH_2CONHC_6H_4Cl$		L
SiA_4	-49.483	(1-3)	--	25	C	S:1,2-Dichloro-ethane A=Cl	79Ka	
SbA_5	-42.0	(1)	--	25	C	S:1,2-Dichloro-ethane A=Cl	76Kb	
B202	BUTANOIC ACID, 3-oxo, amide, N-(4-chlorophenyl)			$C_{10}H_{10}O_2NCl$		$CH_3COCH_2CONHC_6H_4Cl$		
SiA_4	-52.007	(1-3)	--	25	C	S:1,2-Dichloro-ethane A=Cl	79Ka	
SbA_5	-51.9	(1)	--	25	C	S:1,2-Dichloro-ethane A=Cl	76Kb	
B203	BUTANOIC ACID, 3-oxo, amide, N-(4-fluorophenyl)			$C_{10}H_{10}O_2NF$		$CH_3COCH_2CONHC_6H_4F$		L
SiA_4	-57.030	(1-3)	--	25	C	S:1,2-Dichloro-ethane A=Cl	79Ka	
SbA_5	-59.2	(1)	--	25	C	S:1,2-Dichloro-ethane A=Cl	76Kb	
B204	BUTANOIC ACID, 3-oxo, amide, N-(4-iodophenyl)			$C_{10}H_{10}O_2NI$		$CH_3COCH_2CONHC_6H_4I$		L
SiA_4	-46.826	(1-3)	--	25	C	S:1,2-Dichloro-ethane A=Cl	79Ka	
SbA_5	-51.8	(1)	--	25	C	S:1,2-Dichloro-ethane A=Cl	76Kb	

Metal Ion	ΔH, kJ /mol	Log K	ΔS, J /mol K	T,°C	M	Conditions	Ref.	Re-marks
B205	BUTANOIC ACID, 3-oxo, amide, N-(2-methoxyphenyl)	$C_{11}H_{13}O_3N$				$CH_3COCH_2CONHC_6H_4OCH_3$		L
SiA_4	-55.424	(1-3)	--	25	C	S:1,2-Dichloro-ethane A=Cl	79Ka	
SbA_5	-67.1	(1)	--	25	C	S:1,2-Dichloro-ethane A=Cl	76Kb	
B206	BUTANOIC ACID, 3-oxo, amide, N-(3-methoxyphenyl)	$C_{11}H_{13}O_3N$				$CH_3COCH_2CONHC_6H_4OCH_3$		L
SiA_4	-63.172	(1-3)	--	25	C	S:1,2-Dichloro-ethane A=Cl	79Ka	
SbA_5	-103.8	(1)	--	25	C	S:1,2-Dichloro-ethane A=Cl	76Kb	
B207	BUTANOIC ACID, 3-oxo, amide, N-(4-methoxyphenyl)	$C_{11}H_{13}O_3N$				$CH_3COCH_2CONHC_6H_4OCH_3$		L
SiA_4	-55.973	(1-3)	--	25	C	S:1,2-Dichloro-ethane A=Cl	79Ka	
SbA_5	-70.2	(1)	--	25	C	S:1,2-Dichloro-ethane A=Cl	76Kb	
B208	BUTANOIC ACID, 3-oxo, amide, N-(2-methylphenyl)	$C_{11}H_{13}O_2N$				$CH_3COCH_2CONHC_6H_4CH_3$		L
SiA_4	-56.130	(1-3)	--	25	C	S:1,2-Dichloro-ethane A=Cl	79Ka	
SbA_5	-46.2	(1)	--	25	C	S:1,2-Dichloro-ethane A=Cl	76Kb	
B209	BUTANOIC ACID, 3-oxo, amide, N-(3-methylphenyl)	$C_{11}H_{13}O_2N$				$CH_3COCH_2CONHC_6H_4CH_3$		L
SiA_4	-55.982	(1-3)	--	25	C	S:1,2-Dichloro-ethane A=Cl	79Ka	
SbA_5	-64.4	(1)	--	25	C	S:1,2-Dichloro-ethane A=Cl	76Kb	
B210	BUTANOIC ACID, 3-oxo, amide, N-(4-methylphenyl)	$C_{11}H_{13}O_2N$				$CH_3COCH_2CONHC_6H_4CH_3$		L
SiA_4	-56.572	(1-3)	--	25	C	S:1,2-Dichloro-ethane A=Cl	79Ka	
SbA_5	-67.6	(1)	--	25	C	S:1,2-Dichloro-ethane A=Cl	76Kb	

Metal Ion	ΔH, kJ /mol	Log K	ΔS, J /mol K	T, °C	M	Conditions	Ref.	Re-marks
B211	BUTANOIC ACID, 3-oxo, amide, N-(3-nitrophenyl)		$C_{10}H_{10}O_4N_2$			$CH_3C(O)CH_2C(O)NHC_6H_4NO_2$		
SbA_5	-33.5	(1)	--	25	C	S:1,2-Dichloro-ethane A=Cl	76Kb	
B212	BUTANOIC ACID, 3-oxo, amide, N-phenyl		$C_{10}H_{11}O_2N$			$CH_3COCH_2CONHC_6H_5$		L
SiA_4	-54.508	(1-3)	--	25	C	S:1,2-Dichloro-ethane A=Cl	79Ka	
SbA_5	-61.9	(1)	--	25	C	S:1,2-Dichloro-ethane A=Cl	76Kb	
B213	BUTANOIC ACID, 3-oxo-2-bromo, amide, N-(4-methylphenyl)		$C_{11}H_{12}O_2NBr$			$CH_3C(O)CH(Br)C(O)NHC_6H_4CH_3$		
SbA_5	-12.4	(1)	--	25	C	S:1,2-Dichloro-ethane A=Cl	76Kb	
B214	BUTANOIC ACID, 3-oxo-2-chloro, amide, N-(4-chlorophenyl)		$C_{10}H_9O_2NCl_2$			$CH_3C(O)CH(Cl)C(O)NHC_6H_4Cl$		
SbA_5	2.08	(1)	--	25	C	S:1,2-Dichloro-ethane A=Cl	76Kb	
B215	BUTANOIC ACID, 3-oxo-2-chloro, amide, N-(4-methoxyphenyl)		$C_{11}H_{12}O_3NCl$			$CH_3C(O)CH(Cl)C(O)NHC_6H_4OCH_3$		
SbA_5	-12.9	(1)	--	25	C	S:1,2-Dichloro-ethane A=Cl	76Kb	
B216	BUTANOIC ACID, 3-oxo-2-chloro, amide, N-(4-methylphenyl)		$C_{11}H_{12}O_2NCl$			$CH_3C(O)CH(Cl)C(O)NHC_6H_4CH_3$		
SbA_5	-8.06	(1)	--	25	C	S:1,2-Dichloro-ethane A=Cl	76Kb	
B217	BUTANOIC ACID, 3-oxo-2-chloro, amide, N-phenyl		$C_{10}H_{10}O_2NCl$			$CH_3C(O)CH(Cl)C(O)NHC_6H_5$		
SbA_5	0.02	(1)	--	25	C	S:1,2-Dichloro-ethane A=Cl	76Kb	
B218	BUTANOIC ACID, 3-oxo-2-methyl, amide, N-(4-methylphenyl)		$C_{12}H_{15}O_2N$			$CH_3C(O)CH(CH_3)C(O)NHC_6H_4CH_3$		
SbA_5	-82.1	(1)	--	25	C	S:1,2-Dichloro-ethane A=Cl	76Kb	

Metal Ion	ΔH, kJ /mol	Log K	ΔS, J /mol K	T,°C	M	Conditions	Ref.	Remarks
B219	BUTANOIC ACID, 3-thio, amide, N-(4-methylphenyl)		$C_{11}H_{13}ONS$			$CH_3C(S)CH_2C(O)NHC_6H_4CH_3$		
SbA_5	-121.6	(1)	--	25	C	S:1,2-Dichloroethane A=Cl	76Kb	
B220	2-BUTANONE		C_4H_8O			$CH_3COCH_2CH_3$		L
$[NiA_2]_2$	-20.5	1.41(100°,1)	-28	75-125	T	S:Squalane A=3-Trifluoroacetylcamphorate	74Sj	b,gg
SbA_5	-72.92	(1)	--	25	C	Excess $SbCl_5$ CL low, variable, ∿0.02M S:1,2-Dichloroethane A=Cl	64Oa	
TeA_4	-22.6	(1)	--	25	C	S:Benzene A=Cl	72Pg	
I_2	-14.6	0.013(25°,1)	-47.7	5-25	T	S:Heptane	71Td	b
B221	3-BUTANONE, 2-amino-2-methyl-, oxime		$C_5H_{12}ON_2$			$CH_3C(CH_3)(NH_2)C(NOH)CH_3$		
Ni^{2+}	-35.6	6.08	-4	25	C	μ<0.06 R:Ni^{2+}+2L=$[NiL_2-H]^+$+H^+	64Wb	
	-70.7	--	--	25	C	μ<0.06 R:$[NiL_2-H]^+$+$3H^+$=Ni^{2+}+$2LH^+$	64Wb	
Cu^{2+}	-47.3	7.84	-8	25	C	μ<0.06 R:Cu^{2+}+2L=$[CuL_2-H]^+$+H^+	64Wb	
	-59.0	--	--	25	C	μ<0.06 R:$[CuL_2-H]^+$+$3H^+$=Cu^{2+}+$2LH^+$	64Wb	
B222	2-BUTANONE, 3,3-dimethyl-		$C_6H_{12}O$			$CH_3COC(CH_3)_3$		L
SbA_5	-70.92	(1)	--	25	C	Excess $SbCl_5$ CL low, variable ∿0.02M S:1,2-Dichloroethane A=Cl	64Oa	
I_2	-15.1	0.057(25°,1)	-49.4	5-25	T	S:Heptane	71Td	b
B223	3-BUTANONE, 2-ethylamino-2-methyl-, oxime		$C_7H_{16}ON_2$			$CH_3C(CH_3)(NHCH_2CH_3)C(:NOH)CH_3$		L
Ni^{2+}	10.9	--	--	25	C	μ<0.06 R:Ni^{2+}+2L=$[NiL_2-H]^+$+H^+	64Wb	
	-99.6	--	--	25	C	μ<0.06 R:$[NiL_2-H]^+$+$3H^+$=Ni^{2+}+$2HL^+$	64Wb	

Metal Ion	ΔH, kJ /mol	Log K	ΔS, J /mol K	T,°C	M	Conditions	Ref.	Re-marks
B224	2-BUTANONE, 3-methyl-		$C_5H_{10}O$			$CH_3COCH(CH_3)_2$		L
SbA_5	-71.42	(1)	--	25	C	Excess $SbCl_5$ CL low, variable ~0.02M S:1,2-Dichloroethane A=Cl	64Oa	
I_2	-15.1	0.033(25°,1)	-49.8	5-25	T	S:Heptane	71Td	b
B225	3-BUTANONE, 2-methylamino-2-methyl-, oxime		$C_6H_{14}ON_2$			$CH_3C(CH_3)(NHCH_3)C(:NOH)CH_3$		L
Ni^{2+}	-2.9	3.59	58	25	C	μ<0.06 R:Ni^{2+}+ 2L=$[NiL_2\text{-}H]^+$+H^+	64Wb	
	-83.3	--	--	25	C	μ<0.06M R:$[NiL_2\text{-}H]^+$+$3H^+$=Ni^{2+}+$2HL^+$	64Wb	
Cu^{2+}	-36.4	6.96	13	25	C	μ<0.06 R:Cu^{2+}+ 2L=$[CuL_2\text{-}H]^+$+H^+	64Wb	
	-49.6	--	--	25	C	μ<0.06M R:$[CuL_2\text{-}H]^+$+$3H^+$=Cu^{2+}+$2HL^+$	64Wb	
B226	3-BUTANONE, 2-methyl-2-(n-propylamino)-, oxime		$C_8H_{18}ON_2$			$CH_3C(CH_3)(NHCH_2CH_2CH_3)C(:NCH)CH_3$		L
Ni^{2+}	8.4	1.1	50	25	C	μ<0.06M R:Ni^{2+}+ 2L=$[NiL_2\text{-}H]^+$+H^+	64Wb	
	-98.7	--	--	25	C	μ<0.06M R:$[NiL_2\text{-}H]^+$+$3H^+$=Ni^{2+}+$2HL^+$	64Wb	
B227	1-BUTENE		C_4H_8			$CH_3CH_2CH:CH_2$		L
$RhAB_2$	4.2	-1.036(25°)	-7.1	0&25	T	S:Toluene A=2,4-Pentanedionato B=Ethylene R:L+$RhAB_2$=RhABL+B	67Cc	a,b,q: ±5.8
PdA_4^{2-}	0	1.093(20°)	21	5-25	T	μ=4.0 R:PdA_4^{2-}+ L=PbA_3L^-+A^- A=Cl	65Pb	b
	0	0.53(20°)	10.5	5-25	T	μ=4.0 R:PdA_4+ L+H_2O=$PdA_2L(OH)_2$ +2A A=Cl	65Pb	b
Ag^+	-24.89	2.08(1)	-43.35	25	T	μ=0.1M(KNO_3)	72Pf	
	15.5	0.944(25°,1)	-33.5	0-40	T	S:Ethylene glycol	65Ci	b
AgAB	-1.2	-0.301(25°)	-8.4	0&25	T	S:Toluene A=NO_3 B=Ethylene R:L+AgAB= AbAL+B	67Cc	a,b,i: 65Ci

Metal Ion	ΔH, kJ /mol	Log K	ΔS, J /mol K	T,°C	M	Conditions	Ref.	Remarks
B228	2-BUTENE(cis)		C_4H_8			$CH_3CH:CHCH_3$		L
$RhAB_2$	7.5	-2.383(25°)	-20.5	0&25	T	S:Toluene A=2,4-Pentanedionato B=Ethylene R:L+$RhAB_2$=RhABL+B	67Cc	a,b,q: ±3.3
Ag^+	14.2	0.690(25°,1)	-34.3	0-40	T	S:Ethylene glycol	65Ci	b
	-0.4	-0.553(25°)	-9.2	0-25	T	S:Toluene A=NO_3 B=Ethylene R:L+AgAB=AgAL+B	67Cc	a,b,i: 65Ci
B229	2-BUTENE(trans)		C_4H_8			$CH_3CH:CHCH_3$		
$RhAB_2$	7.9	-2.699(25°)	-25.5	0&25	T	S:Toluene A=2,4-Pentanedionato B=Ethylene R:L+$RhAB_2$=RhABL+B	67Cc	a,b,q: ±2.9
Ag^+	10.9	0.204(25°,1)	-32.2	0-40	T	S:Ethylene glycol	65Ci	b
B230	2-BUTENE, 2,3-dimethyl-		C_6H_{12}			$(CH_3)_2C:C(CH_3)_2$		L
Ag^+	7.9	-0.468(25°,1)	-35.6	0-40	T	S:Ethylene glycol	65Ci	b
B231	1-BUTENE, 2-methyl-		C_5H_{10}			$CH_3CH_2C(CH_3):CH_2$		L
Ag^+	14.6	0.644(25°,1)	-36.0	0-40	T	S:Ethylene glycol	65Ci	b
B232	1-BUTENE, 3-methyl-		C_5H_{10}			$CH_3CH(CH_3)CH:CH_2$		
Ag^+	16.3	0.903(25°,1)	-36.8	0-40	T	S:Ethylene glycol	65Ci	b
B233	2-BUTENOIC ACID (trans)		$C_4H_6O_2$			$CH_3CH:CHCO_2H$		
Tl^{1+}	-54.4	1.30(28°,1)	-215	28-48	T	--	76Sa	b
B234	2-BUTEN-1-ol(cis)		C_4H_8O			$CH_3CH:CHCH_2OH$		L
CuA_2B	-13.10	0.992(25°)	-25.1	0-30	T	S:2-Butenyl alcohol A=trans-2-Butenyl alcohol B=H_2O R:CuA_2B^++2L=CuL_2A^++2A	72Ib	b

Metal Ion	ΔH, kJ /mol	Log K	ΔS, J /mol K	T,°C	M	Conditions	Ref.	Re-marks
B235	2-BUTEN-1-OL (trans)		C_4H_8O			$CH_3CH:CHCH_2OH$		
Ag^+	-20.8	0.81(25°,1)	-54.0	25-45	T	μ∿0	74Ha	
	-19.6	-0.08(25°,1)	-63.7	25-45	T	S:Ethanol	77Ha	q:±4.6
	-15.8	0.55(25°,1)	-42.6	25-45	T	S:Methanol	77Ha	q:±1.4
	-17.8	1.47(25°,1)	-31.5	25-45	T	S:Propylene carbonate	77Ha	q:±1.6
B236	3-BUTEN-1-OL		C_4H_8O			$CH_2:CHCH_2CH_2OH$		
Ag^+	-23.8	1.80(25°,1)	-44.9	25-45	T	μ∿0	77Ha	
	-17.7	1.34(25°,1)	-33.5	25-45	T	S:Ethanol	77Ha	
	-18.6	1.67(25°,1)	-30.2	25-45	T	S:Methanol	77Ha	
	-33.2	2.81(25°,1)	-57.4	25-45	T	S:Propylene carbonate	77Ha	
B237	3-BUTEN-2-ONE		C_4H_6O			$CH_2:CHC(O)CH_3$		L
NiAB	-57.3	2.58(27°)	-141.8	24-41	T	S:Tetrahydrofuran A=Dipyridyl B=Tetrahydrofuran R:NiAB+L=NiAL+B	71Ya	b
B238	3-BUTEN-2-ONE, 4-amino-3-methyl-		C_5H_9ON			$CH_3COC(CH_3):CHNH_2$		L^-
Co^{2+}	7.99	-0.493(25°)	17.36	-33-37	T	S:15v% Tetra-methylsilane in $CHCl_3$	68Ea	b,t
B239	2-BUTEN-1-ONE, 3-amino-1-phenyl-		$C_{10}H_{11}ON$			$C_6H_5COCH:C(NH_2)CH_3$		L^-
Co^{2+}	6.86	0.100(25°)	24.94	-38-57	T	S:15v% Tetra-methylsilane in $CHCl_3$	68Ea	b,t
B240	2-BUTEN-1-ONE, 3-methylamino-1-(β-naphthyl)-		$C_{15}H_{15}ON$			β-$C_{10}H_7COCH:C(NHCH_3)CH_3$		L^-
Ni^{2+}	9.623	1.35(25°)	6.53	57-102	T	S:$CDCl_3$	68Ea	b,t
B241	2-BUTEN-1-ONE, 3-methylamino-1-phenyl		$C_{11}H_{13}ON$			$C_6H_5COCH:C(NHCH_3)CH_3$		L^-
Ni^{2+}	20.12	-1.730(25°)	34.39	-33-87	T	S:$CDCl_3$	68Ea	b,t
B242	3-BUTEN-2-ONE, 3-methyl-4-methylamino-		$C_6H_{11}O$			$CH_3COC(CH_3):CHNHCH_3$		L^-
Ni^{2+}	23.5	-3.26(25°)	16.40	67-127	T	S:$CDCl_3$	68Ea	b,t

Metal Ion	ΔH, kJ /mol	Log K	ΔS, J /mol K	T,°C	M	Conditions	Ref.	Remarks
B243	BUTYROHYDROXAMIC ACID, N-phenyl		$C_5H_{11}O_3N$			$CH_3CH_2CH_2CH_2COONOH$		
Zn^{2+}	12.34	7.85(25°,1)	-109.2	25-35	T	μ=0 S:50v% Dioxane	71Aa	b
	17.61	6.40(25°,2)	-61.5	25-35	T	μ=0 S:50v% Dioxane	71Aa	b

C

Metal Ion	ΔH, kJ /mol	Log K	ΔS, J /mol K	T,°C	M	Conditions	Ref.	Remarks
C1	CADMIUMATE(II) ION, tetra-chloro-		Cl_4Cd			$CdCl_4^{2-}$		L^{2-}
Na^+	-30.5	0.30(1)	109	25	T	4M(Li,Na,ClO_4,Cl)	63Mg	
K^+	-28.9	0.15(1)	96	25	T	4M(Li,Na,ClO_4,Cl)	63Mg	
Rb^+	-28.4	0.26(1)	92	25	T	4M(Li,Na,ClO_4,Cl)	63Mg	
Cs^+	-26.3	0.34(1)	84	25	T	4M(Li,Na,ClO_4,Cl)	63Mg	
NH_4^+	-23.0	0.04(25°,1)	-75	15-65	T	μ=1	63Md	b
C2	CAMPHOR(d)		$C_{10}H_{16}O$			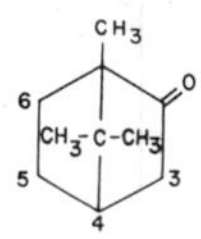		L
[Fe]	36.4	5.47(4.6°,1)	236.0	0-13	T	μ=0.1M(KCl) pH= 7.4 [Fe]=Ferric Cytochrome P-450	72Gg	b
	0	5.72(21.0°,1)	109.6	13-40	T	μ=0.1M(KCl) pH= 7.4 [Fe]=Ferric Cytochrome P-450	72Gg	b
C3	CAMPHOR, thiono		$C_{10}H_{16}S$			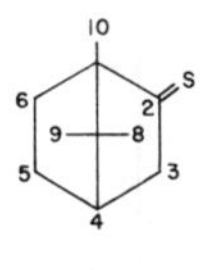		
I_2	-46.0	1.978(25°,1)	-116.3	0-45	T	S:Cyclohexane	66Bc	b,q:±4
C4	CARBAMIC ACID, N,N-benzyl-dithio		$C_{15}H_{15}NS_2$			$(C_6H_5CH_2)_2NC(S)(SH)$		
Cd^{2+}	-20	2.398	-43	30	C	S:Benzene R:$2CdL_2=Cd_2L_4$	76Af	

Metal Ion	ΔH, kJ /mol	Log K	ΔS, J /mol K	T,°C	M	Conditions	Ref.	Re-marks
C5 CARBAMIC ACID, N,N-dibutyl-dithio			$C_9H_{19}NS_2$			$(C_4H_9)_2NC(S)(SH)$		
Cd^{2+}	-15	1.301	-38	30	C	S:Benzene R:$2CdL_2{=}Cd_2L_4$	76Af	
C6 CARBAMIC ACID, N,N-diethyl-dithio			$C_5H_{11}NS_2$			$(C_2H_5)_2NC(S)(SH)$		
Cd^{2+}	-40	2.398	-100	30	C	S:Benzene R:$2CdL_2{=}Cd_2L_4$	76Af	
C7 CARBAMIC ACID, N,N-dimethyl-, choride			C_3H_6ONCl			$(CH_3)_2NCOCl$		L
SbA_2	-72.05	(1)	--	25	C	S:Ethylene chloride A=Cl	67Da	
C8 CARBAMIC ACID, N,N-dimethyl-, ethyl ester			$C_5H_{11}O_2N$			$(CH_3)_2NC(O)OC_2H_5$		L
SbA_2	-93.60	(1)	--	25	C	S:Ethylene chloride A=Cl	67Da	
SbA_5	-83.7	(1)	--	25	C	S:CCl_4 A=Cl	73Da	
C9 CARBAMIC ACID, N-isopropyl-, thiono-, methyl ester			$C_5H_{11}ONS$			$(CH_3)_2CHNHC(S)OCH_3$		L
I_2	-32.6	2.204(25°,1)	-67.4	0-45	T	S:Cyclohexane	66Bc	b,q:±4
C10 CARBAMIC ACID, N-methyl-N-benzyl-dithio-			$C_9H_{11}NS_2$			$C_6H_5CH_2N(CH_3)C(S)SH$		
FeL_3A	-0.63	0.477(25°)	7.1	-54-36	T	S:CH_2Cl_2 A=BF_4 R:Cis→Trans	73Ea	b
C11 CARBAMIC ACID, N-methyl-N-isopropyl-dithio-			$C_5H_{11}ONS_2$			$(CH_3)_2CHN(CH_3)C(O)SH$		
FeL_3	3.01	0.613(25°)	21.8	-54-34	T	S:CH_2Cl_2 R:Cis→Trans	73Ea	b
FeL_3A	0.71	0.462(25°)	11.3	-56-43	T	S:CH_2Cl_2 A=BF_4 R:Cis→Trans	73Ea	b
C12 CARBAMIC ACID, N-methyl, thiono-, methyl ester			C_3H_7ONS			$CH_3NHC(S)OCH_3$		L
I_2	-30.5	3.910(25°,1)	-28.0	0-45	T	S:Cyclohexane	66Bc	b,q:±4

Metal Ion	ΔH, kJ /mol	Log K	ΔS, J /mol K	T,°C	M	Conditions	Ref.	Re-marks
C13 CARBONATE ION			CO_3			CO_3^{2-}		L^{2-}
CrA_3	19.20	-0.046(1)	63.6	25	C	μ=3.0($NaClO_4$) A=Ethylenediamine	73Ua	
	13.81	-0.070(2)	45.2	25	C	μ=3.0($NaClO_4$) A=Ethylenediamine	73Ua	
CoA_3^{3+}	1.05	2.86(1)	57.3	25	C	μ=0 A=Ethylenediamine	76Me	q:±0.2
	3.34	1.47(1)	38.5	25	C	μ=0.1($NaClO_4$) A=Ethylenediamine	76Me	
	3.56	1.25(1)	36.0	25	C	μ=0.2($NaClO_4$) A=Ethylenediamine	76Me	
	3.8	1.06(1)	33.0	25	C	μ=0.3($NaClO_4$) A=Ethylenediamine	76Me	q:±0.4
	4.39	0.92(1)	32.2	25	C	μ=0.4($NaClO_4$) A=Ethylenediamine	76Me	
	4.27	0.84(1)	30.1	25	C	μ=0.5($NaClO_4$) A=Ethylenediamine	76Me	
	-4.81	0.46(1)	-25.1	25	C	μ=3.0M($NaClO_4$+ Na_2L) A=Ethylenediamine	73Md	q:±0.4
	-2.9	1.61(2)	21	25	C	μ=0 A=Ethylenediamine	76Me	q:±0.8
	-2.9	0.59(2)	0	25	C	μ=0.1($NaClO_4$) A=Ethylenediamine	76Me	q:±0.4
	-3.8	0.81(2)	≤ 4	25	C	μ=0.2($NaClO_4$) A=Ethylenediamine	76Me	q:±0.8
	-3.8	0.66(2)	≤ 6	25	C	μ=0.3($NaClO_4$) A=Ethylenediamine	76Me	q:±1.2
	-4.2	0.44(2)	-8	25	C	μ=0.4($NaClO_4$) A=Ethylenediamine	76Me	q:±0.8
	-4.2	0.37(2)	-8	25	C	μ=0.5($NaClO_4$) A=Ethylenediamine	76Me	q:±0.8
	-1.26	0.146(2)	-8	25	C	μ=3.0M($NaClO_4$+ Na_2L) A=Ethylenediamine	73Md	q:±0.4
	0.42	0.88(3)	16.7	25	C	μ=0 A=Ethylenediamine	76Me	q:±0.4
	<2	0.73(3)	~12	25	C	μ=0.2($NaClO_4$) A=Ethylenediamine	76Me	
	0	0.73(3)	~12	25	C	μ=0.3($NaClO_4$) A=Ethylenediamine	76Me	
	0	0.95(3)	~16	25	C	μ=0.4($NaClO_4$) A=Ethylenediamine	76Me	
	0	0.37(3)	~8	25	C	μ=0.5($NaClO_4$) A=Ethylenediamine	76Me	
	6.07	-0.40(3)	29	25	C	μ=3.0M($NaClO_4$+ Na_2L) A=Ethylenediamine	73Md	q:±1.2
CoA_5B^{2+}	4.0	0.40(1)	21	25	C	μ=3.0M($NaClO_4$) A=NH_3 B=NO_2	74Mb	
	1.9	0.12(2)	8.4	25	C	μ=3.0M($NaClO_4$) A=NH_3 B=NO_2	74Mb	

Metal Ion	ΔH, kJ /mol	Log K	ΔS, J /mol K	T,°C	M	Conditions	Ref.	Remarks
C13, continued								
CoA_5B^{2+}, cont.	0.63	0.07(3)	4.2	25	C	μ=3.0M($NaClO_4$) A=NH_3 B=NO_2	74Mb	
	-1.67	0.25(4)	0	25	C	μ=3.0M($NaClO_4$) A=NH_3 B=NO_2	74Mb	
	3.6	0.13(5)	12	25	C	μ=3.0M($NaClO_4$) A=NH_3 B=NO_2	74Mb	
C14 CARBONIC ACID, ethyl ester		$C_3H_6O_3$				$HCO_3CH_2CH_3$		L
SbA_5	-66.86	(1)	--	25	C	S:1,2-Dichloro-ethane A=Cl	63Oa	
C15 CARBONIC ACID, methyl ester		$C_2H_4O_3$				HCO_3CH_3		L
SbA_5	-63.47	(1)	--	25	C	S:1,2-Dichloro-ethane A=Cl	63Oa	
C16 CARBONIC ACID, chloro-, ethyl ester		$C_3H_5O_2Cl$				$ClCO_2CH_2CH_3$		L
SbA_5	-89.5	(1)	--	25	C	Excess $SbCl_5$ CL low, variable ~0.02M S:Ethyl-ene chloride A=Cl	67Oa	
C17 CARBON MONOXIDE		CO				CO		L
$[MoAL_3]_2$	1.84	0.418(25°)	14.14	-15-62	T	S:Acetone A=(h^5-C_5H_5) R:Gauche→Trans Transformation	73Ab	b,q: ±0.63
[Fe]	-88.7	(1)	-161.1	0-30	T	0.001M THAM buffer pH=8.5 [Fe]=Sperm whale myoglobin. Standard state for L = unit mole fraction in organic solvent.	71Ka	b
$RhABC_2$	-8	1.477(20°,1)	-0.4	10-60	T	S:Chlorobenzene A=I B=CO C=Ph_3P	71Vb	b
$IABC_2$	-4	3.653(20°,1)	55.6	10-60	T	S:Chlorobenzene A=I B=CO C=Ph_3P	71Vb	b
C18 CHALCONE, 2'-hydroxy-5'-bromo		$C_{15}H_{11}O_2Br$				$C_6H_5CH{:}CHCO[C_6H_3(Br)(OH)]$		
UO_2^{2+}	-16.90	9.044(25°,1-2)	229.7	25-45	T	S:Ethanol pH=5.6	79Sk	b

Metal Ion	ΔH, kJ /mol	Log *K*	ΔS, J /mol K	T,°C	M	Conditions	Ref.	Remarks
C19 CHLORATE ION			ClO_3			ClO_3^-		L^-
K^+	-42.34	-1.277(1)	-129.3	25	T	μ=0	34La	
Mn^{2+}	-5.21	-0.27(1)	-22.6	25	C	μ=1M	75Ah	
Co^{2+}	-2.73	0.21(1)	-5.1	25	C	μ=1M	75Ah	
Ni^{2+}	-5.30	-0.35(1)	-24.5	25	C	μ=1M	75Ah	
Cu^{2+}	-6.51	-0.34(1)	-28.4	25	C	μ=1M	75Ah	
Zn^{2+}	-4.37	-0.12(1)	-16.9	25	C	μ=1M	75Ah	
C20 CHLORIDE ION			Cl			Cl^-		L^-
Li^+	61.5	12.06(1)	439	25	T	CM≅10^{-4}M S:25w% Butanol - 75w% Hexane	61Rd	
Na^+	-4.263	-1.850(1)	-49.8	25	T	μ=0	34La	
	70.7	12.70(100°)	430.5	25-300	T	S:Dioxane-Water mixtures R:Na^++L^-=NaL+7.8H_2O	72La	b
	29.3	17.31(500°)	369.4	300-800	T	R:Na^++L^-= NaL+10.2H_2O		
K^+	-17.234	-0.885(1)	-74.9	25	T	μ=0	34La	
Sc^{3+}	-6.3	1.08(1)	0	25	T	μ=0.5($NaClO_4$)	62Pa	
	-7.1	1.03(2)	-4	25	T	μ=0.5($NaClO_4$)	62Pa	
Y^{3+}	-1.2	0.36(1)	-4	25	T	μ=0.5($NaClO_4$)	62Pa	
Ce^{3+}	22.6	0.0(1)	75.3	25	C	--	67Ab 62Mi	
Pr^{3+}	11.5	1.66(25°,1)	71	15-45	T	S:n-Propyl alcohol	73Ld	b,q
Sm^{3+}	-6.7	0.13(25°,1)	-20	0-50	T	μ=1M(ClO_4^-)	75Mf	
Eu^{3+}	-0.21	-0.05(1)	-1.7	25	T	μ=1.0 Data in 65Ce recalculated from 63Cb which contains an error in the unitary entropy value	65Ce 63Cb	
	-13.4	0.07(25°,1)	-43.5	0-50	T	μ=1M(ClO_4^-)	75Mf	
Gd^{3+}	-14.6	0.09(25°,1)	-47.3	0-50	T	μ=1M(ClO_4^-)	75Mf	
Tb^{3+}	-7.5	0.11(25°,1)	-23.0	0-50	T	μ=1M(ClO_4^-)	75Mf	q:±0.8
Dy^{3+}	-4.2	0.11(25°,1)	-11.7	0-50	T	μ=1M(ClO_4^-)	75Mf	q:±0.8
VAB_4^{2+}	31.4	-1.92(27°)	68.2	20-130	T	μ=0 A=O B=H_2O R:VAB_4^{2+}+L^-= VAB_3L^++B	71Zb	b

Metal Ion	ΔH, kJ /mol	Log K	ΔS, J /mol K	T, °C	M	Conditions	Ref.	Remarks
C20, continued								
VAB_3L^+	10.9	-3.00(27°)	-20.9	--	T	μ=0 A=0 B=H_2O R:VAB_3L^++L^-= VAB_2L_2+B	71Zb	b
Cr^{3+}	26.3	(1)	--	25	C	0.5M($HClO_4$)	76Dd	
	-1.8	-0.056(25°,1)	-7.1	10-50	T	μ=1.0($HClO_4$)	68Mf	b,q: ±0.96
	25.5	-0.710(1)	72.0	25	T	μ≅4.4 [Cl^-]=0	58Ga	
	18.8	-0.982(1)	44.3	25	T	μ≅4.4 [Cl^-]=4.4	58Ga	
	27.6	-0.66(1)	79	25	C	μ=5.1($NaClO_4$)	67Ab 58Se	
	18.0	-1.54(2)	31.4	25	T	μ≅4.4	58Ga	
	20.9	(2)	--	25	C	μ=5.1($NaClO_4$)	58Se	
$CrA_3{}^{3+}$	22.32	1.62(35°)	103.51	25-45	T	μ=0 A=2-Guanidinobenzimidazole C_6H_4-NH-C(=N)-NH-C(=NH)-NH_2	79Mf	b
$CrA_6{}^{3+}$	31.0	-0.745(43.6°,1)	83.7	30-60	T	0M%CH_3OH in water at μ= 0.418 A=H_2O	64Bb	b
	31.0	-0.623(43.6°,1)	85.8	30-60	T	0.048M%CH_3OH in water at μ= 0.418 A=H_2O	64Bb	b
	31.0	-0.426(43.6°,1)	90.0	30-60	T	0.128M%CH_3OH in water at μ= 0.418 A=H_2O	64Bb	b
	31.0	-0.267(43.6°,1)	92.5	30-60	T	0.190M%CH_3OH in water at μ= 0.418 A=H_2O	64Bb	b
	31.0	-0.011(43.6°,1)	96.6	30-60	T	0.306M%CH_3OH in water at μ= 0.418 A=H_2O	64Bb	b
	31.0	0.193(43.6°,1)	101.2	30-60	T	0.391M%CH_3OH in water at μ= 0.418 A=H_2O	64Bb	b
	31.0	0.415(43.6°,1)	105.8	30-60	T	0.505M%CH_3OH in water at μ= 0.418 A=H_2O	64Bb	b
	31.0	0.785(43.6°,1)	113.0	30-60	T	0.625M%CH_3OH in water at μ= 0.418 A=H_2O	64Bb	b
	31.0	1.04(43.6°,1)	117.6	30-60	T	0.706M%CH_3OH in water at μ= 0.418 A=H_2O	64Bb	b
	25.1	-0.27(60°)	65.7	40-80.5	T	0.98M($HClO_4$) R:$CrA_6{}^{3+}$+L^-= CrA_5L^{2+}+A A=H_2O	67Ha	b
	23.4	-0.054(60°)	65.7	40-80.5	T	4.00M($HClO_4$) R:$CrA_6{}^{3+}$+L^-= CrA_5L^{2+}+A A=H_2O		

Metal Ion	ΔH, kJ /mol	Log K	ΔS, J /mol K	T,°C	M	Conditions	Ref.	Re-marks
C20, continued								
CrA_6^{3}, cont.	15.1	0.26(60°)	54.4	40-80.5	T	6.70M($HClO_4$) R:CrA_6^{3+}+L^-= CrA_5L^{2+}+A A=H_2O	67Ha	b
	13.0	0.54(60°)	57.7	40-80.5	T	10.0M($HClO_4$) CrA_6^{3+}+L^-= CrA_5L^{2+}+A A=H_2O	67Ha	b
$CrAL_2$	-34.3	--	--	--	C	0.1M($CaCl_2$) R:$CrAL_2$+HL= CrL_3+H_2O A=OH	10Ba	
	-40.2	--	--	--	T	0.1M($CaCl_2$) R:$CrAL_2$+HL= CrL_3+H_2O A=OH	10Ba	
CrA_2L	-33.723	--	--	--	T	0.1M($CaCl_2$) R:CrA_2L+HL= $CrAL_2$+H_2O A=OH	10Ba	
$HCrO_4^-$	4.73	1.05	36.0	15-35	T	μ=1.0 R:H^++L^-+ $HCrO_4^-$=CrO_3L^-+ H_2O	66Tb	b
	-21.76	1.515(25°)	-48.62	26&38	T	μ=2.0M($NaClO_4$) R:$HCrO_4^-$+2H^++L= $HCrO_3L$+H_2O	71Yc	b
CrA_5B	25	0.030	79	45	T	μ=0.106M A=NH_3 B=H_2O R:CrA_5B+ L^-=CrA_5L^{2+}+B	67Df	
U^{4+}	-30.1	0.025(25°,1)	-100	16-56	T	μ=4($HClO_4$,HCl)	74Bi	
	-28.9	0.02(25°,1)	-87.9	16-56	T	μ=4M($HClO_4$,HCl)	74Bj	
UO_2^{2+}	15.9	-0.06(1)	50.2	25	T	μ=2.0($NaClO_4$)	54Db	
NpO_2^{2+}	-36.4	-0.72(0°,1)	-134.7	0-9.84	T	μ=3.0 Difference in ΔH due to mechanism assumed to obtain rate constant	55Cb	b
	-28.9	0.51(0°,1)	-88	0-9.84	T	μ=3.0 Difference in ΔH due to mechanism assumed to obtain rate constant	55Cb	b
	14.6	1.02(0°,2)	67	0-9.84	T	μ=3.0	55Cb	b
Pu^{3+}	19.2	0.57(1)	75	25	C	μ=0.1	58Ma	
Pu^{4+}	7.9	-0.25(1)	21	25	T	1M(HCl)	55Ra	
PuO_2^{2+}	13.8	-0.25(20.2°,1)	41.4	2.4-29.6	T	μ=2.0	61Ra	b
Mn^{2+}	0	-0.736(25°,1)	-14.1	25-60	T	μ=5($NaClO_4$+$HClO_4$)	76An	
Fe^{2+}	0	-0.648(25°,1)	-12.4	25-60	T	μ=5($NaClO_4$+$HClO_4$)	76An	
Fe^{3+}	19.413	(1)	92.5	10	C	μ=0	67Vb	ΔCp=255

Metal Ion	ΔH, kJ /mol	Log K	ΔS, J /mol K	T,°C	M	Conditions	Ref.	Remarks
C20, continued								
Fe^{3+}, cont.	21.506	(1)	98.7	18	C	μ=0	67Vb	ΔCp=242
	23.430	1.38(1)	105.0	25	C	μ=0	67Vb	ΔCp=242
	35.5	1.48(1)	146	25	T	μ=0	42Ra 52Ua	
	35.6	0.95(1)	137.6	25	T	μ=0	77Nc	
	25.773	(1)	112.5	35	C	μ=0	67Vb	ΔCp=234
	27.949	(1)	119.7	45	C	μ=0	67Vb	ΔCp=226
	41.6	1.10(1)	150.6	50	T	μ=0	77Nc	
	30.125	(1)	126.4	55	C	μ=0	67Vb	ΔCp=222
	47.3	1.16(1)	159.8	70	T	μ=0	77Nc	
	53.1	1.38(1)	172.4	90	T	μ=0	77Nc	
	56.1	1.49(1)	178.6	100	T	μ=0	77Nc	
	64.0	1.80(1)	194.6	125	T	μ=0	77Nc	
	72.0	2.13(1)	211.3	150	T	μ=0	77Nc	
	20.9	0.64(1)	100	25	T	μ=0.5	61Se	
	17.6	0.46(1)	69.4	25	C	μ=1.0($HClO_4$)	62Wa	
	25.1	0.83(25°,1)	100.0	22-45	-	μ=1.382	59Cb	b
	13.807	(1)	56.5	10	C	μ=2.0M($HClO_4$)	67Vb	ΔCp=181
	15.355	(1)	61.5	18	C	μ=2.0M($HClO_4$)	67Vb	ΔCp=172
	16.652	0.53(1)	65.7	25	C	μ=2.0M($HClO_4$)	67Vb	ΔCp=172
	17.824	(1)	71.1	35	C	μ=2.0M($HClO_4$)	67Vb	ΔCp=163
	19.623	(1)	76.1	45	C	μ=2.0M($HClO_4$)	67Vb	ΔCp=159
	21.129	(1)	82.0	55	C	μ=2.0M($HClO_4$)	67Vb	ΔCp=155
	20.3	0.518(22°,1)	78	22-60	T	μ=2($HClO_4$)	76An	
	12.134	(1)	54.8	10	C	μ=3.0M($HClO_4$)	67Vb	ΔCp=159
	13.347	(1)	59.0	18	C	μ=3.0M($HClO_4$)	67Vb	ΔCp=155
	13.933	0.76(1)	62.8	25	C	μ=3.0M($HClO_4$)	67Vb	ΔCp=151
	15.732	(1)	67.4	35	C	μ=3.0M($HClO_4$)	67Vb	ΔCp=142
	15.9	0.73(25°,1)	67	25-45	C	μ=3.0($HClO_4$)	59Ca	b
	17.7	(1)	72.0	45	C	μ=3.0M($HClO_4$)	67Vb	ΔCp=138
	18.744	(1)	76.1	55	C	μ=3.0M($HClO_4$)	67Vb	ΔCp=134
	18.6	0.747(22°,1)	77	22-60	T	μ=3($HClO_4$)	76An	
	10.209	(1)	54.4	10	C	μ=4.0M($HClO_4$)	67Vb	ΔCp=134
	11.056	(1)	58.2	18	C	μ=4.0M($HClO_4$)	67Vb	ΔCp=130
	12.092	1.10(1)	61.1	25	C	μ=4.0M($HClO_4$)	67Vb	ΔCp=126
	13.389	(1)	65.3	35	C	μ=4.0M($HClO_4$)	67Vb	ΔCp=121
	14.602	(1)	69.4	45	C	μ=4.0M($HClO_4$)	67Vb	ΔCp=117
	15.941	(1)	73.2	55	C	μ=4.0M($HClO_4$)	67Vb	ΔCp=113
	14.6	1.037(22°,1)	69	22-60	T	μ=4($HClO_4$)	76An	
	7.698	(1)	54.4	10	C	μ=5.0M($HClO_4$)	67Vb	ΔCp=105
	8.577	(1)	57.3	18	C	μ=5.0M($HClO_4$)	67Vb	ΔCp=100
	8.954	1.50(1)	59.4	25	C	μ=5.0M($HClO_4$)	67Vb	ΔCp=100
	9.749	(1)	62.8	35	C	μ=5.0M($HClO_4$)	67Vb	ΔCp=96
	11.046	(1)	66.1	45	C	μ=5.0M($HClO_4$)	67Vb	ΔCp=96
	12.008	(1)	69.0	55	C	μ=5.0M($HClO_4$)	67Vb	ΔCp=92
	16.8	1.009(22°,1)	76	22-60	T	μ=5(4M $NaClO_4$, 1M $HClO_4$)	76An	
	13.7	1.245(22°,1)	70	22-60	T	μ=5(2M $NaClO_4$, 3M $HClO_4$)	76An	
	10.5	1.464(22°,1)	63	22-60	T	μ=5($HClO_4$)	76An	
	-0.920	(1)	--	0	C	μ=7.0M($HClO_4$)	67Vb	q:±0.12
	-0.008	(1)	46.9	10	C	μ=7.0M($HClO_4$)	67Vb	q:±0.10 ΔCp=96

Metal Ion	ΔH, kJ /mol	Log K	ΔS, J /mol K	T,°C	M	Conditions	Ref.	Re-marks
C20, continued								
Fe^{3+}, cont.	0.950	(1)	50.2	18	C	μ=7.0M($HClO_4$)	67Vb	q:±0.12 ΔCp=92
	1.665	2.46(1)	52.7	25	C	μ=7.0M($HClO_4$)	67Vb	ΔCp=92
	2.314	(1)	54.8	35	C	μ=7.0M($HClO_4$)	67Vb	ΔCp=88
	3.297	(1)	58.2	45	C	μ=7.0M($HClO_4$)	67Vb	ΔCp=84
	4.125	(1)	60.7	55	C	μ=7.0M($HClO_4$)	67Vb	ΔCp=84
	-3.217	3.13(1)	49.0	25	C	μ=8.0M($HClO_4$)	67Vb	q:±0.33
	-9.6	3.44(25°,1)	34.3	25-45	T	μ=8.5($HClO_4$)	59Ca	b
	17.6	0.94(1)	77.0	25	T	μ=0.5 S:D_2O	61Se	
	19.5	-0.46(22°,2)	59	22-60	T	μ=5($HClO_4$+$NaClO_4$)	76An	
Co^{2+}	12.1	-0.770(1)	25.5	27	T	μ=0.5-16M(HCl) R:$Co(H_2O)_6^{2+}$+L^-= $CoL(H_2O)_5^+$+H_2O	68Za	q:±4.2
	2.18	0.137(1)	4.64	25	C	μ=2.0	66Kb	
	4.2	-0.176(25°,1)	10.5	12-40	T	μ=2.0	60Lb	b
	2.09	0.119(1)	4.39	40	C	μ=2.0	66Kb	
	3.8	-0.41(1)	~4	25	C	μ=3.0N($LiClO_4$) R:$Co(H_2O)_6^{2+}$+L^-= $Co(H_2O)_6L^+$	74Bf	
	0	-0.280(25°,1)	-5.4	21-60	T	μ=5($NaClO_4$+$HClO_4$)	76An	
	8.8	-2.770(2)	-23.4	27	T	μ=0.5-16M(HCl) R:$CoL(H_2O)_5^+$+L^-= $CoL_2(H_2O)_2$+$3H_2O$	68Za	q:±4.2
	47.3	-2.507(3)	109.6	27	T	μ=0.5-16M(HCl) R:$CoL_2(H_2O)_2$+L^-= $CoL_3(H_2O)^-$+H_2O	68Za	q:±5.4
	-32.6	17.62(3)	36.4	25	C	S:Acetone 0.01M(LiCl)	80Ba	
	3.35	-2.056(4)	-28.0	27	T	μ=0.5-16M(HCl) R:$CoL_3(H_2O)^-$+L^-= CoL_4^{2-}+H_2O	68Za	q:±1.7
	-9.2	6.41(4)	22.2	25	C	S:Acetone 0.01M(LiCl)	80Ba	
CoA_6^{3+}	18.07	2.59(1)	109	25	T	μ=0 A=NH_3	55Na 53Ea	
	-0.09	0.081(1)	1.3	25	C	0.2M($LiClO_4$,LiCl) A=NH_3	70Me	q:±2
	2.5	(1)	--	25	C	0-0.4M(NaCl) A=NH_3	62Fa 53Ya	
	24.3	-1.185(1)	46	25	T	0-0.9M(NaCl) A=NH_3	62Fa	
	23.4	-0.29(1)	75.3	25	T	1M($NaClO_4$&NaCl) A=NH_3	62Me	
cis-$CoAB_2S$	3.89	4.179	92.5	30	T	μ=0 S:N,N-Di methylformamide R:cis-$CoAB_2S$+L= $CoALB_2^+$+S A=Cl B=Ethylenedi-amine S=Solvent	66La	

Metal Ion	ΔH, kJ /mol	Log *K*	ΔS, J /mol K	T,°C	M	Conditions	Ref.	Re-marks
C20, continued								
cis-$CoAB_2S$, cont.	-15.9	6.083	64.4	30	T	μ=0 S:N,N-Dimethylformamide R:cis-$CoAB_2S$+2L=$CoAL_2B_2$+S A=Cl B=Ethylenediamine S=Solvent	66La	
cis-$CoA_2B_2^+$	-9.6	3.712(25°,1)	39.3	20-30	T	μ=0 S:N,N-Dimethylformamide A=Ethylenediamine B=Cl	66Md	b
CoA_3B^-	-18.0	3.44(25°)	5.5	12-43	T	μ=0.1(Bu_4NClO_4) S:Acetone R:CoA_3B^-+L^-=CoA_3L^{2-}+B A=Cl^- B=Acetone	75Tc	b
	-14.2	2.87(25°)	7.3	12-43	T	μ=0.1(Bu_4NClO_4) S:Acetonitrile R:CoA_3B^-+L^-=CoA_3L^{2-}+B A=Cl^- B=Acetonitrile	75Tc	b
	-20.9	3.53(25°)	-2.5	12-43	T	μ=0.1(Bu_4NClO_4) S:Benzonitrile R:CoA_3B^-+L^-=CoA_3L^{2-}+B A=Cl B=Benzonitrile	75Tc	b
	-7.5	1.50(25°)	3.5	12-43	T	μ=0.1(Bu_4NClO_4) S:Dimethylacetamide R:CoA_3B^-+L^-=CoA_3L^{2-}+B A=Cl^- B=Dimethylacetamide	75Tc	b, q:±1.7
	-5.8	1.20(25°)	3.4	12-43	T	μ=0.1(Bu_4NClO_4) S:Dimethylformamide R:CoA_3B^-+L^-=CoA_3L^{2-}+B A=Cl^- B=Dimethylformamide	75Tc	b, q:±2.1
	-8.8	4.83(25°)	63	12-43	T	μ=0.1(Bu_4NClO_4) S:Nitromethane R:CoA_3B^-+L^-=CoA_3L^{2-}+B A=Cl^- B=Nitromethane	75Tc	b,q:±2
CoA_4L_2	56.78	--	--	25-63	T	μ=0.354M(LiCl) S:CH_3OH A=CH_3OH R:CoL_2A_4+L^-=CoL_3A^-+3A	67Sa	b,aa
	48.95	--	--	25-69	T	μ=8.47M(LiCl) A=H_2O R:CoL_2A_4+L^-=CoL_3A^-+3A	67Sa	b,aa
CoA_5B	1.7	0.176(25°)	6.7	25-86	T	CM≅2x10^{-3}M A=NH_3 B=H_2O	67Lb	b

Metal Ion	ΔH, kJ /mol	Log K	ΔS, J /mol K	T, °C	M	Conditions	Ref.	Re-marks
C20, continued								
CoA_5B, cont.						R: CoA_5B^{3+}---L^- = CoA_5L^{2+}---H_2O (ligand interchange between inner and outer coordination spheres)		
Ni^{2+}	2.09	-0.174(1)	3.64	25	C	μ=2.0	66Kb	
	5.23	0.569(25°,1)	28.4	12-40	T	μ=2.0	60Ld	b
	1.84	-0.155(1)	2.55	40	C	μ=2.0	66Kb	
	9.6	-0.56(1)	21	25	C	μ=3.0N($LiClO_4$)	74Bf	
						R: $Ni(H_2O)_6{}^{2+}$+L^- = $Ni(H_2O)_6{}^{+}$		
	10.5-10.9	≅-0.48(4)	--	19-70	T	3.6M(Ni^{2+})	56Ka	b,r
NiA_2L_2	6.862	-0.394(25°)	15.5	-18-52	T	S:Chloroform A=[(p-C_6H_5)(p-C_6H_5)MeP]	70Pc	b,t
	8.954	-0.626(25°)	18.03	-18-52	T	S:Chloroform A=[(p-C_6H_5)(p-ClC_6H_4)MeP]	70Pc	b,t
	7.657	-0.356(25°)	18.79	-13-54	T	S:Chloroform A=[(p-C_6H_5)(p-MeC_6H_4)MeP]	70Pc	b,t
	7.238	-0.260(25°)	19.37	0-70	T	S:Chloroform A=[(p-C_6H_5)(p-$MeOC_6H_4$)MeP]	70Pc	b,t
	5.899	-0.0088(25°)	19.58	-5-59	T	S:Chloroform A=[(p-C_6H_5)(p-$Me_2NC_6H_4$)MeP]	70Pc	b,t
	7.615	-0.274(25°)	20.33	-7-50	T	S:Chloroform A=[(p-MeC_6H_4)(p-MeC_6H_4)MeP]	70Pc	b,t
	5.774	-0.194(25°)	15.73	-8-59	T	S:Chloroform A=[(p-$MeOC_6H_4$)(p-$MeOC_6H_4$)MeP]	70Pc	b,t
NiA_4L_2	61.25	--	--	40-85	T	μ=2.4M(LiBr) S:C_2H_5OH A=C_2H_5OH R: NiL_2A_4+L^- = NiL_3A^-+3A	67Sa	b,aa
	54	--	--	110-175	T	S:Dimethylsulfoxide A=DMSO R: NiL_2A_4+L^- = NiL_3A^-+3A	75Gi	
	55.6	1.176(100°)	172	63-160	T	S:Dimethyl Sulphoxide A=Dimethyl Sulphoxide R: NiA_4L_2+L^- = $NiAL_3{}^-$+3A	70Gk	b,aa

Metal Ion	ΔH, kJ /mol	Log K	ΔS, J /mol K	T,°C	M	Conditions	Ref.	Remarks
C20, continued								
RuA_2L^+	-9.2	4.137(25°,1)	49.0	25-49	T	S:N,N-Dimethyl- acetamide μ=0 A=Tiphenylphos- phine	74Ja	b
RhA_2BC	2.5	2.699	58	50	T	μ=0.2M A=Ethyl- enediamine B=Cl $C=H_2O$ R:trans- $RhA_2BC^{2+}+L^-=$ trans-RhA_2BL^++C	67Bc	q:±1.7
	-5.0	3.222	46	50	T	μ=0.2M A=Ethyl- enediamine B=Br $C=H_2O$ R:trans- $RhA_2BC^++L^-=$ trans-RhA_2BL^++C	67Bc	q:±2.9
	-6.7	3.000	33	50	T	μ=0.2M A=Ethyl- enediamine B=I $C=H_2O$ R:trans- $RhA_2BC^{2+}+L^-=$ trans-RhA_2BL^++C	67Bc	q:±2.1
RuA_5B^{3+}	7.1	2.272(64°)	63	35-90	T	μ=0.1+0.1M p-toluenesul- fonic acid K values good to ±30% R:RuA_5B +L=RuA_5L+B $A=NH_3$ $B=H_2O$	64Be	b
	14.6	--	113	77- 97.5	T	μ=0.01-0.07M $A=NH_3$ $B=H_2O$ R:$RhA_5B^{3+}+L^-=$ RhA_5L^{2+}+B	68Lb	b
	5.4	--	--	--	-	$A=NH_3$ $B=H_2O$ R:$RhA_5B^{3+}+L^-=$ RhA_5L^{2+}+B	67Pa	u,q: ±2.5
Pd^{2+}	-12.68	4.47(1	43.1	25	C	μ=1M($HClO_4$)	72Re	
	-10.84	3.29(2)	26.8	25	C	μ=1M($HClO_4$)	72Re	
	-10.71	2.41(3)	10.0	25	C	μ=1M($HClO_4$)	72Re	
	-14.27	1.37(4)	-21.8	25	C	μ=1M($HClO_4$)	72Re	
	-11.7	1.42(25°,4)	-12.1	10-45	T	μ=1.0	66Sb 57De	b
	-11.7	1.59(25°,4)	-8.8	15-40	T	μ=2.0($LiClO_4$) CM=$1x10^{-4}$M	68Le	b
	-23.0	(1-4)	--	25	C	μ=0.10	67Ic	
PdA_2L_2	22.2	0.39(32°)	79.9	22- 153	T	S:$CDCl_3$ R:cis-A_2PdL_2= trans-A_2PdL_2 $A=(C_6H_5)_2PCH_3$	73Ra	
	13.0	-0.26(32°)	38	22- 153	T	S:$CDCl_3$ R:cis-A_2PdL_2= trans-A_2PdL_2 $A=(CH_3)_2PC_2H_5$	73Ra	

Metal Ion	ΔH, kJ /mol	Log K	ΔS, J /mol K	T,°C	M	Conditions	Ref.	Remarks
C20, continued								
PdA_2L_2, cont.	33.9	-0.19(32°)	108	22-153	T	S:$C_6H_5NO_2$ R:cis-A_2PdL_2= trans-A_2PdL_2 A=$(C_6H_5)_2PCH_3$	73Ra	
	32.2	-1.12(32°)	84.5	22-153	T	S:$C_6H_5NO_2$ R:cis-A_2PdL_2= trans-A_2PdL_2 A=$(CH_3)_2PC_2H_5$	73Ra	
	23.8	-0.97(32°)	62.3	22-153	T	S:CD_3NO_2 R:cis-A_2PdL_2= trans-A_2PdL_2 A=$(C_6H_5)_2PCH_3$	73Ra	
	18.4	-0.99(32°)	67.8	22-153	T	S:CD_3NO_2 R:cis-A_2PdL_2= trans-A_2PdL_2 A=$(CH_3)_2PC_2H_5$	73Ra	
PdA_2B_2	11.7	-0.28(32°)	33.0	--	T	S:Acetone A=$PhP(CH_3)_2$ B=Cl	73Ra	q:±2.1 u:32,bb
	13.0	-0.26(32°)	37.6	--	T	S:Deuterio-chloroform A=$PhP(CH_3)_2$ B=Cl	73Ra	q:±2.1 u:32,bb
	19.2	1.12(32°)	84.5	--	T	S:Deuterio-chloroform A=Ph_2PCH_3 B=Cl	73Ra	q:±2.1 u:32,bb
	18.4	0.36(32°)	63.2	--	T	S:m=Dichloro-benzene A=$PhP(CH_3)_2$ B=Cl	73Ra	q:±2.1 u:32,bb
	19.2	-0.24(32°)	58.6	--	T	S:o-Dichloro-benzene A=$PhP(CH_3)_2$ B=Cl	73Ra	q:±2.1 u:32,bb
	32.2	-1.12(32°)	84.5	--	T	S:Nitrobenzene A=$PhP(CH_3)_2$ B=Cl	73Ra	q:±2.1 u:32,bb
	33.9	-0.19(32°)	107.5	--	T	S:Nitrobenzene A=Ph_2PCH_3 B=Cl	73Ra	q:±2.1 u:32,bb
	18.4	-0.99(32°)	67.8	--	T	S:Nitromethane A=$PhP(CH_3)_2$ B=Cl	73Ra	q:±2.1 u:32,bb
	23.8	-0.83(32°)	62.3	--	T	S:Nitromethane A=Ph_2PCH_3 B=Cl	73Ra	q:±2.1 u:32,bb
	24.3	-0.81(32°)	64.0	--	T	S:sym-Tetra-chloroethane A=$PhP(CH_3)_2$ B=Cl	73Ra	q:±2.1 u:32,bb
	26.8	-0.46(32°)	79.1	--	T	S:Tetrachloro-ethane A=Ph_2PCH_3 B=Cl	73Ra	q:±2.1 u:32,bb
PtA_2^{2+}	≅0	2.52(23°,1)	48.1	23-54	T	μ=0 S:CH_3OH A=o-Phenylene-bisdimethylar-sine	66Db	b
PtA_3^{4+}	16.3	-0.21(1)	50	25	C	3.0M($NaClO_4$+NaCl) A=Ethylenediamine	72Mf	q:±2.1
	6.7	1.24(25°,1)	46.0	10-40	T	μ→0 A=Ethylene-diamine	63Gb	b

Metal Ion	ΔH, kJ /mol	Log K	ΔS, J /mol K	T,°C	M	Conditions	Ref.	Re-marks
C20, continued								
PtA_2L_2	8.263	1.470(25°)	55.6	20-35	T	S:Benzene A=PPr_3	56Cb	b,bb
	4.937	2.243(25°)	59.4	15-70	T	S:Benzene A=$AsEt_3$	52Cb	b,bb
	10.083	0.281(25°)	39.3	15-35	T	S:Benzene A=$SbEt_3$	52Cb	b,bb
	9.205	0.600(25°)	42.2	20-35	T	S:Benzene A=$SbPr_3$	56Cb	b,bb
Cu^+	76.6	2.94(1)	313	25-40	T	μ=0	74Nc	b
	-15.1	0.31(3)	-46.4	25	C	μ=0	75Vb	
	-15.0	0.22(3)	-46.0	25	C	μ=0	75Vb	
	-14.7	0.14(3)	-45.2	25	C	μ=0	75Vb	
	-12.4	5.64(1-2)	64.8	25	C	μ=0	75Vb	
	-11.6	5.57(1-2)	67.8	25	C	μ=0	75Vb	
	-11.0	5.50(1-2)	69.9	25	C	μ=0	75Vb	
	-17.2	5.845(50°,1-2)	59.0	25-80	T	μ=10M(NH_4NO_3)	69Si	b,q:±2.1
	-36.4	5.447(50°,1-3)	-7.9	25-80	T	μ=10M(NH_4NO_3)	69Si	b,q:±2.1
	-48.5	4.863(50°,1-3)	-55.2	25-80	T	μ=10M(NH_4NO_3)	69Si	b
	-55.6	12.079(50°,2-3)	56.5	25-80	T	μ=10M(NH_4NO_3)	69Si	b,q:±4.2
	-66.9	11.845(50°,2-4)	16.3	25-80	T	μ=10M(NH_4NO_3)	69Si	b
	-73.6	11.602(50°,2-5)	-7.5	25-80	T	μ=10M(NH_4NO_3)	69Si	b
	-70.7	11.301(50°,2-6)	-6.3	25-80	T	μ=10M(NH_4NO_3)	69Si	b
	-89.6	18.176(50°,3-4)	66.5	25-80	T	μ=10M(NH_4NO_3)	69Si	b,q:±6.3
	-93.7	18.041(50°,3-5)	52.7	25-80	T	μ=10M(NH_4NO_3)	69Si	b,q:±6.3
	-111.7	17.699(50°,3-6)	-11.3	25-80	T	μ=10M(NH_4NO_3)	69Si	b
	-139.7	24.477(50°,4-5)	29	25-80	T	μ=10M(NH_4NO_3)	69Si	b,q:±10.5
	-147.3	24.255(50°,4-6)	2.5	25-80	T	μ=10M(NH_4NO_3)	69Si	b,q:±10.5
Cu^{2+}	2.51	0.114(25°,1)	10.5	25.2& 46.9	T	μ=1.0($NaClO_4$)	50Ma 60Ld 52Ua	a,b
	6.61	0.086(1)	23.72	25	C	μ=2.0	66Kb	
	8.28	0.149(1)	29.25	40	C	μ=2.0	66Kb	
	13.0	-0.346(1)	33	25	C	3.0N($LiClO_4$+ LiCl) Values are for formation of inner sphere complexes	70Bf	
	9.29	-0.060(1)	29	25	C	μ=3($NaClO_4$)	74Bg	
	12.1	-0.34(1)	33	25	C	μ=3.0N($LiClO_4$) R:$Cu(H_2O)_5{}^{2+}+L^-=Cu(H_2O)_5L^+$	74Bf	
	9.2	-0.37(1)	25	25	C	μ=3.0N($LiClO_4$) R:$Cu(H_2O)_6{}^{2+}+L^-=Cu(H_2O)_6L^+$	74Bf	
	13.56	-0.0550(1)	44.5	25	C	4N($LiClO_4$)	77Ke	
	12.0	0.013(1)	42	25	C	μ=3($NaClO_4$) S:5v%Ethanol	74Bg	
	13.8	0.155(1)	50	25	C	μ=3($NaClO_4$) S:20v%Ethanol	74Bg	
	11.1	0.623(1)	50	25	C	μ=3($NaClO_4$) S:40v%Ethanol	74Bg	

Metal Ion	ΔH, kJ /mol	Log K	ΔS, J /mol K	T,°C	M	Conditions	Ref.	Re-marks
C20, continued								
Cu^{2+}, cont.	13.6	1.01(1)	65	25	C	μ=3($NaClO_4$) S:60v%Ethanol	74Bg	
	9.41	1.79(1)	67	25	C	μ=3($NaClO_4$) S:70v%Ethanol	74Bg	
	12.1	-0.013(1)	42	25	C	μ=3($NaClO_4$) S:5v%Methanol	74Bg	
	13.4	0.057(1)	46	25	C	μ=3($NaClO_4$) S:20v%Methanol	74Bg	
	11.7	0.589(1)	50	25	C	μ=3($NaClO_4$) S:40v%Methanol	74Bg	
	14.8	1.03(1)	71	25	C	μ=3($NaClO_4$) S:60v%Methanol	74Bg	
	11.1	1.43(1)	65	25	C	μ=3($NaClO_4$) S:75v%Methanol	74Bg	
	75.3	2.91(1)	305	20	C	S:Methanol	76Ra	q:±6.3
	10.0	-0.346(1)	27.2	25	C	Values are for formation of outer sphere complexes	70Bf	
	-63.8	2.38(2)	-168	25-40	T	μ=0	74Nc	b
	53.6	1.90(2)	213	20	C	S:Methanol	76Ra	
	-19.9	-0.119(3)	-69	25	C	μ=5M($NaClO_4$) H^+=0.1M($HClO_4$)	77Ad	
	48.1	1.18(3)	184	20	C	S:Methanol	76Ra	q:±4
	15.40	0.512(3)	61.5	25	T	CL→0	65Ta	
CuA^{2+}	-9.58	0.892(25°,1)	-15.5	25-45	T	μ=0.4M(KNO_3) A=Macrocyclic ligand, rac-5,7,7,12,14,14-hexamethyl-1,4,8,11-tetra-azacyclotetra-decane, (tet b), (blue form)	76Cd	
CuA_2^{2+}	12.05	2.13(1)	81.2	25	C	S:CH_3OH A=Ethyl-enediamine	72Be	
	9.16	2.65(1)	81.6	25	C	S:CH_3OH A=Ethyl-enediamine	72Be	
	8.70	3.01(1)	86.6	25	C	S:CH_3OH A=N,N-dimethylethyl-enediamine	72Be	
	6.82	3.04(1)	81.2	25	C	S:CH_3OH A=N,N-dimethylethyl-enediamine	72Be	
Ag^+	-11.3	3.3(25°,1)	25	15-35	T	μ=0	67Ab	b
	-7.45	2.98(1)	34.7	60	T	μ=0	75Ic	
	-8.37	3.09(1)	35.5	80	T	μ=0	75Ic	
	-9.33	2.81(1)	28.8	100	T	μ=0	75Ic	
	-10.33	2.81(1)	27.4	120	T	μ=0	75Ic	
	-11.42	2.71(1)	24.2	140	T	μ=0	75Ic	
	-12.59	2.63(1)	21.3	160	T	μ=0	75Ic	
	-33.30	1.70(370°,1)	-19.2	370&436	T	μ=0 S:KNO_3	61Dc	a,b

C20, continued

Metal Ion	ΔH, kJ /mol	Log K	ΔS, J /mol K	T,°C	M	Conditions	Ref.	Remarks
Ag^+, cont.	-18.41	1.548(333°,1)	-0.75	370&374	T	μ=0 S:$NaNO_3$-KNO_3 Eutectic	61Dc	a,b
	-3.81	2.13(2)	29.3	60	T	μ=0	75Ic	
	-4.27	1.77(2)	21.7	80	T	μ=0	75Ic	
	-4.81	1.76(2)	20.8	100	T	μ=0	75Ic	
	-5.36	1.74(2)	19.8	120	T	μ=0	75Ic	
	-5.90	1.77(2)	19.6	140	T	μ=0	75Ic	
	-6.44	1.95(2)	22.4	160	T	μ=0	75Ic	
	-20.58	1.23(370°,2)	-8.4	370&436	T	μ=0 S:KNO_3	61Dc	a,b
	-32.34	1.127(333°,2)	-31.8	333&374	T	μ=0 S:$NaNO_3$-KNO_3 Eutectic	61Dc	a,b
	9.1	(3)	--	60	T	μ=0	75Ic	
	10.3	0.15(3)	31.9	80	T	μ=0	75Ic	
	11.5	0.30(3)	36.5	100	T	μ=0	75Ic	
	12.7	0.15(3)	35.2	120	T	μ=0	75Ic	
	14.0	0.47(3)	43.0	140	T	μ=0	75Ic	
	15.4	(3)	--	160	T	μ=0	75Ic	
	-23	-0.788(4)	-92	25	C	μ=5M($NaClO_4$)	77Ad	q:±10
	-16.3	5.3(25°,1-2)	46	15-35	T	μ=0	52Jc	b
	-39.2	6.150(1-3)	-14	25	C	μ=5M($NaClO_4$)	77Ad	q:±4.8
	-58	5.90(25°,1-4)	-83.3	0&25	T	Sat Sol AgCl	41Ea	a,b
Zn^{2+}	13.0	-0.139(25°,1)	41.0	15-65	T	μ=1M($LiClO_4$)	70Fb	b
	6.48	-0.176(25°,1)	18.0	15-65	T	μ=2M($LiClO_4$)	70Fb	b
	5.52	-0.19(1)	15.1	25	C	μ=3M($NaClO_4$)	69Gc	
	14.94	0.00(25°,1)	49.8	15-65	T	μ=3M($LiClO_4$)	70Fb	b
	17.11	0.147(25°,1)	59.4	15-65	T	μ=4M($LiClO_4$)	70Fb	b,q: ±1.9
	0	-0.332(1)	-6.3	25	C	μ=4.5	56Sa	n
	22.3	1.945(1)	112	25	C	μ=1M(NH_4ClO_4) S:DMSO	76Af	q:±3.8
	17.99	1.66(70°,1)	84.1	55-85	T	S:$NH_4NO_3 \cdot 2H_2O$ melt	74Ne	b
	10.5	-0.132(25°,2)	32.2	15-65	T	μ=1M($LiClO_4$)	70Fb	b
	7.4	0.081(25°,2)	25.1	15-65	T	μ=2M($LiClO_4$)	70Fb	b,q: ±0.88
	37.6	-0.42(2)	238	25	C	μ=3M($NaClO_4$)	69Gc	q:±8.4
	10.08	-0.117(25°,2)	33.0	15-65	T	μ=3M($LiClO_4$)	70Fb	b,q: ±3.7
	11.84	0.418(25°,2)	51.9	15-65	T	μ=4M($LiClO_4$)	70Fb	b,q: ±1.4
	0.8	3.890(2)	77	25	C	μ=1M(NH_4ClO_4) S:DMSO	76Af	q:±4.3
	52.63	-0.601(25°,3)	170.3	15-65	T	μ=2M($LiClO_4$)	70Fb	b,q:±7.9
	0	0.75(3)	8	25	C	μ=3M($NaClO_4$)	69Gc	q:±8.4
	17.24	0.476(25°,3)	63.6	15-65	T	μ=3M($LiClO_4$)	70Fb	b,q:±4.2
	-3.43	0.212(25°,3)	-10.9	15-65	T	μ=4M($LiClO_4$)	70Fb	b,q:±7.1
	-10.2	2.260(3)	9.1	25	C	μ=1M(NH_4ClO_4) S:DMSO	76Af	
	21.338	-0.044(1-2)	70.3	25	C	μ=4.5	56Sa	n
	80.332	-0.300(1-3)	263	25	C	μ=4.5	56Sa	n
	0.502	-0.154(1-4)	-1.46	25	C	μ=4.5	56Sa	n
	29.3	(1-4)	--	--	C	S:25w%NH_4Cl	57Ta	

Metal Ion	ΔH, kJ /mol	Log K	ΔS, J /mol K	T,°C	M	Conditions	Ref.	Remarks
C20, continued								
Cd^{2+}	0.669	1.92(1)	39.3	25	C	μ=0	72Ff	q:±0.12
	3.3	(1)	--	25	C	μ=0	70Pd	q:±2.1
	2.51	1.98(1)	46.9	25	T	μ=0.0	53Va 67Ab	
	3.347	1.971(25°,1)	49.0	5-35	T	μ=0	69Sa	b
	-5.4	2.12(40°,1)	24	35-45	T	μ=0	73Mf	b
	0.88	1.44(1)	30.5	25	C	μ=0.25($NaClO_4$)	68Gc 49Ka	
	1.151	1.33(1)	29.3	25	C	μ=0.5M($LiClO_4$& LiCl)	72Ff	q: ±0.08
	1.05	1.37(1)	29.7	25	C	μ=0.50($NaClO_4$)	68Gc 52Ua	
	1.21	1.37(1)	30.1	25	T	μ=0.5($NaClO_4$)	53Va 62Bc	
	0.573	1.37(1)	28.0	25	C	μ=1M($LiClO_4$&LiCl)	72Ff	q:±0.006
	0.4	(1)	--	25	C	μ=1M	70Pd	q:±2.1
	0.54	1.35(1)	27.6	25	C	μ=1.0($NaClO_4$)	68Gc 59Aa	
	0.75	1.37(1)	28.9	25	T	μ=1.0($NaClO_4$)	53Va	
	-0.456	1.46(1)	26.4	25	C	μ=2M($LiClO_4$&LiCl)	72Ff	q:±0.06
	0.00	1.42(1)	42.2	25	C	μ=2.0($NaClO_4$)	68Gc	
	0	1.43(1)	42.2	25	T	μ=2.0($NaClO_4$)	53Va	
	-1.422	1.51(1)	24.3	25	C	μ=3M($LiClO_4$&LiCl)	72Ff	q:±0.10
	-0.42	1.58(1)	28.9	25	C	μ=3.0($NaClO_4$)	68Gc	
	0.42	1.586(1)	28.9	25	C	μ=3.0($NaClO_4$)	66Ga	
	-0.188	1.54(1)	28.9	25	T	μ=3.0($NaClO_4$)	53Va	
	-2.845	1.77(1)	24.3	25	C	μ=4M($LiClO_4$&LiCl)	72Ff	
	0	1.32(1)	25.5	25	T	μ=4.5	57Sc	
	-160	16.94(40°,1)	-188	35-45	T	μ=0 S:20v%Dioxane	73Mf	b,j
	-6.3	3.225(1)	40.6	25	C	μ=1M(NH_4ClO_4) S:DMSO	76Ae	
	-15.5	2.53(40°,1)	-0.8	40-70	T	S:Melt of NH_4NO_3	72Nb	b
	3.933	0.61(2)	24.7	25	C	μ=0	72Ff	q:±0.38
	-2.5	(2)	--	25	C	μ=0	70Pd	q:±4.2
	6.067	0.538(25°,2)	30.5	5-35	T	μ=0	69Sa	b
	2.80	0.42(2)	17.6	25	C	μ=0.5($NaClO_4$)	68Gc	
	3.159	0.44(2)	18.8	25	C	μ=0.5M($LiClO_4$& LiCl)	72Ff	q: ±0.33
	2.05	0.43(2)	15.1	25	C	μ=1($NaClO_4$)	68Gc	
	2.958	0.23(2)	14.2	25	C	μ=1M($LiClO_4$&LiCl)	72Ff	
	2.5	(2)	--	25	C	μ=1M	70Pd	q:±2.1
	1.42	0.53(2)	15.1	25	C	μ=2($NaClO_4$)	68Gc	
	1.351	0.49(2)	13.8	25	C	μ=2M($LiClO_4$&LiCl)	72Ff	q:±0.12
	0.08	0.64(2)	12.6	25	C	μ=3($NaClO_4$)	68Gc	
	-0.08	1.644(2)	12.6	25	C	μ=3.0($NaClO_4$)	66Ga	
	-1.422	0.82(2)	10.9	25	C	μ=3M($LiClO_4$&LiCl)	72Ff	q:±0.21
	-2.678	0.70(2)	4.6	25	C	μ=4M($LiClO_4$&LiCl)	72Ff	
	-1.255	0.902(2)	13.0	25	T	μ=4.5	57Sc	
	15	1.979(2)	88	25	C	μ=1M(NH_4ClO_4) S:DMSO	76Ae	q:±2
	18.0	(3)	--	25	C	μ=0	70Pd	q:±4.2
	-3.849	-0.20(3)	-18.0	25	C	μ=0($LiClO_4$&LiCl)	72Ff	q:±0.4
	7.5	-0.37(3)	18.0	25	C	μ=1($NaClO_4$)	68Gc	

Metal Ion	ΔH, kJ /mol	Log K	ΔS, J /mol K	T,°C	M	Conditions	Fef.	Re-marks
C20, continued								
Cd^{2+}, cont.	0.653	0.10(3)	4.2	25	C	μ=1($LiClO_4$&LiCl)	72Ff	
	12.1	(3)	--	25	C	μ=1M	70Pd	q:±4.2
	8.08	-0.16(3)	23.8	25	C	μ=2($NaClO_4$)	68Gc	
	4.376	0.22(3)	18.8	25	C	μ=2($LiClO_4$&LiCl)	72Ff	q:±0.63
	7.74	0.18(3)	29.3	25	C	μ=3($NaClO_4$)	68Gc	
	-7.74	0.185(3)	29.3	25	C	μ=3.0($NaClO_4$)	66Ga	
	10.209	0.30(3)	39.7	25	C	μ=3($LiClO_4$+LiCl)	72Ff	
	6.276	0.64(3)	33.0	25	C	μ=4($LiClO_4$&LiCl)	72Ff	
	10.88	0.088(3)	38.5	25	T	μ=4.5	57Sc	
	1	2.567(3)	53	25	C	μ=1M(NH_4ClO_4) S:DMSO	76Ae	q:±2
	24.769	-0.55(4)	72.4	25	C	μ=4.0($LiClO_4$& LiCl)	72Ff	
	12.13	0.454(4)	32.2	25	T	μ=4.5	56Sc	
	-12.2	1.748(4)	-7	25	C	μ=1M(NH_4ClO_4) S:DMSO	76Ae	
	5.02	2.70(1-2)	68.6	25	T	μ=0.0	53Va	
	3.26	1.82(1-2)	45.6	25	T	μ=0.5($NaClO_4$)	53Va	
	3.14	1.79(1-2)	44.8	25	T	μ=1.0($NaClO_4$)	53Va	
	2.93	1.96(1-2)	47.3	25	T	μ=2.0($NaClO_4$)	53Va	
	2.72	2.20(1-2)	51.1	25	T	μ=3.0($NaClO_4$)	53Va	
	16.11	2.11(1-3)	93.3	25	T	μ=0.0	53Va	
	13.81	1.49(1-3)	74.9	25	T	μ=1.0($NaClO_4$)	53Va	
	12.64	1.82(1-3)	77.0	25	T	μ=2.0($NaClO_4$)	53Va	
	11.88	2.30(1-3)	84.1	25	T	μ=3.0($NaClO_4$)	53Va	
Hg^{2+}	-20.1	7.32(1)	71	25	C	μ→0	61Ma 67Ab	
	-28.24	7.2(1)	37.6	8	C	μ=0.50	64Cb	
	-23.0	6.7(1)	51.9	25	C	μ=0.50	64Cb	
	-24.7	6.74(1)	46.4	25	C	μ=0.50	60Ga	
	-23.4	6.6(1)	51.0	40	C	μ=0.50	64Cb	
	-23.2	6.72(1)	51	25	C	μ=1M($NaClO_4$)	75Cd	
	-24.22	7.07(1)	54.0	25	C	3M($NaClO_4$)	65Ae	
	-30.08	5.16(70°,1)	10.5	55-85	T	S:$NH_4NO_3\cdot 2H_2O$ melt	74Ne	b
	-30.5	6.7(2)	19.7	8	C	μ=0.50	64Cb	
	-30.33	6.5(2)	22.2	25	C	μ=0.50	64Cb	
	-28.9	6.48(2)	26.4	25	C	μ=0.50	60Ga	
	-28.9	6.2(2)	25.9	40	C	μ=0.50	64Cb	
	-27.7	6.51(2)	32	25	C	μ=1M($NaClO_4$)	75Cd	
	-27.15	6.91(2)	41.4	25	C	3M($NaClO_4$)	65Ae	
	-9.2	0.95(3)	12.6	25	C	μ=0.5	60Ga	
	-1.0	1.00(3)	16.0	25	C	μ=1M($NaClO_4$)	75Ce	q:±0.1
	-4.31	0.75(1.08)(3)	0(6.3)	25	C	3M($NaClO_4$) Log K and ΔS values in parentheses determined calorimetrically	65Ae	
	0.4	1.049(4)	21.3	25	C	μ=0.5	60Ga	
	-7.6	0.97(4)	-7	25	C	μ=1M($NaClO_4$)	75Ce	
	-6.19	1.38(1.09)(4)	5.4(0)	25	C	3M($NaClO_4$) Log K and ΔS values in parentheses	65Ae	

Metal Ion	ΔH, kJ /mol	Log K	ΔS, J /mol K	T,°C	M	Conditions	Ref.	Remarks
C20, continued								
Hg_2^+, cont.						determined calorimetrically		
	-58.78	13.9(1-2)	57.3	8	C	μ=0.50	64Cb	
	-53.35	13.2(1-2)	74.0	25	C	μ=0.50	64Cb	
	-52.30	12.7(1-2)	77.0	40	C	μ=0.50	64Cb	
	-51.38	(1-2)	--	25	C	$3M(NaClO_4)$	65Ae	
	-62.3	15.22(1-4)	81.6	25	C	μ=0.50	60Ga	
	-61.88	(1-4)	--	25	C	$3M(NaClO_4)$	65Ae	
HgA	-25.1	5.219(1)	15.1	20	C	$μ=0.1(KNO_3)$ $A=CH_3$	65Sa	
	-2.1	0.60(25°)	4.6	5-60	T	$μ≅10^{-3}$ A=Br R:$\frac{1}{2}HgL_2+\frac{1}{2}HgA_2=$ HgLA	64Ea	b,r
	-2.9	0.88(25°)	7.1	5-60	T	$μ≅10^{-3}$ A=I R:$\frac{1}{2}HgL_2+\frac{1}{2}HgA_2=$ HgLA	65Ae	b,r
HgA^{2-}	-26.4	3.59(1)	-19.2	25	C	$μ=2M(NaClO_4)$ $A=EDTA^{4-}$	75Bc	
HgL_2	257.7	1.81(42.1°)	854	42& 55.6	T	S:Benzene R:$2(HgL_2)=(HgL_2)_2$	74Ea	a,b
	-41.0	2.03(42.1°)	-92	42& 55.6	T	S:Toluene R:$2(HgL_2)=(HgL_2)_2$	74Ea	a,b
	-31.8	2.65(55.6°)	-46	55&65	T	S:=p-Xylene R:$2(HgL_2)=(HgL_2)_2$	74Ea	a,b
	-17.2	2.92(55.6°)	4	55&65	T	S:Mesitylene R:$2(HgL_2)=(HgL_2)_2$	74Ea	a,b
AlL_4^-	16.7	-4.0(250°)	-29.3	175-275	T	S:$AlCl_3$-LiCl melt R:$2AlL_4^-=$ $L^-+Al_2L_7^-$	72Tc	b
	56.9	-5.7(300°)	-10.0	175-450	T	S:$AlCl_3$-NaCl melt R:$2AlL_4^-=$ $L^-+Al_2L_7^-$	72Tc	b
	95.8	-7.1(300°)	31.4	275-450	T	S:$AlCl_3$-KCl melt R:$2AlL_4^-=$ $L^-+Al_2L_7^-$	72Tc	b
In^{3+}	5.14	2.10(1)	57.3	25	C	$μ=2.00M(NaClO_4)$	69Rd	
	3.26	1.50(2)	39.7	25	C	$μ=2.00M(NaClO_4)$	69Rd	
	3.3	-0.24(3)	109	25	C	$μ=2.00M(NaClO_4)$	59Rd	
Tl^+	-5.8	0.68(1)	-6.7	25	C	μ=0	69Bl	
	-5.98	0.68(1)	-7.1	25	T	μ=0	34La 53Ba 63Kb	
	∿0	0.67(25°,1)	12	0-45	T	μ=0	67Ab	
	-6.36	0.380(25°,1)	-4.2	10-60	T	μ=0	72Fd	b,q: ±0.42
	-0.63	0.107(25°,1)	-1.84	10-60	T	$μ=0.5(LiClO_4)$	72Fd	b,q: ±0.04
	0.08	-0.013(25°,1)	16.7	10-60	T	$μ=1.0(LiClO_4)$	72Fd	b,q: ±0.46

Metal Ion	ΔH, kJ /mol	Log K	ΔS, J /mol K	T,°C	M	Conditions	Ref.	Remarks
C20, continued								
Tl^+, cont.	0.42	-0.076(25°,1)	-2.1	10-60	T	μ=2.0($LiClO_4$)	72Fd	b,q: ±0.17
	1.13	-0.201(25°,1)	24.7	10-60	T	μ=3.0($LiClO_4$)	72Fd	b,q: ±0.12
	0.50	-0.092(25°,1)	19.2	10-60	T	μ=4.0($LiClO_4$)	72Fd	b,q: ±0.25
	-4.6	0.00(25°,1)	-17	15-80	T	4M(LiCl+$LiClO_4$)	67Ab	
	9.6	-0.81(25°,1-2)	17	15-80	T	4M(LiCl+$LiClO_4$)	63Eb	
Tl^{3+}	-25.27	7.49(1)	58.2	25	C	3M($HClO_4$)+ 1M($NaClO_4$)	64Lf	
	-22.80	7.724(1)	60.7	25	C	μ=3.0	64Wa	
	-32.6	7.1(1)	29	25	C	3M(HCl,$HClO_4$, LiCl,$LiClO_4$)	65Kd	
	-16.94	5.78(2)	54.0	25	C	3M($HClO_4$)+ 1M($NaClO_4$)	64Lf	
	-18.41	5.763(2)	42.2	25	C	μ=3.0	64Wa	
	-15.5	3.4(2)	50	25	C	3M(HCl,$HClO_4$, LiCl, $LiClO_4$)	65Kd	
	-4.52	3.39(3)	49.8	25	C	3M($HClO_4$)+ 1M($NaClO_4$)	64Lf	
	-4.6	3.000(3)	52.7	25	C	μ=3.0	64Wa	
	0	3.4(3)	67	25	C	3M(HCl,$HClO_4$, LiCl, $LiClO_4$)	65Kd	
	-21.3	1.470(25°,4)	-42	0&25	T	μ=0	64Nb	a,b
	-0.71	2.78(4)	51.0	25	C	3M($HClO_4$)+ 1M($NaClO_4$)	64Lf	
	-1.2	1.815(4)	37.2	25	C	μ=3.0	64Wa	
	0	3.4(4)	25	25	C	3M(HCl,$HClO_4$, LiCl,$LiClO_4$)	65Kd	
	-48.1	12.5(1-2)	79	25	C	3M(HCl,$HClO_4$, LiCl,$LiClO_4$)	65Kd	
	-48.1	16.0(1-3)	146	25	C	3M(HCl,$HClO_4$, LiCl,$LiClO_4$)	65Kd	
	-48.1	18.5(1-4)	171	25	C	3M(HCl,$HClO_4$, LiCl,$LiClO_4$)	65Kd	
Sn^{2+}	10.878	1.15(1)	58.6	25	T	3.0M($NaClO_4$)	52Va 67Ab	b
	13.389	1.171(1-2)	77.4	25	T	3.0M($NaClO_4$)	52Vb	a
	23.430	1.69(1-3)	110.9	25	T	3.0M($NaClO_4$)	52Vb	a
Pb^{2+}	5.23	(1)	--	25	C	μ=0	59Aa	
	17.99	1.60(1)	92.0	25	T	μ=0	55Na	a
	13.0	1.36(1)	69.4	25	T	μ=0	76Yf	q:±5
	23.2	1.84(1)	97.5	100	T	μ=0	76Yf	
	42.2	2.23(1)	141.8	150	T	μ=0	76Yf	
	4.113	(1)	--	25	C	μ=0.15	59Aa	
	3.66	1.16(1)	34.3	25	T	μ=3.0($LiClO_4$, LiCl)	64Mb	
	3.66	1.158(25°,1)	33	18-65	T	μ=3M($LiClO_4$)	68Fa	b
	-10.5	(1)	--	40-70	T	S:Melt of NH_4NO_3	72Nb	b
	7.9	1.81(2)	63	25	T	μ=3.0($LiClO_4$, LiCl)	64Mb	
	-4.35	0.652(25°,2)	29	18-65	T	μ=3M($LiClO_4$)	68Fa	b

Metal Ion	ΔH, kJ /mol	Log K	ΔS, J /mol K	T,°C	M	Conditions	Ref.	Remarks
C20, continued								
Pb^{2+}, cont.	10.9	1.91(3)	75	25	T	μ=3.0($LiClO_4$, LiCl)	64Mb	
	-2.9	0.103(25°,3)	12	18-65	T	μ=3M($LiClO_4$)	68Fa	b
	2.5	1.20(4)	31.4	25	T	μ=3.0($LiClO_4$, LiCl)	64Mb	
	8.4	-0.711(25°,4)	-42	18-65	T	μ=3M($LiClO_4$)	68Fa	b
Bi^{3+}	16.736	3.42(1)	135.6	25	C	μ=0	67Va	
	0	2.2(25°,1)	42	25-65	T	μ=3(Li,H,ClO_4, Cl)	63Me	b
	-2.648	2.77(1)	43.1	-7.4	C	μ=4.0 ΔCp=149	67Va	
	-9.489	2.75(1)	48.5	0	C	μ=4.0 ΔCp=147	67Va	
	-0.586	2.82(1)	51.5	5	C	μ=4.0 ΔCp=146	67Va	
	0.481	2.71(1)	54.0	10	C	μ=4.0 ΔCp=144	67Va	
	0.916	2.20(1)	45.2	15	C	μ=4.0	76Vd	q:±0.17
	1.431	2.78(1)	57.7	18	C	μ=4.0 ΔCp=142	67Va	
	2.180	2.78(1)	61.1	25	C	μ=4.0 ΔCp=140	67Va	
	2.18	2.13(1)	48.1	25	C	μ=4.0 ΔCp=139.3	76Vd	
	3.72	2.10(1)	52.3	35	C	μ=4.0	76Vd	
	4.318	2.79(1)	66.9	40	C	μ=4.0 ΔCp=136 ΔCp=220.1-0.268T	67Va	
	5.06	2.64(1)	66.5	45	C	μ=4.0	76Vd	
	-3.088	2.94(1)	52.7	25	C	μ=5.0	67Va	
	-1.092	3.12(1)	40.6	25	C	μ=6.0	67Va	
	18.8	3.5(25°,2)	130	25-65	T	μ=3(Li,H,ClO_4,Cl)	63Me	b
	21.64	2.02(2)	113.8	15	C	μ=4.0	76Vd	n
	24.18	2.04(2)	120.1	25	C	μ=4.0 ΔCp=85.8	76Vd	n
	24.98	2.02(2)	119.7	35	C	μ=4.0	76Vd	n
	24.27	1.30(2)	101.2	45	C	μ=4.0	76Vd	n
	23.8	5.8(25°,3)	192	25-65	T	μ=3(Li,H,ClO_4,Cl)	63Me	b
	-43.68	13.97(3)	-115.9	15	C	μ=4.0	76Vd	n
	-45.52	14.12(3)	-117.6	25	C	μ=4.0 ΔCp=72.4	76Vd	n
	-43.93	13.08(3)	-107.9	35	C	μ=4.0	76Vd	n
	-41.88	12.06(3)	-99.2	45	C	μ=4.0	76Vd	n
	18.8	6.7(25°,4)	192	25-65	T	μ=3(Li,H,ClO_4,Cl)	63Me	b
	37.78	1.24(4)	154.8	15	C	μ=4.0	76Vd	n
	36.82	1.31(4)	148.5	25	C	μ=4.0 ΔCp=-239.3	76Vd	n
	34.98	0.96(4)	131.8	35	C	μ=4.0	76Vd	n
	31.30	1.19(4)	121.1	45	C	μ=4.0	76Vd	n
	16.7	7.29(25°,5)	197	25-65	T	μ=3(Li,H,ClO_4,Cl)	63Me	b
	-50.12	17.08(5)	-153.1	15	C	μ=4.0	76Vd	n
	-48.24	15.92(5)	-143.1	25	C	μ=4.0 ΔCp=209.6	76Vd	n
	-45.10	14.40(5)	-129.3	35	C	μ=4.0	76Vd	n
	-44.22	13.73(5)	-123.8	45	C	μ=4.0	76Vd	n
	0	7.33(25°,6)	142	25-65	T	μ=3(Li,H,ClO_4,Cl)	63Me	b
	-32.970	8.29(1-5)	46.0	15	C	μ=4.0($HClO_4$)	72Vb	ΔCp=351
	-29.916	8.19(1-5)	56.5	25	C	μ=4.0($HClO_4$)	72Vb	ΔCp=339
	-26.443	8.03(1-5)	67.8	35	C	μ=4.0($HClO_4$)	72Vb	ΔCp=331
	-23.012	7.91(1-5)	79.1	45	C	μ=4.0($HClO_4$)	72Vb	ΔCp=318
TeA_4	-89.1	--	--	25	C	S:Benzene A=Cl R:TeA_4+GaL_3= TeA_4·GaL_3	73Gb	

Metal Ion	ΔH, kJ /mol	Log K	ΔS, J /mol K	T, °C	M	Conditions	Ref.	Re-marks
C21 CHLORINE			Cl_2			Cl_2		
$IrABC_2$ (trans)	-338.9	(1)	--	25	C	S:Benzene A=Cl B=CO C=PPh_3	79Bf	
C22 CHROMATE ION			O_4Cr			$CrO_4^{=}$		L^{2-}
Th^{4+}	15.36	0.67(25°)	64.4	10-25	T	μ=0.20M($HClO_4$) R:Th^{4+}+HL^-= ThL^{2+}+H^+	72Bn	b
	16.36	0.28(25°)	60.2	10-25	T	μ=1.00M($HClO_4$) R:Th^{4+}+HL^-= ThL^{2+}+H^+	72Bn	b,q: ±1.7
Np^{4+}	4.90	1.81(25°)	51.0	10-25	T	μ=0.20M($HClO_4$) R:Np^{4+}+HL^-= NpL^{2+}+H^+	72Bn	b
Fe^{3+}	12.05	0.45(25°)	48.9	1-25	T	μ=0.084($HClO_4$) R:Fe^{3+}+HL^-= FeL^++H^+	72Bn	b
	12.93	0.29	49.0	1-25	T	μ=0.20M($HClO_4$) R:Fe^{3+}+HL^-= FeL^++H^+	72Bn	b
	8.74	0.11(25°)	31.4	1-25	T	μ=1.00M($HClO_4$) R:Fe^{3+}+HL^-= FeL^++H^+	72Bn	b
C23 CHROMIUM(III) ION, hexaaquo-			$O_6H_{12}Cr$			$Cr(H_2O)_6^{3+}$		L^{3+}
Np^{5+}	-13.8	0.418(25°,1)	-37.6	25-50	T	μ=8.0	64Se	b
C24 CINNAMIC ACID			$C_9H_8O_2$			$C_6H_5CH{:}CHCO_2H$		
Tl^+	-54.9	1.50(28°,1)	-211	28-48	T	--	76Sa	b
C25 CITRIC ACID			$C_6H_8O_7$			$HO_2CCH_2C(OH)(CO_2H)CH_2CO_2H$		L^{3-}
Na^+	2.22	0.88(1)	24.3	25	C	μ=0.05M($NaClO_4$)	80Aa	
	1.67	0.81(1)	20.9	25	C	μ=0.10M($NaClO_4$)	80Aa	
	0.25	0.73(1)	15.1	25	C	μ=0.15M($NaClO_4$)	80Aa	
	5.8	0.22	23.8	25	C	μ=0.05M($NaClO_4$) R:Na^++HL^{2-}=$NaHA^-$	80Aa	
	0.8	0.15	5.4	25	C	μ=0.10M($NaClO_4$) R:Na^++HL^{2-}=$NaHA^-$	80Aa	
	-2.9	0.07	-8.4	25	C	μ=0.15M($NaClO_4$) R:Na^++HL^{2-}=$NaHA^-$	80Aa	

Metal Ion	ΔH, kJ /mol	Log K	ΔS, J /mol K	T,°C	M	Conditions	Ref.	Re-marks
C25, continued								
Na^+, cont.	-1.2	0.07	-2.9	25	C	μ=0.05M($NaClO_4$) R:Na^++H_2L^-=NaH_2L	80Aa	
	-2.1	0.07	-5.4	25	C	μ=0.10M($NaClO_4$) R:Na^++H_2L^-=NaH_2L	80Aa	
	-5.8	0.0	-19.7	25	C	μ=0.15M($NaClO_4$) R:Na^++H_2L^-=NaH_2L	80Aa	
Ca^{2+}	-10.376	3.10(28.2°,1)	24.7	19.6-65.0	T	μ=0.2M(NH_4Cl)	71Re	b
Sr^{2+}	-24.476	2.72(20°,1)	-31.8	10-25	T	μ=0.2M(NH_4Cl)	71Re	b
	-3.908	2.59(40°,1)	61.9	30-55	T	μ=0.2M(NH_4Cl)	71Re	b
Ba^{2+}	-21.171	2.30(20°,1)	-28.9	10-25	T	μ=0.2M(NH_4Cl)	71Re	b
	17.908	2.27(40°,1)	100.4	30-55	T	μ=0.2M(NH_4Cl)	71Re	b
Ra^{2+}	0	2.04(37°,1)	38	7-37	T	μ=0.15(NaCl)	50Sa	b
C26	COBALTATE(II) ION, hexacyano-	C_6H_6Co				$Co(CN)_6{}^{4-}$		L^{4-}
Al^{3+}	17.99	-1.301	35.1	10-40	T	μ=0 R:$Al(OH_2)L$= $AlL+H_2O$ (Outer Sphere)=(Inner Sphere)	71Ke	b,q:±4
C27	COBALTATE(III)ION, hexacyano	C_6N_6Co				$Co(CN)_6{}^{3-}$		L^{3-}
La^{3+}	5.56	3.762(25°,1)	90.8	25	T	μ=0	50Ja	a
C28	CONCANAVALIN A							L
Mn^{2+}	105.4	4.12(15°)	440.2	0-30	T	0.2M(NaCl) 0.05M(Acetate) pH=5.1 Tenfold excess of Ca^{2+} over L	77Ae	b,j, q:9
	57.24	3.21(15°)	262.3	0-30	T	0.2M(NaCl) 0.05M(Acetate) pH=5.1 No Ca^{2+} present	77Ae	b,j
C29	COPPER(II) CHLORIDE		Cl_2Cu			$CuCl_2$		L
LiA	6.7	(2-1)	--	55	T	S:Acetic acid sol A=Cl	67Eb	
	12.6	--	--	55	T	S:Acetic acid sol R:$CuCl_2$+ 2LiA=CuA_2+2LiCl A=Acetate	67Eb	

Metal Ion	ΔH, kJ /mol	Log K	ΔS, J /mol K	T,°C	M	Conditions	Ref.	Remarks
C30	COUMARIN, 8-amino-7-hydroxy-4-methyl		$C_{10}H_9O_3N$			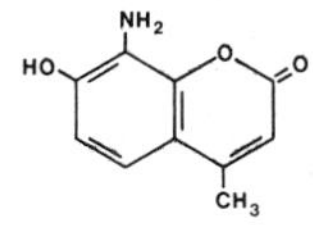		
Fe^{3+}	-15.3	7.51(30°,1-2)	93.1	10-40	T	μ=0.1 S:50v%Ethanol	75Pc	b
C31	6-COUMARINCARBOXYLIC ACID, 7-hydroxy-4-methyl		$C_{11}H_8O_5$			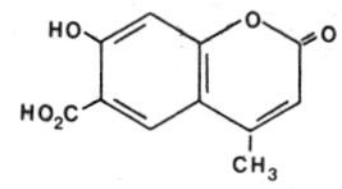		
Fe^{3+}	35.9	8.81(30°,1-2)	304	10-40	T	μ=0.1 S:50v%Ethanol	75Pc	b
C32	12-CROWN-4		$C_8H_{16}O_4$			$CH_2CH_2O(CH_2CH_2O)_2CH_2CH_2O$		
Li^+	-36.8	>4(1)	--	25	C	S:Nitromethane 0.02M($LiClO_4$)	80Sb	
	-19.2	1.6(2)	-34.3	25	T	S:Nitromethane 0.02M($LiClO_4$)	80Sb	
	-16.3	4.25(1)	27.2	25	T	S:Acetonitrile 0.02M($LiClO_4$)	80Sb	
	-10.9	(1)	>41.8	25	C	S:Propylene carbonate 0.02M($LiClO_4$)	80Sb	
	-13.4	1.62(1)	-14.2	25	C	S:Acetone 0.5M($LiClO_4$)	80Sb	
	-3.3	~0(1)	--	25	C	S:Methanol 0.5M($LiClO_4$)	80Sb	
C33	15-CROWN-5		$C_{10}H_{20}O_5$			$CH_2CH_2O(CH_2CH_2O)_3CH_2CH_2O$		
Li^+	-43.5	>4(1)	--	25	C	S:Nitromethane 0.02M($LiClO_4$)	80Sb	
	-21.3	>4(1)	>6.3	25	C	S:Acetonitile 0.02M($LiClO_4$)	80Sb	
	-16.7	>4(1)	>20.9	25	C	S:Propylene carbonate 0.02M($LiClO_4$)	80Sb	
	-19.7	3.59(1)	--	25	C	S:Acetone 0.5M($LiClO_4$)	80Sb	
	-19.2	3.59(1)	4.2	25	T	S:Acetone 0.5M($LiClO_4$)	80Sb	
	-11.3	1.23(1)	-14.2	25	C	S:Methanol 0.5M($LiClO_4$)	80Sb	
Na^+	-6.28	0.70(1)	-7.5	25	C	μ=0.1	76Ia	
	-20.88	3.48(1)	-3.37	25	C	S:Methanol	80La	
K^+	-17.2	0.74(1)	-43.5	25	C	μ=0.1	76Ia	
	-32.2	3.77(1)	-36.06	25	C	S:Methanol	80La	

Metal Ion	ΔH, kJ /mol	Log K	ΔS, J /mol K	T,°C	M	Conditions	Ref.	Re-marks
C33, continued								
Rb^+	-7.95	0.62(1)	-14.6	25	C	μ=0.1	76Ia	
Cs^+	-5.4	0.8(1)	-2.1	25	C	μ=0.1	76Ia	q:±0.8
	-48.9	2.18(1)	-122.1	25	C	S:Methanol	80La	
	-6.07	2.18(1)	21.33	25	C	S:Methanol	80La	
Sr^{2+}	-3.8	1.95(1)	25	25	C	μ=0.1	76Ia	q:±0.42
	-19.62	2.63(1)	-15.4	25	C	S:Methanol	80La	
Ba^{2+}	-4.77	1.71(1)	16.7	25	C	μ=0.1	76Ia	
Ag^+	-13.51	0.94(1)	-27.2	25	C	μ=0.1	76Ia	
	-27.53	3.62(1)	-23.01	25	C	S:Methanol	80La	
Hg^{2+}	-15.1	1.68(1)	-17	25	C	μ=0.1	76Ia	q:±1.2
Tl^+	-16.78	1.23(1)	-32.6	25	C	μ=0.1	76Ia	
Pb^{2+}	-13.64	1.85(1)	-10.5	25	C	μ=0.1	76Ia	
NH_4^+	-1.00	1.71(1)	29.3	25	C	μ=0.1	76Ia	q:±0.17

C34 15-CROWN-5, dibenzo- $C_{18}H_{20}O_5$ L

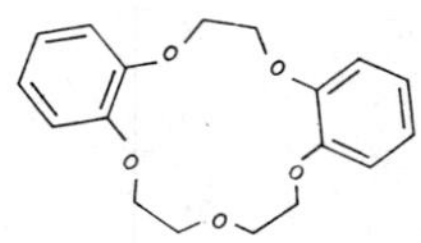

Metal Ion	ΔH, kJ /mol	Log K	ΔS, J /mol K	T,°C	M	Conditions	Ref.	Re-marks
Na^+	-7.40	0.72(1)	-10.9	25	C	μ=0 S:20w%CH_3OH	74Ce	
	-11.00	1.17(1)	-14.6	25	C	μ=0 S:40w%CH_3OH	74Ce	
	-15.82	1.64(1)	-15.5	25	C	μ=0 S:60w%CH_3OH	74Ce	
	-15.98	1.99(1)	-15.5	25	C	μ=0 S:70w%CH_3OH	74Ce	
	-34.82	2.26(1)	-73.6	25	C	μ=0 S:80w%CH_3OH	74Ce	
K^+	-7.5	1.22(1)	-2.1	25	C	μ=0 S:20w%CH_3OH	74Ce	
	-10.50	1.92(1)	1.7	25	C	μ=0 S:40w%CH_3OH	74Ce	
	-14.73	2.54(1)	-0.8	25	C	μ=0 S:60w%CH_3OH	74Ce	
	-15.36	2.93(1)	4.6	25	C	μ=0 S:70w%CH_3OH	74Ce	
	-42.68	2.82(1)	-89.1	25	C	μ=0 S:80w%CH_3OH	74Ce	

C35 18-CROWN-6 $C_{12}H_{24}O_6$ $CH_2CH_2O(CH_2CH_2O)_4CH_2CH_2O$ (ring)

Metal Ion	ΔH, kJ /mol	Log K	ΔS, J /mol K	T,°C	M	Conditions	Ref.	Re-marks
Li^+	-14.6	1.50(1)	-20.5	25	C	S:Acetone 0.5M($LiClO_4$)	80Sb	
	~0	2.34(1)	44.8	25	C	S:Acetonitrile 0.02M($LiClO_4$)	80Sb	
	-11.7	~0(1)	--	25	C	S:Methanol 0.5M($LiClO_4$)	80Sb	
	~12	>4(1)	>0	25	C	S:Nitromethane 0.02M($LiClO_4$)	80Sb	
	-15.9	2.69(1)	-2.5	25	T	S:Propylene carbonate 0.02M($LiClO_4$)	80Sb	
Na^+	-9.41	0.80(1)	-15.5	25	C	μ=0.1	76Ia	
	-35.1	4.36(1)	-33.7	25	C	S:Methanol	80La	

Metal Ion	ΔH, kJ /mol	Log K	ΔS, J /mol K	T,°C	M	Conditions	Ref.	Re-marks
C35, continued								
Na^+, cont.	-34.98	4.36(1)	-33.82	25	C	S:Methanol μ=0.005	80Bb	
K^+	-25.98	2.03(1)	-47.7	25	C	μ=0.1	76Ia	
	-56.11	6.06(1)	-72.13	25	C	S:Methanol	80La	
	-56.11	6.06(1)	-72.13	25	C	S:Methanol μ=0.005	80Bb	
Rb^+	-15.98	1.56(1)	-23.8	25	C	μ=0.1	76Ia	
	-50.58	5.32(1)	-67.78	25	C	S:Methanol	80La	
	-50.58	5.32(1)	-67.78	25	C	S:Methanol μ=0.005	80Bb	
Cs^+	-15.86	0.99(1)	-33.9	25	C	μ=0.1	76Ia	
	-47.24	4.79(1)	-66.66	25	C	S:Methanol	80La	
	-24.3	1.891(25°)	-44.8	-38-24	T	S:Pyridine R:$CsL^+ + L = CsL_2^+$	77Md	
Ca^{2+}	-11.51	3.86(1)	-35.36	25	C	S:Methanol	80La	
Sr^{2+}	-15.10	2.72(1)	1.2	25	C	μ=0.1	76Ia	
	-36.02	>5.5(1)	--	25	C	S:Methanol	80La	
Ba^{2+}	-31.71	3.87(1)	-33.0	25	C	μ=0.1	76Ia	
	-43.55	7.04(1)	-11.23	25	C	S:Methanol	80La	
	-43.56	7.04(1)	-11.23	25	C	S:Methanol μ=0.005	80Bb	
Ag^+	-9.08	1.50(1)	-1.7	25	C	μ=0.1	76Ia	
	-38.28	4.58(1)	-40.7	25	C	S:Methanol	80La	
	-38.28	4.56(1)	-40.70	25	C	S:Methanol μ=0.005	80Bb	
Hg^{2+}	-19.62	2.42(1)	-19.7	25	C	μ=0.1	76Ia	
Tl^+	-18.58	2.27(1)	-18.8	25	C	μ=0.1	76Ia	
	-22.6	2.2(25°,1)	-34.7	5-35	T	μ=0.1M $[(C_2H_5)_4NClO_4]$ pH=2	76Kd	
Pb^{2+}	-21.59	4.27(1)	9.2	25	C	μ=0.1	76Ia	
	-13.0	4.4(25°,1)	40.6	5-35	T	μ=0.1M $[(C_2H_5)_4NClO_4]$ pH=2	76Kd	
NH_4^+	-9.79	1.23(1)	-9.2	25	C	μ=0.1	76Ia	
	-38.78	4.27(1)	-48.27	25	C	S:Methanol μ=0.005	80Bb	

C36 18-CROWN-6, di(4-aminobenzo-) $C_{20}H_{26}O_6N_2$ (cis)

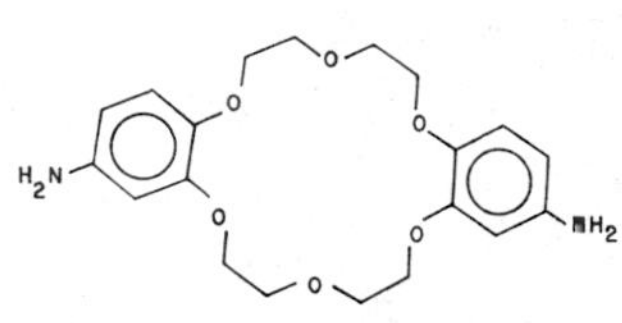

L

Metal Ion	ΔH, kJ /mol	Log K	ΔS, J /mol K	T,°C	M	Conditions	Ref.	Re-marks
Na^+	-25	--	-29	10-40	T	S:Dimethoxy-	73Sp	b

Metal Ion	ΔH, kJ /mol	Log K	ΔS, J /mol K	T,°C	M	Conditions	Ref.	Re-marks
C36, continued								
Na^+, cont.						fluorenyl Salt:NaSCN		
C37	18-CROWN-6, 1,10-diaza		$C_{12}H_{26}O_4N_2$			$HN(CH_2CH_2O)_2CH_2CH_2NH(CH_2CH_2O)_2CH_2CH_2$		
Sr^{2+}	-10.9	2.56(1)	13.0	25	C	μ=0.1[$(CH_3)_4NCl$]	75Ad	
Ba^{2+}	-12.6	2.97(1)	15.1	25	C	μ=0.1[$(CH_3)_4NCl$]	75Ad	
Ag	-38.3	7.8(1)	20.9	25	C	μ=0.1[$(CH_3)_4NCl$]	75Ad	
Cd^2	-2.9	5.25(1)	90.3	25	C	μ=0.1[$(CH_3)_4NCl$]	75Ad	
Hg^{2+}	-71.8	17.85(1)	101.7	25	C	μ=0.1[$(CH_3)_4NCl$]	75Ad	
C38	18-CROWN-6, dibenzo-		$C_{20}H_{24}O_6$			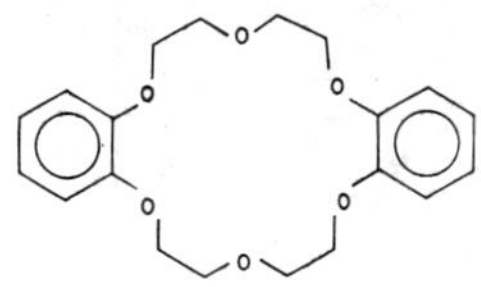		L
Na^+	-16.3	3.728(20°,1)	16.3	10-40	T	S:Dimethoxy-ethane	73Sp	b
	-25	2.778(25°,1)	-29	0-40	T	S:N,N-Dimethyl-formamide CM=10^{-3}M	71Sc	b
	-31.3	4.5(1)	-18.8	25	C	CM=0.1-5x10^{-3}M CL=10^{-3}M S:Methanol	73Fb	
NaA	-14.2	3.684(20°,1)	22.6	10-40	T	S:Dimethoxy-ethane A=Tetra-phenylboron $B(Ph)_4^-$	73Sp	b
K^+	-23.0	2.5(1)	-28.9	25	C	μ=0 S:Dimethyl sulfoxide	74Ce	
	-40.1	5.1(1)	-36.7	25	C	CM=0.1-5x10^{-3}M CL=10^{-3}M S:Methanol	73Fb	
Ba^{2+}	-21.2	4.28(1)	11.1	25	C	CM=0.1-5x10^{-3}M CL=10^{-3}M S:Methanol	73Fb	
C39	18-CROWN-6, dicyclohexano- (cis-syn-cis isomer)		$C_{20}H_{36}O_6$			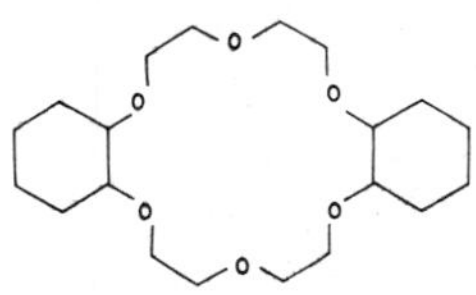		L
Na^+	0.67	1.21(1)	25.5	25	C	μ=0.1	76Ia	q:±0.17
K^+	-17.32	2.15(1)	-20.1	10	C	μ=0	71Ii	
	-16.23	2.02(1)	-15.9	25	C	μ=0	71Ii	

Metal Ion	ΔH, kJ /mol	Log K	ΔS, J /mol K	T,°C	M	Conditions	Ref.	Re-marks
C39, continued								
K^+,	-16.28	2.01(1)	-15.9	25	C	μ=0	69Ic	
cont.	-14.98	1.91(1)	-11.3	40	C	μ=0	71Ii	
	-42.34	(1)	--	25	C	S:Acetone Salt:Kl	71Ag	
KAL	-16.3	0.255(25°)	-49.0	-17-25	T	S:Tetrahydrofuran A=Fluorenyl R:$A^-K^+L=A^-LK^+$	71Zd	b,q: ±0.3
Rb^+	-14.35	1.61(1)	-23.0	10	C	μ=0	71Ii	
	-13.93	1.52(1)	-17.6	25	C	μ=0	71Ii	
	-14.56	1.47(1)	-20.9	25	C	μ=0	69Ic	
	-13.76	1.40(1)	-17.2	40	C	μ=0	71Ii	
Cs^+	-10.04	1.00(1)	-16.3	10	C	μ=0	71Ii	
	-10.08	0.96(1)	-15.5	25	C	μ=0	71Ii	
	-8.37	1.07(1)	-7.5	25	C	μ=0	69Ic	
	-9.96	0.96(1)	-13.4	40	C	μ=0	71Ii	
Sr^{2+}	-15.40	3.43(1)	11.3	10	C	μ=0	71Ii	
	-15.40	3.24(1)	10.5	25	C	μ=0	71Ii	
	-15.48	3.16(1)	10.9	40	C	μ=0	71Ii	
Ba^{2+}	-20.79	3.84(1)	0.0	10	C	μ=0	71Ii	
	-20.58	3.57(1)	0.8	25	C	μ=0	71Ii	
	-20.29	3.47(1)	1.7	40	C	μ=0	71Ii	
Ag^+	0.29	2.36(1)	46.0	25	C	μ=0.1	76Ia	q:±0.04
Hg^{2+}	-2.97	2.75(1)	42.7	25	C	μ=0.1	76Ia	
Hg_2^{2+}	-9.04	1.93(1)	6.7	25	C	μ=0.1	76Ia	
	25.5	1.2(2)	109	25	C	μ=0.1	76Ia	q:±2.9
Tl^+	-15.15	2.44(1)	-4.2	25	C	μ=0.1	76Ia	
Pb^{2+}	-23.35	4.95(1)	16.3	25	C	μ=0.1	76Ia	
	-22.93	5.29(1)	12.6	25	C	μ=0.1	76Ia	
NH_4^+	-9.04	1.33(1)	-5.0	25	C	μ=0	71Ii	
$CH_3NH_3^+$	-3.22	0.82(1)	5.0	25	C	μ=0.1	76Ia	
C40	18-CROWN-6, dicyclohexano- (cis-anti-cis isomer)			$C_{20}H_{36}O_6$		see C39		L
Na^+	-6.57	0.69(1)	-8.8	25	C	μ=0.1	76Ia	
	-10.5	1.7(1)	-2.5	25	C	μ=0 S:Dimethyl sulfoxide	74Ce 71Ii	
	∿-36.0	∿4.5(1)	--	25	C	μ=0 S:Ethanol	74Ce 71Ii	
	-23.4	3.68(1)	-7.9	25	C	μ=0 S:CH_3OH	74Ce 71Ii	
K^+	-24.18	1.79(1)	-51.0	10	C	μ=0	71Ii	
	-21.21	1.63(1)	-40.2	25	C	μ=0	71Ii	
	-21.67	1.60(1)	-42.2	25	C	μ=0	69Ic	
	-17.53	1.50(1)	-27.2	40	C	μ=0	71Ii	
	-46.28	(1)	--	25	C	S:Acetone Salt:Kl	71Ag	

Metal Ion	ΔH, kJ /mol	Log K	ΔS, J /mol K	T,°C	M	Conditions	Ref.	Remarks
C40, continued								
K^+, cont.	-32.2	2.7(1)	-56.5	25	C	μ=0 S:Dimethyl sulfoxide	74Ce 71Ii	
	∿-52.7	>6.0(1)	--	25	C	μ=0 S:Ethanol	74Ce 71Ii	
	-43.9	5.38(1)	-43.9	25	C	μ=0 S:CH_3OH	74Ce 71Ii	
KAL	-13.0	-0.097(25°)	-44.8	-17-25	T	S:Tetrahydrofuran A=Fluorenyl R:$A^-K^+L=A^-LK^+$	71Zd	b,q: ±1.2
Rb^+	-19.2	0.95(1)	-49.8	10	C	μ=0	71Ii	
	-16.61	0.87(1)	-38.9	25	C	μ=0	71Ii	
	-13.81	0.86(1)	-27.6	40	C	μ=0	71Ii	
Sr^{2+}	-14.43	2.80(1)	2.5	10	C	μ=0	71Ii	
	-13.22	2.64(1)	6.3	25	C	μ=0	71Ii	
	-12.18	2.56(1)	10.0	40	C	μ=0	71Ii	
Ba^{2+}	-28.53	3.44(1)	-34.7	10	C	μ=0	71Ii	
	-25.94	3.27(1)	-24.3	25	C	μ=0	71Ii	
	-24.18	3.12(1)	-17.6	40	C	μ=0	71Ii	
Ag^+	-8.74	1.59(1)	1.2	25	C	μ=0	71Ii	
Hg^{2+}	-10.67	2.60(1)	13.8	25	C	μ=0.1	76Ia	
Hg_2^{2+}	-18.16	1.57(1)	-31.0	25	C	μ=0.1	76Ia	
	-23.8	1.1(2)	-58.6	25	C	μ=0.1	76Ia	
Tl^+	-17.95	1.83(1)	-25.1	25	C	μ=0.1	76Ia	
Pb^{2+}	-17.61	4.43(1)	25.9	25	C	μ=0.1	76Ia	
NH_4^+	-14.27	0.80(1)	-32.6	25	C	μ=0	71Ii	
$CH_3NH_3^+$	-3.76	0.66(1)	0	25	C	μ=0.1	76Ia	
C41	18-CROWN-6, dicyclohexano- (Isomer A+B)		$C_{20}H_{36}O_6$			see C39		L
Li^+	-16.19	(1)	--	25	C	S:Acetone Salt:LiBr	71Ag	
	∿0	(1)	--	25	C	S:Dimethyl Sulfoxide	71Ag	
	∿0	(1)	--	25	C	S:Tetrahydrofuran	71Ag	
Na^+	-26.90	(1)	--	25	C	S:Acetone Salt:NaI	71Ag	
	-25.06	(1)	--	25	C	S:Acetone Salt:$NaB(C_6H_5)_4$	71Ag	
	-1.88	(1)	--	25	C	S:Dimethyl Sulfoxide Salt:NaI	71Ag	
	-20.00	(1)	--	25	C	S:Tetrahydrofuran Salt:$NaB(C_6H_5)_4$ Salt added to ligand	71Ag	

Metal Ion	ΔH, kJ /mol	Log K	ΔS, J /mol K	T,°C	M	Conditions	Ref.	Remarks
C41, continued								
Na^+, cont.	-21.04	(1)	--	25	C	S:Tetrahydrofuran Salt:$NaB(C_6H_5)_4$ Ligand added to salt	71Ag	
K^+	-40.63	(1)	--	25	C	S:Acetone Salt:KI	71Ag	
	-38.87	(1)	--	25	C	S:Acetone Salt:$KB(C_6H_5)_4$	71Ag	
	-30.92	(1)	--	25	C	S:Dimethyl sulfoxide Salt:KI Metal added to ligand	71Ag	
	-27.28	(1)	--	25	C	S:Dimethyl sulfoxide Salt:KI Ligand added to metal	71Ag	
	-31.50	(1)	--	25	C	S:Dimethyl sulf-oxide Salt:KNO_3	71Ag	
	-31.17	(1)	--	25	C	S:Dimethyl sulfoxide Salt:$KB(C_6H_5)_4$	71Ag	
Cs^+	-35.14	(1)	--	25	C	S:Acetone Salt:$CsB(C_6H_5)_4$	71Ag	
	-29.04	(1)	--	25	C	S:Dimethyl sulf-oxide Salt:CsI	71Ag	
NH_4^+	-41.76	(1)	--	25	C	S:Acetone Salt:NH_4I	71Ag	
	-17.36	(1)	--	25	C	S:Dimethyl sulf-oxide Salt:NH_4I	71Ag	
C42	18-CROWN-6, 2,6-diketo		$C_{12}H_{20}O_8$			$CH_2CH(=O)O(CH_2CH_2O)_4CH(=O)CH_2O$		
Na^+	-9.50	2.5(1)	15.4	25	C	S:Methanol	80Lb 80Bb	
K^+	-24.56	2.79(1)	-28.91	25	C	S:Methanol	80Lb 80Bb	
Rb^+	-29.25	2.09(1)	-57.5	25	C	S:Methanol	80Lb 80Bb	
Cs^+	-6.36	2.55(1)	28.1	25	C	S:Methanol	80Lb 80Bb	
Ba^{2+}	-1.7	3.1(1)	54.7	25	C	S:Methanol	80Lb 80Bb	
Ag^+	-6.40	2.50(1)	26.38	25	C	S:Methanol	80Lb 80Bb	
C43	18-CROWN-6, 4,13-dithia-7,10-diaza		$C_{10}H_{22}ON_2S_2$			$CH_2CH_2OCH_2CH_2S$- $(CH_2CH_2NH)_2CH_2CH_2S$		H_2L
Cu^{2+}	-52.4	13.26(1)	77.4	25	C	0.1M($NaClO_4$)	79Ab	

Metal Ion	ΔH, kJ /mol	Log K	ΔS, J /mol K	T,°C	M	Conditions	Ref.	Re-marks
C43, continued								
Pb^{2+}	-39.8	6.78(1)	-4.0	25	C	0.1M($NaClO_4$)	79Ab	
C44	18-CROWN-6, (4'-methylbenzo)-	$C_{17}H_{26}O_6$				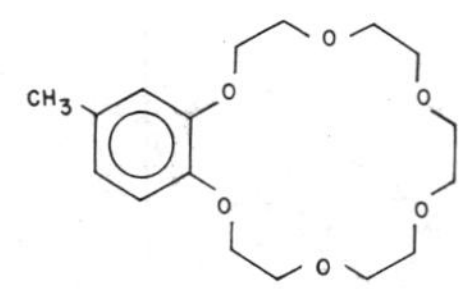		L
KAL	-12.6	-0.032(25°)	-43.5	-17-25	T	S:Tetrahydrofuran A=Fluorenyl R:$A^-K^+L=A^-LK^+$	71Ta	b,q: ±1.2
C45	18-CROWN-6, pyridino	$C_{15}H_{23}O_5N$				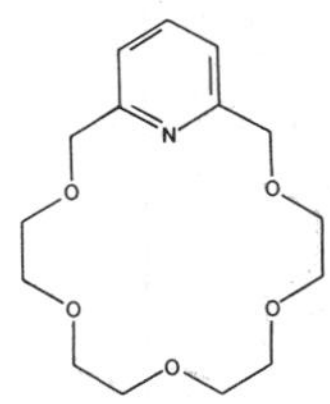		L
Na^+	-22.76	4.09(1)	1.96	25	C	S:Methanol	80La	
K^+	-38.12	5.35(1)	-25.40	25	C	S:Methanol	80La	
Rb^+	-36.4	4.56(1)	-35.1	25	C	S:Methanol	80La	
Ca^{2+}	-12.13	5.26(1)	60.3	25	C	S:Methanol	80La	
Ba^{2+}	-32.30	>5.5(1)	--	25	C	S:Methanol	80La	
Cu^{2+}	-7.03	4.63(1)	65.11	25	C	S:Methanol	80La	
Ag^+	-34.85	>5.5(1)	--	25	C	S:Methanol	80La	
C46	18-CROWN-6, thia	$C_{12}H_{24}O_5S$				$CH_2CH_2O(CH_2CH_2O)_4CH_2CH_2S$		
Na^+	-20.88	2.57(1)	-20.77	25	C	S:Methanol	80La	
K^+	-37.70	3.61(1)	-57.25	25	C	S:Methanol	80La	
Ba^{2+}	-25.5	3.4(1)	-20.35	25	C	S:Methanol	80La	
Ag^+	-51.46	>5.5(1)	--	25	C	S:Methanol	80La	
C47	19-CROWN-6, 2,4-diketo	$C_{13}H_{22}O_8$				$CH_2C(=O)O(CH_2CH_2O)_5C(=O)$		
Na^+	-4.6	1.8(1)	19.6	25	C	S:Methanol	80Lb	
K^+	-33.10	2.55(1)	-62.17	25	C	S:Methanol	80Lb	
Ba^{2+}	-20.42	1.41(1)	-42.1	25	C	S:Methanol	80Lb	

Metal Ion	ΔH, kJ /mol	Log K	ΔS, J /mol K	T,°C	M	Conditions	Ref.	Re-marks
C48	20-CROWN-6,	2,5-diketo	$C_{14}H_{24}O_8$			$CH_2C(=O)O(CH_2CH_2O)_5C(=O)CH_2$		
Na^+	-4.2	1.7(1)	18.2	25	C	S:Methanol	80Lb	
K^+	-23.4	1.94(1)	-42.1	25	C	S:Methanol	80Lb	
Rb^+	-29.3	1.74(1)	-64.6	25	C	S:Methanol	80Lb	
C49	21-CROWN-6,	2,6-diketo	$C_{15}H_{26}O_8$			$CH_2C(=O)O(CH_2CH_2O)_5C(=O)CH_2CH_2$		
K^+	-20.5	1.71(1)	-36.5	25	C	S:Methanol	80Lb	
Rb^+	-27.6	1.63(1)	-61.7	25	C	S:Methanol	80Lb	
Cs^+	-48.1	1.02(1)	-141.7	25	C	S:Methanol	80Lb	
C50	21-CROWN-7		$C_{14}H_{28}O_7$			$CH_2CH_2O(CH_2CH_2O)_5CH_2CH_2O$		
Na^+	-43.39	1.73(1)	-112.3	25	C	S:Methanol	80La	
K^+	-35.94	4.22(1)	-39.71	25	C	S:Methanol	80La	
Rb^+	-40.42	4.86(1)	-42.52	25	C	S:Methanol	80La	
Cs^+	-46.78	5.01(1)	-60.90	25	C	S:Methanol	80La	
Sr^{2+}	-29.71	1.77(1)	-65.82	25	C	S:Methanol	80La	
Ba^{2+}	-28.49	5.44(1)	8.56	25	C	S:Methanol	80La	
Ag^+	-28.9	2.46(1)	-49.1	25	C	S:Methanol	80La	
C51	21-CROWN-7,	2,6-diketo	$C_{14}H_{24}O_9$			$CH_2C(=O)(CH_2CH_2O)_5C(=O)CH_2O$		
K^+	-27.03	2.32(1)	-46.17	25	C	S:Methanol	80Lb	
Rb^+	-41.25	2.21(1)	-94.86	25	C	S:Methanol	80Lb	
Ba^{2+}	-34.89	1.73(1)	-83.92	25	C	S:Methanol	80Lb	
C52	24-CROWN-8,	dibenzo	$C_{24}H_{32}O_8$			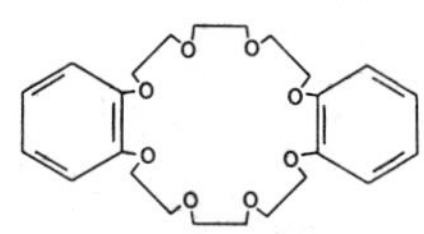		
Na^+	-37.7	4.16(25°,1)	-46.6	10-35	T	S:Propylene carbonate CM=5x10^{-4}M CL=1x10^{-2}M	80Ta	b
	-32.4	1.54(25°,1)	-79.1	10-35	T	S:70w%Methanol CM=5x10^{-4}M CL=5x10^{-3}M	80Ta	b
K^+	-34.5	3.73(25°,1)	-44.4	10-35	T	S:Propylene carbonate CM=5x10^{-4}M CL=1x10^{-2}M	80Ta	b
	-35.8	3.57(25°,1)	-49.4	10-35	T	S:Methanol CM=5x10^{-4}M	80Ta	b

Metal Ion	ΔH, kJ /mol	Log K	ΔS, J /mol K	T,°C	M	Conditions	Ref.	Re-marks
C52, continued								
K^+, cont.						$CL=5x10^{-3}M$		
	-35.7	2.42(25°,1)	-73.6	10-35	T	S:70w%Methanol $CM=5x10^{-4}M$ $CL=5x10^{-3}M$	80Ta	b
Rb^+	-32.9	3.55(25°,1)	-42.3	10-35	T	S:Propylene carbonate $CM=5x10^{-4}M$ $CL=1x10^{-2}M$	80Ta	b
	-39.4	3.86(25°,1)	-57.7	10-35	T	S:Methanol $CM=5x10^{-4}M$ $CL=5x10^{-3}M$	80Ta	b
	-36.5	2.55(25°,1)	-73.6	10-35	T	S:70w%Methanol $CM=5x10^{-4}M$ $CL=5x10^{-3}M$	80Ta	b
Cs^+	-32.3	3.46(25°,1)	-42.3	10-35	T	S:Propylene carbonate $CM=5x10^{-4}M$ $CL=1x10^{-2}M$	80Ta	b
	-36.8	3.84(25°,1)	-49.0	10-35	T	S:Methanol $CM=5x10^{-4}M$ $CL=5x10^{-3}M$	80Ta	b
	-37.4	2.48(25°,1)	-77.8	10-35	T	S:70w%Methanol $CM=5x10^{-4}M$ $CL=5x10^{-3}M$	80Ta	b
C53	30-CROWN-10, dibenzo		$C_{28}H_{40}O_{10}$			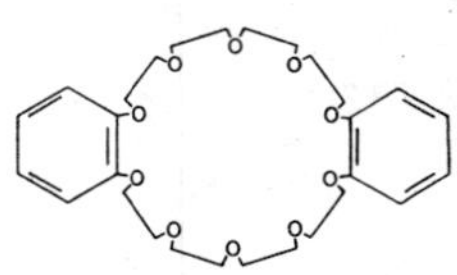		L
Na^+	-17	2.114(25°,1)	-15.9	5-35	T	μ=0.15M(Tetra-butylammonium perchlorate) $S:CH_3OH$	72Cc	b,q:±4
K^+	-48.1	4.568(25°,1)	-74.0	5-35	T	μ=0.15M(Tetra-butylammonium perchlorate) $S:CH_3OH$	72Cc	b,q:±8
Rb^+	-53.1	4.643(25°,1)	-89.5	5-35	T	μ=0.15M(Tetra-burylammonium perchlorate) $S:CH_3OH$	72Cc	b,q:±8
Cs^+	-46.9	4.230(25°,1)	-76.1	5-35	T	μ=0.15M(Tetra-butylammonium perchlorate) $S:CH_3OH$	72Cc	b,q:±8
Tl^+	-46	4.505(25°,1)	-68.2	5-35	T	μ=0.15M(Tetra-butylammonium	72Cc	b,q:±8

Metal Ion	ΔH, kJ /mol	Log K	ΔS, J /mol K	T,°C	M	Conditions	Ref.	Remarks
C53, continued								
Tl^{+}, cont.						perchlorate) S:CH_3OH		
NH_4^{+}	-23.0	2.431(25°,1)	-30.5	5-35	T	μ=0.15M(Tetra-butylammonium perchlorate S:CH_3OH	72Cc	b,q:±4

C54 CRYPTATE, diaza-heptaoxa $C_{20}H_{40}O_7N_2$

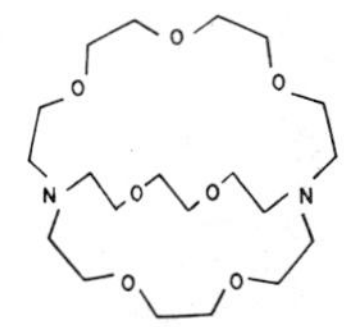

L

Metal Ion	ΔH, kJ /mol	Log K	ΔS, J /mol K	T,°C	M	Conditions	Ref.	Remarks
K^{+}	-25.9	2.1(1)	-46	25	C	4-8x10^{-2}M(LiCl)	74Ce	
	-12	2.20(1)	0	25	C	μ=0.057M pH=10.4	76Ka	
Rb^{+}	-22.6	2.05(1)	-37.6	25	C	4-8x10^{-2}M(LiCl)	74Ce	
	-17.6	2.05(1)	-19.7	25	C	μ=0.057M pH=10.4	76Ka	
Cs^{+}	-27.2	1.8(1)	-58	25	C	4-8x10^{-2}M(LiCl)	74Ce	
	-22.6	1.796(1)	-41.4	25	C	μ=0.057M pH=10.4	76Ka	
Ca^{2+}	0.67	1.98(1)	40.2	25	C	μ=0.057M pH=10.4	76Ka	
Sr^{2+}	-13.8	3.37(1)	18.4	25	C	μ=0.057M pH=10.4	76Ka	q:±1.2
Ba^{2+}	-25.9	6.01(1)	28.0	25	C	μ=0.057M pH=10.4	76Ka	

C55 CRYPTATE, diaza-hexaoxa $C_{18}H_{36}O_6N_2$

L

Metal Ion	ΔH, kJ /mol	Log K	ΔS, J /mol K	T,°C	M	Conditions	Ref.	Remarks
Na^{+}	-31.0	3.88(1)	-29.3	25	C	μ=0.057M pH=10.4	76Ka	
	-24.3	3.90(1)	-8	25	C	4-8x10^{-2}M [$N(CH_3)_4Br$]	74Ce 71Lc	
	-31.0	4.11(1)	-25	25	C	μ=0.1[$(CH_3)_4NCl$]	75Ad	
	-44.4	7.20(1)	-10.9	25	C	μ=0.057M S:95v%Methanol pH=10.4	76Ka	
K^{+}	-79.5	9.73(1)	-80.3	25	C	μ=0.057M S:95v%Methanol pH=10.4	76Ka	
	-46.4	5.30(1)	-54	25	C	4-8x10^{-2}M [$N(CH_3)_4Br$]	74Ce 71Lc	
	-46.0	5.58(1)	-48.1	25	C	μ=0.1[$(CH_3)_4NCl$]	75Ad	
	-47.7	5.28(1)	-59.0	25	C	μ=0.057M S:95v%Methanol pH=10.4	76Ka	

Metal Ion	ΔH, kJ /mol	Log K	ΔS, J /mol K	T, °C	M	Conditions	Ref.	Remarks
C55, continued								
Rb^+	-49.4	4.32(1)	-82.8	25	C	μ=0.057M pH=10.4	76Ka	
	-43.9	4.35(1)	-63	25	C	4-8x10⁻²M [N(CH₃)₄Br]	74Ce 71Lc	
	-49.4	4.06(1)	-87.4	25	C	μ=0.1[(CH₃)₄NCl]	75Ad	
	-82.0	8.38(1)	-114.6	25	C	μ=0.057M S:95v%Methanol pH=10.4	76Ka	
Cs^+	-54.0	4.255(29°,1)	-112.1	24-54	T	S:Acetone	77Me	b
	-23.8	2.193(25°,1)	-46.9	-40-54	T	S:N,N-Dimethyl-formamide	77Me	b
	-49.8	3.53(1)	-99.2	25	C	μ=0.057M S:95v%Methanol pH=10.4	76Ka	
	-36.0	3.826(29°,1)	-57.3	25-105	T	S:Propylene carbonate	77Me	b
	-10.5	0.617(25°)	-23.4	24-54	T	S:Acetone R:CsL^+(exclusive) =CsL^+(inclusive)	77Me	b
	-10.9	0.248(25°)	-31.8	-40-54	T	S:N,N-Dimethyl-formamide R:CsL^+(exclusive) =CsL^+(inclusive)	77Me	n
	-12.1	0.589(25°)	-29.3	25-105	T	S:Propylene carbonate R:CsL^+(exclusive) =CsL^+(inclusive)	77Me	b
Ca^{2+}	-0.84	4.40(1)	81.6	25	C	μ=0.057M pH=10.4	76Ka	
	-0.63	4.40(1)	84	25	C	4-8x10⁻²M [N(CH₃)₄Br]	74Ce 71Lc	
	-0.8	4.57(1)	84.5	25	C	μ=0.1[(CH₃)₄NCl]	75Ad	
	-2.1	4.447(25°,1)	79	5-40	T	μ=0.1M(piperi-dine-piperidine H^+Cl^-) pH~11.5	77Lc	b
	-27.6	7.59(1)	52.7	25	C	μ=0.057M S:95v%Methanol pH=10.4	76Ka	
Sr^{2+}	-43.1	7.99(1)	8.4	25	C	μ=0.057 pH=10.4	76Ka	
	-33.9	8.00(1)	38.	25	C	(4-8)x10⁻²M [N(CH₃)₄Br]	76Ce 71Lc	
	-44.4	8.29(1)	9.2	25	C	μ=0.1[(CH₃)₄NCl]	75Ad	
	-59.0	11.48(1)	21.9	25	C	μ=0.057M S:95v%Methanol pH=10.4	76Ka	
Ba^{2+}	-59.0	9.46(1)	-16.7	25	C	μ=0.057M pH=10.4	76Ka	
	-54.0	9.50(1)	0	25	C	(4-8)x10⁻²M [N(CH₃)₄Br]	74Ce 71Lc	
	-59.8	9.68(1)	-15.5	25	C	μ=0.1[(CH₃)₄NCl]	75Ad	

Metal Ion	ΔH, kJ /mol	Log K	ΔS, J /mol K	T,°C	M	Conditions	Ref.	Remarks
C55, continued								
Ba^{2+}, cont.	-84.1	11.95(1)	-53.3	25	C	μ=0.057M S:95v%Methanol pH=10.4	76Ka	
Ag^{+}	-53.6	9.6(1)	4.2	25	C	$\mu=0.1[(CH_3)_4NCl]$	75Ad	
Cd^{2+}	2.1	6.8(1)	138.1	25	C	$\mu=0.1[(CH_3)_4NCl]$	75Ad	
Hg^{2+}	-66.7	18.2(1)	124.7	25	C	$\mu=0.1[(CH_3)_4NCl]$	75Ad	
Tl^{+}	-55.2	5.5(1)	-61.9	25	C	$\mu=0.1[(CH_3)_4NCl]$	75Ad	
Pb^{2+}	-57.7	12.36(1)	42.2	25	C	$\mu=0.1[(CH_3)_4NCl]$	75Ad	
C56 CRYPTATE, diaza-pentaoxa			$C_{16}H_{32}O_5N_2$			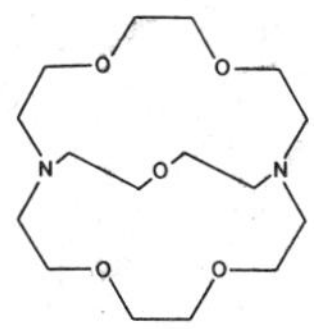		L
Li^{+}	0	2.49(1)	47.7	25	C	μ=0.057M pH=12	76Ka	
	1.7	2.50(1)	54	25	C	$4\text{-}8x10^{-2}M$ $[N(CH_3)_4Br]$	71Lc 74Ce	
Na^{+}	-22.4	5.28(1)	25.9	25	C	μ=0.057M pH=12	76Ka	
	-13.4	5.40(1)	59	25	C	$4\text{-}8x10^{-2}M$ $[N(CH_3)_4Br]$	71Lc	
	-19.7	5.30(1)	33	25	C	$4\text{-}8x10^{-2}M$ $[N(CH_3)_4Br]$	74Ce	
K^{+}	-28.4	3.96(1)	-19.7	25	C	μ=0.057M pH=12	76Ka	
	-24.7	3.95(1)	-8	25	C	$4\text{-}8x10^{-2}M$ $[N(CH_3)_4Br]$	71Lc 74Ce	
Rb^{+}	-22.6	2.53(1)	-27.2	25	C	μ=0.057M pH=12	76Ka	
	-13.4	2.55(1)	4	25	C	$4\text{-}8x10^{-2}M$ $[N(CH_3)_4Br]$	71Lc 74Ce	
	-12.1	6.96(1)	92.0	25	C	μ=0.057M pH=12	76Ka	
Ca^{2+}	-7.1	6.95(1)	109	25	C	$4\text{-}8x10^{-2}M$ $[N(CH_3)_4Br]$	71Lc	
	-11.3	6.95(1)	96	25	C	$4\text{-}8x10^{-2}M$ $[N(CH_3)_4Br]$	74Ce	
	-5.4	6.863(25°,1)	113	5-40	T	μ=0.1M(piperi-dine-piperidine $H^{+}Cl^{-}$) pH∿11.5	77Lc	b
Sr^{2+}	-25.5	7.33(1)	54.8	25	C	μ=0.057M pH=12	76Ka	
	-20.9	7.35(1)	71	25	C	$4\text{-}8x10^{-2}M$ $[N(CH_3)_4Br]$	71Lc 74Ce	
Ba^{2+}	-26.4	6.30(1)	32.2	25	C	μ=0.057M pH=12	76Ka	
	-21.3	6.30(1)	50	25	C	$4\text{-}8x10^{-2}M$ $[N(CH_3)_4Br]$	71Lc 74Ce	

Metal Ion	ΔH, kJ /mol	Log K	ΔS, J /mol K	T, °C	M	Conditions	Ref.	Remarks
C57	CRYPTATE, diaza-tetraoxa		$C_{14}H_{28}O_4N_2$			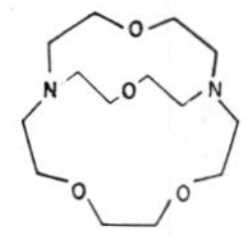		L
Li^+	-21.3	5.50(1)	33	25	C	μ=0.057M pH=12	76Ka	
	-19.2	4.30(1)	17	25	C	4-8x10^{-2}M [N(CH_3)$_4$Br]	71Lc 74Ce	
Na^+	-22.6	3.30(1)	-12	25	C	μ=0.057M pH=12	76Ka	
	-19.7	2.70(1)	-17	25	C	4-8x10^{-2}M [N(CH_3)$_4$Br]	74Ce	
Ca^{2+}	-0.4	2.49(1)	46	25	C	μ=0.057M pH=12	76Ka	q:±0.04
	0	2.80(1)	54	25	C	4-8x10^{-2}M [N(CH_3)$_4$Br]	71Lc 74Ce	
	-1.2	2.301(25°,1)	46	5-40	T	μ=0.1M(piperidine-piperidine H^+Cl^-) pH∿11.5	77Lc	b
C58	CYANAMIDE, dimethyl-		$C_3H_6N_2$			(CH_3)$_2$NCN		L
[Co^{3+}]	-53.1	--	--	24	C	S:Dichloromethane [Co]= Methylcobaloxime R:[Co]$_2$+2L=2[Co]L	73Cf	
I_2	-11.7	0.249(25°,1)	-34.3	13-43	T	S:CCl_4	65Af	b
IA	-23.4	1.202(25°,1)	-55.6	9-40	T	S:CCl_4 A=Br	65Af	b
	-30.5	2.13(25°,1)	-63.2	13-44	T	S:CCl_4 A=Cl	65Af	b
C59	CYANIC ACID		CHON			HOCN		L^-
[Fe^{3+}]	-44.4	2.86(1)	-94.6	25	-	Chironomus hemoglobin	60Gb	
CoA_2BC	-17.6	3.68(25°,1)	13	32.7-48	T	S:DMSO R:trans-CoA_2BC+L^-=trans-CoA_2BL^-+C A=Methylglyoximato B=CH_3 C=DMSO	74Lb	b
C60	CYANIDE ION		CN			CN^-		L^-
Mg^{2+}	18.0	3.77(24.8°)	133	15.2-35	T	μ=0 R:Mg^{2+}+ $Fe(CN)_6^{4-}$= $MgFe(CN)_6^{2-}$	75Ja	q:±2.4
Ca^{2+}	5.8	2.631(25°)	71	15-35	T	μ=0 R:Ca^{2+}+ $Fe(CN)_6^{3-}$= $CaFe(CN)_6^-$	74Ha	b,q: ±1.2
	8.4	3.634(25°)	100	15-35	T	μ=0 R:Ca^{2+}+ $Fe(CN)_6^{4-}$= $CaFe(CN)_6^{2-}$	74Ha	b,q: ±1.2

Metal Ion	ΔH, kJ /mol	Log K	ΔS, J /mol K	T,°C	M	Conditions	Ref.	Re-marks
C60, continued								
Ca^{2+}, cont.	13.0	3.67(24.8°)	115	15.2-35	T	μ=0 R:Ca^{2+}+ $Fe(CN)_6^{4-}$= $CaFe(CN)_6^{2-}$	75Ja	q:±3.8
	6.7	2.631	71	25	C	μ=0.07M R:Ca^{2+}+ $Fe(CN)_6^{3-}$= $CaFe(CN)_6^{-}$	74Ha	q:±0.8
	8.8	3.634	100	25	C	μ=0.07M R:Ca^{2+}+ $Fe(CN)_6^{4-}$= $CaFe(CN)_6^{2-}$	74Ha	q:±0.8
Sr^{2+}	7.8	3.66(24.8°)	96	15.2-35	T	μ=0 R:Sr^{2+}+ $Fe(CN)_6^{4-}$= $SrFe(CN)_6^{2-}$	75Ja	q:±9.2
Ba^{2+}	17.5	3.78(24.8°)	131	4.8-24.8	T	μ=0 R:Ba^{2+}+ $Fe(CN)_6^{4-}$= $BaFe(CN)_6^{2-}$	75Ja	q:±6.1
V^{2+}	-196.6	(1-6)	--	25	C	1M(KCN)	64Gc	
Cr^{2+}	-264.4	(1-6)	--	25	C	1M(KCN)	64Gc	
Mn^{2+}	-144.3	(1-6)	--	25	C	1M(KCN)	64Gc	
Fe^{2+}	-358.86	35.4(1-6)	-526.8	25	C	μ=0	65Wa 64Gc	
	-63.39	--	--	25	C	CFeT=0 R:Fe^{2+}+FeL_6^{3-}= Turnbull's blue	70Ia	
[Fe(II)]	-81.2	4.93(25°,1)	-177.8	15.5-36.5	T	μ=0.10($NaClO_4$) pH=7.0 [Fe(II)]= Iron(II) 2,4-bis-(cysteinylhystidine)dipeptide substituted mesoporphyrin [heme(cys-his)$_2$]	76Gf	
	-36.0	3.90(25°,1)	-46.0	15.5-36.5	T	μ=0.10($NaClO_4$) pH=7.0 [Fe(II)]= Iron(II) 2,4-dicysteine-substituted mesoporphyrin (Cyanoheme C)	76Gf	q:±3.8
	-58.2	5.66(25°,1)	-86.6	15.5-36.5	T	μ=0.1($NaClO_4$) pH=6.5 [Fe(II)]= Iron(II) 2,4-dicysteine-substituted mesoporphyrin (heme c)	76Gf	
Fe^{3+}	-20.1	7.04(1)	67	25	-	--	60Gb	
	-293.5	43.6(1-6)	-149.4	25	C	μ=0	64Gc 65Wa	

Metal Ion	ΔH, kJ /mol	Log K	ΔS, J /mol K	T,°C	M	Conditions	Ref.	Remarks
C60, continued								
Fe^{3+}, cont.	3.01	--	--	25	C	CFeT=0 R:Fe^{3+}+$FeL_6{}^{4-}$= Prussian blue	70Ia	q:±0.75
[Fe(III)]	-35.5	3.89(25°,1)	-44.8	15.5-36.5	T	μ=0.10($NaClO_4$) pH=7.0 [Fe(III)] =Iron(III) 2,4-bis(cysteinyl-histidine)dipeptide substituted mesoporphyrin [hem(cys-his)$_2$]	76Gf	q:±2.9
	-31.4	5.41(25°,1)	-1.7	15.5-36.5	T	μ=0.10($NaClO_4$) pH=6.0 [Fe(III)] =Iron(III) 2,4-dicysteine-substituted mesoporphyrin (hemin c)	76Gf	q:±2.5
	-32.2	4.79(25°,1)	-16.3	15.5-36.5	T	μ=0.10($NaClO_4$) pH=6.0 [Fe(III)] =Iron(III) 2,4-dicysteine-substituted mesoporphyrin (cyanohemin C)	76Gf	
[Fe^{3+}]	-77.8	8.36(1)	-100	25	-	Ferrimyglobin	60Gb 55Ga	
[Fe^{3+}]	4.6	6.10(24.6°,1)	131.0	11.7-30.4	T	Ferricytochrome 0.075M borate buffer	55Ga 52Ga	b
[Fe^{3+}]	-69.04	9.102(1)	-61.1	20	T	Human methaemoglobin A 6.0	68Ab	d:0.05
	-76.15	9.028(1)	-87.03	20	T	6.4	68Ab	
	-81.59	8.968(1)	-106.7	20	T	6.8	68Ab	
	-83.26	8.946(1)	-105.4	20	T	7.0	68Ab	
	-76.94	8.819(1)	-92.5	20	T	7.2	68Ab	
	-60.25	8.610(1)	-40.67	20	T	7.6	68Ab	
	-47.70	8.558(1)	1.13	20	T	7.8	68Ab	
[Fe^{3+}]	-60.67	9.073(1)	-33.26	20	T	Human methaemoglobin C 6.0	68Ab	d:0.05
	-68.20	9.020(1)	-59.8	20	T	6.4	68Ab	
	-71.96	8.924(1)	-74.5	20	T	6.8	68Ab	
	-75.73	8.871(1)	-88.3	20	T	7.0	68Ab	
	-79.50	8.834(1)	-102.1	20	T	7.2	68Ab	
	-88.28	8.633(1)	-136.0	20	T	7.5	68Ab	
	-86.19	8.416(1)	-133.0	20	T	7.8	68Ab	
$FeL_6{}^{3-}$	-63	3.176	-149.8	12-25	T	μ=0.6M(HAC+Ac^-) pH=5.0 A=EDTA R:Fe(III)$L_6{}^{3-}$+ Co(II)A^{2-}=	67Hh	b,r

Metal Ion	ΔH, kJ /mol	Log K	ΔS, J /mol K	T,°C	M	Conditions	Ref.	Re-marks
C60, continued								
FeL_6^{3-}, cont.						$[ACo(III)L-Fe(II)L_5]^{5-}$		
Co^{2+}	-257.3	(1-5)	--	25	C	μ=0	68Ic	
	-311.3	(1-6)	--	25	C	1M(KCN) Product may be $Co(CN)_5H_2O^{3-}$ or a dimer rather than $Co(CN)_6^{4-}$	64Gc	
CoA^{2-}	-36.0	1.580(1)	-90.8	25	T	μ=0.10($NaClO_4$) A=1,2-diamino-cyclohexane-tetraacetate	69Jb	
Ni^{2+}	-12.6	-0.721(25°,5)	-55.6	15-35	T	μ=1.34	60Mb	b
	-189.1	32.2(1-4)	-32.6	10	C	μ=0	71Ih	
	-180.7	30.1(1-4)	-29	25	C	μ=0	63Cc	
				25	C	μ=0	71Ih	ΔCp=151
	-183.7	27.43(1-4)	-61.5	40	C	μ=0	71Ih	
	179.08	(1-4)	--	25	C	1M(NaOH) pH=14	78Fb	
NiA^{2+}	101.34	--	--	25	C	1M(NaOH) pH=14 R:$NiA^{2+}+4L^-=NiL_4^{2-}+A$ A=3,7-Diazanonane, 1,9-diamine	78Fb	
	-129.3	--	--	25	C	R:$NiA^{2+}+4L^-=NiL_4^{2-}+A$ A=1,4,7,10-Tetra-azacyclododecane	80Fa	
	-104.2	--	--	25	C	R:$NiA^{2+}+4L^-=NiL_4^{2-}+A$ A=1,4,8,12-Tetra-azacyclopenta-decane	80Fa	
	-94.06	--	--	25	C	1M(NaOH) pH=14 R:$NiA^{2+}+4L^-=NiL_4^{2-}+A$ A=1,4,8,11-Tetra-azacyclotetra-decane	78Fb	
	-122.6	--	--	25	C	R:$NiA^{2+}+4L^-=NiL_4^{2-}+A$ A=1,4,7,10-Tetra-azacyclotridecane	80Fa	
Pd^{2+}	-0.8	2.8(5)	33	25	C	μ=0	67Ic	
	-386.2	42.2(1-4)	-485	25	C	μ=0	67Ic	
Cu^{+}	-46.4	5.30(3)	-56.1	25	C	μ=0 Log K value from ref 50Va corrected using HCN	67Ib	

Metal Ion	ΔH, kJ /mol	Log K	ΔS, J /mol K	T,°C	M	Conditions	Ref.	Remarks
C60, continued								
Cu^+,						pK=9.21		
cont.	-46.9	1.5(4)	-130	25	C	μ=0 Log K value from ref 50Va corrected using HCN pK=9.21	67Ib	
	-121.8	23.94(1-2)	50	25	C	μ=0 Log K value from ref 50Va corrected using HCN pK=9.21	67Ib	
CuA	-44.4	3.079(1)	-92	25	C	I=0.1 pH=11.5 A=Tetradentate α-amineoxime $(PnAO-H)^+$	78Me	
	-71.1	4.618(1)	-180	25	C	I=0.1 pH=11.5 A=Tetradentate α-amineoxime $(EnAO-2H)^+$	78Me	
Ag^+	-2.5	1.55(3)	21	25	C	μ=0	67Ib	
	-137.6	20.44(1-2)	-67	25	C	μ=0	67Ib	
	-149.4	10.87(1-2)	-77.4	250	T	S:Equimolar molten salt of KNO_3 & $NaNO_3$ K in units of $moles^2/1000gm^2$	56Fa	
Zn^{2+}	-39.4	5.17(3)	-41.4	10	C	μ=0	71Ih	
	-38.5	4.98(3)	-33.5	25	C	μ=0	71Ih	ΔCp=21
	-35.1	4.98(3)	-22.6	25	C	μ=0	65Ia 67Ab	
	-38.9	4.50(3)	-38.1	40	C	μ=0	71Ih	
	-31.0	3.79(4)	-36.8	10	C	μ=0	71Ih	
	-32.2	3.57(4)	-39.7	25	C	μ=0	71Ih	ΔCp=-71
	-36.0	3.57(4)	-51.9	25	C	μ=0	65Ia	
	-33.0	3.10(4)	-59.4	40	C	μ=0	71Ih	
	-48.5	11.47(1-2)	48.1	10	C	μ=0	71Ih	
	-46.0	11.07(1-2)	57.3	25	C	μ=0	71Ih	ΔCp=151
	-45.2	11.067(1-2)	58	25	C	μ=0	65Ia	
	-43.9	10.70(1-2)	64.4	40	C	μ=0	71Ih	
	-80.3	16.050(1-3)	37.6	25	C	μ=0	65Ia	
	-119.2	20.43(1-4)	-30.1	10	C	μ=0	71Ih	
	-116.7	19.62(1-4)	-15.9	25	C	μ=0	71Ih	ΔCp=100
	-116.3	19.62(1-4)	-14.2	25	C	μ=0	65Ia	
	-115.9	18.30(1-4)	-19.7	40	C	μ=0	71Ih	
	-115.9	(1-4)	--	25	C	1M(KCN)	64Gc	
Cd^{2+}	-33.0	6.22(1)	2.5	10	C	μ=0	71Ih	
	-30.5	6.01(1)	12.6	25	C	μ=0	71Ih	ΔCp=176
	-27.82	5.73(1)	20.9	40	C	μ=0	71Ih	
	-30.92	5.48(1)	1.2	25	C	3.0M($NaClO_4$)	66Gb	
	-28.83	5.38(2)	-1.2	10	C	μ=0	71Ih	

Metal Ion	ΔH, kJ /mol	Log K	ΔS, J /mol K	T, °C	M	Conditions	Ref.	Re-marks
C60, continued								
Cd^{2+}, cont.	-23.8	5.11(2)	18.0	25	C	μ=0	71Ih	ΔCp=234
	-21.56	4.90(2)	25.1	40	C	μ=0	71Ih	
	-32.34	5.12(2)	-10.5	25	C	3.0M($NaClO_4$)	66Gb	
	-34.7	4.77(3)	-32.0	10	C	μ=0	71Ih	
	-35.82	4.53(3)	-33.5	25	C	μ=0	71Ih	ΔCp=-29
	-36.0	4.12(3)	-35.6	40	C	μ=0	71Ih	
	-29.79	4.54(3)	-13.0	25	C	3.0M($NaClO_4$)	66Gb	
	-21.8	2.52(4)	-29.3	10	C	μ=0	71Ih	
	-21.3	2.27(4)	-28.9	25	C	μ=0	71Ih	ΔCp=-29
	-22.68	2.12(4)	-31.8	40	C	μ=0	71Ih	
	-29.41	3.16(4)	-38.1	25	C	3.0M($NaClO_4$)	66Gb	
	-118.70	18.89(1-4)	-57.7	10	C	μ=0	71Ih	
	-111.96	17.92(1-4)	-33.5	25	C	μ=0	71Ih	ΔCp=351
	-107.82	16.87(1-4)	-21.3	40	C	μ=0	71Ih	
	-122.6	18.79(1-4)	-51.5	25	C	μ=3	67Ab	
Hg^{2+}	-100.08	17.97(1)	-9.6	10	C	μ=0	71Ih	
	-97.1	17.00(1)	0.0	25	C	μ=0	71Ih	ΔCp=180
	-96.2	17.00(1)	2.9	25	C	μ=0	65Cg	
	-94.6	16.26(1)	9.2	40	C	μ=0	71Ih	
	-100.58	16.74(2)	-34.7	10	C	μ=0	71Ih	
	-97.9	15.75(2)	-26.8	25	C	μ=0	71Ih	ΔCp=155
	-106.7	15.75(2)	-56.1	25	C	μ=0	65Cg	
	-95.8	15.02(2)	-18.4	40	C	μ=0	71Ih	
	-30.79	3.81(3)	-35.6	10	C	μ=0	71Ih	
	-28.62	3.56(3)	-27.6	25	C	μ=0	71Ih	ΔCp=88
	-31.8	3.56(3)	-37.6	25	C	μ=0	65Cg	
	-28.07	3.37(3)	-25.1	40	C	μ=0	71Ih	
	-22.01	2.81(4)	-23.8	10	C	μ=0	71Ih	
	-26.4	2.66(4)	-37.6	25	C	μ=0	71Ih	ΔCp=-251
	-30.1	2.66(4)	-50.6	25	C	μ=0	65Cg	
	-29.71	2.46(4)	-47.7	40	C	μ=0	71Ih	
	-253.51	41.33(1-4)	-104.2	10	C	μ=0	71Ih	
	-250.2	38.97(1-4)	-92.0	25	C	μ=0	71Ih	ΔCp=172
	-248.1	37.11(1-4)	-82.0	40	C	μ=0	71Ih	
HgA^+	-92.5	14.016(1)	-47.7	20	C	μ=0.1(KNO_3) A=CH_3	65Sa	
C61 CYCLOBUTANE, methylene-			C_5H_8			$CH_2CH_2CH_2CH{=}CH_2$		L
Ag^+	-16.90	-1.046(1)	-77.0	25	T	1M(KNO_3)	59Ta	o
I_2	-3.51	0.444(1)	3.3	25	T	S:2,2,4-Trimethylpentane	59Ta	
C62 1,1-CYCLOBUTANEDICARBOXYLIC ACID			$C_6H_8O_4$			$CH_2CH_2CH_2C(COOH)_2$		L^{2-}
Cu^{2+}	10.9	4.993(1)	132.2	25	C	μ=0.1M($NaClO_4$)	77Am	
	2.9	3.434(2)	74.9	25	C	μ=0.1M($NaClO_4$)	77Am	
	-36.4	14.209	150.2	25	C	μ=0.1M($NaClO_4$) R:$Cu^{2+}+L^{2-}+A=CuAL$ A=2,2'-Bipyridyl	77Am	

Metal Ion	ΔH, kJ /mol	Log K	ΔS, J /mol K	T, °C	M	Conditions	Ref.	Re-marks
C63	CYCLOBUTENEDIONE, methyl-hydroxy		$C_5H_4O_3$			O=C-C-OH O=C-C-CH_3		L^-
Eu^{3+}	-11.3	0.653(25°,1)	-26.8	2-51	T	μ=1.0M($NaClO_4$) pH=2	76Yd	b
C64	1,5,9-CYCLODODECATRIENE		$C_{12}H_{18}$			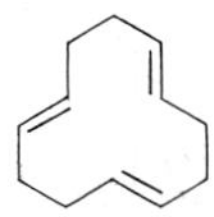		L
PdA_2B_2	-24.7	--	--	25	C	S:CH_2Cl_2 CL=4.6x10^{-3}M CM=2.6x10^{-4}M A=Cl B=C_6H_5CN R:PdA_2B_2+L= PdA_2L+2B	72Pd	
C65	CYCLOHEPTANE, 1-aza-		$C_6H_{13}N$			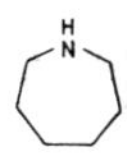		
NiA_2	-32.6	2.037(24.6°,1)	-66.9	8-39.5	T	S:Benzene A=O,O'-Diethyl- dithiophosphate	72Mc	b
	-33.0	2.561(24.6°,1)	-63	8-39.5	T	S:Benzene A=O,O'-Dimethyl- dithiophosphate	72Mc	b
C66	CYCLOHEPTANE, 1,4-diaza		$C_5H_{12}N_2$			$CH_2NHCH_2CH_2NHCH_2CH_2$ (ring)		L
Cu^{2+}	-23.4	7.72(30°,1)	71	10-40	T	μ→0	61Pa	b
	-24.3	5.93(30°,2)	33	10-40	T	μ→0	61Pa	b
C67	CYCLOHEPTANE, methylene-		C_8H_{14}			$CH_2CH_2CH_2CH_2CH_2CH_2C=CH_2$ (ring)		L
Ag^+	-13.43	-1.140(1)	-67.4	25	T	1M(KNO_3)	59Ta	o
I_2	-12.05	0.4265(1)	32.2	25	T	S:2,2,4-Tri- methylpentane	59Ta	
C68	2,4,6-CYCLOHEPTATRIEN-1-one, 2-hydroxy-		$C_7H_6O_2$			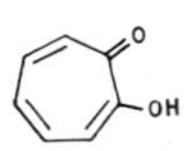		
Y^{3+}	-10.71	7.17(1)	101.3	25	C	μ=0.1M(KNO_3)	70Ca	
La^{3+}	-10.38	6.19(1)	84	25	C	μ=0.1M(KNO_3)	70Ca	
Ce^{3+}	-11.30	6.55(1)	87.9	25	C	μ=0.1M(KNO_3)	70Ca	
Pr^{3+}	-11.80	6.60(1)	86.6	25	C	μ=0.1M(KNO_3)	70Ca	
Nd^{3+}	-11.92	6.77(1)	91.2	25	C	μ=0.1M(KNO_3)	70Ca	
Sm^{3+}	-11.76	6.90(1)	92.9	25	C	μ=0.1M(KNO_3)	70Ca	

Metal Ion	ΔH, kJ /mol	Log K	ΔS, J /mol K	T,°C	M	Conditions	Ref.	Re-marks
C68, continued								
Eu^{3+}	-11.55	7.10(1)	97.1	25	C	μ=0.1M(KNO_3)	70Ca	
Gd^{3+}	-11.38	7.03(1)	96.6	25	C	μ=0.1M(KNO_3)	70Ca	
Tb^{3+}	-11.46	7.14(1)	98.3	25	C	μ=0.1M(KNO_3)	70Ca	
Dy^{3+}	-11.72	7.23(1)	99.2	25	C	μ=0.1M(KNO_3)	70Ca	
Ho^{3+}	-11.55	7.40(1)	102.9	25	C	μ=0.1M(KNO_3)	70Ca	
Er^{3+}	-11.67	7.54(1)	105.4	25	C	μ=0.1M(KNO_3)	70Ca	
Tm^{3+}	-12.30	7.59(1)	104.2	25	C	μ=0.1M(KNO_3)	70Ca	
Yb^{3+}	-13.05	7.84(1)	106.7	25	C	μ=0.1M(KNO_3)	70Ca	
Lu^{3+}	-12.13	7.69(1)	106.7	25	C	μ=0.1M(KNO_3)	70Ca	
C69	2,4,6-CYCLOHEPTATRIEN-1-ONE, 2-hydroxy-7-isopropenyl-		$C_{10}H_{10}O_2$			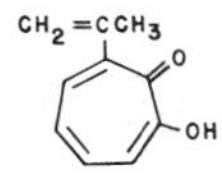		
CoL_3	-3.3	0.518(25°)	-1.3	-56-20	T	S:$CDCl_3$	72Ea	b,q: ±0.33,bb
AlL_3	-7.1	0.623(25°)	-11.3	-35-30	T	S:$C_2H_2Cl_4$	72Ea	b,g: ±0.8,bb
	-0.84	0.602(25°)	9.2	-63-25	T	S:CH_2Cl_2	72Ea	b,q: ±0.08,bb
C70	2,4,6-CYCLOHEPTATRIEN-1-ONE, 2-hydroxy-		$C_{10}H_{11}O_2$			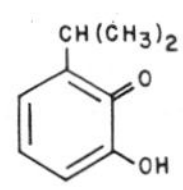		
CoL_3	-3.93	0.623(25°)	-0.8	-55--20	T	S:$CHCl_3$	72Ea	b,q: ±1.7,bb
AlL_3	0.0	0.505(25°)	12.1	-55--20	T	S:$C_2H_2Cl_4$	72Ea	b,q: ±2.1,bb
C71	CYCLOHEPTENE		C_7H_{12}			$CH_2CH_2CH_2CH_2CH_2CH=CH$		L
Ag^+	-27.15	-1.674(1)	-123.0	25	T	1M(KNO_3)	59Ta	o
	-27.66	-1.66(1)	-124.7	25	T	μ=1.0($AgNO_3$)	59Ta	o
I_2	-8.16	0.484(1)	18.0	25	T	S:2,2,4-Tri-methylpentane	59Ta	
C72	1,4-CYCLOHEXADIENE		C_6H_8			$CH=CHCH=CHCH_2CH_2$		L
Cu^+	-29.3	2.53(1)	-48.1	30	T	S:1M($LiClO_4$) in 2-propanol	69Hb	

Metal Ion	ΔH, kJ /mol	Log K	ΔS, J /mol K	T,°C	M	Conditions	Ref.	Remarks
C73	CYCLOHEXANE, 1,2-diamino-(cis)		$C_6H_{14}N_2$			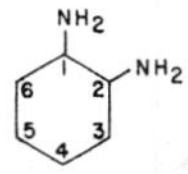		L
Ni^{2+}	-32.2	7.12(30°,1)	29	10-40	T	μ=0	58Ba	b
	-26.3	5.80(30°,2)	25	10-40	T	μ=0	58Ba	b
Cu^{2+}	-48.5	10.72(30°,1)	42	10-40	T	μ=0	58Ba	b
	-50.2	9.40(30°,2)	12.6	10-40	T	μ=0	58Ba	b
Zn^{2+}	-21.8	5.74(30°,1)	37.6	10-40	T	μ=0	58Ba	b
	-17.6	5.38(30°,2)	42	10-40	T	μ=0	58Ba	b
Cd^{2+}	-23.8	5.65(30°,1)	29	10-40	T	μ=0	58Ba	b
	-23.4	4.64(30°,2)	8	10-40	T	μ=0	58Ba	b
C74	CYCLOHEXANE, 1,2-diamino-(trans)		$C_6H_{14}N_2$			$C_6H_{10}(NH_2)_2$		L
Ni^{2+}	-35.6	7.82(30°,1)	33	10-40	T	μ=0	58Ba	b
	-33.0	6.71(30°,2)	21	10-40	T	μ=0	58Ba	b
	-38.1	4.67(30°,3)	38	10-40	T	μ=0	58Ba	b
Cu^{2+}	-56.9	10.96(30°,1)	21	10-40	T	μ=0	58Ba	b
	-51.5	9.54(30°,2)	12	10-4	T	μ=0	58Ba	b
Zn^{2+}	-20.9	6.14(30°,1)	50	10-40	T	μ=0	58Ba	b
	-21.8	5.31(30°,2)	29	10-40	T	μ=0	58Ba	b
Cd^{2+}	-23.4	5.74(30°,1)	33	10-40	T	μ=0	58Ba	b
	-27.6	4.74(30°,2)	0	10-40	T	μ=0	58Ba	b
C75	CYCLOHEXANE, 1,2-diamino-N,N',N'-tetraacetic acid (trans)		$C_{14}H_{22}O_8N_2$			$C_6H_{10}[N(CH_2CO_2H]_2$		L^{4-}
Mg^{2+}	15.90	10.97(1)	264.0	25	C	0.1M(KNO_3)	63Ab	
	6.7	10.3(1)	217.6	25	C	μ=0.1(KNO_3)	65Wd	
	19.20	10.78(1)	270.7	25	C	μ=0.20(KNO_3)	69Ca	
	7.91	10.408(1)	226.4	20	C	μ=1M(KCl) CL=0.01M	69Sh	g:k
Ca^{2+}	-15.48	13.15(1)	198.7	20	C	0.1M(KNO_3)	63Ab	
	-25.9	12.2(1)	146	25	C	μ=0.1(KNO_3)	65Wd	
	-15.15	12.92(1)	196.6	25	C	μ=0.20(KNO_3)	69Ca	
	-23.22	12.712(1)	164.0	20	C	μ=1M(KCl) CL=0.01M	69Sh	g:k
Sr^{2+}	-3.10	10.5(1)	191.2	20	C	0.1M(KNO_3)	63Ab	
	-15.1	9.97(1)	142	25	C	μ=0.1(KNO_3)	65Wd	
	-3.05	10.36(1)	187.9	25	C	μ=0.20(KNO_3)	69Ca	
	-2.34	10.520(1)	193.3	20	C	μ=1M(KCl) CL=0.01M	69Sh	g:k
Ba^{2+}	1.38	8.64(1)	171.1	20	C	0.1M(KNO_3)	63Ab	
	-9.2	7.84(1)	121	25	C	μ=0.1(KNO_3)	65Wd	
	1.17	8.49(1)	165.7	25	C	μ=0.20(KNO_3)	69Ca	

Metal Ion	ΔH, kJ /mol	Log K	ΔS, J /mol K	T,°C	M	Conditions	Ref.	Remarks
C75, continued								
Ba^{2+}, cont.	-3.81	8.112(1)	142.2	20	C	μ=1M(KCl) CL=0.01M	69Sh	g:k
Y^{3+}	17.70	19.41(1)	431.0	25	T	0.1M(KNO_3)	62Mg	
	8.8	19.34(1)	402	20	T	--	72Ya	q:±4
	19	19.42(1)	418	30	T	--	72Ya	q:±4
	28	19.55(1)	460	40	T	--	72Ya	q:±4
La^{3+}	6.69	16.91(1)	346.4	20	C	0.1M(KNO_3)	63Ab	
	14.85	16.35(1)	362.8	25	T	0.1M(KNO_3)	62Mg	
	6.3	16.28(1)	335	20	T	--	72Ya	q:±4
	18	16.35(1)	372	30	T	--	72Ya	q:±4
	29	16.48(1)	410	40	T	--	72Ya	q:±4
Pr^{3+}	20.92	17.23(1)	400.0	25	T	0.1M(KNO_3)	62Mg	
	9.2	17.15(1)	360	20	T	--	72Ya	q:±4
	21	17.24(1)	397	30	T	--	72Ya	q:±4
	33	17.39(1)	439	40	T	--	72Ya	q:±4
Nd^{3+}	20.92	17.69(1)	408.4	25	T	0.1M(KNO_3)	62Mg	
	12	17.62(1)	376	20	T	--	72Ya	q:±4
	22	17.72(1)	410	30	T	--	72Ya	q:±4
	32	17.87(1)	444	40	T	--	72Ya	q:±4
Sm^{3+}	20.92	18.63(1)	427	25	T	0.1M(KNO_3)	62Mg	
	9.2	18.55(1)	389	20	T	--	72Ya	q:±4
	21	18.64(1)	427	30	T	--	72Ya	q:±4
	33	18.79(1)	464	40	T	--	72Ya	q:±4
Eu^{3+}	23.18	18.77(1)	437.2	25	T	0.1M(KNO_3)	62Mg	
	-2.5	17.60(50°,1)	328.4	30-80	T	μ=1.0(KCl) CM=0.1M Complex with structure 1 as determined by absorption spectra	71Kc	b
	22.26	16.97(50°,1)	393.3	30-80	T	μ=1.0(KCl) CM=0.1M Complex with structure 2 as determined by absorption spectra	71Kc	b
	17	18.69(1)	414	20	T	--	72Ya	q:±4
	24	18.81(1)	435	30	T	--	72Ya	q:±4
	30	18.96(1)	460	40	T	--	72Ya	q:±4
Gd^{3+}	24.06	18.80(1)	440.6	25	T	0.1M(KNO_3)	62Mg	
	15	18.73(1)	410	20	T	--	72Ya	q:±4
	23	18.84(1)	435	30	T	--	72Ya	q:±4
	31	18.99(1)	460	40	T	--	72Ya	q:±4
Tb^{3+}	20.92	19.30(1)	439.7	25	T	0.1M(KNO_3)	62Mg	
	13	19.23(1)	414	20	T	--	72Ya	q:±4
	21	19.33(1)	439	30	T	--	72Ya	q:±4
	30	19.47(1)	464	40	T	--	72Ya	q:±4
Dy^{3+}	12.93	19.69(1)	420.5	25	T	0.1M(KNO_3)	62Mg	
	7.1	19.64(1)	402	20	T	--	72Ya	q:±4

Metal Ion	ΔH, kJ /mol	Log K	ΔS, J /mol K	T,°C	M	Conditions	Ref.	Remarks
C75, continued								
Dy^{3+}, cont.	13	19.70(1)	420	30	T	--	72Ya	q:±4
	19	19.79(1)	439	40	T	--	72Ya	q:±4
Ho^{3+}	4.94	19.89(1)	397.5	25	T	0.1M(KNO_3)	62Mg	
	5.8	19.88(1)	397	20	T	--	72Ya	q:±4
	4.2	19.91(1)	393	30	T	--	72Ya	q:±4
	2.9	19.93(1)	389	40	T	--	72Ya	q:±4
Er^{3+}	0.50	20.20(1)	388.3	25	T	0.1M(KNO_3)	62Mg	
	9.2	20.20(1)	448	20	T	--	72Ya	q:±4
	0.8	20.28(1)	389	30	T	--	72Ya	q:±4
	-7.9	20.21(1)	347	40	T	--	72Ya	q:±4
Tm^{3+}	-6.65	20.46(1)	369.4	25	T	0.1M(KNO_3)	62Mg	
Yb^{3+}	-18.79	20.80(1)	335.6	25	T	0.1M(KNO_3)	62Mg	
	-18	20.85(1)	334	20	T	--	72Ya	q:±4
	-18	20.75(1)	335	30	T	--	72Ya	q:±4
	-18	20.64(1)	335	40	T	--	72Ya	q:±4
Lu^{3+}	-20.58	20.91(1)	331.4	25	T	0.1M(KNO_3)	62Mg	
	-34	20.94(1)	276	20	T	--	72Ya	q:±4
	-20	20.77(1)	326	30	T	--	72Ya	q:±4
	-4.6	20.70(1)	381	40	T	--	72Ya	q:±4
Zr^{4+}	-17.61	20.68(1)	336.85	25	T	μ=0.1(KNO_3)	65Hd	
Th^{4+}	-2.22	23.78(1)	448.02	25	T	μ=0.1(KNO_3)	65Hd	
Mn^{2+}	-17.32	17.43(1)	274.47	20	C	0.1M(KNO_3)	63Ab	
	-29.7	16.6(1)	218	25	C	μ=0.1(KNO_3)	65Wd	
Fe^{2+}	-27.6	18.2(1)	255	25	C	μ=0.1(KNO_3)	65Wd	
	3.3	5.04(24.2°)	109	13-35	T	μ=0.1($NaClO_4$) R:Fe(II)L^{2-} = Fe(III)L^-	74Wa	b
	-3.8	9.322(28.1°)	163	18.2-40	T	μ=0.20 R:Fe^{2+}+ HL^{3-}=FeHL$^-$	72Bj	b,q: ±1.2
	-4.6	5.301(28.1°)	84	18.2-40	T	μ=0.20 R:Fe^{2+}+ H_2L^{2-}=FeH_2L	72Bj	b,q: ±3.8
Fe^{3+}	-25.1	29.77(25°,1)	485	13-35	T	μ=0.1($NaClO_4$)	74Wa	b
	-59.4	29.146(28.1°,1)	360	18.2-40	T	μ=0.20	72Kb	b
Co^{2+}	-11.72	19.57(1)	334.7	20	C	0.1M(KNO_3)	63Ab	
	-22.6	18.8(1)	284.5	25	C	μ=0.1(KNO_3)	65Wd	
	-73.2	1.639(25°)	-214.2	17-30	T	μ=0.660(NaAc+HAc) pH=5.0 R:CoL^{2-}+ $Fe(CN)_6{}^{3-}$= [LCoCNFeCN$_5{}^{5-}$]	73Hf	b
Ni^{2+}	-22.47	(1)	--	20	C	0.1M(KNO_3)	63Ab	
	-31.4	19.4(1)	264	25	C	μ=0.1(KNO_3)	65Wd	
Cu^{2+}	-25.40	21.95(1)	333.5	20	C	0.1M(KNO_3)	63Ab	
	-36.8	21.1(1)	280	25	C	μ=0.1(KNO_3)	65Wd	
	-30.04	20.94(25°,1)	300.0	10-40	T	μ=0.1(KNO_3)	67Hg	b
Zn^{2+}	-8.12	19.32(1)	342.3	20	C	0.1M(KNO_3)	63Ab	
	-32.2	18.5(1)	247	25	C	μ=0.1(KNO_3)	65Wd	

Metal Ion	ΔH, kJ /mol	Log K	ΔS, J /mol K	T,°C	M	Conditions	Ref.	Remarks
C75, continued								
Zn^{2+},	-14.98	18.53(25°,1)	304.6	10-40	T	μ=0.1(KNO_3)	67Hg	b
cont.	-8.12	18.99(1)	336.4	25	C	μ=0.20(KNO_3)	69Ca	
Cd^{2+}	-30.96	19.88(1)	274.9	20	C	0.1M(KNO_3)	63Ab	
	-46.9	19.1(1)	209	25	C	μ=0.1(KNO_3)	65Wd	
	-27.74	18.50(25°,1)	261.1	10-40	T	μ=0.1(KNO_3)	67Hg	b
	-29.08	19.54(1)	276.6	25	C	μ=0.20(KNO_3)	69Ca	
Hg^{2+}	-69.45	24.95(1)	246.8	20	C	0.1M(KNO_3)	63Ab 62Mg	
	-79.1	24.3(1)	201	25	C	μ=0.1(KNO_3)	65Wd 65Hd 66Mf	
	-67.15	24.53(1)	244.3	25	C	μ=0.20(KNO_3)	69Ca	
Al^{3+}	44.8	18.73(1)	509.2	25	T	μ=0.1(KNO_3)	66Mf	
Pb^{2+}	-47.53	20.33(1)	226.8	20	C	0.1M(KNO_3)	63Ab	
	-51.9	19.4(1)	197	25	C	μ=0.1(KNO_3)	65Wd	
C76	CYCLOHEXANE, 1,2-dihydroxy-(trans)		$C_6H_{12}O_2$			$C_6H_{10}(OH)_2$		L
$H_5TeO_6^-$	-13.4	1.16	-21	25	T	μ=0.1 R:$H_5TeO_6^-$ +L=$HTeO_4^-$+$2H_2O$	62Ea	
C77	CYCLOHEXANE, 1,4-epoxy-		$C_6H_{10}O$					L
I_2	-19.83	0.517(25°,1)	-56.5	6.7-40	T	S:n-Heptane ΔH determined from visible spectral data	64Ta	b
	-20.67	0.591(25°,1)	-57.7	6.7-40	T	S:n-Heptane ΔH determined from ultraviolet spectral data	64Ta	b
C78	CYCLOHEXANE, methylene		C_7H_{12}			$C_6H_{11}(CH_2)$		L
Ag^+	-13.01	-1.223(1)	-66.9	25	T	1M(KNO_3)	59Ta	o
I_2	-7.45	0.562(1)	14.2	25	T	S:2,2,4-Trimethylpentane	59Ta	
C79	CYCLOHEXANE, 1,3,5-triamino (cis,cis)		$C_6H_{15}N_3$			$C_6H_9(NH_2)_3$		L
Zn^{2+}	-0.21	6.90(1)	129.7	25		μ=0.1M(KCl)	76Fa	q:±2.7
C80	1,1-CYCLOHEXANEDICARBOXYLIC ACID		$C_8H_{12}O_4$			$CH_2CH_2CH_2CH_2CH_2C(CCOH)_2$		L^{2-}
Cu^{2+}	14.2	4.625(1)	136.0	25	C	μ=0.1M($NaClO_4$)	77Am	
	7.1	2.611(2)	74.1	25	C	μ=0.1M($NaClO_4$)	77Am	

Metal Ion	ΔH, kJ /mol	Log K	ΔS, J /mol K	T,°C	M	Conditions	Ref.	Re-marks
C80, continued								
Cu^{2+}, cont.	-35.6	13.018	129.7	25	C	μ=0.1M($NaClO_4$) R:$Cu^{2+}+L^{2-}$+A=CuAL A=2,2'-Bipyridyl	77Am	
C81	1,2-CYCLOHEXANEDIOL(cis)		$C_6H_{12}O_2$			$C_6H_{10}(OH)_2$		L
BA_4^-	-16.74 -41.84	0.000(25°,1) -0.155(25°,1-2)	-42 -167	0-35 0-35	T T	CL=0-0.9M A=OH^- CL=0-0.9M A=OH^-	67Cb	b,q:±0.4 b,q:±10.5
C82	CYCLOHEXANONE		$C_6H_{10}O$			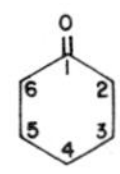		L
TiA_4	-57.40 -56.90	(1) (2)	-- --	25 25	C C	S:CH_2Cl_2 A=Cl S:CH_2Cl_2 A=Cl	72Ec 72Ec	
CuA_2	-36.8	(1)	--	--	C	S:CCl_4 A=Hexafluoro-acetylacetonato	76Mc	u:25°
BF_3	-131.0	(1)	--	25	C	S:CH_2Cl_2	71Pa	
$SiAB_3$	-15.27	1.45(30°,1)	-22.64	20-60	T	S:CCl_4 XL=5x10^{-2} XM=5x10^{-4} A=OH B=C_2H_5	69Lb	b
	-13.68	1.46(30°,1)	-17.28	10-45	T	S:CCl_4 XL=5x10^{-2} XM=5x10^{-4} A=OH B=CH_3	69Lb	b
	-16.48	1.74(30°,1)	-21.00	0-50	T	S:CCl_4 XL=5x10^{-2} XM=4x10^{-4} A=OH B=Phenyl	69Lb	b
I_2	-25.5	1.34(20°,1)	-63	-15-35	T	S:Freon	65Wc	b,k
I_2L	-84	3.95(20°)	-209	-15-35	T	S:Freon R:$2I_2L=(I_2L)_2$	65Wc	b,k
C83	CYCLOHEXANONE, 2,6-dimethyl-		$C_8H_{14}O$			$C_6H_{10}O(CH_3)_2$		L
TiA_4	-40.84 -60.2	(1) (2)	-- --	25 25	C C	S:CH_2Cl_2 A=Cl S:CH_2Cl_2 A=Cl	72Ec 72Ec	
BF_3	-98.3	(1)	--	25	C	S:CH_2Cl_2	71Pa	
C84	CYCLOHEXANONE, 2-methyl-		$C_7H_{12}O$			$C_6H_{10}O(CH_3)$		L
TiA_4	-44.89 -39.16	(1) (2)	-- --	25 25	C C	S:CH_2Cl_2 A=Cl S:CH_2Cl_2 A=Cl	72Ec 72Ec	
BF_3	-104.6	(1)	--	25	C	S:CH_2Cl_2	71Pa	

Metal Ion	ΔH, kJ /mol	Log K	ΔS, J /mol K	T,°C	M	Conditions	Ref.	Re-marks
C85	CYCLOHEXANONE, 3-methyl-		$C_7H_{12}O$			$C_6H_{10}O(CH_3)$		L
TiA_4	-47.86	(1)	--	25	C	S:CH_2Cl_2 A=Cl	72Ec	
	-38.32	(2)	--	25	C	S:CH_2Cl_2 A=Cl	72Ec	
BF_3	-106.7	(1)	--	25	C	S:CH_2Cl_2 A=Cl	71Pa	
C86	CYCLOHEXANONE, 4-methyl-		$C_7H_{12}O$			$C_6H_{10}O(CH_3)$		L
TiA_4	-51.21	(1)	--	25	C	S:CH_2Cl_2 A=Cl	72Ec	
	-37.32	(2)	--	25	C	S:CH_2Cl_2 A=Cl	72Ec	
BF_3	-113.0	(1)	--	25	C	S:CH_2Cl_2	71Pa	
C87	CYCLOHEXENE(cis)		C_6H_{10}			$CH_2CH_2CH_2CH_2CH{=}CH$ (ring)		L
Cu^+	-34.7	2.09(1)	-74.5	30	T	S:1M($LiClO_4$) in 2-propanol	69Hb	
Ag^+	-24.02	-1.735(1)	-113.8	25	T	1M(KNO_3)	59Ta	o
	-23.35	-1.73(1)	-111.3	25	T	1.0M($AgNO_3$)	56Tc	o
	-24.27	-1.735(25°,1)	-114.6	0-25	T	μ=1.0($AgNO_3$,KNO_3)	38Wa	b,o
I_2	-9.12	0.518(1)	20.5	25	T	S:2,2,4-Trimethylpentane	59Ta	
C88	2-CYCLOHEXEN-1-ONE		C_6H_8O			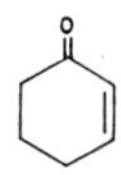		L
I_2	-22.6	1.49(20°,1)	-46	-15-35	T	S:Freon	65Wc	b,k
C89	CYCLOLEUCINE		$C_6H_{13}O_2N$			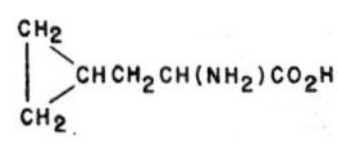		
CoA	-21.3	3.60(25°,1)	28.9	15-55	T	μ=0.1M(KNO_3) A=Methyliminodiacetate	72Ic	b
NiA	-19.7	4.50(25°,1)	20.1	15-55	T	μ=0.1M(KNO_3) A=Methyliminodiacetate	72Ic	b,q: ±3.3
CuA	-31.8	6.11(25°,1)	10.9	15-55	T	μ=0.1M(KNO_3) A=Methyliminodiacetate	72Ic	b,q: ±6.7
CdA	-20.1	3.41(25°,1)	5.0	15-55	T	μ=0.1M(KNO_3) A=Methyliminodiacetate	72Ic	b,q: ±2.1
C90	1,5-CYCLOOCTADIENE(cis,cis)		C_8H_{12}			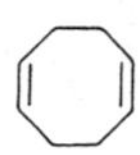		
PdA_2B_2	-54.4	--	--	25	C	S:CH_2Cl_2	72Pd	

Metal Ion	ΔH, kJ /mol	Log K	ΔS, J /mol K	T,°C	M	Conditions	Ref.	Re- marks
C90, continued								
PdA_2B_2, cont.						$CL=5x10^{-3}M$ $CM=1.3x10^{-4}M$ A=Cl $B=C_6H_5CN$ $R:PdA_2B_2+L=$ PdA_2L+2B	73Pe	
C91 1,3,5,7-CYCLOOCTATETRAENE			C_8H_6			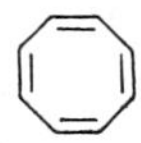		L
Ru_3A_8L	2.9	0.74	-4.2	-56-60	T	S:Hexane R:Isomerization from (a) to (b) form A=CO	79Kb	b,q:±0.4
Ru_3A_8B	2.5	0.50	-8.4	-56-60	T	S:Hexane R:Isomerization from (a) to (b) form A=CO $B=C_8H_5[Si(CH_3)_3]$	79Kb	b,q:±0.4
	3.8	0.83	2.9	-56-60	T	S:Hexane R:Isomerization from (a) to (b) form A=CO $B=C_8H_3[Si(CH_3)_3]$	79Kb	b,q:±0.4
PdA_2B_2	-23.4	--	--	25	C	$S:CH_2Cl_2$ $CL=5x10^{-3}M$ $CM=2x10^{-4}M$ A=Cl $B=C_6H_5CN$ $R:PdA_2B_2+L=$ PdA_2L+2B	72Pd	
C92 CYCLOOCTENE(cis)			C_8H_{14}			$CH_2(CH_2)_5CH=CH$		L
Cu^+	-64.0	3.46(1)	-144.8	30	T	$S:1M(LiClO_4)$ in 2-propanol	69Hb	
Ag^+	-17.20	-2.298(1)	-103.8	25	T	$1M(KNO_3)$	59Ta	o
I_2	-2.01	0.037(1)	5.8	25	T	S:2,2,4-Tri-methylpentane	59Ta	
C93 CYCLOPENTANE, 1,2-diamino-N,N,N',N'-tetraacetic acid (trans)			$C_{13}H_{20}O_8N_2$			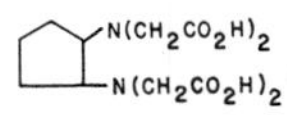		
Mg^{2+}	18.16	(1)	--	20	C	$\mu=0.1M(NaNO_3)$	71If	
	18.28	9.349(1)	241.4	20	C	$\mu=1M(HCl)$ CL=0.01M	69Sh	g:k
Ca^{2+}	-20.38	11.370(1)	148.1	20	C	$\mu=1M(HCl)$ CL=0.01M	69Sh	g:k
Sr^{2+}	-1.34	9.528(1)	177.8	20	C	$\mu=1M(HCl)$ CL=0.01M	69Sh	g:k

Metal Ion	ΔH, kJ /mol	Log K	ΔS, J /mol K	T,°C	M	Conditions	Ref.	Re-marks
C93, continued								
Ba^{2+}	-11.00	8.678(1)	128.4	20	C	μ=1M(HCl) CL=0.01M	69Sh	g:k
La^{3+}	-7.046	17.01(1)	301.7	20	C	μ=0.1M($NaNO_3$)	71If	
	-25.162	--	--	20	C	μ=0.1M($NaNO_3$) R:La^{3+}+MgL^{2-}= LaL^-+Mg^{2+}	71If	
Ce^{3+}	-6.832	17.28(1)	307.5	20	C	μ=0.1M($NaNO_3$)	71If	
	-24.949	--	--	20	C	μ=0.1M($NaNO_3$) R:Ce^{3+}+MgL^{2-}= CeL^-+Mg^{2+}	71If	
Pr^{3+}	-10.364	17.47(1)	299.2	20	C	μ=0.1M($NaNO_3$)	71If	
	-28.480	--	--	20	C	μ=0.1M($NaNO_3$) R:Pr^{3+}+MgL^{2-}= PrL^-+Mg^{2+}	71If	
Nd^{3+}	-11.389	17.72(1)	300.4	20	C	μ=0.1M($NaNO_3$)	71If	
	-29.506	--	--	20	C	μ=0.1M($NaNO_3$) R:Nd^{3+}+MgL^{2-}= NdL^-+Mg^{2+}	71If	
Sm^{3+}	-11.171	18.11(1)	308.8	20	C	μ=0.1M($NaNO_3$)	71If	
	-29.288	--	--	20	C	μ=0.1M($NaNO_3$) R:Sm^{3+}+MgL^{2-}= SmL^-+Mg^{2+}	71If	
Eu^{3+}	-7.761	18.21(1)	322.2	20	C	μ=0.1M($NaNO_3$)	71If	
	-25.878	--	--	20	C	μ=0.1M($NaNO_3$) R:Eu^{3+}+MgL^{2-}= EuL^-+Mg^{2+}	71If	
Gd^{3+}	-5.167	18.24(1)	331.8	20	C	μ=0.1M($NaNO_3$)	71If	
	-23.284	--	--	20	C	μ=0.1M($NaNO_3$) R:Gd^{3+}+MgL^{2-}= GdL^-+Mg^{2+}	71If	
Tb^{3+}	-1.456	18.64(1)	351.9	20	C	μ=0.1M($NaNO_3$)	71If	q:±0.08
	-19.573	--	--	20	C	μ=0.1M($NaNO_3$) R:Tb^{3+}+MgL^{2-}= TbL^-+Mg^{2+}	71If	
Dy^{3+}	3.971	18.94(1)	376.1	20	C	μ=0.1M($NaNO_3$)	71If	
	-14.146	--	--	20	C	μ=0.1M($NaNO_3$) R:Dy^{3+}+MgL^{2-}= DyL^-+Mg^{2+}	71If	
Ho^{3+}	6.640	19.24(1)	391.2	20	C	μ=0.1M($NaNO_3$)	71If	
	-11.477	--	--	20	C	μ=0.1M($NaNO_3$)	71If	

Metal Ion	ΔH, kJ /mol	Log K	ΔS, J /mol K	T,°C	M	Conditions	Ref.	Re-marks
C93, continued								
Ho^{3+}, cont.						R:Ho^{3+}+MgL^{2-}= HoL^{-}+Mg^{2+}		
Er^{3+}	5.034	19.49(1)	390.4	20	C	μ=0.1M($NaNO_3$)	71If	
	-13.083	--	--	20	C	μ=0.1M($NaNO_3$)	71If	
						R:Er^{3+}+MgL^{2-}= ErL^{-}+Mg^{2+}		
Tm^{3+}	-1.100	19.71(1)	373.6	20	C	μ=0.1M($NaNO_3$)	71If	q:±0.08
	-19.217	--	--	20	C	μ=0.1M($NaNO_3$)	71If	
						R:Tm^{3+}+MgL^{2-}= TmL^{-}+Mg^{2+}		
Yb^{3+}	-7.786	19.95(1)	355.2	20	C	μ=0.1M($NaNO_3$)	71If	
	-25.903	--	--	20	C	μ=0.1M($NaNO_3$)	71If	
						R:Yb^{3+}+MgL^{2-}= YbL^{-}+Mg^{2+}		
Lu^{3+}	0.895	20.20(1)	389.9	20	C	μ=0.1M($NaNO_3$)	71If	
	-16.803	--	--	20	C	μ=0.1M($NaNO_3$)	71If	
						R:Lu^{3+}+MgL^{2-}= LuL^{-}+Mg^{2+}		
C94	CYCLOPENTANE, 1,3-diaza- 1,3-dimethyl-2-oxo-2-thia-		$C_4H_{12}ON_2S$			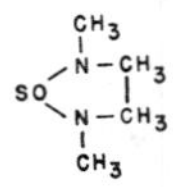		
AlA_3	-101.75	--	--	28	C	S:Hexane A=Cl R:½$[AlA_3]_2$+L= $AlLA_3$	68Hd	
C95	CYCLOPENTANE, methylene-		C_6H_{10}			$CH_2CH_2CH_2CH_2C$=CH_2 (ring)		L
Ag^{+}	-13.85	-1.174(1)	-69.0	25	T	1M(KNO_3)	59Ta	o
I_2	-4.02	0.415(1)	5.4	25	T	S:2,2,4-Tri- methylpentane	59Ta	
C96	1,1-CYCLOPENTANEDICARBOXYLIC ACID		$C_7H_{10}O_4$			$CH_2CH_2CH_2CH_2C(COOH)_2$ (ring)		L^{2-}
Cu^{2+}	13.6	4.888(1)	139.3	25	C	μ=0.1M($NaClO_4$)	77Am	
	6.5	2.803(2)	75.3	25	C	μ=0.1M($NaClO_4$)	77Am	
	-36.0	13.263	130.1	25	C	μ=0.1M($NaClO_4$)	77Am	
						R:Cu^{2+}+L^{2-}+A=CuAL A=2,2'-Bipyridyl		

Metal Ion	ΔH, kJ /mol	Log K	ΔS, J /mol K	T,°C	M	Conditions	Ref.	Re-marks
C97	1,2-CYCLOPENTANEDIOL(cis)		$C_5H_{10}O_2$			$C_5H_8(OH)_2$		L
BA_4^-	-14.77	1.415(25°,1)	-22.6	0-35	T	CL=0-0.9M A=OH^-	67Cc	b,q:±1.7
	-25.36	2.152(25°,1-2)	-46	0-35	T	CL=0-0.9M A=OH^-		b,q:±2.5
C98	1,2,3,4-CYCLOPENTANETETRA-CARBOXYLIC ACID (cis)		$C_9H_{10}O_8$			$C_5H_6(CO_2H)_4$		H_4L
Ce^{3+}	18.74	9.80(30°,1)	249.4	30-50	T	μ=0.20M(ClO_4^-) pH=2.75-3.50	79Nb	
Pr^{3+}	18.79	10.05(30°,1)	254.0	30-50	T	μ=0.20M(ClO_4^-) pH=2.75-3.50	79Nb	
Nd^{3+}	20.96	10.10(30°,1)	262.3	30-50	T	μ=0.20M(ClO_4^-) pH=2.75-3.50	79Nb	
Sm^{3+}	21.76	10.15(30°,1)	266.3	30-50	T	μ=0.20M(ClO_4^-) pH=2.75-3.50	79Nb	
Eu^{3+}	19.16	10.28(30°,1)	260.4	30-50	T	μ=0.20M(ClO_4^-) pH=2.75-3.50	79Nb	
Gd^{3+}	18.95	10.20(30°,1)	257.4	30-50	T	μ=0.20M(ClO_4^-) pH=2.75-3.50	79Nb	
Tb^{3+}	23.05	10.55(30°,1)	277.8	30-50	T	μ=0.20M(ClO_4^-) pH=2.75-3.50	79Nb	
Dy^{3+}	25.31	10.60(30°,1)	286.7	30-50	T	μ=0.20M(ClO_4^-) pH=2.75-3.50	79Nb	
Ho^{3+}	28.45	10.62(30°,1)	297.1	30-50	T	μ=0.20M(ClO_4^-) pH=2.75-3.50	79Nb	
C99	CYCLOPENTANONE		C_5H_8O			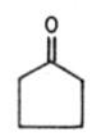		L
TiA_4	-56.82	(1)	--	25	C	S:CH_2Cl_2 A=Cl	72Ec	
	-56.48	(2)	--	25	C	S:CH_2Cl_2 A=Cl	72Ec	
BF_3	-119.2	(1)	--	25	C	S:CH_2Cl_2	71Pa	
I_2	-10.5	(1)	--	0-45	T	S:Cyclohexane CL≅1M	66Bc	b,q:±4
	-19.7	1.34(20°,1)	-42	-15-35	T	S:Freon	65Wc	b,k
I_2L	-67	3.88(20°)	-155	-15-35	T	S:Freon R:$2I_2L=(I_2L)_2$	65Wc	b,k
C100	CYCLOPENTANONE, 2,5-dimethyl-		$C_7H_{12}O$			$C_5H_8O(CH_3)_2$		L
TiA_4	-43.26	(1)	--	25	C	S:CH_2Cl_2 A=Cl	72Ec	
	-20.84	(2)	--	25	C	S:CH_2Cl_2 A=Cl	72Ec	

Metal Ion	ΔH, kJ /mol	Log K	ΔS, J /mol K	T,°C	M	Conditions	Ref.	Re-marks
C101	CYCLOPENTANONE, 2-methyl-		$C_6H_{10}O$			$C_5H_8O(CH_3)$		L
TiA_4	-54.10	(1)	--	25	C	S:CH_2Cl_2 A=Cl	72Ec	
	-42.68	(2)	--	25	C	S:CH_2Cl_2 A=Cl	72Ec	
BF_3	-118.0	(1)	--	25	C	S:CH_2Cl_2	71Pa	
C102	CYCLOPENTANONE, 2,2,5,5-tetramethyl-		$C_9H_{16}O$			$C_5H_8O(CH_3)_4$		L
TiA_4	-39.12	(1)	--	25	C	S:CH_2Cl_2 A=Cl	72Ec	
	-20.17	(2)	--	25	C	S:CH_2Cl_2 A=Cl	72Ec	
BF_3	-92.9	(1)	--	25	C	S:CH_2Cl_2	71Pa	
C103	CYCLOPENTANONE, 2,2,4-trimethyl-		$C_8H_{14}O$			$C_5H_8O(CH_3)_3$		L
BF_3	-110.0	(1)	--	25	C	S:CH_2Cl_2	71Pa	
C104	CYCLOPENTANONE, 2,4,4-trimethyl-		$C_8H_{14}O$			$C_5H_8O(CH_3)_3$		L
TiA_4	-50.00	(1)	--	25	C	S:CH_2Cl_2 A=Cl	72Ec	
	-43.14	(2)	--	25	C	S:CH_2Cl_2 A=Cl	72Ec	
BF_3	-113.0	(1)	--	25	C	S:CH_2Cl_2	71Pa	
C105	CYCLOPENTENE		C_5H_8			$CH_2CH_2CH_2CH=CH$ (ring)		L
Cu^+	-53.1	2.86(1)	-120.1	30	T	S:1M($LiClO_4$) in 2-propanol	69Hb	
Ag^+	18.0	1.009(25°,1)	-41.4	0-40	T	S:Ethylene glycol	65Ci	b
	-27.74	-0.925(1)	-110.9	25	T	1M(KNO_3)	59Ta	o
	-29.41	-0.945(1)	-116.7	25	T	1.0M$AgNO_3$	56Tc	o
I_2	-1.80	0.447(1)	-2.5	25		TS:2,2,4-Trimethylpentane	59Ta	
C106	1,1-CYCLOPROPANEDICARBOXYLIC		$C_5H_6O_4$			$CH_2CH_2C(COOH)_2$ (ring)		L^{2-}
Cu^{2+}	3.8	5.992(1)	127.8	25	C	μ=0.1M($NaClO_4$)	77Am	
	0.3	3.609(2)	70.3	25	C	μ=0.1M($NaClO_4$)	77Am	
	-46.0	15.471	141.8	25	C	μ=0.1M($NaClO_4$) R:$Cu^{2+}+L^{2-}+A=CuAL$ A=2,2'-Bipyridyl	77Am	
C107	CYSTEINE		$C_3H_7O_2NS$			$HSCH_2CH(NH_2)CO_2H$		L^{2-}
Ni^{2+}	-57.3	19.77(25°,1-2)	188	15-40	T	μ=0.2M(KNO_3)	73St	b
Pb^{2+}	-42.4	13.207(25°,1)	110.6	10-40	T	μ=3.00M($NaClO_4$)	76Cf	
	-56.9	17.434(25°)	142.9	10-40	T	μ=3.00M($NaClO_4$) R:$Pb^{2+}+HL=PbHL^{2+}$	76Cf	

Metal Ion	ΔH, kJ /mol	Log K	ΔS, J /mol K	T,°C	M	Conditions	Ref.	Remarks
C107, continued								
Pb^{2+}, cont.	-111.8	27.301(25°)	147.8	10-40	T	μ=3.00M($NaClO_4$) R:Pb^{2+}+L^-+HL=$PbHL_2^+$	76Cf	
C108 CYTIDINE-5'-diphosphoric acid $C_{14}H_{24}O_{15}N_3P_2$								L^{4-}
Mg^{2+}	15.1	3.17(25°,1)	111.3	5-35	T	μ=0.1M(KNO_3)	73Bb	b,g:Na
	16.3	1.64(25°)	86.2	5-35	T	μ=0.1M(KNO_3) R:Mg^{2+}+HL^{3-}=$MgHL^-$	73Bb	b,g:Na
C109 CYTOCHROME C								
[Fe(II)]	-52.7	--	--	25	C	0.05M Buffer pH=6.00 R:Fe(II)L+Fe(III) =Fe(III)L+Fe(II)	69Wa	
	-52.7	--	--	25	C	0.05M Buffer pH=6.50 R:Fe(II)L+Fe(III) =Fe(III)L+Fe(II)	69Wa	
	-52.7	--	--	25	C	0.05M Buffer pH=7.00 R:Fe(II)L+Fe(III) =Fe(III)L+Fe(II)	69Wa	
	-48.5	--	--	25	C	0.05M Buffer pH=7.50 R:Fe(II)L+Fe(III) =Fe(III)L+Fe(II)	69Wa	
	-45.2	--	--	25	C	0.05M Buffer pH=8.00 R:Fe(II)L+Fe(III) =Fe(III)L+Fe(II)	69Wa	
	-37.32	--	--	25	C	0.05M Buffer pH=8.45 R:Fe(II)L+Fe(III) =Fe(III)L+Fe(II)	69Wa	
	-32.64	--	--	25	C	0.05M Buffer pH=8.90 R:Fe(II)L+Fe(III) =Fe(III)L+Fe(II)	69Wa	
	-37.11	--	--	25	C	0.05M Buffer pH=9.00 R:Fe(II)L+Fe(III) =Fe(III)L+Fe(II)	69Wa	
	-7.86	--	--	25	C	0.05M Buffer pH=9.75 R:Fe(II)L+Fe(III) =Fe(III)L+Fe(II)	69Wa	
	2.93	--	--	25	C	0.05M Buffer pH=10.00 R:Fe(II)L+Fe(III) =Fe(III)L+Fe(II)	69Wa	

Metal Ion	ΔH, kJ /mol	Log K	ΔS, J /mol K	T,°C	M	Conditions	Ref.	Remarks
C109, continued								
CoA_3^{3+}	19.4	1.92	$\underline{24.3}$	25	C	0.05MPhosphate buffer pH=6.8 R: CoA_3^{3+}+L(II)= CoL_3^{2+} +L(III) A=1,10-Phenanthroline	78Cb	
	26.4	1.56	$\underline{28.2}$	25	C	0.05MPhosphate buffer pH=6.8 R: CoA_3^{3+}+L(II)= CoL_3^{2+}+L(III) A=2,2'-Bipyridine	78Cb	

D

Metal Ion	ΔH, kJ /mol	Log K	ΔS, J /mol K	T,°C	M	Conditions	Ref.	Remarks
D1 DECANE, 1-amino-			$C_{10}H_{23}N$			$CH_3(CH_2)_9NH_2$		L
VOA_2	-44.31	$(\underline{1})$	--	25	C	μ→0 A:Nitrobenzene A=Acetylacetonate	65Ca	
NiA	-72.0	6.13$(\underline{1-2})$	-123.8	25	T	CM=10^{-4}M S:Benzene A=Diacetylbis-benzoyl hydrazine	60Sb	
D2 DECANE, 4,7-diaza-1,10-diamino- $C_8H_{22}N_4$						$NH_2(CH_2)_3NH(CH_2)_2NH(CH_2)_3NH_2$		L
Ni^{2+}	-80.29	14.69$(\underline{1})$	12.1	25	C	μ=0.5M(KNO_3)	73Bc	
	-50.2	9.75	18.4	25	C	μ=0.5M(KNO_3) R: Ni^{2+}+HL^+=$NiHL^{3+}$	73Bc	
Cu^{2+}	-103.8	21.69$(\underline{1})$	66.1	25	C	μ=0.10M	73Ha	
	-108.32	21.83$(\underline{1})$	54.8	25	C	μ=0.5M(KNO_3)	73Bc	
	-70.3	14.69	45.6	25	C	μ=0.10M R: Cu^{2+}+HL^+=$CuHL^{3+}$	73Ha	
	-75.7	14.73	28.0	25	C	μ=0.5M(KNO_3) R: Cu^{2+}+HL^+=$CuHL^{3+}$	73Bc	
Zn^{2+}	-44.43	11.25$(\underline{1})$	66.5	25	C	μ=0.5M(KNO_3)	73Bc	
	-32.2	7.18	29.3	25	C	μ=0.5M(KNO_3) R: Zn^{2+}+HL^+=$ZnHL^{3+}$	73Bc	

Metal Ion	ΔH, kJ /mol	Log K	ΔS, J /mol K	T,°C	M	Conditions	Ref.	Remarks
D3	DECANE, 4,7-dithio-1,10-diamino		$C_8H_{20}N_2S_2$			$NH_2(CH_2)_3S(CH_2)_2S(CH_2)_3NH_2$		L
Ni^{2+}	-50.3	7.691(1)	-21	25	C	0.5M(KNO_3)	80Ha	
	-20	3.55	-3	25	C	0.5M(KNO_3) R:Ni^{2+}+HL^+=$NiHL^{3+}$	80Ha	
Cu^{2+}	-67.9	10.911(1)	-18.9	25	C	0.5M(KNO_3)	80Ha	
	-35	5.05	-20	25	C	0.5M(KNO_3) R:Cu^{2+}+HL^+=$CuHL^{3+}$	80Ha	
D4	DECANE, 3,8-dithio-1,10-dihydroxy		$C_8H_{18}O_2S_2$			$HOCH_2CH_2SCH_2CH_2CH_2CH_2SCH_2CH_2OH$		
Ag^+	-56.9	5.82(1)	-82.4	20	C	μ=1.0(KNO_3) pH=3	70Wc	
	-38.9	2.38(2)	-87.0	20	C	μ=1.0(KNO_3) pH=3	70Wc	
	-15.5	2.38	-6.3	20	C	μ=1.0(KNO_3) pH=3 R:AgL+Ag=Ag_2L	70Wc	
D5	DECANE, 4,4,9,9-tetramethyl-5,8-diaza-2,11-isonitroso		$C_{14}H_{30}O_2N_4$			$CH_3(NOH)CCH_2C(CH_3)_2NHCH_2$-$CH_2NHC(CH_3)_2CH_2C(NOH)CH_3$		H_2L
Cu^{2+}	-54.8	13.24(1)	70	25	C	μ=0.1M(NaCl) R:Cu^{2+}+H_2L=CuH_2L^{2+}	78Da	
	23.7	-3.24	18	25	C	μ=0.1M(NaCl) R:$Cu(H_2L)^{2+}$= $Cu(HL)^+$+H^+	78Da	
D6	DECANE, 4,4,9,9-tetramethyl-5,8-diaza-2,11-isonitroso-o,o'-dimethyl		$C_{16}H_{34}O_2N_4$			$CH_3(NOCH_3)CCH_2C(CH_3)_2NH$-$CH_2CH_2NHC(CH_3)_2CH_2C$-$(NOCH_3)CH_3$		L
Cu^{2+}	-17.9	9.10(1)	114	25	C	μ=1.0M(NaCl)	78Da	
D7	DECANE, 4,4,9,9-tetramethyl-5,8-diaza-2,11-isonitroso-o-methyl		$C_{15}H_{32}O_2N_4$			$CH_3(NOH)CCH_2C(CH_3)_2NHCH_2$-$CH_2NHC(CH_3)_2CH_2C(NOCH_3)CH_3$		HL
Cu^{2+}	-43	12.11(1)	87	25	C	μ=1.0M(NaCl) R:Cu^{2+}+H_2L=CuH_2L^{2+}	78Da	
	21.9	-6.76	-56	25	C	μ=1.0M(NaCl) R:$Cu(HL)^{2+}$=CuL^++H^+	78Da	
D8	DEOXYRIBONUCLEIC ACID							
Fe^{3+}	≅0	5.114(25°,1)	96	25-45	T	CM<$5x10^{-4}$M pH=3.1	65Kb	b

Metal Ion	ΔH, kJ /mol	Log K	ΔS, J /mol K	T,°C	M	Conditions	Ref.	Remarks
D9 4,7-DIAZADECANE, 1,10-diamino $C_8H_{22}N_4$						$H_2NCH_2CH_2CH_2NHCH_2CH_2NH$-$CH_2CH_2CH_2NH_2$		
Ni^{2+}	18.4	-1.09(25°)	63	20-45	T	μ=1.0M($NaClO_4$) R:$NiL(H_2O)_2^{2+}$ (blue)=NiL^{2+} (yellow)+$2H_2O$	77Aj	
	-67.6	13.98(1)	41.9	15-40	T	μ=0.1M(KCl)	74So	b
Cu^{2+}	-103.8	21.18(1)	58.6	15-40	T	μ=0.1M(KCl)	74So	b
D10 3,7-DIAZANONANE, 1,9-diamino $C_7H_{20}N_4$						$H_2NCH_2CH_2NHCH_2CH_2CH_2NHCH_2CH_2NH_2$		
Ni^{2+}	-81.2	15.8(1)	30	25	C	μ=0.1	74He	
	-81.2	15.8(1)	31	25	C	μ=0.1	74Hf	
	-70.3	15.3(1)	57.7	25	C	μ=0.1 Complex is square-planar	74He	
	-70.3	15.4(1)	58	25	C	μ=0.1 Complex is square-planar	74Hf	
	-77.8	15.8(1)	43.5	25	C	1M(NaOH) pH=14 Equilibrium mixture	78Fb	
	-80.3	16.0(1)	36.4	25	C	1M(NaOH) pH=14 Blue(octahedral) form	78Fb	
	-66.1	15.4(1)	72.8	25	C	1M(NaOH) pH=14 Yellow(square-planar) form	78Fb	
	14.2	-0.540(25°)	38	19-56	T	0.1M($NaClO_4$) R:$NiL(H_2O)_2^{2+}$ (blue)=NiL^{2+} (yellow)+$2H_2O$	77Aj	
Cu^{2+}	-132.6	(1)	--	25	C	0.15M(NEt_4Br) S:26v%Acetonitrile in chloroform	75Df	
CuA_4^{2-}	-230.1	(1)	--	25	C	0.15M(NEt_4Br) S:26v%Acetonitrile in chloroform R:CuA_4^{2-}+L= $CuLA_n^{(2-n)+}$ +(4-n)A^- A=Br	75Df	
Pb^{2+}	-30.5	7.8(25°,1)	46	15-35	T	μ=0.2M($NaClO_4$)	77Kd	
D11 3,6-DIAZAOCTANE, 1,8-diamine $C_6H_{18}N_4$						$H_2NCH_2CH_2NHCH_2CH_2NHCH_2CH_2NH_2$		
Cu^{2+}	-90.4	20.2(25°,1)	81.6	15-35	T	μ=0.2M($NaClO_4$)	78Kd	b
Zn^{2+}	-37.2	12.1(25°,1)	105	15-35	T	μ=0.2M($NaClO_4$)	78Kd	b
Cd^{2+}	-38.5	10.4(25°,1)	79	15-35	T	μ=0.2M($NaClO_4$)	78Kd	b

Metal Ion	ΔH, kJ /mol	Log K	ΔS, J /mol K	T, °C	M	Conditions	Ref.	Remarks
D11, continued								
Hg^{2+}	-126.3	24.5(25°,1)	46	15-35	T	μ=0.2M($NaClO_4$)	78Kd	b
Pb^{2+}	-34.7	10.3(25°,1)	84	15-35	T	μ=0.2M($NaClO_4$)	78Kd	b
	-34.7	10.3(25°,1)	84	15-35	T	μ=0.2M($NaClO_4$)	77Kd	
D12 DICARBONYL, pentahaptocyclopentadienyl iron (dimer)								
Fe	3.76	--	14.6	-90--70	T	S:3:1 v/v CS_2-$C_6H_5CD_3$ mixture R:Complex structure(I)→complex structure(III)	72Bn	b
D13 α-DICYCLOPENTADIENE				$C_{10}H_{12}$		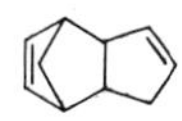		L
PdA_2B_2	-20.5	--	--	25	C	S:CH_2Cl_2 CL=5x10^{-3}M CM=1.9x10^{-4}M A=Cl B=C_6H_5CN R:PdA_2B_2+L= PdA_2L+2B	72Pc	
D14 DIETHYLENE GLYCOL, 3-thio-				$C_4H_{10}OS$		$HOCH_2CH_2SCH_2CH_2OH$		
Ag^+	-31.6	3.53(1)	-38	25	C	μ=0.5($NaClO_4$)	76Mb	
	-34.9	2.28(2)	-73	25	C	μ=0.5($NaClO_4$)	76Mb	
	-8.5	0.87(3)	-12	25	C	μ=0.5($NaClO_4$)	76Mb	
Hg^{2+}	-33.3	6.37(1)	10	25	C	μ=0.5($NaClO_4$)	76Mb	
	-36.3	4.10(2)	-43	25	C	μ=0.5($NaClO_4$)	76Mb	
	7.6	1.03(3)	46	25	C	μ=0.5($NaClO_4$)	76Mb	
	-8.0	2.6(4)	20	25	C	μ=0.5($NaClO_4$)	76Mb	
D15 DIHYDROBIS (3-methyl-1-pyrazolyl) borate ion				$C_8H_{12}N_4B$		$H_2B[C_3H_2N_2(CH_3)]_2$		L^-
Co^{2+}	-93.1	(1-2)	--	22-27	C	CM=0.02M S:Acetonitile	75Jc	
Ni^{2+}	-96.6	(1-2)	--	22-27	C	CM=0.02M S:Acetonitile	75Jc	
Cu^{2+}	-100.1	(1-2)	--	22-27	C	CM=0.02M S:Acetonitile	75Jc	
Zn^{2+}	-118.7	(1-2)	--	22-27	C	CM=0.02M S:Acetonitile	75Jc	

Metal Ion	ΔH, kJ /mol	Log K	ΔS, J /mol K	T,°C	M	Conditions	Ref.	Remarks
D16	DIHYDROBIS (1-pyrazolyl) borate ion		$C_6H_8N_4B$			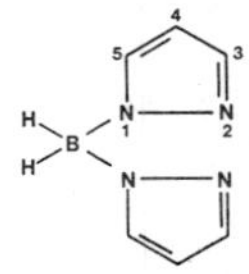		L^-
Co^{2+}	-145.0	(1-3)	--	22-27	C	CM=0.02M S:Acetonitile	75Jc	
Ni^{2+}	-150.6	(1-3)	--	22-27	C	CM=0.02M S:Acetonitile	75Jc	
Cu^{2+}	-119.0	(1-2)	--	22-27	C	CM=0.02M S:Acetonitile	75Jc	
Zn^{2+}	-112.9	(1-2)	--	22-27	C	CM=0.02M S:Acetonitile	75Jc	
D17	2,12-DIMETHYL-3,7,11,17-tetra-azabicyclo [11.3.1] heptadeca-1(17),2,11,13,15-pentaene		$C_{15}H_{22}N_3$			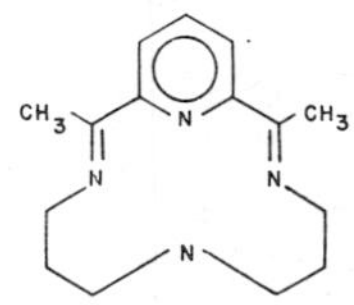		
$NiLA_2$	-17.91	--	-46.0	-43.84	T	S:Acetonitrile $A=PF_6$ R:$NiLA_2$-(diamagnetic)=Ni-LA_2(paramagnetic)	71Rf	b
	-29.16	--	-75.6	-3-84	T	$A=BF_4$ R:$NiLA_2$-(diamagnetic)=Ni-LA_2(paramegnetic)	71Rf	b
	-14.6	--	-28.0	-43-127	T	S:N,N-Dimethyl formamide $A=PF_6$ R:$NiLA_2$(diamagnetic)=$NiLA_2$(paramagnetic)	71Rf	b
	-19.2	--	-38.5	21-127	T	S:Dimethyl sulfoxide $A=PF_6$ R:$NiLA_2$(diamagnetic)=$NiLA_2$-(paramagnetic)	71Rf	b
	-17.8	--	-48.1	-65-60	T	S:CH_3OH $A=BF_4$ R:$NiLA_2$(diamagnetic)=$NiLA_2$-paramagnetic)	71Rf	b
D18	DINACTIN							
Na^+	-27.61	2.9(1)	-36.8	25	C	S:CH_3OH	73Za	
D19	1,4-DIOXANE			$C_4H_8O_2$		$\overline{OCH_2CH_2OCH_2CH_2}$		L
$RhA_2A_4B_2$	∿-2.1	--	--	24	C	S:Benzene	76Pd	

Metal Ion	ΔH, kJ /mol	Log K	ΔS, J /mol K	T, °C	M	Conditions	Ref.	Remarks
D19, continued								
$RhA_2A_4B_2$, cont.						R:$Rh_2A_4B_2$+2L= 2RhA_2BL A=CO B=Cl		
CuA_2	-6.48	-0.155(1)	23.8	30	C	S:Benzene A=N-nitroso-N-Phenyl-hydroxylamine	71Gi	q: ±0.36
HgA_2	-26.9	0.58(1)	-77	30	C	CM=3.9-9.4x10^{-3} S:Benzene A=Cl	74Fb	
	-22.7	0.36(1)	-68	30	C	CM=3.9-8.6x10^{-3} S:Benzene A=Br	74Fb	
Al_2A_6	-87.4	--	--	25	C	S:Benzene CM=0.07M A=Br R:0.5(Al_2A_6)+L= AlA_3L	68Rc	
	-154.8	--	--	25	C	S:Benzene CM=0.07M A=Br R:0.5(Al_2A_6)+2L= AlA_3L_2	68Rc	
I_2	-13.8	0.959(25°,1)	-28.0	--	T	S:CCl_4	57Dd 52Ka	1
	-11.30	0.29(25°,1)	-32.2	17-50	T	CL=0.015 in mole fraction units S:CCl_4	51Ka	b,1
	-9.6	(1)	--	17-50	T	CL=0-0.15 in mole fraction units S:CCl_4	51Ka	b,1
	-14.64	0.969(25°,1)	-30.5	25	T	S:n-Hexane	57Dd 52Ka 54Ka	b,1
	-18.0	1.367(25°,1)	-34.3	--	T	S:n-Hexane	57Dd 52Ka	1
	-15.1	(1)	--	--	C	S:Dioxane Recalculation of data in 50Ha ref	61Aa	
	-13.60	(1)	--	--	C	S:Dioxane	50Ha	
D20	2,4-DIOXO-19-CROWN-6		$C_{13}H_{22}O_8$			$CH_2C(=O)O(CH_2CH_2O)_5C(=O)$		
Na^{+}	-4.6	1.8(1)	19.6	25	C	μ~0 S:Methanol	77Ib	q:±0.8
K^{+}	-33.1	2.55(1)	-62.2	25	C	μ~0 S:Methanol	77Ib	
Ba^{+}	-20.4	1.41(1)	-42.1	25	C	μ~0 S:Methanol	77Ib	
D21	2,6-DIOXO-18-CROWN-6		$C_{12}H_{20}O_8$			$CH_2C(=O)O(CH_2CH_2O)_4C(=O)CH_2O$		
Na^{+}	-9.50	2.5(1)	15.4	25	C	μ~0 S:Methanol	77Ib	q:±0.75
K^{+}	-24.56	2.79(1)	-28.9	25	C	μ~0 S:Methanol	77Ib	
Ba^{2+}	-1.92	3.1(1)	53.3	25	C	μ~0 S:Methanol	77Ib	q:±0.5

Metal Ion	ΔH, kJ /mol	Log K	ΔS, J /mol K	T,°C	M	Conditions	Ref.	Remarks
D22	DISULFIDE, dibutyl-		$C_8H_{18}S_2$			$CH_3(CH_2)_3SS(CH_2)_3CH_3$		
I_2	-18.0	0.544(1)	-50.6	25	C	S:Ethylene chloride	66Nb	
D23	DISULFIDE, di-*tert*-butyl-		$C_8H_{18}S_2$			$(CH_3)_3CSSC(CH_3)_3$		L
I_2	-26.40	0.773(1)	-74.0	25	C	S:CCl_4 CM=0.05M	72Na	
	-23.8	0.832(1)	-64.4	25	C	S:Ethylene chloride	66Nb	
D24	DISULFIDE, diethyl-		$C_4H_{10}S_2$			$C_2H_5SSC_2H_5$		L
I_2	-18.7	0.569(1)	-51.9	25	C	S:CCl_4 CM=0.05M	72Na	
	-24.06	1.559(30°,1)	-49.4	20-40	T	S:CCl_4	61Ga	b
	-19.33	0.750(20°,1)	-51.5	0-30	T	CM=10^{-3}M CL=0.07-0.3M S:Dichloromethane	61Ta	b
	-18.0	0.491(1)	-50.6	25	C	S:Ethylene chloride	66Nb	
	-29.83	1.415(30°,1)	-71.1	20-40	T	S:n-Heptane	61Ga	b
D25	DISULFIDE, bis(diethylthio-carbamyl)-		$C_{10}H_{20}N_2S_4$			$(C_2H_5)_2NC(S)SSC(S)N(C_2H_5)_2$		L
I_2	-26.69	2.243(25°,1)	-47.7	10-30	T	S:CCl_4	69Ge	b,q:±3.3
D26	DISULFIDE, diheptyl-		$C_{14}H_{30}S_2$			$C_7H_{15}SSC_7H_5$		L
I_2	-33.9	2.854(1)	-58.6	25	C	S:Octane	66Ad	
D27	DISULFIDE, diisopropyl-		$C_6H_{14}S_2$			$(CH_3)_2CHSSCH(CH_3)_2$		L
I_2	-19.79	0.639(1)	-54.4	25	C	S:CCl_4 CM=0.05M	72Na	
	-18.4	0.544(1)	-51.9	25	C	S:Ethylene chloride	66Nb	
D28	DISULFIDE, dimethyl-		$C_2H_6S_2$			CH_3SSCH_3		L
I_2	-21.21	0.423(1)	-63.2	25	C	S:CCl_4 CM=0.05M	72Na	
	-18.4	0.380(1)	-54.8	25	C	S:Ethylene chloride	66Nb	
D29	DISULFIDE, bis(dimethyl-arsine)-		$C_4H_{12}S_2As_2$			$(CH_3)_2AsSSAs(CH_3)_2$		L
As	28.32	0.72(89°)	64.4	29-120	T	S:$C_2H_2Cl_4$ CM=0.05-0.10M (Rearrangement reaction) R:	71Zf	

Metal Ion	ΔH, kJ /mol	Log K	ΔS, J /mol K	T,°C	M	Conditions	Ref	Re-marks
D29, continued								
As, cont.						$(CH_3)_2As(S)SAs-(CH_3)_2=(CH_3)_2-AsSSAs(CH_3)_2$		
D30	DISULFIDE, bis(dimethylthio-carbamyl)-		$C_6H_{12}N_2S_4$			$(CH_3)_2NC(S)SSC(S)N(CH_3)_2$		L
I_2	-30.04	2.182(20°,1)	-60.2	10-30	T	S:CCl_4	69Ge	b,q:±2.6
D31	DISULFIDE, dioctyl-		$C_{16}H_{34}S_2$			$H_{17}C_8SSC_8H_{17}$		L
I_2	-31.0	3.000(1)	-46.0	25	C	S:Octane	66Ad	
D32	DISULFIDE, dipropyl-		$C_6H_{14}S_2$			$CH_3CH_2CH_2SSCH_2CH_2CH_3$		L
I_2	-18.8	0.447(1)	-54.8	25	C	S:Ethylene chloride	66Nb	q:±1.2
D33	DISULFIDE, 1-methyl-2-tert-butyl-		$C_5H_{12}S_2$			$CH_3SSC(CH_3)_3$		L
I_2	-20.92	0.559(1)	-59.4	25	C	S:CCl_4 CM=0.05M	72Na	
D34	1,4-DITHIA-9-CROWN-3		$C_6H_{12}OS_2$			$CH_2SCH_2CH_2OCH_2CH_2SCH_2$ (cyclic)		
Hg^{2+}	-52.01	(1)	--	25	C	μ=0 S:70w%Methanol	78Ib	
	-62.55	(2)	--	25	C	μ=0 S:70w%Methanol	78Ib	
D35	1,7-DITHIA-12-CROWN-4		$C_8H_{16}O_2S_2$			$CH_2SCH_2CH_2OCH_2CH_2SCH_2CH_2OCH_2$ (cyclic)		
Ag^+	-70.21	(1)	--	25	C	μ=0 S:70w%Methanol	78Ib	
	-49.33	4.26(2)	-83.9	25	C	μ=0 S:70w%Methanol	78Ib	
Hg^{2+}	-69.83	(1)	--	25	C	μ=0 S:70w%Methanol	78Ib	
	-62.51	(2)	--	25	C	μ=0 S:70w%Methanol	78Ib	
D36	1,4-DITHIA-15-CROWN-5		$C_{10}H_{20}O_3S_2$			$CH_2SCH_2CH_2S(CH_2CH_2O)_3CH_2$ (cyclic)		
Ag^+	-50.71	(1)	--	25	C	μ=0	78Ib	
	-23.43	3.31(2)	-15.2	25	C	μ=0	78Ib	
Hg^{2+}	-47.36	(1)	--	25	C	μ=0	78Ib	
	-32.59	5.1(2)	-11.2	25	C	μ=0	78Ib	
Pb^{2+}	-23.8	1.21(1)	-56.8	25	C	μ=0	78Ib	

Metal Ion	ΔH, kJ /mol	Log K	ΔS, J /mol K	T,°C	M	Conditions	Ref.	Remarks
D37	1,7-DITHIA-15-CROWN-5		$C_{10}H_{20}O_3S_2$			$CH_2SCH_2CH_2OCH_2CH_2S(CH_2CH_2O)_2CH_2$		
Ag^+	-69.33	(1)	--	25	C	μ=0	78Ib	
	-4.18	2.7(2)	37.9	25	C	μ=0	78Ib	
Hg^{2+}	-67.36	(1)	--	25	C	μ=0	78Ib	
	-20.9	2.91(2)	-14.4	25	C	μ=0	78Ib	
Pb^{2+}	-31.80	1.62(1)	-22.6	25	C	μ=0	78Ib	
D38	1,4-DITHIA-18-CROWN-6		$C_{12}H_{24}O_4S_2$			$CH_2SCH_2CH_2S(CH_2CH_2O)_4CH_2$		
Ag^+	-65.69	3.0(1)	-168.4	25	C	μ=0	78Ib	
Hg^{2+}	-118.8	(2)	--	25	C	μ=0	78Ib	
Tl^+	-30.5	1.38(1)	-76.1	25	C	μ=0	78Ib	
Pb^{2+}	-36.94	2.63(1)	-73.5	25	C	μ=0	78Ib	
D39	1,10-DITHIA-18-CROWN-6		$C_{12}H_{24}O_4S_2$			$CH_2S(CH_2CH_2O)_2CH_2CH_2S(CH_2CH_2O)_2CH_2$		
Ag^+	-69.9	(1)	--	25	C	μ=0	78Ib	
Hg^{2+}	-74.48	(1)	--	25	C	μ=0	78Ib	
Tl^+	-46.0	0.93(1)	-136.5	25	C	μ=0	78Ib	
Pb^{2+}	-88.7	3.13(1)	-237.4	25	C	μ=0	78Ib	
D40	1,2-DITHIANE		$C_4H_6S_2$			$CH_2:CHSSCH:CH_2$		L
I_2	-24.02	0.997(1)	-61.5	25	C	S:CCl_4 CM=0.037M	72Na	
D41	1,4-DITHIANE		$C_4H_8S_2$					L
I_2	-19.46	1.064(1)	-45.6	25	C	S:Ethylene chloride	66Nb	
D42	1,4-DITHIA-24-CROWN-8		$C_{16}H_{32}O_6S_2$			$CH_2S(CH_2CH_2O)_6CH_2CH_2SCH_2$		
Ag^+	-59.8	4.5(1)	-115.1	25	C	μ=0	78Ib	
	-8.8	5.0(2)	67.0	25	C	μ=0	78Ib	
Hg^{2+}	-57.3	(1)	--	25	C	μ=0	78Ib	
D43	1,13-DITHIA-24-CROWN-8		$C_{16}H_{32}O_6S_2$			$CH_2S(CH_2CH_2O)_3CH_2CH_2S(CH_2CH_2O)_3CH_2$		
Ag^+	-57.7	(1)	--	25	C	μ=0	78Ib	
Hg^{2+}	-56.5	(1)	--	25	C	μ=0	78Ib	
D44	DITHIODIACETIC ACID		$C_4H_6O_4S_2$			$HO_2CCH_2SSCH_2CO_2H$		L
Ni^{2+}	1.2	1.5(1)	33.5	25	C	μ=1.0($NaNO_3$)	79Ad	q:±0.4

Metal Ion	ΔH, kJ /mol	Log K	ΔS, J /mol K	T,°C	M	Conditions	Ref.	Remarks
D44, continued								
Zn^{2+}	2.9	1.3(1)	34.7	25	C	μ=1.0($NaNO_3$)	79Ad	q:±0.8
Cd^{2+}	3.8	1.6(1)	43.5	25	C	μ=1.0($NaNO_3$)	79Ad	q:±0.8
Al^{3+}	49.79	6.363(25°,1-2)	288.9	25-45	T	μ=0	79Dd	
D45 DI-p-TOLYLTHIOVIOLURIC ACID								
UO_2^{2+}	-35.8	5.09(30°,1)	-21.1	30-50	T	μ=0.100M($NaClO_4$) S:75v%Dioxane	78Md	b
	-47.3	4.37(30°,2)	-71.3	30-50	T	μ=0.100M($NaClO_4$) S:75v%Dioxane	78Md	b
Mn^{2+}	-68.5	5.08(30°,1-2)	-127	30-50	T	μ=0.100M($NaClO_4$)	78Md	b
Fe^{2+}	-39.6	4.81(30°,1)	-38.9	30-50	T	μ=0.100M($NaClO_4$) S:75v%Dioxane	78Md	b
	-51.8	3.94(30°,2)	-92.5	30-50	T	μ=0.100M($NaClO_4$) S:75v%Dioxane	78Md	b
Co^{2+}	-47.7	4.58(30°,1)	-69.9	30-50	T	μ=0.100M($NaClO_4$) S:75v%Dioxane	78Md	b
	-47.7	4.08(30°,2)	-78.6	30-50	T	μ=0.100M($NaClO_4$) S:75v%Dioxane	78Md	b
Ni^{2+}	-51.8	5.13(30°,1)	-73.2	30-50	T	μ=0.100M($NaClO_4$) S:75v%Dioxane	78Md	b
	-25.3	4.08(30°,2)	-4.6	30-50	T	μ=0.100M($NaClO_4$) S:75v%Dioxane	78Md	b
Cu^{2+}	-47.7	5.15(30°,1)	-59.0	30-50	T	μ=0.100M($NaClO_4$) S:75v%Dioxane	78Md	b
	-30.0	4.28(30°,2)	-17	30-50	T	μ=0.100M($NaClO_4$) S:75v%Dioxane	78Md	b
Zn^{2+}	-70.1	4.60(30°,1)	-144	30-50	T	μ=0.100M($NaClO_4$) S:75v%Dioxane	78Md	b
	-12.9	3.59(30°,2)	-26.4	30-50	T	μ=0.100M($NaClO_4$) S:75v%Dioxane	78Md	b
Cd^{2+}	-17.2	3.09(30°,1)	1.6	30-50	T	μ=0.100M($NaClO_4$) S:75v%Dioxane	78Md	b
	-54.4	2.78(30°,2)	-126	30-50	T	μ=0.100M($NaClO_4$) S:75v%Dioxane	78Md	b
Pb^{2+}	-54.8	3.79(30°,1)	-108	30-50	T	μ=0.100M($NaClO_4$) S:75v%Dioxane	78Md	b
	-40.6	4.61(30°,2)	-45.2	30-50	T	μ=0.100M($NaClO_4$) S:75v%Dioxane	78Md	b
D46 DODECANE, 5,8-diaza-, 2,11-diamino			$C_{14}H_{34}N_4$			$H_2NCH(CH_3)CH_2C(CH_3)_2NH(CH_2)_2$ $NHC(CH_3)_2CH_2CH(CH_3)NH_2$		L
Cu^{2+}	-104.01	22.41(1)	80.21	25	C	μ=0.10M(NaCl)	73Hc	

Metal Ion	ΔH, kJ /mol	Log K	ΔS, J /mol K	T,°C	M	Conditions	Ref.	Re-marks
D47	DODECANE, 4,9-diaza-1,12-diamino		$C_{10}H_{30}N_4$		$H_2NCH_2(CH_2)_2NH(CH_2)_4NH(CH_2)_2CH_2NH_2$			
Cu^{2+}	-88.1	14.70(1)	-14	25	C	μ=0.10M(NaCl)	76Gg	
D48	DODECANE, 5,8-diaza-, 2,11-dionedioxime-,4,4,9,9-tetramethyl-		$C_{14}H_{30}O_2N_4$		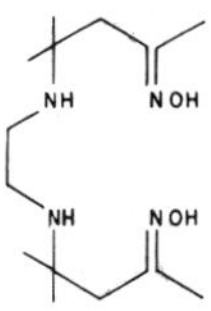			L^{2-}
	-54.81	13.24	69.9	25	C	μ=0.10M R:Cu^{2+}+ H_2L=CuH_2L^{2+}	74Hb	
	23.68	-3.23	17.99	25	C	μ=0.10M R:CuH_2L^{2+} =$Cu(HL)^+$+H^+	74Hb	
D49	DODECANE, 5,8-dithio-		$C_{10}H_{22}S_2$		$CH_3(CH_2)_3S(CH_2)_2S(CH_2)_3CH_3$			L
TiA_4	-105.4	3.66(1)	-283.7	25	C	S:Benzene A=Cl	68Ge 70Gh	
SnA_4	-84.1	(1)	--	--	C	S:Benzene CM=0.05-0.08M A=Br	68Ge	u
	-116.3	4.25(1)	-308.8	25	C	S:Benzene A=Cl	65Gb 68Ge 70Gh	
TeA_4	-53.6	(1)	--	25	C	S:Benzene A=Cl	73Pi	
D50	DODECANE, 2,5,8,11-tetraoxa-		$C_8H_{18}O_4$		$CH_3OCH_2CH_2OCH_2CH_2OCH_2CH_2OCH_3$			L
NaA	-15.1	1.83(25°)	-15.1	-40-25	T	CL<$4x10^{-2}$M S:2-Methyl tetrahydrofuran R:NaA+L=LNaA A=Biphenyl	67Sb	b
	-22.6	1.83(25°)	-40.6	-40-25	T	CL<$4x10^{-2}$M S:2-Methyl tetrahydrofuran R:NaA+L=NaLA A=Biphenyl	67Sb	b
	-22.6	0.954(25°)	-60.7	0-50	T	S:Tetrahydrofuran A=Fluorenyl ion R:NaA+L=NaLA	70Ce	b
	-19.2	1.54(25°)	-35.1	-40-25	T	CL<$4x10^{-2}$M S:Tetrahydropyran R:NaA+L=LNaA A=Biphenyl	67Sb	b
	-29.3	1.39(25°)	-71	-40-25	T	CL<$4x10^{-2}$M S:Tetrahydropyran R:NaA+L=NaLA A=Biphenyl	67Sb	b

Metal Ion	ΔH, kJ /mol	Log K	ΔS, J /mol K	T, °C	M	Conditions	Ref.	Remarks
D51 DUROSEMIQUINONE ION								
Na^+	11.3	9.54(-75°,1)	159	--	T	S:1,2-Dimethoxy-ethane	72Ac	b,u:(-78° --60°)
K^+	14.6	12.64(-75°,1)	209	--	T	S:1,2-Dimethoxy-ethane	72Ac	b,u:(-78° --60°)
Rb^+	23.0	9.76(-75°,1)	201	--	T	S:1,2-Dimethoxy-ethane	72Ac	b,u:(-78° --60°)
Cs^+	31.3	9.87(-75°,1)	230	--	T	S:1,2-Dimethoxy-ethane	72Ac	b,u:(-78° --60°)

E

E1 EICOSANE, 5,16-dithio-			$C_{18}H_{38}S_2$			$CH_3(CH_2)_3S(CH_2)_{10}S(CH_2)_3CH_3$		L
SnA_4	-100.4	(1)	--	--	C	S:Benzene CM=0.05-0.08M A=Cl	65Gb	u
TeA_4	-52.7	3.14(1)	-116.7	25	C	S:Benzene A=Cl	73Pi	
	-30.1	(2)	--	25	C	S:Benzene A=Cl	73Pi	
E2 EPHEDRINE			$C_{10}H_{15}ON$			$C_6H_5CH(OH)CH(CH_3)NHCH_3$		L
Be^{2+}	-52.7	6.57(25°,1)	-54	0-45	T	μ=0.12M(KCl) CM≅2x10^{-3}M CL≅8x10^{-3}M	69Cb	b
	-56.5	5.47(25°,2)	-88	0-45	T	μ=0.12M(KCl) CM≅2x10^{-3}M CL=8x10^{-3}M	69Cb	b

E3 ERIOCHROME BLACK T $C_{20}H_{12}O_7N_3SNa$

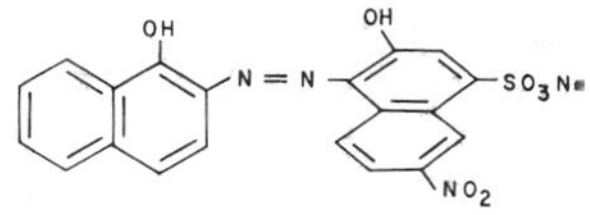

L^{3-}

Mg^{2+}	41.4	3.954	213	10-60	T	μ=0.010 R:Mg^{2+}+ HL^{2-}=MgL^-+H^+	71Kd	b,r
	41.8	3.786	213	10-60	T	μ=0.029 R:Mg^{2+}+ HL^{2-}=MgL^-+H^+	71Kd	b,r
	40.6	3.636	205	10-60	T	μ=0.060 R:Mg^{2+}+ HL^{2-}=MgL^-+H^+	71Kd	b,r
	41.4	3.509	205	10-60	T	μ=0.100 R:Mg^{2+}+ HL^{2-}=MgL^-+H^+	71Kd	b,r

Metal Ion	ΔH, kJ /mol	Log K	ΔS, J /mol K	T,°C	M	Conditions	Ref.	Remarks
E4 ETHANE, 1-amino-			C_2H_7N			$NH_2CH_2CH_3$		L
NiA	-69.0	4.65(1-2)	-139.3	25	T	$CM=10^{-4}M$ S:Benzene A=Di-acetylbisbenzoyl hydrazine	60Sb	
Ag^+	-21	3.46(1)	-14.2	25	C	μ=0	71He	q:±8
	-29	3.90(2)	-28.0	25	C	μ=0	71He	q:±8
	-52.34	7.36(1-2)	-34.7	25	C	μ=0	71He	
	-54.4	7.32(1-2)	-42.2	25	C	--	55Fa	
I_2	-31.4	2.857(20°,1)	-52.3	22	C	low μ S:n-Heptane	60Sd 60Ya	b
	-41.0	3.833(20°,1)	-66.5	20-40	T	$CM=10^{-5}M$ $CL=10^{-4}-10^{-3}M$ S:n-Heptane	60Ya	b
E5 ETHANE, 1-amino-2-diphenyl-methoxy-N,N-dimethyl-			$C_{17}H_{21}ON$			$(C_6H_5)_2CHOCH_2CH_2N(CH_3)_2$		L^-
Be^{2+}	-68.2	6.30(25°,1)	-108.8	0-45	T	μ=0.12M(KCl) $CM \cong 2x10^{-3}M$ $CL \cong 8x10^{-3}M$	69Cb	b
	-10.5	5.04(25°,2)	58.6	0-45	T	μ=0.12M(KCl) $CM \cong 2x10^{-3}M$ $CL \cong 8x10^{-3}M$	69Cb	b
Co^{2+}	-7.5	6.54(25°,1-2)	100.4	0&25	T	0.06M(KCl)	62Aa	a,b
Ni^{2+}	-26.4	6.52(25°,1-2)	36.8	0&25	T	0.06M(KCl)	62Aa	a,b
Cu^{2+}	-32.6	10.50(25°,1-2)	92.0	0&25	T	0.06M(KCl)	62Aa	a,b
Cd^{2+}	-36.0	7.28(1)	17	25	T	μ=0.1(KNO_3)	64Ad	
E6 ETHANE, 1-amino-2-methoxy-			C_3H_9ON			$NHCH_2CH_2OCH_3$		L
Ag^+	-23.0	3.18(30°,1)	-12	25	T	μ=0	59Lb	b
	-22.2	3.37(30°,2)	-8	25	T	μ=0	59Lb	b
E7 ETHANE, 1,2-bis(methylthio)-			$C_4H_{10}S_2$			$CH_3SCH_2CH_2SCH_3$		L
CuA_2	-62.13	(1)	--	--	C	A=Cl	25Ma	
CuA	-24.60	(1)	--	--	C	A=I	25Ma	
ZnA_2	-64.22	(1)	--	--	C	A=Br	25Ma	
	-55.27	(1)	--	--	C	A=Cl	25Ma	
	-59.54	(1)	--	--	C	A=I	25Ma	
CdA_2	-47.15	(1)	--	--	C	A=Br	25Ma	
	-50.88	(1)	--	--	C	A=Cl	25Ma	
	-41.67	(1)	--	--	C	A=I	25Ma	
HgA_2	-47.40	(1)	--	--	C	A=Br	25Ma	
	-5.35	(1)	--	--	C	A=Cl	25Ma	
	-20.21	(1)	--	--	C	A=I	25Ma	

Metal Ion	ΔH, kJ /mol	Log K	ΔS, J /mol K	T,°C	M	Conditions	Ref.	Re-marks
E7, continued								
SnA_4	-81.17	(1)	--	--	C	A=Br	25Ma	
	-117.40	(1)	--	--	C	A=Cl	25Ma	
	-69.37	(1)	--	--	C	A=I	25Ma	
BiA_3	-50.2	(2-3)	--	--	C	A=I	25Ma	
E8 ETHANE, cyano-			C_3H_5N			CH_3CH_2CN		L
TiA_4	-48.1	3.59(1)	-92.5	25	C	S:Benzene A=Cl	70Gh	
	-49.8	3.96(1)	-91.2	25	C	S:Benzene A=Cl	70Gh 68Ge	
SnA_4	-21.3	(1)	--	25	C	S:Benzene A=Cl	70Gh 68Ge	
E9 ETHANE, 1,2-diamino-			$C_2H_8N_2$			$H_2NCH_2CH_2NH_2$		L
Y^{3+}	-83.7	(1)	--	23	C	S:Acetonitrile 0.026M $Y(ClO_4)_3$	69Fc	c
	-78.2	(2)	--	23	C	S:Acetonitrile 0.026M $Y(ClO_4)_3$	69Fc	c
	-54.0	(3)	--	23	C	S:Acetonitrile 0.026M $Y(ClO_4)_3$	69Fc	c
	-43.9	(4)	--	23	C	S:Acetonitrile 0.026M $Y(ClO_4)_3$	69Fc	c
La^{3+}	-72.4	9.5(1)	-63.2	23	C	S:Acetonitrile 0.026M $La(ClO_4)_3$	69Fc	c
	-58.2	(1)	--	23	C	S:Acetonitrile 0.026M $La(NO_3)_3$	69Fc	c
	-64.9	7.5(2)	-76.1	23	C	S:Acetonitrile 0.026M $La(ClO_4)_3$	69Fc	c
	-52.7	(2)	--	23	C	S:Acetonitrile 0.026M $La(NO_3)_3$	69Fc	c
	-57.7	6.2(3)	-77.0	23	C	S:Acetonitrile 0.026M $La(ClO_4)_3$	69Fc	c
	-46.0	3.3(4)	-92.5	23	C	S:Acetonitrile 0.026M $La(ClO_4)_3$	69Fc	c
Pr^{3+}	-78.2	(1)	--	23	C	S:Acetonitrile 0.026M $Pr(ClO_4)_3$	69Fc	c
	-59.8	(1)	--	23	C	S:Acetonitrile 0.026M $Pr(NO_3)_3$	69Fc	c
	-70.3	(2)	--	23	C	S:Acetonitrile 0.026M $Pr(ClO_4)_3$	69Fc	c
	-53.6	(2)	--	23	C	S:Acetonitrile 0.026M $Pr(NO_3)_3$	69Fc	c
	-56.9	(3)	--	23	C	S:Acetonitrile 0.026M $Pr(ClO_4)_3$	69Fc	c
	-45.2	(4)	--	23	C	S:Acetonitrile 0.026M $Pr(ClO_4)_3$	69Fc	c
Nd^{3+}	-78.6	(1)	--	23	C	S:Acetonitrile 0.026M $Nd(ClO_4)_3$	69Fc	c
	-61.5	(1)	--	23	C	S:Acetonitrile	69Fc	c

Metal Ion	ΔH, kJ /mol	Log K	ΔS, J /mol K	T,°C	M	Conditions	Ref.	Re-marks
E9, continued								
Nd^{3+}, cont.						0.026M Nd(NO_3)$_3$		
	-70.7	(2)	--	23	C	S:Acetonitrile 0.026M Nd(ClO_4)$_3$	69Fc	c
	-54.8	(2)	--	23	C	S:Acetonitrile 0.026M Nd(NO_3)$_3$	69Fc	c
	-57.7	(3)	--	23	C	S:Acetonitrile 0.026M Nd(ClO_4)$_3$	69Fc	c
	-45.6	(4)	--	23	C	S:Acetonitrile 0.026M Nd(ClO_4)$_3$	69Fc	c
Sm^{3+}	-80.8	(1)	--	23	C	S:Acetonitrile 0.026M Sm(ClO_4)$_3$	69Fc	c
	-75.3	(2)	--	23	C	S:Acetonitrile 0.026M Sm(ClO_4)$_3$	69Fc	c
	-56.5	(3)	--	23	C	S:Acetonitrile 0.026M Sm(ClO_4)$_3$	69Fc	c
	-41.4	(4)	--	23	C	S:Acetonitrile 0.026M Sm(ClO_4)$_3$	69Fc	c
Eu^{3+}	-82.8	(1)	--	23	C	S:Acetonitrile 0.026M Eu(ClO_4)$_3$	69Fc	c
	-76.6	(2)	--	23	C	S:Acetonitrile 0.026M Eu(ClO_4)$_3$	69Fc	c
	-58.2	(3)	--	23	C	S:Acetonitrile 0.026M Eu(ClO_4)$_3$	69Fc	c
	-40.6	(4)	--	23	C	S:Acetonitrile 0.026M Eu(ClO_4)$_3$	69Fc	c
Gd^{3+}	-81.6	(1)	--	23	C	S:Acetonitrile 0.026M Gd(ClO_4)$_3$	69Fc	c
	-75.3	(2)	--	23	C	S:Acetonitrile 0.026M Gd(ClO_4)$_3$	69Fc	c
	-58.2	(3)	--	23	C	S:Acetonitrile 0.026M Gd(ClO_4)$_3$	69Fc	c
	-39.7	(4)	--	23	C	S:Acetonitrile 0.026M Gd(ClO_4)$_3$	69Fc	c
Tb^{3+}	-83.3	10.4(1)	-82.8	23	C	S:Acetonitrile 0.026M Tb(ClO_4)$_3$	69Fc	c
	-77.8	8.4(2)	-102.5	23	C	S:Acetonitrile 0.026M Tb(ClO_4)$_3$	69Fc	c
	-54.8	6.2(3)	-66.9	23	C	S:Acetonitrile 0.026M Tb(ClO_4)$_3$	69Fc	c
	-37.6	3.2(4)	-66.1	23	C	S:Acetonitrile 0.026M Tb(ClO_4)$_3$	69Fc	c
Dy^{3+}	-83.3	(1)	--	23	C	S:Acetonitrile 0.026M Dy(ClO_4)$_3$	69Fc	c
	-77.0	(2)	--	23	C	S:Acetonitrile 0.026M Dy(ClO_4)$_3$	69Fc	c
	-52.7	(3)	--	23	C	S:Acetonitrile 0.026M Dy(ClO_4)$_3$	69Fc	c
	-38.5	(4)	--	23	C	S:Acetonitrile 0.026M Dy(ClO_3)$_3$	69Fc	c
Ho^{3+}	-83.3	(1)	--	23	C	S:Acetonitrile 0.026M Ho(ClO_4)$_3$	69Fc	c

Metal Ion	ΔH, kJ /mol	Log K	ΔS, J /mol K	T,°C	M	Conditions	Ref.	Re-marks
E9, continued								
Ho^{3+}, cont.	-76.1	(2)	--	23	C	S:Acetonitrile 0.026M $Ho(ClO_4)_3$	69Fc	c
	-53.1	(3)	--	23	C	S:Acetonitrile 0.026M $Ho(ClO_4)_3$	69Fc	c
	-41.8	(4)	--	23	C	S:Acetonitrile 0.026M $Ho(ClO_4)_3$	69Fc	c
Er^{3+}	-84.1	(1)	--	23	C	S:Acetonitrile 0.026M $Er(ClO_4)_3$	69Fc	c
	-78.2	(2)	--	23	C	S:Acetonitrile 0.026M $Er(ClO_4)_3$	69Fc	c
	-54.8	(3)	--	23	C	S:Acetonitrile 0.026M $Er(ClO_4)_3$	69Fc	c
	-48.1	(4)	--	23	C	S:Acetonitrile 0.026M $Er(ClO_4)_3$	69Fc	c
Yb^{3+}	-84.1	11.5(1)	-64.4	23	C	S:Acetonitrile 0.026M $Yb(ClO_4)_3$	69Fc	c
	-78.6	9.3(2)	-88.3	23	C	S:Acetonitrile 0.026M $Yb(ClO_4)_3$	69Fc	c
	-60.2	6.5(3)	-79.5	23	C	S:Acetonitrile 0.026M $Yb(ClO_4)_3$	69Fc	c
	-53.6	3.8(4)	-108.4	23	C	S:Acetonitrile 0.026M $Yb(ClO_4)_3$	69Fc	c
Lu^{3+}	-83.7	(1)	--	23	C	S:Acetonitrile 0.026M $Lu(ClO_4)_3$	69Fc	c
	-77.8	(2)	--	23	C	S:Acetonitrile 0.026M $Lu(ClO_4)_3$	69Fc	c
	-59.8	(3)	--	23	C	S:Acetonitrile 0.026M $Lu(ClO_4)_3$	69Fc	c
	-53.6	(4)	--	23	C	S:Acetonitrile 0.026M $Lu(ClO_4)_3$	69Fc	c
Mn^{2+}	-11.72	2.75(1)	13.4	25	C	1M(KCl)	60Cb 60Ca	
	-25.10	4.87(1-2)	9.2	25	C	1M(KCl)	60Cb	
	-46.23	5.79(1-3)	-43.9	25	C	1M(KCl)	60Cb	
Fe^{2+}	-21.13	4.32(1)	12.1	25	C	1M(KCl)	60Cb 60Ca	
	-43.51	7.66(1-2)	0.8	25	C	1M(KCl)	60Cb	
	-66.32	9.71(1-3)	-36.4	25	C	1M(KCl)	60Cb	
Co^{2+}	-28.9	5.94(1)	16.7	25	C	1M(KCl)	60Cb 60Ca	
	-58.37	10.66(1-2)	8.4	25	C	1M(KCl)	60Cb 54Wa	
	-92.68	13.67(1-3)	49.4	25	C	1M(KCl)	60Cb	
CoA	-33.9	4.36(1)	-30.1	25	C	μ=0.10M(KNO_3) A=Ethylenediamine diacetic acid	70Da	
Ni^{2+}	-39.7	7.27(30°,1)	8	10-40	T	μ=0	58Ba 60Cb	b

Metal Ion	ΔH, kJ /mol	Log K	ΔS, J /mol K	T,°C	M	Conditions	Ref.	Remarks
E9, continued								
Ni^{2+}, cont.	-62.76	7.32(1)	-71	25	C	0.1M(KCl)	54Da 60Ca	
	-38.70	7.49(1)	13.8	25	C	μ=0.30(ClO_4^-)	67Hf	
	-20.5	7.48(25°,1)	75	15-30	T	μ=1.0M($NaClO_4$)	52Ba 74Sm	b,n
	-37.70	7.51(1)	17.2	25	C	1M(KNO_3)	55Pb	
	-37.87	7.51(1)	17	25	C	1M(KNO_3)	54Ba 60Ca	
	-37.95	(1)	--	25	C	μ=1M(KNO_3)	54Wa 68Pa	
	-18.0	7.11(25°,1)	75	15-30	T	μ=1.0M($NaClO_4$) S:1.6mol%Acetone	74Sm	b,n,q: ±2.5
	-12.1	6.82(25°,1)	92	15-30	T	μ=1.0M($NaClO_4$) S:6.0mol%Acetone	74Sm	b,n,q: ±2.5
	-12.6	6.67(25°,1)	84	15-30	T	μ=1.0M($NaClO_4$) S:10.6mol%Acetone	74Sm	b,n,q: ±3.8
	-12.1	6.89(25°,1)	92	15-30	T	μ=1.0M($NaClO_4$) S:21.6mol%Acetone	74Sm	b,n
	-12.1	7.11(25°,1)	96	15-30	T	μ=1.0M($NaClO_4$) S:32.6mol%Acetone	74Sm	b,n,q: ±2.9
	-12.6	7.40(25°,1)	100	15-30	T	μ=1.0M($NaClO_4$) S:43.6mol%Acetone	74Sm	b,n,q: ±2.5
	-31.4	6.11(30°,2)	12	10-40	T	μ=0	58Ba	b
	-38.45	6.44(2)	-5.4	25	C	μ=0.30(ClO_4^-)	55Cc 67Hf	
	-27.2	6.38(25°,2)	29	15-30	T	μ=1.0M($NaClO_4$)	46Ca 74Sm	b,n
	-38.41	6.35(2)	-7.5	25	C	1M(KNO_3)	55Pb	
	-38.66	(2)	--	25	C	μ=1M(KNO_3)	68Pa	
	-24.3	5.86(25°,2)	29	15-30	T	μ=1.0M($NaClO_4$) S:1.6mol%Acetone	74Sm	b,n
	-17.2	5.57(25°,2)	50	15-30	T	μ=1.0M($NaClO_4$) S:6.0mol%Acetone	74Sm	b,n,q: ±1.7
	-10.0	5.42(25°,2)	71	15-30	T	μ=1.0M($NaClO_4$) S:10.6mol%Acetone	74Sm	b,n,q: ±1.2
	-10.9	5.57(25°,2)	71	15-30	T	μ=1.0M($NaClO_4$) S:21.6mol%Acetone	74Sm	b,n,q: ±1.7
	-10.9	5.72(25°,2)	71	15-30	T	μ=1.0M($NaClO_4$) S:32.6mol%Acetone	74Sm	b,n,q: ±1.2
	-12.1	6.08(25°,2)	75	15-30	T	μ=1.0M($NaClO_4$) S:43.6mol%Acetone	74Sm	b,n,q: ±2.5
	-36.4	4.20(30°,3)	-42	10-40	T	μ=0	58Ba	b
	-39.58	4.11(3)	-54.0	25	C	μ=0.30(ClO_4^-)	67Hf	
	-19.2	4.32(25°,3)	17	15-30	T	μ=1.0M($NaClO_4$)	74Sm	b,n
	-40.63	4.42(3)	-51.9	25	C	1M(KNO_3)	55Pb	
	-40.42	(3)	--	25	C	μ=1M(KNO_3)	68Pa	
	-21.3	3.81(25°,3)	4	15-30	T	μ=1.0M($NaClO_4$) S:1.6mol%Acetone	74Sm	b,n,q: ±4
	-19.7	3.52(25°,3)	0	15-30	T	μ=1.0M($NaClO_4$) S:6.0mol%Acetone	74Sm	b,n
	-14.2	3.30(25°,3)	21	15-30	T	μ=1.0M($NaClO_4$) S:10.6mol%Acetone	74Sm	b,n,q: ±1.7
	-15.5	2.84(25°,3)	13	15-30	T	μ=1.0M($NaClO_4$) S:21.6mol%Acetone	74Sm	b,n,q: ±1.7

Metal Ion	ΔH, kJ /mol	Log K	ΔS, J /mol K	T,°C	M	Conditions	Ref.	Remarks
E9, continued								
Ni^{2+}, cont.	-13.4	3.52(35°,3)	21	15-30	T	μ=1.0M($NaClO_4$) S:32.6mol%Acetone	74Sm	b,n,q: ±3.8
	-15.5	3.66(25°,3)	17	15-30	T	μ=1.0M($NaClO_4$) S:43.6mol%Acetone	74Sm	b,n,q: ±5
	-72.17	13.38(1-2)	14.2	25	C	0.1M(KCl)	54Da	
	-68.2	14.5(1-2)	29	0	C	0.50M(KNO_3)	54Bb	
	-74.89	13.80(1-2)	14.6	25	C	1M(KNO_3)	60Ca	
	-117.19	17.48(1-3)	-58.6	25	C	0.1M(KCl)	54Da	
	-104.2	20.1(1-3)	4	0	C	0.50M(KNO_3)	54Bb	
	-116.73	18.28(1-3)	-42.2	25	C	1M(KNO_3)	55Pb	
	-114.64	18.25(1-3)	-35.6	25	C	1M(KNO_3)	60Ca	
NiA	-41.25	6.31(1)	-17.6	25	C	μ=0.10M(KNO_3) A=Ethylenediaminediacetic acid	70Da	
	-39.58	5.51(1)	-27.2	25	C	μ=0.10M(KNO_3) A=Ethylenediaminedipropionic acid	70Da	
NiL_3	24.3	-1.69(25°)	49.4	0.2-75.0		S:DMSO R:δ→λ ring inversion	73Cg	b
	31.8	-1.61(25°)	76.1	20.1-103.5		S:D_2O R:δ→λ ring inversion	73Cg	b
Cu^{2+}	-52.7	10.48(1)	23.8	25	C	μ→0	67Pc 52Ba	
	-53.6	10.36(30°,1)	21	10-40	T	μ=0	58Ba 54Ba	b
	-49.8	10.67(25°,1)	37.6	0-50	T	μ=0.15	55Cc 54Wa	b
	-53.4	10.57(1)	23	25	C	μ=0.2M(KCl)	76Sp 76Sq	
	-56.9	10.60(25°,1)	12.6	15-35	T	μ=0.2M(KNO_3) CM=$2x10^{-3}$M	79Mg	b
	-52.09	10.44(1)	25.1	25	C	μ=0.30(ClO_4^-)	67Hf 48Ba	
	-50.6	10.57(1)	27.2	15-40	T	μ=0.3M($NaClO_4$)	72Md	b
	-52.55	10.582(1)	26.4	25	C	0.5M(KNO_3)	72Bd	
	-54.4	10.72(1)	22.6	25	C	1M(KNO_3)	55Pb	
	-61.1	11.023(1)	5.61	25	T	μ=2	53Sb	
	-52.7	9.07(2)	-2.9	25	C	μ→0	67Pc	
	-51.5	8.93(30°,2)	0	10-40	T	μ=0	59Ma	b
	-47.3	9.10(25°,2)	15.5	0-50	T	μ=0.15	55Cc	b
	-50.7	9.11(2)	4	25	C	μ=0.2M(KCl)	76Sp 76Sq	
	-50.2	9.11(25°,2)	5.8	15-35	T	μ=0.2M(KNO_3) CM=$2x10^{-3}$M	79Mg	b
	-51.38	8.98(2)	-0.4	25	C	μ=0.30(ClO_4^-)	67Hf	
	-52.3	9.19(2)	0.8	15-40	T	μ=0.3M($NaClO_4$)	72Md	b
	-52.88	9.151(2)	-2.1	25	C	0.5M(KNO_3)	72Bd	
	-51.9	9.31(2)	3.8	25	C	1M(KNO_3)	55Pb	
	-105.4	19.56(1-2)	20.9	25	C	μ→0	67Pc	
	-105.27	20.03(1-2)	29.7	25	C	0.1M(KCl)	54Da	

Metal Ion	ΔH, kJ /mol	Log K	ΔS, J /mol K	T,°C	M	Conditions	Ref.	Remarks
E9, continued								
Cu^{2+}, cont.	-102.9	21.3(1-2)	29	0	C	0.5M(KNO_3)	54Bb	
	-107.9	20.07(1-2)	22.8	20	T	μ=1	52Sa	
	-106.3	20.03(1-2)	26.4	25	C	1M(KNO_3)	55Pb	
	-118.8	20.61(1-2)	-2.9	25	C	μ=2	53Sb	
	-2.13	1.42(25°)	20.1	-75&25	T	S:50v%CH_3OH R:Cu+CuL_2=2CuL	73Ya	a,b,q: ±1.0
	-102.9	18.83	15	25	C	μ=0.2M(KCl) R:Cu+A+L=CuAL A=1,3-Diamino- propane	76Sp	
	-98.7	18.55	24	25	C	μ=0.2M(KCl) R:Cu+A+L=CuAL A=1,2-Diamino- N,N-dimethylethane	76Sp	
	-101.3	19.27	29	25	C	μ=0.2M(KCl) R:Cu+A+L=CuAL A=1,3-Diamino- N,N'-dimethyl- ethane	76Sp	
	-79.5	79.5	72	25	C	μ=0.2M(KCl) R:Cu+A+L=CuAL A=Glycine	76Sp	
	-104.6	18.66	6	25	C	μ=0.2M(KCl) R:Cu+A+L=CuAL A=Histamine	76Sp	
CuA	-55.6	6.66(1)	-59.0	25	C	μ=0.1M(KNO_3) A=Ethylenedi- aminediacetic acid	70Da	
	-46.4	9.41(25°,1)	24.7	15-35	T	μ=0.2M(KNO_3) CM=$2x10^{-3}$M A=2,2'-Bipyridyl	79Mg	
	-102.9	18.83(1)	15	25	C	μ=0.2M(KCl) A=1,3-Diamino- propane	76Sq	
	-98.7	18.55(1)	24	25	C	μ=0.2M(KCl) A=1,2-Diamino- N,N-dimethylethane	76Sq	
	-101.3	19.27(1)	29	25	C	μ=0.2M(KCl) A=1,2-Diamino- N,N'-dimethyl- ethane	76Sq	
	-79.5	17.69(1)	72	25	C	μ=0.2M(KCl) A=Glycine	76Sq	
	-104.6	18.66(1)	6	25	C	μ=0.2M(KCl) A=Histamine	76Sq	
CuA_2	-106	(1)	--	30	C	S:Benzene A=1,1,1,5,5,5- Hexafluoropentane- 2,4-dionato	75Gg 76Gh	
	-79	(1)	--	30	C	S:Benzene A=1,1,1-Tri- fluoropentane-	75Gg 76Gh	

Metal Ion	ΔH, kJ /mol	Log K	ΔS, J /mol K	T,°C	M	Conditions	Ref.	Remarks
E9, continued								
CuA_2, cont.						2,4-dione		
	-75	(1)	--	30	C	S:Benzene A=1,1,1-Trifluoro-4-phenylbutane-2,4-dione	75Gg 76Gh	
	-68	(1)	--	30	C	S:Benzene A=1,1,1-Trifluoro-4-thenyl-2,4-dione	75Gg 76Gh	
	-77	(2)	--	30	C	S:Benzene A=1,1,1,5,5,5-Hexafluoropentane-2,4-dionato	75Gg 76Gh	
CuA^{2+}	-46.4	9.41(25°,1)	24.7	15-35	T	μ=0.2M(KNO_3) CM=2x10^{-3}M A=2,2'-Bipyridyl	79Mg	b
CuL_2	-3.60	1.44(20°)	15.27	20&50	T	S:50v%CH_3OH A=L-Alanine R:$CuL_2+CuA_2=2CuAL$	71Yb 71Yc	a,b,q: ±2.5
	0.04	0.78(20°)	15.15	-100-50	T	S:50v%CH_3OH A=β-Alanine R:$CuL_2+CuA_2=2CuAL$	71Yb 71Yc	b,q: ±0.8
	4.27	0.14(20°)	17.24	-100-50	T	S:50v%CH_3OH A=N,N-Dimethylglycine R:$CuL_2+CuA_2=2CuAL$	71Yb 71Yc	q:±1.2
Ag^+	-48.5	5.063(25°,1)	-66	10-40	T	μ=1M(KNO_3)	78Ma	
	-6.7	2.593(25°,2)	27	10-40	T	μ=1M(KNO_3)	78Ma	
	-107.8	13.175(25°)	-109	10-40	T	μ=1M(KNO_3) R:$2Ag^++2L=Ag_2L_2^{2+}$	78Ma	
	-10.8	3.048(25°)	22	10-45	T	μ=1M(KNO_3) R:$AgL^++AgL^+=Ag_2L_2^{2+}$	78Ma	
Zn^{2+}	-20.9	5.55(30°,1)	38	10-40	T	μ=0	58Ba 60Cb	b
	-41.0	5.50(1)	-33	25	C	0.1M(KCl)	54Da 59Ma	
	-27.82	5.79(1)	17.6	25	C	1M(KCl)	60Ca 54Wa	
	-21.8	5.88(1)	40.2	27	C	1M(KNO_3)	61Me	
	-27.6	6.148(1)	25.1	25	T	μ=2	53Sb	
	-21.8	4.89(30°,2)	21	10-40	T	μ=0	58Ba	b
	-21.8	5.12(2)	25.5	27	C	1M(KNO_3)	61Me	
	-7.5	3.22(30°,3)	38	10-40	T	μ=0	58Ba	b
	-21.8	1.85(3)	-36.8	27	C	1M(KNO_3)	61Me	
	-47.91	10.05(1-2)	31.8	25	C	0.1M(KCl)	54Da	
	-57.53	10.55(1-2)	9.2	25	C	1M(KCl)	60Cb	
	-52.3	11.49(1-2)	44.8	25	T	μ=2	53Sb	
	-77.24	11.40(1-3)	-41.0	25	C	0.1M(KCl)	54Da	
	-86.61	12.31(1-3)	-54.8	25	C	1M(KCl)	60Cb	

Metal Ion	ΔH, kJ /mol	Log K	ΔS, J /mol K	T,°C	M	Conditions	Ref.	Remarks
E9, continued								
ZnA	-29.7	4.44(1)	-14.6	25	C	μ=0.10(KNO_3) A=Ethylenediaminediacetic acid	70Da	
	-27.6	4.03(1)	-15.5	25	C	μ=0.10M(KNO_3) A=Ethylenediaminedipropionic acid	70Da	
Cd^{2+}	-25.9	5.23(30°,1)	17	10-40	T	μ=0	58Ba	b
	-5.02	5.2(1)	84	25	C	0.1M(KCl)	54Da	
	-25.9	5.6(1)	21	27	C	1M(KNO_3)	62Md	
	-29.41	5.836(1)	13.0	25	T	μ=2	53Sa	
	-31.4	4.38(30°,2)	-21	10-40	T	μ=0	58Ba	b
	-55.77	9.4(1-2)	-7.1	25	C	0.1M(KCl)	54Da	
	-51.9	10.2(1-2)	21.8	27	C	1M(KNO_3)	62Md	
	-56.5	10.62(1-2)	13.8	25	T	μ=2	53Sa	
	-82.42	11.1(1-3)	-64.0	25	C	0.1M(KCl)	54Da	
	-77.8	12.2(1-3)	-25.1	27	C	1M(KNO_3)	62Md	
	-229.2	10.25(25°)	-564.4	25&35	T	μ=2.0M(KNO_3) R:$Cd^{2+}+A^{2-}+L=CdAL$ A=Hexanedioic acid	79Ga	a,b
	-206.7	10.27(25°)	-483.2	25&35	T	μ=2.0M(KNO_3) R:$Cd^{2+}+2A^{2-}+L=CdLA_2{}^{2-}$ A=Hexanedioic acid	79Ga	a,b
	-79.96	11.00(25°)	-53.56	25&35	T	μ=2.0M(KNO_3) R:$Cd^{2+}+A^{2-}+2L=CdAL_2$ A=Hexanedioic acid	79Ga	a,b
CdA	-29.7	4.33(1)	-16.7	25	C	μ=0.10M(KNO_3) A=Ethylenediaminediacetic acid	70Da	
HgA_2	-43.5	2.58(1)	-94	30	C	CM=1-20x10^{-3} S:Benzene A=C_6F_5	74Fc	
	-137.6	23.10(1-2)	-21	25	T	0.1M(KNO_3)	61Rc	
	-36.23	5.54	-15.5	25	C	μ=0 R:$HgA_2+L=HgAL^++A^-$ A=Cl	66Pd	
	-37.78	4.19	-46.4	25	C	μ=0 R:$HgAL^++L=HgL_2{}^{2+}+A^-$ A=Cl		

E10 ETHANE, 1,2-diamino-N-n-butyl- $C_6H_{16}N_2$ $CH_3CH_2CH_2CH_2NHCH_2CH_2NH_2$ L

Metal Ion	ΔH, kJ /mol	Log K	ΔS, J /mol K	T,°C	M	Conditions	Ref.	Remarks
Ni^{2+}	-32.6	7.25(1)	20.1	0	T	μ=0.5(KNO_3)	52Ba	
	-25.5	5.97(2)	21.3	0	T	μ=0.5(KNO_3)	52Ba	
	-36.8	2.79(3)	-81.2	0	T	μ=0.5(KNO_3)	52Ba	

Metal Ion	ΔH, kJ /mol	Log K	ΔS, J /mol K	T,°C	M	Conditions	Ref.	Remarks
E10, continued								
Cu^{2+}	-33.5	10.47(1)	79	0	T	μ=0.5(KNO_3)	52Ba	
	-34.3	8.82(2)	50	0	T	μ=0.5(KNO_3)	52Ba	
E11	ETHANE, 1,2-diamino-N,N'-diacetic acid		$C_6H_{12}O_4N_2$			$HO_2CCH_2NHCH_2CH_2NHCH_2CO_2H$		L^{2-}
La^{3+}	-4.69	6.955(1)	117.2	25	C	μ=1.00($NaClO_4$)	74Ge	c
	-8.54	4.598(2)	59.8	25	C	μ=1.00($NaClO_4$)	74Ge	
Ce^{3+}	-7.20	(1)	117.2	25	C	μ=1.00($NaClO_4$)	74Ge	
	-5.73	(2)	73.6	25	C	μ=1.00($NaClO_4$)	74Ge	
Pr^{3+}	-8.74	(1)	118.8	25	C	μ=1.00($NaClO_4$)	74Ge	
	-7.15	(2)	75.3	25	C	μ=1.00($NaClO_4$)	74Ge	
Nd^{3+}	-10.50	7.976(1)	117.2	25	C	μ=1.00($NaClO_4$)	74Ge	
	-8.70	5.613(2)	78.6	25	C	μ=1.00($NaClO_4$)	74Ge	
Sm^{3+}	-12.22	8.259(1)	117.2	25	C	μ=1.00($NaClO_4$)	74Ge	
	-12.97	6.180(2)	74.9	25	C	μ=1.00($NaClO_4$)	74Ge	
Gd^{3+}	-8.87	8.245(1)	127.2	25	C	μ=1.00($NaClO_4$)	74Ge	
	-16.07	6.173(2)	66.9	25	C	μ=1.00($NaClO_4$)	74Ge	
Tb^{3+}	-4.77	8.263(1)	141.8	25	C	μ=1.00($NaClO_4$)	74Ge	
	-17.36	6.786(2)	72.0	25	C	μ=1.00($NaClO_4$)	74Ge	
Dy^{3+}	-3.56	(1)	148.1	25	C	μ=1.00($NaClO_4$)	74Ge	
	-16.28	(2)	80.8	25	C	μ=1.00($NaClO_4$)	74Ge	
Ho^{3+}	-2.01	(1)	153.1	25	C	μ=1.00($NaClO_4$)	74Ge	
	-11.97	(2)	106.7	25	C	μ=1.00($NaClO_4$)	74Ge	
Er^{3+}	-1.38	8.508(1)	158.2	25	C	μ=1.00($NaClO_4$)	74Ge	
	-13.01	7.430(2)	98.7	25	C	μ=1.00($NaClO_4$)	74Ge	
Tm^{3+}	-1.63	(1)	159.8	25	C	μ=1.00($NaClO_4$)	74Ge	
	-14.06	(2)	104.2	25	C	μ=1.00($NaClO_4$)	74Ge	
Yb^{3+}	-3.22	8.827(1)	158.2	25	C	μ=1.00($NaClO_4$)	74Ge	
	-15.94	8.110(2)	101.7	25	C	μ=1.00($NaClO_4$)	74Ge	
VO_2^{+}	-59	14.5(25°,1)	79	15-35	T	μ=1.0M($NaClO_4$ & $HClO_4$)	76Yb	
UO_2^{2+}	6.3	11.41(25°,1)	238	20-40	T	μ=0.1M(KNO_3)	69Sg	b
Mn^{2+}	-3.56	7.05(1)	123.0	25	C	μ=0.10M(KNO_3)	70Da	
	-2.9	6.85(1)	122	25	C	μ=0.1M(KNO_3)	79Gc	q:0.4
Co^{2+}	-24.10	11.25(1)	134.7	25	C	μ=0.10M(KNO_3)	70Da	
Ni^{2+}	-41.8	13.65(1)	120.9	25	C	μ=0.10M(KNO_3)	70Da	
Zn^{2+}	-25.5	11.22(1)	129.7	25	C	μ=0.10M(KNO_3)	70Da	
	-24.4	10.99(1)	128	25	C	μ=0.1M(KNO_3)	79Gc	
	60.7	-10.56	2	25	C	μ=0.1M(KNO_3) R: $ZnL+H_2O=ZnLOH^-+H^+$	79Gc	
Cd^{2+}	-22.5	9.40(1)	104.6	25	C	μ=0.10M(KNO_3)	70Da	
	-16.7	9.16(1)	119	25	C	μ=0.1M(KNO_3)	79Gc	

Metal Ion	ΔH, kJ /mol	Log K	ΔS, J /mol K	T,°C	M	Conditions	Ref.	Remarks
E11, continued								
Pb^{2+}	-28.0	10.66(1)	110	25	C	μ=0.1M(KNO_3)	79Gc	
	38.5	-11.02	-82	25	C	μ=0.1M(KNO_3) R: $PbL+H_2O=PbLOH^-+H^+$	79Gc	
E12	ETHANE, 1,2-diamino-N,N'-di-(2-amino ethyl)-		$C_6H_{18}N_4$			$H_2N(CH_2)_2NH(CH_2)_2NH(CH_2)_2NH_2$	L	
Mn^{2+}	-9.62	4.91(1)	63	25	C	0.1M(KCl)	60Sc 61Sa	b
	-17	5.43(30°,1)	47.3	30-40	T	$\mu\cong$2(KCl+KNO_3)	52Ja	b
Fe^{2+}	-25.31	7.69	63	25	C	0.1M(KCl)	60Sc 61Sa	
	-38	8.31(30°,1)	32.6	30-40	T	$\mu\cong$2(KCl+KNO_3)	52Ja	b
Co^{2+}	-44.56	10.92(1)	60.7	25	C	0.1M(KCl)	60Sc 61Sa	
	-38	11.21(30°,1)	90.4	30-40	T	$\mu\cong$2(KCl+KNO_3)	52Ja	b
Ni^{2+}	-58.58	13.78(1)	67	25	C	0.1M(KCl)	60Sc 61Sa	
	-58.2	14.1(1)	75	25	C	μ=0.1(KNO_3)	65Wd	
	-54.4	14.3(30°,1)	93.7	30&40	T	1M(KNO_3)	51Jc	a,b
	-54	14.34(30°,1)	95.4	30&40	T	μ=2(KCl+KNO_3)	52Ja	a,b
	-90.8	20.1(30°,1-2)	85.4	30&40	T	1M(KNO_3)	51Jc	a,b
	-38	5.63(30°,2-3)	-16.7	30&40	T	μ=2(KCl+KNO_3)	52Ja	a,b
	-36.4	5.6(30°,3-2)	-12.6	30&40	T	1M(KNO_3)	51Jc	a,b
Cu^{2+}	-90.17	20.08(1)	81.6	25	C	0.1M(KCl)	60Sc 61Sa	
	-89.5	20.2(1)	88	25	C	μ=0.1(KNO_3)	65Wd	
	-92	20.62(30°,1)	91.2	30&40	T	μ=2(KCl+KNO_3)	52Ja	a,b
	-90.8	20.6(30°,1-2)	95.4	30&40	T	μ=1(KNO_3)	51Jc	a,b
Zn^{2+}	-37.24	12.02(1)	104.6	25	C	0.1M(KCl)	60Sc 61Sa	
	-34.7	11.9(1)	113	25	C	μ=0.1(KNO_3)	65Wd	
	-17	11.94(30°,1)	172	30&40	T	μ=2(KCl+KNO_3)	52Ja	a,b
	-29.7	12.05(1)	131	25	C	μ=0.10M(KNO_3) S:50v%Dioxane	79Mc	
	30	6.90	33	25	C	μ=0.10M(KNO_3) S:50v%Dioxane R: $Zn^{2+}+HL^+=Zn(HL)^{3+}$	79Mc	q:±4
Cd^{2+}	-38.5	10.8(1)	79	25	C	μ=0.1(KNO_3)	65Wd	
	-17	10.92(30°,1)	154.0	30&40	T	μ=2(KCl+KNO_3)	52Ja	a,b
Pb^{2+}	-43.3	10.35(1)	53	25	C	μ=0.10M(KNO_3) S:50v%Dioxane	79Mc	
	15.8	6.07	63	25	C	μ=0.10M(KNO_3) S:50v%Dioxane	79Mc	q:±7.8

Metal Ion	ΔH, kJ /mol	Log K	ΔS, J /mol K	T,°C	M	Conditions	Ref.	Re-marks
E12, continued								
Pb^{2+}, cont.						R: $Zn^{2+}+HL^{+}=Zn(HL)^{3+}$		
E13	ETHANE, 1,2-diamino-N,N-diethyl		$C_6H_{16}N_2$			$NH_2CH_2CH_2N(CH_2CH_3)_2$		
Cu^{2+}	-29.4	8.33(1)	60.7	25	C	μ=0.5M($NaClO_4$)	75Bg	
	-25.9	5.72(2)	21	25	C	μ=0.5M($NaClO_4$)	75Bg	
E14	ETHANE, 1,2-diamino-N,N'-diethyl-		$C_6H_{16}N_2$			$CH_3CH_2NHCH_2CH_2NHCH_2CH_3$		L
Ni^{2+}	-32.6	12.24(1-2)	113	0	C	0.5M(KNO_3)	54Bb	
Cu^{2+}	-31.4	8.01(1)	47.7	15-40	T	μ=0.3M($NaClO_4$)	72Md	b
	-34.3	8.96(1)	56.0	25	C	μ=0.5M($NaClO_4$)	75Bg	
	-26.8	5.70(2)	19.2	15-40	T	μ=0.3M($NaClO_4$)	72Md	b
	-30.5	5.64(2)	8.4	25	C	μ=0.5M($NaClO_4$)	75Bg	
	-73.2	18.6(1-2)	88	0	C	0.5M(KNO_3)	54Bb	
	-64.8	15.62(1-2)	81.6	25	C	μ=0.5M($NaClO_4$)	75Bf	
	-9.83	1.99(25°)	5.0	-75-25	T	S:50v%CH_3OH R: Cu+CuL_2=2CuL	72Ho	a,b,q: ±2.5
E15	ETHANE, 1,2-diamino-N,N'-di(ethyldiphenylphosphine)		$C_{28}H_{30}O_2N_2P_2$			$(C_6H_5)_2P(=O)CH_2NHCH_2$-$CH_2NHCH_2P(=O)(C_6H_5)_2$		L^{IV}
CuA^{+}	37.53	4.58(25°,1)	217	5-45	T	μ<0.005 S:1v Acetone 1v Nitromethane 0.5v Chloroform A=Cl^-	79Ya	b,q:±2
E16	ETHANE, 1,2-diamino-N,N-dimethyl-		$C_4H_{12}N_2$			$NH_2CH_2CH_2N(CH_3)_2$		L
Ni^{2+}	-27.40	(1)	--	--	C	--	79La	μ:25°
	-27.32	(2)	--	--	C	--	79La	
Cu^{2+}	-40.5	9.24(1)	41	25	C	μ=0.2M(KCl)	76Sp 76Sq	
	-41.09	9.280(1)	39.7	25	C	0.5M(KNO_3)	72Bd	
	-40.2	6.96(2)	-2	25	C	μ=0.2M(KCl)	76Sp 76Sq	
	-39.66	7.028(2)	1.7	25	C	0.5M(KNO_3)	72Bd	
	-55.2	13.04(1-2)	64.4	25	C	μ=0.5M($NaClO_4$)	75Bf	
	-98.7	18.55	25	25	C	μ=0.2M(KCl) R: Cu+A+L=CuAL A=1,2-Diamino-ethane	76Sp	
	-90.4	16.98	22	25	C	μ=0.2M(KCl) R: Cu+A+L=CuAL	76Sp	

Metal Ion	ΔH, kJ /mol	Log K	ΔS, J /mol K	T,°C	M	Conditions	Ref.	Remarks
E16, continued								
Cu^{2+}, cont.						A=1,3-Diamino-propane		
	-70.7	16.70	82	25	C	μ=0.2M(KCl) R:Cu+A+L=CuAL A=Glycine	76Sp	
CuA	-98.7	18.55(1)	24	25	C	μ=0.2M(KCl) A=1,2-Diamino-ethane	76Sq	
	-70.7	16.70(1)	82	25	C	μ=0.2M(KCl) A=Glycine	76Sq	
	-90.4	16.98(1)	22	25	C	μ=0.2M(KCl) A=1,3-Diamino-propane	76Sq	
CuA_2	-106	(1)	--	30	C	S:Benzene A=1,1,1,5,5,5-Hexafluoropentane-2,4-dionato	76Gh 75Gg	
	-69	(1)	--	30	C	S:Benzene A=1,1,1-Tri-fluoropentane-2,4-dione	76Gh 75Gg	
	-76	(1)	--	30	C	S:Benzene A=1,1,1-Tri-fluoro-4-phenylbutane-2,4-dione	76Gh 75Gg	
	-79	(1)	--	30	C	S:Benzene A=1,1,1-Tri-fluoro-4-thenyl-2,4-dione	76Gh 75Gg	
	-76	(2)	--	30	C	S:Benzene A=1,1,1,5,5,5-Hexafluoropentane-2,4-dionato	76Gh 75Gg	
	-45	(2)	--	30	C	S:Benzene A=1,1,1-Tri-fluoropentane-2,4-dione	76Gh 75Gg	
	-48	(2)	--	30	C	S:Benzene A=1,1,1-Tri-fluoro-4-phenylbutane-2,4-dione	76Gh 75Gg	
	-57	(2)	--	30	C	S:Benzene A=1,1,1-Tri-fluoro-4-thenyl-2,4-dione	76Gh 75Gg	

Metal Ion	ΔH, kJ /mol	Log K	ΔS, J /mol K	T,°C	M	Conditions	Ref.	Remarks
E17	ETHANE, 1,2-diamino-N,N'-dimethyl-		$C_4H_{12}N_2$			$CH_3NHCH_2CH_2NHCH_3$		
Cu^{2+}	-46.6	9.94(1)	34	25	C	μ=0.2M(KCl)	76Sp 76Sq	
	-46.44	10.06(1)	37.6	25	C	0.5M(KNO_3)	72Be	
	-39.0	7.03(2)	4	25	C	μ=0.2M(KCl)	76Sp 76Sq	
	-40.46	7.151(2)	1.2	25	C	0.5M(KNO_3)	72Be	
	-101.3	19.27	29	25	C	μ=0.2M(KCl) R:Cu+A+L=CuAL A=1,2-Diamino-ethane	76Sp	
	-94.1	17.51	19	25	C	μ=0.2M(KCl) R:Cu+A+L=CuAL A=1,3-Diamino-propane	76Sp	
	-74.9	17.29	80	25	C	μ=0.2M(KCl) R:Cu+A+L=CuAL A=Glycine	76Sp	
	-91.6	16.90	16	25	C	μ=0.2M(KCl) R:Cu+A+L=CuAL A=Histamine	76Sp	
	-112.0	22.66	58	25	C	μ=0.2M(KCl) R:Cu+H+2L=$CuHL_2$	76Sq	
	-7.03	2.10(25°)	16.7	-75-25	T	S:50v%CH_3OH R:Cu+CuL_2=2CuL	73Ya	a,b,q: ±1.8
CuA	-101.3	19.27(1)	29	25	C	μ=0.2M(KCl) A=1,2-Diamino-ethane	76Sq	
	-94.1	17.51(1)	19	25	C	μ=0.2M(KCl) A=1,3-Diamino-propane	76Sq	
	-74.9	17.29(1)	80	25	C	μ=0.2M(KCl) A=Glycine	76Sq	
	-91.6	16.90(1)	16	25	C	μ=0.2M(KCl) A=Histamine	76Sq	
E18	ETHANE, 1,2-diamino-N,N'-dimethyl-N,N'-diacetic acid-		$C_8H_{16}N_2$			$CH_3N(CH_2CO_2H)CH_2CH_2N$-($CH_2CO_2H)CH_3$		L^{2-}
Mg^{2+}	4	5.7(1-2)	130	25	T	μ=0	56Ma	
E19	ETHANE, 1,2-diamino-1,2-dimethyl-N,N,N',N'-tetraacetic acid		$C_{12}H_{20}O_8N_2$			$(HO_2CH_2C)_2NCH(CH_3)CH$-$(CH_3)N(CH_2CO_2H)_2$		L^{4-}
Mg^{2+}	15.90	8.462(1)	216.3	20	C	μ=1M(HCl) CL=0.01M	69Sh	g:k
Cu^{2+}	-14.81	9.200(1)	125.5	20	C	μ=1M(HCl) CL=0.01M	69Sh	g:k
Sr^{2+}	1.55	7.210(1)	132.6	20	C	μ=1M(HCl) CL=0.01M	69Sh	g:k

Metal Ion	ΔH, kJ /mol	Log K	ΔS, J /mol K	T,°C	M	Conditions	Ref.	Re-marks
E19, continued								
Ba^{2+}	-7.66	6.032(1)	89.5	20	C	μ=1M(HCl) CL=0.01M	69Sh	g:k
E20	ETHANE, 1,2-diamino-N-ethyl-	$C_4H_{12}N_2$				$CH_3CH_2NHCH_2CH_2NH_2$		L
Ni^{2+}	-25.5	7.19(1)	44.4	0	T	μ=0.5(KNO_3)	52Ba	
	-30.1	5.78(2)	1.7	0	T	μ=0.5(KNO_3)	52Ba	
	-31.8	2.51(3)	-69.0	0	T	μ=0.5(KNO_3)	52Ba	
Cu^{2+}	-22.6	10.55(1)	121	0	T	μ=0.5(KNO_3)	52Ba	
	-26.8	8.81(2)	71	0	T	μ=0.5(KNO_3)	52Ba	
E21	ETHANE, 1,2-diamino-N-(2-hydroxyethyl)	$C_4H_{12}ON_2$				$H_2NCH_2CH_2NHCH_2CH_2OH$		
Ni^{2+}	-32.6	6.96(1)	24	25	C	μ=0.5M($NaClO_4$)	75Bb	
	-34.8	5.83(2)	-5.0	25	C	μ=0.5M($NaClO_4$)	75Bb	
Cu^{2+}	-46.4	10.33(1)	42.1	25	C	μ=0.5M($NaClO_4$)	75Bb	
	-4.4	7.75(2)	133	25	C	μ=0.5M($NaClO_4$)	75Bb	
E22	ETHANE, 1,2-diamino-N-(2-hydroxyethyl)-N,N',N'-triacetic acid-	$C_{10}H_{18}O_7N_2$				$(CH_2CO_2H)_2NCH_2CH_2N$-$(CH_2CO_2H)CH_2CH_2OH$		L^{3-}
Mg^{2+}	14.2	6.96(1)	180	25	C	μ=0.1	65Wd	
	-15.90	6.92(1)	79	25	T	μ=0.10(KNO_3)	65Ma	
Ca^{2+}	-27.2	8.06(1)	63	25	C	μ=0.1(KNO_3)	61Mf	
	-22.6	8.44(1)	84	25	T	0.1M(KNO_3)	65Wd	
	-22.76	8.12(1)	79	25	T	μ=0.10(KNO_3)	65Ma	
Sr^{2+}	-21.8	6.82(1)	58	25	C	μ=0.1(KNO_3)	65Wd	
	-31.02	6.62(1)	46	25	T	μ=0.10(KNO_3)	65Ma	
Ba^{2+}	-22.6	6.16(1)	42	25	C	μ=0.1(KNO_3)	65Wd	
	-26.94	5.89(1)	21	25	T	μ=0.10(KNO_3)	65Ma	
Y^{3+}	-10.17	14.49(1)	243.5	25	C	μ=0.1(KCl)	68Fe	
	-1.21	14.6(1)	276.5	25	T	0.1M(KNO_3)	61Mf	
La^{3+}	-17.45	13.22(1)	194.6	25	C	μ=0.1(KCl)	68Fe	
	-9.20	13.47(1)	227.6	25	T	0.1M(KNO_3)	61Mf	
Ce^{3+}	-21.80	14.08(1)	196.6	25	C	μ=0.1(KCl)	68Fe	
	-12.80	14.12(1)	227.6	25	T	0.1M(KNO_3)	61Mf	
Pr^{3+}	-23.51	14.39(1)	190.4	25	C	μ=0.1(KCl)	68Fe	
	-18.62	14.62(1)	217.6	25	T	0.1M(KNO_3)	61Mf	
Nd^{3+}	-25.77	14.71(1)	195.4	25	C	μ=0.1(KCl)	68Fe	
	-17.78	14.87(1)	225.1	25	T	0.1M(KNO_3)	61Mf	
Sm^{3+}	-29.41	15.15(1)	191.2	25	C	μ=0.1(KCl)	68Fe	
	-19.46	15.30(1)	227.6	25	T	0.1M(KNO_3)	61Mf	
Eu^{3+}	-29.46	15.21(1)	192.5	25	C	μ=0.1(KCl)	68Fe	
	-20.12	15.36(1)	226.4	25	T	0.1M(KNO_3)	61Mf	

Metal Ion	ΔH, kJ /mol	Log K	ΔS, J /mol K	T,°C	M	Conditions	Ref.	Remarks
E22, continued								
Gd^{3+}	-28.28	15.10(1)	194.1	25	C	μ=0.1(KCl)	68Fe	
	-19.50	15.23(1)	226.4	25	T	0.1M(KNO_3)	61Mf	
Tb^{3+}	-24.77	15.10(1)	205.8	25	C	μ=0.1(KCl)	68Fe	
	-14.18	15.33(1)	246.0	25	T	0.1M(KNO_3	61Mf	
Dy^{3+}	-20.12	15.08(1)	221.3	25	C	μ=0.1(KCl)	68Fe	
	-8.87	15.31(1)	262.8	25	T	0.1M(KNO_3)	61Mf	
Ho^{3+}	-14.31	15.06(1)	240.6	25	C	μ=0.1(KCl)	68Fe	
	-4.77	15.33(1)	277.4	25	T	0.1M(KNO_3)	61Mf	
Er^{3+}	-10.04	15.17(1)	256.9	25	C	μ=0.1(KCl)	68Fe	
	-1.34	15.44(1)	290.4	25	T	0.1M(KNO_3)	61Mf	
Tm^{3+}	-7.57	15.38(1)	269.0	25	C	μ=0.1(KCl)	68Fe	
	3.85	15.60(1)	311.7	25	T	0.1M(KNO_3)	61Mf	
Yb^{3+}	-8.83	15.64(1)	269.9	25	C	μ=0.1(KCl)	68Fe	
	1.51	15.89(1)	309.6	25	T	0.1M(KNO_3)	61Mf	
Lu^{3+}	-9.20	15.79(1)	271.5	25	C	μ=0.1(KCl)	68Fe	
	0.92	15.89(1)	307.1	25	T	0.1M(KNO_3)	61Mf	
Mn^{2+}	-21.8	10.8(1)	134	25	C	μ=0.1(KNO_3)	65Wd	
Fe^{2+}	-25.1	11.6(1)	138	25	C	μ=0.1(KNO_3)	65Wd	
Co^{2+}	-27.2	14.4(1)	184	25	C	μ=0.1(KNO_3)	65Wd	
Ni^{2+}	-43.1	17.1(1)	188	25	C	μ=0.1(KNO_3)	65Wd	
Cu^{2+}	-39.3	17.4(1)	201	25	C	μ=0.1(KNO_3)	65Wd	
	-28.4	17.41(1)	238	25	T	0.1M(KNO_3)	61Mf	
Zn^{2+}	-35.1	14.4(1)	159	25	C	μ=0.1(KNO_3)	65Wd	
Cd^{2+}	-43.1	13.0(1)	105	25	C	μ=0.1(KNO_3)	65Wd	
Hg^{2+}	-84.7	20.0(1)	105	25	C	μ=0.1(KNO_3)	65Wd	
	-29.7	19.38(1)	271.1	25	T	μ=0.1(KNO_3)	66Mf	
Al^{3+}	39.3	12.54(1)	372.8	25	T	μ=0.1(KNO_3)	66Mf	
Pb^{2+}	-52.7	15.5(1)	121	25	C	μ=0.1(KNO_3)	65Wd	
E23 ETHANE, 1,2-diamino-N-methyl- $C_3H_{10}N_2$						$CH_3NHCH_2CH_2NH_2$		L
Ni^{2+}	-38.9	6.97(30°,1)	4	10-40	T	μ=0	59Ma	b
	-36.8	7.95(1)	16.7	0	T	μ=0.5(KNO_3)	52Ba	
	-32.6	5.40(30°,2)	-4	10-40	T	μ=0	59Ma	b
	-25.5	6.15(2)	24.7	0	T	μ=0.5(KNO_3)	52Ba	
	-17.2	2.68(30°,3)	-4	10-40	T	μ=0	59Ma	b
	-25.1	2.41(3)	-46.0	0	T	μ=0.5(KNO_3)	52Ba	
	-71.1	-17.2(1-2)	4	0	C	0.5(KNO_3)	54Bb	
Cu^{2+}	-51.0	10.06(30°,1)	25	10-40	T	μ=0	59Ma	b
							52Ba	
	-48.24	10.417(1)	37.6	25	C	0.5M(KNO_3)	72Bd	
	-52.3	8.38(30°,2)	-12	10-40	T	μ=0	59Ma	b
	-51.0	8.703(2)	-4.2	25	C	0.5M(KNO_3)	72Bd	
	-96.2	20.2(1-2)	33	0	C	0.5(KNO_3)	54Bb	

Metal Ion	ΔH, kJ /mol	Log K	ΔS, J /mol K	T,°C	M	Conditions	Ref.	Remarks
E23, continued								
CuA_2	-108	(1)	--	30	C	S:Benzene A=1,1,1,5,5,5-Hexafluoropentane-2,4-dionato	75Gg 76Gh	
	-78	(2)	--	30	C	S:Benzene A=1,1,1,5,5,5-Hexafluoropentane-2,4-dionato	75Gg 76Gh	
Zn^{2+}	-21.3	5.29(30°,1)	4	10-40	T	μ=0	59Ma	b
	-14.2	4.23(30°,2)	4	10-40	T	μ=0	59Ma	b
E24	ETHANE, 1,2-diamino-N-(1-methylethyl)-	$C_5H_{14}N_2$				$H_2NCH_2CH_2NHCH(CH_3)_2$		L
Ni^{2+}	-28.0	5.62(1)	4.6	0	T	μ=0.5(KNO_3)	52Ba	
	-23.0	3.84(2)	-10.9	0	T	μ=0.5(KNO_3)	52Ba	
Cu^{2+}	-24.3	9.46(1)	92	0	T	μ=0.5(KNO_3)	52Ba	
	-34.3	8.00(2)	29	0	T	μ=0.5(KNO_3)	52Ba	
E25	ETHANE, 1,2-diamino-1-phenyl-	$C_8H_{12}N_2$				$C_6H_5CH(NH_2)CH_2NH_2$		
Cu^{2+}	-54.8	10.25(1)	12.1	15-40	T	μ=0.3M($NaClO_4$)	72Md	b
	-56.5	9.01(2)	-16.7	15-40	T	μ=0.3M($NaClO_4$)	72Md	b
E26	ETHANE, 1,2-diamino-N-n-propyl-	$C_5H_{14}N_2$				$CH_3CH_2CH_2NHCH_2CH_2NH_2$		L
Ni^{2+}	-31.4	7.10(1)	21.3	0	T	μ=0.5(KNO_3)	52Ba	
	-41.4	5.82(2)	-39.7	0	T	μ=0.5(KNO_3)	52Ba	
	-16.3	2.26(3)	-16.7	0	T	μ=0.5(KNO_3)	52Ba	
Cu^{2+}	-31.8	10.49(1)	84	0	T	μ=0.5(KNO_3)	52Ba	
	-33.5	8.70(2)	46	0	T	μ=0.5(KNO_3)	52Ba	
E27	ETHANE, 1,2-diamino-N,N'-(2-pyridylmethyl) (bis)	$C_{14}H_{20}N_4$				$(C_5H_5N)CH_2NHCH_2CH_2NHCH_2(NC_5H_5)$		
Mn^{2+}	-18.8	5.57(1)	43.5	25	C	μ=0.1(KNO_3)	75Ae	
Co^{2+}	-59.4	11.95(1)	29.3	25	C	μ=0.1(KNO_3)	75Ae	
Ni^{2+}	-72.8	14.51(1)	33.5	25	C	μ=0.1(KNO_3)	75Ae	
Cu^{2+}	-76.1	17.00(1)	70.3	25	C	μ=0.1(KNO_3)	75Ae	
Ag^{+}	-110.0	15.9	-64.4	25	C	μ=0.1(KNO_3) R:$2Ag^{+}+2L=Ag_2L_2^{2+}$	75Ae	
Zn^{2+}	-47.3	11.14(1)	54.8	25	C	μ=0.1(KNO_3)	75Ae	
Cd^{2+}	-40.6	9.67(1)	48.9	25	C	μ=0.1(KNO_3)	75Ae	
Hg^{2+}	-92.0	19.13(1)	57.3	25	C	μ=0.1(KNO_3)	75Ae	
	-23.0	5.58(2)	30.1	25	C	μ=0.1(KNO_3)	75Ae	

Metal Ion	ΔH, kJ /mol	Log K	ΔS, J /mol K	T,°C	M	Conditions	Ref.	Remarks
E28	ETHANE, 1,2-diamino-N,N,N',N'-tetraacetic acid		$C_{10}H_{16}O_8N_2$			$(CH_2CO_2H)_2NCH_2CH_2N(CH_2CO_2H)_2$		L^{4-}
Li^+	0.4	2.8(1)	54	25	C	μ≦0.1	54Cb	
	0.8	2.97(1)	59.8	20	C	μ=0.1M(Tetramethylammonium chloride)	76Ah	
	0.42	(1)	54	20-40	T	μ=0.1M(KNO_3)	77Fb	
Na^+	-9.50	2.58(1)	18.8	15	C	μ=0	76Vb	
	-8.66	2.52(1)	19.7	25	C	μ=0 ΔCp=68.6	76Vb	
	-8.16	2.48(1)	20.9	35	C	μ=0	76Vb	
	-5.9	1.7(1)	12	25	C	μ≦0.1	54Cb	
	-5.4	1.85(1)	17	20	C	μ=0.1M(NaCl)	76Ah	
	-10.0	1.40(1)	-6.7	15	C	μ=0.3(KNO_3)	76Vb	
	-9.46	1.34(1)	-5.8	25	C	μ=0.3(KNO_3) ΔCp=49.0	76Vb	
	-9.71	1.34(1)	-7.1	25	C	μ=0.3[$(CH_3)_4NNO_3$]	76Vb	
	-9.08	1.30(1)	-4.6	35	C	μ=0.3(KNO_3)	76Vb	
	-8.70	1.29(1)	-4.2	15	C	μ=0.5(KNO_3)	76Vb	
	-8.58	1.24(1)	-5.0	25	C	μ=0.5(KNO_3) ΔCp=16.3	76Vb	
	-9.37	1.24(1)	-7.5	25	C	μ=0.5[$(CH_3)_4NNO_3$]	76Vb	
	-8.45	1.20(1)	-4.6	35	C	μ=0.5(KNO_3)	76Vb	
	-7.11	1.25(1)	4.2	15	C	μ=1.0(KNO_3)	76Vb	
	-10.8	1.69(1)	-6.3	20	C	μ=1M(Tetramethylammonium chloride)	76Ah	
	-6.90	1.21(1)	0.0	25	C	μ=1.0(KNO_3) ΔCp=17.6	76Vb	
	-8.53	1.21(1)	-5.4	25	C	μ=1.0[$(CH_3)_4NNO_3$]	76Vb	
	-6.61	1.18(1)	1.2	35	C	μ=1.0(KNO_3)	76Vb	
	-5.8	(1)	12	20-40	T	μ=0.1M(KNO_3)	77Fb	
K^+	3.26	2.66(1)	62.3	20	C	μ=1M(Tetramethylammonium chloride)	76Ah	
Be^{2+}	40.6	9.63(25°,1)	318	20-40	T	μ=0.1M(KNO_3)	77Fb	
Mg^{2+}	18.9	10.43(1)	265	15	C	μ=0	75Va	
	21.0	10.54(1)	272	25	C	μ=0 ΔCp=233	75Va	
	23.6	10.68(1)	281	35	C	μ=0	75Va	
	23.0	(1)	--	25	C	CM=10^{-2}M	56Jc, 54Ca	
	13.0	9.09(1)	218	25	C	μ<0.1	54Cb, 56Ma	
	13.0	(1)	218	20-40	T	μ=0.1M(KNO_3)	77Fb	
	14.60	8.69(1)	213.4	20	C	0.1M(KNO_3)	63Ab	
	13.14	8.69(1)	211.3	20	C	μ=0.17	56Ca, 64Ab	
	14.9	8.04(1)	205	15	C	μ=0.3(KNO_3)	75Va	
	16.3	8.15(1)	210	25	C	μ=0.3(KNO_3) ΔCp=158	75Va	
	18.0	8.26(1)	217	35	C	μ=0.3(KNO_3)	75Va	
	13.8	7.98(1)	200	15	C	μ=0.5(KNO_3)	75Va	

Metal Ion	ΔH, kJ /mol	Log K	ΔS, J /mol K	T,°C	M	Conditions	Ref.	Remarks
E28, continued								
Mg^{2+}, cont.	14.9	8.06(1)	204	25	C	μ=0.5(KNO_3) ΔCp=132	75Va	
	16.4	8.16(1)	209	35	C	μ=0.5(KNO_3)	75Va	
	12.0	7.80(1)	191	15	C	μ=1.0(KNO_3)	75Va	
	12.8	7.87(1)	194	25	C	μ=1.0(KNO_3) ΔCp=91.6	75Va	
	13.9	7.95(1)	197	35	C	μ=1.0(KNO_3)	75Va	
Ca^{2+}	-27.15	12.6(1)	146.8	15	C	μ=0	76Va	y
	-25.19	12.42(1)	153.1	25	C	μ=1 ΔCp=200	76Va	
	-12.6	11.0(1)	176	25	T	μ=0	56Ma 54Ca	
	-23.14	12.31(1)	160.7	35	C	μ=0	76Va	
	-23.8	(1)	--	25	C	CM=10^{-2}M	57Jc 64Ab	
	-24.02	11.18(1)	133.5	25	C	μ=0.060	59Yb	
	-23.81	11.06(1)	131.8	25	C	μ=0.075	59Yb	
	-24.3	10.97(1)	130	25	C	μ≦0.1	54Cb	
	-27.40	10.7(1)	111.3	20	C	0.1M(KNO_3)	63Ab	
	-24.3	(1)	130	20-40	T	μ=0.1M(KNO_3)	77Fb	
	-21.84	10.58(1)	129.3	25	C	μ=0.15	59Yb	
	-26.99	10.69(1)	112.5	20	C	μ=0.17(KNO_3)	56Ca	
	-21.34	10.45(1)	128.4	25	C	μ=0.22	59Yb	
	-30.46	10.22(1)	90.0	15	C	μ=0.3(KNO_3)	76Va	
	-29.00	10.03(1)	94.6	25	C	μ=0.3(KNO_3) ΔCp=152	76Va	
	-27.40	9.90(1)	100.4	35	C	μ=0.3(KNO_3)	76Va	
	-30.84	10.15(1)	87.0	15	C	μ=0.5(KNO_3)	76Va	
	-29.46	9.95(1)	91.6	25	C	μ=0.5(KNO_3) ΔCp=136	76Va	
	-28.12	9.81(1)	96.6	35	C	μ=0.5(KNO_3)	76Va	
	-19.75	10.13(1)	127.6	25	C	μ=0.54	59Yb	
	-19.46	10.15(1)	128.9	25	C	μ=0.75	59Yb	
	-20.17	10.16(1)	126.8	25	C	μ=0.80	59Yb	
	-19.79	10.17(1)	128.4	25	C	μ=0.86	58Ya	
	-30.92	9.95(1)	82.8	15	C	μ=1.0(KNO_3)	76Va	
	-30.00	9.75(1)	85.8	25	C	μ=1.0(KNO_3) ΔCp=119	76Va	
	-28.53	9.61(1)	91.2	35	C	μ=1.0(KNO_3)	76Va	
	-20.79	10.29(1)	127.2	25	C	μ=1.08	59Yb	
	-22.05	10.61(1)	129.3	25	C	μ=1.46	59Yb	
	-23.10	10.73(1)	128.0	25	C	μ=1.58	59Yb	
Sr^{2+}	-17	8.72(1)	109	25	T	μ=0	54Ca 56Ma	
	-17.07	8.63(1)	107.1	20	C	0.1M(KNO_3)	63Ab	
	-17.6	8.72(1)	109	25	C	μ≦0.1	54Cb	
	-17.6	(1)	109	20-40	T	μ=0.1M(KNO_3)	77Fb	
	-17.20	8.62(1)	106.3	20	C	μ=0.17	56Ca	
	68.0	7.99(32°,1)	290	32&42	T	--	68Ld	a,b
Ba^{2+}	-17	7.70(1)	92	25	T	μ=0	54Ca 56Ma	
	-21.3	7.72(1)	75	25	C	μ≦0.1	54Cb	
	-20.63	7.76(1)	78.2	20	C	0.1M(KNO_3)	63Ab	

Metal Ion	ΔH, kJ /mol	Log K	ΔS, J /mol K	T,°C	M	Conditions	Ref	Remarks
E28, continued								
Ba^{2+}, cont.	-22.2	7.84(1)	75	25	C	μ=0.1(KNO_3)	65Wd	
	-21.3	(1)	75	20-40	T	μ=0.1M(KNO_3)	77Fb	
	-20.21	7.75(1)	79.5	20	C	μ=0.17	56Ca	
Y^{3+}	1.34	18.085(1)	350.6	20	C	μ=0.1	58Sf	
	-2.460	17.372(1)	324.3	25	C	μ=0.1(KNO_3)	62Ma	
	-6.7	17.11(1)	305	25	C	μ=0.44	65Ka	
La^{3+}	-11.72	15.50(1)	256.9	20	C	0.1M(KNO_3)	63Ab	
	-4.81	15.88(1)	287.8	25	C	μ=0.1	64Ab, 65Fb	
	-12.242	15.19(1)	249.8	25	C	μ=0.1(KNO_3)	59Ba, 62Ma	
	-17.2	-14.51(1)	222	25	C	μ=0.44	62Ya, 65Ka	
	-22.112	--	--	10	C	μ=1.0M($NaClO_4$) R:La^{3+}+MgL^{2-}= LaL^-+Mg^{2+}	73Oc	
	-21.510	--	--	15	C	μ=1.0M($NaClO_4$) R:La^{3+}+MgL^{2-}= LaL^-+Mg^{2+}	73Oc	
	-21.004	--	--	20	C	μ=1.0M($NaClO_4$) R:La^{3+}+MgL^{2-}= LaL^-+Mg^{2+}	73Oc	
	-20.577	--	--	25	C	μ=1.0M($NaClO_4$) R:La^{3+}+MgL^{2-}= LaL^-+Mg^{2+} ΔCp=20.9	73Oc	
	-20.108	--	--	30	C	μ=1.0M($NaClO_4$) R:La^{3+}+MgL^{2-}= LaL^-+Mg^{2+}	73Oc	
	-19.614	--	--	35	C	μ=1.0M($NaClO_4$) R:La^{3+}+MgL^{2-}= LaL^-+Mg^{2+}	73Oc	
	-18.987	--	--	40	C	μ=1.0M($NaClO_4$) R:La^{3+}+MgL^{2-}= LaL^-+Mg^{2+}	73Oc	
Ce^{3+}	-10.17	15.98(1)	271.1	20	C	μ=0.1	58Sf	
	-12.292	15.44(1)	254.4	25	C	μ=0.1(KNO_3)	59Ba, 62Ma	
	-20.1	15.00(1)	222	25	C	μ=0.44	65Ka	
Pr^{3+}	-13.380	15.75(1)	256.9	25	C	μ=0.1(KNO_3)	62Ma	
	-19.7	15.42(1)	230	25	C	μ=0.44	59Ba, 65Ka	
	-18.3	5.18(25°)	37.2	25-45	T	μ=0.2M($NaClO_4$) S:50v%Dioxane R:Pr+L+A=PrLA	76Md	

Metal Ion	ΔH, kJ /mol	Log K	ΔS, J /mol K	T,°C	M	Conditions	Ref.	Re-marks
E28, continued								
Pr^{3+}, cont.	-37.6	5.26(25°)	-26.4	25-45	T	A=3,4-Dihydroxy-benzaldehyde μ=0.2M($NaClO_4$) R:Pr+L+A=PrLA	76Md	
	-30.6	3.40(25°)	-37.6	25-45	T	A-1,2-Dihydroxy-benzene μ=0.2M($NaClO_4$) R:Pr+L+A=PrLA	76Md	
	31.6	4.27(25°)	188.3	25-45	T	A=3,4-Dihydroxy-benzoic acid μ=0.2M($NaClO_4$) R:Pr+L+A=PrLA	76Md	
	-22.824	--	--	10	C	A=1,2,3-Tri-hydroxybenzene μ=1.0M($NaClO_4$) R:Pr^{3+}+MgL^{2-}= PrL^-+Mg^{2+}	73Oc	
	-21.886	--	--	20	C	μ=1.0M($NaClO_4$) R:Pr^{3+} MgL^{2-}= PrL^-+Mg^{2+}	73Oc	
	-21.527	--	--	25	C	μ=1.0M($NaClO_4$) R:Pr^{3+}+MgL^{2-}=PrL^- +Mg^{2+} ΔCp=69	73Oc	
	-21.179	--	--	30	C	μ=1.0M($NaClO_4$) R:Pr^{3+}+MgL^{2-}= PrL^-+Mg^{2+}	73Oc	
	-20.351	--	--	40	C	μ=1.0M($NaClO_4$) R:Pr^{3+}+MgL^{2-}= PrL^-+Mg^{2+}	73Oc	
Nd^{3+}	-12.47	16.60(1)	275.3	20	C	μ=0.1	58Sf 59Ba	
	-15.159	16.04(1)	256.5	25	C	μ=0.1(KNO_3)	62Ma 63Pd	
	-20.1	16.16(1)	230	25	C	μ=0.44	65Ka	
	-24.389	--	--	10	C	μ=1.0M($NaClO_4$) R:Nd^{3+}+MgL^{2-}= NdL^-+Mg^{2+}	73Oc	
	-23.305	--	--	20	C	μ=1.0M($NaClO_4$) R:Nd^{3+}+MgL^{2-}= NdL^-+Mg^{2+}	73Oc	
	-22.920	--	--	25	C	μ=1.0M($NaClO_4$) R:Nd^{3+}+MgL^{2-}=NdL^- +Mg^{2+} ΔCp=74.1	73Oc	
	-22.560	--	--	30	C	μ=1.0M($NaClO_4$)	73Oc	

Metal Ion	ΔH, kJ /mol	Log K	ΔS, J /mol K	T,°C	M	Conditions	Ref.	Re-marks
E28, continued								
Nd^{3+}, cont.						R:Nd^{3+}+MgL^{2-}= NdL^{-}+Mg^{2+}		
	-21.673	--	--	40	C	μ=1.0M($NaClO_4$) R:Nd^{3+}+MgL^{2-}= NdL^{-}+Mg^{2+}	730c	
Sm^{3+}	-14.012	16.52(1)	269.4	25	C	μ=0.1(KNO_3)	62Ma	
	-19.2	16.16(1)	247	25	C	μ=0.44	59Ba 65Ka 63Pd	
	-25.334	--	--	10	C	μ=1.0M($NaClO_4$) R:Sm^{3+}+MgL^{2-}= SmL^{-}+Mg^{2+}	730c	
	-23.945	--	--	20	C	μ=1.0M($NaClO_4$) R:Sm^{3+}+MgL^{2-}= SmL^{-}+Mg^{2+}	730c	
	-23.330	--	--	25	C	μ=1.0M($NaClO_4$) R:Sm^{3+}+MgL^{2-}=SmL^{-} +Mg^{2+} ΔCp=120.9	730c	
	-22.715	--	--	30	C	μ=1.0M($NaClO_4$) R:Sm^{3+}+MgL^{2-}= SmL^{-}+Mg^{2+}	730c	
	-21.355	--	--	40	C	μ=1.0M($NaClO_4$) R:Sm^{3+}+MgL^{2-}= SmL^{-}+Mg^{2+}	730c	
Eu^{3+}	-10.703	16.65(1)	282.8	25	C	μ=0.1(KNO_3)	62Ma	
	-14.6	16.37(1)	264	25	C	μ=0.44	59Ba 65Ka	
	-20.9	15.27(50°,1)	227.2	40-70	T	μ=1.0(KCl) CM=0.1 Complex having Structure 1 as determined by ab-sorption spectra	71Kc	b
	0	15.21(50°,1)	292.5	40-70	T	μ=1.0(KCl) CM=0.1 Complex having Structure 2 as determined by ab-sorption spectra	71Kc	b
	23.14	7.29(50°)	210.9	40-70	T	μ=1.0(KCl) CM=0.1 R:Eu^{3+}+ HL^{3-}=EuHL	71Kc	b
	-23.267	--	--	10	C	μ=1.0M($NaClO_4$) R:Eu^{3+}+MgL^{2-}= EuL^{-}+Mg^{2+}	730c	
	-22.539	--	--	15	C	μ=1.0M($NaClO_4$)	730c	

Metal Ion	ΔH, kJ /mol	Log K	ΔS, J /mol K	T,°C	M	Conditions	Ref.	Remarks
E28, continued								
Eu^{3+}, cont.						R:Eu^{3+}+MgL^{2-}= EuL^{-}+Mg^{2+}		
	-21.882	--	--	20	C	μ=1.0M($NaClO_4$)	730c	
						R:Eu^{3+}+MgL^{2-}= EuL^{-}+Mg^{2+}		
	-21.242	--	--	25	C	μ=1.0M($NaClO_4$)	730c	
						R:Eu^{3+}+MgL^{2-}=EuL^{-} +Mg^{2+} ΔCp=123.0		
	-20.652	--	--	30	C	μ=1.0M($NaClO_4$)	730c	
						R:Eu^{3+}+MgL^{2-}= EuL^{-}+Mg^{2+}		
	-20.01	--	--	35	C	μ=1.0M($NaClO_4$)	730c	
						R:Eu^{3+}+MgL^{2-}= EuL^{-}+Mg^{2+}		
	-19.368	--	--	40	C	μ=1.0M($NaClO_4$)	730c	
						R:Eu^{3+}+Mg^{2-}= EuL^{-}+Mg^{2+}		
Gd^{3+}	-4.64	17.36(1)	316.7	20	C	μ=0.1	58Sf, 59Ba	
	-7.238	16.81(1)	297.9	25	C	μ=0.1(KNO_3)	62Ma, 63Pd	
	-10.5	16.39(1)	276	25	C	μ=0.44	65Ka	
	-20.322	--	--	10	C	μ=1.0M($NaClO_4$)	730c	
						R:Gd^{3+}+MgL^{2-}= GdL^{-}+Mg^{2+}		
	-19.133	--	--	20	C	μ=1.0M($NaClO_4$)	730c	
						R:Gd^{3+}+MgL^{2-}= GdL^{-}+Mg^{2+}		
	-18.652	--	--	25	C	μ=1.0M($NaClO_4$)	730c	
						R:Gd^{3+}+MgL^{2+}=GdL^{-} +Mg^{2+} ΔCp=93.7		
	-18.179	--	--	30	C	μ=1.0M($NaClO_4$)	730c	
						R:Gd^{3+}+MgL^{2+}= GdL^{-}+Mg^{2+}		
	-17.066	--	--	40	C	μ=1.0M($NaClO_4$)	730c	
						R:Gd^{3+}+MgL^{2+}= GdL^{-}+Mg^{2+}		
Tb^{3+}	-4.661	17.31(1)	315.9	25	C	μ=0.1(KNO_3)	62Ma, 59Ba	
	-16.656	--	--	10	C	μ=1.0M($NaClO_4$)	730c	
						R:Tb^{3+}+MgL^{2-}= TbL^{-}+Mg^{2+}		

Metal Ion	ΔH, kJ /mol	Log K	ΔS, J /mol K	T, °C	M	Conditions	Ref.	Re-marks
E28, continued								
Tb^{3+}, cont.	-16.489	--	--	15	C	μ=1.0M($NaClO_4$) R:Tb^{3+}+MgL^{2-}= TbL^-+Mg^{2+}	730c	
	-16.384	--	--	20	C	μ=1.0M($NaClO_4$) R:Tb^{3+}+MgL^{2-}= TbL^-+Mg^{2+}	730c	
	-16.276	--	--	25	C	μ=1.0M($NaClO_4$) R:Tb^{3+}+MgL^{2-}=TbL^- +Mg^{2+} ΔCp=21.3	730c	
	-16.175	--	--	30	C	μ=1.0M($NaClO_4$) R:Tb^{3+}+MgL^{2-}= TbL^-+Mg^{2+}	730c	
	-15.933	--	--	35	C	μ=1.0M($NaClO_4$) R:Tb^{3+}+MgL^{2-}= TbL^-+Mg^{2+}	730c	
	-15.648	--	--	40	C	μ=1.0M($NaClO_4$) R:Tb^{3+}+MgL^{2-}= TbL^-+Mg^{2+}	730c	
Dy^{3+}	-5.067	17.77(1)	323.4	25	C	μ=0.1(KNO_3)	62Ma	
	-12.6	17.32(1)	289	25	C	μ=0.44	59Ba/65Ka	
Ho^{3+}	-5.674	18.038(1)	326.4	25	C	μ=0.1(KNO_3)	62Ma	
	-13.0	17.76(1)	297	25	C	μ=0.44	59Ba/65Ka	
	-16.832	--	--	10	C	μ=1.0M($NaClO_4$) R:Ho^{3+}+MgL^{2-}= HoL^-+Mg^{2+}	730c	
	-16.928	--	--	20	C	μ=1.0M($NaClO_4$) R:Ho^{3+}+MgL^{2-}=HoL^- +Mg^{2+} ΔCp=-14	730c	
	-17.012	--	--	25	C	μ=1.0M($NaClO_4$) R:Ho^{3+}+MgL^{2-}= HoL^-+Mg^{2+} ΔCp=-14	730c	
	-17.050	--	--	30	C	μ=1.0M($NaClO_4$) R:Ho^{3+}+MgL^{2-}= HoL^-+Mg^{2+}	730c	
	-16.800	--	--	40	C	μ=1.0M($NaClO_4$) R:Ho^{3+}+MgL^{2-}= HoL^-+Mg^{2+}	730c	

Metal Ion	ΔH, kJ /mol	Log K	ΔS, J /mol K	T,°C	M	Conditions	Ref.	Re-marks
E28, continued								
Er^{3+}	-7.146	18.368(1)	327.6	25	C	μ=0.1(KNO_3)	62Ma 59Ba	
	-12.6	17.87(1)	301	25	C	μ=0.44	65Ka	
Tm^{3+}	-7.824	18.647(1)	331.0	25	C	μ=0.1(KNO_3)	62Ma 59Ba	
Yb^{3+}	-9.665	18.992(1)	331.4	25	C	μ=0.1(KNO_3)	62Ma 59Ba	
	-11.3	18.53(1)	318	25	C	μ=0.44	65Ka	
	-19.974	--	--	10	C	μ=1.0M($NaClO_4$) R:Yb^{3+}+MgL^{2-}= YbL^-+Mg^{2+}	73Oc	
	-20.004	--	--	20	C	μ=1.0M($NaClO_4$) R:Yb^{3+}+MgL^{2-}= YbL^-+Mg^{2+}	73Oc	
	-20.108	--	--	25	C	μ=1.0M($NaClO_4$) R:Yb^{3+}+MgL^{2-}=YbL^- +Mg^{2+} ΔCp=22	73Oc	
	-20.204	--	--	30	C	μ=1.0M($NaClO_4$) R:Yb^{3+}+MgL^{2-}= YbL^-+Mg^{2+}	73Oc	
	-20.138	--	--	40	C	μ=1.0M($NaClO_4$) R:Yb^{3+}+MgL^{2-}= YbL^-+Mg^{2+}	73Oc	
Lu^{3+}	-10.510	19.138(1)	331.0	25	C	μ=0.1(KNO_3)	62Ma 59Ba	
	-13.4	18.84(1)	314	25	C	μ=0.44	65Ka	
Hf^{4+}	-4.60	33.74(1)	629.7	15	C	μ=0	78Va	
	-0.96	33.71(1)	642.2	25	C	μ=0 ΔCp=385	78Va	
	3.10	33.72(1)	655.6	35	C	μ=0	78Va	
	28.58	--	--	15	C	2N($HClO_4$) R:Hf^{4+}+H_6L^{2+}= HfL+6H^+	78Va	
	26.53	--	--	25	C	2N($HClO_4$) R:Hf^{4+}+H_6L^{2+}= HfL+6H^+	78Va	
	24.89	--	--	35	C	2N($HClO_4$) R:Hf^{4+}+H_6L^{2+}= HfL+6H^+	78Va	
	28.91	--	--	15	C	3N($HClO_4$) R:Hf^{4+}+H_6L^{2+}= HfL+6H^+	78Va	
	26.44	--	--	25	C	3N($HClO_4$)	78Va	

Metal Ion	ΔH, kJ /mol	Log K	ΔS, J /mol K	T, °C	M	Conditions	Ref.	Re-marks
E28, continued								
Hf^{4+}, cont.						$R: Hf^{4+} + H_6L^{2+} = HfL + 6H^+$		
	24.89	--	--	35	C	$3N(HClO_4)$	78Va	
						$R: Hf^{4+} + H_6L^{2+} = HfL + 6H^+$		
	26.15	--	--	25	C	$3.7N(HClO_4)$	78Va	
						$R: Hf^{4+} + H_6L^{2+} = HfL + 6H^+$		
	29.16	--	--	15	C	$4N(HClO_4)$	78Va	
						$R: Hf^{4+} + H_6L^{2+} = HfL + 6H^+$		
	24.43	--	--	35	C	$4N(HClO_4)$	78Va	
						$R: Hf^{4+} + H_6L^{2+} = HfL + 6H^+$		
UO_2^{2+}	13.4	7.40(25°)	188	20-40	T	$\mu=0.1(KNO_3)$	69Sg	b
						$R: UO_2^{2+} + HL = UO_2HL$		
Pu^{3+}	-17.70	18.07(1)	287.0	25	C	$\mu=0.1$	65Fb	
Am^{3+}	-19.54	18.16(1)	282.4	25	C	$\mu=0.1$	65Fb	
Mn^{2+}	-19.08	13.80(1)	199.2	20	C	$0.1M(KNO_3)$	63Ab 64Ab	
	-21.8	12.6(1)	172	25	C	$\mu \leqq 0.1$	54Cb	
	-0.4	6.89	130	25	C	$\mu=0.10M(KNO_3)$	69Bk	q:±1.7
						$R: Mn^{2+} + HL^{3-} = MnHL^-$		
Fe^{2+}	-16.7	14.1(1)	213	25	C	$\mu=0.1(KNO_3)$	65Wd	
	-1.72	6.903(20°)	126	10-40	T	$\mu=0.20$ $R: Fe^{2+} + HL^{3-} = FeHL^-$	72Ej	b,q: ±0.25
	-3.35	3.322(20°)	50	10-40	T	$\mu=0.20$ $R: Fe^{2+} + H_2L^{2-} = FeH_2L$	72Ej	b,q: ±1.7
Fe^{3+}	-51.0	25.447(20°,1)	314	10-40	T	$\mu=0.20$	72Bj	b
Co^{2+}	-16.9	18.14(1)	288.3	15	C	$\mu=0$	76Vc	
	-15.9	18.03(1)	291.6	25	C	$\mu=0$ $\Delta Cp=142$	76Vc	
	-14.1	17.97(1)	298.3	35	C	$\mu=0$	76Vc	
	-17.57	16.31(1)	252.3	20	C	$0.1M(KNO_3)$	63Ab 64Ab	
	-17.6	(1)	--	25	C	--	57Jc	
	-17.2	15.7(1)	243	25	C	$\mu \leqq 0.1$	54Cb	
	-21.1	15.77(1)	228.4	15	C	$\mu=0.3(KNO_3)$	76Vc	
	-20.3	15.64(1)	231.4	25	C	$\mu=0.3(KNO_3)$ $\Delta Cp=92$	76Vc	
	-19.3	15.55(1)	235.1	35	C	$\mu=0.3(KNO_3)$	76Vc	
	-21.7	15.69(1)	225.1	15	C	$\mu=0.5(KNO_3)$	76Vc	
	-21.0	15.55(1)	227.2	25	C	$\mu=0.5(KNO_3)$ $\Delta Cp=84$	76Vc	
	-20.0	15.45(1)	230.5	35	C	$\mu=0.5(KNO_3)$	76Vc	
	-23.7	15.51(1)	214.6	15	C	$\mu=1.0(KNO_3)$	76Vc	

Metal Ion	ΔH, kJ /mol	Log K	ΔS, J /mol K	T,°C	M	Conditions	Ref.	Re-marks
E28, continued								
Co^{2+}, cont.	-22.6	15.36(1)	218.4	25	C	μ=1.0(KNO_3) ΔCp=86	76Vc	
	-22.0	15.25(1)	220.5	35	C	μ=1.0(KNO_3)	76Vc	
	-1.7	9.16	167	25	C	μ=0.1M(KNO_3) R:Co^{2+}+HL^{3}=$CoHL^{-}$	69Bk	q:±1.7
	-69.9	2.920(25°)	-178.2	15-30	T	μ=0.660(NaAc+HAc) pH=5.0 R:CoL^{2-}+ $Fe(CN)_6^{3-}$= [$LCoCNFeCN_5^{5-}$]	73Hf	b
Ni^{2+}	-30.0	20.49(1)	287.8	15	C	μ=0	76Vc	
	-28.0	20.30(1)	294.6	25	C	μ=0 ΔCp=161	76Vc	
	-26.8	20.17(1)	299.2	35	C	μ=0	76Vc	
	-31.0	(1)	--	25	C	CM=10^{-2}M	57Jc 64Ab	
	-31.59	18.62(1)	248.5	20	C	0.1M(KNO_3)	63Ab	
	-31.8	17.6(1)	230	25	C	$\mu\leqq$0.1	54Cb	
	-35.6	18.7(1)	238	25	C	μ=0.1(KNO_3)	65Wd	
	-34.94	18.61(1)	237.2	20	C	μ=0.17(KNO_3)	56Ca	
	-28.78	18.11(1)	250.2	25	C	μ=0.22	59Yb	
	-33.8	18.12(1)	230.1	15	C	μ=0.3(KNO_3)	76Vc	
	-32.5	17.91(1)	233.9	25	C	μ=0.3(KNO_3) ΔCp=107	76Vc	
	-31.6	17.75(1)	237.2	35	C	μ=0.3(KNO_3)	76Vc	
	-34.4	18.04(1)	225.9	15	C	μ=0.5(KNO_3)	76Vc	
	-33.1	17.82(1)	230.1	25	C	μ=0.5(KNO_3) ΔCp=121	76Vc	
	-32.0	17.66(1)	234.3	35	C	μ=0.5(KNO_3)	76Vc	
	-28.45	17.79(1)	245.2	25	C	μ=0.59	59Yb	
	-27.99	17.78(1)	246.4	25	C	μ=0.75	59Yb	
	-35.6	17.86(1)	218.4	15	C	μ=1.0(KNO_3)	76Vc	
	-34.6	17.03(1)	221.3	25	C	μ=1.0(KNO_3) ΔCp=111	76Vc	
	-33.3	17.46(1)	225.9	35	C	μ=1.0(KNO_3)	76Vc	
	-29.66	18.25(1)	249.8	25	C	μ=1.45	59Yb	
	-54.46	18.28(1)	246.4	25	C	μ=1.50	59Yb	
	-15.5	11.58	167	25	C	μ=0.10M(KNO_3) R:Ni^{2+}+HL^{3-}=$NiHL^{-}$	69Bk	q:±1.7
Cu^{2+}	-34.52	20.69(1)	276	15	C	μ=0	77Va	
	-33.18	20.47(1)	281	25	C	μ=0 ΔCp=117	77Va	
	-32.13	20.31(1)	284	35	C	μ=0	77Va	
	-34.3	(1)	--	25	C	CM=10^{-2}M	57Jc 64Ab	
	-34.10	18.80(1)	243.5	20	C	0.1M(KNO_3)	63Ab	
	-34.3	17.88(1)	230	25	C	$\mu\leqq$0.1	54Cb	
	-36.8	18.62(1)	234	25	C	0.1M(KNO_3)	65Wd	
	-36.28	18.79(1)	236.0	20	C	μ=0.17(KNO_3)	56Ca	
	-38.32	18.31(1)	218	15	C	μ=0.3(KNO_3)	77Va	
	-37.95	18.07(1)	219	25	C	μ=0.3(KNO_3) ΔCp=58	77Va	
	-37.15	17.89(1)	222	35	C	μ=0.3(KNO_3)	77Va	
	-39.08	18.24(1)	218	15	C	μ=0.5(KNO_3)	77Va	

Metal Ion	ΔH, kJ /mol	Log K	ΔS, J /mol K	T,°C	M	Conditions	Ref.	Re-marks
E28, continued								
Cu^{2+}, cont.	-38.41	17.99(1)	215	25	C	μ=0.5(KNO_3) ΔCp=67	77Va	
	-37.74	17.80(1)	218	35	C	μ=0.5(KNO_3)	77Va	
	-40.38	18.05(1)	205	15	C	μ=1.0(KNO_3)	77Va	
	-39.96	17.80(1)	207	25	C	μ=1.0(KNO_3) ΔCp=50	77Va	
	-39.37	17.60(1)	209	35	C	μ=1.0(KNO_3)	77Va	
	-18.8	11.51	155	25	C	μ=0.10M(KNO_3) R:Cu^{2+}+HL^{3-}=$CuHL^-$	69Bk	q:±1.7
Ag^+	-23.8	6.96(1)	53.0	25	C	pH=12 CL=7x10^{-3}-1x10^{-2}M CM=1x10^{-3}-3x10^{-3}M	78Rb	
	-31.5	7.32(1)	32.6	20	C	μ=0.1M(KNO_3)	76Ah	
	-17.56	7.28	80.5	15-35	T	-- Data from G.A. Rechnitz and Z.F. Lin, Anal. Chem. 39, 1406 (1967).	78Rb	i
Zn^{2+}	-19.2	(1)	--	25	C	CM=10^{-2}M	57Jc 64Ab	
	-20.29	16.50(1)	246.8	20	C	0.1M(KNO_3)	63Ab	
	-18.8	15.3(1)	230	25	C	μ≦0.1	54Cb	
	-23.4	16.3(1)	238	25	C	μ=0.1(KNO_3)	65Wd	
	105	15.52(32°,1)	483	32&42	T	--	68Ld	a,b
	-23.47	16.49(1)	235.6	20	C	μ=0.17(KNO_3)	56Ca	
	-5.8	9.02	151	25	C	μ=0.10M(KNO_3) R:Zn^{2+}+HL^{3-}=$ZnHL^-$	69Bk	q:±1.7
Cd^{2+}	-38.5	(1)	--	25	C	CM=10^{-2}M	57Jc 64Ab	
	-37.86	16.46(1)	185.8	20	C	0.1M(KNO_3)	63Ab	
	-38.1	15.0(1)	159	25	C	μ≦0.1	54Cb	
	-42.2	16.34(1)	176	25	C	0.1M(KNO_3)	65Wd	
	-42.17	16.45(1)	171.1	20	C	μ=0.17(KNO_3)	56Ca	
	105	15.32(32°,1)	476	32&42	T	--	68Ld	a,b
	-15.9	9.09	121	25	C	μ=0.10M(KNO_3) R:Cd^{2+}+HL^{3-}=$CdHL^-$	69Bk	q:±1.7
Hg^{2+}	-79.10	21.80(1)	148.5	20	C	0.1M(KNO_3)	63Ab 64Ab	
	-80.3	22.1(1)	155	25	C	μ=0.1(KNO_3)	65Wd 66Mf	
	-59.4	14.59	121	25	C	μ=0.10M(KNO_3) R:Hg^{2+}+HL^{3-}=$HgHL^-$	69Bk	
HgA_2	-29.5	1.37(1)	-71	30	C	CM=1-20x10^{-3} S:Benzene A=C_6F_5	74Fc	q:±3
Al^{3+}	10.79	16.10(1)	487.8	20	C	μ=0.1	58Sf	
		16.78(1)	491.2	25	T	μ=0.1(KNO_3)	66Mf	
		15.61(1)	468.6	25	C	μ=0.22	58Ya	

Metal Ion	ΔH, kJ /mol	Log K	ΔS, J /mol K	T,°C	M	Conditions	Ref.	Remarks
E28, continued								
In^{3+}	-30.25	24.94(1)	374.5	20	C	μ=0.1	58Sf	
Tl^{+}	-43.5	6.53(1)	-22.2	20	C	μ=0.1M(KNO_3)	76Ah	
Pb^{2+}	-53.6	(1)	--	25	C	CM=10^{-2}M	57Jc 54Cb	
	-55.23	18.04(1)	156.9	20	C	0.1M(KNO_3)	63Ab 64Ab	
	-54.8	17.3(1)	146	25	C	μ=0.1(KNO_3)	65Wd	
	-58.91	18.02(1)	144.3	20	C	μ=0.17(KNO_3)	56Ca	
E29	ETHANE, 1,2-diamino-N,N,N',N'-tetra(2-aminoethyl)- $C_{10}H_{28}N_2$					$(NH_2CH_2CH_2)_2NCH_2CH_2N(CH_2CH_2NH_2)_2$		L
Mn^{2+}	-37.03	9.30(1)	52.3	25	C	μ=0.1(KCl)	64Sb 60Cc	
Fe^{2+}	-40.38	11.05(1)	77.4	25	C	μ=0.1(KCl)	64Sb 60Cc	
Co^{2+}	-61.71	11.55(1)	14.2	25	C	μ=0.1(KCl)	64Sb 60Cc	
	-58.58	12.40	41.8	25	C	μ=0.1(KCl) R:Co^{2+}+HL^{+}=$CoHL^{3+}$	64Sb	
Ni^{2+}	-82.22	19.05(1)	90.0	25	C	μ=0.1(KCl)	64Sb 60Cc	
	-76.78	15.65	41.8	25	C	μ=0.1(KCl) R:Ni^{2+}+HL^{+}=$NiHL^{3+}$	64Sb	
Cu^{2+}	-102.51	22.15(1)	79.5	25	C	μ=0.1(KCl)	64Sb 60Cc	
	-103.76	20.15	37.6	25	C	μ=0.1(KCl) R:Cu^{2+}+HL^{+}=$CuHL^{3+}$	64Sb	
Zn^{2+}	-60.67	16.05(1)	104.6	25	C	μ=0.1(KCl)	64Sb 60Cc	
	-61.30	14.00	62.8	25	C	μ=0.1(KCl) R:Zn^{2+}+HL^{+}=$ZnHL^{3+}$	64Sb	
E30	ETHANE, 1,2-diamino-N,N,N',N'-tetrakis (2-hydroxyethyl)- $C_{10}H_{24}O_4N_2$					$(HOCH_2CH_2)_2NCH_2CH_2N(CH_2CH_2OH)_2$		
Ni^{2+}	-18.8	6.40(25°,1)	58	10-55	T	μ=0.5($NaClO_4$)	69Pa	b
Cu^{2+}	-27.2	8.60(25°,1)	75	10-55	T	μ=0.5($NaClO_4$)	69Pa	b
Zn^{2+}	-10.5	4.78(25°1)	58	10-55	T	μ=0.5($NaClO_4$)	69Pa	b
E31	ETHANE, 1,2-diamino-N,N,N',N'-tetrakis (2-pyridylmethyl)- $C_{26}H_{16}N_6$					$(C_5H_4NCH_2)_2NCH_2CH_2N(CH_2C_5H_4N)_2$		
Mn^{2+}	-47.7	10.3(1)	33.5	20	C	μ=0.1	70Wb	

Metal Ion	ΔH, kJ /mol	Log K	ΔS, J /mol K	T,°C	M	Conditions	Ref.	Re-marks
E31, continued								
Fe^{2+}	-69.9	14.6(1)	41.4	20	C	μ=0.1	70Wb	
Co^{2+}	-72.0	(1)	--	20	C	μ=0.1	70Wb	
Ni^{2+}	-98.7	18.0(1)	6.7	20	C	μ=0.1	70Wb	
Cu^{2+}	-96.6	20.6(1)	64.8	20	C	μ=0.1	70Wb	
Zn^{2+}	-61.9	18.0(1)	133	20	C	μ=0.1	70Wb	
E32 ETHANE, 1,2-diamino-N,N,N',N'-tetramethyl-		$C_6H_{16}N_2$				$(CH_3)_2NCH_2CH_2(CH_3)_2$	L	
Cu^{2+}	-25.73	7.376(1)	54.8	25	C	0.5M(KNO_3)	72Ad	
	13.8	-0.66	33.5	25	C	0.5M(KNO_3) R:$Cu^{2+}+L+H_2O=CuOHL^{+}+H^{+}$	72Ad	
	62.76	-10.91	1.67	25	C	0.5M(KNO_3) R:$Cu^{2+}+L+2H_2O=Cu(OH)_2L+2H^{+}$	72Ad	
	19.66	2.59	115.5	25	C	0.5M(KNO_3) R:$2Cu^{2+}+2L+2H_2O=[Cu_2(OH)_2L_2]^{2+}+2H^{+}$	72Ad	
	74.0	-8.15	92.5	25	C	0.5M(KNO_3) R:$3Cu^{2+}+2L+4H_2O=[Cu_3(OH)_4L_2]^{2+}+4H^{+}$	72Ad	
CuA_2	-107	(1)	--	30	C	S:Benzene A=1,1,1,5,5,5-Hexafluoropentane-2,4-dionato	76Gh 75Gg	
	-79	(1)	--	30	C	S:Benzene A=1,1,1-Tri-fluoro-4-thenyl-2,4-dione	76Gh 75Gg	
	-75	(1)	--	30	C	S:Benzene A=1,1,1-Tri-fluoro-4-phenylbutane-2,4-dione	76Gh 75Gg	
	-65	(1)	--	30	C	S:Benzene A=1,1,1-Tri-fluoropentane-2,4-dione	76Gh 75Gg	
HgA_2	-94.7	6.60(1)	-187	30	C	CM=0.9-5.1x10^{-3} S:Benzene A=Br	74Fb	
	-90.2	6.00(1)	-182	30	C	CM=1.3-6.1x10^{-3} S:Benzene A=Cl	74Fb	
	-96.3	6.30(1)	-196	30	C	CM=1-4.8x10^{-3} S:Benzene A=I	74Fb	
SnA_2B_2	-28.7	1.666(1)	-63	25	C	S:Benzene A=Bu	77Ge	

Metal Ion	ΔH, kJ /mol	Log K	ΔS, J /mol K	T, °C	M	Conditions	Ref.	Re-marks
E32, continued								
SnA_2B_2, cont.	-26.6	1.663(1)	-56	25	C	B=(OCOMe) S:Benzene A=Bu B=(OCOEt)	77Ge	
	-22.9	1.913(1)	-39	25	C	S:Benzene A=Ph B=(OCOMe)	77Ge	
E33	ETHANE, 1,2-diamino-N,N,N'-trimethyl-		$C_5H_{14}N_2$			$(CH_3)_2NCH_2CH_2NH(CH_3)$		
CuA_2	-106	(1)	--	30	C	S:Benzene A=1,1,1,5,5,5-Hexafluoropentane-2,4-dionato	76Gh 75Gg	
	-80	(1)	--	30	C	S:Benzene A=1,1,1-Tri-fluoro-4-thenyl-2,4-dione	76Gh 75Gg	
	-75	(1)	--	30	C	S:Benzene A=1,1,1-Tri-fluoro-4-phenylbutane-2,4-dione	76Gh 75Gg	
	-66	(1)	--	30	C	S:Benzene A=1,1,1-Tri-fluoropentane-2,4-dione	76Gh 75Gg	
	-76	(2)	--	30	C	S:Benzene A=1,1,1,5,5,5-Hexafluoropentane-2,4-dionato	76Gh 75Gg	
E34	ETHANE, 1,2-di(2-carboxypiperidyl)		$C_{14}H_{24}O_4N_2$			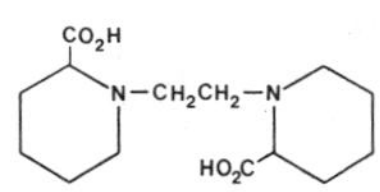		
Ni^{2+}	79.4	10.95(20°,1)	485	5-35	T	0.200M(KCl)	78Ca	q:±12.5
E35	ETHANE, 1,2-bis(dimethyl-amino)-		$C_6H_{16}N_2$			$(H_3C)_2NCH_2CH_2N(CH_3)_2$		L
PdA_2B_2	-114.2	--	--	25	C	S:CH_2Cl_2 CL=$4x10^{-4}$M CM=$2x10^{-5}$M A=Cl B=H_6H_5CN R:PdA_2B_2+L= PdA_2L+2B	72Pd	
E36	ETHANE, 1,2-bis(diphenyl-arsino)-		$C_{26}H_{24}As_2$			$(C_6H_5)_2AsCH_2CH_2As(C_6H_5)_2$		L
PdA_2B_2	-147.3	--	--	25	C	S:CH_2Cl_2	72Pd	

Metal Ion	ΔH, kJ /mol	Log K	ΔS, J /mol K	T, °C	M	Conditions	Ref.	Remarks
E36, continued								
PdA_2B_2, cont.						$CL=1x10^{-4}M$ $CM=2x10^{-4}M$ A=Cl B=C_6H_5CN R:PdA_2B_2+L= PdA_2L+2B		
E37	ETHANE, 1-(diphenylarsino)-2-(diphenylphosphino)		$C_{26}H_{24}PAs$			$(C_6H_5)_2PCH_2CH_2As(C_6H_5)_2$		
HgA_2	-114.0	4.114(1)	-298	30	C	S:Benzene $CM=5-10x10^{-4}M$ A=Cl	76Gb	
	-119.0	3.845(1)	-320	30	C	S:Benzene $CM=6-8x10^{-4}M$ A=Br	76Gb	
	-112.0	3.544(1)	-302	30	C	S:Benzene $CM=8-12x10^{-4}M$ A=I	76Gb	
E38	ETHANE, 1,2-bis(diphenyl-phosphino)-		$C_{26}H_{24}P_2$			$(C_6H_5)_2PCH_2CH_2P(C_6H_5)_2$		L
PdA_2B_2	-211.3	--	--	25	C	S:CH_2Cl_2 $CL=6x10^{-5}M$ $CM=3x10^{-4}M$ A=Cl B=C_6H_5CN R:PdA_2B_2+L= PdA_2L+2B	72Pd	
E39	ETHANE, 1,2-diphosphino-P,P,P',P'-tetramethyl		$C_6H_{16}P_2$			$(CH_3)_2PCH_2CH_2P(CH_3)_2$		
$PtAB_2^+$	-116.3	(1)	--	25	C	S:THF A=CH_3 B=$[P(CH_3)_2(C_6H_5)]$	75Ma	
E40	ETHANE, 1,1-diphosphonic acid		$C_2H_8O_6P_2$			$CH_3CH(PO_3H_2)_2$		L^{4-}
Li^+	-1.51	3.12(1)	54.8	25	C	$0.5M(CH_3)_4NCl$	68Ca	
Na^+	7.1	1.51(1)	52.7	25	C	$0.5M(CH_3)_4NCl$	68Ca	
K^+	3.3	1.20(1)	33.9	25	C	$0.5M(CH_3)_4NCl$	68Ca	
Ca^{2+}	9.2	5.22(1)	130	25	C	$0.5M(CH_3)_4NCl$	68Ca	
E41	ETHANE, 1,2-diylbis-(diphenylarsino)		$C_{26}H_{24}As_2$			$[(C_6H_5)_2As]_2C_2H_4$		
HgA_2	-73.0	2.818(1)	-188	30	C	S:Benzene $CM=5-10x10^{-4}M$ A=Cl	76Gb	
	-72.6	3.255(1)	-177	30	C	S:Benzene	76Gb	

Metal Ion	ΔH, kJ /mol	Log K	ΔS, J /mol K	T,°C	M	Conditions	Ref.	Remarks
E41, continued								
HgA_2, cont.						CM=6-8x10^{-4}M A=Br		
	-57.0	2.535(1)	-140	30	C	S:Benzene CM=8-12x10^{-4}M A=I	76Gb	
E42	ETHANE, 1,2-diylbis-(diphenylphosphino)		$C_{26}H_{24}P_2$			$[(C_6H_5)_2P]_2C_2H_4$		
NiA	-42.7	2.28(1)	-97	30	C	S:Benzene CM=1-2x10^{-3}M A=BF_2-bridged glyoximato R_1=Me R_2=Et 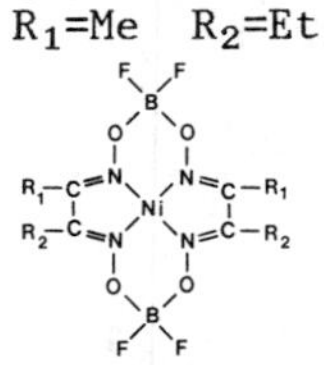	76Gh	
	-38.1	2.30(1)	-82	30	C	S:Benzene CM=1-3x10^{-3}M A=BF_2-bridged glyoximato R_1=Me R_2=Pr	76Gh	
	-42.9	2.08(1)	-102	30	C	S:Benzene CM=1-2x10^{-3}M A=BF_2-bridged glyoximato R_1=Me R_2=iso-Pr	76Gh	
E43	ETHANE, 1-hydroxy, 1,1-diphosphonic acid		$C_2H_8O_7P_2$			$CH_3C(OH)(PO_3H_2)_2$		L^{4-}
Li^+	-2.38	3.35(1)	56.1	25	C	0.5M$(CH_3)_4NCl$	68Ca	
Na^+	1.7	2.07(1)	45.6	25	C	0.5M$(CH_3)_4NCl$	68Ca	
K^+	1.7	1.79(1)	39.3	25	C	0.5M$(CH_3)_4NCl$	68Ca	
Ca^{2+}	-2.18	5.75(1)	102.9	25	C	0.5M$(CH_3)_4NCl$	68Ca	
E44	ETHANE, 1,2-bis(3-mercapto-propanoate)		$C_8H_{14}O_4S_2$			$HSCH_2CH_2C(O)OCH_2CH_2OC(O)$-$CH_2CH_2SH$		
In^{3+}	54	0.30(30°,1)	184	30&40	T	μ=1.0M(KCl)	72Sf	a,b
	-34	1.70(30°,2)	-84	30&40	T	μ=1.0M(KCl)	72Sf	a,b
	-18	1.93(30°,3)	-21	30&40	T	μ=1.0M(KCl)	72Sf	a,b
Tl^+	-6.7	0.740(40°,1)	-7.5	30&40	T	μ=0.5M($NaClO_4$)	72Se	a,b,j
	-18.0	0.903(40°,1-2)	-40.6	30&40	T	μ=0.5M($NaClO_4$)	72Se	a,b,j
	-21.8	2.845(40°,1-3)	-14.6	30&40	T	μ=0.5M($NaClO_4$)	72Se	a,b,j

Metal Ion	ΔH, kJ /mol	Log K	ΔS, J /mol K	T,°C	M	Conditions	Ref.	Remarks
E45	ETHANE, 1,1,2,2-tetramercapto-S,S',S'',S'''-tetraacetic acid		$C_{10}H_{14}O_8S_4$			$(HO_2CCH_2S)_2CHCH(SCH_2CO_2H)_2$		L^{4-}
Cu^{2+}	7.03	5.00(1)	119.2	25	C	μ=0.10(KNO_3)	70Cd	
	10.42	7.33(2-1)	175.3	25	C	μ=0.10(KNO_3)	70Cd	
E46	1,2-ETHANEDIOL		$C_2H_6O_2$			HOCHCHOH		L
BA_4^-	-11.30	0.332(25°,1)	-31.4	0-35	T	CL=0-0.9M A=OH^-	67Cb	b,q:±1.2
	-8.24	0.061(25°,1-2)	-26.4	0-35	T	CL=0-0.9M A=OH^-		b,q:±0.8
H_5TeO_6	-19.2	1.17	-42	25	T	μ=0.1 R:$H_5TeO_6^-$+L=$HTeO_4L^-$+$2H_2O$	62Ea	
E47	1,2-ETHANEDIOL, di(2'-amino-ethyl)ether		$C_6H_{16}O_2N_2$			$H_2N(CH_2CH_2O)_2CH_2CH_2NH_2$		L
Cu^{2+}	-33.0	7.82(30°,1)	42	10-40	T	μ=0	59Lb	b
Ag^+	-55.2	7.71(30°,1)	-33	10-40	T	μ=0	59Lb	b
E48	1,2-ETHANEDIOL, dibutyl ether		$C_{10}H_{22}O_2$			$CH_3(CH_2)_3OCH_2CH_2O(CH_2)_3CH_3$		
TiA_4	-107.5	(1)	--	25	C	S:Benzene CM=0.05-0.08M A=Cl	68Ge 70Gh	
SnA_4	-91.6	4.03(1)	-230.1	25	C	S:Benzene CM=0.05-0.08M A=Cl	68Ge 70Gh	
E49	1,2-ETHANEDIOL, diethyl ether		$C_6H_{14}O_2$			$C_2H_5OCH_2CH_2OC_2H_5$		L
TiA_4	-88.7	(1)	--	25	C	S:Benzene CM=0.05-0.08M A=Br	68Ge 70Gh	
	-99.6	3.96(1)	-258.2	25	C	S:Benzene CM=0.05-0.08M A=Cl	68Ge 70Gh	
SnA_4	-95.4	(1)	--	25	C	S:Benzene CM=0.05-0.08M A=Cl	68Ge 70Gh	
E50	1,2-ETHANEDITHIOL		$C_2H_6S_2$			$HSCH_2CH_2SH$		
Th^{4+}	-35.2	30.64(25°,1-2)	483	25&35	T	μ=0.1M($NaClO_4$) S:25v%Ethanol	74Sc	a,b
UO_2^{2+}	-26.3	13.00(25°,1)	160	25&35	T	μ=0.1M($NaClO_4$) S:25v%Ethanol	74Sc	a,b
Cu^{2+}	-27.0	7.317(25°,1)	48.8	25-45	T	μ=0.1M($NaClO_4$) S:25v%Ethanol	75Sc	

Metal Ion	ΔH, kJ /mol	Log K	ΔS, J /mol K	T,°C	M	Conditions	Ref.	Remarks
E50, continued								
Ag^+	-24.4	7.993(25°,2-1)	71.5	25-45	T	μ=0.1M($NaClO_4$) S:25v%Ethanol	75Sc	
Hg^{2+}	-41.8	8.405(25°,1)	19.0	25-45	T	μ=0.1M($NaClO_4$) S:25v%Ethanol	75Sc	
E51 1-ETHANESULPHONIC ACID, 2-amino-			$C_2H_7O_3NS$			$NH_2(CH_2)_2SO_3H$		L^-
Ag^+	-18.83	2.969(1)	-6.3	25	C	μ=0.5M(KNO_3)	72Va	
	-29.497	3.181(2)	-38.1	25	C	μ=0.5M(KNO_3)	72Va	
E52 ETHANESULFONIC ACID, 2-amino, N-(1-methyl-3-oxo-butyl)			$C_7H_{15}O_4NS$			$CH_3C(O)CH_2CH(CH_3)NH-CH_2CH_2SO_3H$		L^{2-}
UO_2^{2+}	-30.7	15.40(25°,1)	292	25&35	T	S:50v%Dioxane	760a	a,b
Co^{2+}	-18.1	8.20(25°,1)	209	25&35	T	S:50v%Dioxane	760a	a,b
Ni^{2+}	-12.6	10.30(25°,1)	234	25&35	T	S:50v%Dioxane	760a	a,b
Cu^{2+}	-8.4	14.60(25°,1)	256	25&35	T	S:50v%Dioxane	760a	a,b
Zn^{2+}	-18.4	7.80(25°,1)	214	25&35	T	S:50v%Dioxane	760a	a,b
Cd^{2+}	-22.3	5.85(25°,1)	189	25&35	T	S:50v%Dioxane	760a	a,b
E53 ETHANESULFONIC ACID, 2-(furfuralamino)-			$C_7H_9O_4NS$			$(C_4H_3O)CHNCH_2CH_2SO_3H$		
Cr^{2+}	53.93	7.15(25°,1-2)	302.4	25-45	T	μ=0	79Sg	b
Mn^{2+}	48.95	7.65(25°,1-2)	307.6	25-45	T	μ=0	79Sg	b
Fe^{2+}	58.99	8.70(25°,1-2)	357.5	25-45	T	μ=0	79Sg	b
Co^{2+}	29.84	9.80(25°,1-2)	283.8	25-45	T	μ=0	79Sg	b
Ni^{2+}	22.37	10.40(25°,1-2)	272.3	25-45	T	μ=0	79Sg	b
Cu^{2+}	27.21	11.20(25°,1-2)	297.7	25-45	T	μ=0	79Sg	b
E54 ETHANESULFONIC ACID, N-(2-hydroxybenzyl)			$C_9H_{13}O_4NS$			$C_6H_4(OH)(CH_2NHCH_2CH_2SO_3H)$		H_2L
La^{3+}	89.96	10.15(25°,1-2)	488.3	25-45	T	--	79Sf	
Ce^{3+}	104.6	10.68(25°,1-2)	572.4	25-45	T	--	79Sf	
Pr^{3+}	99.75	11.32(25°,1-2)	562.3	25-45	T	--	79Sf	
Nd^{3+}	114.43	12.05(25°,1-2)	621.7	25-45	T	--	79Sf	
Sm^{3+}	86.19	12.88(25°,1-2)	539.7	25-45	T	--	79Sf	
Gd^{3+}	90.04	13.82(25°,1-2)	569.9	25-45	T	--	79Sf	
Tb^{3+}	73.18	14.44(25°,1-2)	526.8	25-45	T	--	79Sf	
Dy^{3+}	62.55	14.91(25°,1-2)	501.7	25-45	T	--	79Sf	

Metal Ion	ΔH, kJ /mol	Log K	ΔS, J /mol K	T,°C	M	Conditions	Ref.	Re-marks
E54, continued								
Ho^{3+}	79.58	15.80(25°,1-2)	568.6	25-45	T	--	79Sf	
E55 ETHANETHIOL, 2-(diisopropyl-amino)-			$C_8H_{19}NS$			$(C_3H_7)_2NC_2H_4SH$		L^-
UO_2^{2+}	68.87	15.75(30°)	536	30-40	T	μ=0.1M(KNO_3) R:UO_2^{2+}+2HL+2OH^-= UO_2L_2+2H_2O	73Fb	a,b
Ag^+	-7.2	10.53(30°,1)	180	30&40	T	μ=0.1M(KNO_3) pH=2.0-3.5	73Se	a,b
E56 ETHANETHIOL, 2-(dimethylamino)-			$C_4H_{11}NS$			$(CH_3)_2NCH_2CH_2SH$		L^-
UO_2^{2+}	64.39	15.20(25°,1-2)	511.16	20&35	T	μ=0.25M(KNO_3)	73Me	a,b
Cu^{2+}	-17.20	19.20(25°,1-2)	309.11	20&35	T	μ=0.25M(KNO_3)	73Me	a,b
Zn^{2+}	43.47	13.44(25°,1-2)	406.02	20&35	T	μ=0.25M(KNO_3)	73Me	a,b
Cd^{2+}	-110.29	15.98(25°,1-2)	-70.08	20&35	T	μ=0.25M(KNO_3)	73Me	a,b
Hg^{2+}	-21.84	20.28(25°,1-2)	314.34	20&35	T	μ=0.25M(KNO_3)	73Me	a,b
Pb^{2+}	-5.73	14.48(25°,1-2)	259.00	20&35	T	μ=0.25M(KNO_3)	73Me	a,b
E57 ETHANOL			C_2H_6O			CH_3CH_2OH		L
[Fe(II)]	-18.4	0.18(25°,1)	-58	7-37	T	S:Benzene [Fe(II)]=Ferrous deuteroporphyrin dimethyl ester	74Bh	b,q:±1.8
SO_2	-15.23	0.387(20°,1)	-45.19	0-30	T	0.4F(NaOH) 0.2F(HCl) S:CCl_4	57Da	b
I_2	-14.6	0.88(23°,1)	-33.9	23-45	T	Dilute solution	56Ab	b
	-8.79	0.61(25°,1)	-17.11	7-25	T	CL=0.06-0.96 mole fraction S:CCl_4	57Dc	b
	-13.4	0.176(25°,1)	-42	10-40	T	S:n-Heptane	77Tb	
	-14.6	-0.022(25°,1)	-49	10-40	T	S:n-Heptane	77Tb	
	-10.9	(1)	--	--	C	S:C_2H_5OH Recalculation of data in Ref 50Ha	61Aa	
	-8.8	(1)	--	--	C	S:C_2H_5OH	50Ha	
E58 ETHANOL, 2-amino-			C_2H_7ON			$NH_2CH_2CH_2OH$		L
Ag^+	-23.0	3.07(30°,1)	-17	10-40	T	μ=0	59Lb 67Cd	b
	-26.8	3.57(30°,2)	-21	10-40	T	μ=0	59Lb	b
Hg^{2+}	-42.7	8.56(1)	20.9	25	C	μ=0	70Ea	
	-34.7	8.77(2)	51.5	25	C	μ=0	70Ea	

Metal Ion	ΔH, kJ /mol	Log K	ΔS, J /mol K	T,°C	M	Conditions	Ref.	Re-marks
E59	ETHANOL, 1,1-dimethyl-		$C_4H_{10}O$			$(CH_3)_3COH$		L
I_2	-14.2	1.05(25°,1)	-28.0	1.6-45.8	T	CL=0.01-0.15 mole fraction S:CCl_4	55Kb	b,m
E60	ETHANOL, 2-mercapto-		C_2H_6OS			$HOCH_2CH_2SH$		L^-
Be^{2+}	-102	11.84(35°,1-2)	-105	25-45	T	μ=0.065M($NaClO_4$)	75Gh	b
Y^{3+}	92.5	10.02(35°,1-2)	492	25-45	T	μ=0.065M($NaClO_4$)	75Gh	b
La^{3+}	153	7.64(35°,1-2)	644	25-45	T	μ=0.065M($NaClO_4$)	75Gh	b
Th^{4+}	-20.6	25.31(20°,1-4)	414.2	10-30	T	μ=0.1M($NaClO_4$) pH=3.2-4.8	77Se	b,j
UO_2^{2+}	-46.4	13.97(35°,1-2)	117	25-45	T	μ=0.065M($NaClO_4$)	75Gh	b
	-95.3	14.53(20°,1-2)	-47.0	10-30	T	μ=0.1M($NaClO_4$) pH=3.3-6.5	77Se	b,j
HgA^+	-82.8	16.104(1)	25.9	20	C	μ=0.1(KNO_3) A=CH_3	65Sa	
Tl^+	-8.58	0.973(25°,1)	-10.1	25&35	T	0.1M(KNO_3)	76Sl	a,b
	0.54	0.902(25°,2)	19.0	25&35	T	0.1M(KNO_3)	76Sl	a,b
E61	ETHENE		C_2H_4			$CH_2:CH_2$		L
PdA_4^{2-}	-6.3	1.182(20°)	0	5-25	T	μ=4.0 R:PdA_4^{2-}+ L=PdA_3L^-+A^- A=Cl	65Pb	b
	-48.1	0.634(20°)	-151	5-25	T	μ=4.0 R:PdA_4^{2-}+ L+H_2O=$PdA_2L(OH)_2$+ $2A^-$ A=Cl	65Pb	b
$IrABC_2$	-50	0.431(20°,1)	-163.2	10-60	T	S:Chlorobenzene A=I B=CO C=Ph_3P	71Vb	b
Ag^+	-23.01	1.929(25°,1)	-37.99	0-25	T	μ=0	70Wa	b
	-30.84	1.973(25°,1)	-65.7	17.5-32.5	T	1M Sodium trifluoroacetate	59Bb	b
	14.6	1.243(25°,1)	-25.1	0-40	T	S:Ethylene glycol	65Ci	b
E62	ETHENE, n-butoxy-		$C_6H_{12}O$			$CH_2:CHOC_4H_9$		L
Ag^+	-15.5	0.890(20°,1)	-36.8	10-30	T	S:Ethyleneglycol	68Fd	b,q:±0.8
E63	ETHENE, chloro-		C_2H_3Cl			$CH_2:CHCl$		L
$RhAB_2$	3.3	-0.770	-3.3	0&25	T	S:Toluene A=2,4-Pentanedionato B=Ethylene R:L+$RhAB_2$=RhABL+B	67Cc	a,b,q: ±3.3
E64	ETHENE, cyano-		C_3H_3N			$CH_2:CHCN$		L
TiA_4	-46.9	3.22(1)	-95.4	25	C	S:Benzene A=Cl	70Gh	

Metal Ion	ΔH, kJ /mol	Log K	ΔS, J /mol K	T,°C	M	Conditions	Ref.	Re-marks
E65	ETHENE, 1,1-dimethyl-2-ethoxy-		$C_6H_{12}O$			$(CH_3)_2C:CHOC_2H_5$		L
Ag^+	-14.6	-0.633(20°,1)	-63	10-30	T	S:Ethyleneglycol	68Fd	b,q:±0.8
E66	ETHENE, 1,2-diphenyl-(cis)		$C_{14}H_{12}$			$C_6H_5CH:CHC_6H_5$		L
I_2	-10.04	0.161(1)	-30.63	25	T	CM=10^{-4}M CL=0.1M S:CCl_4	59Ya	
	-31.55	0.707(1)	-92.5	25	T	CM=10^{-4}M CL=0.1M S:n-Hexane	59Ya	
E67	ETHENE, 1,2-diphenyl-(trans)		$C_{14}H_{12}$			$C_6H_5CH:CHC_6H_5$		L
I_2	-10.63	0.295(1)	-30.04	25	T	CM=10^{-4}M CL=0.1M S:n-Hexane	59Ya	
E68	ETHENE, 1,2-diylbis (diphenylphosphine)		$C_{26}H_{22}P_2$			$[(C_6H_5)_2P]_2C_2H_2$		
HgA_2	-79	(1)	--	30	C	S:Benzene A=Br L is (Z) isomer	76Ga	
	-103	(1)	--	30	C	S:Benzene A=Br L is (E) isomer	76Ga	
	-80	(1)	--	30	C	S:Benzene A=Cl L is (Z) isomer	76Ga	
	-87	(1)	--	30	C	S:Benzene A=Cl L is (E) isomer	76Ga	
	-78	(1)	--	30	C	S:Benzene A=I L is (Z) isomer	76Ga	
	-95	(1)	--	30	C	S:Benzene A=I L is (E) isomer	76Ga	
E69	ETHENE, ethoxy-		C_4H_8O			$CH_2:CHOC_2H_5$		L
Ag^+	-18.4	0.827(20°,1)	-46.0	10-30	T	S:Ethyleneglycol	68Fd	b,q:±0.8
E70	ETHENE, 1-ethoxy-2-ethyl-(cis)		$C_6H_{12}O$			$C_2H_5CH:CHOC_2H_5$		L
Ag^+	-18.8	0.627(20°,1)	-52.7	10-30	T	S:Ethyleneglycol	68Fd	b,q:±0.8
E71	ETHENE, 1-ethoxy-2-ethyl-(trans)		$C_6H_{12}O$			$C_2H_5CH:CHOC_2H_5$		L
Ag^+	-13.8	-0.137(20°,1)	-50.2	10-30	T	S:Ethyleneglycol	68Fd	b,q:±0.8
E72	ETHENE, 1-ethoxy-2-isopropyl-(cis)		$C_7H_{14}O$			$C_3H_7CH:CHOC_2H_5$		L
Ag^+	-19.7	0.649(20°,1)	-54.0	10-30	T	S:Ethyleneglycol	68Fd	b,q:±0.8

Metal Ion	ΔH, kJ /mol	Log K	ΔS, J /mol K	T,°C	M	Conditions	Ref.	Re-marks
E73	ETHENE, 1-ethoxy-2-isopropyl- (trans)		$C_7H_{14}O$			$C_3H_7CH{:}CHOC_2H_5$		L
Ag^+	-14.6	-0.229(20°,1)	-54.0	10-30	T	S:Ethyleneglycol	68Fd	b,q:±0.8
E74	ETHENE, 1-ethoxy-2-methyl- (cis)		$C_5H_{10}O$			$CH_3CH{:}CHOC_2H_5$		L
Ag^+	-16.3	0.496(20°,1)	-46.0	10-30	T	S:Ethyleneglycol	68Fd	b,q:±0.8
E75	ETHENE, 1-ethoxy-2-methyl- (trans)		$C_5H_{10}O$			$CH_3CH{:}CHOC_2H_5$		L
Ag^+	-13.4	-0.268(20°,1)	-50.6	10-30	T	S:Ethyleneglycol	68Fd	b,q:±0.8
E76	ETHENE, fluoro-		C_2H_3F			$CH_2{:}CHF$		L
$RhAB_2$	-6.7	-0.495(25°)	-31.4	0&25	T	S:Toluene A=2,4-Pentanedionate B=Ethylene R:$L+RhAB_2=RhABL+B$	67Cc	a,b,q: ±4.6
E77	ETHENE, isobutoxy-		$C_6H_{12}O$			$CH_2{:}CHOC_4H_9$		L
Ag^+	-16.7	0.834(20°,1)	-41.0	10-30	T	S:Ethyleneglycol	68Fd	b,q:±0.8
E78	ETHENE, 1-isobutoxy-2-methyl- (cis)		$C_7H_{14}O$			$CH_3CH{:}CHOC_4H_9$		L
Ag^+	-15.5	0.456(20°,1)	-43.5	10-30	T	S:Ethyleneglycol	68Fd	b,q:±0.8
E79	ETHENE, 1-isobutoxy-2-methyl- (trans)		$C_7H_{17}O$			$CH_3CH{:}CHOC_4H_9$		L
Ag^+	-11.7	-0.222(20°,1)	-44.4	10-30	T	S:Ethyleneglycol	68Fd	b,q:±0.8
E80	ETHENE, 1-methoxy-1-methyl-		C_4H_8O			$CH_2{:}C(CH_3)OCH_3$		L
Ag^+	-19.7	0.866(20°,1)	-50	10-30	T	S:Ethyleneglycol	68Fd	b,q:±0.8
E81	ETHENE, tetracyano-		C_6N_4			$(CN)_2C{:}C(CN)_2$		L
Na^+	-29.7	--	--	-95--50	T	S:2-Methyltetra-hydrofuran ΔH determined by ESR R:$2ML=M_2L_2$	72Ie	b
	-31.8	4.699(-104°)	-49.4	-119--88	T	S:2-Methyltetra-hydrofuran ΔH determined by ESR R:$2ML=M_2L_2$	72Ie	b

Metal Ion	ΔH, kJ /mol	Log K	ΔS, J /mol K	T,°C	M	Conditions	Ref.	Re-marks
E81, continued								
Na^+, cont.	-23.8	4.556(-104°)	-27.2	-129- -98	T	S:5.4% Tetra-hydrofuran R:2ML=M_2L_2	72Ie	b
	-14.2	4.176(-104°)	-2.1	-126- -104	T	S:9.6% Tetra-hydrofuran R:2ML=M_2L_2	72Ie	b
	-7.5	4.000(-104°)	15.5	-126- -104	T	S:14.5% Tetra-hydrofuran R:2ML=M_2L_2	72Ie	b
K^+	-20.1	--	--	-115- -85	T	S:2-Methyltetra-hydrofuran R:2ML=M_2L_2	72Ie	b
Rb^+	-17.99	--	--	-105- -65	T	S:2-Methyltetra-hydrofuran R:2ML=M_2L_2	72Ie	b
Cs^+	-16.7	--	--	-115- -85	T	S:2-Methyltetra-hydrofuran R:2ML=M_2L_2	72Ie	b
$IrABC_2$ (trans)	-77.8	(1)	--	25	C	S:Benzene A=Br B=CO C=PPh_3	79Bf	
	-83.3	(1)	--	25	C	S:Benzene A=Cl B=CO C=PPh_3	79Bf	
	-87.4	(1)	--	25	C	S:Benzene A=Cl B=CO C=$P(OPh)_3$	79Bf	
	-118.4	(1)	--	25	C	S:Benzene A=Cl B=CO C=PMe_2Ph	79Bf	
	-104.6	(1)	--	25	C	S:Benzene A=Cl B=CO C=$PMePh_2$	79Bf	
	-110.9	(1)	--	25	C	S:Benzene A=Cl B=CO C=PEt_2Ph	79Bf	
	-68.2	(1)	--	25	C	S:Benzene A=Cl B=CO C=$P(t\text{-}Bu)Ph_2$	79Bf	
	-83.3	(1)	--	25	C	S:Benzene A=Cl B=CO C=PPh_3	79Bf	
	-82.4	(1)	--	25	C	S:Benzene A=Cl B=CO C=$AsPh_3$	79Bf	
	-100.0	(1)	--	25	C	S:Benzene A=F B=CO C=PPh_3	79Bf	
	-52.7	(1)	--	25	C	S:Benzene A=I B=CO C=PPh_3	79Bf	
[Se]	-8.8	0.01(20°,1)	-30	10-30	T	S:Dichloromethane [Se]=Phenoxasel-enine 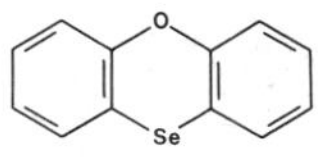	74Hc	b,q:±0.8

Metal Ion	ΔH, kJ /mol	Log K	ΔS, J /mol K	T,°C	M	Conditions	Ref.	Re-marks
E81, continued								
[Te]	-8.4	0(20°,1)	-28	10-30	T	S:Dichloromethane [Te]=Phenoxatellurine 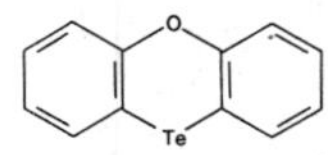	74Hc	b,q:±1.7
[S]	-8.8	0.03(20°,1)	-30	10-30	T	S:Dichloromethane [S]=Phenoxthiin 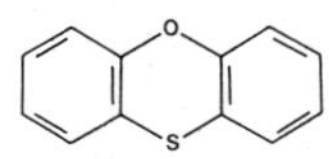	74Hc	b
E82	ETHENE, tetrakis(4-methoxy-phenyl)		$C_{30}H_{28}O_4$			$[C_6H_4(OCH_3)]_2C:[C_6H_4(OCH_3)]_2$		L
Br_2	-40.00	3.94(1)	-58	26	T	CM≅10^{-3}M S:Ethylene chloride	58Bc	
E83	ETHENE, tetraphenyl-		$C_{26}H_{20}$			$(C_6H_5)_2C:C(C_6H_5)_2$		L
LiL	66.11	2.613(30°)	268	0-50	T	S:Tetrahydrofuran R:2LiL=Na_2L+L	63Ea	b
NaL	46.0	3.447(30°	218	0-50	T	S:Tetrahydrofuran R:2NaL=Na_2L+L	63Ea	b
	-54	0.143(0°)	-188	-37-20	T	S:Tetrahydrofuran R:NaL+L^-=NaL^-+L	65Ra	b,g:Na, q:±8
Na_2L	-79	-1.591(0°)	-314	-37-20	T	S:Tetrahydrofuran R:Na_2L+L=2NaL	65Ra	b,g:Na, q:±8
E84	ETHENE, 1,1,2-trimethyl-		C_5H_{10}			$CH_3CH:C(CH_3)_2$		L
Ag^+	-25.14	-2.212(25°)	126.4	0-25	T	μ=1.0($AgNO_3$,KNO_3)	38Wa	b,o
	10.0	0.0043(25°,1)	-32.6	0-40	T	S:Ethyleneglycol	65Ci	b
E85	ETHER, allyl ethyl-		$C_5H_{10}O$			$C_2H_5OCH_2CH:CH_2$		L
	-50.2	(1)	--	25	C	S:Benzene CM=0.05-0.08M A=Cl	68Ge	
E86	ETHER, dibutyl-		$C_8H_{18}O$			$CH_3(CH_2)_3O(CH_2)_3CH_3$		L
TiA_4	-25.1	--	--	25	C	S:Benzene CM=0.05-0.08M A=Cl	70Gh 68Ge	u
AlA_3	-81.6	0.76(1)	-254	28.5	C	CM=0.10-0.88M S:Benzene A=Isobutyl	74Sp	

Metal Ion	ΔH, kJ /mol	Log K	ΔS, J /mol K	T,°C	M	Conditions	Ref.	Remarks
E86, continued								
Al_2A_6	-97.1	--	--	25	C	S:Benzene CM=0.07M A=Br R:0.5(Al_2A_6)+L= AlA_3L	68Rc	
	-90.0	--	--	25	C	S:Cyclohexane CM=0.07M A=Br R:0.5(Al_2A_6)+L= AlA_3L	68Rc	
$Al_2A_4B_2$	-92.9	--	--	25	C	S:Toluene R:0.5$Al_2A_4B_2$+L= (AlA_2BL) A=Ethyl B=Cl	77Pd	
GaA_3	-66.9	3.66(1)	-155	25	C	S:Benzene A=Cl	70Gh	
TeA_4	-6.3	(1)	--	25	C	S:Benzene A=Cl	72Pg	
E87 ETHER, diethyl-			$C_4H_{10}O$			$CH_3CH_2OCH_2CH_3$		L
$TiAB_3$	2.1	-1.42(1-2)	-21	-18	T	S:Toluene $A=CH_3$ B=Cl	69Ta	b
$[NiA_2]_2$	-23	0.96(100°,1)	-42	75-125	T	S:Squalane A=3-Trifluoro-acetylcamphorate	74Sj	b,gg
AlA_3	-84.56	--	--	25-28	C	S:n-Hexane $A=CH_3$	68Hc	
Al_2A_6	-97.1	--	--	25	C	S:Benzene CM=0.07M A=Br R:0.5(Al_2A_6)+L= AlA_3L	68Rc	
$Al_2A_4B_2$	-92.9	--	--	25	C	S:Toluene R:0.5$Al_2A_4B_2$+L= (AlA_2BL) A=Ethyl B=Cl	77Pd	
TeA_4	-17.6	(1)	--	25	C	S:Benzene A=Cl	72Pg	
I_2	-17.6	0.82(25°,1)	-43.1	4-25	T	S:n-Heptane	60Ta 60Bb	b
	-17.99	0.96(25°,1)	-41.25	4-25	T	S:CCl_4	60Ta 60Bb	b
	-17.00	(1)	--	2-25	T	CL=0.00-0.16M S:CCl_4	61Da	
	-17.99	0.96(25°,1)	-41.25	7-25	T	CL=0.06-0.95 mole fraction S:CCl_4	57Dc	b,l
	-3.93	(1)	--	2-25	T	CL=0.00-0.16M S:$CHCl_3$	61Da	
	-15.15	(1)	--	10-25	T	CL=0.00-0.16M S:Cyclohexane	61Da	
	-20.5	(1)	--	--	C	S:$(C_2H_5)_2O$	61Aa	i:50Ha
	-18.4	(1)	--	--	C	S:$(C_2H_5)_2O$	50Ha	
	-11.30	(1)	--	2-25	T	CL=0.00-0.16M S:n-Heptane	61Da	

Metal Ion	ΔH, kJ /mol	Log K	ΔS, J /mol K	T, °C	M	Conditions	Ref.	Remarks
E87, continued								
I_2, cont.	-15.73	(1)	--	2-25	T	CL=0.00-0.16M S:n-Hexane	61Da	
E88	ETHER, diethyl, 2,2'-diamino-	$C_4H_{12}O$				$O(CH_2CH_2NH_2)_2$		L
Ni^{2+}	-29.7	5.54(30°,1)	8	10-40	T	μ=0	59Lb	b
	-28.0	5.895(1)	18.8	25	C	μ=0.5M(KNO_3)	74Bb	
	-31.4	3.19(30°,2)	-42	10-40	T	μ=0	59Lb	b
	-27.2	3.62(2)	-21	25	C	μ=0.5M(KNO_3)	74Bb	
Cu^{2+}	-46.0	8.58(30°,1)	12	10-40	T	μ=0	59Lb	b
	-39.9	8.970(1)	38.1	25	C	μ=0.5M(KNO_3)	74Bb	
	-15.5	4.35(30°,2)	33	10-40	T	μ=0	59Lb	b
	-18.8	3.78(2)	8.4	25	C	μ=0.5M(KNO_3)	74Bb	
Ag^+	-60.7	5.31(30°,1)	-100	10-40	T	μ=0	59Lb	b
Zn^{2+}	-20.1	5.74(1)	41.4	25	C	μ=0.5M(KNO_3)	74Bb	
	-30.5	4.12(2)	-17	25	C	μ=0.5M(KNO_3)	74Bb	
E89	ETHER, diethyl, 2,2'-diamino-N,N,N',N'-tetraacetic acid	$C_{12}H_{20}O_9N_2$				$(CH_2CO_2H)_2N(CH_2)_2O-(CH_2)_2N(CH_2CO_2H)_2$		L^{4-}
Mg^{2+}	14.69	8.32(1)	209.37	20	C	μ=0.1(KNO_3)	64Ab	
Ca^{2+}	-28.66	10.0(1)	93.7	20	C	μ=0.1(KNO_3)	64Ab	
La^{3+}	-14.02	16.6(1)	270.0	20	C	μ=0.1(KNO_3)	64Ab	
Mn^{2+}	-24.7	13.76(1)	190.8	20	C.	μ=0.1(KNO_3)	64Ab	
Co^{2+}	-26.57	15.27(1)	201.7	20	C	μ=0.1(KNO_3)	64Ab	
Ni^{2+}	-19.8	15.07(1)	220.9	20	C	μ=0.1(KNO_3)	64Ab	
Cu^{2+}	-41.09	18.1(1)	205.0	20	C	μ=0.1(KNO_3)	64Ab	
Zn^{2+}	-25.06	15.3(1)	207.5	20	C	μ=0.1(KNO_3)	64Ab	
Cd^{2+}	-39.41	16.2(1)	175.7	20	C	μ=0.1(KNO_3)	64Ab	
Hg^{2+}	-85.8	23.9(1)	149.4	20	C	μ=0.1(KNO_3)	64Ab	
Pb^{2+}	-55.02	15.03(1)	100.0	20	C	μ=0.1(KNO_3)	64Ab	
E90	ETHER, diethyl, 2,2'-dimercapto	$C_4H_{10}OS_2$				$O(CH_2CH_2SH)_2$		
Be^{2+}	-10.5	12.04(25°,1)	181	25&35	T	μ=0.1M($NaClO_4$) S:40v%Ethanol	75Sg 75Sf	a,b
UO_2^{2+}	-8.8	12.68(25°,1)	213	25&35	T	μ=0.1M($NaClO_4$) S:40v%Ethanol	75Sg 75Sf	a,b
Zn^{2+}	-20.4	14.08(25°,1)	200	25-45	T	μ=0.1M($NaClO_4$) S:40v%Ethanol	74Sg	b
Tl^+	-19.9	2.23(20°,1-2)	-25.4	20&30	T	0.1M($NaClO_4$) pH=3.8 S:40v%Ethanol	75Se	a,b

Metal Ion	ΔH, kJ /mol	Log K	ΔS, J /mol K	T,°C	M	Conditions	Ref.	Remarks
E91	ETHER, diethyl, 2,2'-dimethoxy-		$C_6H_{14}O_3$			$CH_3O(CH_2)_2O(CH_2)_2OCH_3$		L
NaA	-18.8	(1)	-36.0	-40-25	T	CL<4x10^{-2}M S:Tetrahydropyran R:NaA+L=LNaA A=Biphenyl	67Sb	r,b
	0	(1)	--	-40-25	T	CL<4x10^{-2}M S:Tetrahydrofuran R:NaA+L=NaLA A=Biphenyl	67Sb	
	-11.7	0.146(25°)	-38	0-50	T	S:Tetrahydrofuran A=Fluorenyl ion R:NaA+L=NaLA	70Ce	b
SnA_4	-84.9	(1)	--	--	C	S:Benzene CM=0.05-0.08M A=Cl	68Ge	u
E92	ETHER, diisopropyl-		$C_6H_{14}O$			$(CH_3)_2CHOCH(CH_3)_2$		L
I_2	-25.6	0.647(25°,1)	--	--	T	S:n-Hexane	57Dd 52Ka	1
E93	ETHER, dimethyl-		C_2H_6O			CH_3OCH_3		L
AlA_3	-84.89	--	--	25-28		S:n-Hexane A=CH_3	68Hc	
E94	ETHER, dioctyl-		$C_{16}H_{34}O$			$C_8H_{17}OC_8H_{17}$		L
TiA_4	-26.4	(1)	--	25	C	S:Benzene CM=0.05-0.08M A=Cl	68Ge 70Gh	
GaA_3	-72.8	3.81(1)	-171.1	25	C	S:Benzene A=Cl	70Gh	
E95	ETHER, diphenyl-		$C_{12}H_{10}O$			$C_6H_5OC_6H_5$		L
AlA_3	-16.32	(1)	-42.68	29-110	T	S:L R:½Al_2A_6+L= AlA_3L A=Et	72Ab	b,r
Al_2A_6	-50.2	--	--	25	C	S:Benzene CM=0.07M A=Br R:0.5(Al_2A_6)+L= AlA_3L	68Rc	
E96	ETHER, diphenyl, 4,4'-dibromo-		$C_{12}H_8OBr_2$			$C_6H_4(Br)OC_6H_4(Br)$		L
Al_2A_6	-39.7	--	--	25	C	S:Benzene CM=0.07M A=Br R:0.5(Al_2A_6)+L= AlA_3L	68Rc	

Metal Ion	ΔH, kJ /mol	Log K	ΔS, J /mol K	T, °C	M	Conditions	Ref.	Remarks
E97	ETHER, diphenyl, 4,4'-dimethyl-		$C_{14}H_{14}O$			$C_6H_4(CH_3)OC_6H_4(CH_3)$		L
Al_2A_6	-54.4	--	--	25	C	S:Benzene CM=0.07M A=Br R:0.5(Al_2A_6)+L= AlA_3L	68Rc	
E98	ETHER, dipropyl		$C_6H_{14}O$			$O(CH_2CH_2CH_3)_2$		
$[NiA_2]_2$	-23	0.54(100°,1)	-51	75-125	T	S:Squalane A=3-Trifluoro-acetylcamphorate	74Sj	b,gg
E99	ETHER, methyl propyl-		$C_4H_{10}O$			$CH_3OCH_2CH_2CH_3$		L
SbA_5	-75.69	(1)	--	25	C	S:Ethylene chloride A=Cl	68Oa	
E100	ETHENE, tetrachloro-		C_2Cl_4			$Cl_2C:CCl_2$		L
I_2A	-0.71	0.364(25°)	4.56	-20-40	T	S:Hexane A=Hexane R:I_2A+L=I_2L+A	73Sn	b,q: ±0.12
E101	ETHYLENETRITHIOCARBONATE		$C_3H_4S_3$			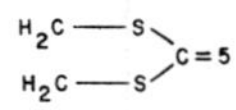		L
I_2	-35.6	1.653(25°,1)	-87.9	0-45	T	S:Cyclohexane	66Bc	b,q:±4
E102	ETHYNE		C_2H_2			CH:CH		L
Cr^{2+}	-8.4	0.29(0°,2-1)	-25	-50-0	T	CL=10^2CM S:Ethanol R:2Cr+L=Cr_2L Solution is saturated with acetylene	71Bl	b,q:±2
$IrABC_2$	-38.9	0.079(20°,1)	-131.4	10-60	T	S:Chlorobenzene A=I B=CO C=Ph_3P	71Vb	b
Ag^+	-55.23	1.60(1)	-154.40	25	T	P:$AgC_2H_2^+$ C_2H_2(g) is passed into soln. containing Ag_2SO_4+ H_2SO_4 $\Delta H_T=\Delta H_{298}$+ 30.0 (T=298)	62Tc	
	28.70	1.38(1)	123.0	25	T	P:$AgC\equiv CH^+$ C_2H_2(g)is passed into soln. containing Ag_2SO_4+	62Tc	

Metal Ion	ΔH, kJ /mol	Log K	ΔS, J /mol K	T,°C	M	Conditions	Ref.	Remarks
E102, continued								
Ag^+, cont.						H_2SO_4 $\Delta H_T=\Delta H_{298}-6.0\ (T-298)$		
E103 ETHYNE, di(methoxycarbonyl)		$C_6H_6O_4$				$CH_3OC(=O)CCC(=O)OCH_3$		
$IrABC_2$ (trans)	-52.7	(1)	--	25	C	S:Benzene A=F B=CO C=PPh_3	79Bf	
	-61.1	(1)	--	25	C	S:Benzene A=Cl B=CO C=PPh_3	79Bf	
	-62.8	(1)	--	25	C	S:Benzene A=Cl B=CO C=$PMePh_2$	79Bf	
	-55.2	(1)	--	25	C	S:Benzene A=Cl B=CO C=$AsPh_3$	79Bf	
	-62.3	(1)	--	25	C	S:Benzene A=Br B=CO C=PPh_3	79Bf	
	-23.8	(1)	--	25	C	S:Benzene A=Br B=CO C=$PMePh_2$	79Bf	
	-64.4	(1)	--	25	C	S:Benzene A=I B=CO C=PPh_3	79Bf	
	-33.5	(1)	--	25	C	S:Benzene A=NCS B=CO C=PPh_3	79Bf	
	-52.7	(1)	--	25	C	S:Benzene A=N_3 B=CO C=PPh_3	79Bf	

F

Metal Ion	ΔH, kJ /mol	Log K	ΔS, J /mol K	T,°C	M	Conditions	Ref.	Remarks
F1 FERRATE(II) ION, hexacyano		C_6N_6Fe				$Fe(CN)_6^{4-}$		L^{4-}
K^+	4.2	2.35(1)	54	25	C	μ=0	67Ea	
La^{3+}	3.56	3.702(1)	82.8	25	C	μ=0	73Su	
Pr^{3+}	3.64	3.636(1)	81.6	25	C	μ=0	73Su	
Nd^{3+}	3.35	3.768(1)	83.3	25	C	μ=0	73Su	
Sm^{3+}	3.81	3.724(1)	84.1	25	C	μ=0	73Su	
Eu^{3+}	4.10	3.651(1)	83.7	25	C	μ=0	73Su	
Gd^{3+}	4.35	3.592(1)	83.3	25	C	μ=0	73Su	
Tb^{3+}	3.93	3.760(1)	85.4	25	C	μ=0	73Su	
Dy^{3+}	4.27	3.680(1)	84.9	25	C	μ=0	73Su	
Ho^{3+}	4.44	3.658(1)	84.9	25	C	μ=0	73Su	
Er^{3+}	4.35	3.680(1)	84.9	25	C	μ=0	73Su	

Metal Ion	ΔH, kJ /mol	Log K	ΔS, J /mol K	T,°C	M	Conditions	Ref.	Remarks
F1, continued								
Tm^{3+}	4.39	3.673(1)	85.9	25	C	μ=0	73Su	
Yb^{3+}	4.35	3.658(1)	84.5	25	C	μ=0	73Su	
Lu^{3+}	4.22	3.695(1)	84.9	25	C	μ=0	73Su	
Tl^{+}	4.6	3.0(1)	72.8	25	T	μ=0	58Pa	
	-7.45	0.82(1)	-9.2	25	C	3M($LiClO_4$)	67Mc	
F2 FERRATE(III) ION, hexacyano			C_6N_6Fe			$Fe(CN)_6^{3-}$		L^{3-}
Na^{+}	-16.7	-0.77(1)	-71	25	C	μ=3	66Me	
K^{+}	2.1	1.46(1)	33	25	C	μ=0	67Ea	
	-22.6	-0.42(1)	-84	25	C	μ=3	66Me	
Mg^{2+}	-14.2	-1.03(1)	-67	25	C	μ=3	66Me	
Ba^{2+}	-15.5	-0.60(1)	-63	25	C	μ=3	66Me	
La^{3+}	8.45	3.73(25°,1)	-100.0	20-35	T	μ=0	52Jb 48Da	b
Tl^{+}	3.43	3.22(1)	72.8	25	T	μ=0	53Ba	
F3 FLUORENE			$C_{13}H_{10}$			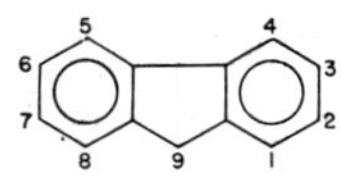		L^{-}
LiL	12.6	--	67	--	T	S:3,4-Dihydropyran R:Solvent separated ion pair→contact ion pair $Li^+SL^- = Li^+L^- + S$	72Si	u
	23.4	--	63	-40-80	T	S:1,2-Dimethoxyethane R:Solvent separated ion pair→contact ion pair $Li^+SLi^- = Li^+L^- + S$	72Gi	b
	12.1	--	67	-55-30	T	S:2,5-Dimethoxytetrahydrofuran R:Solvent separated ion pair→contact ion pair $Li^+SLi^- = Li^+L^- + S$	72Si	
	8.4	--	58	-55-30	T	S:2,5-Dimethyltetrahydrofuran R:Solvent separated ion pair→contact ion pair $Li^+SL^- = Li^+L^- + S$	72Si	
	15.1	--	58	-55-30	T	S:Dioxane-tetrahydrofuran(50%)	72Si	

Metal Ion	ΔH, kJ /mol	Log K	ΔS, J /mol K	T,°C	M	Conditions	Ref.	Re-marks
F3, continued								
LiL, cont.						R:Solvent separated ion pair→ contact ion pair $Li^+SL^-=Li^+L^-+S$		
	14.6	--	7.1	-55-30	T	S:Dioxolane R:Solvent separated ion pair→ contact ion pair $Li^+SL^-=Li^+L^-+S$	72Si	
	31.4	--	113	-55-30	T	S:2-Methyltetrahydrofuran R:Solvent separated ion pair→ contact ion pair $Li^+SL^-=Li^+L^-+S$	72Si	
	17.99	--	67	-55-30	T	S:Oxepane R:Solvent separated ion pair→contact ion pair $Li^+SL^-=Li^+L^-+S$	72Si	
	31.4	--	92	-55-30	T	S:Tetrahydrofuran R:Solvent separated ion pair→ contact ion pair $Li^+SL^-=Li^+L^-+S$	72Si	
	27.6	--	117	-55-30	T	S:Tetrahydropyran R:Solvent separated ion pair→ contact ion pair $Li^+SL^-=Li^+L^-+S$	72Si	
LiA	-1.213	0.613(20°)	7.5	-10-30	T	S:Cyclohexylamine A=2,3-Benzfluorene R:CuA+L=CuL+A	72Sj	b,q: ±0.06
Na^+	5.8	4.32($\underline{1}$)	113	-70	T	S:Tetrahydrofuran	75Ha	b
	34.7	6.12($\underline{1}$)	234	20	T	S:Tetrahydrofuran	75Ha	b
	20.9	8.47($\underline{1}$)	254	-40	T	S:Tetrahydrofuran	75Ha	b
	20.9	9.45($\underline{1}$)	253	20	T	S:Tetrahydrofuran	75Ha	b
NaL	22.6	--	71	-40-80	T	S:1,2-Dimethoxyethane R:Solvent separated ion pair→contact ion pair $Na^+SL^-=Na^+L^-+S$	72Gi	b
	28.0		113	-40-80	T	S:Tetrahydrofuran R:Solvent separated ion pair→	72Gi	b

Metal Ion	ΔH, kJ /mol	Log K	ΔS, J /mol K	T,°C	M	Conditions	Ref.	Re-marks
F3, continued								
NaL, cont.						contact ion pair $Na^+SL^-=Na^+L^-+S$		
	31.8	1.194(24.2°)	129.3	-63.0-24.2	T	S:Tetrahydrofuran R:Solvent separated ion pair→ contact ion pair $Na^+SL^-=Na^+L^-+S$	66Hc	b
NaA^+	0	5.48(1)	105	-60	T	S:Tetrahydrofuran A=4,4'-Dimethyl-dibenzo-18-crown-6	75Ha	b
	0	5.42(1)	105	20	T	S:Tetrahydrofuran A=4,4'-Dimethyl-dibenzo-18-crown-6	75Ha	b
	-7.1	7.60(1)	115	-40	T	S:Tetrahydropyran	75Ha	b
	0	7.42(1)	146	20	T	S:Tetrahydropyran	75Ha	b
KL	17.2	--	67	-40-80	T	S:1,2-Dimethoxy-ethane R:Solvent separated ion pair→contact ion pair $K^+SL^-=L^+L^-+S$	72Gi	b
Sr^{2+}	25.9	(1)	113	-80-20	T	S:$C_2H_5NH_2$	74Pf	b,j
	0	5.59(1-2)	105	-70	T	S:Tetrahydrofuran	75Ha	b
	55.6	6.76(1-2)	318	20	T	S:Tetrahydrofuran	75Ha	b
SrA^{2+}	-10.9	6.56(1-2)	71	-70	T	S:Tetrahydrofuran A=4,4'-Dimethyl-dibenzo-18-crown-6	75Ha	b
	0	6.13(1-2)	117	20	T	S:Tetrahydrofuran A=4,4'-Dimethyl-dibenzo-18-crown-6	75Ha	b
	0	5.94(1-2)	117	-70	T	S:Tetrahydrofuran A=Diaza-hexaoxa cryptate	75Ha	b
	2.9	6.08(1-2)	130	20	T	S:Tetrahydrofuran A=Diaza-hexaoxa cryptate	75Ha	b
Ba^{2+}	0	6.66(1-2)	228	-70	T	S:Tetrahydrofuran	75Ha	b
	0	8.28(1-2)	228	20	T	S:Tetrahydrofuran	75Ha	b
BaA^{2+}	0	6.17(1-2)	119	-70	T	S:Tetrahydrofuran	75Ha	b
	0	6.18(1-2)	119	20	T	S:Tetrahydrofuran	75Ha	b
F4 FLUORENE, 9-tert-butyl-			$C_{17}H_{18}$			$C_{13}H_9[C(CH_3)_3]$	L^-	
LiA	4.770	-1.678(20°)	-15.9	-10-30	T	S:Cyclohexyl-amine A=2,3-Benzfluorene R:LiA+L=LiL+A	72Sj	b
CsA	5.523	-1.086(20°)	2.5	-10-30	T	S:Cyclohexyl-amine A=2,3-	72Sj	b

Metal Ion	ΔH, kJ /mol	Log K	ΔS, J /mol K	T,°C	M	Conditions	Ref.	Remarks
F4, continued								
CsA, cont.						Benzfluorene R:CsA+L=CsL+A		
F5 FLUORENE, 2-fluoro-			$C_{13}H_9F$			$C_{13}H_9(F)$		L^-
Li^+	12.1	5.208(15°,1)	138	25	T	S:1,2-Dimethoxy-ethane	69Ea	
	14.6	5.523(1)	155	25	T	S:Tetrahydrofuran	69Ea	
Na^+	16.3	5.260(1)	155	25	T	S:1,2-Dimethoxy-ethane	69Ea	
	34.7	6.174(1)	234	25	T	S:Tetrahydrofuran	69Ea	
K^+	24.7	5.538(20°,1)	188	25	T	S:1,2-Dimethoxy-ethane	69Ea	
	19.7	6.796(1)	197	25	T	S:Tetrahydrofuran	69Ea	
Cs^+	19.2	6.602(15°,1)	188	25	T	S:1,2-Dimethoxy-ethane	69Ea	
	12.1	7.854(1)	192	25	T	S:Tetrahydrofuran	69Ea	
F6 FLUORENE, 9-ethyl-			$C_{15}H_{14}$			$C_{13}H_9(C_2H_5)$	L	
LiA	-3.222	0.196(20°)	-7.5	-10-30	T	S:Cyclohexyl-amine A=2,3-Benzfluorene R:LiA+L=LiL+A	72Sj	b
CsA	-2.720	0.580(20°)	1.7	-10-30	T	S:Cyclohexyl-amine A=2,3-Benzfluorene R:CsA+L=CsL+A	72Sj	b,q: ±0.21
F7 FLUORENE, 9-isopropyl-			$C_{16}H_{16}$			$C_{13}H_9[CH(CH_3)_2]$		L^-
LiA	0.753	-0.585(20°)	-8.8	-10-30	T	S:Cyclohexyl-amine A=2,3-Benzfluorene R:LiA+L=LiL+A	72Sj	b,q: ±0.17
CsA	1.674	-0.086(20°)	4.2	-10-30	T	S:Cyclohexyl-amine A=2,3-Benzfluorene R:CuA+L=CuL+A	72Sj	b,q: ±0.17
F8 FLUORENE, 9-methyl-			$C_{14}H_{12}$			$C_{13}H_9(CH_3)$		L^-
LiA	-3.264	0.573(20°)	0.0	-10-30	T	S:Cyclohexyl-amine A=2,3-Benzfluorene R:LiA+L=LiL+A	72Sj	b
CsA	-3.096	0.724(20°)	2.93	-10-30	T	S:Cyclohexyl-amine A-2,3-	72Sj	b

Metal Ion	ΔH, kJ /mol	Log K	ΔS, J /mol K	T,°C	M	Conditions	Ref.	Remarks
F8, continued								
						Benzfluorene R:CsA+L=CsL+A		
F9 FLUORENE, 9-(1-methylpentyl)			$C_{19}H_{22}$			$C_{13}H_9[CH(CH_3)C_4H_9]$		L^-
LiL	34.3	--	138	-55-30	T	S:3,4-Dihydropyran R:Solvent separated ion pair→contact ion pair $Li^+SL^-=Li^+L^-+S$	72Si	
	41.8	--	209	-55-30	T	S:2,5-Dimethyltetrahydrofuran R:Solvent separated ion pair→contact ion pair $Li^+SL^-=Li^+L^+S$	72Si	
	46.0	--	151	-55-30	T	S:Dioxane-tetrahydrofuran (28%) R:Solvent separated ion pair→contact ion pair $Li^+SL^-=Li^+L^-+S$	72Si	
	41.0	--	134	-55-30	T	S:2-Methyltetrahydrofuran R:Solvent separated ion pair→contact ion pair $Li^+SL^-=Li^+L^-+S$	72Si	
F10 9-FLUORENONE			$C_{13}H_8O$					L
SnA_4	-31.8	6.6(30°,1)	-63.2	30-40	T	S:Benzene ΔS calculated using ΔG in atm^{-1} A=Cl	64La	b
F11 FLUORIDE ION			F			F^-		L^-
Be^{2+}	-0.8	5.99(1)	113	25	T	μ=0	67Ab	
	-1.7	5.04(1)	92	25	T	μ=0.5	67Ab	
	-1.46	4.900(1)	88.7	25	T	1.0M(NaCl)	69Mb	
	-4.85	3.762(2)	55.6	25	T	1.0M(NaCl)	69Mb	
	-1.21	2.788(3)	49.4	25	T	1.0M(NaCl)	69Mb	
	-1.92	1.426(4)	20.9	25	T	1.0M(NaCl)	69Mb	
Mg^{2+}	17	1.301(25°,1)	79	15-25	T	$\mu=0.50(NaClO_4)$	54Ce 68Tb	b

Metal Ion	ΔH, kJ /mol	Log K	ΔS, J /mol K	T,°C	M	Conditions	Ref	Remarks
F11, continued								
Mg^{2+}, cont.	13.4	1.32(1)	70.3	25	C	μ=1.0($NaClO_4$)	68Tb	
	35	4.556(25°,1)	200	15-25	T	μ=0.05M(Tetraethyl ammonium perchlorate) S:Methanol	78Bb	b
Ca^{2+}	14.6	0.63(1)	62.8	25	C	μ=1.0($NaClO_4$)	68Tb	
	15.9	0.63(1)	66.1	25	T	μ=1.0($NaClO_4$)	68Tb	
Sr^{2+}	17	0.15(1)	58.6	25	C	μ=1.0($NaClO_4$)	68Tb	
	17	0.15(1)	58.6	25	T	μ=1.0($NaClO_4$)	68Tb	
	28	2.612(25°,1)	140	15-25	T	μ=0.05M(Tetraethyl ammonium perchlorate) S:Methanol	78Bb	b
Ba^{2+}	0	-0.22(1)	-4.2	25	T	μ=1.0($NaClO_4$)	68Tb	
Sc^{3+}	2.5	6.16(25°,1)	126	15-35	T	μ=0.5	67Ab	b
	-9.6	3.28(25°)	29	15-35	T	μ=0.5 R:Sc^{3+}+ HF=ScF_2^++H^+	59Kb	b
	-17.2	2.37(25°)	-12	15-35	T	μ=0.5 R:ScF^{2+}+ HF=ScF_2^++H^+	59Kb	b
	-14.2	1.16(25°)	-25	15-35	T	μ=0.5 R:ScF_2^++ HF=ScF_3+H^+	59Kb	b
Y^{3+}	9.6	4.79(1)	126	25	T	μ=0	67Ab 61Fb	
	9.2	3.91(1)	105	25	T	μ=0.5	67Ab	
	5.042	3.91(25°,1)	92.0	15-35	T	μ=0.500($NaClO_4$)	69Ab	b,n
	34.81	3.599(1)	186.2	25	C	μ=1M($NaClO_4$)	67Wa	
	-5.02	3.25(25°,2)	46.0	15-35	T	μ=0.500($NaClO_4$)	69Ab	b,n
La^{3+}	16.74	2.668(1)	107.1	25	C	μ=1M($NaClO_4$)	67Wa	
Ce^{3+}	20.17	2.815(1)	121.3	25	C	μ=1M($NaClO_4$)	67Wa	
Pr^{3+}	24.02	3.013(1)	138.5	25	C	μ=1M($NaClO_4$)	67Wa	
Nd^{3+}	28.58	3.086(1)	154.4	25	C	μ=1M($NaClO_4$)	67Wa	
Sm^{3+}	39.29	3.116(1)	191.6	25	C	μ=1M($NaClO_4$)	67Wa	
Eu^{3+}	38.58	3.189(1)	190.0	25	C	μ=1M($NaClO_4$)	67Wa	
	-17.2	3.13(25°,1)	2.1	0-50	T	μ=1M(ClO_4)	75Mf	
Gd^{3+}	37.24	3.313(1)	188.7	25	C	μ=1M($NaClO_4$)	67Wa	
	-17.2	3.16(25°,1)	2.5	0-50	T	μ=1M(ClO_4)	75Mf	
Tb^{3+}	31.42	3.423(1)	170.7	25	C	μ=1M($NaClO_4$)	67Wa	
	-17.2	3.18(25°,1)	2.9	0-50	T	μ=1M(ClO_4)	75Mf	
Dy^{3+}	29.41	3.460(1)	164.8	25	C	μ=1M($NaClO_4$)	67Wa	
Ho^{3+}	30.38	3.519(1)	169.0	25	C	μ=1M($NaClO_4$)	67Wa	
Er^{3+}	31.09	3.541(1)	172.4	25	C	μ=1M($NaClO_4$)	67Wa	
Tm^{3+}	36.23	3.563(1)	190.0	25	C	μ=1M($NaClO_4$)	67Wa	
Yb^{3+}	40.00	3.585(1)	202.9	25	C	μ=1M($NaClO_4$)	67Wa	

Metal Ion	ΔH, kJ /mol	Log K	ΔS, J /mol K	T, °C	M	Conditions	Ref.	Remarks
F11, continued								
Lu^{3+}	39.87	3.614(1)	202.9	25	C	μ=1M($NaClO_4$)	67Wa	
Th^{4+}	-5.0	8.44(25°,1)	163	5-40	T	μ=0	70Be	b
	-3.3	6.62(25°,2)	130	5-40	T	μ=0	70Be	b
	-3.3	4.75(25°,3)	79	5-40	T	μ=0	70Be	b
	-3.8	3.36(25°,4)	50	5&25	T	μ=0	70Be	a,b
VO^{2+}	7.86	3.371(1)	90.8	25	C	μ=1.0($NaClO_4$)	71Ab	
	6.48	2.369(2)	66.9	25	C	μ=1.0($NaClO_4$)	71Ab	
	5.82	1.554(3)	49.4	25	C	μ=1.0($NaClO_4$)	71Ab	
	6.11	0.785(4)	35.40	25	C	μ=1.0($NaClO_4$)	71Ab	
Cr^{3+}	-1.38	--	--	25	C	0.5M($HClO_4$) R:Cr^{3+}+HF=CrF^{2+}+H^+	76Dd	
$CrA_6{}^{3+}$	5.4	1.346(86°)	25.9	77.2-94.7	T	μ=1 R:$CrA_6{}^{3+}$+HF=CrA_5F^{2+}+H^++A A=H_2O	65Se	b
$UO_2{}^{2+}$	1.699	4.535(1)	92.5	25	C	μ=1.0($NaClO_4$)	71Ab	
	≅-8	4.4(25°,1)	≅54	10-40	T	μ=2 Data recalculated from those given in 54Db	67Ab 54Db	b
	0.402	3.442(2)	67.4	25	C	μ=1.0($NaClO_4$)	71Ab	
	0.251	2.437(3)	47.7	25	C	μ=1.0($NaClO_4$)	71Ab	
	-2.058	1.477(4)	21.38	25	C	μ=1.0($NaClO_4$)	71Ab	
Mn^{2+}	15.1	1.000(1)	69.6	25	C	μ=3.0M($NaClO_4$& NaF) pH=3-5	76Km	
Fe^{3+}	14.2	6.03(25°,1)	163	15-35	T	μ=0	67Ab	b
	9.83	5.17(1)	131.8	25	C	μ=0.5	67Ab	
	13.8	5.210(25°,1)	146.0	0-25	T	μ=0.5($NaClO_4$)	53Ha	b
	3.3	3.949(25°,2)	87.0	0-25	T	μ=0.5($NaClO_4$)	53Ha	b
	-2.72	2.26	34.3	25	T	μ=0.5 R:Fe^{3+}+HF=FeF^{2+}+H^+	56Cd	
	-5.0	1.01	2.5	25	T	μ=0.5 R:FeF^{2+}+HF=$FeF_2{}^+$+H^+	56Cd	
[Fe^{3+}]	-6.3	1.46(1)	7.5	25	-	Ferrimyglobin	55Ga 60Gb	
[Fe^{3+}]	-10.5	1.76(1)	-2.5	25	-	Ferrihaemoglobin	55Ga 60Gb	
[Fe^{3+}]	ΔH, obtained by temperature variation of K, is reported in graphical form as a function of pH in the reactions: Hb(H_2O)+L=HbL+H_2O					Dog methaemoglobin	69Be	
[Fe^{3+}]	ΔH, obtained by temperature variation of K, is reported in graphical form as a function of pH in the reactions: Hb(H_2O)+L=HbL+H_2O					Guinea Pig methaemoglobin	69Be	

Metal Ion	ΔH, kJ /mol	Log K	ΔS, J /mol K	T, °C	M	Conditions	Ref.	Remarks
F11, continued								
[Fe^{3+}]	-5.77	2.01(1)	18.8	20	T	Human methaemoglobin A 6.0	68Ab 69Be	d:0.05
	-9.12	1.94(1)	6.3	20	T	6.5	68Ab	
	-12.09	1.86(1)	-5.4	20	T	7.0	68Ab	
	-9.96	1.77(1)	0	20	T	7.5	68Ab	
	-8.12	1.69(1)	4.6	20	T	8.0	68Ab	
	-5.77	1.68(1)	12.6	20	T	8.5	68Ab	
	-5.19	1.62(1)	13.4	20	T	9.0	68Ab	
[Fe^{3+}]	-5.27	2.13(1)	22.84	20	T	Human methaemoglobin C 6.0	68Ab 69Be	d:0.05
	-8.49	2.06(1)	10.5	20	T	6.5	68Ab	
	-11.17	1.99(1)	0	20	T	7.0	68Ab	
	-13.43	1.89(1)	-9.6	20	T	7.5	68Ab	
	-11.17	1.78(1)	-4.6	20	T	8.0	68Ab	
	-7.20	1.67(1)	7.5	20	T	8.5	68Ab	
	-4.18	1.60(1)	15.86	20	T	9.0	68Ab	
[Fe^{3+}]	ΔH, obtained by temperature variation of K, is reported in graphical form as a function of pH in the reactions: $Hb(H_2O)+L=HbL+H_2O$					Pigeon methaemoglobin	69Be	
[Fe^{3+}]	ΔH, obtained by temperature variation of K, is reported in graphical form as a function of pH in the reactions: $Hb(H_2O)+L=HbL+H_2O$					Sperm whale Metmyoglobin	69Be	
[Fe^{3+}]	-2.9	1.90(1)	26.8	25	-	Chironomus hemoglobin	60Gb	
Co^{2+}	13.8	0.643(1)	58.6	25	C	μ=3.0M($NaClO_4$& NaF) pH=3-5	76Km	
Ni^{2+}	5.8	0.763(1)	33.5	25	C	μ=3.0M($NaClO_4$& NaF) pH=3-5	76Km	
Cu^{2+}	3.8	0.70(1)	25	25	T	μ=0.5	58Cd	
	13.4	1.057(1)	64.8	25	C	μ=3.0M($NaClO_4$& NaF) pH=3-5	76Km	
Ag^{+}	-11.7	-0.167(1)	-42	25	T	μ=0.50	61Cc	q:±4.2
Zn^{2+}	15.90	1.16(1)	75.3	25	C	μ=0	71Ce	
	6.3	0.73(1)	33	25	T	μ=0.5	67Ab	
	8.33	0.748(1)	42.2	25	C	μ=1M($NaClO_4$)	69Gc	
	7.66	0.845(1)	41.8	25	C	μ=3M($NaClO_4$)	69Gc	
	9.2	0.875(1)	48.1	25	C	μ=3.0M($NaClO_4$& NaF) pH=3-5	76Km	
Cd^{2+}	4.27	0.568(1)	25.1	25	C	μ=3($NaClO_4$)	67Ab 66Ga	
Hg^{2+}	≅3.8	1.02(1)	≅33	25	T	μ=0.5	67Ab	
Al^{3+}	4.44	6.61(1)	141.4	25	C	μ=0.07	59Ka	
	4.812	6.14(1)	134	25	C	μ=0.06-0.2	53La	
	3.10	6.114(1)	127.6	25	C	μ=1.0M($NaClO_4$)	71Wa	q:±0.33
	3.85	5.36(2)	115.5	25	C	μ=0.07	59Ka	
	32.64	5.02(2)	109	25	C	μ=0.06-0.2	53La	

Metal Ion	ΔH, kJ /mol	Log *K*	ΔS, J /mol K	T,°C	M	Conditions	Ref.	Re-marks
F11, continued								
Al^{3+}, cont.	0.75	4.06(3)	80.3	25	C	μ=0.07	59Ka	
	0.795	3.85(3)	75	25	C	μ=0.06-0.2	53La	
	0.17	2.68(4)	51.9	25	C	μ=0.07	59Ka	
	1.172	2.74(4)	12	25	C	μ=0.06-0.2	53La	
	-1.51	1.33(5)	20.5	25	C	μ=0.07	59Ka	
	-3.138	1.63(5)	21	25	C	μ=0.06-0.2	53La	
	-6.485	0.469(6)	-12	25	C	μ=0.06-0.2	53La	
	0.418	19.842(1-6)	381	25	C	μ=0.06-0.2	53La	
Ga^{3+}	10.50	4.362(1)	118.8	25	C	μ=1.0M($NaClO_4$)	71Wa	
In^{3+}	9.79	4.44(1)	122	15	C	μ=0	74Vb	
	10.88	4.68(1)	126	25	C	μ=0	74Va	
						ΔCp=134	74Vb	
	14.6	4.62(1)	138	25	T	μ=0	67Ab	
	12.47	4.77(1)	131	35	C	μ=0	74Vb	
	8.49	3.56(1)	100	15	C	μ=0.5($NaClO_4$)	74Vb	
	10.0	3.74(1)	105	25	T	μ=0.5	67Ab	
	9.29	3.75(1)	103	25	C	μ=0.5($NaClO_4$)	74Va	
						ΔCp=96	74Vb	
	10.29	3.83(1)	106	35	C	μ=0.5($NaClO_4$)	74Vb	
	8.25	3.52(1)	98.3	15	C	μ=1.0($NaClO_4$)	74Vb	
	9.20	3.69(1)	101.2	25	C	μ=1.00M($NaClO_4$)	69Rd	
	12.47	3.729(1)	113.8	25	C	μ=1.0M($NaClO_4$)	71Wa	q:±0.63
	8.95	3.71(1)	101	25	C	μ=1.0($NaClO_4$)	74Va	
						ΔCp=105	74Vb	
	10.42	3.75(1)	105	35	C	μ=1.0($NaClO_4$)	74Vb	
	8.03	3.58(1)	98.7	15	C	μ=2.0($NaClO_4$)	74Vb	
	8.74	3.74(1)	101	25	C	μ=2.0($NaClO_4$)	74Va	
						ΔCp=71	74Vb	
	9.41	3.82(1)	103	35	C	μ=2.0($NaClO_4$)	74Vb	
	10.7	3.27(2)	102	15	C	μ=0	74Vb	
	12.4	3.47(2)	108	25	C	μ=0	74Va	
						ΔCp=100	74Vb	
	12.7	3.56(2)	109	35	C	μ=0	74Vb	
	9.04	2.70(2)	84.9	15	C	μ=0.5($NaClO_4$)	74Vb	
	10.5	2.86(2)	90.0	25	C	μ=0.5($NaClO_4$)	74Va	
						ΔCp=84	74Vb	
	10.9	2.93(2)	91.2	35	C	μ=0.5($NaClO_4$)	74Vb	
	8.20	2.68(2)	81.6	15	C	μ=1.0($NaClO_4$)	74Vb	
	7.66	2.83(2)	79.9	25	C	μ=1.00M($NaClO_4$)	69Rd	
	8.28	2.83(2)	82.0	25	C	μ=1.0($NaClO_4$)	74Va	
						ΔCp=38	74Vb	
	8.95	2.90(2)	84.1	35	C	μ=1.0($NaClO_4$)	74Vb	
	6.65	2.74(2)	77.4	15	C	μ=2.0($NaClO_4$)	74Vb	
	6.32	2.84(2)	75.6	25	C	μ=2.0($NaClO_4$)	74Va	
						ΔCp=54	74Vb	
	7.78	2.97(2)	81.6	35	C	μ=2.0($NaClO_4$)	74Vb	
	6.78	2.04(3)	64.0	15	C	μ=0	74Vb	
	6.23	2.16(3)	62.3	25	C	μ=0	74Va	
						ΔCp=58	74Vb	
	7.86	2.20(3)	67.4	35	C	μ=0	74Vb	
	6.28	1.88(3)	59.0	15	C	μ=0.5($NaClO_4$)	74Vb	
	5.48	1.99(3)	56.5	25	C	μ=0.5($NaClO_4$)	74Va	
						ΔCp=8.4	74Vb	

Metal Ion	ΔH, kJ /mol	Log K	ΔS, J /mol K	T,°C	M	Conditions	Ref.	Re-marks
F11, continued								
In^{3+}, cont.	6.15	2.06(3)	59.0	35	C	μ=0.5($NaClO_4$)	74Vb	
	5.52	2.01(3)	59.0	15	C	μ=1.0($NaClO_4$)	74Vb	
	13.8	2.11(3)	86.6	25	C	μ=1.00M($NaClO_4$)	69Rd	
	6.86	2.10(3)	63.2	25	C	μ=1.0($NaClO_4$)	74Va	
						ΔCp=100	74Vb	
	7.32	2.18(3)	65.3	35	C	μ=1.0($NaClO_4$)	74Vb	
	5.52	2.30(3)	64.8	15	C	μ=2.0($NaClO_4$)	74Vb	
	6.48	2.41(3)	67.8	25	C	μ=2.0($NaClO_4$)	74Va	
						ΔCp=4.2	74Vb	
	5.61	2.46(3)	64.8	35	C	μ=2.0($NaClO_4$)	74Vb	
	6.86	1.19(4)	47.3	15	C	μ=0	74Vb	
	8.54	1.26(4)	52.7	25	C	μ=0	74Va	
						ΔCp=113	74Vb	
	9.20	1.32(4)	55.2	35	C	μ=0	74Vb	
	7.86	1.20(4)	51.0	15	C	μ=0.5($NaClO_4$)	74Vb	
	9.58	1.27(4)	56.5	25	C	μ=0.5($NaClO_4$)	74Va	
						ΔCp=159	74Vb	
	11.3	1.29(4)	61.5	35	C	μ=0.5($NaClO_4$)	74Vb	
	9.37	1.19(4)	56.1	15	C	μ=1.0($NaClO_4$)	74Vb	
	9.58	1.27(4)	56.5	25	C	μ=1.0($NaClO_4$)	74Va	
						ΔCp=50	74Vb	
	9.54	1.31(4)	56.1	35	C	μ=1.0($NaClO_4$)	74Vb	
	11.38	0.712(4)	53.6	15	C	μ=2.0($NaClO_4$)	74Vb	
	11.6	1.29(4)	63.6	25	C	μ=2.0($NaClO_4$)	74Va	
						ΔCp=71	74Vb	
	13.0	1.32(4)	67.8	35	C	μ=2.0($NaClO_4$)	74Vb	
	-2.125	0.839	8.8	25	T	μ=0.5 R:In^{3+}+HF=InF_2^++H^+	54Ha	
	4.2	-0.30	8.4	25	T	μ=0.5 R:InF^{2+}+HF=InF_2^++H^+	54Ha	
SiO_2	-153.7	5.29(53°)	-369	53&72	T	0-6M($HClO_4$) CM=0.54-0.62M CL=2M R:SiO_2+4HL=$SiL_4(H_2O)_2$	72Bk	a,b
Sn^{2+}	43.30	8.415(25°,1-3)	306.3	25-60	T	μ=0.85	68Ha	b
F12 FORMIC ACID			CH_2O_2			HCO_2H		L^-
Mg^{2+}	-7.40	1.42(25°,1)	2.5	25-35	T	μ=0	56Na	b
Ca^{2+}	4.10	1.42(25°,1)	42	25-35	T	μ=0	56Na	b
Sr^{2+}	2.47	1.38(25°,1)	34.7	25-35	T	μ=0	56Na	b
Ba^{2+}	-7.91	1.37(25°,1)	-1.2	25-35	T	μ=0	56Na	b
Cr^{2+}	13.22	1.928(25°,1)	79	25&35	T	μ=1.0($HClO_4$)	73Ta	a,b,y:ΔS
	11.38	2.000(35°,1)	75	35&50	T	μ=1.0($HClO_4$)	73Ta	a,b,y:ΔS
	5.52	0.674(25°,2)	33	25&35	T	μ=1.0($HClO_4$)	73Ta	a,b,y:ΔS
	19.66	0.709(35°,2)	75	35&50	T	μ=1.0($HClO_4$)	73Ta	a,b,y:ΔS
	26.86	1.283(25°,3)	113	25&35	T	μ=1.0($HClO_4$)	73Ta	a,b,y:ΔS
	-26.86	1.433(35°,3)	-58	35&50	T	μ=1.0($HClO_4$)	73Ta	a,b,y:ΔS

Metal Ion	ΔH, kJ /mol	Log K	ΔS, J /mol K	T,°C	M	Conditions	Ref.	Remarks
F12, continued								
Mn^{3+}	49.70	-0.046(49.8°)	154.0	40.8-59.6	T	μ=4.0M R:Mn^{3+}+HL=MnL^{2+}+H^+	72Wa	b
[Fe^{3+}]	-26.4	1.46(1)	-61.1	25	-	Hemoglobin	60Gb	
[Fe^{3+}]	-26.36	1.51(1)	-61.1	20	T	Methemoglobin	55Sa	
CoA_5B^{2+}	13.4	-0.48(1)	35.6	25	C	μ=3.0M($NaClO_4$) A=NH_3 B=Br	74Mb	
	13.0	-0.59(1)	33	25	C	μ=3.0M($NaClO_4$) A=NH_3 B=Cl	74Mb	
	-13.4	-0.21(2)	-48.1	25	C	μ=3.0M($NaClO_4$) A=NH_3 B=Br	74Mb	
	-13.0	-0.15(2)	-46	25	C	μ=3.0M($NaClO_4$) A=NH_3 B=Cl	74Mb	
F13 FORMIC ACID, amide, N,N-dimethyl-		C_3H_7ON				$HCON(CH_3)_2$		L
[Fe(II)]	-22.2	0.53(25°,1)	-63	7-37	T	S:Benzene [Fe(II)]=Ferrous deuteroporphyrin dimethyl ester	74Bh	b,q:±2.5
[Co]	-33.0	1.63(0°,1)	-88	-23-23	T	S:Toluene [Co]=Cobalt(II) protoporphyrin IX dimethyl ester	73Sx	b,q: ±2.5
NiA_2B	-15.5	-1.041(22°,1)	-71	-50-50	T	S:N,N-Dimethylformamide (L) A=C_2H_5 B=2,2'-Bipyridine	76Yc	b
NiA	-35.37	1.434(25°,1-2)	-91.2	5-110	T	S:N,N-Dimethylformic acid amide (L) A=1,4,8,11-Tetraazacyclotetradecane	79Vb	
IrA_2BC_2	-74.9	1.42(1)	-226	25	C	S:CH_2Cl_2 A=Cl B=C(O)CF_3 C=P$(C_6H_5)_3$	77La	
I_2	-15.5	0.462(25°,1)	-43.5	8.6-28.2	T	CM=5.5x10^{-3}M CL=4.9x10^{-3}M S:CCl_4	62Dc	b
	-16.86	0.158(10°,1)	-56.5	0-20	T	CM=5x10^{-4}M CL=0.2-0.7M S:CH_2Cl_2	61Ta	b
F14 FORMIC ACID, amide, N,N-dimethyl-thio		C_3H_7ONS				$HSCON(CH_3)_2$		
$Rh_2A_4B_2$	-50.2	--	--	24	C	S:Benzene R:$Rh_2A_4B_2$+2L=	76Pd	

Metal Ion	ΔH, kJ /mol	Log K	ΔS, J /mol K	T, °C	M	Conditions	Ref.	Re-marks
F14, continued								
$Rh_2A_4B_2$, cont.						$2RhA_2BL$ A=CO B=Cl		
F15 FORMIC ACID, amide, thio- N,N-dimethyl-			C_3H_7NS			$HC(S)N(CH_3)_2$		
$Rh_2A_4B_2$	-50.2	--	--	--	C	S:Benzene R:$Rh_2A_4B_2$+2L= $2RhA_2BL$ A=CO B=Cl	76Pd	u:24°
PdAB	-49.4	--	--	24	C	S:Benzene R:$[PdAB]_2$+2L= 2[PdABL] A=(π-MeAllyl) B=Cl	77Lb	
F16 FORMIC ACID, ethyl ester			$C_3H_6O_2$			$HCO_2C_2H_5$		
$[NiA_2]_2$	-17.4	1.05(100°,1)	-26	75-125	T	S:Squalane A=3-Trifluoro-acetylcamphorate	74Sj	b,gg
F17 FORMIC ACID, methyl ester			$C_2H_4O_2$			HCO_2CH_3		
$[NiA_2]_2$	-15.7	1.05(100°,1)	-22	75-125	T	S:Squalane A=3-Trifluoro-acetylcamphorate	74Sj	b,gg
F18 FRUCTOSE			$C_6H_{12}O_6$			$CH_2(CHOH)_3COHCH_2OH$ O		L
[B(III)]	-12.9	3.47(25°,1)	-23.4	0-35	T	CM=0.0272M Cl=0.01-0.6M $[B(III)]=HB_4O_7^-$	70Ci	b
	-8.3	1.62(25°,2)	-2.9	0-35	T	CM=0.0272M Cl=0.01-0.6M $[B(III)]=HB_4O_7^-$	70Ci	b
$H_5TeO_6^-$	-11.3	2.46	8	25	T	μ=0.1 R:$H_5TeO_6^-$+ L=$HTeO_4L^-$+$2H_2O$	62Ea	
F19 FUMARIC ACID			$C_4H_4O_4$			$HO_2CCH:CHCO_2H$		
La^{3+}	11.25	2.74(1)	90.4	25	C	μ=0.10M($NaClO_4$)	73Cc	
Ce^{3+}	13.22	2.80(1)	97.9	25	C	μ=0.10M($NaClO_4$)	73Cc	
Pr^{3+}	13.56	2.84(1)	99.6	25	C	μ=0.10M($NaClO_4$)	73Cc	
Nd^{3+}	15.36	2.74(1)	103.8	25	C	μ=0.10M($NaClO_4$)	73Cc	
Sm^{3+}	14.48	2.83(1)	102.5	25	C	μ=0.10M($NaClO_4$)	73Cc	
Eu^{3+}	14.39	2.86(1)	102.9	25	C	μ=0.10M($NaClO_4$)	73Cc	

Metal Ion	ΔH, kJ /mol	Log K	ΔS, J /mol K	T, °C	M	Conditions	Ref.	Re-marks
F19, continued								
Gd^{3+}	16.07	2.88(1)	108.4	25	C	μ=0.10M($NaClO_4$)	73Cc	
Tb^{3+}	15.23	2.77(1)	104.2	25	C	μ=0.10M($NaClO_4$)	73Cc	
Dy^{3+}	15.94	2.80(1)	106.7	25	C	μ=0.10M($NaClO_4$)	73Cc	
Ho^{3+}	15.10	2.80(1)	104.2	25	C	μ=0.10M($NaClO_4$)	73Cc	
Er^{3+}	16.23	2.80(1)	107.9	25	C	μ=0.10M($NaClO_4$)	73Cc	
Tm^{3+}	14.56	2.81(1)	102.5	25	C	μ=0.10M($NaClO_4$)	73Cc	
Yb^{3+}	15.90	2.80(1)	106.7	25	C	μ=0.10M($NaClO_4$)	73Cc	
Lu^{3+}	15.94	2.81(1)	107.1	25	C	μ=0.10M($NaClO_4$)	73Cc	
F20 FURAN			C_4H_4O			CH:CHCH:CHO (ring)		
$[NiA_2]_2$	-7.3	-0.30(100°,1)	-25	75-125	T	S:Squalane A=3-Trifluoro-acetylcamphorate	74Sj	b,gg
F21 FURAN, 2,5-dimethyl-tetrahydro-			$C_6H_{12}O$			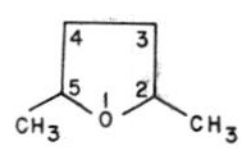		
AlA_3	-96.02	(1)	--	25-28	C	S:n-Hexane A=CH_3	68Hc	
F22 FURAN, 2-methyl-tetrahydro-			C_5H_{10}			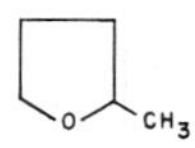		
AlA_3	-95.98	(1)	--	25-28	C	S:n-Hexane A=CH_3	68Hc	
I_2	-25.9	1.35(25°,1)	-61.1	1-25	T	S:n-Heptane	60Bb 60Ta	b
F23 FURAN, tetrahydro-			C_4H_8O			$CH_2CH_2CH_2CH_2O$ (ring)		L
$TiAB_3$	13.4	-0.46(1-2)	-8.4	-18	T	S:Toluene A=CH_3 B=Cl	69Ta	b
[Fe(II)]	-28.4	0.72(25°,1)	-79	7-37	T	S:Benzene [Fe(II)]=Ferrous deuteroporphyrin dimethyl ester	74Bh	b:q±4
$[Co^{3+}]$	-66.1	--	--	24	C	S:Dichloromethane [Co]=Methylco-baloxime R:$[Co]_2$+2L=2[Co]L	73Cf	
NiA_2B	-15.5	-1.638(22°,1)	-84	-50-50	T	S:Tetrahydrofuran (L) A=C_2H_5 B=2,2'-Bipyridine	76Yc	b
$[NiA_2]_2$	-23	2.44(100°,1)	-15	75-125	T	S:Squalane A=3-Trifluoro-acetylcamphorate	74Sj	b,gg

Metal Ion	ΔH, kJ /mol	Log K	ΔS, J /mol K	T,°C	M	Conditions	Ref.	Re-marks
F23, continued								
CuA_2	-17	0.732(1)	-43	20	T	S:Benzene A=Trifluoro- acetylacetone	79Si	b,q:±3
	-38.1	(1)	--	--	C	S:CCl_4 A=Hexafluoro- acetylacetonato	76Mc	u:25°
	-36.0	3.54(1)	--	--	C	S:o-$Cl_2C_6H_4$ A=Hexafluoro- acetylacetonato	76Mc	u:25°
	-14.2	1.38(0°,1)	-25.5	-60-60	T	S:Toluene A=Thenoyltri- fluoroacetonato	78Mb	
HgA_2	-15.10	0.78(1)	-34.98	30	C	S:Benzene A=Cl CM=1-5x10^{-3}M	73Fa	q:±2.5
	-12.51	0.70(1)	-27.99	30	C	S:Benzene A=Br CM=1-5x10^{-3}M	73Fa	q:±2.5
AlA_3	-95.81	(1)	--	25-28	C	S:n-Hexane A=CH_3	68Hc	
SbA_5	-80.8	(1)	--	25	C	S:CCl_4 A=Cl	73Oc	
I_2	-22.2	1.25(25°,1)	-48.5	6-25	T	S:n-Heptane	60Bb 60Ta	b
F24 2-FURANCARBOXYLIC ACID			$C_5H_4O_3$			$C_4H_3O(CO_2H)$		L^-
La^{3+}	6.19	1.478(1)	49.8	25	C	μ=2.0M($NaClO_4$)	76Ye	
Pr^{3+}	6.48	1.590(1)	52.3	25	C	μ=2.0M($NaClO_4$)	76Ye	
Nd^{3+}	6.99	1.605(1)	54.0	25	C	μ=2.0M($NaClO_4$)	76Ye	
Sm^{3+}	6.40	1.675(1)	53.6	25	C	μ=2.0M($NaClO_4$)	76Ye	
Eu^{3+}	6.32	1.673(1)	53.1	25	C	μ=2.0M($NaClO_4$)	76Ye	
Gd^{3+}	6.32	1.640(1)	52.7	25	C	μ=2.0M($NaClO_4$)	76Ye	
Tb^{3+}	7.32	1.599(1)	55.2	25	C	μ=2.0M($NaClO_4$)	76Ye	
Dy^{3+}	8.87	1.514(1)	58.6	25	C	μ=2.0M($NaClO_4$)	76Ye	
Ho^{3+}	8.66	1.504(1)	57.7	25	C	μ=2.0M($NaClO_4$)	76Ye	
Er^{3+}	9.41	1.450(1)	59.4	25	C	μ=2.0M($NaClO_4$)	76Ye	
Tm^{3+}	8.95	1.476(1)	58.2	25	C	μ=2.0M($NaClO_4$)	76Ye	
Yb^{3+}	8.95	1.470(1)	58.2	25	C	μ=2.0M($NaClO_4$)	76Ye	
Lu^{3+}	9.75	1.449(1)	60.7	25	C	μ=2.0M($NaClO_4$)	76Ye	
F25 FURFURAL, oxime(syn)			$C_5H_5O_2N$			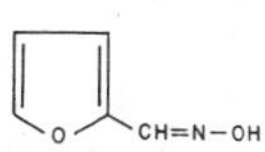		L
Cr^{3+}	-22.38	10.64(25°,1)	128.62	15-35	T	μ=0 S:70v% Dioxane	63Aa	b
	-17.03	10.04(25°,2)	135.06	15-35	T	μ=0 S:70v% Dioxane	63Aa	b

Metal Ion	ΔH, kJ /mol	Log K	ΔS, J /mol K	T, °C	M	Conditions	Ref.	Remarks
F25, continued								
Cr^{3+}, cont.	-24.31	9.42(25°,3)	98.83	15-35	T	μ=0 S:70v% Dioxane	63Aa	b
Fe^{3+}	-95.73	12.64(25°,1)	-79.20	15-35	T	μ=0 S:70v% Dioxane	63Aa	b
	-216.31	10.80(25°,2)	-519	15-35	T	μ=0 S:70v% Dioxane	63Aa	b
	-478.23	8.52(25°,3)	-1,439	15-35	T	μ=0 S:70v% Dioxane	63Aa	b
Ni^{2+}	-85.19	7.60(25°,1)	-140.6	15-35	T	μ=0 S:70v% Dioxane	63Aa	b
	-45.94	7.52(25°,2)	-10.12	15-35	T	μ=0 S:70v% Dioxane	63Aa	b
Cu^{2+}	-41.25	10.28(25°,1)	58.41	15-35	T	μ=0 S:70v% Dioxane	63Aa	b
	-58.49	9.88(25°,2)	-7.15	15-35	T	μ=0 S:70v% Dioxane	63Aa	b

G

Metal Ion	ΔH, kJ /mol	Log K	ΔS, J /mol K	T, °C	M	Conditions	Ref.	Remarks
G1 GALACTOSE			$C_6H_{12}O_6$			$CH_2OHCH(CHOH)_3CHOH$ (CH and CHOH linked through O)		L
[B(III)	-12.8	2.24(25°,1)	0	0-35	T	CM=0.0272M CL=0.01-0.6M $[B(III)]=HB_4O_7^-$	70Ci	b
	-0.9	0.39(25°,2)	-4.2	0-35	T	CM=0.0272M CL=0.01-0.6M $[B(III)]=HB_4O_7^-$	70Ci	b
$H_5TeO_6^-$	-5.8	1.66(1)	12	25	T	μ=0 R:$H_5TeO_6^-$+L =$HTeO_4L^-$+$2H_2O$	62Ea	
G2 GALLION								
Y^{3+}	-22.2	19.30(25°,1-2)	293	25-45	T	μ=0.1M($NaClO_4$)	74Pb	b
G3 GLUCONIC ACID			$C_6H_{12}O_7$			$HOCH_2CH(OH)CH(OH)CH-(OH)CH(OH)CO_2H$		L^-
La^{3+}	-42	10.23(1-2)	50	25	T	--	72Ya	q:±4,y, ee,ff
	-50	9.84(1-2)	29	40	T	--	72Ya	q:±4,y, ee,ff

Metal Ion	ΔH, kJ /mol	Log K	ΔS, J /mol K	T,°C	M	Conditions	Ref.	Re-marks
G3, continued								
La^{3+}, cont.	-54	9.57(1-2)	17	50	T	--	72Ya	q:±4,y, ee,ff
Pr^{3+}	-8	9.22(1-2)	142	25	T	--	72Ya	q:±4,y, ee,ff
	-50	8.97(1-2)	12	40	T	--	72Ya	q:±4,y, ee,ff
	-71	8.64(1-2)	-50	50	T	--	72Ya	q:±4,y, ee,ff
Sm^{3+}	-8	8.82(1-2)	113	25	T	--	72Ya	q:±4,y, ee,ff
	-88	8.48(1-2)	-113	40	T	--	72Ya	q:±4,y, ee,ff
	-138	7.85(1-2)	-276	50	T	--	72Ya	q:±4,y, ee,ff
Eu^{3+}	-25	8.65(1-2)	79	25	T	--	72Ya	q:±4,y, ee,ff
	-88	8.17(1-2)	-125	40	T	--	72Ya	q:±4,y, ee,ff
	-130	7.60(1-2)	-259	50	T	--	72Ya	q:±4,y, ee,ff
Gd^{3+}	-46	8.60(1-2)	96	25	T	--	72Ya	q:±4,y, ee,ff
	-79	8.08(1-2)	-96	40	T	--	72Ya	q:±4,y, ee,ff
	-100	7.62(1-2)	-163	50	T	--	72Ya	q:±4,y, ee,ff
Tb^{3+}	-71	8.20(1-2)	-71	25	T	--	72Ya	q:±4,y, ee,ff
	-71	7.6(1-2)	-79	40	T	--	72Ya	q:±4,y, ee,ff
	-67	7.24(1-2)	-75	50	T	--	72Ya	q:±4,y, ee,ff
Dy^{3+}	-75	7.97(1-2)	-109	25	T	--	72Ya	q:±4,y, ee,ff
	-71	7.35(1-2)	-84	40	T	--	72Ya	q:±4,y, ee,ff
	-67	7.00(1-2)	-67	50	T	--	72Ya	q:±4,y, ee,ff
Er^{3+}	-63	7.25(1-2)	-71	25	T	--	72Ya	q:±4,y, ee,ff
	-50	6.77(1-2)	-29	40	T	--	72Ya	q:±4,y, ee,ff
	-46	6.52(1-2)	-12	50	T	--	72Ya	q:±4,y, ee,ff
Yb^{3+}	-58	6.85(1-2)	-58	25	T	--	72Ya	q:±4,y, ee,ff
	-54	6.38(1-2)	-54	40	T	--	72Ya	q:±4,y, ee,ff
	-54	6.10(1-2)	-50	50	T	--	72Ya	q:±4,y, ee,ff

Metal Ion	ΔH, kJ /mol	Log K	ΔS, J /mol K	T,°C	M	Conditions	Ref.	Remarks
G3, continued								
Lu^{3+}	-38	6.77(1-2)	12	25	T	--	72Ya	q:±4,y, ee,ff
	-58	6.38(1-2)	-58	40	T	--	72Ya	q:±4,y, ee,ff
	-71	6.05(1-2)	-100	50	T	--	72Ya	q:±4,y, ee,ff
G4 GLUCOSE			$C_6H_{12}O_6$			$CH_2OHCH(CHOH)_3CHOH$ (ring: –O–)		L
BA_4^-	-14.73	2.130(25°,1)	-8.4	0-35	T	CL=0-0.9M A=OH^-	67Cb	b,q:±1.7
	-0.63	2.940(25°,1-2)	54	0-35	T	CL=0-0.9M A=OH^-	67Cb	b,q: ±0.08
[B(III)]	-20.6	2.18(25°,1)	-27.2	0-35	T	CM=0.0272M CL=0.01-0.6M [B(III)]=$HB_4O_7^-$	70Ci	b
	5.3	0.70(25°,2)	31.4	0-35	T	CM=0.0272M CL=0.01-0.6M [B(III)]=$HB_4O_7^-$	70Ci	b
$H_5TeO_6^-$	-2.9	1.11(1)	12	25	T	μ=0.1 R:$H_5TeO_6^-$ +L=$HTeO_4L^-$ +$2H_2O$	62Ea	
G5 GLUCOSE, 2-amino-(D,β)			$C_6H_{13}O_5N$					
Co^{2+}	-12.6	4.35(1)	40.6	25	C	μ=0.05(KCl)	72Gd 72Gc	
	-29.3	8.04(1-2)	54	25	C	μ=0.05M(KCl)	72Gd 72Gc	
Ni^{2+}	-17.6	5.39(1)	44.8	25	C	μ=0.05M(KCl)	72Gd 72Gc	
	-36.0	9.85(1-2)	67	25	C	μ=0.05M(KCl)	72Gd 72Gc	
Cu^{2+}	-27.6	8.07(1)	40.6	25	C	μ=0.05M(KCl)	72Gd 72Gc	
	-53.6	14.76(1-2)	109	25	C	μ=0.05M(KCl)	72Gd 72Gc	
Zn^{2+}	-7.5	4.63(1)	62.8	25	C	μ=0.05M(KCl)	72Gd 72Gc	
	-22.6	8.64(1-2)	92	25	C	μ=0.05M(KCl)	72Gc	
G6 GLUCOSE, 1-phosphate			$C_6H_{13}O_9P$			$CH_2OHCH(CHOH)_3CHOPO_3H_2$ (ring: –O–)		L
Mg^{2+}	12.188	2.479(1)	88.3	25	T	μ=0 ΔCp=247	54Cd	
	15.230	2.678(1)	98.3	37	T	μ=0 ΔCp=257.3	54Cd	
Ca^{2+}	10.054	2.495(1)	81.6	25	T	μ=0 ΔCp=264	56Cc	
	13.263	2.573(1)	92.0	37	T	μ=0 ΔCp=272	56Cc	

Metal Ion	ΔH, kJ /mol	Log K	ΔS, J /mol K	T,°C	M	Conditions	Ref.	Re-marks
G7 GLUTAMIC ACID (D)			$C_5H_9O_4$			$HO_2CCH_2CH_2CH(NH_2)CO_2H$		
Ce^{3+}	35.6	5.35(25°,1)	222	25&45	T	μ=0	74Sk	a,b
	27.2	5.28(25°,1)	192	25&45	T	μ=0.1M(KNO_3)	74Sk	a,b
	58.2	4.84(25°,2)	289	25&45	T	μ=0	74Sk	a,b
	77.0	4.47(25°,2)	343	25&45	T	μ=0.1M(KNO_3)	74Sk	a,b
Y^{3+}	-10.0	5.28(25°,1)	67	25&45	T	μ=0.1M(KNO_3)	74Sk	a,b
	-9.2	4.69(25°,2)	58	25&45	T	μ=0.1M(KNO_3)	74Sk	a,b
La^{3+}	-35.40	3.585(30°,1)	-48.2	30&40	T	μ=0.1M(KCl)	76Ab	a,b
Ce^{3+}	-29.96	3.795(30°,1)	-26.2	30&40	T	μ=0.1M(KCl)	76Ab	a,b
Pr^{3+}	-27.78	3.855(30°,1)	-17.86	30&40	T	μ=0.1(KCl)	78Aa	a,b
Nd^{3+}	-25.78	3.956(30°,1)	-9.49	30&40	T	μ=0.1(KCl)	78Aa	a,b
Sm^{3+}	-24.15	4.087(30°,1)	-1.44	30&40	T	μ=0.1(KCl)	78Aa	a,b
Ni^{2+}	-14.6	5.52(1)	59	25	C	μ=0.2(KCl)	75Gc	
	-30.92	9.81(2)	84.9	25	C	μ=0.1M	70Bd 71Be	y:ΔH (70Bd)
	-18.8	4.35(2)	21	25	C	μ=0.2(KCl)	75Gc	
Cu^{2+}	-47.66	14.1(2)	110.9	25	C	μ=0.1M	70Bd 71Be	y:ΔH (70Bd)
Zn^{2+}	15.5	3.86(2)	113	25	C	μ=0.2(KCl)	75Gc	
G8 GLUTAMIC ACID (L)			$C_5H_9O_4N$			$HO_2CCH_2CH_2CH(NH_2)CO_2H$		L^-
Ni^{2+}	-30.79	9.81(2)	84.1	25	C	μ=0.1M	70Bd 71Be	y:ΔH (70Bd)
Cu^{2+}	-47.32	14.1(2)	112.1	25	C	μ=0.1M	70Bd 71Be	y:ΔH (70Bd)
G9 GLUTAMIC ACID (DL)			$C_5H_9O_4N$			$HO_2CCH_2CH_2CH(NH_2)CO_2H$		L^{2-}
Ni^{2+}	-30.71	9.71(1-2)	82.8	25	C	μ=0.1M	70Bd 71Be	y:ΔH (70Bd)
Cu^{2+}	-20.9	8.27(1)	88	25	C	μ=0.2(KCl)	74Na	
	-28.0	6.47(2)	29	25	C	μ=0.2(KCl)	74Na	
	-47.86	14.1(1-2)	110.0	25	C	μ=0.1M	70Bd 71Be	y:ΔH (70Bd)
G10 GLUTAMIC ACID, N(2-hydroxy-benzyl)			$C_{12}H_{15}O_5N$			$C_6H_4(OH)[CH_2NHCH(CO_2H)-CH_2CH_2CO_2H]$		
$CoA_4{}^{3+}$	35.65	3.00(25°,1)	107.24	15-30	T	0.1M(Sodium acetate A=NH_3	79Mb	b
$CoA_5{}^{3+}$	28.12	1.51(25°,1)	123.97	15-30	T	0.1M(Sodium acetate A=NH_3	79Mb	b

Metal Ion	ΔH, kJ /mol	Log K	ΔS, J /mol K	T, °C	M	Conditions	Ref.	Remarks
G11 GLUTAMIC ACID, N(3-methoxy-4-hydroxybenzyl)		$C_{13}H_{17}O_6N$				$C_6H_2(OH)(OCH_3)[CH_2NHCH(CO_2H)-CH_2CH_2CO_2H]$		
CoA_4^{3+}	43.14	2.41(25°,1)	191.08	15-30	T	0.1M(Sodium acetate) A=NH_3	79Mb	b
CoA_5^{3+}	20.63	1.63(25°,1)	100.83	25-45	T	0.1M(Sodium acetate) A=NH_3	79Mb	b
G12 GLUTAMINE			$C_5H_{10}O_3N_2$			$HO_2CCH(NH_2)CH_2CH_2CONH_2$		
Ni^{+}	-14.2	5.08(1)	50	25	C	μ=0.2(KCl)	75Gc	
	-18.8	4.20(2)	17	25	C	μ=0.2(KCl)	75Gc	
Cu^{2+}	-23.6	7.62(1)	66.9	25	C	μ=0.2(KCl)	75Ge	
	-25.5	6.38(2)	36.4	25	C	μ=0.2(KCl)	75Ge	
Zn^{2+}	-6.86	5.199(1)	74.9	5	T	μ=0.1(KCl)	73Rb	q:±3.0
	-6.69	5.113(1)	75.3	25	T	μ=0.1(KCl)	73Rb	q:±3.0
	-6.19	5.042(1)	77.0	45	T	μ=0.1(KCl)	73Rb	q:±3.0
	15.9	3.66(2)	96	25	C	μ=0.2(KCl)	75Gc	
G13 GLUTAMINE SYNTHETASE FROM ESCHERICHIA COLI								
Mn^{2+}	-26.4	--	--	37	C	Hepes (N-2-hydroxyethylpiperazine-N'-2-ethanesulfonic acid) buffer pH=7.14	72Hf	
	-82.4	--	--	37	C	Tris-chloride buffer pH=7.19 R:$Mn^{2+}+H_2L=MnL+2H^+$	72Hf	
G14 GLUTATHIONE			$C_{10}H_{17}O_6N_3S$			$HO_2CCH(NH_2)CH_2CH_2CONHCH(CH_2SH)-CONHCH_2CO_2H$		
Pb^{2+}	-67.6	9.913(25°,1)	-36.8	10-40	T	μ=3.00M($NaClO_4$)	76Cf	q:±5
	-82.2	16.821(25°)	46.4	10-40	T	μ=3.00M($NaClO_4$) R:$Pb^{2+}+HL=PbHL^{2+}$	76Cf	
	-101.4	23.400(25°)	107.9	10-40	T	μ=3.00M($NaClO_4$) R:$Pb^{2+}+L^-+HL=PbHL_2^+$	76Cf	
	-143.5	32.313(25°)	137.6	10-40	T	μ=3.00M($NaClO_4$) R:$Pb^{2+}+2HL=PbHL_2^+$	76Cf	
G15 GLYCEROL			$C_3H_8O_3$			$HOCH_2CH(OH)CH_2OH$		L
BA_4^-	-11.92	1.146(25°,1)	-18.0	0-35	T	CL=0-0.9M A=OH^-	67Cb	b,q:±1.2
	-16.23	1.763(25°,1-2)	-20.5	0-35	T	CL=0-0.9M A=OH^-	67Cb	b,q:±1.7
$H_5TeO_6^-$	-15.9	1.90	-17	25	T	μ=0.1 R:$H_5TeO_6^-+L=HTeO_4L^-+2H_2O$	62Ea	

Metal Ion	ΔH, kJ /mol	Log K	ΔS, J /mol K	T,°C	M	Conditions	Ref.	Re-marks
G16	GLYCEROL, 2-phosphoric acid		$C_3H_9O_6P$			$HOCH_2CH(OPO_3H)CH_2OH$		
Mg^{2+}	14.380	2.486(1)	95.8	25	T	μ=0 ΔCp=56	54Cd	
	17.196	2.700(1)	105.0	37	T	μ=0 ΔCp=58	54Cd	
G17	GLYCINE		$C_2H_5O_2N$			$H_2NCH_2CO_2H$		L^-
Ce^{3+}	13.8	0.531(25°,1)	58	0-55	T	μ=2.0M($NaClO_4$) pH=3.64	68Ta	b,q: ±1.7
Pm^{3+}	14.6	0.672(25°,1)	63	0-55	T	μ=2.0M($NaClO_4$) pH=3.64	68Ta	b,q: ±1.7
Eu^{3+}	9.6	0.699(25°,1)	46	0-55	T	μ=2.0M($NaClO_4$) pH=3.64	68Ta	b,q: ±1.7
Am^{3+}	12.1	0.690(25°,1)	54	0-55	T	μ=2.0M($NaClO_4$) pH=3.64	68Ta	b,q: ±1.7
Cm^{3+}	13.8	0.806(25°,1)	63	0-55	T	μ=2.0M($NaClO_4$) pH=3.64	68Ta	b,q: ±1.7
Mn^{2+}	-1.7	3.23(1)	54	10	C	μ=0	72Ig	
	-1.2	3.21(1)	58	25	C	μ=0	72Ig	q:±0.4
	-1.21	3.17(1)	56.5	25	T	μ=0 ΔCp=67	64Bd	
	-0.17	3.15(1)	58	40	C	μ=0	72Ig	q:±0.54
				10-40	C	μ=0 ΔCp=50	72Ig	
	-21	2.96(1)	-12	30	T	μ=0.09(KCl)	57Mb	
	-5.8	2.60(25°,1)	31.0	10&25	T	μ=0.65(KCl)	64Lg	a,b
	-7.9	1.98(25°,2)	9.6	10&25	T	μ=0.65(KCl)	64Lg	a,b
	-16.7	5.36(25°)	46.4	10&25	T	μ=0.65(KCl) R:$Mn^{2+}+L^-+$ (Pyruvate)$^-$=MnL (Pyruvate)	64Lg	a,b
	-49.4	9.79(25°)	22.6	10&25	T	μ=0.65(KCl) R:$Mn^{2+}+2L^-+$ 2(Pyruvate)$^-$= $MnL_2(Pyruvate)_2^{2-}$	64Lg	a,b
Fe^{2+}	-15.5	4.36(1)	28.4	10	C	μ=0	72Ig	
	-15.23	4.31(1)	31.4	25	C	μ=0	72Ig	
	-14.85	4.28(1)	34.3	40	C	μ=0	72Ig	
				10-40	C	μ=0 ΔCp=29	72Ig	
Co^{2+}	-12.55	5.16(1)	54	10	C	μ=0	72Ig	
	-10.38	4.61(1)	53.6	25	C	μ=0	67Bd	
							66Na	
							64Bd	
							64Sd	
	-11.97	5.07(1)	56.9	25	C	μ=0	72Ig	
	-9.00	4.98(1)	66.5	40	C	μ=0	72Ig	
				10-40	C	μ=0 ΔCp=117	72Ig	
	-12.6	4.66(1)	46	25	C	μ=0.05M(KCl)	70Gc	
							71Gc	
	-8.4	4.76(15°,1)	62.3	15&40	T	μ=0.2	65Sb	a,b
	-16.74	4.07(2)	18.8	10	C	μ=0	72Ig	
	-14.81	4.02(2)	27.2	25	C	μ=0	72Ig	

Metal Ion	ΔH, kJ /mol	Log K	ΔS, J /mol K	T,°C	M	Conditions	Ref.	Remarks
G17, continued								
Co^{2+}, cont.	-10.67	3.97(2)	40.2	25	C	μ=0	67Bd	
	-14.27	3.91(2)	29.3	40	C	μ=0	72Ig	
				10-40	C	μ=0 ΔCp=84		
	-18.4	3.98(2)	12	25	C	μ=0.05M(KCl)	70Gc	
							71Gc	
	-15.1	3.56(15°,2)	15.5	15&40	T	μ=0.2	65Sb	a,b
	-15.19	2.67(3)	-2.5	10	C	μ=0	72Ig	
	-14.27	2.54(3)	0.8	25	C	μ=0	72Ig	
	-12.59	2.45(3)	6.7	40	C	μ=0	72Ig	
				10-40	C	μ=0 ΔCp=88		
	-27.6	8.4(1-2)	67.4	20	C	0.1M(KNO_3)	67Sc	
	-23.4	8.32(15°,1-2)	77.8	15&40	T	μ=0.2	65Sb	a,b
CoA	-20.9	4.03(25°,1)	7.5	15-70	T	μ=0.1M(KNO_3) A=Methylimino-diacetate	72Id	b
Ni^{2+}	-21.8	6.36(1)	44.8	10	C	μ=0	67Aa	
	-19.37	6.28(1)	51.9	10	C	μ=0	72Ig	
	-18.32	6.13(1)	56.1	25	C	μ=0	72Ig	
	-20.5	6.18(1)	49.8	25	C	μ=0	67Aa	
	-17.32	6.18(1)	60.2	25	C	μ=0	67Bd	
	-17.11	6.18(1)	60.7	25	T	μ=0 ΔCp=75	64Bd	
	-18.0	6.09(1)	18.4	40	C	μ=0	67Aa	
	-15.77	6.00(1)	64.4	40	C	μ=0	72Ig	
				10-40	C	μ=0 ΔCp=117	72Ig	
	-15.1	5.77(1)	58	25	C	μ=0.05M(KCl)	70Gc	
							71Gc	
	-20.9	5.79(1)	40.2	25	C	μ=0.1M(KNO_3)	71Ld	
	-17.2	5.73(25°,1)	54	20-30	T	μ=0.1M(KNO_3)	67Gb	b
	-21.0	5.65(1)	38	25	C	μ=0.2M(KCl)	74Sn	
	-18.0	6.04(15°,1)	53.1	15-40	T	μ=0.2	65Sb	b
	-7.9	5.66(25°,1)	82.8	10&25	T	μ=0.65(KCl)	64Lg	
	-17.24	23.822(1)	51.17	25	C	μ=1M($NaClO_4$)	79Ea	
	-22.18	5.14(2)	20.1	10	C	μ=0	72Ig	
	-24.3	5.29(2)	15.5	10	C	μ=0	67Aa	
	-20.79	4.92(2)	24.3	25	C	μ=0	72Ig	
	-19.7	5.07(2)	31.8	25	C	μ=0	67Aa	
	-19.62	4.95(2)	28.9	25	T	μ=0 ΔCp=117	64Bd	
	-19.7	4.92(2)	31.0	40	C	μ=0	67Aa	
	-19.2	4.76(2)	29.7	40	C	μ=0	72Ig	
				10-40	C	μ=0 ΔCp=96	72Ig	
	-20.1	4.88(2)	25	25	C	μ=0.05M(KCl)	70Gc	
							71Gc	
	-18.4	4.78(2)	29.3	25	C	μ=0.1M(KNO_3)	71Ld	
	-22.2	4.83(25°,2)	17	20-30	T	μ=0.1(KNO_3)	67Gb	b
	-20.9	4.75(2)	17	25	C	μ=0.2M(KCl)	74Sn	
	-18.4	4.98(15°,2)	25.1	15-40	T	μ=0.2	65Sb	b
	-23.4	4.85(25°,2)	12.6	10&25	C	μ=0.65(KCl)	64Lg	a,b
	-20.58	20.138(2)	23.09	25	C	μ=1M($NaClO_4$)	79Ea	
	-23.60	3.51(3)	-16.3	10	C	μ=0	72Ig	
	-23.22	3.18(3)	-17.2	25	C	μ=0	72Ig	
	-23.01	3.00(3)	-15.9	40	C	μ=0	72Ig	
				10-40	C	μ=0 ΔCp=21	72Ig	
	-22.64	14.402(3)	-10.02	25	C	μ=1M($NaClO_4$)	79Ea	

Metal Ion	ΔH, kJ /mol	Log K	ΔS, J /mol K	T,°C	M	Conditions	Ref.	Remarks
G17, continued								
Ni^{2+}, cont.	-36.4	(1-2)	--	24.6	C	μ=0.01M($HClO_4$)	72Sh	
	-36.8	10.5(1-2)	75.3	20.0	C	0.1M(KNO_3)	67Sc	
	-36.4	11.02(15°,1-2)	78.2	15-40	T	μ=0.2	65Sb	b
	-58.6	13.22(1-2)	58.6	25	T	μ=0.004 S:45% Dioxane	56Lb	
	-37.64	44.671	78.14	25	C	μ=1M($NaClO_4$) R:Ni^{2+}+L+A=$NiLA^{2+}$ A=α-Alanine	79Ea	
	-57.21	59.031	78.20	25	C	μ=1M($NaClO_4$) R:Ni^{2+}+L+2A= $NiLA_2{}^{2+}$ A=α-Alanine	79Ea	
	-59.94	59.659	71.92	25	C	μ=1M($NaClO_4$) R:Ni^{2+}+2L+A= NiL_2A^{2+} A=α-Alanine	79Ea	
	-10.9	8.09(25°)	119.2	10&25	T	μ=0.65(KCl) R:Ni^{2+}+L^-+ $(Pyruvate)^-$= NiL(Pyruvate)	64Lg	a,b
	-49.8	15.29(25°)	126.4	10&25	T	μ=0.65(KCl) R:Ni^{2+}+2L^-+ 2$(Pyruvate)^-$= $NiL_2(Pyruvate)_2{}^{2-}$	64Lg	a,b
NiA	-10.9	4.91(25°,1)	22.6	15-70	T	μ=0.1M(KNO_3) A=Methylimino- diacetate	72Id	b,q: ±1.2
	-28.4	4.34(25°,1)	-11.7	15-70	T	μ=0.1M(KNO_3) A=Triethylene- tetramine	72Id	b
Cu^{2+}	-26.07	8.85(1)	70.7	10	C	μ=0	72Ig	
	-30.46	8.85(1)	61.9	10	C	μ=0	66Aa 56Lb	
	-24.35	8.57(1)	82.4	25	C	μ=0	72Ig	
	-26.02	8.58(1)	77.0	25	C	μ=0	66Aa	
	-25.1	8.59(20°,1)	79	25	C	μ=0	64Ia	
	-28.28	8.58(1)	69.4	25	C	μ=0	67Bc	
	-24.06	8.42(1)	84.5	40	C	μ=0	66Aa	
	-22.89	8.33(1)	86.2	40	C	μ=0	72Ig	
				10-40	C	μ=0 ΔCp=105		
	-28.4	8.18(1)	58	25	C	μ=0.05M(KCl)	70Gc 71Gc	
	-29.3	8.00(1)	54	30	T	μ=0.09(KCl)	57Mb	
	-25.1	9.56(1)	73.2	25	C	μ=0.1M($NaClO_4$)	72Ia	
	-24.7	8.23(25°,1)	75	20-30	T	μ=0.1(KNO_3)	67Gb	b
	-25.6	8.07(1)	70	25	C	μ=0.2M(KCl)	76Sq 76Sp	
	-24.56	8.07(1)	71	25	C	μ=0.2M(KCl)	73Ga	
	-27.6	8.16(25°,1)	63.6	15-35	T	μ=0.2M(KNO_3) CM=$2.0x10^{-3}$	79Mg	b

Metal Ion	ΔH, kJ /mol	Log K	ΔS, J /mol K	T,°C	M	Conditions	Ref.	Remarks
G17, continued								
Cu^{2+}, cont.	-20.1	8.54(15°,1)	93.7	15-40	T	μ=0.2	65Sb	b
	-30.12	7.52(2)	37.6	10	C	μ=0	72Ig	
	-28.95	7.36(2)	38.5	10	C	μ=0	66Aa	
	-29.12	7.09(2)	38.1	25	C	μ=0	66Aa	
	-26.8	7.24(20°,2)	46	25	C	μ=0	64Ia	
	-28.83	6.94(2)	36.4	25	C	μ=0	67Bd	
	-29.00	7.26(2)	41.8	25	C	μ=0	72Ig	
	-27.57	7.00(2)	46.0	40	C	μ=0	72Ig	
				10-40	C	μ=0 ΔCp=84		
	-30.67	6.85(2)	33.1	40	C	μ=0	66Aa	
	-28.4	7.00(15°,2)	47.7	15-40	T	μ=0.2	65Sb	b
	-25.9	6.87(2)	46	25	C	μ=0.05M(KCl)	70Gc 71Gc	
	-27.6	6.80(2)	37.2	25	C	μ=0.1M($NaClO_4$)	72Ia	
	-27.2	6.96(25°,2)	46	20-30	T	μ=0.1(KNO_3)	67Gb	b
	-28.3	6.77(2)	33	25	C	μ=0.2M(KCl)	76Sp 76Sq	
	-31.92	6.77(2)	25	25	C	μ=0.2M(KCl)	73Ga	
	-28.9	6.82(25°,2)	33.9	15-35	T	μ=0.2M(KNO_3) CM=2.0x10^{-3}	79Mg	b
	-28.4	7.00(15°,2)	47.7	15-40	T	μ=0.2	65Sb	b
	-56.9	(1-2)	--	24.1	C	μ=0.01M($HClO_4$)	72Sh	
	-53.6	11.2(1-2)	105.4	20.5	C	0.1M(KNO_3)	67Sc	
	-48.5	15.54(15°,1-2)	141.4	15-40	T	μ=0.2	65Sb	b
	-51.67	15.81	129.3	25	C	μ=0 A=Alanine R:$Cu^{2+}+L^-+A^-=CuLA$	72Yb	
	-49.16	15.05	123.0	25	C	μ=0.1M(KNO_3) A=α-Alanine R:$Cu^{2+}+L^-+A^-=CuLA$	72Ia	
	-48.07	15.89	143.1	25	C	μ=0 A=α-Amino-isobutric acid R:$Cu^{2+}+L^-+A^-=CuLA$	72Yb	
	-55.6	15.78	117	25	C	μ=0.2(KCl) A=Aspartic acid R:$Cu^{2+}+A^-+L^-=CuLA$	74Na	
	-70.7	16.70	82	25	C	μ=0.2M(KCl) R:Cu+A+L=CuAL A=1,2-Diamino-N,N-dimethyl-ethane	76Sp	
	-74.9	17.30	80	25	C	μ=0.2M(KCl) R:Cu+A+L=CuAL A=1,2-Diamino-N,N'-dimethyl-ethane	76Sp	
	-79.5	17.69	72	25	C	μ=0.2M(KCl) R:Cu+A+L=CuAL A=1,2-Diamino-ethane	76Sp	
	-77.0	16.91	65	25	C	μ=0.2M(KCl) R:Cu+A+L=CuAL	76Sp	

Metal Ion	ΔH, kJ /mol	Log K	ΔS, J /mol K	T,°C	M	Conditions	Ref.	Remarks
G17, continued								
Cu^{2+}, cont.						A=1,3-Diamino-propane		
	-80.8	17.00	54	25	C	μ=0.2M(KCl)	76Sp	
						R:Cu+A+L=CuAL		
						A=Histamine		
	-45.65	15.59	145.6	25	C	μ=0 A=Sarcosine	72Yb	
						R: $Cu^{2+}+L^{-}+A^{-}$=CuLA		
	-49.0	14.66	116.3	25	C	μ=0.1M(KNO_3)	72Ia	
						A=Serine		
						R: $Cu^{2+}+L^{-}+A^{-}$=CuLA		
	-49.54	15.06	122.2	25	C	μ=0.1M(KNO_3)	72Ia	
						A=Valine		
						R: $Cu^{2+}+L^{-}+A^{-}$=CuLA		
CuA^{2+}	-33.5	7.74(25°,1)	36.0	15-35	T	μ=0.2M(KNO_3)	79Mg	b
						CM=2.0x10^{-3}		
						A=2,2'-Bipyridyl		
CuA	-70.7	16.70(1)	82	25	C	μ=0.2M(KCl)	76Sq	
						A=1,2-Diamino-N,N-dimethyl-ethane		
	-74.9	17.29(1)	80	25	C	μ=0.2M(KCl)	76Sq	
						A=1,2-Diamino-N,N'-dimethyl-ethane		
	-79.5	17.69(1)	72	25	C	μ=0.2M(KCl)	76Sq	
						A=1,2-Diamino-ethane		
	-77.0	16.91(1)	65	25	C	μ=0.2M(KCl)	76Sq	
						A=1,3-Diamino-propane		
	-80.8	17.00(1)	54	25	C	μ=0.2M(KCl)	76Sq	
						A=Histamine		
	-33.9	5.92(25°,1)	1.2	15-70	T	μ=0.1M(KNO_3)	72Id	b
						A=Methylimino-diacetate		
Ag^{+}	6.90	3.24(1)	90.8	5	T	μ=0.01 ΔCp=-1008	59Da	
	-3.39	3.36(1)	54.8	15	T	μ=0.01 ΔCp=-1046	59Da	
	-13.97	3.43(1)	18.8	25	T	μ=0.01 ΔCp=-1079	59Da	
	-24.98	3.43(1)	-18.0	35	T	μ=0.01 ΔCp=-1117	59Da	
	-36.28	3.36(1)	-54.0	45	T	μ=0.01 ΔCp=-1155	59Da	
	-48.1	3.24(1)	-90.0	55	T	μ=0.01 ΔCp=-1188	59Da	
	-19.2	3.15(1)	-4.2	25	C	0.5M(KNO_3)	70Va	
	-15.9	3.15(25°,1)	6.7	15-40	T	0.5M(KNO_3)	70Va	b
	-33.68	3.43(2)	-46.9	25	T	μ=0.01	59Da	
	-28.9	4.65(2)	-27.6	25	C	0.5M(KNO_3)	70Va	
	-36.0	4.65(25°,2)	-54.4	15-40	T	0.5M(KNO_3)	70Va	b
Zn^{2+}	-13.14	5.50(1)	59.4	10	C	μ=0	72Ig	
	-11.55	5.38(1)	64.4	25	C	μ=0	72Ig	
	-14.18	5.50(1)	57.7	25	C	μ=0	67Bd	
	-9.29	5.29(1)	71.5	40	C	μ=0	72Ig	

Metal Ion	ΔH, kJ /mol	Log K	ΔS, J /mol K	T,°C	M	Conditions	Ref.	Remarks
G17, continued								
Zn^{2+}, cont.				10-40	C	μ=0 ΔCp=130		
	-8.4	5.06(1)	67	25	C	μ=0.05M(KCl)	70Gc 71Gc	
	-7.82	5.156(1)	70.3	5	T	μ=0.1(KCl)	73Rb	q:±3.0
	-8.70	5.052(1)	67.4	25	T	μ=0.1(KCl)	73Rb	q:±3.0
	-9.58	4.951(1)	64.4	45	T	μ=0.1(KCl)	73Rb	q:±3.0
	-13.8	5.27(15°,1)	53.1	15-40	T	μ=0.2	65Sb	b
	-8.8	4.88(25°,1)	64.4	10&25	T	μ=0.65(KCl)	64Lg	a,b
	-15.61	4.57(2)	32.2	10	C	μ=0	72Ig	
	-13.47	4.43(2)	39.7	25	C	μ=0	72Ig	
	-12.13	4.29(2)	43.5	40	C	μ=0	72Ig	
				10-40	C	μ=0 ΔCp=117	72Ig	
	-13.4	4.38(2)	38	25	C	μ=0.05M(KCl)	70Gc 71Gc	
	-16.7	4.31(15°,2)	25.1	15-40	T	μ=0.2	65Sb	b
	-15.9	4.13(25°,2)	25.1	10&25	T	μ=0.65(KCl)	64Lg	a,b
	-14.77	2.63(3)	-1.7	10	C	μ=0	72Ig	
	-14.90	2.52(3)	-1.7	25	C	μ=0	72Ig	
	-15.23	2.40(3)	-2.5	40	C	μ=0	72Ig	
				10-40	C	μ=0 ΔCp=-17	72Ig	
	-26.4	9.5(1-2)	92.5	20	C	0.1M(KNO_3)	67Sc	
	-30.5	9.58(15°,1-2)	78.2	15-40	T	μ=0.2	65Sb	b
	-7.9	7.53(25°)	118.0	10&25	T	μ=0.65(KCl) A=Pyruvate R:$Zn^{2+}+L^-+A^-=ZnLA$	64Lg	a,b
	-31.4	14.25(25°)	168.6	10&25	T	μ=0.65(KCl) A=Pyruvate R:$Zn^{2+}+2L^-+2A^-=$ $ZnL_2A_2{}^{2-}$	64Lg	a,b
ZnA	-19.2	4.41(25°,1)	27.2	15-70	T	μ=0.1M(KNO_3) A=Methylimino-diacetate	71Ie	b
Cd^{2+}	-9.46	4.73(1)	57.3	10	C	μ=0	72Ig	
	-8.87	4.69(1)	60.2	25	C	μ=0	72Ig	
	-8.16	4.60(1)	61.9	40	C	μ=0	72Ig	q:±0.42
				10-40	C	μ=0 ΔCp=42		
	-15.65	3.76(2)	16.7	10	C	μ=0	72Ig	
	-13.56	3.71(2)	25.5	25	C	μ=0	72Ig	
	-12.34	3.60(2)	29.3	40	C	μ=0	72Ig	
				10-40	C	μ=0 ΔCp=109	72Ig	
	-11.88	2.53(3)	6.3	10	C	μ=0	72Ig	
	-13.43	2.28(3)	-1.2	25	C	μ=0	72Ig	
	-14.18	2.00(3)	-7.1	40	C	μ=0	72Ig	
				10-40	C	μ=0 ΔCp=-75	72Ig	
CdA	-18.4	3.82(25°,1)	11.7	15-70	T	μ=0.1M(KNO_3) A=Methylimino-diacetate	72Ie	b
HgA_2	-12.30	2.61	8.8	25	C	μ=0 R:$HgAL+L^-=$ HgL_2+A^- A=Cl	66Pd	
	-25.52	3.42	-20.1	25	C	μ=0 R:$HgA_2+L^-=$ $HgAL+A^-$ A=Cl	66Pd	

Metal Ion	ΔH, kJ /mol	Log K	ΔS, J /mol K	T,°C	M	Conditions	Ref.	Remarks
G17, continued								
In^{3+}	12.6	2.46(35°,1)	88	25-45	T	μ=0.2M($NaClO_4$)	73Sc	b,q:±2.1
Pb^{2+}	-12.4	5.752(25°,1)	68.6	10-40	T	μ=3.00M($NaClO_4$)	76Cf	q:±3
	-25.1	11.880(25°)	143.2	10-40	T	μ=3.00M($NaClO_4$)	76Cf	q:±4
						R:Pb^{2+}+HL=$PbHL^{2+}$		
	16.5	-1.886(25°)	19.2	10-40	T	μ=3.00M($NaClO_4$)	76Cf	q:±3
						R:Pb^{2+}+L^{2-}=PbL		
G18 GLYCINE, amide-			$C_2H_6ON_2$			$H_2NCH_2C(O)NH_2$		
	-23.0	5.40(1)	26.4	25	C	μ=0.1M(KNO_3)	72Ba	
	-26.8	5.43(1)	14.2	25	C	μ=0.10M($NaClO_4$)	72Tb	
	-23.8	4.22(2)	0.8	25	C	μ=0.1M(KNO_3)	72Ba	
	-25.5	4.10(2)	-7.1	25	C	μ=0.10M($NaClO_4$)	72Tb	
G19 GLYCINE, 2-alanyl-			$C_5H_{10}O_3N_2$			$CH_3CHNH_2COCH(NH_2)CO_2H$		
Cu^{2+}	-28.0	5.34(1)	7.9	25	C	μ=0.1M(KNO_3)	72Ba	
	-25	5.25(1)	15	25	C	μ=0.2M(KCl)	77Gd	q:±2
	3	1.35	35	25	C	μ=0.2M(KCl)	77Gd	q:±1
						ΔH for formation of $[CuLH_{-1}]$		
	47	-8.16	0	25	C	μ=0.2M(KCl)	77Gd	
						ΔH for formation of $[CuLH_{-2}]^-$		
	-21	3.95	4	25	C	μ=0.2M(KCl)	77Gd	
						ΔH for formation of $[CuL_2H_{-1}]^-$		
	38	-4.66	40	25	C	μ=0.2M(KCl)	77Gd	
						ΔH for formation of $[Cu_2L_2H_{-3}]^-$		
	-24	2.60	-31	25	C	μ=0.2M(KCl)	77Gd	
						R:$[CuLH_{-1}]$+L^-= $[CuL_2H_{-1}]^-$		
	-13	4.31	-39	25	C	μ=0.2M(KCl)	77Gd	q:±2
						R:$(OH)^-$+$CuLH_{-1}$= $[CuLH_{-1}(OH)]^-$		
	-11	2.15	4	25	C	μ=0.2M(KCl)	77Gd	q:±2
						R:$[CuLH_{-1}(OH)]^-$+ +$[CuLH_{-1}]$= $[(CuL_2H_{-1})_2(OH)]^-$		
G20 GLYCINE, N-ALANYL			$C_5H_{10}O_3N_2$			$CH_3CH(NH_2)CONHCH_2CO_2H$		
Cu^{2+}	-25.4	5.25(1)	15	25	C	μ=0.2M(KCl)	76Gd	
	-2.8	1.35	35	25	C	μ=0.2M(KCl)	76Gd	
						R:$Cu^{2+}L^-$=CuL^-+H^+		
	46.7	-8.16	0	25	C	μ=0.2M(KCl)	76Gd	
						R:Cu^{2+}+L^-=CuL^-+ $2H^+$		

Metal Ion	ΔH, kJ /mol	Log K	ΔS, J /mol K	T,°C	M	Conditions	Ref.	Re-marks
G20, continued								
Cu^{2+}, cont.	-21.3	3.95	4	25	C	μ=0.2M(KCl) R:$Cu^{2+}2L^{-}=CuL_2^{-}+H^{+}$	76Gd	
	38.5	-4.66	40	25	C	μ=0.2(KCl) R:$2Cu^{2+}+2L^{-}=Cu_2L_2^{-}+3H^{+}$	76Gd	
G21 GLYCINE, N,N-(dihydroxyethyl)		$C_6H_{13}O_4N$				$(HOCH_2CH_2)_2NCH_2CO_2H$		
Th^{4+}	-134	28.56(30°,1-4)	89	30-50	T	μ=0.1M($NaClO_4$)	73Kc	b
VO^{2+}	-134	13.08(30°,1-2)	-194	30-50	T	μ=0.1M($NaClO_4$)	73Kc	b
G22 GLYCINE, N,N,-dimethyl-		$C_4H_9O_2N$				$(CH_3)_2NCH_2CO_2H$		L^{-}
Mg^{2+}	0	1.68(1)	67	25	T	μ=0	56Ma	
CoA	-18.4	3.30(25°,1)	1.2	15-70	T	μ=0.1M(KNO_3) A= Nitrilotriacetate	71Ie	b
NiA	-16.7	4.42(25°,1)	27.2	15-70	T	μ=0.1M(KNO_3) A= Nitrilotriacetate	71Ie	b
CuA	-45.2	5.34(25°,1)	-48.5	15-70	T	μ=0.1M(KNO_3) A= Nitrilotriacetate	71Ie	b
ZnA	-18.8	3.28(25°,1)	-1.2	15-70	T	μ=0.1M(KNO_3) A= Nitrilotriacetate	71Ie	b
CdA	-11.3	2.70(25°,1)	13.4	15-70	T	μ=0.1M(KNO_3) A= Nitrilotriacetate	71Ie	b
G23 GLYCINE, N-glycyl-		$C_4H_8O_3N_2$				$NH_2CH_2CONHCH_2CO_2H$		L^{-}
Mn^{2+}	52.3	3.44(1)	25	30	T	μ=0.09(KCl)	57Mb	p
Ni^{2+}	-20.5	4.09(1)	9.6	25	C	μ=0.10($NaClO_4$)	72Tb	
	-16.3	3.10(2)	4.6	25	C	μ=0.10($NaClO_4$)	72Tb	
	-23.0	2.11(3)	-36.8	25	C	μ=0.10($NaClO_4$)	72Tb	
	-37.99	(1-2)	--	24	C	μ=0.01M($HClO_4$)	72Sh	
Cu^{2+}	-25.5	(1)	--	--	-	μ=0.01M($HClO_4$)	72Sh	r,u
	-25.5	5.55(1)	20.79	25	C	μ=0.10($NaClO_4$)	68Bc	
	-5.0	6.02(25°,1)	98	10-40	T	μ=0.14	72Ph	b
	-29	5.56(1)	8	25	C	μ=0.2M(KCl)	77Gd 76Gd	
	-35.1	7.34(25°,1)	22	10-40	T	μ=0.14 S:45v%Dioxane	72Ph	b
	-5.8	5.04(25°,2)	77	10-40	T	μ=0.14	72Ph	b
	-37.6	6.44(25°,2)	-2.9	10-40	T	μ=0.14 S:45v%Dioxane	72Ph	b
	38.0	-4.51	41	25	C	μ=0.2(KCl) R:$2Cu^{2+}+2L^{-}=Cu_2L_2^{-}+3H^{+}$	76Gd	
	-40.46	--	--	--	-	μ=0.01M($HClO_4$)	72Sh	r,u

Metal Ion	ΔH, kJ /mol	Log K	ΔS, J /mol K	T,°C	M	Conditions	Ref.	Remarks
G23, continued								
Cu^{2+}, cont.	-28.3	4.46	-10	25	C	R:Cu^{2+}+2L^-= CuA+HL A=L^{2-} μ=0.2(KCl)	76Gd	
	2.8	1.33	35	25	C	R:Cu^{2+}+2L^-= CuL_2^-+H^+ μ=0.2(KCl)	76Gd	
	46.2	-8.04	1	25	C	R:Cu^{2+}+L^-=CuL+H^+ μ=0.2(KCl)	76Gd	
	33.0	-7.6	-25	0	T	R:Cu^{2+}+L^-=CuL^-+ $2H^+$ μ=0.09(KCl)	57Mb	
	-3	1.33	35	25	C	R:Cu^{2+}+HL=CuL+$2H^+$ μ=0.2M(KCl)	77Gd	j,q:±1
	46	-8.04	1	25	C	ΔH for formation of $[CuLH_{-1}]$ μ=0.2M(KCl)	77Gd	j
	-28	4.46	-10	25	C	ΔH for formation of $[CuLH_{-2}]^-$ μ=0.2M(KCl)	77Gd	j
	38	-4.51	41	25	C	ΔH for formation of $[CuL_2H_{-1}]^-$ μ=0.2M(KCl)	77Gd	j
	-31	3.13	-45	25	C	ΔH for formation of $[Cu_2L_2H_{-3}]^-$ μ=0.2M(KCl)	77Gd	
	-13	4.45	-14	25	C	R:$[CuLH_{-1}]$+L^-= $[CuL_2H_{-1}]^-$ μ=0.2M(KCl)	77Gd	q:±2
	-11	2.20	5	25	C	R:$(OH)^-$+$CuLH_{-1}$= $[CuLH_{-1}(OH)]^-$ μ=0.2M(KCl)	77Gd	q:±2
						R:$[CuLH_{-1}(OH)]^-$+ $[CuLH_{-1}]$= $[(CuLH_{-1})_2(OH)]^-$		
CuL	-70.7	--	--	--	-	μ=0.01M($HClO_4$)	72Sh	r,u
CuA	-28.0	2.99(1)	-36.8	25	C	R:Cu^{2+}+A^{2-}+L^-= $CuAL^-$ A=L^{2-} μ=0.10($NaClO_4$) A=Glyclyglycine	73Pe	
Zn^{2+}	-13.0	3.44(1)	22.2	25	C	μ=0.1M(KNO_3)	72Bl	
	-27.6	3.59(25°,1)	-25	10-40	T	μ=0.14	72Ph	b
	-32.6	5.02(25°,1)	-13	10-40	T	μ=0.14 S:45v%Dioxane	72Ph	b
	-20.5	2.85(2)	-14.2	25	C	μ=0.1M(KNO_3)	72Bl	
	-27.6	2.97(25°,2)	-33	10-40	T	μ=0.14	72Ph	b
	-25.9	4.03(25°,2)	-9.6	10-40	T	μ=0.14 S:45v%Dioxane	72Ph	b

Metal Ion	ΔH, kJ /mol	Log K	ΔS, J /mol K	T,°C	M	Conditions	Ref.	Remarks
G23, continued								
Cd^{2+}	-10.5	3.08(25°,1)	23.8	10-40	T	μ=0.06	66Vb 65Va	b
	-11.7	2.85(1)	15.5	25	C	μ=0.1M(KNO_3)	72B1	
	-9.2	2.566(25°,2)	18.0	10-40	T	μ=0.06M	65Va	b
	-18.0	2.48(2)	-12.6	25	C	μ=0.1M(KNO_3)	72B1	
	-19.7	5.64(25°,1-2)	42	10-40	T	μ=0.06	66Vc	b
Pb^{2+}	-12.7	3.824(25°,1)	30.5	10-40	T	μ=3.00M($NaClO_4$)	76Cf	q:±5
	-37.8	10.007(25°)	34.3	10-40	T	μ=3.00M($NaClO_4$) R:Pb^{2+}+HL=PbHL	76Cf	
G24 GLYCINE, N-glycylglycyl			$C_6H_{11}O_4N_3$			$NH_2CH_2CONHCH_2CONHCH_2CO_2H$		
Pb^{2+}	-15.0	3.967(25°,1)	25.5	10-40	T	μ=3.00M($NaClO_4$)	76Cf	q:±1
	-27.5	10.618(25°)	111.0	10-40	T	μ=3.00M($NaClO_4$) R:Pb^{2+}+HL=$PbHL^{2+}$	76Cf	q:±2
	22.0	-3.401(25°)	8.7	10-40	T	μ=3.00M($NaClO_4$) R:Pb^{2+}+L^{2-}=PbL	76Cf	
G25 GLYCINE, N(N-glycylglycyl-glycine)			$C_8H_{14}O_5N_4$			$NH_2CH_2CONHCH_2CONHCH_2$-$CONHCH_2CO_2H$		L^-
Cu^{2+}	-25.1	5.13(1)	13.8	25	C	μ=0.1($NaClO_4$)	69Na	
G26 GLYCINE, N(2-hydroxybenzyl)			$C_9H_{11}O_3N$			$C_6H_4(OH)(CH_2NHCH_2CO_2H)$		
$CoA_4{}^{3+}$	53.14	4.49(25°,1)	264.43	15-30	T	0.1M(Sodium acetate) A=NH_3	79Mb	b
$CoA_5{}^{3+}$	38.53	0.47(25°,1)	138.82	25-40	T	0.1M(Sodium acetate) A=NH_3	79Mb	b
G27 GLYCINE, N(3-methoxy-4-hydroxybenzyl)			$C_{10}H_{13}O_4N$			$C_6H_3(OH)(OCH_3)(CH_2NHCH_2CO_2H)$		
$CoA_4{}^{3+}$	10.21	3.62(25°,1)	104.18	15-30	T	0.1M(Sodium acetate) A=NH_3	79Mb	b
$CoA_5{}^{3+}$	21.67	3.33(25°,1)	137.28	15-30	T	0.1M(Sodium acetate) A=NH_3	79Mb	b
G28 GLYCINE, N-valine (DL)			$C_7H_{14}O_3N_2$			$NH_2CH_2C(=O)NHCH[CH(CH_3)_2]COOH$		L
Gd^{3+}	14.510	3.87(25°,1)	122.63	25-45	T	μ=0.15M(KNO_3)	79Sb	b
Dy^{3+}	9.012	3.98(25°,1)	106.52	25-45	T	μ=0.15M(KNO_3)	79Sb	b
Er^{3+}	8.134	4.04(25°,1)	104.60	25-45	T	μ=0.15M(KNO_3)	79Sb	b
Yb^{3+}	8.134	4.08(25°,1)	108.70	25-45	T	μ=0.15M(KNO_3)	79Sb	b

Metal Ion	ΔH, kJ /mol	Log K	ΔS, J /mol K	T,°C	M	Conditions	Ref.	Re-marks
G29	GLYOXAL, dioxime, dimethyl-		$C_4H_8O_2N_2$			HON:C(CH_3)C(CH_3):NOH		L^-
Ni^{2+}	8.8	21.58(25°,1-2)	445.6	25&40	T	S:50v% Dioxane	54Cc	a,b
Cu^{2+}	-58.2	23.30(25°,1-2)	253.1	25&40	T	S:50v% Dioxane	54Cc	a,b
G30	GUANOSINE-5'-diphosphoric acid		$C_{10}H_{15}O_{10}N_5P_2$					L^{3-}
Mg^{2+}	14.27	3.40(1)	112.1	30	C	μ=0.2[(CH_3)$_4$NCl] pH=8.50	73Sa	
G31	GUANOSINE-5'-phosphoric acid		$C_{10}H_{14}O_8N_5P$					
Mg^{2+}	7.20	1.75(1)	57.3	30	C	μ=0.2[(CH_3)$_4$NCl] pH=8.50	73Sa	
G32	GUANOSINE-5'-triphosphoric acid		$C_{10}H_{16}O_{14}N_5P_3$					
Mg^{2+}	3.76	4.98(25°,1)	109	25-45	T	μ=0.10M(KNO_3)	73Kf	b,g:Na
	17.95	3.91(1)	134.3	30	C	μ=0.2[(CH_3)$_4$NCl] pH=8.50	73Sa	
Ca^{2+}	-6.7	4.92(25°,1)	71	25-45	T	μ=0.10M(KNO_3)	73Kf	b,g:Na q:±0.4
Mn^{2+}	-8.8	5.18(25°,1)	67	25-45	T	μ=0.10M(KNO_3)	73Kf	b,g:Na
Co^{2+}	-6.7	5.57(25°,1)	84	25-45	T	μ=0.10M(KNO_3)	73Kf	b,g:Na q:±0.4
Ni^{2+}	-7.9	5.78(25°,1)	84	25-45	T	μ=0.10M(KNO_3)	73Kf	b,g:Na q:±0.4
Cu^{2+}	-18.8	6.55(25°,1)	63	25-45	T	μ=0.10M(KNO_3)	73Kf	b,g:Na
Zn^{2+}	-7.5	5.72(25°,1)	84	25-45	T	μ=0.10M(KNO_3)	73Kf	b,g:Na q:±0.4

Metal Ion	ΔH, kJ /mol	Log K	ΔS, J /mol K	T, °C	M	Conditions	Ref.	Remarks
				H				
H1 HAEMATOPORPHYRIN								L°
Mg^{2+}	4.8	1.35(25°,1)	42	25-40	T	0.05M Tris buffer CM=0.01-0.1M CL=10^{-4}M pH=7.4	73Af	b
	40.9	1.17(25°,1)	160	25-40	T	0.05M Tris buffer CM=0.01-0.1M CL=10^{-4}M pH=8.2	73Af	b
	-39.7	1.10(25°,1)	-112	25-40	T	0.05M Tris buffer CM=0.01-0.1M CL=10^{-4}M pH=9.0	73Af	b
H2 HENDECANE, 5,7-thio-			$C_9H_{20}S_2$			$CH_3(CH_2)_3SCH_2S(CH_2)_3CH_2$		
SnA_4	-60.7	(1)	--	25	C	S:Benzene CM=0.05-0.08M A=Cl	68Ge 65Gb	
H3 HENDECANE, 3,6,9-triaza-1,11-diamino-			$C_8H_{23}N_5$			$NH_2(CH_2CH_2NH)_3CH_2CH_2NH_2$		L
Mn^{2+}	-15.5	6.56(1)	73.2	25	C	μ=0.1(KCl)	64Pa	
	-21.59	7.62(25°,1)	73.2	26.5	C	--	58Jb	b
Fe^{2+}	-36.4	9.86(1)	66.9	25	C	μ=0.1(KCl)	64Pa	
	-39.16	11.40(25°,1)	86.6	26.5	C	--	58Jb	b
Co^{2+}	-57.9	13.30(1)	60.7	25	C	μ=0.1(KCl)	64Pa	
Ni^{2+}	-77.0	17.4(1)	75.7	25	C	μ=0.1(KNO_3)	65Wd	
	-79.08	17.44(1)	69.0	25	C	μ=0.1(KCl)	64Pa	
	-47.49	(1)	--	25	C	μ=0.5 Anion:Cl^-	57Jb	
	-43.64	17.51(1)	-188.7	25	C	μ=0.5 Anion:ClO_4^-	57Jb	
Cu^{2+}	-104.39	22.80(1)	85.8	25	C	μ=0.1(KCl)	64Pa 57Ja	
	-100	23.0(1)	105	25	C	μ=0.1(KNO_3)	65Wd	
Zn^{2+}	-57.95	15.10(1)	94.1	25	C	μ=0.1(KCl)	64Pa	
	-58.6	15.5(1)	100	25	C	μ=0.1(KNO_3)	64Pa	
Cd^{2+}	-53.6	14.1(1)	92	25	C	μ=0.1(KNO_3)	65Wd	
H4 2-HENDECANONE			$C_{11}H_{22}O$			$CH_3(CH_2)_8COCH_3$		
$SiAB_3$	-12.47	1.43(30°,1)	-13.8	10-45	T	S:CCl_4 XL=$5x10^{-2}$ XM=$5x10^{-4}$ A=OH B=CH_3	69Lb	b
	-14.35	1.33(30°,1)	-21.8	20-60	T	S:CCl_4 XL=$5x10^{-2}$ XM=$5x10^{-4}$ A=OH B=C_2H_5	69Lb	b
	-15.56	1.51(30°,1)	-22.38	0-50	T	S:CCl_4 XL=$5x10^{-2}$ XM=$4x10^{-4}$ A=OH B=Phenyl	69Lb	b

Metal Ion	ΔH, kJ /mol	Log K	ΔS, J /mol K	T,°C	M	Conditions	Ref.	Remarks
H5	HEPTANE, 1-amino-		$C_7H_{17}N$			$CH_3(CH_2)_6NH_2$		L
NiA	-70.7	6.12(1-2)	-120.5	25	T	$CM=10^{-4}M$ S:Benzene A=Diacetylbisbenzoyl hydrazine	60Sb	
H6	HEPTANE, 1-amino-3-aza-6-thia		$C_5H_{14}N_2S$			$NH_2CH_2CH_2NHCH_2CH_2SCH_3$		
Ni^{2+}	-36.28	7.063(1)	13.4	25	C	0.5(KNO_3)	77Hd	
	-44.4	5.858(2)	-36.8	25	C	0.5(KNO_3)	77Hd	
Cu^{2+}	-58.49	11.377(1)	21.3	25	C	0.5M(KNO_3)	77Ed	
	-45.27	6.306(2)	-31.4	25	C	0.5M(KNO_3)	77Ed	
H7	HEPTANE, 3-aza-1,7-diamino		$C_6H_{17}N_3$			$NH_2CH_2CH_2NHCH_2CH_2CH_2CH_2NH_2$		L
Cu^{2+}	-64.0	13.44(1)	42.6	25	C	μ=0.10M(NaCl)	76Gg	
	-63.93	13.046(1)	35.3	25	C	μ=0.5M(KNO_3)	76Ba	
	-28.91	4.611(2)	-8.8	25	C	μ=0.5M(KNO_3)	76Ba	
	-44.7	8.94	21	25	C	μ=0.10M(NaCl) R:$Cu^{2+}+HL^{+}=CuHL^{3+}$	76Gg	
	-48.91	9.187	11.8	25	C	μ=0.5M(KNO_3) R:$Cu^{2+}+HL^{+}=[CuHL]^{3+}$	76Ba	
	-51.13	10.096	-29.9	25	C	μ=0.5M(KNO_3) R:$[CuHL]^{3+}+HL^{+}=[Cu(HL)_2]^{4+}$	76Ba	
	-38.74	6.810	0.4	25	C	μ=0.5M(KNO_3) R:$CuL^{2+}+H^{+}=Cu(HL)^{3+}$	76Ba	
	-114.77	20.26	2.9	25	C	μ=0.5M(KNO_3) R:$CuL_2{}^{2+}+2H^{+}=[Cu(HL)_2]^{4+}$	76Ba	
	-97.74	17.43	5.8	25	C	μ=0.5M(KNO_3) R:$Cu^{2+}+HL^{+}+L=[CuL(HL)]^{3+}$	76Ba	
	-58.70	10.44	2.9	25	C	μ=0.5M(KNO_3) R:$CuL_2{}^{2+}+H^{+}=CuL(HL)^{3+}$	76Ba	
H8	HEPTANE, 1,7-diamine-4-thio		$C_6H_{16}N_2S$			$NH_2(CH_2)_3S(CH_2)_3NH_2$		L
Ni^{2+}	-41.9	7.377(1)	0.7	25	C	μ=0.5M(KNO_3)	79Ha	
	-48.6	6.142(2)	-45	25	C	μ=0.5M(KNO_3)	79Ha	
Cu^{2+}	-51.4	9.020(1)	0.2	25	C	μ=0.5M(KNO_3)	79Ha	
	-33.7	5.238(2)	-12	25	C	μ=0.5M(KNO_3)	79Ha	
	-18.0	5.90	52	25	C	μ=0.5M(KNO_3)	79Ha	

Metal Ion	ΔH, kJ /mol	Log K	ΔS, J /mol K	T, °C	M	Conditions	Ref.	Remarks
H8, continued								
Cu^{2+}, cont.						R: $CuL^{2+} + OH^- = Cu(OH)L^+$		
H9 HEPTANE, 4-oxa-			$C_6H_{14}O$			$(CH_3CH_2CH_2)_2O$		L
SbA_5	-68.6	(1)	--	25	C	S: CCl_4 A=Cl	73Oa	
	-74.64	(1)	--	25	C	S: Ethylene chloride A=Cl	68Oa	
H10 HEPTANEDIOIC ACID, 4-thio-			$C_6H_{10}O_4S$			$HO_2CCH_2CH_2SCH_2CH_2CO_2H$		L^{2-}
Ca^{2+}	-8.49	2.99(30°,1-3)	29.29	30&40	T	μ=1.2M(KNO_3)	72Ra	a,b
	-8.03	2.94(40°,1-3)	30.75	40&50	T	μ=1.2M(KNO_3)	72Ra	a,b
Pb^{2+}	3.72	3.27(40°,1-3)	50.6	30-50	T	μ=1.2M(KNO_3)	72Rb	b
Cd^{2+}	-11.34	1.63(40°,1)	-5.02	30-50	T	μ=1.2(KNO_3) P.C.Rowat & C.M. Gupta, J.Inorg. Nucl.Chem., 34, 951 (1973).	74Ma	b,i
	-0.50	0.69(40°,2)	11.72	30-50	T	μ=1.2(KNO_3) P.C.Rowat & C.M. Gupta, J.Inorg. Nucl.Chem., 34, 951 (1973)	74Ma	b,i
	-1.42	0.51(40°,3)	6.48	30-50	T	μ=1.2(KNO_3) P.C.Rowat & C.M. Gupta, J.Inorg. Nucl.Chem., 34, 951 (1973)	74Ma	b,i
H11 3,5-HEPTANEDIONE			$C_7H_{12}O_2$			$CH_3CH_2COCH_2COCH_2CH_3$		
Co^{2+}	-4.18	7.32(25°,1)	126.98	5-45	T	S:50v% Dioxane	73Ak	b
	-6.69	5.84(25°,2)	90.00	5-45	T	S:50v% Dioxane	73Ak	b
H12 3,5-HEPTANEDIONE, 2,6-dimethyl-			$C_9H_{16}O_2$			$CH_3CH(CH_3)COCH_2COCH(CH_3)CH_3$		
NiL_2	-21.00	3.301	-6.07	30	C	S: Benzene R: $3NiL_2 = Ni_3L_6$	74Gb	
H13 6-HEPTANOIC ACID, 3-seleno-			$C_6H_{12}O_2Se$			$CH_3(CH_2)_3SeCH_2CO_2H$		L^-
Ag^+	-32.59	4.57(25°,1)	-21.80	0-40	T	μ=0.20M pH=5.8	71Bd 71Bc	b,q: ±7.9
	-26.99	3.45(25°,2)	-23.43	0-40	T	μ=0.20M pH=5.8	71Bd 71Bc	b,q: ±2.1
H14 4-HEPTANONE, 2,6-dimethyl-			$C_9H_{18}O$			$(CH_3)_2CHCH_2COCH_2CH(CH_3)_2$		L
SnA_4	-56.9	1.529(30°,1)	-132.2	30-40	T	S: Benzene ΔS	64La	b

Metal Ion	ΔH, kJ /mol	Log K	ΔS, J /mol K	T,°C	M	Conditions	Ref.	Remarks
H14, continued								
SnA_4, cont.						calculated using ΔG in units of atm^{-1} at 30° A=Cl		
H15	6-HEPTENOIC ACID, 3-seleno-		$C_6H_{10}O_2Se$			CH_2:$CH(CH_2)_2SeCH_2CO_2H$		L^-
Ag^+	-54.4	5.17(25°,1)	-83.7	0-40	T	μ=0.20M pH=5.8	71Bd 71Bc	b,q: ±7.9
	-100.8	2.78(25°,2)	-284.9	0-40	T	μ=0.20M pH=5.8	71Bc 71Bc	b,q: ±17.8
	-3.01	1.96(25°)	26.82	0-40	T	μ=0.20M R:AgL+ Ag^+=Ag_2L^+ pH=5.8	71Bc	b,q: ±7.9
H16	1-HEPTYNE		C_7H_{12}			$CH_3(CH_2)_4C$:CH		L
Ag^+	-24.7	1.049(25°,1)	-64.0	10-40	T	μ=1M($NaNO_3$) S:52v% Ethanol	73Sz	b,q: ±3.3
	-3.8	1.025(25°,1)	-19.7	10-40	T	μ=1M($NaNO_3$) S:75v% Ethanol	73Sz	b,q: ±0.8
H17	HEXADECANE, 5,12-dithio-		$C_{14}H_{30}S_2$			$CH_3(CH_2)_3S(CH_2)_6S$-$(CH_2)_3CH_3$		L
SnA_4	-114.6	(1)	--	--	C	S:Benzene CM=0.05-0.08M A=Cl	65Gb	u
TeA_4	-51.5	3.00(1)	-115.1	25	C	S:Benzene A=Cl	73Pi	
	-30.5	(2)	--	25	C	S:Benzene A=Cl	73Pi	
H18	1,5-HEXADIENE		C_6H_{10}			CH_2:$CHCH_2CH_2CH$:CH_2		L
PdA_2B_2	-33.5	--	--	25	C	S:CH_2Cl_2 CL=$5x10^{-3}$M CM=$1.7x10^{-4}$M A=Cl B=C_6H_5CN R:PdA_2B_2+L=PdA_2L+2B	72Pd	
H19	HEXAETHYLENE GLYCOL, dimethyl ether		$C_{14}H_{30}O_7$			$CH_3OCH_2(CH_2OCH_2)_5$-CH_2OCH_3		L
NaA	-37.6	2.903(25°)	-71	0-50	T	S:Tetrahydrofuran A=Fluorenyl ion R:NaA+L=NaLA	70Ce	b

Metal Ion	ΔH, kJ /mol	Log K	ΔS, J /mol K	T,°C	M	Conditions	Ref.	Re-marks
H20	HEXAMETHYLENETETRAMINE		$C_6H_{12}N_4$			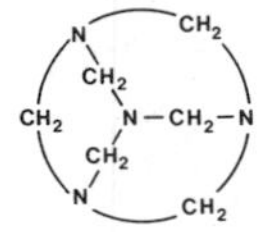		
Cd^{2+}	-0.67	0.924(25°)	15.1	25&40	T	μ=1.0M(KOAc)	78Pb	a,b,j,n
	-25.9	1.144(25°)	-63.4	25&40	T	μ=1.0M(KCl)	78Pb	a,b,j,n
	-2.5	0.0147(25°)	-8.1	25&40	T	μ=1.0M(KNO_3)	78Pb	a,b,j,n
H21	HEXANE, 1-amino-		$C_6H_{15}N$			$CH_3(CH_2)_5NH_2$		L
NiA	-70.3	6.09(1-2)	-119.7	25	T	CM=10^{-4}M S:Benzene A=Diacetylbisbenzoyl hydrazine	60Sb	
CuA_2	-30.5	0.322(25°,1)	-96	10-50	T	S:Methylcyclohexane-n-Hexylamine A=di-n-Butyldithiocarbamate	71Cf	b,q: ±5.8
Ag^+	-25	3.54(1)	-85.4	25	C	μ=0	71He	q:±8
	-29	4.02(2)	-14.2	25	C	μ=0	71He	q:±8
	-53.05	7.56(1-2)	-33.5	25	C	μ=0	71He	
H22	HEXANE, 3-aza-1,6-diamino		$C_5H_{15}N_3$			$H_2NCH_2CH_2NHCH_2CH_2CH_2NH_2$		L
Ni^{2+}	-56.6	11.23(1)	24.7	25	C	0.5M(KNO_3)	74Ba	
	-57.4	9.63(2)	-57.3	25	C	0.5M(KNO_3)	74Ba	
	-32.6	7.99	2.9	25	C	0.5M(KNO_3) R:$Ni^{2+}+HL^+=NiHL^{3+}$	74Ba	
Cu^{2+}	-80.4	22.639(1)	48.1	25	C	0.5M(KNO_3)	74Ba	
	-25.5	4.47(2)	-23.0	25	C	0.5M(KNO_3)	74Ba	
	-48.1	12.30	11.3	25	C	0.5M(KNO_3) R:$Cu^{2+}+HL^+=CuHL^{3+}$	74Ba	
	-24.7	3.42	-34.7	25	C	0.5M(KNO_3) R:$CuL^{2+}+HL^+=[CuHL_2]^{3+}$	74Ba	
Zn^{2+}	-35.2	11.96(1)	50.6	25	C	0.5M(KNO_3)	74Ba	
	-35.4	5.18(2)	-45.6	25	C	0.5M(KNO_3)	74Ba	
H23	HEXANE, 1,6-diamino-N,N,N',N'-tetraacetic acid		$C_{14}H_{24}O_8N_2$			$(CH_2CO_2H)_2N(CH_2)_6N(CH_2CO_2H)_2$		L^{4-}
Ca^{2+}	-5.4	3.69(1)	52.3	25	C	μ=0.50	58Ya	
	-5.8	3.69(1)	50.6	25	C	μ=0.90	58Ya	
Mn^{2+}	3.64	4.8(1)	184.9	20	C	μ=0.1(KNO_3)	64Ab	
Co^{2+}	-19.08	13.05(1)	184.5	20	C	μ=0.1(KNO_3)	64Ab	
Ni^{2+}	-35.6	13.82(1)	143.1	20	C	μ=0.1(KNO_3)	64Ab	

Metal Ion	ΔH, kJ /mol	Log K	ΔS, J /mol K	T,°C	M	Conditions	Ref.	Re-marks
H23, continued								
Zn^{2+}	-16.7	12.68(1)	185.8	20	C	μ=0.1(KNO_3)	64Ab	
Cd^{2+}	-17.82	11.9(1)	166.9	20	C	μ=0.1(KNO_3)	64Ab	
Hg^{2+}	-87.74	21.58(1)	114.2	20	C	μ=0.1(KNO_3)	64Ab	
Pb^{2+}	-31.51	--	--	20	C	μ=0.1(KNO_3)	64Ab	
						R:Pb^{2+}+L+H_2O= HLPbOH		
H24	HEXANE, 1,6-diamino-3-thio		$C_5H_{14}N_2S$			$H_2N(CH_2)_2S(CH_2)_3NH_2$		L
Ni^{2+}	-39.0	5.989(1)	-16.2	25	C	μ=0.5M(KNO_3)	79Ha	
	-21.3	2.75	-18	25	C	μ=0.5M(KNO_3)	79Ha	
						R:Ni^{2+}+HL^+=$NiHL^{3+}$		
	-38.9	3.86(2)	-56	25	C	μ=0.5M(KNO_3)	79Ha	
	-29.7	1.88	-63	25	C	μ=0.5M(KNO_3)	79Ha	
						R:NiL^{2+}+HL^+= $NiHL_2{}^{3+}$		
Cu^{2+}	-59.0	10.035(1)	-5.8	25	C	μ=0.5M(KNO_3)	79Ha	
	-29.3	4.66	-9	25	C	μ=0.5M(KNO_3)	79Ha	
						R:Cu^{2+}+HL^+=$CuHL^{3+}$		
	-32.7	2.86(2)	-54	25	C	μ=0.5M(KNO_3)	79Ha	
	-31.8	2.44	-59	25	C	μ=0.5M(KNO_3)	79Ha	
						R:CuL^{2+}+HL^+= $CuHL_2{}^{3+}$		
	-13.4	4.60	-43	25	C	μ=0.5M(KNO_3)	79Ha	
						R:CuL^{2+}+OH^-= $Cu(OH)L^+$		
H25	HEXANEDIOIC ACID		$C_6H_{10}O_4$			$HO_2C(CH_2)_4CO_2H$		
Cd^{2+}	-229.2	10.21(25°)	-564.4	25&35	T	μ=2.0M(KNO_3)	79Ga	a,b
						R:Cd^{2+}+L^{2-}+A=CdLA A=1,2-Diamino-ethane		
	-206.7	10.27(25°)	-483.2	25&35	T	μ=2.0M(KNO_3)	79Ga	a,b
						R:Cd^{2+}+2L^{2-}+A= $CdAL_2{}^{2-}$ A=1,2-Diaminoethane		
	-79.96	11.00(25°)	-53.56	25&35	T	μ=2.0M(KNO_3)	79Ga	a,b
						R:Cd^{2+}+L^{2-}+2A= $CdLA_2$ A=1,2-Diaminoethane		
	-31.2	2.873(30°)	-47.6	30&40	T	μ=2M($NaClO_4$)	75Ba	a,b
						R:Cd^{2+}+L^{2-}+A^{2-}= $CdLA^{2-}$ A=Pentanedioate		

Metal Ion	ΔH, kJ /mol	Log K	ΔS, J /mol K	T,°C	M	Conditions	Ref.	Remarks
H25, continued								
Cd^{2+}, cont.	-39.4	3.061(30°)	-71.0	30&40	T	μ=2M($NaClO_4$) R:Cd^{2+}+L^{2-}+2A^{2-}= $CdLA_2^{2-}$ A=Pentanedioate	75Ba	a,b
	-10.5	2.727(30°)	17.6	30&40	T	μ=2M($NaClO_4$) R:Cd^{2+}+2L^{2-}+A= CdL_2A^{2-} A=Pentanedioate	75Ba	a,b
H26	2,4-HEXANEDIONE, 5-methyl-1-phenyl-	$C_{13}H_{16}O_2$				$C_6H_5CH_2C(O)CH_2C(O)CH(CH_3)CH_3$		
Gd^{3+}	1.59	0.398(25°)	13.0	-20-50	T	S:Chlorobenzene R:AlL_3(cis)= AlL_3(trans)	71Hj	b,q: ±2.1
Al^{3+}	0.63	0.398(25°)	9.6	-20-80	T	S:Chlorobenzene R:AlL_3(cis)= AlL_3(trans)	71Hj	b,q: ±2.1
H27	HEXANOIC ACID, 2-amino-	$C_6H_{13}O_2N$				$CH_3(CH_2)_3CH(NH_2)CO_2H$		
Co^{2+}	-24.7	7.89(1-2)	67	25	C	μ=0.05M(KCl)	72Gd 72Gc	
Ni^{2+}	-16.7	5.27(1)	45	25	C	μ=0.2M(KCl)	74Sn	
	-16.0	4.34(2)	30	25	C	μ=0.2M(KCl)	74Sn	
	-31.4	9.74(1-2)	79	25	C	μ=0.05M(KCl)	72Gd 72Gc	
H28	HEXANOIC ACID, 6-amino	$C_6H_{13}O_2N$				$H_2NCH_2(CH_2)_3CH_2CO_2H$		
Ag^+	-24.1	3.59(1)	-12.1	25	C	0.5M(KNO_3)	70Va	
	-25.9	3.59(25°,1)	-18.8	15-40	T	0.5M(KNO_3)	70Va	b
	-31.0	5.39(2)	-16.3	25	C	0.5M(KNO_3)	70Va	
	-32.1	5.39(25°,2)	-33.5	15-40	T	0.5M(KNO_3)	70Va	b
H29	3-HEXANONE, 2,2,5,5-tetramethyl-	$C_{10}H_{20}O$				$(CH_3)_3CC(L)CH_2C(CH_3)_3$		L
SbA_5	-49.29	2.423(1)	-145.6	25	C	S:Ethylene chloride A=Cl	67Of	
H30	2-HEXEN-4-ONE, 2-amino-5,5-dimethyl-	$C_8H_{15}ON$				$C(CH_3)_3C(O)CH:C(NH_2)CH_3$		L
Co^{2+}	11.72	-0.385(25°)	31.92	-38-52	T	S:15v% Tetra-methylsilane in $CHCl_3$	68Ea	b,t

Metal Ion	ΔH, kJ /mol	Log K	ΔS, J /mol K	T,°C	M	Conditions	Ref.	Re-marks
H31	2-HEXEN-4-ONE, 5,5-dimethyl-2-methylamino-		$C_9H_{17}ON$			$(CH_3)_3CC(O)CH:C-(NHCH_3)CH_3$		L
Ni^{2+}	21.25	-2.02(25°)	32.64	17-107	T	S:$CDCl_3$	68Ea	b,t
H32	5-HEXEN-1-OL		$C_6H_{12}O$			$CH_2:CHCH_2CH_2CH_2CH_2OH$		
Ag^+	-27.7	2.04(25°,1)	-53.7	25-45	T	μ∿0	77Ha	
	-21.0	1.52(25°,1)	-41.6	25-45	T	S:Methanol	77Ha	
	-19.7	1.04(25°,1)	-46.0	25-45	T	S:Ethanol	77Ha	
	-25.8	2.26(25°,1)	-42.9	25-45	T	S:Propylene carbonate	77Ha	
H33	3-HEXYNE		C_6H_{10}			$CH_3CH_2C:CCH_2CH_3$		L
Ag^+	-18.8	-0.883(25°)	46.4	25-35	T	μ=1.0 CL=$7x10^{-3}M$ CL=$7x10^{-3}M$ R:Ag^++$L_{(liquid)}$=AgL^+	56Hb	b
H34	3-HEXYNE, 2,2,5,5-tetramethyl-		$C_{10}H_{18}$			$CH_3C(CH_3)_2C:CC(CH_3)_2CH_3$		L
Ag^+	-20.9	-2.719(25°)	18.0	25-35	T	μ=1.0 CL=$1.5x10^{-4}M$ R:Ag^++$L_{(liquid)}$=AgL^+	57Hb	b
H35	HISTAMINE		$C_5H_9N_3$			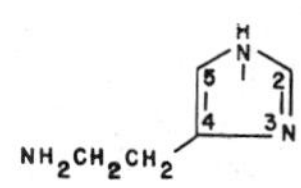		
Be^{2+}	-118.8	7.12(25°,1)	-259	0-45	T	μ=0.12M(KCl) CM≅$2x10^{-3}M$ CL≅$8x10^{-3}M$	69Cb 70Cf	b
	-45.6	5.35(25°,2)	-50	0-45	T	μ=0.12M(KCl) CM≅$2x10^{-3}M$ CL≅$8x10^{-3}M$	69Cb 70Cf	b
Mn^{2+}	-2.1	2.98(15°,1)	49.8	15&40	T	μ=0.2M(KNO_3)	71Rb	a,b
	-2.1	(2)	--	15&40	T	μ=0.2M(KNO_3)	71Rb	a,b
Co^{2+}	-28.9	5.08(30°,1)	-4	10-40	T	μ→0	61Nb	b
	-18.4	5.16(25°,1)	37.2	15-40	T	μ=0.2M(KNO_3)	71Rb	b
	-21	5.34(30°,1)	33	30-50	T	1M(KNO_3)	56Ha	b
	-20.1	3.76(30°,2)	4	10-40	T	μ→0	61Nb	b
	-19.4	3.64(25°,2)	5.0	15-40	T	μ=0.2M(KNO_3)	71Rb	b
	-42	3.75(30°,2)	-67	30-50	T	1M(KNO_3)	56Ha	b
	-29	1.88(30°,3)	-63	30-50	T	1M(KNO_3)	56Ha	b
Ni^{2+}	-29.7	6.69(1)	29	25	T	μ=0	59Sa	
	-45.6	6.77(30°,1)	-21	10-40	T	μ→0	61Nb	b

Metal Ion	ΔH, kJ /mol	Log K	ΔS, J /mol K	T,°C	M	Conditions	Ref.	Remarks
H35, continued								
Ni^{2+}, cont.	-32.2	6.45(1)	21	37	T	μ=0	59Sa	
	-35.1	6.22(1)	12	50	T	μ=0	59Sa	
	-17	6.88(25°,1)	42	0&25	T	μ=0.135	55Ma	a,b,f:3
	-42.7	6.85(1)	-12	25	C	μ=0.2M(KCl)	76Ge	
	-26.4	6.70(25°,1)	40.6	15-40	T	μ=0.2M(KNO_3)	71Rb	b
	-33	6.87(30°,1)	21	30-50	T	μ=0.2M(KNO_3)	71Rb	b
	-34.3	5.01(30°,2)	-17	10-40	T	$\mu\to0$	61Nb	b
	-17	5.03(25°,2)	42	0&25	T	μ=0.135	55Ma	a,b,f:3
	-36.8	5.04(2)	-27	25	C	μ=0.2M(KCl)	76Ge	
	-27.6	5.03(25°,2)	3.8	15-40	T	μ=0.2M(KNO_3)	71Rb	b
	-17	4.96(30°,2)	42	30-50	T	1M(KNO_3)	56Ha	b
	-9.6	3.25(30°,3)	29	10-40	T	$\mu\to0$	61Nb	b
	-17	3.09(25°,3)	42	0&25	T	μ=0.135	55Ma	a,b,f:3
	-25	3.08(30°,3)	-25	30-50	T	1M(KNO_3)	56Ha	b
Cu^{2+}	-43.1	9.47(1)	38	25	T	μ=0	60Se	
	-64.8	9.50(30°,1)	-4	10-40	T	$\mu\to0$	61Nb	b
	-46.9	9.20(1)	25	37	T	μ=0	60Se	
	-50.6	8.80(1)	12	50	T	μ=0	60Se	
	-47.6	9.56(1)	23.0	25	C	μ=0.1M(NO_3^-) pH<6	70Ec	
	-29	9.55(25°,1)	50	0&25	T	μ=0.135	55Ma	a,b
	-56.4	9.56(1)	-5.8	25	C	μ=0.16(KNO_3)	70Md	
	-54.3	9.58(1)	1	25	C	μ=0.2M(KCl)	76Sp 76Sq 76Ge	
	-56.1	9.60(1)	-4	25	C	μ=0.2M(KCl)	73Ga	
	-51.5	9.59(25°,1)	11.7	15-35	T	μ=0.2M(KNO_3) CM=2.0x10^{-3}	79Mg	b
	-45.90	9.53(25°,1)	28.45	15-40	T	μ=0.2M(KNO_3)	71Rb	b
	-43.1	9.47(1)	36.4	25	T	--	56Sa	
	-46.9	9.20(1)	25.5	37	T	--	56Sa	
	-50.6	8.80(1)	12.1	50	T	--	56Sa	
	-48.5	6.45(30°,2)	-38	10-40	T	$\mu\to0$	61Nb	b
	-40.1	6.57(2)	9.2	25	C	μ=0.1M(NO_3^-) pH<6	70Ec	
	-29	6.48(25°,2)	50	0&25	T	μ=0.135	55Ma	a,b
	-45.4	6.37(2)	-30	25	C	μ=0.16(KNO_3)	70Md	
	-41.1	6.48(2)	-14	25	C	μ=0.2M(KCl)	76Sp 76Sq 76Ge	
	-41.8	6.49(2)	-12	25	C	μ=0.2M(KCl)	73Ga	
	-41.0	6.52(25°,2)	-12.6	15-35	T	μ=0.2M(KNO_3) CM=2.0x10^{-3}	79Mg	b
	-42.68	6.21(25°,2)	-24.27	15-40	T	μ=0.2M(KNO_3)	71Rb	b
	-129.0	21.79	-16	25	C	μ=0.2M(KCl) R:Cu+H+2L=$CuHL_2$	76Sq	
	-129.0	21.76	-16	25	C	μ=0.2M(KCl) R:Cu^{2+}+H^++2L= CuL_2H^{3+}	76Ge	
	-45.6	6.99	-19	25	C	μ=0.2M(KCl) R:2Cu^{2+}+2H_2O+2L= $Cu_2L_2(OH)_2^{2+}$+2H^+	76Ge	

Metal Ion	ΔH, kJ /mol	Log K	ΔS, J /mol K	T,°C	M	Conditions	Ref.	Remarks
H35, continued								
Cu^{2+}, cont.	-80.8	17.00	54	25	C	μ=0.2M(KCl) R:Cu+A+L=CuAL A=Glycine	76Sp	
	-104.6	18.66	6	25	C	μ=0.2M(KCl) R:Cu+A+L=CuAL A=1,2-Diamino-ethane	76Sp	
	-91.6	16.90	16	25	C	μ=0.2M(KCl) R:Cu+A+L=CuAL A=1,2-Diamino-N,N'-dimethyl-ethane	76Sp	
CuA^{2+}	-37.2	7.15(25°,1)	12.1	15-35	T	μ=0.2M(KNO_3) CM=2.0x10^{-3} A=2,2'-Bipyridyl	79Mg	b
CuA	-104.6	18.66(1)	6	25	C	μ=0.2M(KCl) A=1,2-Diamino-ethane	76Sq	
	-91.6	16.90(1)	16	25	C	μ=0.2M(KCl) A=1,2-Diamino-N,N'-dimethyl-ethane	76Sq	
	-80.8	17.00(1)	54	25	C	μ=0.2M(KCl) A=Glycine	76Sq	
Zn^{2+}	-16.7	5.15(25°,1)	42.7	15-40	T	μ=0.2M(KNO_3)	71Fb	b
	-23.0	5.23(25°,1)	21	25	T	--	62Kb	
	-15.5	4.84(25°,2)	40.6	15-40	T	μ=0.2M(KNO_3)	71Fb	b
Cd^{2+}	-22.6	4.80(1)	17	25	T	μ=0	59Sa	
	-31.0	4.8(30°,1)	-12	10-40	T	$\mu\to0$	61Nb	b
	-24.7	4.62(1)	8	37	T	μ=0	59Sa	
	-26.8	4.44(1)	4	50	T	μ=0	59Sa	
	-49.4	8.57(1)	4	25	T	μ=0.1(KNO_3)	64Ad	
	-18.4	3.39(30°,2)	4	10-40	T	$\mu\to0$	61Mb	b
H36 HISTIDINE			$C_6H_9O_2N_3$			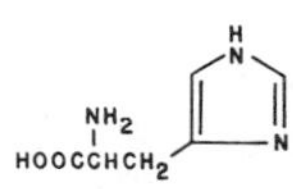		L^-
Be^{2+}	-135.6	5.52(35°,1)	-330	25-40	T	μ=0.12(KCl) CM≅2x10^{-3}M CL≅9x10^{-3}M	70Cf	b
	-35.6	4.50(35°,2)	-29	25-40	T	μ=0.12(KCl) CM≅2x10^{-3}M CL≅9x10^{-3}M	70Cf	b
La^{3+}	-11.51	4.10(1)	37.71	25	C	μ=3.00M($NaClO_4$)	70Jc	
	-9.12	3.40(1)	35.6	37	C	μ=3.00M($NaClO_4$)	71Ja	
	31.0	1.30(2)	129.3	25	C	μ=3.00M($NaClO_4$)	70Jc	
	32.47	3.45(2)	170.7	37	C	μ=3.00M($NaClO_4$)	71Ja	
	-11.59	11.75	185.8	25	C	μ=3.00M($NaClO_4$) R:La^{3+}+HL=$LaHL^{3+}$	70Jc	

Metal Ion	ΔH, kJ /mol	Log K	ΔS, J /mol K	T,°C	M	Conditions	Ref.	Remarks
H36, continued								
La^{3+}, cont.	-8.20	11.07	185.4	37	C	μ=3.00M($NaClO_4$)	71Ja	q:±0.88
						R:La^{3+}+HL=$LaHL^{3+}$		
Pr^{3+}	-10.29	4.36(1)	49.0	25	C	μ=3.00M($NaClO_4$)	70Jc	
	-8.54	3.69(1)	43.1	37	C	μ=3.00M($NaClO_4$)	71Ja	
	31.4	1.84(2)	140.2	25	C	μ=3.00M($NaClO_4$)	70Jc	
	33.93	4.09(2)	187.9	37	C	μ=3.00M($NaClO_4$)	71Ja	
	-12.30	11.77	184.1	25	C	μ=3.00M($NaClO_4$)	70Jc	
						R:Pr^{3+}+HL=$PrHL^{3+}$		
	-8.83	11.04	182.8	37	C	μ=3.00M($NaClO_4$)	71Ja	q:±0.96
						R:Pr^{3+}+HL=$PrHL^{3+}$		
Nd^{3+}	-5.48	3.95(1)	57.7	37	C	μ=3.00M($NaClO_4$)	71Ja	q:±0.63
	-7.82	4.40(1)	58.2	25	C	μ=3.00M($NaClO_4$)	70Jc	
	28.32	2.19(2)	136.8	25	C	μ=3.00M($NaClO_4$)	70Jc	
	30.04	4.17(2)	177.0	37	C	μ=3.00M($NaClO_4$)	71Ja	
	-9.96	11.77	190.8	25	C	μ=3.00M($NaClO_4$)	70Jc	q:±0.92
						R:Nd^{3+}+HL=$NdHL^{3+}$		
	-6.36	11.20	193.7	37	C	μ=3.00M($NaClO_4$)	71Ja	q:±1.05
						R:Nd^{3+}+HL=$NdHL^{3+}$		
Sm^{3+}	-2.30	4.46(1)	77.4	25	C	μ=3.00M($NaClO_4$)	70Jc	q:±0.50
	-0.25	4.37(1)	82.8	37	C	μ=3.00M($NaClO_4$)	71Ja	q:±0.54
	17.40	4.25(2)	140.6	25	C	μ=3.00M($NaClO_4$)	70Jc	
	20.38	4.41(2)	150.2	37	C	μ=3.00M($NaClO_4$)	71Ja	
	-10.50	11.78	190.0	25	C	μ=3.00M($NaClO_4$)	70Jc	
						R:Sm^{3+}+HL=$SmHL^{3+}$		
	-5.02	11.18	197.9	37	C	μ=3.00M($NaClO_4$)	71Ja	q:±0.75
						R:Sm^{3+}+HL=$SmHL^{3+}$		
Gd^{3+}	7.11	4.39(1)	107.9	25	C	μ=3.00M($NaClO_4$)	70Jc	
	9.04	4.94(1)	123.8	37	C	μ=3.00M($NaClO_4$)	71Ja	
	1.51	4.22(2)	86.2	25	C	μ=3.00M($NaClO_4$)	70Jc	q:±0.79
	3.85	4.22(2)	92.9	37	C	μ=3.00M($NaClO_4$)	71Ja	
	-4.14	11.47	205.8	25	C	μ=3.00M($NaClO_4$)	70Jc	q:±0.71
						R:Gd^{3+}+HL=$GdHL^{3+}$		
	-0.75	11.30	213.8	37	C	μ=3.00M($NaClO_4$)	71Ja	q:±0.84
						R:Gd^{3+}+HL=$GdHL^{3+}$		
Dy^{3+}	9.20	4.40(1)	115.1	25	C	μ=3.00M($NaClO_4$)	70Jc	
	11.30	5.09(1)	133.0	37	C	μ=3.00M($NaClO_4$)	71Ja	
	-0.58	4.74(2)	89.1	25	C	μ=3.00M($NaClO_4$)	70Jc	q:±1.00
	1.17	4.76(2)	95.0	37	C	μ=3.00M($NaClO_4$)	71Ja	q:±1.09
	4.10	11.16	228.0	25	C	μ=3.00M($NaClO_4$)	70Jc	q:±0.79
						R:Dy^{3+}+HL=$DyHL^{3+}$		
	5.52	11.56	238.9	37	C	μ=3.00M($NaClO_4$)	71Ja	q:±0.88
						R:Dy^{3+}+HL=$DyHL^{3+}$		
Er^{3+}	7.07	4.49(1)	110.0	25	C	μ=3.00M($NaClO_4$)	70Jc	q:±0.71
	8.49	4.99(1)	123.0	37	C	μ=3.00M($NaClO_4$)	71Ja	q:±0.79
	21.30	4.50(2)	92.9	25	C	μ=3.00M($NaClO_4$)	70Jc	q:±0.79

Metal Ion	ΔH, kJ /mol	Log K	ΔS, J /mol K	T,°C	M	Conditions	Ref	Re-marks
H36, continued								
Er^{3+}, cont.	4.48	4.92(2)	108.4	37	C	μ=3.00M($NaClO_4$)	71Ja	
	1.55	11.18	218.8	25	C	μ=3.00M($NaClO_4$) R:Er^{3+}+HL=$ErHL^{3+}$	70Jc	q:±0.58
	4.98	11.40	234.3	37	C	μ=3.00M($NaClO_4$) R:Er^{3+}+HL=$ErHL^{3+}$	71Ja	q:±0.79
Yb^{3+}	6.61	4.23(1)	102.9	25	C	μ=3.00M($NaClO_4$)	70Jc	q:±0.71
	7.57	4.76(1)	115.5	37	C	μ=3.00M($NaClO_4$)	71Ja	q:±0.84
	-1.51	5.60(2)	102.1	25	C	μ=3.00M($NaClO_4$)	70Jc	q:±0.71
	1.46	5.55(2)	110.9	37	C	μ=3.00M($NaClO_4$)	71Ja	q:±0.79
	1.21	11.40	223.0	25	C	μ=3.00M($NaClO_4$) R:Yb^{3+}+HL=$YbHL^{3+}$	70Jc	q:±0.58
	4.73	11.60	237.2	37	C	μ=3.00M($NaClO_4$) R:Yb^{3+}+HL=$YbHL^{3+}$	71Ja	q:±0.71
Mn^{2+}	-11.34	3.915(1)	36.8	25	C	μ=3.0($NaClO_4$)	70Wd	
	-10.54	2.691(2)	16.3	25	C	μ=3.0($NaClO_4$)	70Wd	
	-5.0	5.71(40°,1-2)	93.7	15&40	T	μ=0.2(KNO_3)	69Ra	a,b
Fe^{2+}	-18.28	5.883(1)	51.5	25	C	μ=3.0($NaClO_4$)	70Wd	
	-19.79	4.543(2)	20.5	25	C	μ=3.0($NaClO_4$)	70Wd	
	-16.7	8.50(40°,1-2)	110.0	15&40	T	μ=0.2(KNO_3)	69Ra	a,b
[Fe(II)]	-66.9	4.36(25°,1-2)	-141.0	15.5-36.5	T	μ=0.10($NaClO_4$) pH=7.0 [Fe(II)]= Iron(II) 2,4-dicysteine-substituted mesoporphyrin (heme c)	76Gf	
[Fe(III)]	-50.6	4.67(25°,1-2)	-80.3	15.5-36.5	T	μ=0.10($NaClO_4$) pH=6.1 [Fe(III)] =Iron(III) 2,4-dicysteine-substituted mesoporphyrin (hemin c)	76Gf	
Co^{2+}	-21	7.32(25°,1)	71	15-35	T	μ=0	79Va	b
	-33.5	7.10(15°,1)	20.33	0-40	T	μ=0.25(KCl)	65Ad	b
	-23.56	7.442(1)	63.6	25	C	μ=3.0($NaClO_4$)	70Wd	
	-4	5.72(25°,2)	92	15-35	T	μ=0	79Va	b
	-52.7	5.62(15°,2)	-75.3	0-40	T	μ=0.25(KCl)	65Ad	b
	-27.70	6.036(2)	22.6	25	C	μ=3.0($NaClO_4$)	70Wd	
	-48.49	(1-2)	--	25	C	μ=0.02M	72Pk	
	-50.42	(1-2)	--	25	C	μ=0.10(KNO_3) (bis complex of D or L-Histidine)	77Pa	
	-40.2	12.84(25°,1-2)	111.3	15-45	T	μ=0.2(KNO_3)	69Ra	b
	-1.39	0.599	6.81	25	C	μ=0.10(KNO_3) R:0.5Co$(L-L)_2$+ 0.5Co$(D-L)_2$= Co(L-L)(D-L) L=Histidine	77Pa	

Metal Ion	ΔH, kJ /mol	Log K	ΔS, J /mol K	T, °C	M	Conditions	Ref.	Remarks
H36, continued								
Ni^{2+}	-25	8.90(25°,1)	87	15-35	T	μ=0	79Va	b
	-33.9	8.69(1)	53	25	C	0.1M(KNO_3)	69Tb	
	-47.7	8.75(15°,1)	2.9	0-40	T	μ=0.25(KCl)	65Ad	b
	-36.11	9.199(1)	54.8	25	C	μ=3.0($NaClO_4$)	70Wd	
	-21	7.00(25°,2)	64	15-35	T	μ=0	79Va	b
	-33.9	6.83(2)	17	25	C	0.1M(KNO_3)	69Tb	
	-57.3	7.08(15°,2)	-64.0	0-40	T	μ=0.25(KCl)	65Ad	b
	-42.93	7.454(2)	-1.2	25	C	μ=3.0($NaClO_4$)	70Wd	
	-69.4	15.5(1-2)	20.5	22	C	0.1M(KNO_3)	67Sc	
	-71.63	(1-2)	--	25	C	μ=0.10(KNO_3) (bis complex of D or L-Histidine)	77Pa	
	-68.91	15.5(1-2)	64.8	25	C	μ=0.1M Histidine(-)	70Bd 71Be	y:ΔH (70Bd)
	-69.54	15.5(1-2)	64.8	25	C	μ=0.1M Histidine(+)(rac)	70Bd	y:ΔH (70Bd)
	-70.92	15.7(1-2)	61.92	25	C	μ=0.1M Histidine(+)(rac)	70Bd	y:ΔH (70Bd)
	-56.9	16.05(25°,1-2)	117.2	15-45	T	μ=0.2(KNO_3)	69Ra	b
	-2.55	0.780	6.37	25	C	μ=0.10(KNO_3) R:0.5Co(L-L)$_2$+ 0.5Co(D-L)$_2$= Co(L-L)(D-L) L=Histidine	77Pa	
Rh_2A_4	-32.7	4.38(1)	-25.9	25	C	A=MeO_2Ac	77Da	
	-31.8	2.79(2)	-53.1	25	C	A=MeO_2Ac	77Da	
	-30.4	4.05(1)	-24.7	25	C	A=OAc	77Da	
	-26.4	2.69(2)	-36.8	25	C	A=OAc	77Da	
	-28.0	4.14(1)	-15.1	25	C	A=Propionate	77Da	
	-25.1	2.71(2)	-32.2	25	C	A=Propionate	77Da	
Cu^{2+}	-48.4	10.01(1)	29	25	C	μ=0.16(KNO_3)	70Md	
	-43.89	10.086(1)	46.0	25	C	μ=3.0($NaClO_4$)	72Wc	
	-40.8	8.01(2)	17	25	C	μ=0.16(KNO_3)	70Md	
	-40.04	8.940(2)	36.90	25	C	μ=3.0($NaClO_4$)	72Wc	
	-89.1	18.8(1-2)	58.2	21.9	C	0.1M(KNO_3)	67Sc	
	-81.58	(1-2)	--	25	C	μ=0.1(KNO_3) (bis complex of D or L-Histidine)	77Pa	
	-83.64	(1-2)	--	25	C	μ=0.1M Histidine(-)	70Bd	y:ΔH (70Bd)
	-82.69	(1-2)	--	25	C	μ=0.1M Histidine(+)(rac)	70Bd	y:ΔH (70Bd)
	-83.64	(1-2)	--	25	C	μ=0.1M Histidine(+)(rac)	70Bd	y:ΔH (70Bd)
	-92.5	19.40(25°,1-2)	60.7	15-40	T	μ=0.2(KNO_3)	69Ra	b
	-114.6	11.2(1-2)	-183.2	0-40	T	μ=0.25(KCl)	65Ad	b
	-66.98	15.615	74.0	25	C	μ=3.0($NaClO_4$) R: $Cu^{2+}+L^-+H^+=CuLH^{2+}$	72Wc	
	-107.61	25.884	134.7	25	C	μ=3.0($NaClO_4$) R: $Cu^{2+}+2L^-+H^+=CuL_2H^+$	72Wc	

Metal Ion	ΔH, kJ /mol	Log K	ΔS, J /mol K	T,°C	M	Conditions	Ref.	Re-marks
H36, continued								
Cu^{2+}, cont.	-129.70	30.750	153.6	25	C	μ=3.0($NaClO_4$) R:Cu^{2+}+2L^-+2H^+= $CuL_2H_2^{2+}$	72Wc	
	-26.61	5.97	25.1	25	C	μ=3.0($NaClO_4$) R:Cu^{2+}+HL=$CuHL^{2+}$	72Wc	
	-48.91	11.5	55.6	25	C	μ=3.0($NaClO_4$) R:Cu^{2+}+2HL= $CuH_2L_2^{2+}$	72Wc	
	-67.5	18.60	129.7	25	C	μ=3.0M($NaClO_4$) R:Cu^{2+}+L^-+A^-=CuLA A=Asparagine	75Bd	
	-88.2	23.32	150.7	25	C	μ=3.0M($NaClO_4$) R:Cu^{2+}+HL+A^-= $CuHLA^+$ A=Asparagine	75Bd	
	2.05	0.302	12.6	25	C	μ=0.1(KNO_3) R:0.5Co(L-L)$_2$+ 0.5Co(D-L)= Co(L-L)(D-L) L=Histidine	77Pa	
	-67.4	18.61	130.3	25	C	μ=3.0M($NaClO_4$) R:Cu^{2+}+L^-+A^-=CuLA A=THreonine	76Bd	
	-103.5	23.42	101.3	25	C	μ=3.0M($NaClO_4$) R:Cu^{2+}+$HL^+$$A^-$= $CuHLA^+$ A=Threonine	75Bd	
CuL	-23.10	5.53	28.4	25	C	μ=3.0($NaClO_4$)	72Wc	
	-23.30	6.16	39.7	25	C	μ=3.0($NaClO_4$) R:CuL^++HL=$CuHL_2^+$	72Wc	
CuL_2	-23.68	6.86	51.9	25	C	μ=3.0($NaClO_4$) R:CuL_2+H^+=$CuHL_2^+$	72Wc	
CuAL	-40.58	10.28(1)	60.2	25	C	μ=3.0($NaClO_4$) A=H	72Wc	
	-22.30	5.50	30.5	25	C	μ=3.0($NaClO_4$) A=H R:$CuAL^{2+}$+AL= $CuA_2L_2^{2+}$	72Wc	
$CuAL_2$	-22.09	5.58	18.8	25	C	μ=3.0($NaClO_4$) A=H R:$CuAL_2^+$+A= $CuA_2L_2^{2+}$	72Wc	
Zn^{2+}	-28.9	6.78(15°,1)	28.9	0-40	T	μ=0.25(KCl)	65Ad	b
	-23.10	7.068(1)	57.7	25	C	μ=3.0($NaClO_4$)	7CWd	
	-26.07	5.673(2)	21.3	25	C	μ=3.0($NaClO_4$)	7CWd	
	-49.0	11.8(1-2)	61.1	21.3	C	0.1M(KNO_3)	67Sc	
	-50.25	(1-2)	--	25	C	μ=0.1(KNO_3)	77Pa	

Metal Ion	ΔH, kJ /mol	Log K	ΔS, J /mol K	T,°C	M	Conditions	Ref.	Re-marks
H36, continued								
Zn^{2+}, cont.						(bis complex of D or L-Histidine)		
	-2.48	0.420	-0.27	25	C	0.1(KNO_3) R:0.5Co(L-L)$_2$+ 0.5Co(D-L)$_2$= Co(L-L)(D-L) L=Histidine	77Pa	
	-47.70	12.05(1-2)	71.1	25	C	μ=0.1M Histidine(-)	70Bd 71Be	y:ΔH (70Bd)
	-47.86	12.05(1-2)	69.9	25	C	μ=0.1M Histidine(+)(rac)	70Bd 71Be	y:ΔH (70Bd)
	-49.20	12.16(1-2)	67.8	25	C	μ=0.1M Histidine(+)(rac)	70Bd 71Be	y:ΔH (70Bd)
	-34.7	12.44(25°,1-2)	122.2	15-40	T	μ=0.2(KNO_3)	69Ra	b
	-47.7	5.55(15°,1-2)	-59.4	0-40	T	μ=0.25(KCl)	65Ad	b
Cd^{2+}	-59.4	10.20(1)	0	25	T	μ=0.1(KNO_3)	64Ad	
H37 HISTIDINE, histidyl-			$C_{12}H_{16}O_4N_6$			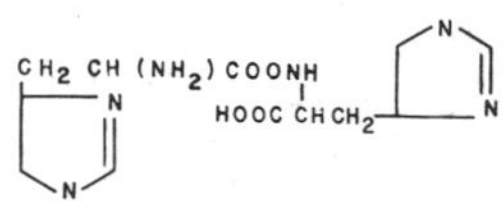		
Be^{2+}	-210.4	4.14(35°,1)	-598	25-45	T	μ=0.12(KCl) CM≅2x10^{-3}M CL≅9x10^{-3}M	70Cf	b
	-84.9	3.45(35°,2)	-209	25-45	T	μ=0.12(KCl) CM≅2x10^{-3}M CL≅9x10^{-3}M	70Cf	b
H38 HISTIDINE, methyl ester			$C_7H_{11}O_2N_3$			$C_3H_3N_2[CH_2CH(NH_2)COOCH_3]$		L^-
Be^{2+}	-59.0	4.50(35°,1)	-105	25-45	T	μ=0.12(KCl) CM≅2x10^{-3}M CL≅9x10^{-3}M	70Cf	b
	-42.2	3.22(35°,2)	-75	25-45	T	μ=0.12(KCl) CM≅2x10^{-3}M CL≅9x10^{-3}M	70Cf	b
Co^{2+}	-66.9	5.00(15°,1)	-136.8	0-40	T	μ=0.25(KCl)	65Ad	b
	-66.9	3.57(15°,2)	-164.0	0-40	T	μ=0.25(KCl)	65Ad	b
Ni^{2+}	-66.9	6.65(15°,1)	-104.6	0-40	T	μ=0.25(KCl)	65Ad	b
	-66.9	5.14(15°,2)	-133.9	0-40	T	μ=0.25(KCl)	65Ad	b
	-72.0	2.76(15°,3)	-197.5	0-40	T	μ=0.25(KCl)	65Ad	b
Cu^{2+}	-51.5	8.52(25°,1)	-9.6	25-50	T	μ=0.1M	71Hf	b,q:±2.9
	-46.2	8.46(1)	7.1	25	C	μ=0.16(KNO_3)	70Md	
	-76.6	9.12(15°,1)	-91.6	0-40	T	μ=0.25(KCl)	65Ad	b
	-37.6	5.98(25°,2)	-12.6	25-50	T	μ=0.1M	71Hf	b
	-34.1	5.92(2)	-1.2	25	C	μ=0.16(KNO_3)	70Md	
	-66.9	6.54(15°,2)	-107.5	0-40	T	μ=0.25(KCl)	65Ad	b

Metal Ion	ΔH, kJ /mol	Log K	ΔS, J /mol K	T,°C	M	Conditions	Ref.	Re- marks
H38, continued								
Zn^{2+}	-38.5	4.82(15°,1)	-40.63	0-40	T	μ=0.25(KCl)	65Ad	b
	-38.5	3.93(15°,2)	-58.2	0-40	T	μ=0.25(KCl)	65Ad	b
Cd^{2+}	-46.9	7.42(1)	-12	25	T	μ=0.1(KNO_3)	64Ab	
H39 HYDANTOIN			$C_3H_4O_2N_2$			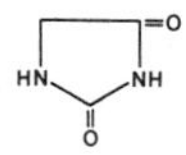		L
Co^{2+}	-9.66	2.93(25°,1)	23.8	25	C	μ=0.5M(KNO_3)	79Bd	
	-8.74	1.31(25°,2)	-4.6	25	C	μ=0.5M(KNO_3)	79Bd	
	-11.72	0.92(25°,3)	-22.2	25	C	μ=0.5M(KNO_3)	79Bd	
Ni^{2+}	-11.67	3.33(25°,1)	24.7	25	C	μ=0.5M(KNO_3)	79Bd	
	-17.20	1.78(25°,2)	-23.8	25	C	μ=0.5M(KNO_3)	79Bd	
	-0.42	0.98(25°,3)	17.4	25	C	μ=0.5M(KNO_3)	79Bd	
Cu^{2+}	-18.83	4.73(25°,1)	27.2	25	C	μ=0.5M(KNO_3)	79Bd	
	-16.74	3.63(25°,2)	13.4	25	C	μ=0.5M(KNO_3)	79Bd	
	-21.34	3.40(25°,3)	-6.4	25	C	μ=0.5M(KNO_3)	79Bd	
Ag^{+}	-24.43	4.42(25°,1)	2.68	25	C	μ=0.5M(KNO_3)	79Bd	
	-28.49	4.63(25°,2)	-6.9	25	C	μ=0.5M(KNO_3)	79Bd	
Zn^{2+}	-8.24	3.09(25°,1)	31.4	25	C	μ=0.5M(KNO_3)	79Bd	
	-8.49	2.46(25°,2)	18.8	25	C	μ=0.5M(KNO_3)	79Bd	
	-10.46	3.25(25°,3)	26.8	25	C	μ=0.5M(KNO_3)	79Bd	
Cd^{2+}	-11.13	3.33(25°,1)	26.4	25	C	μ=0.5M(KNO_3)	79Bd	
	-9.37	2.01(25°,2)	7.1	25	C	μ=0.5M(KNO_3)	79Bd	
	-10.46	1.11(25°,3)	-13.8	25	C	μ=0.5M(KNO_3)	79Bd	
H40 HYDRAZINE, 1,1,2-trimethyl-silyl-			$C_9H_{28}N_2S_3$			$[(CH_3)_3S]_2NNH[S(CH_3)_3]$		
Li^{+}	18.4	1.155(-50°)	105	-80--20	T	S:Diethylether R:$2LiL=(LiL)_2$ (monomer=dimer)	72Wb	b
H41 HYDROAZOIC ACID			HN_3			HN_3		L
Fe^{3+}	8.41	0.233	32.6	25	T	μ=0 R:$Fe^{3+}+L=FeN_3^{2+}+H^{+}$	61Wa	
	8.49	--	--	25	T	1M(HNO_3) R:$Fe^{3+}+L=FeN_3^{2+}+H^{+}$	61Wa	
	7.91	--	--	25	T	1M(HNO_3) 4M($NaNO_3$) R:$Fe^{3+}+L=FeN_3^{2+}+H^{+}$	61Wa	
	8.79	--	--	25	T	2.5M(HNO_3) 2.5M($NaNO_3$)R:$Fe^{3+}+L=FeN_3^{2+}+H^{+}$	61Wa	
	10.12	--	--	25	T	5M(HNO_3) R:$Fe^{3+}+L=FeN_3^{2+}+H^{+}$	61Wa	

Metal Ion	ΔH, kJ /mol	Log K	ΔS, J /mol K	T,°C	M	Conditions	Ref.	Re-marks
H42 HYDROBROMIC ACID			HBr			HBr		L
$IrABC_2$ (trans)	-108.8	(1)	--	25	C	S:Benzene A=Cl B=CO C=PPh_3	79Bf	
I_2	-6.703	1.079(25°,1)	21.3	15-45	T	0-3M(HBr)	36Lb	b
H43 HYDROGEN			H_2			H_2		
$IrABC_2$ (trans)	-62.8	(1)	--	25	C	S:Benzene A=Cl B=CO C=PPh_3	79Bf	
H44 HYDROGEN CHLORIDE			HCl			HCl		L
$IrABC_2$ (trans)	-87.9	(1)	--	25	C	S:Benzene A=Cl B=CO C=PPh_3	79Bf	
	-108.8	(1)	--	25	C	S:Benzene A=Cl B=CO C=$PMePh_2$	79Bf	
I_2	4.339	0.093(25°,1)	16.3	25-45	T	μ=0	37La	b
H45 HYDROGEN PEROXIDE			O_2H_2			H_2O_2		L^-
TiO^{2+}	-44.31	11.96(1)	75.23	15	C	μ=0.5N($HClO_4$)	70Vb	
	-44.18	11.65(1)	74.94	25	C	μ=0.5N($HClO_4$)	70Vb	
	-44.18	11.36(1)	74.01	35	C	μ=0.5N($HClO_4$)	70Vb	
	-44.18	10.98(1)	71.42	45	C	μ=0.5N($HClO_4$)	70Vb	
	-44.64	12.01(1)	75.23	15	C	μ=1N($HClO_4$)	70Vb	
	-43.97	11.52(1)	73.00	25	C	μ=1N($HClO_4$)	70Vb	
	-44.35	11.36(1)	73.64	35	C	μ=1N($HClO_4$)	70Vb	
	-44.43	11.02(1)	71.25	45	C	μ=1N($HClO_4$)	70Vb	
	-43.81	11.37(1)	70.21	25	C	μ=2N($HClO_4$)	70Vb	
	-44.60	11.76(1)	70.29	15	C	μ=3N($HClO_4$)	70Vb	
	-43.72	11.18(1)	67.36	25	C	μ=3N($HClO_4$)	70Vb	
	-44.60	11.21(1)	69.79	35	C	μ=3N($HClO_4$)	70Vb	
	-44.85	10.9(1)	67.78	45	C	μ=3N($HClO_4$)	70Vb	
	-44.35	11.27(1)	66.94	25	C	μ=4N($HClO_4$)	70Vb	
	-44.02	11.32(1)	63.89	15	C	μ=5N($HClO_4$)	70Vb	
	-44.02	11.03	63.60	25	C	μ=5N($HClO_4$)	70Vb	
	-44.14	10.77(1)	63.01	35	C	μ=5N($HClO_4$)	70Vb	
	-44.60	10.65(1)	63.68	45	C	μ=5N($HClO_4$)	70Vb	
Nb^{5+}	-29.2	2.73(25°)	-46	15-45	T	S:63.51m% H_2SO_4 R:Either NbO_2+L= NbO_2L^+ or $NbOSO_4^+$ +L=$NbOSO_4L^+$	71Vb	b,j,q: ±6.3
	-32.6	3.18(25°)	-46	15-45	T	S:78.42m% H_2SO_4 R:Either NbO_2^++L= NbO_2L^+ or $NbOSO_4^+$ +L=$NbOSO_4L^+$	71Vb	b,j,q: ±4.2
	-34.7	3.56(25°)	-46	15-45	T	S:88.71m% H_2SO_4 R:Either NbO_2^++L=	71Vb	b,j,q: ±6.3

Metal Ion	ΔH, kJ /mol	Log K	ΔS, J /mol K	T,°C	M	Conditions	Ref.	Re-marks
H45, continued								
Nb^{5+}, cont.						NbO_2L^+ or $NbOSO_4^+$ $+L=NbOSO_4L^+$		
	-25.1	3.67(25°)	-12	5-45	T	S:94.85m% H_2SO_4 R:Either NbO_2^++L= NbO_2L^+ or $NbOSO_4^+$+ L=$NbOSO_4L^+$	71Vb	b,j,q: ±3.8
TaABn	-17.49	2.40(25°,1)	-4	15-55	T	S:63.5w% H_2SO_4 A=0 B=SO_4	73Vc	b,q: ±4.2
	-18.37	2.80(1)	-7.9	25	C	S:78.4w% H_2SO_4 A=0 B=SO_4	73Vc	
	-18.83	2.73(1)	-8.8	35	C	S:78.4w% H_2SO_4 A=0 B=SO_4	73Vc	
	-18.07	2.60(1)	-7.1	45	C	S:78.4w% H_2SO_4 A=0 B=SO_4	73Vc	
	-18.20	2.80(25°,1)	-7.5	15-55	T	S:78.4w% H_2SO_4 A=0 B=SO_4	73Vc	b,q: ±6.3
	-16.07	3.05(25°,1)	4	15-55	T	S:88.71w% H_2SO_4 A=0 B=SO_4	73Vc	b,q: ±4.2
	-22.2	3.45(25°,1)	-8	15-55	T	S:94.85w% H_2SO_4 A=0 B=SO_4	73Vc	b,q: ±4.2
Fe^{3+}	7.5	9.32(1)	205	20	T	μ=0	52Ua 49Ea	
Co^{3+}	-21	13.95(12.5°,1)	192	0-12.5	T	≅1M($HClO_4$)	57Bb	b
H46 HYDROGEN SULFIDE ION			HS			HS^-		L^-
[Fe^{3+}]	ΔH, obtained by temperature variation of K, is reported in graphical form as a function of pH for the reaction: $Hb(H_2O)+L^-=HbLH+H_2O$					Cow methaemoglobin	69Be	
[Fe^{3+}]	ΔH, obtained by temperature variation of K, is reported in graphical form as a function of pH in the reaction: $Hb(H_2O)+L^-=HbLH+H_2O$					Dog methaemoglobin	69Be	
[Fe^{3+}]	ΔH, obtained by temperature variation of K, is reported in graphical form as a function of pH in the reaction: $Hb(H_2O)+L^-=HbLH+H_2O$					Guinea Pig methaemoglobin	69Be	
[Fe^{3+}]	ΔH, obtained by temperature variation of K, is reported in graphical form as a function of pH in the reaction: $Hb(H_2O)+L^-=HbLH+H_2O$					Human methaemoglobin	69Be	
[Fe^{3+}]	ΔH, obtained by temperature variation of K, is reported in graphical form as a function of pH in the reaction: $Hb(H_2O)+L^-=HbLH+H_2O$					Mouse methaemoglobin	69Be	
[Fe^{3+}]	ΔH, obtained by temperature variation of K, is reported in graphical form as					Pigeon methaemoglobin	69Be	

Metal Ion	ΔH, kJ /mol	Log K	ΔS, J /mol K	T,°C	M	Conditions	Ref.	Remarks
H46, continued								
$[Fe^{3+}]$, cont.	a function of pH in the reaction: $Hb(H_2O)+L^-=HbLH+H_2O$							
$[Fe^{3+}]$	ΔH, obtained by temperature variation of K, is reported in graphical form as a reaction of pH in the reaction: $Hb(H_2O)+L^-=HbLH+H_2O$					Sperm whale metmyoglobin	69Be	
H47	HYDROTRIS (1-pyrazolyl) borate ion		$C_9H_{10}N_6B$			$HB(C_3H_3N_2)_3$	L^-	
Co^{2+}	-175.6	(1-2)	--	22-27	C	CM=0.02M S:Acetonitrile	75Jc	
Ni^{2+}	-192.2	(1-2)	--	22-27	C	CM=0.02M S:Acetonitrile	75Jc	
Cu^{2+}	-135.0	(1-2)	--	22-27	C	CM=0.02M S:Acetonitrile	75Jc	
Zn^{2+}	-153.0	(1-2)	--	22-27	C	CM=0.02M S:Acetonitrile	75Jc	
H48	HYDROXAMIC ACID, benzhydryl		$C_{14}H_{13}O_2N$			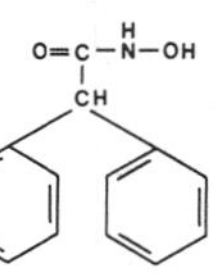	L	
Mn^{2+}	-11.46	4.60(30°,1)	50.12	20-50	T	μ=0.1M S:50v%Ethanol	79Rb	
	-10.63	3.80(30°,2)	37.53	20-50	T	μ=0.1M S:50v%Ethanol	79Rb	
Co^{2+}	-14.35	5.58(30°,1)	59.79	20-50	T	μ=0.1M S:50v%Ethanol	79Rb	
	-14.35	4.40(30°,2)	36.86	20-50	T	μ=0.1M S:50v%Ethanol	79Rb	
Ni^{2+}	-19.92	5.65(30°,1)	42.38	20-50	T	μ=0.1M S:50v%Ethanol	79Rb	
	-19.79	4.45(30°,2)	19.87	20-50	T	μ=0.1M S:50v%Ethanol	79Rb	
Cu^{2+}	-27.32	9.15(30°,1)	84.89	20-50	T	μ=0.1M S:50v%Ethanol	79Rb	
	-24.94	7.70(30°,2)	65.02	20-50	T	μ=0.1M S:50v%Ethanol	79Rb	
Zn^{2+}	-24.94	5.90(30°,1)	30.63	20-50	T	μ=0.1M S:50v%Ethanol	79Rb	
	-23.93	4.60(30°,2)	8.95	20-50	T	μ=0.1M S:50v%Ethanol	79Rb	
Cd^{2+}	-18.37	4.90(30°,1)	33.14	20-50	T	μ=0.1M S:50v%Ethanol	79Rb	

Metal Ion	ΔH, kJ /mol	Log K	ΔS, J /mol K	T,°C	M	Conditions	Ref.	Remarks
H48, continued								
Cd^{2+}, cont.	-16.48	4.15(30°,2)	24.98	20-50	T	μ=0.1M S:50v%Ethanol	79Rb	
H49 HYDROXAMIC ACID, 2-furoyl-N-phenyl		$C_{12}H_9O_4N$				OCH:CHCH:CCOC(=O)N(C_6H_5)OH		
UO_2^{2+}	-15.5	9.67(25°,1)	133	25&35	T	0.001M($HClO_4$) S:50v%Dioxane	76Aa	a,b
	-10.5	8.36(25°,2)	125	25&35	T	0.001M($HClO_4$) S:50v%Dioxane	76Aa	a,b
Mn^{2+}	-10.5	5.08(25°,1)	62	25&35	T	0.001M($HClO_4$) S:50v%Dioxane	76Aa	a,b
	-5.0	4.37(25°,2)	67	25&35	T	0.001M($HClO_4$) S:50v%Dioxane	76Aa	a,b
Co^{2+}	-12.1	5.69(25°,1)	68	25&35	T	0.001M($HClO_4$) S:50v%Dioxane	76Aa	a,b
	-8.8	4.83(25°,2)	63	25&35	T	0.001M($HClO_4$) S:50v%Dioxane	76Aa	a,b
Ni^{2+}	-10.5	5.92(25°,1)	78	25&35	T	0.001M($HClO_4$) S:50v%Dioxane	76Aa	a,b
	-6.7	5.01(25°,2)	73	25&35	T	0.001M($HClO_4$) S:50v%Dioxane	76Aa	a,b
Cu^{2+}	-17.6	8.90(25°,1)	111	25&35	T	0.001M($HClO_4$) S:50v%Dioxane	76Aa	a,b
	-8.8	7.40(25°,2)	112	25&35	T	0.001M($HClO_4$) S:50v%Dioxane	76Aa	a,b
Zn^{2+}	-10.5	6.51(25°,1)	89	25&35	T	0.001M($HClO_4$) S:50v%Dioxane	76Aa	a,b
	-6.7	5.71(25°,2)	87	25&35	T	0.001M($HClO_4$) S:50v%Dioxane	76Aa	a,b
Pd^{2+}	-15.5	9.01(25°,1)	120	25&35	T	0.001M($HClO_4$) S:50v%Dioxane	76Aa	a,b
Pd^{2+}	-13.8	7.64(25°,2)	100	25&35	T	0.001M($HClO_4$) S:50v%Dioxane	76Aa	a,b
H50 HYDROXAMIC ACID, 2-furoyl-N-(p-tolyl)		$C_{13}H_{11}O_4N$				OCH:CHCH:CCOC(=O)N($CH_3C_6H_4$)OH		
UO_2^{2+}	-19.2	9.74(25°,1)	122	25&35	T	0.001M($HClO_4$) S:50v%Dioxane	76Aa	a,b
	-15.5	8.40(25°,2)	109	25&35	T	0.001M($HClO_4$) S:50v%Dioxane	76Aa	a,b
Mn^{2+}	-15.5	5.14(25°,1)	46	25&35	T	0.001M($HClO_4$) S:50v%Dioxane	76Aa	a,b
	-12.1	4.39(25°,2)	43	25&35	T	0.001M($HClO_4$) S:50v%Dioxane	76Aa	a,b
Co^{2+}	-13.8	5.76(25°,1)	64	25&35	T	0.001M($HClO_4$) S:50v%Dioxane	76Aa	a,b

Metal Ion	ΔH, kJ /mol	Log K	ΔS, J /mol K	T,°C	M	Conditions	Ref.	Remarks
H50, continued								
Co^{2+}, cont.	-10.5	4.86(25°,2)	58	25&35	T	0.001M($HClO_4$) S:50v%Dioxane	76Aa	a,b
Ni^{2+}	-13.8	5.99(25°,1)	68	25&35	T	0.001M($HClO_4$) S:50v%Dioxane	76Aa	a,b
	-12.1	5.05(25°,2)	56	25&35	T	0.001M($HClO_4$) S:50v%Dioxane	76Aa	a,b
Cu^{2+}	-13.8	8.97(25°,1)	125	25&35	T	0.001M($HClO_4$) S:50v%Dioxane	76Aa	a,b
	-13.8	7.43(25°,2)	96	25&35	T	0.001M($HClO_4$) S:50v%Dioxane	76Aa	a,b
Zn^{2+}	-13.8	6.56(25°,1)	79	25&35	T	0.001M($HClO_4$) S:50v%Dioxane	76Aa	a,b
	-12.1	5.75(25°,2)	69	25&35	T	0.001M($HClO_4$) S:50v%Dioxane	76Aa	a,b
Pd^{2+}	-17.6	9.16(25°,1)	116	25&35	T	0.001M($HClO_4$) S:50v%Dioxane	76Aa	a,b
	-13.8	7.69(25°,2)	101	25&35	T	0.001M($HClO_4$) S:50v%Dioxane	76Aa	a,b
H51	HYDROXAMIC ACID, methyl		$C_2H_5O_2N$			$CH_3C(=O)NH(OH)$		HL
Fe^{3+}	9.6	2.037(25°)	67	2.5-51.3	T	μ=2.00M($NaClO_4$+ $HClO_4$) R:Fe^{3+}+HL=FeL^{2+}+H^+	79Mh	b,q:±0.8
H52	HYDROXAMIC ACID, methyl-N-methyl		$C_3H_7O_2N$			$CH_3C(=O)N(OH)CH_3$		HL
Fe^{3+}	2.5	2.500(25°)	61.1	14.8-49.3	T	μ=2.00M($NaClO_4$+ $HClO_4$) R:Fe^{3+}+HL=FeL^{2+}+H^+	79Mh	b,q:±4
H53	HYDROXAMIC ACID, methyl-N-phenyl		$C_8H_9O_2N$			$H_3CC(=O)-N(OH)C_6H_5$		HL
Fe^{3+}	5.4	2.352(25°)	65.3	25.0-51.3	T	μ=2.00M($NaClO_4$+ $HClO_4$) R:Fe^{3+}+HL=FeL^{2+}+H^+	79Mh	b
H54	HYDROXAMIC ACID, phenyl		$C_7H_7O_2N$			$H_5C_6C(=O)NHOH$		HL
Th^{4+}	-25.9	10.05(30°)	89	30-50	T	μ=0 R:Th^{4+}+H_2L= $ThL^{2+}$$2H^+$	72Td	b
	13.8	3.15(30°)	9.8	30-50	T	μ=0	72Td	b

Metal Ion	ΔH, kJ /mol	Log K	ΔS, J /mol K	T,°C	M	Conditions	Ref.	Re-marks
H54, continued								
Th^{4+}, cont.						R: $Th^{4+}+H_2L=ThL^{2+}+2H^+$		
U^{6+}	-25.5	9.70(30°,1)	97.1	30-50	T	$\mu=0$ S:1:1 Dioxane-water solution	74Te	b
	-12.3	2.85(30°,2)	12.4	30-50	T	$\mu=0$ S:1:1 Dioxane-water solution	74Te	b
Fe^{3+}	10.9	2.236(25°)	78.7	2.0-50.4	T	$\mu=2.00M(NaClO_4+HClO_4)$ R: $Fe^{3+}+HL=FeL^{2+}+H^+$	79Mh	b
H55	HYDROXAMIC ACID, phenyl-N-methyl		$C_8H_9O_2N$			$H_5C_6C(=O)N(OH)CH_3$		HL
Fe^{3+}	4.2	2.740(25°)	69.9	13.8-45.7	T	$\mu=2.00M(NaClO_4+HClO_4)$ R: $Fe^{3+}+HL=FeL^{2+}+H^+$	79Mh	b,q:±0.8
	4.6	2.425(25°)	61.5	15.3-50.3	T	$\mu=2.00M(NaClO_4+HClO_4)$ R: $Fe^{3+}+HL=FeL^{2+}+H^+$	79Mh	b
H56	HYDROXAMIC ACID, phenyl-N-(m-tolyl)		$C_{14}H_{13}O_2N$			$(O)C(C_6H_5)N(C_6H_4CH_3)OH$		
Mn^{2+}	-15.8	7.40(25°,1)	88	25&35	T	0.1M(KOH) S:70v%Dioxane	75Sj	
	-15.8	6.62(25°,2)	73	25&35	T	0.1M(KOH) S:70v%Dioxane	75Sj	
Ni^{2+}	-17.6	8.53(25°,1)	104	25&35	T	0.1M(KOH) S:70v%Dioxane	75Sj	
	-15.6	6.86(25°,2)	78	25&35	T	0.1M(KOH) S:70v%Dioxane	75Sj	
Cu^{2+}	-21.1	11.70(25°,1)	153	25&35	T	0.1M(KOH) S:70v%Dioxane	75Sj	
	-17.6	9.80(25°,2)	129	25&35	T	0.1M(KOH) S:70v%Dioxane	75Sj	
Zn^{2+}	-8.8	9.25(25°,1)	147	25&35	T	0.1M(KOH) S:70v%Dioxane	75Sj	
	-14.0	7.78(25°,2)	102	25&35	T	0.1M(KOH) S:70v%Dioxane	75Sj	
H57	HYDROXIDE ION		OH			OH^-		L^-
Ca^{2+}	8.4	1.222(1)	51.5	25	C	$\mu=0$	65Hb	
	4.98	1.40(1)	43.5	25	T	$\mu=0$	53Ba	
Sr^{2+}	4.81	0.815(1)	31.8	25	T	$\mu=0$	54Ga	

Metal Ion	ΔH, kJ /mol	Log K	ΔS, J /mol K	T,°C	M	Conditions	Ref.	Re-marks
H57, continued								
Ba^{2+}	7.32	0.640(1)	36.8	25	T	μ=0	54Ga	
Th^{4+}	26.3	-8.60(35°)	-95	30&35	T	μ=0.1M(KNO_3) R:Th+A+L=ThAL A=3-Aza-1,5-diamino-N,N',-N",N"-pentaacetic acid	75Pa	a,b
V^{3+}	-15.5	12.6(1)	190.4	25	-	--	52Ua	
Fe^{2+}	-233.26	-6.63(25°,1)	-653	12-30	T	μ=1.0($NaClO_4$)	60Ba	b,n
Fe^{3+}	5.0	11.70(1)	209	25	T	μ=0	60Gb 42Ra 52Ua	
A_5FeB	-67.8	6.18	-109.2	25	T	μ=1.0(KCl) R:$A_5FeB^{2-}+2L^-$= $A_5FeB^{4-}+H_2O$ A=CN B=NO	66Se	
[Fe^{3+}]	-32.2	5.06(1)	-10.9	25	-	Ferrimyoglobin	55Ga 60Gb	
[Fe^{3+}]	-39.7	5.2(1)	-33.0	25	-	Ferrihaemoglobin	55Ga 60Gb	
[Fe^{3+}]	-40.2	5.86(1)	-20.9	25	-	Chironomus Hemoglobin	60Gb	
Co^{3+}	-14.6	12.16(1)	184	25	T	μ=1.0	56Sd	
CoA_5B	-25.5	--	--	25	C	μ=0.1F(NaOH) A=NH_3 B=Cl R:$CoA_5B^{2+}+L^-$= $CoA_5L^{2+}+B^-$	69Hd 71Hh	
	-23.8	--	--	25	C	μ=0.1F(NaOH) A-NH_3 B=Br R:$CoA_5B^{2+}+L^-$= $CoA_5L^{2+}+B^-$	69Hd 71Hh	
	-15.27	--	--	25	C	μ=0.1F(NaOH) A=NH_3 B=I R:$CoA_5B^{2+}+L^-$= $CoA_5L^{2+}+B^-$	69Hd 71Hh	
	-16.11	--	--	25	C	μ=0.1F(NaOH) A=NH_3 B=NO_3 R:$CoA_5B^{2+}+L^-$= $CoA_5L^{2+}+B^-$	69Hd 71Hh	
$RuA_6{}^{3+}$	15.5	0.732(25°,1)	77.8	5-50	T	μ=0.1M($NaCH_3SO_3$)	79Wa	
	25.5	0.255(25°)	90.4	5-50	T	μ=0.1M($NaCH_3SO_3$) R:Ru_6AL^{3+}= $RuA_5NH_2{}^{2+}$·HL A=NH_3	79Wa	

Metal Ion	ΔH, kJ /mol	Log K	ΔS, J /mol K	T, °C	M	Conditions	Ref.	Re-marks
H57, continued								
RuA_6	21.3	3.36	136.0	25	T	A=NH_3 R:$RuA_6{}^{3+}$+ L^-=$RuA_5NH_2{}^{2+}$+H_2O	71Wc	q:±1.2
Cu^{2+}	-84	14.1	-12	25	C	μ=0.1M($NaNO_3$-$NaClO_4$) R:Cu^{2+}+A+L^-=$CuAL^+$ A=2,2'-Bipyridyl	76Al	
	-133.0	33.1	186.6	25	C	μ=0.1M($NaNO_3$-$NaClO_4$) R:2Cu^{2+}+2A+2L^-= $Cu_2A_2L_2{}^{2+}$ A=2,2'-Bipyridyl	76Al	
	-43.1	17.8(2-2)	197	25	C	3M($NaClO_4$)	68Ae	
	-64.60	16.45(25°)	96	5-35	T	μ=0.5($LiClO_4$) S:50% C_2H_5OH R:Cu^{2+}+2-carboxyphenyl-azo-β-naphthol+L= complex	58Ja	b
	-92.55	19.05(25°)	54	5-35	T	μ=0.5($LiClO_4$) S:50% C_2H_5OH R:Cu^{2+}+L+2-azo-β-napthylamine= complex	58Ja	b
	-156.36	22.34(25°)	-96	5-35	T	μ=0.5($LiClO_4$) S:50% C_2H_5OH R:Cu^{2+}+L+o,o-dihydroxyazobenzene=complex	58Ja	b
CuA	-12.9	4.31(1)	-39	25	C	μ=0.2(KCl) A=N-Alanylglycine (A^{2-})	76Gd	
	-12.8	4.38(1)	-41	25	C	μ=0.2(KCl) A=N-Alanyl-α-alanine (A^{2-})	76Gd	
	-38	6.0(1)	-12	25	C	μ=0.1M($NaNO_3$-$NaClO_4$) A=2,2'-Bipyridyl	76Al	
	-13.3	4.45(1)	-40	25	C	μ=0.2(KCl) A=N-Glycylglycine (A^{2-})	76Gd	
	-12.3	4.33(1)	-42	25	C	μ=0.2(KCl) A=N-Glycyl-α-alanine(A^{2-})	76Gd	
	-51.6	17.64(1-2)	164	25	C	μ=0.5M($NaClO_4$) A=N-(2-hydroxyethyl)ethylenediamine	75Bb	
	-43.9	16.9	175.3	25	C	μ=0.1M($NaNO_3$-$NaClO_4$)	76Al	

Metal Ion	ΔH, kJ /mol	Log K	ΔS, J /mol K	T,°C	M	Conditions	Ref.	Remarks
H57, continued								
CuA, cont.						R: $2CuA^{2+}+2L^{-}=Cu_2A_2L_2^{2+}$ A=2,2'-Bipyridyl		
CuA^{2+}	-10.79	4.35(1)	42.15	25	C	μ=0.5M(KNO_3) A=3-Aza-1,7-diaminoheptane	76Ba	
	-9.53	6.44(1)	58.2	25	C	0.5M(KNO_3) A=3-Aza-1,6-diaminohexane	74Ba	
	-18.0	5.90(1)	52	25	C	μ=0.5(KNO_3) A=1,7-Diamino-4-thioheptane	79Ha	
	-13.4	4.60(1)	43	25	C	μ=0.5(KNO_3) A=1,8-Diamine-4-thiooctane	79Ha	
	-5.4	4.42(1)	66	25	C	μ=0.10M(NaCl) A=3-Aza-1,7-diaminoheptane	76Gg	
	-38.9	5.48(1)	-25.1	25	C	μ=0.5M(KNO_3) A=2,2'-Diamino-diethylether	74Bb	
	-22.6	0.81(1)	29	25	C	μ=0.5M($NaClO_4$) A=1,2-Diamino-N,N'-diethylethane	75Bg	
	-25.9	4.69(1)	29	25	C	μ=0.5M($NaClO_4$) A=1,2-Diamino-N,N'-diethylethane	75Bg	
	-9.54	4.14(1)	47.3	25	C	0.1M(KCl) A=3,3'-Diaminodipropyl amine	66Pa	
	-11.3	4.5(1)	48.1	25	C	μ=0.1M(KCl) A=Diethylenetriamine	68Pa	
	-16.02	3.988(1)	22.6	25	C	0.1M(KCl) A=Tri-(3-aminopropyl) amine	68Va	
	-15.5	4.65(1)	35.6	25	C	μ=0.1M(KCl) A=Triethylene-tetramine	63Pa	
	0	3.2(1)	60.7	25	C	μ=0.1M(KCl) A=Tetraethylene-pentaamine	63Pa	
	-24.7	8.789	84	25	C	0.5M(KNO_3) A=1,2-Diamino-N,-N-dimethylethane	72Bd	
						R: $CuA^{2+}+2L^{-}=Cu(L)_2A$		
	-31.0	8.306	54	25	C	0.5M(KNO_3) A=1,2-Diamino-N,N'-dimethylethane	72Bd	
						R: $CuA^{2+}+2L^{-}=Cu(L)_2A$		

Metal Ion	ΔH, kJ /mol	Log K	ΔS, J /mol K	T,°C	M	Conditions	Ref.	Re-marks
H57, continued								
CuA^{2+}, cont.	-23.4	9.119	96.2	25	C	0.5M(KNO_3) A=1,2-Diamino-N,N,N',N'-tetramethylethane R:CuA^{2+}+2L^-= Cu$(L)_2$A	72Ad 72Bd	
	-46.9	15.416	138	25	C	0.5M(KNO_3) A=1,2-Diamino-N,N-dimethylethane R:2CuA^{2+}+2L^-= $[Cu_2(L)_2A_2]^{2+}$	72Bd	
	-46.9	15.020	130	25	C	0.5M(KNO_3) A=1,2-Diamino-N,N'-dimethylethane R:2CuA^{2+}+2L^-= $[Cu_2(L)_2A_2]^{2+}$	72Bd	
	-39.3	15.255	160.2	25	C	0.5M(KNO_3) A=1,2-Diamino-N,N,N',N'-tetramethylethane R:2CuA^{2+}+2L^-= $[Cu_2(L)_2A_2]^{2+}$	72Ad 72Bd	
	-47.7	0.59	113	25	C	μ=0.5M($NaClO_4$) R:2CuA^{2+}+2L^-= $CuL_2A_2{}^{2+}$ A=1,2-Diamino-N,N'-diethyl-ethane	75Bg	
	-52.3	3.96	109	25	C	μ=0.5M($NaClO_4$) R:2CuA^{2+}+2L^-= $CuL_2A_2{}^{2+}$ A=1,2-Diamino-N,N'-diethyl-ethane	75Bg	
$CuA_2{}^{2+}$	-23.8	4.39(1)	4	25	C	0.5(KNO_3) A=3-Thia-5-amino-1-pentanol	77Hd	
ZnA^{2+}	-0.21	6.80(1)	94.6	25	C	0.5M(KNO_3) A=3-Aza-1,6-diaminohexane	74Ba	
	-15.77	5.20(1)	46.9	25	C	0.1M(KCl) A=3,3'-Diaminodipropyl amine	66Pa	
	1.7	5.11(1)	105	25	C	μ=0.5M(KNO_3) A=2,2'-Diamino-diethyl ether	74Bb	
	-8.4	4.03(1)	49.0	25	C	μ=0.5M(KNO_3) A=4,7-Diazade-cane-1,10-diamine	73Bc	

Metal Ion	ΔH, kJ /mol	Log K	ΔS, J /mol K	T, °C	M	Conditions	Ref.	Remarks
H57, continued								
ZnA^{2+}, cont.	-17.36	5.85(1)	54.4	25	C	μ=0.1M(KCl) A=1,3,5-Triamino-cyclohexane (cis, cis)	76Fa	q:±3.2
HgL	0	0.5	10	25	C	μ=1M($NaClO_4$) R:$HgL^+ + \frac{1}{2}Hg^{2+} = \frac{1}{2}HgL_2$	76Ce	q:±1.6
HgA^+	-35.6	9.394(1)	57.3	20	C	μ=0.1(KNO_3) A=CH_3	65Sa	
HgA	-0.8	0.6	9	25	C	μ=1M($NaClO_4$) R:$HgA^+ + HgL^+ = HgAL + Hg^{2+}$ A=Cl^-	76Ce	q:±1.7
	1	0.3	9	25	C	μ=1M($NaClO_4$) R:$HgA^+ + HgL^+ = HgAL + Hg^{2+}$ A=SCN^-	76Ce	q:±2
HgA_2	-5.19	4.09(1)	61.1	25	C	μ=0 A=Cl	65Pa	
	-5.06	3.77(2)	55.2	25	T	μ=0	65Pa	
	1.6	0.2	9	25	C	μ=1M($NaClO_4$) R:$\frac{1}{2}HgA_2 + \frac{1}{2}HgL_2 = HgAL$ A=Cl^-	76Ce	q:±3
	1	0.35	10	25	C	μ=1M($NaClO_4$) R:$\frac{1}{2}HgA_2 + \frac{1}{2}HgL_2 = HgAL$ A=SCN^-	76Ce	q:±4
BL_3	-42.89	(1)	-52.93	0	T	μ=0 ΔCp=16.3	72Me	b
	-42.34	(1)	-50.96	25	T	μ=0 ΔCp=28.0	72Me	b
	-41.50	(1)	-48.28	50	T	μ=0 ΔCp=39.3	72Me	b
	-40.38	(1)	-44.89	75	T	μ=0 ΔCp=51.0	72Me	b
	-38.95	(1)	-40.96	100	T	μ=0 ΔCp=62.8	72Me	b
	-37.24	(1)	-36.53	125	T	μ=0 ΔCp=74.1	72Me	b
	-35.23	(1)	-31.67	150	T	μ=0 ΔCp=85.8	72Me	b
	-32.97	(1)	-26.44	175	T	μ=0 ΔCp=97.5	72Me	b
	-30.38	(1)	-20.84	200	T	μ=0 ΔCp=108.8	72Me	b
	-27.53	(1)	-14.94	225	T	μ=0 ΔCp=120.5	72Me	b
	-24.35	(1)	-8.74	250	T	μ=0 ΔCp=131.8	72Me	b
	-20.92	(1)	-2.34	275	T	μ=0 ΔCp=143.5	72Me	b
	-17.20	(1)	4.35	300	T	μ=0 ΔCp=157.7	72Me	b
	-43.1	(1)	-50.2	25	T	μ=1M(KCl)	72Me	b
	-38.5	(2-1)	-38.1	25	T	μ=1M(KCl)	72Me	b
	-60.2	(3-1)	-71.5	25	T	μ=1M(KCl)	72Me	b
	-142	(4-2)	-230	25	T	μ=1M(KCl)	72Me	b,q:±25
	-180	(5-3)	-243	25	T	μ=1M(KCl)	72Me	b,q:±17
Al^{3+}	-37.6	(2-2)	255	25-125	T	μ=1M(KCl)	71Mc	b,q:±10.9
	-97.9	(3-4)	464	25-125	T	μ=1M(KCl)	71Mc	b,q:±14.2
	-843.5	(14-34)	4,021	25-125	T	μ=1M(KCl)	71Mc	b
Tl^+	1.55	0.82(1)	21.3	25	T	μ=0	53Ba	
	7.5	0.088(1)	26.8	25	C	μ=3.0($LiClO_4$)	73Kj	q:±0.8
	20.9	-0.953(2)	52.30	25	C	μ=3.0($LiClO_4$)	73Kj	q:±2.5

Metal Ion	ΔH, kJ /mol	Log K	ΔS, J /mol K	T,°C	M	Conditions	Ref.	Remarks
H57, continued								
SiL_4	-4.2	5.6(25°)	92	15-65	T	μ=5.0M(NaCl) R:$SiL_4+L^- = SiOL_3^- + H_2O$	73Pj	b,q: ±2.1
	-6.3	7.1(25°)	117	15-65	T	μ=5.0M(NaCl) R:$SiL_4+2L^- = SiO_2L_2^{2-} + 2H_2O$	73Pj	b,q: ±2.1
	-12	12.0(25°)	188	15-65	T	μ=5.0M(NaCl) R:$4SiL_4+2L^- = Si_4O_6^{4-} + 6H_2O$	73Pj	b,q:±4

H58 HYPOXANTHINE $C_5H_4ON_4$

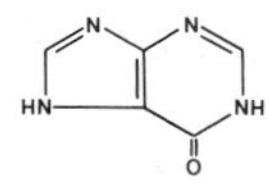

L

Metal Ion	ΔH, kJ /mol	Log K	ΔS, J /mol K	T,°C	M	Conditions	Ref.	Remarks
Mn^{2+}	-87.86	3.50(25°,1)	-227.86	25&35	T	μ=0.1M($KClO_4$)	79Ra	a,b
	28.37	2.30(25°,2)	139.24	25&35	T	μ=0.1M($KClO_4$)	79Ra	a,b
Co^{2+}	-42.17	3.67(25°,1)	-71.17	25&35	T	μ=0.1M($KClO_4$)	79Ra	a,b
	-107.19	3.41(25°,2)	-294.43	25&35	T	μ=0.1M($KClO_4$)	79Ra	a,b
Ni^{2+}	-96.65	4.55(25°,1)	-237.11	25&35	T	μ=0.1M($KClO_4$)	79Ra	a,b
	-133.55	3.71(25°,2)	-377.10	25&35	T	μ=0.1M($KClO_4$)	79Ra	a,b
Cu^{2+}	-110.67	6.22(25°,1)	-252.30	25&35	T	μ=0.1M($KClO_4$)	79Ra	a,b
	-191.50	5.75(25°,2)	-532.20	25&35	T	μ=0.1M($KClO_4$)	79Ra	a,b

I

I1 IMIDAZOLE $C_3H_4N_2$

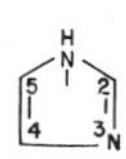

L

Metal Ion	ΔH, kJ /mol	Log K	ΔS, J /mol K	T,°C	M	Conditions	Ref.	Remarks
[Fe(II)]	-63.2	5.11(25°,1-2)	-114.2	15.5-36.5	T	μ=0.10($NaClO_4$) pH=7.0 [Fe(II)]= Iron(II) 2,4-Dicysteine-substituted mesoporphyrin (heme c)	76Gf	

Metal Ion	ΔH, kJ /mol	Log K	ΔS, J /mol K	T,°C	M	Conditions	Ref.	Re-marks
I1, continued								
[Fe(III)]	-45.2	6.00(25°,1-2)	-36.4	15.5-36.5	T	μ=0.10($NaClO_4$) pH=6.0 [Fe(III)]= Iron(III) 2,4-Dicysteine-substituted mesoporphyrin (hemin c)	76Gf	
	-82.8	4.76(30°,1-2)	-184	23-46	T	S:Acetone R:FeAB+2L= $FeAL_2^+ + B^-$ A=Tetraphenyl-porphyrin B=Cl [Fe(III)]=Tetra-phenylporphyrin iron(III)	73Dg	b
	-92.0	5.89(30°,1-2)	-192	30	T	S:Dichloromethane [Fe^{3+}]=Fe(deut)Cl Fe(deut)=Iron(III) deuteroporphyrin (IX) dimethyl ester	74Aa	
[Fe^{3+}]	-5.8	1.27(25°,1)	12.1	14.4-30.9	T	0.1(K-phos) pH=7.0 [Fe^{3+}]= Horse heart ferric cytochrome c	75Ia	
[Fe^{3+}]	4.6	1.462(18°,1)	42.7	18-32	T	0.1M Phosphate Buffer pH=7.4 Ferricytochrome C	69Sf	b
[Fe^{3+}]	-16.94	2.199(1)	-14.6	25	T	μ=0.2 Ferrimyglobin	64Ga	
Co^{2+}	-24.8	2.611(1)	-36	15	T	μ=0	74Lc	
	-20.5	2.473(1)	-21	25	T	μ=0	74Lc	
	-16.0	2.369(1)	-6	35	T	μ=0	74Lc	
	-17.6	2.47(25°,1)	-10.9	10-50	T	μ=0.16(KNO_3)	66Sc	b
	-16.3	1.93(25°,2)	-19.7	10-50	T	μ=0.16(KNO_3)	66Sc	b
	-14.6	1.45(25°,3)	-21.3	10-50	T	μ=0.16(KNO_3)	66Sc	b
	-15.5	1.00(25°,4)	-33.0	10-50	T	μ=0.16(KNO_3)	66Sc	b
	-12.1	0.54(25°,5)	-29.7	10-50	T	μ=0.16(KNO_3)	66Sc	b
	-15.1	-0.04(25°,6)	-50.2	10-50	T	μ=0.16(KNO_3)	66Sc	b
[Co^{2+}]	-29.34	4.608(1)	-10.22	25	C	μ=0.1M(KNO_3) R:[Co^{2+}]·H_2O+L= [Co^{2+}]L+H_2O [Co^{2+}]=Aquo-cobalamin (Vitamin B_{12})	76Ea	
[Co]	-30.12	4.59	-13.4	25	T	μ=0 Aquocabal-amin R:B-Co^+(H_2O) +L=B-Co^+=L+H_2O-	64Hb	

Metal Ion	ΔH, kJ /mol	Log K	ΔS, J /mol K	T,°C	M	Conditions	Ref.	Re-marks
I1, continued								
[Co], cont.						$BCo^{+}(H_2O)$= aquocobalamine		
[Co]	-33.0	4.00(0°,1)	-42	-23-23	T	S:Toluene [Co]=Cobalt(II) protoporphyrin IX dimethyl ester	73Sx	b,q: ±2.5
[Co]	-26.4	4.09	-10.0	25	T	μ=0 Cobalamin Factor B R:NC-$Co^{+}(H_2O)$ + L = NC-Co^{+} - L + H_2O-NC-$Co^{+}(H_2O)$ = cobalamine factor B	66Ha	
Ni^{2+}	-31.8	3.146(1)	-50	15	T	μ=0	74Lc	
	-28.7	2.962(1)	-39	25	T	μ=0	74Lc	
	-25.4	2.808(1)	-28	35	T	μ=0	74Lc	
	-17	2.94(25°,1)	-17	0&25	T	μ=0.135	55Ma	a,b,f:4
	-21.8	3.09(25°,1)	-12.1	10-50	T	μ=0.16(KNO_3)	66Sc	b
	-22.2	3.10(25°,1)	-15.5	20-50	T	CM=0.01M CL=0.100M	65Sd	b
	-15.2	2.621(2)	-3	15	T	μ=0	74Lc	
	-6.3	2.553(2)	26	25	T	μ=0	74Lc	
	2.4	2.540(2)	55	35	T	μ=0	74Lc	
	-17	2.41(25°,2)	-17	0&25	T	μ=0.135	55Ma	a,b,f:4
	-19.2	2.47(25°,2)	-17.2	10-50	T	μ=0.16(KNO_3)	66Sc	b
	-17	1.99(25°,3)	-17	0&25	T	μ=0.135	55Ma	a,b,f:4
	-18.0	2.00(25°,3)	-22.2	10-50	T	μ=0.16(KNO_3)	66Sc	b
	-17	1.3(25°,4)	-17	0&25	T	μ=0.135	55Ma	a,b,f:4
	-15.9	1.54(25°,4)	-22.6	10-50	T	μ=0.16(KNO_3)	66Sc	b
	-13.4	1.06(25°,5)	-24.7	10-50	T	μ=0.16(KNO_3)	66Sc	b
	-11.7	0.45(25°,6)	-30.1	10-50	T	μ=0.16(KNO_3)	66Sc	b
	-19.2	-0.49(25°,1-2)	-15.5	20-50	T	CM=0.01M CL=0.100M	65Sd	b
	-17.6	1.98(25°,1-3)	-20.5	20-50	T	CM=0.01M CL=0.100M	65Sd	b
	-15.5	1.53(25°,1-4)	-23.0	20-50	T	CM=0.01M CL=0.100M	65Sd	b
	-15.1	1.01(25°,1-5)	-31.0	20-50	T	CM=0.01M CL=0.100M	65Sd	b
	-77.0	11.4(1-6)	-42	25	C	μ=0.16(KNO_3)	64Bc	
	-11.3	0.39(25°,1,6)	-31.0	20-50	T	CM=0.10M CL-0.100M	65Sd	b
Rh_2A_4	-31.09	4.021(1)	-27.2	25	C	μ=0.1 pH=7.4 PT=0.038M A=Acetate	76Db 77Da	
	-30.54	3.943(1)	-26.8	25	C	μ=0.1 pH=7.4 PT=0.038M A=Methoxyacetate	76Db 77Da	
	-36.19	4.067(1)	-43.9	25	C	μ=0.1 pH=7.4 PT=0.038M A=Propionate	76Db 77Da	

Metal Ion	ΔH, kJ /mol	Log K	ΔS, J /mol K	T,°C	M	Conditions	Ref.	Re-marks
I1, continued								
Rh_2A_4, cont.	-12.6	2.477(2)	5.0	25	C	μ=0.1 pH=7.4 PT=0.038M A=Acetate	76Db 77Da	q:±1.3
	-12.6	2.398(2)	4.2	25	C	μ=0.1 pH=7.4 PT=0.038M A=Methoxyacetate	76Db 77Da	q:±1.2
	-22.2	2.650(2)	-23.8	25	C	μ=0.1 pH=7.4 PT=0.038M A=Propionate	76Db 77Da	
Cu^{2+}	-45.2	4.352(1)	-73	15	T	μ=0	74Lc	
	-28.6	4.127(1)	-17	25	T	μ=0	74Lc	
	-11.5	4.012(1)	39	35	T	μ=0	74Lc	
	-25	4.20(25°,1)	-29	0&25	T	μ=0.135	55Ma 54Ea 57Nd	a,b, f:4
	-30.1	4.31(1)	-17.6	25	T	μ=0.16(KNO_3)	66Sc	
	-18.8	3.783(2)	7	15	T	μ=0	74Lc	
	-34.4	3.620(2)	-47	25	T	μ=0	74Lc	
	-51.5	3.375(2)	102	35	T	μ=0	74Lc	
	-25	3.47(25°,2)	-29	0&25	T	μ=0.135	55Ma	a,b,f:4
	-22.6	3.53(2)	-7.5	25	T	μ=0.16(KNO_3)	66Sc	
	-208	3.480(3)	-656	15	T	μ=0	74Lc	
	-60	2.661(3)	-148	25	T	μ=0	74Lc	
	94	2.754(3)	360	35	T	μ=0	74Lc	
	-25	2.84(25°,3)	-29	0&25	T	μ=0.135	55Ma	a,b,f:4
	-19.2	2.92(3)	-8.8	25	T	μ=0.16(KNO_3)	66Sc	
	223	2.749(4)	826	15	T	μ=0	74Lc	
	35	3.539(4)	183	25	T	μ=0	74Lc	
	-160	3.191(4)	-460	35	T	μ=0	74Lc	
	-25	2.0(25°,4)	-29	0&25	T	μ=0.135	55Ma	a,b,f:4
	-13.0	2.14(4)	-2.1	25	T	μ=0.16(KNO_3)	66Sc	
	-96.2	12.6(1-4)	-84	25	T	μ=0.16(KNO_3)	64Bc	
Ag^+	-19.00	3.516(1)	-2.01	0	T	μ=0.1 ΔCp=-199.6	66Da	
	-20.96	3.381(1)	-9.2	10	T	μ=0.1 ΔCp=-199.6	66Da	
	-22.97	3.243(1)	-17.2	20	T	μ=0.1 ΔCp=-199.6	66Da	
	-25.00	3.100(1)	-23.85	30	T	μ=0.1 ΔCp=-199.6	66Da	
	-27.95	2.955(1)	-31.8	40	T	μ=0.1 ΔCp=-199.6	66Da	
	-29.96	2.807(1)	-38.9	50	T	μ=0.1 ΔCp=-199.6	66Da	
	-30.5	3.05(1)	-46.0	25	C	μ=1(KNO_3)	64Bc	
	-25.1	4.361(2)	-8.8	0	T	μ=0.1 ΔCp=-1397	66Da	
	-36.8	4.148(2)	-51.5	10	T	μ=0.1 ΔCp=-900	66Da	
	-42	3.893(2)	-72.8	20	T	μ=0.1 ΔCp=-300	66Da	
	-42	3.635(2)	-74.0	30	T	μ=0.1 ΔCp=-301	66Da	
	-36.8	3.409(2)	-54.0	40	T	μ=0.1 ΔCp=-900	66Da	
	-25.1	3.245(2)	-13.8	50	T	μ=0.1 ΔCp=-1598	66Da	
	-35.1	3.83(2)	-46.9	25	C	μ=1(KNO_3)	64Bc	
	-65.3	6.89(1-2)	-88	25	C	μ=1(KNO_3)	64Bc	
Zn^{2+}	-8.1	2.633(1)	22	15	T	μ=0	74Lc	
	-12.8	2.570(1)	6	25	T	μ=0	74Lc	
	-17.6	2.484(1)	-9	35	T	μ=0	74Lc	
	-67.8	9.24(1-4)	-50	25	C	μ=0.16(KNO_3)	64Bc	
Cd^{2+}	-23.3	2.817(1)	-27	15	T	μ=0	74Lc	

Metal Ion	ΔH, kJ /mol	Log K	ΔS, J /mol K	T,°C	M	Conditions	Ref.	Re-marks
I1, continued								
Cd^{2+}, cont.	-26.8	2.665(1)	-38	25	T	μ=0	74Lc	
	-30.3	2.503(1)	-50	35	T	μ=0	74Lc	
	-29.2	3.03(25°,1)	-38	0-45	T	μ=0.1(KNO_3)	64Ac	b
	-21	2.8(1)	16.7	25	T	μ=0.15	53Ta	
	-20.4	2.77(1)	-15	25	C	μ=0.5M($NaClO_4$)	75Mb 78Mc	
	-29.1	2.70(25°,1)	-45.7	5-40	T	μ=1.0($NaClO_4$) pH=9.5	75Jb	b
	-26.7	2.161(2)	-52	15	T	μ=0	74Lc	
	-31.2	1.982(2)	-70	25	T	μ=0	74Lc	
	-36.7	1.786(2)	-87	35	T	μ=0	74Lc	
	-23.4	2.11(25°,2)	-42	0-45	T	μ=0.1(KNO_3)	64Ac	b
	-20.3	2.10(2)	-28	25	C	μ=0.5M(NaClO)	75Mb 78Mc	
	0	2.40(25°,2)	45.7	5-40	T	μ=1.0($NaClO_4$) pH=9.5	75Jb	b
	-18.0	1.34(25°,3)	-38	0-45	T	μ=0.1(KNO_3)	64Ac	b
	-17.3	1.45(3)	-30	25	C	μ=0.5M($NaClO_4$)	75Mb	q:±2
	-18.3	1.53(25°,3)	-31.6	5-40	T	μ=1.0($NaClO_4$) pH=9.5	75Jb	b
	-11.7	0.79(25°,4)	-25	0-45	T	μ=0.1(KNO_3)	64Ac	b
	-2	1.17(4)	16	25	C	μ=0.5M($NaClO_4$)	75Mb	q:±0.5
	-44.0	0.97(25°,4)	-130	5-40	T	μ=1.0($NaClO_4$) pH=9.5 ΔCp=-3160	75Jb	b
	54.8	0.58(25°,5)	196	5-40	T	μ=1.0($NaClO_4$) pH=9.5 ΔCp=-2280	75Jb	b
	-68.1	0.77(25°,6)	-214	5-40	T	μ=1.0($NaClO_4$) pH=9.5 ΔCp=-2910	75Jb	b
	-45.2	7.48(25°,1-4)	9.2	0-35	T	μ=0.15(KNO_3) S:18.8% C_2H_5OH	54La	b
Hg^{2+}	-75	9.18(1)	-77	25	C	μ=0.5M(ClO_4^- or NO_3^-)	78Mc	
	-60.7	9.64(1)	-19	25	C	μ=0.5M($NaClO_4$)	75Mb	
	-32.1	9.00(2)	66	25	C	μ=0.5M(ClO_4^- or NO_3^-)	78Mc	
	-43.3	8.55(2)	18	25	C	μ=0.5M($NaClO_4$)	75Mb	
	-107	18.19(1-2)	-11	25	C	μ=0.5M($NaClO_4$)	77Mc	
	-113.8	21.86	37	25	C	μ=0.5M(ClO_4^- or NO_3^-) R:Hg^{2+}+L=$[Hg(OH)L]^+$	78Mc	j
	-76	21.86	161	25	C	μ=0.5M($NaClO_4$) R:HgL^{2+}+H_2O=$Hg(OH)L^+$+H^+	77Mc	q:±15

I2 IMIDAZOLE, N-acetyl- $C_5H_6ON_2$ $C_3H_3N_2(CH_3CO)$

Metal Ion	ΔH, kJ /mol	Log K	ΔS, J /mol K	T,°C	M	Conditions	Ref.	Re-marks
CoA	-42.2	2.895(25°,1)	-84	20-50	T	S:Toluene A=α,β,γ,δ-Tetra-(p-methoxyphenyl) porphyrin	73Wa	b,q:±4

Metal Ion	ΔH, kJ /mol	Log K	ΔS, J /mol K	T,°C	M	Conditions	Ref.	Remarks
I3	IMIDAZOLE, 4(5)-acetic acid		$C_5H_6O_2N_2$			$C_3H_3N_2(CH_2CO_2H)$		L^-
Co^{2+}	-9.6	4.00(15°,1)	43.5	0-40	T	μ=0.25(KCl)	65Ad	b
	-14.2	3.03(25°,2)	8.8	0-40	T	μ=0.25(KCl)	65Ad	b
Ni^{2+}	-9.6	4.83(15°,1)	59.4	0-40	T	μ=0.25(KCl)	65Ad	b
	-14.2	3.71(15°,2)	21.7	0-40	T	μ=0.25(KCl)	65Ad	b
Cu^{2+}	-14.2	7.34(15°,1)	91.6	0-40	T	μ=0.25(KCl)	65Ad	b
	-9.6	5.81(15°,2)	78.6	0-40	T	μ=0.25(KCl)	65Ad	b
Zn^{2+}	-9.6	3.83(15°,1)	39.3	0-40	T	μ=0.25(KCl)	65Ad	b
	0.0	3.32(15°,2)	64.0	0-40	T	μ=0.25(KCl)	65Ad	b
	-14.2	2.63(15°,3)	21.8	0-40	T	μ=0.25(KCl)	65Ad	b
I4	IMIDAZOLE, 4(5)-(2'-aminoethyl)-	$C_5H_9N_3$				$C_3H_3N_2(CH_2CH_2NH_2)$		L
Ni^{2+}	-32.47	6.84(1)	22.2	25	C	μ=0.30(ClO_4^-)	67Hf	
	-29.29	5.04(2)	-1.7	25	C	μ=0.30(ClO_4^-)	67Hf	
	-25.31	3.15(3)	-24.7	25	C	μ=0.30(ClO_4^-)	67Hf	
Cu^{2+}	-43.05	9.56(1)	38.5	25	C	μ=0.30(ClO_4^-)	67Hf	
	-42.05	6.64(2)	-13.8	25	C	μ=0.30(ClO_4^-)	67Hf	
I5	IMIDAZOLE, 4(5)-aminomethyl-		$C_4H_7N_3$			$C_3H_3N_2(CH_2NH_2)$		L
Ni^{2+}	-36.36	5.85(1)	-10.04	25	C	μ=0.30(ClO_4^-)	67Hf	
	-36.28	4.82(2)	-29.3	25	C	μ=0.30(ClO_4^-)	67Hf	
	-36.19	3.12(3)	-61.9	25	C	μ=0.30(ClO_4^-)	67Hf	
Cu^{2+}	-47.86	8.72(1)	6.7	25	C	μ=0.30(ClO_4^-)	67Hf	
	-47.74	7.72(2)	-12.1	25	C	μ=0.30(ClO_4^-)	67Hf	
I6	IMIDAZOLE, 2-methoxy-		$C_4H_6ON_2$			$C_3H_3N_2(OCH_3)$		
Cu^{2+}	-28.62	4.27(1)	-14.2	25	C	μ=3.00($NaClO_4$)	68Wa	f:4
	-28.62	3.92(2)	-20.9	25	C	μ=3.00($NaClO_4$)	68Wa	f:4
	-28.62	3.13(3)	-36.0	25	C	μ=3.00($NaClO_4$)	68Wa	f:4
	-28.62	2.10(4)	-55.6	25	C	μ=3.00($NaClO_4$)	68Wa	f:4
I7	IMIDAZOLE, 4-methoxy-		$C_4H_6ON_2$			$C_3H_3N_2(OCH_3)$		
Cu^{2+}	-29.46	4.25(1)	-17.6	25	C	μ=3.0($NaClO_4$)	68Wa	f:4
	-29.46	3.90(2)	-24.3	25	C	μ=3.0($NaClO_4$)	68Wa	f:4
	-29.46	3.11(3)	-39.3	25	C	μ=3.0($NaClO_4$)	68Wa	f:4
	-29.46	2.08(4)	-59.0	25	C	μ=3.0($NaClO_4$)	68Wa	f:4
I8	IMIDAZOLE, 1-methyl		$C_4H_6N_2$		$C_3H_3N_2(CH_3)$			L
CoA	-40.2	(1)	--	-45--31	T	S:Toluene A=Protoporphyrin IX dimethyl ester	78Dd	
	-47.7	3.375(25°,1)	-96	20-50	T	S:Toluene A=α,β,-γ,δ-Tetra(p-	73Wa	b,q:±4

Metal Ion	ΔH, kJ /mol	Log K	ΔS, J /mol K	T, °C	M	Conditions	Ref.	Remarks
I8, continued								
CoA, cont.						methoxy-phenyl) porphyrin		
$[Co^{2+}]$	-23.08	4.403(1)	6.85	25	C	μ=0.1M(KNO_3) R:$[Co^{2+}]\cdot H_2O$+L= $[Co^{2+}]\cdot L+H_2O$ $[Co^{2+}]$=Aquo-cobalamin (Vitamin B_{12})	76Ea	
[Co]	-57.7	(1)	--	25	C	S:CH_2Cl_2 [Co]= Methylcobaloxime	74Td	
[Co]	-92.5	--	--	24	C	S:Dichloromethane [Co]=Methylcobal-oxime R:$[Co]_2$+2L= 2[Co]L	73Cf	
[Co]	-44.8	4.40(0°,1)	-79	-23-23	T	S:Toluene [Co]= Cobalt(II) proto-porphyrin IX dimethyl ester	73Sx	b,q: ±2.5
RhAB	-66.9	--	--	24	C	S:Benzene R:[RhAB]+2L= 2RhABL A=Cl B=1,5-Cyclo-octadiene	76Lb	
$Rh_2A_4B_2$	-64.0	--	--	24	C	S:Benzene R:$Rh_2A_4B_2$+2L= $2RhA_2BL$ A=CO B=Cl	76Pd	u:24°
PdAB	-50.6	--	--	24	C	S:Benzene R:$[PdAB]_2$+2L= 2[PdABL] A=(π-MeAllyl) B=Cl	77Lb	
Cu^{2+}	-28.0	3.46(29°,1)	26.4	2.5&29	T	μ=0.16	57Nd	a,b,q:±4
	-26.8	3.24(29°,2)	26.4	2.5&29	T	μ=0.16	57Nd	a,b,q:±12
	-22.2	2.99(29°,3)	15.1	2.5&29	T	μ=0.16	57Nd	a,b,q:±12
CuA_2	-67.4	(1)	--	25	C	S:CCl_4 A=Hexafluoro-acetylacetone	74Td	
	-64.8	7.0(1)	--	--	C	S:CCl_4 A=Hexafluoro-acetylacetonato	76Mc	u:25°
Ag^+	-65.3	6.9(1-2)	-88	25	C	1M(KNO_3)	64Bc	
Zn^{2+}	-11.7	1.83(27°,1)	-4.06	2.5&27	T	μ=0.16	57Nd	a,b,q:±4 kcal
	-13.0	2.33(27°,2)	1.38	2.5&27	T	μ=0.16	57Nd	a,b,q:±12 kcal

Metal Ion	ΔH, kJ /mol	Log K	ΔS, J /mol K	T,°C	M	Conditions	Ref.	Re-marks
I8, continued								
Zn^{2+}, cont.	-13.0	2.80(27°,3)	10.5	2.5&27	T	μ=0.16	57Nd	a,b,q:±12 kcal
ZnA_2	-60.2	(1)	--	25	C	S:Cyclohexane $A=N[Si(CH_3)_3]_2$	75Fc	
	-60.2	(1)	--	25	C	S:n-Hexane $A=NSi(CH_3)_3$	74Td	
I9 IMIDAZOLE, 4(5)-methyl			$C_4H_6N_2$			$C_3H_4N_2(CH_3)$		
$[Co^{2+}]$	-22.12	(1)	--	25	C	$\mu=0.1M(KNO_3)$ R: $[Co^{2+}]\cdot H_2O+L=[Co^{2+}]\cdot L+H_2O$ $[Co^{2+}]$=Aquo-cobalamin (Vitamin B_{12})	76Ea	
I10 IMIDAZOLE, 2-(2-pyridyl)-			$C_8H_7N_3$			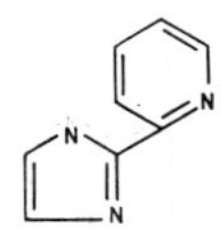		
Fe^{2+}	-81.2	11.60(1-3)	-50.6	25	C	$\mu=0.1M(NO_3^-)$	70Eb	
Co^{2+}	-32.34	5.26(1)	-6.7	25	C	$\mu=0.1M(NO_3^-)$	70Eb	
	-26.28	4.79(2)	12.1	25	C	$\mu=0.1M(NO_3^-)$	70Eb	
	-17.53	3.82(3)	3.8	25	C	$\mu=0.1M(NO_3^-)$	70Eb	
Ni^{2+}	-38.74	6.39(1)	-7.9	25	C	$\mu=0.1M(NO_3^-)$	70Eb	
	-34.31	6.22(2)	-6.3	25	C	$\mu=0.1M(NO_3^-)$	70Eb	
	-21.59	5.19(3)	26.8	25	C	$\mu=0.1M(NO_3^-)$	70Eb	
Cu^{2+}	-44.18	7.94(1)	3.3	25	C	$\mu=0.1M(NO_3^-)$	70Eb	
	-27.57	5.70(2)	16.3	25	C	$\mu=0.1M(NO_3^-)$	70Eb	
	-24.72	3.28(3)	20.5	25	C	$\mu=0.1M(NO_3^-)$	70Eb	
Zn^{2+}	-26.65	4.39(1)	0.4	25	C	$\mu=0.1M(NO_3^-)$	70Eb	
	-28.41	4.57(2)	-13.8	25	C	$\mu=0.1M(NO_3^-)$	70Eb	
	-7.20	3.11(3)	34.7	25	C	$\mu=0.1M(NO_3^-)$	70Eb	
I11 IMIDAZOLE, 4-(2-pyridyl)-			$C_8H_7N_3$			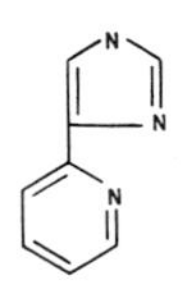		
Fe^{2+}	-92.0	13.76(1-3)	-46.0	25	C	$\mu=0.1M(NO_3^-)$	70Eb	
Ni^{2+}	-42.05	7.20(1)	-3.8	25	C	$\mu=0.1M(NO_3^-)$	70Eb	
	-39.96	6.75(2)	-4.6	25	C	$\mu=0.1M(NO_3^-)$	70Eb	
	-24.48	5.87(3)	30.1	25	C	$\mu=0.1M(NO_3^-)$	70Eb	
Cu^{2+}	-49.00	8.76(1)	2.9	25	C	$\mu=0.1M(NO_3^-)$	70Eb	

Metal Ion	ΔH, kJ /mol	Log K	ΔS, J /mol K	T,°C	M	Conditions	Ref.	Re-marks
I11, continued								
Cu^{2+},	-30.38	6.40(2)	20.1	25	C	μ=0.1M(NO_3^-)	70Eb	
cont.	-85.48	3.25(3)	-1.7	25	C	μ=0.1M(NO_3^-)	70Eb	
Zn^{2+}	-29.62	5.42(1)	4.2	25	C	μ=0.1M(NO_3^-)	70Eb	
	-24.27	4.82(2)	14.6	25	C	μ=0.1M(NO_3^-)	70Eb	
	-15.53	3.60(3)	32.67	25	C	μ=0.1M(NO_3^-)	70Eb	
I12	2-IMIDAZOLIDONE, N,N'-dimethyl		$C_5H_{10}ON_2$			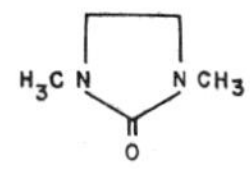		
SbA_5	-115.9	--	--	25	C	S:CH_2Cl_2 A=Cl	71Oa	j
I13	INDENE		C_9H_8			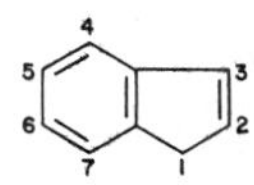		
Li^+	20.9	--	84	-40-80	T	S:1,2-Dimethoxy-ethane R:NaSL=NaL(Solvent sep-arated=Contact ion pair)	72Gi	b
Na^+	14.6	--	63	-40-80	T	S:1,2-Dimethoxy-ethane R:NaSL=NaL(Solvent sep-arated=Contact ion pair)	72Gi	b
I14	INOSINE-5'-triphosphoric acid		$C_{10}H_{15}O_{14}N_4P_3$					
Mg^{2+}	6.3	3.76(25°,1)	92	25-45	T	μ=0.10M(KNO_3)	73Kf	b,g:Na q:±3.3
	18.74	3.91(1)	136.8	30	C	μ=0.2 (Tetrame-thylammonium chloride) pH=8.50	73Sa	
Ca^{2+}	-4.6	3.41(25°,1)	50	25-45	T	μ=0.10M(KNO_3)	73Kf	b,g:Na, q:±0.4
Mn^{2+}	-10	4.45(25°,1)	50	25-45	T	μ=0.10M(KNO_3)	73Kf	b,g:Na
Co^{2+}	-5.0	4.97(25°,1)	79	25-45	T	μ=0.10M(KNO_3)	73Kf	b,g:Na, q:±0.4
Ni^{2+}	-8.4	5.06(25°,1)	79	25-45	T	μ=0.10M(KNO_3)	73Kf	b,g:Na
Cu^{2+}	-25.1	5.76(25°,1)	46	25-45	T	μ=0.10M(KNO_3)	73Kf	b,g:Na
Zn^{2+}	-7.1	4.57(25°,1)	63	25-45	T	μ=0.10M(KNO_3)	73Kf	b,g:Na, q:±0.8
I15	INOSITOL, hexaphosphate		$C_6H_{18}O_{24}P_6$			$C_6H_6[OP(=O)(OH)_2]_6$		
[Fe]	-53.6	6.0(1)	-64	25	C	0.1M[Cl^-]	74Nd	

Metal Ion	ΔH, kJ /mol	Log K	ΔS, J /mol K	T,°C	M	Conditions	Ref.	Re-marks
I15, continued								
[Fe], cont.								
	-48.5	6.0(1)	-48	25	C	0.05M bis-tris buffer pH=7.0 [Fe]=Deoxy-hemoglobin 0.1M[Cl^-]	74Nd	
	-42.2	6.0(1)	-27	25	C	0.05M bis-tris buffer pH=7.2 [Fe]=Deoxy-hemoglobin 0.1M[Cl^-]	74Nd	
	-13.0	~4.8(1)	48	25	C	0.05M bis-tris buffer pH=7.4 [Fe]=Methemo-globin 0.1M[Cl^-]	74Nd	
						0.05M bis-tris buffer pH=7.4 [Fe]=Deoxy-hemoglobin		
I16 INSULIN, bovine								
[Zn]	-72.0	6.38	-121	25	C	μ=0.1 pH=2 [Zn]=Bovine zinc insulin R:2ZnL=$[ZnL]_2$ monomer=dimer	73Lc	
I17 IODATE ION			IO_3			IO_3^-		L^-
Na^+	19.16	1.73(1)	97.5	25	C	μ=0 S:Ethanol	72Kd	
Eu^{3+}	11.09	1.14(25°,1)	59.8	--	T	μ=0.10M($NaClO_4$)	73Cb	b,q: ±1.7,u
Ag^+	21.50	0.801(1)	84.9	35	T	μ=1.0($LiClO_4$)	56Ra	
	-21.76	1.79(1-2)	-35.56	35	T	μ=1.0($LiClO_4$)	56Ra	
I18 IODIDE ION			I			I^-		L^-
Li^+	54.8	11.32(1)	402	25	T	CM=10^{-4}M S:25w% Butanol 75w% Hexane D=2.85	61Rd	
	10.04	15.56(1)	330	25	T	CM=10^{-4}M S:12.65 w% Butanol 87.35 w% Hexane D-2.16	61Rd	
	-19.25	17.83(1)	276	25	T	CM=10^{-4}M S:7w% Butanol 93wt% Hexane	61Rd	
Na^+	4.18	3.96(30°,1)	88	25&35	T	S:2-Methyl-propanl-ol	70Ed	a,b,q: ±0.63
	5.02	3.84(30°,1)	92	25&35	T	S:2-Methylpro-panl-ol	70Ed	a,b,q: ±0.63

Metal Ion	ΔH, kJ /mol	Log K	ΔS, J /mol K	T,°C	M	Conditions	Ref.	Remarks
I18, continued								
Na^+, cont.						Pressure=506 bar $\Delta V=17cm^3$/mole		
	6.32	3.72(30°,1)	92	25&35	T	S:2-Methylpro-panl-ol	70Ed	a,b,q: ±0.63
						Pressure=1013 bar $\Delta V=13cm^3$/mole		
	7.11	3.55(30°,1)	92	25&35	T	S:2-Methylpro-panl-ol	70Ed	a,b,q: ±0.63
						Pressure=2026 bar $\Delta V=10cm^3$/mole		
	7.91	3.45(30°,1)	92	25&35	T	S:2-Methylpro-panl-ol	70Ed	a,b,q: ±0.63
						Pressure=3040 bar $\Delta v=6cm^3$/mole		
Sm^{3+}	-9.2	0.24(25°,1)	-25.9	0-50	T	μ=1M(ClO_4)	75Mf	
Eu^{3+}	-10.5	0.24(25°,1)	-30.1	0-50	T	μ=1M(ClO_4)	75Mf	
Gd^{3+}	-11.3	0.29(25°,1)	-32.2	0-50	T	μ=1M(ClO_4)	75Mf	
Tb^{3+}	-9.6	0.28(25°,1)	-26.8	0-50	T	μ=1M(ClO_4)	75Mf	
Dy^{3+}	-4.2	0.25(25°,1)	-9.2	0-50	T	μ=1M(ClO_4)	75Mf	q:±0.8
$CrA_6{}^{3+}$	31.8	-4.155(25°)	27.6	15-45	T	μ=4.2M(KI-HI) CH^+=0.26M A=H_2O R:$CuA_6{}^{3+}+L^-$= CuA_5L^{2+}+A	68Se	b
$CoA_6{}^{3+}$	9.62	1.95(1)	67	25	T	μ=0 A=NH_3	55Na	
	6.82	1.24(1)	46	25	T	μ=0.054 A=NH_3	53Ea	
	5.10	0.938(1)	33	25	T	μ=0.054 A= Ethylenediamine	53Ea	
NiA_2	0.749	0.438(25°)	10.88	-36-50	T	S:Chloroform A=[(p-C_6H_5) (p-C_6H_5)MeP] R:NiA_2L_2(planar)= NiA_2L_2 (tetrahedral)	70Pc	b
	3.636	0.081(25°)	13.76	-29-50	T	S:Chloroform A=[(p-C_6H_5) (p-ClC_6H_4)MeP] R:NiA_2L_2(planar)= NiA_2L_2 (tetrahedral)	70Pc	b
	1.238	0.484(25°)	13.43	-41-42	T	S:Chloroform A=[(p-C_6H_5) (p-MeC_6H_4)MeP] R:NiA_2L_2(planar)= NiA_2L_2 (tetrahedral)	70Pc	b
	6.276	-0.242(25°)	16.40	-47-30	T	S:Chloroform A[(p-C_6H_5) (p-$CF_3C_6H_4$)MeP]	70Pc	b

Metal Ion	ΔH, kJ /mol	Log K	ΔS, J /mol K	T,°C	M	Conditions	Ref.	Remarks
I18, continued								
NiA_2, cont.						R:NiA_2L_2(planar)= NiA_2L_2 (tetrahedral)		
	6.402	-0.246(25°)	16.74	-50-45	T	S:Chloroform A=[(p-ClC_6H_4) (p-$Cl_3C_6H_4$)MeP] R:NiA_2L_2 (planar)= NiA_2L_2 (tetrahedral)	70Pc	b
	0.0218	0.693(25°)	13.35	-36-40	T	S:Chloroform A=[(p-MeC_6H_4) (p-MeC_6H_4)MeP] R:NiA_2L_2(planar)= NiA_2L_2 (tetrahedral)	70Pc	b
$NiAL_2$	12.64	0.428(27°)	50.6	7-137		S:Nitrobenzene A=$(C_6H_5)_2P(CH_2)_3$-$P(C_6H_5)_2$ R:$NiAL_2$ (planar)=$NiAL_2$ (tetrahedral)	66Va	b
	18.87	0.585(27°)	74.0	-43-57		S:Chloroform A=$(C_6H_5)_2P(CH_2)_3$-$P(C_6H_5)_2$ R:$NiAL_2$ (planar)=$NiAL_2$ (tetrahedral)	66Va	b
$RhA_2B_2^+$	-10.96	0.934(70°)	-14.2	70&90	T	μ=2.0M A=Ethyl-enediamine B=Br R:transRh$A_2B_2^+$+L^- =transRhA_2BL^++B^-	66Bg	a,b,q: ±1.6
	3.8	0.781(85°)	25	85&96	T	μ=0.15M A=Ethyl-enediamine B=Cl R:transRh$A_2B_2^+$+L= transRhA_2BL^++B^-	66Bg	a,b,q: ±7.9
RhA_2BL^+	-25.86	0.757(50°)	-65.3	45-55	T	μ=0.6M A=Ethyl-enediamine B=Cl R:transRhA_2BL^++L^- =transRh$A_2L_2^+$+B^-	66Bg	b,q: ±1.7
	-14.14	0.508(50°)	-33.9	45-55	T	μ=0.5M A=Ethyl-enediamine B=Br R:transRhA_2BL^++L^- =transRh$A_2L_2^+$+B^-	66Bg	b,q: ±1.4
RhA_2BC	-29.7	3.699	4	50	T	μ=0.2M A=Ethyl-diamine B=I C=H_2O R:trans RhA_2BC^{2+}+L^-= transRhA_2BL^++C	67Bc	
RhA_5B	-7.9	--	--	--	-	A=NH_3 B=H_2O R:RhA_5B^{3+}+L^- =	67Pa	q:±1.2, u

Metal Ion	ΔH, kJ /mol	Log K	ΔS, J /mol K	T, °C	M	Conditions	Ref.	Remarks
I18, continued								
ThA_5B, cont.						RhA_5L^{2+}+B		
PdA_2	-23.8	3.63(1)	-10.0	25	T	μ=0 A=o-Phenylenebisdimethylarsine	67Ec 67Ed	
	-24.3	3.348(35°,1)	-13.4	22-55	T	μ=0.003($NaNO_3$) A=o-Phenylenebisdimethylarsine	67Ed	b
$PdAB^+$	-15.5	1.950(27°)	-14.2	24-63	T	μ=0.5 A=Diethylenetriamine- B=Cl^- R:$PdAB^+$+L^-= $PdAL^+$+B^-	67Hd	b
	-10.5	1.480(27°)	-6.7	19-85	T	μ=0.5 A-Diethylenetriamine B=Br^- R:$PdAb^+$+L^-= $PdAL^+$+B^-	67Hd	b
PtA_2	-15.9	5.68(23°,1)	55.2	21-55	T	μ=0 S:CH_3OH A=o-Phenylenebisdimethylarsine	66Db	b
	-22.6	3.279(1)	-13.4	25	T	μ=0 A=o-Phenylenebisdimethylarsine	67Ec 67Ed	
	-23.0	2.969(40°,1)	-17.2	25-55	T	μ=0.003($NaNO_3$) A=o-Phenylenebisdimethylarsine	67Ed	b
PtA_3^{4+}	5.4	1.14(1)	40.2	25	T	μ=0 A=Ethylenediamine	63Gb	
PtA_6^{2-}	-79	18.0	-79	25	T	μ=0.5 R:PtA_6^{2-}+ 6L^-=PtL_6^{2-}+6A^- A=Cl	60Pf	
	-79	--	--	0-44.5	T	μ=0.5 A=Cl R:PtA_6^{2-}+6L^-= PtL_6^{2-}+6A^-	59Pa	
	-96	16.1	12	25	T	μ=0.5 R:PtA_6^{2-}+ 6L^-=PtL_6^{2-}+6A^- A=Br	60Pf	
CuA^{2+}	-5.90	1.029(25°,1)	-0.4	25-45	T	μ=0.4M(KNO_3) A=Macrocyclic ligand, rac-5,7,7,12,14,14-hexamethyl-1,4,8,11-tetraazacyclotetradecane (tet b), (blue form)	76Cd	
	4.69	2.53(1)	64.8	25	C	S:CH_3OH A=NN'-bis(2-aminoethyl)propane-1,3-diamine	72Bc	
	-1.51	3.10(1)	54.0	25	C	S:CH_3OH A=NN'-bis(3-aminopropyl)	72Bc	

Metal Ion	ΔH, kJ /mol	Log K	ΔS, J /mol K	T,°C	M	Conditions	Ref.	Re-marks
I18, continued								
CuA^{2+}, cont.						propane-1,3-diamine		
	6.57	2.83(1)	76.1	25	C	S:CH_3OH A=Triethylenetetramine	72Bc	
CuA_2^{2+}	5.40	2.89(1)	73.2	25	C	S:CH_3OH A=NN'-Dimethylethylenediamine	72Be	
	2.68	3.05(1)	67.4	25	C	S:CH_3OH A=NN'-Dimethylethylenediamine	72Be	
	11.88	1.73(1)	72.8	25	C	S:CH_3OH A=Ethylenediamine	72Bc 72Be	
	4.22	2.62(1)	64.0	25	C	S:CH_3OH A=N-Methylethylenediamine	72Be	
	2.13	2.52(1)	55.2	25	C	S:CH_3OH A=Trimethylenediamine	72Bc	
Zn^{2+}	1.7	-0.47(1)	-3.3	25	C	μ=4($LiClO_4$) pH=1	79Cb	q:±0.4
	0	-2.93(1)	-56.1	25	C	μ=4.5	56Sa	n
	-19.0	-0.701(1)	50	25	C	μ=1M(NH_4ClO) S:DMSO	76Af	q:±7.5
	29.4	1.402(2)	127	25	C	μ=1M(NH_4ClO) S:DMSO	76Af	q:±7.8
	12.7	0.158(3)	45	25	C	μ=1M(NH_4ClO) S:DMSO	76Af	q:±2.3
	-38.1	-0.51(4)	137.6	25	C	μ=4($LiClO_4$) pH=1	79Cb	q:±0.8
	12.13	-1.76(1-2)	7.1	25	C	μ=4.5	56Sa	n
	55.23	-1.83(1-3)	150.2	25	C	μ=4.5	56Sa	n
	-76.57	-2.42(1-4)	-302.9	25	C	μ=4.5	56Sa	n
	33.4	-0.27	107.5	25	C	μ=4($LiClO_4$) R:$ZnL^+ + 2L^- = ZnL_3^-$ pH=1	79Cb	
	18.8	--	--	25	C	S:Dimethylformamide R:ZnL_2+LiL=Li(ZnL_3)	74Ga	
	17.6	--	--	25	C	S:Dimethylformamide R:ZnL_2+NaL=Na(ZnL_3)	74Ga	
	15.1	--	--	25	C	S:Dimethylformamide R:ZnL_2+KL=K(ZnL_3)	74Ga	
	13.4	--	--	25	C	S:Dimethylformamide R:ZnL_2+RbL=Rb(ZnL_3)	74Ga	
	13.8	--	--	25	C	S:Dimethylformamide R:ZnL_2+CsL=Cs(ZnL_3)	74Ga	

Metal Ion	ΔH, kJ /mol	Log K	ΔS, J /mol K	T,°C	M	Conditions	Ref.	Remarks
I18, contnued								
Zn^{2+}, cont.	69.4	--	--	25	C	S:Dimethyl-formamide R:$2ZnL_2+CaL_2=Ca(ZnL_3)_2$	74Ga	
	60.2	--	--	25	C	S:Dimethyl-formamide R:$2ZnL_2+BaL_2=Ba(ZnL_3)_2$	74Ga	
	9.2	--	--	25	C	S:DMSO R:$ZnL_2+LiL=Li(ZnL_3)$	74Ga	
	8.4	--	--	25	C	S:DMSO R:$ZnL_2+NaL=Na(ZnL_3)$	74Ga	
	6.3	--	--	25	C	S:DMSO R:$ZnL_2+KL=K(ZnL_3)$	74Ga	
	10.0	--	--	25	C	S:DMSO R:$ZnL_2+RbL=Rb(ZnL_3)$	74Ga	
	10.0	--	--	25	C	S:DMSO R:$ZnL_2+CsL=Cs(ZnL_3)$	74Ga	
	32.6	--	--	25	C	S:DMSO R:$ZnL_2+CaL_2=Ca(ZnL_4)$	74Ga	
	29.7	--	--	25	C	S:DMSO R:$ZnL_2+BaL_2=Ba(ZnL_4)$	74Ga	
Cd^{2+}	-5.0	2.40(1)	29	25	C	μ=0	74Fd	
	-8.58	(1)	--	15&25	T	μ=0	59Aa, 66Ga	a,b
	-9.71	1.94(1)	4.6	25	C	μ=0.25($NaClO_4$)	68Gc, 54Ya	
	-5.0	1.82(1)	18	25	C	μ=0.5($LiClO_4$)	74Fd	
	-10.17	1.88(1)	2.1	25	C	μ=0.50($NaClO_3$)	68Gc, 60Ca	
	-6.3	1.91(1)	15	25	C	μ=1.0($LiClO_4$)	74Fd	q:±0.8
	-10.25	1.88(1)	1.7	25	C	μ=1($NaClO_4$)	68Gc	
	-7.5	2.00(1)	13	25	C	μ=2.0($LiClO_4$)	74Fd	
	-9.71	1.94(1)	5.0	25	C	μ=2($NaClO_4$)	68Gc	
	-9.6	2.20(1)	9.6	25	C	μ=3.0($LiClO_4$)	74Fd	
	-9.46	2.08(1)	7.9	25	C	μ=3($NaClO_4$)	68Gc	
	-9.6	2.07(1)	7.1	25	C	μ=3($NaClO_4$) pH=1	79Cb	q:±0.8
	-12.1	2.40(1)	5.0	25	C	μ=4.0($LiClO_4$)	74Fd	q:±1.2
	2.4	2.182(1)	50	25	C	μ=1M(NH_4ClO_4) S:DMSO	76Ae	
	-13.0	1.68(2)	-11	25	C	μ=0	74Fd	
	-7.1	0.70(2)	-10.5	25	C	μ=0.25($NaClO_4$)	68Gc	
	-12.6	1.32(2)	-17	25	C	μ=0.5($LiClO_4$)	74Fd	
	-3.89	0.75(2)	1.2	25	C	μ=0.50($NaClO_4$)	68Gc	
	-12.1	1.43(2)	-13	25	C	μ=1.0($LiClO_4$)	74Fd	q:±1.2

Metal Ion	ΔH, kJ /mol	Log K	ΔS, J /mol K	T,°C	M	Conditions	Ref.	Re-marks
I18, continued								
Cd^{2+}, cont.	-2.01	0.78(2)	7.9	25	C	μ=1($NaClO_4$)	68Gc	
	-12.6	1.76(2)	-8.4	25	C	μ=2.0($LiClO_4$)	74Fd	q:±1.7
	-1.21	0.63(2)	7.9	25	C	μ=2($NaClO_4$)	68Gc	
	-12.6	1.50(2)	-13	25	C	μ=3.0($LiClO_4$)	74Fd	q:±1.7
	-0.8	0.70(2)	10.5	25	C	μ=3($NaClO_4$) pH=1	79Cb	q:±2.5
	-0.84	0.70(2)	10.5	25	C	μ=3($NaClO_4$)	68Gc	
	-13.0	1.71(2)	-11	25	C	μ=4.0($LiClO_4$)	74Fd	q:±2.1
	27	1.398(2)	117	25	C	μ=1M(NH_4ClO_4) S:DMSO	76Ae	q:±2
	-2.9	1.03(3)	9.6	25	C	μ=0	74Fd	q:±1.7
	-5.8	1.15(3)	2.9	25	C	μ=0.5($LiClO_4$)	74Fd	q:±2
	-3.14	1.64(3)	20.9	25	C	μ=0.50($NaClO_4$)	68Gc	
	-3.3	1.31(3)	14	25	C	μ=1.0($LiClO_4$)	74Fd	q:±2
	-5.77	1.68(3)	13.0	25	C	μ=1($NaClO_4$)	68Gc	
	-15.5	1.24(3)	19	25	C	μ=2.0($LiClO_4$)	74Fd	q:±2.5
	-5.06	2.10(3)	23.4	25	C	μ=2($NaClO_4$)	68Gc	
	-10.9	2.70(3)	15	25	C	μ=3.0($LiClO_4$)	74Fd	q:±3.3
	-2.9	2.12(3)	31.0	25	C	μ=3($NaClO_4$) pH=1	79Cb	q:±1.2
	-3.05	2.13(3)	30.5	25	C	μ=3($NaClO_4$)	68Gc	
	-9.2	1.93(3)	5.8	25	C	μ=4.0($LiClO_4$)	74Fd	q:±3.3
	-5	2.925(3)	40	25	C	μ=1M($NHClO_4$) S:DMSO	76Ae	q:±2
	-18.8	-0.01(4)	-64	25	C	μ=0	74Fd	q:±2.5
	-22.6	1.02(4)	-57	25	C	μ=0.5($LiClO_4$)	74Fd	q:±4
	-17.70	1.2(4)	-36.4	25	C	μ=0.50($NaClO_4$)	68Gc	
	-17.6	1.21(4)	-36	25	C	μ=1.0(LiClO)	74Fd	q:±4
	-16.65	1.28(4)	-31.4	25	C	μ=1($NaClO_4$)	68Gc	
	-1.2	0.81(4)	-36	25	C	μ=2.0($LiClO_4$)	74Fd	q:±3.3
	-15.90	1.33(4)	-27.6	25	C	μ=2($NaClO_4$)	68Gc	
	-4.6	0.80(4)	-0.4	25	C	μ=3.0($LiClO_4$)	74Fd	q:±3.3
	-15.5	1.60(4)	-21.3	25	C	μ=3($NaClO_4$) pH=1	79Cb	
	-15.94	1.60(4)	-22.6	25	C	μ=3($NaClO_4$)	68Gc	
	-7.1	2.16(4)	18	25	C	μ=4.0($LiClO_4$)	74Fd	q:±4
	-9.54	1.167(4)	-10	25	C	μ=1M(NH_4ClO_4) S:DMSO	76Ae	
	45.2	(1-4)	--	25	C	--	52Ya	
	20.1	--	--	25	C	S:Dimethyl-formamide R:CdL_2+LiL=Li(CdL_3)	74Ga	
	17.2	--	--	25	C	S:Dimethyl-formamide R:CdL_2+NaL=Na(CdL_3)	74Ga	
	19.2	--	--	25	C	S:Dimethyl-formamide R:CdL_2+KL=K(CdL_3)	74Ga	
	18.8	--	--	25	C	S:Dimethyl-formamide R:CdL_2+RbL=Rb(CdL_3)	74Ga	
	16.7	--	--	25	C	S:Dimethyl-formamide	74Ga	

Metal Ion	ΔH, kJ /mol	Log K	ΔS, J /mol K	T,°C	M	Conditions	Ref.	Remarks
I18, continued								
Cd^{2+}, cont.						R: CdL_2+CsL= Cs(CdL_3)		
	-31.4	--	--	25	C	S:Dimethyl- formamide	74Ga	
						R: CdL_2+ZnL_2= Zn(CdL_4)		
	66.1	--	--	25	C	S:Dimethyl- formamide	74Ga	
						R: 2CdL_2+CaL_2= Ca$(CdL_3)_2$		
	56.1	--	--	25	C	S:Dimethyl- formamide	74Ga	
						R: 2CdL_2+BaL_2= Ba$(CdL_3)_2$		
	18.4	--	--	25	C	S:DMSO R: CdL_2+ LiL=Li(CdL_3)	74Ga	
	18.8	--	--	25	C	S:DMSO R: CdL_2+ NaL=Na(CdL_3)	74Ga	
	15.9	--	--	25	C	S:DMSO R: CdL_2+ KL=K(CdL_3)	74Ga	
	15.9	--	--	25	C	S:DMSO R: CdL_2+ RbL=Rb(CdL_3)	74Ga	
	31.0	--	--	25	C	S:DMSO R: CdL_2+ CaL_2=Ca(CdL_4)	74Ga	
	14.6	--	--	25	C	S:DMSO R: CdL_2+ CsL=Cs(CdL_3)	74Ga	
	27.2	--	--	25	C	S:DMSO R: CdL_2+ BaL_2=Ba(CdL_4)	74Ga	
	10.9	--	--	25	C	S:DMSO R: CdL_2+ ZnL_2=Zn(CdL_4)	74Ga	
Hg^{2+}	-73.6	13.5(1)	12	25	C	μ=0	61Ma	
	-79.1	13.53(1)	-22.2	8	C	μ=0.50	64Cb	
	-75.3	12.87(1)	-6.3	25	C	μ=0.50	64Cb	
	-76.1	12.81(1)	-10.0	25	C	μ=0.5($NaClO_4$) pH=1	79Cb	
	-72.4	12.44(1)	6.3	40	C	μ=0.50	64Cb	
	-69.4	13.5(1)	25	25	C	--	54Ya	
	-64.0	11.59(2)	-5.8	8	C	μ=0.50	64Cb	
	-67.8	11.69(2)	-17.6	25	C	μ=0.50	64Cb	
	-68.2	10.91(2)	-20.1	25	C	μ=0.5($NaClO_4$) pH=1	79Cb	
	-72.0	10.64(2)	-26.8	40	C	μ=0.50	64Cb	
	-21.3	3.65(3)	-1.7	25	C	μ=0.5M($NaClO_4$) pH=1	79Cb	
	-26.8	6.1(3)	27	25	C	μ=1M($NaClO_4$) S:Dimethyl sulfoxide	76Am	
	-4.2	2.32(4)	30.1	25	C	μ=0.5($NaClO_4$) pH=1	79Cb	
	-22.8	2.58(4)	-27	25	C	μ=1M($NaClO_4$) S:Dimethyl sulfoxide	76Am	
	-142.2	24.97(1-2)	-28.0	8	C	μ=0.50	64Cb	

Metal Ion	ΔH, kJ /mol	Log K	ΔS, J /mol K	T,°C	M	Conditions	Ref.	Remarks
I18, continued								
Hg^{2+}, cont.	-142.9	23.78(1-2)	-23.8	25	C	μ=0.50	64Cb	
	-144.3	23.08(1-2)	-20.5	40	C	μ=0.50	64Cb	
	-184.9	29.86(1-4)	-56.1	25	C	μ=0.5	60Ga	
	-179.1	1.23(42.1°)	-544	42&55.6	T	S:Benzene R:$2(HgL_2)=(HgL_2)_2$	74Ea	a,b
	64.8	3.08(42.1°)	264	42&55.6	T	S:Benzene R:$3(HgL_2)=(HgL_2)_3$	74Ea	a,b
	-38.9	--	--	25	C	S:Dimethyl-formamide R:HgL_2+LiL= Li(HgL_3)	74Ga	
	-66.1	--	--	25	C	S:Dimethyl-formamide R:HgL_2+2LiL= Li_2(HgL_4)	74Ga	
	-39.7	--	--	25	C	S:Dimethyl-formamide R:HgL_2+NaL= Na(HgL_3)	74Ga	
	-66.9	--	--	25	C	S:Dimethyl-formamide R:HgL_2+2NaL= Na_2(HgL_4)	74Ga	
	-41.4	--	--	25	C	S:Dimethyl-formamide R:HgL_2+KL= K(HgL_3)	74Ga	
	-42.7	--	--	25	C	S:Dimethyl-formamide R:HgL_2+RbL= Rb(HgL_3)	74Ga	
	-69.9	--	--	25	C	S:Dimethyl-formamide R:HgL_2+2RbL= Rb_2(HgL_4)	74Ga	
	-43.5	--	--	25	C	S:Dimethyl-formamide R:HgL_2+CsL= Cs(HgL_3)	74Ga	
	-71.1	--	--	25	C	S:Dimethyl-formamide R:HgL_2+2CsL= Cs(HgL_4)	74Ga	
	-62.8	--	--	25	C	S:Dimethyl-formamide R:2HgL_2+CaL_2= Ca$(HgL_3)_2$	74Ga	
	-44.4	--	--	25	C	S:Dimethyl-formamide R:HgL_2+CaL_2= Ca(HgL_4)	74Ga	
	-70.3	--	--	25	C	S:Dimethyl-formamide R:2HgL_2+BaL_2=	74Ga	

Metal Ion	ΔH, kJ /mol	Log K	ΔS, J /mol K	T,°C	M	Conditions	Ref.	Re-marks
I18, continued								
Hg^{2+}, cont.	-53.6	--	--	25	C	$Ba(HgL_3)_2$ S:Dimethyl-formamide	74Ga	
	-98.3	--	--	25	C	R:HgL_2+BaL_2=$Ba(HgL_4)$ S:Dimethyl-formamide	74Ga	
	-83.7	--	--	25	C	R:HgL_2+ZnL_2=$Zn(HgL_4)$ S:Dimethyl-formamide	74Ga	
	-69.0	--	--	25	C	R:$2HgL_2$+CdL_2=$Cd(HgL_3)_2$ S:Dimethyl-formamide	74Ga	
	-31.0	--	--	25	C	R:HgL_2+2KL=$K_2(HgL_4)$ S:DMSO R:HgL_2+LiL=Li(HgL_3)	74Ga	
	-42.2	--	--	25	C	S:DMSO R:HgL_2+2LiL=$Li_2(HgL_4)$	74Ga	
	-32.2	--	--	25	C	S:DMSO R:HgL_2+NaL=Na(HgL_3)	74Ga	
	-43.5	--	--	25	C	S:DMSO R:HgL_2+2NaL=$Na_2(HgL_4)$	74Ga	
	-31.4	--	--	25	C	S:DMSO R:HgL_2+KL=K(HgL_3)	74Ga	
	-44.8	--	--	25	C	S:DMSO R:HgL_2+2KL=$K_2(HgL_4)$	74Ga	
	-32.2	--	--	25	C	S:DMSO R:HgL_2+RbL=Rb(HgL_3)	74Ga	
	-45.2	--	--	25	C	S:DMSO R:HgL_2+2RbL=$Rb_2(HgL_4)$	74Ga	
	-31.0	--	--	25	C	S:DMSO R:HgL_2+CsL=Cs(HgL_3)	74Ga	
	-45.6	--	--	25	C	S:DMSO R:HgL_2+2RbL=$Rb_2(HgL_4)$	74Ga	
	-31.0	--	--	25	C	S:DMSO R:HgL_2+CsL=Cs(HgL_3)	74Ga	
	-40.2	--	--	25	C	S:DMSO R:HgL_2+CaL_2=Ca(HgL_2)	74Ga	
	-42.7	--	--	25	C	S:DMSO R:HgL_2+BaL_2=Ba(HgL_4)	74Ga	
	-62.8	--	--	25	C	S:DMSO R:HgL_2+ZnL_2=Zn(HgL_4)	74Ga	
	-50.6	--	--	25	C	S:DMSO R:HgL_2+CdL_2=Cd(HgL_4)	74Ga	
	-43.9	--	--	25	C	CM=0.028M S:Dimethyl sulfoxide R:$HgI_2+2I^-=HgI_4^{2-}$	67Be	
	-29.3	2.86(55.6°)	-33	55&65	T	S:Mesitylene	74Ea	a,b

Metal Ion	ΔH, kJ /mol	Log K	ΔS, J /mol K	T,°C	M	Conditions	Ref.	Re-marks
I18, continued								
Hg^{2+},						R:$2(HgL_2)=(HgL_2)_2$		
cont.	-134.3	0.93(42.1°)	-410	42&55.6	T	S:Toluene R:$2(HgL_2)=(HgL_2)_2$	74Ea	a,b
	-348.9	2.39(42.1°)	-1063	42&55.6	T	S:Toluene R:$3(HgL_2)=(HgL_2)_3$	74Ea	a,b
	-390.4	0.0(55.6°)	1188	55&65	T	S:p-Xylene R:$2(HgL_2)=(HgL_2)_2$	74Ea	a,b
	-457.7	3.28(55.6°)	-1330	55.6& 64.9	T	S:p-Xylene R:$3(HgL_2)=(HgL_2)_3$	74Ea	a,b
HgA^{2-}	-65.7	9.30(1)	-42.2	25	C	μ=2M($NaClO_4$) A=EDTA	75Bc	
HgA_2	-46.9	5.76(20°)	-49.0	10-30	T	μ=0 A=Cl R:HgA_2 $+L^-=HgAL+A^-$	72Bg	b
	-97.9	10.62(20°)	-130.5	10-30	T	μ=0 A=Cl R:HgA_2 $+2L^-=HgL_2+2A^-$	72Bg	b
	4.6	0.902(20°)	33.0	10-30	T	μ=0 A=Cl R:HgA_2 $+HgL_2=2HgAL$	72Bg	b
HgAL	-51.5	4.86(20°)	-82.0	10-30	T	μ=0 A=Cl R:HgLA $+L^-=HgL_2+A^-$	72Bg	b
HgL_2	-1.7	0.55(25°)	5.0	5-60	T	μ=10^{-3} R:½HgL_2 +½$HgBr_2$=HgLBr	64Ea	b,r
	-2.9	0.88(25°)	7.1	5-60	T	μ=10^{-3} R:½HgL_2 +½$HgCl_2$=HgLCl	64Ea	b,r
$HgAB^{3-}$	-23.4	4.03	0	25	C	μ=2M($NaClO_4$) R:$HgAB^{3-}+L^-=$ $HgLB^{3-}+A^-$ $A=Br^-$ $B=EDTA^{4-}$	75Bc	
In^{3+}	-3.05	1.03(1)	9.6	25	C	μ=2.00M($NaClO_4$)	69Rd	
	-3.39	1.25(2)	35.1	25	C	μ=2.00M($NaClO_4$)	69Rd	
Tl^+	-73.346	7.284(1)	-106.7	25	T	μ=0	34La	
Pb^{2+}	-1.255	(1)	--	25	C	μ=0	59Aa	
	-1.736	(1)	--	25	C	μ=6.5x10^{-3}	59Aa	
	-49.4	1.68(25°,1)	-130	5-35	T	μ=3M($LiClO_4$+LiI)	72Fh	b
	-4.18	2.29(1)	29	25	C	--	54Ya	
	-4.2	1.32(25°,2)	8	5-35	T	μ=3M($LiClO_4$+LiI)	72Fh	b
	10.9	1.36(25°,3)	63	5-35	T	μ=3M($LiClO_4$+LiI)	72Fh	b
	-67	0.40(25°,4)	-17	5-35	T	μ=3M($LiClO_4$+LiI)	72Fh	b
	-247	6.201(25°,1-4)	-707	0-35	T	1M	49Kb	b
	65.3	(1-4)	--	--	C	--	52Ya	
PbL_2	15.06	0.041(25°)	54	5-25	T	μ=3M($LiClO_4$+LiI) R:$2PbL_2=PbL^++PbL_3^-$	72Fh	b
[N]	34.73	11.56(1)	339	25	T	CM≅10^{-4}M iso $(C_5H_{11})_4N$ S:25w%Butanol 72w%Hexane D=2.85	61Rd	b
[N]	22.59	13.13(1)	326	25	T	CM≅10^{-4}M iso $(C_5H_{11})_4N$	61Rd	

Metal Ion	ΔH, kJ /mol	Log K	ΔS, J /mol K	T,°C	M	Conditions	Ref.	Re-marks
I18, continued								
[N], cont.						S:20w%Butanol 80w%Hexane D=2.51		
[N]	-6.28	15.53(1)	276	25	T	CM≅10^{-4}M iso $(C_5H_{11})_4N$ S:12.65w%Butanol 87.35w%Hexane D=2.16	61Rd	
Bi^{3+}	-36.0	2.91(25°,1)	-64.8	6-45	T	μ=3	72Fe	b,q:±3.3
	-10.9	3.65(25°,2)	33.0	6-45	T	μ=3	72Fe	b
	-0.4	3.34(25°,3)	62.8	6-45	T	μ=3	72Fe	b
	-17.2	2.46(25°,4)	-10.0	6-45	T	μ=3	72Fe	b
	-14.6	2.02(25°,5)	-11.3	6-45	T	μ=3	72Fe	b
	6.3	0.56(25°,6)	32.2	6-45	T	μ=3	72Fe	b
I_2	-13.452	2.69(46.65°,1)	4.6	25-63	T	μ=0 ΔCp=138	52Da	b
	-15.251	(31.69°,1)	--	25-63	T	μ=0 ΔCp=138	52Da	b
	-13.690	(44.01°,1)	--	25-63	T	μ=0 ΔCp=138	52Da	b
	-11.778	(56.35°,1)	--	25-63	T	μ=0 ΔCp=138	52Da	b
	21.34	2.85(25°,1)	17.2	1-39	T	μ=0.05(KCl)	51Aa	b
	-15.06	--	--	20-30	T	R:$I_2+I^-=I_3^-$	55Ka	b
	-30.12	5.58(38.38°)	10.0	25-49.65	T	μ=0 R:$2I^-+2I_2=I_6^{2-}$	52Da	b
I19 IODINE			I_2			I_2		L
$RhABC_2$ (trans)	-143.9	(1)	--	25	C	S:Benzene A=Cl B=CO C=PEt_2Ph	79Bf	
	-174.9	(1)	--	25	C	S:Benzene A=Cl B=CO C=$P(t\text{-}Bu)Ph_2$	79Bf	
	-84.1	(1)	--	25	C	S:Benzene A=Cl B=CO C=$P(Cy)_3$	79Bf	
	-123.8	(1)	--	25	C	S:Benzene A=Cl B=CO C=$AsPh_3$	79Bf	
	-121.3	(1)	--	25	C	S:Benzene A=Cl B=CO C=PPh_3	79Bf	
	-112.5	(1)	--	25	C	S:Benzene A=Br B=CO C=PPh_3	79Bf	
	-95.8	(1)	--	25	C	S:Benzene A=I B=CO C=PPh_3	79Bf	
	-108.4	(1)	--	25	C	S:Benzene A=Cl B=CO C=$P(OMe)_3$	79Bf	
	-146.0	(1)	--	25	C	S:Benzene A=Cl B=CO C=PMe_3	79Bf	
	-148.5	(1)	--	25	C	S:Benzene A=Cl B=CO C=PMe_2Ph	79Bf	
	-88.7	(1)	--	25	C	S:Benzene A=Cl B=CO C=$P(OPh)_2$	79Bf	
	-121.3	(1)	--	25	C	S:Benzene A=Cl B=CO C=$PMePh_2$	79Bf	
	-114.2	(1)	--	25	C	S:Benzene A=Cl B=CS C=PPh_3	79Bf	

Metal Ion	ΔH, kJ /mol	Log K	ΔS, J /mol K	T,°C	M	Conditions	Ref.	Remarks
I19, continued								
$IrABC_2$ (trans)	-146.8	(1)	--	25	C	S:Benzene A=F B=CO L=PPh_3	79Bf	
	-136.4	(1)	--	25	C	S:Benzene A=Cl B=CO L=PPh_3	79Bf	
	-173.6	(1)	--	25	C	S:Benzene A=Cl B=CO C=PMe_2Ph	79Bf	
	-136.8	(1)	--	25	C	S:Benzene A=Cl B=CO C=$P(OPh)_3$	79Bf	
	-160.2	(1)	--	25	C	S:Benzene A=Cl B=CO C=$PMePh_2$	79Bf	
	-170.3	(1)	--	25	C	S:Benzene A=Cl B=CO C=PEt_2Ph	79Bf	
	-136.4	(1)	--	25	C	S:Benzene A=Cl B=CO C=PPh_3	79Bf	
	-111.7	(1)	--	25	C	S:Benzene A=Cl B=CO C=$P(t\text{-}Bu)Ph_2$	79Bf	
	-116.3	(1)	--	25	C	S:Benzene A=Cl B=CO C=$P(Cy)_3$	79Bf	
	-171.5	(1)	--	25	C	S:Benzene A=Cl B=CO C=$AsPh_3$	79Bf	
	-145.6	(1)	--	25	C	S:Benzene A=Cl B=CS C=PPh_3	79Bf	
	-148.5	(1)	--	25	C	S:Benzene A=Br B=CO L=PPh_3	79Bf	
	-120.1	(1)	--	25	C	S:Benzene A=I B=CO L=PPh_3	79Bf	
	-138.5	(1)	--	25	C	S:Benzene A=NCS B=CO L=PPh_3	79Bf	
	-151.0	(1)	--	25	C	S:Benzene A=N_3 B=CO L=PPh_3	79Bf	
	-43.5	(1)	--	25	C	S:Benzene A=C_6F_5 B=CO C=PPh_3	79Bf	q:±4
PbA_4	-136.27	--	--	--	C	S:$CHCl_3$ A=Phenyl R:$PbA_4+2L_2=PbL_2A_2+2LA$	70Bg	u
PbA_3L	-59.24	--	--	--	C	S:$CHCl_3$ A=Phenyl R:$PbA_3L+L_2=PbA_2L_2+LA$	70Bg	u
[N]	-25.1	-2.60(1)	-134	25	C	C_5H_5N S:Benzene	65Ga	
[O]	-13.8	0(1)	-46.0	25	T	$O(C_2H_4)_2O$ S:CCl_4	61Md	
[O]	-24.7	(1)	--	25	C	C_4H_8O S:not stipulated	65Ga	
[S]	-19.2	0.836(1)	-49.0	25	T	CM=$1.6x10^{-3}$-$6.0x10^{-3}$M CL=$2.1x10^{-4}$-$6.9x10^{-4}$M $C_3H_6S_3$(s-Trithiane) S:CCl_4	62Mc	
[S]	-27.74	1.55(25°,1)	63.6	5-25	T	trans-2,3-	62Ta	b

Metal Ion	ΔH, kJ /mol	Log K	ΔS, J /mol K	T,°C	M	Conditions	Ref.	Remarks
I19, continued								
[S], cont.						Butylene sulfide S:n-Heptane		
[S]	-31.21	2.05(25°,1)	65.3	5-25	T	Thiacyclobutane S:n-Heptane	62Ta	b
[S]	-27.49	1.89(25°,1)	56.1	5-25	T	Thiacyclobutane S:CCl_4	62Ta	b
[S]	-32.51	2.42(25°,1)	-62.8	5-25	T	Thiacyclopentane S:n-Heptane	62Ta	b
[S]	-36.0	(1)	--	25	C	Thiacyclopentane S:not stipulated	65Ga	
[S]	-29.7	2.15(1)	-59.0	25	T	CM=1.6×10^{-3}-6.0×10^{-3}M CL=2.1×10^{-4}-6.9×10^{-4}M Thiacyclohexane S:CCl_4	62Mc	
[S]	-29.54	2.20(25°,1)	56.9	5-25	T	Thiacyclohexane S:n-Heptane	62Ta	b
[S]	-36.4	2.27(1)	-36.4	25	T	CM=1.6×10^{-3}-6.0×10^{-3}M CL=2.1×10^{-4}-6.0×10^{-4}M Thiacyclopentane S:CCl_4	62Mc	
[S]	-37.24	2.30(25°,1)	81.2	5-25	T	Diethyl sulfide S:n-Heptane	62Ta	b
[S]	-25.9	1.90(1)	-50.6	25	T	$S(C_2H_4)_2S$ 1,4-Dithiane S:CCl_4	61Md	
[S]	-21.8	1.77(1)	-39.7	25	T	$O(C_2H_4)_2S$ S:CCl_4	57Mb	
[S]	-34.31	(1)	--	25	C	$(C_4H_9)_2S$ S:Octane	65Ga	
[S]	-28.03	(1)	--	25	C	$(C_4H_9)_2S$ S:Benzene	65Ga	
[S]	-9.62	(1)	--	25	C	$(C_4H_9)_2S$ S:Tetrahydrofuran	65Ga	
[Se]	-36.0	2.70(1)	-68.6	25	T	Dilute solution Dimethyl Se $Se(CH_3)_2$ S:CCl_4	60Ma	
[Se]	-32.0	2.13(1)	-65.3	25	T	$Se(C_2H_4)_2O$ S:CCl_4	61Md	
[Se]	-31.0	2.29(1)	-59.4	25	T	$Se(C_2H_4)_2S$ S:CCl_4	61Md	
[Se]	-29.3	2.46(1)	-51.5	25	C	$Se(C_2H_4)_2Se$ S:CCl_4	61Md	
[Se]	-41.0	3.17(1)	-75	25	T	CM=CL=1.6×10^{-4}-2.0×10^{-3}M Selenacyclohexane SeC_5H_{10} S:CCl_4	67Mg	
[Se]	-46.4	3.36(1)	-92	25	T	CM=1.5×10^{-3}-8x	67Mg	

Metal Ion	ΔH, kJ /mol	Log K	ΔS, J /mol K	T,°C	M	Conditions	Ref.	Re-marks
I19, continued								
[Se], cont.						10^{-3}M CL=1.0x 10^{-4}-1.6x10^{-3}M Selenacyclopentane SeC_4H_8 S:CCl_4		
[Se]	-32.0	1.82(1)	-71.5	25	T	Dilute solution 4,4'-Dimethoxy-diphenyl Se S:CCl_4	60Ma	
[Se]	-28.9	1.60(1)	-68.6	25	T	Dilute solution 4,4'-Dimethyl-diphenyl Se S:CCl_4	60Ma	
[Se]	-17.6	0.923(1)	-41.8	25	T	Dilute solution 4,4'-Dichloro-diphenyl Se S:CCl_4	60Ma	
I_2	-8.20	(1)	--	8-40	T	low μ S:CCl_4	56Da	b
I20 ISOHISTAMINE			$C_5H_9N_3$			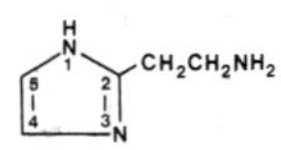		
Cu^{2+}	-48.2	9.85(1)	26.4	25	C	μ=0.1M(NO_3^-) pH<6	70Ec	
	-41.8	7.16(2)	-4.2	25	C	μ=0.1M(NO_3^-) pH<6	70Ec	
I21 ISOLEUCINE			$C_{13}H_{13}O_2N$			$CH_3CH_2CH(CH_3)CH(NH_2)CO_2H$		L^-
Co^{2+}	-18.5	4.590(1)	26	25	C	μ=0.1M($NaNO_3$) L(L form)	78Ia	
	-12.6	4.100(1)	36	25	C	μ=0.1M($NaNO_3$) L(D-allo form)	78Ia	
	-3.5	4.345(2)	71	25	C	μ=0.1M($NaNO_3$) L(L form)	78Ia	
	-4.2	3.364(2)	50	25	C	μ=0.1M($NaNO_3$) L(D-allo form)	78Ia	
Cu^{2+}	-26.2	8.497(1)	75	25	C	μ=0.1M($NaNO_3$) L(L form)	78Ia	
	-24.0	8.094(1)	74	25	C	μ=0.1M($NaNO_3$) L(D-allo form)	78Ia	
	-18.3	7.288(2)	78	25	C	μ=0.1M($NaNO_3$) L(L form)	78Ia	
	-18.0	6.956(2)	73	25	C	μ=0.1M($NaNO_3$) L(D-allo form)	78Ia	

Metal Ion	ΔH, kJ /mol	Log K	ΔS, J /mol K	T,°C	M	Conditions	Ref.	Remarks
I22 ISOQUINOLINE			C_9H_7N			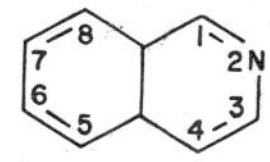		L
CoL_2A_2	-61.5	0.984(1-2)	-191.2	20	C	μ=0 S:Chloroform A=Cl	64Ka	
	-69.4	0.865(1-2)	-220.5	20	C	μ=0 S:Chloroform A=Br	64Ka	
	-72.8	0.36(1-2)	-241.4	20	C	μ=0 S:Chloroform A=I	64Ka	
NiA	-59.12	3.78(1-2)	-125.9	25	T	CM≅CL≅10^{-4}M S:Benzene A=Diacetylbisbenzoyl hydrazine	58Sa	
NiA_2	-87.86	3.52(25°,1-2)	-227.6	20-40	T	S:Benzene A=(p-$CH_3C_6H_4$)CH=NH-N=COC_6H_5	77Ia	
	-92.47	4.18(25°,1-2)	-230.1	20-40	T	S:Benzene A=(m-$OCH_3C_6H_4$)CH=N-N=COC_6H_5	77Ia	
	-86.61	4.00(25°,1-2)	-214.2	20-40	T	S:Benzene A=C_6H_5CH=N-N=COC_6H_5	77Ia	
	-83.14	3.42(25°,1-2)	-213.6	20-40	T	S:Benzene A=(p-$CH_3OC_6H_4$)CH=N-N=COC_6H_5	77Ia	
	-95.14	4.40(25°,1-2)	-235.1	20-40	T	S:Benzene A=(p-ClC_6H_4)CH=N-N=COC_6H_5	77Ia	
I_2	-31.9	2.04(1)	-68	25	C	S:Benzene	80Ua	
	-33.98	2.201(1)	-71.9	25	C	S:CCl_4	79Ba	
	-32.9	2.17(1)	-69	25	C	S:CCl_4	80Ua	
	-37.7	2.41(1)	-80	25	C	S:Chlorobenzene	80Ua	
	-36.6	2.27(1)	-79	25	C	S:Cyclohexane	80Ua	
	-34.8	2.48(25°,1)	-69.4	17-35	T	S:Heptane	68Kb	b
	-34.9	2.444(25°,1)	-70.3	17-35	T	CM=1.1-2.8x10^{-3}M CL=2.7x10^{-4}M S:n=Heptane	75Be	b
	-37.0	2.30(1)	-80	25	C	S:n-Hexane	80Ua	
I23 ISOSERINE			$C_3H_7O_3N$			$H_2NCH_2CH(OH)CO_2H$		L^-
Ni^{2+}	-12.93	4.19(1)	36.8	25	C	μ=0.16	70Lb	
	-15.61	3.66(2)	17.6	25	C	μ=0.16	70Lb	
Cu^{2+}	-24.18	7.31(1)	59.0	25	C	μ=0.16	70Lb	
	-25.77	7.06(2)	48.5	25	C	μ=0.16	70Lb	
I24 ISOTHIOCYANIC ACID, methyl ester			C_2H_3NS			$SCNCH_3$		L
I_2	13.8	-0.149(25°,1)	43.5	25-39.5	T	CM=0.3x10^{-2}M CL=0.36M S:CCl_4	65Wb	b

Metal Ion	ΔH, kJ /mol	Log K	ΔS, J /mol K	T,°C	M	Conditions	Ref.	Remarks
						L		
L1 LEUCINE			$C_6H_{13}O_2N$			$(CH_3)_2CHCH_2CH(NH_2)CO_2H$		
Y^{3+}	33.413	4.26(25°,1)	193.64	25&35	T	μ=0.1(KCl)	73S1	a,b,q: ±2.5
	52.756	3.90(25°,2)	251.58	25&35	T	μ=0.1(KCl)	73S1	a,b
Ce^{3+}	40.442	4.69(25°,1)	225.43	25&35	T	μ=0.1(KCl)	73S1	a,b,q: ±2.5
L2 LEUCINE (L), N-benzoyl			$C_{13}H_{17}O_3N$			$(CH_3)_2CHCH_2CH(NHCOC_6H_5)COOH$		L
Co^{2+}	-3.47	2.82(30°,1)	42.51	20-40	T	μ=0.1M(KNO_3) S:50v%Dioxane	79Sc	
Ni^{2+}	-3.47	2.80(30°,1)	42.09	20-40	T	μ=0.1M(KNO_3) S:50v%Dioxane	79Sc	
Cu^{2+}	-9.50	2.90(30°,1)	24.14	20-40	T	μ=0.1M(KNO_3) S:50v%Dioxane	79Sc	
L3 LIMONENE			$C_{10}H_{16}$			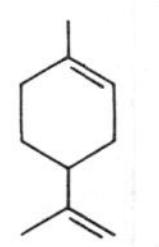		
PdA_2B_2	-28.4	--	--	25	C	S:CH_2Cl_2 CL=3.8x10^{-3}M CM=3.5x10^{-4}M A=Cl B=C_6H_5CN R:PdA_2B_2+L= PdA_2L+2B	72Pd	

Metal Ion	ΔH, kJ /mol	Log K	ΔS, J /mol K	T, °C	M	Conditions	Ref.	Remarks
				M				
M1 MALEIC ACID			$C_4H_4O_4$			$HO_2CCH:CHCO_2H$		L^{2-}
Na^+	-0.29	0.37(1)	5.8	25	C	μ=0.05M($NaClO_4$)	80Aa	
	-0.46	0.29(1)	4.2	25	C	μ=0.10M($NaClO_4$)	80Aa	
	-0.46	0.29(1)	4.2	25	C	μ=0.15M($NaClO_4$)	80Aa	
	-5.4	0.0	-18.4	25	C	μ=0.05M($NaClO_4$)	80Aa	
						R:Na^++HL^-=NaHL		
	-5.4	-0.07	-19.7	25	C	μ=0.10M($NaClO_4$)	80Aa	
						R:Na^++HL^-=NaHL		
	-5.4	-0.07	-19.7	25	C	μ=0.15M($NaClO_4$)	80Aa	
						R:Na^++HL^-=NaHL		
La^{3+}	12.80	3.44(1)	108.8	25	C	μ=0.10M($NaClO_4$)	73Cc	
Pr^{3+}	12.30	3.63(1)	110.9	25	C	μ=0.10M($NaClO_4$)	73Cc	
	10.79	2.814(1)	90.0	25	C	μ=1.0M($NaClO_4$)	73De	
	9.20	1.882(2)	66.9	25	C	μ=1.0M($NaClO_4$)	73De	
Nd^{3+}	12.09	3.66(1)	110.4	25	C	μ=0.10M($NaClO_4$)	73Cc	
Sm^{3+}	12.72	3.82(1)	115.5	25	C	μ=0.10M($NaClO_4$)	73Cc	
	10.67	3.000(1)	93.3	25	C	μ=1.0M($NaClO_4$)	73De	
	7.07	1.906(2)	59.8	25	C	μ=1.0M($NaClO_4$)	73De	
Eu^{3+}	14.22	3.83(1)	120.9	25	C	μ=0.10M($NaClO_4$)	73Cc	
Gd^{3+}	14.94	3.79(1)	122.6	25	C	μ=0.10M($NaClO_4$)	73Cc	
	12.51	2.964(1)	98.7	25	C	μ=1.0M($NaClO_4$)	73De	
	7.32	1.819(2)	59.0	25	C	μ=1.0M($NaClO_4$)	73De	
Tb^{3+}	17.4	3.74(1)	129.7	25	C	μ=0.10M($NaClO_4$)	73Cc	
Dy^{3+}	17.4	3.74(1)	130.1	25	C	μ=0.10M($NaClO_4$)	73Cc	
Ho^{3+}	18.24	3.67(1)	131.4	25	C	μ=0.10M($NaClO_4$)	73Cc	
	15.44	2.886(1)	107.1	25	C	μ=1.0M($NaClO_4$)	73De	
	8.66	1.775(2)	63	25	C	μ=1.0M($NaClO_4$)	73De	
Er^{3+}	18.03	3.64(1)	130.1	25	C	μ=0.10M($NaClO_4$)	73Cc	
Tm^{3+}	18.32	3.62(1)	130.5	25	C	μ=0.10M($NaClO_4$)	73Cc	
Yb^{3+}	18.54	3.64(1)	131.8	25	C	μ=0.10M($NaClO_4$)	73Cc	
	16.32	2.805(1)	108.4	25	C	μ=1.0M($NaClO_4$)	73De	
	10.71	1.843(2)	70.7	25	C	μ=1.0M($NaClO_4$)	73De	
Lu^{3+}	18.45	3.59(1)	130.5	25	C	μ=0.10M($NaClO_4$)	73Cc	
Cu^{2+}	14.4	3.399(1)	113.8	25	C	μ=0.1M($NaClO_4$)	78Ae	
	-34.7	11.879	112.1	25	C	μ=0.1M($NaClO_4$)	78Ae	
						R:Cu^{2+}+L^{2-}+A=CuLA A=2,2'-Bipyridyl		
M2 MALEIC ACID, anhydride			$C_4H_2O_3$			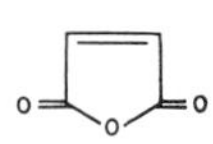		
NiAB	-53.6	4.52(27°)	-92.0	22-47	T	S:Tetrahydrofuran	71Ya	b

Metal Ion	ΔH, kJ /mol	Log K	ΔS, J /mol K	T,°C	M	Conditions	Ref.	Re-marks
M2, continued								
NiAB, cont.						A=Dipyridyl B=Tetrahydrofuran R:NiAB+L=NiAL+B		
M3 MALEIC ACID, ethylene copolymer			$(C_6H_8O_4)_n$			(1:1 Alternating type, MW ∿ 10^5)		L^{2-}
Ba^{2+}	2.5	4.32(1)	92	25	C	0.05M $(CH_3)_4NClO_4$ ΔH is per mole of complex	74Db	
Ni^{2+}	8.4	4.69(1)	117	25	C	0.05M $(CH_3)_4NClO_4$ ΔH is per mole of complex	74Db	
Cu^{2+}	18.4	5.57(1)	167	25	C	0.05M $(CH_3)_4NClO_4$ ΔH is per mole of complex	74Db	
M4 MALEIC ACID - isobutene copolymer			$(C_8H_{12}O_4)_n$			(1:1 Alternating type, MW ∿ 10^5)		L^{2-}
Ba^{2+}	2.9	5.28(1)	113	25	C	0.05M $(CH_3)_4NClO_4$ ΔH is per mole of complex	74Db	
Ni^{2+}	7.5	5.94(1)	138	25	C	0.05M $(CH_3)_4NClO_4$ ΔH is per mole of complex	74Db	
Cu^{2+}	16.7	7.40(1)	51.2	25	C	0.05M $(CH_3)_4NClO_4$ ΔH is per mole of complex	74Db	
M5 MALEIC ACID, methyl			$C_5H_6O_4$			$HO_2CCH{:}C(CH_3)CO_2H$		
Tl^{+}	-54.4	1.84(28°,1)	-216	28-48	T	--	76Sa	b
M6 MALEIC ACID - propylene copolymer			$(C_7H_{10}O_4)_n$			(1:1 Alternating type, MW ∿ 10^5)		L^{2-}
Ba^{2+}	2.5	4.47(1)	92	25	C	0.05M $(CH_3)_4NClO_4$ ΔH is per mole of complex	74Db	
Ni^{2+}	8.4	4.69(1)	117	25	C	0.05M $(CH_3)_4NClO_4$ ΔH is per mole of complex	74Db	
Cu^{2+}	16.7	6.52(1)	180	25	C	0.05M $(CH_3)_4NClO_4$ ΔH is per mole of complex	74Db	

Metal Ion	ΔH, kJ /mol	Log K	ΔS, J /mol K	T,°C	M	Conditions	Ref.	Re-marks
M7 MALONIC ACID			$C_3H_4O_4$			$CH_2(CO_2H)_2$		L^{2-}
Be^{2+}	21.6	2.65(25°,1)	121.3	25&35	T	μ=1.0(ClO_4)	74Tb	a,b
	-0.54	2.67(25°,2)	50.2	25&35	T	μ=1.0(ClO_4)	74Tb	a,b
	13.7	8.50(1-2)	209	?	C	μ∿0.01M	76Ma	u:25°
Mg^{2+}	14.22	2.86(25°,1)	100.0	20-25	T	μ=0	52Ea	b
	13.39	2.86(1)	100.0	25	T	μ=0	54Cd	
	12.97	(1)	--	25-30	T	μ=0	52Ea	b
	13.39	(1)	--	30-35	T	μ=0	52Ea	b
Ca^{2+}	18.62	2.50(25°,1)	110.4	25-40	T	μ=0	70Ge	b
Sc^{3+}	14.27	5.87(1)	160.2	25	C	μ=0.1($NaClO_4$)	69Gf	
	12.26	4.25(2)	122.6	25	C	μ=0.1($NaClO_4$)	69Gf	
	6.11	2.89(3)	75.7	25	C	μ=0.1($NaClO_4$)	69Gf	
La^{3+}	20.1	4.98(25°,1)	163	25&35	T	μ=0	56Ga	a,b
	15.65	3.92(1)	127.6	25	C	μ=0.10M($NaClO_4$)	72Db	
	12.09	3.068(1)	99.2	25	C	μ=1.00M($NaClO_4$)	73Dc	q:±0.71
	16.7	2.56(1)	104.6	25	C	μ=2.00M($NaClO_4$)	72Db	
	8.28	2.085(2)	67.8	25	C	μ=1.00M($NaClO_4$)	73Dc	q:±1.7
	5.19	6.307	138.1	25	C	μ=1.00M($NaClO_4$)	73Dc	q:±0.71
						R:La^{3+}+HL^-=$LaHL^{2+}$		
	17.78	9.180	235.1	25	C	μ=1.00M($NaClO_4$)	73Dc	q:±1.7
						R:La^{3+}+L^{2-}+HL^-= $LaHL_2$		
Ce^{3+}	15.10	4.12(1)	129.3	25	C	μ=0.10M($NaClO_4$)	72Db	
	12.22	3.259(1)	102.9	25	C	μ=1.00M($NaClO_4$)	73Dc	q:±0.79
	6.78	1.980(2)	66.9	25	C	μ=1.00M($NaClO_4$)	73Dc	q:±8.8
	5.31	6.377	140.2	25	C	μ=1.00M($NaClO_4$)	73Dc	q:±0.79
						R:La^{3+}+HL^-=$LaHL^{2+}$		
	20.00	9.303	245.2	25	C	μ=1.00M($NaClO_4$)	73Dc	q:±2.5
						R:La^{3+}+L^{2-}+HL^-= $LaHL_2$		
Pr^{3+}	15.61	4.20(1)	132.6	25	C	μ=0.10M($NaClO_4$)	72Db	
	12.72	3.294(1)	105.8	25	C	μ=1.00M($NaClO_4$)	73Dc	q:±0.84
	8.79	2.313(2)	179.1	25	C	μ=1.00M($NaClO_4$)	73Dc	q:±1.1
	5.31	6.465	141.8	25	C	μ=1.00M($NaClO_4$)	73Dc	q:±0.88
						R:La^{3+}+HL^-=$LaHL^{2+}$		
	19.00	9.28	240.2	25	C	μ=1.00M($NaClO_4$)	73Dc	q:±4.2
						R:La^{3+}+L^{2-}+HL^-= $LaHL_2$		
Nd^{3+}	16.53	4.24(1)	136.4	25	C	μ=0.10M($NaClO_4$)	72Db	
	13.10	3.380(1)	108.8	25	C	μ=1.00M($NaClO_4$)	73Dc	q:±0.71
	7.99	2.540(2)	184.1	25	C	μ=1.00M($NaClO_4$)	73Dc	q:±2.6
	5.02	6.482	141.0	25	C	μ=1.00M($NaClO_4$)	73Dc	q:±0.71
						R:La^{3+}+HL^-=$LaHL^{2+}$		
	19.20	9.443	245.2	25	C	μ=1.00M($NaClO_4$)	73Dc	q:±1.5
						R:La^{3+}+L^{2-}+HL^-= $LaHL_2$		

Metal Ion	ΔH, kJ /mol	Log K	ΔS, J /mol K	T,°C	M	Conditions	Ref.	Remarks
M7, continued								
Sm^{3+}	13.76	4.53(1)	132.6	25	C	μ=0.10M($NaClO_4$)	72Db	
	12.51	3.672(1)	112.1	25	C	μ=1.00M($NaClO_4$)	73Dc	
	14.52	3.20(1)	110.0	25	C	μ=2.00M($NaClO_4$)	72Db	
	8.79	2.400(2)	76.1	25	C	μ=1.00M($NaClO_4$)	73Dc	q:±1.5
	5.61	6.588	144.8	25	C	μ=1.00M($NaClO_4$)	73Dc	q:±0.71
						R:$La^{3+}+HL^{-}=LaHL^{2+}$		
	18.28	9.864	250.2	25	C	μ=1.00M($NaClO_4$)	73Dc	q:±1.4
						R:$La^{3+}+L^{2-}+HL^{-}=LaHL_2$		
Eu^{3+}	13.51	4.62(1)	133.5	25	C	μ=0.10M($NaClO_4$)	72Db	
	12.80	3.721(1)	113.8	25	C	μ=1.00M($NaClO_4$)	73Dc	
	7.82	3.02(2)	84.1	25	C	μ=0.10M($NaClO_4$)	72Db	
	7.20	2.516(2)	72.0	25	C	μ=1.00M($NaClO_4$)	73Dc	q:±3.3
	5.90	6.482	143.9	25	C	μ=1.00M($NaClO_4$)	73Dc	q:±0.79
						R:$La^{3+}+HL^{-}=LaHL^{2+}$		
	16.82	9.986	248.1	25	C	μ=1.00M($NaClO_4$)	73Dc	q:±1.0
						R:$La^{3+}+L^{2-}+HL^{-}=LaHL_2$		
Gd^{3+}	21.3	5.36(25°,1)	176	25&35	T	μ=0	56Ga	a,b
	13.26	4.70(1)	134.3	25	C	μ=0.10M($NaClO_4$)	72Db	
	12.59	3.732(1)	113.8	25	C	μ=1.00M($NaClO_4$)	73Dc	
	7.95	3.00(2)	84.1	25	C	μ=0.10M($NaClO_4$)	72Db	
	7.49	2.50(2)	72.8	25	C	μ=1.00M($NaClO_4$)	73Dc	q:±1.6
	6.48	6.50	146.0	25	C	μ=1.00M($NaClO_4$)	73Dc	q:±0.79
						R:$La^{3+}+HL^{-}=LaHL^{2+}$		
	18.07	9.78	248.1	25	C	μ=1.00M($NaClO_4$)	73Dc	q:±1.5
						R:$La^{3+}+L^{2-}+HL^{-}=LaHL_2$		
Tb^{3+}	13.14	4.74(1)	134.7	25	C	μ=0.10M($NaClO_4$)	72Db	
	12.610	3.816(1)	115.1	25	C	μ=1.00M($NaClO_4$)	73Dc	
	6.86	3.04(2)	81.2	25	C	μ=0.10M($NaClO_4$)	72Db	
	9.41	2.561(2)	81.2	25	C	μ=1.00M($NaClO_4$)	73Dc	
	7.61	6.36	146.8	25	C	μ=1.00M($NaClO_4$)	73Dc	q:±0.71
						R:$La^{3+}+HL^{-}=LaHL^{2+}$		
	17.20	10.09	251.0	25	C	μ=1.00M($NaClO_4$)	73Dc	
						R:$La^{3+}+L^{2-}+HL^{-}=LaHL_2$		
Dy^{3+}	14.22	4.86(1)	140.6	25	C	μ=0.10M($NaClO_4$)	72Db	
	12.828	3.849(1)	117.2	25	C	μ=1.00M($NaClO_4$)	73Dc	
	8.372	3.06(2)	86.6	25	C	μ=0.10M($NaClO_4$)	72Db	
	9.50	2.488(2)	79.1	25	C	μ=1.00M($NaClO_4$)	73Dc	
	0.71	1.19(3)	23.8	25	C	μ=1.00M($NaClO_4$)	73Dc	q:±10
	8.20	6.31	148.1	25	C	μ=1.00M($NaClO_4$)	73Dc	q:±1.2
						R:$La^{3+}HL^{-}=LaHL^{2+}$		
	17.99	9.81	148.1	25	C	μ=1.00M($NaClO_4$)	73Dc	q:±1.7
						R:$La^{3+}+L^{2-}+HL^{-}=LaHL_2$		

Metal Ion	ΔH, kJ /mol	Log K	ΔS, J /mol K	T,°C	M	Conditions	Ref.	Re-marks
M7, continued								
Ho^{3+}	13.510	3.832(1)	118.8	25	C	μ=1.00M($NaClO_4$)	73Dc	
	8.62	2.54(2)	77.0	25	C	μ=1.00M($NaClO_4$)	73Dc	
	7.91	1.30(3)	54.0	25	C	=1.00M($NaClO_4$)	73Dc	q:±3.4
	6.90	6.31	143.9	25	C	μ=1.00M($NaClO_4$	73Dc	q:±1.2
						R:La^{3+}+HL^-=$LaHL^{2+}$		
	17.99	9.95	251.0	25	C	μ=1.00M($NaClO_4$)	73Dc	q:±1.2
						R:La^{3+}+L^{2-}+HL^-= $LaHL_2$		
Er^{3+}	13.01	4.78(1)	135.1	25	C	μ=0.10M($NaClO_4$)	72Db	
	13.39	3.851(1)	118.8	25	C	μ=1.00M($NaClO_4$)	73Dc	
	13.68	3.32(1)	109.6	25	C	μ=2.00M($NaClO_4$)	72Db	
	7.07	3.05(2)	82.0	25	C	μ=0.10M($NaClO_4$)	72Db	
	9.08	2.54(2)	79.1	25	C	μ=1.00M($NaClO_4$)	73Dc	
	5.48	1.23(3)	41.8	25	C	μ=1.00M($NaClO_4$)	73Dc	q:±10
	6.28	6.34	143.1	25	C	μ=1.00M($NaClO_4$)	73Dc	q:±1.5
						R:La^{3+}+HL^-=$LaHL^{2+}$		
	20.5	9.93	259.0	25	C	μ=1.00M($NaClO_4$)	73Dc	q:±2.0
						R:La^{3+}+L^{2-}+HL^-= $LaHL_2$		
Tm^{3+}	13.05	4.80(1)	135.6	25	C	μ=0.10M($NaClO_4$)	72Db	
	14.460	3.846(1)	122.2	25	C	μ=1.00M($NaClO_4$)	73Dc	
	7.95	2.98(2)	83.7	25	C	μ=0.10M($NaClO_4$)	72Db	
	8.740	2.57(2)	79.1	25	C	μ=1.00M($NaClO_4$)	73Dc	
	10.71	1.19(3)	59.0	25	C	μ=1.00M($NaClO_4$)	73Dc	
	5.82	6.27	140.2	25	C	μ=1.00M($NaClO_4$)	73Dc	q:±1.8
						R:La^{3+}+HL^-=$LaHL^{2-}$		
	16.99	9.99	248.1	25	C	μ=1.00M($NaClO_4$)	73Dc	q:±2.2
						R:La^{3+}+L^{2-}+HL^-= $LaHL_2$		
Yb^{3+}	13.22	4.80(1)	136.0	25	C	μ=0.10M($NaClO_4$)	72Db	
	14.238	3.870(1)	12.217	25	C	μ=1.00M($NaClO_4$)	73Dc	
	14.85	3.42(1)	115.1	25	C	μ=2.00M($NaClO_4$)	72Db	
	8.16	3.09(2)	86.2	25	C	μ=0.100M($NaClO_4$)	72Db	
	10.159	2.54(2)	82.8	25	C	μ=1.00M($NaClO_4$)	73Dc	
	8.70	1.37(3)	54.8	25	C	μ=1.00M($NaClO_4$)	73Dc	
	7.91	6.22	146.0	25	C	μ=1.00M($NaClO_4$)	73Dc	q:±1.0
						R:La^{3+}+HL^-=$LaHL^{2+}$		
	21.59	9.76	259.0	25	C	μ=1.00M($NaClO_4$)	73Dc	q:±1.6
						R:La^{3+}+L^{2-}+HL^-= $LaHL_2$		
Lu^{3+}	21.8	5.69(25°,1)	184	25&35	T	μ=0	56Ga	a,b
	13.8	4.74(1)	136.8	25	C	μ=0.10M($NaClO_4$)	72Db	
	14.590	3.879(1)	123.0	25	C	μ=1.00M($NaClO_4$)	73Dc	
	9.08	3.09(2)	89.5	25	C	μ=0.10M($NaClO_4$)	72Db	
	10.79	2.533(2)	84.9	25	C	μ=1.00M($NaClO_4$)	73Dc	
	9.79	1.47(3)	61.1	25	C	μ=1.00M($NaClO_4$)	73Dc	
	7.99	6.25	146.8	25	C	μ=1.00M($NaClO_4$)	73Dc	q:±1.5
						R:La^{3+}+HL^-=$LaHL^{2+}$		

Metal Ion	ΔH, kJ /mol	Log K	ΔS, J /mol K	T,°C	M	Conditions	Ref.	Remarks
M7, continued								
Lu^{3+}, cont.	24.60	9.46	259.8	25	C	μ=1.00M($NaClO_4$) R:La^{3+}+L^{2-}+HL^-=$LaHL_2$	73Dc	q:±3.9
Th^{4+}	11.88	7.477(1)	182.8	25	C	μ=1M($NaClO_4$)	76Df	
	8.49	5.314(2)	130.1	25	C	μ=1M($NaClO_4$)	76Df	
	4.60	3.489(3)	82.4	25	C	μ=1M($NaClO_4$)	76Df	
UO_2^{2+}	5.20	5.417(1)	133.5	25	C	μ=1M($NaClO_4$)	76Df	
	2.72	4.039(2)	87.0	25	C	μ=1M($NaClO_4$)	76Df	
Mn^{2+}	15.40	3.29(1)	114.6	25	C	μ=0	63Mb	
	14.77	3.29(1)	112.1	25	T	μ=0 ΔCp=385	61Na	
Fe^{3+}	11.3	7.497(1)	181.4	25	C	μ=1M($NaClO_4$)	77Db	
	3.1	5.54(2)	117	25	C	μ=1M($NaClO_4$)	77Db	q:±0.7
	-4.6	3.56(3)	53	25	C	μ=1M($NaClO_4$)	77Db	q:±0.7
Co^{2+}	12.13	3.76(1)	112.5	25	C	μ=0	63Mb	
	10.75	3.76(1)	107.9	25	T	μ=0 ΔCp=285	61Na	
	-25.8	0.51(25°)	-75	25-45	T	μ=1.0(ClO_4^-) pH<3.0 R:Cu^{2+}+HL^-=$CuHL^+$	75Tb	b
Ni^{2+}	7.86	4.10(1)	104.6	25	C	μ=0	66Mc 66Na 65Ac	
	7.40	4.10(1)	103.8	25	T	μ=0 ΔCp=192	61Na	
	8.74	4.025(25°,1)	106	15-35	T	μ=0	74Ua	
	12	1.398(25°,1)	67	12-305	T	μ=0 R:$Ni(H_2O)_6^{2+}$+L= [$(H_2O)_5Ni(H_2O)L$] (ion pair)	72Hc	b
	17.4	4.506(25°,1)	145	15-35	T	μ=0 S:2mol%Dioxane	74Ua	
	14.6	4.513(25°,1)	135	15-35	T	μ=0 S:4.2mol%Dioxane	74Ua	
	12.8	4.418(25°,1)	128	15-35	T	μ=0 S:6.4mol%Dioxane	74Ua	
	-11.7	0.51(25°)	-29	25-45	T	μ=1.0(ClO_4^-) pH<3.0 R:Cu^{2+}+HL^-=$CuHL^+$	75Tb	b
Cu^{2+}	11.92	5.64(1)	148.1	25	C	μ=0	66Mc 66Na 65Ac	
	14.64	5.09(25°,1)	146.4	25-40	T	μ=0	70Ge	b
	5.9	5.046(1)	116.7	25	C	μ=0.1M($NaClO_4$)	78Ae	
	4.6	4.81(25°,1)	107.1	15-35	T	μ=0.2M(KNO_3) CM=2.0x10^{-3}	79Mg	b
	9.6	2.66(25°,2)	83.3	15-35	T	μ=0.2M(KNO_3) CM=2.0x10^{-3}	79Mg	b
	27.53	0.90(25°)	109.6	25-40	T	μ=0 R:Cu^{2+}+HL^-= $CuHL^+$	70Ge	b
	-4.94	1.91(25°)	24.3	25-45	T	Potentiometric	75Tb	b

Metal Ion	ΔH, kJ /mol	Log K	ΔS, J /mol K	T,°C	M	Conditions	Ref	Remarks
M7, continued								
Cu^{2+}, cont.						method used μ=1.0(ClO_4^-) pH<3.0 R:Cu^{2+}+HL^-=$CuHL^+$		
	56.0	0.73(25°)	197	25-45	T	Polarography method used μ=1.0(ClO_4^-) pH<3.0 R:Cu^{2+}+HL^-=$CuHL^+$	75Tb	b
	-42.7	13.403	113.8	25	C	μ=0.1M($NaClO_4$) R:Cu^{2+}+L^{2-}+A=CuLA A=2,2'-Bipyridyl	78Ae	
CuA^{2+}	-7.5	5.12(25°,1)	72.8	15-35	T	μ=0.2M(NKO_3) CM=2.0x10^{-3} A=2,2'-Bipyridyl	79Mg	b
Zn^{2+}	13.10	3.83(1)	117.2	25	C	μ=0	66Mc 66Na 51Jb	
	12.47	3.81(1)	114.6	25	T	μ=0 ΔCp=361.9 Equations are given for ΔH as a function of T in 65Na	65Na 65Ac	
	-20.0	0.59(25°)	-54	25-45	T	μ=1.0(ClO_4^-) pH<3.0 R:Cu^{2+}+HL^-=$CuHL^+$	75Tb	b
	-29.9	11.69(25°)	123.4	25&35	T	R:Zn+L+A=ZnLA A=Iminodiacetic acid	78Ka	a,b
Cd^{2+}	-35.1	10.92(25°)	91.0	25&35	T	R:Cd+L+A=CdLA A=Iminodiacetic acid	78Ka	a,b
Al^{3+}	107	3.44(25°,1)	427	25&35	T	μ=1.0(ClO_4)	74Tb	a,b
	-48.9	2.54(25°,2)	-117	25&35	T	μ=1.0(ClO_4)	74Tb	a,b
	12.5	2.62(25°,3)	92	25&35	T	μ=1.0(ClO_4)	74Tb	a,b
M8 MALONIC ACID, benzyl-			$C_{10}H_{10}O_4$			$C_6H_5CH_2CH(CO_2H)_2$		
Mg^{2+}	29.83	2.785(30°,1)	151.5	25-35	T	μ=0	72Pe	b
Cu^{2+}	1.97	5.430(1)	110.4	25	C	μ=0	73Pg	
M9 MANNITOL			$C_6H_{14}O_6$			$HOCH_2[CH(OH)]_4CH_2OH$		
BA_4^-	-33.68	3.041(25°,1)	-54	0-35	T	CL=0-0.9M A=OH^-	67Cb	b,q:±3.3
	-18.62	5.137(25°,1-2)	36.0	0-35	T	CL=0-0.9M A=OH^-	67Cb	b,q:±43.9
[B]	0.033	4.26(1)	79	0	T	0.1M(NaCl) [B]=$H_2BO_3^-$	57Wa	

Metal Ion	ΔH, kJ /mol	Log K	ΔS, J /mol K	T, °C	M	Conditions	Ref.	Remarks
M9, continued								
[B], cont.	-0.096	4.18(1)	79	10	T	0.1M(NaCl) $[B]=H_2BO_3^-$	57Wa	
	-0.243	4.12(1)	75	20	T	0.1M(NaCl) $[B]=H_2BO_3^-$	57Wa	
	-0.410	3.91(1)	75	30	T	0.1M(NaCl) $[B]=H_2BO_3^-$	57Wa	
	-0.598	3.64(1)	67	40	T	0.1M(NaCl) $[B]=H_2BO_3^-$	57Wa	
	-0.808	3.26(1)	58	50	T	0.1M(NaCl) $[B]=H_2BO_3^-$	57Wa	
	-0.393	1.20(2)	21	0	T	0.1M(NaCl) $[B]=H_2BO_3^-$	57Wa	
	-0.251	1.01(2)	17	10	T	0.1M(NaCl) $[B]=H_2BO_3^-$	57Wa	
	-0.088	0.90(2)	21	20	T	0.1M(NaCl) $[B]=H_2BO_3^-$	57Wa	
	-0.105	0.94(2)	17	30	T	0.1M(NaCl) $[B]=H_2BO_3^-$	57Wa	
	0.322	0.98(2)	21	40	T	0.1M(NaCl) $[B]=H_2BO_3^-$	57Wa	
	0.565	1.22(2)	25	50	T	0.1M(NaCl) $[B]=H_2BO_3^-$	57Wa	
M10 MANNOSE			$C_6H_{12}O_6$			$HOCH_2$-C(OH)(H)-C(OH)(H)-C(H)(OH)-C(H)(OH)-CHO		
[B(III)]	-1.5	2.01(25°,1)	33.4	0-35	T	CM=0.0272M CL=0.01-0.6M $[B(III)]=HB_4O_7^-$	70Ci	b
	-16.1	0.65(25°,2)	-41.4	0-35	T	CM=0.0272M CL=0.01-0.6M $[B(III)]=HB_4O_7^-$	70Ci	b
M11 METHANE, amino-			CH_5N			CH_3NH_2		L
NiA	-66.5	3.74(1)	-152.3	25	T	CM=10^{-4}M S:Benzene A=Diacetylbisbenzoyl hydrazine	60Sb	
Ag^+	-12	3.07(1)	14.2	25	C	μ=0	71He	q:±8.4
	-33	3.82(2)	-42.2	25	C	μ=0	71He	q:±8.4
	-60.2	7.48(1-2)	-75.7	2.5	T	μ=0	65Le	b,h: 2.5-45
	-49.04	6.89(1-2)	-32.6	25	C	μ=0	71He	
	-60.2	6.69(1-2)	-74.0	25	T	μ=0	65Le	b,h: 2.5-45
	-60.2	6.18(1-2)	-77.4	35	T	μ=0	65Le	b,h: 2.5-45
	-60.2	5.96(1-2)	-75.3	45	T	μ=0	65Le	b,h: 2.5-45
	-48.1	6.78(1-2)	-31.4	25	C	--	55Fa	

Metal Ion	ΔH, kJ /mol	Log K	ΔS, J /mol K	T,°C	M	Conditions	Ref.	Re-marks
M11, continued								
Ag^+, cont.	-60.2	7.86(1-2)	-68.6	2.5	T	μ=0 S:50%CH_3OH	65Le	b,h: 2.5-45
	-60.2	6.97(1-2)	-68.6	25	T	μ=0 S:50%CH_3OH	65Le	b,h: 2.5-45
	-60.2	6.96(1-2)	-67.8	35	T	μ=0 S:50%CH_3OH	65Le	b,h: 2.5-45
	-60.2	6.36(1-2)	-69.0	45	T	μ=0 S:50%CH_3OH	65Le	b,h: 2.5-45
Cd^{2+}	-29.37	4.808(25°,1)	-6.3	25	T	μ=2	53Sa	
	-57.3	6.55(25°,1-2)	-66.9	25	T	μ=2	53Sb	
HgA_2	-28.4	2.40	-49.0	25	C	μ=0 R:HgA_2+L= $HgAL^+$+A^- A=Cl	66Pd	
	-4.2	2.21	28.0	25	C	μ=0 R:$HgAL^+$+L= L=HgL_2^{2+}+A^- A=Cl	66Pd	
AlA_3	-125.6	(1)	--	25-28	C	S:n-Hexane A=CH_3	68Hc	
I_2	-29.7	2.724(20°,1)	-51.5	20&30	T	S:n-Heptane	60Ya	a,b
M12 METHANE, bis(diphenylarsine)			$C_{25}H_{22}As_2$			$[(C_6H_5)_2As]_2CH_2$		
HgA_2	-50.0	3.496(1)	-98	30	C	S:Benzene CM=5-7x10^{-4}M A=Cl	76Gb	
	-53.0	2.992(1)	-128	30	C	S:Benzene CM=5-8x10^{-4}M A=Br	76Gb	
	-52.0	2.025(1)	-133	30	C	S:Benzene CM=1-2.8x10^{-4}M A=I	76Gb	
M13 METHANE, bis(diphenylarsine)-disulfide			$C_{25}H_{22}S_2As_2$			$[(C_6H_5)_2AsS]_2CH_2$		
HgA_2	-44.9	3.602(1)	-79	30	C	S:Benzene CM=5-8x10^{-4}M A=Cl	76Gb	
	-52.5	3.070(1)	-114	30	C	S:Benzene CM=6-9x10^{-4}M A=Br	76Gb	
M14 METHANE, bis-(diphenylphosphine)			$C_{25}H_{22}P_2$			$[(C_6H_5)_2P]_2CH_2$		
HgA_2	-113	(1)	--	30	C	S:Benzene A=Cl	76Ga	
	-113	(1)	--	30	C	S:Benzene A=Br	76Ga	
	-80	--	--	30	C	S:Benzene A=Cl R:2HgA_2+L= $(HgA_2)_2L$	76Ga	
	-69	--	--	30	C	S:Benzene A=Br R:2HgA_2+L= $(HgA_2)_2L$	76Ga	q:±6

Metal Ion	ΔH, kJ /mol	Log K	ΔS, J /mol K	T, °C	M	Conditions	Ref.	Remarks
M14, continued								
HgA_2, cont.	-44	--	--	30	C	S:Benzene A=I R:$2HgA_2$+L= $(HgA_2)_2L$	76Ga	q:±4
	-96	(1-2)	--	30	C	S:Benzene A=I	76Ga	
M15 METHANE, bis-(diphenylphosphine)disulfide		$C_{25}H_{22}S_2P_2$				$[(C_6H_5)_2PS]_2CH_2$		
HgA_2	-39.7	3.505(1)	-64	30	C	S:Benzene CM=1-1.5x10^{-3}M A=Cl	76Gb	q:±3
	-39.0	3.504(1)	-62	30	C	S:Benzene CM=1-2x10^{-3}M A=Br	76Gb	
	-20.5	3.279(1)	-2	30	C	S:Benzene CM=1.5-2.0x10^{-3}M A=I	76Gb	q:±3
M16 METHANE, 1,1-diphosphonic acid		$CH_6O_6P_2$				$CH_2(PO_3H_2)_2$		
Li^+	2.93	2.48(1)	57.3	25	C	0.5M$(CH_3)_4$NCl	68Ca	
Na^+	6.7	1.13(1)	43.9	25	C	0.5M$(CH_3)_4$NCl	68Ca	
K^+	5.8	1.02(1)	40.2	25	C	0.5M$(CH_3)_4$NCl	68Ca	
Mg^{2+}	18.4	4.396(25°,1)	142	25-50	T	μ=0	62Ia	b
Ca^{2+}	0	4.765(25°,1)	92	25-50	T	μ=0	62Ia	b,q:±8.4
	10.9	4.70(1)	126	25	C	0.5M$(CH_3)_4$NCl	68Ca	
M17 METHANE, (2-furyl)						$CH_3C{:}CHCH{:}CHO$		
Sm^{3+}	-28.3	10.58(25°,1-3)	108	25-45	T	μ=0.1M($NaClO_4$) S:50v%Ethanol	74Sh	b
M18 METHANE, iodo-		CH_3I				CH_3I		L
I_2	-18.8	-2.38(25°,1)	-107.9	5-45	T	S:n-Heptane	57Sa	b
M19 METHANE, mono(thiodibenzoly)		$C_{14}H_{14}S$				$CH_3SCH(C_6H_5)_2$		
Y^{3+}	-27.77	20.38(30°,1-3)	295.4	30-40	T	μ=0.05($NaClO_4$)	79Vc	b,dd
La^{3+}	-25.72	19.02(30°,1-3)	253.2	30-40	T	μ=0.05($NaClO_4$)	79Vc	b,dd
Ce^{3+}	-26.08	19.26(30°,1-3)	251.8	30-40	T	μ=0.05($NaClO_4$)	79Vc	b,dd
Pr^{3+}	-26.90	19.46(30°,1-3)	250.4	30-40	T	μ=0.05($NaClO_4$)	79Vc	b,dd
Nd^{3+}	-29.50	21.14(30°,1-3)	289.8	30-40	T	μ=0.05($NaClO_4$)	79Vc	b,dd
Sm^{3+}	-30.94	23.30(30°,1-3)	330.1	30-40	T	μ=0.05($NaClO_4$)	79Vc	b,dd
Eu^{3+}	-34.46	24.54(30°,1-3)	347.3	30-40	T	μ=0.05($NaClO_4$)	79Vc	b,dd
Gd^{3+}	-31.36	23.94(30°,1-3)	340.2	30-40	T	μ=0.05($NaClO_4$)	79Vc	b,dd

Metal Ion	ΔH, kJ /mol	Log K	ΔS, J /mol K	T,°C	M	Conditions	Ref.	Remarks
M19, continued								
Dy^{3+}	-38.39	25.28(30°,1-3)	345.8	30-40	T	μ=0.05($NaClO_4$)	79Vc	b,dd
Ho^{3+}	-38.51	25.84(30°,1-3)	345.6	30-40	T	μ=0.05($NaClO_4$)	79Vc	b,dd
M20 METHANE, nitro			CH_3O_2N			CH_3NO_2		
$[NiA_2]_2$	-17	0.72(100°,1)	-33	75-175	T	S:Squalane A=3-Trifluoro-acetylcamphorate	74Sj	b,gg
M21 METHANE, tetrachloro-			CCl_4			CCl_4		
I_2A	-0.84	0.188(25°)	2.30	-20-40	T	S:Hexane A=Hexane R:$I_2A+L=I_2L+A$	73Sn	b,q: ±0.21
M22 METHANE, triphenyl-			$C_{19}H_{16}$			$(C_6H_5)_3CH$		
Na^+	2.5	4.41(1)	96	-60	T	S:Tetrahydrofuran	72Cb	
	25.9	5.07(1)	184	25	T	S:Tetrahydrofuran	72Cb	
NaL	34.3	--	117	-40-80	T	S:Tetrahydrofuran R:NaSL=NaL(Solvent separate=Contact ion pair)	72Gi	b
KL	28.0	--	96	-40-80	T	S:Tetrahydrofuran R:KSL=KL(Solvent separated=Contact ion pair)	72Gi	b
	22.2	--	67	-40-80	T	S:1,2-Dimethoxy-ethane R:KSL=KL (Solvent separa-ted=Contact ion pair)	72Gi	b
RbL	14.2	--	54	-40-80	T	S:1,2-Dimethoxy-ethane R:RbSL=RbL(Solvent separated=Contact ion pair)	72Gi	b
M23 METHANETHIOL, (2-furyl)-			C_5H_6OS			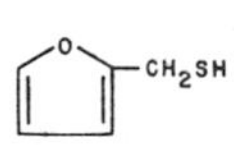		
Be^{2+}	-12.6	13.45(25°,1-2)	216	25-45	T	μ=0.1M($NaClO_4$) S:50v%Ethanol	74Sf	b
La^{3+}	-18.28	8.703(35°,1-3)	105.69	25-45	T	S:50% Ethanol μ=0.1M($NaClO_4$)	73So	b
Ce^{3+}	-18.37	9.518(35°,1-3)	122.67	25-45	T	S:50% Ethanol μ=0.1M($NaClO_4$)	73So	b

Metal Ion	ΔH, kJ /mol	Log K	ΔS, J /mol K	T,°C	M	Conditions	Ref.	Re-marks
M23, continued								
Sm^{3+}	-28.3	10.58(25°,1-3)	108	25-45	T	μ=0.1M($NaClO_4$) S:50v%Ethanol	73Sj	b
Th^{4+}	10.0	17.22(35°,1-4)	368	25-45	T	μ=0.5M($NaClO_4$) S:50v%Ethanol	73Sg	b,j
VO^{2+}	-34.2	16.86(25°,1-2)	206	25-45	T	μ=0.1M($NaClO_4$) S:50v%Ethanol pH=4-5	73Si	b
$UO_2{}^{2+}$	-6.53	15.09(25°,1-2)	265.94	15-35	T	S:50% Ethanol μ=0.1M($NaClO_4$)	73Sh	b
M24 METHANOL			CH_4O			CH_3OH		L
VOA_2	-24.14	-0.26(1)	-86.2	25	C	μ=0 S:Nitrobenzene A=Acetylacetone	65Ca	
CrA_3	15.1	(1)	--	-40-5	T	S:Methanol A=(AcAc)	77Sj	
MoA_2B_2	10.9	(1)	--	-40-5	T	S:Methanol A=0 B=(AcAc)	77Sj	
CoA_2	29.7	(1)	--	-40-5	T	S:Methanol A=(AcAc)	77Sj	
CoAB	6.3	(1)	--	-40-5	T	S:Methanol A=(AcAc) B=CH_3OH	77Sj	
NiA_2	6.3	(1)	--	-40-5	T	S:Methanol A=(AcAc)	77Sj	
	13.4	(1)	--	-40-5	T	S:Methanol A=(CF_3COO)	77Sj	
	11.7	(1)	--	-40-5	T	S:Methanol A=($C_{22}H_{37}COO$)	77Sj	
CuA_2	10.0	(1)	--	-40-5	T	S:Methanol A=(AcAc)	77Sj	
SbA_5	-79.83	(1)	--	25	C	S:Ethylene chloride A=Cl	68Oa	
I_2	-7.95	0.67(25°,1)	-13.89	7-40	T	CL=0.06 mole fraction S:CCl_4	57Dc	b
	-7.9	(1)	--	--	C	S:CCl_4	60Bb	
	-12.6	0.111(25°,1)	-40	10-40	T	Method 1:Constant activity method S:n-Heptane	77Tb	
	-12.6	-0.187(25°,1)	-46	10-40	T	Method 2:Benesi-Hildebrand-Scott method S:n-Heptane	77Tb	
	-8.8	(1)	--	--	C	S:CH_3OH	61Aa	i:50Ha
	-7.1	(1)	--	--	C	S:CH_3OH	50Ha	

Metal Ion	ΔH, kJ /mol	Log K	ΔS, J /mol K	T,°C	M	Conditions	Ref.	Remarks
M25 METHIONINE			$C_5H_{11}O_2NS$			$CH_3SCH_2CH_2CH(NH_2)CO_2H$		L^-
Mn^{2+}	-7.1	2.87(20°,1)	29	20-60	T	μ=0.1(KNO_3)	73Bh	b
	-0.8	2.05(20°,2)	38	20-60	T	μ=0.1(KNO_3)	73Bh	b
Ni^{2+}	-13.2	(1)	--	25	C	μ=0.10M(KNO_3) L=D-Methionine	76Sr	
	-13.1	(1)	--	25	C	μ=0.10M(KNO_3) L=L-Methionine	76Sr	
	-13.3	(1)	--	25	C	μ=0.10M(KNO_3) L=DL-Methionine	76Sr	
	-21.63	5.56(25°,1)	33.9	20-40	T	--	57Pa 60Pc	b
	-16.7	6.84(1)	74.9	25	T	S:44.6% Dioxane	60Pd	
	-12.6	7.3(1)	97.5	25	T	S:59.7% Dioxane	60Pd	
	-21.8	(2)	--	25	C	μ=0.10M(KNO_3) L=D-Methionine	76Sr	
	-22.1	(2)	--	25	C	μ=0.10M(KNO_3) L=L-Methionine	76Sr	
	-22.8	(2)	--	25	C	μ=0.10M(KNO_3) L=DL-Methionine	76Sr	
	-21.63	4.63(25°,2)	16.3	20-40	T	--	60Pc	b
	-25.1	5.91(2)	28.9	25	T	S:44.6% Dioxane	60Pd	
	-16.7	6.55(2)	-69.4	25	T	S:59.7% Dioxane	60Pd	
	-26.86	2.63(25°,3)	-39.7	20-40	T	--	60Pc	b
	-23.0	3.94(3)	-1.7	25	T	S:44.6% Dioxane	60Pd	
	-36.8	10.367	75.1	25	C	μ=0.10M(KNO_3) R:Ni^++(L-Methionine+(D-Methionine)=Ni^+(L-Methionine)(D-Methionine)	76Sr	
M26 METHIONINE (DL), N-acetyl			$C_7H_{18}O_3NS$			$CH_3SCH_2CH_2CH(NHOCCH_3)CO_2H$		
[Fe(II)]	-52.7	2.75(25°,1-2)	-122.6	15.5-36.5	T	μ=0.10($NaClO_4$) pH=6.0 [Fe(II)]= Iron(II) 2,4-dicysteine-substituted mesoporphyrin (heme c)	76Gf	q:±5.2
M27 METHYL, thymol blue			$C_{27}H_{30}O_5S$					
Mn^{2+}	-21.8	7.82(30°,1)	77.9	30&40	T	μ=0	78Sb	a,b,q: ±0.5
Fe^{2+}	-23.6	10.75(30°,1)	128.0	30&40	T	μ=0	78Sb	a,b,q: ±0.5
Co^{2+}	-24.5	12.28(30°,1)	154.2	30&40	T	μ=0	78Sb	a,b,q: ±0.5

Metal Ion	ΔH, kJ /mol	Log K	ΔS, J /mol K	T,°C	M	Conditions	Ref.	Remarks
M27, continued								
Ni^{2+}	-25.4	12.45(30°,1)	154.4	30&40	T	μ=0	78Sb	a,b,q: ±0.5
Cu^{2+}	-32.6	13.84(30°,1)	157.1	30&40	T	μ=0	78Sb	a,b,q: ±0.5
Zn^{2+}	-29.0	13.11(30°,1)	155.2	30&40	T	μ=0	78Sb	a,b,q: ±0.5
M28	MONACTIN		$C_{41}H_{66}O_2$					
Na^+	-22.38	(1)	--	25	C	S:CH_3OH	73Za 71Fe	i:71Fe
	-25.10	2.6(1)	-33.9	25	C	S:CH_3OH H.Diebler, M. Eigen, G. Ilgenfritz, G. Maass, &R. Winkler, Pure Appl. Chem., 20, 93 (1969)	73Za	
M29	MONENSIN		$C_{36}H_{62}O_{11}$					
Na^+	-16.3	6.0(1)	63	25	C	S:CH_3OH	71Li	
	-16.2	6.0(1)	61	25	C	CM=0.1-5x10^{-2}M CL=10^{-3}M 5x10^{-2}M Tributylamine S:Methanol	73Fb	
	-22.89	6.72(25°,1)	51.9	5-45	T	S:Methanol	79Hb 78Ha	b
	-35.6	2.4(25°)	-71	5-45	T	S:Methanol R:HL+Na^+=$NaHL^+$	79Hb	
K^+	-15.6	4.6(1)	35	25	C	CM=0.1-5x10^{-2}M CL=10^{-3}M 5x10^{-2}M Tributylamine S:Methanol	73Fb	
	-16.19	4.477(1)	31.0	25	C	S:CH_3OH	71Fe	
	-15.5	4.6(1)	35.1	25	C	S:CH_3OH	71Li	
M30	MORPHOLINE		C_4H_9ON			$OCH_2CH_2NHCH_2CH_2$ (ring)		L
VOA_2	-39.08	2.48(1)	-83.7	25	C	μ=0 S:Nitrobenzene A=Acetylacetone	65Ca	
NiA_2	-32.2	1.863(24.6°,1)	-73.2	8-39.5	T	S:Benzene A=0,0'-Diethyldithiophosphate	72Mc	b
	-36.4	2.301(24.6°,1)	-78.6	8-39.5	T	S:Benzene A=0,0'-Dimethyldithiophosphate	72Mc	b

Metal Ion	ΔH, kJ /mol	Log K	ΔS, J /mol K	T, °C	M	Conditions	Ref.	Remarks
M31	MORPOHOLINE, n-methyl		$C_5H_{11}ON$			$OCH_2CH_2N(CH_3)CH_2CH_2$		
HgA_2	-37.9	2.43(1)	-79	30	C	CM=0.9-3.9x10^{-3} S:Benzene A=Cl	74Fb	
	-37.1	2.61(1)	-73	30	C	CM=1.1-4.4x10^{-3} S:Benzene A=Br	74Fb	
	-33.6	2.44(1)	-64	30	C	CM=1.2-5.7x10^{-3} S:Benzene A=I	74Fb	
M32	MUREXIDE		$C_8H_8O_6N_6$			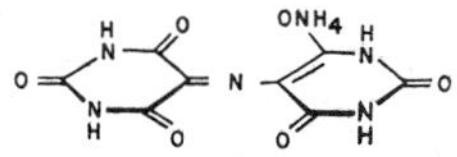		L^{3-}
Ni^{2+}	-6.7	3.218(20°)	39.3	12-32	T	μ=0.1M(NaClO$_4$) R:Ni^{2+}+H_2L^- = NiH_2L^+	71Le	b,q: ±1.2
	-23.4	4.079(20°)	-1.2	12-32	T	S:50v% DMSO μ=0.1M(NaClO$_4$) R:Ni^{2+}+H_2L^- = NiH_2L^+	71Le	b,q: ±2.1
M33	MYOSIN B							
K^+	0	2.903(27°)	55.6	5&27	T	CL=1.5-2.5% pH=7.7 Reaction for binding of K^+ to first set of sites (imidazole groups)	57La	a,b,q: ±13
	-4.1	1.699(27°)	-112.1	5&27	T	CL=1.5-2.5% pH=7.7 Reaction for binding of K^+ to second set of sites (amino groups)	57La	a,b,q: ±10

N

Metal Ion	ΔH, kJ /mol	Log K	ΔS, J /mol K	T, °C	M	Conditions	Ref.	Remarks
N1	1-NAPHTHALDEHYDE, 2-hydroxy-6-nitro-		$C_{11}H_7O_3N$			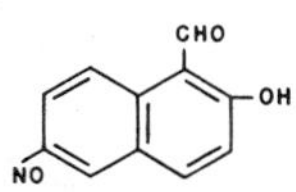		
Cu^{2+}	-22.89	6.9556(25°,1)	55.56	15-50	T	--	68Rb	b
	-20.50	5.6069(25°,2)	38.07	15-50	T	--	68Rb	b

Metal Ion	ΔH, kJ /mol	Log K	ΔS, J /mol K	T,°C	M	Conditions	Ref.	Re-marks
N2 NAPHTHALENE			$C_{10}H_8$			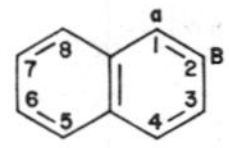		L
Li^+	2.1	-5.027(1)	109	-65	T	S:Tetrahydrofuran	68Nb	q:±1.7
	11.7	-5.420(1)	146	15	T	S:Tetrahydrofuran	68Nb	q:±1.7
Na^+	-19.7	(1)		-65-30	T	S:2-Methyltetra- hydrofuran	61Ab	b
	7.5	6.647(1)	126	-65	T	μ=0 S:Tetra- hydrofuran	66Cb	b
	7.5	(1)	126	-65	T	S:Tetrahydrofuran	68Nb	q:±1.7
	34.3	6.66(1)	243	15	T	μ=0 S:Tetra- hydrofuran	66Cb	b
	34.3	(1)	243	15	T	S:Tetrahydrofuran	68Nb	
	38.5	0.208(-35°,1)	259	-70--35	T	S:Tetrahydrofuran	68He	b
	31.4	(1)	--	-65-30	T	S:Tetrahydrofuran	61Ab	q:±6.3
	10.9	(1)	--	-65-30	T	S:Tetrahydropyran	61Ab	
NaL	-20.1	--	-84	-58-12	T	S:Tetrahydrofuran CL=10^{-5}-10^{-4}M R:Tight ion pair= Loose ion pair	67He	b,q: ±2.1
	-26.8	--	-138	-101--51	T	S:50% Tetra- hydrofuran 50% Diethyl ether CL=10^{-5}-10^{-4}M R:Tight ion pair =Loose ion pair	67He	b,q: ±3.3
VAB_3	-0.711	(1)	--	10-26	T	S:Cyclohexane A=O B=Cl	71Ha	
WA_6	-0.711	(1)	--	-24-26	T	S:CCl_4 A=Cl	71Hd	
OsA_4	-1.674	(1)	--	0-31.5	T	S:CCl_4 A=O	71Hc	
PA_5	-0.293	(1)	--	-30-31	T	S:CCl_4 A=Cl	71Hb	
	-0.460	(1)	--	-30-31	T	S:Dichloromethane A=Cl	71Hb	
I_2	-6.90	(1)	--	1&45	T	CL=0.07-1.76M S:CCl_4	60Dc	a,b
	-6.86	(1)	--	22&45	T	CL=0.07-1.76M S:$CHCl_3$	60Dc	a,b
	-6.53	(1)	--	22&45	T	CL=0.07-1.76M S:Cyclohexane	60Dc	a,b
	-6.82	(1)	--	22&45	T	CL=0.07-1.76M S:n-Hexane	60Dc	a,b
	-7.53	(1)	--	--	T	S:n-Hexane	60Dc	
	-7.53	0.364(1)	-18.0	25	T	S:n-Hexane	54Ka	1
	-5.77	(1)	--	22&45	T	S:n-Heptane	60Dc	a,b
	-7.5	(1)	--	--	-	S:n-Heptane	60Ta	

Metal Ion	ΔH, kJ /mol	Log K	ΔS, J /mol K	T,°C	M	Conditions	Ref.	Remarks
N3 NAPHTHALENE, 2,3-dihydroxy			$C_{10}H_8O_2$			$C_{10}H_6(OH)_2$		
La^{3+}	-72.80	9.81(25°,1)	-54.4	25&45	T	μ=0.1M($NaClO_4$)	78Af	a,b
	-97.30	8.97(25°,2)	-138.1	25&45	T	μ=0.1M($NaClO_4$)	78Af	a,b
Ce^{3+}	-67.36	10.27(25°,1)	-29.3	25&45	T	μ=0.1M($NaClO_4$)	78Af	a,b
	-54.81	9.17(25°,2)	-8.4	25&45	T	μ=0.1M($NaClO_4$)	78Af	a,b
Pr^{3+}	-64.06	10.43(25°,1)	-12.6	25&45	T	μ=0.1M($NaClO_4$)	78Af	a,b
	-55.44	9.39(25°,2)	-4.2	25&45	T	μ=0.1M($NaClO_4$)	78Af	a,b
Nd^{3+}	-68.20	10.68(25°,1)	-8.4	25&45	T	μ=0.1M($NaClO_4$)	78Af	a,b
	-56.65	9.60(25°,2)	-4.2	25&45	T	μ=0.1M($NaClO_4$)	78Af	a,b
Sm^{3+}	-65.77	11.20(25°,1)	-4.2	25&45	T	μ=0.1M($NaClO_4$)	78Af	a,b
	-60.33	10.07(25°,2)	-8.4	25&45	T	μ=0.1M($NaClO_4$)	78Af	a,b
N4 NAPHTHALENE, 1-hydroxy			$C_{10}H_8O$			$C_{10}H_7(OH)$		
PA_3B	-32.8	3.505(25°,1)	-43.1	0-50	T	S:CCl_4 A=$(CH_3)_2N$ B=O	77Kf	
N5 NAPHTHALENE, 2-hydroxy-1-(2-pyridylazo)			$C_{15}H_{11}ON_3$			$C_{10}H_6(OH)(N:NC_5H_4N)$		
Y^{3+}	-30.5	22.59(25°,1-2)	330	25-45	T	μ=0.1M($NaClO_4$) S:50v%Dioxane	74Pb	b
N6 NAPHTHALENE, 2-methoxy-			$C_{11}H_{10}O$			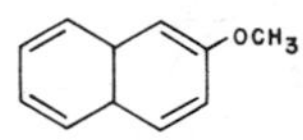		
Al_2A_6	-64.4	--	--	25	C	S:Benzene CM=0.07M A=Br R:0.5(Al_2A_6)+L=AlA_3L	68Rc	
N7 NAPHTHALENE, 1-methyl			$C_{11}H_{10}$			$C_{10}H_7(CH_3)$		L
I_2	-8.79	0.444(1)	-20.9	25	T	S:n-Hexane	54Ka	1
N8 NAPHTHALENE, 2-methyl-			$C_{11}H_{10}$			$C_{10}H_7(CH_3)$		L
I_2	-8.79	0.569(1)	-18.4	25	T	S:n-Hexane	54Ka	1
N9 NAPHTHALENE, octafluoro-			$C_{10}F_8$			$C_{10}H_7(F_8)$		F
SbA_5	0.460	(1)	--	-30-31	T	S:CCl_4 A=Cl	71Hb	
N10 2,7-NAPHTHALENEDISULFONIC ACID, 4,5-dihydroxy-			$C_{10}H_8O_8S_2$			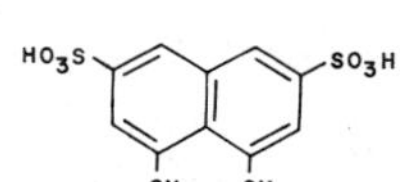		L^{2-}
Y^{+}	-43.01	23.62(30°,1-2)	313.2	20-40	T	μ=0.1M($NaClO_4$)	76Cc	

Metal Ion	ΔH, kJ /mol	Log K	ΔS, J /mol K	T,°C	M	Conditions	Ref.	Remarks
N10, continued								
La^{3+}	-21.22	22.80(30°,1-2)	388.8	20-40	T	μ=0.1M($NaClO_4$)	76Cc	
Pr^{3+}	-27.28	23.50(30°,1-2)	363.2	20-40	T	μ=0.1M($NaClO_4$)	76Cc	
Nd^{3+}	-36.20	23.72(30°,1-2)	337.9	20-40	T	μ=0.1M($NaClO_4$)	76Cc	
Sm^{3+}	-43.01	24.21(30°,1-2)	313.2	20-40	T	μ=0.1M($NaClO_4$)	76Cc	
Ti^{3+}	18	2.86(25°)	110	25	T	μ=0 R:Ti^{3+}+ $2H_2L^{2-}$=$Ti(HL)_2$+$2H^+$	74Pa	
ThA	44.8	--	--	25	C	μ=0.10(KNO_3) A=EDTA R:ThA+ H_2L=$ThAL^{2-}$+$2H^+$	70Kb	g:Na
	47.7	--	--	25	C	μ=0.10(KNO_3) A=1,2-Diaminocyclohexane-N,N,N',N'-tetraacetic acid R:ThA+H_2L= $ThAL^{2-}$+$2H^+$	70Kb	g:Na
Al^{3+}	-19.3	9.57(30°)	120	20-40	T	μ=0.2M(KCl) Al^{3+}+HL^{3-}+OH^-= AlL^-+H_2O	71Ge	b
	-13.3	8.24(30°)	114	20-40	T	μ=0.2M(KCl) AlL^-+HL^{3-}+OH^-= AlL_2^{5-}+H_2O	71Ge	b
	-10.7	6.91(30°)	97	20-40	T	μ=0.2M(KCl) AlL^{5-}+HL^{3-}+OH^-= AlL_3^{9-}+H_2O	71Ge	b
	-46.5	24.72(30°)	320	20-40	T	Al^{3+}+$3HL^{3-}$+3OH= AlL_3^{9-}+$3H_2O$	71Ge	b

N11 2,7-NAPHTHALENEDISULFONIC ACID, $C_{16}H_{12}O_8N_2S_2$
4,5-dihydroxy-3-(phenyl)azo-

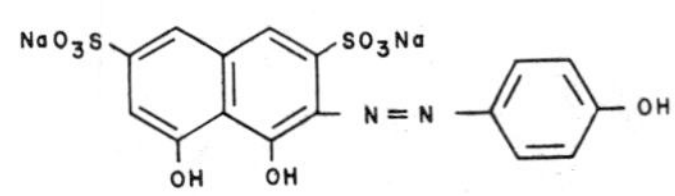

Metal Ion	ΔH, kJ /mol	Log K	ΔS, J /mol K	T,°C	M	Conditions	Ref.	Remarks
Cu^{2+}	-74.98	4.43(25°,1)	-166.61	15-30	T	μ=0.1M($NaClO_4$)	73Df	b
Al^{3+}	5.92	3.837(25°,1)	93	15-35	T	μ=0.1($SO_4^=$) pH=5	72Dc	b

N12 2,7-NAPHTHALENEDISULPHONIC ACID, 4,5-dihydroxy-3-(p-sulphophenyl)azo $C_{16}H_{12}O_{11}N_2S_3$

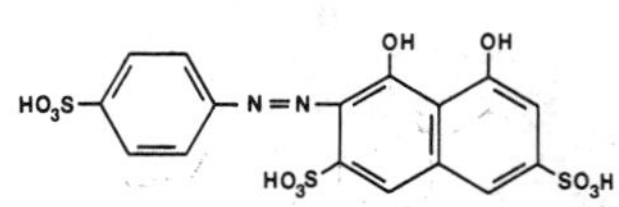

Metal Ion	ΔH, kJ /mol	Log K	ΔS, J /mol K	T,°C	M	Conditions	Ref.	Remarks
Y^{3+}	-15.1	8.12(25°,1)	105	25-45	T	μ=0.1M($NaClO_4$)	74Pb	b
	24.5	15.1(30°,1)	365	30&45	T	0.5M(KCl)	77Mb	a,b,n
	61.2	9.4(30°,2)	379	30&45	T	0.5M(KCl)	77Mb	a,b,n
Ce^{4+}	24.5	10.4(30°,1)	276	30&45	T	0.5M(KCl)	77Mb	a,b,n
	98.0	3.1(30°,2)	382	30&45	T	0.5M(KCl)	77Mb	a,b,n
Zr^{4+}	-73.5	14.9(30°,1)	38	30&45	T	0.5M(KCl)	77Mb	a,b,n
	-24.5	9.4(30°,2)	96	30&45	T	0.5M(KCl)	77Mb	a,b,n

Metal Ion	ΔH, kJ /mol	Log K	ΔS, J /mol K	T,°C	M	Conditions	Ref.	Re-marks
N13	2,7-NAPHTHALENEDISULFONIC ACID, 3-hydroxy-4-nitroso		$C_{10}H_7O_8NS_2$			$C_{10}H_4(OH)(NO)(SO_3H)_2$		L^{2-}
Be^{2+}	-16.1	5.48(30°,1)	50.8	10-50	T	μ=0.10($NaClO_4$)	75Gf	b
	-2.2	4.17(30°,2)	72.4	10-50	T	μ=0.10($NaClO_4$)	75Gf	b
YA	-7.66	3.44(25°,1)	40.6	25-45	T	μ=0.2M(KNO_3) A=Nitrilotriacetic acid	75Pg	b
UO_2^{2+}	-18.8	6.69(30°,1)	65.0	10-50	T	μ=0.10($NaClO_4$)	75Gf	b
	-17.9	5.12(30°,2)	41.5	10-50	T	μ=0.10($NaClO_4$)	75Gf	b
Co^{2+}	-20.1	6.90(30°,1)	64.4	10-50	T	μ=0.10($NaClO_4$)	75Gf	b
	-7.2	6.00(30°,2)	90.3	10-50	T	μ=0.10($NaClO_4$)	75Gf	b
Ni^{2+}	-10.1	6.14(30°,1)	84.2	10-50	T	μ=0.10($NaClO_4$)	75Gf	b
	-11.1	4.00(30°,2)	40.0	10-50	T	μ=0.10($NaClO_4$)	75Gf	b
N14	2-NAPHTHALENESULFONIC ACID, 6,7-dihydroxy		$C_{10}H_7O_5S$			$C_{10}H_5(OH)_2(SO_3)$		
YA	-72.8	8.64(30°,1)	-75.3	30-50	T	μ=0.2(KNO_3) A=Nitrilotriacetic acid	75Pg	b
In^{3+}	-42.2	15.77(25°,1)	159	25-45	T	μ=0	78Sc	
	-37.2	13.07(25°,2)	126	25-45	T	μ=0	78Sc	
	-33.9	9.42(25°,3)	67	25-45	T	μ=0	78Sc	

N15 1-NAPHTHOIC ACID, 2-hydroxy- $C_{11}H_8O_3$

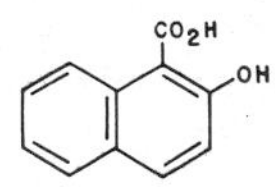

Metal Ion	ΔH, kJ /mol	Log K	ΔS, J /mol K	T,°C	M	Conditions	Ref.	Re-marks
UO_2^{2+}	-10.79	3.52(30°,1)	31.76	20-50	T	μ=0.1	72Pc	b
Fe^{3+}	-30.38	7.83(30°,1-2)	49.79	20-50	T	μ=0.1	72Pc	b

N16 2-NAPHTHOIC ACID, 1-hydroxy- $C_{11}H_8O_3$

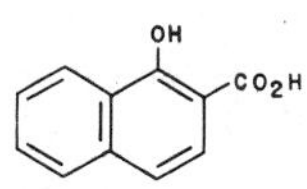

Metal Ion	ΔH, kJ /mol	Log K	ΔS, J /mol K	T,°C	M	Conditions	Ref.	Re-marks
Be^{2+}	469	17.70(30°,1-2)	1890	30-40	T	μ=0.1M(KNO_3) S:70v%Ethanol	75Sa	b
Mg^{2+}	521	8.30(30°,1-3)	1870	30-40	T	μ=0.1M(KNO_3) S:70v%Ethanol	75Sa	b
Ca^{2+}	191	4.75(30°,1)	720	30-40	T	μ=0.1M(KNO_3) S:70v%Ethanol	75Sa	b
UO_2^{2+}	-5.90	3.42(30°,1)	46.06	20-60	T	μ=0.1 pH=4.6	71Wb	b
Fe^{3+}	-23.606	4.20(25°,1)	1.17	20-40	T	μ=0.1M S:50% Ethanol	72Pb	b
Zn^{2+}	460.8	14.34(30°,1-2)	1798	30-40	T	μ=0.1M(KNO_3) S:70v%EtOH	76Sh	b

Metal Ion	ΔH, kJ /mol	Log K	ΔS, J /mol K	T,°C	M	Conditions	Ref.	Re-marks
N16, continued								
Cd^{2+}	382.5	12.43(30°,1-2)	1498	30-40	T	μ=0.1M(KNO_3) S:70v%EtOH	76Sh	b
Hg^{2+}	391.2	6.89(30°,1-2)	1435	30-40	T	μ=0.1M(KNO_3) S:70v%EtOH	76Sh	b
Pb^{2+}	382	12.30(30°,1-2)	1500	30-40	T	μ=0.1M(KNO_3) S:70v%Ethanol	75Sa	b

N17 2-NAPHTHOIC ACID, 3-hydroxy- $C_{11}H_8O_3$

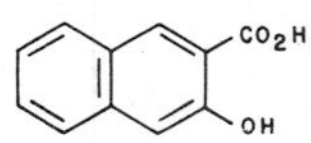

L^{2-}

Metal Ion	ΔH, kJ /mol	Log K	ΔS, J /mol K	T,°C	M	Conditions	Ref.	Re-marks
Ca^{2+}	137.9	4.42(30°,1-2)	539.6	20-40	T	μ=0.1M(KNO_3) S:70v%Dioxane	76Si	
Be^{2+}	506.9	21.56(30°,1-2)	2086	20-40	T	μ=0.1M(KNO_3) S:70v%Dioxane	76Si	
Mg^{2+}	351.3	8.92(30°,1-2)	1330	20-40	T	μ=0.1M(KNO_3) S:70v%Dioxane	76Si	
Y^{3+}	-70.3	16.68(30°,1-2)	87.9	20-40	T	μ=0.1M(KNO_3) S:50v%Dioxane	77Sh	
La^{3+}	-81.6	15.77(30°,1-2)	32.6	20-40	T	μ=0.1M(KNO_3) S:50v%Dioxane	77Sh	
Ce^{3+}	-84.9	16.04(30°,1-2)	26.4	20-40	T	μ=0.1M(KNO_3) S:50v%Dioxane	77Sh	
Pr^{3+}	-45.2	23.49(30°,1-3)	300	20-40	T	μ=0.1M(KNO_3) S:50v%Dioxane	77So	
Nd^{3+}	-75.7	23.71(30°,1-3)	204	20-40	T	μ=0.1M(KNO_3) S:50v%Dioxane	77So	
Gd^{3+}	-56.5	24.43(30°,1-3)	281	20-40	T	μ=0.1M(KNO_3) S:50v%Dioxane	77So	
Dy^{3+}	-21.8	24.53(30°,1-3)	397	20-40	T	μ=0.1M(KNO_3) S:50v%Dioxane	77So	
Er^{3+}	-17.6	24.83(30°,1-3)	418	20-40	T	μ=0.1M(KNO_3) S:v%Dioxane	77Sc	
Yb^{3+}	-18.4	25.72(30°,1-3)	431	20-40	T	μ=0.1M(KNO_3) S:v%Dioxane	77Sc	
UO_2^{2+}	-64.4	11.22(30°,1-2)	2.0	30-40	T	μ=0.1M(KNO_3) S:70v%Dioxane	76Se	b
Fe^{3+}	17.11	4.983(30°,1)	148.5	10-60	T	μ=0.02 pH=2.6	69Gh	b
Co^{2+}	-43.1	14.61(30°,1-2)	137	20-40	T	μ=0.1M(KNO_3) S:50v%Dioxane	78Sa	b
Ni^{2+}	-53.6	13.22(30°,1-2)	75.7	20-40	T	μ=0.1M(KNO_3) S:50v%Dioxane	78Sa	b
Cu^{2+}	55.2	16.48(30°,1-2)	498	20-40	T	μ=0.1M(KNO_3) S:50v%Dioxane	78Sa	b

Metal Ion	ΔH, kJ /mol	Log K	ΔS, J /mol K	T,°C	M	Conditions	Ref.	Remarks
N17, continued								
Zn^{2+}	-25.9	5.26(30°,1-2)	14.6	30-40	T	μ=0.1M(KNO_3) S:70v%Dioxane	76Se	b
Cd^{2+}	-10.5	4.87(30°,1-2)	59.0	30-40	T	μ=0.1M(KNO_3) S:70v%Dioxane	76Se	b
Al^{3+}	39.54	4.772(30°)	218.4	20-60	T	μ=0.02($NaClO_4$) R: $Al^{3+}+HL^-=AlL^++H^+$	67Ge	b
N18 NICOTINAMIDE, adenine dinucleotide phosphate								L^{3-}
Mn^{2+}	5.4	2.568(20°,1)	67	0-20	T	μ=0.22(Cl^-)	79Gb	q:±0.8
N19 NIGERICIN			$C_{40}H_{67}O_{11}$					L^-
Na^+	6.7	3.9(1)	96	25	C	CM=0.1-5x10^{-2}M CL=10^{-3}M 5x10^{-3}M Tributylamine S:CH_3OH	73Fb 71Li	q:±2.9
K^+	-4.10	5.6(1)	92	25	C	CM=0.1-5x10^{-2}M CL=10^{-3} 5x10^{-2}M Tributylamine S:CH_3OH	73Fb 71Li	q:±1.2
N20 NITRATE ION			NO_3			NO_3^-		L^-
Sr^{2+}	-0.13	0.59(1)	10.9	25	C	μ=0.5($LiNO_3$)	64Va	c:Sr $(NO_3)_2$
	-10.13	0.52(1)	-23.8	25	C	μ=0.5M(KNO_3)	64Va	c:Sr $(NO_3)_2$
Ba^{2+}	-7.95	0.94(1)	-8.8	25	C	μ=0.5($LiNO_3$)	64Va	c:Ba $(NO_3)_2$
	-13.47	0.93(1)	-27.2	25	C	μ=0.5M(KNO_3)	64Va	c:Ba $(NO_3)_2$
	-42.68	2.(1-2)	-98.3	25	T	μ=0	34La	
Eu^{3+}	-2.38	0.30(1)	-2.26	25	T	μ=1.0	65Cf	
	-12.1	1.23(25°,1)	-16.7	0-50	T	μ=1M(ClO_4)	75Mf	
Gd^{3+}	-10.0	1.19(25°,1)	-10.9	0-50	T	μ=1M(ClO_4)	75Mf	
Tb^{3+}	-7.9	1.17(25°,1)	-4.2	0-50	T	μ=1M(ClO_4)	75Mf	q:±0.8
Dy^{3+}	-5.0	0.96(25°,1)	1.7	0-50	T	μ=1M(ClO_4)	75Mf	q:±0.8
Cr^{3+}	18.8	-2.01(1)	24.7	25	T	μ=1.0 ΔH changes with NO_3^- conc., K decreases with increasing NO_3^- conc.	67Ac	c

Metal Ion	ΔH, kJ /mol	Log K	ΔS, J /mol K	T,°C	M	Conditions	Ref.	Re-marks
N20, continued								
CrA_3^{3+}	13.67	1.46(35°)	74.42	25-45	T	μ=0 R: $CrA_3^{3+}+L^- = CrA_3L^{2+}$ A=2-Guanidinobenzimidazole C_6H_4-NH-C(=N)-NH-C(=NH)-NH_2	79Mf	b
UO_2^{++}	-18.0	1.5	-32.2	25	T	R: $UO_2^{++}+2L^-+2TBP=UO_2(NO_3)_2(TBP)_2$ where TBP= tributyl phosphate	60Na	
Mn^{2+}	-4.74	-0.15(1)	-18.8	25	C	μ=1M	75Ah	
Co^{2+}	-4.86	-0.15(1)	-19.2	25	C	μ=1M	74Ah	
	-7.1	-0.757(25°,1)	-38.9	15-35	T	μ=1.0(HNO_3,$HClO_4$)	69Me	b,q:±5.0
Ni^{2+}	-3.53	0.08(1)	-10.3	25	C	μ=1M	74Ah	
Cu^{2+}	-4.70	-0.13(1)	-18.3	25	C	μ=1M	75Al	
	3.56	-0.824(1)	-4	25	C	3.0N($LiClO_4$+LiCl) values are for formation of inner sphere complexes	70Bf	
	2.01	-0.398(1)	0	25	C	3.0N($LiClO_4$+LiCl) values are for formation of outer sphere complexes	70Bf 74Bf	
Zn^{2+}	-4.69	-0.12(1)	-18.0	25	C	μ=1M	75Ah	
Cd^{2+}	-21.92	0.31(1)	-67.8	25	C	1M(KNO_3)	62Va 67Va	
	-23.8	0.22(25°,1)	-75	18&25	T	--	62Va	a,b
Tl^+	-2.72	0.33(1)	-4.2	25	T	μ=0	57Na	
	-41.714	1.31(1)	-114.6	25	T	μ=0	34La	
	9.71	-0.30(1)	26.4	25	C	μ=0.5N($LiClO_4$, $LiNO_3$)	75Fb	
	11.2	-0.32(1)	31.8	25	C	μ=1N($LiClO_4$, $LiNO_3$)	75Fb	
	-0.46	-0.37(1)	-8.4	25	C	μ=2N($LiClO_4$, $LiNO_3$)	75Fb	
	-15.5	-0.31(1)	-58.2	25	C	μ=3N($LiClO_4$, $LiNO_3$)	75Fb	
	-25.9	-0.48(1)	-96	25	C	3M($LiClO_4$or$LiNO_3$)	65Ke	
	-27.5	-0.32(1)	-98.3	25	C	μ=4N($LiClO_4$, $LiNO_3$)	75Fb	
Tl^{3+}	0	-0.90(1)	17.2	25	C	2.0M($LiClO_4$) 1.0M($HClO_4$)	67Md	
Pb^{2+}	-2.38	1.19(1)	14.6	25	T	μ=0	55Na	
	-5.48	0.518(1)	-8.4	25	C	μ=3M($LiClO_4$, $LiNO_3$)	72Fg	
	-5.8	0.505(25°,1)	-10.0	2-65	T	μ=3.0M($LiClO_4$, $LiNO_3$)	67Fa	b
	-5.8	0.506(25°,1)	-10.0	18-65	T	μ=3M($LiClO_4$)	67Fa 68Fa	b

Metal Ion	ΔH, kJ /mol	Log K	ΔS, J /mol K	T,°C	M	Conditions	Ref.	Re-marks
N20, continued								
Pb^{2+}, cont.	-3.89	0.58(25°,1)	-2.5	5-65	T	μ=3M $Li(ClO_4,NO_3)$ S:3M Ethanol D=70.1	71Gk	b,q: ±1.2
	-6.94	0.3(25°,1)	-7.5	5-65	T	μ=3M $Li(ClO_4,NO_3)$ S:6M Ethanol D=61.0	71Gk	b,q: ±1.1
	-6.28	0.98(25°,1)	-2.5	5-65	T	μ=3M $Li(ClO_4,NO_3)$ S:9M Ethanol D=51.2	71Gk	b,q: ±1.2
	-7.03	1.23(25°,1)	0.0	5-65	T	μ=3M $Li(ClO_4,NO_3)$ S:12M Ethanol D=41.5	71Gk	b,q: ±2.0
	-2.72	0.72(25°,1)	4.6	5-65	T	μ=3M $Li(ClO_4,NO_3)$ S:6M CH_3OH D=69.0	71Gk	b,q: ±1.2
	-5.48	0.83(25°,1)	-2.5	5-65	T	μ=3M $Li(ClO_4,NO_3)$ S:12M CH_3OH D=58.3	71Gk	b,q: ±0.92
	-8.91	1.15(25°,1)	-7.9	5-65	T	μ=3M $Li(ClO_4,NO_3)$ S:18M CH_3OH D=45.9	71Gk	b,q: ±0.83
	-12.26	1.49(25°,1)	-12.6	5-65	T	μ=3M $Li(ClO_4,NO_3)$ S:24.72M CH_3OH D=31.5	71Gk	b,q: ±2.0
	-1.21	-0.070(2)	-5.4	25	C	μ=3M($LiCLO_4$, $LiNO_3$)	72Fg	
	-5.4	-0.18(25°,2)	-21.8	2-65	T	μ=3.0M($LiClO_4$, $LiNO_3$)	67Fa	b
	-5.4	-0.183(25°,2)	-21.8	18-65	T	μ=3M($LiClO_4$)	68Fa 67Fa	b
	-7.40	0.25(25°,2)	-18.8	5-65	T	μ=5M $Li(ClO_4,NO_3)$ S:3M Ethanol D=70.1	71Gk	b,q: ±2.1
	-0.17	0.21(25°,2)	4.2	5-65	T	μ=5M $Li(ClO_4,NO_3)$ S:6M Ethanol D=61.0	71Gk	b,q: ±2.5
	-0.8	0.28(25°,2)	2.1	5-65	T	μ=5M $Li(ClO_4,NO_3)$ S:9M Ethanol D=51.2	71Gk	b,q: ±2.9
	2.43	0.63(25°,2)	20.5	5-65	T	μ=5M $Li(ClO_4,NO_3)$ S:12M Ethanol D=41.5	71Gk	b,q: ±1.2
	-14.85	0.11(25°,2)	-46.4	5-65	T	μ=3M $Li(ClO_4,NO_3)$ S:6M CH_3OH D=69.0	71Gk	b,q: ±3.8
	-7.91	0.32(25°,2)	-20.1	5-65	T	μ=3M $Li(ClO_4,NO_3)$ S:12M CH_3OH D=58.3	71Gk	b,q: ±1.7
	0.96	0.59(25°,2)	14.2	5-65	T	μ=3 M$Li(ClO_4,NO_3)$ S:18M CH_3OH D=45.9	71Gk	b,q: ±1.7
	-7.82	0.88(25°,2)	-9.6	5-65	T	μ=3M $Li(ClO_4,NO_3)$ S:24.72 CH_3OH D=31.5	71Gk	b,q: ±5.0
	5.4	0.022(25°,3)	14.6	2-65	T	μ=3.0M($LiClO_4$, $LiNO_3$)	67Fa	b

Metal Ion	ΔH, kJ /mol	Log K	ΔS, J /mol K	T,°C	M	Conditions	Ref.	Remarks
N20, continued								
Pb^{2+}, cont.	3.3	-0.194(25°,3)	7.1	2-65	T	μ=3M($LiClO_4$, $LiNO_3$)	72Fg	b,q: ±2.1
	5.4	0.022(25°,3)	18.8	18-65	T	μ=3M($LiClO_4$)	68Fa 67Fa	b
	-27.2	-0.553(25°,4)	-100	2-65	T	μ=3M($LiClO_4$, $LiNO_3$)	72Fg	b,q: ±4.2
Bi^{3+}	-4.14	0.733(1)	0	25	T	μ=3	71Fb	
N21 NITRITE ION			NO_2			NO_2^-		L^-
[Fe^{3+}]	-42	2.4(1)	-92	25	-	Hemoglobin	60Gb	
NiA_2L_2	-9.6	--	-30.1	3-51	T	S:Chloroform CM=0.015M A=Diethylethylenediamine R:$NiA_2(ONO)_2$=$NiA_2(NO_2)_2$ (Nitro=Nitrito equilibrium)	66Ge	b,q: ±2.5
Ag^+	-28.9	2.458(1)	-53	15	T	μ=0 ΔCp=200	72Te	
	-29.7	2.367(1)	-56	20	T	μ=0 ΔCp=203	72Te	
	-31.0	2.276(1)	-60	25	T	μ=0 ΔCp=207	72Te	
	-33.0	2.095(1)	-66	35	T	μ=0 ΔCp=214	72Te	
	-15.1	0.348(2)	-46	15	T	μ=0 ΔCp=105	72Te	
	-15.9	0.301(2)	-50	20	T	μ=0 ΔCp=107	72Te	
	-15.9	0.254(2)	-49	25	T	μ=0 ΔCp=110	72Te	
	-17.2	0.159(2)	-53	35	T	μ=0 ΔCp=214	72Te	
Cd^{2+}	-8.74	1.76(1)	4.6	25	C	3.0M($NaClO_4$)	66Gb	
	-8.83	1.32(2)	-4.2	25	C	3.0M($NaClO_4$)	66Gb	
	-6.61	0.62(3)	-10.0	25	C	3.0M($NaClO_4$)	66Gb	
N22 NITROGEN			N_2			N_2		L
RuA_5	-92	(2-1)	--	25	C	0.1M(H_2SO_4) CM=0.03-0.1M A=NH_3	70Fa	q:±8
RuA_5B	-39.3	4.690	-42	20-35	T	μ=0.10(HCl) A=NH_3 B=H_2O R:RuA_5B^{2+}+L=RuA_5L^{2+}+B K measured as a function of temperature by Kinetic method	70Ab	b,q: ±3.7
	-42.2	4.518	-54	25-45	T	μ=0.10(HCl) A=NH_3 B=H_2O R:RuA_5B^{2+}+L=RuA_5L^{2+}+B K measured as a	70Ab	b,q: ±5.8

Metal Ion	ΔH, kJ /mol	Log K	ΔS, J /mol K	T,°C	M	Conditions	Ref.	Remarks
N22, continued								
RuA_5B, cont.						function of temperature by Static method		
	-46.9	3.863	-84	25-45	T	μ=0.10(HCl) $A=NH_3$ $B=H_2O$ $R:RuA_5B^{2+}+RuA_5L^{2+}$ $=[RuA_5LRuA_5]^{4+}+B$	70Ab	b,q: ±5.8
N23 NITROGEN MONOXIDE			NO			NO		L
CoA_5B	-36.8	4.0	-46	25	-	$R:CoA_5B^{2+}+L=$ $CoA_5L^{2+}+B$ $A=NH_3$ $B=H_2O$	60Gb	
N24 NITROXIDE, di-*tert*-butyl			$C_8H_{18}ON$			$ON[C(CH_3)_3]_2$		
SiA_4	-13.0	0.90(1)	-42	-53	T	$S:SiCl_4$ A=Cl	74Cf	q:±1.2
GeA_4	-12.6	1.70(1)	-25.5	-53	T	$S:GeCl_4$ A=Cl	74Cf	
$SnAB_2C_5xD$	-9.2	--	--	-53	T	S:Toluene A=Cr B=*t*-Butyl C=CO D=Tetrahydrofuran $R+SnAB_2C_5xD+L=$ $SnAB_2C_5xL+D$	74Cf	q:±0.8
N25 NONACTIN			$C_{40}H_{64}O_2$					
Na^+	-27.40	3.27(1)	29.3	25	C	$CM=0.1-5x10^{-2}M$ $CL=10^{-3}M$ S:Ethanol	73Fb 73Za	
	-11.09	2.71(1)	14.6	25	C	$CM=0.1-5x10^{-2}M$ $CL=10^{-3}M$ $S:CH_3OH$	73Fb 73Za	i:71Fe
	-14.18	(1)	--	25	C	$S:CH_3OH$	71Fe	
K^+	-52.22	5.26(1)	-74.5	25	C	$CM=0.1-5x10^{-2}M$ $CL=10^{-3}M$ S:Ethanol	73Fb 73Za	
	-46.0	4.301(1)	-71	25	C	$S:CH_3OH$	71Fe	
	-43.60	4.49(1)	-60.2	25	C	$CM=0.1-5x10^{-2}M$ $CL=10^{-3}M$ $S:CH_3OH$	73Fb 73Za	i:71Fe
N26 NONANE, 1-amino-			$C_9H_{21}N$			$[CH_3(CH_2)_7CH_2NH_2]$		L
NiA	-70.7	6.10(1-2)	-120.5	25	T	$CM=10^{-4}M$ S:Benzene A=Diacetylbisbenzoyl hydrazine	60Sb	

Metal Ion	ΔH, kJ /mol	Log K	ΔS, J /mol K	T,°C	M	Conditions	Ref.	Remarks
N27	NONANE, 3,7-diaza-1,9-diamino		$C_7H_{10}N_4$			$N_2NCH_2CH_2NHCH_2CH_2CH_2$-$NHCH_2CH_2NH_2$		
Hg^{2+}	-114.6	22.1(25°,1)	38.1	15-35	T	μ=0.20M($NaClO_4$)	76Kg	
N28	NONANE, 3,7-dithio-1,9-dihydroxy		$C_7H_{16}O_2S_2$			$OHCH_2CH_2SCH_2CH_2CH_2SCH_2CH_2OH$		
Ag^+	-78.2	7.60(1)	-120	20	C	μ=1.0(KNO_3) pH=3	70Wc	
	-28.0	4.03(2)	-18.8	20	C	μ=1.0(KNO_3) pH=3	70Wc	
	-7.5	2.24	17.2	20	C	μ=1.0(KNO_3) pH=3 R:AgL+Ag=Ag_2L	70Wc	

O

Metal Ion	ΔH, kJ /mol	Log K	ΔS, J /mol K	T,°C	M	Conditions	Ref.	Remarks
O1	OCTANE, 1-amino		$C_8H_{19}N$			$CH_3(CH_2)_7NH_2$		L
NiA	-70.7	6.06(1-2)	-121.8	25	T	CM=10^{-4}M S:Benzene A=Di-acetylbisbenzyl hydrazine	60Sb	
CuA_2	-58	4.30(1)	-116	20	T	S:Benzene A=Trifluoro-acetylacetone	79Si	b
O2	OCTANE, 1,8-diamino-3,6-dithio-		$C_6H_{16}N_2S_2$			$H_2N(CH_2CH_2S)_2CH_2CH_2NH_2$		L
Ni^{2+}	-50.3	7.691(1)	-21	25	C	0.5M(KNO_3)	80Ha	
	-50	7.90(30°,1)	-12	0-50	T	μ=1(KCl or KNO_3)	54Gb	b
	-20	3.55	-3	25	C	0.5M(KNO_3) R:Ni^{2+}+HL^+=$NiHL^{3+}$	80Ha	
Cu^{2+}	-64.8	10.43(30°,1)	12	10-40	T	μ=0	59Ma	b
	-67.9	10.911(1)	-18.9	25	C	0.5M(KNO_3)	80Ha	
	-35	5.05	-20	25	C	0.5M(KNO_3) R:Cu^{2+}+HL^+=$CuHL^{3+}$	80Ha	
O3	OCTANE, 1,8-diamino-3-oxa-6-thio-		$C_6H_{16}ON_2S$			$H_2NCH_2CH_2OCH_2CH_2SCH_2CH_2NH_2$		L
Ni^{2+}	34.3	6.17(30°,1)	4	10-40	T	μ=0	59Lb	b
Cu^{2+}	-54.8	8.86(30°,1)	-12	10-40	T	μ=0	59Lb	b
Ag^+	-60.2	8.12(30°,1)	-42	10-40	T	μ=0	59Lb	b
O4	OCTANE 1,8-diamino-N,N,N',N'-tetraacetic acid		$C_{16}H_{28}O_8N_2$			$(CH_2CO_2H)_2N(CH_2)_8N(CH_2CO_2H)_2$		L^{4-}
Mn^{2+}	2.1	9.01(1)	179.1	20	C	μ=0.1(KNO_3)	64Ab	
Co^{2+}	-19.92	12.91(1)	178.2	20	C	μ=0.1(KNO_3)	64Ab	

Metal Ion	ΔH, kJ /mol	Log K	ΔS, J /mol K	T,°C	M	Conditions	Ref.	Re-marks
04, continued								
Ni^{2+}	-35.6	13.62(1)	138.9	20	C	μ=0.1(KNO_3)	64Ab	
Cu^{2+}	-43.01	15.8(1)	155.6	20	C	μ=0.1(KNO_3)	64Ab	
Zn^{2+}	-18.24	12.66(1)	180.12	20	C	μ=0.1(KNO_3)	64Ab	
Cd^{2+}	-19.20	11.99(1)	164.0	20	C	μ=0.1(KNO_3)	64Ab	
Hg^{2+}	-82.59	21.83(1)	136.0	20	C	μ=0.1(KNO_3)	64Ab	
05 OCTANE, 1,8-diamine-4-thio			$C_7H_{18}N_2S$			$NH_2(CH_2)_4S(CH)_3NH_2$		L
Ni^{2+}	-39.0	5.989(1)	-16.2	25	C	μ=0.5M(KNO_3)	79Ha	
	-38.9	3.86(2)	-56	25	C	μ=0.5M(KNO_3)	79Ha	
	-21.3	2.75	-18	25	C	μ=0.5M(KNO_3) R:Ni^{2+}+HL^{+}=$NiHL^{3+}$	79Ha	
	-29.7	1.88	-63	25	C	μ=0.5M(KNO_3) R:NiL^{2+}+HL^{+}= $NiHL_2{}^{3+}$	79Ha	
Cu^{2+}	-59.0	10.035(1)	-5.8	25	C	μ=0.5M(KNO_3)	79Ha	
	-32.7	2.86(2)	-54	25	C	μ=0.5M(KNO_3)	79Ha	
	-29.3	4.66	-9	25	C	μ=0.5M(KNO_3) R:Cu^{2+}+HL^{+}=$CuHL^{3+}$	79Ha	
	-31.8	2.44	-59	25	C	μ=0.5M(KNO_3) R:CuL^{2+}+HL^{+}= $CuHL_2{}^{3+}$	79Ha	
	-13.4	4.60	43	25	C	μ=0.5M(KNO_3) R:CuL^{2+}+OH^{-}= Cu(OH)L^{-}	79Ha	
06 OCTANE, 3,6-diaza-1,8-diamino			$C_6H_{18}N_4$			$H_2NCH_2CH_2NHCH_2CH_2NHCH_2CH_2NH_2$		
Hg^{2+}	-126.4	24.5(25°,1)	45.2	15-35	T	μ=0.20M($NaClO_4$)	76Kg	
07 OCTANE, 3,6-dioxy-1,8-diamino			$C_6H_{16}O_2N_2$			$H_2NCH_2CH_2OCH_2CH_2OCH_2CH_2NH_2$		
Ag^{+}	-57.5	7.7(1)	-45.6	25	C	μ=0.1[$(CH_3)_4NCl$]	75Af	
Hg^{2+}	-102.5	18.55(1)	11.3	25	C	μ=0.1[$(CH_3)_4NCl$]	75Af	
08 OCTANE, 3,6-dioxa-1,8-diamino-N,N,N',N'-tetraacetic acid			$C_{14}H_{24}O_{10}N_2$			$(CH_2CO_2H)_2N(CH_2CH_2O)_2$-$CH_2CH_2N(CH_2CO_2H)_2$		L^{4-}
Mg^{2+}	21.67	5.2(1)	173.6	20	C	μ=0.1(KNO_3)	64Ab	
	18.4	5.4(1)	167	25	C	μ=0.1(KNO_3)	65Wd	
	22.97	5.28(1)	178.2	25	C	μ=0.1(KClorKNO$_3$)	65Bc	
Ca^{2+}	-35.06	10.97(1)	90.4	20	C	μ=0.1(KNO_3)	65Ab	
	-33.5	10.8(1)	96	25	C	μ=0.1(KNO_3)	65Wd	
	-33.22	10.89(1)	97.1	25	C	μ=0.1(KClorKNO$_3$)	65Bc	

Metal Ion	ΔH, kJ /mol	Log K	ΔS, J /mol K	T,°C	M	Conditions	Ref.	Re-marks
08, continued								
Sr^{2+}	-26.8	8.06(1)	63	25	C	μ=0.1(KNO_3)	65Wd	
	-24.02	8.43(1)	80.8	25	C	μ=0.1(KClorKNO$_3$)	65Bc	
Ba^{2+}	-36.8	7.99(1)	29.7	25	C	μ=0.1(KNO_3)	65Wd	
	-37.61	8.30(1)	32.6	25	C	μ=0.1(KClorKNO$_3$)	65Bc	
La^{3+}	-22.84	15.79(1)	224.3	20	C	μ=0.1(KNO_3)	64Ab	
UO_2^{2+}	10.0	9.49(25°)	213	20-40	T	μ=0.1(KNO_3) R:UO_2+HL=UO_2HL	69Sg	b
Mn^{2+}	-34.14	12.28(1)	90.0	20	C	μ=0.1(KNO_3)	64Ab	
	-36.8	12.2(1)	113	25	C	μ=0.1(KNO_3)	65Wd	
Fe^{2+}	-21.8	11.8(1)	155	25	C	μ=0.1(KNO_3)	65Wd	
Co^{2+}	-11.84	12.28(1)	194.6	20	C	μ=0.1(KNO_3)	64Ab	
	-14.2	12.2(1)	188	25	C	μ=0.1(KNO_3)	65Wd	
Ni^{2+}	-16.02	11.82(1)	171.5	20	C	μ=0.1(KNO_3)	64Ab	
	-20.9	13.6(1)	188	25	C	μ=0.1(KNO_3)	65Wd	
Pd^{2+}	-55.2	11.8(1)	38.1	20	C	μ=0.1(KNO_3)	64Ab	
Cu^{2+}	-46.0	17.71(1)	182.0	20	C	μ=0.1(KNO_3)	64Ab	
	-43.9	17.7(1)	192	25	C	μ=0.1(KNO_3)	65Wd	
Zn^{2+}	-17.91	12.91(1)	185.8	20	C	μ=0.1(KNO_3)	64Ab	
	-15.9	14.4(1)	222	25	C	μ=0.1(KNO_3)	65Wd	
	-21.00	12.86(1)	176.1	25	C	μ=0.1(KClorKNO$_3$)	65Bc	
Cd^{2+}	-61.9	16.1(1)	97.1	20	C	μ=0.1(KNO_3)	64Ab	
	-59.0	16.6(1)	121	25	C	μ=0.1(KNO_3)	65Wd	
	-62.30	16.53(1)	107.5	25	C	μ=0.1(KClorKNO$_3$)	65Bc	
Hg^{2+}	-99.2	23.2(1)	105.4	20	C	μ=0.1(KNO_3)	64Ab	
	-97.5	23.7(1)	130	25	C	μ=0.1(KNO_3)	65Wd	
Pb^{2+}	-52.3	14.6(1)	105	25	C	μ=0.1(KNO_3)	65Wd	
09 OCTANEDIOIC ACID, 3,6-oxy-			$C_6H_{10}O_6$			$HO_2CCH_2OCH_2CH_2OCH_2CO_2H$		L^{2-}
La^{3+}	6.74	4.354(1)	105.8	25	C	μ=1.00($NaClO_4$)	74Ge	
	-10.79	3.360(2)	47.3	25	C	μ=1.00($NaClO_4$)	74Ge	
Ce^{3+}	5.52	4.644(1)	107.1	25	C	μ=1.00($NaClO_4$)	74Ge	
	-8.79	3.243(2)	51.9	25	C	μ=1.00($NaClO_4$)	74Ge	
Pr^{3+}	4.31	4.806(1)	105.8	25	C	μ=1.00($NaClO_4$)	74Ge	
	-6.44	3.085(2)	38.1	25	C	μ=1.00($NaClO_4$)	74Ge	
Nd^{3+}	3.30	4.920(1)	105.0	25	C	μ=1.00($NaClO_4$)	74Ge	
	-2.72	3.035(2)	49.0	25	C	μ=1.00($NaClO_4$)	74Ge	
Sm^{3+}	1.59	5.074(1)	102.1	25	C	μ=1.00($NaClO_4$)	74Ge	
	6.36	2.909(2)	75.7	25	C	μ=1.00($NaClO_4$)	74Ge	
Gd^{3+}	3.56	4.894(1)	105.8	25	C	μ=1.00($NaClO_4$)	74Ge	
	8.66	2.978(2)	86.2	25	C	μ=1.00($NaClO_4$)	74Ge	
Tb^{3+}	7.28	4.753(1)	115.1	25	C	μ=1.00($NaClO_4$)	74Ge	
	6.78	3.145(2)	82.8	25	C	μ=1.00($NaClO_4$)	74Ge	

Metal Ion	ΔH, kJ /mol	Log K	ΔS, J /mol K	T,°C	M	Conditions	Ref.	Re-marks
09, continued								
Dy^{3+}	10.34	4.668(1)	125.9	25	C	μ=1.00($NaClO_4$)	74Ge	
	6.94	3.258(2)	84.9	25	C	μ=1.00($NaClO_4$)	74Ge	
Ho^{3+}	12.97	(1)	132.0	25	C	μ=1.00($NaClO_4$)	74Ge	
	6.44	(2)	84.1	25	C	μ=1.00($NaClO_4$)	74Ge	
Er^{3+}	15.61	4.606(1)	141.0	25	C	μ=1.00($NaClO_4$)	74Ge	
	13.43	3.432(2)	110.0	25	C	μ=1.00($NaClO_4$)	74Ge	
Tm^{3+}	16.74	4.637(1)	144.8	25	C	μ=1.00($NaClO_4$)	74Ge	
	15.06	3.597(2)	119.2	25	C	μ=1.00($NaClO_4$)	74Ge	
Yb^{3+}	15.94	4.847(1)	146.0	25	C	μ=1.00($NaClO_4$)	74Ge	
	15.6	3.829(2)	125.9	25	C	μ=1.00($NaClO_4$)	74Ge	
Lu^{3+}	15.23	(1)	143.1	25	C	μ=1.00($NaClO_4$)	74Ge	
	14.18	(2)	124.7	25	C	μ=1.00($NaClO_4$)	74Ge	
010	OCTANEDIOIC ACID, 3,6-thio-		$C_6H_{10}O_4S_2$			$HO_2CCH_2SCH_2CH_2SCH_2CO_2H$		L^{2-}
Sm^{3+}	12.30	2.320(1)	85.8	25	C	μ=1.00($NaClO_4$)	74Ge	
	8.91	1.171(2)	52.3	25	C	μ=1.00($NaClO_4$)	74Ge	
Dy^{3+}	20.42	2.086(1)	107.9	25	C	μ=1.00($NaClO_4$)	74Ge	
	3.89	1.391(2)	40.2	25	C	μ=1.00($NaClO_4$)	74Ge	
Er^{3+}	19.58	2.009(1)	103.8	25	C	μ=1.00($NaClO_4$)	74Ge	
	13.43	1.137(2)	67.4	25	C	μ=1.00($NaClO_4$)	74Ge	
011	7-OCTENOIC ACID, 3-seleno-		$C_7H_{12}O_2Se$			$CH_2:CH(CH_2)_3SeCH_2CO_2H$		L^-
Ag^+	-38.20	4.62(25°,1)	-40.25	0-40	T	μ=0.20M pH=5.8	71Bd 71Bc	b,q: ±2.1
	-33.01	3.05(25°,2)	-53.64	0-40	T	μ=0.20M pH=5.8	71Bd 71Bc	b,q: ±7.1
	-14.98	1.80(25°)	-16.78	0-40	T	μ=0.20M R:$AgL+Ag^+=Ag_2L^+$ pH=5.8	71Bd	b,q: ±5.0
012	OXALIC ACID		$C_2H_2O_4$			HO_2CCO_2H		L^{2-}
Be^{2+}	33.8	5.91(1-2)	226	?	C	μ~0.01M	76Ma	u:25°
Ca^{2+}	-21.296	8.642	-94.1	25	T	μ=0 R:$Ca^{2+}+L^{2-}+H_2O=CaL\cdot H_2O$	34La	
Er^{3+}	-29.83	8.42(30°,1-3)	63.2	18&30	T	μ=0.10 Signs on ΔG and ΔH are reversed from those reported in the article	69Ka	a,b,q: ±2.8
Ti^{3+}	-6.3	--	--	25	C	μ=0 R:$Ti^{3+}+2H_2L=TiL_2^-+4H^+$	72Pa	
Th^{4+}	-14.14	8.004(25°,1)	105.8	10-40	T	μ=1.0M($HClO_4$)	77Ra	b
CrA_5B	-5.4	0.672(40°,1)	-3.8	40-60	T	μ=1.0M($NaClO_4$)	73Na	b,q:

Metal Ion	ΔH, kJ /mol	Log K	ΔS, J /mol K	T,°C	M	Conditions	Ref.	Remarks
012, continued								
CrA_5B, cont.	-18.0	0.146(40°)	-54.4	40-60	T	$A=NH_3$ $B=H_2O$ μ=1.0M($NaClO_4$) $A=NH_3$ $B=H_2O$ R:$CrA_5B^{3+}+HL^-=$ $CrA_5B^{3+}HL^-$	73Na	±4.2 b,q: ±1.7
Np^{4+}	-19.08	9.220(25°,1)	112.5	10-40	T	μ=1.0M($HClO_4$)	77Ra	b
NpO_2^+	0.0	0.290(25°,1)	5.4	10-47	T	μ=0.5	53Ga	b
Pu^{3+}	-5.439	7.936(20°,1-2)	133.5	20&70	T	μ=1	57Ga 58Gb	b
	-5.021	8.252(20°,1-3)	141.0	20&70	T	μ=1	57Ga 58Gb	b
	-5.439	8.602(20°,1-4)	146.0	20&70	T	μ=1	57Ga 58Gb	b
Pu^{4+}	-2.02	9.734(25°,1)	179.5	10&25	T	μ=1.0M($HClO_4$)	77Ra	a,b
Mn^{2+}	5.94	3.96(25°,1)	95.8	0-45	T	μ=0 ΔCp=272	61Mb	b
	0.84	5.4(1-2)	99.6	25	C	CM=1-2.2M CL=0.25-0.8M	54Yb	
Fe^{3+}	6.3	7.583(1)	166	25	C	μ=1M($NaClO_4$)	77Db	q:±0.9
	-1.2	9.68(1)	180	--	-	--	52Ua	r
	3.7	6.223(2)	277	25	C	μ=1M($NaClO_4$)	77Db	q:±1.8
	0.4	4.853(3)	359	25	C	μ=1M($NaClO_4$)	77Db	q:±1.3
Co^{2+}	2.47	4.79(25°,1)	100.0	0-45	T	μ=0 ΔCp=84	61Mb	b
	3.35	6.7(1-2)	118.0	25	C	CM=1-2.2M CL=0.25-0.80M	54Yb	
Ni^{2+}	0.63	5.17(25°,1)	101.2	0-45	T	μ=0 ΔCp=146	61Mb	b
	5.15	7.2(1-2)	120.5	25	C	CM=1-2.2M CL=0.25-0.80M	54Yb	
Cu^{2+}	-0.21	6.22(1)	118.4	25	C	μ=0	66Mc 67Mf	
	-11.7	5.53(1)	66.5	25	C	μ=1M($NaNO_3$)	74Ad	
	-0.2	6.220(1)	118.4	25	C	μ=0.1M($NaClO_4$)	78Ae	
	2.8	4.01(2)	86.2	25	C	μ=1M($NaNO_3$)	74Ad	
	6.65	7.48(1-2)	121	25	C	CM=1-2.2M CL=0.25-0.80M	54Yb	
Zn^{2+}	6.48	7.33(1-2)	118.8	25	C	CM=1-2.2M CL=0.25-0.80M	54Yb	
	-17.6	6.74(25°)	70.0	25&35	T	R:Zn+L+A=ZnLA A=Iminodiacetic acid	78Ka	a,b
Cd^{2+}	-19.3	6.04(25°)	50.7	25&35	T	R:Cd+L+A=CdLA A=Iminodiacetic acid	78Ka	a,b
SbO^+	249.12	12.30(27°)	594	25-30	T	R:$SbO^++2L^{=}+H_2O=$ $SbL_2^-+2OH^-$	53Na	b

Metal Ion	ΔH, kJ /mol	Log K	ΔS, J /mol K	T,°C	M	Conditions	Ref.	Remarks
013 OXYGEN (aq)			O_2					
[Fe]	-26.8	--	--	20	C	[Fe]=Ox Haemoglobin pH=6.8 Buffer:0.07M Phosphate R:1 mole O_2(solution)+[Fe] (solution)=O_2[Fe] (solution)	35Ra	q:±2.9
	-43.9	--	--	20	C	[Fe]=Ox Haemoglobin pH=9.5 R:1 mole O_2(solution)+[Fe](solution)=O_2[Fe] (solution)	35Ra	q:±8.8
	-74.9	(1)	-138.1	0-30	T	0.001M THAM buffer pH=8.5 [Fe]=Sperm whale myoglobin Standard state for L= unit mole fraction in organic solvent	71Ka	b
CoA_2	-125.9	6.63(2-1)	-293	25	C	μ=0.02M A=Histidinate	72Pk	
	-62.88	4.728(20°,1)	-121	15-31.5	T	S:Pyridine L(gas) A=(acetylacetone) ethylenediamine	70Aa	b
CoAB	-38.9	2.84(-65°,1)	-134	-80--20	T	S:Toluene A=α,β,γ,δ-Tetra-(p-methoxyphenyl) porphyrin B=Pyridine Standard state of 1M O_2	73Wb	b,q: ±4.6
	-36.8	2.95(-65°,1)	-121	-80--20	T	S:Toluene A=α,β,γ,δ-Tetra-(p-methoxyphenyl) porphyrin B=4-Picoline Standard state of 1M O_2	73Wb	b,q: ±3.8
	-38.5	2.81(-65°,1)	-130	-80--20	T	S:Toluene A=α,β,γ,δ-Tetra-(p-methoxyphenyl) porphyrin B=3,4-Lutidine Standard state of 1M O_2	73Wb	b
	-35.6	3.41(-65°,1)	-105	-80--20	T	S:Toluene A=α,β,γ,δ-Tetra-(p-methyoxyphenyl) porphyrin B=4-Dimethylamino	73Wb	b,q: ±3.3

Metal Ion	ΔH, kJ /mol	Log K	ΔS, J /mol K	T,°C	M	Conditions	Ref.	Remarks
013, continued								
CoAB, cont.						pyridine Standard state of 1M O_2		
	-39.7	2.41(-65°,1)	-146	-80--20	T	S:Toluene A=α,β,γ,δ-Tetra-(p-methoxyphenyl) porphyrin B=γ-Collidine Standard state of 1M O_2	73Wb	b,q: ±5.4
	-36.0	2.10(65°,1)	-130	-80--20	T	S:Toluene- A=α,β,γ,δ-Tetra-(p-methoxyphenyl) porphyrin B=5-Chloro-N-methyl-imidazole Standard state of 1M O_2	73Wb	b
	-37.2	3.58(-65°,1)	-109	-80--20	T	S:Toluene A=α,β,γ,δ-Tetra-(p-methoxyphenyl) porphyrin B=N-methylimidazole Standard state of 1M O_2	73Wb	b,q: ±2.1
	-34.3	2.87(-65°,1)	-109	-80--20	T	S:Toluene A=α,β,γ,δ-Tetra-(p-methoxyphenyl) porphyrin B=Piperidine Standard state of 1M O_2	73Wb	b
	-35.6	-2.18(-65°,1)	-207.2	-80--20	T	S:Carbon Disulfide A=α,β,γ,δ-Tetra(p-methoxyphenyl) porphyrin B=Pyridine Standard state of 1M O_2	73Wb	b,q: ±2.1
	-37.6	-2.10(-65°,1)	-216.7	-80--20	T	S:Carbon Disulfide A=α,β,γ,δ-Tetra(p-methoxyphenyl) porphyrin B=3,4-Lutidine Standard state of 1M O_2	73Wb	b,q: ±2.5
$CoAB_2$	-31.4	2.863(-65°,1)	-96	-80--20	T	S:Toluene A=α,β,γ,δ-Tetra-(p-methoxyphenyl) porphyrin B=3,4-Lutidine	73Wb	b,q: ±4.2
	-28.0	2.322(-65°,1)	-88	-80--20	T	S:Toluene A=α,β,γ,δ-Tetra-	73Wb	b,q: ±4.2

Metal Ion	ΔH, kJ /mol	Log K	ΔS, J /mol K	T,°C	M	Conditions	Ref.	Remarks
013, continued								
$CoAB_2$, cont.						(p-methoxyphenyl) porphyrin B= Piperidine		
CoA_2B_2	-123.0	10.84	-205	25	C	μ=1.0M(KCl) A=Ethylenediamine $B=H_2O$ R:$2CoA_2B_2^+ +L=Co_2A_4LOH+H_3O^+$	72Pk	
	-123.0	7.33	-272	25	-	μ=1.0M(KCl) A=Ethylenediamine $B=H_2O$ R:$2CoA_2B_2^+ +L=Co_2A_4LB_2$	72Pk	
[Co]A	-46.0	-2.27(-45°,1)	-247	-63.5- -31	T	S:Toluene [Co]= Cobalt(II) proto-porphyrin IX dimethyl ester A=N,N-Dimethyl-formamide L(gas)	73Sx	b,cc
	-38.5	-2.25(-57.5°,1)	-222	-63.5- -45	T T	S:Toluene [Co]=Cobalt(II) protoporphyrin IX dimethyl ester A=Pyridine L(gas)	72Sk 73Sx	b,cc
	-41.0	-2.77(-45°,1)	-234	-63.5- -37.4	T	S:Toluene [Co]=Cobalt(II) protoporphyrinIX dimethyl ester A=4-tert-butyl-pyridine L(gas)	73Sx 72Sk	b,cc
	-47.3	-1.84(-45°,1)	-243	-57.5- -31	T	S:Toluene [Co]=Cobalt(II) protoporphyrinIX dimethyl ester A=Imidazole L(gas)	73Sx	b,cc
	-49.4	-2.04(-37.4°,1)	-247	-45--31		S:Toluene [Co]=Cobalt(II) protoporphyrinIX dimethyl ester A=1-Methylimida-zole L(gas)	73Sx	b,cc
	-41.4	-2.05(-45°,1)	-222	-63.5- -31		S:Toluene [Co]=Cobalt(II) protoporphyrinIX dimethyl ester A=4-Aminopyridine L(gas)	73Sx	b,cc
	-37.6	-2.34(-45°,1)	-209	-63.5- -31		S:Toluene [Co]=Cobalt(II) protopophyrinIX dimethyl ester	73Sx	b,cc

Metal Ion	ΔH, kJ /mol	Log K	ΔS, J /mol K	T,°C	M	Conditions	Ref.	Re-marks
013, continued								
[Co]A, cont.						A=Piperidine L(gas)		

<u>P</u>

Metal Ion	ΔH, kJ /mol	Log K	ΔS, J /mol K	T,°C	M	Conditions	Ref.	Re-marks
P1 PENTAAMINE COBALT(II)FUMARATE			$C_4H_{18}O_4N_5Co$					
Cu^{1+}	-48.1	3.643(23°,<u>1</u>)	-92	5-40	T	μ=1.0($HClO_4$)	73Hg	b,q:±13.4
P2 PENTAAMINE (PYRAZINE) RUTHENIUM(II)			$C_4H_{19}N_7Ru$			$Ru(NH_3)_5$ 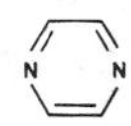		L^{2+}
Ni^{2+}	-25	1.230(25°,<u>1</u>)	-63	5-30	T	μ=1.5M($LiClO_4$)	74Pe	b,q:±3.3
Cu^{2+}	-18.0	1.505(25°,<u>1</u>)	-29	5-30	T	μ=1.0M($LiClO_4$)	74Pe	b,q:±2.1
Zn^{2+}	-18.0	0.491(25°,<u>1</u>)	-50	5-30	T	μ=2.0M($LiClO_4$)	74Pe	b,q:±2.1
P3 1,4,7,11,14-PENTAAZACYCLO-HEPTADECANE			$C_{12}H_{29}N_5$			$CH_2(CH_2CH_2N)_3CH_2CH_2CH_2NCH_2CH_2N$		L
Cu^{2+}	-113.8	23.8(25°,<u>1</u>)	75	15-35	T	μ=0.2M($NaClO_4$)	78Kd 78Kc	b
Zn^{2+}	-53.1	15.8(25°,<u>1</u>)	126	15-35	T	μ=0.2M($NaClO_4$)	78Kd	b
Cd^{2+}	-52.7	15.5(25°,<u>1</u>)	121	15-35	T	μ=0.2M($NaClO_4$)	78Kd	b
Hg^{2+}	-139.7	26.5(25°,<u>1</u>)	38	15-35	T	μ=0.2M($NaClO_4$)	78Kd	b
Pb^{2+}	-41.4	11.6(25°,<u>1</u>)	126	15-35	T	μ=0.2M($NaClO_4$)	78Kd	b
P4 1,4,7,10,13-PENTAAZACYCLO-HEXADECANE			$C_{11}H_{27}N_5$			$CH_2(CH_2CH_2N)_5CH_2$		L
Cu^{2+}	-136.8	27.1(25°,<u>1</u>)	58	15-35	T	μ=0.2M($NaClO_4$)	78Kd 78Kc	b
Zn^{2+}	-56.5	17.9(25°,<u>1</u>)	154	15-35	T	μ=0.2M($NaClO_4$)	78Kd	b
Cd^{2+}	-54.4	18.1(25°,<u>1</u>)	163	15-35	T	μ=0.2M($NaClO_4$)	78Kd	b

Metal Ion	ΔH, kJ /mol	Log K	ΔS, J /mol K	T,°C	M	Conditions	Ref.	Re-marks
P4, continued								
Hg^{2+}	-143.9	27.4(25°,1)	42	15-35	T	μ=0.2M($NaClO_4$)	78Kd	b
Pb^{2+}	-43.9	14.3(25°,1)	126	15-35	T	μ=0.2M($NaClO_4$)	78Kd	b
P5 1,4,7,10,13-PENTAAZACYCLO-PENTADECANE			$C_{10}H_{25}N_5$			$CH_2CH_2N(CH_2CH_2N)_3CH_2CH_2N$		L
Cu^{2+}	-137.6	28.3(25°,1)	92	15-35	T	μ=0.2M($NaClO_4$)	78Kd 78Kc	b
Zn^{2+}	-57.3	19.1(25°,1)	172	15-35	T	μ=0.2M($NaClO_4$)	78Kd	b
Cd^{2+}	-54.4	19.2(25°,1)	184	15-35	T	μ=0.2M($NaClO_4$)	78Kd	b
Hg^{2+}	-136.8	28.5(25°,1)	84	15-35	T	μ=0.2M($NaClO_4$)	78Kd	b
Pb^{2+}	-41.8	17.3(25°,1)	188	15-35	T	μ=0.2M($NaClO_4$)	78Kd	b
P6 PENTADECANE, 2,14-dioxy-5,8,11-trithia			$C_{10}H_{22}O_2S_3$			$CH_3OCH_2(CH_2SCH_2)_3CH_2OCH_3$		L
Ag^+	-73.47	(1)	--	25	C	μ=0	78Ib	
	-29.00	3.75(2)	-7.6	25	C	μ=0	78Ib	
Hg^{2+}	-71.71	(1)	--	25	C	μ=0	78Ib	
	-39.12	4.69(2)	-12.3	25	C	μ=0	78Ib	
P7 PENTADECANE, 5,11-dithio-			$C_{13}H_{28}S_2$			$C_4H_9S(CH_2)_5SC_4H_9$		L
SnA_4	-105.0	(1)	--	--	C	S:Benzene CM=0.05-0.08M A=Cl	65Gb	u
TeA_4	-50.2	2.98(1)	-113.0	25	C	S:Benzene A=Cl	73Pi	
	-20.9	(2)	--	25	C	S:Benzene A=Cl	73Pi	
P8 PENTADECANE, 2,14-dithio-5,8,11-trioxy			$C_{10}H_{22}O_3S_2$			$CH_3SCH_2(CH_2OCH_2)_3CH_2SCH_3$		L
Ag^+	-60.12	4.46(1)	-116.2	25	C	μ=0	78Ib	
Hg^{2+}	-61.30	5.3(1)	-103.8	25	C	μ=0	78Ib	
P9 PENTADECANE, 5,11-dithio-2,8,14-trioxy			$C_{10}H_{22}O_3S_2$			$CH_3OCH_2CH_2SCH_2CH_2OCH_2CH_2$-$SCH_2CH_2OCH_3$		L
Ag^+	-59.12	(1)	--	25	C	μ=0	78Ib	
	-15.40	3.06(2)	6.9	25	C	μ=0	78Ib	
Hg^{2+}	-58.66	(1)	--	25	C	μ=0	78Ib	
	-29.66	3.22(2)	-37.9	25	C	μ=0	78Ib	

Metal Ion	ΔH, kJ /mol	Log K	ΔS, J /mol K	T,°C	M	Conditions	Ref.	Remarks
P10	PENTADECANE, 8-oxa-2,14-diaza-5,11-dithio-		$C_{10}H_{24}ON_2S_2$			$HNCH_2CH_2SCH_2CH_2OCH_2$-$CH_2SCH_2CH_2NH$		H_2L
Cu^{2+}	-40.0	9.15(1)	40.9	25	C	0.1M($NaClO_4$)	79Ab	
Pb^{2+}	-39.9	7.49(1)	9.4	25	C	0.1M($NaClO_4$)	79Ab	
P11	PENTADECANE, 3,6,10,13-tetrathio-1,15-dihydroxy		$C_{11}H_{24}O_2S_4$			$OHCH_2CH_2SCH_2CH_2SCH_2CH_2CH_2$-$SCH_2CH_2SCH_2CH_2OH$		
Ag^+	-33.5	3.65(1)	-42.2	20	C	μ=1.0(KNO_3) pH=3	70Wc	
	-28.4	2.46(2)	-52.7	20	C	μ=1.0(KNO_3) pH=3	70Wc	
P12	PENTAETHYLENE GLYCOL, dimethyl ether		$C_{12}H_{26}O_6$			$CH_3OCH_2(CH_2OCH_2)_4CH_2OCH_3$		L
Na^+	-38.2	1.0(1)	-109	25	C	CM=0.1-5x10^{-3}M CL=10^{-3}M S:Methanol	73Fb	q:±5
NaA	-38.5	2.653(25°)	-77.4	0-50	T	S:Tetrahydrofuran A=Fluorenyl ion R:NaA+L=NaLA	70Ce	b
K^+	-36.4	2.1(1)	-83.1	25	C	CM=0.1-5x10^{-3}M CL=10^{-3}M S:Methanol	73Fb	
Ba^{2+}	-23.2	2.3(1)	-34.2	25	C	CM=0.1-5x10^{-3}M CL=10^{-3}M S:Methanol	73Fb	
P13	PENTANE, 1-amino-		$C_5H_{13}N$			$CH_3(CH_2)_4NH_2$		L
NiA	-17.5	5.96(1-2)	-125.5	25	T	CM=10^{-4}M S:Benzene A=Diacetylbisbenzoyl hydrazine	60Sb	
Ag^+	-21	3.57(1)	-14.2	25	C	μ=0	71He	q:±8
	-29	3.93(2)	-28.0	25	C	μ=0	71He	q:±8
	-51.63	7.50(1-2	-29.7	25	C	μ=0	71He	
P14	PENTANE, 1-amino-5-hydroxy-3-thia-		$C_4H_{11}ONS$			$HOCH_2CH_2SCH_2CH_2NH_2$		L
Cu^{2+}	-21.3	5.26(30°,1)	29	10-30	T	μ=0	59Lb	b
	-24.7	4.85(30°,2)	12	10-30	T	μ=0	59Lb	b
Ag^+	-36.8	4.78(30°,1)	-29	10-40	T	μ=0	59Lb	b
	-24.7	3.95(30°,2)	-8	10-40	T	μ=0	59Lb	b

Metal Ion	ΔH, kJ /mol	Log K	ΔS, J /mol K	T,°C	M	Conditions	Ref.	Re-marks
P15	PENTANE, 3-aza-1,5-diamino-N,N,N',N",N"-pentaacetic acid		$C_{14}H_{23}O_{10}N_3$			$(HO_2CCH_2)_2N(CH_2)_2N$-$(CH_2CO_2H)(CH_2)_2N(CH_2CO_2H)_2$		L^{5-}
Mg^{2+}	12.6	9.09(1)	219.2	20	C	0.1M(KNO_3)	65Ac	
	15.1	9.09(1)	226	25	C	μ=0.1(KNO_3)	65Wd	
	10.9	9.10(1)	21	27	C	μ=0.1(KNO_3,HNO_3)	68Cb	
Ca^{2+}	-24.89	10.9(1)	125.1	20	C	0.1M(KNO_3)	65Ac	
	-25.5	10.7(1)	121	25	C	μ=0.1(KNO_3)	65Wd	
	-23.4	10.6(1)	126	27	C	μ=0.1(KNO_3,HNO_3)	68Cb	
Sr^{2+}	-31.4	9.6(1)	79	25	C	μ=0.1(KNO_3)	65Wd	
	-28.0	9.5(1)	88	27	C	μ=0.1(KNO_3,HNO_3)	68Cb	
Ba^{2+}	-30.5	8.50(1)	58	25	C	μ=0.1(KNO_3)	65Wd	
	-28.0	8.52(1)	88	27	C	μ=0.1(KNO_3,HNO_3)	68Cb	
Y^{3+}	-21.8	22.05(1)	348.9	25	T	0.1M(KNO_3)	62Mh	
	-27.2	22.0(1)	330	27	C	0.1M(KNO_3)	68Cc	
	22	22.21(1)	498	10	T	--	72Ya	q:±4
	-22	22.21(1)	347	20	T	--	72Ya	q:±4
	-66.5	21.95(1)	201	30	T	--	72Ya	q:±4
La^{3+}	-19.7	19.6(1)	307.5	20	C	0.1M(KNO_3)	65Ac	
	-23.8	19.48(1)	292.9	25	T	0.1M(KNO_3)	62Mh	
	-21.8	19.4(1)	301	27	C	0.1M(KNO_3)	68Cc	
	-4.6	19.68(1)	360	10	T	--	72Ya	q:±4
	-24	19.59(1)	293	20	T	--	72Ya	q:±4
	-43.9	19.39(1)	222	30	T	--	72Ya	q:±4
Ce^{3+}	-24.06	(1)	--	27	C	μ=0.10	68Cd	
Pr^{3+}	-29.7	21.07(1)	301.2	25	T	0.1M(KNO_3)	62Mh	
	-26.99	21.0(1)	313.0	27	C	μ=0.10	68Cd	
	-1.2	21.32(1)	402	10	T	--	72Ya	q:±4
	-30	21.23(1)	301	20	T	--	72Ya	q:±4
	-61.9	20.96(1)	197	30	T	--	72Ya	q:±4
Nd^{3+}	-24.3	21.60(1)	332.2	25	T	0.1M(KNO_3)	62Mh	
	-29.7	21.6(1)	313.8	27	C	μ=0.10	68Cd	
	-3.8	21.81(1)	402	10	T	--	72Ya	q:±4
	-25	21.72(1)	330	20	T	--	72Ya	q:±4
	-46.4	21.51(1)	259	30	T	--	72Ya	q:±4
Sm^{3+}	-34.3	22.34(1)	312.5	25	T	0.1M(KNO_3)	62Mh	
	-33.0	22.3(1)	316.7	27	C	μ=0.10	68Cd	
	-10	22.60(1)	397	10	T	--	72Ya	q:±4
	-34	22.46(1)	314	20	T	--	72Ya	q:±4
	-57.7	22.19(1)	234	30	T	--	72Ya	q:±4
Eu^{3+}	-33.9	22.39(1)	315.0	25	T	0.1M(KNO_3)	62Mh	
	-33.0	22.4(1)	318.0	27	C	μ=0.10	68Cd	
	-7.1	22.68(1)	406	10	T	--	72Ya	q:±4
	-34	22.55(1)	314	20	T	--	72Ya	q:±4
	-61.1	22.27(1)	222	30	T	--	72Ya	q:±4
Gd^{3+}	-31.4	22.46(1)	324.7	25	T	0.1M(KNO_3)	62Mh	
	-32.6	22.4(1)	320.9	27	C	μ=0.10	63Cd	
	-12	22.72(1)	393	10	T	--	72Ya	q:±4
	-31	22.59(1)	322	20	T	--	72Ya	q:±4
	-54.4	22.34(1)	247	30	T	--	72Ya	q:±4

P15, continued

Metal Ion	ΔH, kJ /mol	Log K	ΔS, J /mol K	T,°C	M	Conditions	Ref.	Remarks
Tb^{3+}	-32.2	22.71(1)	326.8	25	T	0.1M(KNO_3)	62Mh	
	-32.2	22.6	326.4	27	C	μ=0.10	68Cd	
	6.7	22.97(1)	402	10	T	--	72Ya	q:±4
	-33	22.89(1)	326	20	T	--	72Ya	q:±4
	-72.8	22.58(1)	188	30	T	--	72Ya	q:±4
Dy^{3+}	-33.5	22.82(1)	324.7	25	T	0.1M(KNO_3)	62Mh	
	-33.0	22.8(1)	326.4	27	C	μ=0.10	68Cd	
	2.1	23.10(1)	418	10	T	--	72Ya	q:±4
	-34	23.0(1)	322	20	T	--	72Ya	q:±4
	-71.1	22.69(1)	197	30	T	--	72Ya	q:±4
Ho^{3+}	-31.8	22.78(1)	329.3	25	T	0.1M(KNO_3)	62Mh	
	-31.4	22.7(1)	330.5	27	C	μ=0.10	68Cd	
	3.8	23.04(1)	448	10	T	--	72Ya	q:±4
	-32	22.95(1)	330	20	T	--	72Ya	q:±4
	-69.4	22.65(1)	201	30	T	--	72Ya	q:±4
Er^{3+}	-30.5	22.74(1)	333.0	25	T	0.1M(KNO_3)	62Mh	
	-31.0	22.7(1)	331.8	27	C	μ=0.10	68Cd	
	-2.5	22.99(1)	444	10	T	--	72Ya	q:±4
	-14	22.89(1)	389	20	T	--	72Ya	q:±4
	-62.3	22.62(1)	226	30	T	--	72Ya	q:±4
Tm^{3+}	-23.0	22.72(1)	357.7	25	T	0.1M(KNO_3)	62Mh	
	-27.6	22.7(1)	343.1	27	C	μ=0.10	68Cd	
	10.5	22.90(1)	473	10	T	--	72Ya	q:±4
	-24	22.86(1)	356	20	T	--	72Ya	q:±4
	-58.2	22.62(1)	238	30	T	--	72Ya	q:±4
Yb^{3+}	-23.0	22.62(1)	356.0	25	T	0.1M(KNO_3)	62Mh	
	-25.9	22.6(1)	346.0	27	C	μ=0.10	68Cd	
	4.2	22.81(1)	452	10	T	--	72Ya	q:±4
	-23	22.75(1)	356	20	T	--	72Ya	q:±4
	-51.5	22.53(1)	255	30	T	--	72Ya	q:±4
Lu^{3+}	-19.2	22.44(1)	364.8	25	T	0.1M(KNO_3)	62Mh	
	-21.3	22.4(1)	358.6	27	C	μ=0.10	68Cd	
	17	22.58(1)	489	10	T	--	72Ya	q:±4
	-20	22.57(1)	360	20	T	--	72Ya	q:±4
	-58.2	22.34(1)	234	30	T	--	72Ya	q:±4
Th^{4+}	-28.6	3.81(30°)	-21.4	30&35	T	μ=0.1M(KNO_3) R:Th+A+L=ThAL A=Iminodiacetic acid	75Pa	a,b
	-17.9	3.94(30°)	16.4	30&35	T	μ=0.1M(KNO_3) R:Th+A+L=ThAL A=Iminotriacetic acid	75Pa	a,b
	26.3	-8.60(35°)	-95	30&35	T	μ=0.1M(KNO_3) R:Th+A+L=ThAL A=OH	75Pa	a,b
Mn^{2+}	-30.04	15.60(1)	196.6	20	C	0.1M(KNO_3)	65Ac	
	-31.4	15.5(1)	192	25	C	μ=0.1(KNO_3)	65Wd	
Fe^{2+}	-32.2	16.4(1)	205	25	C	μ=0.1(KNO_3)	65Wd	

Metal Ion	ΔH, kJ /mol	Log K	ΔS, J /mol K	T,°C	M	Conditions	Ref.	Remarks
P15, continued								
Co^{2+}	-39.37	19.26(1)	235.1	20	C	0.1M(KNO_3)	65Ac	
	-39.7	19.1(1)	234	25	C	μ=0.1(KNO_3)	65Wd	
Ni^{2+}	-49.0	20.2(1)	220.5	20	C	0.1M(KNO_3)	65Ac	
	-46.9	20.0(1)	226	25	C	μ=0.1(KNO_3)	65Wd	
Cu^{2+}	-56.9	21.5(1)	218.4	20	C	0.1M(KNO_3)	65Ac	
	-56.1	21.3(1)	222	25	C	μ=0.1(KNO_3)	65Wd	
Zn^{2+}	-36.8	18.54(1)	230.1	20	C	0.1M(KNO_3)	65Ac	
	-44.4	18.8(1)	209	25	C	μ=0.1(KNO_3)	65Wd	
Cd^{2+}	-51.67	19.3(1)	194.1	20	C	0.1M(KNO_3)	65Ac	
	-51.9	18.9(1)	188	25	C	μ=0.1(KNO_3)	65Wd	
Hg^{2+}	-99.2	26.7(1)	172.8	20	C	0.1M(KNO_3)	65Ac 62Mh	
	-98.7	26.9(1)	184	25	C	μ=0.1(KNO_3)	65Wd 66Mf	
Al^{3+}	35.1	18.51(1)	472.8	25	T	μ=0.1(KNO_3)	66Mf	
	31.0	18.6(1)	460	27	C	μ=0.1(KNO_3)	68Cc	
Pb^{2+}	-78.6	18.9(1)	91.2	25	C	0.1M(KNO_3)	65Ac	
	-78.6	18.5(1)	92	27	C	μ=0.1(KNO_3)	65Wd	
P16	PENTANE, 1,5-diamino-3-oxa-N,N',N'-tetraacetic acid		$C_{12}H_{20}O_5N_2$			$O[CH_2CH_2N(CH_2CO_2H)_2]_2$		L^{4-}
Mg^{2+}	15.1	8.4(1)	209	25	C	μ=0.1(KNO_3)	65Wd	
Ca^{2+}	-26.8	9.9(1)	100	25	C	μ=0.1(KNO_3)	65Wd	
Sr^{2+}	-33.9	8.6(1)	50	25	C	μ=0.1(KNO_3)	65Wd	
Ba^{2+}	-27.2	8.1(1)	63	25	C	μ=0.1(KNO_3)	65Wd	
Mn^{2+}	-23.4	13.0(1)	172	25	C	μ=0.1(KNO_3)	65Wd	
Fe^{2+}	-26.8	14.7(1)	192	25	C	μ=0.1(KNO_3)	65Wd	
Co^{2+}	-27.6	14.6(1)	188	25	C	μ=0.1(KNO_3)	65Wd	
Ni^{2+}	-213.4	14.6(1)	209	25	C	μ=0.1(KNO_3)	65Wd	
Cu^{2+}	-40.6	17.5(1)	201	25	C	μ=0.1(KNO_3)	65Wd	
Zn^{2+}	-31.8	15.2(1)	184	25	C	μ=0.1(KNO_3)	65Wd	
Cd^{2+}	-40.6	16.1(1)	172	25	C	μ=0.1(KNO_3)	65Wd	
Hg^{2+}	-81.6	23.0(1)	167	25	C	μ=0.1(KNO_3)	65Wd	
Pb^{2+}	-51.0	14.4(1)	105	25	C	μ=0.1(KNO_3)	65Wd	
P17	PENTANE, 1,3-diamino-N,N'-(2-pyridylmethyl) (bis)		$C_{15}H_{22}N_4$			$(C_5H_5N)CH_2NHCH_2CH_2CH_2$-$NHCH_2(NC_5H_5)$		
Mn^{2+}	-13.4	4.47(1)	40.6	25	C	μ=0.1(KNO_3)	75Ae	
Co^{2+}	-55.2	11.21(1)	29.3	25	C	μ=0.1(KNO_3)	75Ae	
Ni^{2+}	-68.2	14.22(1)	43.5	25	C	μ=0.1(KNO_3)	75Ae	

Metal Ion	ΔH, kJ /mol	Log K	ΔS, J /mol K	T,°C	M	Conditions	Ref.	Re-marks
P17, continued								
Cu^{2+}	-84.1	18.32(1)	68.6	25	C	μ=0.1(KNO_3)	75Ae	
Ag^{+}	-45.6	6.30(1)	-32.2	25	C	μ=0.1(KNO_3)	75Ae	
Zn^{2+}	-40.6	10.34(1)	61.9	25	C	μ=0.1(KNO_3)	75Ae	
Cd^{2+}	-32.6	8.58(1)	54.8	25	C	μ=0.1(KNO_3)	75Ae	
Hg^{2+}	-94.6	18.76(1)	42.2	25	C	μ=0.1(KNO_3)	75Ae	
P18	PENTANE, 1,5-diamino-N,N,N',N'-tetraacetic acid		$C_{13}H_{22}O_8N_2$			$(CH_2CO_2H)_2N(CH_2)_5N-(CH_2CO_2H)_2$		L^{4-}
Mn^{2+}	3.8	8.7(1)	180	20	C	μ=0.1(KNO_3)	64Ab	
Co^{2+}	-13.0	13.38(1)	212.1	20	C	μ=0.1(KNO_3)	64Ab	
Ni^{2+}	-28.0	13.9(1)	170.7	20	C	μ=0.1(KNO_3)	64Ab	
Cu^{2+}	-45.6	16.24(1)	155.6	20	C	μ=0.1(KNO_3)	64Ab	
Zn^{2+}	-11.3	12.67(1)	204.2	20	C	μ=0.1(KNO_3)	64Ab	
Cd^{2+}	-18.66	11.6(1)	156.9	20	C	μ=0.1(KNO_3)	64Ab	
P19	PENTANE, 1,5-diamino-3-thio		$C_4H_{12}N_2S$			$H_2N(CH_2)_2S(CH_2)_2NH_2$		L
Ni^{2+}	-41.9	7.377(1)	0.7	25	C	μ=0.5M(KNO_3)	79Ha	
	-48.6	6.142(2)	-45	25	C	μ=0.5M(KNO_3)	79Ha	
Cu^{2+}	-51.4	9.020(1)	0.2	25	C	μ=0.5M(KNO_3)	79Ha	
	-33.7	5.238(2)	-12	25	C	μ=0.5M(KNO_3)	79Ha	
	-18.0	5.90	52	25	C	μ=0.5M(KNO_3) R: $CuL^{2+}+OH^{-}=Cu(OH)L^{+}$	79Ha	
P20	PENTANEDIOIC ACID		$C_5H_8O_4$			$HO_2C(CH_2)_3CO_2H$		L^{-2}
Cd^{2+}	-31.2	2.873(30°)	-47.6	30&40	T	μ=2M($NaClO_4$) R: $Cd^{2+}+L^{2-}+A^{2-}=CdLA^{2-}$ A=Hexanedioate	75Ba	a,b
	-10.5	2.727(30°)	17.6	30&40	T	μ=2M($NaClO_4$) R: $Cd^{2+}+L^{2-}+2A^{2-}=CdLA_2{}^{2-}$ A=Hexanedioate	75Ba	a,b
	-39.4	3.061(30°)	-71.0	30&40	T	μ=2M($NaClO_4$) R: $Cd^{2+}+2L^{2-}+A^{2-}=CdL_2A^{2-}$ A=Hexanedioate	75Ba	a,b
P21	2,4-PENTANEDIONE		$C_5H_8O_2$			$CH_3COCH_2COCH_3$		L^{-}
Be^{2+}	-8.4	7.8(30°,1)	121	10-40	T	μ=0	55Ia	b
	-28.9	6.7(30°,2)	33.4	10-40	T	μ=0	55Ia	b

Metal Ion	ΔH, kJ /mol	Log K	ΔS, J /mol K	T,°C	M	Conditions	Ref.	Remarks
P21, continued								
Mg^{2+}	-7.5	3.6(30°,1)	46	10-40	T	μ=0	55Ia	b
	-18.0	3.1(30°,2)	10.5	10-40	T	μ=0	55Ia	b
La^{3+}	-1.30	4.809(1)	87.9	25	C	μ=0.2($NaNO_3$)	76Ki	q:±0.7
	-1.42	3.350(2)	59.0	25	C	μ=0.2($NaNO_3$)	76Ki	q:±1.5
	-3.14	2.448(3)	37.2	25	C	μ=0.2($NaNO_3$)	76Ki	q:±0.8
Ce^{3+}	0	5.3(30°,1)	100	10-40	T	μ=0	55Ia	b
	0.8	4.0(30°,2)	75	10-40	T	μ=0	55Ia	b
Pr^{3+}	10.04	4.32(25°,1)	116.3	25-51	T	μ=1M($LiClO_4$) S:50v%Propanol	76Kk	j,q:±1.7
Nd^{3+}	-2.51	5.153(1)	90.0	25	C	μ=0.2($NaNO_3$)	76Ki	q:±0.7
	-2.93	3.995(2)	66.9	25	C	μ=0.2($NaNO_3$)	76Ki	q:±1.5
	-10.0	3.152(3)	27.2	25	C	μ=0.2($NaNO_3$)	76Ki	q:±0.8
Sm^{3+}	-3.26	5.439(1)	93.3	25	C	μ=0.2($NaNO_3$)	76Ki	q:±0.7
	-8.24	4.362(2)	55.6	25	C	μ=0.2($NaNO_3$)	76Ki	q:±1.5
	-8.58	2.852(3)	26.8	25	C	μ=0.2($NaNO_3$)	76Ki	q:±0.8
Gd^{3+}	-4.35	5.747(1)	95.4	25	C	μ=0.2($NaNO_3$)	76Ki	q:±0.7
	-9.04	4.384(2)	53.6	25	C	μ=0.2($NaNO_3$)	76Ki	q:±1.5
	-11.72	3.357(3)	26.8	25	C	μ=0.2($NaNO_3$)	76Ki	q:±0.8
Dy^{3+}	-5.23	5.879(1)	95.0	25	C	μ=0.2($NaNO_3$)	76Ki	q:±0.7
	-9.62	4.574(2)	55.2	25	C	μ=0.2($NaNO_3$)	76Ki	q:±1.5
	-16.5	3.284(3)	8.8	25	C	μ=0.2($NaNO_3$)	76Ki	q:±0.8
Ho^{3+}	-4.73	5.901(1)	97.1	25	C	μ=0.2($NaNO_3$)	76Ki	q:±0.7
	-9.41	4.582(2)	56.1	25	C	μ=0.2($NaNO_3$)	76Ki	q:±1.5
	-16.8	3.350(3)	5.8	25	C	μ=0.2($NaNO_3$)	76Ki	q:±0.8
Yb^{3+}	-4.22	6.033(1)	101.2	25	C	μ=0.2($NaNO_3$)	76Ki	q:±0.7
	-8.79	4.758(2)	61.5	25	C	μ=0.2($NaNO_3$)	76Ki	q:±1.5
	-10.0	3.548(3)	33.9	25	C	μ=0.2($NaNO_3$)	76Ki	q:±0.8
ZrA_2L_2	0.21	0.806(25°)	14.2	7.5-53.1	T	S:Benzene A=Trifluoroacetylacetone R:ZrA_2L_2+ZrL_4=2$ZrAL_3$	66Pf	b,q: ±2.2
	-0.42	0.755(25°)	13.0	18-79.8	T	S:CCl_4 A=Trifluoroacetylacetone R:ZrA_2L_2+ZrL_4=2$ZrAL_3$	66Pf	b,q: ±0.79
	0.08	0.843(25°)	16.3	7.5-63.9	T	S:Benzene A=Trifluoroacetylacetone R:ZrA_2L_2+ZrA_4=2ZrA_3L	66Pf	b,q: ±1.1
	-0.84	0.894(25°)	14.2	-18.5-51.6	T	S:CCl_4 A=Trifluoroacetylacetone R:ZrA_2L_2+ZrA_4=2ZrA_3L	66Pf	b,q: ±1.0
ZrA_3L	-0.21	0.806(25°)	14.6	7.5-77.8	T	S:Benzene A=Trifluoroacetylacetone R:ZrA_3L+	66Pf	b,q: ±0.75

Metal Ion	ΔH, kJ /mol	Log K	ΔS, J /mol K	T,°C	M	Conditions	Ref.	Remarks
P21, continued								
ZrA_3L, cont.	1.05	0.799(25°)	18.8	-18.5-63.4	T	$ZrAL_3$=$2ZrA_2L_2$ S:CCl_4 A=Trifluoroacetylacetone R:ZrA_3L+$ZrAL_3$=$2ZrA_2L_2$	66Pf	b,q: ±0.75
VO^{++}	-7.1	8.65(25°)	142.2	25-40	T	R:VO^{++}+HL=VOL^++H^+	56Td	b
	-28.4	7.04(25°)	40.2	25-40	T	R:VOL^++HL=VOL_2+H^+	56Td	b
UO_2^{2+}	-16.7	5.3(30°,2)	67	10-40	T	μ=0	55Ia	b
NpO_2^+	-26.4	4.08(25°,1)	-9.6	18-32	T	μ=0.1M	72Gb	b,q:±4.2
	-13.4	2.92(25°,2)	10.9	18-32	T	μ=0.1M	72Gh	b,q:±4.2
Mn^{2+}	-10.5	4.2(30°,1)	46	10-40	T	μ=0	55Ia	b
	-6.3	4.07(1)	58	25	C	μ=0.1($NaClO_4$)	68Gg	
	-2.5	4.24(1)	79	0	T	S:1-Propanol-H_2O Mix 1/D=0.0113	71Da	
	-1.7	4.18(1)	73.6	25	T	S:1-Propanol-H_2O Mix 1/D=0.0127	71Da	
	-2.1	4.26(1)	70.7	40	T	S:1-Propanol-H_2O Mix 1/D=0.0137	71Da	
	-2.5	4.48(1)	83.7	0	T	S:1-Propanol-H_2O Mix 1/D=0.0131	71Da	
	-0.4	4.40(1)	80.8	25	T	S:1-Propanol-H_2O Mix 1/D=0.0149	71Da	
	-0.8	4.47(1)	81.2	40	T	S:1-Propanol-H_2O Mix 1/D=0.0159	71Da	
	-1.2	4.80(1)	94.1	0	T	S:1-Propanol-H_2O Mix 1/D=0.0160	71Da	
	0.8	4.76(1)	93.3	25	T	S:1-Propanol-H_2O Mix 1/D=0.0183	71Da	
	0.4	4.82(1)	87.9	40	T	S:1-Propanol-H_2O Mix 1/D=0.0199	71Da	
	-1.2	5.04(1)	100.0	0	T	S:1-Propanol-H_2O Mix 1/D=0.0182	71Da	
	2.1	5.13(1)	101.2	25	T	S:1-Propanol-H_2O Mix 1/D=0.0213	71Da	
	1.7	5.10(1)	98.7	40	T	S:1-Propanol-H_2O Mix 1/D=0.0230	71Da	
	0.4	5.44(1)	110.0	0	T	S:1-Propanol-H_2O Mix 1/D=0.212	71Da	
	2.9	5.50(1)	113.4	25	T	S:1-Propanol-H_2O Mix 1/D=0.0247	71Da	
	3.3	5.51(1)	112.5	40	T	S:1-Propanol-H_2O Mix 1/D=0.0269	71Da	
	1.7	5.84(1)	121.8	0	T	S:1-Propanol-H_2O Mix 1/D=0.0248	71Da	
	4.6	6.01(1)	129.7	25	T	S:1-Propanol-H_2O Mix 1/D=0.0294	71Da	
	5.4	6.00(1)	128.4	40	T	S:1-Propanol-H_2O Mix 1/D=0.0323	71Da	
	3.3	6.40(1)	138.9	0	T	S:1-Propanol-H_2O Mix 1/D=0.0297	71Da	
	7.9	6.60(1)	149.4	25	T	S:1-Propanol-H_2O	71Da	

Metal Ion	ΔH, kJ /mol	Log K	ΔS, J /mol K	T,°C	M	Conditions	Ref.	Re-marks
P21, continued								
Mn^{2+},						Mix 1/D=0.0352		
cont.	9.2	6.70(1)	154.8	40	T	S:1-Propanol-H_2O Mix 1/D=0.0392	71Da	
	-19.7	3.0(30°,2)	-6.3	10-40	T	μ=0	55Ia	b
	-7.5	3.12(2)	29.7	0	T	S:1-Propanol-H_2O Mix 1/D=0.0113	71Da	
	-5.8	3.00(2)	34.3	25	T	S:1-Propanol-H_2O Mix 1/D=0.0127	71Da	
	-5.4	3.00(2)	37.6	40	T	S:1-Propanol-H_2O Mix 1/D=0.0137	71Da	
	-6.7	3.36(2)	36.4	0	T	S:1-Propanol-H_2O Mix 1/D=0.0131	71Da	
	-5.4	3.22(2)	41.0	25	T	S:1-Propanol-H_2O Mix 1/D=0.0149	71Da	
	-4.6	3.14(2)	44.8	40	T	S:1-Propanol-H_2O Mix 1/D=0.0159	71Da	
	-6.3	3.60(2)	48.5	0	T	S:1-Propanol-H_2O Mix 1/D=0.0160	71Da	
	-4.6	3.52(2)	52.7	25	T	S:1-Propanol-H_2O Mix 1/D=0.0183	71Da	
	-1.7	3.49(2)	57.7	40	T	S:1-Propanol-H_2O Mix 1/D=0.0199	71Da	
	-5.0	3.84(2)	56.5	0	T	S:1-Propanol-H_2O Mix 1/D=0.0182	71Da	
	-1.7	3.74(2)	64.4	25	T	S:1-Propanol-H_2O Mix 1/D=0.0213	71Da	
	-0.8	3.77(2)	69.0	40	T	S:1-Propanol-H_2O Mix 1/D=0.0230	71Da	
	-3.3	4.08(2)	66.9	0	T	S:1-Propanol-H_2O Mix 1/D=0.0212	71Da	
	0.0	4.10(2)	76.6	25	T	S:1-Propanol-H_2O Mix 1/D=0.0247	71Da	
	2.5	4.12(2)	82.8	40	T	S:1-Propanol-H_2O Mix 1/D=0.0269	71Da	
	-2.5	4.48(2)	79.9	0	T	S:1-Propanol-H_2O Mix 1/D=0.0248	71Da	
	2.5	4.47(2)	95.0	25	T	S:1-Propanol-H_2O Mix 1/D=0.0294	71Da	
	5.8	4.54(2)	102.1	40	T	S:1-Propanol-H_2O Mix 1/D=0.0323	71Da	
	0.8	4.96(2)	97.9	0	T	S:1-Propanol-H_2O Mix 1/D=0.0297	71Da	
	2.5	5.06(2)	118.0	25	T	S:1-Propanol-H_2O Mix 1/D=0.0353	71Da	
	9.6	5.16(2)	128.9	40	T	S:1-Propanol-H_2O Mix 1/D=0.0392	71Da	
MnA_3	-4.895	5.094	81.2	25	T	R:$MnA_3+2H_2O+H^+=$ $MnA_2(H_2O)_2^++HL$ A=L	52Ca	
	12.678	3.92	-32.6	25	T	R:$MnA_3+2H_2O=$ $MnA_2(H_2O)_2^++A^-$ A=L	52Ca	

Metal Ion	ΔH, kJ /mol	Log K	ΔS, J /mol K	T,°C	M	Conditions	Ref.	Re-marks
P21, continued								
Co^{2+}	-5.0	5.4(30°,1)	88	10-40	T	μ=0	55Ia	b
	-20.9	4.1(30°,2)	10.0	10-40	T	μ=0	55Ia	b
CoL_2AB	-13.0	0.394(29°)	-34.7	29-60	T	S:$CdCl_3$ CM=0.1M A=Pyridine B=N_3^- R:CoL_2AB(trans)= CoL_2AB(cis)	73Hd	b,q: ±8.8
	-23.0	0.570(29°)	-87.0	29-60	T	S:$CDCl_3$ CM=0.1M A=Pyridine B=NCO^- R:CoL_2AB(trans)= CoL_2AB(cis)	73Hd	b,q: ±13.8
	-13.8	-0.226(29°)	-50.2	29-60	T	S:$CDCl_3$ CM=0.1M A=Pyridine B=NO_2^- R:CoL_2AB(trans)= CoL_2AB(cis)	73Hd	b,q: ±3.8
Ni^{2+}	-23.4	6.3(1)	35.1	0	T	--	68Gb	
	-18.8	6.1(1)	53.6	25	T	--	68Gb	
	-15.9	5.9(1)	61.5	40	T	--	68Gb	
	-28.0	5.9(30°,1)	50	10-40	T	μ=0	55Ia	b
	-19.2	5.69(25°,1)	46	15-40	T	μ=0	71Yb	b,q:±2.1
	-14.2	5.72(1)	63	25	C	μ=0.1($NaClO_4$)	68Gg	
	-8.8	8.77(25°,1)	138	15-40	T	μ=0 S:75v% Dioxane	71Yb	b,q:±2.1
	-21.8	6.6(1)	46.0	0	T	0.098 w fraction CH_3OH	68Gb	
	-16.7	6.3(1)	64.4	25	T	0.099 w fraction CH_3OH	68Gb	
	-13.4	6.2(1)	74.9	40	T	0.099 w fraction CH_3OH	68Gb	
	-19.2	6.8(1)	59.8	0	T	0.220 w fraction CH_3OH	68Gb	
	-13.0	6.7(1)	84.1	25	T	0.222 w fraction CH_3OH	68Gb	
	-8.4	6.5(1)	97.5	40	T	0.224 w fraction CH_3OH	68Gb	
	-17.6	7.0(1)	69.0	0	T	0.289 w fraction CH_3OH	68Gb	
	-10.5	6.8(1)	93.4	25	T	0.295 w fraction CH_3OH	68Gb	
	-5.4	6.6(1)	112.5	40	T	0.299 w fraction CH_3OH	68Gb	
	-15.5	7.2(1)	81.2	0	T	0.374 w fraction CH_3OH	68Gb	
	-7.5	7.1(1)	110.9	25	T	0.383 w fraction CH_3OH	68Gb	
	-1.7	7.0(1)	128.4	40	T	0.387 w fraction CH_3OH	68Gb	
	-12.6	7.4(1)	96.6	0	T	0.474 w fraction CH_3OH	68Gb	
	-3.3	7.4(1)	130.5	25	T	0.485 w fraction CH_3OH	68Gb	
	4.2	7.4(1)	154.8	40	T	0.489 w fraction CH_3OH	68Gb	
	-9.2	7.8(1)	115.1	0	T	0.590 w fraction	68Gb	

Metal Ion	ΔH, kJ /mol	Log K	ΔS, J /mol K	T,°C	M	Conditions	Ref.	Remarks
P21, continued								
Ni^{2+},						CH_3OH		
cont.	1.7	7.8(1)	155.6	25	T	0.610 w fraction CH_3OH	68Gb	
	10.5	8.0(1)	185.4	40	T	0.621 w fraction CH_3OH	68Gb	
	-24.3	6.3(1)	32.2	0	T	0 w% 1-propanol	68Gb	
	-22.2	6.0(1)	40.2	25	T	0 w% 1-propanol	68Gb	
	-20.9	5.8(1)	43.1	40	T	0 w% 1-propanol	68Gb	
	-23.0	6.6(1)	41.0	0	T	0.056 w fraction 1-propanol	68Gb	
	-20.5	6.2(1)	50.6	25	T	0.056 w fraction 1-propanol	68Gb	
	-19.2	6.0(1)	54.0	40	T	0.056 w fraction 1-propanol	68Gb	
	-21.3	6.9(1)	53.6	0	T	0.132 w fraction 1-propanol	68Gb	
	-17.2	6.6(1)	68.2	25	T	0.134 w fraction 1-propanol	68Gb	
	-15.5	6.4(1)	74.0	40	T	0.134 w fraction 1-propanol	68Gb	
	-19.2	7.1(1)	64.8	0	T	0.183 w fraction 1-propanol	68Gb	
	-14.2	6.9(1)	83.7	25	T	0.188 w fraction 1-propanol	68Gb	
	-12.6	6.8(1)	90.0	40	T	0.188 w fraction 1-propanol	68Gb	
	-17.2	7.4(1)	79.1	0	T	0.251 w fraction 1-propanol	68Gb	
	-11.3	7.3(1)	101.7	25	T	0.256 w fraction 1-propanol	68Gb	
	-8.4	7.1(1)	109.6	40	T	0.256 w fraction 1-propanol	68Gb	
	-14.2	7.8(1)	97.5	0	T	0.334 w fraction 1-propanol	68Gb	
	-6.3	7.8(1)	127.2	25	T	0.346 w fraction 1-propanol	68Gb	
	-3.3	7.6(1)	136.0	40	T	0.346 w fraction 1-propanol	68Gb	
	-9.6	8.3(1)	124.3	0	T	0.450 w fraction 1-propanol	68Gb	
	-0.4	8.4(1)	160.2	25	T	0.473 w fraction 1-propanol	68Gb	
	4.2	8.3(1)	172.0	40	T	0.473 w fraction 1-propanol	68Gb	
	-28.0	5.1(2)	-7.5	0	T	--	68Gb	
	-26.4	4.7(2)	1.3	25	T	--	68Gb	
	-23.8	4.5(2)	9.2	40	T	--	68Gb	
	-26.4	4.5(30°,2)	-1.7	10-40	T	μ=0	55Ia	b
	-22.2	4.47(25°,2)	12	15-40	T	μ=0	71Yb	b,q:±2.1
	-9.6	7.21(25°,2)	105	15-40	T	μ=0 S:75v%Dioxane	71Yb	b,q:±2.1
	-27.6	5.3(2)	-0.8	0	T	0.098 w fraction CH_3OH	68Gb	
	-24.7	4.8(2)	9.6	25	T	0.099 w fraction CH_3OH	68Gb	

Metal Ion	ΔH, kJ /mol	Log K	ΔS, J /mol K	T,°C	M	Conditions	Ref.	Re-marks
P21, continued								
Ni^{2+}, cont.	-21.8	4.6(2)	18.8	40	T	0.099 w fraction CH_3OH	68Gb	
	-25.9	5.4(2)	7.9	0	T	0.220 w fraction CH_3OH	68Gb	
	-22.6	5.1(2)	20.9	25	T	0.222 w fraction CH_3OH	68Gb	
	-18.8	4.8(2)	32.2	40	T	0.224 w fraction CH_3OH	68Gb	
	-25.1	5.5(2)	13.8	0	T	0.289 w fraction CH_3OH	68Gb	
	-20.9	5.1(2)	28.0	25	T	0.295 w fraction CH_3OH	68Gb	
	-17.6	5.0(2)	40.2	40	T	0.299 w fraction CH_3OH	68Gb	
	-23.8	5.6(2)	19.7	0	T	0.374 w fraction CH_3OH	68Gb	
	-19.7	5.3(2)	35.1	25	T	0.383 w fraction CH_3OH	68Gb	
	-14.6	5.2(2)	52.3	40	T	0.387 w fraction CH_3OH	68Gb	
	-22.2	5.8(2)	29.7	0	T	0.474 w fraction CH_3OH	68Gb	
	-16.7	5.5(2)	49.0	25	T	0.485 w fraction CH_3OH	68Gb	
	-10.9	5.5(2)	6.94	40	T	0.489 w fraction CH_3OH	68Gb	
	-20.1	6.0(2)	41.4	0	T	0.590 w fraction CH_3OH	68Gb	
	-13.8	5.8(2)	64.4	25	T	0.610 w fraction CH_3OH	68Gb	
	-6.7	5.7(2)	88.3	40	T	0.621 w fraction CH_3OH	68Gb	
	-26.4	5.0(2)	0.8	0	T	0 w% 1-propanol	68Gb	
	-24.3	4.5(2)	5.4	25	T	0 w% 1-propanol	68Gb	
	-23.8	4.4(2)	7.1	40	T	0 w% 1-propanol	68Gb	
	-25.9	5.1(2)	4.2	0	T	0.056 w fraction 1-propanol	68Gb	
	-23.0	4.8(2)	14.2	25	T	0.056 w fraction 1-propanol	68Gb	
	-22.6	4.6(2)	15.1	40	T	0.056 w fraction 1-propanol	68Gb	
	-23.8	5.4(2)	15.9	0	T	0.132 w fraction 1-propanol	68Gb	
	-20.5	5.1(2)	28.4	25	T	0.134 w fraction 1-propanol	68Gb	
	-19.7	4.9(2)	31.4	40	T	0.134 w fraction 1-propanol	68Gb	
	-22.6	5.6(2)	24.7	0	T	0.183 w fraction 1-propanol	68Gb	
	-18.0	5.4(2)	41.8	25	T	0.188 w fraction 1-propanol	68Gb	
	-17.6	5.2(2)	43.5	40	T	0.188 w fraction 1-propanol	68Gb	
	-20.9	5.9(2)	37.2	0	T	0.251 w fraction	68Gb	

Metal Ion	ΔH, kJ /mol	Log K	ΔS, J /mol K	T, °C	M	Conditions	Ref.	Remarks
P21, continued								
Ni^{2+},						1-propanol		
cont.	-15.5	5.6(2)	56.9	25	T	0.256 w fraction 1-propanol	68Gb	
	-13.8	5.5(2)	60.2	40	T	0.256 w fraction 1-propanol	68Gb	
	-18.8	6.2(2)	51.5	0	T	0.334 w fraction 1-propanol	68Gb	
	-11.7	6.2(2)	77.8	25	T	0.346 w fraction 1-propanol	68Gb	
	-10.0	6.0(2)	81.6	40	T	0.346 w fraction 1-propanol	68Gb	
	-15.1	6.7(2)	73.2	0	T	0.450 w fraction 1-propanol	68Gb	
	-6.3	6.7(2)	106.7	25	T	0.473 w fraction 1-propanol	68Gb	
	-3.3	6.5(2)	113.8	40	T	0.473 w fraction 1-propanol	68Gb	
	-28.0	2.2(30°,3)	-50	10-40	T	μ=0	55Ia	b
	-31.8	9.66(1-2)	79	25	T	μ=0.1($NaClO_4$)	68Gg	
NiL_2	192.00	6.0	748.1	30	C	S:Benzene CM=0.007-0.070M R:$3NiL_2=Ni_3L_6$	73Ag	q
Pd^{2+}	-75	16.2(30°,1)	58	20-40	T	μ≅0.45	57De	b
	-75	10.9(30°,2)	-46	20-40	T	μ≅0.45	57De	b
Cu^{2+}	-19.7	8.2(30°,1)	92	10-40	T	μ=0	55Ia	b
	-29.7	8.29(25°,1)	58	15-40	T	μ=0.02[$N(CH_3)_4Cl$]	71Yb	b
	-20.1	8.16(1)	88	25	C	μ=0.1($NaClO_4$)	68Gg	
	-17.2	12.06(25°,1)	172	15-40	T	μ=0 S:75v% Dioxane	71Yb	b,q: ±2.1
	-27.6	6.7(30°,2)	38	10-40	T	μ=0	55Ia	b
	-32.2	6.70(25°,2)	21	15-40	T	μ=0.02[$N(CH_3)_4Cl$]	71Yb	b
	-17.6	10.43(25°,2)	142	15-40	T	μ=0 S:75v% Dioxane	71Yb	b,q: ±2.1
	-42.3	14.76(1-2)	142	25	C	μ=0.1($NaClO_4$)	68Gg	
CuL_2	16.7	--	54	0-60	T	S:Pyridine R:CuL_2(low temperature species) =CuL_2(high temperature species)	72Yc	b,q:±4
Zn^{2+}	-7.9	5.0(30°,1)	71	10-40	T	μ=0	55Ia	b
	-5.52	5.36(1)	84	25	C	μ=0.1	67Mc	
	-6.3	4.68(1)	67	25	C	μ=0.1($NaClO_4$)	68Gg	
	-6.3	5.20(1)	81.6	0	T	S:1-Propanol-H_2O Mix 1/D=0.0113	71Da	
	-4.2	5.06(1)	80.3	25	T	S:1-Propanol-H_2O Mix 1/D=0.0127	71Da	
	-2.9	5.02(1)	80.8	40	T	S:1-Propanol-H_2O Mix 1/D=0.0137	71Da	
	-5.0	5.36(1)	91.2	0	T	S:1-Propanol-H_2O Mix 1/D=0.0131	71Da	
	-2.9	5.28(1)	90.4	25	T	S:1-Propanol-H_2O	71Da	

Metal Ion	ΔH, kJ /mol	Log K	ΔS, J /mol K	T,°C	M	Conditions	Ref.	Remarks
P21, continued								
Zn^{2+}, cont.						Mix 1/D=0.0149		
	-1.2	5.30(1)	91.2	40	T	S:1-Propanol-H_2O Mix 1/D=0.0159	71Da	
	-2.9	5.68(1)	105.8	0	T	S:1-Propanol-H_2O Mix 1/D=0.0160	71Da	
	0.0	5.64(1)	107.5	25	T	S:1-Propanol-H_2O Mix 1/D=0.0183	71Da	
	2.9	5.72(1)	102.1	40	T	S:1-Propanol-H_2O Mix 1/D=0.0199	71Da	
	-1.7	5.92(1)	115.9	0	T	S:1-Propanol-H_2O Mix 1/D=0.0182	71Da	
	2.9	5.94(1)	123.4	25	T	S:1-Propanol-H_2O Mix 1/D=0.0213	71Da	
	5.8	6.00(1)	127.2	40	T	S:1-Propanol-H_2O Mix 1/D=0.0230	71Da	
	0.8	6.24(1)	131.4	0	T	S:1-Propanol-H_2O Mix 1/D=0.0212	71Da	
	6.3	6.30(1)	140.2	25	T	S:1-Propanol-H_2O Mix 1/D=0.0247	71Da	
	10.5	6.42(1)	147.3	40	T	S:1-Propanol-H_2O Mix 1/D=0.0269	71Da	
	4.2	6.64(1)	151.0	0	T	S:1-Propanol-H_2O Mix 1/D=0.0248	71Da	
	10.5	6.82(1)	164.0	25	T	S:1-Propanol-H_2O Mix 1/D=0.0294	71Da	
	15.9	6.98(1)	175.3	40	T	S:1-Propanol-H_2O Mix 1/D=0.0323	71Da	
	7.9	7.20(1)	176.1	0	T	S:1-Propanol-H_2O Mix 1/D=0.0297	71Da	
	16.3	7.40(1)	195.0	25	T	S:1-Propanol-H_2O Mix 1/D=0.0352	71Da	
	23.0	7.68(1)	211.7	40	T	S:1-Propanol-H_2O Mix 1/D=0.0392	71Da	
	-5.98	4.49(2)	67	25	C	μ=0.1	67Mc	
	4.2	4.64(2)	115.1	0	T	S:1-Propanol-H_2O Mix 1/D=0.0160	71Da	
	2.9	4.76(2)	97.1	25	T	S:1-Propanol-H_2O Mix 1/D=0.0183	71Da	
	2.1	4.75(2)	92.9	40	T	S:1-Propanol-H_2O Mix 1/D=0.0199	71Da	
	4.6	4.88(2)	109.2	0	T	S:1-Propanol-H_2O Mix 1/D=0.0182	71Da	
	2.1	4.98(2)	100.0	25	T	S:1-Propanol-H_2O Mix 1/D=0.0213	71Da	
	1.7	5.02(2)	96.6	40	T	S:1-Propanol-H_2O Mix 1/D=0.0230	71Da	
	4.2	5.20(2)	120.9	0	T	S:1-Propanol-H_2O Mix 1/D=0.0212	71Da	
	1.2	5.28(2)	104.2	25	T	S:1-Propanol-H_2O Mix 1/D=0.0247	71Da	
	1.2	5.30(2)	102.5	40	T	S:1-Propanol-H_2O Mix 1/D=0.0269	71Da	
	3.3	5.60(2)	125.9	0	T	S:1-Propanol-H_2O Mix 1/D=0.0248	71Da	

Metal Ion	ΔH, kJ /mol	Log K	ΔS, J /mol K	T,°C	M	Conditions	Ref.	Remarks
P21, continued								
Zn^{2+}, cont.	0.0	5.72(2)	109.2	25	T	S:1-Propanol-H_2O Mix 1/D=0.0294	71Da	
	0.0	5.72(2)	109.2	40	T	S:1-Propanol-H_2O Mix 1/D=0.0323	71Da	
	2.5	6.16(2)	133.0	0	T	S:1-Propanol-H_2O Mix 1/D=0.0297	71Da	
	-1.2	6.16(2)	117.2	25	T	S:1-Propanol-H_2O Mix 1/D=0.0353	71Da	
	-0.4	6.28(2)	118.8	40	T	S:1-Propanol-H_2O Mix 1/D=0.0392	71Da	
	-14.2	7.92(1-2)	100	25	C	μ=0.1($NaClO_4$)	68Gg	
Cd^{2+}	-5.8	3.8(30°,1)	54	10-40	T	μ=0	55Ia	b
	-3.97	3.59(1)	54	25	C	μ=0.1	67Mc	
	-7.91	2.55(2)	21	25	C	μ=0.1	67Mc	
AlA_3	-16.292	--	--	25	C	S:50v% Dioxane (0.174 mole fraction dioxane) R:AlA_3+3NaL=AlL_3+ 3NaA A=Cl	66Hb	
GaA_3	1.34	0.413(25°)	12.47	5.7-43	T	S:Benzene A=Dibenzoylmethane R:2/3GaA_3+ 1/3GaL_3=$GaLA_2$	69Fc	b,q: ±1.0
	-17.2	3.049(25°)	0.4	36-98	T	S:Benzene A=Hexafluoroacetylacetone R:2/3GaA_3+ 1/3GaL_3=$GaLA_2$	69Fc	b,q: ±3.3
	0.17	0.393(25°)	7.99	5.7-43	T	S:Benzene A=Dibenzoylmethane R:1/3GaA_3+ 2/3GaL_3=$GaAL_2$	69Fc	b,q: ±1.0
	-18.8	3.086(25°)	-3.8	36-98	T	S:Benzene A=Hexafluoroacetylacetone R:1/3GaA_3+ 2/3GaL_3=$GaAL_2$	69Fc	b,q:±4
P22	2,4-PENTANEDIONE, 3-benzyl-		$C_{12}H_{14}O_2$			$CH_3C(O)CH(CH_2C_6H_5)C(C)CH_3$		
NiL_2	113.8	4.0	452	30	C	S:Benzene CM=0.002-0.044M R:3NiL_2=Ni_3L_6	73Ag	
P23	2,4-PENTANEDIONE, 3-butyl-		$C_9H_{16}O_2$			$CH_3C(O)CH(C_4H_9)C(O)CH_3$		
NiL_2	204.2	2.9	728	30	C	S:Benzene CM=0.006-0.045M R:3NiL_2=Ni_3L_6	73Ag	

Metal Ion	ΔH, kJ /mol	Log K	ΔS, J /mol K	T,°C	M	Conditions	Ref.	Re-marks
P24	2,4-PENTANEDIONE, 3-phenyl		$C_{11}H_{12}O_2$			$CH_3C(O)CH(C_6H_5)C(O)CH_3$		
Ni^{2+}	-43.1	3.59(25°)	-75	10-50	T	S:Toluene R:$3NiL_2$=$[NiL_2]_3$	61Fa	b,q:±4.3
P25	2,4-PENTANEDIONE, 1,1,1-trifluoro-		$C_5H_5O_2F_3$			$CF_3C(O)CH_2C(O)CH_3$		
VL_3	-5.8	0.785	-4.2	25	T	S:$CdCl_3$ R:VL_3(cis)= VL_3(trans)	71Gg	q:±1.2
MnL_3	-4.6	0.643	-2.9	25	T	S:$CDCl_3$ R:MnL_3(cis)= MnL_3(trans)	71Gg	q:±1.7
CoL_3	-1.42	0.650(25°)	7.5	79.1-115		S:Chloroform R:CoL_3(cis)= CoL_3(trans)	63Fa	b,q: ±3.0
RuL_3	-4.6	0.690	2.5	150	T	S:sym-$C_2H_2Cl_4$ R:RuL_3(cis)= RuL_3(trans)	71Gg	q:±2.5
AlL_3	-1.00	0.658(25°)	9.2	-25.5-3		S:Chloroform R:AlL_3(cis) AlL_3(trans)	63Fa	b,q: ±0.79
	-1.97	0.682(25°)	6.3	-25.5-29.5		S:Cloroform R:AlL_3(cis)= AlL_3(trans)	63Fa	b,q: ±0.92
GaL_3	-1.97	0.672	6.3	25	T	S:$CDCl_3$ R:GaL_3(cis)= GaL_3(trans)	71Gg	q: ±0.92
P26	2,4-PENTANEDIONE, 1,1,1-trifluoro-, 4-imine, N-isopropyl		$C_8H_{11}ONF_3$			$CF_3C(O):CHC(CH_3):NCH_2CH_2CH_3$		
Ni^{2+}	-0.008	0.602(25°)	11.3	17&107	T	S:$CDCl_3$ R:NiL_2+ NiA_2=2NiAL Where A=N-iso-propylacetyl-acetoneimine	73La	a,b,q: ±0
P27	1-PENTANESULPHONIC ACID, 5-amino-		$C_5H_{13}O_3NS$			$H_2N(CH_2)_5SO_3H$		L^-
Ag^+	-23.18 -30.96	3.489(1) 3.908(2)	-10.88 -29.3	25 25	C C	μ=0.5M(KNO_3) μ=0.5M(KNO_3)	72Va 72Va	
P28	PENTANOIC ACID, nitrile		C_5H_9N			$CH_3(CH_2)_3CN$		L
I_2	-11.30	-0.303(25°,1)	-43.1	--	T	S:CCl_4 CL=0.2-1.0M CM=$2x10^{-3}$M	72Ma	j,u,q: ±1.5

Metal Ion	ΔH, kJ /mol	Log K	ΔS, J /mol K	T,°C	M	Conditions	Ref	Remarks
P28, continued								
I_2, cont.						Data analyzed by method 1		
	-12.26	-0.303(25°,1)	-45.98	--	T	S:CCl_4 CL=0.2-1.0M CM=$2x10^{-3}$M Data analyzed by method 2	72Ma	j,u
	-12.6	0.009(25°,1)	-41.0	25-55	T	S:Heptane CL=0.2-0.9M CM=$8x10^{-3}$-$1.6x10^{-3}$M	67Ma	b,j,q: ±1.2
	-17.78	0.009(25°,1)	-59.08	--	T	S:n-Heptane CL=0.2-1.0M CM=$2x10^{-3}$M Data analyzed by method 2	72Ma	j,u
P29 PENTANOIC ACID, 2-amino-		$C_5H_{11}O_2N$				$CH_3CH_2CH_2CH(NH_2)CO_2H$		
Mn^{2+}	-13.0	3.30(20°,1)	21	20-60	T	μ=0.1(KNO_3)	73Bh	b
	6.7	1.89(20°,2)	58	20-60	T	μ=0.1(KNO_3)	73Bh	b
Ni^{2+}	-17.9	5.28(1)	41	25	C	μ=0.2M(KCl)	74Sn	
	-17.6	4.34(2)	24	25	C	μ=0.2M(KCl)	74Sn	
Cu^{2+}	-21.42	8.07(1)	84	25	T	μ=0.2M(KCl)	73Ga	
	-28.78	6.75(2)	33	25	C	μ=0.2M(KCl)	73Ga	
P30 PENTANOIC ACID, 5-amino		$C_5H_{11}O_2N$				$H_2NCH_2(CH_2)_2CH_2CO_2H$		
Ag^+	-23.2	3.56(1)	-9.6	25	C	0.5M(KNO_3)	70Va	
	-27.6	3.56(25°,1)	-24.7	15-40	T	0.5M(KNO_3)	70Va	b
	-30.6	5.18(2)	-20.5	25	C	0.5M(KNO_3)	70Va	
	-23.8	5.18(25°,2)	-8.4	15-40	T	0.5M(KNO_3)	70Va	b
P31 1-PENTANOL, 3-thia-5-amino		$C_4H_{11}ONS$				$H_2NCH_2CH_2SCH_2CH_2OH$		
Ni^{2+}	-20.12	3.208(1)	-6.3	25	C	0.5(KNO_3)	77Hd	
	-22.68	2.55(2)	-27.2	25	C	0.5(KNO_3)	77Hd	
Cu^{2+}	-31.63	5.243(1)	-5.8	25	C	0.5(KNO_3)	77Hd	
	-36.07	4.765(2)	-29.7	25	C	0.5(KNO_3)	77Hd	
	-23.8	4.39	4	25	C	0.5(KNO_3) R:CuL_2^{2+}+OH^-= $Cu(OH)L_2^-$	77Hd	
P32 2-PENTANONE		$C_5H_{10}O$				$CH_3CH_2CH_2COCH_3$		L
$[NiA_2]_2$	-19.0	1.46(100°,1)	-23	75-125	T	S:Squalane A=3-Trifluoro-acetylcamphorate	74Sj	b,gg
SbA_5	-67.4	(1)	--	25	C	S:CCl_4 A=Cl	73Oa	
TeA_4	-23.4	(1)	--	25	C	S:Benzene A=Cl	72Pg	

Metal Ion	ΔH, kJ /mol	Log K	ΔS, J /mol K	T,°C	M	Conditions	Ref.	Re-marks
P33	3-PENTANONE		$C_5H_{10}O$			$CH_3CH_2COCH_2CH_3$		L
SbA_5	-69.16	(1)	--	25	C	S:1,2-Dichloro-ethane A=Cl	650a	
P34	3-PENTANONE, 2,2-dimethyl-		$C_7H_{14}O$			$CH_3CH_2COC(CH_3)_3$		L
SbA_5	-61.00	(1)	--	25	C	S:1,2-Dichloro-ethane A=Cl	650a	
P35	3-PENTANONE, 2,4-dimethyl-		$C_7H_{14}O$			$(CH_3)_2CHCOCH(CH_3)_2$		L
SbA_5	-67.78	(1)	--	25	C	S:1,2-Dichloro-ethane A=Cl	650a	
I_2	-10.5	(1)	--	0-45	T	S:Cyclohexane CL≅1M	66Bc	b,q:±4
P36	3-PENTANONE, 2-methyl-		$C_6H_{12}O$			$CH_3CH_2COCH(CH_3)_2$		L
SbA_5	-67.61	(1)	--	25	C	S:1,2-Dichloro-ethane A=Cl	650a	
P37	3-PENTANONE, 2,2,4,4-tetramethyl-		$C_9H_{18}O$			$(CH_3)_3CCOC(CH_3)_3$		L
SbA_5	-43.10	(1)	--	25	C	S:1,2-Dichloro-ethane A=Cl	650a	
	-40.2	2.52(1)	-113.0	--	C	S:Ethylene chloride A=Cl	670f	r
P38	3-PENTANONE, 2,2,4-trimethyl-		$C_8H_{16}O$			$(CH_3)_3CCOCH(CH_3)_2$		L
SbA_5	-47.45	(1)	--	25	C	S:1,2-Dichloro-ethane A=Cl	650a	
	-48.53	3.57(1)	-120.9	--	C	S:Ethylene chloride A=Cl	670f	
P39	1-PENTENE		C_5H_{10}			$CH_3CH_2CH_2CH{:}CH_2$		L
Ag^+	15.1	0.826(25°,1)	-33.9	0-40	T	S:Ethylene glycol	65Ci	b
P40	2-PENTENE(cis)		C_5H_{10}			$CH_3CH_2CH{:}CHCHCH_3$		L
Ag^+	15.1	0.748(25°,1)	-35.6	0-40	T	S:Ethylene glycol	65Ci	b
P41	2-PENTENE(trans)		C_5H_{10}			$CH_3CH_2CH{:}CHCH_3$		L
Ag^+	12.1	0.255(25°,1)	-36.4	0-40	T	S:Ethylene glycol	65Ci	b
P42	4-PENTENOIC ACID		$C_5H_8O_2$			$CH_2{:}CHCH_2CH_2CO_2H$		
Ag^+	-17.5	2.009(25°,1)	-20	2-38	T	μ=0.10M(CH_3COO^-)	75Ib	b

Metal Ion	ΔH, kJ /mol	Log K	ΔS, J /mol K	T,°C	M	Conditions	Ref.	Remarks
P43	4-PENTEN-1-OL		$C_5H_{10}O$			$CH_2:CHCH_2CH_2CH_2OH$		
Ag^+	-26.3	1.96(25°,1)	-50.3	25-45	T	μ~0	77Ha	
	-20.6	1.47(25°,1)	-40.9	25-45	T	S:Methanol	77Ha	
	-20.0	1.04(25°,1)	-47.0	25-45	T	S:Ethanol	77Ha	
	-27.8	2.32(25°,1)	-48.6	25-45	T	S:Propylene carbonate	77Ha	
P44	3-PENTEN-2-ONE, 4-amino-		C_5H_9ON			$CH_3C(NH_2):CHCOCH_3$		L^-
Co^{2+}	2.954	0.492(25°)	19.33	-33-72	5	S:15v% Tetra-methylsilane in $CHCl_3$	68Ea	b,t
P45	3-PENTEN-2-ONE, 4-amino-1,1,1-trifluoro-		$C_5H_6ONF_3$			$CH_3C(NH_2):CHCOCF_3$		L^-
Co^{2+}	~0	0.307	5.90	-43-67	T	S:15v% Tetra-methylsilane in $CHCl_3$	68Ea	b,t
P46	3-PENTEN-2-ONE, 4-methylamino-		$C_6H_{11}ON$			$CH_3C(NHCH_3):CHCOCH_3$		L^-
Ni^{2+}	12.343	-1.30(25°)	16.65	-43-77	T	S:$CDCl_3$	68Ea	b,t
P47	3-PENTEN-2-ONE, 4-methyl-amino-1,1,1-trifluoro-		$C_6H_8ONF_3$			$CH_3C(NHCH_3):CHCOCF_3$		L^-
Ni^{2+}	11.548	1.129(25°)	17.20	17-102	T	S:$CDCl_3$	68Ea	b,t
P48	PEPTIDE (AAHNMA, L-Asp-L-Ala-L-His-$NHCH_3$)							
Cu^{2+}	-37.03	8.39(1)	36.40	25	C	μ=0.15M(NACl)	79Aa	
	-4.94	-0.55	-28.0	25	C	μ=0.15M(NACl) R:$Cu+H_2L=CuL+2H$	79Aa	
P49	PEPTIDE (GGHNMA, GlyGly-L-His$NHCH_3$)							
Cu^{2+}	-5.23	-0.48	-25.9	25	C	μ=0.15M(NaCl) R:$Cu+H_2L=CuL+2H$	79Aa	
P50	PERCHLORATE ION		ClO_4			ClO_4^-		L^-
Ce^{3+}	-49.4	1.9(1)	-130	25	T	μ→0	55Ha	
	-71	0.079(26.5°,1)	-238	17.5-31.4	T	μ=1.14	56Sc	b
	-59	-0.387(26°,1)	-201	17.5-31.4	T	μ=5.11	56Sc	b
[Fe]	-14.6	--	--	25	C	Tris-cacodylate buffer (≅0.1M Tris)[Fe]=G	71Lb	

Metal Ion	ΔH, kJ /mol	Log K	ΔS, J /mol K	T,°C	M	Conditions	Ref.	Remarks
P50, continued								
[Fe], cont.						gouldii Hemerythrin=$HrFe_2O_3$ R:$HrFe_2O_3$+L^- = $HrFe_2O_3L^-$		
$[Fe^{2+}]$	-18.8	--	--	25	C	Tris-cacodylate buffer(≅0.1M Tris) [Fe]=G gouldii Hemerythrin=$HrFe_2$ R:$HrFe_2$+L^- = $HrFe_2L^-$	71Lb	q:Na
PtA_4	-6.19	0.455(1)	-11.7	25	C	μ=1M(NaF) CM=0.040-0.050M A=NH_3	73Ma	q:±0.42
	-5.36	0.484(1)	-8.8	25	C	μ=1M(NaF) CM=0.040-0.050M A=Ethylenediamine	73Ma	q:±0.42
NA_4^+	0.92	0.25(1)	7.9	25	C	μ=0 A=CH_3	69Ia	
	0.79	0.18(1)	6.3	25	C	μ=0 A=CH_3CH_2	69Ia	
	10.5	0.03(1)	35.1	25	C	μ=0 A=$CH_3CH_2CH_2$	69Ia	
P51 PERIODATE ION			IH_2O_6			$H_2IO_6^{3-}$		L^{3-}
CuA_4^{2-}	-31.4	10.468(40°)	99.6	30-50	T	1.3M(ClO_4^-) R:CuA_4^{2-}+$2L^{3-}$= $Cu(IO_6)_2^{7-}$+$4H_2O$ A=OH	53Lb	b
P52 PEROXYDISULPHATE ION			O_8S_2			S_2O_8		L^{2-}
Li^+	-61.5	2.45(25°,1)	-159.8	25&35	T	μ<10^{-2}	77Af	
Na^+	-72.8	2.32(25°,1)	-200.0	25&35	T	μ<10^{-2}	77Af	
Cu^{2+}	-82.0	2.65(25°,1)	-225.1	25&35	T	μ<10^{-2}	77Af	
P53 PERSULFATE ION			O_8S_2			$S_2O_8^{2-}$		L^{2-}
Na^+	18.0	0.580(25°,1)	71.1	25&37	T	μ=0	71Cb	a,b
K^+	6.7	0.914(25°,1)	40.2	25&37	T	μ=0	71Cb	a,b
Rb^+	5.4	1.167(25°,1)	41.0	25&37	T	μ=0	71Cb	a,b
Cs^+	-4.2	1.415(25°,1)	13.4	25&37	T	μ=0	71Cb	a,b

Metal Ion	ΔH, kJ /mol	Log K	ΔS, J /mol K	T,°C	M	Conditions	Ref.	Remarks
P54	PERYLENE		$C_{20}H_{12}$			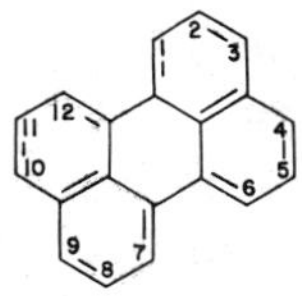		L
Li^+	0.0	5.194(1)	100	-65	T	S:Tetrahydrofuran	68Nb	q:±1.7
	5.8	5.301(1)	121	15	T	S:Tetrahydrofuran	68Nb	q:±1.7
Na^+	0	4.72(1)	90.0	-55	T	μ≅0 S:Dimethoxy ethane	66Cb	
	10.5	5.15(15°,1)	136.0	20	T	μ≅0 S:Dimethoxy ethane	66Cb	
	0.0	4.54(1)	88	-65	T	μ≅0 S:Tetrahydrofuran	66Cb 68Nb	q:±1.7
	9.2	4.75(1)	121	15	T	μ≅0 S:Tetrahydrofuran	66Cb 68Nb	q:±1.7
P55	1,10-PHENANTHROLINE		$C_{12}H_8N_2$			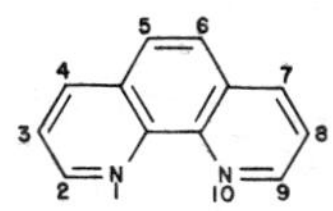		L
Mn^{2+}	-14.6	4.13(1)	28.4	20	C	μ=0.1($NaNO_3$)	63Ac	
	-12	3.8(25°,1)	33	-46-0	T	μ=0.2($NaClO_4$) S:CH_3OH	73Bg	b,q
	-29.3	7.61(1-2)	45.6	20	C	μ=0.1($NaNO_3$)	63Ac	
	-37.6	10.3(1-3	43.5	20	C	μ=0.1($NaNO_3$)	63Ac	
Fe^{2+}	-129.9	(1)	--	25	C	μ∿10^{-3}M	76La	
	-143.8	(1)	--	25	C	μ∿10^{-3}M S:10w%Dioxane	76La	
	-156.5	(1)	--	25	C	μ∿10^{-3}M S:20w%Dioxane	76La	
	-147.6	(1)	--	25	C	μ∿10^{-3}M S:30w%Dioxane	76La	
	-130.8	(1)	--	25	C	μ∿10^{-3}M S:40w%Dioxane	76La	
	-130.9	(1)	--	25	C	μ∿10^{-3}M S:10w%Ethanol	76La	
	-133.2	(1)	--	25	C	μ∿10^{-3}M S:20w%Ethanol	76La	
	-128.4	(1)	--	25	C	μ∿10^{-3}M S:30w%Ethanol	76La	
	-131.04	20.49(1-3)	-47.3	25	C	μ∿0	77Ba	
	-131.0	20.48(1-3)	-47.3	25	T	μ≅10^{-3}	64Lc	
	-138.1	21.3(1-3)	-64.4	20	C	μ=0.1($NaNO_3$)	63Ac	
	-142.88	20.08(1-3)	-95.0	25	C	μ∿0 S:8w%Methanol	77Ba	
	-162.21	19.81(1-3)	-165.1	25	C	μ∿0 S:16.4w%Methanol	77Ba	
	-170.62	19.55(1-3)	-198.3	25	C	μ∿0 S:25.2w%Methanol	77Ba	
	-160.79	19.05(1-3)	-175.1	25	C	μ∿0	77Ba	

Metal Ion	ΔH, kJ /mol	Log K	ΔS, J /mol K	T,°C	M	Conditions	Ref.	Re- marks
P55, continued								
Fe^{2+},						S:34.4w%Methanol		
cont.	-153.18	18.50(1-3)	-159.8	25	C	µ∼0 S:44.1w%Methanol	77Ba	
	-144.68	17.63(1-3)	-148.0	25	C	µ∼0 S:54.2w%Methanol	77Ba	
	-157.65	16.32(1-3)	-216.5	25	C	µ∼0 S:64.7w%Methanol	77Ba	
	-80.6	5.34	-168.2	25	C	µ∼0 R:Fe^{+}+3HL^{+}= $FeL_3{}^{2+}$+3H^{+}	77Ba	
	-79.5	--	--	25	C	µ∼10^{-3}M R:Fe^{+}+ 3HL^{+}=$FeL_3{}^{2+}$+3H^{+}	76La	
	-87.6	--	--	25	C	µ∼10^{-3}M S:10w%Dioxane R:Fe^{2+}+3HL^{+}= $FeL_3{}^{2+}$+3H^{+}	76La	
	-53.0	--	--	25	C	µ∼10^{-3}M S:20w%Dioxane R:Fe^{2+}+3HL^{+}= $FeL_3{}^{2+}$+3H^{+}	76La	
	-88.7	--	--	25	C	µ∼10^{-3}M S:30w%Dioxane R:Fe^{2+}+3HL^{+}= $FeL_3{}^{2+}$+3H^{+}	76La	
	-79.8	--	--	25	C	µ∼10^{-3}M S:45w%Dioxane R:Fe^{2+}+3HL^{+}= $FeL_3{}^{2+}$+3H^{+}	76La	
	-77.6	--	--	25	C	µ∼10^{-3}M S:10w%Ethanol R:Fe^{2+}+3HL^{+}= $FeL_3{}^{2+}$+3H^{+}	76La	
	-76.6	--	--	25	C	µ∼10^{-3}M S:20w%Ethanol R:Fe^{2+}+3HL^{+}= $FeL_3{}^{2+}$+3H^{+}	76La	
	-74.3	--	--	25	C	µ∼10^{-3}M S:30w%Ethanol R:Fe^{2+}+3HL^{+}= $FeL_3{}^{2+}$+3H^{+}	76La	
	-87.77	5.44	-190.3	25	C	µ∼0 S:8w%Methanol R:Fe^{2+}+3HL^{+}= $FeL_3{}^{2+}$+3H^{+}	77Ba	

Metal Ion	ΔH, kJ /mol	Log K	ΔS, J /mol K	T,°C	M	Conditions	Ref.	Remarks
P55, continued								
Fe^{2+}, cont.	-83.51	5.56	-173.9	25	C	μ≃0 S:16.4w%Methanol R:Fe^{2+}+3HL^{+}= FeL_3^{2+}+3H^{+}	77Ba	
	-105.60	5.66	-246.0	25	C	μ≃0 S:25.2w%Methanol R:Fe^{2+}+3HL^{+}= FeL_3^{2+}+3H^{+}	77Ba	
	-103.72	5.37	-251.0	25	C	μ≃0 S:34.4w%Methanol R:Fe^{2+}+3HL^{+}= FeL_3^{2+}+3H^{+}	77Ba	
	-99.29	5.56	-226.5	25	C	μ≃0 S:44.1w%Methanol R:Fe^{2+}+3HL^{+}= FeL_3^{2+}+3H^{+}	77Ba	
	-89.79	5.12	-202.6	25	C	μ≃0 S:54.2w%Methanol R:Fe^{2+}+3HL^{+}= FeL_3^{2+}+3H^{+}	77Ba	
	-106.11	4.32	-273.4	25	C	μ≃0 S:64.7w%Methanol R:Fe^{2+}+3HL^{+}= FeL_3^{2+}+3H^{+}	77Ba	
Fe^{3+}	-41.4	13.79(1-3)	121	25	T	μ≅10^{-3}	64Lc	
[Fe^{3+}]	-32.2	2.43(30°,1)	-63	30	T	S:Dichloromethane [Fe^{3+}]= Fe(deut)$Im_2^{+}$$Cl^{-}$ Fe(deut)=Iron(III) deuterporphyrin (IX) dimethyl ester Im= Imidazole	74Aa	
Co^{2+}	-38.1	7.25(1)	8.8	20	C	μ=0.1($NaNO_3$)	63Ac	
	-66.1	13.95(1-2)	41.4	20	C	μ=0.1($NaNO_3$)	63Ac	
	-99.6	19.9(1-3)	41.4	20	C	μ=0.1($NaNO_3$)	63Ac	
Ni^{2+}	-46.9	8.8(1)	8.8	20	C	μ=0.1($NaNO_3$)	63Ac	
	38.62	6.46(1)	-5.8	25	C	μ=0.1M(KNO_3) S:50w% Ethanol	72Bi	
	-38.32	5.98(2)	-13.8	25	C	μ=0.1M(KNO_3) S:50w% Ethanol	72Bi	
	-34.18	5.24(3)	-14.2	25	C	μ=0.1M(KNO_3) S:50w% Ethanol	72Bi	
	-85.8	17.1(1-2)	34.3	20	C	μ=0.1($NaNO_3$)	63Ac	
	-125.5	24.8(1-3)	46.9	20	C	μ=0.1($NaNO_3$)	63Ac	

Metal Ion	ΔH, kJ /mol	Log K	ΔS, J /mol K	T,°C	M	Conditions	Ref.	Re-marks
P55, continued								
NiA_2	-34.3	5.15(20°,1)	-19.2	10-30	T	S:$CHCl_3$ A=N-(p-Chlorophenyl)-β-mercaptocinnamamide	74Cd	b
	-33.5	5.03(20°,1)	-17.6	10-30	T	S:$CHCl_3$ A=N-(p-Ethoxyphenyl)-β-mercaptocinnamamide	74Cd	b
	-24.7	4.05(20°,1)	-63	10-30	T	S:$CHCl_3$ A=N-Ethyl-β-mercaptocinnamamide	74Cd	b
	-36.8	5.46(20°,1)	-20.9	10-30	T	S:$CHCl_3$ A=Ethyl-β-mercaptothiocinnamate	74Cd	b
	-41.4	5.90(20°,1)	-29.7	10-30	T	S:$CHCl_3$ A=Ethyl-β-mercaptocinnamate	74Cd	b
	-20.1	3.39(20°,1)	-0.8	10-30	T	S:$CHCl_3$ A=N-Methyl-N-phenyl-β-mercaptocinnamamide	74Cd	b
	-32.2	4.99(20°,1)	-14.2	10-30	T	S:$CHCl_3$ A=N-(α-Naphtyl)-β-mercaptocinnamamide	74Cd	b
	-23.0	3.87(20°,1)	-3.8	10-30	T	S:$CHCl_3$ A=Monothiodibenzoylmethane	74Cd	b
	-33.5	5.10(20°,1)	-18.0	10-30	T	S:$CHCl_3$ A=N-Phenyl-β-mercaptocinnamamide	74Cd	b
	-21.8	3.83(20°,1)	-0.4	10-30	T	S:$CHCl_3$ A=Thiobenzoylacetone	74Cd	b
Cu^{2+}	-46.15	9.14(1)	20.1	25	C	μ=0	70Ea	
	-49.0	9.25(1)	10.0	20	C	μ=0.1($NaNO_3$)	63Ac	
	-22.68	6.87(2)	55.2	25	C	μ=0	70Ea	
	-21.3	5.42(3)	4.6	25	C	μ=0	70Ea	
	-76.1	16.00(1-2)	46.9	20	C	μ=0.1($NaNO_3$)	63Ac	
	-110.4	21.35(1-3)	31.4	20	C	μ=0.1($NaNO_3$)	63Ac	
CuA_2	-87.0	(1)	--	30	C	S:Benzene A=1,-1,1,5,5,5-Hexafluoropentane-2,4-dione	74Gc	
	-53.26	(1)	--	30	C	S:Benzene A=1,-1,1-Trifluoropentane-2,4-dione	74Gc	
	-54.4	(1)	--	30	C	S:Benzene A=1,-1,1-Trifluoro-4-phenylbutane-2,4-dione	74Gc	
	-62.3	(1)	--	30	C	S:Benzene A=1,-1,1-Trifluoro-4-(2-thienyl)butane-2,4-dione	74Gc	

Metal Ion	ΔH, kJ /mol	Log K	ΔS, J /mol K	T,°C	M	Conditions	Ref.	Re-marks
P55, continued								
CuA_2, cont.	-35.52	5.40(1)	-15.48	25	C	S:Chloroform A=4,4,4-Trifluoro-1-(2-thienyl)-1,3-butanedione	72Ka	
Zn^{2+}	-31.4	6.17(1)	13.0	25	C	μ=0	70Ea	
	-31.4	6.55(1)	18.4	20	C	μ=0.1($NaNO_3$)	63Ac	
	-27.91	5.97(1)	20.9	25	C	μ=0.1M(KNO_3) S:50w% Ethanol	72Bi	
	-19.92	5.91(2)	46.4	25	C	μ=0	70Ea	
	-30.38	5.50(2)	2.9	25	C	μ=0.1M(KNO_3) S:50w% Ethanol	72Bi	
	-12.1	5.25(3)	59.8	25	C	μ=0	70Ea	
	-19.66	4.97(3)	29.3	25	C	μ=0.1M(KNO_3) S:50w% Ethanol	72Bi	
	-62.8	12.35(1-2)	23.0	20	C	μ=0.1($NaNO_3$)	63Ac	
	-80.8	17.55(1-3)	59.8	20	C	μ=0.1($NaNO_3$)	63Ac	
ZnA_2	-46.36	2.50(1)	-112.26	25	C	S:Chloroform A=4,4,4-Trifluoro-1-(2-thienyl)-1,3-butanedione	72Ka	
Cd^{2+}	-26.4	5.78(1)	20.1	20	C	μ=0.1($NaNO_3$)	63Ac	
	-22.47	5.48(1)	29.3	25	C	μ=0.1M(KNO_3) S:50w% Ethanol	72Bi	
	-24.39	5.31(2)	20.1	25	C	μ=0.1M(KNO_3) S:50w% Ethanol	72Bi	
	-16.65	4.05(3)	21.8	25	C	μ=0.1M(KNO_3) S:50w% Ethanol	72Bi	
	-54.8	10.82(1-2)	20.1	20	C	μ=0.1($NaNO_3$)	63Ac	
	-67.4	14.92(1-3)	55.6	20	C	μ=0.1($NaNO_3$)	63Ac	
HgA_2	-70.8	(1)	--	30	C	CM=0.5-0.6x10^{-3} S:Benzene A=Br	74Fb	
	-68.5	(1)	--	30	C	CM=0.45-0.7x10^{-3} S:Benzene A=I	74Fb	
	-47.4	3.36(1)	-93	30	C	CM=1-20x10^{-3} S:Benzene A=C_6F_5	74Fc	
Tl^{3+}	-49.4	11.6(1)	-54	25	C	μ=1.0	62Kc	
	-98.7	18.3(2)	-21	25	C	μ=1.0	62Kc	
	-36.8	5.4	-21	25	C	μ=1.0 R:Tl^{3+}+ HL^{+}=TlL^{3+}+H^{+}	62Kc	
	-46.0	10.6	50	25	C	μ=1.0 R:TlL^{3+}+ HL^{+}=$TlL_2{}^{3+}$+H^{+}	62Kc	
SnA_2B_2	-37.0	1.371(1)	-96	25	C	S:Benzene A=Bu B=(OCOMe)	77Ge	
	-37.0	1.143(1)	-100	25	C	S:Benzene A=Bu B=(OCOEt)	77Ge	
	-107.0	5.30(1)	-250	30	C	S:Benzene A=Phenyl B=Cl	75Fa	
	-86.0	4.70(1)	-195	30	C	S:Benzene A=Methyl B=Cl	75Fa	
	-84.3	4.70(1)	-189	30	C	S:Benzene	75Fa	

Metal Ion	ΔH, kJ /mol	Log K	ΔS, J /mol K	T,°C	M	Conditions	Ref.	Remarks
P55, continued								
SnA_2B_2, cont.	-84.7	>5(1)	<-184	30	C	A=Butyl B=Cl S:Benzene A=Butyl B=NCS	75Fa	
$SnAB_3$	-128	>5(1)	<-330	30	C	S:Benzene A=Butyl B=Cl	75Fa	
	-144	>6(1)	<-360	30	C	S:Benzene A=Phenyl B=Cl	75Fa	
P56 1,10-PHENANTHROLINE, 2,9-dimethyl-			$C_{14}H_{12}N_2$			$C_{12}H_6N_2(CH_3)_2$		
NiA_2	-46.9	(1)	--	30	C	S:Benzene A=Di-ethyldithio-phosphate	71Dc	q:±5.0
CuA_2	-54.8	(1)	--	30	C	S:Benzene A=1,-1,1,5,5,5-Hexa-fluoropentane-2,4-dione	74Gb	
	-30.00	3.531(1)	-31.00	30	C	S:Benzene A=1,-1,1-Trifluoro-pentane-2,4-dione	74Gb	
	-34.02	3.672(1)	-41.8	30	C	S:Benzene A=1,-1,1-Trifluoro-4-phenylbutane-2,4-dione	74Gb	
	-37.99	(1)	--	30	C	S:Benzene A=1,-1,1-Trifluoro-4-(2-thienyl)butane-2,4-dione	74Gb	
Zn^{2+}	-5.52	3.79(1)	52.3	25	C	μ=0.1M(KNO_3) S:50w% Ethanol	72Ee	
	-18.24	2.58(2)	-10.0	25	C	μ=0.1M(KNO_3) S:50w% Ethanol	72Ee	
Cd^{2+}	-18.4	3.94(1)	13.4	25	C	μ=0.1M(KNO_3) S:50w% Ethanol	72Ee	
	-18.66	2.40(2)	-16.3	25	C	μ=0.1M(KNO_3) S:50w% Ethanol	72Ee	
P57 1,10-PHENANTHROLINE, 4,7-dimethyl-			$C_{14}H_{12}N_2$			$C_{12}H_6N_2(CH_3)_2$		
Ni^{2+}	-38.99	6.54(1)	-5.4	25	C	μ=0.1M(KNO_3) S:50w% Ethanol	72Ee	
	-47.70	5.59(2)	-53.1	25	C	μ=0.1M(KNO_3) S:50w% Ethanol	72Ee	
	-27.82	6.30(3)	28.0	25	C	μ=0.1M(KNO_3) S:50w% Ethanol	72Ee	
Zn^{2+}	-22.76	6.55(1)	49.0	25	C	μ=0.1M(KNO_3) S:50w% Ethanol	72Ee	
	-24.27	6.73(2)	47.3	25	C	μ=0.1M(KNO_3)	72Ee	

Metal Ion	ΔH, kJ /mol	Log K	ΔS, J /mol K	T,°C	M	Conditions	Ref.	Remarks
P57, continued								
Zn^{2+},						S:50w% Ethanol		
cont.	-26.32	5.86(3)	24.3	25	C	μ=0.1M(KNO_3) S:50w% Ethanol	72Ee	
Cd^{2+}	-21.17	6.19(1)	47.7	25	C	μ=0.1M(KNO_3) S:50w% Ethanol	72Ee	
	-19.83	5.92(2)	46.4	25	C	μ=0.1M(KNO_3) S:50w% Ethanol	72Ee	
	-22.8	4.76(3)	14.6	25	C	μ=0.1M(KNO_3) S:50w% Ethanol	72Ee	
P58	1,10-PHENANTHROLINE, 5,6-dimethyl-		$C_{14}H_{12}N_2$			$C_{12}H_6N_2(CH_3)_2$		L
Fe^{2+}	-128.66	21.95(25°,1-3)	-10.8	20-40	T	μ=0	68La	b
Fe^{3+}	-77.15	18.54(25°,1-3)	96.2	20-40	T	μ=0	68La	b
Ni^{2+}	-43.39	5.83(1)	-33.9	25	C	μ=0.1M(KNO_3) S:50w% Ethanol	72Ee	
	-46.48	6.30(2)	-35	25	C	μ=0.1M(KNO_3) S:50w% Ethanol	72Ee	
	-39.20	4.53(3)	-44.8	25	C	μ=0.1M(KNO_3) S:50w% Ethanol	72Ee	
Zn^{2+}	-30.59	5.71(1)	6.7	25	C	μ=0.1M(KNO_3) S:50w% Ethanol	72Ee	
	-38.95	6.19(2)	-12.1	25	C	μ=0.1M(KNO_3) S:50w% Ethanol	72Ee	
	-24.39	5.15(3)	16.7	25	C	μ=0.1M(KNO_3) S:50w% Ethanol	72Ee	
Cd^{2+}	-30.1	5.77(1)	9.2	25	C	μ=0.1M(KNO_3) S:50w% Ethanol	72Ee	
	-30.5	5.34(2)	0	25	C	μ=0.1M(KNO_3) S:50w% Ethanol	72Ee	
	-17.24	4.13(3)	21.3	25	C	μ=0.1M(KNO_3) S:50w% Ethanol	72Ee	
P59	1,10-PHENANTHROLINE, 6-methyl-		$C_{13}H_{10}N_2$			$C_{12}H_7N_2(CH_3)$		L
Fe^{2+}	-141.4	21.01(25°,1-3)	-72.0	15-45	T	μ=0	67La	b
Ni^{2+}	-42.22	6.19(1)	-23.0	25	C	μ=0.1M(KNO_3) S:50w% Ethanol	72Bi	
	-37.11	5.93(2)	-10.9	25	C	μ=0.1M(KNO_3) S:50w% Ethanol	72Bi	
	-46.02	4.93(3)	-60.25	25	C	μ=0.1M(KNO_3) S:50w% Ethanol	72Bi	
Zn^{2+}	-28.70	5.83(1)	15.1	25	C	μ=0.1M(KNO_3) S:50w% Ethanol	72Bi	
	-34.22	5.70(2)	-5.4	25	C	μ=0.1M(KNO_3) S:50w% Ethanol	72Bi	
	-24.39	5.01(3)	14.2	25	C	μ=0.1M(KNO_3) S:50w% Ethanol	72Bi	
Cd^{3+}	-26.19	5.32(1)	14.2	25	C	μ=0.1M(KNO_3)	72Bi	

Metal Ion	ΔH, kJ /mol	Log K	ΔS, J /mol K	T,°C	M	Conditions	Ref.	Re-marks
P59, continued								
Cd^{3+},						S:50w% Ethanol		
cont.	-22.89	5.27(2)	23.8	25	C	μ=0.1M(KNO_3) S:50w% Ethanol	72Bi	
	-20.25	4.43(3)	17.2	25	C	μ=0.1M(KNO_3) S:50w% Ethanol	72Bi	
P60	1,10-PHENANTHROLINE, 6-nitro- $C_{12}H_7O_2N_3$					$C_{12}H_7N_2(NO_2)$		L
Fe^{2+}	-105.4	15.63(1-3	-54.4	25	T	μ=0	64Ld	
	-75.86	5.94	-146.4	25	T	μ=0 R:Fe^{2+}+3HL^+ =$FeL_3{}^{2+}$+3H^+	64Ld	
P61	1,10-PHENANTHROLINE, 5-phenyl- $C_{18}H_{12}N_2$					$C_{12}H_7N_2(C_6H_5)$		
Ni^{2+}	-40.38	5.67(1)	-26.8	25	C	μ=0.1M(KNO_3) S:50w% Ethanol	72Bi	
	-36.86	4.90(2)	-29.7	25	C	μ=0.1M(KNO_3) S:50w% Ethanol	72Bi	
	-40.75	4.66(3)	-47.7	25	C	μ=0.1M(KNO_3) S:50w% Ethanol	72Bi	
Zn^{2+}	-28.62	5.51(1)	9.6	25	C	μ=0.1M(KNO_3) S:50w% Ethanol	72Bi	
	-22.01	5.39(2)	29.3	25	C	μ=0.1M(KNO_3) S:50w% Ethanol	72Bi	
	-19.20	4.08(3)	13.8	25	C	μ=0.1M(KNO_3) S:50w% Ethanol	72Bi	
Cd^{2+}	-23.01	5.12(1)	20.5	25	C	μ=0.1M(KNO_3) S:50w% Ethanol	72Bi	
	-20.92	4.77(2)	21.3	25	C	μ=0.1M(KNO_3) S:50w% Ethanol	72Bi	
P62	PHENOL		C_6H_6O			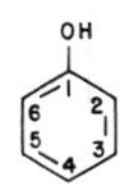		L^-
Fe^{3+}	-0.4	7.84(1)	151	25	T	μ=0.1	67Mi 69Da	
PA_3B	-29.9	3.260(25°,1)	-37.7	0-50	T	S:CCl_4 A=$(CH_3)_2N$ B=O	77Kf	
[P]	-34.3	2.710(25°,1)	-63.6	20-70	T	[P]=$Et_2P(O)OEt$ S:CCl_4	68Aa	b
	-33.0	2.512(25°,1)	-63.2	20-70	T	[P]=$EtP(O)(OEt)_2$ S:CCl_4	68Aa	b
	-33.9	2.338(25°,1)	-69.0	20-70	T	[P]=$(EtO)_3P$=O S:CCl_4	68Aa	b
	-32.6	2.227(25°,1)	-66.5	20-70	T	[P]= 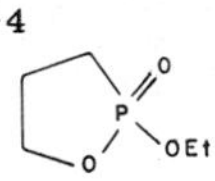 S:CCl_4	68Aa	b

Metal Ion	ΔH, kJ /mol	Log K	ΔS, J /mol K	T,°C	M	Conditions	Ref.	Remarks
P62, continued								
[P], cont.	-31.8	1.981(25°,1)	-68.2	20-70	T	[P]= 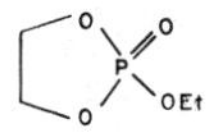 S:CCl_4	68Aa	b
	-31.8	2.487(25°,1)	-59.4	20-70	T	[P]= 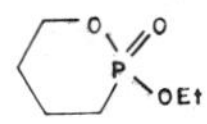 S:CCl_4	68Aa	b
	-31.8	2.257(25°,1)	-63.8	20-70	T	[P]= 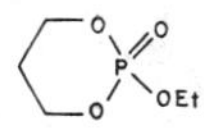 S:CCl_4	68Aa	b
	-34.7	2.520(25°,1)	-68.6	20-70	T	[P]= 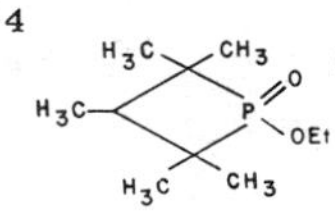 S:CCl_4	68Aa	b
	-40.2	2.797(25°,1)	-81.2	20-70	T	[P]=Ph_3P=O S:CCl_4	68Aa	b
	-39.7	2.834(25°,1)	-80.3	20-70	T	[P]=$Ph_2(CH_3)P$=O S:CCl_4	68Aa	b
	-40.2	2.959(25°,1)	-78.2	20-70	T	[P]=$Ph(CH_3)_2P$=O S:CCl_4	68Aa	b
	-40.2	3.000(25°,1)	-77.4	20-70	T	[P]=Bu_3P=O S:CCl_4	68Aa	b
	-43.9	3.245(25°,1)	-85.8	20-70	T	[P]= 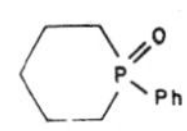 S:CCl_4	68Aa	b
	-53.6	3.332(25°,1)	-115.9	20-70	T	[P]= 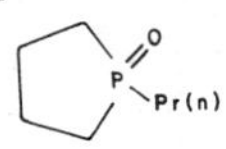 S:CCl_4	68Aa	b
	-37.2	2.281(25°,1)	-81.2	20-70	T	[P]= 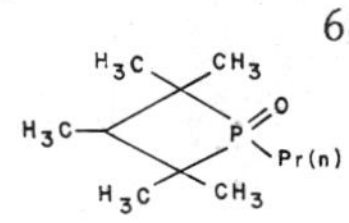 S:CCl_4	68Aa	b
	-31.4	1.806(20°,1)	-72.8	20&50	T	[P]=Pyridine S:CCl_4	68Aa	a,b
P63 PHENOL, 2-amino-			C_6H_7ON			$C_6H_5O(NH_2)$		
La^{3+}	29.7	3.54(25°,1)	230	15-35	T	μ=0.12M($NaClO_4$) S:50% DMSO	69Bd	b
	-8.4	4.50(25°,1)	58	15-35	T	μ=0.12M($NaClO_4$) S:50% Dioxane	69Bd	b
	-8.4	3.59(25°,1)	42	15-35	T	μ=0.12M($NaClO_4$) S:50% Ethanol	69Bd	b
Nd^{3+}	28.9	3.72(25°,1)	218	15-35	T	μ=0.12M($NaClO_4$) S:50% DMSO	69Bd	b
	-12.6	4.62(25°,1)	46	15-35	T	μ=0.12M($NaClO_4$)	69Bd	b

Metal Ion	ΔH, kJ /mol	Log K	ΔS, J /mol K	T,°C	M	Conditions	Ref.	Re-marks
P63, continued								
Nd^{3+}, cont.	-10.9	4.31(25°,1)	46	15-35	T	S:50% Dioxane μ=0.12M($NaClO_4$) S:50% Ethanol	69Bd	b
Gd^{3+}	25.5	3.80(25°,1)	218	15-35	T	μ=0.12M($NaClO_4$) S:50% DMSO	69Bd	b
	-7.5	5.04(25°,1)	71	15-35	T	μ=0.12M($NaClO_4$) S:50% Dioxane	69Bd	b
	-9.6	4.48(25°,1)	54	15-35	T	μ=0.12M($NaClO_4$) S:50% Ethanol	69Bd	b
Tb^{3+}	18.4	3.82(25°,1)	184	15-35	T	μ=0.12M($NaClO_4$) S:50% DMSO	69Bd	b
	-9.2	5.14(25°,1)	67	15-35	T	μ=0.12M($NaClO_4$) S:50% Dioxane	69Bd	b
	-11.3	4.65(25°,1)	50	15-35	T	μ=0.12M($NaClO_4$) S:50% Ethanol	69Bd	b
Dy^{3+}	14.2	3.98(25°,1)	172	15-35	T	μ=0.12M($NaClO_4$) S:50% DMSO	69Bd	b
	-11.3	5.25(25°,1)	63	15-35	T	μ=0.12M($NaClO_4$) S:50% Dioxane	69Bd	b
	-11.3	4.75(25°,1)	54	15-35	T	μ=0.12M($NaClO_4$) S:50% Ethanol	69Bd	b
Ho^{3+}	20.1	4.23(25°,1)	201	15-35	T	μ=0.12M($NaClO_4$) S:50% DMSO	69Bd	b
	-11.7	5.32(25°,1)	63	15-35	T	μ=0.12M($NaClO_4$) S:50% Dioxane	69Bd	b
	-11.7	4.79(25°,1)	54	15-35	T	μ=0.12M($NaClO_4$) S:50% Ethanol	69Bd	b
Er^{3+}	26.8	4.41(25°,1)	238	15-35	T	μ=0.12M($NaClO_4$) S:50% DMSO	69Bd	b
	-10.5	5.52(25°,1)	71	15-35	T	μ=0.12M($NaClO_4$) S:50% Dioxane	69Bd	b
	-12.6	4.89(25°,1)	54	15-35	T	μ=0.12M($NaClO_4$) S:50% Ethanol	69Bd	b
Tm^{3+}	28.9	4.48(25°,1)	205	15-35	T	μ=0.12M($NaClO_4$) S:50% DMSO	69Bd	b
	-9.6	5.65(25°,1)	75	15-35	T	μ=0.12M($NaClO_4$) S:50% Dioxane	69Bd	b
	-13.4	5.01(25°,1)	50	15-35	T	μ=0.12M($NaClO_4$) S:50% Ethanol	69Bd	b
Yb^{3+}	30.5	4.54(25°,1)	259	15-35	T	μ=0.12M($NaClO_4$) S:50% DMSO	69Bd	b
	-10.0	5.82(25°,1)	79	15-35	T	μ=0.12M($NaClO_4$) S:50% Dioxane	69Bd	b
	-13.4	5.12(25°,1)	54	15-35	T	μ=0.12M($NaClO_4$) S:50% Ethanol	69Bd	b

P64 PHENOL, 4-bromo C_6H_5OBr OHC_6H_4Br

Metal Ion	ΔH, kJ /mol	Log K	ΔS, J /mol K	T,°C	M	Conditions	Ref.	Re-marks
PA_3B	-32.9	3.915(25°,1)	-35.5	0-50	T	S:CCl_4 A=$(CH_3)_2N$ B=O	77Kf	

Metal Ion	ΔH, kJ /mol	Log K	ΔS, J /mol K	T,°C	M	Conditions	Ref.	Remarks
P65	PHENOL, 2-chloro-		C_6H_5OCl			$C_6H_5O(Co)$		L^-
Fe^{3+}	0.8	6.08(1)	117	25	T	μ=0.1	69Da	
P66	PHENOL, 3-chloro-		C_6H_5OCl			$C_6H_5O(Cl)$		L^-
Fe^{3+}	-0.8	6.89(1)	130	25	T	μ=0.1	69Da	
P67	PHENOL, 4-chloro-		C_6H_5OCl			$C_6H_5O(Cl)$		L^-
Fe^{3+}	-0.8	7.26(1)	134	25	T	μ=0.1	67Mi 69Da	
PA_3B	-33.5	3.919(25°,1)	-35.2	0-50	T	$S:CCl_4$ $A=(CH_3)_2N$ B=O	77Kf	
P68	PHENOL, 2-cyano-		C_7H_5ON			$C_6H_5O(CN)$		L^-
Fe^{3+}	2.72	5.53(1)	113	25	T	μ=0.1	69Da	
P69	PHENOL, 3-cyano-		C_7H_5ON			$C_6H_5O(CN)$		L^-
Fe^{3+}	3.8	6.30(1)	130	25	T	μ=0.1	69Da	
P70	PHENOL, 4-cyano-		C_7H_5ON			$C_6H_5O(CN)$		L^-
Fe^{3+}	-0.8	5.79(1)	109	25	T	μ=0.1	69Da	
P71	PHENOL, 2,6-di(tert-butyl)		$C_{14}H_{21}O$			$HOC_6H_3[C(CH_3)_3]_2$		
PA_3B	-12.8	0.740(25°,1)	-28.8	0-50	T	$S:CCl_4$ $(CH_3)_2N$ B=O	77Kf	
P72	PHENOL, 2,6-diisopropyl		$C_{12}H_{18}O$			$HOC_6H_3[CH(CH_3)_2]_2$		
PA_3B	-20.3	1.948(25°,1)	-30.8	0-50	T	$S:CCl_4$ $A=(CH_3)_2N$ B=O	77Kf	
P73	PHENOL, 2,6-dimethyl		$C_8H_{10}O$			$HOC_6H_3(CH_3)_2$		
PA_3B	-20.1	2.100(25°,1)	-27.2	0-50	T	$S:CCl_4$ $A=(CH_3)_2N$ B=O	77Kf	
P74	PHENOL, 5-dimethylamino-o-(2-thiazolylazo)		$C_{11}H_{13}ON_4S$			$(CH_3)_2N(C_6H_3OH)N:N(SNC_3H_3)$		
Ho^{3+}	-15.4	11.12(25°,1)	161	15-35	T	$μ=0.1(KNO_3)$	75Ka	b

Metal Ion	ΔH, kJ /mol	Log K	ΔS, J /mol K	T, °C	M	Conditions	Ref.	Remarks
P75	PHENOL, 4-ethyl-o-(2-thiazolylazo)		$C_{11}H_{12}ON_3S$			$H_5C_2(C_6H_3OH)N{:}N(SNC_3H_3)$		
Ho^{3+}	-20.4	9.90(25°,1)	121	15-35	T	μ=0.1(KNO_3)	75Ka	b
P76	PHENOL, 4-fluoro		C_6H_5OF			OHC_6H_4F		
PA_3B	-32.3	3.678(25°,1)	-38.1	0-50	T	S:CCl_4 A=$(CH_3)_2N$ B=O	77Kf	
P77	PHENOL, 3-nitro-		$C_6H_5O_3N$			$C_6H_5O(NO_2)$		L^-
Fe^{3+}	5.0	6.2(1)	138	25	T	μ=0.1	67Mi 69Da	
P78	PHENOL, 4-nitro-		$C_6H_5O_3N$			$C_6H_5O(NO_2)$		L^-
Fe^{3+}	1.2	5.3(1)	105	25	T	μ=0.1	67Mi 69Da	
P79	PHENOL, pentabromo		C_6HOBr_5			OHC_6Br_5		
PA_3B	-24.7	2.624(25°,1)	-32.7	0-50	T	S:CCl_4 A=$(CH_3)_2N$ B=O	77Kf	
P80	PHENOL, pentachloro		C_6HOCl_5			OHC_6Cl_5		
PA_3B	-26.5	3.185(25°,1)	-27.9	0-50	T	S:CCl_4 A=$(CH_3)_2N$ B=O	77Kf	
P81	PHENOL, pentafluoro		C_6HOF_5			OHC_6F_5		
PA_3B	-35.0	4.428(25°,1)	-32.6	0-50	T	S:CCl_4 A=$(CH_3)_2N$ B=O	77Kf	
P82	PHENOL, 2(2-thiazolylazo)-4-chlorophenol		$C_9H_8ON_3ClS$			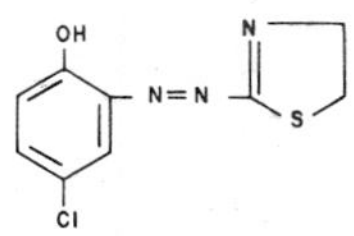		
La^{3+}	-61.00	6.63(25°,1)	-77.78	15-35	T	μ=0.1(KNO_3)	74Ka	b
Pr^{3+}	-51.67	7.48(25°,1)	-30.04	15-35	T	μ=0.1(KNO_3)	74Ka	b
Sm^{3+}	-58.45	7.77(25°,1)	-46.74	15-35	T	μ=0.1(KNO_3)	74Ka	b
Eu^{3+}	-49.29	7.82(25°,1)	-15.6	15-35	T	μ=0.1(KNO_3)	74Ka	b
Gd^{3+}	-42.34	7.94(25°,1)	10.17	15-35	T	μ=0.1(KNO_3)	74Ka	b
Dy^{3+}	-54.22	8.38(25°,1)	-21.92	15-35	T	μ=0.1(KNO_3)	74Ka	b
Ho^{3+}	-49.16	8.24(25°,1)	-7.11	15-35	T	μ=0.1(KNO_3)	74Ka	b

Metal Ion	ΔH, kJ /mol	Log K	ΔS, J /mol K	T,°C	M	Conditions	Ref.	Remarks
P82, continued								
Er^{3+}	-50.84	8.18(25°,1)	-13.76	15-35	T	μ=0.1(KNO_3)	74Ka	b
Yb^{3+}	-38.12	8.43(25°,1)	33.39	15-35	T	μ=0.1(KNO_3)	74Ka	b
P83	PHENOL, 2(2-thiazolylazo)-4-methoxy		$C_{10}H_{11}O_2N_3S$			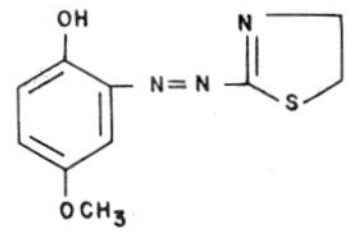		
Pr^{3+}	-26.40	8.52(25°,1)	74.52	15-35	T	μ=0.1(KNO_3)	74Ka	b
Nd^{3+}	-24.10	8.53(25°,1)	82.47	15-35	T	μ=0.1(KNO_3)	74Ka	b
Sm^{3+}	-35.27	8.98(25°,1)	53.60	15-35	T	μ=0.1(KNO_3)	74Ka	b
Eu^{3+}	-27.53	9.09(25°,1)	81.67	15-35	T	μ=0.1(KNO_3)	74Ka	b
Gd^{3+}	-26.44	8.96(25°,1)	82.89	15-35	T	μ=0.1(KNO_3)	74Ka	b
Dy^{3+}	-43.72	9.39(25°,1)	32.93	15-35	T	μ=0.1(KNO_3)	74Ka	b
Ho^{3+}	-35.10	9.63(25°,1)	66.57	15-35	T	μ=0.1(KNO_3)	74Ka	b
Er^{3+}	-30.50	9.49(25°,1)	79.37	15-35	T	μ=0.1(KNO_3)	74Ka	b
Yb^{3+}	-20.17	9.60(25°,1)	116.15	15-35	T	μ=0.1(KNO_3)	74Ka	b
P84	PHENYLALANINE, N-benzoyl(ℓ)		$C_{15}H_{15}O_3N$			$C_6H_5CH_2CH(NHOC_6H_5)CO_2H$		
Cu^{2+}	-19.83	3.11(25°,1)	-7.32	25-45	T	μ=1M(KNO_3) S:50v%Dioxane	79Sd	
Ni^{2+}	-9.20	2.80(25°,1)	23.10	25-45	T	μ=1M(KNO_3) S:50v%Dioxane	79Sd	
Co^{2+}	-11.92	2.90(25°,1)	16.02	25-45	T	μ=1M(KNO_3) S:50v%Dioxane	79Sd	
P85 PHOSPHATE ION			O_4P			PO_4^{3-}		L^{3-}
Li^+	≅25	0.72(25°,1)	≅100	0&25	T	μ=0.2	56Sc	a,b
Na^+	≅25	0.60(25°,1)	≅100	0&25	T	μ=0.2	56Sc	a,b
K^+	≅25	0.49(25°,1)	≅100	0&25	T	μ=0.2	56Sc	a,b
Mg^{2+}	12.205	2.708(25°,1)	92.9	25	T	μ=0 ΔCp=251	54Cd	
	15.230	2.791(37°,1)	102.9	37	T	μ=1 ΔCp=259	54Cd	
	≅21	1.881(25°)	≅105	0&25	T	μ=0.2[n-$(C_3H_7)_4$-N]$^+$ R:Mg^{2+}+HL^{2-}=MgHL	56Sb	a,b
Ca^{2+}	13.0	0.462(25°,1)	167	25&37	T	μ=0	68Ce	a,b,q: ±8.4
	13.8	2.739(25°,1)	96	25&37	T	μ=0 R:Ca^{2+}+HL^{2-}=CaHL	68Ce	a,b,q: ±2.5
	≅21	1.699(25°)	≅105	0&25	T	μ=0.2[n-$(C_3H_7)_4$-	56Sb	a,b

Metal Ion	ΔH, kJ /mol	Log K	ΔS, J /mol K	T,°C	M	Conditions	Ref.	Remarks
P85, continued								
Ca^{2+}, cont.						$_4N]^+$ R: Ca^{2+} + HL^{2-}=CaHL		
	14.2	1.408(25°,1)	75	25&37	T	μ=0 R: Ca^{2+} + H_2L^- = CaH_2L^+	68Cd	a,b,q: ±7.1
Sr^{2+}	≅21	1.518(25°)	≅105	0&25	T	μ=0.2[n-(C_3H_7)- $_4N]^+$ R: Sr^{2+} + HL^{2-}=SrHL	56Sb	a,b
Mn^{2+}	≅21	2.582(25°)	≅105	0&25	T	μ=0.2[n-$(C_3H_7)_4$- $N]^+$ R: Mn^{2+} + HL^{2-}=MnHL	56Sb	a,b
Fe^{2+}	53.6	7.340(30°,1)	314	20&30	T	μ=0	64Lb	a,b,q:±8
Fe^{3+}	-15.77	10.88(25°)	155	25-50	T	μ=0.037 R: Fe^{3+} + HL^{2-} = $FeHL^+$	65La	b
P86 PHOSPHATE ION, diimidotri-			$P_3H_2O_8N_2$			$[P_3O_8(NH)_2]^{5-}$		L^{5-}
Ca^{2+}	-38.1	7.10(1)	7.9	25	T	μ=0	61Ia	
P87 PHOSPHATE ION, dithio- dicyclohexyl ester			$C_{12}H_{22}O_2S_2P$			$(C_6H_{11}O)_2PS_2$		
CdL_2	-12.97	3.40	-11.00	30	C	S:Benzene R: $2CdL_2$ = Cd_2L_4	71Db	q:±2.5
P88 PHOSPHATE ION, dithio-, diisobutyl ester			$C_8H_{18}O_2S_2P$			$[(CH_3)_2CHCH_2O]_2PS_2$		
ZnL_2	-17.28	0.85	-49.0	30	C	S:Benzene R: $2ZnL_2$ = Zn_2L_4	71Db	q:±1.7
CdL_2	-17.11	3.70	-21.00	30	C	S:Benzene R: $2CdL_2$ = Cd_2L_4	71Db	q:±2.9
P89 PHOSPHATE ION, dithio-, diisopropyl ester			$C_6H_{14}O_2S_2P$			$[(CH_3)_2CHO]_2PS_2$		
ZnL_2	-16.48	0.68	-48.1	30	C	S:Benzene R: $2ZnL_2$ = Zn_2L_4	71Db	
CdL_2	-20.00	3.80	-30.1	30	C	S:Benzene R: $2CdL_2$ = Cd_2L_4	71Db	q:±2.9
P90 PHOSPHATE ION, dithio-, dipropyl ester			$C_6H_{14}O_2S_2P$			$(CH_3CH_2CH_2O)PS_2$		
HgL_2	-3.18	0.00	-11.00	30	C	S:Benzene R: $2HgL_2$ = Hg_2L_4	71Db	q:±0.8

Metal Ion	ΔH, kJ /mol	Log K	ΔS, J /mol K	T,°C	M	Conditions	Ref.	Remarks
P91	PHOSPHATE, hexa(meta)		$P_6O_{18}Na_6$			$(NaPO_3)_6$		L^-
Ba^{2+}	-46.69	10.32(35°,3)	46.0	35-45	T	CM=$3x10^{-4}$-$4x10^{-3}$M CL=$1x10^{-3}$-1.5 $x10^{-2}$M $(NaPO_3)_n$	69Bh	b
P92	PHOSPHATE ION, imidodi-		P_2HO_6N			$[P_2O_6(NH)]^{4-}$		L^{4-}
Ca^{2+}	-23.8	6.07(1)	36.4	25	T	μ=0	61Ia	
P93	PHOSPHATE, long chain		P_n			P_n		L
Ca^{2+}	-2.9	7.28	130	25	T	μ=0 R:Ca+P_6=CaP_6	60Ia	
	-5.0	7.23	121	25	T	μ=0 R:Ca+P_{14}= CaP_{14}	60Ia	
	4	7.54	159	25	T	μ=0 R:Ca+P_{60}= CaP_{60}	60Ia	
P94	PHOSPHATE, tri-n-butyl-		$C_{12}H_{27}O_4P$			$PO_4(CH_2CH_2CH_2CH_3)_3$		L
UO_2^{2+}	-18.0	1.5	-32.2	25	T	R:UO_2^{2+}+$2NO_3^-$+2L =$UO_2(NO_3)_2L_2$	60Na	
I_2	-12.30	1.290(25°,1)	-16.69	10-35	T	CM=$0.67x10^{-3}$M CL=0.03-0.12M S:n-Heptane	60Td	b
P95	PHOSPHATE ION, tripoly-		$O_{10}H_5P_3$					L^{5-}
Be^{2+}	19.7	(1)	--	20	C	$0.1M(CH_3)_4NNO_3$	65Aa	
Mg^{2+}	18.16	7.05(1)	197	20	C	$0.1M(CH_3)_4NNO_3$	65Aa	
	-8.83	4.93(25°,1)	126	2-45	T	μ=0.10M(KNO_3)	73Ke	b
	5.8	3.33(25°)	83.3	2-45	T	μ=0.10M(KNO_3) R:Mg^{2+}+HL^{4-}= $MgHL^{2-}$	73Ke	b,q: ±1.7
Ca^{2+}	-13.4	6.90(1)	88	25	T	μ=0	60Ia	
	13.8	6.31(1)	167.4	20	C	$0.1M(CH_3)_4NNO_3$	65Aa	
	-27.6	4.80(25°,1)	39.7	2-45	T	μ=0.10M(KNO_3)	73Ke	b,q: ±12.6
	-3.43	3.25(25°)	61.5	2-45	T	μ=0.10M(KNO_3) R:Ca^{2+}+HL^{4-}= $CaHL^{2-}$	73Ke	b
Sr^{2+}	13.22	5.46(1)	149.4	20	C	$0.1M(CH_3)_4NNO_3$	65Aa	
	-24.94	4.00(25°,1)	20.9	2-45	T	μ=0.10M(KNO_3)	73Ke	b
	-5.8	2.86(25°)	35.1	2-45	T	μ=0.10M(KNO_3) R:Sr^{2+}+HL^{4-}= $SrHL^{2-}$	73Ke	b,q: ±1.2
Y^{3+}	-6.3	9.1(25°,1-2)	151	25-45	T	μ=0.1M(KNO_3)	74Ta	b,q:±0.8
	-18.8	6.7(25°)	63	25-45	T	μ=0.1M(KNO_3)	74Ta	b

Metal Ion	ΔH, kJ /mol	Log K	ΔS, J /mol K	T, °C	M	Conditions	Ref.	Remarks
P95, continued								
Y^{3+}, cont.						R:Y^{3+}+2HL^{8-}= $YH_2L_2{}^{5-}$		
YA^-	-23.4	5.43(25°,1)	33	2-45	T	μ=0.10M(KNO_3) A=$ETAT^{2-}$	73Kg	b,q:Na, q:±4.2
	-14.6	4.02(25°)	29	2-45	T	μ=0.10M(KNO_3) A=$EDTA^{2-}$ R:YA^-+ HL^{4-}=$YAHL^{5-}$	73Kg	b,q:Na, q:±4.2
La^{3+}	-10.5	8.0(25°,1-2)	117	25-45	T	μ=0.1M(KNO_3)	74Ta	q:±0.8
	-19.7	6.1(25°)	50	25-45	T	μ=0.1M(KNO_3) R:La^{3+}+2HL^{8-}= $LaH_2L_2{}^{5-}$	74Ta	
LaA^-	-14.2	3.90(25°,1)	25.1	2-45	T	μ=0.10M(KNO_3)	73Kg	b,g:Na
	-12.1	2.54(25°)	8.8	2-45	T	μ=0.10M(KNO_3) A=$EDTA^{2-}$ R:LaA^- +HL^{4-}=$LaAHL^{5-}$	73Kg	b,g:Na q:±2.1
Ce^{3+}	-10.0	8.2(25°,1-2)	121	25-45	T	μ=0.1M(KNO_3)	74Ta	b,q: ±0.8
	-10.5	6.2(25°)	84	25-45	T	μ=0.1M(KNO_3) R:Ce^{3+}=2HL^{8-}= $CeH_2L_2{}^{5-}$	74Ta	b,q: ±0.8
CeA^-	-12.1	4.25(25°,1)	42	2-45	T	μ=0.10M(KNO_3) A=$EDTA^{2-}$	73Kg	b,g:Na
	-14.2	2.90(25°)	8.4	2-45	T	μ=0.10M(KNO_3) A=$EDTA^{2-}$ R:CeA^- +HL^{4-}=$CeAHL^{5-}$	73Kg	b,g:Na q:±2.5
Pr^{3+}	-10.0	8.3(25°,1-2)	121	25-45	T	μ=0.1M(KNO_3)	74Ta	b,q: ±0.8
	-10.9	6.3(25°)	84	25-45	T	μ=0.1M(KNO_3) R:Pr^{3+}+2HL^{8-}= $PrH_2L_2{}^{5-}$	74Ta	b,q: ±0.8
PrA^-	9.2	4.75(25°,1)	58	2-45	T	μ=0.10M(KNO_3) A=$EDTA^{2-}$	73Kg	b,g:Na
	-15.9	3.10(25°)	6.3	2-45	T	μ=0.10M(KNO_3) A=$EDTA^{2-}$ R:PrA^- +HL^{4-}=$PrAHL^{5-}$	73Kg	b,g:Na q:±4.2
Nd^+	-18.8	18.5(25°,1-2)	96	25-45	T	μ=0.1M(KNO_3)	74Ta	b
	-10.9	6.4(25°)	84	25-45	T	μ=0.1M(KNO_3) R:Nd^{3+}+2HL^{8-}= $NdH_2L_2{}^{5-}$	74Ta	b,q: ±0.8
NdA^-	-10.5	4.90(25°,1)	58	2-45	T	μ=0.10M(KNO_3) A=$EDTA^{2-}$	73Kg	b,g:Na q;±1.7
	-17.2	3.20(25°)	4	2-45	T	μ=0.10M(KNO_3) A=$EDTA^{2-}$ R:NdA^- +HL^{4-}=$NdAHL^{5-}$	73Kg	b,g:Na q:±3.8
Sm^{3+}	-28.4	8.7(25°,1-2)	71	25-45	T	μ=0.1M(KNO_3)	74Ta	b
	-10.9	6.5(25°)	88	25-45	T	μ=0.1M(KNO_3) R:Sm^{3+}+2HL^{8-}= $SmH_2L_2{}^{5-}$	74Ta	b,q: ±0.8

Metal Ion	ΔH, kJ /mol	Log K	ΔS, J /mol K	T,°C	M	Conditions	Ref.	Re-marks
P95, continued								
SmA^-	-8.4	5.10(25°,1)	67	2-45	T	μ=0.10M(KNO_3) A=$EDTA^{2-}$	73Kg	b,g:Na q:±2.9
	-15.1	3.35(25°)	12	2-45	T	μ=0.10M(KNO_3) R:SmA^- +HL^{4-} = $SmAHL^{5-}$	73Kg	b,g:Na q:±2.9
Eu^{3+}	-28.9	8.8(25°,1-2)	71	25-45	T	μ=0.1M(KNO_3)	74Ta	b
	-10.9	6.6(25°)	88	25-45	T	μ=0.1M(KNO_3) R:Eu^{3+}+2HL^{8-} = $EuH_2L_2{}^{5-}$	74Ta	b,q: ±0.8
EuA^-	-9.6	5.24(25°,1)	67	2-45	T	μ=0.10M(KNO_3) A=$EDTA^{2-}$	73Kg	b,g:Na q:±2.9
	-11.7	3.44(25°)	25	2-45	T	μ=0.10M(KNO_3) A=$EDTA^{2-}$ R:EuA^- +HL^{4-}=$EuAHL^{5-}$	73Kg	b,g:Na q:±2.9
Gd^{3+}	-28.9	8.9(25°,1-2)	71	25-45	T	μ=0.1M(KNO_3)	74Ta	b
	-10.9	6.6(25°)	88	25-45	T	μ=0.1M(KNO_3) R:Gd^{3+}+2HL^{8-} = GdH_2L_2	74Ta	b,q: ±0.8
GdA^-	-8.8	5.31(25°,1)	71	2-45	T	μ=0.10M(KNO_3) A=$EDTA^{2-}$	73Kg	b,g:Na q:±2.1
	-9.6	3.58(25°)	38	2-45	T	μ=0.10M(KNO_3) A=$EDTA^{2-}$ R:GdA^- +HL^{4-}=$GdAHL^{5-}$	73Kg	b,g:Na q:±0.8
Tb^{3+}	-28.4	9.1(25°,1-2)	75	25-45	T	μ=0.1M(KNO_3)	74Ta	b
	-19.7	6.8(25°)	63	25-45	T	μ=0.1M(KNO_3) R:Tb^{3+}+2HL^{8-} = $TbH_2L_2{}^{5-}$	74Ta	b
TbA^-	-20.5	5.54(25°,1)	38	2-45	T	μ=0.10M(KNO_3) A=$EDTA^{2-}$	73Kg	b,g:Na q:±1.2
	-13.0	3.95(25°)	33	2-45	T	μ=0.10M(KNO_3) A=$EDTA^{2-}$ R:TbA^- +HL^{4-}=$TbAHL^{5-}$	73Kg	b,g:Na q:±1.7
Dy^{3+}	-20.1	9.2(25°,1-2)	109	25-45	T	μ=0.1M(KNO_3)	74Ta	b
	-20.1	6.9(25°)	63	25-45	T	μ=0.1M(KNO_3) R:Dy^{3+}+2HL^{8-} = $DyH_2L_2{}^{5-}$	74Ta	b
DyA^-	-11.7	5.71(25°,1)	71	2-45	T	μ=0.10M(KNO_3) A=$EDTA^{2-}$	73Kg	b,g:Na q:±1.2
	-13.4	4.18(25°)	33	2-45	T	μ=0.10M(KNO_3) A=$EDTA^{2-}$ R:DyA^- +HL^{4-}=$DyAHL^{5-}$	73Kg	b,g:Na q:±2.5
Ho^{3+}	-28.4	9.3(25°,1-2)	109	25-45	T	μ=0.1M(KNO_3)	74Ta	b
	-28.4	7.1(25°)	38	25-45	T	μ=0.1M(KNO_3) R:Ho^{3+}+2HL^{8-} = $HoH_2L_2{}^{5-}$	74Ta	b
HoA^-	-10.0	6.50(25°,1)	88	2-45	T	μ=0.10M(KNO_3) A=$EDTA^{2-}$	73Kg	b,g:Na q:±1.7
	-7.5	4.43(25°)	58	2-45	T	μ=0.10M(KNO_3) A=$EDTA^{2-}$ R:HoA^-	73Kg	b,g:Na q:±1.7

Metal Ion	ΔH, kJ /mol	Log K	ΔS, J /mol K	T,°C	M	Conditions	Ref.	Remarks
P95, continued								
HoA, cont.						$+HL^{4-}=HoAHL^{5-}$		
Er^{3+}	-28.0	9.5(25°,1-2)	84	25-45	T	μ=0.1M(KNO_3)	74Ta	b
	-19.2	7.2(25°)	71	25-45	T	μ=0.1M(KNO_3)	74Ta	b
						R:$Er^{3+}+2HL^{8-}=$ $ErH_2L_2{}^{5-}$		
ErA^-	-9.2	7.01(25°,1)	105	2-45	T	μ=0.10M(KNO_3) A=$EDTA^{2-}$	73Kg	b,g:Na q:±1.2
	-13.8	4.95(25°)	46	2-45	T	μ=0.10M(KNO_3) A=$EDTA^{2-}$ R:ErA^- $+HL^{4-}=ErAHL^{5-}$	73Kg	b,g:Na q:±8.4
Tm^{3+}	-28.0	9.7(25°,1-2)	88	25-45	T	μ=0.1M(KNO_3)	74Ta	b
	-19.2	7.3(25°)	75	25-45	T	μ=0.1M(KNO_3)	74Ta	b
						R:$Tm^{3+}+2HL^{8-}=$ $TmH_2L_2{}^{5-}$		
TmA^-	-16.3	7.64(25°,1)	92	2-45	T	μ=0.10M(KNO_3) A=$EDTA^{2-}$	73Kg	b,g:Na q:±4.2
	-19.7	5.10(25°)	25	2-45	T	μ=0.10M(KNO_3) A=$EDTA^{2-}$ R:TmA^- $+HL^{4-}=TmAHL^{5-}$	73Kg	b,g:Na q:±3.8
Mn^{2+}	11.7	8.04(1)	194.1	20	C	0.1M$(CH_3)_4NNO_3$	65Aa	
	-25.1	6.20(25°,1)	35.1	2-45	T	μ=0.10M(KNO_3)	73Ke	b
	-0.04	4.30(25°)	82.0	2-45	T	μ=0.10M(KNO_3)	73Ke	b,q: ±0.04
						R:$Mn^{2+}+HL^{4-}=$ $MnHL^{2-}$		
Co^{2+}	18.87	7.95(1)	216.3	20	C	0.1M$(CH_3)_4NNO_3$	65Aa	
	-29.7	6.95(25°,1)	40.6	2-45	T	μ=0.10M(KNO_3)	73Ke	b,q:±9.2
	-0.84	4.05(25°)	74.5	2-45	T	μ=0.10M(KNO_3)	73Ke	b,q: ±0.04
						R:$Co^{2+}+HL^{4-}=CoHL^{2-}$		
Ni^{2+}	20.9	7.8(1)	220.5	20	C	0.1M$(CH_3)_4NNO_3$	65Aa	
	-6.7	7.07(25°,1)	105	2-45	T	μ=0.10M(KNO_3)	73Ke	b,q:±0.4
	0.12	3.95(25°)	76.1	2-45	T	μ=0.10M(KNO_3)	73Ke	b,q: ±0.04
						R:$Ni^{2+}+HL^{4-}=$ $NiHL^{2-}$		
Cu^{2+}	20.5	9.3(1)	247.1	20	C	0.1M$(CH_3)_4NNO_3$	65Aa	
	-34.3	8.20(25°,1)	42	2-45	T	μ=0.10M(KNO_3)	73Ke	b,q:±2.1
	-11.7	5.20(25°)	59.8	2-45	T	μ=0.10M(KNO_3)	73Ke	b,q: ±2.1
						R:$Cu^{2+}+HL^{4-}=$ $CuHL^{2-}$		
Zn^{2+}	26.44	8.35(1)	250.2	20	C	0.1M$(CH_3)_4NNO_3$	65Aa	
	-18.4	6.83(25°,1)	67	2-45	T	μ=0.10M(KNO_3)	73Ke	b,q:±7.9
	-5.8	3.75(25°)	51.9	2-45	T	μ=0.10M(KNO_3)	73Ke	b,q: ±0.1
						R:$Zn^{2+}+HL^{4-}=$ $ZnHL^{2-}$		
Cd^{2+}	11.3	8.1(1)	193.3	20	C	0.1M$(CH_3)_4NNO_3$	65Aa	

Metal Ion	ΔH, kJ /mol	Log K	ΔS, J /mol K	T,°C	M	Conditions	Ref.	Re-marks
P96	PHOSPHINE, bis(dimethylamine) fluoro-		$C_4H_{12}N_2FP$			$[(CH_3)_2N]_2PF$		
CoA_2	-25.1	-0.252(30°,1)	-88	24-66	T	S:CH_2Cl_2 CM=0.37M CL=0.9M A=Cl	72Nc	b
	-25.1	-0.420(28°,1)	-92	9-46	T	S:CH_2Cl_2 CM=0.37M CL=1.0M A=Cl	72Nc	b
	-22.6	0.577(30°,1)	-84	23-43	T	S:CH_2Cl_2 CM=0.37M CL=1.25M A=Cl	72Nc	b
	-18.0	0.471(22°,1)	-54	8-41	T	S:CH_2Cl_2 CM=0.37M CL=1.0M A=Br	72Nc	b
P97	PHOSPHINE, bis(dimethylamine) methyl, oxide		$C_5H_9ON_2P$			$[(CH_3)_2N]_2CH_3PO$		
$SnAB_3$	-52.3	>12(1)	<-75	25	C	S:Benzene $A=C_2H_5$ B=Cl	79Kc	
	-49.0	8.9(2)	<-90	25	C	S:Benzene $A=C_2H_5$ B=Cl	79Kc	
SnA_2B_2	-40.6	7.2(1)	-131.8	25	C	S:Benzene $A=C_2H_5$ B=Cl	79Kc	
	-58	<1.5(2)	>-58	25	C	S:Benzene $A=C_2H_5$ B=Cl	79Kc	q:±17
P98	PHSOPHINE, butyl(diphenyl)		$C_{16}H_{19}P$			$(C_6H_5)_2PC_4H_9$		
HgA_2	90	(1)	--	30	C	S:Benzene A=Cl	76Ga	
	-37	(2)	--	30	C	S:Benzene A=Cl	76Ga	
	-92	(1)	--	30	C	S:Benzene A=Br	76Ga	
	-43	(2)	--	30	C	S:Benzene A=Br	76Ga	q:±3
	-80	(1)	--	30	C	S:Benzene A=I	76Ga	
	-48	(2)	--	30	C	S:Benzene A=I	76Ga	
P99	PHOSPHINE, chlorodiethyl, sulfide		$C_4H_{10}ClSP$			$(CH_3CH_2)_2P(Cl)S$		
I_2	-15.1	0.462(20°,1)	-42	10-40	T	S:n-Heptane Results from UV data	71La	b,q: ±1.7
	-18.4	0.491(20°,1)	-54	10-40	T	S:n-Heptane Results from Visible data	71La	b
P100	PHOSPHINE, cyclohexyldiethyl-		$C_8H_{21}P$			$C_6H_{11}P(C_2H_5)_2$		L
CoL_2A_2	-17.2	0.301(10°,1)	-54	-30-20	T	S:1,2-Dichloro-ethane A=NCS	67Bb	b

Metal Ion	ΔH, kJ /mol	Log K	ΔS, J /mol K	T,°C	M	Conditions	Ref.	Re-marks
P101	PHOSPHINE, dibutoxy (butyl), oxide		$C_{12}H_{27}O_3P$			$(C_4H_9O)_2(C_4H_9)PO$		
$SnAB_3$	-46.0	6.3(1)	-102.5	25	C	S:Benzene $A=C_2H_5$ B=Cl	79Kc	
	-29.7	3.1(2)	-74.5	25	C	S:Benzene $A=C_2H_5$ B=Cl	79Kc	
P102	PHOSPHINE, dibutoxy (methylsulfide), oxide		$C_9H_{21}O_3SP$			$(C_4H_9O)_2(CH_3S)PO$		
$SnAB_3$	-39.7	5.2(1)	-131.8	25	C	S:Benzene $A=C_2H_5$ B=Cl	79Kc	
SnA_2B_2	-37.2	2.7(1)	-102.5	25	C	S:Benzene $A=C_2H_5$ B=Cl	79Kc	
P103	PHOSPHINE, dibutyl (methylsulfide)-, oxide		$C_9H_{21}OSP$			$(CH_3S)(C_4H_9)_2PO$		
$SnCl_4$	-65.7	12.0(1)	-120.9	25	C	S:Benzene A=Cl	79Kc	
	-54.0	9.5(2)	-102.5	25	C	S:Benzene A=Cl	79Kc	
P104	PHOSPHINE, dibutylphenyl		$C_{14}H_{23}P$			$(C_4H_9)_2PC_6H_5$		
HgA_2	-108	(1)	--	30	C	S:Benzene A=Cl	76Ga	
	-49	(2)	--	30	C	S:Benzene A=Cl	76Ga	
	-107	(1)	--	30	C	S:Benzene A=Br	76Ga	
	-44	(2)	--	30	C	S:Benzene A=Br	76Ga	
	-86	(1)	--	30	C	S:Benzene A=I	76Ga	
	-46	(2)	--	30	C	S:Benzene A=I	76Ga	
P105	PHOSPHINE, dichloro(phenyl), oxide		$C_6H_5OCl_2P$			$C_6H_5Cl_2PO$		
SnA_4	-40.6	4.7(1)	-26.8	25	C	S:Benzene A=Cl	79Kc	
	-45.2	2.6(2)	-15.5	25	C	S:Benzene A=Cl	79Kc	
P106	PHOSPHINE, diethoxy(methyl), sulfide		$C_5H_{13}O_2SP$			$(C_2H_5O)_2(CH_3)PS$		
I_2	-17.15	3.4(1)	29.3	25	C	S:CCl_4	79Kc	
P107	PHOSPHINE, diethylphenyl-		$C_8H_{15}P$			$(C_2H_5)_2PC_6H_5$		L
CoL_2A_2	-77.4	3.556(10°,1)	-205	-30-20	T	S:1,2-Dichloro-ethane A=NCS	67Bb	b
NiL_2A_2	-77.8	1.970(1)	-226	20	T	S:$C_2H_4Cl_2$ A=CN	69Rc	
	-67.4	3.120(1)	-172	20	T	S:C_2H_5OH A=CN	69Rc	
	-71.1	(1)	--	20	C	S:C_2H_5OH A=CN	69Rc	

Metal Ion	ΔH, kJ /mol	Log K	ΔS, J /mol K	T,°C	M	Conditions	Ref.	Remarks
P108	PHOSPHINE, (dimethylamine) dichloro-, oxide		$C_2H_6ONCl_2P$			$[(CH_3)_2N]Cl_2PO$		
SnA_4	-47.3	5.3(1)	-115.1	25	C	S:Benzene A=Cl	79Kc	q:±6
	-38.5	3.6(2)	-99.6	25	C	S:Benzene A=Cl	79Kc	
P109	PHOSPHINE, dimethyl (dimethylamine), oxide		$C_4H_{12}ONP$			$(CH_3)_2[(CH_3)_2N]PO$		
SnA_2B_2	-45.2	8.0(1)	-85.7	25	C	S:Benzene A=C_2H_5 B=Cl	79Kc	
	-38	<3(2)	>-100	25	C	S:Benzene A=C_2H_5 B=Cl	79Kc	q:±12
P110	PHOSPHINE, di(4-methyl-naphthyl)phenyl, sulfide		$C_{28}H_{23}SP$			$(C_6H_5)(CH_3C_{10}H_6)_2PS$		
I_2	-25.6	2.32(25°,1)	-41.4	25-45	T	S:$CHCl_3$	66Ta	b
P111	PHOSPHINE, dimethylphenyl		$C_8H_{11}P$			$(CH_3)_2PC_6H_5$		
$PtAB_2^+$	-104.6	(1)	--	25	C	S:THF A=CH_3 B=$[P(CH_3)_2(C_6H_5)]$	76Ma	
HgA_2	-109	(1)	--	30	C	S:Benzene A=Cl	76Ga	
	-37	(2)	--	30	C	S:Benzene A=Cl	76Ga	
	-103	(1)	--	30	C	S:Benzene A=Br	76Ga	
	-43	(2)	--	30	C	S:Benzene A=Br	76Ga	
	-92	(1)	--	30	C	S:Benzene A=I	76Ga	
	-36	(2)	--	30	C	S:Benzene A=I	76Ga	q:±3
P112	PHOSPHINE, di(1-naphthyl) phenyl, sulfide		$C_{26}H_{19}SP$			$(C_6H_5)(C_{10}H_7)_2PS$		
I_2	-28.4	2.03(25°,1)	-56.6	25-45	T	S:$CHCl_3$	66Ta	b
P113	PHOSPHINE, diphenyl (N,N-diethyl-2-amino-ethane), oxide		$C_{18}H_{24}ONP$			$(C_6H_5)_2P(=O)CH_2CH_2N(C_2H_5)_2$		
CuA^+	0	3.88(25°,1)	75	5-45	T	μ<0.005 S:1v Acetone+ 1v Nitromethane+ 0.5v Chloroform A=Cl^-	79Ya	b
P114	PHOSPHINE, diphenylethene		$C_{14}H_{13}P$			$(C_6H_5)_2PCH=CH_2$		
HgA_2	-85	(1)	--	30	C	S:Benzene A=Cl	76Ga	
	-38	~3.38(2)	--	30	C	S:Benzene A=Cl	76Ga	
	-78	(1)	--	30	C	S:Benzene A=Br	76Ga	
	-42	~3.60(2)	--	30	C	S:Benzene A=Br	76Ga	q:±3

Metal Ion	ΔH, kJ /mol	Log K	ΔS, J /mol K	T, °C	M	Conditions	Ref.	Remarks
P114, continued								
HgA_2,	-60	(1)	--	30	C	S:Benzene A=I	76Ga	
cont.	-49	~3.38(2)	--	30	C	S:Benzene A=I	76Ga	
P115	PHOSPHINE, diphenylethyl-		$C_{14}H_{15}P$			$(C_6H_5)_2PC_2H_5$		L
CoL_2A_2	-58.2	1.00(10°,1)	-188	-30-20	T	S:1,2-Dichloro-ethane A=NCS	67Bb	b
NiL_2A_2	-37.2	0.380(1)	-121	20	T	S:$C_2H_4Cl_2$ A=CN	69Rc	
	-28.4	1.508(1)	-67	20	T	S:C_2H_5OH A=CN	69Rc	
Ga_2A_6	-115.1	--	--	25	C	S:Benzene CM= 0.03-0.04M A=Cl R:½Ga_2A_6+L=GaA_3L	73Rd	
P116	PHOSPHINE, diphenylethylene		$C_{14}H_{14}P$			$(C_6H_5)_2PC_2H_4$		
HgA_2	-130	(1)	--	30	C	CM=0.4x10^{-3} S:Benzene A=Cl	74Fb	
	-130.1	~7(1)	-295	30	C	CM=0.4-0.6x10^{-3} S:Benzene A=Br	74Fb	
	-118.8	6.30(1)	-270	30	C	CM=0.5-0.8x10^{-3} S:Benzene A=I	74Fb	
P117	PHOSPHINE, diphenylmethyl		$C_{13}H_{13}P$			$(C_6H_5)_2PCH_3$		
$PtAB_2^+$	-92.5	(1)	--	25	C	S:THF A=CH_3 B=[$P(CH_3)_2(C_6H_5)$]	75Ma	
HgA_2	-92	(1)	--	30	C	S:Benzene A=Cl	76Ga	
	-37	(2)	--	30	C	S:Benzene A=Cl	76Ca	
	-89	(1)	--	30	C	S:Benzene A=Br	76Ga	
	-39	(2)	--	30	C	S:Benzene A=Br	76Ga	q:±3
	-82	(1)	--	30	C	S:Benzene A=I	76Ga	
	-38	(2)	--	30	C	S:Benzene A=I	76Ga	
P118	PHOSPHINE, diphenyl-4-(methylnaphthyl), selenide		$C_{25}H_{18}PSe$			$(C_6H_5)_2(CH_3C_{10}H_6)PSe$		
I_2	-38.1	3.62(25°,1)	-58	5-35	T	S:$CHCl_3$	68Cf	b
P119	PHOSPHINE, diphenyl-4-methyl, sulfide		$C_{13}H_{13}SP$			$(C_6H_5)_2(CH_3)PS$		
I_2	-23.9	2.10(25°,1)	-39.9	25-45	T	S:$CHCl_3$	66Ta	b
P120	PHOSPHINE, diphenyl-1-naphthyl, selenide		$C_{22}H_{17}PSe$			$(C_6H_5)_2(C_{10}H_7)PSe$		
I_2	-41.8	3.71(25°,1)	-69	5-35	T	S:$CHCl_3$	68Cf	b

Metal Ion	ΔH, kJ /mol	Log K	ΔS, J /mol K	T,°C	M	Conditions	Ref.	Re-marks
P121	PHOSPHINE, diphenyl-1-naphthyl, sulfide		$C_{22}H_{17}SP$			$(C_6H_5)_2(C_{10}H_7)PS$		
I_2	-25.1	2.04(25°,1)	-45.2	25-45	T	S:$CHCl_3$	66Ta	b
P122	PHOSPHINE, diphenylpropyl		$C_{15}H_{17}P$			$(C_6H_5)_2PC_3H_7$		
HgA_2	-90	(1)	--	30	C	S:Benzene A=Cl	76Ga	
	-41	(2)	--	30	C	S:Benzene A=Cl	76Ga	
	-90	(1)	--	30	C	S:Benzene A=Br	76Ga	
	-39	(2)	--	30	C	S:Benzene A=Br	76Ga	q:±3
	-79	(1)	--	30	C	S:Benzene A=I	76Ga	
	-45	(2)	--	30	C	S:Benzene A=I	76Ga	
P123	PHOSPHINE, dipropylphenyl-		$C_{12}H_{19}P$			$(CH_3CH_2CH_2)_2PC_6H_5$		
	-100	(1)	--	30	C	S:Benzene A=Cl	76Ga	
	-37	(2)	--	30	C	S:Benzene A=Cl	76Ga	q:±4
	-105	(1)	--	30	C	S:Benzene A=Br	76Ga	
	-48	(2)	--	30	C	S:Benzene A=Br	76Ga	
	-87	(1)	--	30	C	S:Benzene A=I	76Ga	
	-51	(2)	--	30	C	S:Benzene A=I	76Ga	
Al_2A_6	-115.9	--	--	25	C	S:Benzene CM= 0.03-0.04M A=Br R:½Al_2A_6+L=AlA_3L	73Rd	
P124	PHOSPHINE, (dipropylsulfide) ethyl		$C_8H_{19}S_2P$			$(C_3H_7S)_2PC_2H_5$		
SnA_4	-44.6	6.6(1)	-95.0	25	C	S:Benzene A=Cl	79Kc	
P125	PHOSPHINE, tribenzyl-		$C_{21}H_{21}P$			$P(CH_3C_6H_4)_3$		
Ni°	-54	1.204(30°,4)	-155	10-70	T	S:Benzene	70Gi	b,q:±6.3
$PtAB_2^+$	-97.1	(1)	--	25	C	S:THF A=CH_3 B=$[P(CH_3)_2(C_6H_5)]$	75Ma	
P126	PHOSPHINE, tributoxy		$C_{12}H_{27}O_3P$			$(CH_3CH_2CH_2CH_2O)_3P$		
SnA_4	-74.0	9.4(1)	-169.9	25	C	S:Benzene A=Cl	79Kc	
	-58.2	5.9(2)	-146.0	25	C	S:Benzene A=Cl	79Kc	
P127	PHOSPHINE, tri(tert-butoxy)		$C_{12}H_{27}O_3P$			$P[OC(CH_3)_3]_3$		
$PtAB_2^+$	-74.9	(1)	--	25	C	S:THF A=CH_3 B=$[P(CH_3)_2(C_6H_5)]$	75Ma	
P128	PHOSPHINE, tributyl-		$C_{12}H_{27}P$			$P(CH_2CH_2CH_2CH_3)_3$		L
NiA	-41.1	1.84(1)	-100	30	C	S:Benzene CM=2-5x10^{-3}M	76Gh	

Metal Ion	ΔH, kJ /mol	Log K	ΔS, J /mol K	T,°C	M	Conditions	Ref.	Remarks
P128, continued								
NiA, cont.						A=H-bridged glyoximato R_1=Me R_2=iso-Pr		
	-43.5	1.73(1)	-110	30	C	S:Benzene CM=1-4x10^{-3}M A=H-bridged glyoximato R_1=Me R_2=Bu	76Gh	
	-45.4	2.96(1)	-93	30	C	S:Benzene CM=1-2x10^{-3}M A=BF_2-bridged glyoximato R_1=Me R_2=Et	76Gh	
	-39.4	2.74(1)	-78	30	C	S:Benzene CM=2-4x10^{-3}M A=BF_2-bridged glyoximato R_1=Me R_2=Pr	76Gh	
	-43.7	3.01(1)	-87	30	C	S:Benzene CM=1-2x10^{-3}M A=BF_2-bridged glyoximato R_1=Me R_2=iso-Pr	76Gh	
	-42.0	1.75(1)	-105	30	C	S:Benzene CM=2-4x10^{-3}M A=H-bridged glyoximato R_1=Me R_2=Et	76Gh	
	-40.6	1.77(1)	-100	30	C	S:Benzene CM=2-5x10^{-3}M A=H-bridged glyoximato R_1=Me R_2=Pr	76Gh	
	-61.9	5.75(1-2)	-86.2	25	T	CM=10^{-4}M S:Benzene A=Diacetylbisbenzoyl hydrazine	60Sd	
NiL_2A_2	-48.5	0.447(0°,1)	-159	20	T	S:C_2H_5OH A=CN	69Rc	b
	-45.2	0.681(0°,1)	-151	20	T	S:n-Hexane A=CN	69Rc	b
PdA_2	-53	1.00(1)	-156	30	C	S:Benzene	78Gb	
	-53	1.60(2)	-144	30	C	S:Benzene	78Gb	

Metal Ion	ΔH, kJ /mol	Log K	ΔS, J /mol K	T,°C	M	Conditions	Ref.	Remarks
P128, continued								
HgA_2	-123.0	(1)	--	30	C	S:Benzene CM=1-5x10^{-3}M A=Cl	73Fa 76Ga	
	-122.2	(1)	--	30	C	S:Benzene CM=1-5x10^{-3}M A=Br	73Fa 76Ga	
	-97.9	(1)	--	30	C	S:Benzene CM=1-5x10^{-3}M A=I	73Fa 76Ga	
	-54.0	(2)	--	30	C	S:Benzene CM=1-5x10^{-3}M A=Cl	73Fa 76Ga	
	-58.2	(2)	--	30	C	S:Benzene CM=1-5x10^{-3}M A=Br	73Fa 76Ga	
	-69.0	(2)	--	30	C	S:Benzene CM=1-5x10^{-3}M A=I	73Fa 76Ga	
BA_3C	-82.4	--	--	25	C	S:Benzene CM= 0.03-0.04M A=F C=Diethyl ether R:$BA_3C+L=BA_3L+C$	73Rd	
Al_2A_6	-143.1	--	--	25	C	S:Benzene CM= 0.03-0.04M A=Br R:$\frac{1}{2}Al_2A_6+L=AlA_3L$	73Rd	
Ga_2A_6	-146.4	--	--	25	C	S:Benzene CM= 0.03-0.04M A=Cl R:$\frac{1}{2}Ga_2A_6+L=GaA_3L$	73Rd	
SnA_4	-141.8	(1)	--	25	C	S:Benzene A=Cl	79Kc	
	-98.7	(2)	--	25	C	S:Benzene A=Cl	79Kc	
SnA_2B_2	-73.7	4.28(1)	-161	30	C	S:Benzene A=Phenyl B=Cl	75Fa	
	-54.0	3.00(1)	-121	30	C	S:Benzene A=Methyl B=Cl	75Fa	
	-54.8	2.51(1)	-133	30	C	S:Benzene A=Butyl B=Cl	75Fa	
	-78.0	>4(1)	<-182	30	C	S:Benzene A=Butyl B=NCS	75Fa	
	-26.2	1.053(1)	-66	25	C	S:Benzene A=Bu B=(OCOMe)	77Ge	
	-26.2	0.863(1)	-70	25	C	S:Benzene A=Bu B=(OCOEt)	77Ge	
	-38.3	1.260(1)	-102	25	C	S:Benzene A=Bu B=(NCO)	77Ge	
$SnAB_3$	-92.6	4.30(1)	-222	30	C	S:Benzene A=Butyl B=Cl	75Fa	
	-94.3	>5(1)	<-215	30	C	S:Benzene A=Phenyl B=Cl	75Fa	
	-101.3	8.7(1)	-266.9	25	C	S:Benzene A=C_2H_5 B=Cl	79Kc	
	-75.1	3.38(2)	-181	30	C	S:Benzene A=Phenyl B=Cl	75Fa	
	-30.5	(2)	--	25	C	S:Benzene A=C_2H_5 B=Cl	79Kc	

Metal Ion	ΔH, kJ /mol	Log K	ΔS, J /mol K	T,^{o}C	M	Conditions	Ref.	Remarks
P129	PHOSPHINE, tri(tert-butyl)		$C_{12}H_{27}P$			$P(C_4H_9)_3$		
$PtAB_2^+$	-20.1	(1)	--	25	C	S:THF A=CH_3 B=$[P(CH_3)_2(C_6H_5)]$	75Ma	
P130	PHOSPHINE, tributyl, oxide		$C_{12}H_{27}OP$			$(C_4H_9)_3PO$		
$SnAB_3$	-75.7	>11.5(1)	<-159	25	C	S:Benzene A=C_2H_5 B=Cl	79Kc	
	-49.4	>12.0(2)	<-67	25	C	S:Benzene A=C_2H_5 B=Cl	79Kc	
SnA_2B_2	-52.7	8.0(1)	-102.5	25	C	S:Benzene A=CH_3 B=Cl	79Kc	
	-33.5	<3(2)	>-87	25	C	S:Benzene A=CH_3 B=Cl	79Kc	q:±12
	-53.6	7.3(1)	-119.2	25	C	S:Benzene A=C_2H_5 B=Cl	79Kc	
SnA_3B	-25.9	<7(1)	-29	25	C	S:Benzene A=C_2H_5 B=Cl	79Kc	
	-11.7	<5(2)	--	25	C	S:Benzene A=C_2H_5 B=Cl	79Kc	q:±5
I_2	-26.8	4.5(1)	-51.9	25	C	S:CCl_4	79Kc	
P131	PHOSPHINE, tri(2-tert-butylphenoxy)		$C_{30}H_{39}O_3P$			$P[OC_6H_4(C_4H_9)]_3$		
$PtAB_2^+$	-38.1	--	--	25	C	S:THF R:2$PtAB_2$+L=PtABL+B A=CH_3 B=$[P(CH_3)_2(C_6H_5)]$	75Ma	
P132	PHOSPHINE, tributyl, sulfide		$C_{12}H_{27}SP$			$(C_4H_9)_3PS$		
SnA_4	-65.3	11.5(1)	-57.7	25	C	S:Benzene A=Cl	79Kc	
	-54.8	9.0(2)	-109.6	25	C	S:Benzene A=Cl	79Kc	
P133	PHOSPHINE, tri(2-chloroethoxy)		$C_6H_{12}O_3Cl_3P$			$P(OCH_2CH_2Cl)_3$		
$PtAB_2^+$	-110.4	(1)	--	25	C	S:THF A=CH_3 B=$[P(CH_3)_2(C_6H_5)]$	75Ma	
P134	PHOSPHINE, trichloro, oxide		OCl_3P			Cl_3PO		
NbA_5	-31.13	(1)	--	--	C	S:CCl_4 A=Cl	76Na	u:25°
	-24.06	(1)	--	--	C	S:Dichloroethane A=Cl	76Na	u:25°
	9.08	(1)	--	--	C	S:Nitromethane A=Cl	76Na	u:25°

Metal Ion	ΔH, kJ /mol	Log *K*	ΔS, J /mol K	T,°C	M	Conditions	Ref.	Remarks
P135	PHOSPHINE, tri(2-cyanoethyl)		$C_9H_{12}N_3P$			$P(CH_2CH_2CN)_3$		
$PtAB_2^+$	-73.6	(1)	--	25	C	S:THF A=CH_3 B=$[P(CH_3)_2(C_6H_5)]$	75Ma	
P136	PHOSPHINE, tri(2-cyanophenoxy)		$C_{21}H_{12}O_3N_3P$			$P(OC_6H_4CN)_3$		
$PtAB_2^+$	-78.6	(1)	--	25	C	S:THF A=CH_3 B=$[P(CH_3)_2(C_6H_5)]$	75Ma	
P137	PHOSPHINE, tri(cyclohexoxy)		$C_{18}H_{33}P$			$P(C_6H_{11})_3$		
$PtAB_2^+$	-66.5	(1)	--	25	C	S:THF A=CH_3 B=$[P(CH_3)_2(C_6H_5)]$	75Ma	
P138	PHOSPHINE, tricyclohexyl		$C_{18}H_{33}P$			$(C_6H_{11})_3P$		
HgA_2	-119	(1)	--	30	C	S:Benzene A=Cl	76Ga	
	-59	(2)	--	30	C	S:Benzene A=Cl	76Ga	
	-111	(1)	--	30	C	S:Benzene A=Br	76Ga	
	-60	(2)	--	30	C	S:Benzene A=Br	76Ga	
	-102	(1)	--	30	C	S:Benzene A=I	76Ga	
	-58	(2)	--	30	C	S:Benzene A=I	76Ga	
P139	PHOSPHINE, tri(2,5-dimethylphenoxy)		$C_{24}H_{27}O_3P$			$P[OC_6H_3(CH_3)_2]_3$		
$PtAB_2^+$	-2.1	(1)	--	25	C	S:THF A=CH_3 B=$[P(CH_3)_2(C_6H_5)]$	75Ma	
P140	PHOSPHINE, triethoxy		$C_6H_{15}OP$			$P(OC_2H_5)_3$		
$PtAB_2^+$	-105.4	(1)	--	25	C	S:THF A=CH_3 B=$[P(CH_3)_2(C_6H_5)]$	75Ma	
P141	PHOSPHINE, triethyl-		$C_6H_{15}P$			$P(CH_2CH_3)_3$		L
CoL_2A_2	-79.5	3.285(10°,1)	-218	-30-20	T	S:1,2-Dichloroethane A=NCS	67Ed	b
Ni°	-67	1.921(25°,4)	-188	-50-70	T	S:Benzene	74Tc	b,q:±8
NiA	58.2	6.18(1-2)	-76.6	25	T	CM=10^{-4}M S:Benzene A=Diacetylbisbenzoyl hydrazine	60Sd	
NiL_2A_2	-56.1	0.279(0°,1)	-184	20	T	S:$C_2H_4Cl_2$ A=CN	69Fc	
	-44.8	1.127(0°,1)	-130	20	T	S:C_2H_5OH A=CN	69Fc	
$PtAB_2^+$	-101.7	(1)	--	25	C	S:THF A=CH_3 B=$[P(CH_3)_2(C_6H_5)]$	75Ma	
AlA_3	-92.55	(1)	--	25-38	C	S:n-Hexane A=CH_3	68Hc	

Metal Ion	ΔH, kJ /mol	Log K	ΔS, J /mol K	T,°C	M	Conditions	Ref.	Remarks
P142 PHOSPHINE, trihexyl-			$C_{18}H_{39}P$			$P[CH_3(CH_2)_5]_3$		L
Al_2A_6	-138.9	--	--	25	C	S:Benzene CM= 0.03-0.04M A=Br R:½Al_2A_6+L=AlA_3L	73Rd	
P143 PHOSPHINE, triisobutyl, oxide			$C_{12}H_{27}OP$			$(i\text{-}C_4H_9)_3PO$		
SnA_2B_2	-50.2	6.5(1)	-115.1	25	C	S:Benzene $A=C_2H_5$ B=Cl	79Kc	
	-33	<2.5(2)	>-91.2	25	C	S:Benzene $A=C_2H_5$ B=Cl	79Kc	q:±17
P144 PHOSPHINE, tri(isopropoxy)			$C_9H_{21}O_3P$			$P(OC_3H_7)_3$		
$PtAB_2^+$	-104.2	(1)	--	25	C	S:THF $A=CH_3$ $B=[P(CH_3)_2)(C_6H_5)]$	75Ma	
SnA_4	-71.1	9.1(1)	-162.8	25	C	S:Benzene A=Cl	79Kc	
	-25.5	7.0(2)	-28.0	25	C	S:Benzene A=Cl	79Kc	
P145 PHOSPHINE, tri(isopropyl)			$C_9H_{21}P$			$P(C_3H_7)_3$		
$PtAB_2^+$	-53.1	(1)	--	25	C	S:THF $A=CH_3$ $B=[P(CH_3)_2(C_6H_5)]$	75Ma	
P146 PHOSPHINE, tri(2-isopropylphenoxy)			$C_{27}H_{33}O_3P$			$P[OC_6H_4(C_3H_7)]_3$		
$PtAB_2^+$	-71.5	(1)	--	25	C	S:THF $A=CH_3$ $B=[P(CH_3)_2(C_6H_5)]$	75Ma	
P147 PHOSPHINE, trimethoxy			$C_3H_9O_3P$			$P(OCH_3)_3$		
IrA_2BC_2	-87.9	5.04(1)	-197	25	C	S:CH_2Cl_2 A=Cl $B=C(O)CF_3$ $C=P(C_6H_5)_3$	77La	
	-111.7	(1)	--	25	C	S:CH_2Cl_2 A=Cl $B=CH_3$ $C=P(C_6H_5)_3$	77La	
$PtAB_2^+$	-110.4	(1)	--	25	C	S:THF $A=CH_3$ $B=[P(CH_3)_2(C_6H_5)]$	75Ma	
CuA_2	-28.9	2.62(1)	--	--	C	S:o-$Cl_2C_6H_4$ A=Hexafluoro-acetylacetonato	76Mc	u:25°
	-29.3	2.41(1)	--	--	C	S:CH_2Cl_2 A=Hexafluoro-acetylacetonato	76Mc	u:25°

Metal Ion	ΔH, kJ /mol	Log K	ΔS, J /mol K	T,°C	M	Conditions	Ref	Remarks
P148	PHOSPHINE, tri(4-methoxyphenoxy)		$C_{21}H_{21}O_6P$			$P(OC_6H_4OCH_3)_3$		
$PtAB_2^+$	-89.1	(1)	--	25	C	S:THF A=CH_3 B=$[P(CH_3)_2(C_6H_5)]$	75Ma	
P149	PHOSPHINE, trimethoxy, sulfide		C_3H_9OSP			$(CH_3O)_3PS$		
I_2	-16.74	1.68(1)	-42.2	25	C	S:CCl_4	79Kc	
P150	PHOSPHINE, trimethyl-		C_3H_9P			$P(CH_3)_3$		
ThA_4	75	(1-2)	--	--	C	S:Acetone A=NO_3	75Dh	u:25°
	110	(1-3)	--	--	C	S:Acetone A=NO_3	75Dh	u:25°
UA_4	60	--	--	--	C	S:Acetone R:UA_4L_3+2L= UA_2L_5+2A A=NO_3	75Dh	j,u:25°
$PtAB_2^+$	-109.6	(1)	--	25	C	S:THF A=CH_3 B=$[P(CH_3)_2(C_6H_5)]$	75Ma	
AlA_3	-87.95	(1)	--	25-28	C	S:n-Hexane A=CH_3	68Hc	
P151	PHOSPHINE, tri(4-methylnaphthyl), sulfide		$C_{33}H_{27}SP$			$(CH_3C_{10}H_6)_3PS$		
I_2	-25.6	2.16(25°,1)	-44.6	25-45	T	S:$CHCl_3$	66Ta	b
P152	PHOSPHINE, (trimethylol-propane) ester		$C_6H_{15}O_3P$			$CH_3CH_2C(CH_2OH)_2CH_2OPH_2$		L
[Co]	-82.4	--	--	24	C	[Co]=methylcobaloxime S:Dichloromethane R:$[Co]_2$ +2L=2[Co]L	73Cf	
P153	PHOSPHINE, trimethyl-, oxide		C_3H_9OP			$(CH_3)_3PO$		L
NbA_5	-17.6	-0.29	-79	-48	T	S:CH_3CN A=OCH_3 R:$[NbA_5]_2$+2L= $2NbA_5L$	71Hi	q:±1.7
TaA_5	-14.2	1.84	-29	-48	T	S:CH_3CN A=OCH_3 R:$[TaA_5]_2$+2L= $2TaA_5L$	71Hi	q:±1.2
CuA_2	-48.5	3.48(1)	--	--	C	S:CCl_4 A=Hexafluoroacetylacetonato	76Mc	u:25°

Metal Ion	ΔH, kJ /mol	Log K	ΔS, J /mol K	T,°C	M	Conditions	Ref.	Re-marks
P154	PHOSPHINE, tri(2-methylphenoxy)		$C_{21}H_{24}O_3P$			$P(OC_6H_5CH_3)_3$		
$PtAB_2^+$	-80.8	(1)	--	25	C	S:THF A=CH_3 B=$[P(CH_3)_2(C_6H_5)]$	75Ma	
P155	PHOSPHINE, tri(2-methylphenyl)		$C_{21}H_{21}P$			$P[C_6H_4(CH_3)]_3$		
$PtAB_2^+$	-7.9	(1)	--	25	C	S:THF A=CH_3 B=$[P(CH_3)_2(C_6H_5)]$	75Ma	
P156	PHOSPHINE, tri(1-naphthyl), selenide		$C_{30}H_{21}PSe$			$(C_{10}H_7)_3PSe$		
I_2	-43.7	3.96(25°,1)	-71	5-35	T	S:$CHCl_3$	68Cf	b
P157	PHOSPHINE, tri(1-naphthyl), sulfide		$C_{30}H_{21}SP$			$(C_{10}H_7)_3PS$		
I_2	-19.5	1.98(25°,1)	-27.5	25-45	T	S:$CHCl_3$	66Ta	b
P158	PHOSPHINE, trioctyl-		$C_{24}H_{51}P$			$[CH_3(CH_2)_6CH_2]_3P$		L
Ga_2A_6	-151.9	--	--	25	C	S:Benzene CM= 0.03-0.04M A=Cl R:½Ga_2A_6+L=GaA_3L	73Rd	
P159	PHOSPHINE, trioctyl-, sulfide		$C_{24}H_{51}SP$			$[CH_3(CH_2)_6CH_2]_3PS$		
I_2	-47.70	3.929(20°,1)	-88	10-40	T	S:n-Heptane Results from UV data	71La	b
	-48.1	3.944(20°,1)	-88	10-40	T	S:n-Heptane Results from Visible data	71La	b
P160	PHOSPHINE, triphenoxy		$C_{18}H_{15}O_3P$			$P(OC_6H_5)_3$		
$PtAB_2^+$	-89.5	(1)	--	25	C	S:THF A=CH_3 B$[P(CH_3)_2(C_6H_5)]$	75Ma	
P161	PHOSPHINE, triphenyl-		$C_{18}H_{15}P$			$(C_6H_5)_3P$		L
NiA	-22.6	0.602(1)	-63	30	C	S:Acetonitrile CM=1-2x10^{-3}M A=BF_2-bridged glyoximato R_1=Me R_2=Me	76Gh	

Metal Ion	ΔH, kJ /mol	Log K	ΔS, J /mol K	T,°C	M	Conditions	Ref.	Re-marks
P161, continued								
NiA, cont.						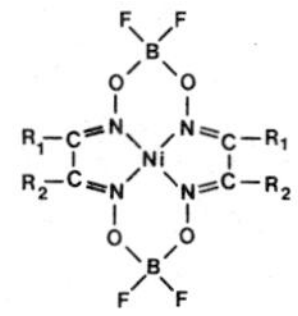		
	-23.0	0.643(1)	-66	30	C	S:Benzene $CM=0.5x10^{-3}M$ $A=BF_2$-bridged glyoximato R_1=Me R_2=Pr	76Gh	
RuA_2L_2	-17.6	2.569(25°,1)	-9.2	25-49	T	μ=0 S:Benzene	74Ja	b,q:±1.2
	-13.0	1.337(25°,1)	-17.2	25-49	T	μ=0 S:N,N-Dimethylacetamide	74Ja	b,q: ±1.2
$RhAB_2$	-114.2	--	--	21-24	C	S:Dichloromethane A=Cl B=1,5-Cyclooctadiene R:$[RhAB]_2$+2L= 2RhABL	73Pe	
	-121.8	--	--	21-24	C	S:Dichloromethane A=B B=1,5-Cyclooctadiene R:$[RhAB]_2$+2L= 2RhABL	73Pe	
	-127.2	--	--	21-24	C	S:Dichloromethane A=I B=1,5-Cyclooctadiene R:$[RhAB]_2$+2L= 2RhABL	73Pe	
PdAB	-72.4	--	--	24	C	S:Benzene R:$[PdAB]_2$+2L= 2[PdABL] A=(π-MeAllyl) B=Cl	77Lb	
PdA_2B_2	-163.2	--	--	21-24	C	S:Dichloromethane A=Cl B=Benzonitrile R:PdA_2B_2 +2L=PdA_2L_2+2B	73Pe	
$PtAB_2^+$	-81.6	(1)	--	25	C	S:THF A=CH_3 B=$[P(CH_3)_2(C_6H_5)]$	75Ma	
Cu^+	-47	6.57(1)	-32	25	C	μ=0.1M(NH_4ClO_4) S:Dimethyl sulfoxide	78Ab	
	-41	3.87(2)	-64	25	C	μ=0.1M(NH_4ClO_4) S:Dimethyl sulfoxide	78Ab	
	-25	1.40(3)	-57	25	C	μ=0.1M(NH_4ClO_4) S:Dimethyl sulfoxide	78Ab	
CuA_2	-11.3	0.886(10°,1)	-23.0	-60-60	T	S:Toluene	78Mb	

Metal Ion	ΔH, kJ /mol	Log K	ΔS, J /mol K	T,°C	M	Conditions	Ref.	Remarks
P161, continued								
CuA_2, cont.						A=Thenoyltri-fluoroacetonato		
AgA	-47.3	(2)	--	25	C	S:CH_2Cl_2 A=Hexa-fluoroacetyl-acetone	73Pf	
AgAB	-55.6	--	--	25	C	S:CH_2Cl_2 A=Hexa-fluoroacetyl-acetone B=Cyclo-hexene R:[AgAB]+L=[AgAL]+B	73Pf	
	-75.3	--	--	25	C	S:CH_2Cl_2 A=Hexa-fluoroacetyl-acetone B=Cyclo-pentene R:[AgAB]+L=[AgAL]+B	73Pf	
	-59.8	--	--	25	C	S:CH_2Cl_2 A=Hexa-fluoroacetyl-acetone B=Cyclo-heptene R:[AgAB]+L=[AgAL]+B	73Pf	
	-41.34	--	--	25	C	S:CH_2Cl_2 A=Hexa-fluoroacetyl-acetone B=Cyclo-octene R:[AgAB]+L=[AgAL]+B	73Pf	
	-49.0	--	--	25	C	S:CH_2Cl_2 A=Hexa-fluoroacetyl-acetone B=1,5-Cyclooctadiene R:[AgAB]+L=[AgAL]+B	73Pf	
	-44.8	--	--	25	C	S:CH_2Cl_2 A=Hexa-fluoroacetyl-acetone B=1,3,-5,7-Cycloocta-tetraene R:[AgAB]+L=[AgAL]+B	73Pf	
ZnA	-32.2	1.429(25°,1)	-81.2	15-35	T	S:Benzene A=α,β,γ,δ-Tetra-phenylporphine	73Ve	b,q: ±2.5
ZnA_2	-15	(1)	--	30	C	S:Acetonitrile A=I	76Ak	q:±3
CdA_2	-40	2.130(1)	-91	30	C	S:Acetonitrile A=I	76Ak	q:±4
Hg^{2+}	-57	11.06(1)	20	25	C	μ=0.1M(NH_4ClO_4) S:Dimethyl sulfoxide	78Ab	
	-51	6.55(2)	-47	25	C	μ=0.1M(NH_4ClO_4) S:Dimethyl sulfoxide	78Ab	

Metal Ion	ΔH, kJ /mol	Log K	ΔS, J /mol K	T,°C	M	Conditions	Ref	Re-marks
P161, continued								
HgA_2	-74.9	(1)	--	30	C	S:Benzene CM= 1-5x10^{-3}M A=Cl	73Fa 76Ga	
	-77.0	(1)	--	30	C	S:Benzene CM= 1-5x10^{-3}M A=Br	73Fa 76Ga	
	-60	>4(1)	<-125	30	C	S:Acetonitrile A=I	76Ak	
	-64.8	(1)	--	30	C	S:Benzene CM= 1-5x10^{-3}M A=I	73Fa 76Ga	
	-33.01	3.712(2)	-38.1	30	C	S:Benzene CM= 1-5x10^{-3}M A=Cl	73Fa 76Ga	q:±3.6
	-41.8	3.550(2)	-71	30	C	S:Benzene CM= 1-5x10^{-3}M A=Br	73Fa 76Ga	q:±5.0
	-25	>3(2)	-35	30	C	S:Acetonitrile A=I	76Ak	
	-40.00	3.204(2)	-71	30	C	S:Benzene CM= 1-5x10^{-3} A=I	73Fa 76Ga	
AlA_3	-73.76	(1)	--	25-28	C	S:n-Hexane A=CH_3	68Hc	
Al_2A_6	-90.4	--	--	25	C	S:Benzene CM= 0.03-0.04M A=Br R:½Al_2A_6+L=AlA_3L	73Rd	
	-94.6	--	--	25	C	S:Benzene CM= 0.03-0.04M A=Br R:Al_2A_6+L= L$(AlA_3)_2$	73Rd	
Ga_2A_6	-102.1	--	--	25	C	S:Benzene CM= 0.03-0.04M A=Cl R:½Ga_2A_6+L=GaA_3L	73Rd	
	-107.5	--	--	25	C	S:Benzene CM=0.03-0.04M A=Cl R:Ga_2A_6+L= L$(GaA_3)_2$	73Rd	
P162	PHOSPHINE, triphenyl-, oxide-			$C_{18}H_{15}OP$		$(C_6H_5)_3PO$		L
VOA_2	-24.73	1.50(1)	-54.4	25	C	μ→0 S:Nitrobenzene A=Acetylacetonate	65Ca	
I_2	-18.8	1.52(1)	-33.5	25	C	CM=6.8-73x10^{-4}M CL=65-173x10^{-4}M S:CCl_4	75Ld	
P163	PHOSPHINE, tri(2-phenylphenoxy)			$C_{36}H_{27}O_3P$		$P[OC_6H_4(C_6H_5)]_3$		
$PtAB_2^+$	-74.5	(1)	25	25	C	S:THF A=CH_3 B=$[P(CH_3)_2(C_6H_5)]$	75Ma	

Metal Ion	ΔH, kJ /mol	Log K	ΔS, J /mol K	T,°C	M	Conditions	Ref.	Remarks
P164	PHOSPHINE, triphenyl, selenide		$C_{18}H_{15}PSe$			$(C_6H_5)_3PSe$		
HgA_2	-45	2.301(1)	-105	30	C	S:Benzene CM=2-4x10^{-4}M A=I	76Gb	
I_2	-38.5	3.45(1)	-18.8	25	C	CM=6-73x10^{-4}M CL=7-40x10^{-4}M S:CCl_4	75Ld	
	-39.0	3.60(25°,1)	-62	5-35	T	S:$CHCl_3$	68Cf	b
P165	PHOSPHINE, triphenyl, sulfide		$C_{18}H_{15}SP$			$(C_6H_5)_3PS$		
HgA_2	-12.9	2.072(1)	-4	30	C	S:Benzene CM=1.2=1.8x10^{-3}M A=Cl	76Gb	q:±1.6
	-13.8	2.260(1)	-3	30	C	S:Benzene CM=1.5-2.5x10^{-3}M A=Br	76Gb	q:±1
	-13.3	1.929(1)	-6	30	C	S:Benzene CM=1.7-3.0x10^{-3}M A=I	76Gb	q:±1.7
I_2	-30.5	2.20(1)	-18.0	25	C	CM=6.8-73x10^{-4}M CL=16-135x10^{-4}M S:CCl_4	75Ld	
P166	PHOSPHINE, tripropyl-		$C_9H_{21}P$			$(C_3H_7)_3P$		L
CoL_2A_2	-74.9	2.505(10°,1)	-218	-30-20	T	S:1,2-Dichloro-ethane A=NCS	67Bb	b
NiA	-57.7	5.75(1-2)	-83.3	25	T	CM=10^{-4}M S:Ben-zene A=Diacetyl-bisbenzoyl hydrazine	60Sd	
NiL_2A_2	-50.2	0.914(0°,1)	-167	20	T	S:C_2H_5OH A=CN	69Rc	b
	-48.5	0.643(0°,1)	-163	20	T	S:n-Hexane A=CN	69Rc	b
P167	PHOSPHINE, tris (dimethylamino)		$C_6H_{18}N_3P$			$[(CH_3)_2N]_3P$		
SnA_4	-123.0	11(1)	-321	25	C	S:Benzene A=Cl	79Kc	
	-9.2	(2)	--	25	C	S:Benzene A=Cl	79Kc	
P168	PHOSPHINE, tris(dimethyl-amino)-, oxide		$C_6H_{18}ON_3P$			$[N(CH_3)_2]_3PO$		L
ThA_4	31	(1-2)	--	--	C	S:Acetone A=NO_3	76Dh	u:25°
	36	3	--	--	C	S:Acetone A=NO_3	76Dh	j,u:25°
NbA_5	-6.7	-0.78	-46	-48	T	S:$C_6H_5CH_3$ A=OCH_3 R:$[NbA_5]_2$+2L=	71Hi	q:±8

Metal Ion	ΔH, kJ /mol	Log K	ΔS, J /mol K	T,°C	M	Conditions	Ref.	Re-marks
P168, continued								
NbA_5, cont.	-5.8	-0.51	-38	-60	T	$2NbA_5L$ S:CH_3CN A=OCH_3 R:$[NbA_5]_2$+2L= $2NbA_5L$	71Hi	q:±1.2
CuA_2	-42.2	4.0(1)	--	--	C	S:CCl_4 A=Hexafluoro- acetylacetonato	76Mc	u:25°
SnA_2B_2	-47.9	8.8(1)	-87.9	25	C	S:Benzene A=C_2H_5 B=Cl	79Kc	
	-33.5	3.3(2)	-85.8	25	C	S:Benzene A=C_2H_5 B=Cl	79Kc	q:±5
P169	PHOSPHINE, tris- (dimethylamino), sulfide		$C_6H_{18}NSP$			$[(CH_3)_2N]_3PS$		
SnA_4	-31.8	7.0(1)	-48.9	25	C	S:Benzene A=Cl	79Kc	
I_2	-34.10	7.2(1)	-54.8	25	C	S:CCl_4	79Kc	
P170	PHOSPHINE, tri- (2,2,2-trichloroethoxy)		$C_6H_6O_3Cl_9P$			$P(OCH_2CCl_3)_3$		
$PtAB_2^+$	-81.2	(1)	--	25	C	S:THF A=CH_3 B=$[P(CH_3)_2(C_6H_5)]$	75Ma	
P171	PHOSPHORIC ACID, di- (2-ethylhexyl)ester		$C_{16}H_{35}O_4P$			$(C_8H_{17}O)_2POOH$		L
Zn^{2+}	10.75	-1.836(25°)	0.33	5-50	T	μ=0.1M($NaClO_4$) S:Benzene R:Zn^{2+}(aq)+ 2L(benzene)= ZnL_2(benzene)+ $2H^+$(aq)	77Na	
P172	PHOSPHORIC ACID, triamide, hexamethyl-		$C_6H_{18}ON_3P$			$[(CH_3)_2N]_3PO$		L
VOA_2	-24.81	2.02(1)	-44.8	25	C	μ→0 S:Nitrobenzene A=Acetylacetonate	65Ca	
ZnA_2	-48.5	(1)	--	25	C	S:Benzene A=$N[Si(CH_3)_3]_2$	75Fc	
	-48.9	(1)	--	25	C	S:Cyclohexane A=$N[Si(CH_3)_3]_2$	75Fc	
CdA_2	-42.7	4.64(1)	-54.4	25	C	S:Cyclohexane A=$N[Si(CH_3)_3]_2$	75Fc	
$SnAB_3$	-42.2	2.584(1)	91.6	26	C	S:Isooctane A=Cl B=CH_3	66Be	

Metal Ion	ΔH, kJ /mol	Log K	ΔS, J /mol K	T,°C	M	Conditions	Ref.	Re-marks
P173	PHOSPHORIC ACID, tributyl ester		$C_{12}H_{27}O_4P$			$(C_4H_9O)_3PO$		
Zn^{2+}	17.32	-4.239(25°)	-23.1	5-50	T	μ=0.1M($NaClO_4$) A:Benzene R:Zn^{2+}(aq)+ 2A(benzene)+ L(benzene)= ZnA_2L(benzene)+ $2H^+$(aq) A=1-Thenoyl- 3,3,3-trifluoro- 2-propanone	77Na	
$ZnA_2{}^{2+}$	-28.24	3.559(25°,1)	-26.2	5-50	T	μ=0.1M($NaClO_4$) S:Benzene A=1-Thenoyl- 3,3,3-trifluoro- 2-propanone	77Na	
$SnAB_3$	-50.6	5.2(1)	-126.4	25	C	S:Benzene A=C_2H_5 B=Cl	79Kc	
P174	PHOSPHORIC ACID, tri(4-chlorophenyl) ester		$C_{18}H_{12}O_3Cl_3P$			$[C_6H_4O(Cl)]_3P$		
Ni°	-88	9.699(25°,4)	-109	-50-70	T	S:Benzene	74Tc	b,q:±7
P175	PHOSPHORIC ACID, triethyl ester		$C_6H_{15}O_4P$			$(C_2H_5O)_3PO$		
RhAB	-55.2	3.613	<u>-116</u>	24	C	S:Benzene R:$[RhAB]_2$+2L= 2RhABL A=Cl B=1,5-Cyclo- octadiene	76Lb	
I_2	-13.4	0.792(20°,1)	-33.0	20&50	T	S:CCl_4	62Ga	a,b
P176	PHOSPHORIC ACID, triisopropyl ester		$C_9H_{21}O_3P$			$(C_3H_7O)_3PO$		
Ni°	-98.3	4.569(25°,4)	-243	-50-70	T	S:Benzene	74Tc	b,q:±6.3
P177	PHOSPHORIC ACID, triphenyl ester		$C_{18}H_{15}O_4P$			$(C_6H_5O)_3PO$		
PtA_2BL	-44.2	--	--	--	C	S:Benzene R:PtA_2BL+L= PtA_2L_2+B A=Cl B=Pyridine	76Pa	u:25°,y
PtA_2BC	-68	--	--	--	C	S:Benzene R:PtA_2BC+L=	76Pa	u:25°

Metal Ion	ΔH, kJ /mol	Log K	ΔS, J /mol K	T,°C	M	Conditions	Ref.	Re- marks
P177, continued								
PtA_2BC, cont.						PtA_2CL+B A=Cl B=cis-Butene C=Pyridine		
	-65	--	--	--	C	S:Benzene R:$PtA_2BC+L=$ PtA_2CL+B A=Cl B=Cycloheptene C=Pyridine	76Pa	u:25°,y
	-82	--	--	--	C	S:Benzene R:$PtA_2BC+L=$ PtA_2CL+B A=Cl B=Cyclohexene C=Pyridine	76Pa	u:25°,y
	-61	--	--	--	C	S:Benzene R:$PtA_2BC+L=$ PtA_2CL+B A=Cl B=cis-Cyclooctene C=Pyridine	76Pa	u:25°,y
	-77	--	--	--	C	S:Benzene R:$PtA_2BC+L=$ PtA_2CL+B A=Cl B=Cyclopentene C=Pyridine	76Pa	u:25°,y
	-58	--	--	--	C	S:Benzene R:$PtA_2BC+L=$ PtA_2CL+B A=Cl B=Ethylene C=Pyridine	76Pa	u:25°
	-82	--	--	--	C	S:Benzene R:$PtA_2BC+L=$ PtA_2CL+B A=Cl B=p-Nitrostyrene C=Pyridine	76Pa	u:25°,y
	-77	--	--	--	C	S:Benzene R:$PtA_2BC+L=$ PtA_2CL+B A=Cl B=Styrene C=Pyridine	76Pa	u:25°,y
	-66	--	--	--	C	S:CH_2Cl_2 R:$PtA_2BC+L=$ PtA_2CL+B A=Cl B=cis-Butene C=Pyridine	76Pa	u:25°,y
	-63	--	--	--	C	S:CH_2Cl_2 R:$PtA_2BC+L=$ PtA_2CL+B A=Cl B=Cycloheptene C=Pyridine	76Pa	u:25°,y
	-74	--	--	--	C	S:CH_2Cl_2 R:$PtA_2BC+L=$ PtA_2CL+B A=Cl B=Cyclopentene C=Pyridine	76Pa	u:25°,y

Metal Ion	ΔH, kJ /mol	Log K	ΔS, J /mol K	T,°C	M	Conditions	Ref.	Remarks
P177, continued								
PtA_2BC, cont.	-58	--	--	--	C	S:Dichloromethane R:PtA_2BC+L= PtA_2CL+B A=Cl B=Ethylene C=Pyridine	76Pa	u:25°
	-77	--	--	--	C	S:CH_2Cl_2 R:PtA_2BC+L= PtA_2CL+B A=Cl B=p-Nitrostyene C=Pyridine	76Pa	u:25°,y
	-75	--	--	--	C	S:CH_2Cl_2 R:PtA_2BC+L= PtA_2CL+B A=Cl B=Styene C=Pyridine	76Pa	u:25°,y
SnA_4	-33.5	7.8(1)	-47.7	25	C	S:Benzene A=Cl	79Kc	
	-29.7	5.00(2)	-57.7	25	C	S:Benzene A=Cl	79Kc	
P178	PHOSPHOROUS ACID, tri(4-chloro-2-methylphenyl) ester		$C_{21}H_{18}O_3Cl_3P$			$[C_6H_3O(Cl)(CH_3)]_3P$		
Ni°	-52.3	2.155(25°,4)	-134	-50-70	T	S:Benzene	74Tc	b,q:±4
P179	PHOSPHOROUS ACID, tri(2,4-dimethylphenyl) ester		$C_{24}H_{27}O_3P$			$[C_6H_3O(CH_3)_2]_3P$		
Ni°	-50	1.509(25°,4)	-138	-50-70	T	S:Benzene	74Tc	b,q:±8
P180	PHOSPHOROUS ACID, trimethyl ester		C_3H_9OP			$(CH_3O)_3P$		L
CoAB	-19.7	--	--	-90-50	T	S:CH_2Cl_2 A=CH_3 B=N,N'-Ethylene-bis(acetylacetoniminato) R:CoAB+L= CoABL	73Ge	b,q:±4
[Co]	-80.3	--	--	25	C	S:Dichloromethane [Co]=Methyl cobaloxime R:$[Co]_2$ +2L=2[Co]L	73Cf	
CuA_2	-25.1	(1)	--	25	C	S:Methylene chloride A=Hexafluoroacetylacetonate	74Dc	
P181	PHOSPHOROUS ACID, triphenyl ester		$C_{18}H_{15}O_3P$			$(C_6H_5O)_3P$		L
RhAB	-99.6	--	--	25	C	S:CCl_4 A=Acetylacetonate B=1,5-	73Pd	

Metal Ion	ΔH, kJ /mol	Log K	ΔS, J /mol K	T,°C	M	Conditions	Ref.	Remarks
P181, continued								
RhAB, cont.						Cyclooctadiene R:RhAB+2L=RhAL$_2$+B		
	-118.4	--	--	25	C	S:CCl$_4$ A=Acetyl-acetonate B=1,3-5,7-Cyclooctate-traene R:RhAB+2L=RhAL$_2$+B	73Pd	
	-89.5	--	--	25	C	S:CH$_2$Cl$_2$ A=Acetylacetonate B=1,5-Cycloocta-diene R:RhAB+2L=RhAL$_2$+B	73Pd	
	-107.9	--	--	25	C	S:CH$_2$Cl$_2$ A=Acetylacetonate B=1,3,5,7-Cyclo-octatetraene R:RhAB+2L=RhAL$_2$+B	73Pd	
	-118.0	--	--	25	C	S:CH$_2$Cl$_2$ A=Cl B=1,5-Cycloocta-diene R:[RbAB]$_2$ +2L=[Rb$_2$A$_2$BL$_2$]+B	73Pd	
	-118.0	--	--	21-24	C	S:Dichloromethane A=Cl B=1,5-Cyclooctadiene R:[RhAB]$_2$+2L=Rh$_2$A$_2$BL$_2$+B	73Pe	
	-115.5	--	--	25	C	S:CH$_2$Cl$_2$ A=Cl B=1,3,5,7-Cyclo-octatetraene R:[RhAB]$_2$+2L=[Rh$_2$A$_2$BP$_2$]+B	73Pd	
	-133.0	--	--	25	C	S:CH$_2$Cl$_2$ A=Cl B=<u>cis</u>-Cyclooctene R:[RhAB]$_2$+2L=[Rh$_2$A$_2$BL$_2$]+B	73Pd	
	-119.2	--	--	25	C	S:CH$_2$Cl$_2$ A=Cl B=Dicyclopenta-diene R:[RhAB]$_2$ +2L=[Rh$_2$A$_2$BL$_2$]+B	73Pd	
	-117.6	--	--	25	C	S:CH$_2$Cl$_2$ A=Cl B=Norbornadiene R:[RhAB]$_2$+2L=[Rb$_2$A$_2$BL$_2$]+B	73Pd	
	-125.9	--	--	21-24	C	S:Dichloromethane A=Br B=1,5-Cyclooctadiene R:[RhAB]$_2$+2L=Rh$_2$A$_2$BL$_2$+B	73Pe	
	-126.4	--	--	21-24	C	S:Dichloromethane A=I B=1,5-Cyclo-octadiene R:[RhAB]$_2$+2L=Rh$_2$A$_2$BL$_2$+B	73Pe	

Metal Ion	ΔH, kJ /mol	Log K	ΔS, J /mol K	T,°C	M	Conditions	Ref.	Remarks
P181, continued								
$RhAL_2$	-96.2	--	--	25	C	S:CH_2Cl_2 A=Cl R:$[RhAL_2]_2$+2L=2$[RhAL_3]$	73Pd	
	-98.7	--	--	21-24	C	S:Dichloromethane A=Cl R:$[RhAL_2]_2$ +2L=2$RhAL_3$	73Pe	
	-97.9	--	--	21-24	C	S:Dichloromethane A=Br R:$[RhAL_2]_2$ +2L=2$RhAL_3$	73Pe	
	-96.2	--	--	21-24	C	S:Dichloromethane A=I R:$[RhAL_2]_2$ +2L=2$RhAL_3$	73Pe	
$Rh_2A_2BL_2$	-140.2	--	--	25	C	S:CH_2Cl_2 A=Cl B=Benzoquinone R:$[Rh_2A_2BL_2]$+2L=$[RhAL_2]_2$+B	73Pd	
	-96.2	--	--	25	C	S:CH_2Cl_2 A=Cl B=1,5-Cycloocta-diene R:$[Rh_2A_2BL_2]$+2L=$[RhAL_2]_2$+B	73Pd	
	-98.3	--	--	21-24	C	S:Dichloromethane A=Cl B=1,5-Cyclo-octadiene R:$Rh_2A_2BL_2$+2L=$[RhAL_2]_2$+B	73Pe	
	-105.4	--	--	25	C	S:CH_2Cl_2 A=Cl B=1,3,5,7-Cyclo-octatetraene R:$[Rh_2A_2BL_2]$+2L=$[RhAL_2]_2$+B	73Pd	
	-126.8	--	--	25	C	S:CH_2Cl_2 A=Cl B=<u>cis</u>-Cyclooctene R:$[Rh_2A_2BL_2]$+2L=$[RhAL_2]_2$+B	73Pd	
	-110.0	--	--	25	C	S:CH_2Cl_2 A=Cl B=Dicyclopenta-diene R:$[Rh_2A_2BL_2]$+2L=$[RhAL_2]_2$+B	73Pd	
	-94.6	--	--	25	C	S:CH_2Cl_2 A=Cl R:Norbornadiene R:$[Rh_2A_2BL_2]$+2L=$[RhAL_2]_2$+B	73Pd	
	-104.6	--	--	21-24	C	S:Dichloromethane A=Br B=1,5-Cyclooctadiene R:$Rh_2A_2BL_2$+2L=$[RhAL_2]_2$+B	73Pe	
	-102.9	--	--	21-24	C	S:Dichloromethane A=I B=1,5-Cyclo-octadiene	73Pe	

Metal Ion	ΔH, kJ /mol	Log K	ΔS, J /mol K	T,°C	M	Conditions	Ref.	Remarks
P181, continued								
RhA_2BL_2, cont.						R:$Rh_2A_2BL_2$+2L= $[RhAL_2]_2$+B		
PdA_2B	-88.7	--	--	21-24	C	S:Dichloromethane A=Cl B=1,5-Cyclooctadiene R:PdA_2B+2L= PdA_2L_2+B	73Fe	
PtA_2BC	-58	--	--	--	C	S:Benzene R:PtA_2BC+L= PtA_2CL+B A=Cl B=Ethylene C=Pyridine	76Pa	u:25°,y
	-58	--	--	--	C	S:CH_2Cl_2 R:PtA_2BC+L= PtA_2CL+B A=Cl B=Ethylene C=Pyridine	76Pa	u:25°,y
	-68	--	--	--	C	S:Benzene R:PtA_2BC+L= PtA_2CL+B A=Cl B=cis-Butene C=Pyridine	76Pa	u:25°,y
P182	PHOSPHOROUS ACID, tri(2-tolyl) ester		$C_{21}H_{21}O_3P$			$(C_7H_7O)_3P$		
Ni°	-96	9.222(25°,4)	-146	-50-70	T	S:Benzene	74Tc	b,q:±12
P183	PHOSPHORIC ACID, thiono-, triethyl ester		$C_6H_{15}O_3SP$			$(C_2H_5O)_3PS$		
I_2	-24.7	0.602(20°,1)	-78.2	20&50	T	S:CS_2	62Ga	a,b
	-24.7	1.000(20°,1)	-69.9	20&50	T	S:CCl_4	62Ga	a,b
P184	PHTHALIC ACID		$C_8H_6O_4$			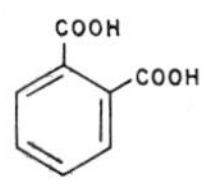		L^{2-}
Ca^{2+}	29.75	2.41(25°,1)	145.6	25-45	T	μ=0	70Ge	b
	13.14	2.64(25°,1)	91.2	20-35	T	μ=0.073M	76Pb	
ThA	23.51	3.10(1)	138.1	25	C	μ=0.10(KNO_3) A=EDTA	70Kb	g:k
	20.71	2.63(1)	120.1	25	C	μ=0.10(KNO_3) A=1,2-Diamino-cyclohexane-N,N,-N',N'-tetraacetic acid	70Kb	g:k
	23.81	--	--	25	C	μ=0.10(KNO_3) A=EDTA R:ThA+HL= $ThAL^-$+H^+	70Kb	g:k
	21.00	--	--	25	C	μ=0.10(KNO_3)	70Kb	g:k

Metal Ion	ΔH, kJ /mol	Log K	ΔS, J /mol K	T,°C	M	Conditions	Ref.	Re-marks
P184, continued								
ThA, cont.						A=1,2-Diamino-cyclohexane-N,N,-N',N'-tetraacetic acid R:ThA+HL=$ThAL^-+H^+$		
Mn^{2+}	10.63	(1)	--	25	C	μ=0	71Wc	
	9.20	2.741(1)	83.3	25	T	μ=0 ΔCp=264	62Da	
	8.91	(1)	--	25	C	μ=0.024M	71Wc	
Co^{2+}	7.82	2.829(1)	80.3	25	T	μ=0 ΔCp=213	62Da	
	-1.2	2.899(25°,1)	52	15-35	T	μ=0	75Lc	b
Ni^{2+}	7.86	(1)	--	25	C	μ=0	71Wc	
	7.385	2.951(1)	284	25	T	μ=0 ΔCp=285	62Da	
	13.4	3.030(25°,1)	103	15-35	T	μ=0	75Lc	b
	5.94	(1)	--	25	C	μ=0.028M	71Wc	
Cu^{2+}	8.41	4.038(1)	105.4	25	C	μ=0	71Wc	
	39.7	3.14(25°,1)	193.3	25-45	T	μ=0	70Ge	b
	11.0	3.999(25°,1)	113	15-35	T	μ=0	75Lc	b
	6.53	(1)	--	25	C	μ=0.034M	71Wc	
	10.3	3.241(1)	96.7	25	C	μ=0.1M($NaClO_4$)	78Ae	
	-33.8	11.861	113.8	25	C	μ=0.1M($NaClO_4$) R:$Cu^{2+}+L^{2-}$+A=CuLA A=2,2'-Bipyridyl	78Ae	
	-43.9	15.313	146.0	25	C	μ=0.1M($NaClO_4$) R:$Cu^{2+}+L^{2-}+A+H^+$ =$[CuLAH]^+$ A=2,2'-Bipyridyl	78Ae	
	23.14	0.99(25°)	96.6	25-45	T	μ=0 R:$Cu^{2+}+HL^-$=$CuHL^+$	70Ge	b
Zn^{2+}	13.22	2.90(1)	99.6	25	T	μ=0 ΔCp=358.6 Equations of ΔH as a function of T are given	65Na	
	-29.9	11.08(25°)	111.8	25&35	T	R:Zn+L+A=ZnLA A=Iminodiacetic acid	78Ka	a,b
Cd^{2+}	-29.9	10.76(25°)	105.6	25&35	T	R:Zn+L+A=ZnLA A=Iminodiacetic acid	78Ka	a,b

P185 PHTHALOCYANINE, 4,4',4'',4'''-tetraoctadecyl-sulfonamido $C_{104}H_{162}O_8$-$N_{12}S_4$

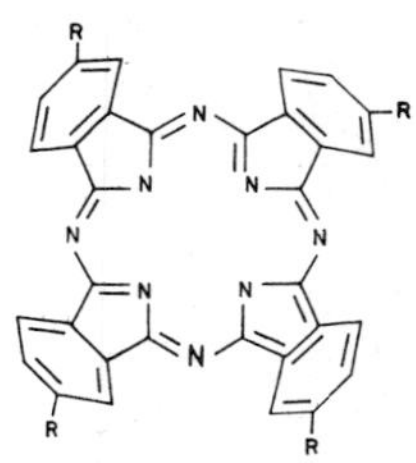

[Cu]	-4.18	4.114	-62.8	25	C	S:Benzene [Cu]=	73Gd	q:±4.2

Metal Ion	ΔH, kJ /mol	Log K	ΔS, J /mol K	T,°C	M	Conditions	Ref.	Re-marks
P185, continued								
[Cu], cont.						4,4',4'',4'''-tetraoctadecyl-sulfonamido copper(II) phthalocyanine R:2[Cu]=$[Cu]_2$ (dimerization)		
P186	PHTHALOCYANINE, 4,4',4'',4'''-tetrasulfo-		$C_{32}H_{18}O_{12}N_8S_4$			see P185, R=SO_3H		L^{2-}
CoL_2	-58	5.280(58°)	-75	38-58	T	μ=0 R:$2CoL_2=(CoL_2)_2$	70Sa	q:±3.8
CuL	-1.92	0.42(25°)	-2.59	25-58	T	R:$2CuL=(CuL)_2$	73Bm	b
	-4.52	0.28(25°)	-9.58	25-58	T	S:20% Ethanol R:$2CuL=(CuL)_2$	73Bm	b
	-4.10	0.35(25°)	-6.90	25-58	T	S:20% Ethanol 10^{-2}M(NaCl) R:$2CuL=(CuL)_2$	73Bm	b
	-2.80	0.37(25°)	-2.38	25-58	T	S:200g/1 Formamide in H_2O R:$2CuL=(CuL)_2$	73Bm	b
	-0.79	0.28(25°)	2.59	25-58	T	S:Formamide 1M(CH_3COOH) R:$2CuL=(CuL)_2$	73Bm	b
	-3.89	0.32(25°)	-7.20	25-58	T	S:20% CH_3OH R:$2CuL=(CuL)_2$	73Bm	b
	-2.38	0.35(25°)	-1.51	25-58	T	S:100g/1 Thiourea in H_2O R:$2CuL=(CuL)_2$	73Bm	b
	-2.18	0.33(25°)	-0.92	25-58	T	S:250g/1 Urea in H_2O R:$2CuL=(CuL)_2$	73Bm	b
	-1.38	0.32(25°)	1.46	25-58	T	S:500g/1 Urea in H_2O R:$2CuL=(CuL)_2$	73Bm	b
	-1.51	0.30(25°)	0.79	25-58	T	S:250g/1 Urea + 100g/1 Thiourea in H_2O R:$2CuL=(CuL)_2$	73Bm	b
P187	PIPERAZINE		$C_4H_{10}N_2$					
Ag^+	-26.82	3.33(1)	-25.9	25	C	μ=0.1M(KNO_3)	73He	
	-32.547	3.42(1)	-43.9	25	C	μ=0.1M(KNO_3) S:50w%Ethanol	73Ed	
	-16.90	2.71(2)	-5.0	25	C	μ=0.1M(KNO_3)	73He	
	-18.422	2.87(2)	-6.7	25	C	μ=0.1M(KNO_3) S:50w% Ethanol	73Ed	

Metal Ion	ΔH, kJ /mol	Log K	ΔS, J /mol K	T,°C	M	Conditions	Ref.	Re-marks
P188	PIPERAZINE, N-(2-aminomethyl)-	$C_6H_{15}N_3$				$CH_2CH_2NHCH_2CH_2N(CH_2CH_2NH_2)$ (ring)		L
Cu^{2+}	-19.2	5.51(20°,1)	42	10-40	T	$\mu\to0$	61Pa	b
	-7.9	3.77(20°,2)	46	10-40	T	$\mu\to0$	61Pa	b
P189	PIPERAZINE, N,N'-dimethyl	$C_6H_{14}N_2$				$CH_3(C_4H_8N_2)CH_3$		
HgA_2	-41.1	3.35(1)	-72	30	C	CM=1.0-2.6x10^{-3} S:Benzene A=I	74Fb	
P190	PIPERAZINE, N,N'-di(methyl-diphenylphosphine)	$C_{30}H_{32}O_2N_2P_2$				$(C_6H_5)_2P(=O)CH_2N(CH_2-CH_2)(CH_2-CH_2)NCH_2P(=O)(C_6H_5)_2$		
CuA^+	31.92	4.13(25°,1)	188	5-45	T	$\mu<0.005$ S:1v Acetone+ 1v Nitromethane+ 0.5v Chloroform A=Cl^-	79Ya	b,q:±2.7
P191	PIPERAZINE, 1-(p-methoxy-phenyl)-2-methyl-	$C_{12}H_{18}ON_2$				$C_4H_8N_2(C_6H_4OCH_3)(CH_3)$		
Ag^+	-26.840	3.23(1)	-28.0	25	C	μ=0.1M(KNO_3) S:52w% Ethanol	73Ed	
	-25.250	1.80(2)	-50.2	25	C	μ=0.1M(KNO_3) S:52w% Ethanol	73Ed	
P192	PIPERAZINE, 1-methyl-	$C_5H_{12}N_2$				$C_4H_9N_2(CH_3)$		
Ag^+	-23.35	2.93(1)	-22.2	25	C	μ=0.1M(KNO_3)	73He	
	-16.61	2.33(2)	-10.9	25	C	μ=0.1M(KNO_3)	73He	
P193	PIPERAZINE, 2-methyl-	$C_5H_{12}N_2$				$C_4H_9N_2(CH_3)$		
Ag^+	-26.28	3.44(1)	-22.2	25	C	μ=0.1M(KNO_3)	73He	
	-32.459	3.53(1)	-41.4	25	C	μ=0.1M(KNO_3) S:52w% Ethanol	73Ed	
	-15.94	2.74(2)	-0.8	25	C	μ=0.1M(KNO_3)	73He	
	-18.920	3.00(2)	-5.8	25	C	μ=0.1M(KNO_3) S:52w% Ethanol	73Ed	
P194	PIPERAZINE, 2-methyl-1-(m-tolyl)-	$C_{12}H_{18}N_2$				$C_4H_8N_2(C_6H_4CH_3)(CH_3)$		
Ag^+	-26.14	3.06(1)	-28.9	25	C	μ=0.1M(KNO_3) S:52w% Ethanol	73Ed	
	-26.100	2.11(2)	-47.3	25	C	μ=0.1M(KNO_3) S:52w% Ethanol	73Ed	

Metal Ion	ΔH, kJ /mol	Log K	ΔS, J /mol K	T,°C	M	Conditions	Ref.	Re-marks
P195	PIPERAZINE, 2-methyl-1-(p-tolyl)-		$C_{12}H_{18}N_2$			$C_4H_8N_2(C_6H_4CH_3)(CH_3)$		
Ag^+	-25.740	3.18(1)	-25.5	25	C	μ=0.1M(KNO_3) S:52w% Ethanol	73Ed	
	-26.526	2.10(2)	-48.5	25	C	μ=0.1M(KNO_3) S:52w% Ethanol	73Ed	
P196	PIPERAZINE, 1-phenyl-		$C_{10}H_{14}N_2$			$C_4H_9N_2(C_6H_5)$		
Ag^+	-22.38	2.97(1)	-19.2	25	C	μ=0.1M(KNO_3)	73He	
	-21.67	2.77(2)	-18.4	25	C	μ=0.1M(KNO_3)	73He	
[Ni]	-10.12	-1.07(34°,1)	-52.7	4-24	T	[Ni]=Nickel mesoporphyrin IX dimethyl ester S:Mixtures of piperidine and tetrahydrofuran	64Ba	b
CuA_2	-31.4	0.591(25°,1)	-92	0-65	T	S:Methylcyclo-hexane-Piperidine A=(di-n-Butyldi-thiocarbamate)	71Cf	b
[Cu]	-9.62	-0.916(34°,1)	-48.5	11-37	T	[Cu]=Copper meso-porphyrin IX dimethyl ester S:Mixtures of piperidine and tetrahydrofuran	64Ea	b
I_2	-43.1	3.973(20°,1)	-67.4	10-30	T	CM=10^{-5}M CL=10^{-4}-10^{-3}M S:n-Heptane	60Ya	b
P197	PIPERIDINE		$C_5H_{11}N$			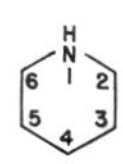		L
VOA_2	-42.09	3.00(1)	-83.7	25	C	μ→0 S:Nitrobenzene A=Acetylacetonate	65Ca	
CoA	-28.4	3.386(25°,1)	-29	20-50	T	S:Toluene A=α β,γ,δ-Tetra(p-methoxyphenyl) porphyrin	73Wa	b
CoAL	-7.1	0.602(-65°,1)	-25	-80--20	T	S:Toluene A=α,-β,γ,δ-Tetra(p-methoxyphenyl) porphyrin	73Wb	b,q: ±4.2
CoA_2	-57.7	2.64(25°,1)	-143.1	25	C	S:Acetone A=Cl	73Lb	
CoAB	-41.0	2.36(1)	-93.3	25	T	S:Cyclohexane A=p-MeOC_6H_4 B=N,N'-bis(acetyl-	77Hc	

Metal Ion	ΔH, kJ /mol	Log K	ΔS, J /mol K	T,°C	M	Conditions	Ref.	Remarks
P197, continued								
CoAB, cont.						acetonato)ethylenediamine		
[Co]	-43.5	4.61(0°,1)	-71	-23-23	T	S:Toluene [Co]=Cobalt (II) protoporphyrin IX dimethyl ester	73Sx	b,q: ±2.5
IrA_2BC_2	-50.2	4.32(1)	-88	25	C	S:CH_2Cl_2 A=Cl B=C(O)CF_3 C=P$(C_6H_5)_3$	77La	
NiA	-34.0	1.94(1)	-75	30	C	S:Acetonitrile CM=1-3x10^{-3}M A=BF_2-bridged glyoximato R_1=Me R_2=iso-Pr	76Gh	
	-37.9	1.86(1)	-89	30	C	S:Acetonitrile CM=1-3x10^{-3}M A=BF_2-bridged glyoximato R_1=Me R_2=Et	76Gh	
	-58.07	5.70(1-2)	-85.8	25	T	CL=10^{-4}M S:Benzene A=Diacetylbisbenzoyl hydrazine	58Sa	
	-31.4	1.68(1)	-71	30	C	S:Benzene CM=1-2x10^{-3}M A=BF_2-bridged glyoximato R_1=Me R_2=Et	76Gh	
	-26.5	1.82(1)	-53	30	C	S:Benzene CM=2-4x10^{-3}M A=BF_2-bridged glyoximato R_1=Me R_2=Pr	76Gh	
	-36.2	1.87(1)	-84	30	C	S:Benzene CM=1-2x10^{-3}M A=BF_2-bridged glyoximato R_1=Me R_2=iso-Pr	76Gh	

Metal Ion	ΔH, kJ /mol	Log K	ΔS, J /mol K	T,°C	M	Conditions	Ref.	Remarks
P197, continued								
NiA_2	-37.2	2.601(24.6°,1)	-75.3	8-39.5	T	S:Benzene A=O,-O'-Diethyldithiophosphate	72Mc	b
	-31.8	3.025(24.6°,1)	-49.4	8-39.5	T	S:Benzene A=O,-O'-Dimethyldithiophosphate	72Mc	b
	-105	2.398(1)	-392	30	C	S:Benzene A=$(SOCNPr_2)$	77Gf	
	-10	1.699(2)	-65	30	C	S:Benzene A=$(SOCNPr_2)$	77Gf	
RhAB	-68.6	5.447	-126	24	C	S:Benzene R:$[RhAB]_2$+2L= 2RhABL A=Cl B=1,5-Cyclooctadiene	76Lb	
$Rh_2A_4B_2$	-65.7	--	--	--	C	S:Benzene R:$Rh_2A_4B_2$+2L= $2RhA_2BL$ A=CO B=Cl	76Pd	u:24°
PdAB	-52.3	5.146	-77	24	C	S:Benzene R:$[PdAB]_2$+2L= 2[PdABL] A=(π-MeAllyl) B=Cl	77Lb	
Ag^+	-22.0	3.12(1)	-13.8	25	C	μ=0.5M(KNO_3)	75Eb	
	-28.2	3.31(2)	-31.4	25	C	μ=0.5M(KNO_3)	75Eb	
ZnA_2	-60.2	(1)	--	25	C	S:Benzene A=$N[Si(CH_3)_3]_2$	75Fc	
	-44.4	2.52(1)	-101	25	C	S:Cyclohexane A=$N[Si(CH_3)_3]_2$	75Fc	
Hg^{2+}	-45	8.70(1)	23	25	C	μ=0.5M(ClO_4^- or NO_3^-)	78Mc 77Mc	q:±15
	-29	8.74(2)	65	25	C	μ=0.5M(ClO_4^- or NO_3^-)	78Mc 77Mc	q:±9
P198	PIPERIDINE, 2-(2-aminoethyl)-	$C_7H_{16}N_2$				$C_5H_{10}N(CH_2CH_2NH_2)$		L
Ni^{2+}	-4	5.28(30°,1)	87.0	10-40	T	μ=0	63Ha	b
	17	4.42(30°,2)	139.7	20-40	T	μ=0	63Ha	b
Cu^{2+}	-33	8.60(30°,1)	42.7	10-40	T	μ=0	63Ha	b
	-33	5.86(30°,2)	1.7	10-40	T	μ=0 Log K_2 values vary slightly with $\bar{n}$	63Ha	b
P199	PIPERIDINE, 2-aminomethyl-	$C_6H_{14}N_2$				$C_5H_{10}N(CH_2NH_2)$		L
Co^{2+}	-12	4.81(30°,1)	50.6	10-40	T	μ=0	63Ha	b
	-29	3.96(30°,2)	-2.1	10-40	T	μ=0	63Ha	b
Ni^{2+}	-29	6.23(30°,1)	22.6	10-40	T	μ=0	63Ha	b

Metal Ion	ΔH, kJ /mol	Log K	ΔS, J /mol K	T,°C	M	Conditions	Ref.	Re-marks
P199, continued								
Ni^{2+},	-29	4.98(30°,2)	-2.9	10-40	T	μ=0	63Ha	b
cont.	-25	2.90(30°,3)	-27.2	10-40	T	μ=0	63Ha	b
Cu^{2+}	-50	9.60(30°,1)	18.0	10-40	T	μ=0	63Ha	b
	-46	7.96(30°,2)	0.4	10-40	T	μ=0	63Ha	b
P200	PIPERIDINE, 2,6-dimethyl		$C_7H_{15}N$			$C_5H_9N(CH_3)_2$		
Ag^+	-24.4	3.96(1)	-5.8	25	C	μ=0.5M(KNO_3)	75Eb	
	-24.0	3.75(2)	-8.4	25	C	μ=0.5M(KNO_3)	75Eb	
P201	PIPERIDINE, 2-methyl-		$C_6H_{13}N$			$C_5H_{10}N(CH_3)$		L
VOA_2	-32.17	1.28(1)	-83.3	25	C	μ→0 S:Nitrobenzene A=Acetylacetonate	65Ca	
Ag^+	-23.0	3.45(1)	-10.9	25	C	μ=0.5M(KNO_3)	75Eb	
	-28.7	3.50(2)	-29.7	25	C	μ=0.5M(KNO_3)	75Eb	
P202	PIPERIDINE, 3-methyl		$C_6H_{13}N$			$C_5H_{10}N(CH_3)$		
Ag^+	-23.2	3.04(1)	-19.7	25	C	μ=0.5M(KNO_3)	75Eb	
	-27.7	3.39(2)	-28.0	25	C	μ=0.5M(KNO_3)	75Eb	
P203	PIPERIDINE, 4-methyl		$C_6H_{13}N$			$C_5H_{10}N(CH_3)$		
Ag^+	-22.6	3.20(1)	-14.4	25	C	μ=0.5M(KNO_3)	75Eb	
	-27.7	3.30(2)	-29.9	25	C	μ=0.5M(KNO_3)	75Eb	
P204	PIPERIDINE, 2,2,6,6-tetramethyl-		$C_9H_{19}N$			$C_5H_7N(CH_3)_4$		
CuA_2	-49.0	(1)	--	25	C	S:Cyclohexane	72Lc	
BF_3	-110.9	(1)	--	25	C	S:1,2-Dichloro-ethane	72Lc	
P205	2-PIPERIDINECARBOXYLIC ACID		$C_7H_{13}O_2N$			$C_5H_9N(CO_2H)(CH_3)$		
Ni^{2+}	-54.3	9.77(20°,1-2)	0	5-35	T	0.200M(KCl)	78Ca	q:±12.5
P206	2-PIPERIDONE, N-methyl		$C_6H_{11}ON$			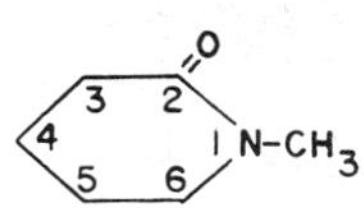		L
I_2	-25.1	1.47(1)	-56.1	25	T	S:n-Heptane	66Kc	

Metal Ion	ΔH, kJ /mol	Log K	ΔS, J /mol K	T, °C	M	Conditions	Ref.	Remarks
P207	PLUMBATE(II) ION, tetrabromo		Br_4Pb			$PbBr_4^{2-}$		L^{2-}
K^+	-20.9	0.0(25°,1)	-71	25&35		4M(Li,K,ClO_4,Cl)	63Mc	a,b
P208	POLYACRYLIC ACID		$(C_3H_4O_2)_n$			$[-CH_2CH(CO_2H)-]_n$		L_n^-
Cu^{2+}	15.5	3.5(25°)	-16.3	10-40	T	1M(KCl) L=Linear polyacrylic acid R:Cu^{2+}+2HL= CuL_2+$2H^+$	55Lb	b
	2.1	5.9(25°,1-2)	117	10-40	T	1M(KCl) L=Linear polyacrylic acid	55Lb	b
	3.3	3.4(25°)	-54	3-40	T	1M(KCl) L=Cross-linked polyacrylic acid R:Cu^{2+}+ 2HL=CuL_2+$2H^+$	55Lb	b
	1.7	7.0(25°,1-2)	138	3-40	T	1M(KCl) L=Cross-linked polyacrylic acid	55Lb	b
P209	POLYADENYLIC ACID		$[C_{10}H_{14}O_7N_5P]_n$					L
Mg^{2+}						R:Mg^{2+}+L(Na^+ form)=MgL+iNa^+ θ=fraction of nucleotide sites with bound Mg^{2+}		
	-16.3	--	--	10-40	T	C_{Na^+}=0.01M θ=0	71Kd	b,q:±6.3
	-14.2	--	--	10-40	T	C_{Na^+}=0.029M θ=0	71Kd	b,q:±6.3
	-7.5	--	--	10-40	T	C_{Na^+}=0.06M θ=0	71Kd	b,q:±6.3
	-2.5	--	--	10-40	T	C_{Na^+}=0.01M θ=0.25	71Kd	b,q:±6.3
	-1.2	--	--	10-40	T	C_{Na^+}=0.029M θ=0.25	71Kd	b,q: ±6.3
	2.9	--	--	10-40	T	C_{Na^+}=0.06M θ=0.25	71Kd	b,q:±6.3
	10.5	--	--	10-40	T	C_{Na^+}=0.01M θ=0.50	71Kd	b,q:±6.3
	12.6	--	--	10-40	T	C_{Na^+}=0.029M θ=0.50	71Kd	b,q: ±6.3
	8.4	--	--	10-40	T	C_{Na^+}=0.06M θ=0.50	71Kd	b,q:±6.3
P210	POLY(ADENYLIC+URIDYLIC) ACID							L
Mg^{2+}						R:Mg^{2+}+L(Na^+ form)=MgL+iNa^+ θ=fraction of nucleotide sites with bound Mg^{2+}		
	10.0	--	--	10-40	T	C_{Na^+}=0.01M θ=0	71Kd	b,q:±6.3
	8	--	--	10-40	T	C_{Na^+}=0.029M θ=0	71Kd	b,q:±6.3
	-2.9	--	--	10-40	T	C_{Na^+}=0.06M θ=0	71Kd	b,q:±6.3
	9.2	--	--	10-40	T	C_{Na^+}=0.01M θ=0.25	71Kd	b,q:±6.3

Metal Ion	ΔH, kJ /mol	Log K	ΔS, J /mol K	T, °C	M	Conditions	Ref.	Remarks
P210, continued								
Mg^{2+}, cont.	4	--	--	10-40	T	C_{Na^+}=0.029M θ=0.25	71Kd	b,q: ±6.3
	2.1	--	--	10-40	T	C_{Na^+}=0.06M θ=0.25	71Kd	b,q:±6.3
	8.8	--	--	10-40	T	C_{Na^+}=0.01M θ=0.50	71Kd	b,q:±6.3
	21	--	--	10-40	T	C_{Na^+}=0.029M θ=0.50	71Kd	b,q: ±6.3
	4	--	--	10-40	T	C_{Na^+}=0.06M θ=0.50	71Kd	b,q: ±6.3
P211 POLY(ADENYLIC+2URIDYLIC) ACID								
Mg^{2+}						R.Mg^{2+}+L(Na^+ form)=MgL+iNa^+ θ=fraction of nucleotide sites with bound Mg^{2+}		
	13.0	--	--	10-40	T	C_{Na^+}=0.01M θ=0	71Kd	b,q:±6.3
	-4	--	--	10-40	T	C_{Na^+}0=.029M θ=0	71Kd	b,q:±6.3
	2.9	--	--	10-40	T	C_{Na^+}=0.06M θ=0	71Kd	b,q:±6.3
	9.6	--	--	10-40	T	C_{Na^+}=0.01M θ=0.25	71Kd	b,q:±6.3
	2.9	--	--	10-40	T	C_{Na^+}=0.029M θ=0.25	71Kd	b,q: ±6.3
	3.3	--	--	10-40	T	C_{Na^+}=0.06M θ=0.25	71Kd	b,q:±6.3
	5.4	--	--	10-40	T	C_{Na^+}=0.01M θ=0.50	71Kd	b,q:±6.3
	8	--	--	10-40	T	C_{Na^+}=0.029M θ=0.50	71Kd	b,q: ±6.3
	4	--	--	10-40	T	C_{Na^+}=0.06M θ=0.50	71Kd	b,q:±6.3
P212 POLYDIOXOLAN						$[-CH_2-O-CH_2-CH_2-O-]_n$		
$Al_2A_4B_2$	-99.2	--	--	25	C	S:Toluene L=$[-CH_2-O-CH_2-CH_2-O-]$ R:0.5$Al_2A_4B_2$+L=(AlA_2BL) A=Ethyl B=Cl	77Pd	
P213 POLY(ETHYLENE OXIDE)						$[-CH_2-CH_2-O-]_n$		
$Al_2A_4B_2$	-97.5	--	--	25	C	S:Toluene L=$[-CH_2-CH_2-O-]$ R=0.5$Al_2A_4B_2$+L=(AlA_2BL) A=Ethyl B=Cl	77Pd	
P214 POLYPHOSPHATE ION, P_6			$P_6O_{19}^{8-}$					
Mg^{2+}	19.7	8.577(1)	230	25	C	μ=0	77Si	
Ca^{2+}	13.4	8.283(1)	205	25	C	μ=0	77Si	

Metal Ion	ΔH, kJ /mol	Log K	ΔS, J /mol K	T,°C	M	Conditions	Ref.	Remarks
P214, continued								
Sr^{2+}	13.0	7.550(1)	188	25	C	μ=0	77Si	
P215 POLYPHOSPHATE ION, P_8			$P_8O_{24}{}^{10-}$					
Mg^{2+}	20.9	9.530(1)	251	25	C	μ=0	77Si	
Ca^{2+}	14.2	8.430(1)	209	25	C	μ=0	77Si	
Sr^{2+}	13.8	7.770(1)	197	25	C	μ=0	77Si	
P216 POLY(PROPYLENE OXIDE)						$[-O-C_6H_4-]_n$		
$Al_2A_4B_2$	-72.8	--	--	25	C	S:Toluene L=$[-O-C_6H_4-]$ R:0.5$Al_2A_4B_2$+L= AlA_2BL A=Ethyl B=Cl	77Pd	
P217 POLYTETRAHYDROFURAN						$[-CH_2-CH_2-CH_2-CH_2-O-]_n$		
$Al_2A_4B_2$	-104.6	--	--	25	C	S:Toluene L=$(-CH_2-CH_2-CH_2-$ $CH_2-O-)$ R:0.5$Al_2A_4B_2$+L= (AlA_2BL) A=Ethyl B=Cl	77Pd	
P218 POLYTRIOXOCAN						$[-CH_2-O-(-CH_2-CH_2-O)_2-]_n$		
$Al_2A_4B_2$	-97.9	--	--	25	C	S:Toluene L=$[-CH_2-O-(-CH_2-$ $CH_2-O)_2-]$ R:0.5$Al_2A_4B_2$+L= (AlA_2BL) A=Ethyl B=Cl	77Pd	
P219 POLYURIDYLIC ACID			$(C_9H_{13}O_9N_2P)_n$					
Mg^{2+}						R:Mg^{2+}+L(Na^+ form)=MgL+iNa^+ θ=fraction of nucleotide sites with bound Mg^{2+}		
	14.2	--	--	10-40	T	C_{Na^+}=0.01M θ=0	71Kd	b,q:±6.3
	7.9	--	--	10-40	T	C_{Na^+}=0.029M θ=0	71Kd	b,q:±6.3
	9.2	--	--	10-40	T	C_{Na^+}=0.01M θ=0.25	71Kd	b,q:±6.3
	7.9	--	--	10-40	T	C_{Na^+}=0.029M θ=0.25	71Kd	b,q: ±6.3
	4.2	--	--	10-40	T	C_{Na^+}=0.01M θ=0.50	71Kd	b,q:±6.3
	7.5	--	--	10-40	T	C_{Na^+}=0.029M θ=0.50	71Kd	b,q: ±6.3

Metal Ion	ΔH, kJ /mol	Log K	ΔS, J /mol K	T,°C	M	Conditions	Ref.	Remarks
P220	PROLINE		$C_5H_9O_2N$			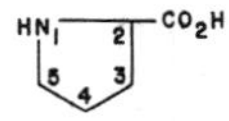		
Y^{3+}	26.376	5.40(25°,1)	191.84	25&35	T	μ=0.1(KCl)	73Sl	a,b,q: ±2.5
	42.204	4.81(25°,1)	233.59	25&35	T	μ=0.1(KCl)	73Sl	a,b,q: ±2.5
Ce^{3+}	12.309	6.00(25°,1)	156.31	25&35	T	μ=0.1(KCl)	73Sl	a,b,q: ±2.5
CoA	-19.7	4.26(25°,1)	15.9	15-55	T	μ=0.1M(KNO_3) A=Methylimino-diacetate	73Ib	b,q: ±1.2
	-23.0	3.85(25°,1)	-3.3	15-70	T	μ=0.1M(KNO_3) A= Nitrilotriacetate	71Ib	b,q: ±1.7
Ni^{2+}	-23.8	5.95(1)	33.9	25	C	μ=0.1M	73Ia	
	-25.5	4.95(2)	8.4	25	C	μ=0.1M	73Ia	
NiA	-25.5	5.18(25°,1)	14.2	15-55	T	μ=0.1M(KNO_3) A=Methylimino-diacetate	73Ib	b,q: ±2.9
	-27.2	4.99(25°,1)	3.8	15-70	T	μ=0.1M(KNO_3) A= Nitrilotriacetate	71Ib	b,q: ±2.1
Cu^{2+}	-31.0	8.84(1)	64.0	25	C	μ=0.1M	73Ia	
	-32.2	7.44(2)	35.1	25	C	μ=0.1M	73Ia	
	-59.8	16.4(1-2)	113.8	25	C	μ=0.1M	71Be	
CuA	-33.4	6.68(25°,1)	8.8	15-55	T	μ=0.1M(KNO_3) A=Methylimino-diacetate	73Ib	b,q: ±3.3
	-45.6	6.24(25°,1)	-50.2	15-70	T	μ=0.1M(KNO_3) A= Nitrilotriacetate	71Ib	b
ZnA	-23.8	3.98(25°,1)	-3.3	15-70	T	μ=0.1M(KNO_3) A= Nitrilotriacetate	71Ib	b
CdA	-17.6	3.95(25°,1)	16.7	15-55	T	μ=0.1M(KNO_3) A=Methylimino diacetate	73Ib	b,q: ±1.2
	-16.3	3.05(25°,1)	3.8	15-70	T	μ=0.1M(KNO_3) A= Nitrilotriacetate	71Ib	b
P221	PROLINE (L)		$C_5H_9O_2N$			see P220		
Cu^{2+}	-59.4	16.4(2)	115.1	25	C	μ=0.1M	71Be	
P222	PROLINE, n-benzyl		$C_{12}H_{15}O_2N$			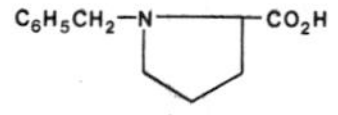		
Cu^{2+}	-36	7.07(25°,1)	12	15-35	T	μ=0.1(KCl)	78Ke	
	-16	6.29(25°,2)	68	15-35	T	μ=0.1(KCl) Complex: Cu(N-benzyl-L-proline)$_2$	78Ke	
	-12	7.29(25°,2)	100	15-35	T	μ=0.1(KCl)	78Ke	

Metal Ion	ΔH, kJ /mol	Log K	ΔS, J /mol K	T,°C	M	Conditions	Ref.	Re-marks
P222, continued								
Cu^{2+}, cont.						L=N-benzyl-DL-proline Complex: Cu(N-benzyl-L-proline) (N-benzyl-D-proline)		
P223	PROLINE, N-glycyl-(L)		$C_7H_{12}O_3N_2$			$C_5H_8O_2N(COCH_2NH_2)$		
Y^{3+}	21.88	3.80(25°,1)	145.48	25-45	T	μ=0.15(KNO_3)	79Se	b
La^{3+}	18.16	3.46(25°,1)	126.90	25-45	T	μ=0.15(KNO_3)	79Se	b
Ce^{3+}	20.92	3.50(25°,1)	138.36	25-45	T	μ=0.15(KNO_3)	79Se	b
Pr^{3+}	20.12	3.59(25°,1)	135.10	25-45	T	μ=0.15(KNO_3)	79Se	b
Nd^{3+}	22.76	3.65(25°,1)	160.79	25-45	T	μ=0.15(KNO_3)	79Se	b
Cu^{2+}	-26.4	6.34(1)	32.2	25	C	μ=0.1M(KNO_3)	72Bl	
	-28.4	5.01(2)	0.0	25	C	μ=0.1M(KNO_3)	72Bl	
P224	PROLINE, 4-hydroxy-		$C_5H_9O_3N$			$C_5H_8O_2N(OH)$		
Y^{3+}	7.033	4.52(1)	110.12	25&35	T	μ=0.1(KCl)	73Sl	a,b,q: ±2.5
	14.067	4.40(2)	131.42	25&35	T	μ=0.1(KCl)	73Sl	a,b,q: ±2.5
La^{3+}	-7.28	3.80(25°,1)	48.3	25&45	T	μ=0.05M	75Si	a,b
	-10.8	3.05(25°,2)	21.9	25&45	T	μ=0.05M	75Si	a,b
	-14.5	7.55(25°,1-2)	95.9	25&45	T	μ=0	75Si	a,b
Ce^{3+}	52.756	4.90(1)	270.75	25&35	T	μ=0.1(KCl)	73Sl	a,b
UO_2^{2+}	-29.04	7.023(25°,1)	37.07	25&45	T	μ=0.05M(KCl)	73Sm	a,b
	-29.04	6.817(25°,2)	33.14	25&45	T	μ=0.05M(KCl)	73Sm	a,b
	-18.28	13.891(25°,1-2)	205.10	25&45	T	μ=0	73Sm	a,b
Co^{2+}	-13.8	4.83(1)	46.0	25	C	μ=0.1M	73Ia	q:±0.8
	-14.06	4.05(2)	30.1	25	C	μ=0.1M	73Ia	
Ni^{2+}	-20.9	5.94(1)	43.5	25	C	μ=0.1M	73Ia	q:±0.8
	-22.93	5.01(2)	19.2	25	C	μ=0.1M	73Ia	
Cu^{2+}	-28.9	8.38(1)	63.6	25	C	μ=0.1M	73Ia	
	-32.6	7.04(2)	25.1	25	C	μ=0.1M	73Ia	
Zn^{2+}	-10.9	5.03(1)	60.2	25	C	μ=0.1M	73Ia	q:±0.8
	-12.1	4.35(2)	43.1	25	C	μ=0.1M	73Ia	
P225	PROPANAL		C_3H_6O			CH_3CH_2CHO		
$[NiA_2]_2$	-17.2	1.19(100°,1)	-23	75-125	T	S:Squalene A=3-Trifluoro-acetylcamphorate	74Sj	b,gg

Metal Ion	ΔH, kJ /mol	Log K	ΔS, J /mol K	T,°C	M	Conditions	Ref.	Remarks
P226	PROPANAL, 3-methyl		C_4H_8O			$(CH_3)_2CHCHO$		
$[NiA_2]_2$	-18	1.19(100°,1)	-25	75-125	T	S:Squalene A=3-Trifluoro-acetylcamphorate	74Sj	b,gg
P227	PROPANE, 1-amino-		C_3H_9N			$CH_3CH_2CH_2NH_2$		L
NiA	-70.7	5.89(1)	-124.3	25	T	CM≅10^{-4}M S:Benzene A=Dibenzoyl hydrazine	60Sb	
Ag^+	-12	3.45(1)	14.2	25	C	μ=0	71He	q:±8
	-38	3.99(2)	-56.1	25	C	μ=0	71He	q:±8
	-53.22	7.44(1-2)	-36.0	25	C	μ=0	71He	
P228	PROPANE, 2-amino-		C_3H_9N			$CH_3CH(NH_2)CH_3$		L
NiA	-72.4	5.20(1)	-143.1	25	T	CM≅10^{-4}M S:Benzene A=Dibenzoyl hydrazine	60Sb	
P229	PROPANE, 2-amino-1,3-dihydroxy-2-hydroxymethyl-		$C_4H_{11}O_3N$			$(HOCH_2)_3CNH_2$		L
Ag^+	-10.0	3.503(1)	31.8	0	T	μ=0.05 ΔCp=-1498	66Da	
	-21.8	3.362(1)	-12.1	10	T	μ=0.05 ΔCp=-1000	66Da	
	-28.9	3.251(1)	-36.0	20	T	μ=0.05 ΔCp=-400	66Da	
	-28.9	3.040(1)	-38.9	30	T	μ=0.05 ΔCp=200	66Da	
	-23.8	2.909(1)	-20.9	40	T	μ=0.05 ΔCp=900	66Da	
	-12.1	2.796(1)	17.2	50	T	μ=0.05 ΔCp=1598	66Da	
	8	2.793(1)	74.9	60	T	μ=0.05 ΔCp=2301	66Da	
	-42	4.042(2)	-77.8	0	T	μ=0.05 ΔCp=300	66Da	
	-38.5	3.766(2)	-66.9	10	T	μ=0.05 ΔCp=300	66Da	
	-36.0	3.529(2)	-54	20	T	μ=0.05 ΔCp=300	66Da	
	-33.0	3.328(2)	-42	30	T	μ=0.05 ΔCp=300	66Da	
	-28.9	3.158(2)	-31.8	40	T	μ=0.05 ΔCp=400	66Da	
	-25.1	3.017(2)	-20.9	50	T	μ=0.05 ΔCp=400	66Da	
	-21.8	2.902(2)	-10.0	60	T	μ=0.05 ΔCp=400	66Da	
Cd^{2+}	-130.1	4.77(25°,1-2)	-345	25&35	T	μ=0.5M(KNO_3)	76Rb	a,b
Pb^{2+}	-41.4	6.29(25°,1-2)	-18.8	25&35	T	μ=0.5M(KNO_3)	76Rb	ab
P230	PROPANE, 1-amino-2-methyl-		$C_4H_{11}N$			$(CH_3)_2CHCH_2NH_2$		L
NiA	-69.4	4.32(1-2)	-150.2	25	T	CM=10^{-4}M S:Benzene A=Diacetyl-bisbenzoyl hydrazine	60Sb	
P231	PROPANE, 2-amino-2-methyl-		$C_4H_{11}N$			$(CH_3)_3CNH_2$		L
NiA	-67.4	3.39(1-2)	-161.1	25	T	CM=10^{-4}M	60Sb	

Metal Ion	ΔH, kJ /mol	Log K	ΔS, J /mol K	T,°C	M	Conditions	Ref.	Remarks
P231, continued								
NiA, cont.						S:Benzene A=Di-acetylbisbenzoyl hydrazine		
ZnA_2	-41.0	2.580(25°,1)	-88	17.8-194	T	S:Toluene $A=[S_2CN(CH_3)_2]$	65Ch	b
P232	PROPANE, 2,2-bis(hydroxy-methyl)-1,3-dihydroxy-		$C_5H_{12}O_4$			$CH_2OHC(CH_2OH)_2CH_2OH$		L
$H_5TeO_6^-$	-18.0	0.49	-50	25	T	μ=0.1 R:$H_5TeO_6^-$ + L=$HTeO_4L^-$ +$2H_2O$	62Ea	
P233	PROPANE, 1,2-diamino		$C_3H_{10}N$			$CH_3CH(NH_2)CH_2NH_2$		L
Ni^{2+}	-34.48	7.30(1)	23.8	25	C	μ=0	73Pl	
	-34.48	6.28(2)	4.6	25	C	μ=0	73Pl	
	-43.1	4.98(3)	-49.0	25	C	μ=0	73Pl	
Cu^{2+}	-50.00	10.322(1)	29.7	25	C	μ→0	67Pc	
	-54.4	10.54(1)	19.2	25	T	μ→0	62Na	
	-59.0	10.59	4.6	15-40	T	μ=0.3M($NaClO_4$)	72Md	b
	-49.79	8.881(2)	3.8	25	C	μ→0	67Pc	
	-55.2	9.06(2)	-11.3	25	T	μ→0	62Na	
	-57.7	9.15(2)	-18.0	15-40	T	μ=0.3M($NaClO_4$)	72Md	b
	-100.0	19.20(1-2)	33.5	25	T	μ→0	67Pc	
Hg^{2+}	-141.4	23.40(1-2)	-25	25	T	0.1M(KNO_3)	61Rc	
P234	PROPANE, 1,3-diamino-		$C_3H_{10}N_2$			$H_2NCH_2CH_2CH_2NH_2$		L
Ni^{2+}	-42.7	6.18(30°,1)	-21	10-40	T	μ=0	58Ba	b
	-36.8	6.39(25°,1)	-1.2	0-50	T	μ=0.15	55Cc	b
	-36.0	5.84(1)	-9	25	C	μ=0.2M(KCl)	76Ge	
	-30.29	6.91(1)	30.5	25	C	μ=0.30(ClO_4^-)	67Hf	
	-32.6	6.39(1)	12.6	25	C	0.1M(KNO_3)	55Pb	
	-34.3	4.28(30°,2)	-33	10-40	T	μ=0	58Ba	b
	-34.3	4.38(25°,2)	31.0	0-50	T	μ=0.15	55Cc	b
	-31.4	4.35(2)	-22	25	C	μ=0.2M(KCl)	76Ge	
	-30.75	5.34(2)	-0.8	25	C	μ=0.30(ClO_4^-)	67Hf	
	-30.1	4.39(2)	-17.2	25	C	0.1M(KNO_3)	55Pb	
	-25	1.5(25°,3)	-54	0-50	T	μ=0.15	55Cc	b
	-29.79	3.01(3)	-42.2	25	C	μ=0.30(ClO_4^-)	67Hf	
	-26.4	1.23(3)	-64.8	25	C	0.1M(KNO_3)	55Pb	
	-89.1	12.01(1-3)	-69.4	25	C	0.1M(KNO_3)	55Pb	
Cu^{2+}	-58.2	9.45(30°,1)	-12	10-40	T	μ=0	58Ba	b
	-52.3	9.68(25°,1)	9.6	0-50	T	μ=0.15	55Cc	b
	-49.5	9.65(1)	19	25	C	μ=0.2M(KCl)	76Ge 76Sp 76Sq	
	-46.11	10.16(1)	40.2	25	C	μ=0.30(ClO_4^-)	67Hf	
	-58	9.62(30°,1)	-12	0-30	T	1M(KNO_3)	56Ha	b
	-51.5	7.12(25°,2)	-36.4	0-50	T	μ=0.15	55Cc	b

Metal Ion	ΔH, kJ /mol	Log K	ΔS, J /mol K	T,°C	M	Conditions	Ref.	Re-marks
P234, continued								
Cu^{2+}, cont.	-46.0	7.17(2)	-17	25	C	μ=0.2M(KCl)	76Ge 76Sp 76Sq	
	-46.11	7.14(2)	-17.6	25	C	μ=0.30(ClO_4^-)	67Hf	
	-54	7.00(30°,2)	-50	0-30	T	0.1M(KNO_3)	56Ha	b
	-95.4	17.17(1-2)	8.4	25	C	0.1M(KNO_3)	55Pb	
	-4.52	2.26(25°)	28.0	-75&25	T	S:50v% CH_3OH R:Cu+CuL_2=2CuL	73Ya	a,b,q: ±2.2
	-90.4	16.98	22	25	C	μ=0.2M(KCl) R:Cu+A+L=CuAL A=1,2-Diamino-N,N-dimethyl-ethane	76Sp	
	-94.1	17.51	19	25	C	μ=0.2M(KCl) R:Cu+A+L=CuAL A=1,2-Diamino-N,N'-dimethyl-ethane	76Sp	
	-102.9	18.83	15	25	C	μ=0.2M(KCl) R:Cu+A+L=CuAL A=1,2-Diamino-ethane	76Sp	
	-77.0	16.91	65	25	C	μ=0.2M(KCl) R:Cu+A+L=CuAL A=Glycine	76Sp	
CuA	-90.4	16.98(1)	22	25	C	μ=0.2M(KCl) A=1,2-Diamino-N,N-dimethyl-ethane	76Sq	
	-94.1	17.51(1)	19	25	C	μ=0.2M(KCl) A=1,2-Diamino-N,N'-dimethyl-ethane	76Sq	
	-77.0	16.91(1)	65	25	C	μ=0.2M(KCl) A=Glycine	76Sq	
	-102.9	18.83(1)	15	25	C	μ=0.2M(KCl) A=1,2-Diamino-ethane	76Sq	
CuA_2	-114	(1)	--	30	C	S:Benzene A=1,1,1,5,5,5-Hexafluoropentane-2,4-dionato	75Gg 75Gh	
	-67	(2)	--	30	C	S:Benzene A=1,1,1,5,5,5-Hexafluoropentane-2,4-dionato	75Gg 75Gh	
Ag^+	-61.1	5.56(30°,1)	-96	10-40	T	μ=0	58Ba	b
Cd^{2+}	-22.2	4.62(25°,1)	14.2	0-50	T	μ=0.15	55Cc	b
	-18.4	3.05(25°,2)	-2.5	0-50	T	μ=0.15	55Cc	b

Metal Ion	ΔH, kJ /mol	Log K	ΔS, J /mol K	T,°C	M	Conditions	Ref.	Re-marks
P235	PROPANE, 1,3-diamino-N-(2-aminoethyl)-		$C_5H_{15}N_3$			$NH_2CH_2CH_2NHCH_2CH_2CH_2NH_2$		
Co^{2+}	-28.0	8.27(25°,1)	63	15-40	T	μ=0.1M(KCl)	73Sr	b
Ni^{2+}	-39.7	11.01(25°,1)	79	15-40	T	μ=0.1M(KCl)	73Sr	b
Cu^{2+}	-68.2	16.30(25°,1)	84	15-40	T	μ=0.1M(KCl)	73Sr	b
Zn^{2+}	-20.5	8.62(25°,1)	96	15-40	T	μ=0.1M(KCl)	73Sr	b
P236	PROPANE, 1,3-diamino-N,N'-bis(2-aminoethyl)		$C_7H_{20}N_4$			$H_2NCH_2CH_2NHCH_2CH_2CH_2NHCH_2CH_2NH_2$		
Co^{2+}	-68.6	14.37(25°,1)	46.0	15-40	T	μ=0.1M(KCl)	73Ss	b
Ni^{2+}	-78.2	16.07(25°,1)	46.0	15-40	T	μ=0.1M(KCl)	73Ss	b
	-74.9	16.4(1)	63.2	25	C	μ=0.5M(KCl)	72Fa	
	-17.2	3.66(2)	12.6	25	C	μ=0.5M(KCl)	72Fa	
Cu^{2+}	-114.2	23.17(25°,1)	60.7	15-40	T	μ=0.1M(KCl)	73Ss	b
	-115.9	23.9(1)	68.6	25	C	μ=0.5M(KCl)	72Fa	
Zn^{2+}	-49.8	12.67(25°,1)	75.3	15-40	T	μ=0.1M(KCl)	73Ss	b
	-49.8	12.8(1)	78.6	25	C	μ=0.5M(KCl)	72Fa	
P237	PROPANE, 1,3-diamino, 2,2-dimethyl-		$C_5H_{14}N_2$			$H_2NCH_2C(CH_3)_2CH_2NH_2$		L
Co^{2+}	-29	4.88(1)	0	30	T	$1M(KNO_3)$	56Ha	
	-25	3.07(2)	-25	30	T	$1M(KNO_3)$	56Ha	
Ni^{2+}	-33	6.59(1)	12	30	T	$1M(KNO_3)$	56Ha	
	-29	4.41(2)	-12	30	T	$1M(KNO_3)$	56Ha	
Cu^{2+}	-50	9.94(1)	29	30	T	$1M(KNO_3)$	56Ha	
	-50	7.45(2)	-29	30	T	$1M(KNO_3)$	56Ha	
Zn^{2+}	-21	5.21(30°,1)	33	0-30	T	$1M(KNO_3)$	56Ha	b
	-21	5.20(30°,1)	33	0-30	T	$1M(KNO_3)$	56Ha	b
P238	PROPANE, 1,2-diamino-2-methyl		$C_4H_{12}N_2$			$CH_3C(NH_2)(CH_3)CH_2NH_2$		L
Ni^{2+}	-29.50	6.47(1)	25.1	25	C	μ=0 Data refer to reactions involving paramagnetic bisdiamine complex	73Pl	
	-35.10	(2)	--	25	C	μ=0 Data refer to reactions involving paramagnetic bisdiamine complex	73Pl	
	-31.00	(2)	--	25	C	μ=0 Data refer to mixed diamagnetic-paramagnetic complex	73Pl	
	-12.59	(3)	--	25	C	μ=0 Data refer	73Pl	q:±1.7

Metal Ion	ΔH, kJ /mol	Log K	ΔS, J /mol K	T,°C	M	Conditions	Ref.	Re- marks
P238, continued								
Ni^{2+}, cont.						to reactions involving paramagnetic bisdiamine complex		
	-16.69	(3)	--	25	C	μ=0 Data refer to mixed diamagnetic-paramagnetic complex	73Pl	q:±1.7
NiL_2	-9.79	0.079	-31.34	25	T	μ=0 R:NiL_2^{2+} (diamagnetic)= NiL_2^{2+} (paramagnetic)	73Pl	
Cu^{2+}	-47.15	9.765(1)	28.9	25	C	μ→0	67Pc	
	-51.30	8.819(2)	-2.9	25	C	μ→0	67Pc	
	-98.3	18.59(1-2)	25.9	25	C	μ→0	67Pc	
P239	PROPANE, 1,3-diamino-N,N,-N',N'-tetraacetic acid		$C_{11}H_{18}O_8N_2$			$(CH_2CO_2H)_2NCH_2CH_2CH_2N(CH_2CO_2H)_2$		L
Mg^{2+}	38.03	6.21(1)	246.8	20	C	0.1M(KNO_3)	64Ab	
	4.18	9.603(1)	197.9	20	C	μ=1M(HCl) CL=0.01M	69Sh	g:k
Ca^{2+}	-7.28	7.28(1)	114.6	20	C	0.1M(KNO_3)	64Ab	
	-12.64	11.250(1)	172.4	20	C	μ=1M(HCl) CL=0.01M	69Sh	g:k
Sr^{2+}	-2.30	9.237(1)	169.0	20	C	μ=1M(HCl) CL=0.01M	69Sh	g:k
Ba^{2+}	-8.12	8.231(1)	129.7	20	C	μ=1M(HCl) CL=0.01M	69Sh	g:k
La^{3+}	15.73	11.23(1)	268.6	20	C	0.1M(KNO_3)	64Ab	
	15.82	11.23(1)	266.5	20	C	μ=0.1(KNO_3)	71Af	
Ce^{3+}	15.06	11.70(1)	275.3	20	C	μ=0.1(KNO_3)	71Af	
Pr^{3+}	16.40	11.99(1)	284.9	20	C	μ=0.1(KNO_3)	71Af	
Nd^{3+}	17.6	12.34(1)	290.8	20	C	μ=0.1(KNO_3)	71Af	
Sm^{3+}	23.10	13.14(1)	330.1	20	C	μ=0.1($KClO_4$)	71Af	
	23.64	13.14(1)	332.2	20	C	μ=0.1(KNO_3)	71Af	
Eu^{3+}	24.31	13.54(1)	341.8	20	C	μ=0.1($KClO_4$)	71Af	
	24.89	13.54(1)	343.9	20	C	μ=0.1(KNO_3)	71Af	
Gd^{3+}	20.71	13.70(1)	331.4	20	C	μ=0.1(KNO_3)	71Af	
Dy^{3+}	19.08	14.65(1)	345.6	20	C	μ=0.1(KNO_3)	71Af	
Er^{3+}	17.61	14.80(1)	337.6	20	C	μ=0.1(KNO_3)	71Af	
Yb^{3+}	15.27	15.42(1)	336.0	20	C	μ=0.1(KNO_3)	71Af	
Lu^{3+}	14.77	15.54(1)	348.9	20	C	μ=0.1(KNO_3)	71Af	
Mn^{2+}	-3.01	9.99(1)	221.3	20	C	0.1M(KNO_3)	64Ab	

Metal Ion	ΔH, kJ /mol	Log K	ΔS, J /mol K	T,°C	M	Conditions	Ref.	Remarks
P239, continued								
Co^{2+}	-10.9	15.55(1)	260.2	20	C	0.1M(KNO_3)	64Ab	
Ni^{2+}	-27.86	18.15(1)	252.2	20	C	0.1M(KNO_3)	64Ab	
Cu^{2+}	-32.38	18.92(1)	251.4	20	C	0.1M(KNO_3)	64Ab	
Zn^{2+}	-9.50	15.26(1)	258.6	20	C	0.1M(KNO_3)	64Ab	
Cd^{2+}	-22.8	13.90(1)	188.3	20	C	0.1M(KNO_3)	64Ab	
Hg^{2+}	-79.1	19.9(1)	111.3	20	C	0.1M(KNO_3)	64Ab	
Pb^{2+}	-26.8	13.78(1)	170.7	20	C	0.1M(KNO_3)	64Ab	
P240	PROPANE, 1,3-diamino-N,N,N',N'-tetramethyl		$C_7H_{18}N_2$			$(CH_3)_2NCH_2CH_2CH_2N(CH_3)_2$		
CuA_2	-74	(1)	--	30	C	S:Benzene A=1,1,1,5,5,5-Hexafluoropentane-2,4-dionato	75Gg 76Gh	
HgA_2	-80.0	4.48(1)	-179	30	C	CM=1.0-5.3x10^{-3} S:Benzene A=Cl	74Fb	
	-84.8	4.84(1)	-184	30	C	CM=1.0-4.9x10^{-3} S:Benzene A=Br	74Fb	
	-83.6	5.00(1)	-180	30	C	CM=1.5-4.0x10^{-3} S:Benzene A=I	74Fb	
	-34.8	0.62(1)	-102	30	C	CM=1-20x10^{-3} S:Benzene A=C_6F_5	74Fc	
P241	PROPANE, 1,3-di-(2-carboxypiperidyl)		$C_{15}H_{26}O_4N_2$			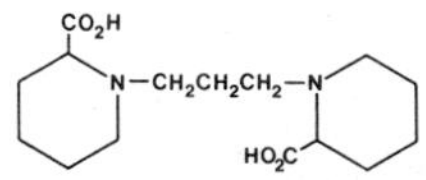		
Ni^{2+}	5.85	7.21(20°,1)	159	5-35	T	0.200M(KCl)	78Ca	q:±6.27
P242	PROPANE, 1,3-diethoxy-		$C_7H_{16}O_2$			$C_2H_5O(CH_2)_3OC_2H_5$		
TiA_4	-97.1	3.74(1)	-254.0	25	C	S:Benzene CM=0.05-0.08M A=Cl	68Ge 70Gh	
SnA_4	-87.4	3.81(1)	-220.5	25	C	S:Benzene CM=0.05-0.08M A=Cl	68Ge 70Gh	
P243	PROPANE, 2,2-diphosphonic acid		$C_3H_{10}O_6P_2$			$CH_3C(PO_3H_2)_2CH_3$		L^{4-}
Li^+	-5.52	3.83(1)	54.8	25	C	0.5M$(CH_3)_4$NCl	68Ca	
Na^+	1.2	2.08(1)	44.8	25	C	0.5M$(CH_3)_4$NCl	68Ca	
K^+	0.0	1.60(1)	33.5	25	C	0.5M$(CH_3)_4$NCl	68Ca	
Ca^{2+}	-4.27	6.34(1)	107.1	25	C	0.5M$(CH_3)_4$NCl	68Ca	

Metal Ion	ΔH, kJ /mol	Log K	ΔS, J /mol K	T,°C	M	Conditions	Ref.	Remarks
P244	PROPANE, 1,3-diylbis (diphenylphosphine)		$C_{27}H_{26}P_2$			$[(C_6H_5)_2P]_2C_3H_6$		
HgA_2	-93	--	--	30	C	S:Benzene A=Cl R:$2HgA_2$+L= $(HgA_2)_2L$	76Ga	
	-78	--	--	30	C	S:Benzene A=Br R:$2HgA_2$+L= $(HgA_2)_2L$	76Ga	
P245	PROPANE, 1,2-epoxy		C_3H_6O			OCH_2CHCH_3 (ring)		
$[NiA_2]_2$	-18.2	1.43(100°,1)	-21	75-125	T	S:Squalane A=3-Trifluoro-acetylcamphorate	74Sj	b,gg
I_2	-15.9	0.81(25°,1)	-37.6	1-25	T	S:n-Heptane	60Ta 60Bb	b
P246	PROPANE, 1,3-epoxy-		C_3H_6O			$CH_2CH_2CH_2$ (ring via O)		L
I_2	-26.8	1.41(25°,1)	-62.8	2.5-25	T	S:n-Heptane	60Bb 60Ta	b
P247	PROPANE, 1-nitro		$C_3H_7O_2N$			$CH_3CH_2CH_2NO_2$		
$[NiA_2]_2$	-21	0.27(100°,1)	-52	75-125	T	S:Squalane A=3-Trifluoro-acetylcamphorate	74Sj	b,gg
P248	1,2-PROPANEDIOL		$C_3H_8O_2$			$CH_3CH(OH)CH_2OH$		L
BA_4^-	-12.64	0.607(25°,1)	-30.5	0-35	T	CL=0-0.9M A=OH^-	67Cb	b,q:±1.2
	-30.21	0.585(25°,1-2)	-92	0-35	T	CL=0-0.9M A=OH^-	67Cb	b,q:±2.9
$H_5TeO_6^-$	-20.9	1.52	-44	25	T	μ=0.1 R:$H_5TeO_6^-$+ L=$HTeO_4L^-$+$2H_2O$	62Ea	
P249	1,3-PROPANEDIOL		$C_3H_8O_2$			$HOCH_2CH_2CH_2OH$		L
BA_4^-	-19.04	0.104(25°,1)	-63	0-35	T	CL=0-0.9M A=OH^-	67Cb	b,q:±2.1
	-25.86	-0.959(25°,1-2)	-67	0-35	T	CL=0-0.9M A=OH^-		b,q:±2.5
$H_5TeO_6^-$	-20.1	0.51	-58	25	T	μ=0.1 R:$H_5TeO_6^-$+ L=$HTeO_4L^-$+$2H_2O$	62Ea	
P250	1,2-PROPANEDIOL, carbonate		$C_4H_6O_3$			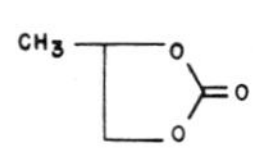		
SbA_5	-58.2	(1)	--	40	C	S:CCl_4 A=Cl	73Oa	

Metal Ion	ΔH, kJ /mol	Log K	ΔS, J /mol K	T,°C	M	Conditions	Ref.	Remarks
P251	1,3-PROPANEDIOL, 2,2-diethyl-	$C_7H_{16}O_2$				$HOCH_2C(CH_2CH_3)_2CH_2OH$		L
$H_5TeO_6^-$	-19.7	0.23	-63	25	T	μ=0.1 R: $H_5TeO_6^-$ + L= $HTeO_4L^-$ + $2H_2O$	62Ea	
P252	1,3-PROPANEDIOL, 2-hydroxymethyl-2-methyl-	$C_5H_{12}O_3$				$OHCH_2C(CH_2OH)(CH_3)CH_2OH$		L
$H_5TeO_6^-$	-14.2	0.21	-50	25	T	μ=0.1 R: $H_5TeO_6^-$ + L= $HTeO_4L^-$ + $2H_2O$	62Ea	
P253	1,2-PROPANEDIOL, 3-methoxy-	$C_4H_{10}O_3$				$CH_3OCH_2CH(OH)CH_2OH$		L
$H_5TeO_6^-$	-15.8	1.48	-25	25	T	μ=0.1 R: $H_5TeO_6^-$ + L= $HTeO_4L^-$ + $2H_2O$	62Ea	
P254	1,3-PROPANEDIONE, 1,3-diphenyl-	$C_{15}H_{12}O_2$				$C_6H_5COCH_2COC_6H_5$		L
LiA	-7.933	5.95(30°,1)	87.9	25.5	C	S:75v% Dioxane CL=0.02M A= NO_3	68Mc	
Ni^{2+}	-14.2	9.93(25°,1)	142	15-40	T	μ=0 S:75v% Dioxane	71Rc	b,q: ±2.1
	-17.2	8.73(25°,2)	109	15-40	T	μ=0 S:75v% Dioxane	71Rc	b,q: ±2.1
Cu^{2+}	-40.6	13.32(25°,1)	117	15-40	T	μ=0 S:75v% Dioxane	71Rc	b
	-38.5	11.79(25°,2)	96	15-40	T	μ=0 S:75v% Dioxane	71Rc	b
TlA_2B	-4.046	6.90(30°)	118.8	25.5	C	S:75v% Dioxane CL=0.02M R: TlA_2B +L= TlA_2L +B A= CH_3 B= ClO_4	68Mc	
	-5.690	7.07(30°)	116.7	25.5	C	S:75v% Dioxane CL=0.02M R: TlA_2B +L= TlA_2L +B A= CH_3CH_2 B= ClO_4	68Mc	
	-5.176	6.96(30°)	116.3	25.5	C	S:75v% Dioxane CL=0.02M R: TlA_2B +L= TlA_2L +B A= CH_3CH_2 B= NO_3	68Mc	b
	-6.979	7.51(30°)	162.8	25.5	C	S:75v% Dioxane CL=0.02M R: TlA_2B +L= TlA_2L +B A= $CH_3CH_2CH_2$ B= NO_3	68Mc	b
	-8.268	8.43(30°)	132.6	25.5	C	S:75v% Dioxane CL=0.02M R: TlA_2B +L= TlA_2L +B A= C_6H_5 B= ClO_4	68Mc	b

Metal Ion	ΔH, kJ /mol	Log K	ΔS, J /mol K	T,°C	M	Conditions	Ref.	Re-marks
P255	1-PROPANESULFONIC ACID, 3-amino-		$C_3H_9O_3NS$			$H_2N(CH_2)_3SO_3H$		L^-
Ag^+	-21.42	3.167(1)	-11.3	25	C	μ=0.5M(KNO_3)	72Va	
	-29.33	3.584(2)	-29.7	25	C	μ=0.5M(KNO_3)	72Va	
P256	1-PROPANESULFONIC ACID, 2,3-dimercapto		$C_3H_8O_3S_3$			$CH_2(SH)CH(SH)CH_2SO_3H$		HL
Pt^{2+}	28.28	10.09(25°,1)	287.4	25-75	T	μ=1M($HClO_4$) CM=1x10^{-4}M CL=40x10^{-4}M	78Kb	b
	9.54	4.44(25°,1)	116.3	25-75	T	μ=1M(KOH) CM=1x10^{-4}M CL=40x10^{-4}M	78Kb	b
	14.31	9.63(25°,2)	231.8	25-75	T	μ=1M(KNO_3) CM=1x10^{-4}M CL=40x10^{-4}M	78Kb	b
P257	PROPANOIC ACID		$C_3H_6O_2$			$CH_3CH_2CO_2H$		L^-
Be^{2+}	83.3	0.301(25°,1)	284	25-30	T	μ=1.0(ClO_4^-)	75Ta	b,n
	82.4	0.544(35°,1)	280	30-40	T	μ=1.0(ClO_4^-)	75Ta	b,n
	-35.1	3.954(25°,2)	75	25-30	T	μ=1.0(ClO_4^-)	75Ta	b,n
	-73.6	3.854(25°,2)	67	30-40	T	μ=1.0(ClO_4^-)	75Ta	b,n
Y^{3+}	16.23	1.61(1)	85.4	25	C	μ=2.0	65Cd	
	8.45	1.20(2)	51.0	25	C	μ=2.0	65Cd	
La^{3+}	10.33	1.53(1)	64.0	25	C	μ=2.0	65Cd	
	6.94	0.89(2)	40.2	25	C	μ=2.0	65Cd	
Ce^{3+}	9.04	1.67(1)	62.3	25	C	μ=2.0	65Cd	
	7.40	1.00(2)	43.9	25	C	μ=2.0	65Cd	
Pr^{3+}	7.91	1.78(1)	60.7	25	C	μ=2.0	65Cd	
	7.03	1.08(2)	44.4	25	C	μ=2.0	65Cd	
Nd^{3+}	7.45	1.93(1)	61.9	25	C	μ=2.0	65Cd	
	6.15	1.15(2)	42.7	25	C	μ=2.0	65Cd	
Sm^{3+}	6.53	2.02(1)	60.7	25	C	μ=2.0	65Cd	
	5.56	1.22(2)	41.8	25	C	μ=2.0	65Cd	
Eu^{3+}	7.36	1.98(1)	62.8	25	C	μ=2.0	65Cd	
	9.71	1.927(1)	69.0	25	C	μ=2.0($NaClO_4$)	71Aj	
	4.14	1.30(2)	38.9	25	C	μ=2.0	65Cd	
	7.61	1.314(2)	51.0	25	C	μ=2.0($NaClO_4$)	71Aj	
Gd^{3+}	9.29	1.84(1)	66.5	25	C	μ=2.0	65Cd	
	3.60	1.33(2)	37.2	25	C	μ=2.0	65Cd	
Tb^{3+}	12.68	1.73(1)	75.7	25	C	μ=2.0	65Cd	
	2.72	1.37(2)	35.6	25	C	μ=2.0	65Cd	
Dy^{3+}	15.36	1.63(1)	82.8	25	C	μ=2.0	65Cd	
	3.47	1.34(2)	37.2	25	C	μ=2.0	65Cd	
Ho^{3+}	16.32	1.62(1)	85.8	25	C	μ=2.0	65Cd	
	6.82	1.23(2)	46.4	25	C	μ=2.0	65Cd	

Metal Ion	ΔH, kJ /mol	Log K	ΔS, J /mol K	T,°C	M	Conditions	Ref.	Remarks
P257, continued								
Er^{3+}	16.57	1.60(1)	86.2	25	C	μ=2.0	65Cd	
	9.71	1.12(2)	54.0	25	C	μ=2.0	65Cd	
Tm^{3+}	16.65	1.61(1)	86.6	25	C	μ=2.0	65Cd	
	11.84	1.05(2)	59.4	25	C	μ=2.0	65Cd	
Yb^{3+}	15.77	1.63(1)	84.1	25	C	μ=2.0	65Cd	
	11.42	1.07(2)	58.6	25	C	μ=2.0	65Cd	
Lu^{3+}	15.98	1.66(1)	85.4	25	C	μ=2.0	65Cd	
	10.46	1.12(2)	56.5	25	C	μ=2.0	65Cd	
Cr^{3+}	46.4	(1-3)	--	25-30	T	μ=1.0(ClO_4^-)	75Ta	b,n
Co^{2+}	4.77	0.46(1)	24.8	25	C	μ=1M($NaNO_3$)	75Ag	
	17.46	-0.12(2)	56.3	25	C	μ=1M($NaNO_3$)	75Ag	
Ni^{2+}	4.73	0.50(1)	25.4	25	C	μ=1M($NaNO_3$)	75Ag	
	11.76	0.004(2)	39.5	25	C	μ=1M($NaNO_3$)	75Ag	
Cu^{2+}	4.23	1.93(1)	51.1	25	C	μ=1M($NaNO_3$)	75Ag	
Al^{3+}	9.6	1.778(25°,1)	63	25-30	T	μ=1.0(ClO_4^-)	75Ta	b,n
	15.1	1.806(35°,1)	84	30-40	T	μ=1.0(ClO_4^-)	75Ta	b,n
	215.9	1.637(25°,2)	4.2	25-30	T	μ=1.0(ClO_4^-)	75Ta	b,n
	54.4	2.236(35°,2)	226	30-40	T	μ=1.0(ClO_4^-)	75Ta	b,n
P258	PROPANOIC ACID, amide, N,N-dimethyl-		$C_5H_{11}ON$			$CH_3CH_2CON(CH_3)_2$		L
I_2	-16.7	0.59(25°,1)	-44.8	16-43	T	CM=4.87x10^{-3}M CL=4.99x10^{-3}M S:CCl_4	62Dc	b
P259	PROPANOIC ACID, chloride		C_3H_5OCl			CH_3CH_2COCl		L
SbA_5	-13.8	1.15(1)	-50	25	C	S:Ethylene chloride A=Cl	67Ob	
P260	PROPANOIC ACID, ethyl ester		$C_5H_{10}O_2$			$CH_3CH_2CO_2C_2H_5$		L
$[NiA_2]_2$	-16.1	1.13(100°,1)	-22	75-125	T	S:Squalane A=3-Trifluoro-acetylcamphorate	74Sj	b,gg
SbA_5	-70.37	(1)	--	25	C	Excess $SbCl_5$ CL low, variable ∿0.02M S:1,2-Dichloroethane A=Cl	64Oa	
	-70.37	5.09(1)	-138.5	25	C	Excess $SbCl_5$ CL low, variable ∿0.02M S:Ethylene chloride A=Cl	67Od	

Metal Ion	ΔH, kJ /mol	Log K	ΔS, J /mol K	T,°C	M	Conditions	Ref.	Remarks
P261	PROPANOIC ACID, nitrile		C_3H_5N			CH_3CH_2CN		
I_2	-9.66	-0.296(25°)	-38.5	--	T	S:CCl_4 CL=0.2-1.0M CM=$2x10^{-3}$M Data analyzed by method 1	72Ma	j,u,q: ±1.9
	-12.30	-0.296(25°)	-47.11	--	T	S:CCl_4 CL=0.2-1.0M CM=$2x10^{-3}$M Data analyzed by method 2	72Ma	j,u
	-16.44	-0.021(25°)	-55.73	--	T	S:n-Heptane CL=0.2-1.0M CM=$2x10^{-3}$M Data analyzed by method 2	72Ma	j,u
P262	PROPANOIC ACID, 2-amino-2-methyl-		$C_4H_9O_2N$			$(CH_3)_2C(NH_2)CO_2H$		L^-
Cu^{2+}	-22.6	8.55(20°,1)	88	25	C	μ=0	64Ia	
	-23.8	7.05(20°,2)	54	25	C	μ=0	64Ia	
P263	PROPANOIC ACID, 3-chloro		$C_3H_5O_2Cl$			$ClCH_2CH_2CO_2H$		L^-
UO_2^{2+}	11.3	2.06(1)	77.4	25	C	μ=1.0M($NaClO_4$)	74Ph	
	9.62	1.52(2)	61.1	25	C	μ=1.0M($NaClO_4$)	74Ph	
	0	1.60(3)	31.0	25	C	μ=1.0M($NaClO_4$)	74Ph	
P264	PROPANOIC ACID, 2,3-diamino-, methyl ester (dl)		$C_4H_{10}O_2N_2$			$H_2NCH_2CH(NH_2)CO_2CH_3$		
Cu^{2+}	-46.9	8.986(25°,1)	15.1	25-50	T	μ=0.1M	71Hg	b
	-46.9	7.763(25°,2)	-8.8	25-50	T	μ=0.1M	71Hg	b
P265	PROPANOIC ACID, 2,2-dimethyl-, ethyl ester		$C_7H_{14}O_2$			$(CH_3)_3CCO_2CH_2CH_3$		L
SnA_4	-23.4	0.623(30°,1)	65.3	30-40	T	S:Benzene A=Cl	64La	b
SbA_5	-54.10	(1)	--	25	C	Excess $SbCl_5$ CL low, variable ~0.02M S:1,2-Dichloroethane A=Cl	64Oa	
	-54.52	3.362(1)	-145.2	25	C	Excess $SbCl_5$ CL low, variable ~0.02M S:Ethylene chloride A=Cl	67Oe	
	-54.52	1.97(1)	-145.2	25	C	Excess $SbCl_5$ CL low, variable ~0.02M S:Ethylene chloride A=Cl	67Od	

Metal Ion	ΔH, kJ /mol	Log K	ΔS, J /mol K	T,°C	M	Conditions	Ref.	Re-marks
P266	PROPANOIC ACID, 2,2-dimethyl-, methyl ester		$C_6H_{12}O_2$			$CH_3C(CH_3)_2CO_2CH_3$		L
SbA_5	-53.68	2.89(1)	-151.5	25	C	Excess $SbCl_5$ low μ CL low variable, ∿0.02M S:Ethylene chloride A=Cl	67Oe	
P267	PROPANOIC ACID, 2,2-dimethyl-, nitrile		C_5H_9N			$CH_3C(CH_3)_2CN$		L
I_2	-8.79	-0.226(25°)	-34.3	--	T	S:CCl_4 CL=0.2-1.0M CM=2x10^{-3}M	72Ma	j,u
P268	PROPANOIC ACID, 2,2-dimethyl-, 2-propyl ester		$C_8H_{16}O_2$			$CH_3C(CH_3)_2CO_2CH(CH_3)_2$		L
SbA_5	-56.5	(1)	--	25	C	Excess $SbCl_5$ low μ CL low, variable, ∿0.02M S:Ethylene chloride A=Cl	67Oe	
P269	PROPANOIC ACID, 2,3-diphosphoric acid		$C_3H_8O_{10}P_2$			$H_2O_3POCH_2CH(OPO_3H_2)CO_2H$		
Ca^{2+}	21.8	2.910(1)	129	25	C	0.1M(KCl)	72Dd	
	24.5	3.496(1)	149	25	C	0.1M(TMABr)	72Dd	
Mg^{2+}	22.5	2.660(1)	126	25	C	0.1M(KCl)	72Dd	
	26.9	3.209(1)	137	25	C	0.1M(TMABr)	72Dd	
Sr^{2+}	15.2	2.694(1)	102	25	C	0.1M(KCl)	72Dd	
	16.6	3.246(1)	118	25	C	0.1M(TMABr)	72Dd	
[Fe]	-41.8	4.9(1)	∿-46	25	C	0.1M(Cl^-) 0.05Mbis-tris buffer pH=6 [Fe]= Deoxyhemoglobin	74Nd	
	-54.4	4.5(1)	∿-96	25	C	0.1M(Cl^-) 0.05Mbis-tris buffer pH=7 [Fe]= Deoxyhemoglobin	74Nd	
	-47.7	4.5(1)	∿-75	25	C	0.1M(Cl^-) 0.05Mbis-tris buffer pH=7.2 [Fe]=Deoxy-hemoglobin	74Nd	
	43.1	4.4(1)	∿-58	25	C	0.1M(Cl^-) 0.05Mbis-tris buffer pH=7.4 [Fe]=Deoxy-hemoglobin	74Nd	

Metal Ion	ΔH, kJ /mol	Log K	ΔS, J /mol K	T,°C	M	Conditions	Ref.	Remarks
P269, continued								
[Fe], cont.	-44.8	3.5(1)	∿-84	25	C	0.1M(Cl^-) CM=10CL 0.05Mbis-tris buffer pH=7.4 [Fe]=Deoxy-hemoglobin	74Nd	
	-36.0	4.0(1)	∿-42	25	C	0.1M(Cl^-) 0.05Mbis-tris buffer pH=6 [Fe]= Methemoglobin	74Nd	
	-31.4	3.9(1)	∿-29	25	C	0.1M(Cl^-) 0.05Mbis-tris buffer pH=6 [Fe]= Oxyhemoglobin	74Nd	
	-45.2	4.301(1)	-79	5	C	[Fe]=Horse oxyhemoglobin CCl-=0M pH(initial)=7.32	72He	q:±3.3
	-46.9	5.342(1)	-67	5	C	[Fe]=Horse oxyhemoglobin CCl-=0.005M pH(initial)=6.50	72He	q:±4.2
	-38.1	3.602(1)	-58	25	C	[Fe]=Horse oxyhemoglobin CCl-=0M pH(initial)=6.93	72He	q:±2.5
	-35.6	2.602(1)	-67	25	C	[Fe]=Horse oxyhemoglobin CCl^-=0.10M pH(initial)=7.05	72He	q:±3.3

P270 PROPIONIC ACID, 3-(N-2-furfuralideneimino) $C_8H_9O_3N$

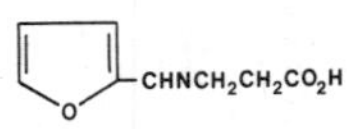

Metal Ion	ΔH, kJ /mol	Log K	ΔS, J /mol K	T,°C	M	Conditions	Ref.	Remarks
Cr^{2+}	52.7	8.30(25°,1-2)	277	25-35	T	μ=0	78Se	
Mn^{2+}	35.1	9.15(25°,1-2)	244	25-35	T	μ=0	78Se	
Fe^{2+}	52.7	9.35(25°,1-2)	356	25-35	T	μ=0	78Se	
Co^{2+}	49.2	10.02(25°,1-2)	357	25-35	T	μ=0	78Se	
Ni^{2+}	52.7	10.50(25°,1-2)	378	25-35	T	μ=0	78Se	
Cu^{2+}	56.2	10.73(25°,1-2)	394	25-35	T	μ=0	78Se	

P271 PROPANOIC ACID, 2-hydroxy- $C_3H_6O_3$ $CH_3CH(OH)CO_2H$ L^-

Metal Ion	ΔH, kJ /mol	Log K	ΔS, J /mol K	T,°C	M	Conditions	Ref.	Remarks
Be^{2+}	50.2	0.398(25°,1)	176	25-30	T	μ=1.0(ClO_4^-)	75Ta	b,n
	156	0.544(35°,1)	502	30-40	T	μ=1.0(ClO_4^-)	75Ta	b,n
	22.2	(2)	--	25-40	T	μ=1.0(ClO_4^-)	75Ta	b,n
Ca^{2+}	45.65	1.55(25°,1)	182.8	25-45	T	μ=0	70Gd	b

Metal Ion	ΔH, kJ /mol	Log K	ΔS, J /mol K	T,°C	M	Conditions	Ref.	Re-marks
P271, continued								
La^{3+}	-6.61	2.27(1)	21.3	25	C	μ=2.0($NaClO_4$)	66Cc	
	-4.6	1.68(2)	16.7	25	C	μ=2.0($NaClO_4$)	66Cc	
	-9.6	1.11(3)	-10.9	25	C	μ=2.0($NaClO_4$)	66Cc	
Ce^{3+}	-7.07	2.29(1)	20.1	25	C	μ=2.0($NaClO_4$)	66Cc	
	-3.47	2.33(1)	33.0	25	T	μ=2($NaClO_4$)	61Ca	
	2.5	1.79(2)	42.2	25	C	μ=2.0($NaClO_4$)	66Cc	
	-22.2	1.01(3)	-54.4	25	C	μ=2.0($NaClO_4$)	66Cc	
Nd^{3+}	-13.51	2.47(1)	2.1	25	C	μ=2.0($NaClO_4$)	66Cc	
	6.3	1.89(2)	57.3	25	C	μ=2.0($NaClO_4$)	66Cc	
Sm^{3+}	-9.75	2.57(1)	16.7	25	C	μ=2.0($NaClO_4$)	65Cc	
	1.88	2.01(2)	44.8	25	C	μ=2.0($NaClO_4$)	65Cc	
	-30.1	1.36(3)	-74.5	25	C	μ=2.0($NaClO_4$)	65Cc	
Eu^{3+}	-8.12	2.54(1)	21.3	25	C	μ=2.0($NaClO_4$)	65Cc	
	-4.31	2.479(1)	33.01	25	C	μ=2.0($NaClO_4$)	71Aj	
	3.3	2.03(2)	50.6	25	C	μ=2.0($NaClO_4$)	65Cc	
	-4.27	2.08(2)	25.02	25	C	μ=2.0($NaClO_4$)	71Aj	
	-18.4	1.36(3)	-36.0	25	C	μ=2.0($NaClO_4$)	66Cc	
	-14.39	1.26(3)	-24.02	25	C	μ=2.0($NaClO_4$)	71Aj	
Gd^{3+}	-8.24	2.60(1)	22.2	25	C	μ=2.0($NaClO_4$)	66Cc	
	-2.26	2.53(1)	41.0	25	T	μ=2($NaClO_4$)	61Ca	
	-0.8	1.89(2)	33.0	25	C	μ=2.0($NaClO_4$)	66Cc	
	-11.7	1.68(3)	-7.1	25	C	μ=2.0($NaClO_4$)	66Cc	
Tb^{3+}	-6.86	2.66(1)	27.6	25	C	μ=2.0($NaClO_4$)	66Cc	
Dy^{3+}	-9.04	2.71(1)	21.8	25	C	μ=2.0($NaClO_4$)	66Cc	
Er^{3+}	-7.99	2.85(1)	27.6	25	C	μ=2.0($NaClO_4$)	66Cc	
	4.2	2.12(2)	54.0	25	C	μ=2.0($NaClO_4$)	66Cc	
	-17.6	1.95(3)	-21.3	25	C	μ=2.0($NaClO_4$)	66Cc	
Yb^{3+}	-9.16	2.89(1)	24.7	25	C	μ=2.0($NaClO_4$)	66Cc	
	-5.8	2.29(2)	24.7	25	C	μ=2.0($NaClO_4$)	66Cc	
	5.8	1.90(3)	16.3	25	C	μ=2.0($NaClO_4$)	66Cc	
Ti^{3+}	-15.4	3.57(30°,1)	17.7	30-50	T	μ=0.1M($NaClO_4$)	75Cb	
	-39.1	10.26(30°,1-3)	68.1	30-50	T	μ=0.1M($NaClO_4$)	75Cb	
Cr^{3+}	74.9	3.301(25°,1)	318	25-30	T	μ=1.0(ClO_4^-)	75Ta	b,n
	-100	3.653(25°,1)	-263	30-40	T	μ=1.0(ClO_4^-)	75Ta	b,n
	66.5	2.398(25°,2)	267	25-30	T	μ=1.0(ClO_4^-)	75Ta	b,n
	-91.6	2.709(35°,2)	238	30-40	T	μ=1.0(ClO_4^-)	75Ta	b,n
	-36.8	2.699(25°,3)	-514	25-30	T	μ=1.0(ClO_4^-)	75Ta	b,n
	82.4	1.680(35°,3)	305	30-40	T	μ=1.0(ClO_4^-)	75Ta	b,n
Al^{3+}	7.9	4.255(25°,1)	67	25-30	T	μ=1.0(ClO_4^-)	75Ta	b,n
	24.7	2.279(35°,1)	113	30-40	T	μ=1.0(ClO_4^-)	75Ta	b,n
	14.2	0.620(25°,2)	0	25-30	T	μ=1.0(ClO_4^-)	75Ta	b,n
	32.2	2.660(35°,2)	167	30-40	T	μ=1.0(ClO_4^-)	75Ta	b,n
Co^{2+}	0.08	1.04(1)	20.2	25	C	μ=1M($NaNO_3$)	75Ag	
	0.54	1.13(2)	23.5	25	C	μ=1M($NaNO_3$)	75Ag	
Ni^{2+}	-0.71	1.33(1)	23.1	25	C	μ=1M($NaNO_3$)	75Ag	
	-1.38	1.25(2)	19.3	25	C	μ=1M($NaNO_3$)	75Ag	

Metal Ion	ΔH, kJ /mol	Log *K*	ΔS, J /mol K	T, °C	M	Conditions	Ref.	Remarks
P271, continued								
Cu^{2+}	37.24	2.36(25°,1)	169.9	25-45	T	μ=0	70Gd	b
	0.67	2.40(1)	48.2	25	C	μ=1M($NaNO_3$)	75Ag	
	-2.6	2.66(1)	42	25	C	2M($NaClO_4$)	78Fc	
	-11.59	1.54(25°,2)	-9.2	25-45	T	μ=0	70Gd	b
	-0.38	1.35(2)	24.6	25	C	μ=1M($NaNO_3$)	75Ag	
	4.4	1.62(2)	46	25	C	2M($NaClO_4$)	78Gc	
	-0.71	1.17(3)	20.0	25	C	μ=1M($NaNO_3$)	75Ag	
	-174	0.61(3)	-570	25	C	2M($NaClO_4$)	78Fc	
In^{3+}	-11.6	3.65(30°,1)	31.8	30-50	T	μ=0.1M($NaClO_4$)	74Cc	b
	13.4	3.21(35°,1)	105	25-45	T	μ=0.2M($NaClO_4$)	73Sc	b,q:±2.1
	10.0	2.66(35°,2)	84	25-45	T	μ=0.2M($NaClO_4$)	73Sc	b,q:±2.1
	-31.2	9.92(30°,1-3)	87.1	30-50	T	μ=0.1M($NaClO_4$)	74Cc	b
Pb^{2+}	-7.5	2.16(1)	16	25	C	2M($NaClO_4$)	78Fc	
	18.3	1.07(2)	82	25	C	2M($NaClO_4$)	78Fc	
	-90	0.44(3)	-290	25	C	2M($NaClO_4$)	78Fc	
P272	PROPANOIC ACID, 3-hydroxy-		$C_3H_6O_3$			$HOCH_2CH_2CO_2H$		
Y^{3+}	19.2	1.431(1)	85.8	25	C	μ=2.00M($NaClO_4$)	69Ja	q:±1.2
La^{3+}	10.75	1.556(1)	65.7	25	C	μ=2.00M($NaClO_4$)	69Ja	
Ce^{3+}	10.67	1.568(1)	65.7	25	C	μ=2.00M($NaClO_4$)	69Ja	
Pr^{3+}	7.86	1.623(1)	57.3	25	C	μ=2.00M($NaClO_4$)	69Ja	q:±1.1
Nd^{3+}	6.28	1.663(1)	158.8	25	C	μ=2.00M($NaClO_4$)	69Ja	q:±1.2
Sm^{3+}	10.33	1.748(1)	68.2	25	C	μ=2.00M($NaClO_4$)	69Ja	
Gd^{3+}	15.90	1.613(1)	84.1	25	C	μ=2.00M($NaClO_4$)	69Ja	q:±1.4
Tb^{3+}	18.24	1.544(1)	90.0	25	C	μ=2.00M($NaClO_4$)	69Ja	q:±1.4
Dy^{3+}	17.61	1.447(1)	87.0	25	C	μ=2.00M($NaClO_4$)	69Ja	q:±1.9
Ho^{3+}	19.45	1.477(1)	94.1	25	C	μ=2.00M($NaClO_4$)	69Ja	
Er^{3+}	20.50	1.322(1)	95.8	25	C	μ=2.00M($NaClO_4$)	69Ja	
Tm^{3+}	21.97	1.447(1)	100.8	25	C	μ=2.00M($NaClO_4$)	69Ja	
Yb^{3+}	18.66	1.505(1)	91.6	25	C	μ=2.00M($NaClO_4$)	69Ja	q:±1.4
Lu^{3+}	18.91	1.398(1)	90.8	25	C	μ=2.00M($NaClO_4$)	69Ja	q:±1.8
Co^{2+}	7.49	0.26(1)	30.1	25	C	μ=1M($NaNO_3$)	75Ag	
Ni^{2+}	4.81	0.54(1)	26.5	25	C	μ=1M($NaNO_3$)	75Ag	
	5.82	0.43(2)	27.8	25	C	μ=1M($NaNO_3$)	75Ag	
Cu^{2+}	3.68	1.53(1)	41.6	25	C	μ=1M($NaNO_3$)	75Ag	
	2.80	1.33(2)	34.9	25	C	μ=1M($NaNO_3$)	75Ag	
In^{3+}	43.9	7.20(35°,1-2)	280	25-45	T	μ=0	72Sa	b,q:±6.3
	36.4	6.98(35°,1-2)	251	25-45	T	μ=0.10M($NaClO_4$)	72Sa	b,q:±6.3
	12.6	3.80(35°,1)	113	25-45	T	μ=0.20M($NaClO_4$)	73Sc	b
Pb^{2+}	0.2	2.10(1)	41	25	C	2M($NaClO_4$)	78Fc	
	2.4	1.07(2)	29	25	C	2M($NaClO_4$)	78Fc	
	-54	0.34(3)	-170	25	C	2M($NaClO_4$)	78Fc	

Metal Ion	ΔH, kJ /mol	Log K	ΔS, J /mol K	T,°C	M	Conditions	Ref.	Re-marks
P273	PROPANOIC ACID, 2-hydroxy-2-methyl-		$C_4H_8O_3$			$(CH_3)_2C(OH)CO_2H$		L^-
Y^{3+}	-5.15	2.82(1)	36.8	25	C	μ=2.0($NaClO_4$)	66Cc	
	-12.1	2.63(2)	9.6	25	C	μ=2.0($NaClO_4$)	66Cc	
	-1.7	1.80(3)	28.4	25	C	μ=2.0($NaClO_4$)	66Cc	
La^{3+}	-8.20	2.28(1)	16.3	25	C	μ=2.0($NaClO_4$)	66Cc	
	-1.7	1.69(2)	26.8	25	C	μ=2.0($NaClO_4$)	66Cc	
	-14.6	1.20(3)	-26.4	25	C	μ=2.0($NaClO_4$)	66Cc	
Ce^{3+}	-6.19	2.41(1)	25.1	25	C	μ=2.0($NaClO_4$)	66Cc	
	-3.47	2.43(1)	35.1	25	T	μ=2($NaClO_4$)	61Ca	
	-1.7	1.93(2)	30.5	25	C	μ=2.0($NaClO_4$)	66Cc	
	-23.8	0.894(3)	-62.3	25	C	μ=2.0($NaClO_4$)	66Cc	
Nd^{3+}	-6.69	2.58(1)	27.2	25	C	μ=2.0($NaClO_4$)	66Cc	
	-7.1	1.98(2)	14.6	25	C	μ=2.0($NaClO_4$)	66Cc	
	-2.9	1.27(3)	14.2	25	C	μ=2.0($NaClO_4$)	66Cc	
Eu^{3+}	-6.32	2.74(1)	31.4	25	C	μ=2.0($NaClO_4$)	66Cc	
	-7.5	2.16(2)	15.9	25	C	μ=2.0($NaClO_4$)	66Cc	
	-6.7	1.69(3)	10.0	25	C	μ=2.0($NaClO_4$)	66Cc	
Gd^{3+}	-3.72	2.82(1)	41.42	25	T	μ=2($NaClO_4$)	61Ca	
Tb^{3+}	-6.15	2.84(1)	33.9	25	T	μ=2.0($NaClO_4$)	66Cc	
	-13.4	2.39(2)	0.58	25	T	μ=2.0($NaClO_4$)	66Cc	
	4.2	1.81(3)	48.1	25	C	μ=2.0($NaClO_4$)	66Cc	
Er^{3+}	-4.10	3.13(1)	46.0	25	C	μ=2.0($NaClO_4$)	66Cc	
	-18.0	2.53(2)	-11.7	25	C	μ=2.0($NaClO_4$)	66Cc	
	6.7	2.24(3)	65.7	25	C	μ=2.0($NaClO_4$)	66Cc	
Yb^{3+}	-2.9	3.57(1)	58	25	C	μ=0.10M($NaClO_4$)	71Bf	
	-4.64	3.04(1)	42.7	25	C	μ=2.0($NaClO_4$)	66Cc	
	-11.7	4.24(1)	42	25	C	μ=0.10M($NaClO_4$) S:40v% DMSO	71Bf	
	-16.7	5.34(1)	46	25	C	μ=0.10M($NaClO_4$) S:DMSO	71Bf	
	-45.2	9.66(1)	33	25	C	μ=0.10M($NaClO_4$) S:Ethanol	71Bf	
	-6.7	4.89(1)	71	25	C	μ=0.10M($NaClO_4$) S:Ethylene glycol	71Bf	
	-6.7	2.92(2)	33	25	C	μ=0.10M($NaClO_4$)	71Bf	
	-21.8	2.84(2)	-18.8	25	C	μ=2.0($NaClO_4$)	66Cc	
	-20.5	3.81(2)	4	25	C	μ=0.10M($NaClO_4$) S:40v% DMSO	71Bf	
	-17.2	4.54(2)	29	25	C	μ=0.10M($NaClO_4$) S:DMSO	71Bf	
	-15.9	8.47(2)	109	25	C	μ=0.10M($NaClO_4$) S:Ethanol	71Bf	
	4.2	2.14(3)	54.8	25	C	μ=2.0($NaClO_4$)	66Cc	
Fe^{3+}	-12.34	3.24(30°,1)	21.25	30&50	T	CL≅CM≅1×10^{-3}M	71Sb	a,b
Cu^{2+}	-0.4	2.99(1)	54	25	C	μ=0.10M($NaClO_4$)	71Bf	
	-15.5	5.99(1)	63	25	C	μ=0.10M($NaClO_4$) S:DMSO	71Bf	
	-9.2	4.67(1)	58	25	C	μ=0.10M($NaClO_4$)	71Bf	

Metal Ion	ΔH, kJ /mol	Log K	ΔS, J /mol K	T,°C	M	Conditions	Ref.	Re-marks
P273, continued								
Cu^{2+},						S:Ethylene glycol		
cont.	-3.3	1.90(2)	25	25	C	μ=0.10M($NaClO_4$)	71Bf	
	-23.4	4.76(2)	12	25	C	μ=0.10M($NaClO_4$)	71Bf	
						S:DMSO		
	-7.5	3.50(2)	42	25	C	μ=0.10($NaClO_4$)	71Bf	
						S:Ethylene glycol		
P274	PROPANOIC ACID, 2-hydroxy-2-phenyl-		$C_9H_{10}O_3$			$C_6H_5C(OH)(CH_3)CO_2H$		L
Fe^{3+}	-11.30	3.45(30°,1)	28.7	30&50	T	CL≅CM≅1x10^{-3}M	71Sb	a,b
P275	PROPANOIC ACID, 2-mercapto-		$C_3H_6O_2S$			$CH_3CH(SH)CO_2H$		L^-
Y^{3+}	5.61	1.699(1)	50.2	--	C	μ=2.00($NaClO_4$)	69Cc	u:25°
La^{3+}	11.04	1.230(1)	56.9	--	C	μ=2.00($NaClO_4$)	69Cc	u:25°
Ce^{3+}	9.66	1.362(1)	59.0	--	C	μ=2.00($NaClO_4$)	69Cc	u:25°
Pr^{3+}	5.94	1.892(1)	56.1	--	C	μ=2.00($NaClO_4$)	69Cc	u:25°
Nd^{3+}	5.61	1.929(1)	54.8	--	C	μ=2.00($NaClO_4$)	69Cc	u:25°
Sm^{3+}	4.10	2.064(1)	51.5	--	C	μ=2.00($NaClO_4$)	69Cc	u:25°
Eu^{3+}	5.06	2.000(1)	55.2	--	C	μ=2.00($NaClO_4$)	69Cc	u:25°
Gd^{3+}	7.40	1.748(1)	55.6	--	C	μ=2.00($NaClO_4$)	69Cc	u:25°
Tb^{3+}	6.57	1.653(1)	48.5	--	C	μ=2.00($NaClO_4$)	69Cc	u:25°
Dy^{3+}	7.24	1.580(1)	53.1	--	C	μ=2.00($NaClO_4$)	69Cc	u:25°
Ho^{3+}	7.82	1.531(1)	55.6	--	C	μ=2.00($NaClO_4$)	69Cc	u:25°
Er^{3+}	10.25	1.531(1)	63.2	--	C	μ=2.00($NaClO_4$)	69Cc	u:25°
Tm^{3+}	10.71	1.477(1)	61.9	--	C	μ=2.00($NaClO_4$)	69Cc	u:25°
Yb^{3+}	13.35	1.431(1)	72.0	--	C	μ=2.00($NaClO_4$)	69Cc	u:25°
Lu^{3+}	6.94	1.505(1)	51.9	--	C	μ=2.00($NaClO_4$)	69Cc	u:25°
Ni^{2+}	-37.66	13.68(30°,1-2)	138.1	20-40	T	μ=0.1M($NaClO_4$)	72Sf	b
	-0.84	13.144(1-2)	247	25	C	μ=0.5M(KNO_3)	75De	ee
	-33.5	30.709(3-4)	113	25	C	μ=0.5M(KNO_3)	75De	ee
Zn^{2+}	-12.6	14.336(1-2)	55	25	C	μ=0.5M(KNO_3)	75De	ee
	-79.5	34.742(3-4)	95	25	C	μ=0.5M(KNO_3)	75De	ee
In^{3+}	-24.3	12.15(35°,1)	155	25-45	T	μ=0.2M($NaClO_4$)	73Sc	b,q:±2.1
	-27.2	10.56(35°,2)	113	25-45	T	μ=0.2M($NaClO_4$)	73Sc	b,q:±2.1
	-31.0	6.37(35°,3)	21	25-45	T	μ=0.2M($NaClO_4$)	73Sc	b,q:±2.1
P276	PROPANOIC ACID, 3-mercapto-		$C_3H_6O_2S$			$HSCH_2CH_2CO_2H$		L^-
Y^{3+}	21.09	1.505(1)	99.2	--	C	μ=2.00($NaClO_4$)	69Cc	u:25°
La^{3+}	11.04	1.544(1)	66.1	--	C	μ=2.00($NaClO_4$)	69Cc	u:25°
Ce^{3+}	10.00	1.568(1)	64.0	--	C	μ=2.00($NaClO_4$)	69Cc	u:25°

Metal Ion	ΔH, kJ /mol	Log K	ΔS, J /mol K	T,°C	M	Conditions	Ref.	Re-marks
P276, continued								
Pr^{3+}	10.84	1.568(1)	65.3	--	C	μ=2.00($NaClO_4$)	69Cc	u:25°
Nd^{3+}	10.25	1.740(1)	67.8	--	C	μ=2.00($NaClO_4$)	69Cc	u:25°
Sm^{3+}	7.24	1.740(1)	57.3	--	C	μ=2.00($NaClO_4$)	69Cc	u:25°
Eu^{3+}	9.20	1.643(1)	63.6	--	C	μ=2.00($NaClO_4$)	69Cc	u:25°
Gd^{3+}	10.84	1.708(1)	69.0	--	C	μ=2.00($NaClO_4$)	69Cc	u:25°
Tb^{3+}	12.26	1.623(1)	72.0	--	C	μ=2.00($NaClO_4$)	69Cc	u:25°
Dy^{3+}	14.81	1.602(1)	80.3	--	C	μ=2.00($NaClO_4$)	69Cc	u:25°
Ho^{3+}	13.51	1.462(1)	73.2	--	C	μ=2.00($NaClO_4$)	69Cc	u:25°
Er^{3+}	19.96	1.415(1)	95.8	--	C	μ=2.00($NaClO_4$)	69Cc	u:25°
Tm^{3+}	19.71	1.398(1)	92.5	--	C	μ=2.00($NaClO_4$)	69Cc	u:25°
Yb^{3+}	16.65	1.431(1)	84.1	--	C	μ=2.00($NaClO_4$)	69Cc	u:25°
Lu^{3+}	20.38	1.491(1)	95.4	--	C	μ=2.00($NaClO_4$)	69Cc	u:25°
VO^{2+}	-38.4	9.47(25°,1-2)	50.9	25-45	T	μ=0.1M(KNO_3) S:25v%Ethanol	76Sj	
$UO_2{}^{2+}$	-42.2	15.05(30°,1-2)	148.5	20-40	T	μ=0.1M(KNO_3)	69Sd	b
Ni^{2+}	-79.5	5.12(1)	-169	25	C	μ=1M($NaNO_3$) L=L^{2-}	77An	q:±10
	-29.41	10.74(30°,1-2)	108.24	20-40	T	μ=0.1M(KNO_3)	69Sc 69Se	b
Zn^{+}	1.2	6.05(1)	120	25	C	μ=1M($NaNO_3$) L=L^{2-}	77An	q:±9
	-9.2	6.14(2)	87.0	25	C	μ=1M($NaNO_3$) L=L^{2-}	77An	q:±13
	-64.7	6.90(25°,1-2)	-87.6	25-45	T	μ=0.1M(KNO_3) S:25v%Ethanol	76Sj	
In^{3+}	-25.1	11.60(35°,1)	142	25-45	T	μ=0.20M($NaClO_4$)	73Sc	b
	-68.6	26.40(35°,1-3)	280	25-45	T	μ=0	72Sa	b,q:±6.3
	-69.87	25.40(35°,1-3)	259	25-45	T	μ=1.0M($NaClO_4$)	72Sa	b,q:±6.3
Tl^{+}	-13.97	2.850(30°,1)	8.41	20-40	T	μ=0.1M(KNO_3)	69Sd	b
Tl^{3+}	-116.3	18.75(30°,1-2)	-25.9	20-40	T	μ=0.1M(KNO_3)	69Sb	b
P277	PROPANOIC ACID, 3-mercapto, methyl ester		$C_4H_8O_2S$			$CH_3OOCCH_2CH_2SH$		
Be^{2+}	-60.4	7.56(25°,1-2)	-57.9	25-45	T	μ=0.1M(KNO_3) S:25v%Ethanol	74Sd	
La^{3+}	-29.5	3.45(25°,1-3)	-32.6	25-45	T	μ=0.1M(KNO_3) S:25v%Ethanol	74Sd	b
Ce^{3+}	-33.3	3.60(25°,1-3)	-42.7	25-45	T	μ=0.1M(KNO_3) S:25v%Ethanol	74Sd	b
Pr^{3+}	-29.9	3.95(25°,1-3)	-24.3	25-45	T	μ=0.1M(KNO_3) S:25v%Ethanol	74Sd	b

Metal Ion	ΔH, kJ /mol	Log K	ΔS, J /mol K	T,°C	M	Conditions	Ref.	Re-marks
P277, continued								
Sm^{3+}	-33.3	4.32(25°,1-3)	-28.9	25-45	T	μ=0.1M(KNO_3) S:25v%Ethanol	74Sd	b
Th^{4+}	-52.6	10.99(30°,1-4)	36.8	30-40	T	μ=0.1M(KNO_3) S:25v%Ethanol	74Se	b
P278 PROPANOIC ACID, methylester $C_4H_8O_2$						$CH_3CH_2CO_2CH_3$		
$[NiA_2]_2$	-16.7	1.01(100°,1)	-25	75-125	T	S:Squalane A=3-Trifluoro-acetylcamphorate	74Sj	b,gg
P280 PROPANOIC ACID, 2-methyl- $C_4H_8O_2$						$(CH_3)_2CHCO_2H$		L^-
Y^{3+}	22.43	1.64(1)	106.7	25	C	μ=2.0	65Cd	
	13.47	1.15(2)	66.9	25	C	μ=2.0	65Cd	
La^{3+}	14.52	1.57(1)	78.6	25	C	μ=2.0	65Cd	
	10.46	0.90(2)	52.3	25	C	μ=2.0	65Cd	
Ce^{3+}	13.93	1.62(1)	77.8	25	C	μ=2.0	65Cd	
	10.67	1.10(2)	56.9	25	C	μ=2.0	65Cd	
Pr^{3+}	12.64	1.80(1)	77.0	25	C	μ=2.0	65Cd	
	10.50	1.11(2)	56.5	25	C	μ=2.0	65Cd	
Nd^{3+}	11.88	1.91(1)	76.6	25	C	μ=2.0	65Cd	
	9.87	1.18(2)	55.6	25	C	μ=2.0	65Cd	
Sm^{3+}	11.13	2.00(1)	75.7	25	C	μ=2.0	65Cd	
	8.7	1.25(2)	53.1	25	C	μ=2.0	65Cd	
Eu^{3+}	12.18	1.98(1)	78.6	25	C	μ=2.0	65Cd	
	7.74	1.31(2)	51.0	25	C	μ=2.0	65Cd	
Gd^{3+}	14.43	1.86(1)	84.1	25	C	μ=2.0	65Cd	
	6.94	1.34(2)	49.0	25	C	μ=2.0	65Cd	
Tb^{3+}	18.32	1.73(1)	94.6	25	C	μ=2.0	65Cd	
	6.11	1.39(2)	47.3	25	C	μ=2.0	65Cd	
Dy^{3+}	21.09	1.56(1)	102.5	25	C	μ=2.0	65Cd	
	7.49	1.32(2)	50.2	25	C	μ=2.0	65Cd	
Ho^{3+}	22.22	1.63(1)	105.8	25	C	μ=2.0	65Cd	
	10.67	1.21(2)	59.0	25	C	μ=2.0	65Cd	
Er^{3+}	22.97	1.61(1)	107.9	25	C	μ=2.0	65Cd	
	14.35	1.11(2)	69.4	25	C	μ=2.0	65Cd	
Tm^{3+}	22.55	1.61(1)	105.7	25	C	μ=2.0	65Cd	
	17.15	1.06(2)	77.8	25	C	μ=2.0	65Cd	
Yb^{3+}	23.0	2.09(1)	117	25	C	μ=0.10M($NaClO_4$)	71Bf	
	22.38	1.62(1)	106.3	25	C	μ=2.0	65Cd	
	15.1	3.04(1)	109	25	C	μ=0.10M($NaClO_4$) S:40v% DMSO	71Bf	
	-23.4	8.04(1)	75	25	C	μ=0.10M($NaClO_4$) S:Ethanol	71Bf	

Metal Ion	ΔH, kJ /mol	Log K	ΔS, J /mol K	T,°C	M	Conditions	Ref.	Re-marks
P280, continued								
Yb^{3+}, cont.	15.5	3.63(1)	121	25	C	μ=0.10M($NaClO_4$) S:Ethylene glycol	71Bf	
	-4.6	3.21(1)	46	25	C	μ=0.10M($NaClO_4$) S:Formamide	71Bf	
	3.3	1.60(2)	42	25	C	μ=0.10M($NaClO_4$)	71Bf	
	16.61	1.05(2)	75.7	25	C	μ=2.0	65Cd	
	4.2	2.54(2)	63	25	C	μ=0.10M($NaClO_4$) S:40v% DMSO	71Bf	
	-28.9	6.81(2)	33	25	C	μ=0.10M($NaClO_4$) S:Ethanol	71Bf	
	12.6	3.05(2))	100	25	C	μ=0.10M($NaClO_4$) S:Ethylene glycol	71Bf	
Lu^{3+}	22.38	1.65(1)	106.7	25	C	μ=2.0	65Cd	
	15.27	1.12(2)	72.8	25	C	μ=2.0	65Cd	
Cu^{2+}	-5.0	6.56(1)	109	25	C	μ=0.10M($NaClO_4$) S:DMSO	71Bf	
	7.1	4.22(1)	105	25	C	μ=0.10M($NaClO_4$) S:Ethylene glycol	71Bf	
	2.5	2.84(1)	63	25	C	μ=0.10M($NaClO_4$) S:Formamide	71Bf	
	-34.3	6.23(2)	4	25	C	μ=0.10M($NaClO_4$) S:DMSO	71Bf	
	-5.4	4.89(2)	75	25	C	μ=0.10M($NaClO_4$) S:Ethylene glycol	71Bf	
	-3.3	2.34(2)	33	25	C	μ=0.10M($NaClO_4$) S:Formamide	71Bf	
P281	PROPANOIC ACID, 2-methyl-, ethyl ester		$C_6H_{12}O_2$			$CH_3CH(CH_3)CO_2C_2H_5$		L
SbA_5	-68.78	(1)	--	25	C	Excess $SbCl_5$ CL low, variable, ~0.02M S:1,2-Dichloroethane A=Cl	64Oa	
	-68.49	4.55(1)	-142.7	25	C	Excess $SbCl_5$ CL low, variable, ~0.02M S:1,2-Dichloroethane A=Cl	67Od	
P282	PROPANOIC ACID, 2-oxo-		$C_3H_4O_3$			$CH_3C(O)CO_2H$		
La^{3+}	-3.72	1.00(1)	6.7	25	C	μ=2.0M($NaClO_4$) pH=2.2	80Ca	
Ce^{3+}	-4.48	1.59(1)	15.5	25	C	μ=2.0M($NaClO_4$) pH=2.2	80Ca	
Pr^{3+}	-4.73	1.72(1)	17.2	25	C	μ=2.0M($NaClO_4$) pH=2.2	80Ca	

Metal Ion	ΔH, kJ /mol	Log K	ΔS, J /mol K	T, °C	M	Conditions	Ref.	Remarks
P282, continued								
Nd^{3+}	-4.73	1.46(1)	12.1	25	C	μ=2.0M($NaClO_4$) pH=2.2	80Ca	
Sm^{3+}	-4.85	1.52(1)	13.0	25	C	μ=2.0M($NaClO_4$) pH=2.2	80Ca	
Eu^{3+}	-4.90	1.88(1)	19.7	25	C	μ=2.0M($NaClO_4$) pH=2.2	80Ca	
	-4.90	1.973(1)	22.01	25	C	μ=2.0($NaClO_4$)	71Aj	
	-10.08	1.344(2)	-10.00	25	C	μ=2.0($NaClO_4$)	71Aj	
Gd^{3+}	-4.94	1.67(1)	15.5	25	C	μ=2.0M($NaClO_4$) pH=2.2	80Ca	
Tb^{3+}	-4.18	1.19(1)	8.8	25	C	μ=2.0M($NaClO_4$) pH=2.2	80Ca	
Dy^{3+}	-4.52	1.27(1)	9.2	25	C	μ=2.0M($NaClO_4$) pH=2.2	80Ca	
Ho^{3+}	-4.22	1.34(1)	11.3	25	C	μ=2.0M($NaClO_4$) pH=2.2	80Ca	
Er^{3+}	-4.69	1.72(1)	17.2	25	C	μ=2.0M($NaClO_4$) pH=2.2	80Ca	
Tm^{3+}	-4.64	1.89(1)	20.5	25	C	μ=2.0M($NaClO_4$) pH=2.2	80Ca	
Yb^{3+}	-5.06	1.56(1)	13.0	25	C	μ=2.0M($NaClO_4$) pH=2.2	80Ca	
Lu^{3+}	-4.02	1.83(1)	21.8	25	C	μ=2.0M($NaClO_4$) pH=2.2	80Ca	
Cu^{2+}	-0.21	1.59(1)	30	25	C	μ=1M($NaNO_3$)	74Ad	
	-4.64	2.395(2)	30	25	C	μ=1M($NaNO_3$)	74Ad	
P283 2-PROPANOL			C_3H_8O			$CH_3CH(OH)CH_3$		
CeA_2B_4	-34.22	0.210(25°,1)	-110.92	25-35	T	μ=1.08M(HNO_3) CM=4x10^{-3}M CL=5x10^{-2}M	71Tb	b
P284 2-PROPANOL, 1,3-diamino-			$C_3H_{10}ON_2$			$H_2NCH_2CH(OH)CH_2NH_2$		L
Ni^{2+}	-24.3	5.42(30°,1)	25	10-40	T	μ=0	58Ba	b
	-20.9	4.16(30°,2)	8	10-40	T	μ=0	58Ba	b
Cu^{2+}	-21.8	3.57(30°)	-4	10-40	T	μ=0 R:Cu^{2+}+HL^+ =CuL^{2+}+H^+	58Bb	b
Ag^+	-55.2	5.31(30°,1)	-79	10-40	T	μ=0	58Ba	b
P285 2-PROPANONE			C_3H_6O			CH_3COCH_3		L
CoA_2B_2	13.4	-3.917(27°)	-27.6	-80-60	T	S:Acetone A=N-Methyl thiourea	71Ea	b,q: ±2.1

Metal Ion	ΔH, kJ /mol	Log K	ΔS, J /mol K	T,°C	M	Conditions	Ref.	Re-marks
P285, continued								
CoA_2B_2, cont.						B=Cl R:CoA_2B_2+S= $CoASB_2$+A		
	14.6	-3.84(27°)	-24.3	-80-60	T	S:Acetone A=Thiourea B=Cl R:CoA_2B_2+S= $CoASB_2$+A	71Ea	b,q: ±1.7
	9.6	-3.686(27°)	-37.2	-80-60	T	S:Acetone A=N,N'-Dimethyl thiourea B=Br R:CoA_2B_2+S= $CoASB_2$+A	71Ea	b
	16.3	-3.889(27°)	-18.8	-80-60	T	S:Acetone A=N-Methyl thiourea B=Br R:CoA_2B_2+S =$CoASB_2$+A	71Ea	b
	16.7	-3.55(27°)	-10.5	-80-60	T	S:Acetone A=Thiourea B=Br R:CuA_2B_2+S= $CoASB_2$+A	71Ea	b,q: ±1.7
	13.8	-3.436(27°)	-21.3	-80-60	T	S:Acetone A=N,N'-Dimethyl thiourea B=I R:CoA_2B_2+S= $CoASB_2$+A	71Ea	b,q: ±3.8
	25.5	-3.41(27°)	20.5	-80-60	T	S:Acetone A=Thiourea B=I R:CoA_2B_2+S= $CoASB_2$+A	71Ea	b,q: ±3.3
	-34.98	-2.74	--	-84-30	T	S:Acetone CM=2.0-0.4M A=Cl R:CoL_2A_2+2L= CoL_4A_2	72Bf	b,l
	15.98	-0.770	--	-84-30	T	S:Acetone CM=0.2-0.4M A=Cl R:2CoL_2A_2+2L= CoL_6^{2+}+CoA_4^{2-}	72Bf	b,l
	11.3	-3.532(27°)	-9.2	-80-60	T	S:Acetone A=N,N'-Dimethyl thiourea B=ClO_4 R:CoA_4B_2+ S=CoA_3SB_2+A	71Ea	b,j,q: ±1.7
	12.1	-2.499(27°)	-7.1	-80-60	T	S:Acetone A=N-Methyl thiourea B=ClO_4 R:CoA_4B_2+ S=CoA_3SB_2+A	71Ea	b,j
	12.1	-3.272(27°)	0.4	-80-60	T	S:Acetone A=Thiourea B=ClO_4 R:CoA_4B_2 +S=CoA_3SB_2+A	71Ea	b,j,q: ±1.7
	19.2	-3.842(27°)	10.9	-80-60	T	S:Acetone A=Thiourea B=NO_3 R:CoA_4B_2+S= CoA_3SB_2+A	71Ea	b,j
$[NiA_2]_2$	-15.3	1.51(100°,1)	-12	75-125	T	S:Squalane A=3-Trifluoro-acetylcamphorate	74Sj	b,gg

Metal Ion	ΔH, kJ /mol	Log K	ΔS, J /mol K	T,°C	M	Conditions	Ref.	Re-marks
P285, continued								
NiA_2B	-13.0	-1.229(22°,1)	-67	-50-50	T	S:Acetone (L) $A=C_2H_5$ B=2,2'-Bipyridine	76Yc	b
$SiAB_3$	-13.60	1.36(30°,1)	-18.91	10-45	T	$S:CCl_4$ $XL=5x10^{-2}$ $XM=5x10^{-4}$ A=OH $B=CH_3$	69Lb	b
	-14.98	1.36(30°,1)	-23.47	20-60	T	$S:CCl_4$ $XL=5x10^{-2}$ $XM=5x10^{-4}$ A=OH $B=C_2H_5$	69Lb	b
	-15.82	1.54(30°,1)	-22.80	0-50	T	$S:CCl_4$ $XL=5x10^{-2}$ $XM=4x10^{-4}$ A=OH B=Phenyl	69Lb	b
$SnAB_3$	-10.92	0.740(1)	-22.6	25	C	S:Benzene $A=C_4H_9$ $B=Cl^-$	71Gf	
SnA_2B_2	-4.73	0.290(1)	-10.0	25	C	S:Benzene $A=C_4H_9$ B=Cl	71Gf	q: ±0.92
SnA_3B	-23.8	-0.377(1)	88	26	C	$S:CCl_4$ $A=CH_3$ B=Cl	66Be	
	-25.1	-0.41(1)	92	26	C	S:Isooctane $A=CH_3$ B=Cl	66Be	
SbA_5	-71.25	(1)	--	25	C	S:1,2-Dichloro-ethane A=Cl	63Oa	
TeA_4	-22.6	(1)	--	25	C	S:Benzene A=Cl	72Pg	
I_2	-10.5	(1)	--	0-45	T	S:Cyclohexane CL≅1M	66Bc	b,q:±4
	-24.3	1.19(20°,1)	-58	-15-35	T	S:Freon	65Wc	b,k
	-15.9	0.094(25°,1)	-50	5-25	T	S:Heptane	71Gl	b
I_2L	-88	2.91(20°)	-238	-15-35	T	S:Freon $R:2I_2L=(I_2L)_2$	65Wc	b,k b,k
P286	2-PROPANONE, 3-(p-methoxy-benzoyl)-		$C_{11}H_{12}O_3$			$(CH_3OC_6H_4CO)CH_2COCH_3$		
Ni^{2+}	-11.3	9.23(25°,1)	138	15-40	T	μ=0 S:70v% Dioxane	71Yb	b,q: ±2.1
	14.2	7.79(25°,2)	100	15-40	T	μ=0 S:70v% Dioxane	71Yb	b,q: ±2.1
Cu^{2+}	-38.5	12.93(25°,1)	117	15-40	T	μ=0 S:75v% Dioxane	71Yb	b
	-35.6	11.24(25°,2)	96	15-40	T	μ=0 S:70v% Dioxane	71Yb	b
P287	1-PROPANONE, 2-methyl-1-phenyl-		$C_{10}H_{12}O$			$C_6H_5C(O)CH(CH_3)_2$		L
I_2	-11.7	-0.14(25°,1)	-42	5-25	T	S:Heptane	71Td	b

Metal Ion	ΔH, kJ /mol	Log K	ΔS, J /mol K	T,°C	M	Conditions	Ref.	Re-marks
P288	1-PROPANONE, 1-phenyl-		$C_9H_{10}O$			$C_6H_5C(O)C_2H_5$		L
I_2	-12.1	-0.10(25°,1)	-42	5-25	T	S:Heptane	71Td	b
P289	2-PROPANONE, 3-thenoyl-1,1,1-trifluoro-		$C_8H_5O_2F_3S$			F_3CCOCH_2CO— 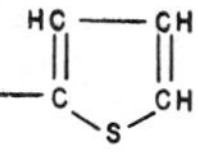		
UO_2^{2+}	-20.63	1.18(50°)	-41.0	30-70	T	S:Benzene pH= 1.6(HCl) A=Dibenzyl sulphoxide R:UO_2^{2+}+2HL+A= $UO_2L_2A+2H^+$	70Sb	b
	-19.16	0.75(50°)	-44.8	30-70	T	S:Benzene pH= 1.6(HCl) A=Diphenyl sulphoxide R:UO_2^{2+}+2HL+A= $UO_2L_2A+2H^+$	70Sb	b
	-21.80	-1.54(50°)	-38.1	30-70	T	S:Benzene pH= 1.6(HCl) A=Tributyl phosphate R:UO_2^{2+}+2HL+A= $UO_2L_2A+2H^+$	70Sb	b
	-15.77	2.56(50°)	0.8	30-70	T	S:Benzene pH= 1.6(HCl) A=Trioctylphosphine oxide R:UO_2^{2+}+ 2HL+A=$UO_2L_2A+2H^+$	70Sb	b
Zn^{2+}	45.56	-7.831(25°)	3.1	5-50	T	μ=0.1M($NaClO_4$) S:Benzene R:Zn^{2+}(aq)+ 2L(benzene)= ZnL_2(benzene)+ $2H^+$(aq)	77Na	
	17.32	-4.239(25°)	-23.1	5-50	T	μ=0.1M($NaClO_4$) S:Benzene R:Zn^{2+}(aq)+ 2L(benzene)+ A(benzene)= ZnL_2A(benzene)+ $2H^+$(aq) A=Tributyl ester phosphoric acid	77Na	
P290	PROPENAL		C_3H_4O			CH_2:CHCHO		L
NiAB	-79.5	3.62(27°)	-196.6	27-42	T	S:Tetrahydrofuran A=Dipyridyl B=Tetrahydrofuran R:NiAB+L=NiAL+B	71Ya	b

Metal Ion	ΔH, kJ /mol	Log K	ΔS, J /mol K	T,°C	M	Conditions	Ref.	Remarks
P290, continued								
NiA_2B	-31.0	0.080(-50°,1)	-130	-70- -18	T	S:Tetrahydrofuran $A=C_2H_5$ B=2,2'-Bipyridyl	76Yc	b
P291	PROPENE		C_3H_6			$CH_3CH:CH_2$		L
$RhAB_2$	5.8	-1.108(25°)	-2.1	0&25	T	S:Toluene A=2,4-Pentanedionato B=Ethylene R:$L+RhAB_2=RhABL+B$	67Cc	a,b,q: ±3.8
PdA_4^{2-}	0	0.90(20°)	17	5-25	T	μ=4.0 R:$PdA_4^{2-}+L=PdA_3L^-+A^-$ A=Cl	65Pb	b
	0	0.66(20°)	12	5-25	T	μ=4.0 R:$PdA_4^{2-}+L+H_2O=PdA_2L(OH)_2+2A^-$ A=Cl	65Pb	b
Cu^+	-42.30	5.018(25°,1)	-45.81	20-50	T	μ=0	70Oa	b
Ag^+	-14.6	0.88(25°,1)	-31.4	0-40	T	S:Ethylene glycol	65Ci	b
AgAB	-0.4	-0.366(25°)	-6.3	--	-	$A=NO_3$ B=Ethylene R:L+AgAB=AgAL+B	67Cc	i:65Cg u
P292	PROPENE, 3-bromo-		C_3H_5Br			$BrCH_2CH:CH_2$		L
I_2A	-4.60	0.745(25°)	-1.21	-20-40	T	S:Hexane A=Hexane R:$I_2A+L=I_2L+A$	73Sn	b,q: ±0.42
P293	PROPENE, 3-chloro-		C_3H_5Cl			$ClCH_2CH:CH_2$		L
CuA_3^{2-}	-71.34	0(1)	-235.6	25	T	CCl→OM A=Cl	65Ta	
	-29.7	0.046	-100.4	25	T	CCl→OM R:$CuA_3^{2-}+L=CuA_2L^-+A^-$ A=Cl	65Ta	
P294	PROPENE, 2-methyl-		C_4H_8			$(CH_3)_2C:CH_2$		L
$RhAB_2$	16.3	-3.456(25°)	-10.9	0&25	T	S:Toluene A=2,4-Pentonedionato B=Ethylene R:$L+RhAB_2=RhABL+B$	67Cc	a,b,q: ±2.9
Cu^+	-29.50	4.495(25°,1)	-12.76	20-50	T	μ=0	70Oa	b
Ag^+	-21.25	1.773(25°,1)	-37.45	20-80	T	μ=0.1M($AgNO_3+KNO_3$)	70Pb	b
P295	PROPENE, 3-sulphonate		$C_3H_5O_3S$			$CH_2:CHCH_2SO_3^-$		L^-
PtA_4	-25.5	3.606(25°)	-17.2	25-45	T	1.90M(NaCl) 0.1M(HCl) A=Cl R:$PtA_4^{2-}+L^-=PrA_3L^{2-}+A^-$	68Me	b,g:k

Metal Ion	ΔH, kJ /mol	Log K	ΔS, J /mol K	T,°C	M	Conditions	Ref.	Remarks
P296	PROPENOIC ACID, amide		C_3H_5ON			CH_2:$CHCONH_2$		L
NiAB	-74.5	3.31(27°)	-184.9	22-46	T	S:Tetrahydrofuran A=Dipyridyl B=Tetrahydrofuran R:NiAB+L=NiAL+B	71Ya	b
P297	PROPENOIC ACID, nitrile		C_3H_3N			CH_2:CHCN		L
NiAB	-66.9	2.82(27°)	-169.4	20-42	T	S:Tetrahydrofuran A=Dipyridyl B=Tetrahydrofuran R:NiAB+L=NiAL+B	71Ya	b
NiA_2B	-26.8	0.38(-50°,1)	-113	-50- -10	T	S:Tetrahydrofuran A=C_2H_5 B=2,2'-Bipyridine	76Yc	b
P298	PROPENOIC ACID, 2-methyl-, nitrile		C_4H_5N			CH_2:$C(CH_3)CN$		L
NiAB	-59.0	2.37(27°)	-151.5	22-43	T	S:Tetrahydrofuran A=Dipyridyl B=Tetrahydrofuran R:NiAB+L=NiAL+B	71Ya	b
P299	2-PROPEN-1-OL		C_3H_6O			CH_2:$CHCH_2OH$		L
PtA_4	-33.9	4.212(25°)	-32.6	25-45	T	1.90M(NaCl) 0.1M(HCl) A=Cl R:PtA_4^{2-}+L= PtA_3L^-+A^-	68Me	b,q: ±2.1
	-33.9	3.860(44.5°)	-31.8	30-60	T	μ=2.0M(NaCl+HCl) A=Cl R:PtA_4^{2-}+L =$PtLA_3^-$+A^-	67Hb	b,q: ±2.1
CuA_2^-	-8.16	0.834(1)	7.66	25	T	CCL→0M A=Cl	65Ta	
CuA_3^{2-}	-25.19	1.333	-59.20	25	T	CCl→0M R:CuA_3^-+ L=CuA_2L^-+A^- A=Cl	65Ta	
Ag^+	-18.6	1.25(25°,1)	-38.2	25-45	T	μ∿0	77Ha	
	-16.0	0.76(25°,1)	-38.9	25-45	T	S:Ethanol	77Ha	
	-17.6	1.09(25°,1)	-38.2	25-45	T	S:Methanol	77Ha	
	-26.3	2.01(25°,1)	-49.6	25-45	T	S:Propylene carbonate	77Ha	
P300	PROPIONIC ACID, 2(4-imidazolyl)		$C_6H_9O_2N_2$			$CH_3CH(C_3H_3N_2)CO_2H$		
Cu^{2+}	-15.3	4.46(1)	34	25	C	μ=0.16(KNO_3)	70Md	
	-21.5	4.03(2)	5.0	25	C	μ=0.16(KNO_3)	70Md	

Metal Ion	ΔH, kJ /mol	Log K	ΔS, J /mol K	T,°C	M	Conditions	Ref.	Remarks
P301	PROPYNE, 3-acetoxy-		$C_5H_6O_2$			$CH_3COOCH_2C{\equiv}CH$		L
Ag^+	-11.3	0.146(25°,1)	-35.6	10-40	T	μ=1M($NaNO_3$) S:52v% Ethanol	73Sz	b,q: ±2.1
P302	PROPYNE, 3-chloro-		C_3H_3Cl			$ClCH_2C{\equiv}CH$		L
Ag^+	-20.5	0.049(25°,1)	-66.9	10-40	T	μ=1M($NaNO_3$) S:52v% Ethanol	73Sz	b,q: ±3.3
	-28.4	0.00(25°,1)	-95.4	10-40	T	μ=1M($NaNO_3$) S:75v% Ethanol	73Sz	b,q: ±1.7
P303	PROPYNE, 3-hydroxy-		C_3H_4O			$HOCH_2C{\equiv}CH$		L
Ag^+	-10.0	0.633(25°,1)	-21	10-40	T	μ=1M($NaNO_3$) S:52v% Ethanol	73Sz	b,q: ±1.7
	-18.0	0.518(25°,1)	-50.6	10-40	T	μ=1M($NaNO_3$) S:75v% Ethanol	73Sz	b,q: ±1.7
P304	PROPYNE, 3-methoxy-		C_4H_6O			$CH_3OCH_2C{\equiv}CH$		L
Ag^+	-10.0	0.431(25°,1)	-26.8	10-40	T	μ=1M($NaNO_3$) S:52v% Ethanol	73Sz	b,q: ±2.1
P305	PURINE, 6-amino-9-cyclohexyl-		$C_{11}H_{14}N_5$			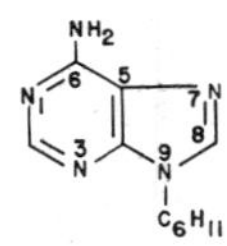		L
I_2	-39.7	2.250(25°,1)	87.4	15-35	T	μ=0 S:CCl_4	74La	b
P306	PYRAN, tetrahydro-		$C_5H_{10}O$			$CH_2CH_2CH_2CH_2CH_2O$ (ring)		L
I_2	-20.5	0.122(25°,1)	-44.8	5.5-25	T	S:n-Heptane	60Ta 60Bb	b
P307	PYRAZINE		$C_4H_4N_2$					L
Ni^{2+}	-13.4	1.013(25°,1)	-25	10-35	T	μ=1.0M($LiClO_4$) pH=4	72Mb	b,q: ±2.1
Ag^+	-17.03	1.38(1)	-30.71	25	C	μ=0.1M(KNO_3)	73Bi	
	-17.0	1.38(1)	-30.7	25	C	μ=0.1M(KNO_3)	74Hg	
	-16.86	1.03(2)	-36.65	25	C	μ=0.1M(KNO_3)	73Bg	
	-16.9	1.03(2)	-36.6	25	C	μ=0.1M(KNO_3)	74Hg	
P308	PYRAZINE, 2-chloro		$C_4H_3N_2Cl$			$C_4H_3N_2Cl$		
Ag^+	-15.9	0.96(1)	-34.7	25	C	μ=0.1M(KNO_3)	74Hg	
	-16.2	0.57(2)	-41.4	25	C	μ=0.1M(KNO_3)	74Hg	

Metal Ion	ΔH, kJ /mol	Log K	ΔS, J /mol K	T,°C	M	Conditions	Ref.	Remarks
P309	PYRAZINE, 2,5-dimethyl		$C_6H_8N_2$			$C_4H_2N_2(CH_3)_2$		
Ag^+	-21.7	1.95(1)	-35.4	25	C	μ=0.1M(KNO_3)	74Hg	
	-18.3	1.52(2)	-32.4	25	C	μ=0.1M(KNO_3)	74Hg	
P310	PYRAZINE, 2,6-dimethyl		$C_6H_8N_2$			$C_4H_2N_2(CH_3)_2$		
Ag^+	-20.9	1.96(1)	-32.6	25	C	μ=0.1M(KNO_3)	74Hg	
	-10.9	1.17(2)	-14.2	25	C	μ=0.1M(KNO_3)	74Hg	
P311	PYRAZINE, 2-methyl-		$C_5H_6N_2$			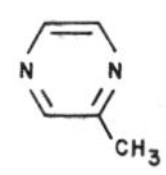		
Ag^+	-20.0	1.65(1)	-35.6	25	C	μ=0.1M(KNO_3)	74Hg	
	-16.8	1.11(2)	-35.1	25	C	μ=0.1M(KNO_3)	74Hg	
I_2	-25.61	1.478(1)	--	30-50	T	S:Cyclohexane K has been corrected for interaction between I_2 and solvent	73Ic	b
	-28.07	1.549(1)	--	30-50	T	S:n-Heptane K has been corrected for interaction between I_2 and solvent	73Ic	b
P312	PYRAZOLE		$C_3H_4N_2$			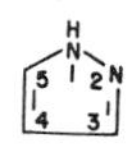		L
Ag^+	-24.9	1.90(30°,1)	-45	25-35	T	μ=0.1M(KNO_3) S:20v%Ethanol	74Pg	b,n,y, dd,ee
	-38.2	1.90(30°,1)	-89.7	25-35	T	μ=0.1M(KNO_3) S:20v%Ethanol	77Pc	b
	-88.4	4.61(30°,2)	3	25-35	T	μ=0.1M(KNO_3) S:20v%Ethanol	74Pg	b,n,p, y,dd,ee
	-38.2	2.41(30°,2)	-80.0	25-35	T	μ=0.1M(KNO_3) S:20v%Ethanol	77Pc	b,e
Cd^{2+}	-16.48	1.50(25°,1)	-26.4	0-45	T	0.1M(KNO_3) CM= 5x10^{-4}M[$Cd(NO_3)_2$] CL=0.05-2.0M	63Ad	b
	-33.89	2.18(25°,1-2)	-84.1	0-45	T	0.1M(KNO_3) CM= 5x10^{-4}M[$Cd(NO_3)_2$] CL=0.05-2.0M	63Ad	b
	-50.79	2.32(25°,1-3)	-125.9	0-45	T	0.1M(KNO_3) CM= 5x10^{-4}M[$Cd(NO_3)_2$] CL=0.05-2.0M	63Ad	b
P313	PYRAZOLE, 3,5-dimethyl		$C_5H_8N_2$			$C_3H_2N_2(CH_3)_2$		
Ag^+	-38.2	2.96(30°,1)	-69.4	25-35	T	μ=0.1M(KNO_3)	77Pc	b

Metal Ion	ΔH, kJ /mol	Log K	ΔS, J /mol K	T,°C	M	Conditions	Ref.	Remarks
P313, continued								
Ag^+, cont.	-40.6	2.42(30°,2)	-87.8	25-35	T	S:20v%Ethanol μ=0.1M(KNO_3) S:20v%Ethanol	77Pc	b,e
P314	PYRAZOLE, 3-methyl		$C_4H_6N_2$			$C_3H_3N_2CH_3$		
Ag^+	-44.2	2.33(30°,1)	-30.0	25-35	T	μ=0.1M(KNO_3) S:20v%Ethanol	77Pc	b
	-38.2	2.28(30°,2)	-32.4	25-35	T	μ=0.1M(KNO_3) S:20v%Ethanol	77Pc	b,e
P315	PYRAZOLE 3,4,5-trimethyl		$C_6H_{10}N_2$			$C_3HN_2(CH_3)_3$		
Ag^+	-52.6	3.74(30°,1)	-101.9	25-35	T	μ=0.1M(KNO_3) S:20v%Ethanol	77Pc	b
	-43.0	2.07(30°,2)	-102.3	25-35	T	μ=0.1M(KNO_3) S:20v%Ethanol	77Pc	b,e
P316	PYRIDAZINE		$C_4H_4N_2$			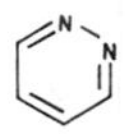		
Ag^+	-30.92	1.48(1)	-75.3	25	C	μ=0.1M(KNO_3)	73Bi	
	-2.80	1.34(2)	16.3	25	C	μ=0.1M(KNO_3)	73Bi	
P317	PYRIDINE		C_5H_5N					L
Be^{2+}	-18.74	2.37(40°,1)	-13.4	35&45	T	--	67Ra	a,b
MgAL	-9.54	-1.024(30°)	-50.2	--	T	S:Piperidine CL=12.36M A= Etioporphyrin II R:MgAL+L=$MgAL_2$	66Sd	u
	-10.67	-1.125(30°)	-57.3	--	T	S:Pyridine-2,6-lutidine CL= 12.36M A=Desoxo-phyllorythrin monomethyl ester R:MgAL+L=$MgAL_2$	66Sd	u
	-16.28	-0.613(30°)	-65.3	--	T	S:Pyridine-2,6-lutidine CL= 12.36M A=Deuter-oporphyrin IX dimethyl ester R:MgAL+L=$MgAL_2$	66Sd	u
	-16.32	-0.548(30°)	-64.4	--	T	S:Pyridine-2,6-lutidine CL= 12.36M A=Di-chlorodeutero-porphyrin IX dimethyl ester	66Sd	u

Metal Ion	ΔH, kJ /mol	Log K	ΔS, J /mol K	T,°C	M	Conditions	Ref.	Remarks
P317, continued								
MgAL, cont.	-12.22	-1.168(30°)	-62.8	--	T	R:MgAL+L=MgAL$_2$ S:Pyridine-2,6-lutidine CL=12.36M A=Etioporphyrin II	66Sd	u
	-15.77	-0.836(30°)	-68.2	--	T	R:MgAL+L=MgAL$_2$ S:Pyridine-2,6-lutidine Cl=12.36M A=Mesoporphyrin IX dicholesteryl ester	66Sd	u
	-16.65	-1.154(30°)	-77.0	--	T	R:MgAL+L=MgAL$_2$ S:Pyridine-2,6-lutidine CL=12.36M A=Mesoporphyrin IX dimethyl ester	66Sd	u
	-17.78	-0.606(30°)	-70.3	--	T	R:MgAL+L=MgAL$_2$ S:Pyridine-2,6-lutidine CL=12.36M A-Protoporphyrin IX dimethyl ester	66Sd	u
	-9.92	-0.779(30°)	-47.7	--	T	R:MgAL+L=MgAL$_2$ S:Pyridine-2,6-lutidine CL=12.36M A=Tetraphenylporphyrin	66Sd	u
	-10.29	-1.024(30°)	-52.7	--	T	R:MgAL+L=MgAL$_2$ S:Pyrrolidine CL=12.36M A=Etioporphyrin II R:MgAL+L=MgAL$_2$	66Sd	u
[Mg]	-10.88	-0.079(30°,2)	-37.6	20-40	T	S:Benzene [Mg]=Magnesium Tetraphenylchlorin	52Ma	b
	-10.88	-0.239(30°,2)	-41.0	20-40	T	S:Benzene [Mg]=Magnesium Tetraphenylporphine	52Ma	b
VOA$_2$	-30.75	1.76(1)	-69.4	25	C	μ→0 S:Nitrobenzene A=Acetylacetonate	65Ca	
	-43.9	1.81(24°,1)	-113	10-73	T	S:Toluene ΔH from plot of lnK vs 1/T A=O,O'-Diethyldithiophosphato	73Sd	
	-50	1.81(24°,1)	-125	10-73	T	S:Toluene ΔH from ESR spectra A=O,O'-	73Sd	

Metal Ion	ΔH, kJ /mol	Log K	ΔS, J /mol K	T,°C	M	Conditions	Ref.	Remarks
P317, continued								
VOA_2, cont.						Diethyldithio- phosphato		
NbA_5	-10.0	-1.4	-71	-54	T	S:$C_6H_5CH_3$ A=OCH_3 R:$[NbA_5]_2$+2L= $2NbA_5L$	71Hi	q:±1.7
	-10.9	-0.68	-63	-48	T	S:CH_3CN A=OCH_3 R:$[NbA_5]_2$+2L= $2NbA_5L$	71Hi	q:±1.7
TaA_5	-14.2	-1.3	-88	-54	T	S:$C_6H_5CH_3$ A=OCH_3 R:$[TaA_5]_2$+2L= $2TaA_5L$	71Hi	q:±2.5
Mn^{2+}	-10.00	1.86(1)	2.1	25	C	1M($NaClO_4$)	63Ae	f:4
	-10.00	1.59(2)	-6.3	25	C	1M($NaClO_4$)	63Ae	f:4
	-10.00	0.90(3)	-16.3	25	C	1M($NaClO_4$)	63Ae	f:4
	-10.00	0.60(4)	-22.2	25	C	1M($NaClO_4$)	63Ae	f:4
[Fe]	-78.6	5.67(25°,1-2)	-115.2	15.5- 36.5	T	μ=0.10($NaClO_4$) pH=7.0 [Fe]= Iron(II) 2,4- dicysteine- substituted mesoporphyrin (heme c)	76Gf	
	-144.3	8.24(25°,1-2)	-326	7-37	T	S:Benzene [Fe]= Iron(II) Deuteroporphyrin dimethyl ester	74Bh	b
	-10.0	2.56(25°,1-2)	17	18-40	T	[Fe]=Iron(II) deuteroporphyrin dimethyl ester S:Benzene	70Ch	b,q: ±0.8
	-14.6	1.76(25°,1-2)	-17	18-40	T	[Fe]=Iron(II) deuteroporphyrin dimethyl ester S:Benzene	71Cd	b,q: ±0.8
	-21.8	2.42(25°,1-2)	-25	18-40	T	[Fe]=Iron(II) mesoporphyrin dimethyl ester S:Benzene	70Ch	b,q: ±5.8
	-13.8	1.32(25°,1-2)	-21	18-40	T	[Fe]=Iron(II) protoporphyrin dimethyl ester S:Benzene	70Ch 71Cd	b,q: ±2.5
	-31.4	1.32(25°,1-2)	134	18-40	T	[Fe]=Iron(II) diacetyldeutero- porphyrin dimethyl ester S:Benzene	70Ch	b,q: ±3.8
	-59.4	2.64(25°,1-2)	-146	18-40	T	[Fe]=Iron(II) deuteroporphyrin dimethyl ester S:CCl_4	71Cd	b,q:

Metal Ion	ΔH, kJ /mol	Log K	ΔS, J /mol K	T,°C	M	Conditions	Ref.	Re-marks
P317, continued								
[Fe], cont.	-66.9	0.44(25°,1-2)	-209	18-40	T	[Fe]=Iron(II) protoporphyrin dimethyl ester S:CCl_4	71Cd	b
	-45.2	0.22(25°,1-2)	-146	18-40	T	[Fe]=Iron(II) protoporphyrin dimethyl ester S:$CHCl_3$	71Cd	b,q: ±6.3
	9.83	0.88(20°)	50.6	5-30	T	pH=11.2 Proto-haemin R:$(Haemin)_2$+2L=$(HaeminL)_2$+$2H_2O$	69Md	b
	-33.5	3.04(25°,1-2)	-54.4	15-5-36.5	T	μ=0.10($NaClO_4$) pH=6.0 [Fe]= Iron(III) 2,4-dicysteine-substituted mesoporphyrin (hemin c)	76Gf	q:±3.3
CoA	-43.5	(1)	--	25	C	S:CH_2Cl_2 A=N,N'-bis(Salicylidene) -o-phenylene-diamino	74Td	
	-35.6	2.686(25°,1)	-67	20-50	T	S:Toluene A=α,β,γ,δ-Tetra-(p-methoxyphenyl) porphyrin S:Toluene	73Wa	b,q:±4
	-37.2	(1)	--	-41- -22	T	S:Toluene A=Protoporphyrin IX dimethyl ester	78Dd	
CoA_2	-21.3	3.15(25°,1)	-11.3	25	C	S:Acetone A=Cl	73Lb	
	-44.8	5.4(1-2)	-47.7	25	C	S:Acetone CM=$2x10^{-3}$M CL=$2x10^{-2}$M A=Cl	71Lg	
	-41.0	5.4(1-2)	-35.1	25	C	S:Acetone CM=$2x10^{-3}$M CL=$2x10^{-2}$M A=Br	71Lg	
	-40.6	5.0(1-2)	-40.6	25	C	S:Acetonitrile CM=$2x10^{-3}$M CL=$2x10^{-2}$M A=Cl	71Lg	
	-41.0	4.8(1-2)	-46.4	25	C	S:Acetonitrile CM=$2x10^{-3}$M CL=$2x10^{-2}$M A=Br	71Lg	
	-26.4	3.4(1-2)	-23.8	25	C	S:n-Butyl alcohol CM=$2x10^{-3}$M CL=$8x10^{-2}$M A=Cl	71Lh	
	-22.2	3.2(1-2)	-14.2	25	C	S:n-Butyl alcohol CM=$2x10^{-3}$M CL=$8x10^{-2}$M A=Br	71Lh	
	-29.3	4.2(1-2)	-18.4	25	C	S:t-Butyl alcohol CM=$2x10^{-3}$M CL=$8x10^{-2}$M A=Cl	71Lh	

Metal Ion	ΔH, kJ /mol	Log K	ΔS, J /mol K	T,°C	M	Conditions	Ref.	Re-marks
P317, continued								
CoA_2, cont.	-24.7	4.5(1-2)	4.2	25	C	S:t-Butyl alcohol CM-2x10^{-3}M CL=8x10^{-2}M A=Br	71Lh	
	-55.6	5.8(1-2)	-75.7	25	C	S:Cyclohexanone CM=2x10^{-3}M CL=2x10^{-2}M A=Cl	71Lg	
	-58.6	5.8(1-2)	-92.9	25	C	S:Cyclohexanone CM=2x10^{-3}M CL=2x10^{-2}M A=Br	71Lg	
	-20.46	4.17(1-2)	11.09	25	C	S:Dimethyl formamide A=Cl	72Ae	
	-32.13	3.22(1-2)	-46.19	25	C	S:Ethanol A=Cl	72Af	
CoAB	-41.0	1.98(1)	-100	25	T	S:Cyclohexane A=p-$MeOC_6H_4$ B=N,N'-bis (acetylacetonato) ethylenediimine	77Hc	
	-59.4	4.167(1)	-119.7	25	T	S:o-Dichlorobenzene A=CF_3 B=N,N'-bis (acetylacetonato) ethylenediimine	77Hc	
	-26.8	1.49(1)	-61.1	25	C	S:o-Dichlorobenzene A=p-$MeOC_6H_4$ B=N,N'-bis (acetylacetonato) ethylenediimine	77Hc	
	-23.0	1.75(1)	-43.9	25	T	S:o-Dichlorobenzene A=CH_3 B=N,N'-bis (acetylacetonato) ethylenediimine	77Hc	n
	-37.6	2.60(1)	-77.0	25	T	S:o-Dichlorobenzene A=CH_2Cl B=N,N'-bis (acetylacetonato) ethylenediimine	77Hc	
	-23.0	1.34(1)	-51.0	25	T	S:o-Dichlorobenzene A=C_6H_5 B=N,N'-bis (acetylacetonato) ethylenediimine	77Hc	
	-23.8	1.76(1)	-46.9	25	T	S:o-Dichlorobenzene A=pFC_6H_4 B=N,N'-bis (acetylacetonato) ethylenediimine	77Hc	
CoL_2A_2	-63.6	1.10(1-2)	-195.8	20	C	$\mu \to 0$ S:$CHCl_3$ A=Cl	62Ka 63Ka	s
	-65.3	0.95(1-2)	-204.2	20	C	$\mu \to 0$ S:$CHCl_3$ A=Br	62Ka 63Ka	s

Metal Ion	ΔH, kJ /mol	Log K	ΔS, J /mol K	T,°C	M	Conditions	Ref.	Remarks
P317, continued								
CoL_2A_2, cont.	-69.4	0.36(1-2)	-229.7	20	C	μ→0 S:$CHCl_3$ A=I	62Ka 63Ka	s
	-57.3	1.36(1-2)	-169.4	20	C	μ→0 S:$CHCl_3$ A=NCO	62Ka 63Ka	s
	-69.4	4.92(1-2)	-142.7	20	C	μ→0 S:$CHCl_3$ A=NCS	62Ka 63Ka	s
	-46.9	-0.146(25°,1-2)	-159.8	27-57	T	μ=0.1(Et_4NClO_4) S:Nitrobenzene A=Br	72Fb	b
	-56.1	1.40(40°,1-2)	-153.6	32-60	T	S:Pyridine A=Cl	61Ka	b
	-56.1	-0.672(25°,1-2)	-201	32-63	T	μ=0.1(Et_4NClO_4) S:Pyridine A=Br	72Fb	b,q: ±5.4
CoA_6	-10.9	2.724	7.5	-30	T	S:CH_3OH A=CH_3OH R:CoA_6^{2+}+L= CoA_5L^{2+}+A	73Pk	
[Co]	-87.0	--	--	24	C	[Co]=Methylcobal- oxime S:Dichloromethane R:$[Co]_2$+2L=2[Co]L	73Cf	
	-28.9	4.03(0°,1)	-25	-23-23	T	[Co]=Cobalt(II) protoporphyrin IX dimethyl ester S:Toluene	73Sx	b,q: ±2.5
Ni^{2+}	-11.09	2.13(1)	3.3	25	C	1M($NaClO_4$)	63Ae	f:4
	-11.09	1.66(2)	-5.4	25	C	1M($NaClO_4$)	63Ae	f:4
	-11.09	1.12(3)	-15.9	25	C	1M($NaClO_4$)	63Ae	f:4
	-11.09	0.65(4)	-24.7	25	C	1M($NaClO_4$)	63Ae	f:4
NiA	-30.0	0.518(1)	-89	30	C	S:Acetonitrile CM=7-15x10^{-3}M A=BF_2-bridged glyoximato R_1=Me R_2=Pr	76Gh	
	-28.0	0.462(1)	-84	30	C	S:Acetonitrile CN=7-15x10^{-2}M A=BF_2-bridged glyoximato R_2=Me R_2=Et	76Gh	
	-25.4	0.398(1)	-76	30	C	S:Acetonitrile CM=10-15x10^{-3}M A=BF_2-bridged glyoximato R_1=Me R_2-Me 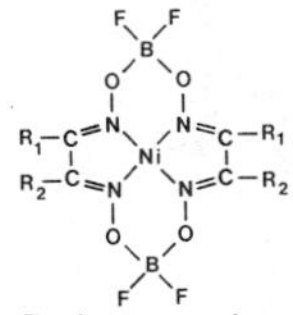	76Gh	
	-19.1	1.502(1)	-34	30	C	S:Acetonitrile CM=2-5x10^{-3}M A=BF_2-bridged glyoximato	76Gh	q:±2.8

Metal Ion	ΔH, kJ /mol	Log K	ΔS, J /mol K	T, °C	M	Conditions	Ref.	Remarks
P317, continued								
NiA, cont.	-20.0	1.107(1)	-45	30	C	R_1=Ph R_2=H S:Acetonitrile CM=2-5x10^{-3}M A=BF_2-bridged glyoximato	76Gh	
	-35.0	0.477(1)	-106	30	C	R_1=Ph R_2=Ph S:Acetonitrile CM=7-15x10^{-3}M A=BF_2-bridged glyoximato	76Gh	
	-62.55	3.78(1-2)	-137.6	25	C	R_1=Me R_2=iso-Pr CM≅CL≅10^{-4}M S:Benzene A=Diacetylbisbenzoyl hydrazine	58Sa	
	-28.0	1.82(25°,1)	-53	10-45	T	S:Benzene A= 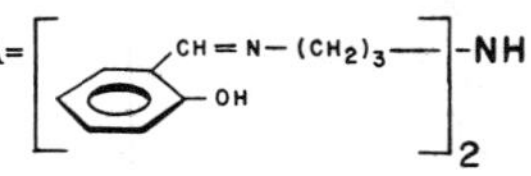	66Sa	b
	-23.0	1.35(25°,1)	-50	10-45	T	S:Benzene A= 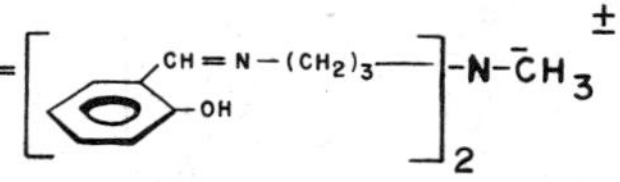	66Sa	b,q: ±1.7
	-31.8	2.25(25°,1)	-63	10-45	T	S:Benzene A= 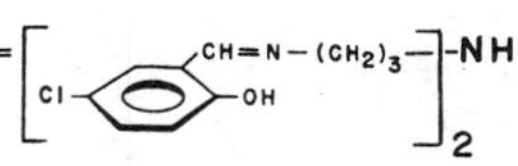	66Sa	b
	-33.0	2.33(25°,1)	-67	10-45	T	S:Benzene A= 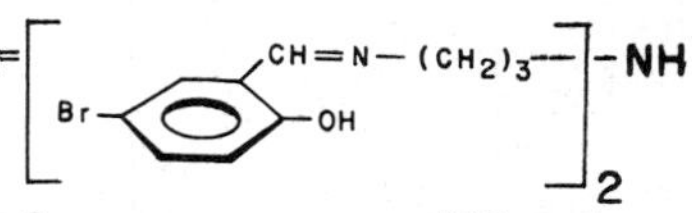	66Sa	b
	-33.9	1.91(25°,1)	-75	10-45	T	S:Benzene A= 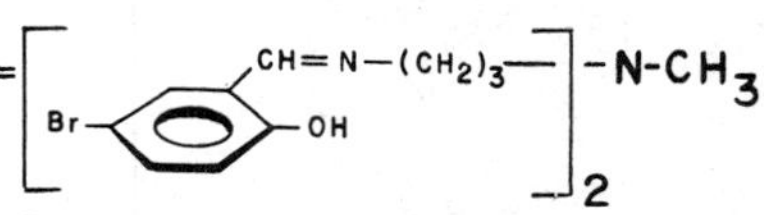	66Sa	b
	-31.8	1.66(25°,1)	-75	10-45	T	S:Benzene A= 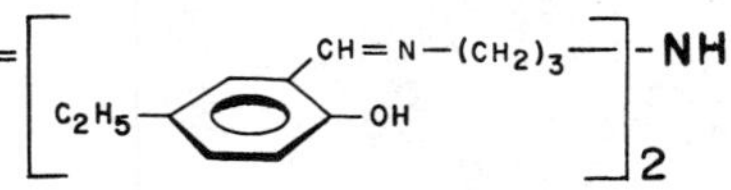	66Sa	b
	-33.0	2.44(25°,1)	-63	10-45	T	S:Benzene A= 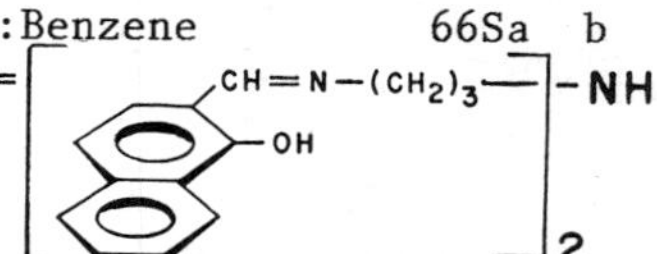	66Sa	b
	-29.7	2.46(25°,1)	-50	10-45	T	S:Benzene A= 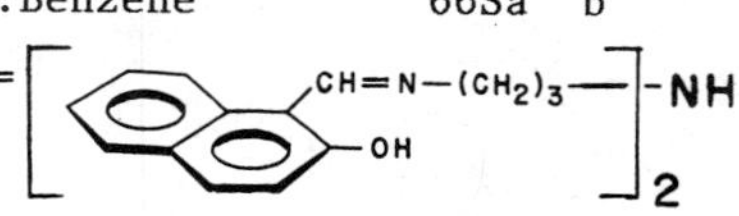	66Sa	b

Metal Ion	ΔH, kJ /mol	Log K	ΔS, J /mol K	T,°C	M	Conditions	Ref.	Re-marks
P317, continued								
NiA_2	-72.8	1.18(1)	-218.0	30	C	S:Benzene A=Diethyldithio-phosphato	71Dc	q:±7.1
	-19.2	0.43(25°,1)	-54	--	T	S:Benzene A=Diethyldithio-phosphate NiA_2= bis(oo'-diethyl-dithiophosphate) nickel(II)	64Ca	b,u
	-29.3	1.46(25°,1)	-71	--	T	S:Benzene A=Diethyldithio-phosphate NiA_2= bis(oo'-diethyl-dithiophosphato) nickel(II)	64Ca	b,u
	-29.3	1.462(25°,1)	-71	--	T	S:Benzene A=oo'-Diethyldi-thiophosphate	64Ca	u
	-71.1	1.18(1)	-210.9	30	C	S:Benzene A=Dipropyldithio-phosphato	71Dc	q:±7.1
	-25.5	2.953(1)	-28.9	25	C	S:Benzene A=2,2,6,6-Tetra-methyl-3,5-heptanedione	73Ce	
	-17	2.00(1)	-18.7	25-30	C	S:Benzene A=Dithiobenzoato	75Af	
	-100	2.301(1)	-374	30	C	S:Benzene A=($SOCNPr_2$)	77Gf	
	-34	2.477(1)	-66	30	C	S:Benzene A=O,O'-Diethyl-thiomalonato	77Ah	
	-15	1.137(1)	-27.7	30	C	S:Butane-2-one A=Dithizonate	78Gc	
	-71	2.000(1)	-196	30	C	S:Butan-2-one A=O,O'-Diethyl-thiomalonato	77Ah	
	-34	2.699(1)	-60.5	30	C	S:Chlorobenzene A=Diphenyldithio-phosphinato	78Ac	
	-18	1.268(1)	-35.1	30	C	S;1,2-Dichloro-ethane A=Dithizonate	78Gc	
	-26	1.342(1)	-63.2	19	C	S:1,2-Dichloro-ethane A=Dithizonate	78Gc	
	-62	1.903(1)	-169	30	C	S:$C_2H_4Cl_2$ A=O,O'-Diethyl-thiomalonato	77Ah	
	-50	2.176(1)	-124	30	C	S:MeCN A=O,O'-Diethyl-thiomalonato	77Ah	
	4.02	1.34(2)	40.2	30	C	S:Benzene A=Diethyldithio-	71Dc	q:±7.1

Metal Ion	ΔH, kJ /mol	Log K	ΔS, J /mol K	T,°C	M	Conditions	Ref.	Remarks
P317, continued								
						phosphato		
NiA_2, cont.	-22	2.000(2)	-34	30	C	S:Benzene A=O,O'-Diethylthiomalonato	77Ah	
	14	2.301(2)	2	30	C	S:Benzene A=$(SOCNPr_2)$	77Gf	
	-36	2.00(2)	-82.4	25-30	C	S:Benzene A=Dithiobenzoato	75Af	
	6	1.699(2)	53	30	C	S:Butan-2-one A=O,O'-Diethylthiomalonato	77Ah	
	-44	1.398(2)	-118.4	30	C	S:Chlorobenzene A=Diphenyldithiophosphinato	78Ac	
	3	1.602(2)	41	30	C	S:$C_2H_4Cl_2$ A=O,O'-Diethylthiomalonato	77Ah	
	7	0.643(2)	36.3	19	C	S:1,2-Dichloroethane A=Dithizonate	78Gc	
	-4	1.875(2)	49	30	C	S:MeCN A=O,O'-Diethylthiomalonato	77Ah	
	-19.2	0.430(25°,2)	-54	--	T	S:Benzene A=oo'-Diethyldithiophosphato	64Ca	u
	2.05	1.43(2)	33.9	30	C	S:Benzene A=Dipropyldithiophosphato	71Dc	q:±7.1
	-54.4	4.301(2)	-100.0	25	C	S:Benzene A=2,2,6,6-Tetramethyl-3,5-heptanedione	73Ce	
	-41.8	0.978(1-2)	-120.1	25	C	S:Benzene A=N-Benzyl-5-chloro-1'-phenylsalicylaldehyde	73Da	q:±4.2
	-35.98	5.5(1-2)	-13.4	30	C	S:Benzene CM=0.002-0.044M A=3-Benzyl-pentane-2,4-dione	73Ag	
	-61.9	1.179(1-2)	-184.9	25	C	S:Benzene A=N=Butyl-5-chloro-1'-phenylsalicylaldehyde	73Da	q:±5.0
	-61.1	0.914(1-2)	-184.9	25	C	S:Benzene A=N-Butyl-5-chloro-1'-phenylsalicylaldehyde	73Da	q:±7.1
	-66.9	2.989(1-2)	-163.2	30	C	S:Benzene A=N-Butyl-5-chloro-salicylaldehyde	73Da	
	-16.99	4.8(1-2)	36.0	30	C	S:Benzene	73Ag	

Metal Ion	ΔH, kJ /mol	Log K	ΔS, J /mol K	T,°C	M	Conditions	Ref.	Re-marks
P317, continued								
NiA$_2$, cont.	-63.8	2.270(1-2)	-168.2	25	C	CM=0.006-0.045M A=3-Butyl-pen-tane-2,4-dione S:Benzene	73Da	
	-59.8	2.114(1-2)	-164.0	30	C	A=N-Butylsalicyl-aldehyde S:Benzene	73Ma	
	-46.9	0.929(1-2)	-138.1	25	C	A=N-Butylsalicyl-aldehyde S:Benzene	73Da	q:±5.0
	-41.8	1.531(1-2)	-110.0	30	C	A=N-(2,4-Di-methylbenzene)-5-chloro-1'-phenyl-salicylaldehyde S:Benzene	73Da	q:±4.2
	-67	7.00(1-2)	-87.0	30	C	A=N-(2,5-Di-methylbenzene)-salicylaldehyde S:Benzene	74Gb	
	-74.9	3.228(1-2)	-185.8	30	C	A=2,6-Dimethyl-heptane-3,5-dione S:Benzene	71Dc	
	-46	3.204(28°,1-2)	-96	--	T	A=Ethylxanthate S:Benzene	64Ca	u
	-36.0	0.949(1-2)	-100.0	25	C	A=Ethylxanthate S:Benzene	73Da	q:±4.2
	-64.0	3.041(1-2)	-154.8	30	C	A=N-Hexyl-5-chloro-1'-phenyl-salicylaldehyde S:Benzene	73Da	
	-61.1	2.176(1-2)	-159.8	30	C	A=N-Isopropyl-5-chlorosalicyl-aldehyde S:Benzene	73Da	
	-41.8	1.204(1-2)	-115.1	30	C	A=N-Isopropyl-salicylaldehyde S:Benzene	73Da	q:±4.2
	-69.0	3.371(1-2	-161.9	30	C	A=N-(2-Methyl-benzene)salicyl-aldehyde S:Benzene	73Da	
	-59.0	2.845(1-2)	-140.2	30	C	A=N-Methyl-5-chlorosalicyl-aldehyde S:Benzene	73Da	
	3.01	6.7(1-2)	138	30	C	A=N-Methylsali-cylaldehyde S:Benzene	73Ag	
	7.1	6.8(1-2)	153.6	30	C	CM=0.007-0.070M A=Pentane-2,4-dione S:Benzene CM-0.006-0.062M	73Ag	

Metal Ion	ΔH, kJ /mol	Log K	ΔS, J /mol K	T,°C	M	Conditions	Ref.	Remarks
P317, continued								
NiA_2, cont.	-56.9	2.623(1-2)	-138.1	25	C	A=1-Phenylbutane-1,3-dione S:Benzene A=N-Phenyl-5-chloro-1'-phenyl-salicylaldehyde	73Da	q:±5.0
	-40.00	3.580(1-2)	-62.8	30	C	S:Benzene A=N-Phenyl-5-chloro-salicylaldehyde	73Da	
	38.00	3.00(1-2)	-67.8	30	C	S:Benzene A=N-Phenylsalicylaldehyde	73Da	q:±4.2
	-61.9	2.00(1-2)	-164.8	30	C	S:Benzene A=N-Propylsalicylaldehyde	73Da	q:±5.0
	-64.8	7.0(1-2)	-79.9	30	C	S:Benzene $CM=0.0004-0.003M$ A=2,2,6,6-Tetramethylheptane-3,5-dione	73Ag	
	-23	0.602(1-2)	-65.6	25-30	C	S:Benzene A=Perthiobenzoato	75Af	q:±4
	-52.30	0.176(1-2)	-172.3	25	C	S:Benzene A=[Ph-C(S)-CH-C(O)-Ph]	78La	j
	-50.54	0.708(1-2)	-159.1	25	C	S:Benzene A=[C(S)-C(Ph)-C(O)]	78La	j
	-47.03	0.544(1-2)	-147.2	25	C	S:Benzene A=[C(S)-CH-C(O)-Ph]	78La	j
	-61.09	2.699(1-2)	-152.3	25	C	S:Benzene A=[Morph-C(S)-N-C(O)-Ph]	78La	j
	-45.61	0.699(1-2)	-139.5	25	C	S:Benzene A=[Morph-C(S)-CH-C(O)-Ph]	78La	j
	-64.02	2.987(1-2)	-157.2	25	C	S:Benzene A=[Ph-C(S)-N-C(O)-Ph]	78La	j
	-55.65	1.666(1-2)	-154.5	25	C	S:Benzene A=[Et_2N-C(S)-N-C(O)-Ph]	78La	j
	-57.32	1.651(1-2)	-160.4	25	C	S:Benzene A=[i-Bu_2N-C(S)-N-C(O)-Ph]	78La	j
	-65.27	2.258(1-2)	-175.4	25	C	S:Benzene A=[MeS-C(S)-CH-C(O)-Ph]	78La	j
	-52.30	1.584(1-2)	-145.0	25	C	S:Benzene A=[Et_2N-C(Se)-N-C(O)-Ph]	78La	j
	-72.8	2.45(1-2)	-193	30	C	$CM=1-4x10^{-3}$ S:Benzene A=$(C_6H_{11})OCS_2$	74Gd	

Metal Ion	ΔH, kJ /mol	Log K	ΔS, J /mol K	T,°C	M	Conditions	Ref.	Remarks
P317, continued								
NiA_2, cont.	-74.9	2.26(1-2)	-204	30	C	CM=1-4x10^{-3} S:Benzene A=2-$CH_3(C_6H_{11})OCS_2$	74Gd	
	-62.4	2.51(1-2)	-158	30	C	CM=1-8x10^{-3} S:Benzene A=$(CH_3)_2CHOCS_2$	74Gd	
	-76.8	3.11(1-2)	-194	30	C	CM=1-8x10^{-3} S:Benzene A=$CH_3(CH_2)_5OCS_2$	74Gd	
	-79.2	3.74(1-2)	-190	30	C	CM=1-6x10^{-3} S:Benzene A=CH_3OCS_2	74Gd	
	-81.4	3.04(1-2)	-210	30	C	CM=1-8x10^{-3} S:Benzene A=$CH_3CH_2CH_2OCS_2$	74Gd	
	-75.0	3.23(1-2)	-186	30	C	CM=4-9x10^{-3} S:Benzene A=$CH_3CH_2OCS_2$	74Gd	
	-77.4	3.08(1-2)	-197	30	C	CM=4-6x10^{-3} S:Benzene A=$(CH_3)_2CHCH_2OCS_2$	74Gd	
	-80.8	3.62(1-2)	-197	30	C	CM=1-2x10^{-3} S:Benzene A=$(C_6H_5)CH_2OCS_2$	74Gd	
	-106.69	4.56(25°,1-2)	-270.7	20-40	T	S:Benzene A=$(p-ClC_6H_4)CH=$ $N-N=COC_6H_5$	77Ia	
	-83.68	3.62(25°,1-2)	-211.3	20-40	T	S:Benzene A=$(p-CH_3OC_6H_4)CH=$ $N-N=COC_6H_5$	77Ia	
	-99.40	4.31(25°,1-2)	-237.6	20-40	T	S:Benzene A=$(m-OCH_3C_6H_4)CH=$ $N-N=COC_6H_5$	77Ia	
	-96.23	4.02(25°,1-2)	-246.0	20-40	T	S:Benzene A=$C_6H_5CH=N-N=$ COC_6H_5	77Ia	
	-20.10	-0.301(1-2)	-74.5	25	C	S:Chloroform A=[Ph-C(S)-CH-C(O)-Ph]	77Ia	j
	-23.0	3.45(1-2)	-8.4	--	C	CM/CL=1/10 S:Dimethylform-amide A=Br^-	76Kj	j,u: 25°,y
	-49.0	1.85(20°,1-2)	-133.5	10-30	T	S:$CHCl_3$ A=Ethyl-β-mercaptocinna-mate	74Cd	b
	-41.4	1.56(20°,1-2)	-113.0	10-30	T	S:$CHCl_3$ A=Ethyl-β-mercaptothio-cinnamate	74Cd	b
	-39.7	0.60(20°,1-2)	-125.5	10-30	T	S:$CHCl_3$ A=N(p-chlorophenyl)-β-mercaptocinnama-mide	74Cd	b
	-36.8	0.35(20°,1-2)	-118.8	10-30	T	S:$CHCl_3$ A=N-	74Cd	b

Metal Ion	ΔH, kJ /mol	Log K	ΔS, J /mol K	T,°C	M	Conditions	Ref.	Remarks
P317, continued								
NiA_2, cont.						Phenyl-β-mercaptocinnamamide		
	-34.3	0.26(20°,1-2)	-114.6	10-30	T	S:$CHCl_3$ A=N-(α-naphtyl)-β-mercaptocinnamamide	74Cd	b
	-29.3	0.03(20°,1-2)	-100.4	10-30	T	S:$CHCl_3$ A=N-(p-ethoxyphenyl)-β-mercaptocinnamamide	74Cd	b
	-20.1	-0.29(20°,1-2)	-74.5	10-30	T	S:$CHCl_3$ A=Monothiodibenzoylmethane	74Cd	b
	-20.5	-0.46(20°,1-2)	-79.5	10-30	T	S:$CHCl_3$ A=Thiobenzoylacetone	74Cd	b
	-19.2	-0.66(20°,1-2)	-78.2	10-30	T	S:$CHCl_3$ A=N-Ethyl-β-mercaptocinnamamide	74Cd	b
	-13.8	-0.93(20°,1-2)	-64.8	10-30	T	S:$CHCl_3$ A=N-methyl-N-phenyl-β-mercaptocinnamamide	74Cd	b
	-83.34	(1-2)	--	25	C	S:Cyclohexane A=2,2,6,6-Tetramethyl-3,5-heptanedione	73Ce	
	-70.3	1.89(25°,1-2)	-199.6	25	C	S:Nitrobenzene A=Diethyldithiophosphate K determined in benzene	70Cb	
	-85.1	3.33(1-2)	-222	25	C	S:Nitrobenzene K determined in benzene A=Ethylxanthato	70Cc	
	-25	--	-79	25-70	T	S:Pyridine A= 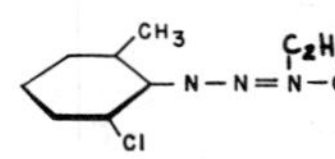 R:NiA_2(planar, S=0)+nL=NiA_2L_n (paramagnetic, S=1)	72Za	b,q:±4
	-46	(1-2)	-163	40-70	T	S:Pyridine A=N-Butyl-2-hydroxyacetophenimine	70Ka	b,q:±4
	-42	(1-2)	-155	40-70	T	S:Pyridine A=N-Butyl-2-hydroxybutyrophenimine	70Ka	b,q:±4
	-33	(1-2)	-138	40-70	T	S:Pyridine A=N-Butyl-2-hydroxypropiophenimine	70Ka	b,q:±4

Metal Ion	ΔH, kJ /mol	Log K	ΔS, J /mol K	T,°C	M	Conditions	Ref.	Re-marks
P317, continued								
NiA_2, cont.	-21	(1-2)	-113	40-70	T	S:Pyridine A= 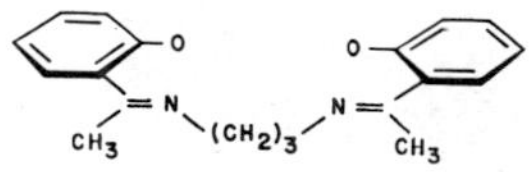	70Ka	b,q:±4
	-17	(1-2)	-105	40-70	T	S:Pyridine A= 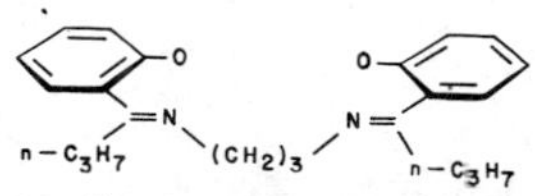	70Ka	b,q:
	-43.51	-1.553(25°,1-2)	-175.7	-10-20	T	S:25v% Pyridine 75v% CH_2Cl_2 A=N,N'-di-n-butyl-dithiocarbamate	72Vd	b
	-28.03	-1.060(25°,1-2)	-114.2	-10-20	T	S:25v% Pyridine 75v% CH_2Cl_2 A=N,N'-diethyl dithiocarbamate	72Vd	b
	-34.52	-1.268(25°,1-2)	-140.2	-10-20	T	S:25v% Pyridine 75v% CH_2Cl_2 A=N,N'-di-n-propyl dithiocarbamate	72Vd	b
	-21	--	-63	27-80	T	S:Pyridine A= 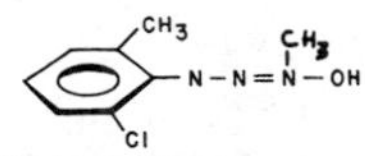 R:NiA_2(planar, S=0)+nL=NiA_2L_n (paramagnetic, S=1)	72Za	b,q:±4
NiL_2A_2	-99.6	4.49(1-2)	-254.0	20	C	μ→0 S:$CHCl_3$ A=I	64Ka	
Ni_2LA_4	-61.9	6.06	-86.2	30	C	S:Benzene A=3-Benzylpentane-2,4-dione R: $Ni_2A_4L+3L=2NiA_2L_2$	73Ag	
	-64.0	5.68	-102.1	30	C	S:Benzene A=3-Butylpentane-2,4-dione R: $Ni_2A_4L+3L=2NiA_2L_2$	73Ag	q: ±10.0
	-84	7.00	-143.9	30	C	S:Benzene A=2,6-Dimethyl-heptane-3,5-dione R:$Ni_2A_4L+3L=2NiA_2L_2$	74Gb	q: ±10.0
	-79.9	5.41	-159.8	30	C	S:Benzene A=Pentane-2,4-dione R:$Ni_2A_4L+3L=2NiA_2L_2$	73Ag	q: ±10.0
	-82.0	6.99	-136.0	30	C	S:Benzene A=1-Phenylbutane-1,3-dione R: $Ni_2A_4L+3L=2NiA_2L_2$	73Ag	q: ±10.0

Metal Ion	ΔH, kJ /mol	Log K	ΔS, J /mol K	T,°C	M	Conditions	Ref.	Remarks
P317, continued								
Ni_2LA_4, cont.	-79.9	7.10	-128.0	30	C	S:Benzene A=2,2,6,6-Tetramethylheptane-3,5-dione R: $Ni_2A_4L+3L=2NiA_2L_2$	73Ag	q: ±10.0
Ni_3A_6	-258.0	6.10	-732	3	C	S:Benzene A=3-Benzylpentane-2,4-dione R: $Ni_3A_6+3L=3Ni_2A_4L$	73Ag	q: ±30.1
	-318.0	5.58	-941	30	C	S:Benzene A=3-Butylpentane-2,4-dione R: $Ni_3A_6+3L=3Ni_2A_4L$	73Ag	q: ±30.1
	-125.9	11.99	185.8	30	C	S:Benzene A=Pentane-2,4-dione R: $Ni_3A_6+3L=3Ni_2A_4L$	73Ag	q: ±30.±
	-156.1	8.06	-359.8	30	C	S:Benzene A=1-Phenylbutane-1,3-dione R: $Ni_3A_6+3L=3Ni_2A_4L$	73Ag	q: ±30.1
	-36.0	4.79	-25.9	30	C	S:Benzene A=2,6-Dimethylheptane-3,5-dione R: $2/3Ni_3A_6+L=Ni_2A_4L$	74Gb	q: ±10.0
[Ni(II)]	-23.0	-1.68(25°,<u>1-2</u>)	-110	6-35	T	S:$CHCl_3$ [Ni(II)]= 2,4-Deuteroporphyrin IX dimethyl ester nickel(II)	68Md	b
	32.6	-0.24(25°,<u>1-2</u>)	-114	6-35	T	S:$CHCl_3$ [Ni(II)]= 2,4-Diacetyldeuteroporphyrin IX dimethyl ester nickel(II)	68Md	b
	-11.7	-2.04(25°,<u>1-2</u>)	-78.6	6-35	T	S:$CHCl_3$ [Ni(II)]= 2,4-Diethyldeuteroporphyrin IX dimethyl ester nickel(II)	68Md	b
	-18.0	-1.71(25°,<u>1-2</u>)	-88.3	6-35	T	S:$CHCl_3$ [Ni(II)]= 2,4-Di(2-ethoxycarbonylcyclopropyl)deuteroporphyrin IX dimethyl ester nickel(II)	68Md	b
	-32.6	0.15(25°,<u>1-2</u>)	-107	6-35	T	S:$CHCl_3$ [Ni(II)]= 2,4-Diformyldeuteroporphyrin IX dimethyl ester nickel(II)	68Md	b

Metal Ion	ΔH, kJ /mol	Log K	ΔS, J /mol K	T,°C	M	Conditions	Ref.	Re-marks
P317, continued								
[Ni(II)], cont.	-21.8	1.39(25°,1-2)	-99.6	6-35	T	S:$CHCl_3$ [Ni(II)]= 2,4-Divinyldeuteroporphyrin IX dimethyl ester nickel(II)	68Md	b
[Ni]	-26.4	3.15(25°,1)	-29	18-40	T	[Ni]=Nickel(II) mesotetraphenyl porphine S:Benzene	72Ce	b
	-17.6	3.00(25°,1)	0	18-40	T	[Ni]=Nickel(II) mesotetraphenyl porphine S:$CHCH_3$	72Ce	b,q:
RhAB	-57.7	3.04	-136	24	C	S:Benzene R:$[RhAB]_2$+2L= 2RhABL A=Cl B=1,5-Cyclooctadiene	76Lb	
	-55.2	3.61	-116	24	C	S:Triethyl phosphate R:$[RhAB]_2$+2L= 2RhABL A=Cl B=1,5-Cyclooctadiene	76Lb	
$RhAB_2$	-80.3	--	--	21-24	C	S:Dichloromethane A=Cl B=1,5-Cyclooctadiene R:$[RhAB]_2$+2L= 2RhABL	73Pe	
	-82.4	--	--	21-24	C	S:Dichloromethane A=Br B=1,5-Cyclooctadiene R: $[RhAB]_2$+2L=2RhABL	73Pe	
	-77.0	--	--	21-24	C	S:Dichloromethane A=I B=1,5-Cyclooctadiene R: $[RhAB]_2$+2L=2RhABL	73Pe	
Rh_2A_4	29.5	4.52(1)	-25.1	25	C	A=MeO_2Ac	77Da	
	-32.6	2.81(2)	-55.6	25	C	A=MeO_2Ac	77Da	
$Rh_2A_4B_2$	-52.3	--	--	24	C	S:Benzene R:$Rh_2A_4B_2$+2L= $2RhA_2BL$ A=CO B=Cl	76Pd	u:24°
PdAB	-43.1	3.362	-80	24	C	S:Benzene R:$[PdAB]_2$+2L= 2[PdABL] A=(π-MeAllyl) B=Cl	77Lb	
	-67.8	--	--	25	C	S:CH_2Cl_2 A=Cl B=cis-Butene R:$(PdA_2B)_2$+2L= $2PdA_2BL$	74Pd 74Pc	

Metal Ion	ΔH, kJ /mol	Log K	ΔS, J /mol K	T,°C	M	Conditions	Ref.	Re-marks
P317, continued								
PdAB, cont.	-67.8	--	--	25	C	S:CH_2Cl_2 A=Cl B=Cycloheptene R:$(PdA_2B)_2$+2L= 2PdA_2BL	74Pd 74Pc	
	-69.4	--	--	25	C	S:CH_2Cl_2 A=Cl B=cis-Cyclooctene R:$(PdA_2B)_2$+2L= 2PdA_2BL	74Pd 74Pc	
	-116.3	--	--	25	C	S:CH_2Cl_2 A=Cl B=Cyclohexene R:$(PdA_2B)_2$+2L= 2PdA_2BL	74Pd 74Pc	
	-90.0	--	--	25	C	S:CH_2Cl_2 A=Cl B=Cyclopentene R:$(PdA_2B)_2$+2L= 2PdA_2BL	74Pd 74Pc	
PdA_2B	-88.3	--	--	25	C	S:CH_2Cl_2 A=Cl B=Ethylene R:$(PdA_2B)_2$+2L= 2PdA_2BL	74Pd 74Pc	
	-95.0	--	--	25	C	S:CH_2Cl_2 A=Cl B=Styrene R:$(PdA_2B)_2$+2L= 2PdA_2BL	74Pd 74Pc	
PdA_2BL	-57.7	--	--	25	C	S:CH_2Cl_2 A=Cl B=Ethylene R:PdA_2BL+L= PdA_2L_2+B	74Pd	
	-51.9	--	--	25	C	S:CH_2Cl_2 A=Cl B=cis-Butene R:PdA_2BL+L= PdA_2L_2+B	74Pd	
	-59.8	--	--	25	C	S:CH_2Cl_2 A=Cl B=Cyclopentene R:PdA_2BL+L= PdA_2L_2+B	74Pd	
	-58.2	--	--	25	C	S:CH_2Cl_2 A=Cl B=Cyclohexene R:PdA_2BL+L= PdA_2L_2+B	74Pd	
	-53.1	--	--	25	C	S:CH_2Cl_2 A=Cl B=Cycloheptene R:PdA_2BL+L= PdA_2L_2+B	74Pd	
	-48.5	--	--	25	C	S:CH_2Cl_2 A=Cl B=cis-Cyclooctene R:PdA_2BL+L= PdA_2L_2=B	74Pd	
	-58.6	--	--	25	C	S:CH_2Cl_2 A=Cl B=Styrene R:PdA_2BL+L= PdA_2L_2+B	74Pd	

Metal Ion	ΔH, kJ /mol	Log K	ΔS, J /mol K	T,°C	M	Conditions	Ref.	Remarks
P317, continued								
PdA_2B_2	-108.8	--	--	21-24	C	S:Dichloromethane A=Cl B=Benzonitrile R:PdA_2B_2+2L=PdA_2L_2+2B	73Pe	
	-108.8	--	--	25	C	S:CH_2Cl_2 A=Cl B=C_6H_5CN R:PdA_2B_2+L=PdA_2L+2B CL=8.3x10^{-3}M CM=1x10^{-4}M	72Pe	
$IrABCD_2$	-31.0	(1)	--	25	C	S:CH_2Cl_2 A=Cl B=NO_2 C=CH_3 D=$P(C_6H_5)_3$	77La	
	-58.2	(1)	--	25	C	S:CH_2Cl_2 A=Cl B=CH_3CO_2 C=CH_3 C=$P(C_6H_5)_3$	77La	
	-76.6	(1)	--	25	C	S:CH_2Cl_2 A=Cl B=NCO C=CH_3 D=$P(C_6H_5)_3$	77La	
	-72.4	(1)	--	25	C	S:CH_2Cl_2 A=Cl B=Br C=CH_3 D=$P(C_6H_5)_3$	77La	
IrA_2BC_2	-61.9	(1)	--	25	C	S:CH_2Cl_2 A=Cl B=$S(O)_2CH_3$ C=$P(C_6H_5)_3$	77La	
	-68.2	(1)	--	25	C	S:CH_2Cl_2 A=Cl B=$C(O)CF_2H$ C=$P(C_6H_5)_3$	77La	
	-73.2	(1)	--	25	C	S:CH_2Cl_2 A=Cl B=CH_3 C=$P(C_6H_5)_3$	77La	
	-74.0	5.40(1)	-146	25	C	S:CH_2Cl_2 A=Cl B=$C(O)CF_3$ C=$P(C_6H_5)_3$	77La	
	-49.4	(1)	--	25	C	S:CH_2Cl_2 A=Cl B=$C(O)CFH_2$ C=$P(C_6H_5)_3$	77La	
	-51.5	(1)	--	25	C	S:CH_2Cl_2 A=Cl B=$S(O)_2CF_3$ C=$P(C_6H_5)_3$	77La	
	-57.7	(1)	--	25	C	S:CH_2Cl_2 A=Cl B=$S(O)_2C_2H_5$ C=$P(C_6H_5)_3$	77La	
	-71.5	5.41(1)	-136	25	C	S:CH_2Cl_2 A=Cl B=$C(O)CF_3$ C=$P(C_6H_5)_3$	77La	
	-79.5	5.76(1)	-156	25	C	S:1,2-$C_2H_4Cl_2$ A=Cl B=$C(O)CF_3$ C=$P(C_6H_5)_3$	77La	
	-74.0	5.46(1)	-144	25	C	S:CH_2Cl_2 A=Cl B=$C(O)CF_3$ C=$P(C_6H_5)_3$	77La	
	-66.9	(1)	--	25	C	S:C_4H_8O	77La	

Metal Ion	ΔH, kJ /mol	Log K	ΔS, J /mol K	T,°C	M	Conditions	Ref.	Re-marks
P317, continued								
IrA_2BC_2, cont.						A=Cl B=C(O)CF_3 C=P$(C_6H_5)_3$		
	-73.6	5.70(1)	-138	25	C	S:C_6H_6 A=Cl B=C(O)CF_3 C=P$(C_6H_5)_3$	77La	
	-68.2	5.75(1)	-119	25	C	S:C_6H_5Cl A=Cl B=C(O)CF_3 C=P$(C_6H_5)_3$	77La	
$PtAB_2^+$	51.0	(1)	--	25	C	S:THF A=CH_3 B=[P$(CH_3)_2(C_6H_5)$]	75Ma	
PtA_2B	-150	--	--	--	C	S:Benzene R:$[PtA_2B]_2$+2L= 2PtA_2BL A=Cl B=Ethylene	76Pa	u:25°,y
	-130	--	--	--	C	S:Benzene R:$[PtA_2B]_2$+2L= 2PtA_2BL A=Cl B=Cyclopentane	76Pa	u:25°,y
	-134	--	--	--	C	S:Benzene R:$[PtA_2B]_2$+2L= 2PtA_2BL A=Cl B=Styrene	76Pa	u:25°,y
	-150	--	--	--	C	S:Benzene R:$[PtA_2B]_2$+2L= 2PtA_2BL A=Cl B=Ethylene	76Pa	u:25°
	-134	--	--	--	C	S:Benzene R:$[PtA_2B]_2$+2L= 2PtA_2BL A=Cl B=Styrene	76Pa	u:25°
	-130	--	--	--	C	S:Benzene R:$[PtA_2B]_2$+2L= 2PtA_2BL A=Cl B=Cyclopentene	76Pa	u:25°
	-130	--	--	--	C	S:Dichloromethane R:$[PtA_2B]_2$+2L= 2PtA_2BL A=Cl B=Cyclopentene	77Pa	u:25°
	-133	--	--	--	C	S:Dichloromethane R:$[PtA_2B]_2$+2L= 2PtA_2BL A=Cl B=cis-Butene	76Pa	u:25°
	-131	--	--	--	C	S:Dichloromethane R:$[PtA_2B]_2$+2L= 2PtA_2BL A=Cl B=Cyclohexene	77Pa	u:25°
	-132	--	--	--	C	S:Dichloromethane R:$[PtA_2B]_2$+2L= 2PtA_2BL A=Cl B=cis-Cyclooctene	76Pa	u:25°
	-130	--	--	--	C	S:Dichloromethane R:$[PtA_2B]_2$+2L= 2PtA_2BL A=Cl	77Pa	u:25°

Metal Ion	ΔH, kJ /mol	Log K	ΔS, J /mol K	T,°C	M	Conditions	Ref.	Re- marks
P317, continued								
PtA_2B, cont.						B=Cycloheptene		
	-151	--	--	--	C	S:Dichloromethane R:$[PtA_2B]_2$+2L= $2PtA_2BL$ A=Cl B=Ethylene	76Pa	u:25°
	-135	--	--	--	C	S:Dichloromethane R:$[PtA_2B]_2$+2L= $2PtA_2BL$ A=Cl B=Styrene	76Pa	u:25°
	-131	--	--	--	C	S:Dichloromethane R:$[PtA_2B]_2$+2L= $2PtA_2BL$ A=Cl B=p-Nitrostyrene	76Pa	u:25°
	-133	--	--	--	C	S:CH_2Cl_2 R:$[PtA_2B]_2$+2L= $2PtA_2BL$ A=Cl B=cis-Butene	76Pa 74Pc	u:25°,y
	-151	--	--	25	C	S:CH_2Cl_2 R:$[PtA_2B]_2$+2L= $2PtA_2BL$ A=Cl B=Ethylene	76Pa 74Pc	u:25°,y
	-130	--	--	25	C	S:CH_2Cl_2 R:$[PtA_2B]_2$+2L= $2PtA_2BL$ A=Cl B=Cycloheptene	76Pa 74Pc	u:25°,y
	-130	--	--	25	C	S:CH_2Cl_2 R:$[PtA_2B]_2$+2L= $2PtA_2BL$ A=Cl B=Cyclopentene	76Pa 74Pc	u:25°,y
	-132	--	--	25	C	S:CH_2Cl_2 R:$[PtA_2B]_2$+2L= $2PtA_2BL$ A=Cl B=cis-Cyclooctene	76Pa 74Pc	u:25°,y
	-131	--	--	25	C	S:CH_2Cl_2 R:$[PtA_2B]_2$+2L= $2PtA_2BL$ A=Cl B=Cyclohexene	76Pa 74Pc	u:25°,y
	-131	--	--	25	C	S:CH_2Cl_2 R:$[PtA_2B]_2$+2L= $2PtA_2BL$ A=Cl B=p-Nitrostyrene	76Pa	u:25°,y
	-135	--	--	25	C	S:CH_2Cl_2 R:$[PtA_2B]_2$+2L= $2PtA_2BL$ A=Cl B=Styrene	76Pa 74Pc	u:25°,y
Cu^{2+}	-16.82	2.50(1)	-8.4	25	C	μ=0	68Ib	
	-18.32	(1)	--	--	-	CL=0.2M CM=$3x10^{-2}$M $CHNO_3$=0.01M	68Wb	f:2.98 u,y
	-12.34	2.46(1)	5.4	25	C	1M($NaClO_4$)	63Ae	f:4
	-18.4	2.58(1)	-11.3	25	C	1.0M(KNO_3)	60La	f:4
	-12.34	1.95(2)	-4.2	25	C	1M($NaClO_4$)	63Ae	f:4

Metal Ion	ΔH, kJ /mol	Log K	ΔS J /mol K	T,^{o}C	M	Conditions	Ref.	Re-marks
P317, continued								
Cu^{2+}, cont.	-18.4	1.85(2)	-25.1	25	C	1.0M(KNO_3)	60La	f:4
	-12.34	1.27(3)	-17.2	25	C	1M($NaClO_4$)	63Ae	f:4
	-18.4	1.41(3)	33.5	25	C	1.0M(KNO_3)	60La	f:4
	-12.34	0.84(4)	-25.1	25	C	1M($NaClO_4$)	63Ae	f:4
	-18.4	0.82(4)	-44.4	25	C	1.0M(KNO_3)	60La	f:4
	>4.6	0.31(5)	8	25	C	1.0M(KNO_3)	66Cb	
	-37.07	4.30(1-2)	-41.8	25	C	μ=0	68Ib	
	-67.4	5.16(1-3)	-128.0	25	C	μ=0	68Ib	
	-90.0	6.04(1-4)	-172	25	C	μ=0	68Ib	
CuA_2	-27.00	0.672(1)	-76.1	30	C	S:Benzene A=Acetylacetone	71Gh	q: ±2.7
	-27.99	0.977(1)	-72.8	30	C	S:Benzene A=Benzoylacetone	71Gh	q: ±2.8
	-26.8	1.312(1)	-64.8	25	C	S:Benzene A=t-Butylacetoacetate	69La	
	-20.1	-0.420(27°,1)	-67	7-47	T	S:Benzene A=Diethyldithio-carbamate	72Fc	b,q: ±2.1
	-12.30	0.298(1)	-35.1	30	C	S:Benzene A=3-Ethylacetyl-acetone	71Gh	q: ±1.2
	-13.30	0.302(1)	-38.1	30	C	S:Benzene A=3-Methylacetyl-acetone	71Gh	q: ±1.3
	-26.40	1.176(1)	-64.8	30	C	S:Benzene A=N-Nitroso-N-Phenyl-hydroxylamine	71Gi	q: ±2.6
	-25.31	0.89(1)	-67.8	25	C	$\mu\cong$0.005 S:Benzene A=2,4-Pentanedionate	63Ma 69Ga	
	-13.85	(1)	--	25	T	$\mu\cong$0.005 S:Benzene A=2,4-Pentanedionate	63Ma	
	-13.18	0.655(1)	-31.0	30	C	S:Benzene A=3-Phenylacetyl-acetone	71Gh	q: ±1.3
	-20.00	0.832(1)	-50.2	30	C	S:Benzene A=3-o-Tolyl-acetylacetone	71Gh	q: ±2.0
	-23.51	0.488(1)	-67.8	30	C	S:Benzene A=3-p-Tolyl-acetylacetone	71Gh	q: ±2.4
	-30.88	2.846(1)	-48.1	30	C	S:Benzene A=1,1,1-Trifluoro-pentane-2,4-dione	74Gc	
	-30.00	2.736(1)	-46.0	30	C	S:Benzene A=1,1,1-Trifluoro-4-phenylbutane-2,4-dione	74Gc	
	-32.80	3.037(1)	-50.2	30	C	S:Benzene A=1,1,1-Trifluoro-4-(2-thienyl) butane-2,4-dione	74Gc	

Metal Ion	ΔH, kJ /mol	Log K	ΔS, J /mol K	T,°C	M	Conditions	Ref.	Re-marks
P317, continued								
CuA_2, cont.	-37.99	(1)	--	30	C	S:Benzene A=1,1,1,5,5,5-Hexafluoropentane-2,4-dione	74Gc	
	-25.1	1.276(1)	-59.8	25	C	S:CCl_4 A=t-Butylacetoacetate Slight ppt	69La	
	-25.5	1.640(1)	-54.4	25	C	S:CCl_4 A=Ethyl-acetoacetate	70Pa	q:±2.1
	-49.4	(1)	--	25	C	S:CCl_4 A=Hexa-fluoroacetyl-acetone	70Pa	
	-19.2	-0.620(25°,1)	-75	-3-57	T	S:Chloroform A=Diethyldithio-carbamate	74Hd	b
	-29.7	1.576(1)	69.0	25	C	S:Cyclohexane A=t-Butylaceto-acetate	69La	
	-56.1	(1)	--	25	C	S:Cyclohexane A=Hexafluoro-acetylacetone	70Pa	
	-21	-0.398(25°,1)	-79	10-30	T	S:Methylcyclo-hexane-Pyridine A=(di-n-Butyldi-thiocarbamate)	71Cf	b,q:±8
	-44.4	(1)	--	25	C	S:Methylene chloride A=Hexafluoro-acetylacetonate	74Dc	
	-20.1	-0.444(25°,1)	-75	-3-57	T	S:Toluene A=Diethyldithio-carbamate	74Hd	b
	-20.1	0.681(1)	-54.4	25	C	S:$CHCl_3$ A=t-Butylacetoacetate	69La	
	-39.5	2.892(2)	-74.9	30	C	S:Benzene A=1,1,1,5,5,5-Hexafluoropentane-2,4-dione	74Gc	
	-26	3.12(25°,1)	-28	15-35	T	S:Benzene A=2-Thenoyltri-fluoroacetonato	76Ag	b,q:±8
	-30.9	3.00(1)	-48	20	T	S:Benzene A=Trifluoro-acetylacetone	79Si	b
	-17.0	0.114(1)	-54	30	C	S:Benzene CM=4-28x10^{-3}M A=	76Ai	
	-18.9	0.301(1)	-57	30	C	S:Benzene CM=6-18x10^{-3}M A=	76Ai	

Metal Ion	ΔH, kJ /mol	Log K	ΔS, J /mol K	T,°C	M	Conditions	Ref.	Remarks
P317, continued								
CuA_2, cont.	-17.7	0.0414(1)	-58	30	C	S:Benzene CM=9-19x10^{-3}M A=	76Ai	q:±1.8
	-8.6	-0.0458(1)	-29	30	C	S:Benzene CM=14-39x10^{-3}M A=	76Ai	
	-19.0	0.623(1)	-51	30	C	S:Benzene CM=9-23x10^{-3}M A=	76Ai	
	-23.3	0.204(1)	-72	30	C	S:Benzene CM=4-12x10^{-3}M A	76Ai	
	-15.2	0.204(1)	-46	30	C	S:Benzene CM=5-25x10^{-3}M A	76Ai	q:±1.5
	-50.6	2.756(2)	-117.2	25	C	S:CCl_4 A=Hexafluoro-acetylacetone	70Pa	
	-49.4	(1)	--	--	C	S:CCl_4 A=Hexafluoro acetylacetonato	76Mc	u:25°
	-21	2.33(25°,1)	-27	5-25	T	S:Chloroform A=2-Thenoyltri-fluoroacetonato	76Ag	b,q:±8
	-33.5	2.86(20°,1)	-59.4	-60-60	T	S:Chloroform A=Thenoyltri-fluoroacetonato	78Mb	
	-22.6	1.16(0°,1)	-60.7	-60-60	T	S:40v%Chloroform 60v%Toluene A=Acetylacetonato	78Mb	
	-25.5	0.991(40°,1)	-63.2	-60-60	T	S:40v%Chloroform 60v%Toluene A=Cupferonato	78Mb	
	-22.2	0.845(40°,1)	-54.0	-60-60	T	S:40v%Chloroform 60v%Toluene A=Dibenzoly-methanato	78Mb	
	-19.2	-0.569(30°,1)	-74.0	-60-60	T	S:40v%Chloroform 60v%Toluene A=Diethyl dithiocarbamato	78Mb	
	-21.3	0.716(40°,1)	-55.2	-60-60	T	S:40v%Chloroform 60v%Toluene A=8-Hydroxy-	78Mb	

Metal Ion	ΔH, kJ /mol	Log K	ΔS, J /mol K	T,°C	M	Conditions	Ref.	Remarks
P317, continued								
CuA_2,						aquinolinato		
cont.	-20.9	-0.356(30°,1)	-76.6	-60-60	T	S:40v%Chloroform	78Mb	
						60v%Toluene		
						A=Piperidyl		
						dithiocarbamato		
	-31.8	2.95(20°,1)	-52.7	-60-60	T	S:40v%Chloroform	78Mb	
						60v%Toluene		
						A=Thenoyltri-		
						fluoroacetonato		
	-12.6	0.954(0°,1)	-24.3	-60-60	T	S:Dimethyl	78Mb	
						sulfoxide		
						A=Cupferonate		
	-20.5	0.38(20°,1)	-60.7	5-40	T	S:Ethylbenzene	73Vd	
						A=Thiolopico-		
						linanilide		
	-52.7	4.53(1)	--	--	C	S:CH_2Cl_2	75Mc	u:25°
						A=Hexafluoro-		
						acetylacetonato		
	-57.8	(1)	--	--	C	S:C_6H_{12}	75Mc	u:25°
						A=Hexafluoro-		
						acetylacetonato		
	-33.5	3.06(20°,1)	-54.4	-60-60	T	S:Toluene	73Mb	
						A=Thenoyltri-		
						fluoroacetonato		
	-26	3.06(25°,1)	-28	15-35	T	S:Toluene	75Ag	b,q:±8
						A=2-Thenoyltri-		
						fluoroacetonato		
Ag^+	-19.2	2.04(1)	-25.9	25	C	μ=0	63Ib	
	-19.96	2.24(1)	-23.8	25	C	μ=0	63Ba	
	-20.21	2.00(1)	-29.3	25	C	0.5M(KNO_3)	66Pc	
	-20.21	2.00(1)	-29.3	25	C	μ=0.5M(KNO_3)	66Pb	
	-19.79	2.063(1)	-26.86	25	C	μ=0.5M(KNO_3)	72Ed	
							74Be	
	-18.4	2.08(1)	-22.6	25	C	μ=0.5M(KNO_3)	72Bh	
						Thermodynamic		
						quantities were		
						evaluated by two		
						different mathe-		
						matical treatments		
						of the data		
	-28.28	1.95(2)	-57.7	25	C	μ=0	63Ba	
	-27.24	2.11(2)	-51.0	25	C	0.5M(KNO_3)	66Pc	
	-27.24	2.11(2)	-51.0	25	C	μ=0.5M(KNO_3)	66Pb	
	-18.03	2.04(2)	-21.3	25	C	μ=0.5M(KNO_3)	72Bh	
						Thermodynamic		
						quantities were		
						evaluated by two		
						different mathe-		
						matical treatments		
						of the data		
	-21.34	2.123(2)	-49.87	25	C	μ=0.5M(KNO_3)	72Ed	
	-26.99	2.123(2)	-49.9	25	C	μ=0.5M(KNO_3)	74Be	
	-26.4	2.05(2)	-48.5	25	C	μ=0.5M(KNO_3)	72Bh	

Metal Ion	ΔH, kJ /mol	Log K	ΔS, J /mol K	T,°C	M	Conditions	Ref.	Remarks
P317, continued								
Ag^+, cont.						Thermodynamic quantities were evaluated by two different mathematical treatments of the data		
	-27.53	2.08(2)	-52.3	25	C	μ=0.5M(KNO_3) Thermodynamic quantities were evaluated by two different mathematical treatments of the data	72Bh	
	-47.07	4.09(1-2)	-79.5	25	C	μ=0	68Ib	
	-48.24	4.19(1-2)	-81.6	25	C	μ=0	63Ba	
	-47.45	4.11(1-2)	-80.3	25	C	0.5M(KNO_3)	66Pc	
Zn^{2+}	-10.79	2.08(1)	3.8	25	C	1M($NaClO_4$)	63Ae	f:4
	-9.58	1.05(25°,1)	-11.7	25-40	T	--	79Da	b
	-10.79	1.69(2)	-3.8	25	C	1M($NaClO_4$)	63Ae	f:4
	-11.67	0.36(25°,2)	-32.6	25-40	T	--	79Da	b
	-10.79	1.03(3)	-16.7	25	C	1M($NaClO_4$)	63Ae	f:4
	-10.79	0.64(4)	-23.8	25	C	1M($NaClO_4$)	63Ae	f:4
ZnA	-38.5	3.780(25°,1)	-58.6	15-35	T	S:Benzene A=$\alpha,\beta,\gamma,\delta$-Tetraphenylporphine	73Ve	b,i: 52Ma
	-41.8	4.400(25°,1)	-56.5	15-35	T	S:Cyclohexane A=$\alpha,\beta,\gamma,\delta$-Tetraphenylporphine	77Vb	
ZnA_2	-26	(1)	--	30	C	S:Acetonitrile A=Cl	76Ak	
	-26	(1)	--	30	C	S:Acetonitrile A=Br	76Ak	
	-28	(1)	--	30	C	S:Acetonitrile A=I	76Ak	
	-32.6	1.626(1)	-76	30	C	S:Benzene CM=3.6-65x10^{-3}M A=[$(Bu)_2NCS_2$]	76Aj	
	-33.7	1.786(1)	-77	30	C	S:Benzene CM=0.9-14.6x10^{-3}M A[$(Et)_2NCS_2$]	76Aj	q:±2.7
	-31.4	2.193(1)	-62	30	C	S:Benzene CM=1-5.3x10^{-3}M A=[$(PhCH_2)_2NCS_2$]	76Aj	
	-33.81	2.550(1)	-62.8	30	C	S:Benzene A=Acetylacetone CM=(9.9-18.3) x10^{-3}M	73Db	q:±3.3
	-31.88	3.015(1)	-48.1	30	C	S:Benzene A=Benzoylacetone CM=(2.9-14.3) x10^{-3}M	73Db	q:±3.3
	-37.78	3.291(1)	-54.8	30	C	S:Benzene	73Db	q:±3.8

Metal Ion	ΔH, kJ /mol	Log K	ΔS, J /mol K	T,°C	M	Conditions	Ref.	Re-marks
P317, continued								
ZnA_2, cont.						A=Dibenzoyl-methane CM=(2.8-12.3)x10^{-3}M		
	-42.2	3.95(1)	-63	30	C	S:Benzene A=Di-isobutyldithio-phosphato	71Db	
	-42.7	3.53(1)	-71	30	C	S:Benzene A=Diisopropyldi-thiophosphato	71Db	
	-15.40	2.705(1)	1.00	30	C	S:Benzene A=2,2,6,6-Tetra-methylheptane-3,5-dione CM=(10.7-18.9)x10^{-3}M	73Db	a:±3.6
	-55.6	3.00(1)	-129	25	C	S:Cyclohexane A=N[Si$(CH_3)_3$]$_2$	75Fc	
	-23	(2)	--	30	C	S:Acetonitrile A=Cl	76Ak	
	-22	(2)	--	30	C	S:Acetonitrile A=Br	76Ak	
	-24	(2)	--	30	C	S:Acetonitrile A=I	76Ak	q:±2
	-13	(3)	--	30	C	S:Acetonitrile A=Cl	76Ak	
	-12	(3)	--	30	C	S:Acetonitrile A=Br	76Ak	
	-13	(3)	--	30	C	S:Acetonitrile A=I	76Ak	q:±3
	-21.8	0.34(2)	-66.9	30	C	S:Benzene A=Diisobutyldi-thiophosphato	71Db	q:±2.9
	-25.1	0.83(2)	-66.1	30	C	S:Benzene A=Diisopropyldi-thiophosphato	71Db	q:±2.9
	-31.8	1.906(25°,1)	-71	18.5-65	T	S:Toluene A=[$S_2CN(CH_3)_2$]	65Ch	b
[Zn]	-36.8	3.59(25°,1)	-54	18-40	T	S:Benzene [Zn]= Zinc(II) Mesotet-raphenylporphine	72Ce	b,q: ±5.0
	-37.66	3.722(30°,1)	-54	20-40	T	S:Benzene [Zn]= Zinc Tetraphenyl-chlorin	52Ma	b
	-38.49	3.570(30°,1)	-58	20-40	T	S:Benzene [Zn]= Zinc Tetraphenyl-porphine	52Ma	b
	-16.7	2.78(25°,1)	0	18-40	T	S:Chloroform [Zn]= Zinc(II) Mesotetraphenyl-porphine	72Ce	b,q: ±1.7
Cd^{2+}	-19.16	1.330(25°,1)	-38.70	18-35	T	μ=0.1	65Cc	b
	-51.30	1.813(25°,1-2)	-137.40	18-35	T	μ=0.1	65Cc	b
	-52.42	1.886(25°,1-3)	-139.79	18-35	T	μ=0.1	65Cc	b
	-49.79	1.477(25°,1-4)	-138.78	18-35	T	μ=0.1	65Cc	b

Metal Ion	ΔH, kJ /mol	Log K	ΔS, J /mol K	T,°C	M	Conditions	Ref.	Remarks
P317, continued								
CdA_2	-21	2.903(1)	-14	30	C	S:Acetonitrile A=I	76Ak	q:±5
	-40	2.845(1)	-78	30	C	S:Benzene CM=1.0-6.4x10^{-3}M A=[$(PhCH_2)_2$)NCS_2]	76Aj	
	-51	2.398(1)	-122	30	C	S:Benzene CM=0.9-4.1x10^{-3}M A=[$(Et)_2NCS_2$]	76Aj	
	-31	1.875(1)	-65	30	C	S:Benzene CM=6.2-9.5x10^{-3}M A=[$(Bu)_2NCS_2$]	76Aj	
	-45.6	2.34(1)	-108	25	C	S:Cyclohexane A=$N[Si(CH_3)_3]_2$	75Fc	
	-24	2.301(2)	-35	30	C	S:Acetonitrile A=I	76Ak	q:±5
	-40.2	0.58(2)	-123.0	30	C	S:Benzene A=Dicyclohexyl-dithiophosphato	71Db	q:±5.0
	-45.2	0.41(2)	-141.0	30	C	S:Benzene A=Diisobutyldi-thiophosphato	71Db	q:±4
Cd_2A_4	-26.99	2.46(1-2)	-41.8	30	C	S:Benzene A=Dicyclohexyl-dithiophosphato	71Db	q:±2.9
	-25.82	2.70(1-2)	-33.0	30	C	S:Benzene A=Diisobutyl-dithiophosphato	71Db	q:±2.9
[Cd]	-36.40	3.430(30°,1)	-54	20-40	T	[Cd]=Cadmium Tetraphenyl-porphine S:Benzene	52Ma	b
Hg^{2+}	-23	6.83(1)	54	25	C	μ=0.5M($NaClO_4$)	75Mb	q:±4
	-36	5.1(1)	-24	25	C	μ=0.5M(ClO_4^- or NO_3^-)	78Mc	q:±5
	-28	3.17(2)	-34	25	C	μ=0.5M($NaClO_4$)	75Mb	q:±10
	-33	4.9(2)	-17	25	C	μ=0.5M(ClO_4^- or NO_3^-)	78Mc	q:±34
	-29	0.38(3)	-90	25	C	μ=0.5M($NaClO_4$)	75Mb	q:±49
	-37	10.00(1-2)	67	25	C	μ=0.5M($NaClO_4$)	77Mc	q:±15
HgA_2	-51	1.740(1)	-132	30	C	S:Acetonitrile A=I	76Ak	
	-11.8	0.32(1)	-33	30	C	CM=1-20x10^{-3} S:Benzene A=C_6F_5	74Fc	q:±1.8
	-63.6	2.336(1)	-166.1	30	C	S:Benzene A=Cl CM=1-5x10^{-3}M	73Fa	
	-67.8	2.354(1)	-179.9	30	C	S:Benzene A=Br CM=1-5x10^{-3}M	73Fa	
	-64.0	2.120(1)	-171.0	30	C	S:Benzene A=I CM=1-5x10^{-3}M	73Fa	
	-18.07	0.322(1)	-54.4	25	C	S:C_6H_6 A=C_6F_5	73Pm	

Metal Ion	ΔH, kJ /mol	Log K	ΔS, J /mol K	T,°C	M	Conditions	Ref.	Re-marks
P317, continued								
HgA_2, cont.	-22.59	1.303(1)	-47.7	25	C	S:CCl_4 A=C_6F_5	73Pm	
	≅-12.1	≅-0.52(1)	≅-50	25	C	S:CCl_4 A=C_6H_5	73Pm	
[Hg]	-20.5	1.083(30°,1)	-46	20-40	T	[Hg]=Mercury Tetraphenylporphine S:Benzene	52Ma	b
BA_3	-105	(1)	--	25	C	S:Nitrobenzene A=F	55Bb	
	-64.0	(1)	--	25	C	S:Nitrobenzene A=CH_3	56Ba	
BA_3C	-54.0	(1)	--	25	C	S:Benzene A=F C=$(CH_3CH_2)_2O$	68Rd	
	-51.0	(1)	--	25	C	S:Nitrobenzene A=F C=Tetrahydropyran	55Bb	
	-59.0	(1)	--	25	C	S:Nitrobenzene A=F C=$(C_2H_5)_2O$	55Bb	
	-66.5	(1)	--	25	C	S:Nitrobenzene A=F C=n-Butyl ether	55Bb	
	-104.6	--	--	25	C	S:Nitrobenzene R:$BA_3C+L=BA_3L+C$ A=F C=$C_5H_5NO_2$	56Bc	
	-128.9	--	--	25	C	S:Nitrobenzene R:$BA_3C+L=BA_3L+C$ A=Cl C=$C_5H_5NO_2$	56Bc	
	-133.9	--	--	25	C	S:Nitrobenzene R:$BA_3C+L=BA_3L+C$ A=Br C=$C_5H_5NO_2$	56Bc	
AlA_3	-115.31	(1)	--	25-28	C	S:n-Hexane A=CH_3	68Hc	
Al_2A_6	-131.4	--	--	25	C	S:Benzene CM=0.07M A=Br R:$0.5(Al_2A_6)+L=AlA_3L$	68Rc	
$SiAB_3$	-22.76	1.90(30°,1)	-38.87	10-45	T	S:CCl_4 XL=$1x10^{-2}$ XM=$5x10^{-4}$ A=OH B=CH_3	69Lb	b
	-23.64	1.76(30°,1)	-44.31	20-60	T	S:CCl_4 XL=$1x10^{-2}$ CM=$5x10^{-4}$ A=OH B=C_2H_5	69Lb	b
	-25.02	2.28(30°,1)	-38.95	0-50	T	S:CCl_4 XL=$1x10^{-2}$ XM=$4x10^{-4}$ A=OH B=Phenyl	69Lb	b
$SnAB_3$	-139	>6(1-2)	<-340	30	C	S:Benzene A=Phenyl B=Cl	75Fa	
	-116	5.60(1-2)	∿-280	30	C	S:Benzene A=Butyl B=Cl	75Fa	
SnA_2B_2	-21.6	0.491(1)	-62	25	C	S:Benzene A=Bu B=(OCOMe)	77Ge	
	-21.2	0.415(1)	-62	25	C	S:Benzene A=Bu B=(OCOEt)	77Ge	

Metal Ion	ΔH, kJ /mol	Log K	ΔS, J /mol K	T,°C	M	Conditions	Ref.	Remarks
P317, continued								
SnA_2B_2, cont.	-25.4	1.539(1)	-57	25	C	S:Benzene A=Bu B=(NCO)	77Ge	
	-45	~4(1)	-73	30	C	S:Benzene A=Butyl B=NCS	75Fa	q:±5
	-40	0.78(1)	-117	30	C	S:Benzene A=Butyl B=Cl	75Fa	q:±10
	-45	1.23(1)	-125	30	C	S:Benzene A=Methyl B=Cl	75Fa	q:±10
	-27.6	1.286(1)	-67	25	C	S:Benzene A=Ph B=(OCOMe)	77Ge	
	-48	1.90(1)	-122	30	C	S:Benzene A=Phenyl B=Cl	75Fa	q:±10
	-35	2.00(2)	-77	30	C	S:Benzene A=Butyl B=NCS	75Fa	q:±5
	-22	0.70(2)	-53	30	C	S:Benzene A=Butyl B=Cl	75Fa	q:±10
	-28	1.11(2)	-71	30	C	S:Benzene A=Methyl B=Cl	75Fa	q:±10
	-57	1.78(2)	-155	30	C	S:Benzene A=Phenyl B=Cl	75Fa	q:±10
SnA_3B	-41.0	0.053(1)	-134	30	C	S:Benzene A=Phenyl B=Cl	75Fa	
	-27.2	0.267(1)	-86.2	26	C	S:CCl_4 A=CH_3 B=Cl	66Be	
PA_5	-92.0	(1)	--	30	C	S:Nitrobenzene A=Cl	63Hb	
PA_4B	-113.0	(1)	--	30	C	S:Nitrobenzene A=Cl B=F	63Hb	
PA_3B_2	-147.7	(1)	--	30	C	S:Nitrobenzene A=Cl B=F	63Hb	
SbA_5	-142.2	(1)	--	25	C	S:$C_2H_4Cl_2$ A=Cl	72Lb	
	-118.4	(1)	--	30	C	S:Nitrobenzene A=Cl	63Hb	
Br_2	-43.1	2.436(1)	-98.3	25	C	S:1,2-Dichloro-ethane	73Tb	
I_2	-25.1	2.346(1)	-39.3	25	C	S:Benzene	66Ad	
	-33.5	1.99(1)	-75.7	22	C	low C S:CCl_4	64Sc	b
	-33.0	2.017(1)	-72.4	25	C	S:CCl_4	73Bj	
	-32.72	2.033(1)	-71.4	25	C	S:CCl_4	79Ba	
	-31.25	2.02(25°,1)	-66.1	22-43	T	S:CCl_4	62Ka	
	-31.25	2.021(25°,1)	-66.1	--	T	S:CCl_4 CL<3x 10^{-2}M CM<10^{-4}M	69Ma	u
	-32.72	1.890(25°,1)	-73.2	--	T	S:Chloroform CL<3x10^{-2}M CM<10^{-4}M	69Ma	u
	-35.1	2.107(1)	-77.0	25	C	S:Cyclohexane	73Bj	
	-32.6	(1)	--	25	C	S:1,2-Dichloro-ethane	73Tb	
	-32.51	2.276(25°,1)	-65.3	--	T	S:1,2-Dichloro-ethane CL<3x	69Ma	u

Metal Ion	ΔH, kJ /mol	Log K	ΔS, J /mol K	T,°C	M	Conditions	Ref.	Re-marks
P317, continued								
I_2, cont.	-35.94	2.179(25°,1)	-79.1	--	T	10^{-2}M CM<10^{-4}M S:Dichloromethane	69Ma	u
	-32.6	2.5(1)	-64.8	17	T	CL<3x10^{-2}M CM<10^{-4}M S:Heptane	54Ra	
	-32.6	2.46(1)	-63.6	22	C	low C S:n-Heptane	64Sc	
	-34.14	2.196(25°,1)	-72.4	--	T	S:n-Heptane CL< 3x10^{-2}M CM<10^{-4}M	69Ma	u
	-31.2	2.22(25°,1)	-62.3	17-35	T	S:Heptane	68Kb	b
	-33.0	(1)	--	--	C	S:Pyridine	61Aa	i:50Ha
	-33.26	(1)	--	--	C	S:Pyridine	50Ha	
P318	PYRIDINE, 3-acetyl-		C_7H_7ON			$C_5H_4N(OCCH_3)$		L
CuA_2	-7.36	0.99(1)	-5.77	25	C	μ≅0.005 S:Benzene A= 2,4-Pentanedione	63Ma	
	-7.24	(1)	--	25	T	μ≅0.005 S:Benzene A= 2,4-Pentanedione	63Ma	
P319	PYRIDINE, 4-acetyl-		C_7H_7ON			$C_5H_4N(OCCH_3)$		L
NiA_2	-78.62	(1-2)	--	25	C	S:Benzene A=2,2,-6,6-Tetramethyl-3,5-heptanedione	73Ce	
P320	PYRIDINE, 2-aldehyde-2'-pyridylhydrazone-		$C_{11}H_{10}N_4$			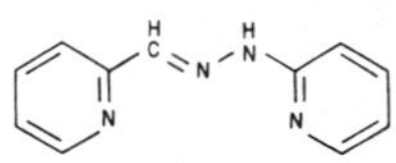		L
Fe^{2+}	-105.8	16.57(25°,1-2)	-38	5-60	T	μ=0	68Gf	b
Zn^{2+}	-46.0	5.82(25°,1)	-42	5-60	T	μ=0	68Gf	b
	-58.2	11.08(25°,2)	17	5-60	T	μ=0	68Gf	b
P321	PYRIDINE, 2-aldoxime-		$C_6H_6N_2$			$C_5H_4N(CH:NOH)$		L
Fe^{2+}	-84	24.85(1-3)	197	25	T	μ=0.045	62Hb	
P322	PYRIDINE, 2-amino-		$C_5H_6N_2$			$C_5H_4N(NH_2)$		L
Be^{2+}	-28.53	4.31(40°,1)	18.0	35&45	T	--	67Rb	a,b,p
Mn^{2+}	-38.1	2.19(25°,1)	-85.4	25-40	T	μ=0.1M(KNO_3& HNO_3)	76Bb	
	-6.7	3.24(25°,2)	39.7	25-40	T	μ=0.1M(KNO_3& HNO_3)	76Bb	
Co^{2+}	-5.0	3.06(25°,1)	41.8	25-40	T	μ=0.1M(KNO_3& HNO_3)	76Bb	
	-15.5	1.81(25°,2)	-16.7	25-40	T	μ=0.1M(KNO_3& HNO_3)	76Bb	

Metal Ion	ΔH, kJ /mol	Log K	ΔS, J /mol K	T,°C	M	Conditions	Ref.	Remarks
P322, continued								
CoA_2	-35.1	7.53(1-2)	25	25	C	S:Acetone A=Cl	74Dd	
Ni^{2+}	-9.6	3.20(25°,1)	29.3	25-40	T	μ=0.1M(KNO_3& HNO_3)	76Bb	
	-52.3	2.16(25°,2)	-133.9	25-40	T	μ=0.1M(KNO_3& HNO_3)	76Bb	
NiA_2	-13.0	1.96(1-2)	-4.2	--	C	CM/CL=1/10 S:Dimethylform-amide A=Br	76Kj	j,u: 25°,y
Cu^{2+}	-6.3	5.43(25°,1)	83.3	25-40	T	μ=0.1M(KNO_3& HNO_3)	76Bb	
	-53.6	4.18(25°,2)	-99.2	25-40	T	μ=0.1M(KNO_3& HNO_3)	76Bb	
Zn^{2+}	-10.9	2.72(25°,1)	15.9	25-40	T	μ=0.1M(KNO_3& HNO_3)	76Bb	
	-21.3	1.92(25°,2)	-34.7	25-40	T	μ=0.1M(KNO_3& HNO_3)	76Bb	
Cd^{2+}	-19.2	2.54(25°,1)	-15.5	25-40	T	μ=0.1M(KNO_3& HNO_3)	76Bb	
	1.7	2.70(25°,2)	57.3	25-40	T	μ=0.1M(KNO_3& HNO_3)	76Bb	
P323	PYRIDINE, 3-amino		$C_5H_6N_2$			$C_5H_4N(NH_2)$		
CoA_2	-60.4	8.85(1-2)	-33	25	C	S:Acetone A=Cl	74Dd	
Ag^{+}	-22.97	2.24(1)	-34.10	25	C	μ=0.5M(KNO_3)	76Be	
	-28.91	2.16(2)	-55.56	25	C	μ=0.5M(KNO_3)	76Be	
P324	PYRIDINE, 4-amino-		$C_5H_6N_2$			$C_5H_4N(NH_2)$		
[Fe^{2+}]	-67.4	5.64(25°,1-2)	-117	18-40	T	[Fe]=Iron(II) protoporphyrin dimethyl ester S:Benzene	70Ch	b,q: ±9.2
CoA_2	-61.0	9.35(1-2)	-25	25	C	S:Acetone A=Cl	74Dd	
[Co^{2+}]	-33.9	4.48(0°,1)	-38	-23-23	T	[Co]=Cobalt(II) protoporphyrin IX dimethyl ester S:Toluene	73Sx	b,q: ±2.5
NiA_2	-74.9	7.78(1-3)	-100	--	C	CM/CL=1/10 S:Dimethylform-amide A=Br	76Kj	j,u: 25°,y
[Ni^{2+}]	-31.8	2.57(25°,1)	-58.6	18-40	T	[Ni]=Nickel(II) mesotetraphenyl-porphine S:$CHCl_3$	72Ce	b,q: ±2.9
	-27.6	3.08(25°,1)	-33	18-40	T	[Ni]=Nickel(II) mesotetraphenyl-porphine S:Benzene	72Ce	b,q: ±5.8

Metal Ion	ΔH, kJ /mol	Log K	ΔS, J /mol K	T,°C	M	Conditions	Ref.	Re-marks
P324, continued								
[Zn^{2+}]	-21.3	3.52(25°,1)	-4	18-40	T	[Zn]=Zinc(II) mesotetraphenyl-porphine S:$CHCl_3$	72Ce	b
	-58.2	4.62(25°,1)	-109	18-40	T	[Zn]=Zinc(II) mesotetraphenyl-porphine S:Benzene	72Ce	b
P325	PYRIDINE, N-(4-aminobutane)- $C_9H_{15}N$					$C_5H_5N[(CH_2)_4NH_2]$		L^+
Ag^+	-22.72	3.240(1)	-14.2	25	C	μ=0.5M(KNO_3)	72Va	
	-33.64	3.365(2)	-48.1	25	C	μ=0.5M(KNO_3)	72Va	
P326	PYRIDINE, N-(2-aminoethane)- $C_7H_{11}N$					$C_5H_5N[(CH_2)_2NH_2]$		L^+
Ag^+	-23.26	2.096(1)	-38.1	25	C	μ=0.5M(KNO_3)	72Va	
	-23.18	2.361(2)	-32.2	25	C	μ=0.5M(KNO_3)	72Va	
P327	PYRIDINE, 2-(2'-aminoethyl)- $C_7H_{10}N_2$					$C_5H_4N(CH_2CH_2NH_2)$		L
Ni^{2+}	-29.87	5.14(30°,1)	0	10-40	T	μ=0	59Gb	b
	-16.99	3.20(30°,1)	5.4	10-40	T	μ=0	59Gb	b
	-28.95	5.42(1)	6.7	25	C	μ=0.30(ClO_4^-)	67Hf	
	-29.71	5.37(1)	3.3	25	C	μ=0.5M(KNO_3)	71Gd	
	-29.29	3.39(2)	-33.5	25	C	μ=0.30(ClO_4^-)	67Hf	
	-25.52	3.37(2)	-20.9	25	C	μ=0.5M(KNO_3)	71Gd	
	-29.25	2.35(3)	-53.1	25	C	μ=0.30(ClO_4^-)	67Hf	
Cu^{2+}	-36.02	7.48(30°,1)	21	10-40	T	μ=0	59Gb	b
	-37.00	7.54(1)	20.5	25	C	μ=0.30(ClO_4^-)	67Hf	
	-39.24	7.64(1)	15.1	25	C	μ=0.5M(KNO_3)	71Gd	
	-37.03	5.64(2)	-16.3	25	C	μ=0.30(ClO_4^-)	67Hf	
	-34.64	5.59(2)	-9.2	25	C	μ=0.5M(KNO_3)	71Gd	
P328	PYRIDINE, 2-aminomethyl- $C_6H_8N_2$					$C_5H_4N(CH_2NH_2)$		L
Co^{2+}	-28.33	5.41(30°,1)	10.0	10-40	T	μ=0	59Gb	b
	-31.04	5.54(1)	2.1	25	C	μ=0.5M(KNO_3)	71Gd	
	-69.79	10.39(1)	-35.10	25	C	μ=0.1(KNO_3) S:50w% Dioxane	66Wa	
	-29.83	4.52(30°,2)	-11.7	10-40	T	μ=0	59Gb	b
	-31.25	4.79(2)	-13.0	25	C	μ=0.5M(KNO_3)	71Gd	
	-25.69	3.33(30°,3)	-20.9	10-40	T	μ=0	59Gb	b
	-28.66	3.50(3)	-29.3	25	C	μ=0.5M(KNO_3)	71Gd	
	-115.6	10.62(25°,1-2)	-185	20-40	T	μ=0.10 pH=8.8-9.6	74Hh	b
Ni^{2+}	-34.94	7.09(30°,1)	20.5	10-40	T	μ=0	59Gb	b
	-36.90	7.18(1)	18.0	25	C	μ=0.30(ClO_4^-)	67Hf	
	-41.46	7.11(1)	-2.1	25	C	μ=0.5M(KNO_3)	71Gd	
	-85.86	13.15(1)	-36.11	25	C	μ=0.1(KNO_3) S:50w% Dioxane	66Wa	
	-38.70	6.08(30°,2)	-11.3	10-40	T	μ=0	59Gb	b
	-35.73	5.96(2)	-5.8	25	C	μ=0.30(ClO_4^-)	67Hf	

Metal Ion	ΔH, kJ /mol	Log K	ΔS, J /mol K	T,°C	M	Conditions	Ref.	Remarks
P328, continued								
Ni^{2+}, cont.	-39.45	6.41(2)	-9.2	25	C	μ=0.5M(KNO_3)	71Gd	
	-35.27	4.95(30°,3)	-21.3	10-40	T	μ=0	59Gb	b
	-35.69	4.41(3)	-35.1	25	C	μ=0.30(ClO_4^-)	67Hf	
	-40.46	5.14(3)	-36.8	25	C	μ=0.5M(KNO_3)	71Gd	
	-40.4	13.31(25°,1-3)	119	15-45	T	μ=0.3	75Hb	b
Cu^{2+}	-40.21	9.45(30°,1)	48.1	10-40	T	μ=0	59Gb	b
	-41.50	9.34(1)	39.7	25	C	μ=0.30(ClO_4^-)	67Hf	
	-90.25	15.69(1)	-2.30	25	C	μ=0.1(KNO_3) S:50w% Dioxane	66Wa	
	-37.57	7.80(30°,2)	25.5	10-40	T	μ=0	59Gb	b
	-41.67	7.92(2)	11.7	25	C	μ=0.30(ClO_4^-)	67Hf	
	-111.8	12.80(25°,1-2)	-129	20-40	T	μ=0.10 pH=4.7-6.8	74Hh	b
Zn^{2+}	-28.49	5.17(30°,1)	5.0	10-40	T	μ=0	59Gb	b
	-27.11	5.37(1)	12.1	25	C	μ=0.5M(KNO_3)	71Gd	
	-52.05	9.84(1)	13.47	25	C	μ=0.1(KNO_3) S:50w% Dioxane	66Wa	
	-30.46	4.21(30°,2)	-20.1	10-40	T	μ=0	59Gb	b
	-26.90	4.47(2)	-5.4	25	C	μ=0.5M(KNO_3)	71Gd	
	-24.48	3.07(30°,3)	-21.8	10-40	T	μ=0	59Gb	b
	-28.87	2.80(3)	-37.2	25	C	μ=0.5M(KNO_3)	71Gd	
Cd^{2+}	-24.60	4.59(30°,1)	6.7	10-40	T	μ=0	59Gb	b
	-24.31	4.76(1)	-9.6	25	C	μ=0.5M(KNO_3)	71Gd	
	-52.68	9.68(1)	8.58	25	C	μ=0.1(KNO_3) S:50w% Dioxane	66Wa	
	-25.69	3.82(30°,2)	-11.7	10-40	T	μ=0	59Gb	b
	-23.89	3.94(2)	-5.4	25	C	μ=0.5M(KNO_3)	71Gd	
	-25.90	2.59(3)	-37.2	25	C	μ=0.5M(KNO_3)	71Gd	

P329 PYRIDINE, 2-aminomethyl-6-methyl- $C_7H_{10}N_2$ $C_5H_3N(CH_2NH_2)(CH_3)$ L

Metal Ion	ΔH, kJ /mol	Log K	ΔS, J /mol K	T,°C	M	Conditions	Ref.	Remarks
Co^{2+}	-44.14	7.00(1)	-14.02	25	C	μ=0.1(KNO_3) S:50w% Dioxane	66Wa	
Ni^{2+}	-24.52	4.71(30°,1)	9.2	10-40	T	μ=0	60Wa	b
	-37.74	8.43(1)	34.81	25	C	μ=0.1(KNO_3) S:50w% Dioxane	66Wa	
	-20.88	3.02(30°,2)	-10.9	10-40	T	μ=0	60Wa	b
Cu^{2+}	-28.70	6.81(30°,1)	35.6	10-40	T	μ=0	60Wa	b
	-63.93	12.69(1)	28.53	25	C	μ=0.1(KNO_3) S:50w% Dioxane	66Wa	
	-30.25	5.65(30°,2)	8.4	10-40	T	μ=0	60Wa	b
Ag^{+}	-39.08	3.97(30°,1)	-53.1	10-40	T	μ=0	60Wa	b
	-4.48	3.45(30°,2)	51.0	10-40	T	μ=0	60Wa	b
Zn^{2+}	-22.34	7.80(1)	74.35	25	C	μ=0.1(KNO_3) S:50w% Dioxane	66Wa	
Cd^{2+}	-18.91	4.01(30°,1)	14.2	10-40	T	μ=0	60Wa	b
	-24.98	7.07(1)	51.55	25	C	μ=0.1(KNO_3) S:50w% Dioxane	66Wa	
	2.18	2.69(30°,2)	59.0	10-40	T	μ=0	60Wa	b

Metal Ion	ΔH, kJ /mol	Log K	ΔS, J /mol K	T,°C	M	Conditions	Ref.	Re- marks
P330	PYRIDINE, N-(5-aminopentane)- $C_{10}H_{17}N$					$C_5H_5N[(CH_2)_5NH_2]$		L^+
Ag^+	-24.31	3.489(1)	-14.6	25	C	μ=0.5M(KNO_3)	72Va	
	-32.26	3.732(2)	-36.8	25	C	μ=0.5M(KNO_3)	72Va	
P331	PYRIDINE, N-(3-aminopropane)- $C_8H_{13}N$					$C_5H_5N[(CH_2)_3NH_2]$		L^+
Ag^+	-21.71	2.910(1)	-16.7	25	C	μ=0.5M(KNO_3)	72Va	
	-32.2	3.101(2)	-49.0	25	C	μ=0.5M(KNO_3)	72Va	
P332	PYRIDINE, 2-(azo-p-dimethyl-aniline)- $C_{13}H_{14}N_4$					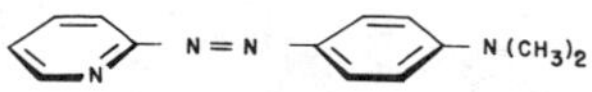		
Co^{2+}	-25.1	(1)	21	15-50	T	μ=0.1M($NaNO_3$)	72Ca	b
Ni^{2+}	-27.2	(1)	12	15-50	T	μ=0.1M($NaNO_3$)	72Ca	b
ZnA_4^{2+}	-7.32	2.360(25°)	20.1	25&35	T	μ=0.15($NaNO_3$) A=H_2O R:ZnA_4^{2+} +L=ZnA_2L^{2+}+2A	53Kb	a,b
P333	PYRIDINE, 4-benzoyl- $C_{12}H_9ON$					$C_5H_4N(COC_6H_5)$		L
CuA_2	-5.94	1.23(1)	3.64	25	C	μ≅0.005 S:Benzene A=2,4-Pentanedione	63Ma	
	-5.44	(1)	--	--	T	μ≅0.005 S:Benzene A=2,4-Pentanedione	63Ma	
	-24.10	2.376(1)	-34.02	30	C	S:Benzene A=1,1,1-Trifluoropentane-2,4-dione	74Gc	
P334	PYRIDINE, 2-benzoyl-, oxime $C_{12}H_{10}ON_2$					$C_5H_4N[C(NOH)(C_6H_5)]$		L
Ni^{2+}	-9.66	9.21(30°,1)	140.04	20-40	T	μ=0 S:40w% Acetone	65Sc	b
	-1.76	7.05(30°,2)	125.31	20-40	T	μ=0 S:40w% Acetone	65Sc	b
Zn^{2+}	-26.36	8.40(30°,1)	71.63	20-40	T	μ=0 S:40w% Acetone	65Sc	b
	-2.64	6.95(30°,2)	120.62	20-40	T	μ=0 S:40w% Acetone	65Sc	b
Cd^{2+}	-51.84	6.95(30°,1)	-36.82	20-40	T	μ=0 S:40w% Acetone	65Sc	b
	-17.57	6.44(30°,2)	63.34	20-40	T	μ=0 S:40w% Acetone	65Sc	b
Pb^{2+}	-29.00	7.52(30°,1)	46.86	20-40	T	μ=0 S:40w% Acetone	65Sc	b
	-16.69	7.11(30°,2)	78.58	20-40	T	μ=0 S:40w% Acetone	65Sc	b

Metal Ion	ΔH, kJ /mol	Log K	ΔS, J /mol K	T,°C	M	Conditions	Ref.	Remarks
P335	PYRIDINE, 4-benzyl-		$C_{12}H_{11}N$			$C_5H_4N(CH_2C_6H_5)$		L
CuA_2	-31.09	3.075(1)	-43.9	30	C	S:Benzene A=1,-1,1,-Trifluoro-pentane-2,4-dione	74Gc	
P336	PYRIDINE, 2-[benzyl(2-dimethylaminoethyl)amino]-		$C_{16}H_{21}N_3$			$C_5H_4N[N(CH_2CH_2N(CH_3)_2)-(CH_2C_6H_5)]$		L
Be^{2+}	-41.4	5.56(25°,1)	-29	0-45	T	μ=0.12M(KCl) CM≅2x10^{-3}M CL≅8x10^{-3}M	69Cb	b
	-4.2	4.55(25°,2)	71	0-45	T	μ=0.12M(KCl) CM≅2x10^{-3}M CL≅8x10^{-3}M	69Cb	b
Co^{2+}	31.4	6.34(25°,1-2)	225.9	0&25	T	0.06M(KCl)	62Aa	a,b
Ni^{2+}	68.6	6.94(25°,1-2)	363.2	0&25	T	0.06M(KCl)	62Aa	a,b
Cu^{2+}	-10.5	8.80(25°,1-2)	133.9	0&25	T	0.06M(KCl)	62Aa	a,b
P337	PYRIDINE, 1-benzyl-2-formamido-		$C_{13}H_{13}ON_2$			$C_5H_4N(CH_2C_6H_5)(NHCOH)$		
Fe(III)A_5	-11.88	3.302(25°,1)	23.30	--	T	μ=0.05 pH=8 A=CN	72Hb	u
P338	PYRIDINE, 3-bromo-		C_5H_4NBr			$C_5H_4N(Br)$		L
CoA_2	-17.86	2.646(1)	-9.2	25	C	S:Acetone CM=0.002M A=Cl	73Dh	
	-6.78	0.66(1)	10.12	25	C	μ≅0.005 S:Benzene A=2,4-Pentanedione	63Ma	
	-9.08	(1)	--	25	T	μ≅0.005 S:Benzene A=2,4-Pentanedione	63Ma	
NiA_2	-7.1	2.40(1-2)	-20.9	--	C	CM/CL=1/10 S:Dimethylform-amide A=Br	76Kj	j,u: 25°,y
Ag^+	-17.15	1.66(1)	-25.82	25	C	μ=0.5M(KNO_3)	76Be	
	-22.18	1.66(2)	-42.68	25	C	μ=0.5M(KNO_3)	76Be	
Br_2	-33.5	1.243(1)	-88.7	25	C	S:1,2-Dichloro-ethane	73Tb	
P339	PYRIDINE, 4-bromo-1-oxide-		C_5H_4ONBr			$C_5H_4N(Br)(\rightarrow O)$		
VAB_2	-25.40	(1)	--	25	T	S:CH_2Cl_2 A=O B=2,4-Pentanedione	69Pd	

Metal Ion	ΔH, kJ /mol	Log K	ΔS, J /mol K	T,°C	M	Conditions	Fef.	Re-marks
P340	PYRIDINE, 4-butanoic acid		$C_{10}H_{13}O_2N$			$C_5H_4N[(CH_2)_4CO_2H]$		
[Fe]	-28.0	3.52(25°,1-2)	-29	18-40	T	[Fe]=Iron(II) deuteroporphyrin dimethyl ester S:Benzene	70Ch	b,q: ±4.2
	-29.3	3.66(25°,1-2)	176	18-40	T	[Fe]=Iron(II) diacetyldeutero-porphyrin dimethyl ester S:Benzene	70Ch	b
	-37.2	3.15(25°,1-2)	-63	18-40	T	[Fe]=Iron(II) mesoporphyrin dimethyl ester S:Benzene	70Ch	b,q: ±4.2
	-42.2	2.93(25°,1-2)	-84	18-40	T	[Fe]=Iron(II) protoporphyrin dimethyl ester S:Benzene	70Ch	b
P341	PYRIDINE, 4-tert-butyl-		$C_9H_{13}N$			$C_5H_4N[C(CH_3)_3]$		
[Co]	-37.6	4.40(0°,1)	-54	-23-23	T	[Co]=Cobalt(II) protoporphyrin IX dimethyl ester S:Toluene	73Sr	b,q: ±2.5
CuA_2	-32.6	1.836(1)	-74.9	25	C	S:Cyclohexane A=t-Butylaceto-acetate	69La	
	-33.1	1.876(25°,1)	-74.9	25-40	C	S:Cyclohexane A=t-Butylaceto-acetate	69Ga	b
P342	PYRIDINE, 4-carbomethoxy-1-oxide-		$C_7H_7O_3N$			$C_5H_4N(CO_2CH_3)(\rightarrow O)$		
VAB_2	-15.40	(1)	--	25	T	S:CH_2Cl_2 A=O B= 2,4-Pentanedione	69Pd	
	-3.56	(1)	--	25	C	S:$CHCl_3$ A=O B= 2,4-Pentanedione	69Pd	
P343	PYRIDINE, 2-chloro		C_5H_4NCl			C_5H_4NCl		
I_2	-12.9	0.70(25°,1)	-29.7	17-35	T	S:Heptane	68Kb	b
P344	PYRIDINE, 3-chloro-		C_5H_4NCl			$C_5H_4N(Cl)$		L
CoA_2	-20.71	3.196(1)	-8.4	25	C	S:Acetone CM=0.002M A=Cl	73Dh	
Ag^+	-16.32	1.59(1)	-24.27	25	C	μ=0.5M(KNO_3)	76Be	
	-21.30	1.43(2)	-44.06	25	C	μ=0.5M(KNO_3)	76Be	
I_2	-22.2	1.38(25°,1)	-48.1	17-35	T	S:Heptane	68Kb	b

Metal Ion	ΔH, kJ /mol	Log K	ΔS, J /mol K	T,°C	M	Conditions	Ref.	Remarks
P345	PYRIDINE, 4-chloro-		C_5H_4NCl			$C_5H_4N(Cl)$		L
CuA_2	-2.76	0.73(1)	4.69	25	C	μ≅0.005 S:Benzene A=2,4-Pentanedione	63Ma	
	-8.83	(1)	--	25	C	μ≅0.005 S:Benzene A=2,4-entanedione	63Ma	
P346	PYRIDINE-2-[4-chloro-α-(2-dimethylaminoethyl)benzyl]-		$C_{16}H_{19}N_2Cl$			$C_5H_4N[CH(CH_2CH_2N(CH_3)_2)-(C_6H_4Cl)]$		L
Cu^{2+}	-103.8	9.26(25°,1-2)	-218	0&25	T	0.06M(KCl)	62Aa	a,b
P347	PYRIDINE, 4-chloro-1-oxide		C_5H_4ONCl			$C_5H_4N(Cl)(\rightarrow O)$		L
VAB_2	-23.93	(1)	--	25	C	S:CH_2Cl_2 A=O B= 2,4-Pentanedione	69Pd	
	-9.12	(1)	--	25	C	S:$CHCl_3$ A=O B= 2,4-Pentanedione	69Pd	
I_2	-19.46	1.564(25°,1)	-35.6	18-32	T	S:CCl_4	68Ga	b
P348	PYRIDINE, 3-cyano		$C_6H_4N_2$			$C_5H_4N(CN)$		L
Ag^+	-38	1.99(1-2)	-91	25	C	μ=0.5M($NaClO_4$)	77Mc	
	-54	2.37	-134	25	C	μ=0.5M($NaClO_4$) R:AgL^++Ag=Ag_2L^{2+}	77Mc	
P349	PYRIDINE, 4-cyano-1-oxide		$C_6H_4ON_2$			$C_5H_4N(CN)(\rightarrow O)$		L
VAB_2	-12.93	(1)	--	25	T	S:CH_2Cl_2 A=O B= 2,4-Pentanedione	69Pd	
P350	PYRIDINE, 3,5-dichloro		$C_5H_3NCl_2$			$C_5H_3N(Cl)_2$		
$Rh_2A_4B_2$	-32.2	--	--	24	C	S:Benzene R:$Rh_2A_4B_2$+2L= $2RhA_2BL$ A=CO B=Cl	76Pd	u:24°
P351	PYRIDINE, 1,3-dimethyl-		$C_7H_{10}N$			$C_5H_4N(CH_3)_2$		L^+
[Fe]	-13.0	5.8(20°)	56.1	5-30	T	pH=11.2 Protohaemin R:$(Haemin)_2$+2L= $(HaeminL)_2$+H_2O	69Md	b,w:I
P352	PYRIDINE, 2,3-dimethyl-		C_7H_9N			$C_5H_3N(CH_3)_2$		L
[Fe]	0	6.2(20°)	118.4	5-30	T	pH=11.2 Protohaemin	69Md	b

Metal Ion	ΔH, kJ /mol	Log K	ΔS, J /mol K	T,°C	M	Conditions	Ref.	Re-marks
P352, continued								
[Fe], cont.						R:(Haemin)$_2$+2HL$^+$ +2OH$^-$=(HaeminL)$_2$ +2H$_2$O		
NiA$_2$	-22.84	-2.386(35°,1-2)	-121	25-85	T	S:L A=Salicyl-aldoximate	71Ma	b
Ag$^+$	-26.1	2.456(1)	-40.6	25	C	μ=0.5M(KNO$_3$)	74Be	
	-22.2	2.331(2)	-29.7	25	C	μ=0.5M(KNO$_3$)	74Be	
P353	PYRIDINE, 2,4-dimethyl-		C$_7$H$_9$N			C$_5$H$_3$N(CH$_3$)$_2$		L
NiA	-57.07	1.02(1-2)	-172.4	25	T	CM≅CL≅10^{-4}M S:Benzene A=Di-acetylbisbenzoyl hydrazine	58Sa	
	-23.22	-1.851(35°,1-2)	-109	25-95	T	S:L A=Salicyl-aldoximate	71Ma	b
Ag$^+$	-29.8	2.544(1)	-51.5	25	C	μ=0.5M(KNO$_3$)	74Be	
	-23.7	2.536(2)	-31.0	25	C	μ=0.5M(KNO$_3$)	74Be	
Br$_2$	-49.4	3.090(1)	-107.5	25	C	S:1,2-Dichloro-ethane	73Tb	
I$_2$	-36.0	2.36(1)	-76	25	C	S:Benzene	80Ua	
	-36.6	2.45(1)	-76	25	C	S:CCl$_4$	80Ua	
	-41.0	2.76(1)	-85	25	C	S:Chlorobenzene	80Ua	
	-38.3	2.56(1)	-80	25	C	S:Cyclohexane	80Ua	
	-37.7	2.59(1)	-77	25	C	S:n-Hexane	80Ua	
P354	PYRIDINE, 2,5-dimethyl-		C$_7$H$_9$N			C$_5$H$_3$N(CH$_3$)$_2$		L
NiA	-52.63	0.54(25°,1-2)	-166.1	25	C	CM≅CL≅10^{-4}M S:Benzene A=Di-acetylbisbenzoyl hydrazine	58Sa	
NiA$_2$	-17.82	-1.955(35°,1-2)	-96	25-95	T	S:L A=Salicyl-aldoximate	71Ma	b
Ag$^+$	-26.2	2.565(1)	-38.9	25	C	μ=0.5M(KNO$_3$)	74Be	
	-21.8	2.353(2)	-28.1	25	C	μ=0.5M(KNO$_3$)	74Be	
P355	PYRIDINE, 2,6-dimethyl-		C$_7$H$_9$N			C$_5$H$_3$N(CH$_3$)$_2$		L
VOA$_2$	-30.79	0.78(1)	-88.7	25	C	μ→0 S:Nitrobenzene A=Acetylacetonate	65Ca	
[Fe]	13.0	2.04(20°)	83.7	5-30	T	pH=11.2 Protohaemin R:(Haemin)$_2$+2L=(HaeminL)$_2$+2H$_2$O	69Md	b
CoA$_2$	-30.29	4.73(1-2)	-11.09	25	C	S:Actone A=Cl	72Ae	
	-45.94	4.10(1-2)	-75.69	25	C	S:Acetonitrile	72Ae	

Metal Ion	ΔH, kJ /mol	Log K	ΔS, J /mol K	T,°C	M	Conditions	Ref.	Re-marks
P355, continued								
CoA_2,						A=Cl		
cont.	-18.07	2.99(1-2)	-3.51	25	C	S:n-Butanol A=Cl	72Af	
	-46.15	4.83(1-2)	-62.47	25	C	S:Cyclohexanone A=Cl	72Ae	
	-15.56	3.46(1-2)	14.06	25	C	S:Dimethylformamide A=Cl	72Ae	
	-20.33	2.59(1-2)	-21.46	25	C	S:Ethanol A=Cl	72Af	
	-17.28	3.20(1-2)	3.22	25	C	S:2-Methyl propan-2-ol A=Cl	72Af	
	-0.67	1.54(1-2)	27.24	25	C	S:Ethylene glycol A=Cl R:CoA_2+2L+2S=$[CoL_2S_2]A_2$	72Af	
NiA	-57.07	1.05(1-2)	-171.5	25	T	CM≅CL≅10^{-4}M S:Benzene A=Diacetylbisbenzoyl hydrazine	58Sa	
NiA_2	-12.1	1.69(1-2)	-8.4	--	C	CM/CL=1/10 S:Dimethylformamide A=Br	76Kj	j,u: 25°,y
	-30.46	-1.114(35°,1-2)	-117	25-95	T	S:L A=Salicylaldoximate	71Ma	b
Ag^+	-30.1	2.683(1)	-49.4	25	C	μ=0.5M(KNO_3)	74Be	
	-20.9	2.382(2)	-24.7	25	C	μ=0.5M(KNO_3)	74Be	
CuA_2	-40	0.978(1)	-118	20	T	S:Benzene A=Trifluoroacetylacetone	79Si	b,q:±3
BA_3	-19.83	(1)		25	C	S:Nitrobenzene CL≅0.18M CM≅0.30M A=F	56Bb	
AlA_3	-83.14	(1)	--	25-28	C	S:n-Hexane A=CH_3	68Hc	
Br_2	-41.4	2.666(1)	-88.3	25	C	S:1,2-Dichloroethane	73Tb	
I_2	-28.6	1.62(1)	-65	25	C	S:Benzene	80Ua	
	-32.06	1.763(1)	-74.0	25	C	S:CCl_4	79Ba	
	-32.5	1.76(1)	-75	25	C	S:CCl_4	80Ua	
	-35.6	1.93(1)	-82	25	C	S:Chlorobenzene	80Ua	
	-34.5	1.88(1)	-80	25	C	S:Cyclohexane	80Ua	
	-34.8	1.86(1)	-81	25	C	S:n=Hexane	80Ua	
P356	PYRIDINE, 3,4-dimethyl-		C_7H_9N			$C_5H_3N(CH_3)_2$		L
VOA_2	-30.58	2.12(1)	-61.9	25	C	μ→0 S:Nitrobenzene A=Acetylacetonate	65Ca	
CoA	-37.6	3.035(25°,1)	-67	20-50	T	S:Toluene A=α,β,γ,δ-Tetra-(p-methoxyphenyl) porphyrin	73Wa	b,q:±4

Metal Ion	ΔH, kJ /mol	Log K	ΔS, J /mol K	T,°C	M	Conditions	Ref.	Remarks
P356, continued								
CoAL	-7.5	-0.097(-65°,1)	-42	-80--20	T	S:Toluene A=α,β,γ,δ-Tetra-(p-methoxyphenyl) porphyrin	73Wb	b,q: ±4.2
CoL_2A_2	-70.3	4.255(1-2)	-159.0	20	C	S:$CHCl_3$ A=NCS	66Ca	s
CuA_2	-21.8	-0.013(25°,1)	-71	-3-57	T	S:Benzene A=Diethyldithio-carbamate	74Hd	b
Ag^+	-25.1	2.434(1)	-37.6	25	C	μ=0.5M(KNO_3)	74Be	
	-30.7	2.419(2)	-56.5	25	C	μ=0.5M(KNO_3)	74Be	
P357	PYRIDINE, 3,5-dimethyl-		C_7H_9N			$C_5H_3N(CH_3)_2$		L
VOA_2	-28.28	2.01(1)	-56.5	25	C	μ→0 S:Nitrobenzene A=Acetylacetonate	65Ca	
CoL_2A_2	-61.1	3.70(1-2)	-138.1	20	C	S:$CHCl_3$ A=NCS	66Ca	s
NiA_2	-80.83	(1-2)	--	25	C	S:Benzene A=2,2,-6,6-Tetramethyl-3,5-heptanedione	73Ce	
	-85.5	3.78(1-2)	-214	25	C	S:Nitrobenzene K determined in benzene A=Ethylxanthato	70Cc	
Br_2	-49.0	3.021(1)	-106.3	25	C	S:1,2-Dichloro-ethane	73Tb	
I_2	-25.1	(1)	--	25	C	S:1,2-Dichloro-ethane	73Tb	
P358	PYRIDINE, 4-(dimethylamino)-		$C_7H_{10}N_2$			$C_5H_4N[N(CH_3)_2]$		L
NiA_2	-88.37	(1-2)	--	25	C	S:Benzene A=2,2,-6,6-Tetramethyl-3,5-heptanedione	73Ce	
P359	PYRIDINE, 2-[α-(2-dimethyl-aminoethyl)benzyl]-		$C_{16}H_{20}N_2$			$C_5H_4N[CH(CH_2CH_2N(CH_3)_2)-(C_6H_5)]$		L
Cu^{2+}	-68.6	10.04(25°,1-2)	-38.5	0&25	T	0.06M(KCl)	62Aa	a,b
P360	PYRIDINE, 4-(dimethylamino)-, 1-oxide-		$C_7H_{10}ON_2$			$C_5H_4N[N(CH_3)_2](\to O)$		
VAB_2	-43.9	(1)	--	25	T	S:CH_2Cl_2 A=O B= 2,4-Pentanedione	69Pd	

Metal Ion	ΔH, kJ /mol	Log K	ΔS, J /mol K	T,°C	M	Conditions	Ref.	Re-marks
P361	PYRIDINE, 2,3-dimethyl-, 1-oxide-		C_7H_9ON			$C_5H_4N(CH_3)_2(\rightarrow O)$		
VAB_2	-19.04	(1)	--	25	C	S:CH_2Cl_2 A=O B= 2,4-Pentanedione	69Pd	
P362	PYRIDINE, 2,4-dimethyl-, 1-oxide-		C_7H_9ON			$C_5H_3N(CH_3)_2(\rightarrow O)$		
VAB_2	-22.47	(1)	--	25	C	S:CH_2Cl_2 A=O B= 2,4-Pentanedione	69Pd	
P363	PYRIDINE, 2,6-dimethyl-, 1-oxide		C_7H_9ON			$C_5H_3N(CH_3)_2(\rightarrow O)$		
VAB_2	-3.64	(1)	--	25	C	S:CH_2Cl_2 A=O B= 2,4-Pentanedione	69Pd	
I_2	-27.99	2.18(25°,1)	-52.3	10-40	T	S:CCl_4	57Ka	b
	-23.4	(1)	--	--	-	S:CCl_4	60Bb	
P364	PYRIDINE, 3,4-dimethyl-, 1-oxide-		C_7H_9ON			$C_5H_3N(CH_3)_2(\rightarrow O)$		L
VAB_2	-29.71	(1)	--	25	C	S:CH_2Cl_2 A=O B= 2,4-Pentanedione	69Pd	
P365	PYRIDINE, 3,5-dimethyl-, 1-oxide-		C_7H_9ON			$C_5H_3N(CH_3)_2(\rightarrow O)$		
VAB_2	-28.12	(1)	--	25	C	S:CH_2Cl_2 A=O B= 2,4-Pentanedione	69Pd	
P366	PYRIDINE, 2-ethyl-		C_7H_9N			$C_5H_4N(CH_2CH_3)$		
Ag	-24.02	2.29(1)	-36.78	25	C	μ=0.5M(KNO_3)	76Be	
	-17.57	2.26(2)	-15.69	25	C	μ=0.5M(KNO_3)	76Be	
BA_3	-37.6	(1)	--	25	C	S:Nitrobenzene A=CH_3	56Ba	
Br_2	-43.9	2.686(1)	-95.8	25	C	S:1,2-Dichloro-ethane	73Tb	
I_2	-32.6	(1)	--	25	C	S:1,2-Dichloro-ethane	73Tb	
P367	PYRIDINE, 3-ethyl-		C_7H_9N			$C_5H_4N(H_2CH_3)$		L
CoL_2A_2	-55.2	2.31(1-2)	-180.3	20	C	S:$CHCl_3$ A=Cl	66Ca	s
	-64.0	4.34(1-2)	-135.1	20	C	S:$CHCl_3$ A=NCS	66Ca	s
CuA_2	-31.8	1.643(1)	-74.9	25	C	S:Cyclohexane A=t-Butylaceto-acetate	69La	

Metal Ion	ΔH, kJ /mol	Log K	ΔS, J /mol K	T,°C	M	Conditions	Ref.	Remarks
P367, continued								
Ag^+	-21.84	2.23(1)	-30.58	25	C	μ=0.5M(KNO_3)	76Be	
	-28.20	2.30(2)	-50.54	25	C	μ=0.5M(KNO_3)	76Be	
Br_2	-46.4	2.700(1)	-105.4	25	C	S:1,2-Dichloro-ethane	73Tb	
P368	PYRIDINE, 4-ethyl-		C_7H_9N			$C_5H_4N(CH_2CH_3)$		L
CoL_2A_2	-66.9	1.053(1-2)	-207.9	20	C	S:$CHCl_3$ A=Cl	66Ca	s
	-69.0	4.89(1-2)	-141.4	20	C	S:$CHCl_3$ A=NCS	66Ca	s
CuA_2	-33.0	1.785(1)	-76.1	25	C	S:Cyclohexane A=T-Butylaceto-acetate	69La 69Ga	
Ag^+	-25.56	2.34(1)	-40.96	25	C	μ=0.5M(KNO_3)	76Be	
	-27.36	2.33(2)	-47.15	25	C	μ=0.5M(KNO_3)	76Be	
SbA_5	-128.4	(1)	--	30	C	S:Nitrobenzene	63Hb	
Br_2	-47.3	2.867(1)	-103.8	25	C	S:1,2-Dichloro-ethane	73Tb	
I_2	-37.6	(1)	--	25	C	S:1,2-Dichloro-ethane	73Tb	
P369	PYRIDINE, 2-ethyl, 1-oxide-		C_7H_9ON			$C_5H_4N(CH_2CH_3)(\rightarrow O)$		
VAB_2	-15.27	(1)	--	25	C	S:CH_2Cl_2 A=O B= 2,4-Pentanedione	69Pd	
P370	PYRIDINE, 2-(2-hydroxyethyl)-		C_7H_9ON			$C_5H_4N(CH_2CH_2OH)$		L
Cu^{2+}	-57.82	18.54(1-2)	161.1	25	C	$\mu=1.5\times10^{-4}$ S:50 v% 1,2-Dioxane	64Le	
	-57.11	18.54(25°,1-2)	163.6	0-25	T	$\mu=4.5\times10^{-3}$ S:50 v% 1,4-Dioxane	64Le	b
P371	PYRIDINE, 2-(2-hydroxyethyl)-6-methyl-		$C_8H_{11}ON$			$C_5H_3N(CH_2CH_2OH)(CH_3)$		L
Cu^{2+}	-52.01	18.12(1-2)	172.4	25	C	$\mu=1.5\times10^{-4}$ S:50 v% 1,4-Dioxane	64Le	
	-51.04	18.12(25°,1-2)	175.7	0-25	T	$\mu=4.5\times10^{-3}$ S:50 v% 1,4-Dioxane	64Le	b
P372	PYRIDINE, 2-hydroxymethyl-		C_6H_7ON			$C_5H_4N(CH_2OH)$		L
Cu^{2+}	-99.62	19.15(1-2)	32.6	25	C	$\mu=7\times10^{-5}$ S:50 v% 1,4-Dioxane	64Le	
	-97.57	19.15(1-2)	39.3	25	C	$\mu=1.5\times10^{-4}$ S:50 v% 1,4-Dioxane	64Le	
	-61.04	19.15(25°,1-2)	161.9	0-25	T	$\mu=4.5\times10^{-3}$ S:50 v% 1,4-Dioxane	64Le	b

Metal Ion	ΔH, kJ /mol	Log K	ΔS, J /mol K	T,°C	M	Conditions	Ref.	Re-marks
P373	PYRIDINE, 2-hydroxymethyl-6-methyl-		C_7H_9ON			$C_5H_3N(CH_2OH)(CH_3)$		L
Cu^{2+}	-60.75	18.35(25°,1-2)	147.7	0-25	T	μ=4.5x10^{-3} S:50 v% 1,4-Dioxane	64Le	b
P374	PYRIDINE, 2-(2-propyl)-		$C_8H_{11}N$			$C_5H_4N[CH(CH_3)_2]$		
BA_3	-31.8	(1)	--	25	C	S:Nitrobenzene A=CH_3	56Ba	
P375	PYRIDINE, 4-(2-propyl)-		$C_8H_{11}N$			$C_5H_4N[CH(CH_3)_2]$		L
P375, continued								
SbA_5	-125.9	(1)	--	30	C	S:Nitrobenzene A=Cl	63Hb	
P376	PYRIDINE, 4-methoxy-, 1-oxide-		$C_6H_7O_2N$			$C_5H_4N(OCH_3)(\rightarrow O)$		
VAB_2	-33.60	(1)	--	25	C	S:CH_2Cl_2 A=O B= 2,4-Pentanedione	69Pd	
	-19.87	(1)	--	25	C	S:$CHCl_3$ A=O B= 2,4-Pentanedione	69Pd	
I_2	-30.1	3.118(25°,1)	-41.0	18-32	T	S:CCl_4	68Ga	b
P377	PYRIDINE, 1-methyl-		C_5H_6N			$C_5H_5N(H)$		L^+
[Fe]	-16.3	5.7(20°)	53.1	5-30	T	pH=11.2 Protohaemin R:$(Haemin)_2+2L=(HaeminL)_2+2H_2O$	69Md	b,w:I
P378	PYRIDINE, 2-methyl-		C_6H_7N			$C_5H_4N(CH_3)$		L
Be^{2+}	-20.84	3.52(35°,1)	-0.12	35&45	T	--	67Rb	a,b
CoA_2	-33.60	4.83(1-2)	-20.38	25	C	S:Acetone A=Cl	72Ae	
	-39.41	4.19(1-2)	-51.96	25		S:Acetonitrile A=Cl	72Ae	
	-23.51	3.75(1-2)	-7.15	25	C	S:n-Butanol A=Cl	72Af	
	-51.04	5.14(1-2)	-72.93	25	C	S:Cyclohexanone A=Cl	72Ae	
	-11.38	3.79(1-2)	34.27	25	C	S:Dimethylformamide A=Cl	72Ae	
	-28.45	2.69(1-2)	-43.93	25	C	S:Ethanol A=Cl	72Af	
	-25.14	3.81(1-2)	-11.51	25	C	S:2-Methyl propan-2-ol A=Cl	72Af	
	-2.55	1.76	24.98	25	C	S:Ethylene glycol A=Cl R:$CoA_2+2L+2S=[CoL_2S_2]A_2$	72Af	
NiA	-53.68	0.48(1-2)	-171.1	25	T	CM=CL=10^{-4}M S:Benzene A=Diacetylbisbenzoyl hydrazine	58Sa	

Metal Ion	ΔH, kJ /mol	Log K	ΔS, J /mol K	T,°C	M	Conditions	Ref.	Re- marks
P378, continued								
NiA_2	-34	1.114(1)	-90.8	30	C	S:Chlorobenzene A=Diphenyldithio-phosphinato	78Ac	
	-22.84	-2.558(35°,1-2)	-121	25-95	T	S:L A=Salicyl-aldoximate	71Ma	b
CuA_2	-13.0	-1.301(25°,1)	-71	-3-57	T	S:Benzene A=Diethyldithio-carbamate	74Hd	b
	-33	1.919(1)	-75	20	T	S:Benzene A=Trifluoro-acetylacetone	79Si	b,q:±4
	-26.4	0.964(1)	-69.9	25	C	S:Cyclohexane A=t-Butylaceto-acetate	69La	
	-22.68	0.633(1)	-63.2	30	C	S:Benzene A=N-Nitroso-N-phenyl-hydroxylamine	71Gi	q: ±2.3
Ag^+	-24.31	2.333(1)	-36.86	25	C	μ=0.5M(KNO_3)	74Be 72Ed	
	-18.28	2.333(2)	-16.23	25	C	μ=0.5M(KNO_3)	74Be 72Ed	
CdA_2	-33	1.176(1)	-87	30	C	S:Benzene CM=16-61x10^{-3}M A=[$(Bu)_2NCS_2$]	76Aj	
HgA_2	-76.6	1.839(1)	-207.9	30	C	S:Benzene A=Cl CM=1-5x10^{-3}M	73Fa	
	-72.4	1.778(1)	-205.0	30	C	S:Benzene A=Br CM=1-5x10^{-3}M	73Fa	
	-62.8	1.415(1)	-179.9	30	C	S:Benzene A=I CM=1-5x10^{-3}M	73Fa	
BA_3	-41.8	(1)	--	25	C	S:Nitrobenzene A=CH_3	56Ba	
	78.6	-2.398(70°,1)	183.2	-50-100	T	Equimolar amounts of L and BA_3 A=C_3H_5	71Bj	b
	-54.8	1.217(70°,1)	182.8	-50-100	T	Equimolar amounts of L and BA_3 A=C_3H_7	71Bj	b
AlA_3	-109.33	(1)	--	25-28	C	S:n-Hexane A=CH_3	68Hc	
$SiAB_3$	-21.17	1.85(30°,1)	-34.39	10-45	T	S:CCl_4 XL=1x10^{-2} XM=5x10^{-4} A=OH B=CH_3	69Lb	b
	-24.98	1.76(30°,1)	-48.74	20-60	T	S:CCl_4 XL=1x10^{-2} XM=5x10^{-4} A=OH B=C_2H_5	69Lb	b
	-25.52	2.31(30°,1)	-40.04	0-50	T	S:CCl_4 XL=1x10^{-2} XM=4x10^{-4} A=OH B=Phenyl	69Lb	b
$SnAB_3$	-80	4.00(1)	∿-188	30	C	S:Benzene A=Butyl B=Cl	75Fa	

Metal Ion	ΔH, kJ /mol	Log K	ΔS, J /mol K	T,°C	M	Conditions	Ref.	Remarks
P378, continued								
$SnAB_3$, cont.	-34	1.48(2)	-84	30	C	S:Benzene A=Butyl B=Cl	75Fa	q:±5
SnA_2B_2	-56	~4(1)	~-109	30	C	S:Benzene A=Butyl B=NCS	75Fa	
	-45	1.01(2)	-130	30	C	S:Benzene A=Butyl B=NCS	75Fa	
	-79	0.65(1-2)	-248	30	C	S:Benzene A=Phenyl B=Cl	75Fa	
	-90	2.52(2-1)	-248	30	C	S:Benzene A=Methyl B=Cl	75Fa	
SbA_5	-109.6	(1)	--	30	C	S:Nitrobenzene A=Cl	63Hb	
I_2	-35.20	2.185(1)	-76.5	25	C	S:CCl_4	79Ba	
	-33.3	2.37(25°,1)	-66.1	17-35	T	S:Heptane	68Kb	b
P379 PYRIDINE, 3-methyl-			C_6H_7N			$C_5H_4N(CH_3)$		L
Be^{2+}	-17.07	2.94(35°,1)	-0.75	35&45	T	--	67Rb	a,b
VOA_2	-33.89	1.81(1)	-79.1	25	C	μ→0 S:Nitrobenzene A=Acetylacetonate	65Ca	
CoA_2	-43.60	4.90(1-2)	-52.51	25	C	S:Acetone A=Cl	72Ae	
	-40.92	4.62(1-2)	-48.87	25	C	S:Acetonitrile A=Cl	72Ae	
	-27.86	4.10(1-2)	-15.15	25	C	S:n-Butanol A=Cl	72Af	
	-55.52	5.64(1-2)	-78.32	25	C	S:Cyclohexanone A=Cl	72Ae	
	-19.00	4.06(1-2)	14.06	25	C	S:Dimethylforma-mide A=Cl	72Ae	
	-31.76	2.85(1-2)	-51.96	25	C	S:Ethanol A=Cl	72Af	
	-31.38	4.47(1-2)	-19.79	25	C	S:2-Methyl pro-pan-2-ol A=Cl	72Af	
	-7.66	2.57(1-2)	23.43	25	C	S:Ethylene glycol A=Cl R:CoA_2+2L+2S→$[CoL_2S_2]A_2$	72Af	
CoL_2A_2	-56.1	0.348(1-2)	-184.5	20	C	S:$CHCl_3$ A=Cl	66Ca	s
	-43.5	0.785(1-2)	-132.6	20	C	S:$CHCl_3$ A=NCO	66Ca	s
	-63.6	4.204(1-2)	-136.4	20	C	S:$CHCl_3$ A=NCS	66Ca	s
NiA	-63.47	3.90(25°,1-2)	-138.5	25	T	CM≅CL≅10^{-4}M S:Benzene A=Di-acetylbisbenzoyl hydrazine	58Sa	
NiA_2	-100.83	3.98(25°,1-2)	-260.7	20-40	T	S:Benzene A=$(p-CH_3C_6H_4)CH=N-N=COC_6H_5$	77Ia	
	-101.25	4.59(25°,1-2)	-251.9	20-40	T	S:Benzene A=$(m-OCH_3C_6H_4)CH=N-N=COC_6H_5$	77Ia	
	-94.47	4.23(25°,1-2)	-236.0	20-40	T	S:Benzene	77Ia	

Metal Ion	ΔH, kJ /mol	Log K	ΔS, J /mol K	T,°C	M	Conditions	Ref.	Remarks
P379, continued								
NiA_2, cont.						A=C_6H_5CH= N-N=COC_6H_5		
	-99.58	3.70(25°,<u>1-2</u>)	-263.2	20-40	T	S:Benzene A=(p-$CH_3OC_6H_4$)CH= N-N=COC_6H_5	77Ia	
	-107.95	4.71(25°,<u>1-2</u>)	-272.1	20-40	T	S:Benzene A=p-ClC_6H_4CH= N-N=COC_6H_5	77Ia	
	-15.9	-0.44(20°,<u>1-2</u>)	-64.4	10-30	T	S:$CHCl_3$ A=N-Ethyl-β-mercaptocinnamamide	74Cd	b
	-51.9	2.05(20°,<u>1-2</u>)	-138.1	10-30	T	S:$CHCl_3$ A=Ethyl-β-mercaptocinnamate	74Cd	b
	-48.1	1.92(20°,<u>1-2</u>)	-126.4	10-30	T	S:$CHCl_3$ A=Ethyl-β-mercaptothiocinnamate	74Cd	b
	-19.2	-0.35(20°,<u>1-2</u>)	-72.0	10-30	T	S:$CHCl_3$ A=Monothiodibenzoylmethane	74Cd	b
	-39.7	0.37(20°,<u>1-2</u>)	-129.3	10-30	T	S:$CHCl_3$ A=N-Phenyl-β-mercaptocinnamamide	74Cd	b
	-24.3	3.78(<u>1-2</u>)	-8.4	--	C	CM/CL=1/10 S:Dimethylformamide A=Br	76Kj	j,u: 25°,y
	-82.6	3.55(<u>1-2</u>)	<u>-209</u>	25	C	S:Nitrobenzene K determined in benzene A=Ethylxanthato	70Cc	
NiL_2A_2	-90.8	3.91(<u>1-2</u>)	-233.5	20	C	μ≅0 S:$CHCl_3$ A=I	65Nc	
CuA_2	-31.4	1.618(<u>1</u>)	-74.0	25	C	S:Cyclohexane A=t-Butylacetoacetate	69La	
	-21.8	-0.310(25°,<u>1</u>)	-79	-3-57	T	S:Benzene A=Diethyldithiocarbamate	74Hd	b
	-33.64	0.68(<u>1</u>)	-100.0	25	C	μ≅0.005 S:Benzene A=2,4-Pentanedione	63Ma	
	-14.14	(<u>1</u>)	--	25	T	μ≅0.005 S:Benzene A=2,4-Pentanedione	63Ma	
	-27.1	1.364(<u>1</u>)	-63.2	30	C	S:Benzene A=N-Nitroso-N-Phenyl hydroxylamine	71Gi	q: ±2.7
Ag^+	-21.63	2.253(<u>1</u>)	-29.3	25	C	μ=0.5M(KNO_3)	74Be 72Ed	
	-28.3	2.233(<u>2</u>)	-52.3	25	C	μ=0.5M(KNO_3)	74Be 72Ed	
Zn^{2+}	-10.63	1.25(25°,<u>1</u>)	-11.7	25-40	T	--	79Da	b
	-11.88	0.65(25°,<u>2</u>)	-27.2	25-40	T	--	79Da	b

Metal Ion	ΔH, kJ /mol	Log K	ΔS, J /mol K	T,°C	M	Conditions	Ref.	Re-marks
P379, continued								
BA_3	-65.3	(1)		25	C	S:Nitrobenzene A=CH_3	56Ba	
Br_2	-45.6	2.644(1)	-103.3	25	C	S:1,2-Dichloro-ethane	73Tb	
I_2	-35.74	2.253(1)	-76.9	25	C	S:CCl_4	79Ba	
	-36.4	(1)	--	25	C	S:1,2-Dichloro-ethane	73Tb	
	-34.8	2.49(25°,1)	-69.1	17-35	T	S:Heptane	68Kb	b
P380	PYRIDINE, 4-methyl-		C_6H_7N			$C_5H_4N(CH_3)$		L
Be^{2+}	-20.84	3.53(35°,1)	0	35&45	T	--	67Rb	a,b
VOA_2	-36.82	2.14(1)	-82.4	25	C	μ≅0 S:Nitrobenzene A=Acetylacetonate	65Ca	
[Fe]	9.83	1.74(20°)	66.9	5-30	T	pH=11.2 Protohaemin R:$(Haemin)_2$+2L=$(HaeminL)_2$+$2H_2O$	69Md	b
	3.3	2.42(25°,1-2)	54	18-40	T	[Fe]=Iron(II) deuteroporphyrin dimethyl ester S:Benzene	70Ch 71Cd	b,q: ±2.5
	-10.9	1.10(25°,1-2)	-17	18-40	T	[Fe]=Iron(II) diacetyldeutero-porphyrin dimethyl ester S:Benzene	70Ch	b,q: ±2.9
	-36.4	3.74(25°,1-2)	-50.2	18-40	T	[Fe]=Iron(II) mesoporphyrin dimethyl ester S:Benzene	70Ch	b,q: ±5.0
	-58.6	6.60(25°,1-2)	-67	18-40	T	[Fe]=Iron(II) protoporphyrin dimethyl ester S:Benzene	70Ch 71Cd	b
	-31.4	2.78(25°,1-2)	-50	18-40	T	[Fe]=Iron(II) deuteroporphyrin dimethyl ester C:CCl_4	71Cd	b,q: ±7.9
	-102.9	2.78(25°,1-2)	-293	18-40	T	[Fe]=Iron(II) protoporphyrin dimethyl ester S:CCl_4	71Cd	b,q: ±10.9
	-51.0	1.10(25°,1-2)	-151	18-40	T	[Fe]=Iron(II) protoporphyrin dimethyl ester S:$CHCl_3$	71Cd	b,q: ±4.2
CoA	30.5	2.900(25°,1)	-46	20-50	T	S:Toluene A=α,β,γ,δ-Tetra-(p-methoxyphenyl) porphyrin	73Wa	b,q:±4

Metal Ion	ΔH, kJ /mol	Log K	ΔS, J /mol K	T,°C	M	Conditions	Ref.	Re-marks
P380, continued								
CoA_2	-46.74	5.14(1-2)	-58.41	25	C	S:Acetone A=Cl	73Ai	
	-45.65	4.97(1-2)	-57.99	25	C	S:Acetonitrile A=Cl	73Ai	
	-34.81	4.22(1)	-36.07	25	C	S:n-Butyl alcohol A=Cl	73Aj	q:±3.0
	-39.41	4.51(1)	-45.90	25	C	S:t-Butyl alcohol A=Cl	73Aj	q:±3.1
	-63.34	5.70(1-2)	-103.47	25	C	S:Cyclohexanone A=Cl	73Ai	
	-27.03	4.12(1-2)	-11.80	25	C	S:Dimethylforma- mide A=Cl	73Ag	
	-8.54	2.62(1)	21.46	25	C	S:Ethylene glycol A=Cl	73Aj	q: ±0.58
	-33.70	2.88(1)	-57.57	25	C	S:Ethyl alcohol A=Cl	73Aj	q:±2.1
CoL_2A_2	-65.7	1.049(1-2)	-204.2	20	C	S:$CHCl_3$ A=Cl	66Ca	s
	-55.6	1.182(1-2)	-167.4	20	C	S:$CHCl_3$ A=NCO	66Ca	s
	-69.9	4.892(1-2)	-143.1	20	C	S:$CHCl_3$ A=NCS	66Ca	s
NiA	-28.1	0.580(1)	-82	30	C	S:Acetonitrile CM=10-15x10^{-3}M A=BF_2-bridged glyoximato R_1=Me R_2=Me	76Gh	q:±3.8
	-44.0	0.556(1)	-135	30	C	S:Acetonitrile CM=2x10^{-3}M A=BF_2-bridged glyoximato R_1-R_2= -$(CH_2)_4$-	76Gh	
	-34.0	0.698(1)	-99	30	C	S:Acetonitrile CM=7-15x10^{-3}M A=BF_2=bridged glyoximato R_1=Me R_2=Pr	76Gh	
	-35.0	0.724(1)	-102	30	C	S:Acetonitrile CM=7-15x10^{-3}M A=BF_2=bridged glyoximato R_1=Me R_2=iso-Pr	76Gh	
	-25.0	0.799(1)	-67	30	C	S:Acetonitrile CM=7-15x10^{-3}M A=BF_2-bridged glyoximato R_1=Me R_2=Et	76Gh	
	-30.4	1.55(1)	-71	30	C	S:Acetonitrile CM=2-5x10^{-3}M A=BF_2-bridged glyoximato	76Gh	

Metal Ion	ΔH, kJ /mol	Log K	ΔS, J /mol K	T,°C	M	Conditions	Ref.	Re-marks
P380, continued								
NiA, cont.	-23.0	1.45(1)	-48	30	C	R_1=Ph R_2=H S:Acetonitrile CM=2-5x10^{-3}M A=BF_2-bridged glyoximato R_1=Ph R_2=Ph	76Gh	
	-65.48	4.45(25°,1-2)	-135.1	25	T	CM≅CL≅10^{-4}M S:Benzene A=Diacetylbisbenzoyl hydrazine	58Sa	
	-42.34	2.062(1)	-102.59	25	C	S:Toluene A=bis-(salicylidene-γ-iminopropyl) methylamine	73Kh	
	-28.16	1.082(1)	-73.76	25	C	S:CH_2Cl_2 A=bis-(salicylidene-γ-iminopropyl) methylamine	73Kh	
NiA_2	-40	2.699(1)	-79	30	C	S:Benzene A=O,O'-Diethylthiomalonato	77Ah	
	-82.8	1.48(1)	-248.1	30	C	S:Benzene A=Diethyldithiophosphato	71Dc	q:±8.4
	7.1	1.88(2)	59.0	30	C	S:Benzene A=Diethyldithiophosphato	71Dc	q:±7.1
	-85.86	(1-2)	--	25	C	S:Benzene A=2,2,-6,6-Tetramethyl-3,5-heptanedione	73Ce	
	-100	2.477(1)	-377	30	C	S:Benzene A=(SOCNPr_2)	77Gf	
	-110	2.477(1)	-410	30	C	S:Benzene A=(SOCNBu_2)	77Gf	
	-62	2.301(1)	-161	30	C	S:Butan-2-one A=O,O'-Diethylthiomalanato	77Ah	
	-74	2.301(1)	-201	30	C	S:CCl_4 A=O,O'-Diethylthiomalanato	77Ah	
	-52.7	2.10(20°,1-2)	-139.3	10-30	T	S:$CHCl_3$ A=Ethyl-β-mercaptocinnamate	74Cd	b
	-19.2	-0.28(20°,1-2)	-72.8	10-30	T	S:$CHCl_3$ A=N-Ethyl-β-mercaptocinnamamide	74Cd	b
	-50.6	2.07(20°,1-2)	-132.2	10-30	T	S:$CHCl_3$ A=Ethyl-β-mercaptocinnamate	74Cd	b
	-25.1	-0.18(25°,1-2)	-89.1	10-30	T	S:$CHCl_3$ A=Monothiodibenzoylmethane	74Cd	b
	-42.7	0.58(20°,1-2)	-133.9	10-30	T	S:$CHCl_3$	74Cd	b

Metal Ion	ΔH, kJ /mol	Log K	ΔS, J /mol K	T,°C	M	Conditions	Ref.	Remarks
P380, continued								
NiA_2, cont.						A=N-Phenyl-β-mercaptocinnamamide		
	-35	3.00(1)	-58.0	30	C	S:Chlorobenzene A=Diphenyldithiophosphinato	78Ac	
	-76	2.602(1)	-202	30	C	S:Cyclohexane A=O,O'-Diethylthiomalonato	77Ah	
	-64	2.000(1)	-171	30	C	S:$C_2H_4Cl_2$ A=O,O'-Diethylthiomalonato	77Ah	
	-22	1.544(1)	-43	30	C	S:1,2-Dichloroethane A=Dithizonate	78Gc	
	-26	1.544(1)	-59	19	C	S:1,2-Dichloroethane A=Dithizonate	78Gc	
	-61	2.301(1)	-157	30	C	S:MeCN A=O,O'-Diethylthiomalonato	77Ah	
	15	2.477(2)	2	30	C	S:Benzene A=$(SOCNBu_2)$	77Gf	
	7	2.477(2)	-24	30	C	S:Benzene A=$(SOCNPr_2)$	77Gf	
	-22	2.000(2)	-34	30	C	S:Benzene A=O,O'-Diethylthiomalonato	77Ah	
	-1	2.000(2)	42	30	C	S:Butan-2-one A=O,O'-Diethylthiomalonato	77Ah	
	7	2.000(2)	61	30	C	S:CCl_4 A=O,O'-Diethylthiomalonato	77Ah	
	-41	1.699(2)	-102.7	30	C	S:Chlorobenzene A=Diphenyldithiophosphinato	78Ac	
	-26	2.301(2)	-42	30	C	S:Cyclohexane A=O,O'-Diethylthiomalonato	77Ah	
	1	2.000(2)	42	30	C	S:$C_2H_4Cl_2$ A=O,O'-Diethylthiomalonato	77Ah	
	7	0.845(2)	40	19	C	S:1,2-Dichloroethane A=Dithizonate	78Gc	
	3	2.000(2)	48	30	C	S:MeCN A=O,O'-Diethylthiomalonato	77Ah	
	-103.76	4.60(25°,1-2)	-260.2	20-40	T	S:Benzene A=C_6H_5CH=N-N=COC_6H_5	77Ia	
	-94.14	4.89(25°,1-2)	-222.2	20-40	T	S:Benzene A=$(m\text{-}OCH_3C_6H_4)CH$=N-N=COC_6H_5	77Ia	

Metal Ion	ΔH, kJ /mol	Log K	ΔS, J /mol K	T,°C	M	Conditions	Ref.	Remarks
P380, continued								
NiA_2, cont.	-104.18	5.00(25°,1-2)	-254.0	20-40	T	S:Benzene A=(p-ClC_6H_4)CH= N-N=COC_6H_5	77Ia	
	-93.30	4.03(25°,1-2)	-236.0	20-40	T	S:Benzene A=(p-$CH_3OC_6H_4$)CH= N-N=COC_6H_5	77Ia	
	-92.05	4.38(25°,1-2)	-225.1	20-40	T	S:Benzene A=(p-$CH_6C_6H_4$)CH= N-N=COC_6H_5	77Ia	
	-70.6	3.21(1-2)	-171	30	C	CM=1.6x10^{-3} S:Benzene A=$(CH_3)_2CHOCS_2$	74Gd	
	-79.4	4.41(1-2)	-178	30	C	CM=1-4x10^{-3} S:Benzene A=CH_3OCS_2	74Gd	
	-64	-1.70(1-2)	-246	25-30	C	S:Benzene A=Dibenzyldithio- carbamato	75Af	q:±2.2
	-6.5	(1-2)	--	25-30	C	S:Benzene A=Dibutyldithio- carbamato	75Af	q:±0.9
	-3.7	(1-2)	--	25-30	C	S:Benzene A=Dipropyldithio- carbamato	75Af	q:±0.5
	-27.6	4.09(1-2)	-12.5	--	C	CM/CL=1.10 S:Dimethylform- amide A=Br	76Kj	j,u: 25°,y
	-80.7	3.89(1-2)	-196	25	C	S:Nitrobenzene K determined in benzene A=Ethylxanthato	70Cc	
RhAB	-60.7	3.67	-134	24	C	S:Benzene R:$[RhAB]_2$+2L= 2RhABL A=Cl B=1,5-Cycloocta- diene	76Lb	
	-57.7	2.94	-138	24	C	S:Tetrahydrofuran R:$[RhAB]_2$+2L= 2RhABL A=Cl B=1,5-Cycloocta- diene	76Lb	
$Rh_2A_4B_2$	-52.3	--	--	24	C	S:Benzene R:$Rh_2A_4B_2$+2L= $2RhA_2BL$ A=CO B=Cl	76Pd	u:24°
PdAB	-44.4	3.954	-73	24	C	S:Benzene R:$[PdAB]_2$+2L= 2[PdABL] A=(π-MeAllyl) B=Cl	77Lb	
IrA_2BC_2	-77.8	5.92(1)	-146	25	C	S:CH_2Cl_2 A=Cl	77La	

Metal Ion	ΔH, kJ /mol	Log K	ΔS, J /mol K	T,°C	M	Conditions	Ref.	Re-marks
P380, continued								
IrA_2BC_2, cont.						$B=C(O)CF_3$ $C=P(C_6H_5)_3$		
Cu^{2+}	-19.66	(1)	--	--	-	CL=0.2M $CM=3x10^{-2}M$ $CHNO_3=0.01M$	68Wb	f:3.44 u,y
CuA_2	-23.8	-0.208(25°,1)	-84	-3-57	T	S:Benzene A=Diethyldithio-carbamate	74Hd	b
	-48.1	0.99(1)	-143.1	25	C	μ≅0.005 S:Benzene A= 2,4-Pentanedione	63Ma 69Ga	
	-11.34	(1)	--	25	T	μ≅0.005 S:Benzene A= 2,4-Pentanedione	63Ma	
	-28.41	1.412(1)	-66.9	30	C	S:Benzene A=N-Nitro-N-Phenyl-hydroxylamine	71Gi	q:±2.8
	-31.8	1.735(1)	-73.6	25	C	S:Cyclohexane A=t-Butylaceto-acetate	69La	
	-30.71	3.175(1)	-48.99	30	C	S:Benzene A=1,-1,1-Trifluoro-pentane-2,4-dione	74Gc	
	-25.3	0.0791(1)	-82	30	C	S:Benzene $CM=10-15x10^{-3}M$ A=	76Ai	q:±3.7
	-25.0	0.0414(1)	-80	30	C	S:Benzene $CM=9-17x10^{-3}M$ A=	74Ai	q:±5
	-21.4	0.518(1)	-61	30	C	S:Benzene $CM=5-16x10^{-3}M$ A=	76Ai	
	-31.1	0.556(1)	-92	30	C	S:Benzene $CM=9-10x10^{-3}M$ A=	76Ai	
	-27.6	0.114(1)	-89	30	C	S:Benzene $CM=1-12x10^{-3}M$ A=	76Ai	q:±2.2

Metal Ion	ΔH, kJ /mol	Log K	ΔS, J /mol K	T,°C	M	Conditions	Ref.	Re- marks
P380, continued								
CuA_2, cont.	-25.9	0.322(1)	-79	30	C	S:Benzene CM=4-15x10^{-3}M	76Ai	
						A=		
	-30.7	3.33(1)	-41	20	T	S:Benzene A=Trifluoro- acetylacetone	79Si	b
Ag^+	-25.48	2.183(1)	-43.68	25	C	μ=0.5M(KNO_3)	74Be 72Vc	
	-28.07	2.464(2)	-46.99	25	C	μ=0.5M(KNO_3)	74Be 72Vc	
Zn^{2+}	-12.55	0.81(25°,1)	-26.4	25-40	T	--	79Da	b
	-11.30	1.40(25°,2)	-10.9	25-40	T	--	79Da	b
ZnA_2	-28	(1)	--	30	C	S:Acetonitrile A=Cl	76Ai	
	-32	(1)	--	30	C	S:Acetonitrile A=Br	76Ai	q:±3
	-30	(1)	--	30	C	S:Acetonitrile A=I	76Ai	
	-35.8	2.132(1)	-77	30	C	S:Benzene CM=7.7-45x10^{-3}M A=[$(Bu)_2NCS_2$]	76Aj	
	-35.8	2.456(1)	-73	30	C	S:Benzene CM=2.7-41x10^{-3}M A=[$(PhCH_2)NCS_2$]	76Aj	
	-58.2	2.46(1)	-148	25	C	S:Cyclohexane A=$N[Si(CH_3)_3]_2$	75Fc	
	-40.2	2.336(25°,1)	-88	25&46	T	S:Toluene A=$[S_2CN(CH_3)_2]$	65Ch	a,b
	-26	(2)	--	30	C	S:Acetonitrile A=Cl	76Ak	
	-23	(2)	--	30	C	S:Acetonitrile A=Br	76Ak	q:±3
	-26	(2)	--	30	C	S:Acetonitrile A=I	76Ak	q:±2
	-7	(3)	--	30	C	S:Acetonitrile A=Cl	76Ak	
	-6	(3)	--	30	C	S:Acetonitrile A=Br	76Ak	q:±3
	-11	(3)	--	30	C	S:Acetonitrile A=I	76Ak	q:±3
CdA_2	-20	3.301(1)	-3	30	C	S:Acetonitrile A=I	76Ak	q:±5
	-31	2.230(1)	-59	30	C	S:Benzene CM=5-54x10^{-3}M A=[$(Bu)_2NCS_2$]	76Aj	
	-35	3.398(1)	-51	30	C	S:Benzene CM=1.4-9x10^{-3}M A=[$(PhCH_2)_2NCS_2$]	76Aj	

Metal Ion	ΔH, kJ /mol	Log K	ΔS, J /mol K	T,°C	M	Conditions	Ref.	Re-marks
P380, continued								
CdA_2, cont.	-46.4	2.58(1)	-106	25	C	S:Cyclohexane $A=N[Si(CH_3)_3]_2$	75Fa	
	-30	2.602(2)	-47	30	C	S:Acetonitrile A=I	76Ak	q:±5
HgA_2	-54	1.799(1)	-144	30	C	S:Acetonitrile A=I	76Ak	
	-73.6	2.412(1)	-197.1	30	C	S:Benzene A=Cl $CM=1\text{-}5x10^{-3}M$	73Fa	
	-72.8	2.579(1)	-190.8	30	C	S:Benzene A=Br $CM=1\text{-}5x10^{-3}M$	73Fa	
	-67.8	2.425(1)	-177.8	30	C	S:Benzene A=I $CM=1\text{-}5x10^{-3}M$	73Fa	
	-12.6	0.67(1)	-29	30	C	$CM=1\text{-}20x10^{-3}$ S:Benzene $A=C_6F_5$	74Fc	
BA_3	-66.5	(1)	--	25	C	S:Nitrobenzene $A=CH_3$	56Ba	
$SnAB_3$	-125	6.0(1-2)	~-300	--	-	S:Benzene A=Butyl B=Cl	75Fa	
	-140	>6(1-2)	<-350	30	C	S:Benzene A=Phenyl B=Cl	75Fa	
SnA_2B_2	-25.7	0.672(1)	-74	25	C	S:Benzene A=Bu B=(OCOMe)	77Ge	
	-25.5	0.544(1)	-74	25	C	S:Benzene A=Bu B=(OCOEt)	77Ge	
	-27.5	1.670(1)	-59	25	C	S:Benzene A=Bu B=(NCO)	77Ge	
	-62	1.08(1)	-184	30	C	S:Benzene A=Butyl B=Cl	75Fa	q:±10
	-18	1.00(2)	-40	30	C	S:Benzene A=Butyl B=Cl	75Fa	q:±10
	-52	~4(1)	-96	30	C	S:Benzene A=Butyl B=NCS	75Fa	q:±5
	-46	2.11(2)	-111	30	C	S:Benzene A=Butyl B=NCS	75Fa	q:±5
	-60	1.60(1)	-167	30	C	S:Benzene A=Methyl B=Cl	75Fa	q:±10
	-18	1.48(2)	-34	30	C	S:Benzene A=Methyl B=Cl	75Fa	q:±10
	-30.8	1.694(1)	-69	25	C	S:Benzene A=Ph B=(OCOMe)	77Ge	
	-50	2.30(1)	-121	30	C	S:Benzene A=Phenyl B=Cl	75Fa	q:±10
	-56	2.18(2)	-143	30	C	S:Benzene A=Phenyl B=Cl	75Fa	q:±10
	-44.8	0.365(1)	-141	30	C	S:Benzene A=Phenyl B=Cl	75Fa	
SbA_5	-129.7	(1)	--	30	C	S:Nitrobenzene A=Cl	63Hb	
Br_2	-47.3	2.867(1)	-103.8	25	C	S:1,2-Dichloro-ethane	73Tb	

Metal Ion	ΔH, kJ /mol	Log K	ΔS, J /mol K	T,°C	M	Conditions	Ref.	Re-marks
P380, continued								
I_2	-37.30	2.233(1)	-81.9	25	C	S:CCl_4	79Ba	
	-37.4	2.58(25°,1)	-75.1	17-35	T	S:Heptane	68Kb	b
P381	PYRIDINE, 4-(methylacetate)-		$C_8H_9O_2N$			$C_5H_4N(CH_2CO_2CH_3)$		
[Ni]	6.3	3.15(25°,1)	25	18-40	T	[Ni]=Nickel(II) mesotetraphenyl-porphine S:$CHCl_3$	72Ce	b,q: ±2.9
[Zn]	-19.7	3.22(25°,1)	-4	18-40	T	[Zn]=Zinc(II) mesotetraphenyl-porphine S:$CHCl_3$	72Ce	b,q: ±2.1
P382	PYRIDINE, 2-(methylamino-ethyl)-		$C_8H_{12}N_2$			$C_5H_4N(CH_2CH_2NHCH_3)$		
Ni^{2+}	-26.36	4.65(1)	0.8	25	C	μ=0.5M(KNO_3)	71Gd	
P383	PYRIDINE, 2-(methylamino-methyl)-		$C_7H_{10}N_2$			$C_5H_4N(CH_2NHCH_3)$		L
Co^{2+}	-27.57	5.10(30°,1)	6.7	10-40	T	μ=0	59Gb	b
	-28.28	5.52(1)	5.0	25	C	μ=0.5M(KNO_3)	71Gd	
	-70.62	(1)	--	25	C	μ=0.1(KNO_3) S:50w% Dioxane	66Wa	
	-23.47	3.84(30°,2)	-3.85	10-40	T	μ=0	59Gb	b
	-25.02	3.98(2)	-7.5	25	C	μ=0.5M(KNO_3)	71Gd	
	-11.09	2.63(30°,3)	1.38	10-40	T	μ=0	59Gb	b
Ni^{2+}	-37.82	6.66(30°,1)	2.9	10-40	T	μ=0	59Gb	b
	-35.56	6.91(1)	13.4	25	C	μ=0.5M(KNO_3)	71Gd	
	-70.75	(1)	--	25	C	μ=0.1(KNO_3) S:50w% Dioxane	66Wa	
	-35.31	5.30(30°,2)	-15.1	10-40	T	μ=0	59Gb	b
	-35.56	5.53(2)	-13.4	25	C	μ=0.5M(KNO_3)	71Gd	
	-15.73	2.79(30°,3)	1.7	10-40	T	μ=0	59Gb	b
Cu^{2+}	-32.09	9.27(30°,1)	71.5	10-40	T	μ=0	59Gb	b
	-44.73	9.07(1)	24.3	25	C	μ=0.5M(KNO_3)	71Gd	
	-80.50	14.60(1)	9.54	25	C	μ=0.1(KNO_3) S:50w% Dioxane	66Wa	
	-41.17	6.55(30°,2)	-10.5	10-40	T	μ=0	59Gb	b
	-34.02	6.71(2)	14.6	25	C	μ=0.5M(KNO_3)	71Gd	
Zn^{2+}	-22.01	4.73(30°,1)	18.0	10-40	T	μ=0	59Gb	b
	-24.18	4.96(1)	13.8	25	C	μ=0.5M(KNO_3)	71Gd	
	-46.02	(1)	--	25	C	μ=0.1(KNO_3) S:50w% Dioxane	66Wa	
	-18.32	3.62(2)	7.9	25	C	μ=0.5M(KNO_3)	71Gd	
Cd^{2+}	-20.46	4.30(30°,1)	14.6	10-40	T	μ=0	59Gb	b
	-21.84	4.49(1)	12.1	25	C	μ=0.5M(KNO_3)	71Gd	
	-41.92	8.54(1)	22.93	25	C	μ=0.1(KNO_3) S:50w% Dioxane	66Wa	
	-20.04	3.35(2)	0.8	25	C	μ=0.5(KNO_3)	71Gd	

Metal Ion	ΔH, kJ /mol	Log K	ΔS, J /mol K	T,°C	M	Conditions	Ref.	Re-marks
P384	PYRIDINE, 2-(methylamino-methyl)-6-methyl-		$C_8H_{12}N_2$			$C_5H_3N(CH_2NHCH_3)(CH_3)$		L
Co^{2+}	-16.15	3.41(30°,1)	12	10-40	T	μ=0	61Rb	b
	-18.28	3.57(1)	7.1	25	C	μ=0.5M(KNO_3)	71Gd	
	-47.86	7.87(1)	-9.83	25	C	μ=0.1(KNO_3) S:50w% Dioxane	66Wa	
	-9.92	1.41(2)	-6.3	25	C	μ=0.5M(KNO_3)	71Gd	
Ni^{2+}	-29.71	4.58(30°,1)	-12	10-40	T	μ=0	61Ra	b
	-24.98	4.74(1)	7.1	25	C	μ=0.5M(KNO_3)	71Gd	
	-31.46	7.57(1)	39.37	25	C	μ=0.1(KNO_3) S:50w% Dioxane	66Wa	
	-19.41	2.52(2)	-16.7	25	C	μ=0.5M(KNO_3)	71Gd	
Cu^{2+}	-29.20	6.48(30°,1)	29	10-40	T	μ=0	61Rb	b
	-29.12	6.88(1)	34.3	25	C	μ=0.5M(KNO_3)	71Gd	
	-50.25	12.26(1)	66.11	25	C	μ=0.1(KNO_3) S:50w% Dioxane	66Wa	
	-18.28	4.94(30°,2)	33	10-40	T	μ=0	61Rb	b
	-27.49	5.44(2)	12.1	25	C	μ=0.5M(KNO_3)	71Gd	
Zn^{2+}	-16.40	7.76(1)	93.55	25	C	μ=0.1(KNO_3) S:50w% Dioxane	66Wa	
P385	PYRIDINE, 2-methyl-, 1-oxide		C_6H_7ON			$C_5H_4N(\rightarrow O)(CH_3)$		L
VAB_2	-16.99	(1)	--	25	C	S:CH_2Cl_2 A=O B= 2,4-Pentanedione	69Pd	
I_2	-25.19	2.05(25°,1)	-45.2	9-42	T	S:CCl_4	57Ka	b
	-25.1	(1)	--	--	-	S:CCl_4	60Bb	
P386	PYRIDINE, 3-methyl-, 1-oxide		C_6H_7ON			$C_5H_4N(\rightarrow O)(CH_3)$		L
VAB_2	-23.30	(1)	--	25	C	S:CH_2Cl_2 A=O B= 2,4-Pentanedione	69Pd	
I_2	-25.48	2.13(25°,1)	-44.8	11-38	T	S:CCl_4	57Ka	b
	-25.5	(1)	--	--	-	S:CCl_4	60Bb	
P387	PYRIDINE, 4-methyl-, 1-oxide		C_6H_7ON			$C_5H_4N(\rightarrow O)(CH_3)$		L
VAB_2	-30.21	(1)	--	25	C	S:CH_2Cl_2 A=O	69Pd	
	-14.6	(1)	--	25	C	S:$CHCl_3$ A=O B= 2,4-Pentanedione	69Pd	q:±0.84
SbA_5	-151.9	(1)	--	25	C	S:$C_2H_4Cl_2$ A=Cl	72Lb	
I_2	-25.73	2.14(25°,1)	-45.2	12-38	T	S:CCl_4	57Ka	b
	-26.4	2.312(25°,1)	-44.4	25&32	T	S:CCl_4	68Ga	a,b
	-25.9	(1)	--	--	-	S:CCl_4	60Bb	
P388	PYRIDINE, 4-nitro-, 1-oxide		$C_5H_4O_3N_2$			$C_5H_4N(NO_2)(\rightarrow O)$		L
VAB_2	-9.37	(1)	--	25	C	S:CH_2Cl_2 A=O B= 2,4-Pentanedione	69Pd	

Metal Ion	ΔH, kJ /mol	Log K	ΔS, J /mol K	T,°C	M	Conditions	Ref.	Remarks
P389	PYRIDINE, 1-oxide		C_5H_5ON			$C_5H_5N(\rightarrow O)$		L
VAB_2	-25.98	(1)	--	25	C	S:CH_2Cl_2 A=O B= 2,4-Pentanedione	69Pd	
	-10.75	(1)	--	25	C	S:$CHCl_3$ A=O B= 2,4-Pentanedione	69Pd	
NiA_2	-23.76	2.456(1)	-32.6	25		S:Benzene A=2,2,-6,6-Tetramethyl-3,5-heptanedione	73Ce	
	-38.9	0.699(2)	-117.2	25	C	S:Benzene A=2,2,-6,6-Tetramethyl-3,5-heptanedione	73Ce	
$Rh_2A_4B_2$	-27.2	5.898	22	--	C	S:Benzene R:$Rh_2A_4B_2$+2L= $2RhA_2BL$ A=CO B=Cl	76Pd	u:24°
CuA_2	-21	2.30(25°,1)	-27	15-35	T	S:Benzene A=2-Thenoyltri-fluoroacetonato	76Ag	b,q:±8
	-10	1.30(25°,1)	-9	5-25	T	S:Chloroform A=2-Thenoyltri-fluoroacetonato	76Ag	b,q:±8
	-22	2.35(25°,1)	-29	5-35	T	S:Toluene A=2-Thenoyltri-fluoracetonato	76Ag	b,q:±8
SnA_2B_2	-32	2.699(1)	-52	25	C	S:Benzene A=Bu B=Cl	77Ge	
	-9	1.699(2)	2.3	25	C	S:Benzene A=Bu B=Cl	77Ge	
	-42	4.301(1)	2.3	25	C	S:Benzene A=Bu B=(NCS)	77Ge	
	-30	2.301(2)	-57	25	C	S:Benzene A=Bu B=(NCS)	77Ge	
	-16.8	1.491(1)	-27	25	C	S:Benzene A=Bu B=(OCOMe)	77Ge	
	-15.3	1.369(1)	-24	25	C	S:Benzene A=Bu B=(OCOEt)	77Ge	
	-37	3.00(1)	-67	25	C	S:Benzene A=Me B=Cl	77Ge	
	-21	1.602(2)	-40	25	C	S:Benzene A=Me B=Cl	77Ge	
	-30.8	1.845(1)	-66	25	C	S:Benzene A=Ph B=(OCOMe)	77Ge	
	-39	∞(1)	--	25	C	S:Benzene A=Ph B=Cl	77Ge	
	-42	1.929(2)	-104	25	C	S:Benzene A=Ph B=Cl	77Ge	
$SnAB_3$	-65	∞(1)	--	25	C	S:Benzene A=Ph B=Cl	77Ge	
	-65	∞(2)	--	25	C	S:Benzene A=Ph B=Cl	77Ge	
I_2	-24.06	1.891(23°,1)	-44.64	7-33	T	S:CCl_4	68Ga	b

Metal Ion	ΔH, kJ /mol	Log K	ΔS, J /mol K	T,°C	M	Conditions	Ref.	Remarks
P389, continued								
I_2, cont.	-24.48	1.892(25°,1)	-46.0	7-33	T	CL=10^{-4}-10^{-3}M S:CCl_4	65Kc	b
	-23.39	1.85(25°,1)	-43.1	10-38	T	S:CCl_4	57Ka	b
	-23.4	(1)	--	--	-	S:CCl_4	60Bb	
P390	PYRIDINE, 2-propyl		$C_8H_{11}N$			$C_5H_4N(CH_2CH_2CH_3)$		
Ag^+	-23.35	2.118(1)	-37.7	25	C	μ=0.5M(KNO_3)	76Be	
	-17.32	1.986(2)	-20.1	25	C	μ=0.5M(KNO_3)	76Be	
P391	PYRIDINE, 4-propyl-		$C_8H_{11}N$			$C_5H_4N(CH_2CH_2CH_3)$		L
CoL_2A_2	-66.1	1.07(1-2)	-205.0	20	C	S:$CHCl_3$ A=Cl	66Ca	s
	-68.6	4.90(1-2)	-140.6	20	C	S:$CHCl_3$ A=NCS	66Ca	s
P392	PYRIDINE, 1,2,4,6-tetramethyl-		$C_9H_{14}N$			$C_5H_2N(CH_3)_4$		L^+
[Fe]	0	6.4(20°)	122.6	5-30	T	pH=11.2 Protohaemin R:$(Haemin)_2$+2L=$(HaeminL)_2$+$2H_2O$	69Md	b,w:I
P393	PYRIDINE, 1,2,4-trimethyl-		$C_8H_{12}N$			$C_5H_3N(CH_3)_3$		L^+
[Fe]	0	6.8(20°)	130.5	5-30	T	pH=11.2 Protohaemin R:$(Haemin)_2$+2L=$(HaeminL)_2$+$2H_2O$	69Md	b,w:I
P394	PYRIDINE, 2,4,6-trimethyl-		$C_8H_{11}N$			$C_5H_2N(CH_3)_3$		L
CuA_2	-14.81	0.415(1)	-41.0	30	C	S:Benzene A=N-Nitroso-N-phenyl-hydroxylamine	71Gi	
BA_3	-23.76	(1)	--	25	C	S:Nitrobenzene Cl≅0.20M CM≅0.25M A=F	56Bb	
Br_2	-47.7	2.725(1)	-108.4	25	C	S:1,2-Dichloro-ethane	73Tb	
I_2	-35.40	1.756(1)	-85.3	25	T	S:CCl_4	79Ba	
	-31.0	(1)	--	25	C	S:1,2-Dichloro-ethane	73Tb	
IA_3B	-20.9	0.95(25°,1)	-52.7	-30-40	T	CL=0-0.48 S:Cyclopentane A=F B=CH_3	65Ld 65Lc	b,1

Metal Ion	ΔH, kJ /mol	Log K	ΔS, J /mol K	T,°C	M	Conditions	Ref.	Re-marks
P395	PYRIDINE, 2,4,6-trimethyl-1-oxide-		$C_8H_{11}ON$			$C_5H_2N(CH_3)_3(\rightarrow O)$		L
VAB_2	-8.62	(1)	--	25	C	S:CH_2Cl_2 A=O B= 2,4-Pentanedione	69Pd	
P396	PYRIDINE, 4-vinyl-		C_7H_7N			$C_5H_4N(CH:CH_2)$		
[Fe]	-6.7	2.86(25°,1-2)	33	18-40	T	[Fe]=Iron(II) deuteroporphyrin dimethyl ester S:Benzene	70Ch	b,q: ±3.8
	-15.1	2.64(25°,1-2)	0	18-40	T	[Fe]=Iron(II) diacetyldeutero-porphyrin dimethyl ester S:Benzene	70Ch	b,q: ±3.8
	-42.7	3.66(25°,1-2)	-67	18-40	T	[Fe]=Iron(II) mesoporphyrin dimethyl ester S:Benzene	70Ch	b,q: ±4.2
	-31.0	3.74(25°,1-2)	-33	18-40	T	[Fe]=Iron(II) protoporphyrin dimethyl ester S:Benzene	70Ch	b,q: ±4.6
P397	4-PYRIDINECARBOXALDEHYDE		C_6H_5ON			$C_5H_4N(COH)$		
[Ni]	25.1	2.27(25°,1)	130	18-40	T	[Ni]=Nickel(II) mesotetraphenyl-porphine S:$CHCl_3$	72Ce	b,q: ±4.2
Cu^{2+}	-18.49	(1)	--	--	-	CL=0.2M $CM=3x10^{-2}M$ $CHNO_3$=0.01M	68Wb	f:3.06 u,y
P398	2-PYRIDINECARBOXYLIC ACID		$C_6H_5O_2N$			$(C_5H_4N)CO_2H$		
YA	-3.8	3.28(25°,1)	50	25-45	T	μ=0.2M(KNO_3) A=EDTA	76Pe	b
	-4.6	3.46(25°,1)	50	25-45	T	μ=0.2M(KNO_3) A=N-Hydroxyethyl-enediaminetri-acetic acid	76Pe	b
	-25.1	3.69(25°,1)	-12	25-45	T	μ=0.2M(KNO_3) A=Iminotriacetic acid	76Pe	b
P399	3-PYRIDINECARBOXYLIC ACID, nitrile		$C_6H_4N_2$			$C_5H_4N(CN)$		L
CuA_2	-10.46	0.79(1)	-19.96	25	C	μ≅0.005 S:Benzene A=2,4-Pentanedione	63Ma	
	-12.22	(1)	--	25	T	μ≅0.005	63Ma	

Metal Ion	ΔH, kJ /mol	Log K	ΔS, J /mol K	T, °C	M	Conditions	Ref.	Re-marks
P399, continued								
CuA_2, cont.						S:Benzene A=2,4-Pentanedione		
Ag^+	-28.4	0.970(1)	-77	25	C	μ=0.02M($NaClO_4$)	75Mb	q:±1.2
	-34.6	0.343(2)	-111	25	C	μ=0.02M($NaClO_4$)	75Mb	q:±15
P400	4-PYRIDINECARBOXYLIC ACID, amide		$C_6H_6ON_2$			$C_5H_4N(CONH_2)$		
Cu^{2+}	-12.89	(1)	--	--	-	CL=0.2M CMx10^{-2}M $CHNO_3$=0.01M	68Wb	f:2.20 u,y
P401	4-PYRIDINECARBOXYLIC ACID, butyl ester		$C_{10}H_{13}O_2N$			$C_5H_4N(CO_2HCH_2CH_2CH_2CH_3)$		L
CuA_2	-25.19	2.258(1)	-40.00	30	C	S:Benzene A=1,1,1-Trifluoro-pentane-2,4-dione	74Gc	
P402	4-PYRIDINECARBOXYLIC ACID methyl ester		$C_7H_7O_2N$			$C_5H_4N(CO_2CH_3)$		
NiA_2	-82.63	(1-2)	--	25	C	S:Cyclohexane A=2,2,6,6-Tetra-methyl-3,5-heptanedione	73Ce	
Cu^{2+}	-12.64	(1)	--	--	-	CL=0.2M CM=3x 10^{-2}M $CHNO_3$=0.01M	68Wb	f:2.20 u,y
P403	4-PYRIDINECARBOXYLIC ACID, nitrile		$C_6H_4N_2$			$C_5H_4N(CN)$		L
[Fe]	-16.3	3.8(20°)	16.7	5-30	T	pH=11.2 R:$(Haemin)_2$+2L=$(HaeminL)_2$+$2H_2O$	69Md	b
	-28.9	3.52(25°,1-2)	-29	18-40	T	[Fe]=Iron(II) deuteroporphyrin dimethyl ester S:Benzene	70Ch 71Cd	b,q: ±4.6
	31.4	1.25(25°,1-2)	134	18-40	T	[Fe]=Iron(II) diacetyldeutero-porphyrin dimethyl ester S:Benzene	70Ch	b
	-49.0	2.78(25°,1-2)	-100	18-40	T	[Fe]=Iron(II) mesoporphyrin dimethyl ester S:Benzene	70Ch	b,q: ±4.6
	-37.6	3.00(25°,1-2)	-67	18-40	T	[Fe]=Iron(II) protoporphyrin dimethyl ester S:Benzene	70Ch 71Cd	b,q: ±4.2

Metal Ion	ΔH, kJ /mol	Log K	ΔS, J /mol K	T,°C	M	Conditions	Ref.	Remarks
P403, continued								
[Fe], cont.	-72.8	3.08(25°,1-2)	-184	18-40	T	[Fe]=Iron(II) deuteroporphyrin dimethyl ester S:CCl_4	71Cd	b,q: ±9.6
	-107.1	3.44(25°,1-2)	-293	18-40	T	[Fe]=Iron(II) protoporphyrin dimethyl ester S:CCl_4	71Cd	b,q: ±12.6
	-50.2	1.54(25°,1-2)	-138	18-40	T	[Fe]=Iron(II) deuteroporphyrin dimethyl ester S:$CHCl_3$	71Cd	b
	-80.8	2.71(25°,1-2)	-218	18-40	T	[Fe]=Iron(II) protoporphyrin dimethyl ester S:$CHCl_3$	71Cd	b,q: ±15.1
[Co]	-41.8	3.88(0°,1)	-79	-23-23	T	[Co]=Cobalt(II) protoporphyrin IX dimethyl ester S:Toluene	73Sx	b,q: ±2.5
NiA_2	-34.3	-0.88(20°,1-2)	-103.3	10-3	T	S:$CHCl_3$ A=Ethyl-β-mercaptocinnamate	74Cd	b
	-32.2	0.52(20°,1-2)	-98.3	10-30	T	S:$CHCl_3$ A=Ethyl-β-mercaptothiocinnamate	74Cd	b
	-33.4	0.06(20°,1-2)	-110.4	10-30	T	S:$CHCl_3$ A=N-Phenyl-β-mercaptocinnamamide	74Cd	b
[Ni]	-18.0	2.86(25°,1)	-4	18-40	T	[Ni]=Nickel(II) mesotetraphenylporphine S:Benzene	72Ce	b
	29.3	2.56(25°,1)	146	18-40	T	[Ni]=Nickel(II) mesotetraphenylporphine S:$CHCl_3$	72Ce	b,q: ±6.3
Cu^{2+}	-10.42	(1)	--	--	-	CL=0.2M CM=3x10^{-2}M $CHNO_3$=0.01M	68Wb	f:1.34 u,y
CuA_2	-18.28	0.716(i)	-48.9	30	C	S:Benzene A=N-nitroso-N-Phenylhydroxylamine	71Gi	q:±1.8
	-15.31	1.892(1)	-14.02	30	C	S:Benzene A=1,1,-1-Trifluoropentane-2,4-dione	74Gc	
[Zn]	-18.0	2.20(25°,1)	-4	18-40	T	[Zn]=Zinc(II) mesotetraphenylporphine S:Benzene	72Ce	b,q: ±2.1
	-20.1	2.05(25°,1)	-4	18-40	T	[Zn]=Zinc(II) mesotetraphenylporphine S:$CHCl_3$	72Ce	b,q: ±1.7

Metal Ion	ΔH, kJ /mol	Log K	ΔS, J /mol K	T,°C	M	Conditions	Ref.	Remarks
P404	2,6-PYRIDINEDICARBOXYLIC ACID $C_7H_5O_4N$					$C_5H_3N(CO_2H)_2$		L^{2-}
Sc^{3+}	-16.69	11.2(1)	158.6	25	C	μ=0.50($NaClO_4$)	69Gf	
	-28.16	7.64(2)	51.5	25	C	μ=0.50($NaClO_4$)	69Gf	
Y^{3+}	-6.016	8.44(1)	141.4	25	C	μ=0.5	63Gc	
	-22.221	15.66(1-2)	225.5	25	C	μ=0.5	63Gc	
	-51.195	21.18(1-3)	233.9	25	C	μ=0.5	63Gc	
La^{3+}	-13.075	7.99(1)	107.9	25	C	μ=0.5	63Gc	
	-26.782	13.71(1-2)	172.8	25	C	μ=0.5	63Gc	
	-36.112	17.96(1-3)	222.6	25	C	μ=0.5	63Gc	
Ce^{3+}	-14.841	8.30(1)	109.2	25	C	μ=0.5	63Gc	
	-29.874	14.33(1-2)	174.0	25	C	μ=0.5	63Gc	
	-43.078	18.68(1-3)	213.0	25	C	μ=0.5	63Gc	
Pr^{3+}	-16.372	8.58(1)	109.2	25	C	μ=0.5	63Gc	
	-32.924	15.00(1-2)	177.0	25	C	μ=0.5	63Gc	
	-47.572	19.79(1-3)	219.2	25	C	μ=0.5	63Gc	
Nd^{3+}	-16.786	8.73(1)	110.9	25	C	μ=0.5	63Gc	
	-33.928	15.39(1-2)	180.7	25	C	μ=0.5	63Gc	
	-49.718	20.14(1-3)	223.8	25	C	μ=0.5	63Gc	
Sm^{3+}	-17.920	8.80(1)	108.4	25	C	μ=0.5	63Gc	
	-37.535	15.77(1-2)	176.1	25	C	μ=0.5	63Gc	
	-55.501	21.08(1-3)	217.6	25	C	μ=0.5	63Gc	
Eu^{3+}	-17.041	8.79(1)	111.3	25	C	μ=0.5	63Gc	
	-38.171	15.86(1-2)	175.7	25	C	μ=0.5	63Gc	
	-57.455	21.32(1-3)	215.5	25	C	μ=0.5	63Gc	
Gd^{3+}	-14.987	8.70(1)	116.3	25	C	μ=0.5	63Gc	
	-36.530	15.96(1-2)	182.8	25	C	μ=0.5	63Gc	
	-56.982	21.67(1-3)	223.8	25	C	μ=0.5	63Gc	
Tb^{3+}	-11.251	8.65(1)	127.6	25	C	μ=0.5	63Gc	
	-33.501	16.01(1-2)	194.1	25	C	μ=0.5	63Gc	
	-57.785	21.86(1-3)	224.7	25	C	μ=0.5	63Gc	
Dy^{3+}	-9.075	8.67(1)	135.6	25	C	μ=0.5	63Gc	
	-29.949	16.10(1-2)	207.9	25	C	μ=0.5	63Gc	
	-57.823	21.97(1-3)	226.8	25	C	μ=0.5	63Gc	
Ho^{3+}	-8.142	8.68(1)	138.9	25	C	μ=0.5	63Gc	
	-25.799	15.99(1-2)	219.2	25	C	μ=0.5	63Gc	
	-56.722	21.91(1-3)	229.3	25	C	μ=0.5	63Gc	
Er^{3+}	-7.740	8.75(1)	162.3	25	C	μ=0.5	63Gc	
	-24.066	16.32(1-2)	231.8	25	C	μ=0.5	63Gc	
	-56.091	21.97(1-3)	232.2	25	C	μ=0.5	63Gc	
Tm^{3+}	-7.673	8.82(1)	143.1	25	C	μ=0.5	63Gc	
	-22.891	16.48(1-2)	238.5	25	C	μ=0.5	63Gc	
	-53.848	21.89(1-3)	238.5	25	C	μ=0.5	63Gc	
Yb^{3+}	-8.054	8.83(1)	142.2	25	C	μ=0.5	63Gc	
	-24.200	16.54(1-2)	235.6	25	C	μ=0.5	63Gc	
	-54.002	21.58(1-3)	231.8	25	C	μ=0.5	63Gc	
Lu^{3+}	-9.167	9.01(1)	141.8	25	C	μ=0.5	63Gc	

Metal Ion	ΔH, kJ /mol	Log K	ΔS, J /mol K	T,°C	M	Conditions	Ref.	Re-marks
P404, continued								
Lu^{3+},	-25.083	16.73(1-2)	236.0	25	C	μ=0.5	63Gc	
cont.	-52.961	21.34(1-3)	231.0	25	C	μ=0.5	63Gc	
Cu^{2+}	-11.5	9.15(1)	137	25	C	μ=0.1M($NaClO_4$)	79Be	
	-29.3	7.22(2)	39	25	C	μ=0.1M($NaClO_4$)	79Be	
	-14	11.2	168	25	C	μ=0.1M($NaClO_4$) R:$Cu^{2+}+L^{2-}+H^+=CuHL^+$	79Be	
	-74.1	16.11	60	25	C	μ=0.1M($NaClO_4$) R:$Cu^{2+}+L^{2-}+A=CuAL$ A=2,2'-Bipyridyl	79Be	

P405 PYRIDINO-18-CROWN-6 $C_{15}H_{23}O_5N$

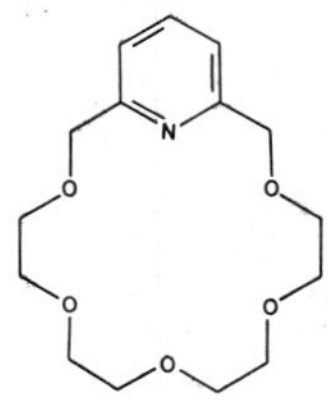

Metal Ion	ΔH, kJ /mol	Log K	ΔS, J /mol K	T,°C	M	Conditions	Ref.	Re-marks
Na^+	-22.76	4.09(1)	1.96	25	C	S:Methanol μ=0.005	80Bb	
K^+	-38.12	5.35(1)	-25.40	25	C	S:Methanol μ=0.005	80Bb	
Rb^+	-36.48	4.56(1)	-35.08	25	C	S:Methanol μ=0.005	80Bb	
Ba^{2+}	-32.30	>5.5(1)	--	25	C	S:Methanol μ=0.005	80Bb	
Ag^+	-34.85	>5.5(1)	--	25	C	S:Methanol μ=0.005	80Bb	

P406 PYRIDINO-18-CROWN-6, chloro, diester $C_{15}H_{18}O_7NCl$

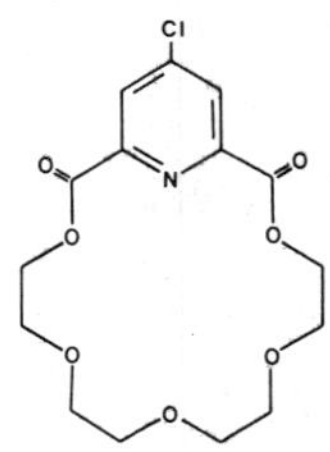

Metal Ion	ΔH, kJ /mol	Log K	ΔS, J /mol K	T,°C	M	Conditions	Ref.	Re-marks
Na^+	-25.23	4.14(1)	-5.6	25	C	S:Methanol μ=0.005	80Bb	
K^+	-33.35	4.73(1)	-21.0	25	C	S:Methanol μ=0.005	80Bb	
Rb^+	-38.62	3.56(1)	-61.7	25	C	S:Methanol μ=0.005	80Bb	

Metal Ion	ΔH, kJ /mol	Log K	ΔS, J /mol K	T,°C	M	Conditions	Ref.	Remarks
P406, continued								
Ag^+	-33.64	3.76(1)	-40.7	25	C	S:Methanol μ=0.005	80Bb	
NH_4^+	-28.37	2.86(1)	-40.7	25	C	S:Methanol μ=0.005	80Bb	
P407	PYRIDINO-18-CROWN-6, diester		$C_{15}H_{19}O_7N$			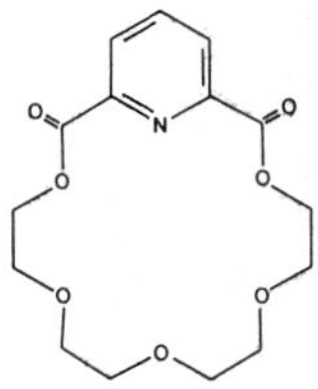		
Na^+	-25.90	4.29(1)	-4.77	25	C	S:Methanol	80Lb 80Bb	
K^+	-38.9	4.66(1)	-42.1	25	C	S:Methanol	80Lb 80Bb	
Rb^+	-37.95	4.24(1)	-46.3	25	C	S:Methanol	80Lb 80Bb	
Ba^{2+}	-25.23	4.34(1)	-1.54	25	C	S:Methanol	80Lb 80Bb	
Ag^+	-32.76	4.88(1)	-16.42	25	C	S:Methanol	80Lb 80Bb	
NH_4^+	-32.43	2.93(1)	-53.3	25	C	S:Methanol	80Lb 80Bb	
P408	PYRIMIDINE		$C_4H_4N_2$					
Ag^+	-17.61	1.61(1)	-28.20	25	C	μ=0.1M(KNO_3)	73Bi	
	-17.32	1.37(2)	-31.84	25	C	μ=0.1M(KNO_3)	73Bi	
P409	PYRIMIDINE, 2-amino-N-(dimethylaminoethyl)-N-		$C_{16}H_{22}ON_4$			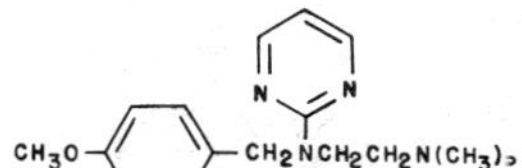		L
Co^{2+}	-18.4	6.30(25°,1-2)	58	0&25	T	0.06M(KCl)	62Aa	a,b
Ni^{2+}	67.8	8.32(25°,1-2)	386.6	0&25	T	0.06M(KCl)	62Aa	a,b
P410	PYROCATECHOL VIOLET							
Y^{3+}	-47.7	6.81(25°,1)	-29	25-45	T	μ=0.1M($NaClO_4$)	74Pb	b
YA	-22.2	4.25(25°,1)	8	25-45	T	μ=0.2M(KNO_3) A=EDTA	76Pe	b
	-33.0	4.84(25°,1)	-17	25-45	T	μ=0.2M(KNO_3)	76Pe	b

Metal Ion	ΔH, kJ /mol	Log K	ΔS, J /mol K	T,°C	M	Conditions	Ref.	Remarks
P410, continued								
YA, cont.						A=N-Hydroxyethyl-enediaminetri-acetic acid		
	-18.0	5.85(25°,1)	50	25-45	T	μ=0.2M(KNO_3) A=Iminotriacetic acid	76Pe	b

P411 4-PYRONE, 5-hydroxy-2-hydroxymethyl $C_6H_6O_4$

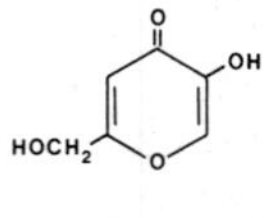

Metal Ion	ΔH, kJ /mol	Log K	ΔS, J /mol K	T,°C	M	Conditions	Ref.	Remarks
LaA	-8.8	(1)	--	27	C	μ=0.1M(KNO_3) A=EDTA	79Ma	j
PrA	-10.5	4.05(1)	10.3	27	C	μ=0.1M(KNO_3) A=EDTA	79Ma	j
SmA	-16.0	4.12(1)	6	27	C	μ=0.1M(KNO_3) A=EDTA	79Ma	j
GdA	-16.4	4.05(1)	5.7	27	C	μ=0.1M(KNO_3) A=EDTA	79Ma	j
HoA	-16.4	4.20(1)	6.0	27	C	μ=0.1M(KNO_3) A=EDTA	79Ma	j
TmA	-18.5	4.20(1)	4.3	27	C	μ=0.1M(KNO_3) A=EDTA	79Ma	j
LuA	-17.6	4.20(1)	5.0	27	C	μ=0.1M(KNO_3) A=EDTA	79Ma	j
Mn^{2+}	-5.4	3.67(25°,1)	52	20-40	T	μ=2M($NaClO_4$)	78Ga	
	-4.6	3.66(25°,1)	55	20-40	T	μ=2M($NaClO_4$)	75Gb	q:±0.4
	-9.4	3.00(25°,2)	26	20-40	T	μ=2M($NaClO_4$)	78Ga	
	1.3	2.99(25°,2)	53	20-40	T	μ=2M($NaClO_4$)	75Gb	q:±0.8
	-10.9	1.82(25°,3)	-1	20-40	T	μ=2M($NaClO_4$)	78Ga	
Co^{2+}	-11.7	4.55(25°,1)	48	20-40	T	μ=2M($NaClO_4$)	78Ga	
	-11.7	4.55(25°,1)	48	20-40	T	μ=2M($NaClO_4$)	75Gb	
	-9.2	3.70(25°,2)	39	20-40	T	μ=2M($NaClO_4$)	78Ga	
	-9.6	3.71(25°,2)	38	20-40	T	μ=2M($NaClO_4$)	75Gb	
	-14.2	2.45(25°,3)	0	20-40	T	μ=2M($NaClO_4$)	78Ga	
	-15.5	2.44(25°,3)	-4	20-40	T	μ=2M($NaClO_4$)	75Gb	
Ni^{2+}	-9.6	4.86(25°,1)	60	20-40	T	μ=2M($NaClO_4$)	78Ga	
	-10.0	4.86(25°,1)	60	20-40	T	μ=2M($NaClO_4$)	75Gb	
	-10.5	3.95(25°,2)	41	20-40	T	μ=2M($NaClO_4$)	78Ga	
	-9.6	3.95(25°,2)	43	20-40	T	μ=2M($NaClO_4$)	75Gb	
	-16.7	2.81(25°,3)	0	20-40	T	μ=2M($NaClO_4$)	78Ga	
	-17.2	2.81(25°,3)	-4	20-40	T	μ=2M($NaClO_4$)	75Gb	
Zn^{2+}	-8.6	5.01(25°,1)	67	20-40	T	μ=2M($NaClO_4$)	78Ga	
	-7.9	5.03(25°,1)	69	20-40	T	μ=2M($NaClO_4$)	75Gb	q:±0.8
	-10.8	4.19(25°,2)	44	20-40	T	μ=2M($NaClO_4$)	78Ga	
	-6.3	4.31(25°,2)	60	20-40	T	μ=2M($NaClO_4$)	75Gb	q:±1.3
	-28.9	2.58(25°,3)	-48	20-40	T	μ=2M($NaClO_4$)	78Ga	

Metal Ion	ΔH, kJ /mol	Log K	ΔS, J /mol K	T,°C	M	Conditions	Ref.	Re-marks
P412	4-PYRONE, 3-hydroxy-2-methyl	$C_6H_6O_3$				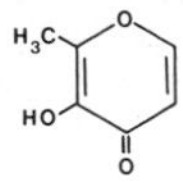		L^-
Mn^{2+}	-6.4	4.19(25°,1)	59	20-40	T	μ=2M($NaClO_4$)	78Ga	
	-8.5	3.30(25°,2)	35	20-40	T	μ=2M($NaClO_4$)	78Ga	
	-8.0	1.83(25°,3)	8	20-40	T	μ=2M($NaClO_4$)	78Ga	
Co^{2+}	-14.5	5.12(25°,1)	49	20-40	T	μ=2M($NaClO_4$)	78Ga	
	-14.7	4.07(25°,2)	29	20-40	T	μ=2M($NaClO_4$)	78Ga	
	-23.0	2.40(25°,3)	-29	20-40	T	μ=2M($NaClO_4$)	78Ga	
Ni^{2+}	-14.1	5.41(25°,1)	57	20-40	T	μ=2M($NaClO_4$)	78Ga	
	-16.5	4.23(25°,2)	26	20-40	T	μ=2M($NaClO_4$)	78Ga	
	-18.8	2.52(25°,3)	-16	20-40	T	μ=2M($NaClO_4$)	78Ga	
Zn^{2+}	-10.9	5.56(25°,1)	70	20-40	T	μ=2M($NaClO_4$)	78Ga	
	-9.2	5.37(25°,1)	72	20-40	T	μ=2M($NaNO_3$)	75Ga	
	-13.4	4.75(25°,2)	46	20-40	T	μ=2M($NaClO_4$)	78Ga	
	-7.9	4.59(25°,2)	61	20-40	T	μ=2M($NaNO_3$)	75Ga	
	-30.1	2.27(35°,3)	-59	20-40	T	μ=2M($NaClO_4$)	78Ga	
	-12.1	2.26(25°,3)	~2	20-40	T	μ=2M($NaNO_3$)	75Ga	
P413	PYROPHOSPHATE ION		O_7P_2			$P_2O_7^{4-}$		L^{4-}
Li^+	4.31	3.702(1)	85.4	25	C	μ=0	73Va	q:±0.38
	1.63	(1)	--	5	C	μ=0.50[$(CH_3)_4$-NNO_3]	73Va	g:$(CH_3)_4$-N^+,q: ±0.17
	1.67	(1)	--	15	C	μ=0.50[$(CH_3)_4$-NNO_3]	73Va	g:$(CH_3)_4$-N^+,q: ±0.17
	2.01	(1)	--	25	C	μ=0.50[$(CH_3)_4$-NNO_3]	73Va	g:$(CH_3)_4$-N^+,q: ±0.17
	1.21	(1)	--	5	C	μ=0.75[$(CH_3)_4$-NNO_3]	73Va	g:$(CH_3)_4$-N^+,q: ±0.17
	1.34	(1)	--	15	C	μ=0.75[$(CH_3)_4$-NNO_3]	73Va	g:$(CH_3)_4$-N^+,q: ±0.17
	1.55	(1)	--	25	C	μ=0.75$(CH_3)_4$ NNO_3]	73Va	g:$(CH_3)_4$-N^+,q: ±0.17
	1.00	(1)	--	5	C	μ=1.0[$(CH_3)_4NNO_3$]	73Va	g:$(CH_3)_4$-N^+,q: ±0.17
	1.00	(1)	--	15	C	μ=1.0[$(CH_3)_4NNO_3$]	73Va	g:$(CH_3)_4$-N^+,q: ±0.17
	1.09	2.390(1)	50.2	25	C	μ=1.0[$(CH_3)_4NNO_3$]	73Va	g:$(CH_3)_4$-N^+,q: ±0.17
	1.21	2.001	42.2	25	C	μ=0 R:Li^++HL^{3-}=$LiHL^{2-}$	73Va	q:±0.33

Metal Ion	ΔH, kJ /mol	Log K	ΔS, J /mol K	T,°C	M	Conditions	Ref.	Remarks
P413, continued								
Li^+, cont.	-0.75	--	--	25	C	μ=0.75[$(CH_3)_4$-NNO_3] R:Li^++HL^{3-}=$LiHL^{2-}$	73Va	g:$(CH_3)_4$-N^+,q: ±0.25
	-0.88	1.026(1)	16.7	25	C	μ=1.0[$(CH_3)_4NNO_3$] R:Li^++HL^{3-}=$LiHL^{2-}$	73Va	g:$(CH_3)_4$-N^+,q: ±0.25
	-1.26	(1)	--	25	C	μ=1.50[$(CH_3)_4$-NNO_3] R:Li^++HL^{3-}=$LiHL^{2-}$	73Va	g:$(CH_3)_4$-N^+,q: ±0.25
	-1.26	(1)	--	25	C	μ=1.75[$(CH_3)_4$-NNO_3] R:Li^++HL^{3-}=$LiHL^{2-}$	73Va	g:$(CH_3)_4$-N^+,q: ±0.25
Na^+	5.73	2.309(1)	63.6	25	C	μ=0	73Va	q:$(CH_3)_4$-N^+,q: ±0.54
	2.929	(1)	--	25	C	μ=0.50[$(CH_3)_4$-NNO_3]	73Va	g:$(CH_3)_4$-N^+,q: ±0.25
	2.22	(1)	--	25	C	μ=0.75[$(CH_3)_4$-NNO_3]	73Va	g:$(CH_3)_4$-N^+,q: ±0.25
	1.92	0.997(1)	25.5	25	C	μ=1.00[$(CH_3)_4$-NNO_3]	73Va	g:$(CH_3)_4$-N^+,q: ±0.25
	2.22	--	--	25	C	R:4Na^++H_2L^{2-}= Na_4L_4+2H^+	67Wc	
K^+	0.73	2.111(1)	64.8	25	C	μ=0	73Va	g:$(CH_3)_4$-N^+,q: ±0.63
	3.81	(1)	--	15	C	μ=0.5M[$(CH_3)_4$-NNO_3]	73Va	g:$(CH_3)_4$-N^+,q: ±0.4
	4.18	(1)	--	25	C	μ=0.5M[$(CH_3)_4$-NNO_3]	73Va	g:$(CH_3)_4$-N^+,q: ±0.4
	4.64	(1)	--	35	C	μ=0.5M[$(CH_3)_4$-NNO_3]	73Va	g:$(CH_3)_4$-N^+,q: ±0.4
	2.55	(1)	--	15	C	μ=1.00M[$(CH_3)_4$-NNO_3]	73Va	g:$(CH_3)_4$-N^+,q: ±0.4
	2.93	0.799(1)	25.1	25	C	μ=1.00M[$(CH_3)_4$-NNO_3]	73Va	g:$(CH_3)_4$-N^+,q: ±0.4

Metal Ion	ΔH, kJ /mol	Log K	ΔS, J /mol K	T,°C	M	Conditions	Ref.	Remarks
P413, continued								
K^+, cont.	3.47	(1)	--	35	C	μ=1.00M[$(CH_3)_4$-NNO_3]	73Va	g:$(CH_3)_4$-N^+,q: ±0.4
	1.76	(1)	--	15	C	μ=1.5M[$(CH_3)_4$-NNO_3]	73Va	g:$(CH_3)_4$-N^+,q: ±0.4
	1.80	(1)	--	25	C	μ=1.5M[$(CH_3)_4$-NNO_3]	73Va	g:$(CH_3)_4$-N^+,q: ±0.4
	2.22	(1)	--	35	C	μ=1.5M[$(CH_3)_4$-NNO_3]	73Va	g:$(CH_3)_4$-N^+,q: ±0.4
Mg^{2+}	12.1	5.7(1)	149.8	25	T	--	57Va	
Ca^{2+}	19.2	5.6(1)	192	25	T	μ=0	60Ia	
Mn^{2+}	12.1	6.03(1)	159.0	5	T	μ=0.2M(Tetra-methylammonium bromide) ΔCp=2280	79Md	b,q:±2.1
	32.2	6.18(1)	230.1	15	T	μ=0.2M(Tetra-methylammonium bromide) ΔCp=1820	79Md	b
	48.1	6.47(1)	285.3	25	T	μ=0.2M(Tetra-methylammonium bromide) ΔCp=1360	79Md	b
	46.0	6.47(1)	277.8	25	C	μ=0.2M(Tetra-methylammonium bromide)	79Md	
	61.1	6.79(1)	327.2	37	T	μ=0.2M(Tetra-methylammonium bromide) ΔCp=795	79Md	b
Co^{2+}	6.7	6.16(1)	141.8	5	T	μ=0.2M(Tetra-methylammonium bromide) ΔCp=2021	79Md	b,q: ±1.7
	25.1	6.28(1)	207.1	15	T	μ=0.2M(Tetra-methylammonium bromide) ΔCp=1644	79Md	b,q: ±2.5
	39.3	6.52(1)	256.9	25	T	μ=0.2M(Tetra-methylammonium bromide) ΔCp=1184	79Md	b
	41.8	6.52(1)	265.2	25	C	μ=0.2M(Tetra-methylammonium bromide)	79Md	
	49.8	6.77(1)	290.4	37	T	μ=0.2M(Tetra-methylammonium bromide) ΔCp=536	79Md	b
Ni^{2+}	17.61	5.82(1)	170.7	25	C	CL=0.025-0.18M	79Md	b,q:
	1.7	6.40(1)	128.4	5	T	μ=0.2M(Tetra-methylammonium bromide) ΔCp=1570	56Ya	±0.4
	15.5	6.45(1)	177.4	15	T	μ=0.2M(Tetra-	79Md	b,q:±2

Metal Ion	ΔH, kJ /mol	Log K	ΔS, J /mol K	T, °C	M	Conditions	Ref.	Re-marks
P413, continued								
Ni^{2+}, cont.						methylammonium bromide) ΔCp=1140		
	25.1	6.59(1)	210.4	25	T	μ=0.2M(Tetra-methylammonium bromide) ΔCp=699	79Md	b,q: ±2.5
	27.2	6.59(1)	217.5	25	C	μ=0.2M(Tetra-methylammonium bromide)	79Md	
	30.5	6.76(1)	228.0	25	T	μ=0.2M(Tetra-methylammonium bromide) ΔCp=117	79Md	b
	-9.25	1.37(2)	-5.0	25	C	CL=0.025-0.18M	56Ya	
	8.37	7.19(1-2)	165.7	25	C	CL=0.025-0.18M	56Ya	
Cu^{2+}	-2.80	9.00(1-2)	163.6	25	C	CL=0.025-0.18M	56Ya	
Zn^{2+}	11.04	6.50(1-2)	161.5	25	C	CL=0.025-0.18M	56Ya	
Pb^{2+}	-4.22	6.24(1-2)	105.4	25	C	CL=0.025-0.18M	56Ya	
P414 PYRROLE			C_4H_5N			CH:CHCH:CHNH		
$[NiA_2]_2$	-26.2	0.46(100°,1)	-62	75-125	T	S:Squalane A=3-Trifluoro-acetylcamphorate	74Sj	b,gg
P415 PYRROLIDINE			C_4H_9N			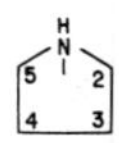		L
VOA_2	-43.81	(1)	--	25	C	μ→0 S:Nitrobenzene A=Acetylacetonate	65Ca	
NiA	-62.51	6.84(1-2)	-78.6	25	T	CM≅CL≅10^{-4}M S:Benzene A=Di-acetylbisbenzoyl hydrazine	58Sa	
NiA_2	-33	3.196(24.6°,1)	-50	8-39.5	T	S:Benzene A=O,O'-Diethyl-dithiophosphate	72Mc	b,q:±4
	-20.5	3.472(24.6°,1)	-2.1	8-39.5	T	S:Benzene A=O,O'-Dimethyl-dithiophosphate	72Mc	b
P416 2-PYRROLIDONE, N-methyl-			C_5H_9ON			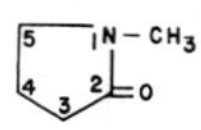		L
I_2	-23.4	1.28(1)	-54	25	T	S:n-Heptane	66Kc	

Metal Ion	ΔH, kJ /mol	Log K	ΔS, J /mol K	T,°C	M	Conditions	Ref.	Remarks

Q

Q1 QUINOLINE C_9H_7N 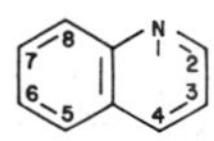 L

Metal Ion	ΔH, kJ /mol	Log K	ΔS, J /mol K	T,°C	M	Conditions	Ref.	Remarks
VOA_2	-20.21	0.26(1)	-62.8	25	C	$\mu\to0$ S:Nitrobenzene A=Acetylacetonate	65Ca	
CoA	0	1.201(25°,1)	-25	20-50	T	S:Toluene A=α,β,γ,δ-Tetra-(p-methoxyphenyl) porphyrin	73Wa	b,q:±4
CoA_2	-10.0	2.42(25°,1)	15.5	25	C	S:Acetone A=Cl	73Lb	q:±1.2
	-33.0	4.3(1-2)	-28.0	25	C	S:Acetone CM=2x10^{-3}M CL=4x10^{-2}M A=Cl	71Lg	
	-31.0	4.8(1-2)	-12.6	25	C	S:Acetone CM=2x10^{-3}M CL=4x10^{-2}M A=Br	71Lg	
	-23.0	4.1(1-2)	1.2	25	C	S:Acetonitrile CM=2x10^{-3}M CL=4x10^{-2}M A=Cl	71Lg	
	-15.9	3.7(1-2)	18.4	25	C	S:Acetonitrile CM=2x10^{-3}M CL=4x10^{-2}M A=Br	71Lg	
	-6.7	3.5(1-2)	44.8	25	C	S:n-Butyl alcohol CM=2x10^{-3}M CL=0.1M A=Cl	71Lh	
	-5.8	3.4(1-2)	46.4	25	C	S:n-Butyl alcohol CM=2x10^{-3}M CL=0.1M A=Br	71Lh	
	-48.1	4.6(1-2)	-73.2	25	C	S:Cyclohexanone CM=2x10^{-3}M CL=4x10^{-2}M A=Cl	71Lg	
	-46.4	4.8(1-2)	-63.2	25	C	S:Cyclohexanone CM=2x10^{-3}M CL=4x10^{-2}M A=Br	71Lg	
NiA	-66.11	2.08(25°,1-2)	-182.0	25	T	CM≅CL≅10^{-4}M S:Benzene A=Diacetylbisbenzoyl hydrazine	58Sa	
Cu^{2+}	-74.18	16.66(25°,1)	71	5-35	T	μ=0.5($LiClO_4$) S:50w% C_2H_5OH R:$Cu^{2+}+A^{2-}+L=CuAL$ A=2-Carboxyphenyl-azo-β-naphthol	58Ja	b

Metal Ion	ΔH, kJ /mol	Log K	ΔS, J /mol K	T,°C	M	Conditions	Ref.	Remarks
Q1, continued								
Cu^{2+}, cont.	-109.29	19.21(25°,1)	0.00	5-35	T	μ=0.5($LiClO_4$) S:50w% C_2H_5OH R:Cu^{2+}+A^{2-}+L= CuAL A=2-Carboxyphenyl-azo-β-naphthylamine	58Ja	b
	-166.73	22.52(25°,1)	-130	5-35	T	μ=0.5($LiClO_4$) S:50w% C_2H_5OH R:Cu^{2+}+A^{2-}+L= CuAL A=o,o-Dihydroxyazobenzene	58Ja	b
CuA_2	-29.00	2.064(1)	-55.98	30	C	S:Benzene A=1,-1,1-Trifluoropentane-2,4-dione	74Gc	
Al_2A_6	-125.9	--	--	25	C	S:Benzene CM=0.07M A=Br R:0.5(Al_2A_6)+2L= AlA_3L	68Rc	
$SiAB_3$	-22.34	1.80(30°,1)	-39.37	10-45	T	S:CCl_4 XL=1x10^{-2} XM=5x10^{-4} A=OH B=CH_3	69Lb	b
	-23.14	1.73(30°,1)	-43.22	20-60	T	S:CCl_4 XL=1x10^{-2} XM=5x10^{-4} A=OH B=C_2H_5	69Lb	b
	-24.73	2.27(30°,1)	-38.12	0-50	T	S:CCl_4 XL=1x10^{-2} XM=4x10^{-4} A=OH B=Phenyl	69Lb	b
I_2	-23.4	2.477(1)	-31.0	25	C	S:Benzene	66Ad	
	-31.88	1.996(1)	-69.0	25	C	S:CCl_4	79Ba	
	-30.2	2.07(25°,1)	-61.7	17-35	T	S:Heptane	68Kb	b
	-30.1	2.070(25°,1)	-61.1	17-35	T	CM=1.1-2.8x10^{-3}M CL=2.7x10^{-4}M S:n-Heptane	75Be	b
ICl	-63.9	3.94(1)	-139	25	C	S:CCl_4	73Bk	
IBr	-51.2	4.10(1)	-90	25	C	S:CCl_4	73Bk	
Q2 QUINOLINE, 6-chloro			C_9H_6NCl			(Cl)C_9H_6N		
I_2	-27.0	1.799(25°,1)	-55.8	17-35	T	CM=1.1-2.8x10^{-3}M CL=2.9x10^{-4}M S:n-Heptane	75Be	
	-26.8	1.83(25°,1)	-54.8	17-35	T	S:Heptane	68Kb	b
Q3 QUINOLINE, 1,2-dimethyl-			$C_{11}H_{12}N$			$C_9H_6N(CH_3)_2$		L^+
[Fe]	-55.6	9.0(25°)	-18.4	5-30	T	pH=11.2 Protohaemin R:$(Haemin)_2$+2L= $(HaeminL)_2$+$2H_2O$	69Md	b,w:I

Metal Ion	ΔH, kJ /mol	Log K	ΔS, J /mol K	T,°C	M	Conditions	Ref.	Remarks
Q4 QUINOLINE, 8-hydroxy-			C_9H_7ON			$C_9H_6N(OH)$		L^-
Pr^{3+}	-8.37	5.81(25°,1)	83.3	25-51	T	μ=1M($LiClO_4$) S:50v%Propanol	76Ek	j,q: ±1.7
Mn^{2+}	-14.6	7.30(1)	92	25	C	μ=0.1($NaClO_4$) S:50v% Dioxane	68Gg	
	-43.9	13.44(2)	113	25	C	μ=0.1($NaClO_4$) S:50v% Dioxane	68Gg	
Co^{2+}	-30.1	9.65(1)	84	25	C	μ=0.1($NaClO_4$) S:50v% Dioxane	68Gg	
	-20.5	(1)	--	25	C	ΔH value cal. by two different methods S:50v% Dioxane	59Fa	
	-26.8	(1)	--	25	C	ΔH value cal. by two different methods S:50v% Dioxane	59Fa	
	-64.4	18.05(1-2)	130	25	C	μ=0.1($NaClO_4$) S:50v% Dioxane	68Gg	
	-90.4	19.6(1-2)	88	25	C	ΔH value cal. by two different methods S:50v% Dioxane	59Fa	
	-85.4	19.6(1-2)	88	25	C	ΔH value cal. by two different methods S:50v% Dioxane	59Fa	
	-46.4	19.6(25°,1-2)	218	0-40	T	S:50v% Dioxane	59Fa	b
Ni^{2+}	-38.9	10.50(1)	71	25	C	μ=0.1($NaClO_4$) S:50v% Dioxane	68Gg	
	-27.6	(1)	--	25	C	ΔH value cal. by two different methods S:50v% Dioxane	59Fa	
	-35.6	(1)	--	25	C	ΔH value cal. by two different methods S:50v% Dioxane	59Fa	
	-80.8	20.27(1-2)	117	25	C	μ=0.1($NaClO_4$) S:50v% Dioxane	68Gg	
	-89.1	21.5(1-2)	121	--	-	ΔH value cal. by two different methods S:50v% Dioxane	59Fa	
	-87.0	21.5(1-2)	121	--	-	ΔH value cal. by two different methods S:50v% Dioxane	59Fa	
	-79.5	21.5(25°,1-2)	146	0-40	T	S:50v% Dioxane	59Fa	b
Cu^{2+}	-42.7	13.29(1)	113	25	C	μ=0.1($NaClO_4$) S:50v% Dioxane	68Gg	
	-43.9	(1)	--	25	C	S:50v% Dioxane	59Fa	

Metal Ion	ΔH, kJ /mol	Log K	ΔS, J /mol K	T,°C	M	Conditions	Ref.	Re-marks
Q4, continued								
Cu^{2+}, cont.	-81.38	17.67(25°,1)	67	5-35	T	μ=0.5($LiClO_4$) S:50w% C_2H_5OH R:Cu^{2+}+A^{2-}+L= CuAL A=2-Carboxylphenyl-azo β-naphthol	58Ja	b
	113.30	20.22(25°,1)	8	5-35	T	μ=0.5($LiClO_4$) S:50w% C_2H_5OH R:Cu^{2+}+A^{2-}+L= CuAL A=2-Carboxyphenyl-azo-β-naphthylamine	58Ja	b
	-174.72	23.51(25°,1)	-138	5-35	T	μ=0.5($LiClO_4$) S:50w% C_2H_5OH R:Cu^{2+}+A^{2-}+L= CuAL A=o,o-Dihydroxy-azo-benzene	58Ja	b
	-82.0	25.90(1-2)	222	25	C	μ=0.1($NaClO_4$) S:50v% Dioxane	68Gg	
	-86.2	25.8(1-2)	201	25	C	ΔH value cal. by two different methods S:50v% Dioxane	59Fa	
	-87.0	25.8(1-2)	201	25	C	ΔH value cal. by two different methods S:50v% Dioxane	59Fa	
	-76.1	25.8(25°,1-2)	222	0-40	T	S:50v% Dioxane	59Fa	b
Zn^{2+}	-24.7	9.45(1)	96	25	C	μ=0.1($NaClO_4$) S:50v% Dioxane	68Gg	
	-40.2	18.15(1-2)	213	25	C	μ=0.1($NaClO_4$) S:50v% Dioxane	68Gg	
Cd^{2+}	-23.0	8.22(1)	79	25	C	μ=0.1($NaClO_4$) S:50v% Dioxane	68Gg	
	-48.5	15.22(1-2)	130	25	C	μ=0.1($NaClO_4$) S:50v% Dioxane	68Gg	
Pb^{2+}	-27.6	10.03(1)	100	25	C	μ=0.1($NaClO_4$) S:50v% Dioxane	68Gg	
	-63.2	17.34(1-2)	121	25	C	μ=0.1($NaClO_4$) S:50v% Dioxane	68Gg	
Q5 QUINOLINE, 8-hydroxy-2-methyl- $C_{10}H_9ON$						$C_9H_5N(OH)(CH_3)$		L^-
Mn^{2+}	-13.8	6.81(1)	84	25	C	μ=0.1($NaClO_4$) S:50v% Dioxane	68Gg	
	-27.6	14.00(25°,1-2)	176	0-40	T	μ=0.005	53Fa 54Ja	b
	-26.4	13.10(1-2)	163	25	C	μ=0.1($NaClO_4$) S:50v% Dioxane	68Gg	

Metal Ion	ΔH, kJ /mol	Log K	ΔS, J /mol K	T,°C	M	Conditions	Ref.	Re-marks
Q5, continued								
Co^{2+}	-17.6	8.59(1)	105	25	C	μ=0.1($NaClO_4$) S:50v% Dioxane	68Gg	
	-19.2	(1)	--	25	C	S:50v% Dioxane	59Fa	
	-43.1	18.44(25°,1-2)	209	0-40	T	μ=0.005	53Fa 54Ja	b
	-57.7	17.38(1-2)	138	25	C	μ=0.1($NaClO_4$) S:50v% Dioxane	68Gg	
	-48.5	18.50(1-2)	188	25	C	S:50v% Dioxane	59Fa	
Ni^{2+}	-20.9	8.96(1)	100	25	C	μ=0.1($NaClO_4$) S:50v% Dioxane	68Gg	
	-27.2	(1)	--	25	C	S:50v% Dioxane	59Fa	
	-44.8	17.85(25°,1-2)	192	0-40	T	μ=0.005	53Fa 54Ja	b
	-50.6	16.94(1-2)	155	25	C	μ=0.1($NaClO_4$) S:50v% Dioxane	68Gg	
	-43.1	17.8(1-2)	197	25	C	ΔH value cal. two different ways S:50v% Dioxane	59Fa	
	-44.4	17.8(1-2)	197	25	C	ΔH value cal. two different ways S:50v% Dioxane	59Fa	
Cu^{2+}	-33.5	11.92	117	25	C	μ=0.1($NaClO_4$) S:50v% Dioxane	68Gg	
	-36.4	(1)	--	25	T	ΔH value cal. two different ways S:50v% Dioxane	59Fa	
	-27.2	(1)	--	25	T	ΔH value cal. two different ways S:50v% Dioxane	59Fa	
	-72.4	23.85(25°,1-2)	209	0-40	T	μ=0.005	53Fa 54Ja	b
	-64.8	22.82(1-2)	222	25	C	μ=0.1($NaClO_4$) S:50v% Dioxane	68Gg	
	-64.8	23.8(1-2)	218	25	T	S:50v% Dioxane	59Fa	
Zn^{2+}	-15.9	9.06(1)	121	25	C	μ=0.1($NaClO_4$) S:50v% Dioxane	68Gg	
	-43.1	18.72(25°,1-2)	213	0-40	T	μ=0.005	53Fa 54Ja	b
	-46.4	17.90(1-2)	188	25	C	μ=0.1($NaClO_4$) S:50v% Dioxane	68Gg	
Pb^{2+}	-26.4	9.97(1)	105	25	C	μ=0.1($NaClO_4$) S:50v% Dioxane	68Gg	
	-57.3	17.18(1-2)	138	25	C	μ=0.1($NaClO_4$) S:50v% Dioxane	68Gg	
Q6 QUINOLINE, 8-hydroxy-4-methyl- $C_{10}H_9ON$						$C_9H_5N(OH)(CH_3)$		L^-
Mn^{2+}	-17.2	7.74(1)	92	25	C	μ=0.1($NaClO_4$) S:50v% Dioxane	68Gg	
	-48.5	15.50(25°,1-2)	134	0-40	T	μ=0.005	53Fa 54Ja	b

Metal Ion	ΔH, kJ /mol	Log K	ΔS, J /mol K	T,°C	M	Conditions	Ref.	Re-marks
A6, continued								
Mn^{2+}, cont.	-26.8	14.81(1-2)	192	25	C	μ=0.1($NaClO_4$) S:50v% Dioxane	68Gg	
Co^{2+}	-28.4	9.95(1)	96	25	C	μ=0.1($NaClO_4$) S:50v% Dioxane	68Gg	
	-29.7	(1)	--	25	C	S:50v% Dioxane	59Fa	
	-84.9	20.00(25°,1-2)	96	0-40	T	μ=0.005	53Fa 54Ja	b
	-74.5	18.92(1-2)	113	25	C	μ=0.1($NaClO_4$) S:50v% Dioxane	68Gg	
	-104.6	20.0(1-2)	33	25	C	S:50v% Dioxane	59Fa	
Ni^{2+}	-39.3	10.56(1)	71	25	C	μ=0.1($NaClO_4$) S:50v% Dioxane	68Gg	
	-28.4	(1)	--	25	C	S:50v% Dioxane	59Fa	
	-108.8	22.28(25°,1-2)	63	0-40	T	μ=0.005	53Fa 54Ja	b
	-76.6	20.47(1-2)	134	25	C	μ=0.1($NaClO_4$) S:50v% Dioxane	68Gg	
	-107.9	22.3(1-2)	63	25	C	S:50v% Dioxane	59Fa	
Cu^{2+}	-45.2	14.04(1)	117	25	C	μ=0.1($NaClO_4$) S:50v% Dioxane	68Gg	
	-38	(1)	--	25	C	S:50v% Dioxane	59Fa	
	-142.7	27.55(25°,1-2)	50	0-40	T	μ=0.005	53Fa 54Ja	b
	-90.4	26.96(1-2)	213	25	C	μ=0.1($NaClO_4$) S:50v% Dioxane	68Gg	
	-102.1	27.6(1-2)	184	25	C	S:50v% Dioxane	59Fa	
Zn^{2+}	-23.4	9.76(1)	109	25	C	μ=0.1($NaClO_4$) S:50v% Dioxane	68Gg	
	-82.0	20.25(25°,1-2)	109	0-40	T	μ=0.05	53Fa 54Ja	b
	-53.6	18.96(1-2)	184	25	C	μ=0.1($NaClO_4$) S:50v% Dioxane	68Gg	
Pb^{2+}	-28.4	10.46(1)	105	25	C	μ=0.1($NaClO_4$) S:50v% Dioxane	68Gg	
	-64.4	18.55(1-2)	138	25	C	μ=0.1($NaClO_4$) S:50v% Dioxane	68Gg	

Q7 QUINOLINE, 8-hydroxy-5-sulfo- $C_9H_8O_4NS$

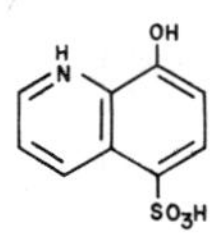

VO^{2+}	-7.1	11.79(25°,1)	188	25-35	T	μ=0.10(KNO_3)	66Mg	b,q:±2.1

Q8 QUINOLINE, 8-mercapto- C_9H_7NS $C_9H_6N(SH)$ L^-

Mn^{2+}	-14.6	6.74(1)	79	25	C	μ=0.1($NaClO_4$) S:50v% Dioxane	68Gg	
Co^{2+}	-64.4	7.9(1)	-63	25	C	μ=0.1($NaClO_4$) S:50v% Dioxane	68Gg	

Metal Ion	ΔH, kJ /mol	Log K	ΔS, J /mol K	T, °C	M	Conditions	Ref.	Re-marks
Q8, continued								
Ni^{2+}	-47.7	11.0(1)	50	25	C	μ=0.1($NaClO_4$) S:50v% Dioxane	68Gg	
Cu^{2+}	-61.5	12.7(1)	38	25	C	μ=0.1($NaClO_4$) S:50v% Dioxane	68Gg	
Zn^{2+}	-30.1	11.0(1)	109	25	C	μ=0.1($NaClO_4$) S:50v% Dioxane	68Gg	
Pb^{2+}	-42.2	11.85(1)	84	25	C	μ=0.1($NaClO_4$) S:50v% Dioxane	68Gg	
Q9 QUINOLINE, 2-methyl			$C_{10}H_9N$			$(CH_3)C_9H_6N$		
I_2	-28.47	1.613(1)	-64.8	25	C	S:CCl_4	79Ba	
Q10 QUINOLINE, 6-methyl			$C_{10}H_9N$			$(CH_3)C_9H_6N$		
I_2	-32.2	2.288(25°,1)	-64.0	17-35	T	CM=1.1-2.8x10^{-3}M CL=2.4x10^{-4}M S:n-Heptane	75Be	b
	-32.5	2.26(25°,1)	-65.7	17-35	T	S:Heptane	68Kb	b
Q11 QUINOLINE, 2-methyl-8-thiol-			$C_{10}H_9NS$			$C_9H_5N(CH_3)(SH)$		L^-
Co^{2+}	-23.8	9.6(1)	105	25	C	μ=0.1($NaClO_4$) S:50v% Dioxane	68Gg	
Ni^{2+}	-33.9	9.2(1)	63	25	C	μ=0.1($NaClO_4$) S:50v% Dioxane	68Gg	
Cu^{2+}	-85.8	11.7(1)	-67	25	C	μ=0.1($NaClO_4$) S:50v% Dioxane	68Gg	
Zn^{2+}	-33.9	11.1(1)	100	25	C	μ=0.1($NaClO_4$) S:50v% Dioxane	68Gg	
Q12 5-QUINOLINE SULFONIC ACID, 7-bromo-8-hydroxy-			$C_9H_6O_4NBrS$			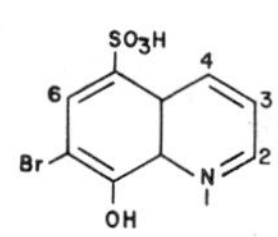		
Pd^{2+}	172	12.41(30°,1-2)	798	20-40	T	μ=0.1M($NaClO_4$) pH=4.5	71Ah	b
Ga^{3+}	32.55	13.6(20°,1-3)	367.4	20-45	T	μ=0.2(NaAc&HAc) pH=4	72Ah	b,n,v: ±0.8

Metal Ion	ΔH, kJ /mol	Log K	ΔS, J /mol K	T,°C	M	Conditions	Ref.	Remarks
Q13	7-QUINOLINE SULFONIC ACID, 5-bromo-8-hydroxy-		$C_9H_6O_4NBrS$			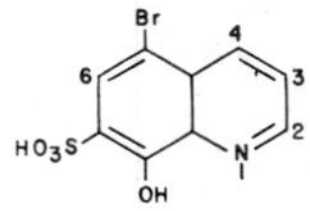		
UO_2^{2+}	14.6	8.623(25°,1-2)	213.8	--	T	--	72Ba	u
Q14	5-QUINOLINE SULPHONIC ACID, 7-chloro-8-hydroxy-		$C_9H_6O_4NClS$			$C_9H_4N(Cl)(OH)(SO_3H)$		
UO_2^{2+}	-36.40	9.43(30°,1-2)	53.76	20-45	T	μ=0.1($NaClO_4$)	71Ai	b,n,v: ±1.5
Cu^{2+}	-58.58	11.6(20°,1-2)	28.472	20-45	T	μ=0.1($NaClO_4$)	70Ad	b,n,v: ±1.2
Ga^{3+}	6.02	13.5(20°,1-3)	273.6	20-50	T	μ=0.2(NaAc&HAc) pH=4	72Ah	b,n,v: ±0.8
Q15	7-QUINOLINE SULFONIC ACID, 5-chloro-8-hydroxy-		$C_9H_6O_4NClS$			$C_9H_4N(Cl)(OH)(SO_3H)$		
UO_2^{2+}	-17.2	8.50(35°,1-2)	218.8	25-45	T	μ=0.02	70Bb	
	13.4	8.556(25°,1-2)	206.7	--	T	--	72Ba	u
Cu^{2+}	-15.31	9.2(25°,1-2)	125.5	25-45	T	μ=0.02 Value of ΔS was reported as 221.3	69Bc	b
Q16	5-QUINOLINE SULFONIC ACID, 8-hydroxy-		$C_9H_7O_4NS$			$C_9H_5N(OH)(SO_3H)$		L^{2-}
Mg^{2+}	-16.7	5.01(1)	42	0	T	μ=0	57Ua	
	-9.2	4.79(1)	63	25	T	μ=0	57Ua	
	-2.9	4.74(1)	96	50	T	μ=0	57Ua	
Y^{3+}	-23.60	22.39	348.1	20	C	μ=0.1M(KNO_3) R:Y^{3+}+$EDTA^{4-}$+ L^{2-}=Y(EDTA)L^{3-}	71Ga	x:62Ma
	-79.9	17.16(25°,1-3)	63	25-45	T	μ=0.2M($NaClO_4$)	73Pb	b
YA	-21.13	4.31(1)	10.5	20	C	μ=0.1M(KNO_3) A=EDTA	71Ga	
	-14.6	5.77(30°,1)	62.8	30-50	T	μ=0.2M(KNO_3) A=Nitrilotri-acetic acid	75Pg	b
La^{3+}	-10.0	5.63(1)	75	25	T	μ=0	58Fa	
	-17.6	4.5(2)	25	25	T	μ=0	58Fa	
	-18.8	3.7(3)	2	25	T	μ=0	58Fa	
	-24.85	18.87	276.1	20	C	μ=0.1M(KNO_3) R:La^{3+}+$EDTA^{4-}$+ L^{2-}=La(EDTA)L^{3-}	71Ga	x:62Ma

Metal Ion	ΔH, kJ /mol	Log K	ΔS, J /mol K	T,°C	M	Conditions	Ref.	Re- marks
Q16, continued								
LaA	-12.59	3.36(1)	21.3	20	C	μ=0.1M(KNO_3) A=EDTA	71Ga	
Ce^{3+}	-15.5	6.05(1)	63	25	T	μ=0	58Fa	
	-17.6	5.0(2)	38	25	T	μ=0	58Fa	
	-19.2	3.9(3)	12	25	T	μ=0	58Fa	
	-31.04	19.77	272.4	20	C	μ=0.10M(KNO_3) R:Ce^{3+}+$EDTA^{4-}$+L^{2-}=Ce(EDTA)L^{3-}	71Ga	x:62Ma
CeA	-18.74	3.80(1)	8.8	20	C	μ=0.1M(KNO_3) A=EDTA	71Ga	
Pr^{3+}	-14.2	6.17(1)	71	25	T	μ=0	58Fa	
	-11.7	5.20(2)	58	25	T	μ=0	58Fa	
	-20.1	4.3(3)	17	25	T	μ=0	58Fa	
	-33.60	20.41	276.1	20	C	μ=0.1M(KNO_3) R:Pr^{3+}+$EDTA^{4-}$+L^{2-}=Pr(EDTA)L^{3-}	71Ga	x:62Ma
PrA	-20.21	4.01(1)	7.9	20	C	μ=0.1M(KNO_3) A=EDTA	71Ga	
Nd^{3+}	-12.6	6.3(1)	79	25	T	μ=0	58Fa	
	-11.7	5.3(2)	63	25	T	μ=0	58Fa	
	-11.7	4.4(3)	46	25	T	μ=0	58Fa	
	-37.53	20.68	267.8	20	C	μ=0.1M(KNO_3) R:Nd^{3+}+$EDTA^{4-}$+L^{2-}=Nd(EDTA)L^{3-}	71Ga	x:62Ma
NdA	-22.38	4.07(1)	16.7	20	C	μ=0.1M(KNO_3) A=EDTA	71Ga	
Sm^{3+}	-12.1	6.58(1)	84	25	T	μ=0	58Fa	
	-15.1	5.70(2)	58	25	T	μ=0	58Fa	
	-15.9	4.76(3)	38	25	T	μ=0	58Fa	
	-37.86	21.50	282.0	20	C	μ=0.1M(KNO_3) R:Sm^{3+}+$EDTA^{4-}$+L^{2-}=Sm(EDTA)L^{3-}	71Ga	x:62Ma
SmA	-23.85	4.35(1)	1.7	20	C	μ=0.1M(KNO_3) A=EDTA	71Ga	
Eu^{3+}	-34.85	21.78	297.9	20	C	μ=0.1M(KNO_3) R:Eu^{3+}+$EDTA^{4-}$+L^{2-}=Eu(EDTA)L^{3-}	71Ga	x:62Ma
EuA	-24.14	4.41(1)	2.1	20	C	μ=0.1M(KNO_3) A=EDTA	71Ga	
Gd^{3+}	-14.6	6.64(1)	79	25	T	μ=0	58Fa	
	-18.0	5.73(2)	50	25	T	μ=0	58Fa	
	-19.7	4.9(3)	29	25	T	μ=0	58Fa	
	-32.76	21.75	304.2	20	C	μ=0.1M(KNO_3) R:Gd^{3+}+$EDTA^{4-}$+L^{2-}=Gd(EDTA)L^{3-}	71Ga	x:62Ma

Metal Ion	ΔH, kJ /mol	Log K	ΔS, J /mol K	T,°C	M	Conditions	Ref.	Remarks
Q16, continued								
GdA	-25.52	4.36(1)	-3.8	20	C	μ=0.1M(KNO_3) A=EDTA	71Ga	
Tb^{3+}	-27.70	22.43	334.7	20	C	μ=0.1M(KNO_3) R:Tb^{3+}+$EDTA^{4-}$+L^{2-}=Tb(EDTA)L^{3-}	71Ga	x:62Ma
TbA	-23.05	4.50(1)	7.5	20	C	μ=0.1M(KNO_3) A=EDTA	71Ga	
Dy^{3+}	-28.08	22.87	341.8	20	C	μ=0.1M(KNO_3) R:Dy^{3+}+$EDTA^{4-}$+L^{2-}=Dy(EDTA)L^{3-}	71Ga	x:62Ma
DyA	-23.0	4.56(1)	8.8	20	C	μ=0.1M(KNO_3) A=EDTA	71Ga	
Ho^{3+}	-29.41	23.40	347.7	20	C	μ=0.1M(KNO_3) R:Ho^{3+}+$EDTA^{4-}$+L^{2-}=Ho(EDTA)L^{3-}	71Ga	x:62Ma
HoA	-23.72	4.65(1)	8.4	20	C	μ=0.1M(KNO_3) A=EDTA	71Ga	
Er^{3+}	-22.6	7.16(1)	63	25	T	μ=0	58Fa	
	-23.8	6.18(2)	38	25	T	μ=0	58Fa	
	-23.4	5.22(3)	21	25	T	μ=0	58Fa	
	32.13	23.52	340.6	20	C	μ=0.1M(KNO_3) R:Er^{3+}+$EDTA^{4-}$+L^{2-}=Er(EDTA)L^{3-}	71Ga	x:62Ma
ErA	-24.98	4.66(1)	4.2	20	C	μ=0.1M(KNO_3) A=EDTA	71Ga	
Tm^{3+}	-33.10	24.07	347.7	20	C	μ=0.1M(KNO_3) R:Tm^{3+}+$EDTA^{4-}$+L^{2-}=Tm(EDTA)L^{3-}	71Ga	x:62Ma
TmA	-25.27	4.76(1)	4.6	20	C	μ=0.1M(KNO_3) A=EDTA	71Ga	
Yb^{3+}	-36.15	24.35	342.7	20	C	μ=0.1M(KNO_3) R:Yb^{3+}+$EDTA^{4-}$+L^{2-}=Yb(EDTA)L^{3-}	71Ga	x:62Ma
YbA	-26.48	4.82(1)	2.1	20	C	μ=0.1M(KNO_3) A=EDTA	71Ga	
Lu^{3+}	-36.86	24.65	346.0	20	C	μ=0.1M(KNO_3) R:Lu^{3+}+$EDTA^{4-}$+L^{2-}=Lu(EDTA)L^{3-}	71Ga	x:62Ma
LuA	-26.36	4.82(1)	2.1	20	C	μ=0.1M(KNO_3) A=EDTA	71Ga	
ThA	-18.49	6.71(1)	66.1	25	C	μ=0.10(KNO_3) A=1,2-Diaminocy- clohexane-N,N,-	70Kb	

Metal Ion	ΔH, kJ /mol	Log K	ΔS, J /mol K	T, °C	M	Conditions	Ref.	Re-marks
Q16, continued								
ThA, cont.						N',N'-tetraacetic acid		
	-6.78	6.99(1)	110.9	25	C	μ=0.10(KNO_3) A=EDTA	70Kb	
	20.00	--	--	25	C	μ=0.10(KNO_3) A=1,2-Diamino-cyclohexane, N,N,-N',N'-tetraacetic acid R:ThA+H_2L=$ThAL^{2-}$+$2H^+$	70Kb	
	31.80	--	--	25	C	μ=0.10(KNO_3) A=EDTA R:ThA+H_2L=$ThAL^{2-}$+$2H^+$	70Kb	
Mn^{2+}	-13.4	7.05(1)	92	25	C	μ=0.1($NaClO_4$) S:50v% Dioxane	68Gg	
	-28.0	13.18(2)	192	25	C	μ=0.1($NaClO_4$) S:50v% Dioxane	68Gg	
Co^{2+}	-26.4	7.38(1)	92	25	C	μ=0.1($NaClO_4$) S:50v% Dioxane	68Gg	
	-60.7	17.55(1-2)	134	25	C	μ=0.1($NaClO_4$) S:50v% Dioxane	68Gg	
Ni^{2+}	-26.8	9.11(1)	84	25	C	μ=0.1($NaClO_4$)	68Gg	
	-30.5	10.22(1)	92	25	C	μ=0.1($NaClO_4$) S:50v% Dioxane	68Gg	
	-61.9	17.34(1-2)	126	25	C	μ=0.1($NaClO_4$)	68Gg	
	-66.5	19.25(1-2)	146	25	C	μ=0.1($NaClO_4$) S:50v% Dioxane	68Gg	
	-107.1	23.23(1-3)	84	25	C	μ=0.1($NaClO_4$)	68Gg	
	-106.3	25.55(1-3)	134	25	C	μ=0.1($NaClO_4$) S:50v% Dioxane	68Gg	
Cu^{2+}	-33.9	12.45(1)	126	25	C	μ=0.1($NaClO_4$)	68Gg	
	-37.6	13.16(1)	126	25	C	μ=0.1($NaClO_4$) S:50v% Dioxane	68Gg	
	-73.6	22.99(1-2)	192	25	C	μ=0.1($NaClO_4$)	68Gg	
	-77.8	25.55(1-2)	226	25	C	μ=0.1($NaClO_4$) S:50v% Dioxane	68Gg	
Zn^{2+}	-21.3	7.95(1)	79	25	C	μ=0.1($NaClO_4$)	68Gg	
	-21.8	9.23(1)	105	25	C	μ=0.1($NaClO_4$) S:50v% Dioxane	68Gg	
	-40.2	14.97(1-2)	151	25	C	μ=0.1($NaClO_4$)	68Gg	
	-43.1	17.56(1-2)	192	25	C	μ=0.1($NaClO_4$) S:50v% Dioxane	68Gg	
Q17	7-QUINOLINE SULFONIC ACID, 8-hydroxy		$C_9H_7O_4NS$			$C_9H_5N(OH)(SO_3H)$		
UO_2^{2+}	-22.2	9.04(35°,1-2)	247	25-45	T	μ=0.02 pH=5.8	70Bc	b
	19.2	8.929(25°,1-2)	236.4	--	T	--	72Ba	u
Pd^{2+}	23.0	11.20(25°,1-2)	293.21	25-45	T	μ=0	70Ba	b
Cu^{2+}	-22.2	9.8(25°,1-2)	115.1	25-45	T	μ=0.02	69Bb	b

Metal Ion	ΔH, kJ /mol	Log K	ΔS, J /mol K	T,°C	M	Conditions	Ref.	Remarks
Q17, continued								
Tl^{3+}	-35.334	6.550(32°,1-3)	-35.303	23.2-38.8	T	CL<1x10^{-4}M CM<2x10^{-4}M	71Ca	b
Q18 5-QUINOLINE SULFONIC ACID, 8-hydoxy-7-iodo-			$C_9H_6O_4NIS$			$C_9H_4N(OH)(I)(SO_3H)$		L^{2-}
Mg^{2+}	-8.8	4.21(1)	50	0	T	μ=0	57Ua	
	-1.2	4.12(1)	75	25	T	μ=0	57Ua	
	9.6	4.17(1)	109	50	T	μ=0	57Ua	
Y^{3+}	-6.4	5.80(30°,1)	79.9	30-50	T	μ=0.1($NaClO_4$)	74Kb	b
	-10.2	4.96(30°,2)	61.1	30-50	T	μ=0.1($NaClO_4$)	74Kb	b
	-12.1	4.08(30°,3)	37.6	30-50	T	μ=0.1($NaClO_4$)	74Kb	b
	-23.8	15.15(25°,1-3)	209	25-45	T	μ=0.2M($NaClO_4$)	73Sf	b
YA	-30.4	5.23(25°,1)	-1.7	25-45	T	μ=0.2M(KNO_3) A=Nitrilotri-acetic acid	75Pg	b
La^{3+}	-6.5	4.52(30°,1)	64.0	30-50	T	μ=0.1($NaClO_4$)	74Kb	b
	-14.0	3.64(30°,2)	23	30-50	T	μ=0.1($NaClO_4$)	74Kb	b
	-12.1	3.00(30°,3)	17	30-50	T	μ=0.1($NaClO_4$)	74Kb	b
Nd^{3+}	-10.2	5.01(30°,1)	63.1	30-50	T	μ=0.1($NaClO_4$)	74Kb	b
	-10.2	4.10(30°,2)	44.4	30-50	T	μ=0.1($NaClO_4$)	74Kb	b
	-12.1	3.02(30°,3)	17.6	30-50	T	μ=0.1($NaClO_4$)	74Kb	b
Gd^{3+}	-17.7	5.34(30°,1)	43.5	30-50	T	μ=0.1($NaClO_4$)	74Kb	b
	-11.2	4.42(30°,2)	47.3	30-50	T	μ=0.1($NaClO_4$)	74Kb	b
	-14.9	3.21(30°,3)	12.1	30-50	T	μ=0.1($NaClO_4$)	74Kb	b
Dy^{3+}	-5.6	5.17(30°,1)	80.3	30-50	T	μ=0.1($NaClO_4$)	74Kb	b
	-13.0	4.11(30°,2)	35.6	30-50	T	μ=0.1($NaClO_4$)	74Kb	b
	-14.0	3.00(30°,3)	10.9	30-50	T	μ=0.1($NaClO_4$)	74Kb	b
Fe^{3+}	8.66	3.50(30°,1)	95.8	20-40	T	--	72Pm	b
	9.83	4.88(30°,2)	126.4	20-40	T	--	72Pm	b
	6.32	4.61(30°,3)	110.4	20-40	T	--	72Pm	b
Tl^{3+}	-39.476	7.149(32°,1-3)	-63.358	23.2-38.8	T	CL<1x10^{-4}M CM<2x10^{-4}M	71Ca	b
Q19 7-QUINOLINE SULFONIC ACID, 8-hydroxy-5-iodo-			$C_9H_6O_4NIS$			$C_9H_4N(OH)(I)(SO_3H)$		L^{2-}
UO_2^{2+}	16.74	8.80(1-2)	225.85	25	T	μ=0.05	72Ba	
	16.65	8.85(1-2)	225.64	30	T	μ=0.05	72Ba	
	16.82	8.89(1-2)	225.68	35	T	μ=0.05	72Ba	
	16.78	8.94(1-2)	225.77	40	T	μ=0.05	72Ba	
	16.69	8.98(1-2)	225.77	45	T	μ=0.05	72Ba	
Cu^{2+}	19.12	9.71(35°,1-2)	247.3	25-45	T	μ=0.02 pH=3	68Ba	
Q20 5-QUINOLINE SULFONIC ACID, 8-hydroxy-7-nitro-			$C_9H_6O_6N_2S$			$C_9H_4N(OH)(NO_2)(SO_3H)$		L^{2-}
Mg^{2+}	-6.3	3.32(1)	38	0	T	μ=0	57Ua	

Metal Ion	ΔH, kJ /mol	Log K	ΔS, J /mol K	T,°C	M	Conditions	Ref.	Re-marks
Q20, continued								
Mg^{2+},	2.9	3.28(1)	71	25	T	μ=0	57Ua	
cont.	16.3	3.41(1)	109	50	T	μ=0	57Ua	
Y^{3+}	-16.78	4.60(30°,1)	32.7	30-50	T	μ=0.1($HClO_4$)	77Ka	b
	-17.78	4.04(30°,2)	18.6	30-50	T	μ=0.1($HClO_4$)	77Ka	b
	-18.66	3.47(30°,3)	4.8	30-50	T	μ=0.1($HClO_4$)	77Ka	b
La^{3+}	-8.37	4.10(30°,1)	50.8	30-50	T	μ=0.1($HClO_4$)	77Ka	b
	-9.33	3.16(30°,2)	29.7	30-50	T	μ=0.1($HClO_4$)	77Ka	b
	-10.25	2.27(30°,3)	9.5	30-50	T	μ=0.1($HClO_4$)	77Ka	b
Pr^{3+}	-9.33	4.30(30°,1)	51.5	30-50	T	μ=0.1($HClO_4$)	77Ka	b
	-10.21	3.55(30°,2)	34.2	30-50	T	μ=0.1($HClO_4$)	77Ka	b
	-11.21	2.87(30°,3)	17.9	30-50	T	μ=0.1($HClO_4$)	77Ka	b
Nd^{3+}	-12.13	4.73(30°,1)	50.5	30-50	T	μ=0.1($HClO_4$)	77Ka	b
	-13.10	3.69(30°,2)	27.3	30-50	T	μ=0.1($HClO_4$)	77Ka	b
	-14.02	3.14(30°,3)	13.9	30-50	T	μ=0.1($HClO_4$)	77Ka	b
Gd^{3+}	-14.94	4.86(30°,1)	43.8	30-50	T	μ=0.1($HClO_4$)	77Ka	b
	-15.86	3.78(30°,2)	20.0	30-50	T	μ=0.1($HClO_4$)	77Ka	b
	-16.82	3.25(30°,3)	6.6	30-50	T	μ=0.1($HClO_4$)	77Ka	b
Tb^{3+}	-16.78	5.16(30°,1)	43.3	30-50	T	μ=0.1($HClO_4$)	77Ka	b
	-17.66	4.05(30°,2)	19.2	30-50	T	μ=0.1($HClO_4$)	77Ka	b
	-18.62	3.50(30°,3)	5.5	30-50	T	μ=0.1($HClO_4$)	77Ka	b
Dy^{3+}	-18.66	5.00(30°,1)	30.1	30-50	T	μ=0.1($HClO_4$)	77Ka	b
	-19.62	4.20(30°,2)	15.6	30-50	T	μ=0.1($HClO_4$)	77Ka	b
	-20.46	3.81(30°,3)	5.4	30-50	T	μ=0.1($HClO_4$)	77Ka	b
Ho^{3+}	-21.46	5.08(30°,1)	26.4	30-50	T	μ=0.1($HClO_4$)	77Ka	b
	-22.43	4.35(30°,2)	9.2	30-50	T	μ=0.1($HClO_4$)	77Ka	b
	-22.93	4.21(30°,3)	4.9	30-50	T	μ=0.1($HClO_4$)	77Ka	b

Q21 5-QUINOLINE SULFONIC ACID, 8-hydroxy-7-(4-nitrophenyl)- $C_{15}H_{10}O_6N_2S$ $C_9H_4N(OH)(C_6H_4NO_2)(SO_3H)$ L^{2-}

Metal Ion	ΔH, kJ /mol	Log K	ΔS, J /mol K	T,°C	M	Conditions	Ref.	Re-marks
Mg^{2+}	-4.2	4.02(1)	58	0	T	μ=0	57Ua	
	-2.9	3.96(1)	67	25	T	μ=0	57Ua	
	-1.7	3.93(1)	75	50	T	μ=0	57Ua	
Ca^{2+}	-2.9	3.27(1)	50	0	T	μ=0	57Ua	
	-0.4	3.23(1)	58	25	T	μ=0	57Ua	
Ba^{2+}	-3.8	2.51(1)	63	25	T	μ=0	57Ua	
	2.9	2.56(1)	75	50	T	μ=0	57Ua	
Mn^{2+}	-8.8	5.31(1)	71	25	T	μ=0	57Ua	
	0	5.25(1)	100	50	T	μ=0	57Ua	
Co^{2+}	-25.9	6.49(1)	38	0	T	μ=0	57Ua	
	-15.9	6.14(1)	67	25	T	μ=0	57Ua	
	-0.8	6.02(1)	105	50	T	μ=0	57Ua	
Ni^{2+}	-17.6	6.91(1)	75	25	T	μ=0	57Ua	
	-5.4	6.75(1)	109	50	T	μ=0	57Ua	
Cu^{2+}	-24.7	9.17(1)	84	0	T	μ=0	57Ua	

Metal Ion	ΔH, kJ /mol	Log K	ΔS, J /mol K	T,°C	M	Conditions	Ref.	Re-marks
Q21, continued								
Cu^{2+},	-24.3	8.77(1)	88	25	T	μ=0	57Ua	
cont.	-23.0	8.44(1)	92	50	T	μ=0	57Ua	
Zn^{2+}	-22.6	6.43(1)	42	0	T	μ=0	57Ua	
	-15.9	6.11(1)	63	25	T	μ=0	57Ua	
	-5.8	5.96(1)	95	50	T	μ=0	57Ua	
Cd^{2+}	-17.2	5.94(1)	50	0	T	μ=0	57Ua	
	-13.4	5.69(1)	63	25	T	μ=0	57Ua	
Q22	5-QUINOLINE SULFONIC ACID, 8-hydroxy-7-phenylazo		$C_{15}H_{11}O_4N_3S$			$C_9H_4N(OH(C_6H_5N{\equiv}N)(SO_3H)$		L^{2-}
Mg^{2+}	-9.2	4.04(1)	42	0	T	μ=0	57Ua	
	-3.3	3.94(1)	63	25	T	μ=0	57Ua	
	5.4	3.95(1)	92	50	T	μ=0	57Ua	
Ba^{2+}	2.93	2.16(1)	50	0	T	μ=0	57Ua	
	5.0	2.22(1)	58	25	T	μ=0	57Ua	
	7.9	2.29(1)	71	50	T	μ=0	57Ua	
Mn^{2+}	-16.3	5.57(1)	46	0	T	μ=0	57Ua	
	-9.2	5.36(1)	71	25	T	μ=0	57Ua	
Ni^{2+}	-25.1	7.57(1)	54	0	T	μ=0	57Ua	
	-20.5	7.21(1)	71	25	T	μ=0	57Ua	
	-13.0	6.97(1)	92	50	T	μ=0	57Ua	
Cu^{2+}	-30.1	9.50(1)	71	0	T	μ=0	57Ua	
	-27.6	9.03(1)	79	25	T	μ=0	57Ua	
	-22.2	8.96(1)	96	50	T	μ=0	57Ua	
Zn^{2+}	-23.8	6.86(1)	42	0	T	μ=0	57Ua	
	-16.7	6.53(1)	71	25	T	μ=0	57Ua	
	-6.3	6.36(1)	100	50	T	μ=0	57Ua	
Q23	QUINUCLIDINE		$C_7H_{13}N$			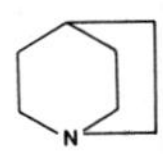		
RhAB	-55.6	1.255	-163	24	C	S:Benzene R:$[RhAB]_2$+2L= 2RhABL A=Cl B=1,5-Cycloocta-diene	76Lb	
PdAB	-45.6	--	--	24	C	S:Benzene R:$[PdAB]_2$+2L= 2[PdABL] A=(π-MeAllyl) B=Cl	77Lb	

Metal Ion	ΔH, kJ /mol	Log K	ΔS, J /mol K	T,°C	M	Conditions	Ref.	Remarks

R

R1 RIBOFLAVIN $C_{17}H_{20}O_6N_4$ L

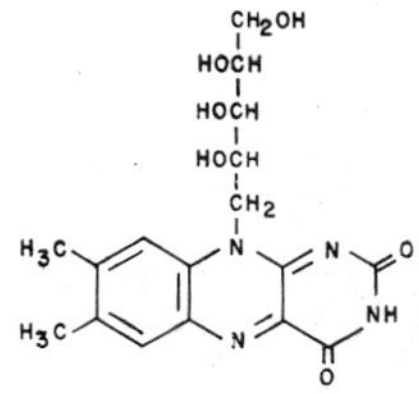

Metal Ion	ΔH, kJ /mol	Log K	ΔS, J /mol K	T,°C	M	Conditions	Ref.	Remarks
Y^+	293	17.95(35°,1-3)	1304	35&45	T	μ=0.05M(KNO_3)	74S1	a,b
La^{3+}	339	15.83(35°,1-3)	1404	35&45	T	μ=0.05M(KNO_3)	74S1	a,b
Ce^{3+}	-138	19.52(35°,1-3)	-76	35&45	T	μ=0.05M(KNO_3)	74S1	a,b
Co^{2+}	21	4.3(25°,2-1)	218	10-40	T	CM=0.01M CL=0.05M	59Ha	b,p
Ni^{2+}	12	4.2(25°,2-1)	192	10-40	T	CM=0.01M CL=0.05M	59Ha	b,p
Cu^{2+}	-12	5.9(25°,2-1)	184	10-40	T	CM=0.01M CL=0.05M	59Ha	b,p
Zn^{2+}	4	4.9(25°,2-1)	197	10-40	T	CM=0.01M CL=0.05M	59Ha	b,p

R2 RIBONUCLEASE A $C_{587}H_{909}N_{171}O_{197}S_{12}$

Metal Ion	ΔH, kJ /mol	Log K	ΔS, J /mol K	T,°C	M	Conditions	Ref.	Remarks
[P]	-64.0	4.556(1)	-128.0	25	C	[P]=3'-Cytidine monophosphate μ=0.05(NaAc) pH=5.5	71Bk	
	-66.1	4.580(1)	-123.4	25	C	[P]=3'-Cytidine monophosphate μ=0.05(KCl) pH=5.5	71Bk	
	-55.2	4.591(1)	-98.3	30	C	[P]=3'-Cytidine monophosphate μ=0.15(NaAc) pH=5.5	71Bk	
	-38.5	3.724(1)	-58.6	25	C	[P]=3'-Cytidine monophosphate μ=0.50(NaAc) pH=5.5	71Bk	
	-41.0	3.800(1)	-64.4	25	C	[P]=3'-Cytidine monophosphate μ=0.50(KCl) pH=5.5	71Bk	
	-31.4	3.176(1)	-44.8	25	C	[P]=3'-Cytidine monophosphate μ=1.20(NaAc) pH=5.5	71Bk	
	-25.5	3.398(1)	-20.9	30	C	[P]=3'-Cytidine monophosphate μ=1.20(NaAc) pH=5.5	71Bk	
	-29.3	3.505(1)	-47.3	25	C	[P]=3'-Cytidine monophosphate μ=1.20(KCl)	71Bk	

Metal Ion	ΔH, kJ /mol	Log K	ΔS, J /mol K	T,°C	M	Conditions	Ref.	Re-marks

R2, continued

Metal Ion	ΔH, kJ /mol	Log K	ΔS, J /mol K	T,°C	M	Conditions	Ref.	Re-marks
[P], cont.	-25.1	3.322(1)	-20.5	30	C	pH=5.5 [P]=3'-Cytidine monophosphate μ=3.00(NaAc) pH=5.5	71Bk	

S

S1 SALICYLALDIMINE, N-n-butyl- $C_{11}H_{15}ON$

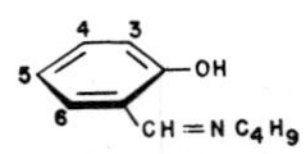

Metal Ion	ΔH, kJ /mol	Log K	ΔS, J /mol K	T,°C	M	Conditions	Ref.	Re-marks
Co^{2+}	-50.2	10.0(25°,1-2)	21	--	T	μ=0.1 S:80v% CH_3OH	72Kc	b,u
Cu^{2+}	-58.2	12.51(25°,1-2)	42	--	T	μ=0.1 S:80v% CH_3OH	72Kc	b,u

S2 SALICYLALDIMINE, N-tert-butyl- $C_{11}H_{15}ON$ $C_7H_6ON(C_4H_9)$

Metal Ion	ΔH, kJ /mol	Log K	ΔS, J /mol K	T,°C	M	Conditions	Ref.	Re-marks
NiA_2	5.4	-0.208(25°)	14.6	17-107	T	S:C_2Cl_4 R:NiL_2+NiA_2=2NiAL A=N-diethylmethyl salicylaldimine	73La	b,q: ±0.08
	-2.80	1.114(25°)	12.1	17-107	T	S:C_2Cl_4 R:NiL_2+NiA_2=2NiAL A=N-isopropylacetylacetoneimine	73La	b,q: ±0.29
	25.1	1.114(25°)	105	--	T	S:C_2Cl_4 R:NiL_2+NiA_2=2NiAL A=N-isopropylacetylacetoneimine	71Lf	b,u
	15.1	-0.796(25°)	38.1	17-107	T	S:C_2Cl_4 R:NiL_2+NiA_2=2NiAL A=N-isopropyl-3-methoxysalicylaldimine	73La	b,q: ±0.21
	5.0	0.079(25°)	18.8	17-107	T	S:C_2Cl_4 R:NiL_2+ NiA_2=2NiAL A=N-isopropyl-4-methylsalicylaldimine	73La	b,q: ±0.21
	-2.22	0.914(25°)	10.0	17-107	T	S:$CDCl_3$ R:NiL_2+ NiA_2=2NiAL where A=N-Isopropylbenzoylacetoneimine	73La	b,q: ±0.21

Metal Ion	ΔH, kJ /mol	Log K	ΔS, J /mol K	T,°C	M	Conditions	Ref.	Re-marks
S2, continued								
NiA_2, cont.	5.4	0.158(25°)	20.9	17-107	T	S:$CDCl_3$ R:NiL_2+ NiA_2=2NiAL where A=N-isopropyl-salicylaldimine	73La	b,q: ±0.21
	28.0	-0.886	77.0	17-107	T	S:$CDCl_3$ R:NiL_2+ NiA_2=2NiAL where A=N-phenylsalicyl-aldimine	73La	b,q: ±0.4
S3	SALICYLALDIMINE, N-n-butyl-3-methoxy-		$C_{12}H_{17}O_2N$			$C_7H_5ON(C_4H_9)(OCH_3)$		
Co^{2+}	-50.2	10.45(25°,1-2)	29	--	T	μ=0.1 S:80v% CH_3OH	72Kc	b,u
Cu^{2+}	-58.6	13.47(25°,1-2)	58	--	T	μ=0.1 S:80v% CH_3OH	72Kc	b,u
S4	SALICYLALDIMINE, N-t-butyl-5-methyl-		$C_{12}H_{17}ON$			$C_7H_5ON(C_4H_9)(CH_3)$		
Ni^{2+}	0.8	0.556(25°)	13.0	17-107	T	S:C_2Cl_4 R:NiL_2+NiA_2=2NiAL A=N-isopropylace-tylacetoneimine	73La	b,q: ±2.9
S5	SALICYLALDIMINE, 3-chloro-N-isopropyl-		$C_{10}H_{12}ONCl$			$C_7H_5ON(Cl)[CH(CH_3)_2]$		
NiL_2	12.6	-0.194(120°)	29	80-170	T	S:Bibenzyl	64Sa	b,t
S6	SALICYLALDIMINE, 5-chloro-N-n-propyl-		$C_{10}H_{12}ONCl$			$C_7H_5ON(Cl)[CH(CH_3)_2]$		
NiL_2	21.8	-1.501(120°)	25	80-170	T	S:Bibenzyl	64Sa	b,t
S7	SALICYLALDIMINE, 5-chloro-N-isopropyl-		$C_{10}H_{12}ONCl$			$C_7H_5ON(Cl)[CH(CH_3)_2]$		
NiL_2	8.4	-0.089(120°)	17	80-170	T	S:Bibenzyl	64Sa	b,t
S8	SALICYLALDIMINE, N-isobutyl-		$C_{11}H_{15}ON$			$C_7H_6ON[CH_2CH(CH_3)_2]$		
Co^{2+}	-48.1	10.08(25°,1-2)	29	--	T	μ=0.1 S:80v% CH_3OH	72Kc	b,u
Cu^{2+}	-55.2	13.13(25°,1-2)	71	--	T	μ=0.1 S:80v% CH_3OH	72Kc	b,u

Metal Ion	ΔH, kJ /mol	Log K	ΔS, J /mol K	T,°C	M	Conditions	Ref.	Remarks
S9	SALICYLALDIMINE, N-isobutyl-3-methoxy-		$C_{12}H_{17}O_2N$			$C_7H_5ON[CH_2CH(CH_3)_2](OCH_3)$		
Co^{2+}	-52.3	11.34(25°,1-2)	33	--	T	μ=0.1 S:80v% CH_3OH	72Kc	b,u
Cu^{2+}	-53.6	14.45(25°,1-2)	92	--	T	μ=0.1 S:80v% CH_3OH	72Kc	b,u
S10	SALICYLALDIMINE, N-isopentane		$C_{12}H_{17}ON$			$C_7H_6ON[CH(C_2H_5)_2]$		
NiL_2	10.9	-0.086(25°)	35.1	17-107	T	S:C_2Cl_4 R:NiL_2+NiA_2=2NiAL A=N-Isopropylace-ylacetoneimine	73La	b,q: ±1.6
	-0.8	0.556(25°)	7.1	17-107	T	S:$CDCl_3$ R:NiL_2+NiA_2=2NiAL A=N-Isopropylben-zoylacetoneimine	73La	b,q: ±3.0
S11	SALICYLALDIMINE, N-isopropyl-		$C_{10}H_{12}ON$			$C_7H_6ON[CH(CH_3)_2]$		
NiA_2	13.0	-0.097	42	17-107	T	S:C_2Cl_4 R:NiL_2+NiA_2=2NiAL A=N-isopropylace-tylacetoneimine	73La	b,q: ±3.0
	18.8	-0.276(25°)		--	T	S:C_2Cl_4 R:NiL_2+NiA_2=2NiAL A=N-isopropylace-tylacetoneimine	71Lf	
	18.0	0.121(25°)	63	17-107	T	S:C_2Cl_4 R:NiL_2+NiA_2=2NiAL A=N-n-propyl-salicylaldimine	73La	b,q: ±9.2
NiL_2	13.4	0.289(120°)	42	80-170	T	S:Bibenzyl	64Sa	b,t
S12	SALICYLALDIMINE, 5-methyl-N-isopropyl-		$C_{11}H_{15}ON$			$C_7H_5ON(CH_3)[CH(CH_3)_2]$		
NiL_2	10.5	-0.111(120°)	25	80-170	T	S:Bibenzyl R:NiL_2(planar)= NiL_2(tetrahedral)	64Sa	b,t
S13	SALICYLALDIMINE, 5-methyl-N-n-propyl-		$C_{11}H_{15}ON$			$C_7H_5ON(CH_3)[O_3H_7]$		
NiL_2	19.2	-1.557(120°)	17	80-170	T	S:Bibenzyl	64Sa	b,t
S14	SALICYLALDIMINE, N-phenyl-		$C_{13}H_{11}ON$			$C_7H_6ON(C_6H_5)$		
NiL_2	38.1	-1.060(25°)	107.1	17-107	T	S:$CDCl_3$ R:NiL_2+NiA_2=2NiAL	73La	b

Metal Ion	ΔH, kJ /mol	Log K	ΔS, J /mol K	T,°C	M	Conditions	Ref.	Re-marks
S14, continued								
NiL_2, cont.						A=N-isopropylace-tylacetoneimine		
S15 SALICYLALDIMINE, N-n-propyl-			$C_{10}H_{13}ON$			$C_7H_6ON(C_3H_7)$		
NiL_2	19.2	-1.612(120°)	17	80-170	T	S:Bibenzyl	64Sa	b,t
S16 SARCOSINE			$C_3H_7O_2N$			$CH_3NHCH_2CO_2H$		L^-
CoA	-17.6	3.13(25°,1)	2.1	15-70	T	μ=0.1M(KNO_3) A= Nitrilotriacetate	71Ie	b
NiA	-20.9	4.23(25°,1)	11.7	15-70	T	μ=0.1M(KNO_3) A= Nitrolotriacetate	71Ie	b
Cu^{2+}	-19.2	8.16(20°,1)	92	25	C	μ=0	64Ia	b
	-22.6	6.89(20°,2)	54	25	C	μ=0	64Ia	b
CuA	-50.2	5.15(25°,1)	-66.5	15-70	T	μ=0.1M(KNO_3) A= Nitrilotriacetate	71Ie	b
ZnA	-20.9	3.22(25°,1)	-7.9	15-70	T	μ=0.1M(KNO_3) A= Nitrilotriacetate	71Ie	b
CdA	-11.7	2.64(25°,1)	12.1	15-70	T	μ=0.1M(KNO_3) A= Nitrilotriacetate	71Ie	b
S17 SARCOSINE, N-glycyl-			$C_5H_{10}O_3N_2$			$H_2NCH_2CON(CH_3)CH_2CO_2H$		
Cu^{2+}	-25.1	6.41(1)	38.5	25	C	μ=0.1M(KNO_3)	72Bl	
	-22.6	5.01(2)	20.1	25	C	μ=0.1M(KNO_3)	72Bl	
S18 SELENATE ION			O_4Se			SeO_4^{2-}		L^{2-}
Mn^{2+}	14.68	2.430(25°,1)	95.8	0-45	T	μ=0	70Gf	b,q:±1.2
Co^{2+}	12.18	2.696(25°,1)	92.9	0-45	T	μ=0	70Gf	b
Ni^{2+}	14.68	2.669(25°,1)	95.8	0-45	T	μ=0	70Gf	b,q: ±1.7
S19 SELENITE ION			HO_3Se			$HSeO_3^-$		L^-
CrA_3^{3+}	23.39	-0.021	77.8	25	C	μ=3.0($NaClO_4$) A=Ethylenediamine R:$CrA_3^{3+}+L^{2-}=CrA_3L^+$	73Ua	
	-12.59	-0.126	44.4	25	C	μ=3.0($NaClO_4$) A=Ethylenediamine R:$CrA_3L^++L^{2-}=CrA_3L_2^-$	73Ua	
Fe^{3+}	16.90	3.45(25°)	123.0	20-40	T	0.5-1.0M($HClO_4$)	65Ha	b

Metal Ion	ΔH, kJ /mol	Log K	ΔS, J /mol K	T, °C	M	Conditions	Ref.	Remarks
S19, continued								
Fe^{3+}, cont.						R: Fe^{3+}+HL= FeL^{2+}+H^+		
CoA_3	5.8	0.556(1)	29	25	C	μ=3.0M($NaClO_4$+ Na_2L) A=Ethylenediamine	73Mc	q:±0.8
	1.7	0.146(2)	8	25	C	μ=3.0M($NaClO_4$+ Na_2L) A=Ethylenediamine	73Mc	q:±0.8
	0.8	-0.15(3)	0	25	C	μ=3.0M($NaClO_4$+ Na_2L) A=Ethylenediamine	73Mc	q:±0.8
	-8.4	-0.52(4)	-29	25	C	μ=3.0M($NaClO_4$+ Na_2L) A=Ethylenediamine	73Mc	q:±2.5
CoA_5B^{2+}	10.2	0.48(1)	42	25	C	μ=3.0M($NaClO_4$) A=NH_3 B=NO_2	74Mb	
	-4.2	0(2)	-12	25	C	μ=3.0M($NaClO_4$) A=NH_3 B=NO_2	74Mb	
	4.8	0.06(3)	17	25	C	μ=3.0M($NaClO_4$) A=NH_3 B=NO_2	74Mb	
	3.1	0.34(4)	17	25	C	μ=3.0M($NaClO_4$) A=NH_3 B=NO_2	74Mb	
S20	SELENOCYANATE ION		CNSe			$SeCN^-$		L^-
Fe^{3+}	-23.4	2.1(1)	-38	25	-	Hemoglobin	60Gb	
Fe^{3+}	-49.0	3.2(1)	-103.3	25	-	Chironomus Hemoglobin	60Gb	
Co^{2+}	-6.35	1.49(1)	7.21	25	C	μ=0.08-0.15 (ClO_4^-)	75Sb	
Ni^{2+}	-12.8	0.987(1)	-24.1	25	C	μ=1.0M($NaClO_4$)	74Kh	
	-8.70	1.66(1)	2.52	25	C	μ=0.08-0.15 (ClO_4−)	75Sb	
	-12	0.268(2)	-35	25	C	μ=1.0M($NaClO_4$)	74Kh	q:±3
PdA_2	-18.8	3.125(1)	-3.8	25	T	μ=0 A=o-Phenylenebis-dimethylarsine	67Ec 67Ed	
	-19.2	2.873(40°,1)	-7.1	25-55	T	μ=0.003($NaNO_3$) A=o-Phenylenebis-dimethylarsine	67Ed	b
PtA_2	-18.0	2.701(35°,1)	-6.3	25-55	T	μ=3x10^{-3}($NaNO_3$) A=o-Phenylenebis-dimethylarsine	67Ec	b,q: ±1.7
	-17.6	2.934(1)	-2.9	25	T	μ=0 A=o-Phenylenebis-dimethylarsine	67Ec	q:±1.7
Zn^{2+}	-5.95	0.44(1)	-11.6	25	C	μ=1.0M($NaClO_4$)	74Ac	
	-0.4	0.20(2)	3	25	C	μ=1.0M($NaClO_4$)	74Ac	q:±2.5

Metal Ion	ΔH, kJ /mol	Log K	ΔS, J /mol K	T,°C	M	Conditions	Ref.	Remarks
S20, continued								
Cd^{2+}	-10.05	1.47(1)	-5.6	25	C	μ=1.0M($NaClO_4$)	74Ac	
	-11.1	1.98(1)	0.75	25	C	μ=0.08-0.15 (ClO_4^-)	75Sb	
	-16.3	0.82(2)	-39	25	C	μ=1.0M($NaClO_4$)	74Ac	
	23	0.54(3)	90	25	C	μ=1.0M($NaClO_4$)	74Ac	q:±7
	-37	1.19(4)	-100	25	C	μ=1.0M($NaClO_4$)	74Ac	q:±7
Hg^{2+}	-194.6	29.6(1-4)	-87.02	25	T	μ=0	56Ta	
	-192.6	29.73(1-4)	-96	25	C	μ=1.0M($NaClO_4$)	74Ki	
Pb^{2+}	-149.0	(1-6)	--	20-30	T	--	59Gd	b
S21 SEMICARBAZIDE			CH_5ON_3			$H_2NCONHNH_2$		L
Ni^{2+}	-59.4	4.785(1-2)	-105	30	C	μ=0.1M(KNO)	77Aa	
	-8.4	1.868(3)	8.4	30	C	μ=0.1M(KNO_3)	77Aa	
Cu^{2+}	-24.3	4.000(1)	-4	30	C	μ=0.1(HNO_3+KNO_3)	71Ae	
	-25.94	2.940(2)	-29	30	C	μ=0.1(HNO_3+KNO_3)	71Ae	
Ag^+	-9.2	1.979(1)	8	30	C	μ=0.1M(KNO_3)	73Ad	q:±2.1
	-43.10	0.733(2)	-130	30	C	μ=0.1M(KNO_3)	73Ad	q:±0.4
Hg^{2+}	≅-67	11.6(1-2)	1.2	30	C	μ=0.1	69Gc	
	≅-75	15.2(1-3)	42.7	30	C	μ=0.1	69Gc	
Cd^{2+}	-11.7	1.246(1)	-17	30	C	μ=0.1M(KNO_3)	73Ad	q:±2.1
	-9.2	1.613(2)	0	30	C	μ=0.1M(KNO_3)	73Ad	q:±1.7
Pb^{2+}	-14.6	2.126(1)	-8	30	C	μ=0.1M(KNO_3)	73Ad	
	-14.2	0.733(2)	-33	30	C	μ=0.1M(KNO_3)	73Ad	
S22 SEMICARBAZIDE, 1,1-diacetic acid-			$C_5H_9O_5N_3$			$H_2NCONHN(CH_2CO_2H)_2$		L^{2-}
Mg^{2+}	41.0	1.4(1)	163	30	C	μ=0.1(KNO_3)	67Gd	
Sr^{2+}	-2.9	2.3(1)	38	30	C	μ=0.1(KNO_3)	67Gd	
Mn^{2+}	13.4	2.6(1)	92	30	C	μ=0.1(KNO_3)	67Gd	
Co^{2+}	-2.9	5.9(1)	121	30	C	μ=0.1(KNO_3)	67Gd	
Ni^{2+}	-4.6	6.6(1)	109	30	C	μ=0.1(KNO_3)	67Gd	
Cu^{2+}	-2.1	8.4(1)	155	30	C	μ=0.1(KNO_3)	67Gd	
Zn^{2+}	-1.7	6.6(1)	121	30	C	μ=0.1(KNO_3)	67Gd	
S23 SEMICARBAZIDE, 4-methyl, thiono			$C_2H_7N_3S$			$H_2NCSNHNHCH_3$		
Cd^{2+}	-19.85	2.745(1)	-14.0	25	C	μ=1.0($NaNO_3$) pH=4.0	79Ta	
	-20.79	2.341(2)	-24.9	25	C	μ=1.0($NaNO_3$) pH=4.0	79Ta	
	-15.55	1.216(3)	-28.9	25	C	μ=1.0($NaNO_3$) pH=4.0	79Ta	

Metal Ion	ΔH, kJ /mol	Log K	ΔS, J /mol K	T,°C	M	Conditions	Ref.	Re-marks
S24	SEMICARBAZIDE, seleno		CH_5N_3Se			$NH_2CSeNHNH_2$		L
Ni^{2+}	-8.8	1.778(3)	4.2	30	C	μ=0.1M(KNO_3)	77Aa	
	-52.7	3.301(1-2)	-109	30	C	μ=0.1M(KNO_3)	77Aa	
Cu^{2+}	-46.0	5.544(1)	-46	30	C	μ=0.1(HNO_3+KNO_3)	71Ae	
	-57.3	5.279(2)	-88	30	C	μ=0.1(HNO_3+KNO_3)	71Ae	
Ag^{+}	-118.0	13.85(1-2)	-130	30	C	μ=0.01M(KNO_3)	73Ad	
	-128.9	15.83(1-3)	-126	30	C	μ=0.01M(KNO_3)	73Ad	
Hg^{2+}	≅-184	30.4(1-3)	≅-29	30	C	μ=0.1	69Gd	
	≅-255	32.4(1-4)	≅-222	30	C	μ=0.1	69Gd	
Cd^{2+}	-13.0	2.93(1)	12	30	C	μ=0.01M(KNO_3)	73Ad	q:±1.7
	-34.7	2.20(1)	-71	30	C	μ=0.01M(KNO_3)	73Ad	
Pd^{2+}	-20.1	2.27(1)	-25	30	C	μ=0.01M(KNO_3)	73Ad	q:±0.84
S25	SEMICARBAZIDE, thiono		CH_5N_3S			$H_2NCSNHNH_2$		L
Ni^{2+}	-42.2	2.405(3)	-92	30	C	μ=0.1M(KNO_3)	77Aa	
	-44.8	4.799(1-2)	-54	30	C	μ=0.1M(KNO_3)	77Aa	
Cu^{2+}	-41.00	6.114(1)	-18.0	30	C	μ=0.1(HNO_3+KNO_3)	71Ae	
	-33.47	5.477(2)	-5.8	30	C	μ=0.1(HNO_3+KNO_3)	71Ae	
Ag^{+}	-93.93	7.498(1)	-171	25	C	μ=0.1M($NaClO_4$)	76Bd	
	-15.31	3.907(2)	23	25	C	μ=0.1M($NaClO_4$)	76Bd	
	-12.26	1.884(3)	-5	25	C	μ=0.1M($NaClO_4$)	76Bd	
	-92.0	10.70(1-2)	-100	30	C	μ=0.03M(KNO_3)	73Ad	
	-113.0	12.68(1-3)	-134	30	C	μ=0.03M(KNO_3)	73Ad	
	-75.3	12.9(25°,1-3)	-7.1	20-50	T	μ=0.80($NaNO_3$)	60Tb	b
	-199.9	21.58(2-3)	-253	25	C	μ=0.1M($NaClO_4$)	76Bd	
	-99.33	10.790	-126	25	C	μ=0.1M($NaClO_4$) R:Ag+1.5L=$AgL_{1.5}$	76Bd	
Cd^{2+}	-18.0	2.27(1)	-17	30	C	μ=0.03M(KNO_3)	73Ad	q:±2.1
	-19.74	2.600(1)	-16.4	25	C	μ=1.0($NaNO_3$) pH=4.0	79Ta	
	-19.2	2.20(2)	-21	30	C	μ=0.03M(KNO_3)	73Ad	q:±2.1
	-19.23	2.080(2)	-24.7	25	C	μ=1.0($NaNO_3$) pH=4.0	79Ta	
	-15.25	1.174(3)	-28.7	25	C	μ=1.0($NaNO_3$) pH=4.0	79Ta	
Hg^{2+}	-147.7	22.5(1-2)	-57.3	30	C	μ=0.1	69Gd	
	-166.9	24.8(1-3)	-75.7	30	C	μ=0.1	69Gd	
	-172	26.1(25°,1-4)	-76.1	2-50	T	μ=0.80($NaNO_3$)	60Tb	b
Pd^{2+}	-17.6	1.90(1)	-21	30	C	μ=0.03M(KNO_3)	73Ad	
S26	SEMICARBAZIDE, thiono, 1,1-diacetic acid-		$C_5H_9O_4N_3S$			$H_2NCSNHN(CH_2CO_2H)_2$		L^{2-}
Mg^{2+}	-5.8	0.7(1)	-4	30	C	μ=0.1(KNO_3)	67Gd	
Sr^{2+}	-0.4	2.2(1)	42	30	C	μ=0.1(KNO_3)	67Gd	
Mn^{2+}	30.1	2.2(1)	138	30	C	μ=0.1(KNO_3)	67Gd	

Metal Ion	ΔH, kJ /mol	Log K	ΔS, J /mol K	T,°C	M	Conditions	Ref.	Re- marks
S26, continued								
Co^{2+}	10.9	5.5(1)	138	30	C	μ=0.1(KNO_3)	67Gd	
Ni^{2+}	-3.3	5.9(1)	100	30	C	μ=0.1(KNO_3)	67Gd	
Cu^{2+}	-10.9	8.1(1)	117	30	C	μ=0.1(KNO_3)	67Gd	
Zn^{2+}	13.0	5.8(1)	155	30	C	μ=0.1(KNO_3)	67Gd	
S27 SERINE			$C_3H_7O_3N$			$HOCH_2CH(NH_2)CO_2H$		L^-
Mn^{2+}	-8.8	3.91(20°,1)	46	20-60	T	μ=0.1(KNO_3)	73Bh	b
	1.2	2.40(20°,2)	46	20-60	T	μ=0.1(KNO_3)	73Bh	b
	-3.3	4.00(15°,1-2)	63	15&40	T	0.2M(KNO_3)	68Ra	a,b
Fe^{2+}	-6.28	6.45(15°,1-2)	100	15&40	T	0.2M(KNO_3)	68Ra	a,b
Co^{2+}	-11.3	4.38(1)	46	25	C	μ=0.05M(KCl)	72Gb 72Ga	
	-13.22	4.38(25°,1)	42	20-35	T	μ=0.05M(KCl)	72Gb 72Ga	b
	-9.2	3.62(2)	38	25	C	μ=0.05M(KCl)	72Gb 72Ga	
	-10.27	3.62(25°,2)	25	20-35	T	μ=0.05M(KCl)	72Gb 72Ga	b
	-16.7	7.66(25°,1-2)	92	15-40	T	0.2M(KNO_3)	68Ra	b
Ni^{2+}	-15.90	5.43(1)	50	25	C	μ=0.05M(KCl)	72Gb 72Ga	
	-14.2	5.43(25°,1)	58	20-35	T	μ=0.05M(KCl)	72Gb 72Ga	b
	-15.73	5.45(1)	51.5	25	C	μ=0.16	70Lb	
	-21.84	5.44(25°,1)	31.0	10-30	T	--	57Pa 60Pc	b
	-19.2	4.53(2)	25	25	C	μ=0.05M(KCl)	72Gb 72Ga	
	-19.2	4.53(25°,2)	21	20-35	T	μ=0.05M(KCl)	72Gb 72Ga	b
	-17.86	4.53(2)	26.8	25	C	μ=0.16	70Lb	
	-23.72	4.50(25°,2)	6.7	10-30	T	--	57Pa 60Pc	b
	-22.09	3.54(3)	-6.3	25	C	μ=0.16	70Lb	
	-26.23	3.3(25°,3)	-30.1	10-30	T	--	57Pa 60Pc	b
	-33.4	10.1(1-2)	79.5	22.0	C	0.1M(KNO_3)	67Sc	
	-32.6	9.76(25°,1-2)	79	15-40	T	0.2M(KNO_3)	68Ra	b
Cu^{2+}	-28.4	7.93(1)	54	25	C	μ=0.05M(KCl)	70Gc 71Gc	
	-23.0	7.93(1)	75	25	C	μ=0.05M(KCl)	72Gb 72Ga	
	-18.0	7.93(25°,1)	92	20-35	T	μ=0.05M(KCl)	72Gb 72Ga	b
	-23.0	7.92(1)	74.5	25	C	μ=0.1M(KNO_3)	72Ia	
	-23.05	7.85(1)	73.2	25	C	μ=0.16	70Lb	
	-23.0	6.74(2)	67	25	C	μ=0.05M(KCl)	70Gc 71Gc	

Metal Ion	ΔH, kJ /mol	Log K	ΔS, J /mol K	T,°C	M	Conditions	Ref.	Remarks
S27, continued								
Cu^{2+}, cont.	-26.4	6.55(2)	33	25	C	μ=0.05M(KCl)	72Gb 72Ga	
	-28.9	6.55(25°,2)	29	20-35	T	μ=0.05M(KCl)	72Gb 72Ga	b
	-25.5	6.65(2)	41.8	25	C	μ=0.1M(KNO_3)	72Ia	
	-25.65	6.65(2)	41.4	25	C	μ=0.16	70Lb	
	-54.4	(1-2)	--	24	C	μ=0.01M($HClO_4$)	72Sh	
	-59.0	14.6(1-2)	79	21.9	C	0.1M(KNO_3)	67Sc	
	-53.59	14.42(1-2)	96.3	25	C	μ=0.10M(KNO_3)	76Pc	
	-54.8	(1-2)	--	25	C	0.1M(KNO_3)	69Tb	
	-40.2	14.40(25°,1-2)	142	16-40	T	0.2M(KNO_3)	68Ra	b
	-33.5	16.05	195	25	C	μ=0.50($NaClO_4$) R:$Cu^{2+}+L^{-}+A^{3-}=CuAL^{2-}$ A=2-hydroxy-5-sulfobenzoic acid	75Mc	
	-1.59	1.50(25°)	33.9	-75&25	T	S:50v% CH_3OH R:$Cu+CuL_2=2CuL$	73Ya	a,b,q: ±1.5
Zn^{2+}	-9.6	4.65(1)	54	25	C	μ=0.05M(KCl)	72Gb 72Ga	
	-12.6	4.65(25°,1)	46	20-35	T	μ=0.05M(KCl)	72Gb 72Ga	b
	-10.9	4.03(2)	42	25	C	μ=0.05M(KCl)	72Gb 72Ga	
	-13.0	4.03(25°,2)	38	20-35	T	μ=0.05M(KCl)	72Gb 72Ga	b
	-28.0	(1-2)	--	22.0	C	0.1M(KNO_3)	67Sc	
	-18.0	8.38(25°,1-2)	100.4	15-40	T	0.2M(KNO_3)	68Ra	b
S28 SILICIC ACID			SiH_4O_4			$Si(OH)_4$		
Fe^{3+}	15.61	-0.602(25°)	41.13	18-32	T	CM=5x10^{-4}M CL=5x10^{-2}M pH=0.5-3.0 R:$Fe^{3+}+L=FeOSi(OH)_3^{2+}+H^+$	73Ob	b
S29 SODIUM DEXTRAN SULFATE								L
$FeAB_2^+$	4.56	4.45	100	21 or 25	C	A=2,2':6',2":6"2'''-quater-pyridyl B=OH^- R:trans-$[FeAB_2]^+$ +L=complex	80Pa	
	3.18	3.76	79	21 or 25	C	A=N,N'-bis (2-methylpyridyl) ethylenediamine B=OH^- R:cis$[FeAB_2]^+$+ L=complex	80Pa	

Metal Ion	ΔH, kJ /mol	Log K	ΔS, J /mol K	T,°C	M	Conditions	Ref.	Re-marks
S29, continued								
$FeABC^{2+}$	-6.44	4.30	54	21 or 25	C	A=N,N'-bis(2-methylpyridyl)-ethylenediamine B=H_2O C=OH^- R:cis$[FeABC]^{2+}$+ L=complex	80Pa	
S30 SODIUM TRIPHOSPHATE			$O_{10}P_3Na_3$			$Na_5P_3O_{10}$		L^{5-}
Ca^{2+}	-14.2	6.4(1)	171	25	C	μ~0.02M	77Rb	c
	-6.3	3.0	38	25	C	μ~0.02M R:Ca^{2+}+CaL^{3-}= Ca_2L^-	77Rb	c
S31 SOLOCHROME, Violet R								L^{3-}
Cr^{3+}	-25	17.05(85°,2)	251	75-100	T	μ=0	62Cf	b,q:±8.4
S32 STIBINE, triethyl			$C_{18}H_{15}Sb$			$Sb(C_6H_5)_3$		
$PtAB_2^+$	-25.5	(1)	--	25	C	S:THF A=CH_3 B=$[P(CH_3)_2(C_6H_5)]$	75Ma	
S33 STIBENE, triphenyl			$C_{18}H_{15}Sb$			$(C_6H_5)_3Sb$		
Cu^+	-11	1.25(1)	-13	25	C	μ=0.1M(NH_4ClO_4) S:Dimethyl sulfoxide	78Ab	
	-25	0.54(2)	-74	25	C	μ=0.1M(NH_4ClO_4) S:Dimethyl sulfoxide	78Ab	
S34 STYRENE			C_8H_8			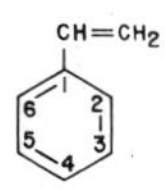		L
PdAB	-27.78	-0.175(25°)	95.8	0-39	T	S:$CDCl_3$ A=6-Acetoxynorbornenyl B=Benzoyltri-fluoroacetonate R:PdAB+L=PdCBL C=5-Acetoxynor-tricyclenyl	73Ba	q:1.9
Ag^+	-8.83	1.25(25°,1)	-5.65	0-40	T	μ=0	65Fa	b
S35 STYRENE, 4-chloro-			C_8H_7Cl			$C_8H_7(Cl)$		L
Ag^+	-9.16	1.03(25°,1)	-10.96	0-40	T	μ=0	65Fa	b

Metal Ion	ΔH, kJ /mol	Log K	ΔS, J /mol K	T, °C	M	Conditions	Ref.	Remarks
S36	STYRENE, 4-dimethylamine		$C_{10}H_{13}N$			$C_8H_7[N(CH_3)_2]$		L
PdAB	-43.1	0.788(25°)	125.9	0-39	T	S:$CDCl_3$ A=6-Acetoxynorbornenyl B=Benzoyltrifluoroacetonate R:PdAB+L=PdCBL C=5-Acetoxynortricyclenyl	73Ba	b
S37	STYRENE, 4-fluoro-		C_8H_7F			$C_8H_7(F)$		L
PdAB	-25.98	-0.263(25°)	92	0-39	T	S:$CDCl_3$ A=6-Acetoxynorbornenyl B=Benzoyltrifluoroacetonate R:PdAB+L=PdCBL C=5-Acetoxynortricyclenyl	73Ba	b,q: ±1.9
S38	STYRENE, 4-methoxy-		$C_9H_{10}O$			$C_8H_7(OCH_3)$		L
PdAB	-33.01	0.175(25°)	105.0	0-39	T	S:$CDCl_3$ A=6-Acetoxynorborenyl B=Benzoyltrifluoroacetonate R:PdAB+L=PdCBL C=5-Acetoxynortricyclenyl	73Ba	b
S39	STYRENE, 4-methyl-		C_9H_{10}			$C_8H_7(CH_3)$		L
Ag^+	-7.86	1.38(25°,1)	0.12	0-40	T	μ=0	65Fa	b
S40	STYRENE, 4-nitro-		$C_8H_7O_2N$			$C_8H_7(NO_2)$		L
PdAB	-11.72	-0.578(25°)	50	0-39	T	S:$CDCl_3$ A=6-Acetoxynorbornenyl B=Benzoyltrifluoroacetonate R:PdAB+L=PdCBL C=5-Acetoxynortricyclenyl	73Ba	b,q: ±1.9
S41	SUCCINIC ACID		$C_4H_6O_4$			$HO_2CCH_2CH_2CO_2H$		L^{2-}
Be^{2+}	11.9	2.48(25°,1)	87.9	25&35	T	μ=1.0(ClO_4)	74Tb	a,b
	46.0	2.41(25°,2)	201	25&35	T	μ=1.0(ClO_4)	74Tb	a,b
Mn^{2+}	12.64	2.27(1)	85.8	25	C	μ=0	67Mf	
	12.34	2.27(25°,1)	84.9	0-45	T	μ=0.2 ΔCp=62	61Mc	b
Co^{2+}	13.18	2.22(1)	86.6	25	C	μ=0	67Mf	
	11.76	2.22(25°,1)	82.0	0-45	T	μ=0.2 ΔCp=72	61Mc	b

Metal Ion	ΔH, kJ /mol	Log K	ΔS, J /mol K	T,°C	M	Conditions	Ref.	Re-marks
S41, continued								
Ni^{2+}	10.29	2.35(1)	79.5	25	C	μ=0	67Mf	
	9.33	2.35(25°,1)	76.1	0-45	T	μ=0.2 ΔCp=42	61Mc	b
Cu^{2+}	19.08	3.24(1)	125.9	25	C	μ=0	67Mf	
	11.3	2.610(1)	87.9	25	C	μ=0.1M($NaClO_4$)	78Ae	
	-42.7	10.915	65.7	25	C	μ=0.1M($NaClO_4$) R:Cu^{2+}+L^{2-}+A= CuLA A=2,2'-Bipyridyl	78Ae	
Al^{3+}	-10.7	3.83(25°,1)	58.6	25&35	T	μ=1.0(ClO_4)	74Tb	a,b
	92.8	2.09(25°,2)	180	25&35	T	μ=1.0(ClO_4)	74Tb	a,b
	12.9	3.05(25°,3)	75.3	25&35	T	μ=1.0(ClO_4)	74Tb	a,b
Cu^{2+}	-33.0	15.979	195.8	25	C	μ=0.1M($NaClO_4$) R:Cu^{2+}+L^{2-}+A+H^+= $[CuLAH]^+$ A=2,2'-Bipyridyl	78Ae	
Zn^{2+}	18.37	2.47(1)	108.8	25	C	μ=0	67Mf	
S42	SUCCINIC ACID, 2-hydroxy-		$C_4H_6O_5$			$HO_2CCH_2CH(OH)CO_2H$		
Ca^{2+}	36.02	3.01(25°,1)	178.6	25-40	T	μ=0	75Dc	b
Y^{3+}	8.4	7.74(25°,1-2)	176	25-45	T	μ=0.2M($NaClO_4$)	75Pb	b
Cu^{2+}	45.93	4.53(25°,1)	240.9	25-40	T	μ=0	75Dc	b
	36.98	2.44(25°)	170.7	25-40	T	μ=0 R:Cu^{2+}+HL^-=$CuHL^+$	75Dc	b
In^{3+}	17.6	4.60(35°,1)	146	25-45	T	μ=0.20M($NaClO_4$)	73Sc	b
	43.56	8.56(35°,1-2)	305	25-45	T	μ=0	72Sb	b
S43	SUCCINIC ACID, imide		$C_4H_5O_2N$			$CH_2C(=O)NHC(=O)CH_2$ (ring)		L
Ni^{2+}	-9.07	2.93(1)	25.69	25	C	0.5M(KNO_3)	79Bc	
	-9.3	1.56(2)	-0.6	25	C	0.5M(KNO_3)	79Bc	
	-6.3	2.14(3)	18.8	25	C	0.5M(KNO_3)	79Bc	
Cu^{2+}	-16.44	4.61(1)	19.08	25	C	0.5M(KNO_3)	79Bc	
	-17.15	4.14(2)	35.7	25	C	0.5M(KNO_3)	79Bc	
Ag^+	-23.43	4.36(1)	5.0	25	C	0.5M(KNO_3)	79Bc	
	-27.20	5.28(2)	9.7	25	C	0.5M(KNO_3)	79Bc	
Cd^{2+}	-8.24	3.31(1)	35.6	25	C	0.5M(KNO_3)	79Bc	
	-8.12	2.19(2)	14.6	25	C	0.5M(KNO_3)	79Bc	
	-6.6	1.42(3)	5.0	25	C	0.5M(KNO_3)	79Bc	
S44	SUCCINIC ACID, 2-mercapto-		$C_4H_6O_4S$			$HO_2CCH_2CH(SH)CO_2H$		L^{3-}
Y^{3+}	10.5	5.21(25°,1-2)	134	25-45	T	μ=0.2M($NaClO_4$)	75Pb	b
Mn^{2+}	-8	4.87(45°,1)	67	35-55	T	μ=0.2M($NaClO_4$)	69Rb	b,q:±8

Metal Ion	ΔH, kJ /mol	Log K	ΔS, J /mol K	T,°C	M	Conditions	Ref.	Re-marks
S44, continued								
Fe^{3+}	21.3	3.446(25°)	138	10-25	T	μ=1.0M($NaClO_4$) R:Fe^{3+}+HL^-= FeL^++H^+	73Eb	b,q: ±6.3
	21.3	0.406(25°)	84	10-25	T	μ=1.0M($NaClO_4$) R:Fe^{3+}+H_2L= FeL^++$2H^+$	73Eb	b,q: ±5.8
FeA^{2+}	-20.1	6.230(25°)	50	10-25	T	μ=1.0M($NaClO_4$) A=OH^- R:FeA^{2+}+ HL^-=FeL^++HA	73Eb	b,q: ±5.8
	-20.9	3.643(25°)	-10.5	10-25	T	μ=1.0M($NaClO_4$) A=OH^- R:FeA^{2+}+ H_2L=FeL^++H^++HA	73Eb	b,q: ±6.3
Ni^{2+}	-29	7.38(45°,1)	46	35-55	T	μ=0.2M($NaClO_4$)	69Rb	b,q:±8
	-25	6.23(45°,2)	42	35-55	T	μ=0.2M($NaClO_4$)	69Rb	b,q:±8
	-36.69	14.15(1-2)	147.95	25	T	0.1M(KNO_3)	68Sb	
Ag^+	-31.4	6.95(1)	29.3	30	T	μ=0.1M(KNO_3)	68Sa	b
Zn^{2+}	-22.2	15.60(30°,1-2)	225.1	25-35	T	μ=0.1M(KNO_3)	68Sc	b
In^{3+}	-25.5	14.46(35°,1)	188	25-45	T	μ=0.20M($NaClO_4$)	73Sc	
	-51.13	26.76(35°,1-2)	346.4	25-45	T	μ=0	72Sb	b
Tl^+	-8	3.86(45°,1)	46	35-55	T	μ=0.2M($NaClO_4$)	69Rb	b,q:±8
S45 SUCCINIC ACID, methylene			$C_5H_6O_4$			CH_2:C(CO_2H)CH_2CO_2H		
Tl^{1+}	-54.6	1.66(28°,1)	-213	28-48	T	--	76Sa	b
S46 SULFATE ION			O_4S			$SO_4{}^{2-}$		L^{2-}
Li^+	≅0	(1)	--	25	C	μ=0	62Ac	
	≅0	(1)	--	25	C	μ=0.025	62Ac	
Na^+	-2.05	0.65(1)	-5.4	25	C	μ=0	69Ia	
	4.686	(1)	--	25	C	μ=0	62Ac	
	3.933	(1)	--	25	C	μ=0.025	62Ac	
	∿0	--	--	25	C	μ=0.75M	80Ja	n
K^+	4.22	0.75(1)	0.0	25	C	μ=0	69Ia	
Be^{2+}	5.8	2.00(25°,1)	58.6	5-45	T	μ=0 R:Be^{2+}+L^{2-}= Be(H_2O)L (outer-sphere complex)	74Kf	b,q:±3.3
	9.6	1.81(25°,1)	67.8	5-45	T	μ=0 R:Be^{2+}+L^{2-}=BeL (inner-sphere complex)	74Kf	b
Mg^{2+}	17.07	2.129(1)	103.3	0	T	μ=0 ΔCp=-32	67Mb	
	2.13	2.23(1)	50.2	25	C	μ=0	69Ib	

Metal Ion	ΔH, kJ /mol	Log K	ΔS, J /mol K	T,°C	M	Conditions	Ref.	Remarks
S46, continued								
Mg^{2+},	6.48	(1)	--	25	C	μ=0	73Hb	
cont.	5.31	2.31(1)	61.5	25	C	μ=0	70La	i:59Lc
	16.78	2.399(1)	102.1	25	T	μ=0 ΔCp=11.7	67Mb	
	23.85	2.34(1)	129.7	25	T	μ=0	54Cd	
	8.54	2.130(25°,1)	69.4	0-40	T	μ=0	73Kd	b
	17.66	2.631(1)	105.0	50	T	μ=0 ΔCp=56	67Mb	
	19.58	2.846(1)	110.9	75	T	μ=0 ΔCp=100	67Mb	
	22.64	3.057(1)	119.2	100	T	μ=0 ΔCp=146	67Mb	
	26.82	3.274(1)	130.1	125	T	μ=0 ΔCp=188	67Mb	
	32.09	3.501(1)	142.7	150	T	μ=0 ΔCp=234	67Mb	
	38.49	3.743(1)	157.3	175	T	μ=0 ΔCp=276	67Mb	
	46.0	4.002(1)	173.6	200	T	μ=0 ΔCp=322	67Mb	
	64.4	4.58(1)	210.4	250	T	μ=0 ΔCp=410	67Mb	
	87.0	5.23(1)	251.9	300	T	μ=0 ΔCp=498	67Mb	
	114.2	5.96(1)	297.1	350	T	μ=0 ΔCp=586	67Mb	
	126.4	6.27(1)	316.3	370	T	μ=0 ΔCp=623	67Mb	
	4.8-5.7	(1)	--	30	C	μ=0 Value of ΔH depends on estimated relative apparent enthalpies of free ions	75Lb	
	13.01	2.644(1)	87.0	50	T	μ=0	73Ac	q:±4.2
	0.7-3.3	1.08-0.48(1)	16-27	25	C	μ=0.75M	80Ja	n
Ca^{2+}	3.39	2.43(1)	57.3	25	C	μ=0	69Ib	
	6.28	2.31(1)	65.3	25	C	μ=0	70La	i:59Lc
	6.40	2.309(1)	66.1	25	T	μ=0	73Ac	q:±4.2
	6.90	2.31(1)	67.4	25	T	μ=0	53Ba	
	10.29	2.49(1)	79.1	40	T	μ=0 ΔCp=260	73Ac	q:±2.1
	19.7	2.27(25°,1)	110	15-40	T	μ=0 P=1	71Ia	
	22.6	2.27(25°,1)	119	15-40	T	μ=0 P=193	71Ia	
	23.8	2.27(25°,1)	123	15-40	T	μ=0 P=387 ΔV(15°C)=12.9 ΔV(25°C)=10.2 ΔV(40°C)=5.7	71Ia	
	24.3	2.20(25°,1)	123	15-40	T	μ=0 P=580	71Ia	
	28.4	2.20(25°,1)	138	15-40	T	μ=0 P=774	71Ia	
	29.7	2.12(25°,1)	140	15-40	T	μ=0 P=967	71Ia	
	31.0	2.12(25°,1)	144	15-40	T	μ=0 P=1160	71Ia	
Sc^{3+}	26.40	4.04(1)	165.7	25	C	μ=0	69Ib	
	16.57	1.66(2)	87.9	25	C	μ=0	69Ib	
Y^{3+}	15.10	3.34(1)	114.6	25	C	μ=0	69Ib	
	13.72	3.468(1)	112.5	25	C	μ=0	69Fa	
	16.90	1.23(1)	80.3	25	C	μ=2.0	67Da	
	3.05	2.00(2)	48.5	25	C	μ=0	69Ib	
	6.3	0.45(2)	29.7	25	C	μ=2.0	67Da	
La^{3+}	13.56	3.50(1)	112.5	25	C	μ=0	69Ib 52Jb	
	14.64	3.620(1)	118.4	25	C	μ=0	69Fa	
	10.5	1.4(1)	63	25	T	μ=1.0($NaClO_4$)	53Nb	
	15.56	1.29(1)	77.0	25	C	μ=2.0	67Da	
	5.27	1.85(2)	53.1	25	C	μ=0	69Ib	

Metal Ion	ΔH, kJ /mol	Log K	ΔS, J /mol K	T,°C	M	Conditions	Ref.	Re-marks
S46, continued								
Ce^{3+}	14.48	3.48(1)	115.5	25	C	μ=0	69Ib	
	15.82	3.585(1)	121.8	25	C	μ=0	69Fa	
	19.62	1.23(1)	130.5	25	T	μ=0	53Nb	
	15.23	1.23(1)	74.5	25	T	μ=1.0($NaClO_4$)	53Nb	
	17.86	1.24(1)	83.7	25	C	μ=2.0	67Da	
	17.57	1.32(1)	83.7	25	T	μ=2.0	67Da	
	6.69	1.75(2)	56.1	25	C	μ=0	69Ib	
Pr^{3+}	14.60	3.58(1)	118.0	25	C	μ=0	69Ib	
	16.40	3.620(1)	124.3	25	C	μ=0	69Fa	
	16.48	1.27(1)	79.5	25	C	μ=2.0	67Da	
	4.81	1.86(2)	51.9	25	C	μ=0	69Ib	
Nd^{3+}	15.15	3.43(1)	116.3	25	C	μ=0	69Ib	
	17.36	3.638(1)	128.0	25	C	μ=0	69Fa	
	17.49	1.25(1)	82.8	25	C	μ=2.0	67Da	
	6.74	1.74(2)	56.1	25	C	μ=0	69Ib	
Pm^{3+}	16.3	1.32(1)	79	25	C	μ=2.0	67Da	
Sm^{3+}	15.48	3.52(1)	119.7	25	C	μ=0	69Ib	
	18.16	3.658(1)	131.0	25	C	μ=0	69Fa	
	17.86	1.30(1)	82.4	25	C	μ=2.0	67Da	
	8.8	1.67(2)	61.5	25	C	μ=0	69Ib	
Eu^{3+}	15.23	3.54(1)	118.8	25	C	μ=0	69Ib	
	17.28	3.658(1)	128.0	25	C	μ=0	69Fa	
	22.6	3.671(1)	146	25	T	μ=0 ΔCp=192	72Ha	
	25.1	3.850(1)	155	38.5	T	μ=0 ΔCp=205	72Ha	
	28.0	4.040(1)	163	51.6	T	μ=0 ΔCp=222	72Ha	
	31.0	4.225(1)	172	65.1	T	μ=0 ΔCp=259	72Ha	
	17.2	2.757(1)	109	25	T	μ=0.045M ΔCp=247	72Ha	
	20.5	2.900(1)	121	38.5	T	μ=0.045M ΔCp=276	72Ha	
	24.3	3.053(1)	134	51.6	T	μ=0.045M ΔCp=310	72Ha	
	28.9	3.227(1)	146	65.1	T	μ=0.045M ΔCp=343	71Ha	
	16.23	1.37(1)	80.8	25	C	μ=2.0	67Da	
	6.32	1.78(2)	55.2	25	C	μ=0	69Ib	
	10.0	0.71(2)	47.3	25	C	μ=2.0	67Da	
Gd^{3+}	15.02	3.48(1)	117.2	25	C	μ=0	69Ib	
	17.15	3.658(1)	127.6	25	C	μ=0	69Fa	
	16.53	1.33(1)	80.8	25	C	μ=2.0	67Da	
	6.28	1.73(2)	54.0	25	C	μ=0	69Ib	
	7.1	0.42(2)	31.8	25	C	μ=2.0	67Da	
Tb^{3+}	15.06	3.47(1)	117.2	25	C	μ=0	69Ib	
	16.82	3.635(1)	125.9	25	C	μ=0	69Fa	
	17.74	1.28(1)	84.10	25	C	μ=2.0	67Da	
	3.22	1.90(2)	47.3	25	C	μ=0	69Ib	
	6.7	0.62(2)	34.3	25	C	μ=2.0	67Da	
Dy^{3+}	15.44	3.43(1)	117.6	25	C	μ=0	69Ib	
	14.98	3.611(1)	119.2	25	C	μ=0	69Fa	
	18.45	1.23(1)	85.4	25	C	μ=2.0	67Da	
	5.73	1.75(2)	52.3	25	C	μ=0	69Ib	
	5.4	0.49(2)	27.6	25	C	μ=2.0	67Da	
Ho^{3+}	15.31	3.38(1)	116.3	25	C	μ=0	69Ib	

Metal Ion	ΔH, kJ /mol	Log K	ΔS, J /mol K	T,°C	M	Conditions	Ref.	Remarks
S46, continued								
Ho^{3+}, cont.	14.81	3.585(1)	118.4	25	C	μ=0	69Fa	
	17.66	1.24(1)	82.8	25	C	μ=2.0	67Da	
	6.69	1.60(2)	53.1	25	C	μ=0	69Ib	
	7.1	0.53(2)	33.9	25	C	μ=2.0	67Da	
Er^{3+}	15.10	3.41(1)	115.9	25	C	μ=0	69Ib	
	14.18	3.585(1)	116.3	25	C	μ=0	69Fa	
	17.61	1.22(1)	82.4	25	C	μ=2.0	67Da	
	5.73	1.78(2)	53.1	25	C	μ=0	69Ib	
	6.3	0.48(2)	30.1	25	C	μ=2.0	67Da	
Tm^{3+}	15.06	3.41(1)	115.9	25	C	μ=0	69Ib	
	13.18	3.585(1)	113.0	25	C	μ=0	69Fa	
	17.66	1.15(1)	81.2	25	C	μ=2.0	67Da	
	3.60	1.80(2)	44.8	25	C	μ=0	69Ib	
	4.2	0.45(2)	22.6	25	C	μ=2.0	67Da	
Yb^{3+}	15.06	3.33(1)	114.2	25	C	μ=0	69Ib	
	12.13	3.585(1)	109.2	25	C	μ=0	69Fa	
	17.32	1.15(1)	80.3	25	C	μ=2.0	67Da	
	4.06	1.72(2)	46.4	25	C	μ=0	69Ib	
	5.0	0.45(2)	25.5	25	C	μ=2.0	67Da	
Lu^{3+}	13.76	3.49(1)	114.6	25	C	μ=0	69Ib	
	14.81	3.585(1)	118.4	25	C	μ=0	69Fa	
	17.78	1.08(1)	80.3	25	C	μ=2.0	67Da	
	4.35	1.8(2)	49.4	25	C	μ=0	69Ib	
	6.3	0.53(2)	31.0	25	C	μ=2.0	67Da	
Th^{4+}	-3.8	2.27(25°)	31	10-40	T	μ=2.0(ClO_4^-,$SO_4^=$) $[H^+]$=2.0M R:$Th^{4+}+HL^-=ThL^{2+}+H^+$	72Pf	b
	-2.26	2.220	35.1	25	C	μ=2.0($KClO_4$) R:$Th^{4+}+HL^-=ThL^{2+}+H^+$	59Za	
	-3.72	1.336	13.4	25	C	μ=2.0($KClO_4$) R:$ThL^{2+}+HL^-=ThL_2+H^+$	59Za	
	0	1.29(25°)	25	10-40	T	μ=2.0(ClO_4^-,$SO_4^=$) $[H^+]$=2.0M R:$ThL^{2+}+HL^-=ThL_2+H^+$	72Pf	b
VO^{2-}	17.24	2.48(1)	105.4	25	C	μ=0	71Ba	
Cr^{3+}	31.8	(1)	--	25	C	0.5M($HClO_4$)	76Dd	
	30.1	1.50(48.2°,1)	122.6	48.2 &82	T	μ=1	62Fb 64Ha	a,b
	30.1	0.76(56°,1)	105.8	56&65	T	μ=2	62Fb	a,b
CrA_3	9.2	0.146(1)	33.5	25	C	3N($LiClO_4+Li_2SO_4$) A=Ethylenediamine	72Mg	
	6.3	-0.143(2)	18.8	25	C	3N($LiClO_4+Li_2SO_4$)	72Mg	q:±0.8

Metal Ion	ΔH, kJ /mol	Log K	ΔS, J /mol K	T, °C	M	Conditions	Ref.	Remarks
S46, continued								
CrA_3, cont.	-15.5	-0.155(3)	-54	25	C	A=Ethylenediamine 3N($LiClO_4$+Li_2SO_4) A=Ethylenediamine	72Mg	q:±1.7
$HCrO_4^-$	-19.58	1.276(26°)	-41.0	26&38	T	μ=2.0M($NaClO_4$) R:$HCrO_4^-$+$2H^+$+SO_4^- =$HCrO_3SO_4^-$+H_2O	71Yc	b
U^{4+}	-13.4	2.5(1)	2.9	25	T	μ=2.0	55Da	
	-9.6	3.9(1-2)	42	25	T	μ=2.0	55Da	
UO_2^{2+}	20.84	2.72(1)	121.8	25	C	μ=0	71Ba	
	9.83	2.73(1)	85.4	25	T	μ→0	60Lc	
	21.34	3.262(35°,1)	131.8	25-50	T	μ=0	67Wb	q:±1.8
	38.49	3.32(1)	182.8	50	T	μ→0	60Lc	
	95.8	4.45(1)	342.2	100	T	μ→0	60Lc	
	153.1	6.51(1)	485.6	150	T	μ→0	60Lc	
	210.4	8.87(1)	615.0	200	T	μ→0	60Lc	
	18.230	1.806(1)	95.69	25	C	μ=1.0($NaClO_4$)	71Ac	
	9.6	1.9(1)	67	25	T	μ=2.0($NaClO_4$)	54Db	
	3.89	1.48(2)	41.4	25	T	μ→0	60Lc	
	7.86	1.101(35°,2)	46.4	25-50	T	μ=0	67Wb	b,q: ±2.0
	4.64	1.52(2)	43.5	50	T	μ→0	60Lc	
	6.11	1.66(2)	48.1	100	T	μ→0	60Lc	
	7.61	1.77(2)	51.9	150	T	μ→0	60Lc	
	9.08	1.88(2)	55.2	200	T	μ→0	60Lc	
	16.88	0.950(2)	74.81	25	C	μ=1.0($NaClO_4$)	71Ac	
	5.8	2.8(2)	75	25	T	μ=2.0($NaClO_4$)	54Db	
Np^{4+}	16.7	3.51(25°,1)	123.4	10.2- 35.3	T	μ=2.0($HClO_4$)	54Sa	b
	28.2	3.56(25°,1)	163	10-40	T	μ=2.0 [H^+]=2.0M	73Ph	b
	26.02	2.49(25°,2)	171.5	10.2- 35.3	T	μ=2.0($HClO_4$)	54Sa	b
	5.2	2.49(25°)	65	10-40	T	μ=2.0 [H^+]=2.0M R:Np^{4+}+HL^-=$NpHL^{3+}$	73Ph	b
	-4.90	2.43(25°)	30.1	10.2- 35.3	T	μ=2.0($HClO_4$) R:Np^{4+}+HL^-= NpL^{2+}+H^+	54Sa	b
	15.23	1.41(25°)	78.2	10.2- 35.3	T	μ=2.0($HClO_4$) R:NpL^{2+}+HL^-= NpL_2^++H^+	54Sa	b
Pu^{4+}	16.5	3.83(25°,1)	129	10-40	T	μ=2.0 [H^+]=2.0M	73Ph	b
	-6.6	2.76(25°)	30	10-40	T	μ=2.0 [H^+]=2.0M R:Pu^{4+}+HL^-=$PuHL^{3+}$	73Ph	b
Am^{3+}	18.4	1.5(1)	38	25	C	μ=2.0	67Da	
Cm^{3+}	17.2	1.3(1)	34	25	C	μ=2.0	67Da	
Cf^{3+}	18.8	1.4(1)	38	25	C	μ=2.0	67Da	
Mn^{2+}	2.55	2.86(1)	63.6	25	C	μ=0	69Ib	

Metal Ion	ΔH, kJ /mol	Log K	ΔS, J /mol K	T,°C	M	Conditions	Ref.	Re-marks
S46, continued								
Mn^{2+}, cont.	9.08	(1)	--	25	C	μ=0	73Hb	
	14.10	2.25(25°,1)	94.6	0-40	T	μ=0	59Na	b
	13.4	2.1(25°,1)	88	20-30	T	μ=0	67Af	b
	9.6	0.16(1)	29	25	C	μ=3.0N($LiClO_4$)	74Bf	
Fe^{3+}	17.6	3.854(1)	134	20	T	μ=0	60Kb	
	2.34	2.20(1)	50.2	25	C	μ=0	69Ib	
	25.9	4.2(1)	163	25	T	μ=0.5($NaClO_4$+ $HClO_4$)	63Wa	
	17.6	1.8(1)	96	20	T	μ=1.0	60Kb	
Co^{2+}	2.09	2.69(1)	58.2	25	C	μ=0	69Ib	
	6.15	(1)	--	25	C	μ=0	53Tb 73Hb	i:59Lc
	7.28	2.35(25°,1)	69.4	0-40	T	μ=0	59Na 66Gd	b
	2.6	0.23(1)	12	25	C	μ=3.0N($LiClO_4$) R:$Co(H_2O)_6^{2+}+L^{2-}=Co(H_2O)_6L$	74Bf	
CoA_3^{3+}	-2.1	3.30(1)	56.1	25	C	μ=0 A=Ethylenediamine	75Me	q:±0.4
	0.29	1.99(1)	37.2	25	C	μ=0.1($NaClO_4$, Na_2SO_4) A=Ethylenediamine	75Me	q:±0.04
	0.96	1.69(1)	36.8	25	C	μ=0.2($NaClO_4$, Na_2SO_4) A=Ethylenediamine	75Me	q:±0.04
	1.46	1.60(1)	35.6	25	C	μ=0.3($NaClO_4$, Na_2SO_4) A=Ethylenediamine	75Me	q:±0.04
	1.88	1.30(1)	31.4	25	C	μ=0.4($NaClO_4$, Na_2SO_4) A=Ethylenediamine	75Me	
	2.18	1.20(1)	30.3	25	C	μ=0.5($NaClO_4$, Na_2SO) A=Ethylenediamine	75Me	
	6.90	0.556(1)	33.5	25	C	3N($NaCl_4$+Na_2SO_4) A=Ethylenediamine	72Mg	q:±0.4
	-2.5	3.22(1)	52.7	25	C	μ=0 A=Propylene-diamine	75Me	q:±0.4
	0	1.95(1)	37.6	25	C	μ=0.1($NaClO_4$, Na_2SO_4) A=Propyl-enediamine	75Me	q:±0.08
	0.71	1.62(1)	33.5	25	C	μ=0.2($NaClO_4$, Na_2SO_4) A=Propyl-enediamine	75Me	q:±0.08
	1.26	1.43(1)	31.6	25	C	μ=0.3($NaClO_4$, Na_2SO_4) A=Propyl-enediamine	75Me	
	1.88	1.30(1)	31.4	25	C	μ=0.4($NaClO_4$, Na_2SO_4) A=Propyl-enediamine	75Me	
	2.34	1.19(1)	30.8	25	C	μ=0.5($NaClO_4$,	75Me	

Metal Ion	ΔH, kJ /mol	Log K	ΔS, J /mol K	T,°C	M	Conditions	Ref.	Remarks
S46, continued								
$CoA_3{}^{3+}$, cont.						Na_2SO_4) A=Propylenediamine		
	8.58	0.204(1)	32.6	25	C	3N(Na_2SO_4+$NaClO_4$) A=Propylenediamine	72Mh	
	-1.05	1.80(2)	31.0	25	C	μ=0 A=Ethylenediamine	75Me	q:±0.8
	-0.21	1.50(2)	28.0	25	C	μ=0.1($NaClO_4$, Na_2SO_4) A=Ethylenediamine	75Me	q:±0.12
	0.46	1.06(2)	22.2	25	C	μ=0.2($NaClO_4$, Na_2SO_4) A=Ethylenediamine	75Me	q:±0.2
	0.88	1.04(2)	23.0	25	C	μ=0.3($NaClO_4$, Na_2SO_4) A=Ethylenediamine	75Me	q:±0.17
	1.26	0.97(2)	22.6	25	C	μ=0.4($NaClO_4$, Na_2SO_4) A=Ethylenediamine	75Me	q:±0.17
	1.05	0.89(2)	20.7	25	C	μ=0.5($NaClO_4$, Na_2SO_4) A=Ethylenediamine	75Me	q:±0.17
	3.3	0.146(2)	14.6	25	C	3N($NaClO_4$+Na_2SO_4) A=Ethylenediamine	72Mg	q:±0.8
	-1.5	1.80(2)	53	25	C	μ=0 A=Propylenediamine	75Me	q:±0.8
	-0.29	1.19(2)	22	25	C	μ=0.2($NaClO_4$, Na_2SO_4) A=Propylenediamine	75Me	q:±0.2
	0.42	1.10(2)	22	25	C	μ=0.3($NaClO_4$, Na_2SO_4) A=Propylenediamine	75Me	q:±0.17
	1.25	1.03(2)	24	25	C	μ=0.4($NaClO_4$, Na_2SO_4) A=Propylenediamine	75Me	q:±0.17
	3.56	0.95(2)	23	25	C	μ=0.5($NaClO_4$, Na_2SO_4) A=Propylenediamine	75Me	q:±0.17
	5.0	-0.081(2)	15.1	25	C	3N(Na_2SO_4+$NaClO_4$) A=Propylenediamine	72Mh	
	0	0.73(3)	14.2	25	C	μ=0 A=Ethylenediamine	75Me	q:±1.2
	-1.4	0.88(3)	12.1	25	C	μ=0.2($NaClO_4$, Na_2SO_4) A=Ethylenediamine	75Me	q:±0.63
	-2.1	0.88(3)	5.8	25	C	μ=0.3($NaClO_4$, Na_2SO_4) A=Ethylenediamine	75Me	q:±0.8
	-1.05	0.70(3)	10.0	25	C	μ=0.4($NaClO_4$, Na_2SO_4) A=Ethylenediamine	75Me	q:±0.4
	-1.46	0.67(3)	7.9	25	C	μ=0.5($NaClO_4$, Na_2SO_4) A=Ethylenediamine	75Me	q:±0.4
	1.7	-0.398(2)	≅2.1	25	C	3N($NaClO_4$+Na_2SO_4)	72Mg	q:±0.8

Metal Ion	ΔH, kJ /mol	Log K	ΔS, J /mol K	T,°C	M	Conditions	Ref.	Remarks
S46, continued								
CoA_3^{3+}, cont.	0.63	2.5(3)	10.5	25	C	A=Ethylenediamine μ=0 A=Propylenediamine	75Me	q:±1.2
	-0.42	0.56(3)	9.6	25	C	μ=0.2($NaClO_4$, Na_2SO_4) A=Propylenediamine	75Me	q:±0.4
	-1.7	0.66(3)	7.1	25	C	μ=0.3($NaClO_4$, Na_2SO_4) A=Propylenediamine	75Me	q:±0.8
	0.42	0.71(3)	15.1	25	C	μ=0.4($NaClO_4$, Na_2SO_4) A=Propylenediamine	75Me	q:±0.4
	0.42	0.66(3)	13.8	25	C	μ=0.5($NaClO_4$, Na_2SO_4) A=Propylenediamine	75Me	q:±0.3
	11.3	0.041(3)	38.5	25	C	3N(Na_2SO_4+$NaClO_4$) A=Propylenediamine	72Mh	
CoA_5^{3+}	-17.2	1.54(40°,1)	-25.9	40-55	T	CM=$5x10^{-3}$M CL=$40-60x10^{-3}$M A=NH_3	74Nb	b
CoA_6^{3+}	-2.1	3.44(1)	59	25	C	μ=0 A=NH_3	75Me	q:±0.4
	1.67	3.32(1)	69.4	25	T	μ=0 A=NH_3	56Pa	
	0.25	2.10(1)	39	25	C	μ=0.1($NaClO_4$, Na_2SO_4) A=NH_3	75Me	q:±0.04
	-0.42	1.06(2)	22	25	C	μ=0.1($NaClO_4$, Na_2SO_4) A=NH_3	75Me	q:±0.2
CoA_5B^{+}	6.3	2.42(1)	66.9	25	C	μ=0 A=NH_3 B=$SO_4^=$	75Md	q:±1.2
	11.7	1.90(1)	75.3	25	C	μ=0.1($NaClO_4$, Na_2SO_4) A=NH_3 B=$SO_4^=$	75Md	q:±1.2, ff
	15.1	1.58(1)	79.5	25	C	μ=0.2($NaClO_4$, Na_2SO_4) A=NH_3 B=$SO_4^=$	75Md	q:±1.2, ff
	18.4	1.43(1)	87.9	25	C	μ=0.3($NaClO_4$, Na_2SO_4) A=NH_3 B=$SO_4^=$	75Md	q:±1.7, ff
	22.8	1.21(1)	98.3	25	C	μ=0.4($NaClO_4$, Na_2SO_4) A=NH_3 B=$SO_4^=$	75Md	q:±1.7, ff
	26.4	1.21(1)	109	25	C	μ=0.5($NaClO_4$, Na_2SO_4) A=NH_3 B=$SO_4^=$	75Md	q:±2.1, ff
	4.2	0.59(2)	25	25	C	μ=0 A=NH_3 B=$SO_4^=$	75Md	q:±1.7
	5.8	1.32(2)	44.8	25	C	μ=0.1($NaClO_4$, Na_2SO_4) A=NH_3 B=$SO_4^=$	75Md	q:±2,ff
	6.7	0.99(2)	41.8	25	C	μ=0.2($NaClO_4$, Na_2SO_4) A=NH_3 B=$SO_4^=$	75Md	q:±3.3, ff
	9.6	0.66(2)	46.0	25	C	μ=0.3($NaClO_4$, Na_2SO_4)	75Md	q:±2,ff

Metal Ion	ΔH, kJ /mol	Log K	ΔS, J /mol K	T,°C	M	Conditions	Ref.	Remarks
S46, continued								
CoA_5B^{+},						A=NH_3 B=$SO_4^{=}$		
cont.	10.0	0.66(2)	45.0	25	C	μ=0.4($NaClO_4$, Na_2SO_4) A=NH_3 B=$SO_4^{=}$	75Md	q:±4.2, ff
	11.3	0.59(2)	50.2	25	C	μ=0.5($NaClO_4$, Na_2SO_4) A=NH_3 B=$SO_4^{=}$	75Md	q:±4.2, ff
	-10.5	-0.11(3)	-37.6	25	C	μ=0 A=NH_3 B=$SO_4^{=}$	75Md	q:±2.9
	-16.7	0.88(3)	-37.6	25	C	μ=0.1($NaClO_4$, Na_2SO_4) A=NH_3 B=$SO_4^{=}$	75Md	q:±8.4, ff
	-20.9	0.73(3)	-66.9	25	C	μ=0.2($NaClO_4$, Na_2SO_4) A=NH_3 B=$SO_4^{=}$	75Md	q:±12.5, ff
	-29.3	0.44(3)	-87.9	25	C	μ=0.3($NaClO_4$, Na_2SO_4) A=NH_3 B=$SO_4^{=}$	75Md	q:±8.4, ff
	-33.5	0.37(3)	-105	25	C	μ=0.4($NaClO_4$, Na_2SO_4) A=NH_3 B=$SO_4^{=}$	75Md	q:±12.5, ff
	-37.6	0.29(3)	-117	25	C	μ=0.5($NaClO_4$, Na_2SO_4) A=NH_3 B=$SO_4^{=}$	75Md	q:±12.5, ff
CoA_5B^{2+}	3.3	2.89(1)	66.5	25	C	μ=0 A=NH_3 B=NO_2^{-}	75Md	q:±0.6
	5.2	1.89(1)	53.6	25	C	μ=0.1($NaClO_4$, Na_2SO_4) A=NH_3 B=NO_2^{-}	75Md	ff
	5.8	1.66(1)	51.0	25	C	μ=0.2($NaClO_4$, Na_2SO_4) A=NH_3 B=NO_2^{-}	75Md	ff
	6.48	1.44(1)	49.4	25	C	μ=0.3($NaClO_4$, Na_2SO_4) A=NH_3 B=NO_2^{-}	75Md	ff
	7.1	1.27(1)	48.1	25	C	μ=0.4($NaClO_4$, Na_2SO_4) A=NH_3 B=NO_2^{-}	75Md	ff
	7.74	1.21(1)	49.0	25	C	μ=0.5($NaClO_4$, Na_2SO_4) A=NH_3 B=NO_2^{-}	75Md	ff
	8.16	0.32(1)	33	25	C	μ=3.0M($NaClO_4$) A=NH_3 B=NO_2^{-}	74Mb	
	3.56	2.87(1)	66.9	25	C	μ=0 A=NH_3 B=Cl^{-}	75Md	
	5.4	1.94(1)	55.6	25	C	μ=0.1($NaClO_4$, Na_2SO_4) A=NH_3 B=Cl^{-}	75Md	q:±0.4, ff
	5.8	1.65(1)	51.0	25	C	μ=0.2($NaClO_4$, Na_2SO_4) A=NH_3 B=Cl^{-}	75Md	q:±0.4, ff
	6.3	1.44(1)	48.5	25	C	μ=0.3($NaClO_4$, Na_2SO_4) A=NH_3 B=Cl^{-}	75Md	ff

Metal Ion	ΔH, kJ /mol	Log K	ΔS, J /mol K	T, °C	M	Conditions	Ref.	Remarks
S46, continued								
CoA_5B^+, cont.	6.7	1.28(1̲)	46.0	25	C	μ=0.4($NaClO_4$, Na_2SO_4) $A=NH_3$ $B=Cl^-$	75Md	ff
	7.9	1.17(1̲)	49.0	25	C	μ=0.5($NaClO_4$, Na_2SO_4) $A=NH_3$ $B=Cl^-$	75Md	ff
	2.9	2.81(1̲)	63.6	25	C	μ=0 $A=NH_3$ $B=Br^-$	75Md	q:±0.6
	5.2	1.90(1̲)	54.4	25	C	μ=0.1($NaClO_4$, Na_2SO_4) $A=NH_3$ $B=Br^-$	75Md	q:±0.42, ff
	5.8	1.68(1̲)	52.3	25	C	μ=0.2($NaClO_4$, Na_2SO_4) $A=NH_3$ $B=Br^-$	75Md	ff
	6.48	1.52(1̲)	50.2	25	C	μ=0.3($NaClO_4$, Na_2SO_4) $A=NH_3$ $B=Br^-$	75Md	ff
	7.1	1.41(1̲)	50.2	25	C	μ=0.4($NaClO_4$, Na_2SO_4) $A=NH_3$ $B=Br^-$	75Md	ff
	7.9	1.31(1̲)	52.3	25	C	μ=0.5($NaClO_4$, Na_2SO_4) $A=NH_3$ $B=Br^-$	75Md	ff
	2.1	1.32(2̲)	32.2	25	C	μ=0 $A=NH_3$ $B=NO_2^-$	75Md	q:±1.2
	2.5	1.61(2̲)	38.9	25	C	μ=0.1($NaClO_4$, Na_2SO_4) $A=NH_3$ $B=NO_2^-$	75Md	q:±0.84, ff
	3.3	1.17(2̲)	33.5	25	C	μ=0.2($NaClO_4$, Na_2SO_4) $A=NH_3$ $B=NO_2^-$	75Md	q:±0.84, ff
	3.6	0.95(2̲)	30.1	25	C	μ=0.3($NaClO_4$, Na_2SO_4) $A=NH_3$ $B=NO_2^-$	75Md	q:±0.84, ff
	4.2	0.84(2̲)	30.1	25	C	μ=0.4($NaClO_4$, Na_2SO_4) $A=NH_3$ $B=NO_2^-$	75Md	q:±0.84, ff
	4.6	0.77(2̲)	30.1	25	C	μ=0.5($NaClO_4$, Na_2SO_4) $A=NH_3$ $B=NO_2^-$	75Md	q:±0.84, ff
	-4.4	0.13(2̲)	-12	25	C	μ=3.0M($NaClO_4$) $A=NH_3$ $B=NO_2$	74Mb	
	0	1.47(2̲)	28.0	25	C	μ=0 $A=NH_3$ $B=Cl^-$	75Md	q:±1.3
	1.5	1.32(2̲)	29.3	25	C	μ=0.1($NaClO_4$, Na_2SO_4) $A=NH_3$ $B=Cl^-$	75Md	q:±0.8, ff
	1.88	1.10(2̲)	27.2	25	C	μ=0.2($NaClO_4$, Na_2SO_4) $A=NH_3$ $B=Cl^-$	75Md	q:±0.8, ff
	3.14	1.06(2̲)	31.4	25	C	μ=0.3($NaClO_4$, Na_2SO_4) $A=NH_3$ $B=Cl^-$	75Md	q:±0.8, ff
	3.8	0.84(2̲)	29.3	25	C	μ=0.4($NaClO_4$, Na_2SO_4)	75Md	q:±0.8, ff

Metal Ion	ΔH, kJ /mol	Log K	ΔS, J /mol K	T, °C	M	Conditions	Ref.	Remarks
S46, continued								
CoA_5B^{2+},						$A=NH_3$ $B=Cl^-$		
cont.	4.6	0.66(2)	29.3	25	C	μ=0.4($NaClO_4$, Na_2SO_4) $A=NH_3$ $B=Cl^-$	75Md	q:±0.8, ff
	1.9	1.17(2)	29.3	25	C	μ=0 $A=NH_3$ $B=Br^-$	75Md	q:±1.2
	2.7	1.14(2)	31.4	25	C	μ=0.1($NaClO_4$, Na_2SO_4) $A=NH_3$ $B=Br^-$	75Md	q:±0.8, ff
	2.9	0.92(2)	27.2	25	C	μ=0.2($NaClO_4$, Na_2SO_4) $A=NH_3$ $B=Br^-$	75Md	q:±0.8, ff
	3.1	0.66(2)	23.0	25	C	μ=0.3($NaClO_4$, Na_2SO_4) $A=NH_3$ $B=Br^-$	75Md	q:±0.8, ff
	3.8	0.40(2)	20.9	25	C	μ=0.4($NaClO_4$, Na_2SO_4) $A=NH_3$ $B=Br^-$	75Md	q:±0.8, ff
	4.6	0.15(2)	18.8	25	C	μ=0.5($NaClO_4$, Na_2SO_4) $A=NH_3$ $B=Br^-$	75Md	q:±0.8, ff
	-5.4	0.58(3)	-6.3	25	C	μ=0 $A=NH_3$ $B=NO_2^-$	75Md	q:±2.1
	-9.2	0.73(3)	-16.7	25	C	μ=0.2($NaClO_4$, Na_2SO_4) $A=NH_3$ $B=NO_2^-$	75Md	q:±2.1, ff
	-10.0	0.73(3)	-20.9	25	C	μ=0.3($NaClO_4$, Na_2SO_4) $A=NH_3$ $B=NO_2^-$	75Md	q:±2.1, ff
	-11.3	0.73(3)	-25.1	25	C	μ=0.4($NaClO_4$, Na_2SO_4) $A=NH_3$ $B=NO_2^-$	75Md	q:±2.1, ff
	-12.6	0.51(3)	-31.4	25	C	μ=0.5($NaClO_4$, Na_2SO_4) $A=NH_3$ $B=NO_2^-$	75Md	q:±2.1, ff
	-3.7	0.09(3)	-31	25	C	μ=3.0M($NaClO_4$) $A=NH_3$ $B=NO_2$	74Ma	
	-3.3	0.15(3)	-8.4	25	C	μ=0 $A=NH_3$ $B=Cl^-$	75Md	q:±1.7
	-9.6	0.88(3)	-17	25	C	μ=0.3($NaClO_4$, Na_2SO_4) $A=NH_3$ $B=Cl^-$	75Md	q:±2,ff
	-10.5	0.59(3)	-25	25	C	μ=0.4($NaClO_4$, Na_2SO_4) $A=NH_3$ $B=Cl^-$	75Md	q:±2,ff
	-12.6	0.59(3)	-31	25	C	μ=0.5($NaClO_4$, Na_2SO_4) $A=NH_3$ $B=Cl^-$	75Md	q:±2.5, ff
	-4.8	0(3)	-16.7	25	C	μ=0 $A=NH_3$ $B=Br^-$	75Md	q:±2.1
	-8.8	0.73(3)	-16.7	25	C	μ=0.2($NaClO_4$, Na_2SO_4) $A=NH_3$ $B=Br^-$	75Md	q:±2.1, ff
	-9.6	0.51(3)	-23.0	25	C	μ=0.3($NaClO_4$, Na_2SO_4) $A=NH_3$ $B=Br^-$	75Md	q:±2.1, ff

Metal Ion	ΔH, kJ /mol	Log K	ΔS, J /mol K	T,°C	M	Conditions	Ref.	Remarks
S46, continued								
CoA_5B^{2+}, cont.	-10.9	0.29(3)	-31.4	25	C	μ=0.4($NaClO_4$, Na_2SO_4) A=NH_3 B=Br^-	75Md	q:±2.5, ff
	-12.5	0.07(3)	-41.8	25	C	μ=0.5($NaClO_4$, Na_2SO_4) A=NH_3 B=Br^-	75Md	q:±2.1, ff
Ni^{2+}	1.72	2.81(1)	59.0	25	C	μ=0	69Ib	
	6.36	2.30(1)	65.7	25	C	μ=0	70La	i:59Lc
	5.31	2.273(25°,1)	61.5	0-40	T	μ=0	73Kd	b
	13.85	2.32(25°,1)	90.8	0-45	T	μ=0	59Na	b
	2.4	0.26(1)	12	25	C	μ=3.0N($LiClO_4$) R:$Ni(H_2O)_6^{2+}+L^{2-}$= $Ni(H_2O)_6L$	74Bf	
RhA_3	7.5	0.146(1)	28.0	25	C	3N($NaClO_4$+Na_2SO_4) A=Ethylenediamine	72Mg	q:±0.8
	4.2	0.114(2)	16.3	25	C	3N($NaClO_4$+Na_2SO_4) A=Ethylenediamine	72Mg	q:±1.2
Cu^{2+}	5.10	2.26(1)	61.1	25	C	μ=0	69Ib	
	7.20	2.36(1)	69.4	25	C	μ=0	70La	i:59Lc
	10.17	(1)	--	25	C	μ=0	73Hb	
	40.166	2.398(25°,1)	54.4	0&25	T	μ=0	72Pj	a,b
	4.8	0.66(1)	29	25	C	μ=3.0N($LiClO_4$) R:$Cu(H_2O)_6^{2+}+L^{2-}$= $Cu(H_2O)_6L$	74Bf	
	1.59	0.161(1)	8.41	25	C	4N($LiClO_4$)	77Ke	
Ag^+	6.3	1.3(1)	46	25	C	μ=0	65Hc 34La 33La 60Sg	
Zn^{2+}	2.64	2.49(1)	56.5	25	C	μ=0	69Ib	
	5.69	2.32(1)	63.6	25	C	μ=0	70La	i:59Lc
	6.65	(1)	--	25	C	μ=0	73Hb	
	16.78	2.38(25°,1)	102.1	0-45	T	μ=0	58Nb	b
	3.18	0.923(1)	28.4	25	C	μ=0.5N($LiClO_4$) CM=0.1M	76Fb	q:±0.5
	2.26	0.887(1)	24.7	25	C	μ=1.0N($LiClO_4$) CM=0.1M	76Fb	q:±0.6
	1.88	0.755(1)	20.9	25	C	μ=2.0N($LiClO_4$) CM=0.1M	76Fb	q:±0.7
	2.30	0.696(1)	20.9	25	C	μ=3.0N($LiClO_4$) CM=0.1M	76Fb	q:±0.7
	3.0	0.21(1)	15	25	C	μ=3.0N($LiClO_4$) R:$Zn(H_2O)_6^{2+}+L^{2-}$= $Zn(H_2O)_6L$	74Bf	
	2.38	0.601(1)	19.7	25	C	μ=4.0N($LiClO_4$) CM=0.1M	76Fb	q:±0.6
	-1.88	1.041(2)	13.4	25	C	μ=5.0N($LiClO_4$) CM=0.1M	76Fb	
	1.17	0.337(2)	10.5	25	C	μ=1.0N($LiClO_4$) CM=0.1M	76Fb	
	0.17	0.417(2)	8.4	25	C	μ=2.0N($LiClO_4$)	76Fb	

Metal Ion	ΔH, kJ /mol	Log K	ΔS, J /mol K	T,^{o}C	M	Conditions	Ref.	Re-marks
S46, continued								
						CM=0.1M		
Zn^{2+}, cont.	0.08	-0.007(2)	0.4	25	C	μ=3.0N($LiClO_4$) CM=0.1M	76Fb	
	0.79	0.278(2)	7.9	25	C	μ=4.0N($LiClO_4$) CM=0.1M	76Fb	
	8.03	-0.059(3)	25.9	25	C	μ=2.0N($LiClO_4$) CM=0.1M	76Fb	
	3.05	0.205(3)	14.2	25	C	μ=3.0N($LiClO_4$) CM=0.1M	76Fb	
	-2.09	0.146(3)	-4.2	25	C	μ=4.0N($LiClO_4$) CM=0.1M	76Fb	
Cd^{2+}	4.10	2.55(1)	62.8	25	C	μ=0	69Ib	
	5.27	1.078(1)	38.4	25	C	μ=5.0N($LiClO_4$) CM=0.1M	76Fb	q:±0.6
	4.52	0.945(1)	33.5	25	C	μ=1.0N($LiClO_4$) CM=0.1M	76Fb	q:±0.5
	3.85	0.858(1)	29.3	25	C	μ=2.0N($LiClO_4$) CM=0.1M	76Fb	q:±0.6
	4.39	0.718(1)	28.4	25	C	μ=3.0N($LiClO_4$) CM=0.1M	76Fb	q:±0.75
	4.18	0.733(1)	28.0	25	C	μ=4.0N($LiClO_4$) CM=0.1M	76Fb	q:±0.67
	9.00	2.31(2)	74.5	25	C	μ=0	70La	i:59Lc
	-0.42	0.880(2)	15.1	25	C	μ=5.0N($LiClO_4$) CM=0.1M	76Fb	
	-0.46	0.608(2)	10.0	25	C	μ=1.0N($LiClO_4$) CM=0.1M	76Fb	
	-0.63	0.447(2)	6.7	25	C	μ=2.0N($LiClO_4$) CM=0.1M	76Fb	
	-0.63	0.125(2)	0.4	25	C	μ=3.0N($LiClO_4$) CM=0.1M	76Fb	
	0.16	0.381(2)	7.9	25	C	μ=4.0N($LiClO_4$) CM=0.1M	76Fb	
	9.79	0.198(3)	36.4	25	C	μ=1.0N($LiClO_4$) CM=0.1M	76Fb	
	8.70	0.293(3)	34.7	25	C	μ=2.0N($LiClO_4$) CM=0.1M	76Fb	
	7.11	0.315(3)	29.7	25	C	μ=3.0N($LiClO_4$) CM=0.1M	76Fb	
	3.05	0.037(3)	10.9	25	C	μ=4.0N($LiClO_4$) CM=0.1M	76Fb	
	-5.82	-0.051(4)	-20.5	25	C	μ=4.0N($LiClO_4$) CM=0.1M	76Fb	
Al^{3+}	9.58	3.01(1)	89.5	25	C	μ=0	69Ib	
	3.26	1.89(2)	47.3	25	C	μ=0	69Ib	
Ga^{3+}	19.71	2.77(1)	118.8	25	C	μ=0	69Ib	
	-7.24	2.29(2)	20.1	25	C	μ=0	69Ib	
In^{3+}	29.08	3.04(1)	155.6	25	C	μ=0	69Ib	
	-7.32	1.96(2)	13.0	25	C	μ=0	69Ib	
Tl^{+}	-0.92	1.36(1)	23.0	25	T	μ=0	53Ba	

Metal Ion	ΔH, kJ /mol	Log K	ΔS, J /mol K	T, °C	M	Conditions	Ref.	Re-marks
S46, continued								
Tl^+, cont.	-23.8	-0.484(1)	-96	25	C	μ=3.0[Li(ClO_4, SO_4)]	69Sj	p
	18.4	-0.484(25°,1)	50	25-45	T	μ=3.0M[Na(ClO_4, SO_4)]	69Sj	b,p
	2.99	0.65(1)	21	25	C	μ=0.5N($LiClO_4$, $LiNO_3$)	75Fb	
	5.27	0.40(1)	25	25	C	μ=1N($LiClO_4$, $LiNO_3$)	75Fb	
	6.78	0.08(1)	25	25	C	μ=2N($LiClO_4$, $LiNO_3$)	75Fb	
	2.5	-0.12(1)	-12	25	C	μ=3N($LiClO_4$, $LiNO_3$)	75Fb	
	-20.0	-0.24(1)	-71	25	C	μ=4N($LiClO_4$, $LiNO_3$)	75Fb	
	-12.9	0.71(2)	-29	25	C	μ=0.5N($LiClO_4$, $LiNO_3$)	75Fb	
	-7.51	0.44(2)	-17	25	C	μ=1N($LiClO_4$, $LiNO_3$)	75Fb	
	4.22	0.26(2)	17	25	C	μ=2N($LiClO_4$, $LiNO_3$)	75Fb	
	-5.60	-0.34(2)	-8	25	C	μ=3N($LiClO_4$, $LiNO_3$)	75Fb	
	-14.7	-0.44(2)	-58	25	C	μ=4N($LiClO_4$, $LiNO_3$)	75Fb	
Tl^{3+}	-11.3	2.26(1)	4.6	25	C	2.0M($LiClO_4$) 1.0(HClO)	67Md	
S47 SULFIDE, (2-aminoethyl) (tert-butyl)		$C_6H_{15}NS$				$(CH_3)_3CSCH_2CH_2NH_2$		
Ag^+	-82.42	9.992(1-2)	-85.4	25	C	μ=0.5M(KNO_3)	77Ta	
	-69.4	7.873	-82.0	25	C	μ=0.5M(KNO_3) R: $2Ag^+ + L = Ag_2L^{2+}$	77Ta	
	-131.0	14.170	-168.6	25	C	μ=0.5M(KNO_3) R: $2Ag^+ + 2L = Ag_2L_2{}^{2+}$	77Ta	
	-35.06	3.350	-52.43	25	C	μ=0.5M(KNO_3) R: $Ag^+ + HL^+ = AgHL^{2+}$	77Ta	
	-61.38	4.985	-110.50	25	C	μ=0.5M(KNO_3) R: $Ag^+ + 2HL^+ = Ag(HL)_2{}^{3+}$	77Ta	
	-74.9	8.416	-90.8	25	C	μ=0.5M(KNO_3) R: $Ag^+ + HL^+ + L = [AgL(HL)]^{2+}$	77Ta	
S48 SULFIDE, (2-aminoethyl) (2'-hydroxyethyl)		$C_4H_{11}NS$				$HOCH_2CH_2SCH_2CH_2NH_2$		
Ag^+	-80.00	9.207(1-2)	-92.0	25	C	μ=0.5M(KNO_3)	77Ta	

Metal Ion	ΔH, kJ /mol	Log K	ΔS, J /mol K	T,°C	M	Conditions	Ref.	Remarks
S48, continued								
Ag^+, cont.	-66.9	7.052	-89.5	25	C	μ=0.5M(KNO_3) R:$2Ag^+$+L=Ag_2L^{2+}	77Ta	
	-118.0	12.821	-150.2	25	C	μ=0.5M(KNO_3) R:$2Ag^+$+2L=$Ag_2L_2{}^{2+}$	77Ta	
	-28.11	2.551	-45.6	25	C	μ=0.5M(KNO_3) R:Ag^++HL^+=$AgHL^{2+}$	77Ta	
	-46.44	4.105	-77.0	25	C	μ=0.5M(KNO_3) R:Ag^++2HL^+= $Ag(HL)_2{}^{3+}$	77Ta	
	-70.3	7.536	-90.8	25	C	μ=0.5M(KNO_3) R:Ag^++HL^++L= $[AgL(HL)]^{2+}$	77Ta	
S49 SULFIDE, (2-aminoethyl)propyl		$C_5H_{13}NS$				$CH_3CH_2CH_2SCH_2CH_2NH_2$		
Ag^+	-84.5	9.706(1-2)	-97.9	25	C	μ=0.5M(KNO_3)	77Ta	
	-72.8	7.404	-102.9	25	C	μ=0.5M(KNO_3) R:$2Ag^+$+L=Ag_2L^{2+}	77Ta	
	-130.1	13.584	-177.0	25	C	μ=0.5M(KNO_3) R:$2Ag^+$+2L=$Ag_2L_2{}^{2+}$	77Ta	
	-31.4	2.947	-49.0	25	C	μ=0.5M(KNO_3) R:Ag^++HL^+=$AgHL^{2+}$	77Ta	
	-59.0	4.611	-109.6	25	C	μ=0.5M(KNO_3) R:Ag^++2HL^+= $Ag(HL)_2{}^{3+}$	77Ta	
	-77.8	8.078	-106.7	25	C	μ=0.5M(KNO_3) R:Ag^++HL^++L= $[AgL(HL)]^{2+}$	77Ta	
S50 SULFIDE, benzylmethyl-		$C_8H_{10}S$				$CH_3SCH_2C_6H_5$		
PtA_2L_2	14.2	0.120(20°)	50	30-60	T	S:Chloroform A=Cl	73Re	b,bb
	13.4	-0.409(21°)	38	30-60	T	S:Dichloromethane =Cl	73Re	b,bb
S51 SULFIDE, bis(dimethylthio-carbamyl)-		$C_6H_{12}N_2S_3$				$(CH_3)_2NC(S)SC(S)N(CH_3)_2$		
I_2	-30.54	2.636(20°,1)	-54.4	10-30	T	S:CCl_4	69Ge	b,q: ±2.8
S52 SULFIDE, diallyl-		$C_6H_{10}S$				$(CH_2:CHCH_2)_2S$		
I_2	-19.25	2.699(1)	-12.6	25	C	S:Octane	66Ad	

Metal Ion	ΔH, kJ /mol	Log K	ΔS, J /mol K	T,°C	M	Conditions	Ref.	Remarks
S53 SULFIDE, dibenzyl-			$C_{14}H_{14}S$			$(C_6H_5CH_2)_2S$		
PtA_2L_2	28.0	-0.796(40°)	75	30-60	T	S:Chloroform A=Cl	73Re	b,bb
	20.1	-0.319(36°)	58	30-60	T	S:Chloroform A=Br	73Re	b,bb
	8.4	0.531(33°)	38	30-60	T	S:Chloroform A=I	73Re	b,bb
SnA_4	-15.9	(1)	--	25	C	S:Benzene CM=0.05-0.08M A=Cl	65Gb 68Ge	
I_2	-20.1	2.788(1)	-13.8	25	C	S:Octane	66Ad	
S54 SULFIDE, dibutyl-			$C_8H_{18}S$			$(C_4H_9)_2S$		
Al_2A_6	-73.6	--	--	25	C	S:Benzene CM=0.07M A=Br R:$0.5(Al_2A_6)+L=AlA_3L$	68Rc	
GaA_3	-79.5	4.03(1)	-189.5	25	C	S:Benzene A=Cl	70Gh	
SnA_4	-52.7	(1)	--	25	C	S:Benzene CM=0.05-0.08M A=Cl	68Ge 65Gb 70Gh	
	-48.5	(2)	--	25	C	S:Benzene CM=0.05-0.08M A=Cl	68Ge 65Gb	
TeA_4	-46.9	2.24(1)	-114.2	25	C	S:Benzene A=Cl	72Pg 73Pi	
I_2	-31.8	2.585(1)	-56.9	25	C	S:Octane	66Ad	
S55 SULFIDE, dicylohexyl-			$C_{12}H_{22}S$			$(C_6H_{11})_2S$		L
Al_2A_6	-70.7	--	--	25	C	S:Benzene CM=0.07M A=Br R:$0.5(Al_2A_6)+L=AlA_3L$	68Rc	
I_2	-43.9	2.638(1)	-96.6	25	C	S:Octane	66Ad	
S56 SULFIDE, diethyl-			$C_4H_{10}S$			$(C_2H_5)_2S$		
TiA_4	-48.1	(1)	--	25	C	S:Benzene CM=0.05-0.08M A=Cl	68Ge	
	-48.1	(2)	--	25	C	S:Benzene CM=0.05-0.08M A=Cl	68Ge	
$[NiA_2]_2$	-26	1.74(100°,1)	-38	75-125	T	S:Squalane A=3-Trifluoro-acetylcamphorate	74Sj	b,gg

Metal Ion	ΔH, kJ /mol	Log K	ΔS, J /mol K	T,°C	M	Conditions	Ref.	Remarks
S56, continued								
AlA_3	-70.08	(1)	--	28.5	C	S:Hexane A=CH_3	67Hc	
I_2	-35.60	3.064(30°,1)	-58.6	20-40	T	S:CCl_4 K's determined from spectral data taken in visible region	61Ga	b
	-33.68	3.212(30°,1)	-49.0	20-40	T	S:CCl_4 K's determined from spectral data taken in ultra-violet region	61Ga	b
	-32.6	2.322(25°,1)	-66.5	15-35	T	S:Cyclohexane	63Va	b
	-32.72	2.322(20°,1)	-66.5	0-30	T	CM=5x10^{-5}M CL=0.001-0.05M S:CH_2Cl_2	61Ta	b
	-37.57	2.961(30°,1)	-66.73	20-40	T	S:n-Heptane K's determined from spectral data taken in visible region	61Ga	b
	-33.68	3.158(30°,1)	-50.2	20-40	T	S:n-Heptane K's determined from spectral data taken in ultra-violet region	61Ga	b
S57	SULFIDE, diethyl, 2-amino		$C_4H_{11}NS$			$CH_3CH_2SCH_2CH_2NH_2$		
Ag^+	-84.9	9.654(1-2)	-99.6	25	C	μ=0.5M(KNO_3)	77Ta	
	-72.0	7.418	-99.2	25	C	μ=0.5M(KNO_3) R:$2Ag^++L=Ag_2L^{2+}$	77Ta	
	-127.2	13.657	-165.3	25	C	μ=0.5M(KNO_3) R:$2Ag^++2L=Ag_2L_2{}^{2+}$	77Ta	
	-31.04	2.998	-46.9	25	C	μ=0.5M(KNO_3) R:$Ag^++HL^+=AgHL^{2+}$	77Ta	
	-58.2	4.655	-106.7	25	C	μ=0.5M(KNO_3) R:$Ag^++2HL^+=Ag(HL)_2{}^{3+}$	77Ta	
	-77.0	8.093	-102.5	25	C	μ=0.5M(KNO_3) R:$Hg^++HL^++L=[HgL(HL)]^{2+}$	77Ta	
S58	SULFIDE, (2-diethylamino-ethyl)methyl		$C_7H_{17}NS$			$CH_3SCH_2CH_2N(C_2H_5)_2$		
Ag^+	-63.72	8.665(1-2)	-48.1	25	C	μ=0.5M(KNO_3)	77Ta	
	-34.9	6.165	0.8	25	C	μ=0.5M(KNO_3) R:$2Ag^++L=Ag_2L^{2+}$	77Ta	

Metal Ion	ΔH, kJ /mol	Log K	ΔS, J /mol K	T,°C	M	Conditions	Ref.	Remarks
S58, continued								
Ag^{+}, cont.	-64.31	11.22	-0.8	25	C	μ=0.5M(KNO_3) R:$2Ag^{+}+2L=Ag_2L_2{}^{2+}$	77Ta	
	-27.28	2.588	-42.2	25	C	μ=0.5M(KNO_3) R:$Ag^{+}+HL^{+}=AgHL^{2+}$	77Ta	
	-53.6	3.980	-103.8	25	C	μ=0.5M(KNO_3) R:$Ag^{+}+2HL^{+}=Ag(HL)_2{}^{3+}$	77Ta	
	-57.7	7.001	-59.8	25	C	μ=0.5M(KNO_3) R:$Ag^{+}+HL^{+}+L=[AgL(HL)]^{2+}$	77Ta	
S59 SULFIDE, diethyl-2,2'-diamino-		$C_4H_{12}N_2S$				$(NH_2CH_2CH_2)_2S$		L
Co^{2+}	-29	5.09(30°,1)	0	0-50	T	μ=0.1(KClorKNO$_3$)	54Gb	b
	-33	3.92(30°,2)	-38	0-50	T	μ=0.1(KClorKNO$_3$)	54Gb	b
	-63	9.01(30°,1-2)	-38	0-50	T	μ=0.1(KClorKNO$_3$)	54Gb	b
Ni^{2+}	-42	7.27(30°,1)	0	0-50	T	μ=0.1(KClorKNO$_3$)	54Gb	b
	-50	6.10(30°,2)	-46	0-50	T	μ=0.1(KClorKNO$_3$)	54Gb	b
	-92	13.37(30°,1-2)	-46	0-50	T	μ=0.1(KClorKNO$_3$)	54Gb	b
S60 SULFIDE, diethyl-, 2,2'-diamino-N,N,N',N'-tetraacetic acid		$C_{12}H_{20}O_8N_2S$				$(CH_2CO_2H)_2NCH_2CH_2SCH_2$-$CH_2N(CH_2CO_2H)_2$		L^{4-}
Mg^{2+}	17.28	4.61(1)	147.15	20	C	μ=0.1(KNO_3)	64Ab	
Ca^{2+}	-10.5	6.21(1)	97.1	20	C	μ=0.1(KNO_3)	64Ab	
La^{3+}	-0.8	12.8(1)	242.09	20	C	μ=0.1(KNO_3)	64Ab	
Mn^{2+}	-6.40	10.07(1)	175.14	20	C	μ=0.1(KNO_3)	64Ab	
Co^{2+}	-19.37	13.99(1)	201.7	20	C	μ=0.1(KNO_3)	64Ab	
Ni^{2+}	-32.2	15.7(1)	190.5	20	C	μ=0.1(KNO_3)	64Ab	
Cu^{2+}	-38.20	16.57(1)	186.82	20	C	μ=0.1(KNO_3)	64Ab	
Zn^{2+}	-15.5	13.44(1)	204.39	20	C	μ=0.1(KNO_3)	64Ab	
Cd^{2+}	-34.3	14.38(1)	158.2	20	C	μ=0.1(KNO_3)	64Ab	
Hg^{2+}	-95.4	23.09(1)	132.00	20	C	μ=0.1(KNO_3)	64Ab	
Pb^{2+}	-54.4	13.86(1)	79.79	20	C	μ=0.1(KNO_3)	64Ab	
S61 SULFIDE, diethyl-, 2,2'-dihydroxy-		$C_4H_{10}O_2S$				$(CH_2CH_2OH)_2S$		
Hg^{2+}	36.82	3.00(30°,1)	179.1	30-80	T	μ=1.5M($HClO_4$) CM=0.01M	71Me	b,q: ±4.2
	-93.3	5.08(30°,2)	-210.4	30-80	T	μ=1.5M($HClO_4$) CM=0.01M	71Me	b,q: ±8.4

Metal Ion	ΔH, kJ /mol	Log K	ΔS, J /mol K	T,°C	M	Conditions	Ref.	Re-marks
S61, continued								
Hg^{2+}, cont.	-6.7	1.37(30°,3)	4.2	30-80	T	μ=1.5M($HClO_4$) CM=0.01M	71Me	b,q: ±8.4
S62 SULFIDE, diheptyl-			$C_{14}H_{30}S$			$(C_7H_{15})_2S$		
GaA_3	-80.3	3.88(1)	-195.0	25	C	S:Benzene A=Cl	70Gh	
TeA_4	-47.7	2.38(1)	-114.6	25	C	S:Benzene A=Cl	73Pi	
	-47.7	(1)	--	25	C	S:Benzene A=Cl	72Pg	
S63 SULFIDE, diisoheptyl (sec)-			$C_{14}H_{30}S$			$(C_7H_{15})_2S$		L
Al_2A_6	-71.5	--	--	25	C	S:Benzene CM=0.07M A=Br R:0.5(Al_2A_6)+L= AlA_3L	68Rc	
S64 SULFIDE, diisooctyl (sec)-			$C_{16}H_{34}S$			$(C_8H_{17})_2S$		L
Al_2A_6	-67.4	--	--	25	C	S:Benzene CM=0.07M A=Br R:0.5(Al_2A_6)+L= AlA_3L	68Rc	
SnA_4	-1.7	(1)	--	25	C	S:Benzene CM=0.05-0.08M A=Cl	68Ge 65Gb	
S65 SULFIDE, diisopentyl-			$C_{10}H_{22}S$			$(C_5H_{11})_2S$		L
TeA_4	-48.5	2.18(1)	-120.9	25	C	S:Benzene A=Cl	73Pi	
S66 SULFIDE, diisopropyl-			$C_6H_{14}S$			$(CH_3)_2CHSCH(CH_3)_2$		L
Al_2A_6	-76.1	--	--	25	C	S:Benzene CM=0.07M A=Br R:0.5(Al_2A_6)+L= AlA_3L	68Rc	
S67 SULFIDE, (2-diisopropylamino-ethyl)methyl			$C_9H_{21}NS$			$CH_3SCH_2CH_2N[CH(CH_3)_2]_2$		
Ag^+	-29.50	2.610	-48.95	25	C	μ=0.5M(KNO_3) R:Ag^++HL^+=$AgHL^{2+}$	77Ta	
	-44.18	4.398	-64.0	25	C	μ=0.5M(KNO_3) R:Ag^++2HL^+=$Ag(HL)_2{}^{3+}$	77Ta	
S68 SULFIDE, dimethyl-			C_2H_6S			CH_3SCH_3		L
$[NiA_2]_2$	-24.0	1.91(100°,1)	-28	75-125	T	S:Squalane	74Sj	b,gg

Metal Ion	ΔH, kJ /mol	Log K	ΔS, J /mol K	T,°C	M	Conditions	Ref.	Remarks
S68, continued								
$[NiA_2]_2$, cont.						A=3-Trifluoro-acetylcamphorate		
PtA_2L_2	7.9	0.806(29°)	42	30-60	T	S:Chloroform A=Cl	73Re	b,q: ±0.8,bb
	9.6	0.274(28°)	38	30-60	T	S:Dichloro-methane A=Cl	73Re	b,bb
AlA_3	-69.83	(1)	--	28.5	C	S:Hexane A=CH_3	67Hc	
S69 SULFIDE, (2-dimethylamino-ethyl)methyl		$C_5H_{13}NS$				$CH_3SCH_2CH_2N(CH_3)_2$		
Ag^+	-69.4	8.423(1-2)	-72.4	25	C	μ=0.5M(KNO_3)	77Ta	
	-61.9	5.94	-92	25	C	μ=0.5M(KNO_3) R:$2Ag^+ + L = Ag_2L^{2+}$	77Ta	
	-90.4	11.377	-84	25	C	μ=0.5M(KNO_3) R:$2Ag^+ + 2L = Ag_2L_2{}^{2+}$	77Ta	
	-26.40	2.522	-40.3	25	C	μ=0.5M(KNO_3) R:$Ag^+ + HL^+ = AgHL^+$	77Ta	
	-52.34	3.841	-102.1	25	C	μ=0.5M(KNO_3) R:$Ag^+ + 2HL^+ = Ag(HL)_2{}^{3+}$	77Ta	
	-66.1	6.678	-93.7	25	C	μ=0.5M(KNO_3) R:$Ag^+ + HL^+ + L = [AgL(HL)]^{2+}$	77Ta	
S70 SULFIDE, N,N-dimethylthiono-carbamyl methyl-		$C_4H_9NS_2$				$(CH_3)_2NC(S)SCH_3$		
I_2	-31.50	2.260(20°,1)	-64.0	10-30	T	S:CCl_4	69Ge	b,q: ±1.5
S71 SULFIDE, dinonyl		$C_{18}H_{38}S$				$(C_9H_{19})_2S$		L
TeA_4	-42.2	2.10(1)	-101.7	25	C	S:Benzene A=C	73Pi	
S72 SULFIDE, dioctyl-		$C_{16}H_{34}S$				$(C_8H_{17})_2S$		L
Al_2A_6	-71.55	--	--	25	C	S:Benzene CM=0.07M A=Br R:0.5(Al_2A_6)+L= AlA_3L	68Rc	
SnA_4	-49.8	(1)	--	25	C	S:Benzene CM=0.05-0.08M A=Cl	68Ge 65Gb	
	-46.0	(2)	--	25	C	S:Benzene CM=0.05-0.08M A=Cl	68Ge 65Gb	u

Metal Ion	ΔH, kJ /mol	Log K	ΔS, J /mol K	T,°C	M	Conditions	Ref.	Re-marks
S72, continued								
I_2	-31.0	2.873(1)	-48.5	25	C	S:Octane	66Ad	
S73 SULFIDE, dipentyl-			$C_{10}H_{22}S$			$(C_5H_{11})_2S$		L
Al_2A_6	-75.7	--	--	25	C	S:Benzene CM=0.07M A=Br R:0.5(Al_2A_6)+L= AlA_3L	68Rc	
TeA_4	-48.5	2.18(1)	-120.9	25	C	S:Benzene A=Cl	73Pi	
I_2	-29.3	2.556(1)	-49.0	25	C	S:Octane	66Ad	
S74 SULFIDE, diphenyl-			$C_{12}H_{10}S$			$(C_6H_5)_2S$		L
Al_2A_6	-42.7	--	--	25	C	S:Benzene CM=0.07M A=Br R:0.5(Al_2A_6)+L= AlA_3L	68Rc	
I_2	-23.8	0.591(25°,1)	-69.4	17-35	T	S:Cyclohexane CM=5x10^{-4}M	74Sa	b
	-1.2	(1)	--	25	C	S:Octane	66Ad	
S75 SULFIDE, diphenyl-, 2,2'-dimethyl-			$C_{14}H_{14}S$			$(C_6H_4CH_3)_2S$		L
I_2	-36.8	0.763(25°,1)	-109.6	17-35	T	S:Cyclohexane CM=5x10^{-4}M	74Sa	b
S76 SULFIDE, diphenyl-, 4,4'-dimethyl-			$C_{14}H_{14}S$			$(C_6H_4CH_3)_2S$		L
I_2	-36.4	0.785(25°,1)	-104.6	17-35	T	S:Cyclohexane CM=5x10^{-4}M	74Sa	b
S77 SULFIDE, diphenyl-, 4-mehoxy-			$C_{13}H_{12}OS$			$(C_6H_5)S(C_6H_4OCH_3)$		L
I_2	-33.9	0.799(25°,1)	-98.7	17-35	T	S:Cyclohexane CM=5x10^{-4}M	74Sa	b
S78 SULFIDE, diphenyl-, 4-methyl-			$C_{13}H_{12}S$			$(C_6H_5)S(C_6H_4CH_3)$		L
I_2	-28.4	0.740(25°,1)	-80.8	17-35	T	S:Cyclohexane CM=5x10^{-4}M	74Sa	b
S79 SULFIDE, diphenyl-, 2,2,2',2'-tetramethyl-			$C_{16}H_{20}S$			$[C_6H_4(CH_3)_2]_2S$		L
I_2	-42.7	0.568(25°,1)	-134.7	17-35	T	S:Cyclohexane CM=5x10^{-4}M	74Sa	b

Metal Ion	ΔH, kJ /mol	Log K	ΔS, J /mol K	T,°C	M	Conditions	Ref.	Re-marks
S80 SULFIDE, dipropyl-			$C_6H_{14}S$			$(C_3H_7)_2S$		L
Al_2A_6	-74.0	--	--	25	C	S:Benzene CM=0.07M A=Br R:0.5(Al_2A_6)+L= AlA_3L	68Rc	
GaA_3	-82.0	3.81(1)	-202.1	25	C	S:Benzene A=Cl	70Gh	
SnA_4	-54.0	(1)	--	25	C	S:Benzene CM=0.05-0.08M A=Cl	68Ge 65Gb	
	-49.4	(2)	--	25	C	S:Benzene CM=0.05-0.08M A=Cl	68Ge 65Gb	
TeA_4	-55.6	1.88(1)	-147.7	25	C	S:Benzene A=Cl	72Pg 73Pi	
I_2	-33.9	2.468(1)	-66.1	25	C	S:Octane	66Ad	
S81 SULFIDE, ethylmethyl, 2-amino			C_3H_9NS			$CH_3SCH_2CH_2NH_2$		L
Ag^+	-82.0	9.288(1-2)	-96.6	25	C	μ=0.5M(KNO_3)	77Ta	
	-62.13	6.861	-77.0	25	C	μ=0.5M(KNO_3) R:$2Ag^+$+L=Ag_2L^{2+}	77Ta	
	-122.6	13.012	-162.3	25	C	μ=0.5M(KNO_3) R:$2Ag^+$+2L=$Ag_2L_2{}^{2+}$	77Ta	
	-27.28	2.639	-41.0	25	C	μ=0.5M(KNO_3) R:Ag^++HL^+=$AgHL^{2+}$	77Ta	
	-55.6	4.061	-109.2	25	C	μ=0.5M(KNO_3) R:Ag^++$2HL^+$= $Ag(HL)_2{}^{3+}$	77Ta	
	-73.6	7.565	-102.5	25	C	μ=0.5M(KNO_3) R:Ag^++HL^++L= $[AgL(HL)]^{2+}$	77Ta	
S82 SULFIDE, ethyl phenyl			$C_8H_{10}S$			$C_6H_5SC_2H_5$		L
Al_2A_6	-59.8	--	--	25	C	S:Benzene CM=0.07M A=Br R:0.5(Al_2A_6)+L= AlA_3L	68Rc	
SnA_4	-8.4	(1)	--	25	C	S:Benzene CM=0.05-0.08M A=Cl	65Gb 68Ge	
S83 SULFIDE, methyl propyl, 3-amino			$C_4H_{11}NS$			$CH_3SCH_2CH_2CH_2NH_2$		
Ag^+	-72.8	7.822(1-2)	-94.6	25	C	μ=0.5M(KNO_3)	77Ta	
	-60.7	7.206	-67	25	C	μ=0.5M(KNO_3)	77Ta	

Metal Ion	ΔH, kJ /mol	Log K	ΔS, J /mol K	T,°C	M	Conditions	Ref.	Re-marks
S83, continued								
Ag^+, cont.	-123.8	13.694	-154.0	25	C	R:$2Ag^+$+L=Ag_2L^{2+} μ=0.5M(KNO_3)	77Ta	
	-31.59	3.328	-42.2	25	C	R:$2Ag^+$+2L=$Ag_2L_2^{2+}$ μ=0.5M(KNO_3)	77Ta	
	-59.0	5.600	-91.2	25	C	R:Ag^++HL^+=$AgHL^{2+}$ μ=0.5M(KNO_3)	77Ta	
	-69.4	7.133	-96.6	25	C	R:Ag^++2HL^+= $Ag(HL)_2^{3+}$ μ=0.5M(KNO_3) R:Ag^++HL^++L= $[AgL(HL)]^{2+}$	77Ta	
S84	SULFIDE, α-naphthyl phenyl-		$C_{16}H_{12}S$			$C_{10}H_7SC_6H_5$		L
Al_2A_6	-41.0	--	--	25	C	S:Benzene CM=0.07M A=Br R:0.5(Al_2A_6)+L= AlA_3L	68Rc	
S85	SULFIDE, pentamethylene-		$C_5H_{10}S$			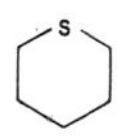		L
AlA_3	-71.09	(1)	--	28.5	C	S:Hexzne A=CH_3	67Hc	
S86	SULFIDE, propylene		C_3H_6S			$H_2CSCHCH_2$		L
$[NiA_2]_2$	-20.3	1.42(100°,1)	-28	75-125	T	S:Squalane A=3-Trifluoro-acetylcamphorate	74Sj	b,gg
AlA_3	-63.76	(1)	--	28.5	C	S:Hexane A=CH_3	67Hc	
S87	SULFIDE, trimethylene-		C_3H_6S			$(CH_2)_3S$		L
$[NiA_2]_2$	-23	1.74(100°,1)	-27	75-125	T	S:Squalane A=3-Trifluoro-acetylcamphorate	74Sj	b,gg
AlA_3	-67.11	(1)	--	28.5		S:Hexane A=CH_3	67Hc	
S88	SULFITE ION		O_3S			SO_3^{2-}		L^{2-}
CrA_3	16.69	0.079(1)	57.3	25	C	μ=3.0($NaClO_4$) A=Ethylenediamine	73Ua	
	2.89	0.042(2)	10.5	25	C	μ=3.0($NaClO_4$) A=Ethylenediamine	73Ua	q:±0.21

Metal Ion	ΔH, kJ /mol	Log K	ΔS, J /mol K	T,°C	M	Conditions	Ref.	Re-marks
S88, continued								
CrA_3, cont.	-3.30	0.042(3)	-10.5	25	C	μ=3.0($NaClO_4$) A=Ethylenediamine	73Ua	q:±0.50
CoA_3	7.9	0.61(1)	38	25	C	μ=3.0M($NaClO_4$+ Na_2L) A=Ethylenediamine	73Mc	q:±0.4
	4.2	0.18(2)	17	25	C	μ=3.0M($NaClO_4$+ Na_2L) A=Ethylenediamine	73Mc	q:±1.2
	2.1	-0.046(3)	8	25	C	μ=3.0M($NaClO_4$+ Na_2L) A=Ethylenediamine	73Mc	q:±1.2
	-10.5	-0.40(4)	-42	25	C	μ=3.0M($NaClO_4$+ Na_2L) A=Ethylenediamine	73Mc	q:±2.9
Ag^+	-20.71	8.326(20.15°, 1-2)	88.7	2-50	T	0.1-0.6M(Na_2SO_3) Done under 1 atm of N_2	51Ja	b
	-22.64	8.332(25°,1-2)	84	20-60	T	0.1-1.4M(Na_2SO_3)	56Kb	b
Hg^{2+}	-91	23.34(1-2)	42	25	C	μ=0.5M($NaClO_4$)	76Mb 75Mb	q:±7,y
S89 SULFITE, diethyl-			$C_4H_{10}O_3S$			$(CH_3CH_2)_2SO_3$		L
I_2	-9.2	-0.503(32.6°,1)	-39.7	23.5-44.1	T	S:CCl_4	63Da	b
S90 SULFONE, tetramethylene-			$C_4H_8O_2S$			$CH_2CH_2CH_2CH_2SO_2$ (cyclic)		L
I_2	-9.2	-0.172(32.1°,1)	-33.5	25-45.5	T	S:CCl_4	63Da	b
S91 SULFOXIDE, bis(dimethylamino)-			$C_4H_{12}ON_2S$			$[(CH_3)_2N]_2SO$		L
AlA_3	-102.76	--	--	28	C	S:Hexane A=CH_3 R:½$[AlA_3]_2$+L= $AlLA_3$	68Hd	
S92 SULFOXIDE, dibenzyl-			$C_{14}H_{14}OS$			$C_6H_5CH_2SOCH_2C_6H_5$		L
UA_2B_2	-16.07	2.994(40°)	5.8	30-60	T	S:Benzene A=O B=Benzoyltri-fluoroacetone R:$UA_2B_2nH_2O$+L= UA_2B_2L+nH_2O	73Sy	b
	-20.5	3.316(40°)	-1.97	30-60	T	S:Benzene A=O B=Thenoyltri-fluoroacetone R:$UA_2B_2nH_2O$+L= UA_2B_2L+nH_2O	73Sy	b

Metal Ion	ΔH, kJ /mol	Log K	ΔS, J /mol K	T,°C	M	Conditions	Ref.	Re-marks
S93	SULFOXIDE, didecyl-		$C_{20}H_{42}OS$			$(C_{10}H_{21})_2SO$		L
UA_2B_2	-12.84	2.785(40°)	12.1	30-60	T	S:Benzene A=0 B=Benzoylacetone R:$UA_2B_2nH_2O$+L= UA_2B_2L+nH_2O	73Sy	b
	-19.16	3.581(40°)	7.5	30-60	T	S:Benzene A=0 B=Benzoyltri-fluoroacetone R:$UA_2B_2 \cdot nH_2O$+L= UA_2B_2L+nH_2O	73Sy	b
	-12.76	3.797(40°)	31.8	30-60	T	S:Benzene A=0 B=Thenoyltri-fluoroacetone R:$UA_2B_2 \cdot nH_2O$+L= UA_2B_2L+nH_2O	73Sy	b
S94	SULFOXIDE, dihexyl-		$C_{12}H_{26}OS$			$(C_6H_{11})_2SO$		L
UA_2B_2	-21.04	2.841(40°)	-13.0	30-60	T	S:Benzene A=0 B=Benzoylacetone R:$UA_2B_2 \cdot nH_2O$+L= UA_2B_2L+nH_2O	73Sy	b
	-15.94	3.678(40°)	19.7	30-60	T	S:Benzene A=0 B=Benzoyltri-fluoroacetone R:$UA_2B_2 \cdot nH_2O$+L= UA_2B_2L+nH_2O	73Sy	b
	-14.73	3.986(40°)	29.3	30-60	T	S:Benzene A=0 B=Thenoyltri-fluoroacetone R:$UA_2B_2 \cdot nH_2O$+L= UA_2B_2L+nH_2O	73Sy	b
TeA_4	-44.8	(1)	--	25	C	S:Benzene A=Cl	72Pg	
S95	SULFOXIDE, diisopentyl		$C_{10}H_{22}OS$			$(C_5H_{11})_2SO$		L
UA_2B_2	-10.88	2.715(40°)	17.2	30-60	T	S:Benzene A=0 B=Benzoylacetone R:$UA_2B_2 \cdot nH_2O$+L= UA_2B_2L+nH_2O	73Sy	b
	-18.66	3.623(40°)	9.6	30-60	T	S:Benzene A=0 B=Benzoyltri-fluoroacetone R:$UA_2B_2 \cdot nH_2O$+L= UA_2B_2L+nH_2O	73Sy	b
	-15.94	3.832(40°)	22.6	30-60	T	S:Benzene A=0 B=Thenoyltri-fluoroacetone R:$UA_2B_2 \cdot nH_2O$+L= UA_2B_2L+nH_2O	73Sy	b

Metal Ion	ΔH, kJ /mol	Log K	ΔS, J /mol K	T,°C	M	Conditions	Ref.	Remarks
S96 SULFOXIDE, dimethyl-			C_2H_6OS			$(CH_3)_2SO$		L
CoAB	-15.5	1.73(1)	-19.7	25	T	S:o-Dichloro- benzene A=CF_3 B=N,N'-bis (acetylacetonato) ethylenediimine	77Hc	
NiA	-17.53	0.328(25°,1-2)	-52.7	5-110	T	S:Dimethyl sulfoxide A=1,4,8,11-Tetra- azacyclotetra- decane	79Vb	
NiA_2	-18.83	1.760(1)	-29.3	25	C	S:Benzene A=2,2,- 6,6-Tetramethyl- 3,5-heptanedione	73Ce	
	-2.55	(2)	--	25	C	S:Benzene A=2,2,- 6,6-Tetramethyl- 3,5-heptanedione	73Ce	
RhAB	-42.2	1.881	-106	24	C	S:Benzene R:$[RhAB]_2$+2L= 2RhABL A=Cl B=1,5-Cycloocta- diene	76Lb	
$IrABC_2$ (trans)	-53.6	(1)	--	25	C	S:Benzene A=F B=CO C=PPh_3	79Bf	
	-47.7	(1)	--	25	C	S:Benzene A=Cl B=CO C=PPh_3	79Bf	
	-48.1	(1)	--	25	C	S:Benzene A=Br B=CO C=PPh_3	79Bf	
	-34.7	(1)	--	25	C	S:Benzene A=I B=CO C=PPh_3	79Bf	
	-59.0	(1)	--	25	C	S:Benzene A=Cl B=CO C=PMe_2Ph	79Bf	
	-57.3	(1)	--	25	C	S:Benzene A=Cl B=CO C=$PMePh_2$	79Bf	
	-53.6	(1)	--	25	C	S:Benzene A=Cl B=CO C=PEt_2Ph	79Bf	
IrA_2BC_2	-77.0	0.037(1)	-259	25	C	S:CH_2Cl_2 A=Cl B=C(O)CF_3 C=P$(C_6H_5)_3$	77La	
CuA_2	-35.6	(1)	--	25	C	S:CCl_4 A=Hexafluoro- acetylacetone	76Mc 70Pa	u:25°
	-30.1	3.549(1)	-33.0	25	C	S:o-Dichloro- benzene A=Hexafluoro- acetylacetone	72Nd	q:±5.4
	-26.4	(1)	--	25	C	S:Methylene chloride A=Hexafluoro- acetylacetonate	74Dc	
	-18.0	1.58(20°,1)	-28.9	-60-60	T	S:Toluene	78Mb	

Metal Ion	ΔH, kJ /mol	Log K	ΔS, J /mol K	T,°C	M	Conditions	Ref.	Remarks
S96, continued								
CuA_2, cont.						A=Thenoyltri-fluoroacetonato		
	-12.6	1.58(20°,1)	-24.3	-60-60	T	S:Toluene A=Cupferonato	78Mb	
CuA_2B	-2.1	--	--	25	C	S:CCl_4 A=Hexafluoro-acetylacetone B= Dimethyl acetate R:CuA_2B+L=CuA_2L+B	72Nd	
	-10.9	--	--	25	C	S:CCl_4 A=Hexafluoro-acetylacetone B=Ethylacetate R:CuA_2B+L=CuA_2L+B	72Nd	
	-2.5	--	--	25	C	S:o-$Cl_2C_6H_4$ A-Hexafluoro-acetylacetone B= Dimethyl acetate R:CuA_2B+L=CuA_2L+B	72Nd	
	-10.5	--	--	25	C	S:o-$Cl_2C_6H_4$ A=Hexafluoro-acetylacetone B=Ethyl acetate R:CuA_2B+L=CuA_2L+B	72Nd	
ZnA	-36.4	3.582(25°,1)	-53.6	15-35	T	S:Cyclohexane A=α,β,γ,δ-Tetra-phenylporphine	77Vb	
AlA_3	-119.83	--	--	28	C	S:5% Benzene in Hexane R:½[AlA_3]$_2$ +L=$AlLA_3$ A=CH_3	68Hd	
$SnAB_3$	-34.3	0.959(1)	-96	26	C	S:CCl_4 A=Cl B=CH_3	66Be	
I_2	-18.4	0.964(35.4°,1)	-41.0	23.4-46.1	T	S:CCl_4	63Da	b
S97 SULFOXIDE, dimethylamino, methyl-		C_3H_9ONS				$CH_3SON(CH_3)_2$		L
AlA_3	-109.91	--	--	28	C	S:Hexane R:½[AlA_3]$_2$+L= $AlLA_3$ A=CH_3	68Hd	
S98 SULFOXIDE, diphenyl-		$C_{12}H_{10}OS$				$(C_6H_5)_2SO$		L
UA_2B	-17.95	2.520(40°)	-9.2	30-60	T	S:Benzene A=O B=Benzoyltri-fluoroacetone R:$UA_2B_2 \cdot nH_2O$+L= UA_2B_2L+nH_2O	73Sy	b

Metal Ion	ΔH, kJ /mol	Log K	ΔS, J /mol K	T,°C	M	Conditions	Ref.	Remarks
S98, continued								
UA_2B_2	-19.2	2.876(40°)	-6.3	30-60	T	S:Benzene A=O B=Thenoyltri-fluoroacetone R:$UA_2B_2 \cdot nH_2O+L=UA_2B_2L+nH_2O$	73Sy	b
TeA_4	-46.4	(1)	--	25	C	S:Benzene A=Cl	72Pg	
S99 SULFOXIDE, phenyl propyl-			$C_9H_{12}OS$			$CH_3CH_2CH_2SOC_6H_5$		L
TeA_4	-46.9	(1)	--	25	C	S:Benzene A=Cl	72Pg	
S100 SULFOXIDE, tetramethylene			C_4H_8OS			$CH_2CH_2CH_2CH_2S$:O (ring)		L
AlA_3	-118.20	--	--	28	C	S:5% Benzene in Hexane A=CH_3 R:$\frac{1}{2}[AlA_3]_2+L=AlLA_3$	68Hd	
I_2	-18.4	1.070(35°,1)	-39.7	24.8-45	T	S:CCl_4	63Da	b
S101 SULFUR DIOXIDE			O_2S			SO_2		L
$RhABC_2$	-33.0	1.875(30°,1)	-71	10-60	T	S:Chlorobenzene A=Cl B=CO C=Ph_3P	71Vb	b
	-39.3	(1)	--	10-60	T	S:Chlorobenzene A=I B=CO C=Ph_3P	71Vb	b
$IrABC_2$ (trans)	-47.7	(1)	--	25	C	S:Benzene A=Cl B=CO C=PPh_3	79Bf	
	-25.1	(1)	--	25	C	S:Benzene A=Cl B=CO C=P(t-Bu)Ph_2	79Bf	
$IrABC_2$	-43.9	3.100(30°,1)	-88	10-60	T	S:Chlorobenzene A=Cl B=CO C=Ph_3P	71Vb	b
	-105	(1)	--	10-60	T	S:Chlorobenzene A=I B=CO C=Ph_3P	71Vb	b

Metal Ion	ΔH, kJ /mol	Log K	ΔS, J /mol K	T,°C	M	Conditions	Ref.	Remarks
			T					
T1 TARTARIC ACID			$C_4H_6O_6$			$HO_2CCH(OH)CH(OH)CO_2H$		L^{2-}
Na^+	-0.50	0.66(1)	10.9	25	C	μ=0.05M($NaClO_4$)	80Aa	
	-0.75	0.44(1)	5.8	25	C	μ=0.10M($NaClO_4$)	80Aa	
	-0.96	0.44(1)	5.0	25	C	μ=0.15M($NaClO_4$)	80Aa	
	-2.1	0.07	-5.4	25	C	μ=0.05M($NaClO_4$) R:Na^++HL^{-1}=NaHL	80Aa	
	-2.1	0.07	-5.4	25	C	μ=0.10M($NaClO_4$) R:Na^++HL^{-1}=NaHL	80Aa	
	-2.1	0.0	-7.1	25	C	μ=0.15M($NaClO_4$) R:Na^++HL^{-1}=NaHL	80Aa	
Ca^{2+}	-14.68	1.84(1)	-14.2	25	-	--	53Ka	
La^{3+}	0.0	4.604(25°,1)	95.0	15-35	T	μ=0	72De	b
	-9.2	4.58(25°,1)	58	9-25	T	μ=0	75Ya	q:±3.3
	44.8	2.986(25°,2)	20.0	15-35	T	μ=0	72De	b
	5.61	2.48(25°)	66.11	15-35	T	μ=0 R:La^{3+}+ HL^-=$LaHL^{2+}$	72De	b
Nd^{3+}	-13.0	4.66(25°,1)	46	9-25	T	μ=0	75Ya	q:±3.3
Sm^{3+}	-11.3	5.21(25°,1)	63	9-25	T	μ=0	75Ya	q:±3.3
Gd^{3+}	-12.1	4.75(25°,1)	50	9-25	T	μ=0	75Ya	q:±3.3
Dy^{3+}	5.0	5.12(25°,1)	113	9-25	T	μ=0	75Ya	q:±3.3
Ho^{3+}	3.8	5.14(25°,1)	113	9-25	T	μ=0	75Ya	q:±3.3
Tm^{3+}	11.7	5.30(25°,1)	142	9-25	T	μ=0	75Ya	q:±3.3
Lu^{3+}	8.4	5.30(25°,1)	126	9-25	T	μ=0	75Ya	q:±3.3
CoA_5B^{2+}	2.3	0.45(1)	17	25	C	μ=3.0M($NaClO_4$) A=NH_3 B=NO_2	74Mb	
	1.0	0.08(2)	4	25	C	μ=3.0M($NaClO_4$) A=NH_3 B=NO_2	74Mb	
	0.4	0.015(3)	2	25	C	μ=3.0M($NaClO_4$) A=NH_3 B=NO_2	74Mb	
	0.4	0.015(4)	2	25	C	μ=3.0M($NaClO_4$) A=NH_3 B=NO_2	74Mb	
ZnA_3^+	-43.1	6.070(1)	28.41	25	C	CM=0.01M CL=0.01M A=OH^-	73Ki	
	-38.1	6.070(1)	28.41	5-35	T	CM=0.01M CL=0.01M A=OH^-	73Ki	b
T2 TELLURATE ION			O_6H_2Te			$H_2TeO_6^{4-}$		L^{4-}
CuA_4^-	-84	11.947(50°)	-29.7	40-60	T	1.2M(ClO_4^-) R:CuA_4^-+$2L^{4-}$= $Cu(HTeO_6)_2^{7-}$+ $2OH^-$+$2H_2O$ A=OH^-	53Lb	b

Metal Ion	ΔH, kJ /mol	Log K	ΔS, J /mol K	T,°C	M	Conditions	Ref.	Re-marks
T3 TELLURITE ION			O_3Te			TeO_3^{2-}		L^{2-}
CrA_3^{3+}	20.50	-0.046(1)	67.8	25	C	μ=3.0($NaClO_4$) A=Ethylenediamine	73Ua	
	-14.60	-0.126(2)	-51.5	25	C	μ=3.0($NaClO_4$) A=Ethylenediamine	73Ua	
CoA_3^{3+}	4.6	0.301(1)	20	25	C	μ=3.0M($NaClO_4$+ Na_2L) A=Ethylenediamine	73Mc	q:±0.4
	2.1	0.041(2)	8	25	C	μ=3.0M($NaClO_4$+ Na_2L) A=Ethylenediamine	73Mc	q:±0.8
	-6.7	-0.30(3)	-25	25	C	μ=3.0M($NaClO_4$+ Na_2L) A=Ethylenediamine	73Mc	q:±1.2
CoA_5B^{2+}	8.6	0.42(1)	38	25	C	μ=3.0M($NaClO_4$) A=NH_3 B=NO_2	74Mb	
	2.9	0.02(2)	8.4	25	C	μ=3.0M($NaClO_4$) A=NH_3 B=NO_2	74Mb	
	0.63	0.06(3)	4	25	C	μ=3.0M($NaClO_4$) A=NH_3 B=NO_2	74Mb	
	-8.6	0.37(4)	-21	25	C	μ=3.0M($NaClO_4$) A=NH_3 B=NO_2	74Mb	
T4 TELLURIUM TETRACHLORIDE			Cl_4Te			$TeCl_4$		L
Al_2A_6	-111.3	--	--	25	C	S:Benzene A=Br R:0.5[Al_2A_6]+L= [AlA_3L]	72Pg	
Ga_2A_6	-44.8	--	--	25	C	S:Benzene A=Cl R:0.5[Ga_2A_6]+L= [GaA_3L]	72Pg	
	-70.3	--	--	25	C	S:Benzene A=Br R:0.5[Ga_2A_6]+L= [GaA_3L]	72Pg	
T5 2,2':6',2"-TERPYRIDINE			$C_{15}H_{12}N_3$			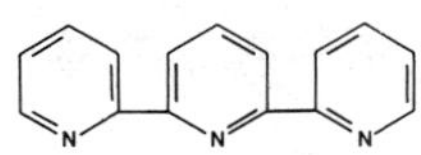		L
Cu^{2+}	-58.7	12.28(1)	38	25	C	μ=0.1($NaNO_3$)	79Ca	
	-29.5	6.82(2)	32	25	C	μ=0.1($NaNO_3$)	79Ca	
	6.0	3.50	87	25	C	μ=0.1($NaNO_3$) R:2CuL(OH)= [CuL(OH)]$_2$	79Ca	
T6 1,4,7,10-TETRAAZACYCLODODECANE			$C_8H_{20}N_4$			$CH_2NH(CH_2CH_2NH)_3CH_2$		L^4
Ni^{2+}	-49.8	(1)	--	25	C	ΔH for formation of high-spin complex	80Fa	
Cu^{2+}	-95.0	8.73(1)	151.5	25	C	μ=0	77Ai	

Metal Ion	ΔH, kJ /mol	Log K	ΔS, J /mol K	T, °C	M	Conditions	Ref.	Remarks
T6, continued								
Cu^{2+}, cont.	-76.6	24.8(25°,1)	213	15-35	T	μ=0.2M($NaClO_4$)	78Kd 76Ke 76Kb	b
	-95.0	24.8(1)	151.5	25	C	1M(NaOH) pH=14	78Ad	
Zn^{2+}	-33.0	16.2(25°,1)	197	15-35	T	μ=0.2M($NaClO_4$)	78Kd 77Kd	b
	-60.7	(1)	--	25	C	1M(NaOH) pH=14	78Ad 77Ai	
Cd^{2+}	-34.3	14.3(25°,1)	159	15-35	T	μ=0.2M($NaClO_4$)	78Kd 77Kd	b
Hg^{2+}	-98.7	25.5(25°,1)	159	15-35	T	μ=0.2M($NaClO_4$)	78Kd 76Kg	b
Pb^{2+}	-27.6	15.9(25°,1)	213	15-35	T	μ=0.2M($NaClO_4$)	78Kd 77Kd	b,q: ±2.1
T7 1,4,8,12-TETRAAZACYCLOPENTA-DECANE			$C_{11}H_{26}N_4$			$CH_2NH(CH_2CH_2CH_2NH)_3CH_2$ (cyclic)		L^7
Ni^{2+}	-74.9	(1)	--	25	C	ΔH for formation of high-spin complex	80Fa	
Cu^{2+}	-110.9	14.46(1)	95.0	25	C	μ=0	77Ai	
	-110.9	24.4(25°,1)	96	15-35	T	μ=0.2M($NaClO_4$)	78Kd 76Kh	b
	-110.9	24.4(1)	95.0	25	C	1M(NaOH) pH=14	78Ad	
Zn^{2+}	-69.0	(1)	--	25	C	μ=0	77Ai	
	-34.3	15.0(25°,1)	171	15-35	T	μ=0.2M($NaClO_4$)	78Kd 77Kd	b
	-69.0	(1)	--	25	C	1M(NaOH) pH=14	78Ad	
Hg^{2+}	-103.3	23.7(25°,1)	109	15-35	T	μ=0.2M($NaClO_4$)	78Kd 76Kg	b
T8 1,4,7,11-TETRAAZACYCLOTETRA-DECANE			$C_{10}H_{20}N_4$			$CH_2NHCH_2CH_2NH(CH_2CH_2NH)_2CH_2$ (cyclic)		L
Ni^{2+}	22.2	0.190(25°)	78.2	21-44.9	T	0.1M($NaClO_4$), R: $NiL(H_2O)_2^{2+}$ (octahedral, blue) = NiL^{2+}(planar, yellow)+$2H_2O$	79Sa	
Cu^{2+}	-116.3	(1)	--	25	C	μ=0	79Fb 78Fa	
	-74.10	(1)	--	25	C	μ=1.0(NaOH)	78Fa	

Metal Ion	ΔH, kJ /mol	Log K	ΔS, J /mol K	T,°C	M	Conditions	Ref.	Re- marks
T9	1,4,7,11-TETRAAZACYCLOTETRA-DECANE, 2,6-dimethyl, pyridino		$C_{15}H_{23}N_4$			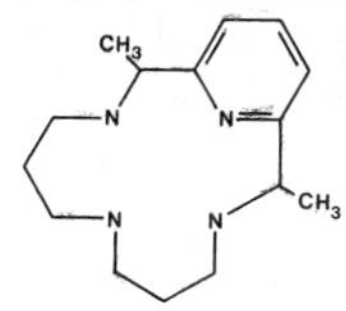		L
Ni^{2+}	-50.2	(1)	--	25	C	ΔH for formation of octahedral high-spin complex	80Fb	
	-31.8	(1)	--	25	C	ΔH for formation of planar low-spin complex	80Fb	
Cu^{2+}	100.0	(1)	--	25	C	μ=0	79Fb 78Fa	
	54.98	(1)	--	25	C	μ=1.0(NaOH)	78Fa	
T10	1,4,8,11-TETRAAZACYCLOTETRA-DECANE		$C_{10}H_{24}N_4$			$CH_2CH_2CH_2NHCH_2CH_2NH$- $CH_2CH_2CH_2NHCH_2CH_2NH$		L[6]
Ni^{2+}	-129.7	22.2(1)	-8	25	C	μ=0.1	74Hf 74He	
	-100.8	21.9(1)	81.6	25	C	1M(NaOH) pH=14 Blue(octahedral) form	78Fb	
	-78.2	22.3(1)	165.3	25	C	1M(NaOH) pH=14 Yellow(square-planar) form	78Fb	
	-84.9	22.2(1)	140.6	25	C	1M(NaOH) pH=14 Equilibrium mixture	78Fb	
	-60.2	(1)	--	25	C	ΔH for formation of planar low-spin complex	80Fb	
	-82.4	(1)	--	25	C	ΔH for formation of octahedral high-spin complex	80Fb	
	22.6	0.390(25°)	84	19-49	T	$0.1M(NaClO_4)$ R:$NiL(H_2O)_2^{2+}$ (blue)=NiL^{2+} (yellow)+$2H_2O$	77Aj	
Cu^{2+}	-135.6	(1)	--	25	C	μ~0	77Ak	
	-127.2	27.2(25°,1)	92	15-35	T	$μ=0.2M(NaClO_4)$	78Ke 77Kc	b
	-135.6	(1)	--	25	C	1M(NaOH) pH=14	78Ad	
	-179.5	(1)	--	25	C	$0.15M(NEt_4Br)$ S:26v%Acetonitrile in Chloroform	75Df	
CuA_4^{2-}	-259.0	(1)	--	25	C	$0.15M(NEt_4Br)$ S:26v%Acetonitrile in Chloroform	75Df	

Metal Ion	ΔH, kJ /mol	Log K	ΔS, J /mol K	T,°C	M	Conditions	Ref.	Remarks
T10, continued								
CuA_4^{2-}, cont.						R: CuA_4^{2-}+L= $CuLA_n^{(2-n)+}$+ (4-n)A^- A=Br		
Zn^{2+}	-61.9	(1)	--	25	C	μ~0	77Ak 77Ai	
	-31.8	15.5(25°,1)	192	15-35	T	μ=0.2M($NaClO_4$)	78Kd 77Kd	b
	-61.9	(1)	--	25	C	1M(NaOH) pH=14	78Ad	
Hg^{2+}	-137.6	23.0(25°,1)	-20.5	15-35	T	μ=0.20M($NaClO_4$)	78Kd 76Kg	b
T11	1,4,8,11-TETRAAZACYCLOTETRADECANE, 5,12-dimethyl(meso)		$C_{12}H_{28}N_4$			$CH_2CH(CH_3)NHCH_2CH_2NHCH_2$- $CH_2C(CH_3)NHCH_2CH_2NHCH_2$		
Ni^{2+}	-117.2	21.9(1)	33	25	C	μ=0.1	74He	
T12	1,4,8,11-TETRAAZACYCLOTETRADECANE, meso-5,5,7,12,12,14-hexamethyl		$C_{16}H_{36}N_4$			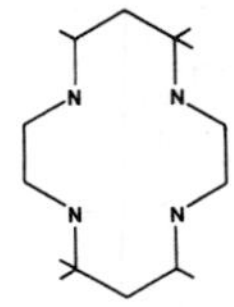		
ZnA_4	-115.5	--	--	25	C	0.15M(NEt_4Br) S:26w%Acetonitrile 74w% Chloroform R: ZnA_4^{2-}+L= $ZnLA_n^{(2-n)+}$+ (4-n)A^- A=Br^-	78Db	
	-70.3	--	--	25	C	0.15M(NEt_4Br) S:50w%Methanol, 50w%Acetone R: ZnA_4^{2-}+L= $ZnLA_n^{(2-n)+}$+ (4-n)A^- A=Br^-	78Db	
T13	1,4,8,11-TETRAAZACYCLOTETRADECANE-2,4,9,11-TETRAENE, 5,5,7,12,12,14-hexamethyl		$C_{16}H_{28}N_4$			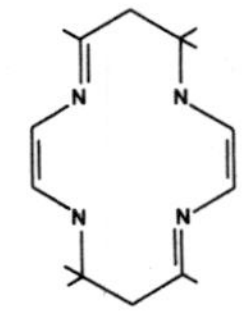		
ZnA_4	-44.8	--	--	25	C	0.15M(NEt_4Br) S:26w%Acetonitrile, 74w%Chloroform R: ZnA_4^{2-}+L=	78Db	

Metal Ion	ΔH, kJ /mol	Log K	ΔS, J /mol K	T, °C	M	Conditions	Ref.	Re-marks
T13, continued								
ZnA_4, cont.						$ZnLA_n^{(2-n)+}$ + $(4-n)A^-$ A=Br^-		
	-25.9	--	--	25	C	0.15(NEt_4Br) S:50w%Methanol, 50w%Acetone R:ZnA_4^{2-}+L= $ZnLA_n^{(2-n)+}$+ $(4-n)A^-$ A=Br^-	78Db	
T14 1,4,7,10-TETRAAZACYCLOTRIDECANE		$C_9H_{22}N_4$				$CH_2CH_2NH(CH_2CH_2NH)_3CH_2$ (ring)		L^5
Ni^{2+}	-52.3	(1)	--	25	C	ΔH for formation of low-spin complex	80Fa	
	-83.7	(1)	--	25	C	ΔH for formation of high-spin complex	80Fa	
	31.4	0.836(25°)	125.5	21-48.5	T	0.1M($NaClO_4$) R:$[NiL(OH_2)_2]^{2+}$ (blue,high spin) =NiL^{2+}+$2H_2O$ (yellow,low spin)	79Fa	
Cu^{2+}	-107.1	(1)	--	25	C	μ=0	79Fb 78Fa	
	-122.2	29.1(25°,1)	142	15-35	T	μ=0.2M($NaClO_4$)	78Kd 76Kf	b
Zn^{2+}	-32.6	15.6(25°,1)	188	15-35	T	μ=0.2M($NaClO_4$)	78Kd 77Kd	b
Hg^{2+}	-103.3	25.3(25°,1)	139.3	15-35	T	μ=0.20M($NaClO_4$)	76Kg	
T15 TETRACYCLINE, (hydrochloride)		$C_{22}H_{24}O_8N_2$				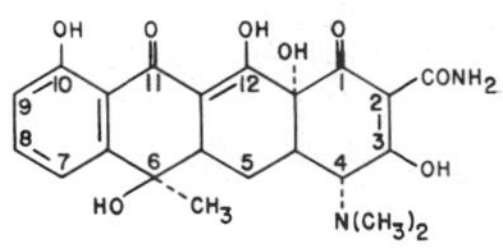		HL^-
Cu^{2+}	-31.8	7.80(1)	42.7	25	C	μ=0.01	66Bb	
	-7.53	5.34(2)	77.0	25	C	μ=0.01	66Bb	
T16 TETRACYCLINE, 7-chloro (hydrochloride)		$C_{22}H_{23}O_8N_2Cl$				$C_{22}H_{23}N_2O_8(Cl)$		HL^-
Cu^{2+}	-49.0	7.32(1)	-16.7	25	C	μ=0.01	66Bb	
	-7.53	5.09(2)	72.4	25	C	μ=0.01	66Bb	

Metal Ion	ΔH, kJ /mol	Log K	ΔS, J /mol K	T,°C	M	Conditions	Ref.	Remarks
T17	TETRACYCLINE, 7-chloro-6,6-dimethyl-(hydrochloride)		$C_{23}H_{25}O_7N_2Cl$			$C_{22}H_{22}N_2O_7(Cl)(CH_3)_2$		HL^-
Cu^{2+}	-51.5	7.85(1)	-22.6	25	C	μ=0.01	66Bb	
	-10.88	5.46(2)	68.2	25	C	μ=0.01	66Bb	
T18	TETRACYCLINE, 7-chloro-4-epi-(hydrochloride)		$C_{22}H_{23}O_8N_2Cl$			$C_{22}H_{24}N_2O_8(Cl)$		HL^-
Cu^{2+}	-48.1	7.63(1)	-15.5	25	C	μ=0.01	66Bb	
	-8.37	5.07(2)	69.0	25	C	μ=0.01	66Bb	
T19	TETRADECANE, 5,10-dithio-		$C_{12}H_{26}S_2$			$CH_3(CH_2)_3S(CH_2)_4S(CH_2)_3CH_3$		L
TiA_4	-95.8	3.37(1)	-256.9	25	C	S:Benzene A=Cl	70Gh	
SnA_4	-85.8	(1)	--	--	C	S:Benzene CM=0.05-0.08M A=Cl	65Gb	u
TeA_4	-50.2	3.34(1)	-104.6	25	C	S:Benzene A=Cl	73Pi	
	-18.4	(2)	--	25	C	S:Benzene A=Cl	73Pi	
T20	TETRADECANE, 3,6,9,12-tetra-aza-1,14-diamino		$C_{10}H_{24}N_6$			$H_2N(CH_2)_2N(CH_2)_2N(CH_2)_2N(CH_2)_2N(CH_2)_2NH_2$		L
Cu^{2+}	-82.09	25.13(25°,1)	206.3	30-40	T	0.1M(KCl)	57Ja	b
T21	TETRADECANE, 3,6,9,12-tetrathio-1,14-dihydroxy		$C_{10}H_{22}O_2S_4$			$OHCH_2CH_2SCH_2CH_2SCH_2-CH_2SCH_2CH_2SCH_2CH_2OH$		
Ag^+	-50.6	5.67(1)	-64.0	20	C	μ=1.0(KNO_3) pH=3	70Wc	
	-55.6	4.62(2)	-100	20	C	μ=1.0(KNO_3) pH=3	70Wc	
T22	TETRAETHYLENE GLYCOL, di(2-benzoic acid) ether		$C_{22}H_{28}O_9$			$(C_6H_5CO_2H)OCH_2(CH_2OCH_2)_3-CH_2O(C_6H_5CO_2H)$		
Li^+	-41	3.45(25°,1)	-70	25	T	μ=0.08M[$N(CH_3)_4$Br] S:Methanol ΔCp=1100	79Tb	b
Na^+	-68	3.41(25°,1)	-161	25	T	μ=0.08M[$N(CH_3)_4$Br] S:Methanol ΔCp=3800	79Tb	b
K^+	-33.3	3.24(25°,1)	-22	25	T	μ=0.08M[$N(CH_3)_4$Br] S:Methanol ΔCp=6700	79Tb	b
Rb^+	-25.1	3.19(25°,1)	-23	25	T	μ=0.08M[$N(CH_3)_4$Br] S:Methanol ΔCp=600	79Tb	b

Metal Ion	ΔH, kJ /mol	Log K	ΔS, J /mol K	T,°C	M	Conditions	Ref.	Remarks
T22, continued								
Cs^+	-24	1.9(25°,1)	-40	25	T	μ=0.08M[N(CH_3)$_4$ Br] S:Methanol ΔCp=1300	79Tb	b
T23	TETRAETHYLENE GLYCOL, dimethyl ether	$C_{10}H_{22}O_5$				$CH_3OCH_2(CH_2OCH_2)_3CH_2OCH_3$		L
NaA	-29.7	2.230(25°)	-58	0-50	T	S:Tetrahydrofuran A=Fluorenyl ion R:NaA+L=NaLA	70Ce	b
Pb^{2+}	-13.4	0.5(25°,1)	-34.7	5-35	T	μ=0.1M([(C_2H_5)$_4$-NClO$_4$] pH=2	76Kd	
	-13.4	1.1(25°,2)	-25.1	5-35	T	μ=0.1M([(C_2H_5)$_4$-NClO$_4$] pH=2	76Kd	
T24	TETRAETHYLENE GLYCOL, di(2-nitrobenzene) ether	$C_{20}H_{24}O_9N_2$				$(C_6H_4NO_2)OCH_2(CH_2OCH_2)_3$-$CH_2O(C_6H_4NO_2)$		
K^+	-29	1.59(25°,1)	-67	25	T	μ=0.08M[N(CH_3)$_4$-Br] S:Methanol	79Tb	b
T25	TETRAETHYLENE GLYCOL, di(8-quinolyl) ether	$C_{26}H_{28}O_5N_2$				$(C_9H_6N)OCH_2(CH_2OCH_2)_3$-$CH_2O(C_9H_6N)$		
Li^+	-63	2.37(25°,1)	-170	25	T	μ=0.08M[N(CH_3)$_4$-Br] S:Methanol ΔCp=400	79Tb	b
Na^+	-36	3.22(25°,1)	-59	25	T	μ=0.08M[N(CH_3)$_4$-Br] S:Methanol ΔCp=1200	79Tb	b
K^+	-21	3.51(25°,1)	-3	25	T	μ=0.08M[N(CH_3)$_4$-Br] S:Methanol	79Tb	b
Rb^+	-20	3.06(25°,1)	-7	25	T	μ=0.08M[N(CH_3)$_4$-Br] S:Methanol	79Tb	b
Cs^+	-25	2.6(25°,1)	-33	25	T	μ=0.08M[N(CH_3)$_4$-Br] S:Methanol	79Tb	b
T26	TETRAETHYLENE GLYCOL monophenyl ether-, mono(8-quinolyl)ether	$C_{23}H_{27}O_5N$				$(C_6H_5)OCH_2(CH_2OCH_2)_3$-$CH_2O(C_9H_6N)$		
K^+	-59	1.87(25°,1)	-160	25	T	μ=0.08M[N(CH_3)$_4$-Br] S:Methanol	79Tb	b
T27	TETRAKIS (1-pyrazolyl) borate ion	$C_{12}H_{12}N_8B$				$B(C_3H_3N_2)_4$		L^-
Co^{2+}	-137.3	(1-2)	--	22-27	C	CM=0.02M S:Acetonitrile	75Jc	

Metal Ion	ΔH, kJ /mol	Log K	ΔS, J /mol K	T,°C	M	Conditions	Ref.	Remarks
T27, continued								
Ni^{2+}	-130.2	(1-2)	--	22-27	C	CM=0.02M S:Acetonitrile	75Jc	
Cu^{2+}	-114.9	(1-2)	--	22-27	C	CM=0.02M S:Acetonitrile	75Jc	
Zn^{2+}	106.8	(1-2)	--	22-27	C	CM=0.02M S:Acetonitrile	75Jc	
T28 TETRANACTIN			$C_{44}H_{71}O_{12}$					
Na^{+}	-41.4	--	--	25	C	S:Methanol	79Ua	j
K^{+}	-64.4	--	--	25	C	S:Methanol	79Ua	j
Rb^{+}	-51.9	--	--	25	C	S:Methanol	79Ua	j
Cs^{+}	-48.1	--	--	25	C	S:Methanol	79Ua	j
T29 TETRAPHENYLBORIDE ION			$C_{24}H_{20}B$			$(C_6H_5)_4B^-$		L^-
Na^{+}	7.5	4.27(25°,1)	107.1	-70-25	T	S:Dimethoxyethane	65Cb	b,q:±1.2
	8.4	4.266(20°,1)	109	10-40	T	S:Dimethoxyethane	73Sp	b
	5.4	4.06(25°,1)	95.8	-70-25	T	S:Tetrahydrofuran	63Cb	b,q:±1.2
NaA^{+}	10.5	4.222(20°,1)	117	10-40	T	S:Dimethoxyethane A=Dibenzo-18-crown-6	73Sp	b
NAB_3^{+}	9.6	4.45(25°,1)	117.6	-70-25	T	S:Dimethoxyethane A=Butyl B=Amyl	65Cb	b,q: ±1.2
	5.4	4.25(25°,1)	99.6	-70-25	T	S:Tetrahydrofuran A=Butyl B=Amyl	65Cb	b,q: ±1.2
Cs^{+}	6.48	5.80(25°,1))	132.6	-70-26	T	S:Tetrahydrofuran	65Cb	b,q:±1.2
	14.6	4.55(25°,1)	136.4	-70-25	T	S:Dimethoxyethane	65Cb	b,q:±1.2
T30 1,2,3,4-TETRAZOLE, 5-trifluoromethyl			$C_2HN_4F_3$			F_3C-C=N N=N NH		
PdA_2	7.1	0.81(32°)	39	22-153	T	S:$CDCl_3$ R:cis-A_2PdL_2= trans-A_2PdL_2 A=$(C_6H_5)_2PCH_3$	74Rc	b
	15.1	-0.65(32°)	51	22-153	T	S:$C_6H_5NO_2$ R:cis-A_2PdL_2= trans-A_2PdL_2 A=$(C_6H_5)_2PCH_3$	74Rc	
	20.1	-0.87(32°)	49.4	22-153	T	S:CD_3NO_2 R:cis-A_2PdL_2= trans-A_2PdL_2 A=$(C_6H_5)_2PCH_3$	74Rc	
	19.2	-0.63(32°)	50.6	22-153	T	S:$C_6H_5NO_2$ R:cis-A_2PdL_2= trans-A_2PdL_2 A=$(CH_3)_2PC_6H_5$	74Rc	

Metal Ion	ΔH, kJ /mol	Log K	ΔS, J /mol K	T, °C	M	Conditions	Ref.	Remarks
T30, continued								
PdA_2, cont.	27.2	0.82(32°)	68.2	22-153	T	S:CD_3NO_2 R:cis-A_2PdL_2=trans-A_2PdL_2 A=$(CH_3)_2PC_6H_5$	74Rc	
T31 THIA-9-CROWN-3			$C_6H_{12}O_2S$			$CH_2S(CH_2CH_2O)_2CH_2$		
Hg^{2+}	-57.62	(1-2)	--	25	C	μ=0 S:70w%Methanol	78Ib	
T32 THIA-12-CROWN-4			$C_8H_{16}O_3S$			$CH_2S(CH_2CH_2O)_3CH_2$		
Ag^+	-42.80	2.71(1)	-91.6	25	C	μ=0	78Ib	
	-43.47	(1)	--	25	C	μ=0 S:70w%Methanol	78Ib	
	-44.10	3.68(2)	-77.5	25	C	μ=0 S:70w%Methanol	78Ib	
Hg^{2+}	-50.21	(1)	--	25	C	μ=0 S:70w%Methanol	78Ib	
	-52.05	(2)	--	25	C	μ=0 S:70w%Methanol	78Ib	
Pb^{2+}	-24.52	0.94(1)	-64.3	25	C	μ=0	78Ib	
T33 THIA-15-CROWN-5			$C_{10}H_{20}O_4S$			$CH_2S(CH_2CH_2O)_4CH_2$		
Ag^+	-39.20	5.0(1)	-35.1	25	C	μ=0	78Ib	
	-14.52	2.45(2)	-1.8	25	C	μ=0	78Ib	
Hg^{2+}	-70.58	(1-2)	--	25	C	μ=0	78Ib	
Tl^+	-32.22	0.80(1)	-27.6	25	C	μ=0	78Ib	
Pb^{2+}	-21.51	1.65(1)	-40.6	25	C	μ=0	78Ib	
T34 THIA-18-CROWN-6, diester			$C_{12}H_{20}O_7S$			$CH_2SCH_2C(=O)O(CH_2CH_2O)_4C(=O)$		
Ag^+	-29.16	3.05(1)	-39.43	25	C	S:Methanol	80Lb	
T35 THIA-21-CROWN-7, diester			$C_{14}H_{24}O_8S$			$CH_2SCH_2C(=O)O(CH_2CH_2O)_5C(=O)$		
K^+	-16.07	2.09(1)	-13.89	25	C	S:Methanol	80Lb	
Rb^+	-22.97	2.52(1)	-28.77	25	C	S:Methanol	80Lb	
Ca^+	-12.68	1.91(1)	-5.89	25	C	S:Methanol	80Lb	
T36 THIAZOLE, 2-amino			$C_3H_4N_2S$			SCH:CHN:C(NH_2)		
Mn^{2+}	-58.2	1.57(25°,1)	-164.8	25-40	T	μ=0.1(HNO_3)	78Ba	b,n
	45.2	3.77(25°,2)	223.8	25-40	T	μ=0.1(HNO_3)	78Ba	b,n
Co^{2+}	-41.8	1.99(25°,1)	-102.1	25-40	T	μ=0.1(HNO_3)	78Ba	b,n
	28.0	2.99(25°,2)	151.5	25-40	T	μ=0.1(HNO_3)	78Ba	b,n

Metal Ion	ΔH, kJ /mol	Log K	ΔS, J /mol K	T,°C	M	Conditions	Ref.	Re-marks
T36, continued								
Ni^{2+}	22.6	2.19(25°,1)	33.5	2540	T	μ=0.1(HNO_3)	78Ba	b,n
	8.8	2.67(25°,2)	21.8	2540	T	μ=0.1(HNO_3)	78Ba	b,n
Cu^{2+}	18.4	2.42(25°,1)	15.1	2540	T	μ=0.1(HNO_3)	78Ba	b,n
	7.1	2.68(25°,2)	27.6	2540	T	μ=0.1(HNO_3)	78Ba	b,n
Zn^{2+}	15.1	1.76(25°,1)	16.7	2540	T	μ=0.1(HNO_3)	78Ba	b,n
	35.1	2.42(25°,2)	71.1	2540	T	μ=0.1(HNO_3)	78Ba	b,n
Cd^{2+}	-7.9	1.56(25°,1)	3.3	2540	T	μ=0.1(HNO_3)	78Ba	b,n
	17.2	2.63(25°,2)	6.7	2540	T	μ=0.1(HNO_3)	78Ba	b,n

T37 THIAZOLE, 2-amino-4 (p-methoxyphenethyl) $C_{12}H_{14}ON_2S$

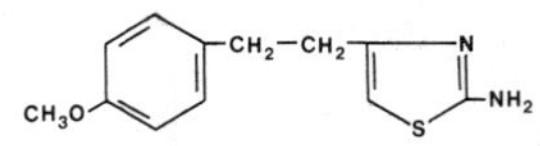

Metal Ion	ΔH, kJ /mol	Log K	ΔS, J /mol K	T,°C	M	Conditions	Ref.	Re-marks
Cu^{2+}	24.27	2.67(25°,1)	30.96	2540	T	μ=0.1(KNO_3) S:70v%Dioxane	79Db	
	17.57	2.54(25°,2)	10.88	2540	T	μ=0.1(KNO_3) S:70v%Dioxane	79Db	
Ni^{2+}	-27.20	2.49(25°,1)						
	-43.51	25-40	T	μ=0.1(KNO_3)		S:70v%Dioxane	79Db	
	-13.39	2.52(25°,2)	1.67	25-40	T	μ=0.1(KNO_3) S:70v%Dioxane	79Db	

T38 2-THIAZOLIDINETHIONE $C_3H_5NS_2$

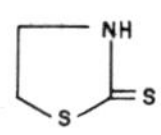

Metal Ion	ΔH, kJ /mol	Log K	ΔS, J /mol K	T,°C	M	Conditions	Ref.	Re-marks
I_2	41.8	3.164(25°,1)	79.9	0-45	T	S:Chloroform	66Bc	b,q:±4

T39 THIOCYANIC ACID CHNS HSCN L^-

Metal Ion	ΔH, kJ /mol	Log K	ΔS, J /mol K	T,°C	M	Conditions	Ref.	Re-marks
Nd^{3+}	-22.89	0.81(1)	-61.5	25	T	μ=0.1($NaClO_4$)	65Ce	
Eu^{3+}	-3.5	0.134(30°,1)	-8.8	15-45	T	μ=1M(NH_4ClO_4, NH_4SCN)	74Kc	q:±0.8
	3.3	0.428(25°,1)	19.2	10-55	T	μ=5.0(ClO_4+SCN)	74Ke	b
	-3.3	0.152(30°,2)	-8.4	15-45	T	μ=1M(NH_4ClO_4, NH_4SCN)	74Kc	q:±1.3
Tm^{3+}	-1.55	0.07(1)	-3.8	30	T	μ=1.0(NH_4ClO_4)	80Ka	
	-4.89	0.35(2)	-9.6	30	T	μ=1.0(NH_4ClO_4)	80Ka	
V^{2+}	-21.8	1.431(25°,1)	-46	11-45	T	μ=0.84($LiClO_4$)	68Ma	b,q:±2.9
V^{3+}	-14.6	2.068(25°,1)	-9.2	5-37	T	μ=1.0($HClO_4$)	67Ba	b,q:±1.2
	-15.1	2.0(25°,1)	-12.1	25-50	T	μ=2.6	51Fa	b
VO^{2+}	1.7	0.92(25°,1)	23.0	25-50	T	μ=2.6	51Fa	b
$CrA_2B_2^-$	5.0	0.653(25°,1)	30.5	25-55	T	μ=1.00F($NaNO_3$) A=C_2O_4 B=H_2O R:cis-$CrA_2B_2^-$+L= cis-CrA_2LB^{2-}+B	72Ag	b

Metal Ion	ΔH, kJ /mol	Log K	ΔS, J /mol K	T,°C	M	Conditions	Ref.	Remarks
T39, continued								
CrA_6^{3+}	-8.91	3.09	28.9	25	T	μ=0 R:$CrA_6^{3+}+L^- = CrA_5L^{2+}+A$ $A=H_2O$	55Pa 57Pb	
	11.55	3.09	90.8	95	T	μ=0 R:$CrA_6^{3+}+L^- = CrLA_5^{2+}+A$ $A=H_2O$	55Pa	
	-5.8	1.87	3.3	25	T	1M($NaClO_4$) R:$CrA_6^{3+}+L^- = CrLA_5^{2+}+A$ $A=H_2O$	54Pb	
CrA_5B	-21	2.233	-25	45	T	μ=0.106M $A=NH_3$ $B=H_2O$ R:$CrA_5B^{3+}+L^-=CrA_5L^{2+}+B$	67Df	
U^{4+}	-23.8	1.9(1)	-42	25	T	μ=2.0	55Da	
	-15.9	2.86(1-2)	-1.2	25	T	μ=2.0	55Da	
UO_2^{2+}	-3.222	0.748(1)	3.51	25	C	μ=1.0($NaClO_4$)	71Ac	
	4.81	1.0(25°,1)	35.6	20-40	T	μ=0 S:20w%CH_3OH	57Ba	b
	-0.569	-0.032(2)	-19.79	25	C	μ=1.0($NaClO_4$)	71Ac	
	0.289	0.460(3)	18.0	25	C	μ=1.0($NaClO_4$)	71Ac	
Pu^{3+}	0.25	0.338(30°,1)	7.1	15-45	T	μ=1M(NH_4ClO_4, NH_4SCN)	74Kc	q:±0.75
Am^{3+}	-18.24	0.51(1)	-51.5	25	T	μ=0.1($NaClO_4$)	65Ce	
	6.7	0.34(25°,1)	29	17-45	T	μ=1.0($NaClO_4$)	72Hd	b,q:±2.9
	2.5	0.167(30°,1)	10.5	15-45	T	μ=1M(NH_4ClO_4), NH_4SCN)	74Kc	q:±2.3
	11.8	0.596(25°,1)	50.2	10-55	T	μ=5.0(ClO_4+SCN)	74Ke	b,q:±2
	-7.4	0.455(30°,2)	14.6	15-45	T	μ=1M(NH_4ClO_4, NH_4SCN)	74Kc	q:±3.7
	-25	0.14(25°,1-3)	-85	17-45	T	μ=1.0($NaClO_4$)	72Hd	b,q:±16.7
	-33	-020(25°)	-113	17-45	T	μ=1.0($NaClO_4$) R:$AmL^{2+}+2L^-=AmL_3$	72Hd	b,q:±20.9
Cm^{3+}	-2.80	0.23(1)	-5.0	30	T	μ=1.0(NH_4ClO_4)	80Ka	
	11.8	0.616(25°,1)	50.6	10-55	T	μ=5.0(ClO_4+SCN)	74Ke	b,q:±2.2
	0.08	0.26(2)	5.4	30	T	μ=1.0(NH_4ClO_4)	80Ka	
Bk^{3+}	2.8	0.858(25°,1)	25.9	10-55	T	μ=5.0(ClO_4+SCN)	74Ke	b,q:±1.8
Cf^{3+}	-8.62	0.18(1)	-25.1	30	T	μ=1.0(NH_4ClO_4)	80Ka	
	2.97	0.58(2)	20.9	30	T	μ=1.0(NH_4ClO_4)	80Ka	
Mn^{2+}	-3.85	1.23(1)	10.5	25	C	μ=0	67Na	
	-13.76	1.15(25°,1)	-24.3	25-45	T	μ=0	71Dd	b
Fe^{3+}	-6.3	3.029(1)	36.4	25	T	μ=0	56La 55La	
	-2.18	3.1(1)	50	25	T	μ=0	61Ya 60Gb	
	-6.7	2.937(1)	33	25	T	μ=0	53Bb	
	-9.2	2.17(1)	8	25	T	0.1M(Li^+,H^+,NO_3^-, NO_3^-,SCN^-) at $[H^+]$=0.1M	65Mb	
	-6.3	2.143(1)	20.1	25	T	μ=0.5	56La	
	-10.0	2.02(1)	4	25	T	0.6M(Li^+,H^+, NO_3^-,SCN^-) at $[H^+]$=0.1M	65Mb	

Metal Ion	ΔH, kJ /mol	Log K	ΔS, J /mol K	T,°C	M	Conditions	Ref.	Remarks
T39, continued								
Fe^{3+}, cont.	-6.3	1.91(1)	17	25	T	1.1M(Li^+,H^+, NO_3^-,SCN^-) at $[H^+]$=0.1M	65Mb	
	-6.7	2.057(1)	17.2	25	T	μ=1.28	53Bb	
	-10.9	2.08(1)	4	25	T	3.1M(Li^+,H^+, NO_3^-,SCN^-) at $[H^+]$=0.1M	65Mb	
	-9.2	2.26(1)	12	25	T	6.1M(Li^+,H^+, NO_3^-,SCN^-) at $[H^+]$=0.1M	65Mb	
	-4.39	3.00(1)	42.7	25	C	μ=0.50(HNO_3) S:70v%C_2H_5OH	76Ve	q:±0.33
	-0.50	2.66(1)	49.4	25	C	μ=0.50(HNO_3) S:70v%$(CH_3)_2CHOH$	76Ve	q:±0.5
	-3.43	3.23(1)	50.4	25	C	μ=0.50(HNO_3) S:70v%CH_3OH	76Ve	q:±0.8
	-1.2	1.31(2)	20.9	25	T	μ=0.5	56La	
	-20.9	3.60(1-2)	-1.2	25	T	0.61M(Li^+,H^+, NO_3^-,SCN^-) at $[H^+]$=0.1M	65Mb	
	-21.8	2.99(1-2)	-12	25	T	1.1M(Li^+,H^+, NO_3^-,SCN^-) at $[H^+]$=0.1M	65Mb	
	-24.3	3.93(1-2)	-8	25	T	6.1M(Li^+,H^+, NO_3^-,SCN^-) at $[H^+]$=0.1M	65Mb	
FeA_5^{2-}	-36.8	2.41(25°,1)	-78.2	18.3-34.6	T	μ=1.00M A=CN	72Ef	b,q: ±2.5
[Fe]	-49.0	3.37(1)	-100.0	--	-	Chironmous hemoglobin	60Gb	
	-36.8	2.20(1)	-82.4	--	-	Hemoglobin	60Gb	
$[Fe^{3+}]$	-60.2	3.602	-136.0	25	C	[Fe]=G. gouldii Hemerythrin=$HrFe_2$ Tris-cacodylate buffer (≅0.1M Tris) R:$HrFe_2(H_2O)_m$+L= $HrFe_2L$+mH_2O	71Lb	g:Na
$[Fe^{3+}]$	-20.88	2.83(1)	16.99	20	T	6.0 Human methaemoglobin A	68Ab	d:0.05
	-26.07	2.76(1)	-35.98	20	T	6.5	68Ab	
	-31.25	2.66(1)	-55.6	20	T	7.0	68Ab	
	-26.65	2.53(1)	-42.7	20	T	7.5	68Ab	
	-16.78	2.43(1)	-10.9	20	T	8.0	68Ab	
	-10.84	2.38(1)	-8.8	20	T	8.5	68Ab	
$[Fe^{3+}]$	-17.57	2.94(1)	-3.72	20	T	6.0 Human methaemoglobin C	68Ab	d:0.05

Metal Ion	ΔH, kJ /mol	Log K	ΔS, J /mol K	T,°C	M	Conditions	Ref.	Re-marks
T39, continued								
$[Fe^{3+}]$,	-23.26	2.93(1)	-23.26	20	T	6.5	68Ab	
	-26.02	2.87(1)	-33.81	20	T	7.0	68Ab	
	-32.64	2.75(1)	-58.6	20	T	7.5	68Ab	
	-26.02	2.53(1)	-40.38	20	T	8.0	68Ab	
	-12.55	2.31(1)	1.42	20	T	8.5	68Ab	
	-6.78	2.24(1)	19.83	20	T	9.0	68Ab	
Co^{2+}	-6.82	1.72(1)	9.2	25	C	μ=0	67Na	
	-18.62	1.87(25°,1)	-26.8	25-45	T	μ=0	71Dd	b,q: ±1.9
CoA_2BC	-15	2.10(25°,1)	-10	32.7-48	T	S:DMSO R:trans-CoA_2BC+L^- =trans-CoA_2BL^-+C A=Methylglyoximato B=CH_3 C=DMSO	74Lb	b
$CoLA_5^+$	-26.4	--	--	35-56	T	S:Dimethyl sulfoxide R:$CoLA_5^++3L^-=$ $CoL_4^{2-}+5A$ A=Dimethyl sulfoxide	75Gj	y
	-28.2	--	--	41-62	T	S:Dimethyl sulfoxide R:$CoLA_5^++3L^-=$ $CoL_4^{2-}+5A$ A=Dimethyl sulfoxide	75Gj	y
	-35.4	--	--	113-144	T	S:Dimethyl sulfoxide R:$CoLA_5^++3L^-=$ $CoL_4^{2-}+5A$ A=Dimethyl sulfoxide	75Gj	y
CoA_6^{2+}	-26.8	(1-4)	--	35-56	T	S:Dimethyl sulfoxide R:$Co(A_6)^{2+}+4L^-=$ $CoL_4^{2-}+6A$ A=Dimethyl sulfoxide	75Gj	y
	-28.5	(1-4)	--	41-62	T	S:Dimethyl sulfoxide R:$Co(A_6)^{2+}+4L^-=$ $CoL_4^{2-}+6A$ A=Dimethyl sulfoxide	75Gj	y
	-35.0	(1-4)	--	113-144	T	S:Dimethyl sulfoxide R:$Co(A_6)^{2+}+4L^-=$ $CoL_4^{2-}+6A$	75Gj	y

Metal Ion	ΔH, kJ /mol	Log K	ΔS, J /mol K	T,°C	M	Conditions	Ref.	Remarks
T39, continued								
CoA_6^{2+}, cont.						A=Dimethyl sulfoxide		
Co^{3+}	-26.4	2.27(25°)	-78	25-55	T	CM=0.001-0.01M S:Chloroform R:$CoA_2B(SCN)$= $CoA_2B(NCS)$ ΔH for difference of linkage isomers S- and N-thiocyanato A=Dimethylglyoximato B=Pyridine	75Na	b
Ni^{2+}	-9.46	1.76(1)	2.1	25	C	μ=0	67Na 61Mg	
	-21.55	1.98(25°,1)	-34.3	25-45	T	μ=0	71Dd	b,q: ±2.2
	-14.4	1.24(25°,1)	-24.3	15-35	T	μ=0.62-0.68 ($HClO_4$)	68Mb	b
	-12.02	1.14(1)	-18.5	25	C	μ=1.0M($NaClO_4$)	74Kg	
	-2.5	1.44(20°,1)	4.6	20-30	T	μ=0 S:Dimethyl sulfoxide	75Ce	d
						R:$Ni(H_2O)_6^{2+}+L^-$= =$Ni(H_2O)_6L^+$ (outer sphere complex)		
	-20.9	2.06(1)	-31.4	25	C	CM=$7x10^{-3}$M CL=0-0.3M S:Methanol	75Ra	
	-8.9	0.44(2)	-21.5	25	C	μ=1.0M($NaClO_4$)	74Kg	q:±1
	13.8	1.04(2)	64.8	25	C	CM=$7x10^{-3}$M CL=0-0.3M S:Methanol	75Ra	
	-8.2	0.022(3)	-27	25	C	μ=1.0M($NaClO_4$)	74Kg	q:±1
	42.7	0.778(3)	156.9	25	C	CM=$7x10^{-3}$M CL=0-0.3M S:Methanol	75Ra	
PdA_2	-18.0	2.336(40°,1)	-12	25-55	T	μ=0.003($NaNO_3$) A=o-Phenylene-bisdimethylarsine	67Ed	b
	-17.6	2.60(1)	-9.2	25	T	μ=0 A=o-Phenylenebis-dimethylarsine	67Ec	
$PdAB^{2+}$	-19.7	2.230(27°)	-22.6	10.6-	T	μ=0.5 A=Di-ethylenetriamine B=Br R:$PdAB^++L^-$ =$PdAL^++B^-$	67Hd	b
	-18.4	2.021(27°)	-18.4	20-82	T	μ=0.5 A=1,1,7,7-Tetraethyldi-ethylenetriamine B-Br R:$PdAB^++L^-$ =$PdAL^++B^-$	67Hd	b

Metal Ion	ΔH, kJ /mol	Log K	ΔS, J /mol K	T,°C	M	Conditions	Ref.	Remarks
T39, continued								
PtA_2	-17.6	2.292(1)	-14.6	25	T	μ=0 A=o-Phenylenebis-dimethylarsine	67Ec	q:±1.7
	-18.0	2.037(40°,1)	-18.0	25-55	T	μ=3x10^{-3}($NaNO_3$) A=o-Phenylenebis-dimethylarsine	67Ec	b,q ±1.7
	-9.6	0.199(1)	-34.3	25	C	μ=1M(NaF) CM=0.040-0.050M A=Ethylenediamine	73Ma	q:±1.2
	-9.41	3.66(23°,1)	38.9	23-56	T	μ=0 S:CH_3OH A=o-Phenylenebis-dimethylarsine	66Eb	b
	-5.4	4.170(35°,1)	61.1	25-55	T	μ=0 S:CH_3OH A=o-Phenylenebis-dimethylarsine	67Ec	b,q: ±1.7
PtA_4	-12.6	0.041(1)	-41.4	25	C	μ=1M(NaF) CM-0.040-0.050M A=NH_3	73Ma	q:±2.5
Cu^{2+}	-12.55	2.33(1)	2.5	26	C	μ=0	67Na	
	-12.65	1.74(1)	-9.1	25	C	μ=1.0M($NaClO_4$)	74Kg	
	2.5	3.50(1)	100	20	C	S:Methanol	76Ra	q:±0.4
	-13.1	1.00(2)	-25	25	C	μ=1.0M($NaClO_4$)	74Kg	q:±1.5
	1.7	2.62(2)	54	20	C	S:Methanol	76Ra	q:±0.4
	-27.2	0.95(3)	-71	20	C	S:Methanol	76Ra	q:±4
CuA^{2+}	3.43	2.73(1)	63.6	25	C	S:CH_3OH A=NN'-bis(2-aminoethyl) propane-1,3-diamine	72Bc	
	-3.05	3.44(1)	55.6	25	C	S:CH_3OH A=NN'-bis(3-aminopropyl) propane-1,3-diamine	72Bc	
	2.93	3.22(1)	71.1	25	C	S:CH_3OH A=Triethylene-tetraamine	72Bc	
$CuA_2{}^{2+}$	-3.89	2.76(1)	39.7	25	C	S:CH_3OH A=NN'-Dimethyl-ethylenediamine	72Be	
	-3.47	3.93(1)	63.6	25	C	S:CH_3OH A=NN'-Dimethyl-ethylenediamine	72Be	
	7.45	2.07(1)	64.4	25	C	S:CH_3OH A=Ethylenediamine	72Bc	
	2.38	2.79(1)	61.5	25		S:CH_3OH A=N-methylethylene-diamine	72Be	
	-0.08	2.70(1)	51.0	25	C	S:CH_3OH A=Tri-methylenediamine	72Bc	
Ag^{+}	-92.4	12.00(25°,1)	-80.3	15-55	T	μ<0.06	75La	b
	-61.71	7.21(25°,1)	-72.4	5-30	T	μ=1M(KNO_3)	80Da	b
	-14.77	1.39(25°,2)	-18.8	5-30	T	μ=1M(KNO_3)	80Da	b

Metal Ion	ΔH, kJ /mol	Log K	ΔS, J /mol K	T,°C	M	Conditions	Ref.	Remarks
T39, continued								
Ag^{+},	19.71	10.71(25°,3)	91.6	5-30	T	μ=1M(KNO_3)	80Da	b,n
cont.	-46.06	-8.37(25°,4)	-142.7	5-30	T	μ=1M(KNO_3)	80Da	b,n
	-117	11.18(25°,1-4)	-180	0-45	T	0.15	49Ja 56Gb	b
AgA^{-}	-65.81	10.38(25°,1)	-19.7	5-30	T	μ=1M(KNO_3) A=Thiosulfate	80Da	b,n
	18.74	1.43(25°,2)	86.2	5-30	T	μ=1M(KNO_3) A=Thiosulfate	80Da	b
AgA_2^{3-}	-65.73	13.94(25°,1)	37.6	5-30	T	μ=1M(KNO_3) A=Thiosulfate	80Da	b,n
AgA_4^{+}	-8.2	0.49(20°,1)	∿37	-40-20	T	S:NH_3(liquid) R:$AgA_4^{+}+L^{-}$= AgA_3L+NH_3 A=NH_3	77Ga	b
Zn^{2+}	0.92	1.85(1)	38.5	25	C	μ=0	67Na	
	-11.17	1.50(25°,1)		25-45	T	μ=0	71Dd	b,q: ±2.3
	-5.77	0.708(1)	-5.8	25	C	μ=1.0($NaClO_4$)	71Ad	
	5.5	1.384(1)	45	25	C	μ=1M(NH_4ClO_4) S:DMSO	76Af	
	-1.80	0.334(2)	0	25	C	μ=1.0($NaClO_4$)	71Ad	
	23.5	1.419(2)	105	25	C	μ=1M(NH_4ClO_4) S:DMSO	76Af	
	-0.79	0.135(3)	0	25	C	μ=1.0($NaClO_4$)	71Ad	
	-17.8	2.400(3)	-13	25	C	μ=1M(NH_4ClO_4) S:DMSO	76Af	
	-7.70	0.329(4)	-18.8	25	C	μ=1.0($NaClO_4$)	71Ad	
	-10.7	1.647(4)	-4	25	C	μ=1M(NH_4ClO_4) S:DMSO	76Af	
	-23.8	2.014(20°,1-4)	-42.68	20&40	T	--	66Gb 58Gc	a,b
Cd^{2+}	-2.93	2.51(1)	38.5	25	C	μ=0	67Na 57Gb	
	-14.35	2.15(25°,1)		25-45	T	μ=0	71Dd	b,q: ±2.0
	-9.37	1.43(1)	-4.2	25	C	μ=0.25($NaClO_4$)	68Gd	
	-9.37	1.36(1)	-5.4	25	C	μ=0.5($NaClO_4$)	68Gd	
	-9.58	1.32(1)	-6.7	25	C	μ=1($NaClO_4$)	68Gd	
	-9.29	1.34(1)	-5.4	25	C	μ=2($NaClO_4$)	68Gd	
	-8.12	1.41(1)	-0.4	25	C	μ=3($NaClO_4$)	68Gd	
	-8.12	1.42(1)	0.0	25	C	3.0M($NaClO_4$)	68Gc	
	-2.97	1.806(1)	24.6	25	C	μ=1M(NH_4ClO_4) S:DMSO	76Ae	
	-11.17	0.67(2)	-25.1	25	C	μ=0.25($NaClO_4$)	68Gd	
	-10.25	0.68(2)	-20.9	25	C	μ=0.5($NaClO_4$)	68Gd	
	-8.08	0.67(2)	-14.2	25	C	μ=1($NaClO_4$)	68Gd	
	-7.74	0.70(2)	-12.6	25	C	μ=2($NaClO_4$)	68Gd	
	-7.24	0.84(2)	-8.4	25	C	μ=3($NaClO_4$)	68Gd	
	-7.24	0.83(2)	-8.4	25	C	3.0M($NaClO_4$)	68Gc	
	-2.8	0.9111(2)	8	25	C	μ=1M(NH_4ClO_4) S:DMSO	76Ae	q:±0.2
	-9.2	0.04(3)	-30.1	25	C	μ=1($NaClO_4$)	68Gd	

Metal Ion	ΔH, kJ /mol	Log K	ΔS, J /mol K	T,°C	M	Conditions	Ref.	Remarks
T39, continued								
Cd^{2+}, cont.	-9.2	0.20(3)	-27.2	25	C	μ=2($NaClO_4$)	68Gd	
	-6.57	0.23(3)	-17.6	25	C	μ=3($NaClO_4$)	68Gd	
	-4.841	0.23(3)	-17.6	25	C	3.0M($NaClO_4$)	68Gc	
	4.2	0.210(3)	18	25	C	μ=1M(NH_4ClO_4) S:DMSO	76Ae	q:±0.6
	-4.35	0.00(4)	-14.6	25	C	μ=3($NaClO_4$)	68Gd	
	-4.35	0.00(4)	-14.6	25	C	3.0M($NaClO_4$)	68Gc	
Hg^{2+}	-49.70	9.079(1)	7.1	25	C	μ=1.0($NaClO_4$)	71Ad	
	-51.2	9.08(1)	2	25	C	μ=1M($NaClO_4$)	75Cd	
	-50.42	7.778(2)	-20.1	25	C	μ=1.0($NaClO_4$)	71Ad	
	-51.2	7.78(2)	-23	25	C	μ=1M($NaClO_4$)	75Cd	
	-20.42	2.842(3)	-13.8	25	C	μ=1.0($NaClO_4$)	71Ad	
	-18.0	2.80(3)	-7	25	C	μ=1M($NaClO_4$)	75Cd	
	-21.00	1.973(4)	-33.0	25	C	μ=1.0($NaClO_4$)	71Ad	
	-18.0	2.0(4)	-22	25	C	μ=1M($NaClO_4$)	75Cd	
	-126	17.3(1-2)	-84	25	T	μ=0	62Tb 66Gb	
	-121	19.9(1-3)	-25	25	T	μ=0	62Tb	
	-151	21.7(1-4)	-84	25	T	μ=0	62Tb	
	-146	21.3(1-4)	-84	25	T	μ=0	56Ta	
	-146.4	21.23(25°,1-4)	-85.8	20-30	T	μ≅0.3	56Ta	b
	-145.3	21.67(1-4)	-72	25	C	μ=1.0M($NaClO_4$)	74Ki	
HgA^+	-46.9	6.039(1)	-44.4	20	C	μ=0.1(KNO_3) A=CH_3	65Sa	
In^{3+}	-6.94	2.56(1)	25.1	25	C	μ=2.00M($NaClO_4$)	69Rd	
	-15.9	0.98(2)	-35.1	25	C	μ=2.00M($NaClO_4$)	69Rd	
	10.0	1.06(3)	53.1	25	C	μ=2.00M($NaClO_4$)	69Rd	
Tl^+	-12.38	0.80(1)	-26.8	25	T	μ=0	53Ba	
	-12.09	0.556(25°,1)	-30.1	10-60	T	μ=0($LiClO_4$)	72Fd	b,q: ±1.7
	-7.61	0.243(25°,1)	-20.5	10-60	T	μ=0.50($LiClO_4$)	72Fd	b,q: ±2.3
	-9.41	0.152(25°,1)	-29.7	10-60	T	μ=1.0($LiClO_4$)	72Fd	b,q: ±1.7
	-11.38	0.117(25°,1)	-36.8	10-60	T	μ=2.0($LiClO_4$)	72Fd	b,q: ±2.0
	-5.36	0.083(25°,1)	-16.3	10-60	T	μ=3.0($LiClO_4$)	72Fd	b,q: ±2.6
	-11.30	0.130(25°,1)	-36.0	10-60	T	μ=4.0($LiClO_4$)	72Fd	b,q: ±4.2
	0.88	0.255(25°)	9.2	10-60	T	μ=0 R:Tl^++Cl^-+L=TlClL	72Fd	b,q: ±3.8
	9.2	-0.155(25°)	27.6	10-60	T	μ=0.50($LiClO_4$) R:Tl^++Cl^-+L=TlClL	72Fd	b,q: ±2.9
	0.04	-0.022(25°)	2.1	10-60	T	μ=1.0($LiClO_4$) R:Tl^++Cl^-+L=TlClL	72Fd	b,q: ±10.0
	0.33	-0.180(25°)	-1.2	10-60	T	μ=2.0($LiClO_4$) R:Tl^++Cl^-+L=TlClL	72Fd	b,q: ±2.8
	-8.54	-0.201(25°)	-33.0	10-60	T	μ=3.0($LiClO_4$) R:Tl^++Cl^-+L=TlClL	72Fd	b,q: ±4.1
	-3.8	-0.222(25°)	-16.7	10-60	T	μ=4.0($LiCl_4$) R:Tl^++Cl^-+L=TlClL	72Fd	b,q: ±4.2

Metal Ion	ΔH, kJ /mol	Log K	ΔS, J /mol K	T, °C	M	Conditions	Ref.	Remarks
T39, continued								
Pb^{2+}	1.26	1.09(1)	25.1	25	C	μ=0	67Na 56Gc	
	-17.2	0.418(25°,1)	-50	15-65	T	μ=3M($LiClO_4$)	69Fb	b,q:±6.3
	4.6	0.293(25°,2)	21	15-65	T	μ=3M($LiClO_4$)	69Fb	b,q: ±10.0
	-3.8	0.205(25°,3)	-8	15-65	T	μ=3M($LiClO_4$)	69Fb	b,q:±5.8
	2.1	-0.337(25°,4)	0	15-65	T	μ=3M($LiClO_4$)	69Fb	b,q:±7.5
	-25.9	0.396(25°,5)	-75	15-65	T	μ=3M($LiClO_4$)	69Fb	b,q:±7.1
Bi^{3+}	33.9	1.28(25°,1)	142	15-55	T	μ=3M	71Fa	b
	7.9	1.39(25°,2)	63	15-55	T	μ=3M	71Fa	b,q:±4
	4.2	1.07(25°,3)	29	15-55	T	μ=3M	71Fa	b,q:±8
	-17	1.46(25°,4)	-42	15-55	T	μ=3M	71Fa	b,q:±17
I_2	-33.9	1.94(1)	-77.0	7.6- 16.5	T	μ=1-5 ΔH & ΔG were extrapolated to 25°C	62Lb	b
T40	THIOCYANIC ACID, allyl ester		C_4H_5NS			CH_2:$CHCH_2SCN$		L
I_2	-15.1	0.015(20°,1)	-50	9-43.6	T	S:CCl_4 CM=1.2x10^{-3}M	65Pc	b,q: ±3.3
T41	THIOCYANIC ACID, ethyl ester		C_3H_5NS			C_2H_5SCN		L
I_2	-16.3	0.045(20°,1)	-54	10.5- 44.2	T	S:Heptane CM=8.0x10^{-4}M	65Pc	b,q: ±3.3
T42	THIOCYANIC ACID, methyl ester		C_2H_3NS			CH_3SCN		L
I_2	9.2	0.176(25°,1)	34.3	25-41.7	T	S:CCl_4 CM=0.5x 10^{-2}M CL=0.35M	65Wb	b
T43	THIODIACETIC ACID		$C_2H_6O_4S$			$S(CH_2COOH)_2$		
Al^{3+}	55.94	6.795(25°,1-2)	319.6	25-45	T	μ=0	79Dd	
T44	THIODIGLYCOLIC ACID		$C_4H_6O_4S$			$S(CH_2CO_2H)_2$		$L^{=}$
Ce^{3+}	12.80	2.654(1)	93.7	25	C	μ=1.00M($NaClO_4$)	73Dd	
	7.49	1.832(2)	60.2	25	C	μ=1.00M($NaClO_4$)	73Dd	
	8.49	5.359	131.0	25	C	μ=1.00M($NaClO_4$) R:$M^{3+}+L^{2-}+H^{+}=MHL^{2+}$	73Dd	q:±1.2
	7.20	2.295	69.0	25	C	μ=1.00M($NaClO_4$) R:$MHL^{2+}+L^{2-}=MHL_2$	73Dd	q:±2.5
Ni^{2+}	1.80	3.7(1)	77.0	25	C	μ=1($NaNO_3$)	78Ah	
Cu^{2+}	1.21	4.0(1)	80.8	25	C	μ=1($NaNO_3$)	78Ah	
Ag^{+}	-31.6	3.52(1)	-38	25	C	μ=0.5M($NaClO_4$)	75Mb	
	-34.9	2.30(2)	-73	25	C	μ=0.5M($NaClO_4$)	75Mb	
	-8.5	0.86(3)	-12	25	C	μ=0.5M($NaClO_4$)	75Mb	q:±7

Metal Ion	ΔH, kJ /mol	Log K	ΔS, J /mol K	T,°C	M	Conditions	Ref.	Remarks
T44, continued								
Zn^{2+}	6.32	2.6(1)	71.1	25	C	μ=1($NaNO_3$)	78Ah	
Cd^{2+}	6.69	2.3(1)	66.5	25	C	μ=1($NaNO_3$)	78Ah	
Hg^{+}	-33.3	6.38(1)	10	25	C	μ=0.5M($NaClO_4$)	75Mb	
	-36.3	4.10(2)	-43	25	C	μ=0.5M($NaClO_4$)	75Mb	
	7.6	1.02(3)	46	25	C	μ=0.5M($NaClO_4$)	75Mb	q:±19
	-8	2.61(4)	20	25	C	μ=0.5M($NaClO_4$)	75Mb	
T45 3,3'-THIODIPROPIONIC ACID			$C_6H_{10}O_4S$			$S(CH_2CH_2CO_2H)_2$		L
Ni^{2+}	2.1	1.3(1)	32.2	25	C	μ=1.0($NaNO_3$)	79Ad	q:±1.2
Cu^{2+}	11.7	2.7(1)	91.2	25	C	μ=1.0($NaNO_3$)	79Ad	q:±2.9
Zn^{2+}	6.7	1.3(1)	47.7	25	C	μ=1.0($NaNO_3$)	79Ad	q:±2
Cd^{2+}	6.3	1.7(1)	53.6	25	C	μ=1.0($NaNO_3$)	79Ad	q:±2
Al^{3+}	33.30	7.199(25°,1-2)	248.6	25-45	T	μ=0	79Dd	b
T46 THIOPHENE			C_4H_4S			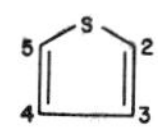		L
Al_2A_6	-77.0	--	--	25	C	S:Benzene CM=0.07M A=Br R:0.5(Al_2A_6)+L= AlA_3L	68Rc	
I_2	-1.7	(1)	--	25	C	S:Octane	66Ad	
T47 THIOPHENE, 2,5-dimethyl-			C_6H_8S			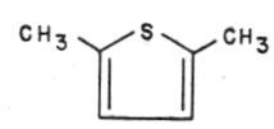		L
Al_2A_6	-79.1	--	--	25	C	S:Benzene CM=0.07M A=Br R:0.5(Al_2A_6)+L= AlA_3L	68Rc	
T48 THIOPHENE, tetrahydro-			C_4H_8S			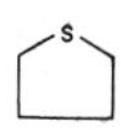		L
TiA_4	-50.2	(1)	--	25	C	S:Benzene CM=0.05-0.08M A=Cl	68Ge	
CoA	-25.1	(1)	--	-46- -23	T	S:Toluene A=Protoporphyrin IX dimethyl ester	78Dd	
CoAB	-18.0	1.23(1)	-36.8	25	T	S:o-Dichloro- benzene A=CF_3 B=N,N'-bis	77Hc	

Metal Ion	ΔH, kJ /mol	Log K	ΔS, J /mol K	T, °C	M	Conditions	Ref.	Remarks
T48, continued								
CoAB, cont.						(acetylacetonato) ethylenediimine		
[Co]	-64.0	--	--	25	C	[Co]=Methyl cobaloxime S:Dichloromethane R:$[Co]_2$+2L=2[Co]L	73Cf	
NiA	-12.0	0.431(1)	-31	30	C	S:Benzene CM=$0.5x10^{-3}$M A=BF_2-bridged glyoximato R_1=Me R_2=Pr	76Gh	
RhAB	-48.1	1.380	-135	24	C	S:Benzene R:$[RhAB]_2$+2L= 2RhABL A=Cl B=1,5-Cycloocta-diene	76Lb	
$Rh_2A_4B_2$	-31.8	--	--	--	C	S:Benzene R:$Rh_2A_4B_2$+2L= $2RhA_2BL$ A=CO B=Cl	76Pd	u:24°
PdAB	-36.4	2.919	-66	24	C	S:Benzene R:$[PdAB]_2$+2L= 2[PdABL] A=(π-MeAllyl) B=Cl	77Lb	
CuA_2	-33.0	2.26(1)	--	--	C	S:CCl_4 A=Hexafluoro-acetylacetonato	76Mc	u:25°
	-22.2	1.90(1)	--	--	C	S:CH_2Cl_2 A=Hexafluoro-acetylacetonato	76Mc	u:25°
	-19.2	2.00(1)	--	--	C	S:o-$Cl_2C_6H_4$ A=Hexafluoro-acetylacetonato	76Mc	u:25°
	-19.7	(1)	--	25	C	S:Methylene chloride A=Hexafluoro-acetylacetonate	74Da	
ZnA	-28.0	1.975(25°,1)	-56.1	15-35	T	S:Cyclohexane A=α,β,γ,δ-Tetra-phenylporphine	77Vb	

Metal Ion	ΔH, kJ /mol	Log K	ΔS, J /mol K	T,°C	M	Conditions	Ref.	Remarks
T48, continued								
HgA_2	-66.5	1.699(1)	-187.9	30	C	S:Benzene A=Cl CM=1-5x10^{-3}M	73Fa	
	-68.2	1.826(1)	-190.0	30	C	S:Benzene A=Br CM=1-5x10^{-3}M	73Fa	
	-58.2	1.580(1)	-161.9	30	C	S:Benzene A=I CM=1-5x10^{-3}M	73Fa	
AlA_3	-70.92	(1)	--	28.5	C	S:Hexane A=CH_3	67Hc	
I_2	-41.8	(1)	--	25	C	S:Octane	66Ad	
T49 2-THIOPHENECARBOXYLIC ACID			$C_5H_4O_2S$			CH:CHSC(CO_2H):CH		
Be^{2+}	-4.56	4.21(30°,1-2)	65.7	30-50	T	μ=0.2M(HNO_3+ KNO_3)	76Sd	b
Mg^{2+}	72.9	1.95(30°,1)	271	30-50	T	μ=0.2M($NaClO_4$)	76Sf	
Ca^{2+}	18.2	2.01(30°,1)	98.2	30-50	T	μ=0.2M($NaClO_4$)	76Sf	
Sr^{2+}	31.0	2.01(30°,1)	139	30-50	T	μ=0.2M($NaClO_4$)	76Sf	
Ba^{2+}	41.0	2.06(30°,1)	171	30-50	T	μ=0.2M($NaClO_4$)	76Sf	
UO_2^{2+}	-7.74	2.20(30°,1)	15.9	30-50	T	μ=0.2M(HNO_3+ KNO_3)	76Sd	b
Mn^{2+}	23.3	1.98(30°,1)	116	30-50	T	μ=0.2M($NaClO_4$)	76Sd	b
Co^{2+}	15.5	2.05(30°,1)	91.0	30-50	T	μ=0.2M($NaClO_4$)	76Sg	
Ni^{2+}	9.1	2.10(30°,1)	70.4	30-50	T	μ=0.2M($NaClO_4$)	76Sg	
	0.13	2.04(1)	39.5	25	C	μ=1.0($NaNO_3$)	79Ac	q:±0.33
Cu^{2+}	10.0	2.15(30°,1)	73.5	30-50	T	μ=0.2M($NaClO_4$)	76Sg	
	1.42	2.09(1)	44.8	25	C	μ=1.0($NaNO_3$)	79Ac	
Zn^{2+}	7.28	2.08(30°,1)	63.6	30-50	T	μ=0.2M($NaClO_4$)	76Sd	b
	0.50	2.02(1)	40.3	25	C	μ=1.0($NaNO_3$)	79Ac	
Cd^{2+}	3.65	2.10(30°,1)	52.3	30-50	T	μ=0.2M($NaClO_4$)	76Sd	b
	0.38	2.04(1)	40.3	25	C	μ=1.0($NaNO_3$)	79Ac	q:±0.54
Pb^{2+}	13.7	2.03(30°,1)	84.6	30-50	T	μ=0.2M($NaClO_4$)	76Sf	
T50 THIOSULFATE ION			O_3S_2			$S_2O_3^{2-}$		L^{2-}
Ba^{2+}	10.9	2.21(25°,1)	353.4	25&35	T	μ=0	49Da 52Ua	a,b
CrA_3	7.28	0.543(1)	35.1	25	C	μ=3.0($NaClO_4$) A=Ethylenediamine	73Ua	q:±0.25
	3.30	0.145(2)	13.8	25	C	μ=3.0($NaClO_4$) A=Ethylenediamine	73Ua	q:±0.25
	1.21	0.042(3)	5.0	25	C	μ=3.0($NaClO_4$) A=Ethylenediamine	73Ua	q:±0.50
Fe^{3+}	37.2	2.10(25°,1)	197.5	1-30	T	μ=0	54Pa	b
	41.0	2.10(25°,1)	177.8	1-30	T	μ=0.47	54Pa	b
CoA_3^{3+}	-0.21	3.37(1)	57.3	25	C	μ=0	76Me	q:±0.4

Metal Ion	ΔH, kJ /mol	Log K	ΔS, J /mol K	T,°C	M	Conditions	Ref.	Re- marks
T50, continued								
CoA_3^{3+},						A=Ethylenediamine		
cont.	-0.25	2.12(1)	39.7	25	C	μ=0.1($NaClO_4$) A=Ethylenediamine	76Me	
	-0.21	1.83(1)	34.7	25	C	μ=0.2($NaClO_4$) A=Ethylenediamine	76Me	q:±0.04
	-0.21	1.61(1)	31.0	25	C	μ=0.3($NaClO_4$) A=Ethylenediamine	76Me	q:±0.04
	-0.21	1.43(1)	26.8	25	C	μ=0.4($NaClO_4$) A=Ethylenediamine	76Me	q:±0.04
	0.17	1.36(1)	25.5	25	C	μ=0.5($NaClO_4$) A=Ethylenediamine	76Me	q:±0.04
	3.3	0.72(1)	25.1	25	C	μ=3.0M($NaClO_4$+ NaL) A=Ethylenediamine	73Mb	q:±0.4
	-0.84	1.83(2)	32.2	25	C	μ=0 A=Ethylenediamine	76Me	q:±0.4
	-0.29	1.17(2)	20.9	25	C	μ=0.1($NaClO_4$) A=Ethylenediamine	76Me	q:±0.12
	-0.29	1.17(2)	20.9	25	C	μ=0.2($NaClO_4$) A=Ethylenediamine	76Me	q:±0.12
	-0.42	1.17(2)	20.5	25	C	μ=0.3($NaClO_4$) A=Ethylenediamine	76Me	q:±0.2
	-0.42	1.10(2)	18.8	25	C	μ=0.4($NaClO_4$) A=Ethylenediamine	76Me	q:±0.2
	-0.62	0.84(2)	13.8	25	C	μ=0.5($NaClO_4$) A=Ethylenediamine	76Me	q:±0.2
	1.7	0.23(2)	10.5	25	C	μ=3.0M($NaClO_4$+ NaL) A=Ethylenediamine	73Mb	q:±0.8
	0.84	0.51(3)	12.6	25	C	μ=0 A=Ethylenediamine	76Me	q:±0.8
	0.84	0.59(3)	12.6	25	C	μ=0.4($NaClO_4$) A=Ethylenediamine	76Me	q:±0.4
	0.84	0.073(3)	4.2	25	C	μ=0.5($NaClO_4$) A=Ethylenediamine	76Me	q:±0.4
	0.8	0.00(3)	4	25	C	μ=3.0M($NaClO_4$+ NaL) A=Ethylenediamine	73Mb	q:±0.4
	-5.8	-0.30(4)	-25	25	C	μ=3.0M($NaClO_4$+ NaL) A=Ethylenediamine	73Mb	q:±0.8
CoA_5B^{2+}	1.9	0.63(1)	5.4	25	C	μ=3.0M($NaClO_4$) A=NH_3 B=NO_2	74Mb	
	2.8	0.19(2)	-5.7	25	C	μ=3.0M($NaClO_4$) A=NH_3 B=NO_2	74Mb	
	1.5	0.24(3)	-0.5	25	C	μ=3.0M($NaClO_4$) A=NH_3 B=NO_2	74Mb	
	1.7	0.029(4)	-5.8	25	C	μ=3.0M($NaClO_4$) A=NH_3 B=NO_2	74Mb	
CoA_6^{3+}	4.6	3.2(25°,1)	75	15-35	T	μ=0 A=NH_3	55Gb	b
PdA_2	-11.7	1.724(40°,1)	4	25-55	T	μ=0.02($NaNO_3$)	67Ed	b

Metal Ion	ΔH, kJ /mol	Log K	ΔS, J /mol K	T,°C	M	Conditions	Ref.	Remarks
T50, continued								
PdA_2, cont.						A=o-Phenylenebis-dimethylarsine		
	-10.9	2.305(1)	10.5	25	T	μ=0	67Ed	
						A=o=Phenylenebis-dimethylarsine		
	6.3	0.097(1)	23.0	25	C	3N($NaClO_4$+ $Na_2S_2O_3$)	71Bi	
						A=Ethylenediamine Values are for formation of inner-sphere complexes		
	3.3	0.230(1)	14.6	25	C	3N($NaClO_4$+ $Na_2S_2O_3$)	71Bi	
						A=Ethylenediamine Values are for formation of outer-sphere complexes		
Ag^+	-32.2	(1)	--	5-45	T	μ=0.01	31Ca	b
	-59.4	9.17(20°,1)	-27.2	10-30	T	μ=1M(KNO_3)	75Dg	b
	-59.0	9.21(25°,1)	-21.3	10-30	T	μ=1M(KNO_3)	74Bd	b,q:±5.4
	-63.30	9.81(25°,1)	-34.7	5-30	T	μ=1M(KNO_3)	80Da	b
	-72.0	9.12(20°,1)	-71.1	10-30	T	μ=1M(KNO_3) S:5v%Dioxane	75Dg	b
	-85.1	9.08(20°,1)	-117	10-30	T	μ=1M(KNO_3) S:10v%Dioxane	75Dg	b
	-54.8	8.82(20°,1)	-18.0	20-70	T	--	55Ca	b
	-5.77	2.69(25°,2)	41.0	5-30	T	μ=1M(KNO_3)	80Da	b
	-21.3	3.27(25°,2)	-10.0	10-30	T	μ=1M(KNO_3)	74Bd	b,q:±2.1
	-15.6	3.60(20°,2)	14.6	10-30	T	μ=1M(KNO_3)	75Dg	b
	8.0	3.81(20°,2)	98.3	10-30	T	μ=1M(KNO_3) S:5v%Dioxane	75Dg	b
	30.9	4.01(20°,2)	179	10-30	T	μ=1M(KNO_3) S:10v%Dioxane	75Dg	b
	-91.88	14.36(1-2)	-59.4	2	T	μ=0	65Le 59Sb	h:2-45°
	-91.88	13.25(1-2)	-54.8	25	T	μ=0	65Le	h:2-45°
	-91.88	12.68(1-2)	-55.6	35	T	μ=0	65Le	h:2-45°
	-91.88	12.21(1-2)	-55.2	45	T	μ=0	65Le	h:2-45°
	-79	12.99(25°,1-2)	-32.2	0-45	T	μ≅0.15	49Ja	b
	-79.70	13.45(1-2)	-9.6	25	T	--	57Cb	
	-73.6	13.46(20°,1-2)	6.3	20-70	T		55Ca	b
	-79.70	14.54(25°,1-2)	-9.6	25-70	T		57Ca	b
	-91.88	15.40(1-2)	-39.3	2	T	μ=0 S:50% CH_3OH	65Le	h:2-45°
	-91.88	14.19(1-2)	-36.8	25	T	μ=0 S:50% CH_3OH	65Le	h:2-45°
	-91.88	13.71(1-2)	-36.0	35	T	μ=0 S:50% CH_3OH	65Le	h:2-45°
	-91.88	13.16(1-2)	-36.8	45	T	μ=0 S:50% CH_3OH	65Le	h:2-45°
	-72.3	12.19(20°)	-10	10-30	T	μ=1M(KNO_3) R:Ag+L+A=AgLA A=2-Thiourea	75Dg	b
	-79.1	11.95(25°)	-35.6	10-30	T	μ=1M(KNO_3)	74Ba	b,q:±12

Metal Ion	ΔH, kJ /mol	Log K	ΔS, J /mol K	T,°C	M	Conditions	Ref.	Remarks
T50, continued								
Ag^+, cont.	-60.4	12.37(20°)	33	10-30	T	$R:Ag^++L^{2-}+A=AgLA^-$ A=2-Thiourea μ=1M(KNO_3) S:5v%Dioxane	75Dg	b
	-40.2	12.56(20°)	105	10-30	T	R:Ag+L+A=AgLA A=2-Thiourea μ=1M(KNO_3) S:10v%Dioxane	75Dg	b
	-90.2	14.59(20°)	-29	10-30	T	R:Ag+L+A=AgLA A=2-Thiourea μ=1M(KNO_3)	75Dg	b
	-112	14.15(25°)	-107.5	10-30	T	$R:Ag+2L+A=AgL_2A$ A=2-Thiourea μ=1M(KNO_3)	74Ba	b,q:±12
	-86.8	14.72(20°)	-17	10-30	T	$R:Ag^++2L^{2-}+A=AgL_2A^{3-}$ A=2-Thiourea μ=1M(KNO_3) S:5v%Dioxane	75Dg	b
	-82.8	14.85(20°)	0	10-30	T	$R:Ag+2L+A=AgL_2A$ A=2-Thiourea μ=1M(KNO_3) S:10v%Dioxane	75Dg	b
	-78.8	13.22(20°)	-12.6	10-30	T	$R:Ag+2L+A=AgL_2A$ A=2-Thiourea μ=1M(KNO_3)	75Dg	b
	-90.0	13.36(25°)	-40.6	10-30	T	$R:Ag+L+2A=AgLA_2$ A=2-Thiourea μ=1M(KNO_3)	74Bd	b,q:±29
	-88.5	13.62(20°)	-43.9	10-30	T	$R:Ag^++L^{2-}+2A=AgLA_2^-$ A=2-Thiourea μ=1M(KNO_3) S:5v%Dioxane	75Dg	b
	-97.6	14.03(20°)	-73.2	10-30	T	$R:Ag+L+2A=AgLA_2$ A=2-Thiourea μ=1M(KNO_3) S:10v%Dioxane	75Dg	b
	27.2	15.48(20°)	385	10-30	T	$R:Ag+L+2A=AgLA_2$ A=2-Thiourea μ=1M(KNO_3)	75Dg	b
	-62.6	15.60(20°)	85.8	10-30	T	$R:Ag+L+3A=AgLA_3$ A=2-Thiourea μ=1M(KNO_3) S:5v%Dioxane	75Dg	b
	-79.1	15.73(20°)	31.4	10-30	T	$R:Ag+L+3A=AgLA_3$ A=2-Thiourea μ=1M(KNO_3) S:10v%Dioxane $R:Ag+L+3A=AgLA_3$ A=2-Thiourea	75Dg	b
Hg^{2+}	-118	28.00(1)	141	25	C	μ=0.5M($NaClO_4$)	75Mb	

T50, continued

Metal Ion	ΔH, kJ /mol	Log K	ΔS, J /mol K	T,°C	M	Conditions	Ref.	Remarks
Hg^{2+},	-26	2.80(2)	-34	25	C	μ=0.5M($NaClO_4$)	75Mb	
cont.	-160.9	33.26(1-3)	98	25	C	μ=1.0M($NaClO_4$)	74Ki	
Zn^{2+}	13.0	3.2(25°,1)	92	15-35	T	μ=0	55Gb	b
	14.6	1.24(1)	71	25	C	μ=1	57Ya	
Cd^{2+}	5.4	3.96(25°,1)	105	15-35	T	μ=0	55Gb	b
	0	2.71(1)	50	25	C	μ=1	57Ya	
	-6.3	2.51(2)	25	25	C	μ=1	57Ya	
	-7.9	1.03(3)	-4	25	C	μ=1	57Ya	
	-6.3	5.20(1-2)	75	25	C	μ=1	57Ya	
	-14.2	6.28(1-3)	71	25	C	μ=1	57Ya	
HgA^+	-49.0	10.885	41.4	20	C	μ=0.1(KNO_3) A=CH_3	65Sa	

T51 THREONINE $C_4H_9O_3N$ $CH_3CH(OH)CH(NH_2)CO_2H$ L^-

Metal Ion	ΔH, kJ /mol	Log K	ΔS, J /mol K	T,°C	M	Conditions	Ref.	Remarks
La^{3+}	-31.8	4.05(25°,1)	-29	25&45	T	μ=0.05M	75Si	a,b
	-4.5	3.00(25°,2)	42	25&45	T	μ=0.05M	75Si	a,b
	-41.8	7.67(25°,1-2)	6.9	25&45	T	μ=0	75Si	a,b
UO_2^{2-}	-20.88	6.348(25°,1)	51.50	25&45	T	μ=0.05M(KCl)	73Sm	a,b
	-20.88	6.486(25°,2)	47.74	25&45	T	μ=0.05M(KCl)	73Sm	a,b
	-30.84	12.682(25°,1-2)	139.45	25&45	T	μ=0	73Sm	a,b
Mn^{2+}	-3.3	3.98(15°,1-2)	63	15&40	T	0.2M(KNO_3)	68Ra	a,b
Fe^{2+}	-8.4	6.62(15°,1-2)	96	15&40	T	0.2M(KNO_3)	68Ra	a,b
Co^{2+}	-4.6	4.38(1)	67	25	C	μ=0.05M(KCl)	72Gb 72Ga	
	-7.5	4.38(25°,1)	58	20-35	T	μ=0.05M(KCl)	72Gb 72Ga	b
	-8.2	3.995(1)	49	25	C	μ=0.1M($NaNO_3$) L(L-allo form)	78Ia	
	-10.8	4.258(1)	45	25	C	μ=0.1M($NaNO_3$) L(L or DL form)	78Ia	
	-18.4	3.63(2)	8	25	C	μ=0.05M(KCl)	72Gb 72Ga	
	-17.2	3.63(25°,2)	12	20-35	T	μ=0.05M(KCl)	72Gb 72Ga	b
	-4.8	3.206(2)	45	25	C	μ=0.1M($NaNO_3$) L(L-allo doem)	78Ia	
	-8.1	3.925(2)	48	25	C	μ=0.1M($NaNO_3$) L(L or DL form)	78Ia	
	-18.0	7.84(25°,1-2)	92	15-40	T	0.2M(KNO_3)	68Ra	b
Ni^{2+}	-13.4	5.42(1)	58	25	C	μ=0.05M(KCl)	72Gb 72Ga	
	-14.2	5.42(25°,1)	54	20-35	T	μ=0.05M(KCl)	72Gb 72Ga	b
	-15.94	5.46(1)	51.0	25	C	μ=0.16	70Lb	
	-22.2	4.53(2)	12	25	C	μ=0.05M(KCl)	72Gb 72Ga	
	-18.0	4.53(25°,2)	29	20-35	T	μ=0.05M(KCl)	72Gb 72Ga	b

Metal Ion	ΔH, kJ /mol	Log K	ΔS, J /mol K	T,°C	M	Conditions	Ref.	Re-marks
T51, continued								
Ni^{2+}, cont.	-18.24	4.51(2)	25.1	25	C	μ=0.16	70Lb	
	-21.76	3.45(3)	-6.7	25	C	μ=0.16	70Lb	
	-33.9	9.94(25°,1-2)	79	15-40	T	0.2M(KNO_3)	68Ra	b
Cu^{2+}	-22.2	8.44(20°,1)	88	25	C	μ=0	64Ia	
	-23.0	8.02(1)	75	25	C	μ=0.05M(KCl)	72Gb 72Ga 70Gc 71Gc	
	-25.5	8.02(25°,1)	67	20-35	T	μ=0.05M(KCl)	72Gb 72Ga	b
	-23.1	7.464(1)	65	25	C	μ=0.1M($NaNO_3$) L(L-allo form)	78Ia	
	-25.6	8.042(1)	68	25	C	μ=0.1M($NaNO_3$) L(L or DL form)	78Ia	
	-23.26	7.95(1)	74.0	25	C	μ=0.16	70Lb	
	-18.0	8.60(1)	104.2	25	C	μ=3.0M($NaClO_4$)	75Bd	q:±1.5
	-25.5	6.96(20°,2)	46	25	C	μ=0	64Ia	
	-25.5	6.70(2)	42	25	C	μ=0.05M(KCl)	72Gb 72Ga 70Gc 71Gc	
	-28.0	6.70(25°,2)	33	20-35	T	μ=0.05M(KCl)	72Gb 72Ga	b
	-18.8	6.483(2)	61	25	C	μ=0.1M($NaNO_3$) L(L-allo form)	78Ia	
	-20.7	6.886(2)	62	25	C	μ=0.1M($NaNO_3$) L(L or DL form)	78Ia	
	-25.48	6.74(2)	43.5	25	C	μ=0.16	70Lb	
	-29.0	7.44(2)	45.2	25	C	μ=3.0M($NaClO_4$)	75Bd	
	-52.91	14.60(1-2)	102.0	25	C	μ=0.10M(KNO_3) L=DL-Threonine	76Pc	
	-53.13	14.60(1-2)	101.3	25	C	μ=0.10M(KNO_3) L=L-Threonine	76Pc	
	-41.4	14.69(25°,1-2)	142	15-40	T	0.2M(KNO_3)	68Ra	b
	-50.0	16.476	147.6	25	C	μ=3.0M($NaClO_4$) R:$Cu^{2+}+L^-+A^-$=CuLA A=Asparagine	75Bd	
	-67.4	18.610	130.1	25	C	μ=3.0M($NaClO_4$) R:$Cu^{2+}+L^-+A^-$=CuLA A=Histidine	75Bd	
	-103.5	23.421	101.3	25	C	μ=3.0M($NaClO_4$) R:$Cu^{2+}+L^-+HA$= $CuLHA^+$ HA=Histidine	75Bd	
Zn^{2+}	-10.5	4.67(1)	54	25	C	μ=0.05M(KCl)	72Gb 72Ga	
	-12.1	4.67(25°,1)	50	20-35	T	μ=0.05M(KCl)	72Gb 72Ga	b
	-11.7	3.99(2)	38	25	C	μ=0.05M(KCl)	72Gb 72Ga	

Metal Ion	ΔH, kJ /mol	Log K	ΔS, J /mol K	T,°C	M	Conditions	Ref.	Re-marks
T51, continued								
Zn^{2+}, cont.	-10.9	3.99(25°,2)	38	20-35	T	μ=0.05M(KCl)	72Gb 72Ga	b
	-18.8	8.51(25°,1-2)	100	15-40	T	0.2M(KNO_3)	68Ra	b
T52 TOLUENE			C_7H_8			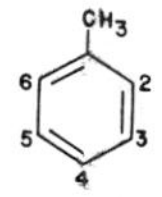		L
Ag^+	-11.38	0.076(1)	-36.69	25	T	μ=0.5 S:Equimolar H_2O-CH_3OH solution	56Oa	b
AgA	14.2	0.599(20°,1)	59.8	14&20	T	S:CH_3OH A=ClO_4^-	64Tb	a,b
I_2	-7.20	0.233(25°,1)	-19.6	25-60	T	S:CCl_4	76Ko	b
	-7.5	(1)	--	--	-	S:Cyclohexane	6CBb	
	-7.531	0.350(1)	-18.4	25	T	S:n-Hexane	54Ka	1
	-8.8	(1)	--	--	C	S:Toluene	61Aa	i:50Ha
	-6.07	(1)	--	--	C	S:Toluene	50Ha	
T53 TOLUENE, 2-amino-			C_7H_9N			$C_6H_4CH_3(NH_2)$		L
CoA_2	-14.06	3.548(1-2)	20.92	25	C	S:Acetone CM=1x10^{-3}M CL=0.2M A=Cl	71Zc	q:±2.5
T54 TOLUENE, 3-amino-			C_7H_9N			$C_6H_4CH_3(NH_3)$		
CoA_2	-40.88	4.039(1-2)	-59.75	25	C	S:Acetone CM=1x10^{-3}M CL=0.2M A=Cl	71Zc	
T55 TOLUENE, 4-amino-			C_7H_9N			$C_6H_4CH_3(NH_2)$		
CoA_2	-38.07	3.768(1-2)	-55.60	25	C	S:Acetone CM=1x10^{-3}M CL=0.2M A=Cl	71Zc	
T56 TOLUENE, 2-amino-N,N-dimethyl-			$C_9H_{13}N$			$C_6H_4CH_3[N(CH_3)_2]$		L
I_2	-9.6	0.38(17°,1)	-25.9	17&32	T	S:n-Heptane CL=0.05-0.13M CM=3x10^{-5}M	60Tc	a,b
T57 TOLUENE, 4-amino-N,N-dimethyl-			$C_9H_{13}N$			$C_6H_4CH_3[N(CH_2)_2]$		L
I_2	-34.7	1.62(21.3°,1)	-87.0	12-33	T	S:n-Heptane CL=0.005-0.03M CM=3x10^{-5}M	60Tc	b

Metal Ion	ΔH, kJ /mol	Log K	ΔS, J /mol K	T,°C	M	Conditions	Ref.	Remarks
T58	TOLUENE, 4-chloro-		C_7H_7Cl			$C_6H_4CH_3(Cl)$		L
IA	-6.99	0.55(25°,1)	-12.93	1.6-25	T	S:CCl_4 CM=10^{-4}-10^{-3}M CL=0.1-1M A=Cl	550a	b,k
	-7.32	0.55(25°,1)	-14.02	25-45.8	T	S:CCl_4 CM=10^{-4}-10^{-3}M CL=0.1-1M A=Cl	550a	b,k
T59	TOLUENE, 4-hydroxy		C_7H_8O			$CH_3C_6H_4OH$		
PA_3B	-29.9	3.158(25°,1)	-40.0	0-50	T	S:CCl_4 A=$(CH_3)_2N$ B=O	77Kf	
T60	TOLUENE, (ℓ)α-hydroxy-α-(aminomethyl)-3,4-dihydroxy-		$C_8H_{11}O_3N$			$(OH)_2C_6H_3CH(OH)CH_2NH_2$		L
Co^{2+}	-79.5	7.36(25°,1-2)	-125	0&25	T	0.06M(KCl)	62Aa	a,b
Ni^{2+}	-93.7	8.00(25°,1-2)	-160.7	0&25	T	0.06M(KCl)	62Aa	a,b
Cu^{2+}	-126.8	16.32(25°,1-2)	-112.1	0&25	T	0.06M(KCl)	62Aa	a,b
T61	TOLUENE, (d)-α-hydroxy-3,4-dihydroxy-α-(methylaminomethyl)-		$C_9H_{13}O_3N$			$(OH)_2C_6H_3CH(OH)CH_2NHCH_3$		L
Co^{2+}	28.9	10.06(25°,1-2)	289.5	0&25	T	0.06M(KCl)	62Aa	a,b
Ni^{2+}	16.3	9.26(25°,1-2)	232.6	0&25	T	0.06M(KCl)	62Aa	a,b
Cu^{2+}	-115.9	18.44(25°,1-2)	-35.1	0&25	T	0.06M(KCl)	62Aa	a,b
T62	TOLUENE, (ℓ)-α-hydroxy-3,4-dihydroxy-α-(methylaminomethyl)-		$C_9H_{13}O_3N$			$(OH)_2C_6H_3CH(OH)CH_2NHCH_3$		L
Co^{2+}	-84.5	8.94(25°,1-2)	-112.1	0&25	T	0.06M(KCl)	62Aa	a,b
Ni^{2+}	-31.0	8.40(25°,1-2)	56.9	0&25	T	0.06M(KCl)	62Aa	a,b
Cu^{2+}	-83.3	17.72(25°,1-2)	60.25	0&25	T	0.06M(KCl)	62Aa	a,b
T63	TOLUENE, α-hydroxy-α-[1-(methylamino)ethyl]-		$C_{10}H_{15}ON$			$C_6H_5CH(OH)CH(CH_3)NHCH_3$		L
Cd^{2+}	-45.2	6.49(1)	-33	25	T	μ=0.1(KNO_3)	64Ad	
T64	TOLUENE, 2-hydroxy-α(2-thiazolylazo)-		$C_{10}H_9ON_3S$			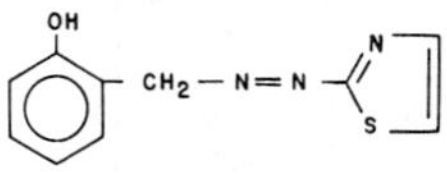		
La^{3+}	-17.78	8.17(25°,1)	94.77	15-35	T	μ=0.1M(KNO_3)	73Ka	b
Pr^{3+}	-15.27	8.91(25°,1)	116.40	15-35	T	μ=0.1M(KNO_3)	73Ka	b
Sm^{3+}	-24.56	9.41(25°,1)	92.88	15-35	T	μ=0.1M(KNO_3)	73Ka	b
Gd^{3+}	-25.40	9.61(25°,1)	94.18	15-35	T	μ=0.1M(KNO_3)	73Ka	b

Metal Ion	ΔH, kJ /mol	Log K	ΔS, J /mol K	T,°C	M	Conditions	Ref.	Re-marks
T64, continued								
Dy^{3+}	-16.94	9.71(25°,1)	125.60	15-35	T	μ=0.1M(KNO_3)	73Ka	b
Ho^{3+}	-26.28	9.61(25°,1)	93.39	15-35	T	μ=0.1M(KNO_3)	73Ka	b
Er^{3+}	-18.66	9.56(25°,1)	119.28	15-35	T	μ=0.1M(KNO_3)	73Ka	b
Ni^{2+}	-64.02	11.70(25°,1)	10.25	15-35	T	μ=0.1M(KNO_3)	73Ka	b
Zn^{2+}	-8.37	7.85(25°,1)	122.13	15-35	T	μ=0.1M(KNO_3)	73Ka	b
Cd^{2+}	-14.64	7.72(25°,1)	98.53	15-35	T	μ=0.1M(KNO_3)	73Ka	b
Hg^{2+}	-14.64	16.07(25°,1)	258.36	15-35	T	μ=0.1M(KNO_3)	73Ka	b
T65 TRANSFER RIBONUCLEIC ACID, yeast phenylalanine								L
Mg^{2+}	0.0	--	--	25	C	CM=1x10^{-3}M Approximately 20 moles of Mg^{2+} bound per mole of L	72Rc	q:±0.4
T66 1,4,7-TRIAZACYCLODECANE		$C_7H_{17}N_3$				$CH_2NH(CH_2CH_2NH)_2CH_2CH_2$ (ring)		L^2
Cu^{2+}	-62.3	14.4(25°,1)	67	15-35	T	μ=0.2M($NaClO_4$)	78Kd	b
Zn^{2+}	-28.0	10.3(25°,1)	105	15-35	T	μ=0.2M($NaClO_4$)	78Kd	b
Cd^{2+}	-30.5	7.8(25°,1)	46	15-35	T	μ=0.2M($NaClO_4$)	78Kd	b
Pb^{2+}	-30.5	8.8(25°,1)	67	15-35	T	μ=0.2M($NaClO_4$)	78Kd	b
T67 1,4,7-TRIAZACYCLONONANE		$C_6H_{15}N_3$				$CH_2NH(CH_2CH_2NH)_2CH_2$ (ring)		L^1
Cu^{2+}	-59.4	15.54(1)	97.9	25	C	0.1M(KNO_3)	77Fa	
	-54.4	16.2(25°,1)	130	15-35	T	μ=0.2M($NaClO_4$)	78Kd 77Kc	b
Zn^{2+}	-49.8	11.58(1)	55.2	25	C	0.1M(KNO_3)	77Fa	
	-30.5	11.3(25°,1)	113	15-35	T	μ=0.2M($NaClO_4$)	78Kd	b
Cd^{2+}	-31.8	9.2(25°,1)	71	15-35	T	μ=0.2M($NaClO_4$)	78Kd	b
Pb^{2+}	-34.3	10.8(25°,1)	112	15-35	T	μ=0.2M($NaClO_4$)	78Kd	b
T68 3,6,9-TRIAZAUNDECANE, 1,11-diamine		$C_8H_{23}N_5$				$H_2N(CH_2CH_2NH)_3CH_2CH_2NH_2$		L^{12}
Cu^{2+}	-104.6	22.8(25°,1)	88	15-35	T	μ=0.2M($NaClO_4$)	78Kd	b
Zn^{2+}	-58.6	15.1(25°,1)	96	15-35	T	μ=0.2M($NaClO_4$)	78Kd	b
Cd^{2+}	-53.6	14.1(25°,1)	92	15-35	T	μ=0.2M($NaClO_4$)	78Kd	b
Hg^{2+}	-139.7	24.8(25°,1)	4	15-35	T	μ=0.2M($NaClO_4$)	78Kd	b
Pb^{2+}	-38.0	9.9(25°,1)	63	15-35	T	μ=0.2M($NaClO_4$)	78Kd	b

Metal Ion	ΔH, kJ /mol	Log K	ΔS, J /mol K	T,°C	M	Conditions	Ref.	Remarks
T69	TRIDECANE, 5,9-dithio-		$C_{11}H_{24}S_2$			$CH_3(CH_2)_3S(CH_2)_3S(CH_2)_3CH_3$		
TiA_4	-91.2	3.96(1)	-227.2	25	C	S:Benzene A=Cl	70Gh 68Ge	
	-85.8	3.88(1)	-213.4	25	C	S:Benzene A=Br	70Gh	
	-86.2	(1)	--	25	C	S:Benzene CM=0.05-0.08M A=Br	70Gh 68Ge 65Gb	
SnA_4	-105.8	3.00(1)	-295.2	25	C	S:Benzene CM=0.05-0.08M A=Cl	68Ge 70Gh	
	-89.1	3.08(1)	-223.0	25	C	S:Benzene A=Br	70Gh 68Ge	
TeA_4	-56.9	3.44(1)	-125.1	25	C	S:Benzene A=Cl	73Pi	
	-10.0	(2)	--	25	C	S:Benzene A=Cl	73Pi	
T70	TRIDECANE, 1,5,9,13-tetraaza-		$C_9H_{24}N_4$			$NH_2(CH_2)_3NH(CH_2)_3NH(CH_2)_3NH_2$		
Ni^{2+}	-55.2	10.48(1)	15.5	25	C	0.1M($NaNO_3$)	72Bb	
Cu^{2+}	-81.38	17.05(1)	53.6	25	C	0.1M($NaNO_3$)	72Bb	
Zn^{2+}	-30.75	9.317(1)	75.3	25	C	0.1M($NaNO_3$)	72Bb	
T71	TRIGLYCINE		$C_6H_{11}O_4N_3$			$H_2NCH_2CONHCH_2CONHCH_2CO_2H$ L^-		
Mg^{2+}	8	1.44(1)	58	30	T	μ=0.09(KCl)	57Mb	
Ni^{2+}	-20.9	3.69(1)	2.5	25	C	μ=0.1M(KNO_3)	71Ld	
	-17.6	3.04(2)	-2.1	25	C	μ=0.1M(KNO_3)	71Ld	
Cu^{2+}	-24.7	(1)	--	24	C	μ=0.01M($HClO_4$)	72Sh	
	-26.4	5.04(1)	7.9	25	C	μ=0.10($NaClO_4$)	68Bc	
T72	2,7,12-TRIMETHYL-3,7,11,17-tetra-azabicyclo [11.3.1] heptadeca-1(17),2,11,13,15-pentane		$C_{16}H_{24}N_4$			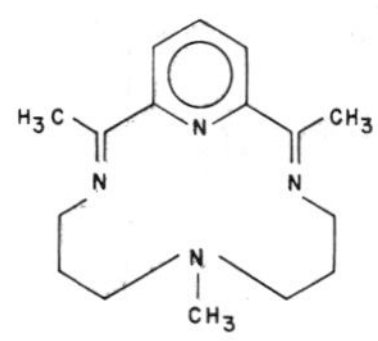		
NiA_2L	-19.16	--	-54.0	--	T	A:BF_4 R:$NiLA_2$-(diamagnetic)=Ni-LA_2(paramagnetic)	71Rf	u
	-16.99	--	-38.1	--	T	S:N,N-Dimethyl-formamide A=ClO_4 R:$NiLA_2$(diamagnetic)=$NiLA_2$(paramagnetic)	71Rf	u
T73	TRINACTIN							
Na^+	-30.50	3.2(1)	-41.0	25	C	S:CH_3OH	73Za	

Metal Ion	ΔH, kJ /mol	Log K	ΔS, J /mol K	T,°C	M	Conditions	Ref.	Re-marks
T74	2,2',2''-TRIPYRIDINE		$C_{15}H_{11}N_3$			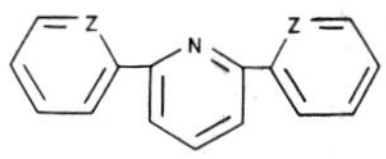		
Mn^{2+}	-23.8	4.4(25°,1)	4.4	9-25	T	CM=0.025 CL=0.20	66Hd	b
	-4	5.0(25°,1)	79	-46-90	T	μ=0.2($NaClO_4$) S:CH_3OH	73Bg	b
Fe^{2+}	-33.5	7.1(25°,1)	23.6	2-25	T	CM=0.003-0.005 CL=0.05-0.10	66Hd	b
Co^{2+}	-44.6	9.496(1)	32	25	C	μ=0.1M(KNO_3)	77Kb	
	-44.8	8.4(25°,1)	10.6	0.2-25	T	CM=0.005-0.01 CL=0.096-0.120	66Hd	b
	92	9.49(25°,1)	-126	8-32	T	μ=0.24(NaCl+HCl)	69Pe	b
	-49.0	9.111(2)	10	25	C	μ=0.1M(KNO_3)	77Kb	
	-51.0	9.10(25°,2)	0	8-32	T	μ=0.24(NaCl+HCl)	69Pe	b
	-60.7	1.49(25°)	-176	8-32	T	μ=0.24(NaCl+HCl) R:$Co^{2+}+H_2L^{2+}=CoL^{2+}+2H^+$	69Pe	b
	-19.7	1.09(25°)	-46	8-32	T	μ=0.24(NaCl+HCl) R:$CoL^{2+}=H_2L^{2+}=CoL_2{}^{2+}+2H^+$	69Pe	b
Ni^{2+}	-39.7	10.7(25°,1)	71.7	6.8-25	T	CM=0.005 CL=0.05-0.14	66Hd	b
	-54.1	10.705(1)	23	25	C	μ=0.1M(KNO_3)	77Kb	
	-59.8	11.091(2)	12	25	C	μ=0.1M(KNO_3)	77Kb	
Cu^{2+}	-66.1	12.983(1)	~27	25	C	μ=0.1M(KNO_3)	77Kb	
CuA_2	-77.8	(1)	--	30	C	S:Benzene A=1,1,1,5,5,5-Hexafluoropentane-2,4-dione	74Gc	
	-34.98	1.898(1)	-79.1	30	C	S:Benzene A=1,-1,1-Trifluoro-pentane-2,4-dione	74Gc	
	-41.8	2.021(1)	-99.2	30	C	S:Benzene A=1,1,1-Tri-fluoro-4-phenyl-butane-2,4-dione	74Gc	
	-40.00	2.468(1)	-84.5	30	C	S:Benzene A=1,-1,1-Trifluoro-4-(2-thienyl) butane-2,4-dione	74Gc	
Zn^{2+}	-42.2	6.0(25°,1)	-26.7	0.2-25	T	CM=0.002 CL=0.03-0.05	66Hd	b
Cd^{2+}	-23.0	5.1(25°,1)	20.5	0.2-25	T	CM=0.002 CL=0.01	66Hd	b

Metal Ion	ΔH, kJ /mol	Log K	ΔS, J /mol K	T,°C	M	Conditions	Ref.	Re-marks
T75 TRYPTOPHAN			$C_{11}H_{12}O_2N_2$			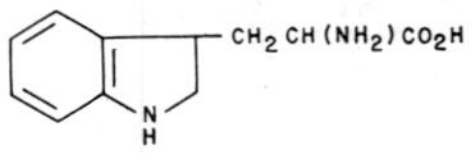		
La^{3+}	-34.7	4.4(25°,1)	-32.3	15-55	T	μ=0.1M(KCl)	76Bc	q:±2
Ce^{+}	-30.5	4.56(25°,1)	-15.4	15-55	T	μ=0.1M(KCl)	76Bc	q:±2.5
Pr^{3+}	-30.5	4.52(25°,1)	-15.4	15-55	T	μ=0.1M(KCl)	76Bc	q:±2.5
Nd^{3+}	-30.5	4.5(25°,1)	-16.8	15-55	T	μ=0.1M(KCl)	76Bc	
Sm^{3+}	-33.9	4.75(25°,1)	-22.4	15-55	T	μ=0.1M(KCl)	76Bc	
Eu^{3+}	-33.0	4.82(25°,1)	-18.2	15-55	T	μ=0.1M(KCl)	76Bc	
Gd^{3+}	-36.0	4.8(25°,1)	-29.5	15-55	T	μ=0.1M(KCl)	76Bc	
Tb^{3+}	-32.2	4.8(25°,1)	-16.8	15-55	T	μ=0.1M(KCl)	76Bc	q:±2
Dy^{3+}	-31.0	4.83(25°,1)	-11.2	15-55	T	μ=0.1M(KCl)	76Bc	
Ho^{3+}	-30.1	4.86(25°,1)	-8.4	15-55	T	μ=0.1M(KCl)	76Bc	
Er^{3+}	-33.0	5.0(25°,1)	-15.4	15-55	T	μ=0.1M(KCl)	76Bc	
Yb^{3+}	-31.0	5.11(25°,1)	-5.6	15-55	T	μ=0.1M(KCl)	76Bc	
Lu^{3+}	-31.8	5.1(25°,1)	-8.4	15-55	T	μ=0.1M(KCl)	76Bc	
Mn^{2+}	-5.4	2.88(20°,1)	38	20-60	T	μ=0.1(KNO_3)	73Bh	b
	-4.52	2.840(1)	39.3	25	C	μ=3.0($NaClO_4$)	70Wd	
	-4.98	2.312(2)	27.6	25	C	μ=3.0($NaClO_4$)	70Wd	
Fe^{2+}	-9.29	3.923(1)	43.9	25	C	μ=3.0($NaClO_4$)	70Wd	
	-9.54	3.471(2)	34.3	25	C	μ=3.0($NaClO_4$)	70Wd	
Co^{2+}	-11.63	4.581(1)	48.5	25	C	μ=3.0($NaClO_4$)	70Wd	
	-11.97	4.321(2)	42.7	25	C	μ=3.0($NaClO_4$)	70Wd	
	-11.72	3.344(3)	24.7	25	C	μ=3.0($NaClO_4$)	70Wd	
CoA	-10.9	3.86(25°)	37.6	15-55	T	μ=0.1M(KNO_3) A=Methylimino-diacetate R:NiA+HL=$NiAL^-$+H^+	71Id	b,q: ±1.2
Ni^{2+}	-23.93	5.757(1)	30.1	25	C	μ=3.0($NaClO_4$)	70Wd	
	-25.52	5.227(2)	14.2	25	C	μ=3.0($NaClO_4$)	70Wd	
	-20.88	4.474(3)	15.9	25	C	μ=3.0($NaClO_4$)	70Wd	
NiA	-22.2	4.80(25°)	17.6	15-55	T	μ=0.1M(KNO_3) A=Methylimino-diacetate R:NiA+HL=$NiAL^-$+H^+	71Id	b,q: ±4.2
Cu^{2+}	-32.09	8.711(1)	59.0	25	C	μ=3.0($NaClO_4$)	70Wd	
	-32.38	7.953(2)	43.5	25	C	μ=3.0($NaClO_4$)	70Wd	
CuA	-35.1	6.38(25°)	3.8	15-55	T	μ=0.1M(KNO_3) A=Methylimino-diacetate R:CuA+HL=$CuAL^-$+H^+	71Id	b,q: ±6.7
Zn^{2+}	-17.15	5.013(1)	38.5	25	C	μ=3.0($NaClO_4$)	70Wd	
	-17.28	4.766(2)	33.0	25	C	μ=3.0($NaClO_4$)	70Wd	

Metal Ion	ΔH, kJ /mol	Log K	ΔS, J /mol K	T,°C	M	Conditions	Ref.	Remarks
T75, continued								
CdA	-13.4	3.66(25°)	25.1	15-55	T	μ=0.1M(KNO_3) A=Methylimino- diacetate R:NiA+HL=NiAL$^-$+H$^+$	71Id	b,q: ±3.3
T76	TYROSINE		$C_9H_{11}O_3N$			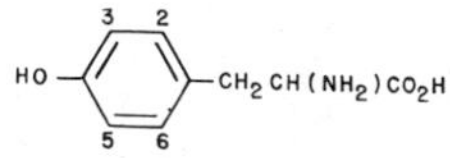		
Y^{3+}	158.2	8.48(25°,1-2)	691.2	25-45	T	μ=0.1M(KNO_3) CL/CM=5	77Sa	
La^{3+}	54.4	3.82(25°,1)	257.3	25-45	T	μ=0.1M(KNO_3) CL/CM=5	77Sa	
Ce^{3+}	70.7	4.17(25°,1)	370.7	25-45	T	μ=0.1M(KNO_3) CL/CM=5	77Sa	
Pr^{3+}	103.3	8.210(25°,1-2)	503.8	25-45	T	μ=0.1M(KNO_3) pH=6.5-8.0	76Sc	b
Nd^{3+}	95.4	8.577(25°,1-2)	483.2	25-45	T	μ=0.1M(KNO_3) pH=6.5-8.0	76Sc	b
Sm^{3+}	80.8	8.797(25°,1-2)	440.2	25-45	T	μ=0.1M(KNO_3) pH=6.5-8.0	76Sc	b
Gd^{3+}	98.7	8.80(25°,1-2)	417	25-45	T	μ=0.1M(KNO_3)	76Sb	b
Dy^{3+}	103.3	9.04(25°,1-2)	519	25-45	T	μ=0.1M(KNO_3)	76Sb	b
Er^{3+}	110.9	9.34(25°,1-2)	550	25-45	T	μ=0.1M(KNO_3)	76Sb	b
Yb^{3+}	118.0	9.70(25°,1-2)	581	25-45	T	μ=0.1M(KNO_3)	76Sb	b
Co^{2+}	-2.5	3.93(25°,1)	84	27	C	μ=0.05M(KCl)	71Gb	
	-0.8	3.93(25°,1)	79	20-35	T	μ=0.05M(KCl)	71Gb	b
	-7.1	3.45(25°,2)	46	27	C	μ=0.05M(KCl)	71Gb	
	-11.3	3.45(25°,2)	25	20-35	T	μ=0.05M(KCl)	71Gb	b
Ni^{2+}	-8.4	5.02(25°,1)	67	27	C	μ=0.05M(KCl)	71Gb	
	-7.1	5.02(25°,1)	71	20-35	T	μ=0.05M(KCl)	71Gb	b
	-11.7	4.25(25°,2)	42	27	C	μ=0.05M(KCl)	71Gb	
	-13.0	4.25(25°,2)	38	20-35	T	μ=0.05M(KCl)	71Gb	
Cu^{2+}	-24.7	7.75(25°,1)	67	27	C	μ=0.05M(KCl)	71Gb	
	-16.7	7.75(25°,1)	88	20-35	T	μ=0.05M(KCl)	71Gb	b
	-23.8	7.68(1)	66.9	25	C	μ=0.2M(KCl)	75Gd	
	-30.5	9.17(1)	71.1	25	C	μ=0.2M(KCl) S:50w%Dioxane	75Gd	
	-27.6	6.82(25°,2)	42	27	C	μ=0.05M(KCl)	71Gb	
	-25.1	6.82(25°,2)	46	20-35	T	μ=0.05M(KCl)	71Gb	b
	-31.0	6.81(2)	25.1	25	C	μ=0.2M(KCl)	75Gd	
	-37.2	7.54(2)	20.9	25	C	μ=0.2M(KCl) S:50w%Dioxane	75Gd	
Zn^{2+}	-5.8	4.16(25°,1)	58	27	C	μ=0.05M(KCl)	71Gb	
	6.3	4.16(25°,1)	100	20-35	T	μ=0.05M(KCl)	71Gb	b
	-10.9	4.11(25°,2)	25	27	C	μ=0.05M(KCl)	71Gb	
	-14.6	4.11(25°,2)	29	20-35	T	μ=0.05M(KCl)	71Gb	b

Metal Ion	ΔH, kJ /mol	Log K	ΔS, J /mol K	T,°C	M	Conditions	Ref.	Remarks

U

U1 UNADENYLYLATED GLUTAMINE SYNTHETASE from Escherichia coli

Metal Ion	ΔH, kJ /mol	Log K	ΔS, J /mol K	T,°C	M	Conditions	Ref.	Remarks
Mn^{2+}	≅12	6.27	≅159	37	C	pH=7.2 Values are per mole of Mn^{2+} bound to ligand	72Hg	

U2 UNDECANE, 3,9-dithio-1,11-dihydroxy $C_9H_{20}O_2S_2$ $HOCH_2CH_2SCH_2(CH_2)_3CH_2SCH_2CH_2OH$

Metal Ion	ΔH, kJ /mol	Log K	ΔS, J /mol K	T,°C	M	Conditions	Ref.	Remarks
Ag^+	-51.0	5.07(1)	-77.4	20	C	μ=1.0(KNO_3) pH=3	70Wc	
	-28.9	2.98(2)	-40.6	20	C	μ=1.0(KNO_3) pH=3	70Wc	
	-8.4	2.46	13.4	20	C	μ=1.0(KNO_3) pH=3 R:AgL+Ag=Ag_2L	70Wc	

U3 UNDECANE, 3,6,9-trithio-1,11-dihydroxy $C_8H_{18}O_2S_3$ $HOCH_2CH_2SCH_2CH_2SCH_2CH_2SCH_2CH_2OH$

Metal Ion	ΔH, kJ /mol	Log K	ΔS, J /mol K	T,°C	M	Conditions	Ref.	Remarks
Ag^+	-53.1	5.37(1)	-79.1	20	C	μ=1.0(KNO_3) pH=3	70Wc	
	-43.5	2.68(2)	-97.1	20	C	μ=1.0(KNO_3) pH=3	70Wc	
	1.7	2.38	51.5	20	C	μ=1.0(KNO_3) pH=3 R:AgL+Ag=Ag_2L	70Wc	

U4 URACIL $C_4H_4O_2N_2$

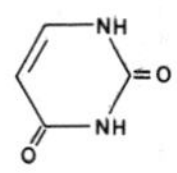

Metal Ion	ΔH, kJ /mol	Log K	ΔS, J /mol K	T,°C	M	Conditions	Ref.	Remarks
Be^{2+}	-32.6	6.52(30°,1)	17	20-40	T	μ=0.1M($NaClO_4$)	78Si	b
	-24.3	5.50(30°,2)	25	20-40	T	μ=0.1M($NaClO_4$)	78Si	b
Hg^{2+}	-27.2	7.30(30°,1)	50	20-40	T	μ=0.1M($NaClO_4$)	78Si	b,q:±3.3
	-30.1	6.29(30°,2)	21	20-40	T	μ=0.1M($NaClO_4$)	78Si	b

U5 URACIL, 5-methyl $C_5H_6O_2N_2$

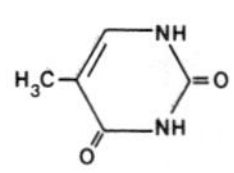

Metal Ion	ΔH, kJ /mol	Log K	ΔS, J /mol K	T,°C	M	Conditions	Ref.	Remarks
Be^{2+}	-19.2	7.01(30°,1)	71	20-40	T	μ=0.1M($NaClO_4$)	78Si	b,q:±3.3
	-21.8	6.20(30°,2)	46	20-40	T	μ=0.1M($NaClO_4$)	78Si	b
Hg^{2+}	-35.6	7.93(30°,1)	21	20-40	T	μ=0.1M($NaClO_4$)	78Si	b
	-28.0	6.77(30°,2)	38	20-40	T	μ=0.1M($NaClO_4$)	78Si	b

Metal Ion	ΔH, kJ /mol	Log K	ΔS, J /mol K	T,°C	M	Conditions	Ref.	Remarks
U6	6-URACILCARBOXYLIC ACID		$C_5H_4O_4N_2$			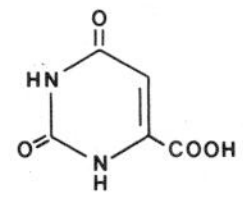		L^{2-}
Mn^{2+}	-31.8	4.39(1)	-25	20	T	CL=5x10^{-3} CM=5x10^{-3}	79Dc	b
	-31.8	4.30(1)	-25	25	T	CL=5x10^{-3} CM=5x10^{-3} ΔCp=0	79Dc	b
	-31.8	4.21(1)	-25	30	T	CL=5x10^{-3} CM=5x10^{-3}	79Dc	b
Ni^{2+}	-19.7	7.25(1)	71	20	T	CL=5x10^{-3} CM=5x10^{-3}	79Dc	b,q:±1.3
	-16.7	7.00(1)	79	25	T	CL=5x10^{-3} CM=5x10^{-3} ΔCp=1260	79Dc	b,q:±1.3
	-12.5	6.82(1)	92	30	T	CL=5x10^{-3} CM=5x10^{-3}	79Dc	b,q:±1.3
U7	UREA, 1-allyl-2-thio		$C_4H_8N_2S$			$H_2NCSNHCH_2CH:CH_2$		
Cd^{2+}	-37.2	3.41(25°,1-3)	-59	25&45	T	μ=0.5M(KNO_3)	74Rb	a,b
Pb^{2+}	-33.9	1.45(25°,1-2)	-82	25&45	T	μ=0.5M(KNO_3)	74Rb	a,b
U8	UREA, 1,3-di-tert-butyl-2-thio-		$C_9H_{20}N_2S$			$(CH_3)_3CNHC(S)NHC(CH_3)_3$		
I_2	-77.8	4.611(25°,1)	-166.5	0-35	T	S:Chloroform CL=4.3x10^{-5}-3.9x10^{-4}M	66Bd	b
U9	UREA, 1,3-diethyl-2-thio-		$C_5H_{12}N_2S$			$CH_3CH_2NHC(S)NHCH_2CH_3$		
NiL_6A_2	-18.4	--	-92	-94-37	T	S:Acetone R:Square planar= octahedral A=ClO_4	75Ea	b,q: ±2.5
	-14.2	--	-75	-94-37	T	S:Acetone R:Square planar= octahedral A=BF_4	75Ea	b,q: ±1.7
Ag^+	-5.4	6.009(1)	96.6	25	C	μ=0.5M(KNO_3)	73Ec	q:±0.8
	-84.47	4.306(2)	-200.8	25	C	μ=0.5M(KNO_3)	73Ec	
	-1.21	3.035(3)	54.0	25	C	μ=0.5M(KNO_3)	73Ec	
	-37.82	0.817(4)	-111.08	25	C	μ=0.5M(KNO_3)	73Ec	
I_2	-60.2	4.191(25°,1)	-118.0	0-35	T	S:Chloroform CL=3.6x10^{-5}-3.7x10^{-4}M	66Bd	b
U10	UREA, 1,3-diisopropyl-2-thio-		$C_7H_{16}N_2S$			$(CH_3)_2CHNHC(S)NHCH(CH_3)_2$		
I_2	-47.3	4.216(25°,1)	-73.6	0-35	T	S:Chloroform	66Bd	b

Metal Ion	ΔH, kJ /mol	Log K	ΔS, J /mol K	T,°C	M	Conditions	Ref.	Re-marks
U10, continued								
I_2, cont.						CL=3.7×10^{-5}-3.7×10^{-4}M		
U11 UREA, 1,3-dimethyl-			$C_3H_8ON_2$			$CH_3NHCONHCH_3$		
SbA_5	-123.8	(1)	--	25	C	S:CH_2Cl_2 A=Cl	71Oa	j
U12 UREA, 1,3-dimethyl-2-thio-			$C_3H_8N_2S$			$CH_3NHC(S)NHCH_2$		
NiL_6A_2	-14.2	--	-71	-94-37	T	S:Acetone R:Square planar= octahedral A=ClO_4	75Ea	b,q: ±2.5
Ag^+	0	6.89(1)	116.52	25	C	μ=0.5M(KNO_3)	73Ec	
	-91.21	4.072(2)	-227.82	25	C	μ=0.5M(KNO_3)	73Ec	
	-1.46	2.598(3)	44.8	25	C	μ=0.5M(KNO_3)	73Ec	
	-39.58	1.116(4)	-111.3	25	C	μ=0.5M(KNO_3)	73Ec	
U13 UREA, 1,3-diphenyl-2-thio-			$C_{13}H_{12}N_2S$			$C_6H_5NHC(S)NHC_6H_5$		
I_2	-63.6	2.794(29°,1)	-156.9	25-32	T	S:Carbon Tetrachloride	72Aa	b
U14 UREA, ethylene-2-thio			$C_3H_6N_2S$			$CH_2NHC(S)NHCH_2$ (ring)		
NiL_6A_2	-14.2	--	-67	-94-37	T	S:Acetone R:Square planar octahedral A=ClO_4	75Ea	b,q: ±1.7
U15 UREA, 1-ethylidene-2-thio-			$C_3H_6N_2S$			$NH_2C(S)NHCH:CH_2$		
I_2	-33.9	2.747(29°,1)	-60.2	25-32	T	S:Carbon tetrachloride	72Aa	b
U16 UREA, 1-methyl-2-thio-			$C_2H_6N_2S$			$NH_2C(S)NHCH_3$		
Ag^+	-6.7	6.710(1)	105.8	25	C	μ=0.5M(KNO_3)	73Ec	q:±1.2
	-86.86	3.916(2)	-216.3	25	C	=0.5M(KNO_3)	73Ec	
	-1.72	2.434(3)	41.0	25	C	μ=0.5M(KNO_3)	73Ec	
	-40.33	1.021(4)	-115.9	25	C	μ=0.5M(KNO_3)	73Ec	
U17 UREA, 1-naphthyl-2-thio			$C_{11}H_{10}N_2s$			$H_2NC(S)NH(C_{10}H_7)$		
NiL_2A_2	10.0	--	54	-94-37		S:Acetone R:Square planar= tetrahedral A=Br	75Ea	b,q: ±1.7
	10.5	--	50	-94-37		S:Acetone R:Square planar= tetrahedral A=I	75Ea	b,q: ±1.7

Metal Ion	ΔH, kJ /mol	Log K	ΔS, J /mol K	T,°C	M	Conditions	Ref.	Remarks
U18	UREA, seleno		CH_4N_2Se			NH_2CSeNH_2		L
Hg^{2+}	-127.6	24.0(1-2)	38.5	30	C	μ=0.1	69Gd	
	-156.5	30.2(1-3)	-61.9	30	C	μ=0.1	69Gd	
	-226.4	32.9(1-4)	-115.9	30	C	μ=0.1	69Gd	
U19	UREA, 1,1,3,3-tetramethyl-		$C_5H_{12}ON_2$			$(CH_3)_2NCON(CH_3)_2$		L
ZnA	-27.2	1.858(25°,1)	-55.6	15-35	T	S:Benzene A=α,β,γ,δ-Tetraphenylporphine	73Ve	b
SbA_5	-124.01	(1)	--	25	C	CT<.01M S:Ethylene chloride A=Cl	64Ob	
I_2	-18.20	0.813(25°,1)	-45.6	24-43	T	$S:CCl_4$ $CM=1.6x10^{-3}M$ $CL=3.8x10^{-2}M$	64Ma	b
	-13.8	0.663(20°,1)	-33	10-30	T	S:Dichloromethane	68Lc	b,q:±2.5
	-20.9	1.279(20°,1)	-46	10-30	T	S:n-Heptane	68Lc	b,q:±1.7
U20	UREA, 1,1,3,3-tetramethyl-2-thio-		$C_5H_{12}N_2S$			$(CH_3)_2NC(S)N(CH_3)_2$		L
ZnA	-23.4	1.079(25°,1)	-57.7	15-35	T	S:Benzene A=α,β,γ,δ-Tetraphenylporphine	73Ve	b
I_2	-43.9	3.903(25°,1)	-72.8	25-40	T	$S:CCl_4$ $CM=10^{-3}M$ $CL=2x10^{-4}M$	64Na	b
	-39.7	4.132(25°,1)	-51.5	0-35	T	S:Chloroform $CL=3.7x10^{-5}$-$3.7x10^{-4}M$	66Bd	b
	-37.6	4.690(20°,1)	-38	10-30	T	S:Dichloromethane	68Lc	b,q:±6.7
	-41.42	4.114(20°,1)	-63	10-30	T	S:n-Heptane	68Lc	b,q:±5.0
U21	UREA, 2-thio-		CH_4N_2S			$H_2NC(S)NH_2$		L
NiL_6A_2	20.9	--	105	-94-37	T	S:Acetone R:Square planar=tetrahedral A=Br	75Ea	b,q: ±1.6
	24.7	--	100	-94-37	T	S:Acetone R:Square planar=tetrahedral A=I	75Ea	b,q: ±1.6
	-10.9	--	-63	-94-37	T	S:Acetone R:Square planar=octahedral $A=ClO_4$	75Ea	b,q: ±2.5
	-8.4	--	-42	-94-37	T	S:Acetone R:Square planar=octahedral $A=BF_4$	75Ea	b,q: ±1.7
PdA_2	-19.2	0.996(1)	-44.8	25	T	μ=0 A=o-Phenylenebis-dimethylarsine	67Ec 67Ed	
	-19.2	0.845(40°,1)	-44.8	25-55	T	$μ=0.0021(NaNO_3)$ A=o-Phenylenebis-	67Ed	b

Metal Ion	ΔH, kJ /mol	Log K	ΔS, J /mol K	T,°C	M	Conditions	Ref.	Remarks
U21, continued								
PdA_2, cont.	-15.1	1.322(40°,1)	-22.6	25-55	T	dimethylarsine $\mu=7\times10^{-4}$ S:CH_3OH A=o-Phenylenebis-dimethylarsine	67Ec	b,q: ±1.7
PtA_2	-20.1	0.676(1)	-54.0	25	T	$\mu=0$ A=o-Phenylenebis-dimethylarsine	67Ec	q:±1.7
	-19.2	1.30(23°,1)	-39.7	23-54	T	$\mu=0$ S:CH_3OH A=o-Phenylenebis-dimethylarsine	66Db	b
	-20.1	0.574(35°,1)	-54.0	25-55	T	$\mu=7\times10^{-4}$($NaNO_3$) A=o-Phenylenebis-dimethylarsine	67Ec	b,q: ±1.7
Ag^+	-154.4	7.30(25°,1)	-376	25-45	T	$\mu=0$	73Bf	
	-81	6.46(1)	-150	25	C	$\mu=0.02M(NaClO_4)$	76Mb 75Mb	
	-10.9	7.050(1)	98.3	25	C	$\mu=0.5M(KNO_3)$	73Ec	q:±1.2
	0	7.043(25°,1)	134	10-30	T	$\mu=0.5M(KNO_3)$	69Bg	b
	-149.2	8.40(20°,1)	-395	10-30	T	$\mu=1M(KNO_3)$	75Dg	b
	-161.9	7.68(25°,1)	-397	10-30	T	$\mu=1M(KNO_3)$	74Bd	b,q:±25
	-101.17	8.77(25°,1)	-179.5	5-30	T	$\mu=1M(KNO_3)$	80Da	b
	-124.0	8.36(20°,1)	-264	10-30	T	$\mu=1M(KNO_3)$ S:5v%Dioxane	75Dg	b
	-98.3	8.32(20°,1)	-172	10-30	T	$\mu=1M(KNO_3)$ S:10v%Dioxane	75Dg	b
	48.5	3.3(25°,2)	226	25-45	T	$\mu=0$	73Bf	
	-32	4.44(2)	-23	25	C	$\mu=0.02(NaClO_4)$	76Mb 75Mb	
	-80.88	3.575(2)	-202.9	25	C	$\mu=0.5M(KNO_3)$	73Ec	
	-69.0	3.571(25°,2)	-161.1	10-30	T	$\mu=0.5M(KNO_3)$	69Bg	b,q: ±16.7
	41.0	2.55(20°,2)	234	10-30	T	$\mu=1M(KNO_3)$	75Dg	b
	41.4	2.78(25°,2)	191	10-30	T	$\mu=1M(KNO_3)$	74Bd	b,q:±4
	-34.85	2.52(25°,2)	-70.3	5-30	T	$\mu=1M(KNO_3)$	80Da	b
	8.4	2.71(20°2)	79	10-30	T	$\mu=1M(KNO_3)$ S:5v%Dioxane	75Dg	b
	-25.4	2.88(20°,2)	-38	10-30	T	$\mu=1M(KNO_3)$ S:10v%Dioxane	75Dg	b
	-3.3	2.2(25°,3)	29	25-45	T	$\mu=0$	73Bf	
	-14	1.98(3)	-10	25	C	$\mu=0.02M(NaClO_4)$	76Mb 75Mb	
	-4.73	2.233(3)	27.2	25	C	$\mu=0.5M(KNO_3)$	73Ec	
	0	2.101(25°,3)	39.54	10-30	T	$\mu=0.5M(KNO_3)$	69Bg	b
	-10.0	2.18(20°,3)	8.4	10-30	T	$\mu=1M(KNO_3)$	75Dg	b
	11.3	2.40(25°,3)	86.6	10-30	T	$\mu=1M(KNO_3)$	74Bd	
	54.43	2.63(25°,3)	236.4	5-30	T	$\mu=1M(KNO_3)$	80Da	b,n
	-12.4	2.20(20°,3)	0	10-30	T	$\mu=1M(KNO_3)$ S:5v%Dioxane	75Dg	b
	-14.8	2.22(20°,3)	-84	10-30	T	$\mu=1M(KNO_3)$ S:10v%Dioxane	75Dg	b
	-33.5	0.92(25°,4)	-92	25-45	T	$\mu=0$	73Bf	
	-37.36	0.781(4)	-110.4	25	C	$\mu=0.5M(KNO_3)$	73Ec	
	-31.4	0.778(25°,4)	-90.46	10-30	T	$\mu=0.5M(KNO_3)$	69Bg	b,q:±8.4

Metal Ion	ΔH, kJ /mol	Log K	ΔS, J /mol K	T,°C	M	Conditions	Ref.	Re-marks
U21, continued								
Ag^+, cont.	-88	10.613(25°,1-2)	-91.2	0-45	T	μ≅0.15(AgCl)	49Ja	b
	-96	12.716(25°,1-3)	-78.6	0-45	T	μ≅0.15(AgBr)	49Ja	b
	-83.7	13.1(25°,1-3)	-30.88	20-50	T	μ=0.80($NaNO_3$)	60Tb	b
	-217	20.72(2-3)	-332	25	C	μ=0.02M($NaClO_4$)	76Mb 75Mb	
	-79.1	11.95(25°)	-35.6	10-30	T	μ=1M(KNO_3) R:Ag^++A^{2-}+L=$AgAL^-$ A=Thiosulfate	74Bd	b,q: ±12
	-72.3	12.19(20°)	-10	10-30	T	μ=1M(KNO_3) R:Ag+A+L=AgAL A=Thiosulfate	75Dg	b
	-60.4	12.37(20°)	33	10-30	T	μ=1M(KNO_3) S:5v%Dioxane R:Ag+A+L=AgAL A=Thiosulfate	75Dg	b
	-40.2	12.56(20°)	105	10-30	T	μ=1M(KNO_3) S:10v%Dioxane R:Ag+A+L=AgAL A=Thiosulfate	75Dg	b
	-90.0	13.36(25°)	-40.6	10-30	T	μ=1M(KNO_3) R:Ag^++A^{2-}+2L= $AgAL_2^-$ A=Thiosulfate	74Bd	b,q: ±29
	-78.8	13.22(20°)	-12.6	10-30	T	μ=1M(KNO_3) R:Ag+A+2L=$AgAL_2$ A=Thiosulfate	75Dg	b
	-88.5	13.62(20°)	-43.9	10-30	T	μ=1M(KNO_3) S:5v%Dioxane R:Ag+A+2L=$AgAL_2$ A=Thiosulfate	75Dg	b
	-97.6	14.03(20°)	-73.2	10-30	T	μ=1M(KNO_3) S:10v%Dioxane R:Ag+A+2L=$AgAL_2$ A=Thiosulfate	75Dg	b
	-112	14.15(25°)	-107.5	10-30	T	μ=1M(KNO_3) R:Ag^++2A^{2-}+L= AgA_2L^{3-} A=Thiosulfate	74Bd	b,q: ±12
	-90.2	14.59(20°)	-29	10-30	T	μ=1M(KNO_3) R:Ag+2A+L=AgA_2L A=Thiosulfate	75Dg	b
	-86.8	14.72(20°)	-17	10-30	T	μ=1M(KNO_3) S:5v%Dioxane R:Ag+2A+L=AgA_2L A=Thiosulfate	75Dg	b
	-82.8	14.85(20°)	0	10-30	T	μ=1M(KNO_3) S:10v%Dioxane R:Ag+2A+L=AgA_2L A=Thiosulfate	75Dg	b
	27.2	15.48(20°)	385	10-30	T	μ=1M(KNO_3) R:Ag+A+3L=$AgAL_3$ A=Thiosulfate	75Dg	b

Metal Ion	ΔH, kJ /mol	Log K	ΔS, J /mol K	T,°C	M	Conditions	Ref.	Re-marks
U21, continued								
Ag^+, cont.	-62.6	15.60(20°)	85.8	10-30	T	μ=1M(KNO_3) S:5v%Dioxane R:Ag+A+3L=$AgAL_3$ A=Thiosulfate	75Dg	b
	-79.1	15.73(20°)	31.4	10-30	T	μ=1M(KNO_3) S:10v%Dioxane R:Ag+A+3L=$AgAL_3$ A=Thiosulfate	75Dg	b
AgA	-86.44	13.89(25°,1)	-100	5-30	T	μ=1M(KNO_3) A=Thiocyanic acid	80Da	b,n
	-115.23	12.40(25°,1-2)	-146.4	5-30	T	μ=1M(KNO_3) A=Thiocyanic acid	80Da	b,n
AgA^-	-75.44	11.77(25°,1)	-26.4	5-30	T	μ=1M(KNO_3) A=Thiosulfate	80Da	b,n
	-64.89	2.08(25°,2)	-187.0	5-30	T	μ=1M(KNO_3) A=Thiosulfate	80Da	b,n
$AgAB^{2-}$	-69.58	13.03(25°,1)	10.0	5-30	T	μ=1M(KNO_3) A=Thiocyanic acid B=Thiosulfate	80Da	b,n
	-90.12	15.20(25°,1-2)	-16.3	5-30	T	μ=1M(KNO_3) Thiocyanic acid B=Thiosulfate	80Da	b,n
AgA_2^+	-140.58	11.24(25°,1)	-258.2	5-30	T	μ=1M(KNO_3) A=Thiocyanic acid	80Da	b,n
AgA_2^{3-}	-97.65	10.42(25°,1)	-52.7	5-30	T	μ=1M(KNO_3) A=Thiosulfate	80Da	b,n
AgA_2	-67.45	13.20(25°,1-2)	29	5-30	T	μ=1M(KNO_3) A=Thiocyanic acid	80Da	b,n
AgA_2B^{3-}	-64.35	13.69(25°,1)	54.4	5-30	T	μ=1M(KNO_3) A=Thiocyanic acid B=Thiosulfate	80Da	b,n
AgA_3^{2-}	-44.98	12.44(25°,1)	87.9	5-30	T	μ=1M(KNO_3) A=Thiocyanic acid	80Da	b,n
Cd^{2+}	-19.3	1.32(1)	-40	25	C	μ=0.5M($NaClO_4$)	76Mb 75Mb	
	-21.7	0.713(2)	-58	25	C	μ=0.5M($NaClO_4$)	76Mb 75Mb	
Hg^{2+}	-74	11.39(1)	-30	25	C	μ=0.5M($NaClO_4$)	76Mb 75Mb	
	-70	10.72(2)	-30	25	C	μ=0.5M($NaClO_4$)	76Mb 75Mb	
	-44	3.04(3)	-87	25	C	μ=0.5($HClO_4$)	76Mb 75Mb	
	-15	1.95(4)	-13	25	C	μ=0.5($HClO_4$)	76Mb 75Mb	
	-138	21.26(1-2)	-54.0	30	C	μ=0.1	69Gd	
	-144.8	24.22(1-3)	-13.8	30	C	μ=0.1	69Gd	
	-194.1	25.88(1-4)	-144.8	30	C	μ=0.1	69Gd	
	-184	26.3(25°,1-4)	-115.1	20-50	T	μ=0.80($NaNO_3$)	60Tb	b

Metal Ion	ΔH, kJ /mol	Log K	ΔS, J /mol K	T, °C	M	Conditions	Ref.	Remarks
U21, continued								
Hg^{2+},	-192	26.50(25°,1-4)	-137.6	20-30	T	1M($NaNO_3$)	56Tb	b
cont.	-210	36.0(2-3)	-17	25	C	μ=0.5($HClO_4$)	76Mb 75Mb	
HgA_2	-6.28	1.97(1)	16.7	25	C	μ=0 A=CN	68Ia	
	-10.0	2.03(1)	5.4	25	C	μ=0 A=CN S:20.0w% C_2H_5OH	68Ia	
	-18.87	1.94(1)	-26.4	25	C	μ=0 A=CN S:40.0w% C_2H_5OH	68Ia	
	-24.81	1.86(1)	-46.4	25	C	μ=0 A=CN S:60.0w% C_2H_5OH	68Ia	
	-24.94	2.11(1)	-43.1	25	C	μ=0 A=CN S:80.0w% C_2H_5OH	68Ia	
	-25.77	2.28(1)	-43.1	25	C	μ=0 A=CN S:92.3w% C_2H_5OH	68Ia	
	-5.8	2.08(1)	20.1	25	C	μ=0 S:20w% Formamide A=CN	71Ig	q:±0.4
	-7.1	1.95(1)	13.4	25	C	μ=0 S:40w% Formamide A=CN	71Ig	
	-7.1	2.01(1)	14.6	25	C	μ=0 S:60w% Formamide A=CN	71Ig	
	-6.7	2.02(1)	15.9	25	C	μ=0 S:80w% Formamide A=CN	71Ig	
	-6.3	2.02(1)	16.7	25	C	μ=0 S:100w% Formamide A=CN	71Ig	
	-33.0	0.58(2)	-99.6	25	C	μ=0 A=CN	68Ia	
	-36.0	0.77(2)	-106.7	25	C	μ=0 A=CN S:20.0w% C_2H_5OH	68Ia	
	-34.3	1.04(2)	-95.0	25	C	μ=0 A=CN S:40.0w% C_2H_5OH	68Ia	
	-21.80	1.26(2)	-48.1	25	C	μ=0 A=CN S:60.0w% C_2H_5OH	68Ia	
	-20.9	1.30(2)	-46.4	25	C	μ=0 A=CN S:80.0w% C_2H_5OH	68Ia	
	-20.1	1.34(2)	-42.2	25	C	μ=0 A=CN S:92.3w% C_2H_5OH	68Ia	
	-38.1	0.56(2)	-117	25	C	μ=0 S:20w% Formamide A=CN	71Ig	
	-38.5	0.57(2)	-117	25	C	μ=0 S:40w% Formamide A=CN	71Ig	
	-36.4	0.60(2)	-109	25	C	μ=0 S:60w% Formamide A=CN	71Ig	
	-32.6	0.64(2)	-96	25	C	μ=0 S:80w% Formamide A=CN	71Ig	
	-31.0	0.64(2)	-92	25	C	S:100w% Formamide A=CN	71Ig	
I_2	-40.2	3.643(25.5°,1)	-61.5	0-35	T	S:Chloroform	66Bd	b
	-40.2	3.929(20°1)	-61.9	4-22	T	CM=2.5×10^{-5}M CL=10^{-4}-4×10^{-4}M S:CH_2Cl_2	62La	b

Metal Ion	ΔH, kJ /mol	Log K	ΔS, J /mol K	T,°C	M	Conditions	Ref.	Re-marks
U22	UREA, 1,1,3-trimethyl-		$C_4H_{10}ON_2$			$H_3CNHC(O)N(CH_3)_2$		
SbA_5	-122.6	(1)	--	25	C	S:CH_2Cl_2 A=Cl	71Oa	j
U23	URIDINE-5'-diphosphoric acid		$C_9H_{14}O_{12}N_2P_2$					L^{3-}
Mg^{2+}	13.47	3.43(1)	110.4	30	C	μ=0.2(Tetra-methylammonium chloride) pH=8.50	73Sa	
U24	URIDINE-5'-phosphoric acid		$C_9H_{13}O_9N_2P$					
Mg^{2+}	7.61	1.69(1)	57.3	30	C	μ=0.2(Tetra-methylammonium chloride) pH=8.50	73Sa	
U25	URIDINE-5'-triphosphoric acid		$C_9H_{15}O_{15}N_2P_3$					
Mg^{2+}	18.41	4.13(1)	139.7	30	C	μ=0.2(Tetra-methylammonium chloride) pH=8.50	73Sa	

V

V1	VALINE		$C_5H_{11}O_2N$			$(CH_3)_2CHCH(NH_2)CO_2H$		L^-
Y^{3+}	-19.355	4.79(25°,1)	26.28	25&35	T	μ=0.1(KCl)	73Sl	a,b,q: ±2.5
	36.928	4.27(25°,2)	205.60	25&35	T	μ=0.1(KCl)	73Sl	a,b,q: ±2.5
La^{3+}	-32.3	3.73(25°,1)	-37.1	15-55	T	μ=0.1(KCl)	74Bc	b
Ce^{3+}	35.171	5.02(25°,1)	214.05	25&35	T	μ=0.1(KCl)	73Sl	a,b,q: ±2.5
	-29.4	3.96(25°,1)	-23.2	15-55	T	μ=0.1(KCl)	74Bc	b
Pr^{3+}	-27.9	3.92(25°,1)	-18.7	15-55	T	μ=0.1(KCl)	74Bc	b
Nd^{3+}	-22.9	3.88(25°,1)	-2.8	15-55	T	μ=0.1(KCl)	74Bc	b
Sm^{3+}	-19.1	3.90(25°,1)	10.4	15-55	T	μ=0.1(KCl)	74Bc	b
Eu^{3+}	-18.4	3.85(25°,1)	11.9	15-55	T	μ=0.1(KCl)	74Bc	b
Gd^{3+}	-19.1	3.90(25°,1)	10.4	15-55	T	μ=0.1(KCl)	74Bc	b
Tb^{3+}	-19.1	4.00(25°,1)	12.3	15-55	T	μ=0.1(KCl)	74Bc	
Dy^{3+}	-17.0	4.03(25°,1)	18.7	15-55	T	μ=0.1(KCl)	74Bc	

Metal Ion	ΔH, kJ /mol	Log K	ΔS, J /mol K	T,°C	M	Conditions	Ref.	Re-marks
V1, continued								
Ho^{3+}	-18.2	4.10(25°,1)	17.4	15-55	T	μ=0.1(KCl)	74Bc	
Er^{3+}	-17.2	4.19(25°,1)	22.5	15-55	T	μ=0.1(KCl)	74Bc	
Yb^{3+}	-18.2	4.30(25°,1)	21.3	15-55	T	μ=0.1(KCl)	74Bc	
Lu^{3+}	-17.0	4.30(25°,1)	25.1	15-55	T	μ=0.1(KCl)	74Bc	
Ni^{2+}	-18.12	5.45(25°,1)	43.5	10-30	T	--	60Pb 57Pa	b
	-15.69	6.87(1)	69.4	25	T	S:44.6w% Dioxane	60Pd	
	-20.9	7.38(1)	71.5	25	T	S:59.7w% Dioxane	60Pd	
	-30.00	8.40(1)	60.2	25	T	S:69w% Dioxane	60Pd	
	-22.68	4.27(25°,2)	5.8	10-30	T	--	60Pb 57Pa	b
	-13.60	5.56(2)	60.2	25	T	S:44.6w% Dioxane	60Pd	
	-34.52	6.31(2)	5.4	25	T	S:59.7w% Dioxane	60Pd	
	-35.56	7.04(2)	15.5	25	T	S:69w% Dioxane	60Pd	
	-18.83	2.48(25°,3)	-15.90	10-30	T	--	60Pb 57Pa	b
	-15.69	3.58(3)	15.5	25	T	S:44.6w% Dioxane	60Pd	
	-21.63	4.30(3)	9.6	25	T	S:69w% Dioxane	60Pd	
	-60.7	9.53(1-2)	-23.0	25	T	μ=0.15	56Lb	
Cu^{2+}	-23.26	8.11(1)	77.4	25	C	μ=0.1M(KNO_3)	72Ia	
	-48.53	15.05(2)	125.1	25	C	μ=0.1M (±)Valine (L)	71Be	
	-49.0	15.0(2)	123.8	25	C	μ=0.1M (±)Valine (D)	71Be	
	-49.0	15.0(2)	125.1	25	C	μ=0.1M (-)Valine (L)	71Be	
	-85.8	14.76(1-2)	0	25	T	μ=0.15	56Lb	
	-48.74	14.84	120.5	25	C	μ=0.1M(KNO_3) R:Cu+L+Serine= [CuL(Serine)]	72Ia	
	-27.2	17.08	236	25	C	μ=0.50($NaClO_4$) R:$Cu^{2+}+L^{-}+A^{3-}$= $CuLA^{2-}$ A=2-Hydroxy-5-sulfobenzoic acid	75Mc	
V2 VALINOMYCIN			$C_{54}H_{90}O_{18}N_6$					
K^+	-37.2	6.079(1)	-9.04	25	C	S:Ethanol CL= 10^{-4}M CM=10^{-5}M	71Mf	
	-23.93	(1)	--	25	C	S:CH_3OH	71Fe	
V3 VITAMIN B_6			$C_8H_{11}O_3N$					
YbA_3	-19.46	(1-3)	--	25	C	A=Cl^-	78Za	

Metal Ion	ΔH, kJ /mol	Log K	ΔS, J /mol K	T,°C	M	Conditions	Ref.	Re-marks

W

W1 WATER H_2O H_2O

Be^{2+}	20.9	-3.30(25°)	5.8	0-60	T	μ=1M(NaCl) R:$2Be^{2+}$+L= $Be_2(OH)^{3+}$+H^+	67Mh	b
	18.54	-3.212	0.8	25	C	3M($NaClO_4$) R:$2Be^{2+}$+L= Be_2OH^{3+}+H^+	62Cd	
	66.9	-8.94(25°	64.0	0-60	T	μ=1M(NaCl) R:$3Be^{2+}$+3L= $Be_3(OH)_3{}^{3+}$+$3H^+$	67Mh	b
	63.51	-8.664	47.28	25	C	3M($NaClO_4$) R:$3Be^{2+}$+3L= $Be_3(OH)_3{}^{3+}$+$3H^+$	62Cd	
	189.5	-25.51(25°)	147.3	·0-60	T	μ=1M(NaCl) R:$5Be^{2+}$+7L= $Be_5(OH)_7{}^{3+}$+$7H^+$	67Mh	b
MgA^{2-}	-14.2	-2.34	-92	25	T	μ=0 ATP^{4-} R: MgA^{2-}+L=Mg(ADP)$^-$ +$HPO_4{}^{2-}$+H^+	63Ga	i:59Bc
$ScA_6{}^{3+}$	26.8	-5.1	-2.9	10	T	μ=1($NaClO_4$) R:$ScA_6{}^{3+}$+L= ScA_5OH^{2+}+H^+ A=H_2O ΔCp=795	54Kb	
	38.9	-4.8	38	25	T	μ=1($NaClO_4$) R:$ScA_6{}^{3+}$+L= ScA_5OH^{2+}+H^+ A=H_2O ΔCp=837	54Kb	
	50	-4.4	79	40	T	μ=1($NaClO_4$) R:$ScA_6{}^{3+}$+L= ScA_5OH^{2+}+H^+ A=H_2O ΔCp=879	54Kb	
	67	-6.6	113	10	T	μ=1($NaClO_4$) R:$2ScA_6{}^{3+}$+2L= $[ScA_5OH]_2{}^{4+}$ A=H_2O ΔCp=-184	54Kb	
	63	-6.0	105	25	T	μ=1($NaClO_4$) R:$2ScA_6{}^{3+}$+2L= $[ScA_5OH]_2{}^{4+}$ A=H_2O ΔCp=-190	54Kb	

Metal Ion	ΔH, kJ /mol	Log K	ΔS, J /mol K	T,°C	M	Conditions	Ref.	Remarks
W1, continued								
ScA_6^{3+}, cont.	63	-5.5	92	40	T	μ=1($NaClO_4$) R:$2ScA_6^{3+}$+2L= $[ScA_5OH]_2^{4+}$ A=H_2O ΔCp=-200	54Kb	
Ce^{4+}	67.4	0.806(24.5°)	241.4	7.6-30.8	T	μ=1.0($NaClO_4$) R: Ce^{4+}+L=$CeOH^{3+}$+H^+	71Eb	b,q: ±5.4
	-64.8	2.27(25°)	231.8	5-35	T	μ=2.0 R:Ce^{4+} +L=$CeOH^{3+}$+H^+	51Ha	b
CeA	59.4	-1.681(25°)	166.9	2.4-25	T	μ=1.0($NaClO_4$) A=H_3AsA_3 R:CeA^{4+} +H_2O=$CeOH^{3+}$+A+H^+	71Eb	b,q: ±10.9
	-9.2	-2.164(25°)	-72.4	2.4-25	T	μ=1.0($NaClO_4$) A=$H_2AsO_3^-$ R:CeA^{3+} +H_2O=$CeOH^{3+}$+HA	71Eb	b,q: ±7.9
	31.4	-0.921(24.5°)	87.4	7.6 30.8	T	μ=1.($NaClO_4$) A=OH^- R:CeA^{3+} +H_2O=$Ce(A)_2^{2+}$+H^+	71Eb	b,q: ±4.2
EuA	-18.0	0.104(20°)	-58	2-81	T	CM=0.01M R:$EuA(L)_{x-1}$+L= $EuA(L)_x$ A=EDTA	69Gb	b,q:±4
Th^{4+}	24.7	-4.10	3.8	25	T	μ=1M($NaClO_4$) R: Th^{4+}+L=$ThOH^{3+}$+H^+ ΔCp=937	65Ba	
	58.2	-7.84	46.0	25	T	μ=1M($NaClO_4$) R:Th^{4+}+2L= $Th(OH)_2^{2+}$+$2H^+$ ΔCp=1272	65Ba	
	61.9	-4.62	118.8	25	T	μ=1M($NaClO_4$) R:$2Th^{4+}$+$2H_2O$= $Th_2(OH)_2^{6+}$+$2H^+$ ΔCp=4	65Ba	
	241.4	-18.99	446.0	25	T	μ=1M($NaClO_4$) R:$4Th^{4+}$+$8H_2O$= $Th_4(OH)_8^{8+}$+$8H^+$ ΔCp=452	65Ba	
	454.0	-36.73	818.8	25	T	μ=1M($NaClO_4$) R:$6Th^{4+}$+$15H_2O$= $Th_6(OH)_{15}^{9+}$+$15H^+$ ΔCp=1000	65Ba	
V^{3+}	46.4	-2.9(25°)	84	25-50	T	μ=1 R:V^{3+}+L= VOH^{2+}+H^+	50Fa	b

Metal Ion	ΔH, kJ /mol	Log K	ΔS, J /mol K	T,°C	M	Conditions	Ref.	Re-marks
W1, continued								
V^{5+}	385	--	--	1	T	R:$HV_{10}O_{28}^{5-}$+2L= $10VO_3^-$+5H^+	61Sb	
$MoAB^{2-}$	-25.1	1.60	-50	25	T	CM=3.7x10^{-4}M 5M(HCl) R:2$MoAB^{2-}$+L= $(MoA)_2O^{4-}$+2H^++2B^- A=ClO_4 B=Cl	62Ha	
MoO_4^{2-}	230.5	--	--	25	C	μ=1M(LiCl) R:7MoO_4^{2-}+4L= Mo_7O_{24}+8OH^-	74Kd	
$Mo_7O_{24}^{6-}$	116.3	--	--	25	C	μ=1M(LiCl) R:$Mo_7O_{24}^{6-}$+2L= $H_2Mo_7O_{24}^{4-}$+2OH^-	74Kd	
	237.2	--	--	25	C	μ=2M(NaCl) R:$Mo_7O_{24}^{6-}$+4L= 7MoO_4^{2-}+8H^+	74Kd	
	230.5	--	--	25	C	μ=3M($NaClO_4$) R:$Mo_7O_{24}^{6-}$+4L= 7MoO_4^{2-}+8H^+	74Kd	
	234	-57.7	-318	25	C	3M($NaClO_4$) R:$Mo_7O_{24}^{6-}$+4L= 7MoO_4^{2-}+8H^+	70Ac	
H_2Mo_7-O_{24}^{4-}	515.5	--	--	25	C	μ=1M(LiCl) R:8$H_2Mo_7O_{24}^{4-}$+L= 7$HMo_8O_{26}^{3-}$+11OH^-	74Kd	
HMo_8-O_{26}^{3-}	164.4	--	--	25	C	μ=1M(LiCl) R:$HMo_8O_{26}^{3-}$+3L= $H_4Mo_8O_{26}$+3OH^-	74Kd	
H_4Mo_8-O_{26}	413.4	--	--	25	C	μ=3M($NaClO_4$) R:7$H_4Mo_8O_{26}$+10L= 8$H_2Mo_7O_{24}$+32H^+	74Kd	
$W_6O_{21}^{6-}$	238.9	-52.5	-205	25	C	3M($NaClO_4$) R:$W_6O_{21}^{6-}$+3L= 6WO_4^{2-}+6H^+	70Ac 69Aa	
$HW_6O_{21}^{5-}$	269.9	65.2	343	25	C	μ≅0.3 R:$HW_6O_{21}^{5-}$ +3L=6WO_4^{2-}+7H^+	62Dd	
	197	65(25°)	586	17-35	T	μ≅0.3 R:$HW_6O_{21}^{5-}$ +3L=6WO_4^{2-}+7H^+	62Dd	b
	261.5	-60.8	-285	25	C	3M($NaClO_4$) R:$HW_6O_{21}^{5-}$+3L= 6WO_4^{2-}+7H^+	70Ac 69Aa	
$W_{12}O_{41}^{10-}$	7.9	-1.69	-5.4	25	C	μ=3M($NaClO_4$) R:$W_{12}O_{41}^{10-}$+L=	69Aa	

Metal Ion	ΔH, kJ /mol	Log K	ΔS, J /mol K	T,°C	M	Conditions	Ref.	Re-marks
W1, continued								
$W_{12}O_{41}^{10-}$, cont.	530.9	-123.2	-577	25	C	$2HW_6O_{21}^{5-}$ 3M($NaClO_4$) R:$W_{12}O_{41}^{10-}$+7L= $12WO_4^{2-}$+$14H^+$	70Ac	
U^{4+}	49.0	-0.678	151	25	T	μ=0 R:U^{4+}+L=UOH^{3+}+H^+	55Kc	
	46.0	-0.66	138	25	T	μ=0 R:U^{4+}+L=UOH^{3+}+H^+	55Ba	
	44.8	-1.168	<u>172</u>	25	T	μ=0.19 R:U^{4+}+L=UOH^{3+}+H^+	55Ba	
	46.9	-1.466	130	25	T	μ=0.5 R:U^{4+}+L=UOH^{3+}+H^+	55Kc	
	45.93	--	--	0-10	T	0.1M($HClO_4$) pH<2.5 CM=0.2M R:U^{4+}+$2Fe^{3+}$+2L= UO_2^{2+}+$2Fe^{2+}$+$4H^+$	77Gb	
	126.23	-11.66(10°)	211.53	0-10	T	0.1M($HClO_4$) pH<2.5 CM=0.2M R:U^{4+}+2L=UO_2^{2+}+ $4H^+$+$2e^-$	77Gb	
UO_2^{2+}	87.0	-5.9	180	25	T	μ=0.0347 R:UO_2^{2+} +L=UO_2OH^++H^+	57Ha	
	46.0	-5.7(25°)	46	25&94	T	μ=0.5 R:UO_2^{2+}+L= UO_2OH^++H^+	62Ba	a,b
	28.0	-6.2	-25	25	T	μ=0.0347 R:$2UO_2^{2+}$ +L=$UO_2UO_3^{2+}$+$2H^+$	57Ha	
	42.7	-5.92(25°)	29.7	25&94	T	μ=0.5 R:$2UO_2^{2+}$+ 2L=$(UO_2)_2(OH)_2^{2+}$ +$2H^+$	62Ba	a,b
	39.7	-6.02	18.0	25	C	3M($NaClO_4$) R:$2UO_2^{2+}$+2L= $(UO_2)_2(OH)_2^{2+}$+$2H^+$	68Ag	
	39.7	-6.01	18.0	25	C	3M($NaClO_4$) R:$2UO_2^{2+}$+2L= $(UO_2)_2(OH)_2^{2+}$+$2H^+$	70Ac	i:62Sa
	39.505	-6.030	17.11	25	C	3.0M($NaClO_4$) R:UO_2^{2+}+2L= $(UO_2)_2(OH)_2^{2+}$+$2H^+$	62Sa	
	75	-13.20	0	25	C	3.0M($NaClO_4$) R:$3UO_2^{2+}$+4L= $(UO_2)_3(OH)_4$+$4H^+$	62Sa	
	105.0	-16.2(25°)	42	25&94	T	μ=0.5 R:$3UO_2^{2+}$ +5L=$(UO_2)_3(OH)_5^+$ +$5H^+$	62Ba	a,b
	102.1	-16.54	25	25	C	3M($NaClO_4$)	68Ae	

Metal Ion	ΔH, kJ /mol	Log K	ΔS, J /mol K	T,°C	M	Conditions	Ref.	Remarks
W1, continued								
UO_2^{2+}, cont.	102.1	-16.6	25	25	C	R:$3UO_2^{2+}+5L=(UO_2)_3(OH)_5^++5H^+$ 3M($NaClO_4$)	70Ac	i:62Sa
	105	-16.55	33	25	C	R:$3UO_2^{2+}+5L=(UO_2)_3(OH)_5^++5H^+$ 3.0M($NaClO_4$)	62Sa	
	100	-19.42	-33	25	C	R:$3UO_2^{2+}+5L=(UO_2)_3(OH)_5^++5H^+$ 3.0M($NaClO_4$)	62Sa	
						R:$4UO_2^{2+}+6L=(UO_2)^4(OH)_6+6H^+$		
Pu^{4+}	35.6	-1.733(25°)	8.4	15.4& 25	T	μ=2.0($HClO_4$) R: $Pu^{4+}+L=PuOH^{3+}+H^+$	60Ra	a,b
MnA_6^{3+}	20.1	-0.056(25°)	65.7	1-35	T	μ=4.0 R:$MnA_6^{3+}+L=MnA_5OH^-+H^+$ A=H_2O	65Wb 65Wd	b
Re(IV)	100	-1.41(25°)	335	5-25	T	1-6M(HCl) R:$[Re_2OCl_{10}]^{4-}+L$ $2[ReOHCl_5]^{2-}$	65Ja	b
Fe^{2+}	50.2	-6.74(25°)	39.7	20-40	T	μ=0.505-2.19 R: $Fe^{2+}+L=FeOH^++H^+$	63Bc	b
[Fe]	102.5	--	--	25	C	[Fe]=G. gouldii Hemerythrin=$HrFe_2$ Tris-cacodylate buffer (≅0.1M Tris) R:$HrFe_2+mL+H_2=HrFe_2L_m+2H^+$	71Lb	
Fe^{3+}	43.5	-2.17	104.6	25	T	μ=0 R:$Fe^{3+}+L=Fe(OH)^{2+}+H^+$	57Ma	
	35.6	-2.609(25°)	69.5	25-80	T	μ=0.1M($NaClO_4$) R:$Fe^{3+}+L=FeOH^{2+}+H^+$	77Sd	
	41.4	-2.32(35.3°)	90.0	18.2& 35.5	T	μ=0.15 R:$Fe^{3+}+L=FeOH^{2+}+H^+$	60Rb	a,b
	76.1	-1.959(25°)	218.0	20&25	T	$CClO_4^-$=0.015M CM=0.005M R: $Fe^{3+}+L=FeOH^{2+}+H^+$	44La	a
	35.6	-2.757(25°)	66.6	25-80	T	μ=0.5M($NaClO_4$) R:$Fe^{3+}+L=FeOH^{2+}+H^+$	77Sd	
	34.3	-2.833(25°)	60.8	25-80	T	μ=1.0M($NaClO_4$) R:$Fe^{3+}+L=FeOH^{2+}+H^+$	77Sd	
	42.7	-2.78	88	25	T	μ=1 R:$Fe^{3+}+L=Fe(OH)^{2+}+H^+$	57Ma	
	32.2	-2.588(45°)	51.7	25-80	T	μ=2.67($NaClO_4$) R:$Fe^{3+}+L=FeOH^{2+}+H^+$	77Sd	

Metal Ion	ΔH, kJ /mol	Log K	ΔS, J /mol K	T,°C	M	Conditions	Ref.	Re-marks
W1, continued								
Fe^{3+}, cont.	32.2	-2.936(25°)	<u>51.8</u>	25-80	T	μ=2.67M($NaClO_4$) R:Fe^{3+}+L=$FeOH^{2+}$+H^+	77Sd	
	46.0	-3.046	96	25	C	3M($NaClO_4$) R: Fe^{3+}+L=$FeOH^{2+}$+H^+	68Af	i:62Sb
	46.0	-3.08	96	25	C	3M($NaClO_4$) R: Fe^{3+}+L=$FeOH^{2+}$+H^+	70Ac	i:62Sb
	56.5	-2.86	134	25	T	μ=0 R:2Fe^{2+}+2L= $Fe_2(OH)_2^{4+}$+2H^+	57Ma	
	51.0	-2.71	117	25	T	μ=1 R: 2Fe^{2+}+2L =$Fe_2(OH)_2^{4+}$+2H^+	57Ma	
	41.8	-2.959	84	25	C	3M($NaClO_4$) R:2Fe^{3+}+2L= $Fe_2(OH)_2^{4+}$+2H^+	68Af	i:62Sb
	41.8	-2.93	83.7	25	C	3M($NaClO_4$) F:2Fe^{3+}+2L= $Fe_2(OH)_2^{4+}$+2H^+	70Ac	i:62Sb
	41.0	-2.914(25°)	<u>81.6</u>	15-51	T	μ=3($NaClO_4$) R:2Fe^{3+}+2L= $Fe_2(OH)_2^{4+}$+2H^+	55Mb	b
	41.0	--	--	--	T	3M($NaClO_4$) CM=0.04 R:2Fe^{3+}+2L= $Fe_2(OH)_2^{4+}$+2H^+	54Ma	
	59.8	-5.770	92	25	C	3M($NaClO_4$) R:3Fe^{3+}+4L= $Fe_3(OH)_4^{5+}$+4H^+	68Af	i:62Sb
	59.8	-5.79	92	25	C	3M($NaClO_4$) R:3Fe^{3+}+4L= $Fe_3(OH)_4^{3+}$+4H^+	70Ac	i:62Sb
FeA^-	41.8	-9.3(25°)	-38	13.7-42.4	T	μ=0.1(KCl) FeA^-+L=$FeOHA^{2-}$ +H^+ A=Cyclohexanediaminetetraacetate	63Gd	b
	42	-7.55(25°)	-8	13-42.4	T	μ=1(KCl) R:FeA^- +L=$FeOHL^{2-}$+H^+ A=EDTA	63Gd	b
	19.7	-12.2(25°)	-167	13-42.4	T	μ=1(KCl) R:2FeA^{2-}+2L= $(FeOHL)_2^{4-}$+2H^+ A=EDTA	63Gd	b
FeA_5B	51.5	-3.52	105.4	25	T	μ=0.1(NaCl) R:FeA_5B+L=FeA_5L^{3-} +B^- A=CN B=NO_2	66Se	

Metal Ion	ΔH, kJ /mol	Log K	ΔS, J /mol K	T,°C	M	Conditions	Ref.	Remarks
W1, continued								
Co^{2+}	34.3	9.82(25°)	302.9	15-40	T	μ=0.327($NaClO_4$) R:Co^{2+}+H_2O= $CoOH^+$+H^+	62Bb 63Be	b,q:±8
CoA_2	-11.92	1.76(1)	-6.3	25	C	S:1-Butanol A=ClO_4	68Hb	
	-19.66	3.08(1-2)	-7.1	25	C	S:1-Butanol A=ClO_4	68Hb	
	-22.38	4.10(1-3)	3.3	25	C	S:1-Butanol A=ClO_4	68Hb	
	-23.43	4.64(1-4)	10.0	25	C	S:1-Butanol A=ClO_4	68Hb	
	-24.3	5.6(1-5)	25	25	C	S:1=Butanol A=ClO_4	68Hb	
Co^{3+}	42	-1.8	105	25	T	μ=0.1 R:Co^{3+}+L= $CoOH^{2+}$+H^+	56Sd	
CoA_2B_2	36.19	4.747(80°)	11.7	75-90	T	μ=0.01 R:CoA_2B_2 +L=CoA_2BL+B A=Dimethylglyoxi- mate B=Thiocyanate	70Fc	b
Co(III)	-18.4	-0.271(25°,1)	-67	-60-50	T	S:CD_3OD A=Corrin ring B=Methyl	68Fc	b
	-18.8	-0.088(25°,1)	-64.8	-60-50	T	S:CD_3OD A=Corrin ring B=Vinyl	68Fc	b
Ni^{2+}	32.47	-9.76(25°)	-77.8	25-50	T	0.03-0.1M($NaClO_4$) R:Ni^{2+}+L=$NiOH^+$+H^+	63Bd	b
	179.9	-28.48	58.6	25	C	3M(NaCl) R:4Ni^{2+}+4L= $Ni_4(OH)_4^{4+}$+4H^+	68Ad	
	179.9	-28.52	58	25	C	3M($NaClO_4$) R:4Ni^{2+}+4L= $Ni_4(OH)_4^{4+}$+4H^+	70Ac	
NiA^{2+}	-20.1	(1-2)	--	12-58	T	CM=0.1M A=1,4,8,11-Tetra- azaundecane	72If	b
	-25.1	1.0(1-2)	-64.4	25	T	μ=0.5M($NaClO_4$) A=1,5,8,12-Tetra- azadodecane	74Fa	
	-22.2	0.5(1-2)	-64.4	25	T	μ=0.5M($NaClO_4$) A=1,4,8,11-Tetra- azaundecane	74Fa	
	-28.9	1.7(1-2)	-64.4	25	T	μ=0.5M($NaClO_4$) A=Triethylene- tetramine	74Fa	

Metal Ion	ΔH, kJ /mol	Log K	ΔS, J /mol K	T,°C	M	Conditions	Ref.	Remarks
W1, continued								
NiA	-13.88	-0.366(25°,<u>1-2</u>)	-53.6	5-80	T	S:Water, L A=1,4,8,11-Tetraazacyclo-tetradecane	79Vb	
	-11.53	-0.420(25°,<u>1-2</u>)	-46.9	5-80	T	0.10M($NaClO_4$) S:Water, L A=1,4,8,11-Tetraazacyclo-tetradecane	79Vb	
NiA_2	-12.76	1.80(<u>1</u>)	-8.4	25	C	Variable μ S:1-Butanol A=ClO_4	68Hb	
	-20.1	3.23(<u>1-2</u>)	-5.4	25	C	Variable μ S:1-Butanol A=ClO_4	68Hb	
	-21.97	4.36(<u>1-3</u>)	9.6	25	C	Variable μ S;1-Butanol A=ClO_4	68Hb	
	-22.8	5.20(<u>1-4</u>)	23.0	25	C	Variable μ S:1-Butanol A=ClO_4	68Hb	
	-23.8	5.8(<u>1-5</u>)	33	25	C	Variable μ S:1-Butanol A=ClO_4	68Hb	
RhA_2BC	17.2	-2.70(50°)	0.4	55-81	T	μ=0.2M A=Ethyl-enediamine B=OH C=I R:trans-RhA_2BC^++L=trans-RhA_2BL^{2+}+C^-	72Fl	b,q: ±0.8
RhA_3B_3	-20.1	-4.25(25°)	<u>-149.0</u>	25-85	T	0.1F($HClO_4$) R:RhA_3B_3+2L=$RhAB_3L_2^{2+}$+2A^- A=Cl B=H_2O	66Ba	b
RhA_5B	-5.4	-2.249	-59.8	50	T	μ=0.2M A=NH_3 B=Cl R:RhA_5B^{2+}+L=RhA_5L^{3+}+B^-	67Pb 67Pa	q:±2.5
	-2.9	-2.159	-49.4	50	T	μ=0.2M A=NH_3 B=Br R:RhA_5B^{2+}+L=RhA_5L^{3+}+B^-	67Pb 67Pa	q:±1.7
	7.5	-2.682	-26.8	50	T	μ=0.2M A=NH_3 B=I R:RhA_5B^{2+}+L=RhA_5L^{3+}+B^-	67Pb 67Pa	q:±1.2
$PtAB_2$	17	-2.66(25°)		15-35	T	μ=0.318M A=Ethylenediamine B=Cl R:$PtAB_2$+L=$PtALB^+$+B^-	73Cd	b,q:±4
$PtALB^+$	-4	-3.84(25°)		15-35	T	μ=0.318M A=Ethylenediamine B=Cl R:$PtALB^+$+L=$PtAL_2^{2+}$+B^-	73Cd	b,q:±8
PtA_4^{2-}	21	<u>-1.8</u>	38	25	T	μ→0 R:PtA_4^{2-}+L=PtA_3L^-+A^- A=Cl	55Gc	

Metal Ion	ΔH, kJ /mol	Log K	ΔS, J /mol K	T, °C	M	Conditions	Ref.	Remarks
W1, continued								
PtA_4^{2-}, cont.	18.4	1.89(25°)	25.1	15-35	T	$0.5M(HClO_4)$ R: $PtA_4^{2-}+L=PtA_3L^-+A^-$ A=Cl	67De	b
$PtAB_3^-$	8.8	-1.8	-4	25	T	μ→0 R: $PtAB_3^-+L=PtAB_2L+A^-$ $A=NH_3$ B=Cl	58Ea	
Cu^{2+}	-20.5	5.65	39.3	25	C	$\mu=0.1M(NaNO_3-NaClO_4)$ R: $2Cu^{2+}+2A+2L=Cu_2A_2(OH)_2^{2+}+2H^+$ A=2,2'-Bipyridyl	76Al	
	-27.2	0.4	84	25	C	$\mu=0.1M(NaNO_3-NaClO_4)$ R: $Cu^{2+}+A+L=CuA(OH)^++H^+$ A=2,2'-Bipyridyl	76Al	
	46-50	-8	≅0	25	C	$3M(NaClO_4)$ ΔH values set assuming ΔS=0 R: $Cu^{2+}+L=CuOH^++H^+$	68Ae	
	66.1	-10.6	18.4	25	C	$3M(NaClO_4)$ R: $2Cu^{2+}+2L=Cu_2(OH)_2^{2+}+2H^+$	68Ae	
	66.1	-10.63	18.4	25	C	$3M(NaClO_4)$ R: $2Cu^{2+}+2L=Cu_2(OH)_2^{2+}+2H^+$	70Ac	
CuAB	4	-1.39(25°,1)	-21	0.3-42.5	T	$\mu=0.1M(KNO_3)$ A=OH B=N,N'-Dihydroxyethylenediamine R: $(CuAB)_2^{2+}+2L=2CuABL^+$	59Ge	b,q:±29
	-4.2	-3.81(25°)	-88	0.3-42.5	T	$\mu=0.1M(KNO_3)$ A=OH B=N,N'-Dimethylethylenediamine R: $(CuAB)_2^{2+}+2L=2CuABL^+$	59Ge	b,q: ±1.7
	-17	-2.20(25°)	-96	0.3-42.5	T	$\mu=0.1M(KNO_3)$ A=OH B=N-Hydroxyethylenediamine R: $(CuAB)_2^{2+}+2L=2CuABL^+$	59Ge	b,q:±12
	-12	-3.88(25°)	-117	0.3-42.5	T	$\mu=0.1M(KNO_3)$ A=OH B=N,N,N',N'-	59Ge	b,q:±12

Metal Ion	ΔH, kJ /mol	Log K	ΔS, J /mol K	T,°C	M	Conditions	Ref.	Re-marks
W1, continued								
CuAB, cont.						Tetramethylethylenediamine $R:(CuAB)_2^{2+}+2L=2CuABL^+$		
CuA^{2+}	17.6	-7.7	88	25	C	μ=0.1M($NaNO_3$-$NaClO_4$) $R:CuA^{2+}+L=CuL(OH)^+ + H^+$ A=2,2'-Bipyridyl	76A1	
	40.4	-8.08	-19	25	C	μ=0.1($NaNO_3$) $R:CuA^{2+}+L=CuAOH^+ + H^+$ A=2,2':6',2"-Terpyridine	79Ca	
	73.4	-18.47	-107	25	C	μ=0.1($NaNO_3$) $R:CuA^{2+}+2L=[CuA(OH)_2]+2H^+$ 2,2':6',2"-Terpyridine	79Ca	
	68.6	-10.6	26.8	25	C	μ=0.1M($NaNO_3$-$NaClO_4$) $R:2CuA^{2+}+2L=Cu_2A_2(OH)_2^{2+}+2H^+$ A=2,2'-Bipyridyl	76A1	
Zn^{2+}	56.1	-8.96(25°)	<u>17</u>	15-42	T	μ=0 $R:Zn^{2+}+L=ZnOH^+ + H^+$	62Pb	b
ZnA	60.7	-10.56	2	25	C	μ=0.1M(KNO_3) $R:ZnA+L=ZnAOH^- + H^+$ A=Ethylenediamine-N,N'-diacetic acid	79Gc	
Cd^{2+}	54.8	-10.2	-113	25	C	3M($LiClO_4$) $R:Cd^{2+}+L=CdOH^+ + H^+$	67Ae 70Ac	
	45.6	-9.1	-21.3	25	C	3M($LiClO_4$) $R:2Cd^{2+}+L=Cd_2OH^{3+}+H^+$	67Ae 70Ac	
	169.9	-31.8	-39.3	25	C	3M($LiClO_4$) $R:4Cd^{2+}+4H_2O=Cd_4(OH)_4^{4+}+H^+$	67Ae 70Ac	
CdA_2^{2+}	8.54	-1.53(<u>1</u>)	-0.8	25	C	S:56.17w% Ethanol A=H_2O	59Va	
CdA_3^{2+}	8.54	-1.67(<u>1</u>)	-3.3	25	C	S:56.17w% Ethanol A=H_2O	59Va	
CdA_4^{2+}	8.54	-1.70(<u>1</u>)	-3.8	25	C	S:56.17w% Ethanol A=H_2O	59Va	

Metal Ion	ΔH, kJ /mol	Log K	ΔS, J /mol K	T, °C	M	Conditions	Ref.	Re-marks
W1, continued								
Hg^{2+}	-20.7	-3.65	0	25	C	μ=1M($NaClO_4$) R:Hg^{2+}+L=$HgOH^+$+H^+	75Cd	
	30.25	-3.56	33.26	25	C	3M($NaClO_4$) R:Hg^{2+}+L=$HgOH^+$+H^+	67Ad	
	30.1	-3.59	33.0	25	C	3M($NaClO_4$) R:Hg^{2+}+L=$HgOH^+$+H^+	70Ac	
	-20.7	-2.64	19	25	C	μ=1M($NaClO_4$) R:$HgOH^+$+H_2O= $Hg(OH)_2$+H^+	75Cd	
	45.145	-6.22	35.1	20	T	μ=0 R:Hg^{2+}+2L= $Hg(OH)_2$+2H^+	58Aa	
	41.17	-6.21	19.20	25	C	3M($NaClO_4$) R:Hg^{2+}+2L= $Hg(OH)_2$+2H^+	67Ad	
	41.0	-6.23	19.2	25	C	3M($NaClO_4$) R:Hg^{2+}+2L= $Hg(OH)_2$+2H^+	70Ac	
	12.80	-2.70	-8.70	25	C	3M($NaClO_4$) R:2Hg^{2+}+L= $Hg_2(OH)^{3+}$+H^+	67Ad	
	13.0	-2.71	-8.8	25	C	3M($NaClO_4$) R:2Hg^{2+}+L= $Hg_2(OH)^{3+}$+H^+	70Ac	
	53.1	-5.20	78.6	25	C	3M($NaClO_4$) R:2Hg^{2+}+2L= $Hg_2(OH)_2{}^{2+}$+2H^+	70Ac	
	53.14	-5.20	78.6	25	C	3M($NaClO_4$) R:2Hg^{2+}+2L= $Hg_2(OH)_2{}^{2+}$+2H^+	67Ad	
HgA^{2+}	-76	21.86	161	25	C	μ=0.5M($NaClO_4$) R:HgA^{2+}+L= $Hg(OH)A^+$+H^+ A=Imidazole	77Mc	q:±15
HgA	20	-3.05	8	25	C	μ=1M($NaClO_4$) R:HgA^++L=HgA(OH) +H^+ A=Cl^-	76Ce	
	20.7	-2.6	19	25	C	μ=1M($NaClO_4$) R:HgA^++L=HgA(OH) +H^+ A=OH^-	76Ce	
	21.9	-3.4	7	25	C	μ=1M($NaClO_4$) R:HgA^++L=HgA(OH) +H^+ A=SCN^-	76Ce	

Metal Ion	ΔH, kJ /mol	Log K	ΔS, J /mol K	T, °C	M	Conditions	Ref.	Remarks
W1, continued								
$B(A)_3$	16.3	-9.0	-118	25	C	μ=2M(KCl) R: $B(A)_3)+L=B(A)_4^-$ $+H^+$ A=OH	73Bd	
Al^{3+}	76.6	-7.45(25°)	113	25-125	T	μ=1M(KCl) R: $2Al^{3+}+2L=$ $Al_2(OH)_2^{4+}+2H^+$ ΔCp=50	71Tb	b,q: ±10.9
	129.3	-13.36(25°)	176	25-125	T	μ=1M(KCl) R: $3Al^{3+}+4L=$ $Al_3(OH)_4^{5+}+4H^+$ ΔCp=377	71Tc	b,q: ±14.2
	1,099.6	-110.45(25°)	1,577	25-125	T	μ=1M(KCl) R: $14Al^{3+}+34L=$ $Al_{14}(OH)_{34}^{8+}+34H^+$ ΔCp=3067	71Tc	b
$AlAB_2$	-51.46	2.65(28°,1̲)	-100	28-130	T	S:Heptane A=Ethyl B=Cl	78Pa	b
In^{3+}	20.29	-4.42	-16.7	25	C	3.0M($NaClO_4$) R: $In^{3+}+L=$ $In(OH)^{2+}+H^+$	61Sc	
	58	-8.32	38	25	C	3.0M($NaClO_4$) R: $In^{3+}+2L=$ $In(OH)_2^{+}+2H^+$	61Sc	
	n(10.18)	0.3($10^{-4.69n}$)	-2.4+ n(12.7)	25	C	3.0M($NaClO_4$) R: $(n+1)In^{3+}+2nL=$ $In[In(OH)]_n^{3+n}+$ $2nH^+$	61Sc	
Tl^{3+}	87.9	-1.19(1̲)	272	25	C	μ=3M($HClO_4$)	76Kl	
	77.0	-1.87(1̲)	223	25	C	μ=3M($HClO_4$) S:0.023 mole fraction DMSO in water	76Kl	
	58.6	-2.73(1̲)	144	25	C	μ=3M($HClO_4$) S:0.049 mole fraction DMSO in water	76Kl	
	104.6	-0.55(1̲)	353	25	C	μ=3M($HClO_4$) S:0.346 mole fraction Aceto- nitrile in water	76Kl	
	119.2	-1.44(2̲)	372	25	C	μ=3M($HClO_4$)	76Kl	
	94.6	-1.28(2̲)	292	25	C	μ=3M($HClO_4$) S:0.023 mole fraction DMSO in water	76Kl	

Metal Ion	ΔH, kJ /mol	Log K	ΔS, J /mol K	T,°C	M	Conditions	Ref.	Remarks
W1, continued								
Tl^{3+}, cont.	137.6	-0.67(2)	437	25	C	μ=3M($HClO_4$) S:0.346 mole fraction Acetonitrile in water	76Kl	
	36.0	-0.90(2)	103	25	C	μ=3M($HClO_4$) S:0.049 mole fraction DMSO in water	76Kl	
	100	-1.41	310	25	C	μ=3.0M($LiClO_4$) R:Tl^{3+}+L=$Tl(OH)^{2+}$+H^+	73Kj	
	88	-1.180	293	25	C	μ=3M($LiClO_4$, $HClO_4$) R:Tl^{3+}+HL=TlL^{2+}+H^+	67Ya	q:±25
	109	-1.15	343	25	C	μ=3.0M($LiClO_4$) A=OH R:$Tl(A)^{2+}$+L=$Tl(A)_2{}^+$+H^+	73Kj	q:±12
	88	-1.415	272	25	C	μ=3M($LiClO_4$, $HClO_4$) A=OH R:TlA^{2+}+HL=$TlA_2{}^+$+H^+	67Fa	q:±100
$GeA_6{}^{2-}$	39.7	-5.28(25°)	33	0-25	T	R:$GeA_6{}^{=}$+2L=[GeA_4LOH]$^-$+HA+A^- A=F	64Rb	a,b
	-2.47	-3.72	-79	25	T	μ=0.4(HCl+NaCl) R:$GeA_6{}^{2-}$+L=GeA_5L^-+A^- A=F	64Ra	
SnA_5	-101.7	(1)	--	25	C	S:Ethylene chloride A=Cl	67Oc	
Pb^{2+}	110.9	-22.87	-66.1	25	C	3M($NaClO_4$) R:3Pb^{2+}+4L=$Pb_3(OH)_4{}^{2+}$+4H^+	62Cc	
	83.97	-19.25	-87.0	25	C	3M($NaClO_4$) R:4Pb^{2+}+4L=$Pb_4(OH)_4{}^{4+}$+4H^+	62Cc	
	206.86	-42.12	-113.0	25	C	3M($NaClO_4$) R:6Pb^{2+}+8L=$Pb_6(OH)_8{}^{4+}$+8H^+	62Cc	
PbA	38.5	-11.02	-82	25	C	μ=0.1M(KNO_3) R:ZnA+L=ZnAOH+H^+ A=Ethylenediamine-N,N'-diacetic acid	79Gc	
Bi^{3+}	17	-1.58	27	25	C	μ=3M($NaClO_4$) R:Bi^{3+}+L=$BiOH^{2+}$+H^+	75Oa	q:±6

Metal Ion	ΔH, kJ /mol	Log K	ΔS, J /mol K	T,°C	M	Conditions	Ref.	Re-marks
W1, continued								
Bi^{3+}, cont.	133	0.33	450	25	C	μ=3M($NaClO_4$) R:$6Bi^{3+}$+12L= $Bi_6(OH)_{12}{}^{6+}$+$12H^+$	75Oa	
IO_4^-	61.9	1.46(25°,<u>1-2</u>)	234	2-18	T	μ=0 <u>R</u>:IO_4^-+2L= $H_4IO_6^-$	68Ka	b
W2 WATER, deuterated			D_2O			D_2O		L
Pu^{4+}	81.2	-1.94(25°)	<u>48.1</u>	15.4& 25	T	μ=2.0($HClO_4$) S:D_2O R:Pu^{4+}+L= $PuOD^{3+}$+D^+	60Ra	a,b

<u>X</u>

Metal Ion	ΔH, kJ /mol	Log K	ΔS, J /mol K	T,°C	M	Conditions	Ref.	Re-marks
X1 XANTHOGENIC ACID, <u>iso</u>-butyl			$C_5H_{10}OS_2$			C_4H_9OCSSH		L
Ni^{2+}	5.73	(<u>1-2</u>)	--	30	C	μ=0	74Cb	g:K, q:±0.46
X2 XANTHOGENIC ACID, <u>n</u>-butyl			$C_5H_{10}S_2$			C_4H_9OCSSH		L^-
Ni^{2+}	3.51	(<u>1-2</u>)	--	30	C	μ=0	74Cb	g:K
X3 XANTHOGENIC ACID, ethyl-			$C_3H_6OS_2$			C_2H_5OCSSH		L^-
Ni^{2+}	4.81	(<u>1-2</u>)	--	30	C	μ=0	74Cb	g:K
X4 XANTHOGENIC ACID, methyl-			$C_2H_4OS_2$			CH_3OCSSH		L^-
Ni^{2+}	10.63	(<u>1-2</u>)	--	30	C	μ=0	74Cb	g:K
X5 XANTHOGENIC ACID, propyl-			$C_4H_8OS_2$			C_3H_7OCSSH		L^-
Ni^{2+}	4.90	(<u>1-2</u>)	--	30	C	μ=0	74Cb	g:K
X6 XANTHOSINE			$C_{10}H_{12}O_6N_2$			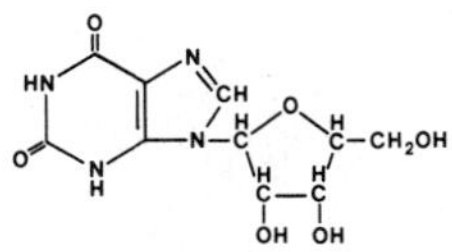		
Mg^{2+}	15.5	2.22(25°,<u>1</u>)	95.4	25-45	T	μ=0.10(KNO_3)	78Ra	b
Ca^{2+}	-6.7	2.37(25°,<u>1</u>)	22.6	25-45	T	μ=0.10(KNO_3)	78Ra	b

Metal Ion	ΔH, kJ /mol	Log K	ΔS, J /mol K	T, °C	M	Conditions	Ref.	Remarks
X6, continued								
Mn^{2+}	-2.1	2.48(25°,1)	45.6	25-45	T	μ=0.10(KNO_3)	78Ra	b
Co^{2+}	-2.1	2.23(25°,1)	36.0	25-45	T	μ=0.10(KNO_3)	78Ra	b
Ni^{2+}	-7.5	2.88(25°,1)	29.7	25-45	T	μ=0.10(KNO_3)	78Ra	b
Cu^{2+}	-7.9	2.91(25°,1)	28.9	25-45	T	μ=0.10(KNO_3)	78Ra	b
Zn^{2+}	-9.2	2.26(25°,1)	13.0	25-45	T	μ=0.10(KNO_3)	78Ra	b
X7 XENON			Xe					
[Fe]	-37.6	2.12(25°,1)	-84	20-30	T	CL=10% Solution of L [Fe]=Cyano-metmyoglobin	70Ee	b,j
	-30.1	2.12(25°,1)	-58	20-30	T	CL=10% Solution of L [Fe]=Met-myoglobin	70Ee	b,j
	-21.3	1.98(25°,1)	-33.5	20-30	T	CL=10% Solution of L [Fe]=Myoglobin	70Ee	b,j
X8 XYLENOL ORANGE								
Mn^{2+}	-18.12	10.00(30°,1)	133.5	20-40	T	μ=0	79Sh	b
Fe^{2+}	-21.76	11.10(30°,1)	145.9	20-40	T	μ=0	79Sh	b
Co^{2+}	-22.68	12.41(30°,1)	166.5	20-40	T	μ=0	79Sh	b
Ni^{2+}	-29.00	14.00(30°,1)	178.2	20-40	T	μ=0	79Sh	b
Cu^{2+}	-32.64	14.81(30°,1)	182.1	20-40	T	μ=0	79Sh	b
	18.0	9.86(25°)	268	15-30	T	μ=0.1M($NaClO_4$) pH=3.5-5.5 R:$2Cu^{2+}+LH_2^{4-}=Cu_2LH_2$	76Ya	b
Zn^{2+}	-23.56	12.95(30°,1)	175.5	20-40	T	μ=0	79Sh	b

EMPIRICAL FORMULA INDEX

EMPIRICAL FORMULA INDEX

EMPIRICAL FORMULA INDEX

ELEMENT INDEX

ELEMENT INDEX

SYNONYM INDEX

REFERENCE INDEX

1980

80Aa F. J. Andres and J. M. Merino, Bull. Soc. Chim. (France), No. 1-2, I61-I65 (1980).

80Ab G. Arena, R. Cali, M. Grasso, S. Musumeci, and S. Sammartano, Thermochimica Acta, 36, 329-342 (1980).

80Ba U. Biader Ceipidor, V. Carunchio, G. Dascenzo, and M. Tomassetti, Thermochimica Acta, 35, 197-200 (1980).

80Bb J. S. Bradshaw, G. E. Maas, J. D. Lamb, R. M. Izatt, and J. J. Christensen, J. Am. Chem. Soc., 102, 467-474 (1980).

80Ca G. R. Choppin and R. E. Cannon, Inorg. Chem., 19, 1889-1892 (1980).

80Cb B. G. Cox, D. Knop, and H. Schnelder, J. Phys. Chem., 84, 320-323 (1980).

80Da D. DeMarco, A. Bellomo, and A. De Robertis, J. Inorg. Nucl. Chem., 42, 559-609 (1980).

80Fa L. Fabbrizzi, M. Micheloni, and P. Paoletti, Inorg. Chem., 19, 535-538 (1980).

80Fb L. Fabbrizzi, M. Micheloni, and P. Paoletti, J.C.S. Dalton, 134-136 (1980).

80Ha G. G. Herman, A. M. Goeminne, C. T. Huys, J. Schaubroeck, and Z. Eeckhaut, Thermochimica Acta, 37, 301-305 (1980).

80Ja O. Johansson, I. Persson, and M. Wedborg, Marine Chem., 8, 191-198 (1980).

80Ka P. K. Khopkar and J. N. Mathur, Thermochimica Acta, 37, 71-78 (1980).

80La J. D. Lamb, R. M. Izatt, C. S. Swain, and J. J. Christensen, J. Am. Chem. Soc., 102, 475-479 (1980).

80Lb J. D. Lamb, R. M. Izatt, C. S. Swain, J. S. Bradshaw, and J. J. Christensen, J. Am. Chem. Soc., 102, 479-482 (1980).

80Pa B. Pispisa and S. Paoletti, J. Phys. Chem., 84, 24-28 (1980).

80Sa J. P. Shukla, Thermochimica Acta, 36, 189-195 (1980).

80Sb A. M. Smetana and A. I. Popov , J. Solution Chem., 9, 183-196 (1980).

80Ta Y. Takeda and H. Yano, Bull. Chem. Soc. Japan, 53, 1720-1722 (1980).

80Ua I. Uruska and H. Inerowicz, J. Solution Chem., 9, 97-108 (1980).

1979

79Aa G. Arena, E. Rizzarelli, and B. Sarkar, Inorg. Chim. Acta, 37, L555-L557 (1979).

79Ab F. Arnaud-Neu, M. J. Schwing-Weill, R. Louis, and R. Weiss, Inorg. Chem., 18, 2956-2961 (1979).

REFERENCE INDEX

79Ac R. Aruga, Aust. J. Chem., 32, 709-713 (1979).

79Ad R. Aruga, J. Inorg. Nucl. Chem., 41, 849-851 (1979).

79Ba A. E. Beezer, M. Orban, and H. J. V. Tyrrell, Acta Chim. Acad. Sci. Hung., 99, 415-419 (1979).

79Bb B. C. Bhuyan and S. N. Dubey, Acta Ciencia Indica, 5, 14-19 (1979).

79Bc M. J. Blais, O. Enea, and G. Berthon, Thermochimica Acta, 30, 37-44 (1979).

79Bd M. J. Blais, O. Enea, and G. Berthon, Thermochimica Acta, 30, 45-53 (1979).

79Be R. P. Bonomo, R. Cali, F. Riggi, E. Rizzarelli, S. Sammartano, and G. Siracusa, Inorg. Chem., 18, 3417-3422 (1979).

79Bf N. E. Burke, A. Singhal, M. J. Hintz, J. A. Ley, H. Hui, L. R. Smith, and D. M. Blake, J. Am. Chem. Soc., 101, 74-79 (1979).

79Ca R. Cali, E. Rizzarelli, S. Sammartano, and G. Siracusa, Transition Metal Chem., 4, 328-332 (1979).

79Cb U. B. Ceipidor, G. Dascenzo, and M. Tomassetti, Thermochimica Acta, 33, 231237 (1979).

79Da H. C. Dasgupta and D. K. Pathak, Indian J. Chem., 17A, 427-428 (1979).

79Db H. C. Dasgupta and V. V. Pd. Singh, J. Indian Chem. Soc., 16, 954-957 (1979).

79Dc N. K. Davidenko and N. N. Zinich, Russ. J. Inorg. Chem., (English Trans.), 24, 1352-1354 (1979). (Russian Page No. 2439-2443).

79Dd S. N. Dubey, A. Singh, H. L. Kalra, and D. M. Puri, J. Indian Chem. Soc., 56, 451-454 (1979).

79Ea O. Enea and G. Berthon, Thermochimica Acta, 33, 311-322 (1979).

79Fa L. Fabbrizzi, J.C.S. Dalton, 1857-1861 (1979).

79Fb L. Fabbrizzi, M. Micheloni, and P. Paoletti, J.C.S. Dalton, 1581-1584 (1979).

79Ga J. N. Gaur, S. C. Baghel, and R. S. Sharma, J. Indian Chem. Soc., 56, 255-256 (1979).

79Gb M. K. Green and G. Kotowycz, Can. J. Chem., 57, 995-999 (1979).

79Gc R. J. Gualtieri, W. A. E. McBryde, and H. K. J. Powell, Can. J. Chem., 57, 113-118 (1979).

79Gd A. Gupta and B. S. Pannu, Thermochimica Acta, 31, 101-108 (1979).

79Ha G. G. Herman, A. M. Goeminne, and H. F. De Brabander, Thermochimica Acta, 32, 27-31 (1979).

79Hb J. G. Hoogerheide and A. I. Popov, J. Solution Chem., 8, 83-95 (1979).

79Ka A. Kettrup and K. H. Ohrbach, Thermochimica Acta, 29, 273-276 (1979).

79Kb S. A. R. Knox, R. J. McKinney, V. Riera, F. G. A. Stone, and A. C. Szary, J.C.S. Dalton, 1801-1811 (1979).

79Kc L. L. Kucheruk, I. P. Goldshtein, E. S. Shcherbakova, E. N. Guryanova, I. Ya. Kuramshin, and A. N. Pudovik, J. General Chem., U.S.S.R. (English Trans.) 49, 643-648 (1979). (Russian Page No. 741-747).

REFERENCE INDEX

79La A. B. P. Lever, P. Paoletti, and L. Fabbrizzi, Inorg. Chem., 18, 1324-1329 (1979).

79Ma N. Mahadevan and R. M. Sathe, Thermochimica Acta, 30, 391-393 (1979).

79Mb W. U. Malik, R. Bembi, and Sushila, J. Indian Chem. Soc., 56, 776-778 (1979).

79Mc W. A. E. McBryde and H. K. J. Powell, Can. J. Chem., 57, 1785-1791 (1979).

79Md M. R. Melardi, G. Ferroni, and J. Galea, Revue Chimie Minerale, 16, 19-29 (1979).

79Me M. R. Melardi, J. Galea, G. Ferroni, A. Belaich, and M. Ragot, Bull. Soc. Chim. Belg., 88, 1015-1017 (1979).

79Mf A. Mishra and J. P. Singh, J. Indian Chem. Soc., 56, 771-775 (1979).

79Mg M. S. Mohan, D. Bancroft, and E. H. Abbott, Inorg. Chem., 18, 344-346 (1979).

79Mh B. Monzyk and A. L. Crumbliss, J. Am. Chem. Soc., 101, 6203-6213 (1979).

79Na S. Nagpal, S. N. Dubey, H. L. Kalra, and D. M. Puri, Indian J. Chem., 18A, 270-271 (1979).

79Nb P. K. R. Nair and K. Srinivasulu, J. Inorg. Nucl. Chem., 41, 251-252 (1979).

79Pa Y. Z. Pathak and G. B. Joshi, J. Indian Chem. Soc., 56, 210-212 (1979).

79Ra S. Randhawa, B. S. Pannu, and S. L. Chopra, Thermochimica Acta, 32, 111-117 (1979).

79Rb V. V. S. Reddy, B. Sethuram, and T. N. Rao, Indian J. Chem., 17A, 199-201 (1979).

79Sa L. Sabatini and L. Fabbrizzi, Inorg. Chem., 18, 438-444 (1979).

79Sb R. S. Sandhu and R. K. Kalia, Thermochimica Acta, 30, 351-354 (1979).

79Sc R. S. Sandhu and R. Kumar, Thermochimica Acta, 28, 191-193 (1979).

79Sd R. S. Sandhu and R. Kumar, Thermochimica Acta, 33, 391-393 (1979).

79Se R. S. Sandhu, R. Kumar, and R. K. Kalia, Thermochimica Acta, 30, 355-358, (1979).

79Sf N. K. Sankhla, C. P. Gupta, and R. K. Mehta, Current Sci., 48, 111-112 (1979).

79Sg N. K. Sankhla, C. P. Gupta, and R. K. Mehta, Z. Phys. Chemie, Leipzig, 260, 1188-1192 (1979).

79Sh I. P. Saraswat, C. L. Sharma, and A. Sharma, J. Indian Chem. Soc., 56, 257-259 (1979).

79Si Y. Sasaki, M. Sakurada, M. Matsui, and T. Shigematsu, Bull. Chem. Soc. Japan, 52, 2295-2298 (1979).

79Sj R. S. Saxena and M. C. Saxena, J. Indian Chem. Soc., 56, 149-151 (1979).

79Sk H. Singh and U. K. Rathi, J. Indian Chem. Soc., 56, 1028-1030 (1979).

79Ta B. Trinderup, Acta Chem. Scand., 33, 7-10 (1979).

79Tb B. Tummler, G. Maass, F. Vogtle, H. Sieger, U. Heimann, and E. Weber, J. Am. Chem. Soc., 101, 2588-2598 (1979).

REFERENCE INDEX

79Ua M. Ueno and H. Kishimoto, Chemistry Letters, Japan, 1487-1490 (1979).

79Va V. P. Vasil'ev, G. A. Zaitseva, and N. V. Provorova, Zhurnal Obshchei Khimii, 49, 2608-2614 (1979).

79Vb G. S. Vigee, C. L. Watkins, and H. F. Bowen, Inorg. Chim. Acta, 35, 255-259 (1979).

79Vc G. Viswanath, K. P. Menon, and A. S. A. Krishnan, J. Inorg. Nucl. Chem., 41, 717-719 (1979).

79Wa D. Waysbort and G. Navon, Inorg. Chem., 18, 9-13 (1979)

79Ya K. B. Yatsimirskii, M. I. Kabachnik, E. I. Sinyavskaya, T. Ya. Medved, F. I. Belskii, Yu. M. Polikarpov, M. A. Konstantinovskaya, and B. K. Shcherbakov, Russ. J. Inorg. Chem., (English Trans.), 24, 63-66 (1979). (Russian Page No. 115-120).

1978

78Aa R. C. Agarwal, J. Indian Chem. Soc., 55, 220-221 (1978).

78Ab S. Ahrland, T. Berg, and P. Blauenstein, Acta Chem. Scand., 32, 933-941 (1978).

78Ac V. T. P. Anh, D. P. Graddon, and V. A. K. Ng, Aust. J. Chem., 31, 2417-2422 (1978).

78Ad A. Anichini, L. Fabbrizzi, and P. Paoletti, J.C.S. Dalton, 577-583 (1978).

78Ae G. Arena, R. Cali, E. Rizzarelli, S. Sammartano, R. Barbucci, and M. J. M. Campbell, J.C.S. Dalton, 1090-1094 (1978).

78Af S. Arora, H. L. Kalra, S. N. Dubey, and D. M. Puri, Acta Ceiencia Indica, 4, 16-20 (1978).

78Ag R. Aruga, Atti Accad. Science Torino, 112, 345-350 (1978).

78Ah R. Aruga, J. Inorg. Nucl. Chem., 40, 1077-1080 (1978).

78Ba A. K. Basak and D. Banerjea, J. Indian Chem. Soc., 55, 853-856 (1978).

78Bb G. Biswas and S. Aditya, J. Inorg. Nucl. Chem., 40, 1069-1071 (1978).

78Bc J. W. Bixler and A. M. Bond, Inorg. Chem., 17, 3684-3689 (1978).

78Ca D. E. Caddy and J. H. P. Utley, J. Inorg. Nucl. Chem., 40, 1103-1106 (1978).

78Cb A. Ciana and V. Crescenzi, Gazz. Chim. Ital., 108, 519-522 (1978).

78Da E. A. Daniel, F. C. March, H. Kipton, J. Powell, W. T. Robinson, and J. M. Russell, Aust. J. Chem., 31, 723-736 (1978).

78Db A. Dei, L. Fabbrizzi, and A. Gallorini, Inorg. Chim. Acta, 26, L45-L48 (1978).

78Dc P. Di Bernardo, P. Zanello, D. Curto, and R. Portanova, Inorg. Chim. Acta, 29, L185-L186 (1978).

78Dd R. S. Drago, T. Beugelsdijk, J. A. Breese, and J. P. Cannady, J. Am. Chem. Soc., 100, 5374-5382 (1978).

78Fa L. Fabbrizzi, M. Micheloni, and P. Paoletti, J.C.S. Chem. Comm., 833-834 (1978).

78Fb L. Fabbrizzi, P. Paoletti, and R. M. Clay, Inorg. Chem., 17, 1042-1046 (1978).

78Fc I. Filipovic and B. Bach-Dragutinovic, Thermochimica Acta, 27, 151-154 (1978).

78Ga C. Gerard and R. P. Hugel, J. Chem. Research, 10, 404-405 (1978).

78Gb D. P. Graddon and I. A. Siddiqi, Inorg. Chim. Acta, 29, L245-L246 (1978).

78Gc D. P. Graddon and I. A. Siddiqi, J. Inorg. Nucl. Chem., 40, 1561-1564 (1978).

78Ha J. G. Hoogerheide and A. I. Popov, J. Solution Chem., 7, 357-372 (1978).

78Ia N. Ivicic and V. Simeon, Thermochimica Acta, 25, 299-305 (1978).

78Ib R. M. Izatt, R. E. Terry, L. D. Hansen, A. G. Avondet, J. S. Bradshaw, N. K. Dalley, T. E. Jensen, and J. J. Christensen, Inorg. Chim. Acta, 30, 1-8 (1978)

78Ja A. K. Jain, V. Kumari, and G. K. Chaturvedi, J. Indian Chem. Soc., 55, 1251-1253 (1978).

78Ka K. Kapoor and G. K. Chaturvedi, Indian J. Chem., 16A, 452-453 (1978).

78Kb Yu. Ya. Kharitonov, U. I. Sholtyrova, and Kh. K. Ospanov, Russ. J. Inorg. Chem., (English Trans.), 23, 1510-1533 (1978). (Russian Page No. 2728-2733).

78Kc M. Kodama and E. Kimura, J.C.S. Dalton, 104-110 (1978).

78Kd M. Kodama and E. Kimura, J.C.S. Dalton, 1081-1085 (1978).

78Ke A. A. Kurganov, L. Ya, Zhuchkova, and V. A. Davankov, J. Incrg. Nucl. Chem., 40, 1081-1083 (1978).

78La J. Lerchner, L. Beyer, and E. Hoyer, Z. Anorg. Allg. Chem., 445, 13-19 (1978).

78Ma B. Magyar and G. Schwarzenbach, Acta Chem. Scand., 32, 943-955 (1978).

78Mb I. N. Marov, O. M. Petrukhin, V. V. Zhukov, and N. B. Kalinichenko, Russ. J. Inorg. Chem. (English Trans.), 23, 1495-1501 (1978). (Russian Page No. 2702-2711).

78Mc F. Marsicano and R. D. Hancock, J.C.S. Dalton, 228-234 (1978).

78Md P. Mathur, D. P. Goel, and R. P. Singh, J. Indian Chem. Soc., 55, 879-881 (1978).

78Me R. K. Murmann and B. Monzyk, Inorg. Chem., 17, 2461-2469 (1978).

78Pa V. D. Petrova, N. N. Rzhevskaya, N. V. Shcherbakova, Yu. Ya. Nelkenbaum, Yu. A. Sangalov, and K. S. Minsker, J. Gen. Chem., U.S.S.R., (English Trans.), 48, 1692-1697 (1978). (Russian Page No. 1854-1859).

78Pb D. R. Prasad and K. Saraswathi, Indian J. Chem., 16A, 631-633 (1978).

78Ra P. R. Reddy, K. V. Reddy, and M. M. Taqui Khan, J. Inorg. Nucl. Chem., 40, 1265-1267 (1978).

78Rb J. S. Redinha and M. L. P. Leitao, Rev. Port. Quim., 20, 3-9 (1978).

78Sa S. S. Sandhu, J. N. Kumaria, and R. S. Sandhu, J. Indian Chem. Soc., 55, 670-674 (1978).

78Sb I. P. Saraswat, C. L. Sharma, and A. Sharma, J. Indian Chem. Soc., 55, 757-758 (1978).

78Sc R. Sarin and K. N. Munshi, Indian J. Chem., 16A, 455-457 (1978).

78Sd R. S. Saxena and G. L. Khandelwal, Electrochimica Acta, 23, 953-955 (1978).

78Se D. C. Sehgal, C. P. Gupta, and R. K. Mehta, Indian J. Chem., 16A, 910-913 (1978).

78Sf C. L. Sharma and P. K. Jain, J. Indian Chem. Soc., 55, 892-893 (1978).

78Sg S. K. Srivastava, Chemical Era, 14, 283-289 (1978).

78Sh S. K. Srivastava, Indian J. Chem., 16A, 89-91 (1978).

78Si R. C. Srivastava and M. N. Srivastava, J. Inorg. Nucl. Chem., 40, 1439-1442 (1978).

78Sj J. P. N. Srivastava and M. N. Srivastava, J. Inorg. Nucl. Chem., 40, 2076-2078 (1978).

78Va V. P. Vasilev, V. P. Lymar, and A. I. Lytkin, Russ. J. Inorg. Chem., (English Trans.), 23, 683-685 (1978). (Russian Page No. 1238-1242).

78Za V. I. Zelenov, Yu. A. Afanasev, L. A. Aslanyan, and V. T. Panyushkin, Russ. J. Inorg. Chem., (English Trans.), 23, 1404-1405 (1978). (Russian Page No. 2538-2540).

1977

77Aa J. A. Adejumobi and D. R. Goddard, J. Inorg. Nucl. Chem., 39, 912-913 (1977).

77Ab Y. K. Agrawal, Monatshefte fur Chemie, 108, 713-723 (1977).

77Ac Y. K. Agrawal, Thermochimica Acta, 18, 245-249 (1977).

77Ad S. Ahrland, B. Tagesson, and D. Tuhtar, Acta Chem. Scand., A31, 625-634 (1977).

77Ae G. M. Alter, E. R. Pandolfino, D. J. Christie, and J. A. Magnuson, Biochemistry, 16, 4034-4038 (1977).

77Af J. Ananthaswamy, B. Sethuram, and T. Navaneeth Rao, Indian J. Chem., 15A, 177-180 (1977).

77Ag G. Anderegg, E. Hubmann, N. G. Podder, and F. Wenk, Helv. Chim. Acta, 60, 123-140 (1977).

77Ah L. T. Ang, D. P. Graddon, and V. A. K. Ng, Aust. J. Chem., 30, 495-501 (1977).

77Ai A. Anichini, L. Fabbrizzi, P. Paoletti, and R. M. Clay, Inorg. Chim. Acta, 22, 25-27 (1977).

77Aj A. Anichini, L. Fabbrizzi, P. Paoletti, and R. M. Clay, Inorg. Chim. Acta, 24, L21-L23 (1977).

77Ak A. Anichini, L. Fabbrizzi, P. Paoletti, and R. M. Clay, J.C.S. Chem. Comm., 244-245 (1977).

77Al A. Anichini, L. Fabbrizzi, R. Barbucci, and A. Mastroianni, J.C.S. Dalton, 2224-2228 (1977).

REFERENCE INDEX

77Am G. Arena, R. Cali, E. Rizzarelli, S. Sammartano, R. Barbucci, and M. J. M. Campbell, J.C.S. Dalton, 581-585 (1977).

77An R. Aruga, Chim. Analitica, 3, 555-561 (1977).

77Ao R. Aruga, J. Inorg. Nucl. Chem., 39, 2159-2162 (1977).

77Ba A. Bhattacharyya and S. C. Lahiri, J. Indian Chem. Soc., 54, 1137-1142 (1977).

77Bb G. Biswas, S. Aditya, and S. C. Lahiri, Thermochimica Acta, 19, 55-62 (1977).

77Da K. Das, E. L. Simmons, and J. L. Bear, Inorg. Chem., 16, 1268-1271 (1977).

77Db I. Dellien, Acta Chem. Scand., 31, 473-479 (1977).

77Dc C. Dragulescu, S. Policec, and T. Simonesou, Rev. Roumaine Chim., 22, 55-59 (1977).

77Fa L. Fabbrizzi and L. J. Zompa, Inorg. Nucl. Chem. Letters, 13, 287-290 (1977).

77Fb J. J. R. Frausto da Silva, M. Candida, and T. Abreu Vaz, J. Inorg. Nucl. Chem., 39, 613-619 (1977).

77Ga P. Gans, J. B. Gill, and M. Griffin, J.C.S. Faraday, 74, 432-439 (1977).

77Gb K. H. Gayer and P. Kothari, Z. Phys. Chemie, Leipzig, 258, 657-665 (1977).

77Gc A. Gergely and T. Kiss, J. Inorg. Nucl. Chem., 39, 109-114 (1977).

77Gd A. Gergely and I. Nagypal, J.C.S. Dalton, 1104-1108 (1977).

77Ge D. P. Graddon and B. A. Rana, J. Organometal. Chem., 136, 19-24 (1977).

77Gf D. P. Graddon and I. A. Siddiqi, Aust. J. Chem., 30, 2133-2141 (1977).

77Ha F. R. Hartley, G. W. Searle, R. M. Alcock, and D. E. Rogers, J.C.S. Dalton, 469-477 (1977).

77Hb Y. Hasegawa and G. R. Choppin, Inorg. Chem., 16, 2931-2934 (1977).

77Hc W. D. Hemphill and D. G. Brown, Inorg. Chem., 16, 766-769 (1977).

77Hd G. G. Herman, A. M. Goeminne, and Z. Eeckhaut, J. Coord. Chem., 7, 53-59 (1977).

77Ia M. F. Iskander and S. Saddeck, Inorg. Chim. Acta, 22, 141-147 (1977).

77Ib R. M. Izatt, J. D. Lamb, G. E. Maas, R. E. Asay, J. S. Bradshaw, and J. J. Christensen, J. Am. Chem. Soc., 99, 2365-2366 (1977).

77Ka Y. D. Kane, D. M. Joshi, and G. S. Natarajan, Proc. Indian Acad. Sci., 85, 303-306 (1977).

77Kb K. Y. Kim and G. H. Nancollas, J. Phys. Chem., 81 948-952 (1977).

77Kc M. Kodama and E. Kimura, J.C.S. Dalton, 1473-1478 (1977).

77Kd M. Kodama and E. Kimura, J.C.S. Dalton, 2269-2276 (1977).

77Ke T. N. Koneva, V. A. Fedorov, and G. L. Trofimov, Russ. J. Phys. Chem., (English Trans.), 51, 470-472 (1977). (Russian Page No. 793-795).

77Kf R. Kuopio, Acta Chem. Scand., 31, 369-374 (1977).

77La T. W. Leung, M. J. Hintz, L. T. Chan, A D. Sherry, and M. D. Blake, Inorg. Chem., 16, 2606-2611 (1977).

REFERENCE INDEX

77Lb M. P. Li, R. S. Drago, and A. J. Pribula, J. Am. Chem. Soc., 99, 6900-6905 (1977).

77Lc V. M. Loyola, R. Pizer, and R. G. Wilkins, J. Am. Chem. Soc., 99, 7185-7188 (1977).

77Ma S. D. Makhijani, and S. P. Sangal, Indian J. Chem., 15, 565-566 (1977).

77Mb W. U. Malik, R. Bembi, R. N. Goyal, and R. K. Paliwal, Proc. Indian Acad. Sci., 85, 531-536 (1977).

77Mc F. Marsicano and R. D. Hancock, Johannesburg, National Institute for Metallurgy, Report No. 1889, 1-12 (1977).

77Md E. Mei, J. L. Dye, and A. I. Popov, J. Am. Chem. Soc., 99, 5308-5311 (1977).

77Me E. Mei, A. I. Popov, and J. L. Dye, J. Am. Chem. Soc., 99, 6532-6536 (1977).

77Mf A. Mudaliar, Y. K. Agrawal, and A. S. Kapur, Thermochimica Acta, 19, 136-139 (1977).

77Na K. L. Nash and G. R. Chcppin, J. Inorg. Nucl. Chem., 39, 131-135 (1977).

77Nb N. M. Nikovaeva, Russ. J. Inorg. Chem. (English Trans.), 22, 1323-1326 (1977). (Russian Page No. 2447-2452).

77Nc N. M. Nikolaeva and L. D. Tsvelodub, Russ. J. Inorg. Chem., (English Trans.), 22, 205-209 (1977). (Russian Page No. 380-385).

77Pa L. D. Pettit and J. L. M. Swash, J.C.S. Dalton, 588-594 (1977).

77Pb L. D. Pettit and J. L. M. Swash, J.C.S. Dalton, 697-701 (1977).

77Pc S. N. Poddar, S. Ghosh, S. M. Bhattacharyya, and S. R. Chaudhuri, J. Inorg. Nucl. Chem., 39, 739-744 (1977).

77Pd A. A. Popov, E. F. Veinshtein, and S. G. Entelis, J. Macromol. Sci. Chem., A11, 859-868 (1977).

77Ra A. Ramanujam, V. V. Ramakrishna, and S. K. Patil, Radiochem. Radioanal. Letters, 31, 119-128 (1977).

77Rb S. J. Rehfeld, H. F. Loken, and D. J. Eatough, Thermochimica Acta, 18, 265271 (1977).

77Sa R. S. Sandhu, Monatshefte fur Chemie, 108, 51-55 (1977).

77Sb S. S. Sandhu, J. N. Kumaria, and R. S. Sandhu, Thermochimica Acta, 18, 329331 (1977).

77Sc S. S. Sandhu, J. N. Kumaria, and R. S. Sandhu, Monatshefte fur Chemie, 108, 1105-1112 (1977).

77Sd R. S. Sapleszko, R. C. Patel, and E. Malijevic, J. Phys. Chem., 81, 1061-1068 (1977).

77Se R. S. Saxena and G. L. Khandelwal, Monatshefte fur Chemie, 108, 533-539 (1977).

77Sf B. R. Singhvi and R. K. Mehta, Chemia Analityczna, 22, 235-239 (1977).

77Sg J. P. N. Srivastava and M. N. Srivastava, Current Science, 46, 443-445 (1977).

77Sh J. P. N. Srivastava and M. N. Srivastava, Revue de Chimie Minerale, 14, 263-268 (1977).

77Si T. Sugano and K. Kubo, Nippon Kaguku Kaishi, 4, 500-503 (1977).

77Sj R. B. Svitych, N. N. Rzhevskaya, O. P. Yablonskii, A. L. Buchachenko, A. A. Petrokhov, and V. A. Belyaev, Kinetics and Catalysis, (English Trans.), 18, 76-80 (1977). (Russian Page No. 76-80).

77Ta J. J. Tombeux, A. M. Goeminne, and J. Schaubroeck, Thermochimica Acta, 19, 327-334 (1977).

77Tb H. C. Tse and M. Tamres, J. Phys. Chem., 81, 1376-1382 (1977).

77Va V. P. Vasilev and A. K. Belonogova, Russ. J. Inorg. Chem., (English Trans.), 22, 1303-1306 (1977). (Russian Page No. 2407-2412).

77Vb G. C. Vogel and J. R. Stahlbush, Inorg. Chem., 16, 950-953 (1977).

1976

76Aa S. A. Abbasi, J. Electroanal. Chem., 73, 115-118 (1976).

76Ab R. C. Agarwal, R. P. Khandelwal, and P. C. Singhal, J. Indian Chem. Soc., 53, 977-979 (1976).

76Ac Y. K. Agrawal and V. P. Khare, J. Inorg. Nucl. Chem., 38, 1663-1667 (1976).

76Ad S. Ahrland and E. Avsar, Acta Chem. Scand., A30, 15-23 (1976).

76Ae S. Ahrland and N. O. Bjork, Acta Chem. Scand., A30, 257-264 (1976).

76Af S. Ahrland, N. O. Bjork, and R. Portanova, Acta Chem. Scand., A30, 270-276 (1976).

76Ag N. S. Al-Niaimi, and I. K. Alzand, Bull. Coll. Sci. Univ. Baghdad, 17, 125138 (1976).

76Ah G. Anderegg, Z. Naturforsch, 31b, 786-789 (1976).

76Ai L. T. Ang and D. P. Graddon, Aust. J. Chem., 29, 565-571 (1976).

76Aj L. T. Ang and D. P. Graddon, Aust. J. Chem., 29, 1429-1434 (1976).

76Ak L. T. Ang and D. P. Graddon, J. Inorg. Nucl. Chem., 38, 2279-2283 (1976).

76Al G. Arena, R. Cali, E. Rizzarelli, and S. Sammartano, Thermochimica Acta, 17, 155-164 (1976).

76Am R. Arnek and D. Poceva, Acta Chem. Scand., A30, 59-62 (1976).

76An K. G. Ashurst, National Institute for Metallurgy, Johannesburg, South Africa, Report No. 1820, (1976).

76Ba R. Barbucci and M. Budini, J.C.S. Dalton, 1321-1325 (1976).

76Bb A. K. Basak and D. Banerjea, Indian J. Chem., 14A, 184-188 (1976).

76Bc I. M. Batyaev and R. S. Fogileva, Russ. J. Inorg. Chem., (English Trans.), 21, 653-654 (1976). (Russian Page No. 1199-1201).

78Bd G. Berthon and T. Berg, J. Chem. Thermodynamics, 8, 1145-1152 (1976).

76Be G. Berthon, O. Enea, and Y. Fuseau, Thermochimica Acta, 16, 323-338 (1976).

REFERENCE INDEX

76Bf A. Bokare, B. K. Deshmukh, and R. B. Kharat, Indian J. Chem., 14A, 541-542 (1976).

76Ca A. K. Chaudhury and S. P. Ghosh, J. Indian Chem. Soc., 53, 973-976 (1976).

76Cb A. T. Chaudhury and S. P. Sangal, Chemical Era., 17-19, Jan. (1976).

76Cc A. T. Chaudhary and S. P. Sangal, Chemical Era, 17-19, Feb. (1976).

76Cd C. S. Chung and S. T. Huang, Chinese Chem. Soc., (Taipei), 23, 139-150 (1976).

76Ce L. Ciavatta, M. Grimaldi, and R. Palombari, J. Inorg. Nucl. Chem., 38, 823-826 (1976).

76Cf A. M. Corrie and D. R. Williams, J.C.S. Dalton, 1068-1072 (1976).

76Da D. R. Dakternieks, J. Inorg. Nucl. Chem., 38, 141-143 (1976).

76Db K. Das and J. L. Bear, Inorg. Chem., 15, 2093-2095 (1976).

76Dc E. Dazzi and M. T. Falqui, Rendicouti Del Seminario Della Facolta Di Scienze Della Universita Di Cagliari, 46, 225-233 (1976).

76Dd I. Dellien and L. G. Hepler, Can. J. Chem., 54, 1383-1387 (1976).

76De P. Di Bernardo, A. Bismondo, R. Portanova, O. Traverso, and L. Magon, Inorg. Chim. Acta, 18, 47-50 (1976).

76Df P. Di Bernardo, A. Cassol, and L. Magon, Chimica E L'Industria, (Milan), 58, 732-733 (1976).

76Ea W. J. Eilbeck and M. S. West, J.C.S. Dalton, 274-278 (1976).

76Fa L. Fabbrizzi, M. Micheloni, and P. Paoletti, Inorg. Chem., 15, 1451-1452 (1976).

76Fb V. A. Fedorov, L. I. Kiprin, and V. E. Mironov, Russ. J. Phys. Chem. (English Trans.), 50, 1658-1660 (1976). (Russian Page No. 2777-2781).

76Ga M. J. Gallagher, D. P. Graddon, and A. R. Sheikh, Aust. J. Chem., 29, 759-765 (1976).

76Gb M. J. Gallagher, D. P. Graddon, and A. R. Sheikh, Aust. J. Chem., 29, 24092413 (1976).

76Gc A. Gergely and T. Kiss, Magyar Kemiai Folyoirat, 82, 89-94 (1976).

76Gd A. Gergely and I. Nagypal, Magyar Kemiai Folyoirat, 82, 442-447 (1976).

76Ge A. Gergely and I. Sovago, Inorg. Chim. Acta, 20, 19-25 (1976).

76Gf H. Goff and L. O. Morgan, Inorg. Chem., 15, 2069-2076 (1976).

76Gg M. Gold and H. K. J. Powell, J.C.S. Dalton, 230-233 (1976).

76Gh D. P. Graddon and I. A. Siddiqi, Aust. J. Chem., 29, 1201-1207 (1976).

76Ha R. C. Hancock and F. Marsicano, J.C.S. Dalton, 1832-1837 (1976).

76Ia R. M. Izatt, R. E. Terry, B. L. Haymore, L. D. Hansen, N. K. Dalley, A. G. Avondet, and J. J. Christensen, J. Am. Chem. Soc., 98, 7620-7626 (1976).

76Ka E. Kauffmann, J. M. Lehn, and J. P. Sauvage, Helv. Chim. Acta, 59, 1099-1111 (1976).

76Kb A. Kettrup and K. Striegler, Thermochimica Acta, 15, 147-158 (1976).

REFERENCE INDEX

76Kc P. V. Khadikar and P. S. Deshmukh, J. Indian Chem. Soc., 53, 322-323 (1976).

76Kd M. Kodama and E. Kimura, Bull. Chem. Soc. Japan, 49, 2465-2468 (1976).

76Ke M. Kodama and E. Kimura, J.C.S. Dalton, 116-120 (1976).

76Kf M. Kodama and E. Kimura, J.C.S. Dalton, 1720-1724 (1976).

76Kg M. Kodama and E. Kimura, J.C.S. Dalton, 2335-2338 (1976).

76Kh M. Kodama and E. Kimura, J.C.S. Dalton, 2341-2345 (1976).

76Ki B. E. Kozer and G. A. Prik, Russ. J. Phys. Chem., (English Trans.), 50, 367-369 (1976). (Russian Page No. 626-629).

76Kj N. A. Kramareva, V. I. Dulova, and E. P. Artyukhova, Russ. J. Phys. Chem., (English Trans.), 50, 306-307 (1976). (Russian Page No. 526-527).

76Kk A. I. Krutous and I. M. Batyaev, Russ. J. Inorg. Chem., (English Trans.), 21, 807-810 (1976). (Russian Page No. 1478-1482).

76Kl F. Y. Kulba, Y. B. Yakovlev, and D. A. Zenchenko, Russ. J. Phys. Chem., (English Trans.), 50, 1332-1334 (1976). (Russian Page No. 2216-2218).

76Km L. A. Kulvinova, V. V. Blokhin, and V. E. Mironov, Russ. J. Phys. Chem., (English Trans.), 50, 773-774 (1976). (Russian Page No. 1287-1288).

76Kn Y. Kuriki, J. Halsey, R. Biltonen, and E. Racker, Biochem., 15, 4956-4961 (1976).

76Ko O. C. Kwun and H. Hartmann, Zeitschrift Physikalische Chemie Neue Folge, 100, 139-148 (1976).

76La S. C. Lahiri, A. K. Roy, and S. Aditya, Thermochimica Acta, 16, 277-282 (1976).

76Lb M. P. Li and R. S. Drago, J. Am. Chem. Soc., 98, 5129-5135 (1976).

76Ma N. Mahadevan and R. M. Sathe, J. Indian Chem. Soc., 53, 97-99 (1976).

76Mb F. Marsicano and R. D. Hancock, J. Coord. Chem., 6, 21-29 (1976).

76Mc D. R. McMillin, R. S. Drago, and J. A. Nusz, J. Am. Chem. Soc., 98, 3120-3126 (1976).

76Md T. H. Mhaske and K. N. Munshi, Indian J. Chem., 14A, 421-423 (1976).

76Me V. E. Mironov, A. K. Pyartman, and N. P. Kolobov, Russ. J. Phys. Chem., (English Trans.), 50, 1186-1187 (1976). (Russian Page No. 1967-1970).

76Na I. L. Nadelyaeva, A. V. Suvorov, and Yu. V. Kondratev, Russ. J. Gen. Chem., (English Trans.), 46, 1601 (1976). (Russian Page No. 1647-1648).

76Nb C. Neveu, G. Folcher, and A. M. Laurent, J. Inorg. Nucl. Chem., 38, 1223-1226 (1976).

76Oa D. D. Ozha and R. K. Mehta, Indian J. Chem., 14A, 452-453 (1976).

76Pa W. Partenheimer, J. Am. Chem. Soc., 98, 2779-2784 (1976).

76Pb S. Parthasarathy and S. Ambujavalli, Inorg. Chim. Acta, 17, 213-216 (1976).

76Pc L. D. Pettit and J. L. M. Swash, J.C.S. Dalton, 2416-2419 (1976).

76Pd A. J. Pribula and R. S. Drago, J. Am. Chem. Soc., 98, 2784-2788 (1976).

76Pe A. M. Pujari and K. N. Munshi, Indian J. Chem., 14A, 192-194 (1976).

REFERENCE INDEX

76Ra G. K. Ragulin, N. E. Aleksandrova, and D. P. Dobychin, Russ. J. Inorg. Chem., (English Trans.), 21, 1499-1501 (1976). (Russian Page No. 2724-2727).

76Rb P. C. Rawat and C. M. Gupta, Indian J. Chem., 14A, 716-718 (1976).

76Sa K. N. Sahu, J. Indian Chem. Soc., 53, 657-659 (1976).

76Sb R. S. Sandhu, Indian J. Chem., 14A, 1020-1021 (1976).

76Sc R. S. Sandhu, Thermochimica Acta, 16, 398-401 (1976).

76Sd S. S. Sandhu, J. N. Kumaria, and R. S. Sandhu, Indian J. Chem., 14A, 817-818 (1976).

76Se S. S. Sandhu, J. N. Kumaria, and R. S. Sandhu, J. Indian Chem. Soc., 53, 435-436 (1976).

76Sf S. S. Sandhu, R. S. Sandhu, and J. N. Kumaria, Indian J. Chem., 14A, 366-367 (1976).

76Sg S. S. Sandhu, R. S. Sandhu, J. N. Kumaria, J. Singh, and N. S. Sekhou, J. Indian Chem. Soc., 53, 114-117 (1976).

76Sh S. S. Sandhu, R. S. Sandhu, and J. N. Kumaria, Thermochimica Acta, 15, 244-247 (1976).

76Si R. S. Sandhu and K. Singh, Thermochimica Acta, 17, 325-328 (1976).

76Sj R. S. Saxena and S. K. Bhatia, Indian J. Chem., 14A, 445-446 (1976).

76Sk R. S. Saxena and S. K. Bhatia, J. Indian Chem. Soc., 53, 636-637 (1976).

76Sl R. S. Saxena and G. L. Khandelwal, J. Indian Chem. Soc., 53, 870-871 (1976).

76Sm R. S. Saxena and M. C. Saxena, Indian J. Chem., 14A, 628-629 (1976).

76Sn R. S. Saxena and M. C. Saxena, J. Indian Chem. Soc., 53, 633-635 (1976).

76So B.R. Singhvi and R.K. Mehta, Current Science, 49, 712-713 (1976).

76Sp I. Sovago and A. Gergely, Inorg. Chim. Acta, 20, 27-32 (1976).

76Sq I. Sovago and A. Gergely, Magyar Kemiai Folyoirat, 82, 290-295 (1976).

76Sr J. L. M. Swash and L. D. Pettit, Inorg. Chim. Acta, 19, 19-21 (1976).

76Va V. P. Vasilev and A. K. Belonogova, Russ. J. Inorg. Chem., (English Trans.), 21, 31-33 (1976). (Russian Page No. 55-60).

76Vb V. P. Vasilev and A. K. Belonogova, Russ. J. Inorg. Chem., (English Trans.), 21, 350-353 (1976).

76Vc V. P. Vasilev and A. K. Belonogova, Russ. J. Inorg. Chem., (English Trans.), 21, 1646-1648 (1976). (Russian Page No. 2982-2986).

76Vd V. P. Vasilev, N. K. Grechina, and A. A. Ikonnikov, Russ. J. Inorg. Chem., (English Trans.), 21, 667-668 (1976). (Russian Page No. 1225-1227).

76Ve V. P. Vasilev, V. D. Korableva, and P. S. Mukhina, Russ. J. Gen. Chem., (English Trans.), 46, 1191-1193 (1976). (Russian Page No. 1207-1209).

76Ya M. Yamada, Bull. Chem. Soc. Japan, 49, 1023-1026 (1976).

76Yb S. Yamada, H. Nagase, S. Funahashi, and M. Tanaka, J. Inorg. Nucl. Chem., 38, 617-621 (1976).

REFERENCE INDEX

76Yc T. Yamamoto, Y. Nakamura, and A. Yamamoto, Bull. Chem. Soc. Japan, 49, 191197 (1976).

76Yd S. S. Yun and G. R. Choppin, J. Inorg. Nucl. Chem., 38, 332-333 (1976).

76Ye S. S. Yun, G. R. Choppin, and D. Blakeway, J. Inorg. Nucl. Chem., 38, 587-589 (1976).

76Yf E. N. Yurchenko, G. R. Kononin, G. P. Shironosova, and T. P. Aksenova, Russ. J. Inorg. Chem., (English Trans.), 21, 1682-1685 (1976). (Russian Page No. 3050-3054).

1975

76Aa S. Ahrland and E. Avsar, Acta Chem. Scand., A29, 881-889 (1975).

76Ab S. Ahrland and E. Avsar, Acta Chem. Scand., A29, 890-894 (1975).

76Ac A. E. Alegria, R. Concepcion, and G. R. Stevenson, J. Phys. Chem., 79, 361365 (1975).

75Ad G. Anderegg, Helv. Chim. Acta., 58, 1218-1225 (1975).

75Ae G. Anderegg, N. G. Podder, P. Blauenstein, M. Hangartner, and H. Stunzi, J. Coord. Chem., 4, 267-275 (1975).

75Af L. T. Ang, D. P. Graddon, L. F. Lindoy, and S. Prakash, Aust. J. Chem., 28, 1005-1010 (1975).

75Ag R. Aruga, Accademia Della Science De Torino, 109, 361-369 (1975).

75Ah R. Aruga, J.C.S. Dalton, 2534-2538 (1975).

75Ba S. C. Baghel, K. K. Choudhary, and J. N. Gaur, J. Inorg. Nucl. Chem., 37, 2513-2516 (1975).

75Bb R. Barbucci, Inorg. Chim. Acta, 12, 113-117 (1975).

75Bc M. Barres, G. Lesgards, J. P. Dubes, and H. Tachoire, Recue. Chimie Minerale, 12, 527-536 (1975).

75Bd A. C. Baxter and D. R. Williams, J.C.S. Dalton, 1757-1761 (1975).

75Be B. B. Bhowmik, D. C. Mukherjee, and P. K. Sriman, Indian J. Chem., 13, 13141316 (1975).

75Bf C. Bianchini, L. Fabbrizzi, P. Paoletti, and A. B. P. Lever, Inorg. Chem., 14, 197-199 (1975).

75Bg C. Bianchini, L. Fabbrizzi, and P. Paoletti, J.C.S. Dalton, 1036-1041 (1975).

75Ca P. B. Chakraborty and H. N. Sharma, J. Indian Chem. Soc., 52, 1211-1212 (1975).

75Cb P. B. Chakrawarti, M. V. M. Bhopal, and H. N. Sharma, Vijnana Parishad Anusandhan Patrika, 18, 169-172 (1975).

75Cc B. K. Choudhary and B. Prasad, J. Indian Chem. Soc., 52, 679-684 (1975).

75Cd L. Ciavatta, M. Grimaldi, and R. Palombari, J. Inorg. Nucl. Chem., 37, 1685-1692 (1975).

75Ce J. F. Coetzee and E. Hsu, J. Solution Chem., 4, 45-52 (1975).

75Da A. R. Das and V. S. K. Nair, J. Inorg. Nucl. Chem., 37, 991-993 (1975).

75Db A. R. Das and V. S. K. Nair, J. Inorg. Nucl. Chem., 37, 995-997 (1975).

75Dc A. R. Das and V. S. K. Nair, J. Inorg. Nucl. Chem., 37, 2121-2123 (1975).

75Dd A. R. Das and V. S. K. Nair, J. Inorg. Nucl. Chem., 37, 2125-2126 (1975).

75De H. F. De Brabander, A. M. Goeminne, and L. C. Van Poucke, J. Inorg. Nucl. Chem., 37, 799-803 (1975).

75Df A. Dei and R. Gori, Inorg. Chim. Acta, 14, 157-160 (1975).

75Dg D. De Marco, A. Bellomo, and A. De Robertis, Atti Soc. Pel. Sc. Fis. Mat. Nat., 21, 41-48 (1975).

75Dh J. G. H. du Preez and C. P. J. van Vuuren, J.C.S. Dalton, 1548-1552 (1975).

75Ea D. R. Eaton and K. Zaw, Can. J. Chem., 53, 633-643 (1975).

75Eb O. Enea, M. J. Blais, and G. Berthon, Thermochimica Acta, 12, 29-37 (1975).

75Fa Y. Farhangi and D. P. Graddon, J. Organometal. Chem., 87, 67-82 (1975).

75Fb V. A. Fedorov, A. M. Robov and I. D. Isaev, Russ. J. Phys. Chem., (English Trans.), 49, 1841-1843 (1975). (Russian Page No. 3115-3118).

76Fc K. J. Fisher and R. S. Drago, Inorg. Chem., 14, 2804-2808 (1975).

75Ga C. Gerard and C. Ducauze, Bull. Soc. Chim., France, 1955-1958 (1975).

75Gb C. Gerard and R. Hugel, Bull. Soc. Chim., France, 2404-2408 (1975).

75Gc A. Gergely and E. Farkas, Magyar Kemiai Folyoirat, 81, 471-477 (1975).

75Gd A. Gergely and T. Kiss, Magyar Kemiai Folyoirat, 81, 15-21 (1975).

75Ge A. Gergely, I. Nagypal, and E. Farkas, J. Inorg. Nucl. Chem., 37, 551-555 (1975).

78Gf P. K. Govil and S. K. Banerji, Analysis, (France), 3, 268-270 (1975).

75Gg D. P. Graddon and W. K. Ong, J. Inorg. Nucl. Chem., 37, 469-472 (1975).

75Gh S. Grewal, B. S. Sekhon and S. L. Chopra, Thermochimica Acta, 11, 315-321 (1975).

75Gi T. R. Griffiths and P. J. Potts, J.C.S. Dalton, 344-352 (1975).

75Gj T. R. Griffiths and P. J. Potts, J. Inorg. Nucl. Chem., 37, 521-530 (1975).

75Ha T. E. Hogen-Esch and J. Smid, J. Phys. Chem., 79, 233-238 (1975).

75Hb T. M. Hseu, Y. H. Tsai, and C. W. Cheng, J. Chinese Chem. Soc., 22, 299-308 (1975).

75Ia M. Ikeda-Saito and T. Iizuka, Biochim. Biophys. Acta, 393, 335-342 (1975).

75Ib M. Israeli and L. D. Pettit, J.C.S. Dalton, 414-417 (1975).

75Ic V. V. Ivanenko and L. A. Pamfilova, Geochem. International, 4, 566-575 (1975).

75Ja F. A. Jackman and M. W. Lister, J. Solution Chem., 4, 1023-1031 (1975).

75Jb J. B. Jensen, Acta Chem. Scand., A29, 250-254 (1975).

75Jc J. R. Jezorek and W. H. McCurdy, Inorg. Chem., 14, 1939-1944 (1975).

75Ka F. Kai, Y. Sadakane, N. Tanaka, and T. Matsuda, J. Inorg. Nucl. Chem., 37, 1311-1313 (1975).

75Kb M. Kodama and E. Kimura, J.C.S. Chem. Comm., 326-327 (1975).

75La S. C. Lal and B. Prasad, Indian J. Chem., 13, 372-374 (1975).

75Lb W. H. Leung and F. J. Millero, J. Solution Chem., 4, 145-159 (1975).

75Lc P. Lumme and E. Kari, Acta Chem. Scand., A29, 125-135 (1975).

75Ld F. Lux, R. Paetzold, J. Danel, and L. Sobczyk, J.C.S. Faraday II, 9, 1610-1616 (1975).

75Ma L. E. Manzer and C. A. Tolman, J. Am. Chem. Soc., 97, 1955-1956 (1975).

75Mb F. Marsicano and R. D. Hancock, National Institute for Metallurgy, Johannesburg, South Africa, Report No. 1764 (1975).

75Mc P. K. Migal, A. P. Gerbeleu, and G. G. Muntyanu, Russ. J. Inorg. Chem., (English Trans.), 20, 1320-1322 (1975). (Russian Page No. 2383-2387).

75Md V. E. Mironov, A. K. Pyartman, N. P. Kolobov, and E. G. Simonov, Russ. J. Phys. Chem., (English Trans.), 49, 1830-1832 (1975). (Russian Page No. 3094-3097).

75Me V. E. Mironov, A. K. Pyartman, E. G. Simonov, N. P. Kolobov, and V. M. Fadeev, Russ. J. Phys. Chem., (English Trans.), 49, 1828-1830 (1975). (Russian Page No. 3090-3093).

75Mf N. Moulin, M. Hussonnois, L. Brillard, and R. Guillaumont, J. Inorg. Nucl. Chem., 37, 2521-2524 (1975).

75Na A. H. Norbury and S. Raghunathan, J. Inorg. Nucl. Chem., 37, 2133-2134 (1975).

75Oa A. Olin, Acta Chem. Scand., A29, 907-910, (1975).

75Pa O. P. Pachauri and J. P. Tandon, J. General Chem. U.S.S.R., (English Trans.), 47, 398-401 (1977). (Russian Page No. 433-436).

75Pb S. P. Pande and K. N. Munshi, Indian J. Chem., 13, 91-92 (1975).

75Pc H. K. Paruthi, J. S. Tyagi, and A. D. Taneja, J. Chinese Chem. Soc., 22, 367-373 (1975).

75Pd S. Yu. Pavlov, O. P. Yablonskii, E. G. Chekalova, and S. G. Kuznetsov, Russ. J. Phys. Chem., (English Trans.), 49, 831-833 (1975). (Russian Page No. 1409-1412).

75Pe R. Portanova, P. Di Bernardo, O. Traverso, G. A. Mazzocchin, and L. Magon, J. Inorg. Nucl. Chem, 37, 2177-2179 (1975).

75Pf J. Pouradier and A. Pailliotet, C.R. Acad. Sci., (Paris), 280, Ser. C, 10491051 (1975).

75Pg A. M. Pujari and K. N. Munshi, Indian J. Chem., 13, 397-399 (1975).

75Ra G. K. Ragulin and N. E. Aleksandrova, Russ. J. Inorg. Chem., (English Trans.), 20, 1192-1193 (1975). (Russian Page No. 2144-2147).

75Sa S. S. Sandhu, J. Singh, J. N. Kumaria, and R. Singh, J. Chem. Sci., (India), 1, 28-34 (1975).

75Sb D. Satyanarayana and G. Sahu, J.C.S. Dalton, 2236-2239 (1975).

75Sc R. S. Saxena and S. K. Bhatia, J. Indian Chem. Soc., 52, 785-787 (1975).

75Sd R. S. Saxena and S. K. Bhatia, J. Inorg. Nucl. Chem., 37, 310-311 (1975).

75Se R. S. Saxena and S. S. Sheelwant, J. Indian Chem. Soc., 52, 495-497 (1975).

75Sf R. S. Saxena and S. S. Sheelwant, J. Indian Chem. Soc., 52, 934-935 (1975).

75Sg R. S. Saxena and S. S. Sheelwant, Monatshefte fur Chemie, 106, 1081-1084 (1975).

75Sh F. Secco and M. Venturini, Inorg. Chem., 14, 1978-1981 (1975).

75Si B. S. Sekhon and S. L. Chopra, Annales Chimie, 10, 21-23 (1975).

75Sj T. P. Sharma and Y. K. Agrawal, J. Inorg. Nucl. Chem., 37, 1830-1831 (1975).

75Ta P. H. Tedesco and V. E. De Rumi, J. Inorg. Nucl. Chem., 37, 1833-1834 (1975).

75Tb P. H. Tedesco and J. A. Gonzalez Quintana, J. Inorg. Nucl. Chem., 37, 17981799 (1975).

75Tc W. Tschebull, V. Gutmann, and U. Mayer, Z. Anorg. Allg. Chem., 416, 323-340 (1975).

75Va V. P. Vasilev and A. K. Belonogova, Russ. J. Inorg. Chem., (English Trans.), 20, 16-18 (1975). (Russian Page No. 30-34).

75Vb V. P. Vasilev and B. T. Kunin, Russ. J. Inorg. Chem., (English Trans.), 20, 1050-1053 (1975). (Russian Page No. 1881-1886).

75Ya S. S. Yun and J. L. Bear, J. Inorg. Nucl. Chem., 37, 1757-1759 (1975).

1974

74Aa E. H. Abbott and P. A. Rafson, J. Am. Chem. Soc., 96, 7378-7379 (1974).

74Ab R. C. Agarwal, J. Indian Chem. Soc., 51, 772-773 (1974).

74Ac S. Ahrland, E. Avsar, and L. Kullberg, Acta Chem. Scand., A28, 855-865 (1974).

74Ad R. Aruga, Ann. Chim., 64, 659-665 (1974).

74Ba R. Barbucci, L. Fabbrizzi, and P. Paoletti, J.C.S. Dalton, 2403-2406 (1974).

74Bb R. Barbucci and A. Vacca, J.C.S. Dalton, 2363-2367 (1974).

74Bc I. M. Batyaev and R. S. Fogileva, Russ. J. Inorg. Chem., (English Trans.), 19, 363-365 (1974). (Russian Page No. 670-673).

74Bd A. Bellomo, D. De Marco, A. Casale, and A. De Robertis, Annali Chimica, (Rome), 64, 295-304 (1974).

74Be G. Berthon and O. Enea, Bull. Soc. Chim. France, 2793-2798 (1974).

74Bf V. V. Blokhin, L. I. Razmyslova, Yu. A. Makashev, and V. E. Mironov, Russ. J. Phys. Chem., (English Trans.), 48, 82-83 (1974). (Russian Page No. 151152).

REFERENCE INDEX

74Bg V. V. Blokhin, L. I. Razmyslova, M. I. Shalaevskaya, Yu. A. Makashev, and V. E. Mironov, Russ. J. Phys. Chem., (English Trans.), 48, 275-276 (1974). (Russian Page No. 469-471).

74Bh D. Brault and M. Rougee, Biochemistry, 13, 4591-4597 (1974).

74Bi F. T. Bunus, J. Inorg. Nucl. Chem., 36, 917-920 (1974).

74Bj F. Bunus, Revista Chimie, 25, 367-370 (1974).

74Ca R. D. Cannon and J. Gardiner, Inorg. Chem., 13, 390 (1974).

74Cb K. J. Cavell, J. O. Hill, and R. J. Magee, Thermochimica Acta, 8, 447 (1974).

74Cc P. B. Chakravarti and H. N. Sharma, Vijnana Parishad Anusandhan Patrika, 17, 53-56 (1974).

74Cd M. Chikuma, A. Yokoyama, and H. Tanaka, J. Inorg. Nucl Chem., 36, 1243 (1974).

74Ce J. J. Christensen, D. J. Eatough, and R. M. Izatt, Chem. Reviews, 74, 351 (1974).

74Cf A. H. Cohen and B. M. Hoffman, Inorg. Chem., 13, 1484 (1974).

74Da H. F. De Brabander, L. C. Van Poucke, and Z. Eeckhaut, Anal. Chim. Acta, 70, 401-410 (1974).

74Db F. Delben and S. Paoletti, J. Phys. Chem., 78, 1486-1489 (1974).

74Dc R. S. Drago, J. A. Nusz, and R. C. Courtright, J. Am. Chem. Soc., 96, 2082 (1974).

74Dd V. I. Dulova and N. A. Skidan, Russ. J. Inorg. Chem., (English Trans.), 19, 725-727 (1974). (Russian Page No. 1333-1336).

74Ea I. Eliezer and G. Algavish, Inorg. Chim. Acta, 9, 257-262 (1974).

74Fa L. Fabbrizzi and P. Paoletti, Gazzetta Chim. Italiana, 104, 929-935 (1974).

74Fb Y. Farhangi and D. P. Graddon, Aust. J. Chem., 27, 2103-2107 (1974).

74Fc Y. Farhangi and D. P. Graddon, J. Organometallic Chem., 71, 17-21 (1974).

74Fd V. A. Fedorov, L. I. Kiprin, N. S. Shchekina, N. P. Samsonova, M. Ya. Kutuzova, and V. E. Mironov, Russ. J. Inorg. Chem., (English Trans.), 19, 474-477 (1974). (Russian Page No. 872-877).

74Fe V. A. Fedorov, A. M. Robov, I.D. Isaev, and A. A. Alekseeva, Russ. J. Inorg. Chem., (English Trans.), 19, 798-800 (1974). (Russian Page No. 1466-1469).

74Ga E. Ya. Gorenbein, M. N. Vainshtein, and E. P. Skorobogatko, Russ. J. General Chem., (English Trans.), 44, 4-7 (1974). (Russian Page No. 7-10).

74Gb D. P. Graddon and T. T. Nyein, Aust. J. Chem., 27, 407 (1974).

74Gc D. P. Graddon and W. K. Ong, Aust. J. Chem., 27, 741 (1974).

74Gd D. P. Graddon and S. Prakash, Aust. J. Chem., 27, 2099-2102 (1974).

74Ge I. Grenthe and G. Gardhammar, Acta Chem. Scand., 28, 125 (1974).

74Ha G. I. H. Hanania and S. A. Israelian, J. Solution Chem., 3, 57 (1974).

74Hb G. R. Hedwig and H. K. J. Powell, J. C. S. Dalton, 47 (1974).

74Hc C. A. Heller, R. A. Zingaro, and E. A. Meyers, Can. J. Chem., 52, 3814-3824 (1974).

74Hd F. G. Herring and R. L. Tapping, J. Phys. Chem., 78, 316 (1974).

74He F. P. Hinz and D. W. Margerum, Inorg. Chem., 13, 2941-2949 (1974).

74Hf F. P. Hinz and D. W. Margerum, J. Am. Chem. Soc., 96, 4993-4994 (1974).

74Hg K. Houngbossa, O. Enea, and G. Berthon, Thermochimica Acta, 10, 415-424 (1974).

74Hh T. M. Hseu and Y. H. Tsai, J. Chinese Chem. Soc., 21, 211-221 (1974).

74Ja B. R. James and L. D. Markham, Inorg. Chem., 13, 97 (1974).

74Ka F. Kai and Y. Sadakane, J. Inorg. Nucl. Chem., 36, 1404 (1974).

74Kb Y. D. Kane and D. M. Jcshi, Current Sci., (India), 43, 332-334 (1974).

74Kc P. K. Khopkar and J. N. Mathur, J. Inorg. Nucl. Chem., 36, 3819-3825 (1974).

74Kd N. Kiba and T. Takeuchi, J. Inorg. Nucl. Chem., 36, 847 (1974).

74Ke W. F. Kinard and G. R. Choppin, J. Inorg. Nucl. Chem., 36, 1131-1134 (1974).

74Kf W. Knoche, C. A. Firth, and D. Hess, Adv. Molecular Relaxation Proc., 6, 1-17 (1974).

74Kg L. Kullberg, Acta Chem. Scand., A28, 829-838 (1974).

74Kh L. Kullberg, Acta Chem. Scand., A28, 897-902 (1974).

74Ki L. Kullberg, Acta Chem. Scand., A28, 979-985 (1974).

74La L. N. Lin, S. D. Christian, and J. D. Childs, J. Am. Chem. Soc., 96, 2727 (1974).

74Lb S. T. D. Lo and D. W. Watts, Inorg. Chim. Acta, 9, 221-225 (1974).

74Lc P. Lumme and P. Virtanen, Acta Chem. Scand., A28, 1055-1067 (1974).

74Ma M. H. Mihailov, V. Ts. Mihailova, and V. A. Khalkin, J. Inorg. Nucl. Chem., 36, 145 (1974).

74Mb V. E. Mironov, N. P. Kolobov, A. V. Komarova, and I. F. Mavrin, Russ. J. Phys. Chem., (English Trans.), 48, 991-992 (1974). (Russian Page No. 16751677).

74Na I. Nagypal, A. Gergely, and E. Farkas, J. Inorg. Nucl. Chem., 36, 699 (1974).

74Nb N. C. Naik and R. K. Nanda, J. Inorg. Nucl. Chem., 36, 3793-3800 (1974).

74Nc G. S. Natarajan and K. A. Venkatachalam, Indian J. Tech., 12, 192-194 (1974).

74Nd D. P. Nelson, W. D. Miller, and L. A. Kiesow, J. Bio. Chem., 249, 4770-4775 (1974).

74Ne R. M. Nikolic and I. J. Gal, J. C. S. Dalton, 985 (1974).

74Pa K. Pan and Y. L. Liang, J. Chinese Chem. Soc., 21, 63-67 (1974).

74Pb S. P. Pande and K. N. Munshi, J. Indian Chem. Soc., 51, 339-340 (1974).

74Pc W. Partenheimer and B. Durham, Inorg. Nucl. Chem. Letters, 10, 1143-1146 (1974).

74Pd W. Partenheimer and B. Durham, J. Am. Chem. Soc., 96, 3800 (1974).

74Pe M. S. Pereira and J. M. Malin, Inorg. Chem., 13, 386 (1974).

74Pf C. Pillot, J. P. Pascault, and J. Gole, Bull. Soc. Chim. France, No. 34, 357 (1974).

74Pg S. N. Poddar, S. M. Bhattacharyya, and S. R. Chowdhury, Indian J. Chem., 12, 754-756 (1974).

74Ph R. Portanova, P. Di Bernardo, A. Cassol, E. Tondello, and L. Magon, Inorg. Chim. Acta, 8, 233-240 (1974).

74Ra E. Raitanen, Acta Polytechnica Scandinavica, 116, 9-64 (1974).

74Rb P. C. Rawat and C. M. Gupta, Indian J. Chem., 12, 174-175 (1974).

74Rc D. A. Redfield, J. H. Nelson, R. A. Henry, D. W. Moore, and H. B. Jonassen, J. Am. Chem. Soc., 96, 6298-6309 (1974).

74Sa S. Santini, G. Reichenback, and U. Mazzucato, J. C. S. Perkin II, 494 (1974).

74Sb J. C. Sari, M. Dellamico, and M. Bourdeaux, Bull. Assoc. Diplomes Microbiologie Fac. Pharm., (France), 134, 27-31 (1974).

74Sc R. S. Saxena and S. K. Bhatia, J. Indian Chem. Soc., 51, 660-662 (1974).

74Sd R. S. Saxena and S. K. Bhatia, J. Indian Chem. Soc., 51, 856-858 (1974).

74Se R. S. Saxena and S. K. Bhatia, Labdev. J. Sci. Tech.,(India), 12A, 183-185 (1974).

74Sf R. S. Saxena and S. S. Sheelwant, Bull. L'academie Polonaise Sci., 22, 689-694 (1974).

74Sg R. S. Saxena and S. S. Sheelwant, J. Inst. Chemists, (India), 46, 66-70 (1974).

74Sh R. S. Saxena and S. S. Sheelwant, Labdev. J. Sci. Tech., (India), 12A, 155-158 (1974).

74Si R. S. Saxena and P. Singh, Monatshefte fur Chemie, 105, 1142-1147 (1974).

74Sj V. Schurig, R. C. Chang, A. Zlatkis, and B. Feibush, J. Chromatography, 99, 147-171 (1974).

74Sk B. S. Sekhon and S. L. Chopra, Israel J. Chem., 12, 917-923 (1974).

74Sl B. S. Sekhon and S. L. Chopra, Z. Naturforsch, 29, 339-342 (1974).

74Sm V. A. Shormanov, G. A. Krestov, and E. A. Trupikov, Russ. J. Inorg. Chem., (English Trans.), 19, 1007-1009 (1974). (Russian Page No. 1848-1851).

74Sn I. Sovago, A. Gergely, and J. Posta, Magyar Kemiai Folyoirat, 80, 539-542 (1974).

74So S. K. Srivastava and H. B. Mathur, Indian J. Chem., 12, 1289-1293 (1974).

74Sp E. V. Stovbun, E. F. Vainshtein, and S. G. Entelis, Russ. J. General Chem., (English Trans.), 44, 1589-1591 (1974). (Russian Page No. 1622-1624).

74Ta M. M. Taqui Khan and P. Rabindra Reddy, J. Inorg. Nucl. Chem., 36, 607 (1974).

74Tb P. H. Tedesco and J. G. Quintana, J. Inorg. Nucl. Chem., 36, 2628-2630 (1974).

74Tc C. A. Tolman, W. C. Seidel, and L. W. Gosser, J. Am. Chem. Soc., 96, 53 (1974).

74Td B. S. Tovrog and R. S. Drago, J. Am. Chem. Soc., 96, 2743 (1974).

74Te S. N. Tripathi and O. C. Saxena, Egypt. J. Chem., 17, 345-356 (1974).

74Ua D. R. Underdown, S. S. Yun, and J. L. Bear, J. Inorg. Nucl. Chem., 36, 2043-2046 (1974).

74Va V. P. Vasilev and E. V. Kozlovskii, Russ. J. Inorg. Chem., (English Trans.), 19, 807-809 (1974). (Russian Page No. 1481-1484).

74Vb V. P. Vasilev and E. V. Kozlovskii, Russ. J. Inorg. Chem., (English Trans.), 19, 971-972 (1974). (Russian Page No. 1781-1784).

74Wa W. H. Woodruff and D. W. Margerum, Inorg. Chem., 13, 2578-2585 (1974).

1973

73Aa R. Abu-Eittah and G. Arafa, Z. Anorg. Allg. Chem., 399, 244 (1973).

73Ab R. D. Adams and F. A. Cotton, Inorg. Chim. Acta, 7, 153 (1973).

73Ac R. G. Ainsworth, Trans. Faraday Soc., 69, 1028 (1973).

73Ad S. O. Ajayi and D. R. Goddard, J.C.S. Dalton, 1751 (1973).

73Ae P. E. M. Allen and R. M. Lough, Trans. Faraday Soc., I, 69 (5), 849 (1973).

73Af V. D. Anand and W. R. Carper, Bull. Chem. Soc. Japan, 46, 3077-3078 (1973).

73Ag L. T. Ang and D. P. Graddon, Aust. J. Chem., 26, 1901 (1973).

73Ah A. C. I. Anusiem and R. Lumry, J. Am. Chem. Soc., 95, 904 (1973).

73Ai E. P. Artyukhova and V. I. Dulova, Russ. J. Inorg. Chem., (English Trans.), 18, 385 (1973). (Russian Page No. 735.)

73Aj E. P. Artyukhova and V. I. Dulova, Russ. J. Phys. Chem., (English Trans.), 47, 317 (1973). (Russian Page No. 559).

73Ak J. L. Ault and J. H. Harries, J.C.S. Dalton, 1095 (1973).

73Ba E. M. Ban, R. P. Hughes, and J. Powell, J.C.S. Chem. Commun., 591 (1973).

73Bb J. L. Banyasz and J. E. Stuehr, J. Am. Chem. Soc., 95, 7226 (1973).

73Bc R. Barbucci, L. Fabbrizzi, P. Paoletti, and A. Vacca, J.C.S. Dalton, 1763 (1973).

73Bd M. Barres, J. P. Redoute, R. Romanetti, H. Tachoire, and C. Zahra, C. R. Acad. Sc., (Paris), 276, 363-366 (1973).

73Be J. K. Beattie and N. Sutin, J. Am. Chem. Soc., 95, 2052 (1973).

73Bf A. Bellomo, D. De Marco, and A. De Robertis, Talanta, 20, 1225-1228 (1973).

73Bg D. J. Benton and P. Moore, J.C.S. Dalton, 399 (1973).

REFERENCE INDEX

73Bh L. P. Berezina, V. G. Samoilenko and A. I. Pozigun, Russ. J. Inorg. Chem., (English Trans.), 18, 205 (1973). (Russian Page No. 393.)

73Bi G. Berthon and O. Enea, Thermochimica Acta, 6, 57 (1973).

73Bj G. L. Bertrand, D. E. Oyler, U. G. Eichelbaum, and L. G. Hepler, Thermochimica Acta, 7, 87 (1973).

73Bk C. Biselx, J. J. Schaer, and D. Janjic, Helv. Chim. Acta, 56, 2098-2100 (1973).

73Bl J. R. Blackborow and J. C. Lockhart, J. Chem. Thermodynamics, 5, 603 (1973).

73Bm R. J. Blagrove and L. C. Gruen, Aust. J. Chem., 26, 225 (1973).

73Ca R. Cetina, J. Gomez-Lara, and R. Contreras, J. Inorg. Nucl. Chem., 35, 4217 (1973).

73Cb G. R. Choppin and S. L. Bertha, J. Inorg. Nucl. Chem., 35, 1309 (1973).

73Cc G. R. Choppin, A. Dadgar, and R. Stampfli, J. Inorg. Nucl. Chem., 35, 875 (1973).

73Cd R. F. Coley and D. S. Martin, Inorg. Chim. Acta, 7, 573 (1973).

73Ce M. J. Collins and H. F. Henneike, Inorg. Chem., 12, 2983 (1973).

73Cf R. L. Courtright, R. S. Drago, J. A. Nusz, and M. S. Nozari, Inorg. Chem., 12, 2809 (1973).

73Cg R. E. Cramer and R. L. Harris, Inorg. Chem., 12, 2575 (1973).

73Da D. R. Dakternieks and D. P. Graddon, Aust. J. Chem., 26, 2379 (1973).

73Db D. R. Dakternieks and D. P. Graddon, Aust. J. Chem., 26, 2537 (1973).

73Dc I. Dellien, Acta Chem. Scand., 27, 733 (1973).

73Dd I. Dellien, I. Grenthe, and G. Hessler, Acta Chem. Scand., 27, 2431 (1973).

73De I. Dellien and L. Malmsten, Acta Chem. Scand., 27, 2877 (1973).

73Df S. C. Dhupar, K. C. Srivastava, and S. K. Banerji, J. Indian Chem. Soc., 20, 19 (1973).

73Dg J. M. Duclos, Bioinorg. Chem., 2, 263-274 (1973).

73Dh V. I. Dulova and T. A. Zhuravel, Russ. J. Phys. Chem., (English Trans.), 47, 1510 (1973). (Russian Page No. 2685.)

73Ea B. L. Edgar, D. J. Duffy, M. C. Palazzotto and L. H. Pignolet, J. Am. Chem. Soc., 95, 1125 (1973).

73Eb K. J. Ellis and A. McAuley, J.C.S. Dalton, 1533 (1973).

73Ec O. Enea and G. Berthon, Thermochimica Acta, 6, 47 (1973).

73Ed O. Enea, K. Houngbossa, and G. Berthon, Thermochimica Acta, 6, 309 (1973).

73Fa Y. Farhang and D. P. Graddon, Aust. J. Chem., 26, 983 (1973).

73Fb P. U. Fruh and W. Simon, Protides Biol. Fluids Colloquium, 20, 505-510 (1973).

73Ga A. Gergely and I. Sovago, J. Inorg. Nucl. Chem., 35, 4355 (1973).

REFERENCE INDEX

73Gb I. P. Goldshtein, E. N. Guryanova, M. E. Peisakhova, and R. R. Shifrina, J. Gen. Chem. USSR, (English Trans.), 43, 2332 (1973). (Russian Page No. 2347.)

73Gc P. K. Govil and S. K. Banerji, J. Inorg. Nucl. Chem., 35, 3932-3935 (1973).

73Gd R. C. Graham, G. H. Henderson, E. M. Eyring, and E. M. Woolley, J. Chem. Eng. Data, 18, 277 (1973).

73Ge R. J. Guschl and T. L. Brown, Inorg. Chem., 12, 2815 (1973).

73Ha G. R. Hedwig and H. K. J. Powell, J.C.S. Dalton, 793 (1973).

73Hb G. R. Hedwig and H. K. J. Powell, J.C.S. Dalton, 798 (1973).

73Hc G. R. Hedwig and H. K. J. Powell, J.C.S. Dalton, 1942 (1973).

73Hd D. R. Herrington and L. J. Boucher, Inorg. Chem., 12, 2378 (1973).

73He K. Houngbossa, G. Berthon, and O. Enea, Thermochimica Acta, 6, 215 (1973).

73Hf D. H. Huchital and R. J. Hodges, Inorg. Chem., 12, 998 (1973).

73Hg J. K. Hurst and R. H. Lane, J. Am. Chem. Soc., 95, 1703 (1973).

73Ia T. P. I, E. J. Burke, J. L. Meyer, and G. H. Nancollas, Thermochimica Acta, 5, 463 (1973).

73Ib J. Israeli and R. Volpe, Bull. Soc. Chim. France, No. 1, 43 (1973).

73Ic P. B. Iyer and M. R. Padhye, Indian J. Chem., 11, 166 (1973).

73Ka F. Kai, Y. Sadakand, H. Yokoi, and H. Aburada, J. Inorg. Nucl. Chem., 35, 2128 (1973).

73Kb R. C. Kapoor and B. S. Aggarwal, Indian J. Chem., 11, 71 (1973).

73Kc R. C. Kapoor and B. S. Aggarwal, Proc. Nat. Acad. Sci. India, 43A, 69-76 (1973).

73Kd S. Katayama, Bull. Chem. Soc. Japan, 46, 106 (1973).

73Ke M. M. T. Khan and P. R. Reddy, J. Inorg. Nucl. Chem., 35, 179 (1973).

73Kf M. M. T. Khan and P. R. Reddy, J. Inorg. Nucl. Chem., 35, 2813 (1973).

73Kg M. M. T. Khan and P. R. Reddy, J. Inorg. Nucl. Chem., 35, 2821 (1973).

73Kh N. K. Kildahl and R. S. Drago, J. Am. Chem. Soc., 95, 6245 (1973).

73Ki Ts. B. Konunova and L. S. Kachkar, Russ. J. Inorg. Chem., (English Trans.), 18, 805 (1973). (Russian Page No. 1527.)

73Kj F. Ya. Kulba, E. A. Kopylov, and Yu. B. Yakovlev, Russ. J. Inorg. Chem., (English Trans.), 18, 38 (1973). (Russian Page No. 76.)

73La J. C. Lockhart and W. J. Mossop, J.C.S. Dalton, 19 (1973).

73Lb V. A. Logachev, V. I. Dulova, and N. R. Mochanova, Russ. J. Inorg. Chem., (English Trans.), 18, 1361 (1973). (Russian Page No. 2565.)

73Lc R. S. Lord, F. Gubensek, and J. A. Rupley, Biochem., 12, 4385 (1973).

73Ld L. N. Lugina, N. K. Davidenko, and K. B. Yatsimirskii, Russ. J. Inorg. Chem., (English Trans.), 18, 1451-1453 (1973).

73Ma V. E. Mironov and N. N. Knyazeva, Russ. J. Phys. Chem., (English Trans.), 47, 736 (1973). (Russian Page No. 1301.)

73Mb V. E. Mironov, G. K. Ragulin, N. P. Kolobov, V. M. Fadeev, and Yu. B. Solovev, Russ. J. Phys. Chem., (English Trans.), 47, 456 (1973). (Russian Page No. 802.)

73Mc V. E. Mironov, G. K. Ragulin, N. P. Kolobov, V. M. Fadeev, and Yu. B. Solovev, Russ. J. Phys. Chem., (English Trans.), 47, 458 (1973). (Russian Page No. 806.)

73Md V. E. Mironov, G. K. Ragulin, Yu. B. Solovev, V. M. Fadeev, and N. P. Kolobov, Russ. J. Phys Chem., (English Trans.), 47, 301 (1973). (Russian Page No. 530.)

73Me M. I. Mittal, R. S. Saxena, and A. V. Pandey, J. Inorg. Nucl. Chem., 35, 1691 (1973).

73Mf S. C. Mohanty, U. C. Mishra, K. C. Singh, and P. K. Das, J. Indian Chem. Soc., 50, 302-306 (1973).

73Na I. Nagypal, A. Gergely, and E. Farkas, Magyar Kemiai Folyoirat, 79, 303-309 (1973).

73Nb O. Nor and A. G. Sykes, J.C.S. Dalton, 1232 (1973).

73Oa G. Olofsson and I. Olofsson, J. Am. Chem. Soc., 95, 7231 (1973).

73Ob L. L. Olson, J. Inorg. Nucl. Chem., 35, 1977 (1973).

73Oc H. Ots, Acta Chem. Scand., 27, 2344 (1973).

73Pa V. A. Palkin, T. A. Kuzina, and N. N. Kuzmina, Russ. J. Inorg. Chem., (English Trans.), 18, 406 (1973).

73Pb S. P. Pande and K. N. Munshi, Indian J. Chem., 11, 1322 (1973).

73Pc A. N. Pant, R. N. Soni, and S. L. Gupta, J. Inorg. Nucl. Chem., 35, 1390 (1973).

73Pd W. Partenheimer and E. F. Hoy, J. Am. Chem. Soc., 95, 2840 (1973).

73Pe W. Partenheimer and E. F. Hoy, Inorg. Chem., 12, 2805 (1973).

73Pf W. Partenheimer and E. H. Johnson, Inorg. Chem., 12, 1274 (1973).

73Pg S. Parthasarathy and S. Ambujavalli, Thermochimica Acta, 7, 225 (1973).

73Ph S. K. Patil, V. V. Ramakrishna, Radiochim., 19, 27-30 (1973).

73Pi M. E. Peisakhova, I. P. Goldshtein, E. N. Guryanova, and E. S. Shcherbakova, J. General Chem. USSR, (English Trans.), 43, 157 (1973). (Russian Page No. 159.)

73Pj I. Z. Pevzner, N. I. Eremin, N. N. Knyazeva, Ya. B. Rozen, and V. E. Mironov, Russ. J. Inorg. Chem., (English Trans.), 18, 596 (1973). (Russian Page No. 1129.)

73Pk K. Plotkin, J. Copes, and J. R. Vriesenga, Inorg. Chem., 12, 1494 (1973).

73Pl H. K. J. Powell and N. F. Curtis, Aust. J. Chem., 26, 977 (1973).

73Pm W. H. Puhl and H. F. Henneike, J. Phys. Chem., 77, 558 (1973).

73Ra D. A. Redfield and J. H. Nelson, Inorg. Chem., 12, 15 (1973).

73Rb G. Reinhard, R. Dreyer, and R. Munze, Z. Phys. Chemie, Leipzig, 254, 226 (1973).

73Rc F. H. Reynolds, R. K. Burkhard, and D. D. Mueller, Biochem., 12, 359 (1973).

REFERENCE INDEX

73Rd I. P. Romm, E. M. Sadykova, E. N. Guryanova, and I. D. Kolli, J. Gen. Chem. USSR, (English Trans.), 43, 727 (1973). (Russian Page No. 728.)

73Re R. Roulet and C. Barbey, Helv. Chim. Acta, 56, 2179 (1973).

73Sa J. C. Sari and J. P. Belaich, J. Am. Chem. Soc., 95, 7491 (1973).

73Sb R. Sarin and K. N. Munshi, J. Indian Chem. Soc., 1, 307 (1973).

73Sc R. Sarin and K. N. Munshi, J. Inorg. Nucl. Chem., 35, 201 (1973).

73Sd M. Sato and T. Kwan, Bull. Chem. Soc. Japan, 46, 3745-3748 (1973).

73Se R. S. Saxena and U. S. Chaturvedi, J. Inst. Chemists, (India), 45, 130-134 (1973).

73Sf R. S. Saxena and U. S. Chaturvedi, Monatshefte fur Chemie, 104, 1208 (1973).

73Sg R. S. Saxena and S. S. Sheelwant, J. Indian Chem. Soc., 50, 455-457 (1973).

73Sh R. S. Saxena and S. S. Sheelwant, J. Inorg. Nucl. Chem., 35, 941 (1973).

73Si R. S. Saxena and S. S. Sheelwant, J. Inst. Chemists, (India), 45, 164-167 (1973).

73Sj R. S. Saxena and S. S. Sheelwant, Labdev. J. Sci. Tech., (India), 11A, 84-87 (1973).

73Sk J. J. Schaer, B. Torre, and D. Janjic, Helv. Chim. Acta, 56, 2101 (1973).

73Sl B. S. Sekhon and S. L. Chopra, Thermochimica Acta, 7, 151 (1973).

73Sm B. S. Sekhon and S. L. Chopra, Thermochimica Acta, 7, 311 (1973).

73Sn G. B. Sergeev, V. Y. Smirnov, and M. S. Sytilin, Russ. J Phys. Chem., (English Trans.), 47, 396 (1973). (Russian Page No. 704.)

73So R. S. Saxena and S. S. Sheelwant, J. Inorg. Nucl. Chem., 35, 3963 (1973).

73Sp E. Shchori and J. Jagur-Grodzinski, Israel J. Chem., 11, 243 (1973).

73Sq H. K. Sinha and B. Prasad, J. Indian Chem. Soc., 1, 177 (1973).

73Sr S. K. Srivastava and H. B. Mathur, Indian J. Chem., 11, 936 (1973).

73Ss S. K. Srivastava and H. B. Mathur, Indian J. Chem., 11, 1293-1296 (1973).

73St S. K. Srivastava, E. V. Raju, and H. B. Mathur, J. Inorg. Nucl. Chem., 35, 253 (1973).

73Su R. Stampfli and G. R. Choppin, J. Inorg. Nucl. Chem., 34, 205 (1973).

73Sv G. R. Stevenson and A. E. Alegria, J. Phys. Chem., 77, 3100 (1973).

73Sw G. R. Stevenson and L. Echegoyen, J. Phys. Chem., 77, 2339 (1973).

73Sx D. V. Stynes, H. C. Stynes, B. R. James, and J. A. Ibers, J. Am. Chem. Soc., 95, 1796 (1973).

73Sy M. S. Subramanian and S. A. Pai, Aust. J. Chem., 26, 77 (1973).

73Sz T. G. Sukhova, O. N. Temkin, L. I. Kuleshova and R. M. Flid, Russ. J. Inorg. Chem., (English Trans.), 18, 172 (1973). (Russian Page No. 325.)

73Ta P. H Tedesco, V. B. DeRumi, J. A. Gonzalez Quintana, J. Inorg. Nucl. Chem., 35, 285 (1973).

73Tb B. Tilquin and L. Lamberts, Z. Phys. Chemie N. F., 84, 84 (1973)

73Ua I. E. Umova, B. I. Lobov, Yu. I. Rutkovskii, and V. E. Mironov, Russ. J. Inorg. Chem., (English Trans.), 18, 438 (1973). (Russian Page No. 836.)

73Va V. P. Vasilev and S. A Aleksandrova, Russ. J. Inorg. Chem., (English Trans.), 18, 1089 (1973). (Russian Page No. 2055.)

73Vb V. N. Vasileva, V. P. Vasilev, T. K. Korbut, and L. I. Bukoyazova, Russ. J. Inorg. Chem., (English Trans.), 18, 974 (1973). (Russian Page No. 1843.)

73Vc V. P. Vasilev and G. A. Zaitseva, Russ. J. Inorg. Chem., (English Trans.), 18, 70 (1973). (Russian Page No. 139.)

73Vd V. G. Vinogradova and K. I. Zamaraev, Teoreticheskaya i Eksperimental'naya Khimiya, 9, 278-281 (1973).

73Ve G. D. Vogel and L. A. Searby, Inorg. Chem., 12, 936 (1973).

73Wa F. A. Walker, J. Am. Chem. Soc., 95, 1150 (1973).

73Wb F. A. Walker, J. Am. Chem. Soc., 95, 1154 (1973).

73Ya H. Yokoi, M. Otagiri, and T. Isobe, Bull. Chem. Soc. Japan, 46, 442 (1973).

73Za Ch. U. Zust, P. U. Fruh, and W. Simon, Helv. Chim. Acta, 56, 495 (1973).

1972

72Aa R. Abu-Eittah and A. El-Kourashy, J. Phys. Chem., 76, 2405 (1972).

72Ab P. E. M. Allen, A. E. Byers, and R. M. Lough, J.C.S. Dalton, No. 4, 479 (1972).

72Ac R. D. Allendoerfer and R. J. Papez, J. Phys. Chem., 76, 1012 (1972).

72Ad E. Arenare, P. Paoletti, A. Dei, and A. Vacca, J.C.S. Dalton, No. 6, 736 (1972).

72Ae E. P. Artyukhova and V. I. Dulova, Russ. J. Inorg. Chem., (English Trans.), 17, 456 (1972). (Russian Page No. 873.)

72Af E. P. Artyukhova and V. I. Dulova, Russ. J. Phys. Chem., (English Trans.), 46, 295 (1972). (Russian Page No. 510.)

72Ag K. R. Ashley and S. Kulprathipanja, Inorg. Chem., 11, 444 (1972).

72Ah B. K. Avinashi and S. K. Banerji, J. Indian Chem. Soc., 49, 693 (1972).

72Ba K. Balachandran and S. K. Banerji, J. Indian Chem. Soc., 49, 543 (1972).

72Bb R. Barbucci, L. Fabbrizzi, and P. Paoletti, J.C.S. Dalton, 745 (1972).

72Bc R. Barbucci, L. Fabbrizzi, and P. Paoletti, J.C.S. Dalton, 1099 (1972).

72Bd R. Barbucci, L. Fabbrizzi, P. Paoletti and A. Vacca, J.C.S. Dalton, 740 (1972).

72Be R. Barbucci, P. Paoletti, and L. Fabbrizzi, J.C.S. Dalton, 2593 (1972).

72Bf G. Beech and K. Miller, J.C.S. Dalton, 801 (1972).

72Bg V. I. Belevantsev, B. I. Peshchevitskii, and Zh. O. Badmaeva, Russ. J. Inorg. Chem., (English Trans.), 17, 1522 (1972). (Russian Page No. 2897.)

72Bh G. Berthon and O. Enea, Thermochimica Acta, 5, 107 (1972).

REFERENCE INDEX

72Bi G. Berthon, O. Enea, and Y. Bokra, Thermochimica Acta, 4, 441 (1972).

72Bj O. K. Borggaard, Acta Chem. Scand., 26, 393 (1972)

72Bk P. M. Borodin and N. K. Zao, Russ. J. Inorg. Chem., (English Trans.), 17, 959-962 (1972).

72Bl A. P. Brunetti, E. J. Burke, M. C. Lim, and G. H. Nancollas, J. Solution Chem., 1, 153 (1972).

72Bm J. G. Bullitt, F. A. Cotton, and T. J. Marks, Inorg. Chem., 11, 671 (1972).

72Bn M. J. Burkhart and R. C. Thompson, J. Am. Chem. Soc., 94, 2999 (1972).

72Ca E. F. Caldin, M. W. Grant, B. B. Hasinoff, Trans. Faraday Soc., 68, 2247 (1972).

72Cb L. L. Chan and J. Smid, J. Phys. Chem., 76, 695 (1972).

72Cc P. B. Chock, Proc. Nat. Acad. Sci., USA, 69, 1939 (1972).

72Cd G. R. Choppin and G. Degischer, J. Inorg. Nucl. Chem., 34, 3473 (1972).

72Ce S. J. Cole, G. C. Curthoys, E. A. Magnusson and J. N. Phillips, Inorg. Chem., 11, 1024 (1972).

72Da A. Dadgar and G. R. Choppin, J. Inorg. Nucl. Chem., 34, 1297 (1972).

72Db G. Degischer and G. R. Choppin, J. Inorg. Nucl. Chem., 34, 2823 (1972).

72Dc S. C. Dhupar, K. C. Srivastava, S. K. Banerji, J. Inst. Chemists, (India), 44, 206-209 (1972).

72Dd P. Dietsch and P. Siegmund, Z. Naturforsch, 27, 509-513 (1972).

72De H. S. Dunsmore and D. Midgley, J.C.S. Dalton, 1138 (1972).

72Ea S. S. Eaton, J. R. Hutchison, R. H. Holm, and E. L. Muetterties, J. Am. Chem. Soc., 94, 6411 (1972).

72Eb S. H. Eberle and C. S. Sabau, Radiochem. Radioanal. Letters, 11, 77-87 (1972).

72Ec L. Elegant, A. Pagliardini, G. Torri, and M. Azzaro, Bull. Soc. Chim. France, 4422 (1972).

72Ed O. Enea and G. Berthon, Comptes Rendus, C, 274, 1968 (1972).

72Ee O. Enea, G. Berthon, and Y. Bokra, Thermochimica Acta, 4, 449 (1972).

72Ef J. H. Espenson and S. G. Wolenuk, Inorg. Chem., 11, 2034 (1972).

72Fa L. Fabbrizzi, R. Barbucci, and P. Paoletti, J.C.S. Dalton, 1529 (1972).

72Fb R. D. Farina and J. H. Swinehart, Inorg. Chem., 11, 645 (1972).

72Fc J. B. Farmer, F. G. Herring, and R. L. Tapping, Can. J. Chem., 50, 2079 (1972).

72Fd V. A. Fedorov, I. D. Isaev, A. M. Robov, A. V. Vertiprakhov, and V. E. Mironov, Russ. J. Inorg. Chem., (English Trans.), 17, 495 (1972). (Russian Page No. 951.)

72Fe V. A. Fedorov, T. N. Kalosh, L. I. Shmyd'ko, and V. E. Mironov, Russ. J. Inorg. Chem., (English Trans.), 17, 1086 (1972). (Russian Page No. 2089.)

72Ff V. A. Fedorov, L. I. Kiprin, and V. E. Mironov, Russ. J. Inorg. Chem., (English Trans.), 17, 641 (1972). (Russian Page No. 1233.)

72Fg V. A. Fedorov, A. M. Robov, T. I. Grigor and V. E. Mironov, Russ. J. Inorg. Chem., (English Trans.), 17, 990 (1972). (Russian Page No. 1909.)

72Fh V. A. Fedorov, N. P. Samsonova, and V. E. Mironov, Russ. J. Inorg. Chem., English Trans.), 17, 674 (1972). (Russian Page No. 1301.)

72Ga A. Gergely, J. Mojzes and Z. Kassai-Bazsa, J. Inorg. Nucl. Chem., 34, 1277 (1972).

72Gb A Gergely, J. Mojzes, and Z. Kassai-Bazsa, Magyar Kemiai Folyoirat, 78, 76 (1972).

72Gc A. Gergely and I. Sovago, Acta Chim. Acad. Sci., Hungary, 74, 273 (1972).

72Gd A. Gergely and I. Sovago, Magyar Kemiai Folyoirat, 78, 274 (1972).

72Ge I. Grenthe and G. Gardhammar, Acta Chem. Scand., 26, 3207 (1972).

72Gf I. Grenthe and H. Ots, Acta Chem. Scand., 26, 1229 (1972).

72Gg B. W. Griffin and J. A. Peterson, Biochem., 11, 4740 (1972).

72Gh J. Gross and C. Keller, J. Inorg. Nucl. Chem., 34, 725 (1972).

72Gi J. B. Grutzner, J. M. Lawlor, and L. M. Jackman, J. Am. Chem. Soc., 94, 2306 (1972).

72Ha C. F. Hale and F. H. Spedding, J. Phys. Chem., 76, 1887 (1972).

72Hb V. Hankonyi, V. Ondrusek, V. Karas-Gasparec and Z. Binenfeld, Z. Phys. Chemie, (Leipzig), 251, 280 (1972).

72Hc S. Harada, K. Amidaiji, and T. Yasunaga, Bull. Chem. Soc. Japan, 45, 1752 (1972).

72Hd H. D. Harmon and J. R. Peterson, J. Inorg. Nucl. Chem., 34, 1381 (1972).

72He B. Hedlund, C. Danielson, and R. Lovrien, Biochem., 11, 4660 (1972).

72Hf J. B. Hunt, P. D. Ross, and A. Ginsburg, Biochem., 11, 3716 (1972).

72Hg J. B. Hunt, P. D. Ross, and A. Ginsburg, Biochem., 11, 3716 (1972).

72Ia T. P. I and G. H. Nancollas, Inorg. Chem., 11, 2414 (1972).

72Ib Y. Ishino, T. Ogura, K. Noda, T, Hirashima, and O. Manabe, Bull. Chem. Soc. Japan, 45, 150 (1972).

72Ic J. Israeli and R. Volpe, Bull. Soc. Chim. France, 1277 (1972).

72Id J. Israeli and R. Volpe, Inorg. Chim. Acta, 6, 5 (1972).

72Ie M. Itoh, Bull. Chem. Soc. Japan, 45, 1947 (1972).

72If K. J. Ivin, R. Jamison, and J. J. McGarvey, J. Am. Chem. Soc., 94, 1763 (1972).

72Ig R. M. Izatt, H. D. Johnston, and J. J. Christensen, J.C.S. Dalton, 1152 (1972).

72Ja R. F. Jameson and M. F. Wilson, J.C.S. Dalton, 2614 (1972).

72Jb R. F. Jameson and M. F. Wilson, J.C.S. Dalton, 2617 (1972).

72Ka E. F. Kassierer and A. S. Kertes, J. Inorg. Nucl. Chem., 34, 3209 (1972).

72Kb A. S. Kertes and E. F. Kassierer, Inorg. Chem., 11, 2108 (1972).

72Kc V. A. Kogan, N. I. Dorokhova, O. A. Osipov, and S. G. Kochin, Russ. J. Phys. Chem., (English Trans.), 46, 120 (1972) (Russian Page No. 205.)

REFERENCE INDEX

72Kd B. S. Krumgal'z, V. V. Subbotina, and K. M. Mishchenko, Russ. J. Phys. Chem., (English Trans.), 46, 165 (1972). (Russian Page No. 272.)

72La T. H. Leong and L. A. Dunn, J. Phys. Chem., 76, 2294 (1972).

72Lb Y. Y. Lim and R. S. Drago, Inorg. Chem., 11, 202 (1972).

72Lc Y. Y. Lim and R. S. Drago, Inorg. Chem., 11, 1334 (1972).

72Ma J. A. Maguire, J. J. Banewicz, R. C. T. Hung, and K. L. Wright, Inorg. Chem., 11, 3059 (1972)

72Mb J. M. Malin and R. E. Shepherd, J. Inorg. Nucl. Chem., 34, 3203 (1972).

72Mc T. Marshall and Q. Fernando, Anal. Chem., 44, 1346 (1972).

72Md J. N. Mathur, Indian J. Chem., 10, 299 (1972).

72Me R. E. Mesmer, C. F. Baes, and F. H. Sweeton, Inorg. Chem., 11, 537 (1972).

72Mf V. E. Mironov, L. A. Nikulina, and N. P. Kolobov, Russ. J. Phys. Chem., (English Trans.), 46, 1370 (1972). (Russian Page No. 2395.)

72Mg V. E. Mironov, G. K. Ragulin, I. E. Umova, Y. B. Solovev, V. P. Mikhailova, and F. T. Dong, Russ. J. Phys. Chem., (English Trans.), 46, 155 (1972). (Russian Page No. 257.)

72Mh V. E. Mironov, Y. B. Solovev, V. M. Fadeev, and N. P. Kolobov, Russ. J. Phys. Chem., (English Trans.), 46, 926 (1972). (Russian Page No. 1612.)

72Mi R. Morassi and A. Dei, Inorg. Chim. Acta, 6, 314 (1972).

72Na B. Nelander and I. Noren, Acta Chem. Scand., 26, 809 (1972).

72Nb R. M. Nikolic and I. J. Gal, J.C.S. Dalton, 162 (1972).

72Nc T. Nowlin and K. Cohn, Inorg. Chem., 11, 560 (1972).

72Nd M. S. Nozari and R. S. Drago, Inorg. Chem., 11, 280 (1972).

72Pa K. Pan, S. C. Hsin, and T. S. Huang, J. Chinese Chem. Soc., 19, 1-6 (1972).

72Pb A. N Pant, R. N. Soni, and S. L. Gupta, J. Inorg. Nucl. Chem., 34, 763 (1972).

72Pc A. N. Pant, R. N. Soni, and S. L. Gupta, J Inorg. Nucl. Chem., 34, 2951 (1972).

72Pd W. Partenheimer, Inorg. Chem., 11, 743 (1972).

72Pe S. Parthasarathy and S. Ambujavalli, Electrochimica Acta, 17, 1219 (1972).

72Pf S. K. Patil and V. V. Ramakrishna, Radiochim. Acta, 18, 190-192 (1972).

72Pg M. E. Peisakhova, I. P. Goldshtein, E. N. Guryanova, and K. A. Kocheshkov, Doklady Chem. Proc. Acad. Sci. USSR, (English Trans.), 203, 372 (1972). (Russian Page No. 1316.)

72Ph S. Pelletier, J. Chim. Phys., 69, 751-753 (1972).

72Pi A. N. Petrov, O. N Temkin, and M. I. Bogdanov, Russ. J. Phys. Chem., (English Trans.), 46, 1043 (1972). (Russian Page No. 1822.)

72Pj K. S. Pitzer, Trans. Faraday Soc., 68, 101 (1972).

72Pk H. K. J. Powell and G. H. Nancollas, J. Am. Chem. Soc., 94, 2664 (1972).

72Pl A. J. Poe and C. P. J. Vuik, J.C.S. Dalton, 2250 (1972).

72Pm C. S. G. Prasad, K. Balchandran, and S. K. Banerji, J. Indian Chem. Soc., 49, 3 (1972).

72Ra P. C. Rawat and C. M. Gupta, J. Inorg. Nucl. Chem., 34, 951 (1972).

72Rb R. C. Rawat and C. M. Gupta, J. Inorg. Nucl. Chem., 34, 1621 (1972).

72Rc G. Rialdi, J. Levy, and R. Biltonen, Biochem., 11, 2472 (1972).

72Rd I. P. Romm, E. N. Kharlamova, and E. N. Guryanova, J. General Chem. USSR, (English Trans.), 42, 2242 (1972). (Russian Page No. 2246.)

72Re T. Ryhl, Acta Chem. Scand., 26, 2961 (1972).

72Sa R. Sarin and K. N. Munshi, Aust. J. Chem., 25, 929 (1972).

72Sb R. Sarin and K. N. Munshi, J. Inorg. Nucl. Chem., 34, 581 (1972).

72Sc R. S. Saxena and U. S. Chaturvedi, J. Indian Chem. Soc., 49, 321 (1972).

72Sd R. S. Saxena and U. S. Chaturvedi, J. Inorg. Nucl. Chem., 34, 913 (1972).

72Se R. S. Saxena and U. S. Chaturvedi, J. Inorg. Nucl. Chem., 34, 2964 (1972).

72Sf R. S. Saxena and U. S. Chaturvedi, J. Inorg. Nucl. Chem., 34, 3272-3275 (1972).

72Sg R. S. Saxena and P. Singh, J. Indian Chem. Soc., 49, 325 (1972).

72Sh H. A. Skinner and E. W. Tipping, Rev. Chim. Minerale, 9, 51 (1972).

72Si J. Smid, Angew. Chem. Internat. Edt., 11, 112 (1972).

72Sj A. Streitwieser, C. J. Chang, and D. M. E. Reuben, J. Am. Chem. Soc., 94, 5730 (1972).

72Sk H. C. Stynes and J. A. Ibers, J. Am. Chem. Soc., 94, 1559 (1972).

72Ta A. D. Taneja and K. P. Srivastava, Indian J. Chem., 10, 1029 (1972).

72Tb E. W. Tipping and H. A. Skinner, Trans. Faraday Soc., 68, 1764 (1972).

72Tc G. Torsi and G. Mamantov, Inorg. Chem., 11, 1439 (1972).

72Td S. N. Tripathi and S. A. I. Rizvi, Philippine J. Sci., 101, 55-70 (1972).

72Te J. Tummavuori and P. Lumme, Suomen Kemistilehti B, 45, 21-27 (1972).

72Va L. C. Van Poucke, G. F. Thiers, and Z. Eeckhaut, Bull. Soc. Chim. Belges, 81, 357 (1972).

72Vb V. P. Vasilev and A. A. Ikonnikov, Russ. J. Inorg. Chem., (English Trans.), 17, 1700 (1972). (Russian Page No. 3232.)

72Vc L. S. Vasilev, M. M. Vartanyan, V. S. Bogdanov, V. G. Kiselev, and B. M. Mikhailov, Russ. J. General Chem., (English Trans.), 42, 1533 (1972). (Russian Page No. 1540.)

72Vd G. S. Vigee and C. L. Walkins, J. Inorg. Nucl. Chem., 34, 3936 (1972).

72Wa C. F. Wells and D. Whatley, Trans. Faraday Soc., 68, 434 (1972).

72Wb R. West and B. Bichlmeir, J. Amer. Chem. Soc., 94, 1649 (1972).

72Wc D. R. Williams, J.C.S. Dalton, 790 (1972).

72Wd D. R. Williams and P. A. Yeo, J.C.S. Dalton, 1988 (1972).

72Ya K. B. Yatsimirskii and N. A. Kostromina, Teoreticheskaya i Eksperimental'naya Khimiya, (English Trans.), 8, 104-108 (1972).

REFERENCE INDEX

72Yb A. Yingst, R. M Izatt, and J. J. Christensen, J.C.S. Dalton, 1199 (1972).

72Yc H. Yokoi, M. Sai, and T. Isobe, Bull. Chem. Soc. Japan, 45, 1100 (1972).

72Za P. S. Zacharias and A. Chakrovorty, Inorg. Chim. Acta, 6, 623 (1972).

72Zb T. V. Zhurba and V. I. Dulova, Russ. J. Inorg. Chem., (English Trans.), 17, 1203-1204 (1972).

1971

71Aa Y. K. Agrawal and J. P. Shukla, J. Indian Chem. Soc., 48, 771 (1971).

71Ab S. Ahrland and L. Kullberg, Acta Chem. Scand., 25, 3471 (1971).

71Ac S. Ahrland and L. Kullberg, Acta Chem. Scand., 25, 3677 (1971).

71Ad S. Ahrland and L. Kullberg, Acta Chem. Scand., 25, 3692 (1971).

71Ae S. O. Ajayi and D. R. Goddard, J. Chem. Soc., (A), 2673 (1971).

71Af G. Anderegg and F. Wenk, Helv. Chim. Acta, 54, 216 (1971).

71Ag E. M. Arnett and T. C. Moriarity, J. Am. Chem. Soc., 93, 4908 (1971).

71Ah B. K. Avinashi and S. K. Banerji, Indian J. Chem., 9, 496-497 (1971).

71Ai B. K. Avinashi and S. K. Banerji, J. Indian Chem. Soc., 48, 174 (1971).

71Aj A. Aziz, S. J. Lyle, and J. E. Newbery, J. Inorg. Nucl. Chem., 33, 1757 (1971).

71Ba A. R. Bailey and J. W. Larson, J. Phys. Chem., 75, 2368 (1971).

71Bb J. R. Baker, A. Hermann, and R. M. Wing, J. Am. Chem. Soc., 93, 6486 (1971).

71Bc D. S. Barnes, G. J. Ford, L. D. Pettit, and C. Sherrington, J.C.S. Chem. Commun., 690 (1971).

71Bd D. S. Barnes, G. J. Ford, L. D. Pettit, and C. Sherrington, J. Chem. Soc., (A), 2883 (1971).

71Be D. S. Barnes and L. D. Pettit, J. Inorg. Nucl. Chem., 33, 2177 (1971).

71Bf J. L. Bear and M. E. Clark, J. Inorg. Nucl. Chem., 33, 3805 (1971).

71Bg J. T. Bell, R. D. Baybarz, and D. M. Helton, J. Inorg. Nucl. Chem., 33, 3077 (1971).

71Bh B. D. Berezin, L. V. Klopova, and A. N. Drobysheva, Russ. J. Inorg. Chem., (English Trans.), 16, 1096 (1971). (Russian Page No. 2053.)

71Bi V. V. Blokhin, V. F. Anufrienko, Yu. A. Makashev and V. E. Mironov, Russ. J. Phys. Chem., (English Trans.), 45, 1062 (1971). (Russian Page No. 1860.)

71Bj V. S. Bogdanov, T. K. Baryshnikova, V. G. Kiselev, and B. M. Mikhailov, Russ. J. General Chem., (English Trans.), 41, 1537 (1971). (Russian Page No. 1533.)

71Bk D. W. Bolen, M. Flogel, and R. Biltonen, Biochem., 10, 4136 (1971).

71Bl Yu. G. Borodko, O. N. Efimov, V. B. Panov, and A. E. Shilov, Russ. J. Phys. Chem., (English Trans.), 45, 139-140 (1971).

71Ca A. Cecal and I. A. Schneider, Inorg. Chim. Acta, 5, 623 (1971).

REFERENCE INDEX

71Cb R. W. Chlebek and M. W. Lister, Can. J. Chem., 49, 2943 (1971).

71Cc G. Ciullo, C. Furlani, L. Sestili, and A. Sgamellotti, Inorg. Chim. Acta, 5, 489 (1971).

71Cd S. J. Cole, G. C. Curthoys, and E. A. Magnusson, J. Am. Chem. Soc., 93, 2153 (1971).

71Ce R. O. Cook, A. Davies, and L. A. K. Staveley, J. Chem. Thermodynamics, 3, 907 (1971).

71Cf B. J. Corden and P. H. Rieger, Inorg. Chem., 10, 263 (1971)

71Da A. Dadgar and R. Delorenzo, J. Inorg. Nucl. Chem., 33, 4155 (1971).

71Db D. R. Dakternieks and D. P. Graddon, Aust. J. Chem., 24, 2077 (1971).

71Dc D. R. Dakternieks and D. P. Graddon, Aust. J. Chem., 24, 2509 (1971).

71Dd R. C. Das, A. C. Dash, D. Satyanarayan, and U. N. Dash, Thermochimica Acta, 2, 435 (1971).

71De N. I. Dorokhova, V. A. Kogan, and O. A. Osipov, Russ. J. Phys. Chem., (English Trans.), 45, 537 (1971). (Russian Page No. 962.)

71Ea D. R. Eaton and K. Zaw, Can. J. Chem., 49, 3315 (1971).

71Eb K. G. Everett and D. A. Skoog, Anal. Chem., 43, 1541 (1971.

71Fa V. A. Fedorov, T. N. Kalosh, G. E. Chernikova, and V. E. Mironov, Russ. J. Phys. Chem., (English Trans.), 45, 775 (1971). (Russian Page No. 1364.)

71Fb V. A. Fedorov, T. N. Kalosh, and V. E. Mironov, Russ. J. Inorg. Chem., (English Trans.), 16, 539 (1971). (Russian Page No. 1014.)

71Fc J. H. Forsberg, T. M. Kubik, T. Moeller and K. Gucwa, Inorg. Chem., 10, 2656 (1971).

71Fd J. H. Forsberg and C. A. Wathen, Inorg. Chem., 10, 1379 (1971).

71Fe P. U. Fruh, J. T. Clerc, and W. Simon, Helv. Chim. Acta, 54, 1445 (1971).

71Ga G. Geier and U. Karlen, Helv. Chim. Acta, 54, 135 (1971).

71Gb A. Gergely, I. Nagypal and B. Kiraly, Acta Chim. Acad. Sci., Hungary, 68, 285 (1971).

71Gc A. Gergely, I. Nagypal, and I. Sovago, Acta Chim. Acad. Sci., Hungary, 67, 241 (1971).

71Gd A. M. Geominne and Z. Eeckhaut, Bull. Soc. Chim. Belges, 80, 605 (1971).

71Ge T. Goina, L. Bocaniciu, and M. Olariu, Revue Roumaine Chimie, 12, 1849-1854 (1971).

71Gf I. P. Goldshtein, E. N. Guryanova, and T. E. Perepelkova, Theor. and Exp. Chemistry, (English Trans.), 7, 343 (1971). (Russian Page No. 410.)

71Gg J. G. Gordon, M. J. O'Connor, and R. H. Holm, Inorg. Chim. Acta, 5, 381 (1971).

71Gh D. P. Graddon and K. B. Heng, Aust. J. Chem., 24, 1781 (1971).

71Gi D. P. Graddon and C. Y. Hsu, Aust. J. Chem., 24, 2267 (1971).

71Gj H. Grasdalen and I. Svare, Acta Chem. Scand., 25, 1089 (1971).

71Gk T. I. Grigor, V. A. Fedorov, and V. E. Mironov, Russ. J. Inorg. Chem., (English Trans.), 16, 337 (1971). (Russian Page No. 633.)

REFERENCE INDEX

71Gl G. Guiheneuf, C. Laurence, and B. Wojtkowiak, Bull Soc. Chim. France, 1157 (1971).

71Ha P. R. Hammond and R. R. Lake, J. Chem. Soc. (A), 3800 (1971).

71Hb P. R. Hammond and R. R. Lake, J. Chem. Soc. (A), 3806 (1971).

71Hc P. R. Hammond and R. R. Lake, J. Chem. Soc. (A), 3819 (1971).

71Hd P. R. Hammond and W. S. McEwan, J. Chem. Soc. (A), 3812 (1971).

71He L. D. Hansen and D. J. Temer, Inorg. Chem., 10, 1439 (1971).

71Hf R. W. Hay and P. J. Morris, J. Chem. Soc. (A), 1518 (1971).

71Hg R. W. Hay and P. J. Morris, J. Chem. Soc. (A), 3562 (1971).

71Hh D. A. House and H. K. J. Powell, Inorg. Chem., 10, 1583 (1971).

71Hi L G. Hubert-Pfalzgraf, J. Guion, and J. G. Riess, Bull. Soc. Chim. France, No. 11, 3855 (1971).

71Hj J. R. Hutchison, J. G. Gordon, and R. H. Holm, Inorg. Chem., 10, 1004 (1971).

71Ia E. Inada, K. Shimizu, J. Osugi, Nippon Kagaku Zasshi, 92, 1096-1101 (1971).

71Ib J. Israeli and J. R. Cayouette, Can. J. Chem., 49, 199 (1971).

71Ic J. Israeli and R. Volpe, Bull. Soc. Chim. France, No. 9, 3119 (1971).

71Id J. Israeli and R. Volpe, Inorg. Nucl. Chem. Letters, 7, 1183 (1971).

71Ie J. Israeli and R. Volpe, J. Inorg. Nucl. Chem., 33, 4358 (1971).

71If N. Ivicic and V. Simeon, Thermochimica Acta, 2, 345 (1971).

71Ig R. M. Izatt, C. H. Bartholomew, C. E. Morgan, D. J. Eatough, and J. J. Christensen, Thermochimica Acta, 2, 313 (1971).

71Ih R. M. Izatt, H. D. Johnston, D. J. Eatough, J. W. Hansen, and J. J. Christensen, Thermochimica Acta, 2, 77 (1971.)

71Ii R. M. Izatt, D. P. Nelson, J. H. Rytting, B. L. Haymore, and J. J. Christensen, J. Am. Chem. Soc., 93, 1619 (1971).

71Ja A. D. Jones and D. R. Williams, J. Chem. Soc. (A), 3159 (1971).

71Ka M. H. Keyes, M. Falley, and R. Lumry, J. Am. Chem. Soc., 93, 2035 (1971).

71Kb V. A. Kogan, N. I. Dorokhova, and O. A. Osipov, Russ. J. Inorg. Chem., (English Trans.), 16, 94 (1971). (Russian Page No. 179.)

71Kc N. A. Kostromina and N. N. Tananaeva, Russ. J. Inorg. Chem., (English Trans.), 16, 1256 (1971). (Russian Page No. 2356.)

71Kd H. Krakauer, Biopolymers, 10, 2459 (1971).

71Ke C. Kuehn and W. Knoche, J.C.S. Trans. Faraday Soc., 67, 2101 (1971).

71La R. P. Lang, J. Am. Chem. Soc., 93, 5047 (1971).

71Lb N. Langerman and J. M. Sturtevant, Biochem., 10, 2809 (1971).

71Lc J. M. Lehn and J. P. Sauvage, J.C.S. Chem. Commun., 440 (1971).

71Ld M. C. Lim and G. H. Nancollas, Inorg. Chem., 10, 1957 (1971).

71Le C. T. Lin and J. L. Bear, J. Phys. Chem., 75, 3705 (1971).

REFERENCE INDEX

71Lf J. C. Lockhart and W. J. Mossop, J.C.S. Chem. Commun., 61 (1971).

71Lg V. A. Logachev and V. I. Dulova, Russ. J. Inorg. Chem., (English Trans.), 16, 145 (1971). (Russian Page No. 275.)

71Lh V. A. Logachev and V. I. Dulova, Russ. J. Phys. Chem., (English Trans.), 45, 1311 (1971). (Russian Page No. 2325.)

71Li W. K. Lutz, P. U. Fruh, and W. Simon, Helv. Chim. Acta, 54, 2767 (1971).

71Ma F. Maggio and V. Romano, J. Inorg. Nucl. Chem., 33, 3993 (1971).

71Mb D. L. Mathur and G. V. Bakore, J. Indian Chem. Soc, 48, 7 (1971).

71Mc R. E. Mesmer and C. F. Baes, Inorg. Chem., 10, 2290 (1971).

71Md B. M. Mikhailov, M. N. Bochkareva, V. S. Bogdanov, O. G. Boldyreva, and V. A. Dorokhov, J. Gen. Chem. USSR, (English Trans.), 41, 1550 (1971). (Russian Page No. 1546.)

71Me A. I. Mogilyanskii, T. G. Sukhova, O. N. Temkin, and R. M. Flid, Russ. J. Inorg. Chem., (English Trans.), 16, 1074 (1971). (Russian Page No. 2017.)

71Mf H. J. Moschler, H. G. Weder, and R. Schwyzer, Helv. Chim. Acta, 54, 1437 (1971).

71Oa G. Olofsson, Acta Chem. Scand., 25, 691 (1971).

71Pa A. Pagliardini, G. Torri, L. Elegant, and M. Azzaro, Bull. Soc. Chim. France, 54 (1971).

71Pb A. N. Pant, R. N. Soni, and S. L. Gupta, J. Inorg. Nucl. Chem., 33, 3202 (1971).

71Pc S. Parthasarathy and V. S. K. Nair, Thermochimica Acta, 2, 69 (1971).

71Ra S. Rani and S. K. Banerji, Indian J. Chem., 9, 495-496 (1971).

71Rb B. Rao and H. B. Mathur, J. Inorg. Nucl. Chem., 33, 809 (1971).

71Rc B. Rao and H. B. Mathur, J. Inorg. Nucl. Chem., 33, 2919 (1971).

71Rd P. V. S. Rao and P. Kamannarayana Z. Phys. Chemie, (Leipzig), 248, 267 (1971).

71Re N. A. Rumbaut, Bull. Soc. Chim. Belges, 80, 63 (1971).

71Rf L. Rusnak and R. B. Jordan, Inorg. Chem., 10, 2199 (1971).

71Sa R. S. Saxena and U. S. Chaturvedi, J. Inorg. Nucl. Chem. 33, 3597 (1971).

71Sb K. K. SenGupta and A. K. Chatterjee, Z. Anorg. Allg. Chem., 384, 280 (1971).

71Sc E. Shchori, J. Jagur-Grodzinski, Z. Luz, and M. Shporer, J. Am. Chem. Soc., 93, 7133 (1971).

71Sd L. O. Spreer and E. L. King, Inorg. Chem., 10, 916 (1971).

71Ta U. Takaki, T. E. Hogen Esch, and J. Smid, J. Am. Chem. Soc., 93, 6760 (1971).

71Tb A. D. Taneja and K. P. Srivastava, J. Inorg. Nucl. Chem., 33, 2694 (1971).

71Tc C. P. Trivedi, P. N. Mathur, and O. P. Sunar, J. Indian Chem. Soc., 48, 270 (1971).

71Td C. P. Trivedi and O. P. Sunar, J. Indian Chem. Soc., 48, 803 (1971).

REFERENCE INDEX

71Ua J. Ulstrup, Acta Chem. Scand., 25, 3397 (1971).

71Va V. P. Vasilev, G. A. Zaitseva, and V. G. Malakhova, Russ. J. Inorg. Chem., (English Trans.), 16, 1144 (1971). (Russian Page No. 2124.)

71Vb L. Vaska, Inorg. Chim. Acta, 5, 295 (1971).

71Wa J. B. Walker, C. R. Twine, and G. R. Choppin, J. Inorg. Nucl. Chem., 33, 1813 (1971).

71Wb J. L. Wardell, J. Chem. Soc. (A), 2628 (1971).

71Wc D. Waysbort and G. Navon, J.C.S. Chem. Commun., 1410 (1971).

71Ya T. Yamamoto, A. Yamamoto, and S. Ikeda, J. Am. Chem. Soc., 93, 3360 (1971).

71Yb H. Yokoi, M. Otagiri, and T. Isobe, Bull. Chem. Soc. Japan, 44, 1445 (1971).

71Yc H. Yokoi, M. Otagiri, and T. Isobe, Bull. Chem. Soc. Japan, 44, 2402 (1971).

71Za P. S. Zacharias and A. Chakrovorty Inorg. Chem., 10, 1961 (1971).

71Zb A. H. Zeltmann and L. O. Morgan, Inorg. Chem., 10, 2739 (1971).

71Zc T. V. Zhurba and V. I. Dulova, Russ. J. Inorg. Chem., (English Trans.), 16, 1168 (1971). (Russian Page No. 2190.)

71Zd T. V. Zhurba and V. I. Dulova, Russ. J. Inorg. Chem., (English Trans.), 17, 1203 (1972). (Russian Page No. 2306.)

71Ze T. V. Zhurba and V. I. Dulova, Russ. J. Inorg. Chem., (English Trans.), 16, 1309 (1971). (Russian Page No. 2454.)

71Zf R. A. Zingaro, K. J. Irgolic, D. H. O'Brien, and L. J. Edmonson, J. Am. Chem. Soc., 93, 5677 (1971).

1970

70Aa G. Amiconi, M. Brunori, E. Antonini, G. Tauzher, and G. Costa, Nature, 228, 549 (1970).

70Ab J. N. Armor and H. Taube, J. Am. Chem. Soc., 92, 6170 (1970).

70Ac R. Arnek, Arkiv Kemi, 32, 55 (1970)

70Ad B. K. Avinashi and S. K. Banerji, Indian J. Chem., 8, 1140 (1970).

70Ba K. Balachandran and S. K. Banerji, Indian J. Chem., 8, 1136 (1970).

70Bb K. Balachandran and S. K. Banerji, J. Indian Chem. Soc., 47, 343 (1970).

70Bc K. Balachandran and S. K. Banerji, J. Indian Chem. Soc., 47, 353 (1970).

70Bd D. S. Barnes and L. D. Pettit, J.C.S. Chem. Commun., 1000 (1970).

70Be E. W. Baumann, J. Inorg. Nucl. Chem., 32, 3823 (1970).

70Bf V. V. Blokhin, G. K. Ragulin, V. F. Anufrienko, Yu. A. Makashev, and V. E. Mironov, Russ. J. Phys. Chem., (English Trans.), 44, 1512 (1970). (Russian Page No. 2654.)

70Bg R. A. Butler, A. S. Carson, P. G. Laye, and W. V. Steele, J. Organometallic Chem., 24, C11 (1970).

REFERENCE INDEX

70Ca D. L. Campbell and T. Moeller, J. Inorg. Nucl. Chem., 32, 945 (1970).

70Cb R. L. Carlin and D. B. Losee, Inorg. Chem., 9, 2087 (1970).

70Cc R. L. Carlin and A. E. Siegel, Inorg. Chem., 9, 1587-1590 (1970).

70Cd A. S. Carson, P. G. Laye, and W. V. Steele, J. Chem. Thermodynamics, 2, 757 (1970).

70Ce L. L. Chan, K. H. Wong, and J. Smid, J. Am. Chem. Soc., 92, 1955 (1970).

70Cf I. D. Chawla and A. C. Andrews, J. Inorg. Nucl. Chem., 32, 91 (1970).

70Cg G. R. Choppin and J. K. Schneider, J. Inorg. Nucl. Chem., 32, 3283 (1970).

70Ch S. J. Cole, G. C. Curthoys, and E. A. Magnusson, J. Am. Chem. Soc., 92, 2991 (1970).

70Ci J. M. Conner, J. Inorg. Nucl. Chem., 32, 3545-3548 (1970).

70Da G. Degischer and G. H. Nancollas, Inorg. Chem., 9, 1259 (1970).

70Ea D. J. Eatough, Anal. Chem., 42, 635 (1970).

70Eb W. J. Eilbeck, F. Holmes, and G. Phillips, J. Chem. Soc. (A), 729 (1970).

70Ec W. J. Eilbeck, F. Holmes, and T. W. Thomas, J. Chem. Soc., (A) 2062-2066 (1970).

70Ed A. H. Ewald and J. A. Scudder, Aust. J. Chem., 23, 1939 (1970).

70Ee G. J. Ewing and S. Maestas, J. Phys. Chem., 74, 2341 (1970).

70Fa E. L. Farquhar, L. Rusnock, and S. J. Gill, J. Am. Chem. Soc., 92, 416 (1970).

70Fb V. A. Fedorov, G. E. Chernikova, and V. E. Mironov, Zhurnal Neorganicheskoi Khimii, 15, 2100 (1970).

70Fc Z. Finta, Cs. Varhelyi, and J. Zsako, J. Inorg. Nucl. Chem., 32, 3013 (1970).

70Ga A. Gergely, B. Kiraly, I. Nagypal, and J. Mojzes, Magyar Kemiai Folyoirat, 76, 452 (1970).

70Gb A. Gergely, B. Kiraly, I. Nagypal, and J. Mojzes, Acta Chim. Acad. Sci., Hungary, 67, 133 (1971).

70Gc A. Gergely, I. Nagypal, and I. Sovago, Magyar Kemiai Folyoirat, 76, 550 (1970).

70Gd R. Ghosh and V. S. K. Nair, J. Inorg. Nucl. Chem., 32, 3025 (1970).

70Ge R. Ghosh and V. S. K. Nair, J. Inorg. Nucl. Chem., 32, 3033 (1970).

70Gf R. Ghosh and V. S. K. Nair, J. Inorg. Nucl. Chem., 32, 3041 (1970).

70Gg A. Y. Girgis and R. C. Fay, J. Am. Chem. Soc., 92, 7061 (1970).

70Gh I. P. Gol'dshtein, E. N. Gur'yanova, and E. S. Scherbakova, Russ. J. Gen. Chem., (English Trans.), 40, 166 (1970). (Russian Page No. 183.)

70Gi L. W. Gosser and C. A. Tolman, Inorg. Chem., 9, 2350 (1970).

70Gj A. F. Grand and M. Tamres, J. Phys. Chem., 74, 208 (1970).

70Gk T. R. Griffiths and R. K. Scarrow, J. Chem. Soc., (A), 827 (1970).

70Ia R. M. Izatt, G. D. Watt, C. H. Bartholomew, and J. J. Christensen, Inorg. Chem., 9, 2019 (1970).

70Ja J. M. Jallon and M. Cohn, Biochim. Biophys. Acta, 222, 542 (1970).

70Jb G. R. Janig, G. Gerber, K. Ruckpaul, S. Rapoport, and F. Jung, Eur. J. Biochem., 17, 441 (1970).

70Jc A. D. Jones and D. R. Williams, J. Chem. Soc. (A), 3138 (1970).

70Ka T. S. Kannan and A. Chakravorty, Inorg. Chem., 9, 1153 (1970).

70Kb G. C. Kugler and G. H. Carey, Talanta, 17, 907 (1970).

70La J. W. Larson, J. Phys. Chem., 74, 3392 (1970).

70Lb J. E. Letter and J. E. Bauman, J. Am. Chem. Soc., 92, 437 (1970).

70Ma J. W. Macklin and R. A. Plane, Inorg. Chem., 9, 821 (1970).

70Mb A. Martin and E. Uhlig, Z. Anorg. Allgem. Chemie, 375, 166 (1970).

70Mc A. Martin and E. Uhlig, Z. Anorg. Allgem. Chemie, 376, 282 (1970).

70Md J. L. Meyer and J. E. Bauman, J. Am. Chem. Soc., 92, 4210-4216 (1970).

70Me V. E. Mironov, K. N. Lyubomirova, and G. K. Ragulin, Russ. J. Phys. Chem., (English Trans.), 44, 230-232 (1970).

70Oa A. L. Oleinikova, O. N. Temkin, M. I. Bogdanov, and R. M. Flid, Russ. J. Phys. Chem., (English Trans.), 44, 1373 (1970). (Russian Page No. 2418.)

70Pa W. Partenheimer and R. S. Drago, Inorg. Chem., 9, 47 (1970).

70Pb A. N. Petrov, O. N. Temkin, and M. I. Bogdanov, Russ. J. Phys. Chem., (English Trans.), 44, 1574 (1970). (Russian Page No. 2766.)

70Pc L. H. Pignolet, W. DeW. Horrocks, and R. H. Holm, J. Am. Chem. Soc., 92, 1855 (1970).

70Pd R. H. Provost and C. A. Wulff, J. Chem. Thermodynamics, 2, 793 (1970).

70Sa Z. A. Schelly, D. J. Harward, P. Hemmes, and E. M. Eyring, J. Phys. Chem., 74, 3040 (1970).

70Sb M. S. Subramanian and S. A. Pai, J. Inorg. Nucl. Chem., 32, 3677 (1970).

70Va L. C. Van Poucke, G. F. Thiers, M. A. Herman, and Z. Eeckhaut, Bull. Soc. Chim. Belges, 79, 611-616 (1970).

70Vb V. P. Vasilev and P. N. Vorobev, Russ. J. Phys. Chem., (English Trans.), 44, 657 (1970). (Russian Page No. 1181.)

70Wa S. P. Wasik and W. Tsang, J. Phys. Chem., 74, 2970 (1970).

70Wb F. Wenk and G. Anderegg, Chimia, 24, 427-431 (1970).

70Wc M. Widmer and G. Schwarzenbach, Chimia, 24, 447-448 (1970).

70Wd D. R. Williams, J. Chem. Soc. (A), 1550 (1970).

1969

69Aa R. Arnek, Acta Chem. Scand., 23, 1986 (1969).

69Ab A. Aziz and S. J. Lyle, Anal. Chim. Acta, 47, 49-56 (1969).

69Ba J. E. Bailey, J. G. Beetlestone, and D. H. Irvine, J. Chem. Soc. (A), 241 (1969).

69Bb K. Balachandran and S. K. Banerji, J. Indian Chem. Soc., 46, 198 (1969).

REFERENCE INDEX

69Bc K. Balachandran and S. K. Banerji, Z. Phys. Chem., (Leipzig), 241, 267 (1969).

69Bd J. L. Bear and M. E. Clark, J. Inorg. Nucl. Chem., 31, 1517 (1969).

69Be J. G. Beetlestone and D. H. Irvine, J. Chem. Soc. (A), 735 (1969).

69Bf J. P. Belaich and J. C. Sari, Proc. Nat. Acad. Sci., U.S., 64, 763 (1969).

69Bg G. Berthon and C. Luca, Bull. Soc. Chim. France, No. 2, 432 (1969).

69Bh R. M. Bhatnagar and A. C. Chatterji, Z. Phys. Chem., (Leipzig), 241, 130 (1969).

69Bi U. C. Bhattacharyya, S. C. Lahiri, and S. Aditya, J. Indian Chem. Soc., 46, 247 (1969).

69Bj U. C. Bhattacharyya, S. C. Lahiri, and S. Aditya, J. Indian Chem. Soc., 46, 715 (1969).

69Bk A. P. Brunetti, G. H. Nancollas, and P. N. Smith, J. Am. Chem. Soc., 91, 4680-4683 (1969).

69Bl G. R. Bruce and M. H. Panckhurst, Aust. J. Chem., 22, 469 (1969).

69Ca A. S. Carson, P. G. Laye, and W. V. Steele, J. Chem. Thermodynamics, 1, 515 (1969).

69Cb I. D. Chawla and A. C. Andrews, J. Inorg. Nucl. Chem., 31, 3809 (1969).

69Cc G. R. Choppin and L. A. Martinez-Perez, Inorg. Chem., 7, 2657 (1969).

69Da A. G. Desai and R. M. Milburn, J. Am. Chem. Soc., 91, 1958 (1969).

69Ea T. Ellingsen and J. Smid, J. Phys. Chem., 73, 2712 (1969).

69Fa D. P. Fay and N. Purdie, J. Phys. Chem., 73, 3462 (1969).

69Fb V. A. Fedorov, N. P. Samsonova, and V. E. Mironov, Zhurnal Neorganicheskoi Khimii, 14, 3264 (1969).

69Fc J. H. Forsberg and T. Moeller, Inorg. Chem., 8, 889 (1969).

69Ga A. F. Garito and B. B. Wayland, J. Am. Chem. Soc., 91, 866 (1969).

69Gb G. Geier, U. Karlen, and A. V. Zelewsky, Helv. Chim. Acta, 52, 1967-1975 (1969).

69Gc P. Gerding, Acta Chem. Scand., 23, 1695 (1969).

69Gd D. R. Goddard, B. D. Lodam, S. . Ajayi, and M. J. Campbell, J. Chem. Soc. (A), 506 (1969).

69Ge A. F. Grand and M. Tamres, Inorg. Chem., 8, 2495 (1969).

69Gf I. Grenthe and E. Hansson, Acta Chem. Scand., 23, 611 (1969).

69Gg S. L. Gupta and R. N. Soni, J. Indian Chem. Soc., 46, 561 (1969).

69Gh S. L. Gupta and R. N. Soni, J. Indian Chem. Soc., 46, 761 (1969).

69Gi S. L. Gupta and R. N. Soni, J. Indian Chem. Soc., 46, 763 (1969).

69Ha H. J. Harries and G. Wright, J. Inorg. Nucl. Chem., 31, 3149 (1969).

69Hb J. M. Harvilchuck, D. A. Aikens, and R. C. Murray, Inorg. Chem., 8, 539 (1969).

69Hc R. W. Henkens, G. D. Watt, and J. M. Sturtevant, Biochem., 8, 1874 (1969).

REFERENCE INDEX

69Hd D. A. House and H. K. J. Powell, J.C.S. Chem. Commun., 382 (1969).

69Ia R. M. Izatt, D. J. Eatough, J. J. Christensen, and C. H. Bartholomew, J. Chem. Soc. (A), 45 (1969).

69Ib R. M. Izatt, D. J. Eatough, J. J. Christensen, and C. H. Bartholomew, J. Chem. Soc. (A), 47 (1969).

69Ic R. M. Izatt, J. H. Rytting, D. P. Nelson, B. L. Haymore, and J. J. Christensen, Science, 164, 443 (1969).

69Ja A. D. Jones and G. R. Choppin, J. Inorg. Nucl. Chem., 31, 3523 (1969).

69Jb J. P. Jones and D. W. Margerum, Inorg. Chem., 8, 1486 (1969).

69Ka I. V. Kolosov and Z. F. Andreeva, Russ. J. Inorg. Chem., (English Trans.), 14, 346 (1969). (Russian Page No. 664.)

69La B. L. Libutti, B. B. Wayland, and A. F. Garito, Inorg. Chem., 8, 1510 (1969).

69Lb M. Lindheimer, J. Rouviere, A. Monnier, and J. Salvinien, C. R. Acad. Sci. (Paris), 269, 364 (1969).

69Ma W. J. McKinney and A. I. Popov, J. Am. Chem. Soc., 91, 5215 (1969).

69Mb R. E. Mesmer and C. F. Baes, Jr., Inorg. Chem., 8, 618 (1969).

69Mc P. M. Milyukov and N. V. Polenova, Khimiia i Khimicheskaia Tekhnologiia, 12, 409 (1969).

69Md P. Mohr and W. Scheler, Eur. J. Biochem., 8, 444 (1969).

69Me D. F. C. Morris and l. Vandersyde, Radiochem. Radioanal. Letters, 1, 123 (1969).

69Na G. H. Nancollas and D. J. Poulton, Inorg. Chem., 8, 680 (1969).

69Pa K. H. Pearson, J. R. Baker, and J. D. Goodrich, Anal. Letters, 2, 577 (1969).

69Pb B. I. Peshchevitskii and V. I. Belevantsev, Russ. J. Inorg. Chem., (English Trans.), 14, 1256 (1969). (Russian Page No. 2393.)

69Pc T. J. Pinnavaia and S. O. Nweke, Inorg. Chem., 8, 639 (1969).

69Pd C. J. Popp, J. H. Nelson, and R. O. Ragsdale, J. Am. Chem. Soc., 91, 610 (1969).

69Pe J. Prasad and N. C. Peterson, Inorg. Chem., 8, 1622-1625 (1969).

69Ra E. V. Raju and H. B. Mathur, J. Inorg. Nucl. Chem., 31, 425 (1969).

69Rb M. V. Reddy and P. K. Bhattacharya, J. Indian Chem. Soc., 46, 1058 (1969).

69Rc P. Rigo, G. Guastalla, and A. Turco, Inorg. Chem., 8, 375 (1969).

69Rd T. Ryhl, Acta Chem. Scand., 23, 2667 (1969).

69Sa G. Sahu and B. Prasad, J. Indian Chem. Soc., 46, 233 (1969).

69Sb R. S. Saxena and K. C. Gupta, J. Indian Chem. Soc., 46, 190 (1969).

69Sc R. S. Saxena and K. C. Gupta, J. Indian Chem. Soc., 46, 258 (1969).

69Sd R. S. Saxena and K. C. Gupta, J. Indian Chem. Soc., 46, 303 (1969).

69Se R. S. Saxena and K. C. Gupta, Z. Phys. Chem., Leipzig, 241, 169 (1969).

69Sf A. Schejter and I. Aviram, Biochem., 8, 149 (1969).

REFERENCE INDEX

69Sg J. J. R. F. Silva and M. L. S. Simoes, Rev. Port. Quim., 11, 54 (1969).

69Sh V. Simeon, B. Svigir, and N. Paulic, J. Inorg. Nucl. Chem., 31, 2085 (1969).

69Si T. G. Sukhova, O. N. Temkin, and R. M. Flid, Russ. J. Incrg. Chem., (English Trans.), 14, 483 (1969). (Russian Page No. 928.)

69Sj V. T. Sushchenko, I. F. Mavrin, R. Ya. Kulba, and V. E. Mironov, Uch. Zap. Leningrad Gos. Pedagog. Inst. in A. I. Gertsena, 385, 74 (1969).

69Ta K. H. Thiele and W. Grahlert, Wissenschaftliche Zeitschrift Technischen Hchschule Otto von Guericke Magdeburg, 13, 177-180 (1969).

69Tb A. C. R. Thornton and H. A. Skinner, J.C.S. Trans. Faraday Soc., 65, 2044-2052 (1969).

69Ua E. Uhlig and A. Martin, J. Inorg. Nucl. Chem., 31, 2781 (1969).

69Wa G. D. Watt and J. M. Sturtevant, Biochem., 8, 4567 (1969).

1968

68Aa G. Aksnes and P. Albriktsen, Acta Chem. Scand., 22, 1866 (1968).

68Ab A. C. Anusiem, J. G. Beetlestone, and D. H. Irvine, J. Chem. Soc. (A), 960 (1968).

68Ac A. C. Anusiem, J. G. Beetlestone, and D. H. Irvine, J. Chem. Soc. (A), 1337 (1968).

68Ad R. Arnek, Acta Chem. Scand., 22, 1102 (1968).

68Ae R. Arnek and C. C. Patel, Acta Chem. Scand., 22, 1097 (1968).

68Af R. Arnek and K. Schlyter, Acta Chem. Scand., 22, 1327 (1968).

68Ag R. Arnek and K. Schlyter, Acta Chem. Scand., 22, 1331 (1968).

68Ba K. Balachandran and S. K. Banerji, Curr. Sci. (India), 37, 528 (1968).

68Bb D. G. Brown, R. S. Drago, and T. F. Bolles, J. Am. Chem. Soc., 90, 5706 (1968).

68Bc A. P. Brunetti, M. C. Lim, and G. H. Nancollas, J. Am. Chem. Soc., 90, 5120 (1968).

68Ca R. L. Carroll and R. R. Irani, J. Inorg. Nucl. Chem., 30, 2971 (1968).

68Cb A. S. Carson, P. G. Laye, and P. N. Smith, J. Chem. Soc. (A), 141 (1968).

68Cc A. S. Carson, P. G. Laye, and P. N. Smith, J. Chem. Soc. (A), 527 (1968).

68Cd A. S. Carson, P. G. Laye, and P. N. Smith, J. Chem. Soc. (A), 1384 (1968).

68Ce A. Chughtai, R. Marshall, and G. H. Nancollas, J. Phys. Chem., 72, 208 (1968).

68Cf B. R. Condray, R. A. Zingaro, and M. V. Kudchadkar, Inorg. Chim. Acta, 2, 309-312 (1968).

68Ea G. W. Everett, Jr. and R. H. Holm, Inorg. Chem., 7, 776 (1968).

68Fa V. A. Fedorov, A. M. Robov, N. P. Samsonova, and M. Ya. Kutuzova, Problemy Sovremennoi Khimii Koordinatsionnykh, 2, 217 (1968).

REFERENCE INDEX

68Fb V. A. Fedorov, N. P. Samsonova, and V. E. Mironov, Russ. J. Inorg. Chem., (English Trans.), 13, 198 (1968). (Russian Page No. 382.)

68Fc R. A. Firth, H. A. O. Hill, B. E. Mann, J. M. Pratt, R. G. Thorp, and R. J. P. Wiliams, J. Chem. Soc. (A), 2419 (1968).

68Fd T. Fueno, O. Kajimoto, T. Okuyama, and J. Furukawa, Bull. Chem. Soc. Japan, 41, 785 (1968).

68Fe J. Fuger, E. Merciny, and G. Duyckaerts, Bull. Soc. Chim. Belges, 77, 455 (1968).

68Ga R. C. Gardner and R. O. Ragsdale, Inorg. Chim. Acta, 2, 139 (1968).

68Gb P. S. Gentile and A. Dadgar, J. Chem. Eng. Data, 13, 367 (1968).

68Gc P. Gerding and I. Jönsson, Acta Chem. Scand., 22, 2247 (1968).

68Gd P. Gerding and B. Johansson, Acta Chem. Scand., 22, 2255 (1968).

68Ge I. P. Gol'dshtein, E. N. Kharlamova, and E. N. Guryanova, J. Gen. Chem. USSR, (English Trans.), 38, 1925 (1968). (Russian Page No. 1984.)

68Gf R. W. Green and W. G. Goodwin, Aust. J. Chem., 21, 1165 (1968).

68Gg G. Gutnikov and H. Freiser, Anal. Chem., 40, 39 (1968).

68Ha F. M. Hall and S. J. Slater, Aust. J. Chem., 21, 2663 (1968).

68Hb P. C. Harris and T. E. Moore, Inorg. Chem., 7, 656 (1968).

68Hc C. H. Henrickson, D. Duffy, and D. P. Eyman, Inorg. Chem., 7, 1047 (1968).

68Hd C. H. Henrickson, K. M. Nykerk, and D. P. Eyman, Inorg. Chem., 7, 1028 (1968).

68He N. Hirota, R. Carraway, and W. Schook, J. Am. Chem. Soc., 90, 3611 (1968).

68Ia R. M. Izatt, D. J. Eatough, and J. J. Christensen, J. Phys. Chem., 72, 2720 (1968).

68Ib R. M. Izatt, D. J. Eatough, J. J. Christensen, and R. L. Snow, J. Phys. Chem., 72, 1208 (1968).

68Ic R. M. Izatt, G. D. Watt, C. H. Bartholomew, and J. J. Christensen, Inorg. Chem., 7, 2236 (1968).

68Ka R. M. Kren, H. W. Dodgen, and C. J. Nyman, Inorg. Chem., 7, 446-451 (1968).

68Kb V. G. Krishna and B. B. Bhowmik, J. Am. Chem. Soc., 90, 1700-1705 (1968).

68La S. C. Lahiri and S. Aditya, J. Inorg. Nucl. Chem., 30, 2487 (1968).

68Lb G. C. Lalor and G. W. Bushnell, J. Chem. Soc. (A), 2520 (1968).

68Lc R. P. Lang, J. Phys. Chem., 72, 2129 (1968).

68Ld S. Laxmi, S. Prakash, and S. Prakash, Indian J. Chem., 6, 31-33 (1968).

68Le O. G. Levanda, Russ. J. Inorg. Chem., (English Trans.), 13, 1707 (1968). (Russian Page No. 3311.)

68Ma J. M. Malin and J. H. Swinehart, Inorg. Chem., 7, 250 (1968).

68Mb O. N. Malyavinskaya and Ya. I. Turyan, Russ. J. Phys. Chem., (English Trans.), 42, 144 (1968).

68Mc D. F. Martin and W. J. Randall, J. Inorg. Nucl. Chem., 30, 2909 (1968).

68Md B. D. McLees and W. S. Caughey, Biochemistry, 7, 642-652 (1968).

68Me R. M. Milburn and L. M. Venanzi, Inorg. Chim. Acta, 2, 97 (1968).

68Mf D. F. C. Morris and S. D. Hammond, Electrochimica Acta, 13, 545 (1968).

68Na G. H. Nancollas and A. C. Park, Inorg. Chem., 7, 58 (1968).

68Nb D. Nicholls, C. Sutphen, and M. Szwarc, J. Phys. Chem., 72, 1021 (1968).

68Oa G. Olofsson, Acta Chem. Scand., 22, 1352 (1968).

68Pa E. Peltonen and P. Kivalo, Suomen Kemistilehti, B41, 187 (1968).

68Pb R. Phillips and P. George, Biochim. Biophys. Acta, 162, 73 (1968).

68Pc N. G. Podder, Current Sci., (India), 37, 48 (1968).

68Ra E. V. Raju and H. B. Mathur, J. Inorg. Nucl. Chem., 30, 2181 (1968).

68Rb C. M. Rodriguez, R. Cetina, and J. Gomez Lara, Boletin Institute Quimica Univ. Nacional Autonoma Mexico, 20, 56 (1968).

68Rc I. P. Romm and E. N. Guryanova, J. General Chem. USSR, (English Trans.), 38, 1872 (1968). (Russian Page No. 1927.)

68Rd I. P. Romm, T. G. Sevastyanova, E. N. Guryanova, I. D. Kolli, and R. A. Rodionov, J. General Chem., USSR (English Trans.), 38, 1881 (1968). (Russian Page No. 1938.)

68Sa R. S. Saxena, K. C. Gupta, and M. L. Mittal, J. Inorg. Nucl. Chem., 30, 189 (1968).

68Sb R. S. Saxena, K. C. Gupta, and M. L. Mittal, Aust. J. Chem., 21, 641 (1968).

68Sc R. S. Saxena, K. C. Gupta, and M. L. Mittal, Can. J. Chem., 46, 311 (1968).

68Sd R. S. Saxena, K. C. Gupta, and M. L. Mittal, Monatshefte fur Chemie, 99, 1779 (1968).

68Se T. W. Swaddle and G. Guastalla, Inorg. Chem., 7, 1915 (1968).

68Ta S. P. Tanner and G. R. Choppin, Inorg. Chem., 7, 2046 (1968).

68Tb S. P. Tanner, J. B. Walker, and G. R. Choppin, J. Inorg. Nucl. Chem., 30, 2067 (1968).

68Va A. Vacca and P. Paoletti, J. Chem. Soc. (A), 2378 (1968).

68Wa D. R. Williams, J. Chem. Soc. (A), 2965 (1968).

68Wb P. T. T. Wong and D. G. Brewer, Can. J. Chem., 46, 131 (1968).

68Za A. H. Zeltmann, N. A. Matwiyoff, and L. O. Morgan, J. Phys. Chem., 72, 121 (1968).

1967

67Aa K. P. Anderson, W. O. Greenhalgh, and E. A. Butler, Inorg. Chem., 6, 1056 (1967).

67Ab S. Ahrland, Helv. Chim. Acta, 50, 306 (1967).

67Ac M. Ardon and N. Sutin, Inorg. Chem., 6, 2268 (1967).

REFERENCE INDEX

67Ad R. Arnek and W. Kakolowicz, Acta Chem. Scand., 21, 1449 (1967).

67Ae R. Arnek and W. Kakolowicz, Acta Chem. Scand., 21, 2180 (1967).

67Af G. Atkinson and S. K. Kor, J. Phys. Chem., 71, 673 (1967).

67Ba B. R. Baker, N. Sutin, and T. J. Welch, Inorg. Chem., 6, 1948 (1967).

67Bb T. Boschi, P. Rigs, C. Pecile, and A. Turco, Gass. Chim. Ital., 97, 1391 (1967).

67Bc H. L. Bott and A. J. Poe, J. Chem. Soc. (A), 205 (1967).

67Bd S. Boyd, J. R. Brannan, H. S. Dunsmore, and G. H. Nancollas, J. Chem. Eng. Data, 12, 601 (1967).

67Be A. Buckingham and R. P. H. Gasser, J. Chem. Soc., (A), 1964-1965 (1967).

67Ca D. K. Cabbiness and E. S. Amis, Bull. Chem. Soc. Japan, 40, 435 (1967).

67Cb J. M. Conner and V. C. Bulgrin, J. Inorg. Nucl. Chem., 29, 1953 (1967).

67Cc R. Cramer, J. Am. Chem. Soc., 89, 4621 (1967).

67Cd G. Curthoys, J. Inorg. Nucl. Chem., 29, 1176 (1967).

67Da R. G. de Carvalho and G. R. Choppin, J. Inorg. Nucl. Chem, 29, 737 (1967)

67Db R. G. Denning, F. R. Hartley, and L. M. Venanzi, J. Chem. Soc. (A), 324 (1967).

67Dc R. G. Denning, F. R. Hartley, and L. M. Vananzi, J. Chem. Soc. (A), 328 (1967).

67Dd R. G. Denning and L. M. Vananzi, J. Chem. Soc. (A), 336 (1967).

67De L. Drougge, L. I. Elding, and L. Gustafson, Acta Chem. Scand., 21, 1647 (1967).

67Df N. V. Duffy and J. E. Earley, J. Am. Chem. Soc., 89, 272 (1967).

67Ea W. A. Eaton, P. George, and G. I. H. Hanania, J. Phys. Chem., 71, 2016 (1967).

67Eb R. P. Eswein, E. S. Howald, R. A. Howald, and D. P. Keeton, J. Inorg. Nucl. Chem., 29, 437 (1967).

67Ec R. Ettorre, G. Dolcetti, and A. Peloso, Gazz. Chim. Ital., 97, 1681 (1967).

67Ed R. Ettorre, A. Peloso, and G. Dolcetti, Gazz. Chim. Ital., 97, 968 (1967).

67Fa V. A. Fedorov, A. M. Robov, and V. E. Mironov, Russ. J. Inorg. Chem., (English Trans.), 12, 1750 (1967). (Russian Page No. 3307.)

67Ga P. Gerding, Acta Chem. Scand., 21, 2015 (1967).

67Gb A. Gergely, I. Nagypal, and J. Mojzes, Acta Chimica Acad. Sci., (Hungary), 51, 381-392 (1967).

67Gc D. R. Goddard, S. I. Nwankwo, and L. A. K. Staveley, J. Chem. Soc. (A), 1376 (1967).

67Gd I. Grenthe and D. R. Williams, Acta Chem. Scand., 21, 347 (1967).

67Ge S. L. Gupta and R. N. Soni, J. Indian Chem. Soc., 44, 195 (1967).

67Ha C. F. Hale and E. L. King, J. Phys. Chem., 71, 1779 (1967).

67Hb F. R. Hartley and L. M. Venanzi, J. Chem. Soc. (A), 330 (1967).

REFERENCE INDEX

67Hc C. H. Henrickson and O. P. Eyman, Inorg. Chem., 6, 1461 (1967).

67Hd D. J. Hewkin and A. J. Poe, J. Chem. Soc. (A), 1884 (1967).

67He N. Hirota, J. Phys. Chem., 71, 127 (1967).

67Hf F. Holmes and D. R. Williams, J. Chem. Soc. (A), 1702 (1967).

67Hg T. M. Hseu and S. C. Wu, J. Chin. Chem. Soc., (Taipei), 14, 56 (1967).

67Hh D. H. Huchital and R. G. Wilkins, Inorg. Chem., 6, 1022 (1967).

67Hi J. U. Hwang nd Y. J. Jun, J. Korean Chem. Soc., 11, 63 (1967).

67Ia S. S. Ivanov and L. P. Gavryuchenkova, Polymer Sci. USSR, (English Trans.), A9, No. 1, 112 (1967). (Russian Page No. 103.)

67Ib R. M. Izatt, H. D. Johnston, G. D. Watt, and J. J. Christensen, Inorg. Chem., 6, 132 (1967).

67Ic R. M. Izatt, G. D. Watt, D. J. Eatough, and J. J. Christensen, J. Chem. Soc. (A), 1304 (1967).

67Ka M. M. T. Kahn and A. E. Martell, J. Am. Chem. Soc., 89, 5585 (1967).

67La S. C. Lahiri and S. Aditya, Z. Phys. Chem., (Frankfurt am Main), 55, 6 (1967).

67Lb C. H. Langford and W. R. Muir, J. Phys. Chem, 71, 2602 (1967).

67Ma J. A. Maguire, A. Bramley, and J. J. Banewicz, Inorg. Chem., 6, 1752 (1967).

67Mb W. L. Marshall, J. Phys. Chem., 71, 3584 (1967).

67Mc G. Matsubayashi, Y. Kawasaki, and T. Tanaka, Nippon Kagaku Zasshi, 88, 1251 (1967).

67Md I. F. Mavrin, F. Ya. Kul'ba, and V. E. Mironov, Russ. J. Inorg. Chem., (English Trans.), 12, 167 (1967). (Russian Page No. 324.)

67Me I. F. Mavrin, F. Ya. Kul'ba, and V. E. Mironov, Russ. J. Inorg. Chem., (English Trans.), 41, 886 (1967). (Russian Page No. 1659.)

67Mf A. McAuley, G. H. Nancollas, and K. Torrance, Inorg. Chem., 6, 136 (1967).

67Mg J. D. McCullough and A. Bunner, Inorg. Chem., 6, 1251 (1967).

67Mh R. E. Mesmer and C. F. Baes, Inorg. Chem., 6, 1951 (1967).

67Mi R. M. Milburn, J. Am. Chem. Soc., 89, 54 (1967).

67Na G. H. Nancollas and K. Torrance, Inorg. Chem., 6, 1567 (1967).

67Oa G. Olofsson, Acta Chem. Scand., 21, 93 (1967).

67Ob G. Olofsson, Acta Chem. Scand., 21, 1114 (1967).

67Oc G. Olofsson, Acta Chem. Scand., 21, 1887 (1967).

67Od G. Olofsson, Acta Chem. Scand., 21, 1892 (1967).

67Oe G. Olofsson, Acta Chem. Scand., 21, 2143 (1967).

67Of G. Olofsson, Acta Chem. Scand., 21, 2415 (1967).

67Pa A. J. Poë, and K. Shaw, J.C.S. Chem. Commun., 52 (1967).

67Pb A. J. Poë, K. Shaw, and M. J. Wendt, Inorg. Chim. Acta, 1, 371 (1967).

67Pc H. K. J. Powell and N. F. Curtis, J. Chem. Soc. (A), 1441 (1967).

67Ra V. H. Rau and K. H. Thiele, Z. Anorg. Allg. Chem., 335, 253 (1967).

67Rb P. S. Relan and P. K. Bhattacharya, J. Indian Chem. Soc., 44, 536 (1967).

67Sa D. E. Scaife and K. P. Wood, Inorg. Chem., 6, 358 (1967).

67Sb R. V. Slates and M. Szwarc, J. Am. Chem. Soc., 89, 6043 (1967).

67Sc W. F. Stack and H. A. Skinner, J.C.S. Trans. Faraday Soc., 63, 1136 (1967).

67Sd M. A. Suwyn and R. E. Hamm, Inorg. Chem., 6, 2150 (1967).

67Va V. P. Vasil'ev and G. A. Lobanov, Russ. J. Inorg. Chem., (English Trans.), 12, 463 (1967). (Russian Page No. 878.)

67Vb V. P. Vasil'ev and G. A. Lobanov, Russ. J. Phys. Chem., (English Trans.), 41, 1053 (1967). (Russian Page No. 1969.)

67Wa J. B. Walker and G. R. Choppin, in Advances in Chemistry Series, 71, 127 (1967). Amer. Chem. Soc., Washington, D.C., R. F. Gould, editor.

67Wb R. M. Wallace, J. Phys. Chem., 71, 1271 (1967).

67Wc C. Wu, R. J. Witonsky, P. George, and R. J. Rutman, J. Am. Chem. Soc., 89, 1987 (1967).

67Ya Yu. B. Yakovlev, F. Ya. Kulba, and V. E. Mironov, Russ. J. Phys. Chem., (English Trans.), 41, 618 (1967). (Russian Page No. 1168.)

1966

66Aa K. P. Anderson, W. O. Greenhalgh, and R. M. Izatt, Inorg. Chem., 5, 2106 (1966).

66Ab K. P. Anderson, D. A. Newell, and R. M. Izatt, Inorg. Chem., 5, 62 (1966).

66Ac A. C. Anusiem, J. G. Beetlestone, and D. H. Irvine, J. Chem. Soc. (A), 357 (1966).

66Ad I. G. Arzamonova and E. N. Guryanova, J. General Chem USSR, (English Trans.), 36, 1171 (1966). (Russian Page No. 1157.)

66Ba A. V. Belyaev and B. V. Ptitsyn, Russ. J. Inorg. Chem., (English Trans.), 11, 836 (1966). (Russian Page No. 1565.)

66Bb L. Z. Benet and J. E. Goyan, J. Pharm. Sci., 55, 1184 (1966).

66Bc K. R. Bhaskar, S. N. Bhat, A. S. N. Murthy, and C. N. R. Rao, J.C.S. Trans. Faraday Soc., 62, 78 (1966).

66Bd K. R. Bhaskar, R. K. Gosavi, and C. N. R. Rao, J.C.S. Trans. Faraday Soc., 62, 29 (1966).

66Be T. F. Bolles and R. S. Drago, J. Am. Chem. Soc., 88, 3921 (1966).

66Bf T. F. Bolles and R. S. Drago, J. Am. Chem. Soc., 88, 5730 (1966)

66Bg E. J. Bounsall and A. J. Poë, J. Chem. Soc. (A), 286 (1966).

66Ca J. de O. Cabral, H. C. A. King. S. M. Nelson, T. M. Shepherd, and E. Koros, J. Chem. Soc. (A), 1348 (1966).

66Cb P. Chang, R V. Slates, and M. Szwarc, J. Phys. Chem., 70, 3180 (1966).

66Cc G. R. Choppin and H. G. Friedman, Inorg. Chem. 5, 1599 (1966).

REFERENCE INDEX

66Da S. P. Datta and A. K. Grzybowski, J. Chem. Soc. (A), 1059 (1966).

66Db G. Dolcetti, A. Peloso, and L. Sindellari, Gazz. Chim. Ital., 96, 1648 (1966).

66Dc R. S. Drago, T. F. Bolles, and R. J. Niedzielski, J. Am. Chem. Soc., 88, 2717 (1966).

66Ga P. Gerding, Acta Chem. Scand., 20, 79 (1966).

66Gb P. Gerding, Acta Chem. Scand., 20, 2771 (1966).

66Gc J. Gohring and B. P. Susz, Helv. Chim. Acta, 49, 486 (1966).

66Gd R. N. Goldberg, R. G. Riddell, M. R. Wingard, H. P. Hopkins, C. A. Wulff, and L. G. Hepler, J. Phys. Chem., 70, 706 (1966).

66Ge D. M. L. Goodgame and M. A. Hitchman, Inorg. Chem., 5, 1303 (1966).

66Ha G. I. H. Hanania, D. H. Irving, and M. V. Irving, J. Chem. Soc. (A), 296 (1966).

66Hb J. O. Hill and R. J. Irving, J. Chem. Soc. (A), 971 (1966).

66Hc T. E. Hogen-Esch, and J. Smid, J. Am. Chem. Soc., 88, 307 (1966).

66Hd R. H. Holyer, C. D. Hubbard, S. F. A. Kettle, and R. C. Wilkins, Inorg. Chem., 5, 622-625 (1966).

66Ka M. M. T. Kahn and A. E. Martell, J. Am. Chem. Soc., 88, 668 (1966).

66Kb M. B. Kennedy and M. W. Kister, Can. J. Chem., 44, 1709 (1966).

66Kc N. I. Kulevsky, J. Chem. Eng. Data., 11, 492 (1966).

66La I. R. Lantzke and D. W. Watts, Aust. J. Chem., 19, 969 (1966).

66Ma R. Matejec, Ber. Busenges. Phys. Chem., 70, 703 (1966).

66Mb G. Matsubayashi, Y. Kawasaki, T. Tanaka, and R. Okawara, J. Inorg. Nucl. Chem., 28, 2937 (1966).

66Mc A. McAuley, G. H. Nancollas, and K. Torrance, J. Inorg. Nucl. Chem., 28, 917 (1966).

66Md W. A. Millen and D. W. Watts, Aust. J. Chem., 19, 43 (1966).

66Me V. E. Mironov and Yu. I. Rutkovskii, Russ. J. Inorg. Chem., (English Trans.), 11 958 (1966). (Russian Page No. 1792.)

66Mf T. Moeller and S. Chu, J. Inorg. Nucl. Chem., 28, 153 (1966).

66Mg G. E. Mont and A. E. Martell, J. Am. Chem. Soc., 88, 1387 (1966).

66Na G. H. Nancollas, Interactions in Electrolyte Solutions, Elsevier, New York, NY, 1966.

66Nb B. Nelander, Acta Chem. Scand., 20, 2289 (1966).

66Pa P. Paoletti, F. Nuzzi, and A. Vacca, J. Chem. Soc. (A), 1385 (1966).

66Pb P. Paoletti, A. Vacca, and D. Arenare, Coordin. Chem. Rev., 1, 280 (1966).

66Pc P. Paoletti, A. Vacca, and D. Arenare, J. Phys. Chem., 70, 193 (1966).

66Pd J. A. Partridge, J. J. Christensen, and R. M. Izatt, J. Am. Cnem. Soc., 8, 1649 (1966).

66Pe R. C. Phillips, P. George, and R. J. Rutman, J. Am. Chem. Soc., 88, 2631 (1966).

66Pf T. J. Pinnavaia and R. C. Fay, Inorg. Chem., 5, 233 (1966).

66Sa L. Sacconi and I. Bertini, J. Am. Chem. Soc., 88, 5180 (1966).

66Sb V. I. Sklenskaya and A. A. Biryukov, Russ. J. Inorg. Chem., (English Trans.), 11, 28 (1966). (Russian Page No. 54.)

66Sc E. V. Sklenskaya and M. Kh. Karapet'yants, Russ. J. Inorg. Chem., (English Trans.), 11, 1102 (1966). (Russian Page No. 2061.)

66Sd C. B. Storm, A. H. Corwin, R. R. Arellano, M. Martz, and R. Weintraub, J. Am. Chem. Soc. 88, 2525 (1966).

66Se J. H. Swinehart and P. A. Rock, Inorg. Chem., 5, 573 (1966).

66Ta W. Tefteller and R. A. Zingaro, Inorg. Chem., 5, 2151-2156 (1966).

66Tb J.Y. Tong and R. L. Johnson, Inorg. Chem., 5, 1902 (1966).

66Va G. R. Van Hecke and W. DeW. Horrocks, Inorg. Chem., 5, 1968 (1966).

66Vb J. Vaissermann, C. R. H. Acad. Sci., Series C, 262, 692 (1966).

66Vc J. Vaissermann and M. Quintin, J. Chim. Phys. Physicochim. Biol., 63, 731 (1966).

66Wa J. L. Walter and S. M. Rosalie, J. Inorg. Nucl. Chem., 28, 2969 (1966).

66Wb B. B. Wayland and R. H. Gold, Inorg. Chem., 5, 154 (1966).

1965

65Aa G. Anderegg, Helv. Chim. Acta, 48, 1712 (1965).

65Ab G. Anderegg, Helv. Chim. Acta, 48, 1718 (1965).

65Ac G. Anderegg, Helv. Chim. Acta, 48, 1722 (1965).

65Ad A. C. Andrews and D. M. Zebolsky, J. Chem. Soc. (London), 742 (1965).

65Ae R. Arnek, Ark. Kemi, 24, 531 (1965).

65Af E. Augdahl and P. Klaeboe, Acta Chem. Scand., 19, 807 (1965).

65Ba C. F. Baes, N. J. Meyer, and C. E. Roberts, Inorg. Chem., 4, 518 (1965).

65Bb T. F Bolles and R. S. Drago, J. Am. Chem. Soc., 87, 5015 (1965).

65Bc S. Boyd, A. Bryson, G. H. Nancollas, and K. Torrance, J. Chem. Soc. (London), 7353 (1965).

65Ca R. L. Carlin and F. A. Walker, J. Am. Chem. Soc., 87, 2128 (1965).

65Cb C. Carvajal, K. J. Tolle, J. Smid, and M. Szwarc, J. Am. Chem. Soc., 87, 5548 (1965).

65Cc L. Chi, T. Ts'ao, C. Lo, and K. Huang, K'o Hsueh T'ung Pao, 16, 352 (1965).

65Cd G. R. Choppin and A. J. Graffeo, Inorg. Chem., 4, 1254 (1965).

65Ce G. R. Choppin and J. Ketels, J. Inorg. Nucl. Chem., 27, 1335 (1965).

65Cf G. R. Choppin and W. F. Strazik, Inorg. Chem., 4, 1250 (1965).

65Cg J. J. Christensen, R. M. Izatt, and D. J. Eatough, Inorg. Chem., 4, 1278 (1965).

REFERENCE INDEX

65Ch E. Coates, B. Rigg, B. Saville, and D. Skelton, J. Chem. Soc. (London), 5613 (1965).

65Ci R. J. Cvetanovic, F. J. Duncan, W. E. Falconer, and R. S. Irwin, J. Am. Chem. Soc., 87, 1827 (1965).

65Da R. C. Das, Indian J. Chem., 3, 179 (1965).

65Db R. L. Davies and K. W. Dunning, J. Chem. Soc. (London), 4168 (1965).

65Fa T. Fueno, T. Okuyama, T. Deguchi, and J. Furukawa, J. Am. Chem. Soc., 87, 170 (1965).

65Fb J. Fuger and B. B. Cunningham, J. Inorg. Nucl. Chem., 27, 1079 (1965).

65Ga I. P. Gol'dshtein, E. N. Gur'yanova, and K. A. Kocheshkov, Proc. Doklady Akad. Nauk Chem., (English Trans.), 39, 491 (1965). (Russian Page No. 932.)

65Gb I. P. Gol'dshtein, E. N. Gur'yanova, and K. A. Kocheshkov, Proc. Doklady Akad. Nauk Chem., (English Trans.), 161, 233 (1965). (Russian Page No. 111.)

65Ha S. Hamada, Y. Ishikawa, and T. Shirai, Nippon Kagaku Zasshi, 86, 1042 (1965).

65Hb H. P. Hopkins and C. A. Wulff, J. Phys. Chem., 69, 6 (1965).

65Hc H. P. Hopkins and C. A. Wulff, J. Phys. Chem., 69, 9 (1965).

65Hd T. M. Hseu, S. C. Wu, and T. Chuang, J. Inorg. Nucl. Chem., 27, 1655 (1965).

65Ia R. M. Izatt, J. J. Christensen, J W. Hansen, and G. D. Watt, Inorg. Chem., 4., 718 (1965).

65Ib R. M. Izatt, J. W. Wrathall, and K. P. Anderson, J. Phys. Chem., 65, 1914 (1965).

65Ja B. Jezowska-Trzebiatowska, W. Wojciechowski, and J. Mrozinski, Roczniki Chem., 39, 1187-1191 (1965).

65Ka V. A. Korobova and G. A. Prik, Russ. J. Inorg. Chem., (English Trans.), 10, 456 (1965). (Russian Page No. 992.)

65Kb E. E. Kriss and K. B. Yatsimirskii, Russ. J. Inorg. Chem., (English Trans.), 10, 1326 (1965). (Russian Page No. 2436.)

65Kc T. Kubota, J. Am. Chem. Soc., 87, 458 (1965).

65Kd F. Ya. Kul'ba, V. E. Mironov, and I. F. Mavrin, Russ. J. Phys. Chem., (English Trans.), 39, 1387 (1965). (Russian Page No. 2595.)

65Ke F. Ya. Kul'ba, V. E. Mironov, I. F. Mavrin, and Y. B. Yakovlev, Russ. J. Inorg. Chem., (English Trans.), 10, 1117 (1965). (Russian Page No. 2053).

65La S. C. Lahiri, J Indian Chem. Soc., 42, 715 (1965).

65Lb D. W. Larson and A. L. Allred, J. Am. Chem. Soc., 87, 1216 (1965).

65Lc D. W. Larsen and A. L. Allred, J. Am. Chem. Soc., 87, 1219 (1965).

65Ld D. W. Larsen and A. L. Allred, J. Phys. Chem., 69, 2400 (1965).

65Le C. Luca, V. Magearu, and G. Popa, Analele Univ. "C. I. Parhon" Bucuresti, Ser. Stiint. Nat., 14, 109 (1965).

65Ma J. B. Martinez, R. Ferus, F. Puerta, and L. L. Mateo, An. Real Soc. Expan. Fis. Quim., Series B, 61, 1227 (1965).

65Mb V. E. Mironov and Y. I. Rutkovskii, Russ. J. Inorg. Chem., (English Trans.), 10, 1450 (1965). (Russian Page No. 2670.)

65Na V. S. K. Nair, J. Chem. Soc. (London), 1450 (1965).

65Nb R. N. Nanda, R. C. Das and R. K. Nanda, Indian J. Chem., 3, 278 (1965).

65Nc S. M. Nelson and T. M. Shepherd, J. Chem. Soc. (London), 3284 (1965).

65Oa G. Olofsson, Acta Chem. Scand., 19, 2155 (1965).

65Pa J. A. Partridge, R. M. Izatt, and J. J. Christensen, J. Chem. Soc. (London), 4231 (1965).

65Pb S. V. Pestrikov, I. I. Moiseev, and T. N. Romanova, Russ. J. Inorg. Chem., (English Trans.), 10, 1199 (1965). (Russian Page No. 2203.)

65Pc E. Plahte, J. Grundnes and P. Klaeboe, Acta Chem. Scand., 19, 1897 (1965).

65Ra R. C. Roberts and M. Szwarc, J. Am. Chem. Soc., 87, 5542 (1965).

65Sa G. Schwarzenbach and M. Schellenberg, Helv. Chim. Acta, 48, 28 (1965).

65Sb V. S. Sharma and H. B. Mathur, Indian J. Chem., 3, 475 (1965).

65Sc D. C. Shuman and B. Sen, Anal. Chim. Acta, 38, 487 (1965).

65Sd E. V. Sklenskaya and M. Kh. Karapet'yants, Tr. Mosk. Khim. Tekhnol. Inst., No. 48, 11 (1965).

65Se T. W. Swaddle and E. L. King, Inorg. Chem., 4, 532 (1965).

65Ta Y. A. Treger, R. M. Flid, L. V. Antonova, and S. S. Spektor, Russ. J. Phys. Chem., (English Trans.), 39, 1515 (1965). (Russian Page No. 2831).

65Va J. Vaissermann and M. Quintin, J. Chim. Phys., 63, 731 (1965).

65Wa G. D. Watt, J. J. Christensen, and R. M. Izatt, Inorg. Chem., 4, 220 (1965).

65Wb C. F. Wells, and G. Davies, Nature (London), 205, 692 (1965).

65Wc D. Wobschall and D. A. Norton, J. Am. Chem. Soc., 87, 3559 (1965).

65Wd D. L. Wright, J. H. Holloway, and C. N. Reilley, Anal. Chem., 37, 884 (1965).

1964

64Aa G. Anderegg, Experientia, Supplementum, 9, 75 (1964).

64Ab G. Anderegg, Helv. Chim. Acta, 47, 1801 (1964).

64Ac A. C. Andrews and J. K. Romary, Kansas Acad. Sci., Trans., 67, 630 (1964).

64Ad A. C. Andrews and J. K. Romary, J. Chem. Soc. (London), 405 (1964).

64Ba E. W. Baker, M. S. Brookhart, and A. H. Corwin, J. Am. Chem. Soc., 86, 4587 (1964).

64Bb R. J. Baltisberger and E. L. King, J. Am. Chem. Soc., 86, 795 (1964).

64Bc J. E. Bauman and J. C. Wang, Inorg. Chem., 3, 368 (1964).

REFERENCE INDEX

64Bd J. R. Brannan, H. S. Dunsmore, and G. H. Nancollas, J. Chem. Soc. (London), 304 (1964).

64Be J. A. Broomhead, F. Basolo, and R. G. Pearson, Inorg. Chem., 3, 826 (1964).

64Ca R. L. Carlin, J. S. Dubnoff, and W. T. Huntress, Proc. Chem. Soc., 228 (1964).

64Cb J. J. Christensen, R. M. Izatt, L. D. Hansen, and J. D. Hale, Inorg. Chem., 3, 130 (1964).

64Da R. C. Das and S. Aditya, J. Indian Chem. Soc., 41, 765 (1964).

64Db P. L. E. De La Praudiere and L. A. K. Staveley, J. Inorg. Nucl. Chem., 26, 1713 (1974).

64Ea I. Eliezer, J. Phys. Chem., 68, 2722 (1964).

64Eb E. Eyal and A. Treinin, J. Am. Chem. Soc., 86, 4287 (1964).

64Ga P. George, G. I. H. Hanania, D. H. Irvine, and I. Abu-Issa, J. Chem. Soc. (London), 5689 (1964).

64Gb I. Grenthe, Acta Chem. Scand., 18, 283 (1964).

64Gc F. H. Guzzetta and W. B. Hadley, Inorg. Chem., 3, 259 (1964).

64Ha G. P. Haight, D. C. Richardson, and N. H. Coburn, Inorg. Chem., 3, 1777 (1964).

64Hb G. I. H. Hanania and D. H. Irvine, J. Chem. Soc. (London), 5694 (1964).

64Hc J. A. Hull, R. H. Davies, and L. A. K. Staveley, J. Chem. Soc. (London), 5422 (1964).

64Ia R. M. Izatt, J. J. Christensen, and V. Kothari, Inorg. Chem., 3, 1565 (1964).

64Ja D. S. Jain and J. N. Gaur, Indian J. Chem., 2, 503 (1964).

64Ka H. C. A. King, E. Koros, and S. M. Nelson, J. Chem. Soc. (London), 4832 (1964).

64La J. Laane and T. L. Brown, Inorg. Chem., 3, 148 (1964).

64Lb S. C. Lahiri and S. Aditya, J. Indian Chem. Soc., 41, 517 (1964).

64Lc S. C. Lahiri and S. Aditya, Z. Phys. Chem., (Frankfurt am Main), 41, 173 (1964).

64Ld S. C. Lahiri and S. Aditya, Z. Phys. Chem., (Frankfurt am Main), 43, 282 (1964).

64Le T. J. Lane, A. J. Kandathil, and S. M. Rosalie, Inorg. Chem., 3, 487 (1964).

64Lf I. Leden and T. Ryhl, Acta Chem. Scand., 18, 1196 (1964).

64Lg D. L. Leussing and D. C. Shultz, J. Am. Chem. Soc., 86, 4846 (1964).

64Ma R. L. Middaugh, R. S. Drago, and R. J. Niedzielski, J. Am. Chem. Soc., 86, 388 (1964).

64Mb V. E. Mironov, F. Ya. Kul'ba, and V. A. Fedorov, Russ. J. Inorg. Chem., (English Trans.), 9, 888 (1964). (Russian Page No. 1641.)

64Mc V. E. Mironov, F. Ya. Kul'ba, A. V. Fokina, V. S. Golubeva, and V. A. Nazarov, Russ. J. Inorg. Chem., (English Trans.), 9, 1152 (1964). (Russian Page No. 2133.)

64Na R. J. Niedzielski, R. S. Drago, and R. L. Middaugh, J. Am. Chem. Soc., 86, 1694 (1964).

64Nb G. Nord and J. Ulstrup, Acta Chem. Scand., 18, 307 (1964).

64Oa G. Olofsson, Acta Chem. Scand., 18, 11 (1964).

64Ob G. Olofsson, Acta Chem. Scand., 18, 1022 (1964).

64Pa P. Paoletti and A. Vacca, J. Chem. Soc. (London), 5051 (1964).

64Pb V. K. Potapov, Y. S. Shabarov, and R. Ya. Levina, Russ. J. General Chem., (English Trans.), 34, 2535 (1964). (Russian Page No. 2512.)

64Ra I. G. Ryss and N. F. Kulish, Russ. J. Inorg. Chem., (English Trans.), 9, 603 (1964). (Russian Page No. 1103.)

64Rb I. G. Ryss and N. F. Kulish, Russ. J. Inorg. Chem., (English Trans.), 9, 752 (1964). (Russian Page No. 1382.)

64Sa L. Sacconi, M. Ciampolini, and N. Nardi, J. Am. Chem. Soc., 86, 819 (1964).

64Sb L. Sacconi, P. Paoletti, and M. Ciampolini, J. Chem. Soc. (London), 5046 (1964).

64Sc C. D. Schmulbach and D. M. Hart, J. Am. Chem. Soc., 86, 2347 (1964).

64Sd V. S. Sharma, H. B. Mathur, and A. B. Biswas, J. Inorg. Nucl. Chem., 26, 382 (1964).

64Se J. C. Sullivan, Inorg. Chem., 3, 315 (1964).

64Ta M. Tamres, S. Searles, and J. M. Goodenow, J. Am. Chem. Soc., 86, 3934 (1964).

64Tb B. G. Torre-Mori, D. Janjic, and B. P. Susz, Helv. Chim. Acta, 47, 1172 (1964).

64Va V. P. Vasil'ev, Russ. J. Inorg. Chem., (English Trans.), 9, 354 (1964). (Russian Page No. 641.)

64Wa M. J. M. Woods, P. K. Gallagher, Z. Z. Hugus, and E. L. King, Inorg. Chem., 3, 1313 (1964).

64Wb J. C. Wang, J. E. Bauman, and R. K. Murmann, J. Phys. Chem., 68, 2296 (1964).

1963

63Aa K. M. J. Al-Komser and B. Sen, Inorg. Chem., 2, 1219 (1963).

63Ab G. Anderegg, Helv. Chim. Acta, 46, 1833 (1963).

63Ac G. Anderegg, Helv. Chim. Acta, 46, 2813 (1963).

63Ad A. C. Andrews and J. K. Romary, Inorg. Chem., 2, 1060 (1963).

63Ae G. Atkinson and J. E. Bauman, Inorg. Chem., 2, 64 (1963).

63Ba F. Becker, J. Barthel, N. G. Schmahl, and H. M. Luschow, Z. Phys. Chem. (Frankfurt am Main), 37, 52 (1963).

REFERENCE INDEX

63Bb M. Bjorkman and L. G. Sillen, Trans. Roy. Inst. Technol., Stockholm, No. 199, (1963).

63Bc J. A. Bolzan and A. J. Arvia, Electrochimica Acta, 8, 375 (1963).

63Bd J. A. Bolzan, E. A. Jauregui, and A. J. Arvia, Electrochimica Acta, 8, 841 (1963).

63Be J. A. Bolzan, J. J. Podesta, and A. J. Arvia, Anales Asociacion Quimica, Argentina, 51, 43 (1963).

63Ca R. L. Carlson and R. S. Drago, J. Am. Chem. Soc., 85, 505 (1963).

63Cb G. R. Choppin and P. J. Unrein, J. Inorg. Nucl. Chem., 25, 387 (1963).

63Cc J. J. Christensen, R. M. Izatt, J. D. Hale, R. T. Pack, and G. D. Watt, Inorg. Chem., 2, 337 (1963).

63Da R. S. Drago, B. B. Wayland, and R. L. Carlson, J. Am. Chem. Soc., 85, 3125 (1963).

63Db J. E. Dubois and H. Herzog, Bull. Soc. Chim. France, 57 (1963).

63Ea A. G. Evans and B. J. Tabner, J. Chem. Soc., 4613 (1963).

63Fa R. C. Fay and T. S. Piper, J. Am. Chem. Soc., 85, 500 (1963).

63Ga P. George, R. C. Phillips, S. J. Rutman, and R. J. Rutman, Biochem., 2, 508 (1963).

63Gb D. C. Giedt and C. J. Nyman, J. Phys. Chem., 67, 2491 (1963).

63Gc I. Grenthe, Acta Chem. Scand., 17, 2487 (1963).

63Gd R. L. Gustafson and A. E. Martell, J. Phys. Chem., 67, 576 (1963).

63Ha R. P. Held and D. E. Goldberg, Inorg. Chem., 2, 585 (1963).

63Hb R. R. Holmes, W. P. Gallagher, and R. P. Carter, Inorg. Chem., 2, 437 (1963).

63Ia H. Irving and J. J. R. F. da Silva, J. Chem. Soc. (London), 448 (1963).

63Ka H. C. A. King, E. Körös, and S. M. Nelson, J. Chem. Soc. (London), 5449 (1963).

63Kb F. Ya. Kul'ba, V. E. Mironov, V. A. Fedorov, and V. A. Baevskii, Russ. J. Inorg. Chem., (English Trans.), 8, 1012 (1963). (Russian Page No. 1945.)

63La T. J. Lane and H. B. Durham, Inorg. Chem., 2, 632 (1963).

63Ma W. R. May and M. M. Jones, J. Inorg. Nucl. Chem., 25, 507 (1963).

63Mb A. McAuley and G. H. Nancollas, J. Chem. Soc. (London), 989 (1963).

63Mc V. E. Mironov, Russ. J. Inorg.Chem., (English Trans.), 8, 388 (1963). (Russian Page No. 764.)

63Md V. E. Mironov, V. A. Fedorov, and V. A. Nazarov, Russ. J. Inorg. Chem., (English Trans.), 8, 1102 (1963). (Russian Page No. 2109.)

63Me V. E. Mironov, F. Ya. Kul'ba, V. A. Fedorov, and T. F. Nikitenko, Russ. J. Inorg. Chem., (English Trans.), 8, 964 (1963). (Russian Page No. 1852.)

63Mf V. E. Mironov, F. Ya. Kul'ba, V. A. Fedorov, and T. F. Nikitenko, Russ. J. Inorg. Chem., (English Trans.), 8, 1215 (1963). (Russian Page No. 2318.)

63Mg V. E. Mironov and V. A. Nazarov, Russ. J. Inorg. Chem., (English Trans.), 8, 967 (1963). (Russian Page No. 1857.)

REFERENCE INDEX

63Oa G. Olofsson, I. Lindquist, and S. Sunner, Acta Chem. Scand., 17, 259 (1963).

63Pa P. Paoletti and M. Ciampolini, Ric. Sci., 3, 399 (1963).

63Pb P. Paoletti, M. Ciampolini, and L. Sacconi, J. Chem. Soc. (London), 3589 (1963).

63Pc W. B. Person, W. C. Golton, and A. I. Popov, J. Am. Chem. Soc., 85, 891 (1963).

63Pd G. A. Prik, Russ. J. Inorg. Chem., (English Trans.), 8, 1097 (1963). (Russian Page No. 2099.)

63Sa A. Ya. Sychev, A. P. Gerbeleu, and P. K. Migal, Russ. J. Inorg. Chem., (English Trans.), 8, 1081 (1963). (Russian Page No. 2070.)

63Va J. VanderVeen and W. Stevens, Rec. Trav. Chim., 82, 287 (1963).

63Wa R. L. S. Willix, J.C.S. Trans. Faraday Soc., 59, 1315 (1963).

1962

62Aa A. C. Andrews, T. D. Lyons, and T. D. O'Brien, J. Chem. Soc. (London), 1776 (1962).

62Ab G. Atkinson and J. E. Bauman, Inorg. Chem., 1, 900 (1962).

62Ac J. M. Austin and A. D. Mair, J. Phys. Chem., 66, 519 (1962).

62Ba C. F. Baes and N. J. Meyer, Inorg. Chem., 1, 780 (1962).

62Bb J. A. Bolzan and A. J. Arvia, Electrochimica Acta, 7, 589 (1962).

62Bc O. D. Bonner, H. Dolyniuk, C. F. Jordan, and G. B. Hanson, J. Inorg. Nucl. Chem., 24, 689 (1962).

62Ca S. Cabani and M. Landucci, J. Chem. Soc. (London), 278 (1962).

62Cb S. Cabani, G. Moretti, and E. Scrocco, J. Chem. Soc. (London), 88 (1962).

63Cc B. Carell and A. Olin, Acta Chem. Scand., 16, 2350 (1962).

63Cd B. Carell and A. Olin, Acta Chem. Scand., 16, 2357 (1962).

63Ce R. L. Carlson and R. S. Drago, J. Am. Chem. Soc., 84, 2320 (1962).

62Cf E. Coates and B. Rigg, J.C.S. Trans. Faraday Soc. 58, 88 (1962).

62Da I. R. Desai and V. S. K. Nair, J. Chem. Soc. (London), 2360 (1962).

63Db R. S. Drago and D. A. Wenz, J. Am. Chem. Soc., 84, 526 (1962).

62Dc R. S. Drago, D. A. Wenz, and R. L. Carlson, J. Am. Chem. Soc., 84, 1106 (1962).

62Dd J. F. Duncan and D. L. Kepert, J. Chem. Soc. (London), 205 (1962).

62Ea H. R. Ellison, J. O. Edwards, and E. L. Healy, J. Am. Chem. Soc., 84, 1820 (1962).

62Fa V. A. Fedorov, V. E. Mironov, and F. Ya. Kul'ba, Russ. J. Inorg. Chem., (English Trans.), 7, 1311 (1962). (Russian Page No. 2528.)

62Fb N. Fogel, J. M. J. Tai, and J. Yarborough, J. Am. Chem. Soc., 84, 1145 (1962).

62Ga T. Gramstad and W. J. Fuglevik, Acta Chem. Scand., 16, 2368 (1962).

62Ha G. P. Haight, J. Inorg. Nucl. Chem., 24, 663 (1962).

62Hb G. I. H. Hanania and D. H. Irvine, J. Chem. Soc. (London), 2745 (1962).

62Ia R. R. Irani and K. Moedritzer, J. Phys. Chem., 66, 1349 (1962).

62Ka H. C. A. King, E. Koros, and S. M. Nelson, Nature (London), 196, 573 (1962).

62Kb P. Kivalo and S. N. R. V. Schalien, Suom. Kemistilehti B, 35, 7 (1962).

62Kc F. Ya. Kul'ba and Yu. A. Makashev, Russ. J. Inorg. Chem., (English Trans.), 7, 661 (1962). (Russian Page No. 1280.)

62La R. P. Lang, J. Am. Chem. Soc., 84, 1185 (1962).

62Lb C. Lewis and D. A. Skoog, J. Am. Chem. Soc., 84, 1101 (1962).

62Ma J. L. Mackey, J. E. Powell, and F. H. Spedding, J. Am. Chem. Soc., 84, 2047 (1962).

62Mb W. R. May and M. M. Jones, J. Inorg. Nucl. Chem., 24, 511 (1962).

62Mc J. D. McCullough and I. C. Zimmermann, J. Phys. Chem., 66, 1198 (1962).

62Md P. M. Milyukov, Izv. Vyssh. Ucheb. Zaved., Khim. Khim. Tekhnol., 5, 249 (1962).

62Me V. E. Mironov and V. A. Fedorov, Russ. J. Inorg. Chem., (English Trans.), 7, 1309 (1962) (Russian Page No. 2524.)

62Mf T. Moeller and R. Ferrus, Inorg. Chem., 1, 49 (1962).

62Mg T. Moeller and T. M. Hseu, J. Inorg. Nucl. Chem., 24, 1635 (1962).

62Mh T. Moeller and L. C. Thomson, J. Inorg. Nucl. Chem., 24, 499 (1962).

62Mi R. L. Montgomery, U.S. Bur. Mines Rep. Invest., 6146 (1962).

62Na R. Näsänen, P. Merliainen, and M. Koskinen, Suom. Kemistilehti B, 35, 59 (1962).

62Nb J. Nixon and R. A. Plane, J. Am. Chem. Soc., 84, 4445 (1962).

62Pa A. D. Paul, J. Phys. Chem., 66, 1248 (1962).

62Pb D. D. Perrin, J. Chem. Soc. (London), 4500 (1962).

1962 (continued)

62Sa K. Schlyter Trans. Roy. Inst. Technol., Stockholm, No. 195 (1962).

62Sb K. Schlyter, Trans. Roy. Inst. Technol., Stockholm, No. 196 (1962).

62Sc A. Ya. Sychev and A. P. Gerbeleu, Russ. J. Inorg. Chem., (English Trans.), 7, 138 (1962). (Russian Page No. 269.)

62Ta M. Tamres and S. Searles, J. Phys. Chem., 66, 1099 (1962).

62Tb N. Tanaka, K. Ebata, and T. Marayama, Bull. Chem. Soc. Japan, 35, 124 (1962).

62Tc O. N. Temkin, R. M. Flid, and A. I. Malakhov, Kinet Katal., (English Trans.), 3, 799 (1962). (Russian Page No. 915.)

62Va V. P. Vasil'ev, Russ. J. Inorg. Chem., (English Trans.), 7, 233 (1962). (Russian Page No. 555.)

62Wa M. J. M. Woods, P, K. Gallagher, and E. L. King, Inorg. Chem., 1, 55 (1962).

62Ya K. B. Yatsimirskii and G. A. Prik, Russ. J. Inorg. Chem., (English Trans.), 7, 30 (1962). (Russian Page No. 62.)

1961

61Aa L. J. Andrews and R. M. Keefer, in Advances in Inorganic Chemistry and Radiochemistry, (H. J. Emeleus and A. G. Sharpe, Eds.), Vol. 3, Academic Press, New York, NY, 1961.

61Ab N. M. Atherton and S. I. Weissman, J. Am. Chem. Soc., 83, 1330 (1961).

61Ba D. Bunn, F. S. Dainton, and S. Duckworth, J.C.S. Trans. Faraday Soc., 57, 1131 (1961).

61Ca G. R. Choppin and J. A. Chopoorian, J. Inorg. Nucl. Chem., 22, 97 (1961).

61Cb M. Ciampolini, P. Paoletti, and L. Sacconi, J. Chem. Soc. (London), 2994 (1961).

61Cc R. E. Connick and A. D. Paul, J. Phys. Chem., 65, 1216 (1961).

61Da P. A. D. De Maine and P. Carapellucci, J. Mol. Spectrosc., 7, 83 (1961).

61Db R. S. Drago, R. L. Carlson, N. J. Rose and D. A. Wenz, J. Am. Chem. Soc., 83, 3572 (1961).

61Dc F. R. Duke and H. M. Garfinkel, J. Phys. Chem., 65, 461 (1961).

61Fa J. P. Fackler and F. A. Cotton, J. Am. Chem. Soc., 83, 3775-3778 (1961).

61Ga M. Good, A. Major, J. Nag-Chaudhuri, and S. P. McGlynn, J. Am. Chem. Soc., 83, 4329 (1961).

61Ia R. R. Irani and C. F. Callis, J. Phys. Chem., 65, 296 (1961).

61Ib R. M. Izatt, J. W. Wrathall, and K. P. Anderson, J. Phys. Chem., 65, 1914 (1961).

61Ja G. Junghähnel, Z. Chem., 1, 91 (1961).

61Ka L. I. Katzin, J. Chem. Phys., 35, 467 (1961).

61Kb G. A. Krestov and K. B. Yatsimirskii, Russ. J. Inorg. Chem., (English Trans.), 6, 1170 (1961). (Russian Page No. 2304.)

61La S. C. Levy and J. E. Powell, U.S. Atomic Energy Comm., IS-421 (1961).

61Ma G. N. Malcolm, H. N. Parton, and I. D. Watson, J. Phys. Chem., 65, 1900 (1961).

61Mb A. McAuley and G. H. Nancollas, J. Chem. Soc. (London), 2215 (1961).

61Mc A. McAuley and G. H. Nancollas, J. Chem. Soc. (London), 4458 (1961).

61Md J. D. McCullough and I. C. Zimmermann, J. Phys. Chem., 65, 888 (1961).

61Me P. M. Milyukov, Izv. Vyssh. Ucheb. Zaved, Khim. Khim. Tekhnol., 4, 212 (1961).

61Mf T. Moeller and R. Ferrus, J. Inorg. Nucl. Chem. 20, 261 (1961).

61Mg P. P. Mohapatra, R. C. Das, and S. Aditya, J. Indian Chem. Soc., 38, 845 (1961).

61Na V. S. K. Nair and G. H. Nancollas, J. Chem. Soc. (London), 4367 (1961).

61Nb W. C. Nicholas and W. C. Fernelius, J. Phys. Chem. 65, 1047 (1961).

61Pa J. M. Pagano, D. E. Goldberg, and W. C. Fernelius, J. Phys. Chem., 65, 1062 (1961).

61Pb A. D. Paul, L. S. Gallo, and J. B. Van Camp, J. Phys. Chem., 65, 441 (1961).

61Ra S. W. Rabideau and B. J. Masters, J. Phys. Chem., 65, 1256 (1961).

61Rb R. E. Reichard and W. C. Fernelius, J. Phys. Chem., 65, 380 (1961).

61Rc D. K. Roe, D. B. Masson, and C. J. Nyman, Anal. Chem., 33, 1464 (1961).

61Rd E. M. Ryzhkov and A. M. Sukhotin, Russ. J. Phys. Chem., (English Trans.), 35, 648 (1961). (Russian Page No. 1327.)

61Sa L. Sacconi, P. Paoletii, and M. Ciampolini, J. Chem. Soc. (London), 5115 (1961).

61Sb K. Schiller and E. Thilo, Z. Anorg. Allg. Chem., 310, 261 (1961).

61Sc K. Schlyter, Trans. Roy. Inst. Technol., Stockholm, No. 182 (1961).

61Sd S. A. Shchukarev and O. A. Lobaneva, Russ. J. Inorg. Chem., (English Trans.), 6, 410 (1961). (Russian Page No. 804.)

61Se N. Sutin, J. K. Rowley, and R. W. Dodson, J. Phys. Chem., 65, 1248 (1961).

61Ta H. Tsubomura and R. P. Lang, J. Am. Chem. Soc., 83, 2085 (1961).

61Wa R. M Wallace and E. K. Dukes, J. Phys. Chem., 65, 2094 (1961).

61Ya R. G. Yalman, J. Am. Chem. Soc., 83, 4142 (1961).

1960

60Aa G. Aksnes and T. Gramstad, Acta Chem. Scand., 14, 1485 (1960).

60Ba J. A. Bolzan, Rev. Fac. Cienc. Quim., Univ. Nac. La Plata, 33, 67 (1960).

60Bb M. Brandon, M. Tamres, and S. Searles, J. Am. Chem. Soc., 82, 2129 (1960).

60Ca M. Ciampolini, P. Paoletti, and L. Sacconi, J. Chem. Soc. (London), 4553 (1960).

60Cb M. Ciampolini, P. Paoletti, and L. Sacconi, Nature, (London), 186, 880 (1960).

60Cc M. Ciampolini, P. Paoletti, and L. Sacconi, Ric. Sci., 30, 2148 (1960).

60Da G. Daniele, Gazz. Chim. Ital., 90, 1082 (1960).

60Db G. Daniele, Gazz. Chim. Ital., 90, 1585 (1960).

60Dc P. A. D. De Maine and J. Peone, J. Mol. Spectrosc., 4, 262 (1960).

60Ea J. H. Espenson and E. L. King, J. Phys. Chem., 64, 380 (1960).

60Ga P. K. Gallagher and E. L. King, J. Am. Chem. Soc, 82, 3510 (1960).

60Gb P. George, D. H. Irvine, and S. C. Glauser, Ann. N. Y. Acad. Sci., 88, 393 (1960).

60Ia R. R. Irani, and C. F. Callis, J. Phys. Chem., 64, 1398 (1960).

60Ja J. Jordan, J. Meier, E. J. Billingham, and J. Pendergrast, Nature, (London), 187, 318 (1960).

60Ka T. Kuge and S. Ono, Bull. Chem. Soc. Japan, 33, 1273 (1960).

60Kb T. Kumai, Nippon Kagaku Zasshi, 81, 1687 (1960).

60La D. L. Leussing and P. K. Gallagher, J. Phys. Chem., 64, 1631 (1960).

60Lb D. L. Leussing, R. E. Laramy, and G. S. Alberts, J. Am. Chem. Soc., 82, 4826 (1960).

60Lc M. H. Lietzke and R. W. Stoughton, J. Phys. Chem., 64, 816 (1960).

60Ld M. W. Lister and P. Rosenblum, Can. J. Chem., 38, 1827 (1960).

60Ma J. D. McCullough and I. C. Zimmermann, J. Phys. Chem., 64, 1084 (1960).

60Mb R. L. McCullough, L. H. Jones, and R. A. Penneman, J. Inorg. Nucl. Chem., 13, 286 (1960).

60Na K. Naito, Bull. Chem. Soc. Japan, 33, 363 (1960).

60Pa P. Paoletti, M Ciampolini, and L. Sacconi, Ric. Sci, 30, 1791 (1960).

60Pb S. Pelletier, J. Chim. Phys. Pysiocochim. Biol., 57, 295 (1960).

60Pc S. Pelletier, J. Chim. Phys. Physicochim. Biol., 57, 301 (1960).

60Pd S. Pelletier, J. Chim. Phys. Physicochim. Biol., 57, 311 (1960).

60Pe S. Pelletier, J. Chim. Phys. Physicochim. Biol., 57, 318 (1960).

60Pf A. J. Poe and M. S. Vaidya, J. Chem. Soc. (London), 3431 (1960).

60Ra S. W. Rabideau and R. J. Kline, J. Phys. Chem., 64, 680 (1960).

60Rb D. H. Richards and K. W. Sykes, J. Chem. Soc. (London), 3626 (1960).

60Sa L. Sacconi and G. Lombardo, J. Am. Chem. Soc., 82, 6266 (1960).

60Sb L. Sacconi, G. Lombardo, and R. Ciofalo, J. Am. Chem. Soc., 82, 4182 (1960).

60Sc L. Sacconi, P. Paoletti, and M. Ciampolini, Ric. Sci., 30, 1610 (1960).

60Sd L. Sacconi, G. Lombardo, and P. Paoletti, J. Am. Chem. Soc., 82, 4185 (1960).

60Se S. N. R. v. Schalien, Suom. Kemistilehti B, 33, 5 (1960).

60Sf C. D. Schmulbach and R. S. Drago, J. Am. Chem. Soc., 82, 4484 (1960).

60Sg R. W. Stoughton and M. H. Lietzke, J. Phys. Chem., 64, 133 (1960).

60Ta M. Tamres and M. Brandon, J. Am. Chem. Soc., 82, 2134 (1960).

60Tb V. F. Toropova and L. S. Kirillova, Russ. J. Inorg. Chem., (English Trans.), 5, 276 (1960). (Russian Page No. 575.)

60Tc H. Tsubomura, J. Am. Chem. Soc., 82, 40 (1960).

60Td H. Tsubomura and J. M. Kliegman, J. Am. Chem. Soc., 82, 1314 (1960).

60Wa H. R. Weimer and W. C. Fernelius, J. Phys. Chem., 64, 1951 (1960).

60Ya H. Yada, J. Tanaka, and S. Nagakura, Bull. Chem. Soc. Japan, 33, 1660 (1960).

REFERENCE INDEX

1959

59Aa J. M. Austin, R. A. Matheson, and H. N. Parton, in The Structure of Electrolytic Solutions, (W. J. Hamer, Ed.), J. Wiley & Sons, Inc., New York, NY, 1959. Chapter 24, pp. 365-379.

59Ba R. H. Betts and O. F. Dahlinger, Can. J. Chem., 37, 91 (1959).

59Bb P. Brandt, Acta Chem. Scand., 13, 1639 (1959).

59Bc K. Burton, Biochem. J., 71, 388 (1959).

59Ca H. Coll, R. V. Nauman, and P. W. West, J. Am. Chem. Soc., 81, 1284 (1959).

59Cb R. E. Connick and C. P. Coppel, J. Am. Chem. Soc., 81, 6389 (1959).

59Da S. P. Datta and A. K. Grzybowski, J. Chem. Soc. (London), 1091 (1959).

59Fa D. Fleischer and H. Freiser, J. Phys. Chem., 63, 260 (1959).

59Ga P. George, G. I. H. Hanania, and D. H. Irvine, J. Chem. Soc. (London), 2548 (1959).

59Gb D. E. Goldberg and W. C. Fernelius, J. Phys. Chem., 63, 1246 (1959).

59Gc D. E. Goldberg and W. C. Fernelius, J. Phys. Chem., 63, 1328 (1959).

59Gd A. M. Golub and Yu. V. Kosmatyi, Russ. J. Inorg. Chem., (English Trans.), 4 606 (1959). (Russian Page No. 1347.)

59Ge R. L. Gustafson and A. E. Martell, J. Am. Chem. Soc., 81, 525 (1959).

59Ha T. R. Harkins and H. Freiser, J. Phys. Chem., 63, 309 (1959).

59Ka E. L. King and P. K. Gallagher, J. Phys. Chem., 63, 1073 (1959).

59Kb J. W. Kury, A. D. Paul, L. G. Hepler, and R. E. Connick, J. Am. Chem. Soc., 81, 4185 (1959).

59La T. J. Lane and J. M. Daly, J. Am. Chem. Soc., 81, 2953 (1959).

59Lb J. R. Lotz, B. P. Block, and W. C. Fernelius, J. Phys. Chem., 63, 541 (1959).

59Lc E. Lange, in The Structure of Electrolyte Solutions, (W. J. Hamer, Ed.), J. Wiley & Sons, Inc. New York, NY 1959.

59Ma G. H. McIntyre, B. P. Block, and W. C. Fernelius, J. Am. Chem. Soc., 81, 529 (1959).

59Na V. S. K. Nair and G. H. Nancollas, J. Chem. Soc. (London), 3934 (1959).

59Pa A. J. Poe and M. S. Vaidya, Nature, (London), 184, 1139 (1959).

59Sa S. N. R. v. Schalien, Suom Kemistilehti B, 32, 148 (1959).

59Sb R. L. Seth and A. K. Day, Z. Phys. Chem., (Leipzig), 210, 108 (1959).

59Ta J. G. Traynham and J. . Olechowski, J. Am. Chem. Soc., 81, 571 (1959).

59Va V. P. Vasil'ev, Termodin. Str. Rastvorov, Tr. Soveshch. M., 140 (1958), (Published 1959).

59Ya S. Yamashita, Bull. Chem. Soc. Japan, 32, 1212 (1959).

59Yb K. B. Yatsimirskii and G. A. Karacheva, Russ. J. Inorg. Chem., (English Trans.), 4, 127 (1959). (Russian Page No. 294.)

59Za A. J. Zielen, J. Am. Chem. Soc., 81, 5022 (1959).

REFERENCE INDEX

1958

58Aa G. Anderegg, G. Schwarzenbach, M. Padmoyo, and O. F. Borg, Helv. Chim. Acta, 41, 988 (1958).

58Ba C. R. Bertsch, W. C. Fernelius, and B. P. Block, J. Phys. Chem., 62, 444 (1958).

58Bb C. R. Bertsch, W. C. Fernelius, and B. P. Block, J. Phys. Chem., 62, 503 (1958).

58Bc R. E. Buckles and W. D. Womer, J. Am. Chem. Soc., 80, 5055 (1958).

58Ca S. Cabani and E. Scrocco, Ann Chim., (Rome), 48, 85 (1958).

58Cb S. Cabani and E. Scrocco, Ann. Chim., (Rome), 48, 99 (1958).

58Cc S. Cabani and E. Scrocco, J. Inorg. Nucl. Chem., 8, 332 (1958).

58Cd R. E. Connick and A. D. Paul, J. Am. Chem. Soc., 80, 2069 (1958).

58Ea T. S. Elleman, J. W. Reishus, and D. S. Martin, J. Am. Chem. Soc., 80, 536 (1958).

58Fa B. R. Freasier, A. G. Oberg, and W. W. Wendlandt, J. Phys. Chem., 62, 700 (1958).

58Ga H. S. Gates and E. L. King, J. Am. Chem. Soc., 80, 5011 (1958).

58Gb A. D. Gelman, N. N. Matorina, and A. I. Moskvin, Proc. Acad. Sci. USSR, Chem. Sect., (English Trans.), 117, 973 (1958). (Russian Page No. 88.)

58Gc A. M. Golub and G. D. Ivanchenko, Russ. J. Inorg. Chem., (English Trans.), 3, No. 2, 130 (1958). (Russian Page No. 333.)

58Ja H. B. Jonassen and J. R. Oliver, J. Am. Chem. Soc., 80, 2347 (1958).

58Jb H. B. Jonassen, A. Schaafsma, and L. Westerman, J. Phys. Chem., 62, 1022 (1958).

58Ma P. E. Martin and A. G. White, J. Chem. Soc. (London), 2490 (1958).

58Na S. Nagakura, J. Am. Chem. Soc., 80, 520 (1958).

58Nb V. S. K. Nair and G. H. Nancollas, J. Chem. Soc., (London), 3706 (1958).

58Pa M. H. Panckhurst and K. G. Woolmington, Proc. Roy. Soc., Series A, 244, 124 (1958).

58Sa L. Sacconi, G. Lombardo, and P. Paoletti, J. Chem. Soc., (London), 848 (1958).

58Sb L. Sacconi, G. Lombardo, and P. Paoletti, J. Inorg. Nucl. Chem., 8, 217-226 (1958).

58Sc D. B. Scaife and H. J. V. Tyrrell, J. Chem. Soc., (London), 386 (1958).

58Sd D. B. Scaife and H. J. V. Tyrrell, J. Chem. Soc., (London), 392 (1958).

58Se K. Schug and E. L. King, J. Am. Chem. Soc., 80, 1089 (1958).

58Sf L. A. K. Staveley and T. Randall, J.C.S. Discuss. Faraday Soc., 26, 157 (1958).

58Ya K. B. Yatsimirskii and G. A. Karacheva, Izv. Vyssh. Ucheb. Zaved., Khim. Khim. Tekhnol., 1, 13 (1958).

REFERENCE INDEX

1957

57Ba W. D. Bale, E. W. Davies, D. B. Morgans, and C. B. Monk, J.C.S. Discuss. Faraday Soc., 24, 94 (1957).

57Bb J. H. Baxendale and C. F. Wells, J.C.S. Trans. Faraday Soc., 53, 800 (1957).

57Ca H. Chateau, B. Hervier, and J. Pouradier, J. Chimie Physique, 54, 246 (1957).

57Cb H. Chateau, B. Hervier, and J. Pouradier, J. Phys. Chem., 61, 250 (1957).

57Da P. A. D. De Maine, J. Chem. Phys., 26, 1036 (1957).

57Db P. A. D. De Maine, J. Chem. Phys., 26, 1042 (1957).

57Dc P. A. D. De Maine, J. Chem. Phys., 26, 1192 (1957).

57Dd P. A. D. De Maine, J. Chem. Phys., 26, 1199 (1957).

57De H. A. Droll, B. P. Block, and W. C. Fernelius, J. Phys. Chem., 61, 1000 (1957).

57Ga A. D. Gelman, N. N. Matorina, and A. I. Moskvin, Akademiia Nauk USSR, Leningrad Comptes Rendus (Doklady), (English Trans.), 117, 973 (1957). Russian Page No. 88).

57Gb A. M. Golub and O. G. Bilyk, Russ. J. Inorg. Chem., (English Trans.), 2, 58 (1957). (Russian Page No. 2723.)

57Gc A. M. Golub and V. M. Samoilenko, Ukr. Khim. Zh, 23, 17 (1957).

57Ha J. A. Hearne and A. G. White, J. Chem. Soc., (London), 2168 (1957).

57Hb G. K. Helmkamp, F. L. Carter, and H. J. Lucas, J. Am. Chem. Soc., 79, 1306 (1957).

57Ja H. B. Jonassen, J. A. Bertrand, F. R. Groves, Jr., and R. I. Stearns, J. Amer. Chem. Soc., 79, 4279 (1957).

57Jb H. B. Jonassen and L. Westerman, J. Phys. Chem., 61, 1006 (1957).

57Jc J. Jordan and T. G. Alleman, Anal. Chem., 29, 9 (1957).

57Ka T. Kubota, Nippon Kagaku Zasshi, 78, 196 (1957).

57La M. S. Lewis and H. A. Saroff, J. Am. Chem. Soc., 79, 2112 (1957).

57Ma R. M. Milburn, J. Am. Chem. Soc., 79, 537 (1957).

57Mb C. B. Murphy and A. E. Martell, J. Biol. Chem., 226, 37 (1957).

57Na V. S. K. Nair and G. H. Nancollas, J. Chem. Soc., (London), 318 (1957).

57Nb T. W. Nakagawa, L. J. Andrews, and R. M. Keefer, J. Am. Chem. Soc., 61, 1007 (1957).

57Nc L. B. Nanninga, J. Phys. Chem., 61, 1144 (1957).

57Nd Y. Nozaki, F. R. Gurd, R. F. Chen, and J. T. Edsall, J. Am. Chem. Soc., 79, 2123 (1957).

57Pa S. Pelletier, C.R.H. Acad. Sci., 245, 160 (1957).

57Pb A. L. Phipps and R. A. Plane, J. Am. Chem. Soc., 79, 2458 (1957).

57Sa H. Schmied and R. W. Fink, J. Chem. Phys., 27, 1034 (1957).

57Sb R. L. Seth and A. K. Dey, Proc. Nat. Acad. Sci., (India), Sect. A, 26, 312 (1957).

57Sc S. A. Shchukarev, L. S. Lilich, and V. A. Latysheva, Uch. Zap. Leningrad Gos. Univ., Ser. Khim. Nauk, 15, 17 (1957).

57Ta T. Takahashi and K. Sasaki, Denki Kagaku, 25, 118 (1957).

57Ua E. Uusitalo, Ann. Acad. Sci. Fenn., Series A, 2, No. 87 (1957).

57Va V. P. Vasil'ev, Zh. Fiz. Khim., 31, 692 (1957).

57Wa S. Wilska, Suomen Kemistilehti B, 30, 185 (1957).

57Ya K. B. Yatsimirskii and L. V. Gus'kova, Russ. J. Inorg. Chem., (English Trans.), 2, No. 9, 91 (1957). (Russian Page No. 2039.)

57Yb K. B. Yatsimirskii and P. M. Milyukov, Russ. J. Inorg. Chem., (English Trans.), 2, No. 5, 98 (1957). (Russian Page No. 1046.)

57Yc K. B. Yatsimirskii and P. M. Milyukov, Zh. Fiz. Khim., 31, 842 (1957).

1956

56Aa A. Agren, Sv. Kem. Tidskr., 68, 181 (1956).

56Ab Y. Amako, Sci. Rep. Tohoku Univ., Series 1, 40, 147 (1956).

56Ba H. C. Brown and D. Gintis, J. Am. Chem. Soc., 78, 5378 (1956).

56Bb H. C. Brown, D. Gintis, and Harold Podall, J. Am. Chem. Soc., 78, 5375 (1956).

56Bc H. C. Brown and R. R. Holmes, J. Am. Chem. Soc., 78, 2173 (1956).

56Ca R. A. Care and L. A. K. Staveley, J. Chem. Soc. (London), 4571 (1956).

56Cb J. Chatt and R. G. Wilkins, J. Chem. Soc. (London), 525 (1956).

56Cc H. B. Clarke and S. P. Datta, Biochem. J., 64, 604 (1956).

56Cd R. E. Connick, L. G. Hepler, Z. Z. Hugus, J. W. Kury, W. M. Latimer, and M. A. Tsao, J. Am. Chem. Soc., 78, 1827 (1956).

56Da P. A. D. De Maine, J. Chem. Phys., 24, 1091 (1956).

56Fa S. N. Flengas and E. Rideal, Proc. Roy. Soc, Series A, 233, 443 (1956).

56Ga E. Gelles and G. H. Nancollas, J.C.S. Trans. Faraday Soc., 52, 680 (1956).

56Gb A. M. Golub, J. Gen. Chem., USSR, (English Trans.), 26, 2049 (1956). Russian Page No. 1837.)

56Ha G. B. Hares, W. C. Fernelius, and B. E. Douglas, J. Am. Chem. Soc., 78, 1816 (1956).

56Hb V. L. Hughes and A. E. Martell, J. Am. Chem. Soc., 78, 1319 (1956).

56Ka E. V. Kiseleva and S. M. Khodeeva, Tr. Mosk. Khim. Tekhnol. Inst., 22, 89 (1956).

56Kb E. Klein, Z. Elektrochem., 60, 1003 (1956).

56La G. S. Lawrence, J.C.S. Trans. Faraday Soc., 52, 236 (1956).

56Lb N. C. Li, J. M. White, and R. L. Yoest, J. Am. Chem. Soc., 78, 5218 (1956).

REFERENCE INDEX

56Ma A. E. Martell, Rec. Trav. Chim. Pays-Bas, 75, 781 (1956).

56Na G. H. Nancollas, J. Chem. Soc. (London), 744 (1956).

56Oa N. Ogimachi, L. J. Andrews, and R. M. Keefer, J. Am. Chem. Soc., 78, 2210 (1956).

56Pa F. A. Posey and H. Taube, J. Am. Chem. Soc., 78, 15 (1956).

56Ra J. J. Renier and D. S. Martin, J. Am. Chem. Soc., 78, 1833 (1956).

56Sa S. A. Shchukarev, L. S. Lilich, and V. A. Latysheva, Russ. J. Inorg. Chem., (English Trans.), 1, No. 2, 36 (1956). (Russsian Page No. 225.)

56Sb R. M. Smith and R. A. Alberty, J Am. Chem. Soc., 78, 2376 (1956).

56Sc R. M. Smith and R. A. Alberty, J. Phys. Chem., 60, 180 (1956).

56Sd L. H. Sutcliffe and J. R. Weber, J.C.S. Trans. Faraday Soc., 52, 1225 (1956).

56Ta V. F. Toropova, Zh. Neorg. Khim., 1, 243 (1956).

56Tb V. F. Toropova, Zh. Neorg. Khim., 1, 930 (1956).

56Tc J. G. Traynham and M. F. Sehnert, J. Am. Chem. Soc., 78, 4024 (1956).

56Td R. Trujillo and F. Brito, Real Soc. Espan. Fis. Quim., Series B, 52, 4024 (1956).

56Ya K. B. Yatsimirskii and V. P. Vasil'ev, Zh. Fiz. Khim., 30, 901 (1956).

1955

55Ba R. H. Betts, Can. J. Chem., 33, 1775 (1955).

55Bb H. C. Brown and R. H. Horowitz, J. Am. Chem. Soc., 77, 1730 (1955).

55Ca H. Chateau and J. Pouradier, Comptes Rendus, (France), 240, 1882 (1955).

55Cb D. Cohen, J. C. Sullivan, and J. C. Hindman, J. Am. Chem. Soc., 77, 4964 (1955).

55Cc F. A. Cotton and F. E. Harris, J. Phys. Chem., 59, 1203 (1955).

55Da R. A. Day, R. N. Wilhoit, and F. D. Hamilton, J. Am. Chem. Soc., 77, 3180 (1955).

55Fa W. S. Fyfe, J. Chem. Soc. (London), 1347 (1955).

55Ga P. George and G. I. H. Hanania, Nature (London), 175, 1034 (1955).

55Gb F. G. R. Gimblett and C. B. Monk, J.C.S. Trans. Faraday Soc., 51, 793 (1955).

55Gc L. F. Grantham, T. S. Elleman, and D. S. Martin, J. Am. Chem. Soc., 77, 2965 (1955).

55Ha L. J. Heidt and J. Berestecki, J. Am. Chem. Soc., 77, 2049 (1955).

55Ia R. M. Izatt, W. C. Fernelius, and B. P. Block, J. Phys. Chem., 59, 235 (1955).

55Ka L. I. Katzin and E. Gebert, J. Am. Chem. Soc., 77, 5814 (1955).

55Kb R. M. Keefer and L. J. Andrews, J. Am. Chem. Soc., 77, 2164 (1955).

55Kc K. A. Kraus and F. Nelson, J. Am. Chem. Soc., 77, 3721 (1955).

55La M. W. Lister and D. E. Rivington, Can. J. Chem., 33, 1572 (1955).

55Lb E. M. Loebl, L. B. Luttinger, and H. P. Gregor, J. Phys. Chem., 59, 559 (1955).

55Ma B. L. Mickel and A. C. Andrews, J. Am. Chem. Soc., 77, 5291 (1955).

55Mb L. N. Mulay and P. W. Selwood, J. Am. Chem. Soc., 77, 2693 (1955).

55Na G. H. Nancollas, J. Chem. Soc. (London), 1458 (1955).

55Oa N. Ogimachi, L. J. Andrews, and R. M. Keefer, J. Am. Chem. Soc., 77, 4202 (1955).

55Pa C. Postmus and E. L. King, J. Phys. Chem., 59, 1208 (1955).

55Pb I. Poulsen and J. Bjerrum, Acta Chem. Scand., 9, 1407 (1955).

55Ra S. W. Rabideau and H. D. Cowan, J. Am. Chem. Soc., 77, 6145 (1955).

55Sa W. Scheler, A. Salewski, and F. Jung, Biochem. Z., 326, 288 (1955).

1954

54Ba F. Basolo, Y. T. Chen, and R. K. Murmann, J. Am. Chem. Soc., 76, 956 (1954).

54Bb F. Basolo and R. K. Murmann, J. Am. Chem. Soc., 76, 211 (1954).

54Ca F. F. Carini and A. E. Martell, J. Am. Chem. Soc., 76, 2153 (1954).

54Cb R. G. Charles, J. Am. Chem. Soc., 76, 5854 (1954).

54Cc R. G. Charles and H. Frieser, Anal. Chim. Acta, 11, 101 (1954).

54Cd H. B. Clarke, D. C. Cusworth, and S. P. Datta, Biochem. J., 58, 146 (1954).

56Ce R. E. Connick, and M. S. Tsao, J. Am. Chem. Soc., 76, 5311 (1954).

54Da T. Davies, S. S. Singer, and L. A. K. Staveley, J. Chem. Soc. (London), 2304 (1954).

54Db R. A. Day and R. M. Powers, J. Am. Chem. Soc., 76, 3895 (1954).

54Ea J. T. Edsall, G. Felsenfeld, D. S. Goodman, and F. R. N. Gurd, J. Am. Chem. Soc., 76, 3054 (1954).

54Ga F. G. R. Gimblett and C. B. Monk, J.C.S. Trans. Faraday Soc., 50, 965 (1954).

54Gb E. Gonick, W. C. Fernelius, and B. E. Douglas, J. Am. Chem. Soc., 76, 4671 (1954).

54Ha L. G. Hepler, J. W. Kury, and Z. Z. Hugus, J. Phys. Chem., 58, 26 (1954).

54Ja W. D. Johnston and H. Freiser, Anal. Chim. Acta, 11, 201 (1954).

54Ka J. A. A. Ketelaar, J. Phys. Radium, 15, 197 (1954).

54Kb M. Kilpatrick and L. Pokras, J. Electrochem. Soc., 101, 39 (1954).

54La N. C. Li, J. M. White, and E. Doody, J. Am. Chem. Soc., 76, 6219 (1954).

54Ma L. N. Mulay and P. W. Selwood, J. Am. Chem. Soc., 76, 6207 (1954).

54Pa F. M. Page, J.C.S. Trans. Faraday Soc., 50, 120 (1954).

54Pb K. G. Poulsen, J. Bjerrum, and I. Poulsen, Acta Chem. Scand., 8, 921 (1954).

54Pc J. Pouradier, A. M. Venet, and H. Chateau, J. Chim. Phys. Physicochim. Biol., 51, 375 (1954).

54Ra C. Reid and R. S. Mulliken, J. Am. Chem. Soc., 76, 3869 (1954).

54Sa J. C. Sullivan and J. C. Hindman, J. Am. Chem. Soc., 76, 5931 (1954).

54Wa R. J. P. Williams, J. Phys. Chem., 58, 121 (1954).

54Ya K. B. Yatsimirskii and A. A. Shutov, Zh. Fiz. Khim., 28, 30 (1954).

54Yb K. B. Yatsimirskii and E. K. Zolotarev, Zh. Fiz. Khim., 28, 1292 (1954).

1953

53Ba R. P. Bell and J. H. B. George, J.C.S. Trans. Faraday Soc., 49, 619 (1953).

53Bb R. H. Betts and F. S. Dainton, J. Am. Chem. Soc., 75, 5721 (1953).

53Ea M. G. Evans and G. H. Nancollas, J.C.S. Trans. Faraday Soc., 49, 363 (1953).

53Fa H. Freiser, Rec. Chem. Progr., 14, 199 (1953).

53Ga D. N. Gruen and J. J. Katz, J. Am. Chem. Soc., 75, 3772 (1953).

53Ha J. Hudis and A. L. Wahl, J. Am. Chem. Soc., 75, 4153 (1953).

53Ka E. L. King, J. Chem. Educ., 30, 71 (1953).

53Kb I. M. Klotz and W. C. Loh Ming, J. Am. Chem. Soc., 75, 4159 (1953).

53La W. M. Latimer and W. L. Jolly, J. Am. Chem. Soc., 75, 1548 (1953).

53Lb M. W. Lister, Can. J. Chem., 31, 638 (1953).

53Na A. C. Nanda and S. Pani, Curr. Sci., (India), 22, 300 (1953).

53Nb T. W. Newton and G. M. Arcand, J. Am. Chem. Soc., 75, 2449 (1953).

53Sa C. G. Spike and R. W. Parry, J. Am. Chem. Soc., 75, 2726 (1953).

53Sb C. G. Spike and R. W. Parry, J. Am. Chem. Soc., 75, 3770 (1953).

53Ta C. Tanford and M. L. Wagner, J. Am. Chem. Soc., 75, 434 (1953).

53Tb H. Taube and F. A. Posey, J. Am. Chem. Soc., 75, 1463 (1953).

53Va C. E. Vanderzee and H. J. Dawson, J. Am. Chem. Soc., 75, 5659 (1953).

53Ya R. G. Yalman and A. B. Lamb, J. Am. Chem. Soc., 75, 1521 (1953).

1952

52Ba F. Basolo and R. K. Murmann, J. Am. Chem. Soc., 74, 5243 (1952).

52Ca G. H. Cartledge, J. Am. Chem. Soc., 74, 6015 (1952).

52Cb J. Chatt and R. G. Wilkins, J. Chem. Soc., (London), 4300 (1952).

52Da M. Davies and E. Gwynne, J. Am. Chem. Soc., 74, 2748 (1952).

52Ea J. I. Evans and C. B. Monk, J.C.S. Trans. Faraday Soc., 48, 934 (1952).

52Fa W. S. Fyfe, J. Chem. Soc., (London), 2023 (1952).

52Ga P. George and C. L. Tsou, Biochem. J., 50, 440 (1952).

52Ja H. B. Jonassen, G. G. Hurst, R. B. LeBlanc, and A. E. Meibohn, J. Phys. Chem., 56, 16 (1952).

52Jb H. W. Jones and C. B. Monk, J.C.S. Trans. Faraday Soc., 48, 929 (1952).

52Jc J. H. Jonte and D. S. Martin, J. Am. Chem. Soc., 74, 2052 (1952).

52Ka J. A. A. Ketelaar, C. Van De Stolpe, A. Gousmit, and W. Dzcubas, Rec. Trav. Chim. Pays-Bas, 71, 1104 (1952).

52Ma J. R. Miller and G. D. Dorough, J. Am. Chem. Soc., 74, 3977 (1952).

52Sa G. Schwarzenbach, Helv. Chim. Acta, 35, 2344-2363 (1952).

52Ua N. Uri, Chem. Rev., 50, 375 (1952).

52Va C. E. Vanderzee, J. Am. Chem. Soc., 74, 4806 (1952).

52Vb C. E. Vanderzee and D. E. Rhodes, J. Am. Chem. Soc., 74, 3552 (1952).

52Ya K. B. Yatsimirskii and A. A. Astasheva, Zh. Fiz. Khim., 26, 239 (1952).

1951

51Aa A. D. Awtrey and R. E. Connick, J. Am. Chem. Soc., 73, 1842 (1951).

51Fa S. C. Furman and C. S. Garner, J. Am. Chem. Soc., 73, 4528 (1951).

51Ha T. J. Hardwick and E. Robertson, Can. J. Chem., 29, 818 (1951).

51Ja E. L. Jahn and H. Staude, Z. Naturforsch. A, 6, 385 (1951).

51Jb J. C. James, J. Chem. Soc., (London), 153 (1951).

51Jc H. B. Jonassen and A. W. Meibohn, J Phys. Colloid Chem., 55, 726 (1951).

51Ka J. A. A. Ketelaar, C. Van De Stolpe, and H. R. Gersmann, Rec. Trav. Chim. Pays-Bas, 70, 499 (1951).

1950

50Ba J. H. Baxendale and P. George, J.C.S. Trans. Faraday Soc., 46, 55 (1950).

50Bb J. H. Baxendale and P. George, J.C.S. Trans. Faraday Soc., 46, 736 (1950).

50Ca T. M. Cromwell and R. L. Scott, J. Am. Chem. Soc., 72, 3825 (1950).

50Fa S. C. Furman and C. S. Garner, J. Am. Chem. Soc., 72, 1785 (1950).

50Ha K. Hartley and H. A. Skinner, J.C.S. Trans. Faraday Soc., 46, 621 (1950).

50Ja J. C. James and C. B. Monk, J.C.S. Trans. Faraday Soc., 46, 1041 (1950).

50Jb H. B. Jonassen, R. B. LeBlanc, and R. M. Rogan, J. Am. Chem. Soc., 72, 4968 (1950).

50Ma H. McConnell and N. Davidson, J. Am. Chem. Soc., 72, 3164 (1950).

50Sa J. Schubert, E. R. Russell, and L. S. Myers, J. Biol. Chem., 185, 387 (1950).

50Va M. G. Vladimirova and I. A. Kakovskii, J. Appl. Chem. (Leningrad), (English Trans.), 23, 615 (1950). (Russian Page No. 580.)

1949

49Da C. W. Davies and P. A. H. Wyatt, J.C.S. Trans. Faraday Soc., 45, 770 (1949).

49Db M. H. Dilke and D. D. Eley, J. Chem. Soc., (London), 2601 (1949).

49Ea M. G. Evans, P. George, and N. Uri, J.C.S. Trans. Faraday Soc., 45, 230 (1949).

49Ha J. Z. Hearon, D. Burk, and A. L. Schade, J. Nat. Cancer Inst., 9, 337 (1949).

49Ja W. Jaenicke and K. Hauffe, Z. Naturforsch. A, 4, 363 (1949)

49Ka E. L. King, J. Am. Chem. Soc., 71, 319 (1949).

49Kb I. A. Korshunov and V. A. Osipova, J. General Chem., USSR, (English Trans.), 19, 269 (1949). (Russian Page No. 1816.)

49Kc P. Krumholz, J. Am. Chem. Soc., 71, 3654 (1949).

49Ma J. D. McCullough and M. K. Barsh, J. Am. Chem. Soc., 71, 3029 (1949).

1948

48Ba J. Bjerrum and E. J. Nielsen, Acta Chem. Scand., 2, 297 (1948).

48Da C. W. Davies and J. C. James, Proc. Roy. Soc. Series A, 195, 116 (1948).

48Ka I. M. Klotz and H. G. Curme, J. Am. Chem. Soc., 70, 939 (1948).

1946

46Ca M. Calvin and R. H. Bailes, J. Am. Chem. Soc., 68, 949 (1946).

1945

45Ma J. D. McCullough and B. A. Eckerson, J. Am. Chem. Soc., 67, 707 (1945).

1944

44La F. Lindstrand, Svensk Kemisk Tidskrift, 56, 251 (1944).

1942

42Na K. Nozaki and R. A. Ogg, J. Am. Chem. Soc., 64, 697 (1942).

42Ra E. Rabinowitch and W. H. Stockmayer, J. Am. Chem. Soc., 64, 335 (1942).

REFERENCE INDEX

1941

41Ea W. Erber and A. Schühly, J. Prakt. Chem., 158, 176 (1941).

1938

38Oa B. B. Owen and S. R. Brinkley, J. Am. Chem. Soc., 60, 2233 (1938).

38Wa S. Winstein and H. J. Lucas, J. Am. Chem. Soc., 60, 836 (1938).

1937

37La F. H. Li and H. C. Chen, Naking Journal, 7, 353 (1937).

37Sa W. V. Smith, O. L. I. Brown, and K. S. Pitzer, J. Am. Chem. Soc., 59, 1213 (1937).

1936

36Ca M. Chatelet, J. Chim. Phys. Physicochim. Biol., 33, 313 (1936).

36La W. M. Latimer, Chem. Rev., 18, 349 (1936).

36Lb F. H. Lee and K. H. Lee, J. Chin. Chem. Soc., (Peiping), 4, 126 (1936).

1935

35Ra F. J. W. Roughton, Biochem. J., 29, 2604 (1935).

1934

34La W. M. Latimer, P. W. Shutz, and J. F. G. Hicks, J. Chem. Phys., 2, 82 (1934).

1933

33La W. M. Latimer, J. F. G. Hicks, and P. W. Schultz, J. Chem. Phys., 1, 424 (1933).

1931

31Ca E. Carriere and E. Raulet, Comptes Rendus, (Paris), 192, 423 (1931).

1925

25Ma G. T. Morgan, S. R. Carter, and W. F. Harrison, J. Chem. Soc., (London), 127, 1917 (1925).

REFERENCE INDEX

1910

10Ba N. Bjerrum, Z. Phys. Chem., (Leipzig), 73, 724 (1910).

1909

09Ha J. H. Hildebrand and B. L. Glascock, J. Am. Chem. Soc., 31, 26 (1909).